# Lecture Notes in Computer Science 13025

More information about this subseries at http://www.springer.com/series/7407

Chi-Yeh Chen · Wing-Kai Hon · Ling-Ju Hung ·
Chia-Wei Lee (Eds.)

# Computing and Combinatorics

## 27th International Conference, COCOON 2021
## Tainan, Taiwan, October 24–26, 2021
## Proceedings

*Editors*
Chi-Yeh Chen
National Cheng Kung University
Tainan, Taiwan

Wing-Kai Hon
National Tsing Hua University
Hsinchu, Taiwan

Ling-Ju Hung
National Taipei University of Business
Taoyuan, Taiwan

Chia-Wei Lee
National Taitung University
Taitung, Taiwan

ISSN 0302-9743 ISSN 1611-3349 (electronic)
Lecture Notes in Computer Science
ISBN 978-3-030-89542-6 ISBN 978-3-030-89543-3 (eBook)
https://doi.org/10.1007/978-3-030-89543-3

LNCS Sublibrary: SL1 – Theoretical Computer Science and General Issues



This Springer imprint is published by the registered company Springer Nature Switzerland AG
The registered company address is: Gewerbestrasse 11, 6330 Cham, Switzerland

# Preface

The 27th International Computing and Combinatorics Conference (COCOON 2021) was held during October 24–26, 2021. COCOON 2021 provided an excellent venue for researchers working in the area of algorithms, theory of computation, computational complexity, and combinatorics related to computing. The technical program of the conference included 56 regular papers selected by the Program Committee from 131 full submissions received in response to the call for papers. All the papers were peer reviewed by at least three (3.10 on average) Program Committee members or external reviewers. Papers of high quality will be invited to special issues of Algorithmica, Theoretical Computer Science (TCS), the Journal of Combinatorial Optimization (JOCO), and the International Journal of Computer Mathematics: Computer Systems Theory (IJCM:CST), respectively.

The conference also included four invited presentations, delivered by Ding-Zhu Du (University of Texas at Dallas), Takeshi Tokuyama (Kwansei Gakuin University), Ralf Klasing (CNRS and University of Bordeaux), and Tony Q.S. Quek (Singapore University of Technology and Design). Abstracts of their talks are included in this volume. We thank everyone who made this meeting possible: the authors for submitting papers, the Program Committee members, and external reviewers for volunteering their time to review conference papers. We thank Springer for publishing the proceedings in the Lecture Notes in Computer Science series. We would also like to extend special thanks to the other chairs and the conference Organizing Committee for their work in making COCOON 2021 a successful event.

September 2021

Chi-Yeh Chen
Wing-Kai Hon
Ling-Ju Hung
Chia-Wei Lee

# Preface

# Organization

## Honorary Chairs

| | |
|---|---|
| Richard Chia-Tong Lee | National Tsing Hua University, Taiwan |
| Huey-Jen Jenny Su | National Cheng Kung University, Taiwan |

## General Co-chairs

| | |
|---|---|
| Chuan-Yi Tang | Providence University, Taiwan |
| Sun-Yuan Hsieh | National Cheng Kung University, Taiwan |

## Program Co-chairs

| | |
|---|---|
| Wing-Kai Hon | National Tsing Hua University, Taiwan |
| Ling-Ju Hung | National Taipei University of Business, Taiwan |
| Chia-Wei Lee | National Taitung University, Taiwan |

## Program Committee

| | |
|---|---|
| Zhipeng Cai | Georgia State University, USA |
| Jou-Ming Chang | National Taipei University of Business, Taiwan |
| Chi-Yeh Chen | National Cheng Kung University, Taiwan |
| Ho-Lin Chen | National Taiwan University, Taiwan |
| Eddie Cheng | Oakland University, USA |
| Kai-Min Chung | Academia Sinica, Taiwan |
| Henning Fernau | Universität Trier, Germany |
| Travis Gagie | Dalhousie University, Canada |
| Arnab Ganguly | University of Wisconsin–Whitewater, USA |
| Hong Gao | Harbin Institute of Technology, China |
| Wing-Kai Hon | National Tsing Hua University, Taiwan |
| Sun-Yuan Hsieh | National Cheng Kung University, Taiwan |
| Ling-Ju Hung | National Taipei University of Business, Taiwan |
| Mong-Jen Kao | National Yang-Ming Chiao-Tung University, Taiwan |
| Ralf Klasing | CNRS and University of Bordeaux, France |
| Dominik Köppl | Tokyo Medical and Dental University, Japan |
| Van Bang Le | Universität Rostock, Germany |
| Chia-Wei Lee | National Taitung University, Taiwan |
| Chung-Shou Liao | National Tsing Hua University, Taiwan |
| Limei Lin | Fujian Normal University, China |
| Hsiang-Hsuan Liu | Utrecht University, The Netherlands |

| | |
|---|---|
| Rolf Niedermeier | TU Berlin, Germany |
| Martin Nöllenburg | Vienna University of Technology, Austria |
| Vangelis Paschos | LAMSADE, Université Paris-Dauphine, France |
| M. Sohel Rahman | Bangladesh University of Engineering and Technology, Bangladesh |
| C. Pandu Rangan | Indian Institute of Technology, Madras, India |
| Peter Rossmanith | RWTH Aachen University, Germany |
| Kunihiko Sadakane | The University of Tokyo, Japan |
| Rahul Shah | Louisiana State University, USA |
| Sharma V. Thankachan | University of Central Florida, USA |
| Takeshi Tokuyama | Tohoku University, Japan |
| Meng-Tsung Tsai | Academia Sinica, Taiwan |
| Shi-Chun Tsai | National Yang-Ming Chiao-Tung University, Taiwan |
| Hung-Lung Wang | National Taiwan Normal University, Taiwan |
| Prudence Wong | University of Liverpool, UK |
| Weili Wu | University of Texas at Dallas, USA |
| Hsu-Chun Yen | National Taiwan University, Taiwan |
| Siu-Ming Yiu | The University of Hong Kong, China |
| Christos Zaroliagis | CTI and University of Patras, Greece |
| Guochuan Zhang | Zhejiang University, China |
| Louxin Zhang | National University of Singapore, Singapore |

## Local Organizing Committee Co-chairs

| | |
|---|---|
| Sheng-Lung Peng | National Taipei University of Business, Taiwan |
| Sun-Yuan Hsieh | National Cheng Kung University, Taiwan |

## Publication Chair

| | |
|---|---|
| Chi-Yeh Chen | National Cheng Kung University, Taiwan |

## Registration Chair

| | |
|---|---|
| Yu-Chih Kuo | National Cheng Kung University, Taiwan |

## Web Chair

| | |
|---|---|
| Chih-Wei Hsu | National Cheng Kung University, Taiwan |

## Additional Reviewers

Abedin, Paniz
Abu-Khzam, Faisal
Ahmed, Abu Reyan
Ahmed, Shehab
Alagesan, Mallika
Banik, Aritra
Bazgan, Cristina
Bevern, René Van
Bhore, Sujoy
Bhuiyan, Mohammad Tawhidul Hasan
Bousquet, Nicolas
Burjons, Elisabet
Chakraborty, Dibyayan
Chan, T.-H. Hubert
Chang, Ching-Lueh
Chang, Jou-Ming
Chang, Shun-Chieh
Chang, Yi-Jun
Chao, Kun-Mao
Chen, Hua
Chen, Li-Hsuan
Chen, Po-An
Chen, Yen Hung
Chen, Yu Han
Cheng, Baolei
Cheng, Dongqin
Chou, Chi-Ning
Diwan, Ajit
Dorbec, Paul
Elbassioni, Khaled
Fan, Jianxi
Feng, Qilong
Fici, Gabriele
Frigioni, Daniele
Gerards, Marco E. T.
Gibney, Daniel
Glück, Roland
Gronemann, Martin
Gu, Mei-Mei
Guan, D. J.
Guo, Litao
Habib, Mursalin
Hansberg, Adriana
Hasan, Md. Manzurul
Heeger, Klaus
Hoffmann, Stefan
Hoi, Gordon
Hooshmand, Sahar
Hsu, Tsan-Sheng
Huang, Guan-Shieng
Huang, Shang-En
Huang, Yao-Ting
Hung, Ruo-Wei
Hunkenschröder, Christoph
Iršič, Vesna
Iwama, Kazuo
Jansson, Jesper
Jia-Jie, Liu
Jo, Seungbum
Kanellopoulos, Panagiotis
Kao, Mong-Jen
Kao, Shih-Shun
Kawase, Yasushi
Kellerer, Hans
Kindermann, Philip
Knop, Dušan
Koana, Tomohiro
Kociumaka, Tomasz
Kung, Tzu-Liang
Kunz, Pascal
Kuo, Jian-Jhih
Kuo, Tung-Wei
Kuželka, Ondřej
Le, Hung
Lee, Chuan-Min
Li, Guangping
Li, Pingshan
Li, Xiaoyan
Liang, Ya-Chun
Liedloff, Mathieu
Lin, Ching-Chi
Lin, Chuang-Chieh
Liu, Fu-Hong
Liu, Zhenwei
Lotze, Henri
Maglione Mathey, German

Mao, Yuchen
Melnichenko, Anna
Mock, Daniel
Molitor, Louise
Morales Ponce, Oscar
Muller, Haiko
Mömke, Tobias
Narboni, Jonathan
Naser Anjum, Naser
Nichterlein, André
Nishimura, Naomi
Oh, Eunjin
Ouvrard, Paul
Pai, Kung-Jui
Patel, Dhrumil
Peng, Sheng-Lung
Pighizzini, Giovanni
Puppis, Gabriele
Rao, Michael
Raptopoulos, Christoforos
Rashid, Syed Md. Mukit
Renault, Marc
Rongxia, Hao
Roshanbin, Elham
Rychlak, Ryan
Sakaue, Shinsaku
Sharma, Gokarna
Shieh, Min-Zheng
Shioura, Akiyoshi
Sin'ya, Ryoma
Spirkl, Sophie
Stamatiou, Yannis
Tan, Tony
Tang, Shyue-Ming
Tang, Zhihao Gavin
Tavakoli, Neda
Traub, Vera
Tsichlas, Kostas
Tu, Deng Yao
Tu, Hai-Lun
Tzeng, Wen-Guey
Unger, Walter
van Zuylen, Anke
Villedieu, Anaïs
Wang, Biing-Feng
Wang, Haitao
Wasa, Kunihiro
Wei, Hao-Ting
Wu, Chenchen
Wu, Hsin-Lung
Wu, Ro-Yu
Yang, Jinn-Shyong
Yen, Chung-Kung
Zehavi, Meirav
Zerovnik, Janez
Zhang, Mingzu
Zhang, Peng
Zhou, Shuming

# Abstracts of Invited Talks

# Coupon Allocation in Social Market: Robust and Machine Learning

Ding-Zhu Du

University of Texas at Dallas, USA

**Abstract.** In this talk, we consider the coupon allocation problem in marketing. It has been reported that 40% of consumers will share an email offer with their friend and 28% of consumers will share deals via social media platforms. What does this mean for a business? Essentially discounts should not just be treated as short term solutions to attract individual customer, instead, allocating coupon to a small fraction of users (called seed users) may trigger a large cascade in a social market. This motivates us to study the influence maximization coupon allocation problem: given a social network and budget, we need to decide to which initial set users should offer the coupon, and how much should the coupon be worth. Our goal is to maximize the number of customers who finally adopt the target product. The talk is based on recent research paper of Jianxiong Guo et al.

# Discrepancy Theory in Combinatorics, Geometry and Computation

Takeshi Tokuyama

Kwansei Gakuin University, Japan

**Abstract.** Discrepancy theory investigates uniformity, and appears in several aspects of mathematics and computer science. Consider a range space consisting of a set of $n$ points $P$ in the unit square $[0, 1] \times [0, 1]$ (in general, d-dimensional unit cube) and a family $R$ of subregions in the square. For a region $R \in R$ with area Area($R$), let $D(P, R) = |n\,Area\,(R) - |P \cap R||$. If $P$ is ideally uniformly distributed, $D(P, R)$ should be small for each $R$, and we define $D(P, \{\in R\}) = \sup_{R \in R} D(P, R)$. The geometric discrepancy of $n$ points with respect to $R$ is $D(n, R) = \inf_{P, |P|=n} D(P, R)$, which gives the limit of uniformity of point distribution with respect to $R$. A classical result is that $D(n, R) = \Theta logn$ if $R$ is the set of all axis-parallel rectangular regions. There are some other related discrepancies defined on hypergraphs. In this talk, discrepancy theory and its applications including recent results on consistent digital curved rays will be discussed.

# Learning Graphs with Topology Properties

Tony Q. S. Quek

Singapore University of Technology and Design, Singapore

**Abstract.** Graphs are mathematical tools, consisting of nodes (vertices) and links (edges), used in various fields to represent, process, visualize, and analyze structured data. In many cases, datasets consist of an unstructured list of samples, and the underlying graph topology (representing the structural relations between samples) is unknown. It is thus desirable to learn the graph from data. Typically, graph learning is an ill-posed problem since multiple solutions may exist associating a graph with the data. In this talk, we show how constraints can be imposed directly on the learned graphs so as to enforce certain topology properties that can best fit the data. Specifically, inspired by a specific application domain (e.g., community detection), we develop a graph learning method that learns a graph with overlapping community structure. Our method encompasses and leverages the community structure information, along with attributes such as sparsity and signal smoothness to capture the intrinsic relationships between data entities, such that the estimated graph can optimally fit the data. Furthermore, we extend to more complex datasets with heterogeneous graph signals. In summary, our methods can incorporate topology properties in graph learning, which makes it possible to capture complex and non-typical behavior of graph signals that cannot be explicitly handled just by observed data.

# Bamboo Garden Trimming Problem
## (Perpetual Maintenance of Machines with Different Urgency Requirements)

Ralf Klasing

CNRS and University of Bordeaux, France

**Abstract.** A garden $G$ is populated by $n \geq 1$ bamboos $b_1, b_2, ..., b_n$ with the respective daily growth rates $h_1 \geq h_2 \geq \cdots \geq h_n$. It is assumed that the initial heights of bamboos are zero. The robotic gardener maintaining the garden regularly attends bamboos and trims them to height zero according to some schedule. The Bamboo Garden Trimming Problem (BGT) is to design a perpetual schedule of cuts to maintain the elevation of the bamboo garden as low as possible. The bamboo garden is a metaphor for a collection of machines which have to be serviced, with different frequencies, by a robot which can service only one machine at a time. The objective is to design a perpetual schedule of servicing which minimizes the maximum (weighted) waiting time for servicing. We consider two variants of BGT. In discrete BGT the robot trims only one bamboo at the end of each day. In continuous BGT the bamboos can be cut at any time, however, the robot needs time to move from one bamboo to the next. For discrete BGT, we show a simple 4-approximation algorithm and, by exploiting relationship between BGT and the classical Pinwheel Scheduling Problem, we derive a 2-approximation for the general case and a tighter approximation when the growth rates are balanced. For continuous BGT, we propose approximation algorithms which achieve approximation ratios $O(\log(h_1/h_n))$ and $O(\log n)$.

# Contents

## Algorithms



## Approximation Algorithms

**Automata**



**Computational Geometry**

**Fault Tolerant Computing and Fault Diagnosis**



**Graph Algorithms**



**Graph Theory and Applications**

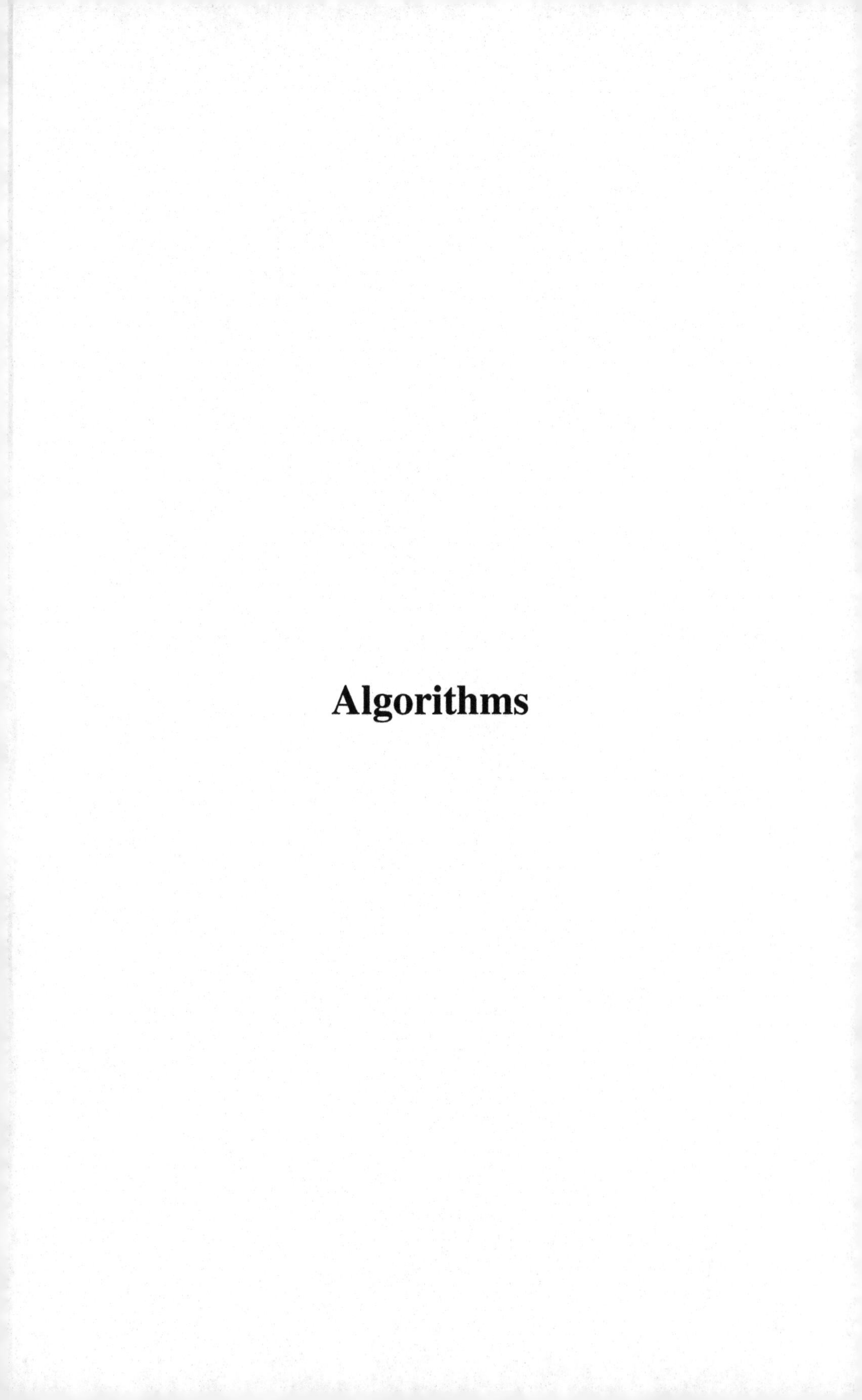

# Algorithms

# Limitations of the Impagliazzo–Nisan–Wigderson Pseudorandom Generator Against Permutation Branching Programs

Edward Pyne(✉) and Salil Vadhan

Harvard University, Cambridge, USA
epyne@college.harvard.edu, salil_vadhan@harvard.edu

**Abstract.** The classic Impagliazzo–Nisan–Wigderson (INW) pseudorandom generator (PRG) (STOC '94) for space-bounded computation uses a seed of length $O(\log n \cdot \log(nw/\varepsilon) + \log d)$ to fool ordered branching programs of length $n$, width $w$, and alphabet size $d$ to within error $\varepsilon$. A series of works have shown that the analysis of the INW generator can be improved for the class of *permutation* branching programs or the more general *regular* branching programs, improving the $O(\log^2 n)$ dependence on the length $n$ to $O(\log n)$ or $\tilde{O}(\log n)$. However, when also considering the dependence on the other parameters, these analyses still fall short of the optimal PRG seed length $O(\log(nwd/\varepsilon))$.

In this paper, we prove that any "spectral analysis" of the INW generator requires seed length

$$\Omega\left(\log n \cdot \log\log(\min\{n, d\}) + \log n \cdot \log(w/\varepsilon) + \log d\right)$$

to fool ordered permutation branching programs of length $n$, width $w$, and alphabet size $d$ to within error $\varepsilon$. By "spectral analysis" we mean an analysis of the INW generator that relies only on the spectral expansion of the graphs used to construct the generator; this encompasses all prior analyses of the INW generator. Our lower bound matches the upper bound of Braverman–Rao–Raz–Yehudayoff (FOCS 2010, SICOMP 2014) for regular branching programs of alphabet size $d = 2$ except for a gap between their $O(\log n \cdot \log\log n)$ term and our $O(\log n \cdot \log\log\min\{n, d\})$ term. It also matches the upper bounds of Koucký–Nimbhorkar–Pudlák (STOC 2011), De (CCC 2011), and Steinke (ECCC 2012) for constant-width ($w = O(1)$) permutation branching programs of alphabet size $d = 2$ to within a constant factor.

To fool permutation branching programs in the measure of *spectral norm*, we prove that any spectral analysis of the INW generator requires a seed of length $\Omega(\log n \cdot \log\log n + \log n \cdot \log(1/\varepsilon) + \log d)$ when the width is at least polynomial in $n$ ($w = n^{\Omega(1)}$), matching the recent upper bound of Hoza–Pyne–Vadhan (ITCS '21) to within a constant factor.

**Keywords:** Pseudorandomness · Space-bounded computation · Spectral graph theory

Full Version [PV21a]: https://eccc.weizmann.ac.il/report/2021/108/.


C.-Y. Chen et al. (Eds.): COCOON 2021, LNCS 13025, pp. 3–12, 2021.
https://doi.org/10.1007/978-3-030-89543-3_1

# 1 Introduction

Starting with the work of Babai, Nisan, and Szegedy [BNS92], there has been three decades of work of constructing and analyzing pseudorandom generators for space-bounded computation, motivated by obtaining unconditional derandomization (e.g. seeking to prove that BPL = L) and a variety of other applications (e.g. [Ind06, Siv02, HVV06, HHR11]). Although we still remain quite far from having pseudorandom generators that suffice for a full derandomization of space-bounded computation, there has been substantial progress on pseudorandom generators for restricted models of space-bounded computation. In particular, a series of works has shown that the analysis of the classic Impagliazzo–Nisan–Wigderson (INW) generator [INW94] can be significantly improved for restricted models (e.g. "permutation branching programs"), but these analyses have not matched the parameters of an optimal pseudorandom generator. In this work, we show that there are inherent limitations to the analysis of the INW generator for these restricted models, proving lower bounds that nearly match the known upper bounds.

## 1.1 Pseudorandom Generators for Space-Bounded Computation

Like previous work, we will work with the following nonuniform model of space-bounded computation.

**Definition 1.** *An* ***ordered branching program (OBP)*** *$B$ of* ***length*** *$n$,* ***width*** *$w$ and* ***alphabet size*** *$d$ computes a function $B : [w] \times [d]^n \to [w]$. On an input $\sigma \in [d]^n$, the branching program computes as follows. It starts at a fixed start state $v_0 \in [w]$. Then for $t = 1, \ldots, n$, it reads the next symbol $\sigma_t$ and updates its state according to a transition function $B_t : [w] \times [d] \to [w]$ by taking $v_t = B_t(v_{t-1}, \sigma_t)$. Note that the transition function $B_t$ can differ at each time step.*

*Moreover, there is a set of accept states $V_e \subseteq [w]$. Let $u$ be the final state of the branching program on input $\sigma$. If $u \in V_e$ the branching program accepts, denoted $B(\sigma) = 1$, and otherwise the program rejects, denoted $B(\sigma) = 0$.*

An ordered branching program can be viewed as a layered digraph, consisting of $(n + 1)$ layers of $w$ vertices each, where for every $t = 1, \ldots, n$ and $v \in [w]$, the $v$'th vertex in layer $t - 1$ has $d$ outgoing edges, going to the vertices $B_t(v, 1), B_t(v, 2), \ldots, B_t(v, d) \in [w]$ in layer $t$.

An ordered branching program corresponds to a streaming algorithm, in that the $n$ input symbols from $[d]$ are each read only once, and in a fixed order. This is the relevant model for derandomization of space-bounded computation because a randomized space-bounded algorithm processes its random bits in a streaming fashion. Specifically, if on an input $x$, a randomized algorithm $A$ uses space $s$ and $n$ random bits $\sigma$, the function $B_x(\sigma) = A(x; \sigma)$ can be computed by an ordered branching program of length $n$, width $w = 2^s$ and alphabet size 2. In particular, if $A$ is a randomized logspace algorithm (i.e. a BPL algorithm), then $n = w = \text{poly}(|x|)$.

The standard definition of pseudorandom generator is as follows.

**Definition 2.** *Let $\mathcal{F}$ be a class of functions $f : [d]^n \to \{0,1\}$. An $\varepsilon$-**pseudorandom generator** ($\varepsilon$-**PRG**) for $\mathcal{F}$ is a function* $\mathrm{GEN} : \{0,1\}^s \to [d]^n$ *such that for every $f \in \mathcal{F}$,*

$$\left| \mathop{\mathbb{E}}_{x \leftarrow U_{[d]^n}} [f(x)] - \mathop{\mathbb{E}}_{x \leftarrow U_{\{0,1\}^s}} [f(\mathrm{GEN}(x))] \right| \leq \varepsilon,$$

*where $U_S$ is the uniform distribution over the set $S$. The value $s$ is the **seed length** of the PRG. We say a generator* GEN *is **explicit** if the $i$th symbol of output is computable in space $O(s)$. We say that* GEN *$\varepsilon$-**fools** $\mathcal{F}$ if it is an $\varepsilon$-PRG for $\mathcal{F}$.*

By the Probabilistic Method, it can be shown that there exist (non-explicit) $\varepsilon$-PRGs for the class of ordered branching programs of length $n$, width $w$, and alphabet size $d$ with seed length $s = O(\log(nwd/\varepsilon))$, and it can be shown that this is optimal up to a constant factor (provided that $2^n \geq w$, $n, d, w \geq 2$, $d$ is even, and $\varepsilon \leq 1/3$). An explicit construction with such a seed length (even for $d = 2$ and $\varepsilon = 1/3$) would suffice to fully derandomize logspace computation (i.e. prove BPL=L).

The classic constructions of Nisan [Nis92] and Impagliazzo, Nisan, and Wigderson [INW94] give explicit PRGs with seed length $s = O(\log n \cdot \log(nw/\varepsilon) + \log d)$. For the case corresponding to derandomizing general logspace computation, where $d$ and $\varepsilon$ are constant and $w$ is polynomially related to $n$, we have $s = O(\log^2 n)$, quadratically worse than the optimal seed length of $s = O(\log n)$. Brody and Verbin [BV10] showed that these classic pseudorandom generators require seed length $\Omega(\log^2 n)$ even for width $w = 3$. Meka, Reingold, and Tal [MRT19] recently gave a completely different explicit construction of pseudorandom generator for width $w = 3$ with seed length $s = \tilde{O}(\log n \cdot \log(1/\varepsilon)))$, but for width $w = 4$ no explicit constructions with seed length $o(\log^2 n)$ are known.

### 1.2 Permutation Branching Programs

Motivated by the lack of progress on the general ordered branching program model, there has been extensive research on restricted models:

**Definition 3.** *An **(ordered) regular branching program** of length $n$, width $w$, and alphabet size $d$ is an ordered branching program where the associated layered digraph consists of regular bipartite graphs between every pair of consecutive layers. Equivalently, for every $t = 1, \ldots, n$ and every $v \in [w]$, there are exactly $d$ pairs $(u, \sigma) \in [w] \times [d]$ such that $B_t(u, \sigma) = v$.*

**Definition 4.** *An **(ordered) permutation branching program** of length $n$, width $w$, and alphabet size $d$ is an ordered branching program where for all $t \in [n]$ and $\sigma \in [d]$, $B_t(\cdot, \sigma)$ is a permutation on $[w]$.*

Every ordered permutation branching program is a regular branching program, but not conversely.

A series of works has shown that the Impagliazzo–Nisan–Wigderson (INW) pseudorandom generator can be instantiated with smaller seed length for regular or permutation branching programs. First, Rozenman and Vadhan [RV05] analyzed the INW generator for carrying out random walks on $d$-regular $w$-vertex graphs, which correspond to regular branching programs in which all of the transition functions $B_t$ are the same. They showed that if the graph is consistently labelled (equivalently, that we have a permutation branching program), then a seed length of $s = O(\log(nwd/\varepsilon))$ suffices for the random walk to get within distance $\varepsilon$ of the uniform distribution on vertices, provided that the length $n$ of the pseudorandom walk is polynomially larger than the mixing time of a truly random walk. (This "pseudo-mixing" property is nonstandard but has applications, including giving a simpler proof of Reingold's Theorem that Undirected Connectivity is in deterministic logspace [Rei08] and the construction of almost $k$-wise independent permutations [KNR05].)

Next, Braverman, Rao, Raz, Yehudayoff [BRRY10] analyzed the INW generator for regular branching programs of alphabet size $d = 2$, and achieved seed length $s = O(\log n \cdot \log\log n + \log n \cdot \log(w/\varepsilon))$, thereby improving the dependence on the length $n$ from $O(\log^2 n)$ to $\tilde{O}(\log n)$ for the standard pseudorandomness property. For the case of *permutation* branching programs of *constant width* $w = O(1)$ and alphabet size $d = 2$, Koucký and Nimbhorkar and Pudlák [KNP11] further improved the seed length to $s = O_w(\log n \cdot \log(1/\varepsilon))$. The hidden constant in the $O_w(.)$ depended exponentially on the width $w$, but was subsequently improved to a polynomial by De [De11] and Steinke [Ste12].

Recently, Hoza, Pyne, and Vadhan [HPV21] turned their attention to permutation branching programs of *unbounded width*, and showed that the INW generator fools such programs in "spectral norm" with seed length $s = O(\log n \cdot \log\log n + \log n \cdot \log(1/\varepsilon) + \log d)$. Here, fooling in spectral norm means that the $w \times w$ matrix of probabilities of going from each initial state to each final state under the generator has distance at most $\varepsilon$ in spectral norm from the same matrix under truly random inputs. $\varepsilon$-fooling in spectral norm can be shown to imply the standard notion of pseudorandomness for programs with a single accept state. Surprisingly, the seed length of [HPV21] even beats the Probabilistic Method; indeed they show that a random function requires seed length $\Omega(n)$ to fool permutation branching programs of unbounded width and a single accept vertex with high probability.

We summarize the aforementioned analyses of the INW generator in Table 1. Let us elaborate on how all of these results are instantations of the INW generator. Specifically, the INW generator can be viewed as a template for a recursive construction of a PRG, where a PRG $\mathrm{INW}_{i-1}$ generating $n_{i-1} = 2^{i-1}$ output symbols is used to construct a PRG $\mathrm{INW}_i$ generating $n_i = 2^i$ output symbols, by running $\mathrm{INW}_{i-1}$ twice on a pair of correlated seeds. The pair of seeds are chosen according to a random edge in an auxiliary expander graph $H_i$:

$$\mathrm{INW}_i(e) = \mathrm{INW}_{i-1}(x) \cdot \mathrm{INW}_{i-1}(y) \text{ for each edge } e = (x, y) \text{ of } H_i, \quad (1)$$

**Table 1.** Spectral analyses of the INW generator

| Model | Seed length | Pseudorandomness | Reference |
|---|---|---|---|
| General | $O(\log n \cdot \log(nwd/\varepsilon))$ | Standard | [INW94] |
| Perm., same trans. | $O(\log(nwd/\varepsilon))$ | Pseudo-mixing | [RV05] |
| Regular, $d = 2$ | $O(\log n \cdot \log\log n +$ $\log n \cdot \log(w/\varepsilon))$ | Standard | [BRRY10] |
| Permutation, $d = 2$ | $O_w(\log n \cdot \log(1/\varepsilon))$ | Standard | [KNP11, De11, Ste12] |
| Permutation | $O(\log n \cdot \log\log n +$ $\log n \cdot \log(1/\varepsilon) + \log d)$ | Spectral | [HPV21] |

where $\cdot$ denotes concatenation. Thus different choices of the sequence of graphs $H_1, H_2, \ldots, H_{\log n}$ yield different instantiations of the INW generator. In all of the aforementioned works,[1] the pseudorandomness property of the generator is proven using only the spectral expansion properties of the graphs $H_i$, namely requiring that all of the nontrivial normalized eigenvalues of $H_i$ have absolute value at most some value $\lambda_i$ for $i = 1, \ldots, \log n$. We call such an analysis a *spectral analysis* of the INW generator. Given a spectral analysis of the INW generator, the degrees of the expanders $H_i$ are then determined by the optimal relationship between expansion and degree $d_i = \text{poly}(1/\lambda_i)$, which in turn determines the seed length of the final generator, namely

$$s = \Theta\left(\log\left(d \cdot d_1 \cdot d_2 \cdots d_{\log n}\right)\right) = \Omega\left(\log d + \sum_{i=1}^{\log n} \log(1/\lambda_i)\right). \tag{2}$$

### 1.3 Our Results

Given the improved analyses of the INW generator described in Table 1, it is natural to wonder how much further these analyses can be pushed. In particular, can the INW generator $\varepsilon$-fool permutation branching programs of length $n$, width $w$, and alphabet size $d$ with seed length matching the optimal seed length of $O(\log(nwd/\varepsilon))$? Our main result is that the answer is no:

**Theorem 1 (informally stated).** *Any spectral analysis of the INW generator for $\varepsilon$-fooling permutation branching programs of length $n$, width $w$, and alphabet size $d$ requires seed length*

$$s = \Omega\left(\log n \cdot \log\log(\min\{n, d\}) + \log n \cdot \log(w/\varepsilon) + \log d\right).$$

Notice that this lower bound nearly matches the upper bounds in Table 1. In particular, we match the upper bound of [BRRY10] for regular branching programs, except that we get a $\log n \cdot \log\log n$ term only when $d = n^{\Omega(1)}$ while

[1] Braverman et al. [BRRY10] analyze the INW generator constructed with *randomness extractors* [NZ96], but the extractor parameters they use follow from spectral expansion properties of the underlying graphs [GW97].

they have such a term even when $d = 2$. We also match the upper bounds of [KNP11,De11,Ste12] for permutation branching programs of alphabet size $d = 2$ and constant width $w = O(1)$.

For fooling with respect to spectral norm, we can get a lower bound of $\log n \cdot \log\log n$ whenever $w = n^{\Omega(1)}$, in particular matching the result of [HPV21] for unbounded-width permutation branching programs:

**Theorem 2 (informally stated).** *For $\varepsilon$-fooling in spectral norm, any spectral analysis of the INW generator for permutation branching programs of length $n$, width $w$, and alphabet size $d$ requires seed length*

$$s = \Omega\left(\log n \cdot \log\log(\min\{n, w\}) + \log n \cdot \log(1/\varepsilon) + \log d\right).$$

While our theorems are quite close to the upper bounds, they leave a few regimes where a spectral analysis of the INW generator could potentially yield an improved seed length. In particular, a couple of open questions stand out regarding the $\log n \cdot \log\log n$ in terms in the bounds:

- Can we achieve seed length $O(\log n \cdot \log(w/\varepsilon))$ for permutation (or even regular) branching programs of alphabet size $d = 2$? When the alphabet size is $d = 2$, the $\log\log(\min\{n, d\})$ term disappears in Theorem 1. However, the upper bound of [BRRY10] for regular branching programs still has an $O(\log n \cdot \log\log n)$ term, and the upper bounds of [KNP11,De11,Ste12] only achieve a polynomial dependence on the width $w$.
- Can we achieve seed length $O(\log n \cdot \log(1/\varepsilon))$ for permutation branching programs with a single accept vertex, alphabet size $d = 2$, and width $w = n$ (or even unbounded width)? The best upper bound for this model is [HPV21], which has a additional $O(\log n \cdot \log\log n)$ term. This term is necessary for fooling in spectral-norm by Theorem 2 but may not be necessary for the easier task of fooling programs with a single accept vertex.

A second opportunity for improvement is to go beyond spectral analysis of the INW generator, and exploit graphs $H_i$ with additional properties. To indicate that there is some hope for this, we include an observation showing that there *exists* an instantiation of the INW generator that achieves optimal seed length, even against more general ordered branching programs:

**Theorem 3.** *For all $n, w, d \in \mathbb{N}$ and $\varepsilon > 0$, there exists a sequence of graphs $\mathcal{H}$ such that the INW generator constructed with this sequence $\varepsilon$-fools ordered branching programs of length $n$, width $w$ and alphabet size $d$ and has seed length $O(\log(nwd/\varepsilon))$.*

This is an application of the Probabilistic Method, and so does not give an explicit PRG.

Our lower bounds also say nothing about constructions that deviate from the template of the INW generator, and better seed lengths can potentially be obtained by modifying the INW generator or using it as a tool in more involved constructions. Examples include the pseudorandom generator for width

3 ordered branching programs [MRT19], which combines the INW generator with pseudorandom restrictions, and [BCG18, CL20, CDR+21, PV21b, Hoz21], which construct "weighted pseudorandom generators" with a better dependence on the error by taking linear combinations of the INW generator (or blends of the Nisan and INW generator).

## 1.4 Techniques

Theorem 1 is really three separate lower bounds, which we state as separate theorems here to discuss the proof ideas separately. (The lower bound of $s = \Omega(\log d)$ is very simple.)

**Theorem 4 (informally stated).** *Any spectral analysis of the INW generator for $(1 - 1/w^{\Omega(1)})$-fooling permutation branching programs of length $n$, width $w$, and alphabet size $d = 2$ requires seed length $s = \Omega(\log n \cdot \log w)$.*

Note that the lower bound holds for a very large error parameter, namely $\varepsilon = 1 - 1/w^{\Omega(1)}$. In fact, it holds even for obtaining a *hitting-set generator*, where we Definition 2 is relaxed to only require that $\mathbb{E}_{x \leftarrow U_{[d]^n}}[f(x)] > \varepsilon$ implies that $\mathbb{E}_{x \leftarrow U_{\{0,1\}^s}}[f(\mathrm{GEN}(x))] > 0$.

To prove this Theorem 4, we show that most of the $\lambda_i$'s parameterizing the INW generator must have $\lambda_i < 1/w^{\Omega(1)}$, which implies the seed-length lower bound by Eq. (2). If that is not the case for some value of $i$, we construct an auxiliary graph $H_i$ to use in the INW generator (with $\lambda(H_i) \leq \lambda_i$) such that a permutation branching program only needs width $\mathrm{poly}(1/\lambda_i) \leq w$ in order to perfectly distinguish a random edge in $H_i$ from a pair of vertices in $H_i$ that are disconnected. Specifically, we can take $H_i$ to be an expander with degree $c_i = \mathrm{poly}(1/\lambda_i)$ and $c_i^2$ vertices. To be able to use such a graph in most levels in the INW generator, we may need to pad the number of vertices to a value larger than $c_i$. We do this by taking a tensor product with a complete graph, which retains both the expansion of $H_i$ and the ability of a width $w$ permutation branching program to distinguish edges and non-edges. We use complete graphs (with an appropriate edge labelling) for the remaining graphs $H_j$ in the INW generator, and argue a permutation branching program of width $w$ can still distinguish the output from uniform.

**Theorem 5 (informally stated).** *Any spectral analysis of the INW generator for $\varepsilon$-fooling permutation branching programs of length $n$, width $w = 2$, and alphabet size $d = 2$ requires seed length $s = \Omega(\log n \cdot \log(1/\varepsilon))$.*

To prove Theorem 5 we use a construction from [RV05] used to show that the tightness of their analysis of the "derandomized square" operation on graphs. (Composing the INW generator with a permutation branching programs amounts to performing $\log n$ iterated derandomized square operations on the graph of the branching program.) Specifically, in order to show that each $\lambda_i$ satisfies $\lambda_i = O(\varepsilon)$, we consider a graph $H_i$ that has a self-loop probability of

$\lambda_i$ but has $\lambda(H_i) \leq \lambda_i$. When the self-loop is taken, it means that two consecutive subsequences of the output of the INW generator of length $2^{i-1}$ are equal to each other, by Eq. (1). Thus the permutation branching program of width 2 that computes the parity of the input bits on the union of those two subsequences will distinguish the output of the INW generator from uniform with advantage $\Omega(\lambda_i)$.

**Theorem 6 (informally stated).** *Any spectral analysis of the INW generator for .1-fooling permutation branching programs of length $n$, width $w = 2$, and alphabet size $d$ requires seed length $s = \Omega(\log n \cdot \log\log(\min\{n, d\}))$.*

To prove Theorem 6, we want to show that most of the $\lambda_i$'s must satisfy $\lambda_i \leq O(1/\log n)$, where we assume without loss of generality that $d = n$. It suffices to prove that $\sum_{i=1}^{\log n} \lambda_i \leq O(1)$. To do this, we again consider graphs $H_i$ that have a self-loop probability of $\lambda_i$, but rather than considering only one such graph, we use all of them in the INW generator. Intuitively, we want to show that the errors of $\Omega(\lambda_i)$ accumulate to lead to an overall error of $\Omega(\sum_i \lambda_i) > \varepsilon$. We consider a permutation branching program that corresponds to a random walk on a graph $G$ with $w = 2$ vertices that has a self-loop probability of $1-1/d = 1-1/n$. A truly random walk of length $n$ on $G$ will end at its start vertex with probability at most $1 - n \cdot (1/n) \cdot (1 - 1/n)^{n-1} < .64$. We show that a pseudorandom walk using the INW generator with the graphs $H_i$ will end at its start vertex with probability at least .75. Specifically, we choose our edge and vertex labellings carefully so that the self-loops in the graphs $H_i$ cause random walks to backtrack with a high constant probability, so that it is as if we are typically doing random walks on $G$ of length at most $n/4$.

Turning to Theorem 7, the only part of the lower bound that does not follow from the same arguments as above is the following:

**Theorem 7 (informally stated).** *For $1/3$-fooling in spectral norm, any spectral analysis of the INW generator for permutation branching programs of length $n$, width $w$, and alphabet size $d = 2$ requires seed length $s = \Omega(\log n \cdot \log\log(\min\{n, w\}))$.*

The proof of Theorem 7 is similar to that of Theorem 6, but instead of considering random walks on a 2-vertex graph $G$ with large degree $d$, we use an graph $G$ of degree 2 and a large number of vertices. Specifically we take $G$ to be the undirected cycle on $w = \Theta(\sqrt{n})$ vertices. The key point is that the truly random walk on the cycle mixes in $n = \Theta(w^2)$ steps in spectral norm. So a truly random walk of length $n$ will differ from complete mixing by at most, say $1/3$, in spectral norm, but due to backtracking, the pseudorandom walks using the INW generator will differ from complete mixing by at least $2/3$ in spectral norm.

**Acknowledgements.** We thank Ronen Shaltiel for asking a question at ITCS 2021 that prompted us to write this paper. S.V. thanks Omer Reingold and Luca Trevisan for discussions many years ago that provided some of the ideas in this paper, in particular the tensor product construction used in the proof of Theorem 4 and the probabilistic existence proof in Theorem 3.

## References

[BCG18] Braverman, M., Cohen, G., Garg, S.: Hitting sets with near-optimal error for read-once branching programs. In: Diakonikolas, I., Kempe, D., Henzinger, M. (eds.) Proceedings of the 50th Annual ACM SIGACT Symposium on Theory of Computing, STOC 2018, Los Angeles, CA, USA, 25–29 June 2018, pp. 353–362. ACM (2018)

[BNS92] Babai, L., Nisan, N., Szegedy, M.: Multiparty protocols, pseudorandom generators for logspace, and time-space trade-offs. J. Comput. Syst. Sci. **45**(2), 204–232 (1992). Twenty-first Symposium on the Theory of Computing (Seattle, WA, 1989)

[BRRY10] Braverman, M., Rao, A., Raz, R., Yehudayoff, A.: Pseudorandom generators for regular branching programs. In: FOCS [IEE10], pp. 40–47 (2010)

[BV10] Brody, J., Verbin, E.: The coin problem and pseudorandomness for branching programs. In: FOCS [IEE10], pp. 30–39 (2010)

[CDR+21] Cohen, G., Doron, D., Renard, O., Sberlo, O., Ta-Shma, A.: Error reduction for weighted PRGs against read once branching programs. In: Kabanets, V. (ed.) 36th Computational Complexity Conference (CCC 2021). Leibniz International Proceedings in Informatics (LIPIcs), Dagstuhl, Germany, vol. 200, pp. 22:1–22:17. Schloss Dagstuhl - Leibniz-Zentrum für Informatik (2021)

[CL20] Chattopadhyay, E., Liao, J.-J.: Optimal error pseudodistributions for read-once branching programs. In: Saraf, S. (ed.) 35th Computational Complexity Conference, CCC 2020, Saarbrücken, Germany, 28–31 July 2020, (Virtual Conference). LIPIcs, vol. 169, pp. 25:1–25:27. Schloss Dagstuhl - Leibniz-Zentrum für Informatik (2020)

[De11] De, A: Pseudorandomness for permutation and regular branching programs. In: IEEE Conference on Computational Complexity, pp. 221–231. IEEE Computer Society (2011)

[GW97] Goldreich, O., Wigderson, A.: Tiny families of functions with random properties: a quality-size trade-off for hashing. Random Struct. Algorithms **11**(4), 315–343 (1997)

[HHR11] Haitner, I., Harnik, D., Reingold, O.: On the power of the randomized iterate. SIAM J. Comput. **40**(6), 1486–1528 (2011)

[Hoz21] Hoza, W.: Better pseudodistributions and derandomization for space-bounded computation. ECCC preprint TR21-019 (2021)

[HPV21] Hoza, W.M., Pyne, E., Vadhan, S.P.: Pseudorandom generators for unbounded-width permutation branching programs. In: Lee, J.R. (ed.) 12th Innovations in Theoretical Computer Science Conference, ITCS 2021, 6–8 January 2021, Virtual Conference. LIPIcs, vol. 185, pp. 7:1–7:20. Schloss Dagstuhl - Leibniz-Zentrum für Informatik (2021)

[HVV06] Healy, A., Vadhan, S., Viola, E.: Using nondeterminism to amplify hardness. SIAM J. Comput. **35**(4), 903–931 (electronic) (2006)

[IEE10] IEEE. 51th Annual IEEE Symposium on Foundations of Computer Science, FOCS 2010, Las Vegas, Nevada, USA, 23–26 October 2010. IEEE Computer Society (2010)

[Ind06] Indyk, P.: Stable distributions, pseudorandom generators, embeddings, and data stream computation. J. ACM **53**(3), 307–323 (2006)

[INW94] Impagliazzo, R., Nisan, N., Wigderson, A.: Pseudorandomness for network algorithms. In: Proceedings of the Twenty-Sixth Annual ACM Symposium on the Theory of Computing, Montréal, Québec, Canada, 23–25 May 1994, pp. 356–364 (1994)

[KNP11] Koucký, M., Nimbhorkar, P., Pudlák, P.: Pseudorandom generators for group products: extended abstract. In: Fortnow, L., Vadhan, S.P. (eds.) STOC, pp. 263–272. ACM (2011)

[KNR05] Kaplan, E., Naor, M., Reingold, O.: Derandomized constructions of $k$-Wise (almost) independent permutations. In: Chekuri, C., Jansen, K., Rolim, J.D.P., Trevisan, L. (eds.) APPROX/RANDOM -2005. LNCS, vol. 3624, pp. 354–365. Springer, Heidelberg (2005). https://doi.org/10.1007/11538462_30

[MRT19] Meka, R., Reingold, O., Tal, A.: Pseudorandom generators for width-3 branching programs. In: Proceedings of the 51st Annual ACM SIGACT Symposium on Theory of Computing, pp. 626–637. ACM (2019)

[Nis92] Nisan, N.: Pseudorandom generators for space-bounded computation. Combinatorica **12**(4), 449–461 (1992)

[NZ96] Nisan, N., Zuckerman, D.: Randomness is linear in space. J. Comput. Syst. Sci. **52**(1), 43–52 (1996)

[PV21a] Pyne, E., Vadhan, S.: Limitations of the Impagliazzo-Nisan-Wigderson Pseudorandom Generator against Permutation Branching Programs. ECCC preprint TR21-108 (2021)

[PV21b] Pyne, E., Vadhan, S.: Pseudodistributions that beat all pseudorandom generators (extended abstract). In: Kabanets, V (ed.) 36th Computational Complexity Conference (CCC 2021). Leibniz International Proceedings in Informatics (LIPIcs), Dagstuhl, Germany, vol. 200, pp. 33:1–33:15. Schloss Dagstuhl - Leibniz-Zentrum für Informatik (2021)

[Rei08] Reingold, O.: Undirected connectivity in log-space. J. ACM **55**(4), 24, Article no. 17 (2008)

[RV05] Rozenman, E., Vadhan, S.: Derandomized squaring of graphs. In: Chekuri, C., Jansen, K., Rolim, J.D.P., Trevisan, L. (eds.) APPROX/RANDOM 2005. LNCS, vol. 3624, pp. 436–447. Springer, Heidelberg (2005). https://doi.org/10.1007/11538462_37

[Siv02] Sivakumar, D.: Algorithmic derandomization via complexity theory. In: Proceedings of the Thirty-Fourth Annual ACM Symposium on Theory of Computing, (electronic), pp. 619–626. ACM, New York (2002)

[Ste12] Steinke, T.: Pseudorandomness for permutation branching programs without the group theory. Technical Report TR12-083, Electronic Colloquium on Computational Complexity (ECCC), July 2012

# All-to-All Broadcast in Dragonfly Networks

Dong Xiang(✉) and Yunzhou Ju

School of Software, Tsinghua University, No. 1, Tsinghua Garden Street, Beijing 100084, China
dxiang@tsinghua.edu.cn

**Abstract.** New deadlock-free unicast-based all-to-all broadcast algorithms are proposed for dragonfly networks. An all-to-all broadcast delivers a message from each router to all routers. Two different all-to-all broadcast algorithms GFA2A and RFA2A using the previous group-first and router-first one-to-all broadcast schemes are presented. A new all-to-all broadcast algorithm named A2A is presented by collecting all messages from all routers in the same group to a single router first and combining them, which are forwarded to all routers in the same group. Each router forwards messages to all other routers in the same groups after receiving all messages from other groups. The proposed algorithms can be implemented with the unicast hardware, that is, each input port is assigned two indistinguishable buffers.

## 1 Introduction

Economical optical signaling enables high-radix topologies with long channels, which are less expensive than the short electrical channels. The use of high-radix routers attracts more interests than those of low-radix ones by maintaining a small number of ports and increasing the bandwidth per port. Dragonfly networks [4,6,8,11,21,22] have been popular in the past decade. It consists of a number of router groups, where each group contains completely connected routers for 1D router groups. Any pair of groups are connected by at least one global channel. Each router group can also be connected as a 2D flattened butterfly [3,15]. Each router has multiple global links that are connected to other groups.

Collective communication has received considerable attention since it places a high demand on network bandwidth and has a great impact on algorithms execution time. Multicast is one of the most useful collective communication operations. Multicast can be easily implemented with no hardware overhead by serially sending a multicast message with $d$ destinations $d$ times and it is delivered to a particular destination for each time, which causes a significant amount of traffic, and introduces an intolerably long delay at the injection point.

Broadcast is frequently used by many important applications such as parallel search, parallel graph and matrix algorithms. One-to-all broadcast is used to implement more complex communication operations, such as, personalized all-to-all broadcast, gather, and barrier synchronization [5,13,17].


C.-Y. Chen et al. (Eds.): COCOON 2021, LNCS 13025, pp. 13–24, 2021.
https://doi.org/10.1007/978-3-030-89543-3_2

Collective communication is classified as follows: (1) tree-based [9,10], (2) path-based [1,14], and (3) unicast-based [12,19–22]. The unicast-based scheme in [12] is purely software-based, which does not need any modification on the hardware.

A multicast message is delivered to all destinations by using a recursive unicast scheme [12] unlike the serial unicast scheme stated earlier, where each multicast message must be delivered to the neighbors of the source $\ln(d+1)$ times from the injection port and $d$ is the number of destinations. The unicast-based multicast scheme in [12] does not duplicate the message to any intermediate node. This unicast-based multicast scheme is promising and cost-effective (Fig. 1).

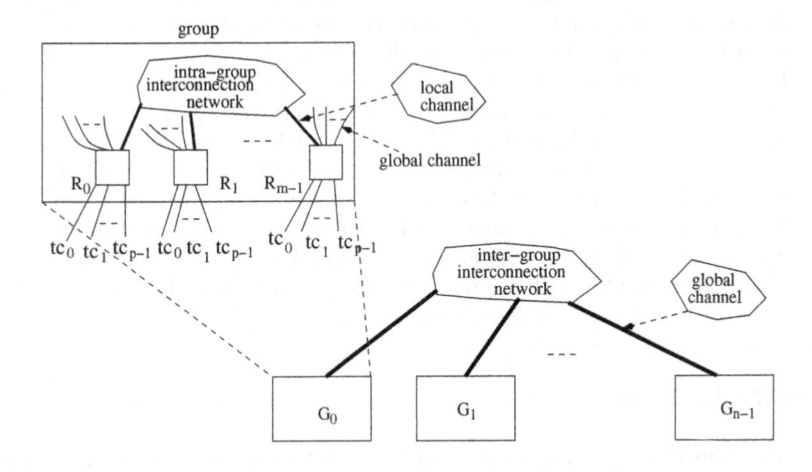


**Fig. 1.** The dragonfly interconnection network.

Collective communication in dragonfly networks [2,22] is very important. Hardware support multicast for tree-based or path-based techniques is very difficult to implement, which makes the switch too complex. McKinley, et al. [12] presented efficient algorithms to implement multicast communication in wormhole-routed meshes and hypercubes in the absence of hardware multicast support, by exploiting the properties of the switching technology. Suh and Yalamanchili [16] used message combining to minimize message start-ups at the expense of larger message sizes. Juurlink, et al. [7] optimized the trade-off between contributions due to start-ups and those due to the bounded capacity of the connections. Up to now, we still do not have any work on all-to-all broadcast in dragonfly networks. It is essential to propose an efficient all-to-all broadcast algorithm for dragonfly networks.

We presented two all-to-all broadcast algorithms using the one-to-all broadcast schemes in [22]. Each of all routers delivers a one-to-all broadcast message separately for both schemes. In the group-first (GFA2A) algorithm, each of all routers delivers a message to all groups in the network first from the source router, which is forwarded to all routers in each group after that. In the router-first all-to-all broadcast (RFA2A) algorithm, each of all routers sends a message to all routers in the source group first, which is delivered from separate routers in the source group to all groups concurrently. The message is finally forwarded to all routers in each group. Both GFA2A and RFA2A algorithms produce quite different global channel traffic, and sequential global channel traversals.

The third all-to-all broadcast algorithm (A2A) contains four separate phases: (1) packet collection inside a router group, (2) packet scattering in the group, (3) delivery of all packets at the router to all directly connected routers in the other groups, and (4) each router scattering all packets received from all $g = m/2$ directly connected routers in other groups to all routers in its own group.

In the rest of this paper, the preliminaries are presented in Sect. 2. The all-to-all broadcast algorithms based on the one-to-all broadcast schemes are presented in Sect. 3. The message combining based all-to-all broadcast algorithm A2A is presented in Sect. 4. The paper is concluded in Sect. 5.

## 2 Preliminaries

Connection of the global links has an impact on the performance of the network. We use the same scheme to connect the global channels for the dragonfly networks as presented in [22]. Let each group have $m$ routers $R_0$, $R_1$, ..., $R_{m-1}$. The groups are labeled as $G_0, G_1, G_2, \ldots, G_{n-1}$, while for $i \in \{0, 1, 2, \ldots, n-2\}$, the last router of $G_i$ is connected to the first router of $G_{i+1}$ for $0 < i < n-1$. Figure 2 presents the global connections between any pairs of adjacent groups for a system with nine groups and each group contains four routers.

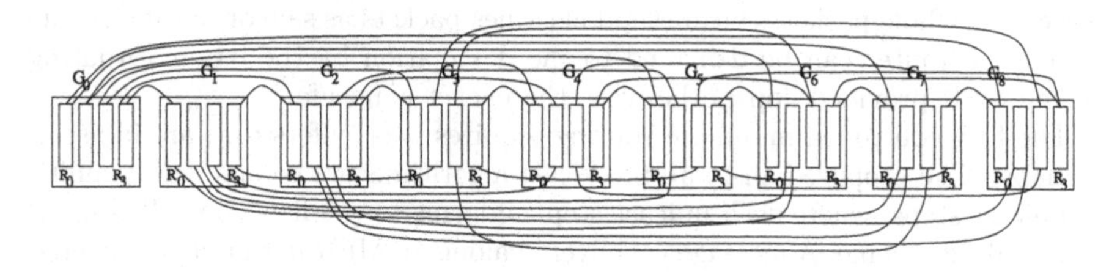

**Fig. 2.** Connecting the global links for the dragonfly networks.

There are $n$ router groups $G_0$, $G_1$, ..., $G_{n-1}$, where each group contains $m$ routers $R_0$, $R_1$, ..., $R_{m-1}$. Let each router be connected to $g = m/2$ global links, therefore, each group has $m \cdot g$ global links. Our method connects the first group $G_0$ to any other group $G_j$ $(j \geq 2)$ from $j = 2$ to $n-1$ in the following way: the router in $G_0$ with the highest ID, that has an available slot, is connected to the router in $G_j$ with the lowest ID and an available slot. The group $G_1$ is then connected with all other groups $G_j$ $(j \geq 3)$ in the same way. This process continues until $G_{n-3}$ has connected a global link with $G_{n-1}$ finally.

For each pair of groups that are not adjacent, $R_v \in G_i$ and $R_{v'} \in G_j$ with $i+1 < j$, $R_v$ is connected to $R_{v'}$. Let $R_v \in G_i$ be connected to $R_{v'} \in G_j$. We have the following equations for the IDs of the routers $R_v$ and $R_{v'}$.

$$v = m - 1 - \lfloor \frac{j - i - 2}{g} \rfloor, \tag{1}$$

$$v^{'} = \lceil \frac{i + 2}{g} \rceil - 1. \tag{2}$$

Figure 2 presents the global link connections in a dragonfly network with nine groups, where each group contains four routers. Each router can be connected to up to $g = 2$ separate groups. Figure 2 presents the global link connections between any pair of groups. All local connections in the same group are not presented. The scheme to connect global channels for 1D dragonfly networks can be applied to dragonfly networks with each group established by a 2D flattened butterfly, such as, the Cray Cascade system [3] and Slingshot [15].

We give each unidirectional link a unique label plus or minus by the identities (IDs) of source and sink. Each router can be represented by a 2-element tuple $(a, b)$, where $a$ and $b$ are the group ID, and router ID, respectively. A link $(s, d)$ with $s$ $(a_1, b_1)$, and $d$ $(a_2, b_2)$ in the same group is *plus* if $b_1 < b_2$, otherwise it is *minus*. A global link $(s, d)$, that connects nodes $s \in G_{a_1}$ and $d \in G_{a_2}$, is *plus* if $a_1 < a_2$; otherwise, the global link $(s, d)$ is *minus*.

The minus-first routing (MFR) algorithm was proposed in [18,22] for dragonfly networks without any VCs. Assume that the dragonfly network is connected as stated above. The main idea of the MFR algorithm is: any packet cannot be delivered across a plus hop unless all minus hops have been traversed.

The MFR algorithm is a partially adaptive routing scheme, which can be enhanced to a minimum routing by using a simple flow control scheme. Each input port for local link contains two indistinguishable buffers, which is enough to keep the whole packet. Our method classifies packets as safe or unsafe: A safe packet at a router can be delivered to the destination by the baseline routing scheme MFR provided hops; otherwise, the packet is unsafe.

The flow controlled minimum routing requires two buffers that are indistinguishable. The proposed all-to-all broadcast algorithms are completed by multiple unicast steps, where each unicast step conforms to the flow controlled minimum routing scheme. A message is delivered along an MFR path if it is delivered across one or more minus hops first, followed by one or more plus hops.

An unsafe packet is kept in a separate buffer. Each input port keeps two buffers which can be written and read directly, however, at most one packet kept in the same input port can be delivered via any output channel. At most one unsafe packet can be kept at the same input port. The flow controlled minimum routing algorithm can forward a safe packet to the next hop that conforms to the MFR algorithm if one of the following conditions can be satisfied: (1) the input port of the next hop contains one empty buffer, and the next hop conforms to the MFR algorithm, (2) two empty buffers at the input port are available, (3) the input port contains one empty buffer and one safe packet no matter whether the next hop conforms to the MFR algorithm. More details on the baseline routing algorithm MFR and the flow controlled minimum routing algorithm can be found in [21].

**Algorithm 1.** group-first-all-to-all()

**Input:**
The dragonfly network, the normalized applied load;

**Output:**
Each processor deliver the message to all processors;

1: Order the groups into an MFR chain $G$.
2: For each router $r \in G$, call *deliver*$(r, G)$ to deliver a message from $r$ to all groups in $G$.
3: **for** each group $G_i \in G$ with router set $D_i$ **do**
4: call *forward*$(v_i, D_i)$ to multicast the message from $v_i$ to all routers $D_i$ in the group, where $v_i$ is the router that received the message in the group-level stage.
5: **end for**

**Algorithm 2.** deliver$(c, G)$

**Input:**
Coordinates of the source node $s$;

**Output:**
Deliver the message to all processors in the network;

1: **if** $|G| = 2$ **then**
2: deliver the message to the remaining group via an MFR path, exit;
3: **end if**
4: divide $G$ into two equal subsets $G'$ and $G''$.
5: **if** $c$ is in the lower half $G''$ **then**
6: deliver the message from $c$ to a router $c_1$ in the group in the upper half $G'$ with the lowest group label; call *deliver*$(c, G')$ at $c$, and call *deliver*$(c_1, G'')$ at $c_1$.
7: **end if**
8: **if** $c$ is in the upper half $G^{'}$ of $G$ **then**
9: deliver the message from $c$ to the router $c_2$ in the group with the highest label in the lower half $G''$;
10: call *deliver*$(c, G')$ at $c$, and call *deliver*$(c_2, G'')$ at $c_2$.
11: **end if**

## 3 All-to-All Broadcast Using One-to-All Broadcast Schemes

We proposed two all-to-all broadcast algorithms using the one-to-all broadcast schemes in [22]: (1) group-first, and (2) router-first. The group-first broadcast scheme directly delivers the message from the source router to all groups in the network first, which is forwarded to all routers in the same group. The router-first broadcast scheme sends the broadcast message from the source to all routers in the source group first, which is delivered from separate routers in the source group to all groups concurrently. The message is finally forwarded to all routers in each group.

Unicast-based broadcast algorithms require no extra hardware support if we implement broadcast by using the baseline routing scheme. The partially adaptive routing scheme called minus-first routing (MFR) is used as the baseline

**Algorithm 3.** forward($v, D$)

**Input:**
Coordinates of the current node $c$;
Coordinates of the destination list $D$;
**Output:**
Deliver the message to all processors in $D$;
1: **if** $|D| = 2$ **then**
2: Forward the message to the other router in $D$.
3: **end if**
4: Divide $D$ into two equal subsets $D_1$ and $D_2$, exit.
5: **if** $v$ is in the first half $D_1$ **then**
6: deliver the message from $v$ to the first router $v_2$ in the latter half $D_2$; call *forward*($v_2, D_2$) at $v_2$, and call *forward*($v, D_1$) at $v$;
7: **end if**
8: **if** $v$ is in the latter half $D_2$ **then**
9: deliver the message from $v$ to the last node $v_1$ in the first half $D_1$; call *forward*($v_1, D_1$) at $v_1$, and call *forward*($v, D_2$) at $v$.
10: **end if**

routing scheme, which can be replaced by a fully adaptive routing algorithm or the flow-controlled minimum routing one in [21]. However, we think that the MFR algorithm is enough for the router-first all-to-all broadcast algorithm and the message combining all-to-all broadcast algorithm because all paths to deliver packets conforms the minimum MFR paths.

Just like the original group-first and router-first broadcast schemes, the all-to-all broadcast algorithms based on the group-first and router-first broadcast schemes are also implemented by unicast steps, which requires no hardware support. They have the following features: (1) all routers receive the message via minimum feasible paths, (2) no router receives the message more than once. Our all-to-all broadcast algorithms are implemented by unicast steps.

The general framework of the group-first all-to-all broadcast algorithm (GFA2A, for short) is presented in Algorithm 1. The groups in the network are ordered into an MFR chain first. For each router $r$, the procedure *deliver*($r, G$) forwards a message from $r$ to all groups in $G$.

The procedure *deliver*($r, G$) as presented in Algorithm 2 is a recursive one. The group set $G$ is equally partitioned into two subsets $G'$ and $G$", where the broadcast message is recursively delivered inside $G'$ and $G$". There may exist some paths for a unicast step that do not conform to the MFR algorithm. However, the length of a unicast path for the procedure *deliver*($r, G$) is at most two in the group-first based all-to-all broadcast algorithm.

After the broadcast message from a single router has reached all groups. Assume that a router $v$ receives the message, and $D$ contains all routers in the same group. Algorithm 3 recursively delivers the message at $v$ to all routers in the same group. The path for each unicast step is always an MFR path because all routers are directly connected.

**Fig. 3.** Delivering a packet inside a group for the router-first A2A broadcast algorithm: (a)-(h) scattering from each of the routers in the same group.

The general framework for the router-first based all-to-all broadcast algorithm (RFA2A, for short) is presented in Algorithm 4. For each router $r$ in the network, call *forward*($r, G_r$) as presented in Algorithm 3 to deliver the message from $r$ to all other routers in the router group $G_r$. Figure 3 presents the separate multicast schemes for different routers in the same group. As shown in Fig. 3(a) presents the multicast scheme to deliver a packet to all eight routers. It takes 3 unicast steps to complete the multicast.

Figure 3(e) shows the multicast details for the router $R_4$. The packet is delivered to the last router $R_3$ of the first half in the first unicast step. The second half of the router set is further equally partitioned, and the packet is delivered to the first router of the second half from $R_4$ in the second unicast step. The packet is simultaneously delivered to $R_1$ from $R_3$ in the second unicast step. The packet is delivered to $R_0$, $R_2$, $R_5$, and $R_7$ simultaneously from $R_1$, $R_3$, $R_4$, and $R_6$, respectively in the third unicast step.

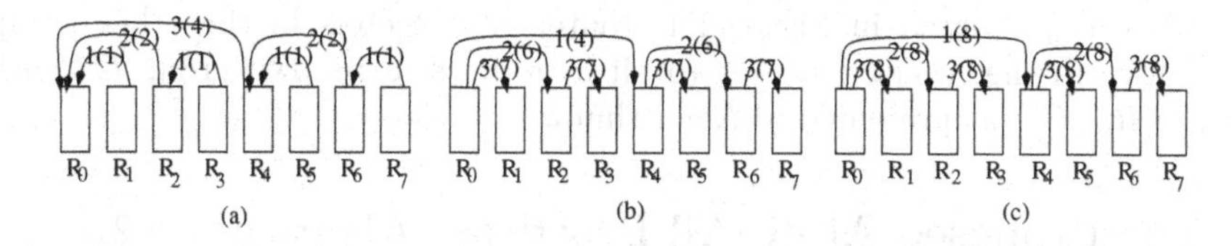


**Fig. 4.** Collection and scattering inside a router group for the proposed A2A algorithm: (a) collection, (b) scattering, and (c) multicasting after receiving packets from other groups.

**Algorithm 4.** router-first-A2A()

**Input:**
Coordinates of the source node $v$;
**Output:**
Deliver the message to all processors in the network;
1: For each router $r$ in the network, call *forward*$(r, G_r)$ as presented in Alg. 3 to deliver the message from $r$ to all other routers in the router group $G_r$.
2: **for** each router $v$ in the network **do**
3: Call *urouter*$(v, D)$ to multicast the $m$ messages from $R_v$ to all routers in the groups that are connected to $v$ via global channels as shown in Fig. 3.
4: **end for**
5: **for** each router $R_{v'} \in G_j$ **do**
6: Call *forward*$(R_{v'}, G_j)$ to multicast each of the $g \cdot m$ messages, that $R_{v'}$ receives, to all other routers in $G_j$.
7: **end for**

**Algorithm 5.** *urouter*$(s,D)$

**Input:**
Coordinates of $s$ and $g$ routers directly connected to $s$ by global channels;
**Output:**
Deliver the message to the $g$ routers in $D$;
1: Let $|D| = 1$, deliver the message from $s$ to the single router via a minimum MFR path;
2: $|D| \geq 2$, divide the $g$ routers in $D$ into two subsets $D_1$ and $D_2$ with $||D_1|-|D_2|| \leq 1$, which are routers in $g/2$ groups with the least IDs and $g/2$ groups with greater IDs;
3: Let $c_2$ be the router in the group with the least ID in $D_2$. Deliver the message from $s$ to $c_2$, and call *urouter*$(c_2, D_2 - \{c_2\})$ at $c_2$;
4: Let $c_1$ be the router in the group with the greatest ID in $D_1$. Deliver the message from $s$ to the node $c_1$, call *urouter*$(c_1, D_1 - \{c_1\})$ at $c_1$.

The RFA2A algorithm delivers the packet from $r$ to all $g = m/2$ routers in other groups after receiving a packet from any other router in the same group by using the recursively procedure *urouter*$(r, D)$ as shown in Algorithm 5, where $D$ contains the $g$ routers, that are directly connected to router $r$, in other groups. All the unicast steps for the recursive procedure *urouter*$(r, D)$ follows the MFR algorithm as presented in Fig. 5. The router $R_{v'}$ receives in the other group $G_j$ multicasts the received packet to all $m$ routers in its own group $G_j$ with *forward*$(R_{v'}, G_j)$ as presented in Algorithm 3.

## 4 The Proposed All-to-All Broadcast Algorithm A2A

The general framework of the proposed all-to-all broadcast algorithm A2A is presented in Algorithm 6. The A2A contains four separate phases: (1) packet collection inside a router group, (2) packet scattering in the group, (3) delivery of all packets at the router to all directly connected routers in the other groups,

**Algorithm 6.** A2A()

**Input:**
The dragonfly network;
**Output:**
Each processor deliver a message to all processors;
1: For each router $r \in G_i$, call *collect*$(r, G_i)$ to collect messages at $r$ for all routers in $G_i$ as shown in Fig. 4(a).
2: **for** each group $G_i$ with router set $D_i$ **do**
3: call *scatter*$(r, D_i)$ to scatter the messages at the selected router $r$, that collected in step 1, to all routers in $D_i$ as presented in Fig. 4(b).
4: **end for**
5: **for** each router in the network **do**
6: Deliver the $m$ messages at $r$ received from all routers in the its group by urouter($r$,$G$) to all $g = m/2$ groups via the global channels;
7: **end for**
8: **for** each router $r \in G_j$ **do**
9: Forward all $m$ messages received at $r$ to all other routers in the group $G_j$ as presented in Fig. 4(c).
10: **end for**

and (4) each router scattering all packets received from all $g = m/2$ directly connected routers in other groups to all routers in its own group.

The packet collection phase collects all $m$ packets along the selected tree and keeps them at the root router of the tree. The packet scattering phase forwards all packets at the root of the tree to each router in the group. The third phase for each router in the network forwards all packets received from all routers inside its group to all routers in other groups, which are directly connected to it. In phase four, each router scatters all $g \cdot m$ received packets to all routers in its own group. Totally, each router receives $g \cdot m \cdot m + m - 1$ packets.

As shown in Fig. 4(a), we give an example for a group with eight routers to collect packets. Each leaf in the multicast tree delivers the packet to its predecessor. The predecessor delivers the packet and its own packet to its own predecessor. The process continues until the root has been reached. As shown in Fig. 4(a), each data contains two number $a(b)$, where $a$ and $b$ represent the unicast step, and the number of packets. The leaves $R_1$, $R_3$, $R_5$, and $R_7$ deliver a single packet to their predecessors $R_0$, $R_2$, $R_4$, and $R_6$ in unicast step 1, respectively. $R_2$ and $R_6$ sends two packets to their predecessors $R_0$ and $R_4$ in unicast step 2, respectively. In unicast step 3, $R_4$ sends four packets to $R_0$. Finally, $R_0$ receives eight packets.

The root router $R_0$ scatters all $m$ (it is eight in Fig. 4) packets to all the $m$ routers in the same group. The root router $R_0$ delivers four packets to $R_4$ as shown in Fig. 4(b) in the first unicast step, where the four packets are received from $R_0$-$R_3$ in the process of packet collection. Router $R_0$ delivers six packets to $R_2$ in the second unicast step, which include the packets from $R_4$-$R_7$, $R_0$ and $R_1$. Simultaneously, router $R_4$ delivers six packets to $R_6$ in the second unicast step, which includes four packets from $R_0$-$R_3$, $R_4$ and $R_5$. In the third unicast step, routers $R_0$, $R_2$, $R_4$, and $R_6$ deliver seven packets to $R_1$, $R_3$, $R_5$, and $R_7$, respectively. The seven packets include packets from all other routers.

After all $m$ packets have been delivered to all $m$ routers in the same group, each router keeps $m$ separate packets. It is required that each router deliver the $m$ packets to the $g = m/2$ directly connected routers in other groups. For each router $r$, the procedure urouter($r$,$G$) is used to deliver the $m$ packets to all $g = m/2$ routers in other groups via the global channels.

As shown in Fig. 5, we present the case when $g = 4$. Figure 5(a) shows the case when all groups connected to the router $r$ have lower group labels than $G_s$. Figure 5(b) shows the case when $g/2$ groups connected to the router $r$ have lower group labels than $G_s$, and $g/2$ groups connected to the router $r$ have higher group labels than $G_s$. Figure 5(c) shows the case when all groups connected to the router $r$ have higher group labels than $G_s$. In all cases in Fig. 5, $m$ packets are packetized into a single big packet, which saves $m - 1$ start-up latency at the source and $m - 1$ receipt latency at the destination compared to the GFA2A and RFA2A algorithms.

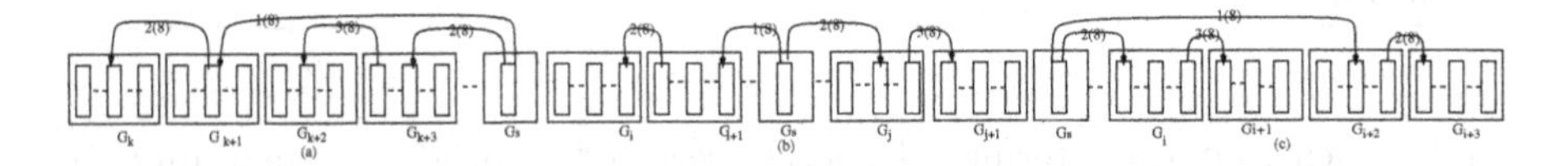

**Fig. 5.** The A2A algorithm: (a) all groups are before the source group, (b) the source group is in the middle, and (c) all groups are after the source group.

In all cases in Fig. 5 for two consecutive unicast steps, all packets can be delivered along MFR paths. As presented in Fig. 5(a), the eight packets are delivered together from $G_s$ to $G_{k+1}$ in the first unicast step, which are delivered from $G_{k+1}$ to $G_k$ in the second unicast step. In both cases, the packets are delivered along MFR paths.

For each router $r$ in the network, it multicasts all $m$ packets to all routers in its own group after it has received all $m$ packets as shown in Fig. 5(c). Just like Fig. 3, each router multicasts the received packets along a separate multicast tree. As shown in Fig. 5(c), our method multicasts all $m$ packets together in $\ln m$ unicast steps to all $m$ routers in the same group. It is shown that all $m$ packets are delivered together in all unicast steps. This can reduce the number of start-ups from $m$ to one, and receipts from $m$ to one for each unicast step compared to the RFA2A algorithm. Based on the proposed A2A algorithm, each router multicasts all $m$ packets to all routers in the same group immediately after received them from any other group. It is not necessary to wait until all $g \cdot m$ packets have been received from other groups.

## 5 Conclusions

New deadlock-free unicast-based all-to-all broadcast algorithms were proposed for dragonfly networks without any special hardware support. The proposed all-to-all broadcast algorithms work under the unicast environment, that is, each input port requires two indistinguishable buffers. An all-to-all broadcast delivers a message from each router to all other routers. The group-first and router-first one-to-all broadcast algorithms [22] were used to implement all-to-all broadcast, named GFA2A and RFA2A, respectively. A new all-to-all broadcast algorithm named A2A is proposed by using message combining. It collects all messages from all the routers in the same group to a single router first, which are forwarded to all the routers in the same group. Packets at each router are directly delivered to routers in other groups in a recursive way. Each router in the network forwards messages to all other routers in the same groups after receiving all messages from other groups.

## References

1. Boppana, R.V., Chalasani, S., Raghavendra, C.S.: Resource deadlocks and performance of wormhole multicast routing algorithms. IEEE Trans. Parallel Distrib. Syst. **9**(6), 535–549 (1998)
2. Dorier, M., Mubarak, M., Ross, R., Li, J.K., Carothers, C.D., Ma, K.L.: Evaluation of topology-aware broadcast algorithms for dragonfly networks. In: Proceedings of International Conference on Cloud Computing, pp. 40–49 (2016)
3. Faanes, G., et al.: Cray cascade: a scalable HPC system based on a dragonfly network. In: Proceedings of International Conference on for High-Performance Computing, Networking, Storage and Analysis, Article no. 103, December 2012
4. Garcia, M., et al.: On-the-fly adaptive routing in high-radix hierarchical networks. In: Proceedings of International Conference on Parallel Processing, pp. 280–288 (2012)
5. Hoefler, T., Mehlan, T., Mietke, F., Rehm, W.: A survey of barrier algorithms for coarse grained supercomputers. Chemnitzer Informatik Berichte, vol. 04, no. 03, presented in Chemnitz, Germany, December 2004. ISSN 0947-5152
6. Jiang, N., Kim, J., Dally, W.J.: Indirect adaptive routing on large scale interconnection networks. In: Proceedings of International Symposium on Computer Architecture, pp. 220–231 (2009)
7. Jiang, N., Kim, J., Dally, W.J.: Gossiping on meshes and tori. IEEE Trans. Parallel Distrib. Syst. **9**(6), 513–525 (1998)
8. Kim, J., Dally, W.J., Scott, S., Abts, D.: Technology-driven, highly-scalable dragonfly topology. In: Proceedings of International Symposium on Computer Architecture, pp. 77–88 (2008)
9. Lin, X., Ni, L.M.: Multicast communication in multicomputer networks. IEEE Trans. Parallel Distrib. Syst. **4**(10), 1105–1117 (1993)
10. Lin, X., McKinley, P.K., Ni, L.M.: Deadlock-free multicast wormhole routing in 2D mesh multicomputers. IEEE Trans. Parallel Distrib. Syst. **5**(8), 793–804 (1994)
11. Maglione-Mathey, G., Yebenes, P., Escudero-Sahuquillo, J., Garcia, P.J., Quiles, F.J., Zahavi, E.: Scalable deadlock-free deterministic minimal-path routing engine for infiniband-based dragonfly networks. IEEE Trans. Parallel Distrib. Syst. **29**(1), 183–197 (2018)
12. McKinley, P.K., Xu, H., Esfahanian, A.-H., Ni, L.M.: Unicast-based multicast communication in wormhole-routed networks. IEEE Trans. Parallel Distrib. Syst. **5**(12), 1252–1265 (1994)

13. Navaridas, J., Miguel-Alonso, J., Ridruejo, F.J.: On synthesizing workloads emulating MPI applications. In: Proceedings of International Symposium on Parallel and Distributed Processing (2008). https://doi.org/10.1109/IPDPS.2008.4536473
14. Panda, D.K., Singal, S., Kesavan, R.: Multidestination message passing in wormhole passing in wormhole $k$-ary $n$-cube networks with base routing conformed paths. IEEE Trans. Parallel Distrib. Syst. **10**(1), 76–96 (1999)
15. De Sensi, D., Di Girolamo, S., McMahon, K.H., Roweth, D., Hoefler, T.: An in-depth analysis of the Slingshot interconnect. In: Proceedings of International Conference on for High-Performance Computing, Networking, Storage and Analysis (2020). https://doi.org/10.1109/SC41405.2020.00039
16. Suh, Y.-J., Valamanchili, S.: All-to-all communication with minimum start-up costs in 2D/3D tori and meshes. IEEE Trans. Parallel Distrib. Syst. **9**(5), 442–458 (1998)
17. Thakur, R., Gropp, W.D.: Improving the performance of collective operations in MPICH. In: Dongarra, J., Laforenza, D., Orlando, S. (eds.) EuroPVM/MPI 2003. LNCS, vol. 2840, pp. 257–267. Springer, Heidelberg (2003). https://doi.org/10.1007/978-3-540-39924-7_38
18. Xiang, D., Zhang, Y., Shan, S., Xu, Y.: A fault-tolerant routing algorithm design for on-chip optical networks. In: Proceedings of Symposium on Reliable Distributed Systems, pp. 1–9 (2013)
19. Xiang, D., Zhang, Y.: Cost-effective power-aware core testing in NoCs based on a new unicast-based multicast scheme. IEEE Trans. Comput.-Aided Des. Integr. Circuits Syst. **30**(1), 135–147 (2011)
20. Xiang, D., Chakrabarty, K., Fujiwara, H.: Multicast-based testing and thermal-aware test scheduling for 3D ICs with a stacked network-on-chip. IEEE Trans. Comput. **65**(9), 2767–2779 (2016)
21. Xiang, D., Li, B., Fu, Y.: Fault-tolerant adaptive routing in dragonfly networks. IEEE Trans. Dependable Secur. Comput. **16**(2), 259–271 (2019)
22. Xiang, D., Liu, X.: Deadlock-free broadcast routing in dragonfly networks without virtual channels. IEEE Trans. Parallel Distrib. Syst. **27**(9), 2520–2532 (2016)

# An Efficient Algorithm for Enumerating Longest Common Increasing Subsequences

Chun Lin(✉), Chao-Yuan Huang, and Ming-Jer Tsai

National Tsing Hua University, Hsinchu, Taiwan
s108062571@m108.nthu.edu.tw

**Abstract.** The longest common increasing subsequence (LCIS) problem is the combination of two classic problems in algorithms: the longest increasing subsequence (LIS) problem and the longest common subsequence (LCS) problem. In this paper, we propose an algorithm that finds every LCIS of two sequences $a, b$ of length $n$ in $O(n + \sigma + I_a)$ time and space, where $\sigma$ denotes the size of the alphabet set and $I_a$ the total number of increasing subsequences contained in $a$ (thus, the running time is output-sensitive). Our algorithm employs the trie and some simple data structures, and thus is implementation-wise simple. In addition, it can be proved that our algorithm is optimal in time complexity when $\sigma \leq \log_2 n$.

**Keywords:** LCIS · Trie · Data structure

## 1 Introduction

The longest common increasing subsequence (LCIS) problem can be formulated as follows: Given a sequence $a = a_1, a_2, \cdots, a_n$, a sequence $a_{i_1}, a_{i_2}, \cdots, a_{i_k}$ is a subsequence of $a$ if $1 \leq i_j < i_{j+1} \leq n$ for all $1 \leq j < k$. And, given two sequences $a, b$ of length $n$, the LCIS problem asks for a longest common subsequence of $a, b$ that is strictly increasing.

This problem can be seen as a combination of the longest increasing subsequence (LIS) problem and the longest common subsequence (LCS) problem, and was first introduced by Yang et al. [6] and then applied to the whole genome alignment by Chan et al. [1] in 2005. Yang et al. and Chan et al. proposed algorithms of $O(n^2)$ and $O(min(r \log \sigma, n\sigma + r) \log \log n + Sort_n)$ time, respectively, where $Sort_n$ denotes the time required to sort input sequences $a, b$, and $r$ the number of ordered pairs $(i, j)$ such that $a_i = b_j$. In 2006, Sakai presented a linear-space and $O(n^2)$-time algorithm using a divide-and-conquer approach [5]. In 2011, Kutz et al. designed an algorithm of $O(n)$ space and $O(nl \log \log \sigma + Sort_n)$ time [3], where $l$ denotes the length of the LCIS of $a, b$. And, for small alphabet set, algorithms of $O(n)$ and $O(n \log \log n)$ time were proposed for $\sigma = 2$ and $\sigma = 3$, respectively. In 2016, Zhu et al. proposed an


C.-Y. Chen et al. (Eds.): COCOON 2021, LNCS 13025, pp. 25–36, 2021.
https://doi.org/10.1007/978-3-030-89543-3_3

$O(n^2)$-time and linear-space algorithm [7]. Recently, in 2020, Lo et al. proposed an algorithm of $O(n + l(n-l)\log\log\sigma)$ time and $O(n)$ space [4], and Duraj presented the first algorithm of subquadratic time [2].

The rest of this paper is organized as follows: In Sect. 2, the proposed algorithm is presented. In Sect. 3, the correctness and complexity are analyzed. Finally, we conclude this paper in Sect. 4.

## 2 The Proposed Algorithm

In this section, three assumptions are first introduced. Subsequently, we outline the proposed algorithm, followed by a step-by-step explanation along with the pseudocode. Finally, an example is given.

### 2.1 Assumptions

**Input Format.** Given the size of the alphabet set $\sigma$, we assume the alphabet set consists of integers $0, 1, ..., \sigma - 1$, i.e., each integer in the input sequences $a, b$ is in $\{0, 1, ..., \sigma - 1\}$.

**Fast Computation.** We assume that the bitwise shift (or bitwise OR) on one (or two) binary encoded data of no more than $\sigma$ bits can be done in $O(1)$ time.

**Constant Space.** We assume that a bitstring of length up to $\sigma$ takes $O(1)$ space.

Remark that when the desired input format is not satisfied, one can map integers in $a, b$ to the integers in $[0, \sigma - 1]$ without changing the order of integers in $O(n \log \sigma)$ time using a balanced binary search tree.

### 2.2 Algorithm Overview

The main procedure of the proposed algorithm (Algorithm 1) involves building a trie $T$ containing the information of every increasing subsequence in $a$. Let $IS_u$ denote the *increasing subsequence* with *binary encoding* $u$, i.e., the $i$-th bit in $u$ is 1 if and only if $i$ is contained in $IS_u$. For example, $IS_{10100100}$ denotes the increasing subsequence $[2, 5, 7]$ as $\sigma = 8$. Then, $T$ has the following properties:

1. A node $T_u$ associated with the length $l_u$ of $IS_u$ exists in $T$ to denote an *increasing subsequence* $IS_u$ if and only if $IS_u$ is found in $a$. Also, the binary encoding $u$ of $IS_u$ is stored in $T_u$ to help retrieval of sequence information.
2. A directed edge associated with $x \in \{0, 1, ..., \sigma - 1\}$ from $T_u$ to $T_v$, denoted by the tuple $(T_u, T_v, x)$, exists in $T$ if and only if $IS_v$ is the concatenation of $IS_u$ and $x$.

After building $T$, a similar trie-building process is run for sequence $b$; but instead of building a new trie for $b$, we walk along the nodes of $T$ that denote the common increasing subsequences of $a, b$, and meanwhile record all the found longest common increasing subsequences of $a, b$.

For the complexity of Algorithm 1, the initialization step takes $O(\sigma+n)$ time. Building $T$ takes $O(n + I_a)$ time by using additional data structures that take $O(\sigma+I_a)$ space. And, walking on $T$ takes $O(n+I_a)$ time. To sum up, Algorithm 1 has space and time complexity of $O(n + \sigma + I_a)$.

### 2.3 Detailed Description

See Algorithm 1 for the pseudocode. Algorithm 1 consists of 5 parts as follows.

**Input/Output.** Algorithm 1 takes two sequences $a, b$, the length $n$ of $a, b$, and the size of the alphabet set $\sigma$ as the inputs, and outputs a list $L$ containing the binary encoding of every LCIS of $a, b$.

**Initialization for First Loop (Lines 2–19).** Firstly, build an array $Cnt$, where $Cnt[i]$ is the frequency of $i$ in $a$ for all $i \in \{0, 1, ..., \sigma - 1\}$. This can be done in $O(n)$ time by simply scanning $a$ once. Secondly, build a doubly linked list $K$ of nodes to store every integer $i$ with $Cnt[i] > 0$ in an increasing order (from $i = 0$ to $i = \sigma - 1$). Then, a pointer array $M$ of size $\sigma$ is created. And, for each integer $i$ stored in $K$, the pointer to the node containing $i$ in $K$ is stored in $M[i]$ so that the node can be removed from $K$, if necessary, in $O(1)$ time. Thirdly, build the root node $T_0$ of $T$, which denotes an empty sequence, and set $l_0$ to 0. Then, for every $i$ with $Cnt[i] > 0$, create a trie node $T_{2^i}$ containing the binary encoding of the sequence $[i]$, set $l_{2^i} = 1$, and add an edge $(T_0, T_{2^i}, i)$ from $T_0$ to $T_{2^i}$. Finally, build $\sigma$ queues $Next_0, \cdots, Next_{\sigma-1}$, where each queue supports $O(1)$ push and pop (for our purpose, one can also use different data structures such as stacks or dynamic arrays, as long as they support push and pop in $O(1)$ time). Let $A_u$ be the address of $T_u$. Then, for all $i$, queue $Next_i$ initially contains $A_{2^i}$ if $Cnt[i] > 0$, and is left empty otherwise.

**First Loop (Lines 20–31).** Algorithm 1 iterates the following two steps when sequence $a$ is scanned one by one from left to right. Firstly, for the $i$-th integer $a_i$ in $a$, we decrease $Cnt[a_i]$ by 1. Secondly, in the inner loop, Algorithm 1 iterates the following two substeps until queue $Next_{a_i}$ is empty. First pop $A_u$ from queue $Next_{a_i}$ and get $u$ from $T_u$. Then, in the (yet deeper) inner loop, for each integer $x$ with $a_i < x < \sigma$ and $Cnt[x] > 0$ (every such $x$ can be found efficiently using $K$), first create a new node $T_v$ of $T$, set $l_v$ to $l_u + 1$, add an edge $(T_u, T_v, x)$ from $T_u$ to $T_v$, where $v = u + 2^x$, and then push $A_v$ into queue $Next_x$. At last, at the end of the $i$-th iteration, remove the node containing $a_i$ from $K$ if $Cnt[a_i]$ has become 0.

**Algorithm 1: LCIS**

```
   Input: (n, σ, a, b): the length of each sequence, the size of the alphabet set,
          the two sequences
   Output: L: a list containing the binary code of every LCIS of (a, b)
 1 begin
 2     Cnt ← new 1D integer array of size σ;
 3     M ← new 1D pointer array of size σ;
 4     for i ← 0 to σ − 1 do
 5         Cnt[i] ← 0;
 6     for i ← 1 to n do
 7         Cnt[a_i] ← Cnt[a_i] + 1;
 8     K ← new doubly linked list;
 9     create trie node T_0;
10     l_0 ← 0;
11     for i ← 0 to σ − 1 do
12         Next_i ← new queue;
13         if Cnt[i] > 0 then
14             add the node K_i containing i to K;
15             M[i] ← address of K_i;
16             create trie node T_{2^i};
17             l_{2^i} ← 1;
18             add edge (T_0, T_{2^i}, i);
19             push A_{2^i} into Next_i;
20     for i ← 1 to n do
21         Cnt[a_i] ← Cnt[a_i] − 1;
22         for A_u ∈ Next_{a_i} do
23             pop A_u from Next_{a_i} and get u from T_u;
24             for x ← a_i + 1 to σ − 1 in K do
25                 v ← u + 2^x;
26                 create trie node T_v;
27                 l_v ← l_u + 1;
28                 add edge (T_u, T_v, x);
29                 push A_v into Next_x;
30         if Cnt[a_i] = 0 then
31             remove the node containing a_i from K;
32     len ← 0;
33     L ← new list;
34     insert 0 into L;
35     for i ← 0 to σ − 1 do
36         if trie node T_{2^i} exists then
37             push A_{2^i} into Next_i;
38     ...(continued in next page)
```

```
37
38   for i ← 1 to n do
39       for A_u in Next_{b_i} do
40           pop A_u from Next_{b_i} and get u and l_u from T_u;
41           if l_u > len then
42               len ← l_u;
43               empty L;
44           if len = l_u then
45               insert u into L;
46           for every edge (T_u, T_v, x) from T_u do
47               push A_v into Next_x;
48   return the list L;
```

**Initialization for Second Loop (Lines 32–37).** Firstly, set $len$, denoting the length of LCIS of $a, b$ currently found, to 0. Secondly, build a list $L$ to store the binary encoding of every common increasing subsequence (CIS) of length $len$ of $a, b$, where $L$ contains only 0 (the binary encoding of the empty sequence) initially. Thirdly, reuse $Next$ queues and for each queue $Next_i$, push $A_{2^i}$ into queue $Next_i$ if $T_{2^i}$ exists in $T$.

**Second Loop (Lines 38–47).** Algorithm 1 iterates the following step when sequence $b$ is scanned one by one from left to right. For the $i$-th integer $b_i$ in $b$, Algorithm 1 iterates the following substeps until queue $Next_{b_i}$ is empty in the inner loop. First pop one $A_u$ from queue $Next_{b_i}$. Then, since $IS_u$ is a newly found CIS of $a, b$, we may need to update $len$ and $L$ accordingly: 1) if $l_u > len$ (i.e., the length of $IS_u$ is greater than that of any CIS of $a, b$ currently found), empty $L$ and update $len$ to $l_u$, and 2) if $l_u = len$, add $u$ into $L$. Finally, in the (yet deeper) inner loop, push $A_v$ into queue $Next_x$ for each edge $(T_u, T_v, x)$ from $T_u$.

### 2.4 Example

Figures 1a and 1b show the statuses of $K$, $Next$, and $T$ on the termination of the initialization and iteration 1, respectively, of the first loop of Algorithm 1 for $a = [1, 4, 1, 0, 3]$ and $\sigma = 5$. During the execution of the initialization, $Cnt[0] = Cnt[3] = Cnt[4] = 1$, $Cnt[1] = 2$, and $Cnt[2] = 0$ since the frequencies of integers $0, 1, 2, 3, 4$ are $1, 2, 0, 1, 1$, respectively. And, since $Cnt[i] > 0$ for $i = 0, 1, 3, 4$, the nodes storing integers $0, 1, 3, 4$ are doubly linked in sequence in $K$, the nodes $T_1$, $T_{10}$, $T_{1000}$, and $T_{10000}$ (which contains the binary encodings of integers $0, 1, 3, 4$, respectively) are created in $T$, and $A_1$, $A_{10}$, $A_{1000}$, and $A_{10000}$ (which are the addresses of $T_1$, $T_{10}$, $T_{1000}$, and $T_{10000}$, respectively) are contained in $Next_0$, $Next_1$, $Next_3$, and $Next_4$, respectively. In iteration 1, $a_1 = 1$. Thus, $Cnt[1]$ is

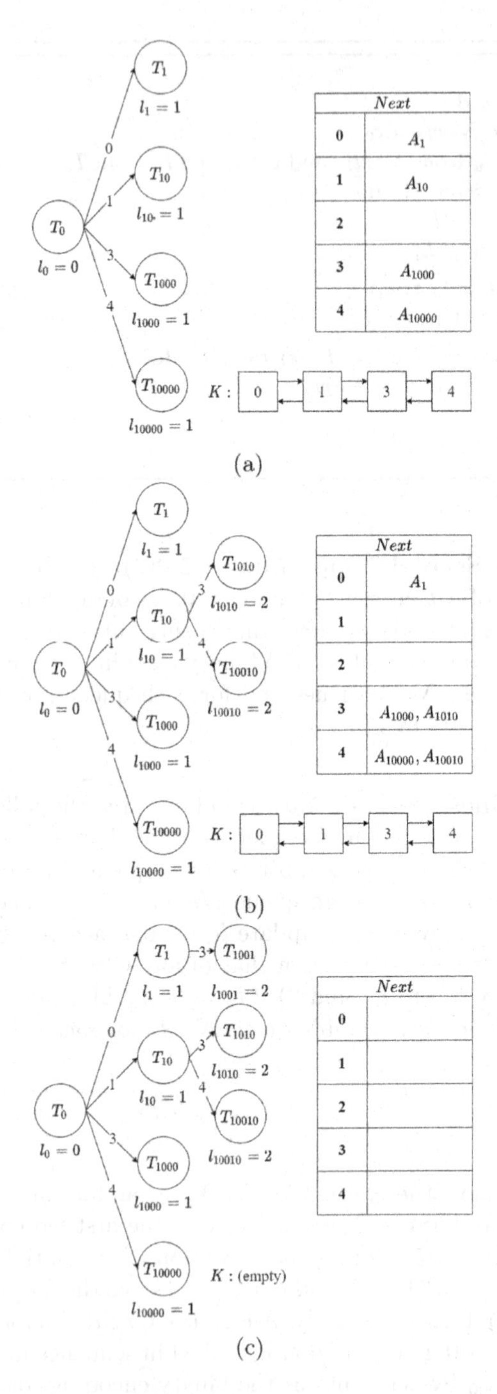


**Fig. 1.** The statuses of $K$, $Next$, and $T$ on the termination of (a) the initialization, (b) iteration 1, (c) the last iteration of the first loop of Algorithm 1 as the input sequence $a$ is $[1, 4, 1, 0, 3]$.

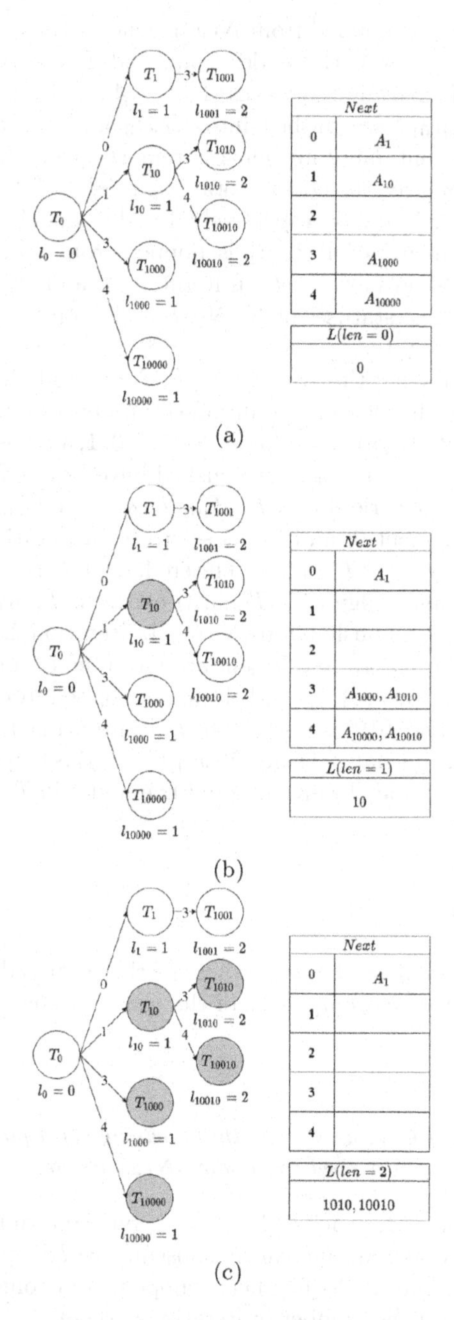


**Fig. 2.** The statuses of $L$, $Next$, and the encountered trie nodes in $T$ on the termination (a) the initialization, (b) iteration 1, (c) the last iteration of the second loop of Algorithm 1 as the input sequence $b$ is $[1, 4, 3, 1, 3]$, where the encountered trie nodes in $T$ are shown in grey.

decreased to 1 and $A_{10}$ is popped from $Next_1$. Due to that $a_1 = 1 < x < 5 = \sigma$ and $Cnt[x] > 0$ for $x = 3, 4$, the nodes $T_{1010}$ and $T_{10010}$ (which contains the binary encodings of increasing sequences $[1, 3], [1, 4]$, respectively) are added to $T$, and $A_{1010}$ and $A_{10010}$ are pushed into $Next_3$ and $Next_4$, respectively. In iteration 2, the node containing integer 4 is removed from $K$ since $a_2 = 4$ and $Cnt[4]$ becomes 0. Similarly, the nodes containing integers 1, 0, and 3 are removed from $K$ in iterations 3, 4, and 5, respectively. In addition, $A_{10000}$ and $A_{10010}$ are popped from $Next_4$ in iteration 2, $A_1$ is popped from $Next_0$, $T_{1001}$ is added to $T$, and $A_{1001}$ is pushed into $Next_3$ in iteration 4, and $A_{1001}$ is popped from $Next_3$ in iteration 5. The statuses of $K$, $Next$, and $T$ on the termination of the first loop is shown in Fig. 1c.

Figures 2a and 2b show the statuses of $L$, $Next$ and the encountered trie nodes in $T$ on the termination of the initialization and iteration 1, respectively, of the second loop of Algorithm 1 for $b = [1, 4, 3, 1, 3]$. For the second loop, initially, $A_1$, $A_{10}$, $A_{1000}$, and $A_{10000}$ are pushed into $Next_0$, $Next_1$, $Next_3$, and $Next_4$, respectively, since trie nodes $T_1$, $T_{10}$, $T_{1000}$, and $T_{10000}$ exist in $T$; also, $L$ contains a single element 0, and $len$ is set to 0. In iteration 1, since $b_1 = 1$, $A_{10}$ is popped from $Next_1$, $T_{10}$ is encountered, and $L$ is updated to contain 10 only. Meanwhile, since edge $(T_{10}, T_{1010}, 3)$ exists in $T$, $A_{1010}$ is pushed into $Next_3$. Similarly, $A_{10010}$ is pushed into $Next_4$. In iteration 2, $A_{10000}$ and $A_{10010}$ is popped from $Next_4$, $T_{10000}$ and $T_{10010}$ are encontered, and $L$ is updated to contain 10010. In iteration 3, $A_{1000}$ and $A_{1010}$ are popped from $Next_3$, $T_{1000}$ and $T_{1010}$ are encontered, and 1010 is inserted to $L$. In iteration 4 (or 5), the statuses of $L$ and $Next$ remains unchanged since $Next_1$ ($Next_3$) is empty. Figure 2c shows the statuses of $L$, $Next$ and the encountered trie nodes in $T$ on the termination of the second loop.

## 3 The Analysis

In this section, we first show the correctness of the proposed algorithm. Subsequently, the time and space complexity of the proposed algorithm is studied.

### 3.1 Correctness

**Lemma 1.** *In the first loop, a non-empty increasing subsequence $IS_u$ of $a$ exists if and only if $A_u$ has been popped from some $Next$ queue.*

*Proof.* It suffices to show for each $i$ $(1 \le i \le n)$, on the termination of iteration $i$ of the first loop, a non-empty increasing subsequence $IS_u$ exists in $a_1, a_2, ..., a_i$ (a prefix of $a$) if and only if $A_u$ has been popped from some $Next$ queue. We show it by induction on the number of iterations executed.

Clearly, on the termination of iteration 1, $[a_1]$ is the only one non-empty increasing subsequence of $a$. Besides, in the initialization of the first loop, Algorithm 1 pushes $A_{2^u}$ into queue $Next_u$ once for each integer $u$ that exists in $a$. Since Algorithm 1 pops all items in queue $Next_{a_1}$ in iteration 1, only $A_{2^{a_1}}$ has been popped on the termination of iteration 1. Thus, we have a basis. We then assume the induction hypothesis: on the termination of iteration $k$, a non-empty increasing subsequence $IS_u$ exists in $a_1, a_2, ..., a_k$ if and only if $A_u$ has been popped from some $Next$ queue. To complete the proof, we only need to show the induction step: on the termination of iteration $k+1$, a non-empty increasing subsequence $IS_u$ exists in $a_1, a_2, ..., a_{k+1}$ if and only if $A_u$ has been popped from some $Next$ queue.

For the *if* part, if $A_u$ is popped from some $Next$ queue before iteration $k+1$, $IS_u$ exists in $a_1, a_2, ..., a_k$ by induction hypothesis, and thus $IS_u$ exists in $a_1, a_2, ..., a_{k+1}$. So, we only need to consider the case where $A_u$ is popped from some $Next$ queue in iteration $k+1$. Note that Algorithm 1 pops all items in queue $Next_{a_{k+1}}$ in iteration $k+1$. Also note that Algorithm 1 only pushes $A_u$ into queue $Next_{a_{k+1}}$ when $IS_u$ ends with $a_{k+1}$. Let $IS_u$ be the concatenation of $IS_v$ and $a_{k+1}$. Clearly, if $IS_v$ is an empty sequence, $IS_u = a_{k+1}$ is an increasing subsequence of $a$. Otherwise, let $IS_v$ end with $a_j$; then, $A_v$ has been popped from some $Next$ queue on the termination of iteration $k$ and $a_j < a_{k+1}$ because otherwise, $A_u$ is not in $Next$ queues in iteration $k+1$ by Algorithm 1. By induction hypothesis, $IS_v$ is an increasing subsequence in $a_1, a_2, ..., a_k$. This implies $IS_u$ is an increasing subsequence in $a_1, a_2, ..., a_{k+1}$, completing the proof of the *if* part.

For the *only if* part, if $IS_u$ does not end with $a_{k+1}$, $IS_u$ exists in $a_1, a_2, ..., a_k$, and thus $A_u$ has been popped from some $Next$ queue on the termination of iteration $k$ by the induction hypothesis. So, we only need to consider the case where $IS_u$ ends with $a_{k+1}$. Since Algorithm 1 pops all items in queue $Next_{a_{k+1}}$ in iteration $k+1$, we only need to show $A_u$ is in queue $Next_{a_{k+1}}$ on the termination of iteration $k$. Let $IS_u$ be the concatenation of $IS_v$ and $a_{k+1}$. Then, if $IS_v$ is an empty sequence, $A_u$ is pushed into queue $Next_{a_{k+1}}$ in the initialization of the first loop. Otherwise, $IS_v$ is a non-empty increasing subsequence in $a_1, a_2, ..., a_k$. Let $IS_v$ end with $a_j$. Since $IS_u$ is an increasing subsequence, we have $a_j < a_{k+1}$. Then, on the termination of iteration $k$, $A_v$ has been popped from some $Next$ queue by induction hypothesis, and then $A_u$ has been pushed into queue $Next_{a_{k+1}}$ due to $a_j < a_{k+1}$ and $Cnt[a_{k+1}] > 0$, completing the *only if* part.

**Theorem 1.** *In the first loop, $IS_u$ is a non-empty increasing subsequence of $a$ if and only if a trie node $T_u$ is created in $T$.*

*Proof.* Note that Algorithm 1 creates a trie node $T_u$ right before $A_u$ is pushed into some $Next$ queue in the first loop. Thus, a trie node $T_u$ is created in $T$ if and only if $A_u$ has been pushed into some $Next$ queue. Besides, $IS_u$ is a non-empty increasing subsequence of $a$ if and only if $A_u$ has been popped from some $Next$ queue by Lemma 1. Thus, to complete the proof, we only need to show $A_u$ has been pushed into some $Next$ queue if and only if $A_u$ has been popped from some $Next$ queue. Clearly, $A_u$ has been pushed into some $Next$ queue if $A_u$ has been popped from some $Next$ queue. On the other hand, suppose $A_u$ is pushed into some $Next$ queue, say $Next_x$, in iteration $j$. Then, $Cnt[x] > 0$ in iteration $j$. This implies $x$ exists in $a_{j+1}, a_{j+2}, \cdots, a_n$. Let $a_k$ be the first integer in $a_{j+1}, a_{j+2}, \cdots, a_n$ such that $x = a_k$. Then, $A_u$ is popped from queue $Next_x$ in iteration $k$. This completes the proof.

**Lemma 2.** *In the second loop, a non-empty common increasing subsequence $IS_u$ of $a, b$ exists if and only if $A_u$ has been popped from some $Next$ queue.*

*Proof.* It suffices to show for each $i$ $(1 \leq i \leq n)$, on the termination of iteration $i$ of the second loop, a non-empty common increasing subsequence $IS_u$ of $a, b$ exists in $b_1, b_2, ..., b_i$ (a prefix of $b$) if and only if $A_u$ has been popped from some $Next$ queue. We show it by induction on the number of iterations executed. In iteration 1, Algorithm 1 pops all items from queue $Next_{b_1}$. Let $u$ be the binary encoding of $b_1$. Then, $b_1$ exists in $a_1, a_2, ..., a_n$ if and only if trie node $T_u$ exists in $T$ by Theorem 1, and thus $b_1$ exists in $a_1, a_2, ..., a_n$ if and only if $A_u$ is pushed into queue $Next_{b_1}$ in the initialization of the second loop. This implies that a non-empty common increasing subsequence $IS_u$ of $a, b$ exists in $b_1$ if and only if $A_u$ has been popped from queue $Next_{b_1}$ in iteration 1. Thus, we have a basis. We omit the induction hypothesis and the proof of the induction step due to their similarities to that of Lemma 1.

**Theorem 2.** *By Algorithm 1, the list L contains exactly the binary encoding of every LCIS of $a, b$.*

*Proof.* By Lemma 2, for every non-empty CIS $IS_u$ of $a, b$, $A_u$ has been popped from some $Next$ queue in the second loop. In Algorithm 1, when $A_u$ is popped from some $Next$ queue, the binary encoding of $IS_u$ is added to $L$ if the length of $IS_u$ is equal to that of the CIS of $a, b$ in $L$, and $L$ is updated to contain only the binary encoding of $IS_u$ if the length of $IS_u$ is greater than that of the CIS of $a, b$ in $L$. This ensures the binary encoding of every LCIS of $a, b$ is contained in $L$.

### 3.2 Complexity

**Lemma 3.** *Every $A_u$ for a non-empty increasing subsequence $IS_u$ of $a$ is pushed into $Next$ queues at most once in (a) the first loop and (b) the second loop.*

*Proof.* The proof of (b) is omitted due to its similarity to that of (a). We show (a) by contradiction. Suppose that $A_u$ is the first one to be pushed into $Next$ queues more than once. Then, $|IS_u|$ must be greater than 1; otherwise, $A_u$ is pushed into $Next$ queues only once in the initialization step.

Let $IS_u$ be the concatenation of $IS_v$ and $x$ (i.e., $IS_u$ ends with $x$). Note that $A_u$ is pushed into queue $Next_x$ right after $A_v$ is popped from some queue. Thus, $A_u$ is pushed into $Next$ queues the second time right after $A_v$ is popped from some queue the second time. This implies $A_v$ is pushed into some queue twice before $A_u$ is pushed into some queue twice. This constitutes a contradiction.

**Theorem 3.** *The time and space complexity of Algorithm 1 is $O(n + \sigma + I_a)$.*

*Proof.* Apart from the input sequences $a, b$ themselves, $K$ and $Cnt$ require $O(\sigma)$ space, and $T$ and $Next$ queues require $O(I_a)$ space by Lemma 3, so the space complexity is $O(n + \sigma + I_a)$. Note that the space complexity can be reduced to $O(n + I_a)$ through removing integers that do not exist in both of $a$ and $b$ from the alphabet set by additional preprocessing before Algorithm 1.

For time complexity, the initialization steps of the first and second loops require $O(n + \sigma)$ time. In the first and second loops, there are $O(I_a)$ queue and trie node operations by Lemma 3. Each queue operation requires $O(1)$ time. And, since the address of $T_u$ is pushed into $Next$ queues, each trie node operation can be achieved in $O(1)$ time. Since Algorithm 1 uses $Cnt$ and $K$ to avoid iterations without queue or trie node operation, the time complexity is $O(n + \sigma + I_a)$.

Remark that due to that $I_a \leq 2^\sigma$, the time complexity of Algorithm 1 is $O(n)$, which is optimal for the LCIS problem, as $\sigma \leq \log_2 n$.

## 4 Conclusion and Discussion

In this paper, we present an algorithm of $O(n+\sigma+I_a)$ time and space complexity to find every LCIS of sequences $a, b$ of length $n$. If the proposed algorithm is run on two computers in parallel, the time complexity can be improved to $O(n + \sigma + min(I_a, I_b))$. When the alphabet set is small, an algorithm of $O(n \log\log n)$ time complexity was proposed in the literature for $\sigma = 3$ [3]. By contrast, the proposed algorithm has $O(n)$ time and space complexity as $\sigma \leq \log_2 n$. For the LCIS problem of $k$ ($k > 2$) sequences, an algorithm of $O(n + \sigma + kI_a)$ time complexity can be obtained through slight modification of the proposed algorithm by just running the second loop for each sequence other than $a$ and keeping track of how many times each node in $T$ is encountered. Whether the proposed algorithm can be modified to better adapt to the cases of more than 2 sequences may be worthy of discussion.

## References

1. Chan, W.T., Zhang, Y., Fung, S.P.Y., Ye, D., Zhu, H.: Efficient algorithms for finding a longest common increasing subsequence. J. Comb. Optim. **13**(3), 277–288 (2006). https://doi.org/10.1007/s10878-006-9031-7
2. Duraj, L.: A sub-quadratic algorithm for the longest common increasing subsequence problem. arXiv:1902.06864 [cs], January 2020
3. Kutz, M., Brodal, G.S., Kaligosi, K., Katriel, I.: Faster algorithms for computing longest common increasing subsequences. J. Discret. Algorithms **9**(4), 314–325 (2011). https://doi.org/10.1016/j.jda.2011.03.013
4. Lo, S.F., Tseng, K.T., Yang, C.B., Huang, K.S.: A diagonal-based algorithm for the longest common increasing subsequence problem. Theor. Comput. Sci. **815**, 69–78 (2020). https://doi.org/10.1016/j.tcs.2020.02.024. https://www.sciencedirect.com/science/article/pii/S0304397520301158
5. Sakai, Y.: A linear space algorithm for computing a longest common increasing subsequence. Inf. Process. Lett. **99**(5), 203–207 (2006). https://doi.org/10.1016/j.ipl.2006.05.005

6. Yang, I.H., Huang, C.P., Chao, K.M.: A fast algorithm for computing a longest common increasing subsequence. Inf. Process. Lett. **93**(5), 249–253 (2005). https://doi.org/10.1016/j.ipl.2004.10.014
7. Zhu, D., Wang, L., Wang, T., Wang, X.: A simple linear space algorithm for computing a longest common increasing subsequence. arXiv:1608.07002 [cs], August 2016

# On Singleton Congestion Games with Resilience Against Collusion

Bugra Caskurlu[1](✉), Özgün Ekici[2], and Fatih Erdem Kizilkaya[1]

[1] TOBB University of Economics and Technology, Ankara, Turkey
{bcaskurlu,f.kizilkaya}@etu.edu.tr
[2] Özyeğin University, Istanbul, Turkey
ozgun.ekici@ozyegin.edu.tr

**Abstract.** We study the subclass of singleton congestion games in which there are identical resources with increasing cost functions. In this domain, we prove that there always exists an outcome that is resilient to weakly-improving deviations by singletons (i.e., the outcome is a Nash equilibrium), by the grand coalition (i.e., the outcome is Pareto efficient), and by coalitions with respect to an *a priori* given partition coalition structure (i.e., the outcome is a partition equilibrium). To our knowledge, this is the strongest existence guarantee in the literature on congestion games when weakly-improving deviations are considered. Our proof technique gives the false impression of a potential function argument but it is a novel application of proof by contradiction.

## 1 Introduction

Game forms are useful mathematical abstractions to study strategic behavior in real-life situations. There is a long tradition of using game forms to analyze real-life multi-agent systems. It goes at least as far back to Wardrop [14], who used a game form to model traffic flow in transportation networks. Ever since, the literature has grown, and today game forms are extensively used to model strategic behavior in a variety of real-life settings where agents interact.

In this paper, we contribute to this line of research by studying the following simple problem: There are $n$ agents and $m$ resources. Each agent has access to a subset (possibly all) of resources. From within her accessible set, the agent's goal is to utilize the "least-crowded" resource, i.e., the one that is used by the smallest number of agents. This game form is a subset of the more general class of singleton congestion games [11]. The additional restriction here is that resources have identical (and increasing) congestion cost functions. Therefore, in the following, we refer to this class of games as *identical singleton congestion games* (ISCGs).

This game form captures the essence of real-life interactions of agents wherein there is collision or interference, as is typical in many settings. For instance, in the

This work is supported by The Scientific and Technological Research Council of Turkey (TÜBİTAK) through grant 118E126.


C.-Y. Chen et al. (Eds.): COCOON 2021, LNCS 13025, pp. 37–48, 2021.
https://doi.org/10.1007/978-3-030-89543-3_4

field of multi-agent navigation, the need for autonomous control of potentially a large number of robots running in the same workspace makes collision avoidance a fundamental issue; see [3]. Figure 1 below presents a motivating toy example. In the figure, robots navigate on a 2-dimensional workspace. At each iteration, a robot either stays put or moves to a neighboring cell. The robot's goal is to avoid crowded cells where collision is more likely, and therefore, at each iteration, the robots play an ISCG. More recently, singleton congestion games have been used to model blockchain competition between miners: With more miners, a miner's likelihood of becoming the first to solve a mining puzzle decreases, and miners' available strategies can change due to political and economic restrictions [1].

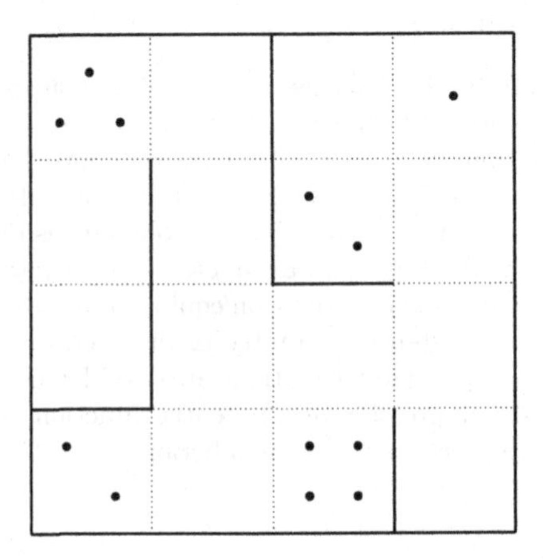

**Fig. 1.** A simple example of a multi-agent navigation problem where dots represent robots and bold lines represent obstacles.

Although they are a restricted form of singleton congestion games, ISCGs draw attention for their simplicity and applications. In this paper, we present another reason for why this game form is noteworthy. The main result of our paper, detailed below, shows that ISCGs give rise to "very stable" outcomes: In an ISCG, there always exists an outcome that is resilient to "weakly-improving deviations" by singletons, the grand coalition, and an *a priori* given partition coalition structure. To our knowledge, this existence guarantee is the strongest one in the literature on congestion games when weakly-improving deviations are considered. Due to a prior finding by Caskurlu et al. [5], we also know that our existence result cannot be extended to the singleton congestion games in general, or even to their restricted form in which every resource is accessible to every agent.

On a technical side, we should note that our proof technique is a novel application of proof by contradiction, although it gives the false sense of a proof by the potential function argument. To ease reading our fairly complicated proof, we should elaborate: In an existence proof, on a finite domain $X$, one needs to show that there exists $x \in X$ such that $x$ satisfies a set of properties, $P$. A proof by the potential function argument proceeds as follows:

- A potential function $f$ is defined on the domain $X$. It is shown that for $x_1 \in X$, if $x_1$ does not satisfy $P$, then there exists $x_2 \in X$ such that $f(x_2) > f(x_1)$. This proves that for $x$ such that $x$ is $f$-maximal in $X$, $x$ satisfies $P$.

Our proof technique in contrast follows the following outline:

- We introduce an asymmetric and transitive relation. For ease in exposition, let it be $\rhd$. As in the potential function argument, we show that $P$ is satisfied by the $\rhd$-maximal elements in $X$. But we show this by way of contradiction and as follows: For $x$ such that $x$ is $\rhd$-maximal in $X$, we suppose that $x$ does not satisfy $P$. Then, we update the set $X$ (we make it smaller) and we show that in the updated set $X$, $x$ is still $\rhd$-maximal and $x$ still does not satisfy $P$. (In our case, $x$ not satisfying $P$ is a non-trivial observation because the set of properties $P$ is a function of $X$.) We iterate and keep on updating the set $X$. Eventually, we obtain that $x$ being $\rhd$-maximal in $X$ leads to a contradiction. This proves that for $x$ such that $x$ is $\rhd$-maximal in $X$, $x$ satisfies $P$.

***Our Results.*** A coalition of agents deviates (i.e., they jointly change their strategies) from an outcome of the game if the deviation results in an outcome where coalition members are better off in the Pareto sense. In other words, we consider "weakly-improving deviations," which lead to no coalition member becoming worse off and at least one coalition member becoming better off. Ideally, we would be interested in a stable outcome that is resilient to weakly-improving deviations by every coalition of agents. An outcome satisfying this property is known as a super-strong equilibrium. However, a super-strong equilibrium does not exist in most game forms, including ISCGs [7]. Instead, we prove the existence of outcomes simultaneously satisfying the following three properties:

**(P1)** Resilience to weakly-improving deviations by singletons (i.e., the outcome should be a **Nash equilibrium**)
**(P2)** Resilience to weakly-improving deviations by the grand coalition (i.e., the outcome should be **Pareto efficient**)
**(P3)** Resilience to weakly-improving deviations by an *a priori* given partition coalition structure (i.e., the outcome should be a **partition equilibrium**)

The property **P1**, requiring that an outcome be a Nash equilibrium, is the simplest notion of stability one can consider, and it is commonly used in game-theoretic studies. In a similar vein, the property **P2**, requiring that an outcome be Pareto efficient, is the most natural notion of efficiency, and it is commonly used in the economics literature. A few words are in order, however, on property **P3**.

Since a super-strong equilibrium does not always exist in most game forms, a growing trend in the recent literature is to study equilibrium outcomes under various restrictions on coalition formation. The kind of coalitions that agents may form can be specified in the form of an *a priori* given coalition structure. In resource selection games (RSGs), Feldman and Tennenholtz [7] considered a partition coalition structure: i.e., the set of viable coalitions is a partition of the set of agents. And on the basis of a partition coalition structure, they introduced the notion of a partition equilibrium. Also related to this notion, a major theme in multi-agent systems is how autonomous agents come together and form coherent groupings to pursue their individual or collective goals more effectively. This problem is known as the coalition structure generation problem; see [12]. How a coalition structure is formed is beyond the scope of our paper. In our paper, we proceed under the assumption that a partition coalition structure is *a priori* given.

In singleton congestion games, even in the special case when every resource is accessible by everyone, there may not exist an outcome that satisfies the properties **P1**, **P2**, **P3**; see [5]. However, we show (in Theorem 1) that in ISCGs, there always exists an outcome that satisfies the above three properties. In other words, we show that in ISCGs, one need not sacrifice efficiency even if attention is confined to Nash and partition equilibrium outcomes.

***Other Related Studies.*** RSGs mentioned above are similar to ISCGs in that they also involve a set of agents selecting from a set of resources to utilize. But in RSGs, it is assumed that every resource is accessible to everyone, and in this sense, they are more restrictive than ISCGs. On the other hand, in RSGs, cost functions need not be identical, and in this sense, they are more general than ISCGs. Singleton congestion games generalize both these game forms: Resources need not be accessible to everyone, and resource cost functions may be non-identical (and non-monotonic). For related studies on these game forms, see [6,7,9]. It is also worth mentioning that ISCGs are equivalent to the subclass of project games with identical rewards and agent weights [4].

Note that every finite game admits an outcome that satisfies property **P2** since the Pareto dominance relation is asymmetric and transitive. In singleton congestion games, an outcome that satisfies property **P1** also always exists, since it is a subclass of congestion games, for which the existence of a Nash equilibrium is guaranteed; see [13]. In RSGs with increasing cost functions, there always exists an outcome that satisfies properties **P1** and **P3** [2]. However, as mentioned earlier, there does not always exist an outcome that also satisfies property **P2** [5].

It is worth emphasizing that in our analysis, we consider "weakly-improving deviations," not "improving deviations." An improving deviation makes *every coalition member* better off. An outcome resilient to improving deviations by all coalitions is known as a strong equilibrium. We should note that a strong equilibrium always exists in singleton congestion games with monotone cost functions [10].

The remainder of the paper is organized as follows: In Sect. 2, we formally define ISCGs, and then we present three lemmas that become useful in showing our main result. In Sect. 3, we present our main result (Theorem 1), and then we discuss various aspects of its proof with some examples.

## 2 The Model and the Preliminaries

An *identical singleton congestion game* (ISCG) is a triplet $\langle N, M, f\rangle$ where:

- $N = \{1, 2, \cdots, n\}$ is the set of agents.
- $M = \{1, 2, \cdots, m\}$ is the set of resources.
- $f = (f_i)_{i \in M}$ is a sequence, called the *feasibility constraint*, such that for each $i \in M$, $f_i \subseteq N$, and for each $j \in N$, $j \in f_i$ for some $i \in M$.

An *allocation* (or an outcome) is an ordered sequence $a = (a_i)_{i \in M}$ such that $a_i \cap a_{\bar{i}} = \emptyset$ for all $i, \bar{i} \in M$, $i \neq \bar{i}$, and $a_1 \cup \cdots \cup a_m = N$. An allocation $a$ is *feasible* if for each $i$, $a_i \subseteq f_i$. Let $\mathcal{A}$ be the domain of allocations. Let $\mathcal{A}^f \subseteq \mathcal{A}$ be the domain of feasible allocations.

The interpretation of the game is as follows: Under allocation $a$, the agents in $a_i$ are served by resource $i$. The congestion level at resource $i$ is $|a_i|$. And agents try to avoid congested resources.

A *coalition* $c \subseteq N$ is a nonempty subset of agents. We say that coalition $c$ *blocks* an allocation $a \in \mathcal{A}^f$ if there exists an allocation $\bar{a} \in \mathcal{A}^f$ such that:

- for each resource $i \in M$, $a_i \setminus c = \bar{a}_i \setminus c$,
- for each $(j, i_1, i_2) \in (c \times M \times M)$ such that $j \in a_{i_1}$ and $j \in \bar{a}_{i_2}$, $|\bar{a}_{i_2}| \leq |a_{i_1}|$,
- for some $(j, i_1, i_2) \in (c \times M \times M)$ such that $j \in a_{i_1}$ and $j \in \bar{a}_{i_2}$, $|\bar{a}_{i_2}| < |a_{i_1}|$.

In simpler terms: Coalition $c$ blocks allocation $a$ if by changing their utilized resources, the coalition can become better off in the Pareto sense. We refer to the allocation that results (above, $\bar{a}$) "the allocation induced when $c$ blocks $a$".

Let $\mathcal{P}_{\geq 1}(N)$ be the domain of coalitions, i.e., $\mathcal{P}_{\geq 1}(N) = \mathcal{P}(N) \setminus \{\emptyset\}$, where $\mathcal{P}(N)$ is the power set of $N$. A *coalition structure* $C \subseteq \mathcal{P}_{\geq 1}(N)$ is a potential set for viable coalitions. We say that an allocation $a \in \mathcal{A}^f$ is *C-stable* if there exists no $c \in C$ such that $c$ blocks $a$. Note that defined this way, a *super-strong equilibrium* is a $\mathcal{P}_{\geq 1}(N)$-stable allocation. That is, a super-strong equilibrium is a feasible allocation $a$ such that there exists no coalition that blocks $a$.

Though the super-strong equilibrium is a very appealing notion, it does not always exist even for the very restricted instances of ISCGs.[1] Therefore, in this paper we consider less demanding conditions. Specifically, we are interested in the existence of feasible allocations that are Pareto efficient, a Nash equilibrium, and a partition equilibrium. We define them next.

Let $\mathcal{P}_{=1}(N) = \{\{1\}, \{2\}, \cdots, \{n\}\}$. Notice that under the coalition structure $\mathcal{P}_{=1}(N)$, the only viable coalitions are singletons. Also, note that when expressed using our "*C-stable*" terminology: An allocation $a \in \mathcal{A}^f$ is a *Nash equilibrium* if it is $\mathcal{P}_{=1}(N)$-stable, and it is *Pareto efficient* if it is $\{N\}$-stable.

We also consider the situation where the set of viable coalitions is a partition of the set of agents. Formally, a *partition coalition structure* $C$ is such that for all $c, \bar{c} \in C, c \neq \bar{c}, c \cap \bar{c} = \emptyset$, and $\bigcup_{c \in C} = N$. Given a partition coalition structure $C$, we refer to an allocation $a \in \mathcal{A}^f$ as a *partition equilibrium* if $a$ is $C$-stable.

[1] See Feldman and Tennenholtz [7] for a very simple ISCG instance, with only three agents and two resources, for which a super-strong equilibrium does not exist.

The main result of our paper is that in an ISCG, for any given partition coalition structure $C$, there exists a feasible allocation that is $\mathcal{P}_{=1}(N)$-stable, $\{N\}$-stable, and $C$-stable. That is, we show that there always exists an outcome that satisfies the properties **P1, P2, P3** given in Sect. 1. We present this result in Theorem 1 in Sect. 3. The key to our proof of this result is what we call the "kernel values" of an allocation $a$. Therefore, in the remainder of this section, we define kernel values, and we present three lemmas related to kernel values that become useful in showing Theorem 1. To ease understanding, we will illustrate the concepts that we introduce by referring to the following example.

*Example 1.* In an ISCG with four resources and eight agents, consider the coalition $c = \{1, 2, 6\}$, and the allocation $a$ such that: $a_1 = \{1, 2\}, a_2 = \{3, 4, 5\}, a_3 = \{6, 7, 8\}, a_4 = \emptyset$. ◊

Given an allocation $a$, let $\omega : M \rightarrow M$ be a bijection such that $|a_{\omega(1)}| \geq |a_{\omega(2)}| \geq \cdots \geq |a_{\omega(m)}|$. That is, the bijection $\omega$ orders resources, in order of the number of agents assigned to them under $a$. For instance, for $a$ in Example 1, $\omega$ may be as follows: $\omega(1) = 2$, $\omega(2) = 3$, $\omega(3) = 1$, $\omega(4) = 4$. Looking at $\omega$, note that under $a$, the cardinality is maximal for resource 2 and minimal for resource 4.

† The *kernel* of allocation $a$, denoted by $k(a)$, is the ordered list: $(|a_{\omega(1)}|, |a_{\omega(2)}|, \cdots, |a_{\omega(m)}|)$. For instance, for $a$ in Example 1, we have $k(a) = (3, 3, 2, 0)$.

Given an allocation $a$, for some coalition $c$, let $\omega^c : M \rightarrow M$ be a bijection such that $|c \cap a_{\omega^c(1)}| \geq |c \cap a_{\omega^c(2)}| \geq \cdots \geq |c \cap a_{\omega^c(m)}|$. That is, the bijection $\omega^c$ orders resources, in order of the number of members of coalition $c$ assigned to them under $a$. For instance, for $a$ and $c$ in Example 1, $\omega^c$ may be as follows: $\omega^c(1) = 1$, $\omega^c(2) = 3$, $\omega^c(3) = 2$, $\omega^c(4) = 4$. Looking at $\omega^c$, note that under $a$, it is resource 1 which is assigned the maximum number of members of coalition $c$, which is followed by resource 3. The resources 2 and 4 are ordered at the end under $\omega^c$ since no member of coalition $c$ is assigned to these them under $a$.

†† The *c-kernel* of allocation $a$, denoted by $k(c, a)$, is the ordered list: $(|c \cap a_{\omega^c(1)}|, |c \cap a_{\omega^c(2)}|, \cdots, |c \cap a_{\omega^c(m)}|)$. For instance, for $a$ and $c$ in Example 1, we have $k(c, a) = (2, 1, 0, 0)$.

Given an allocation $a$, for coalition $c$, let $c_1, \cdots, c_n$ be its partition such that for each $j \in c_s$, if $j \in a_i$ then $|a_i| = s$. That is, under $a$, the agents in $c_s$ are those coalition members assigned to resources with cardinality $s$. For instance, for $a$ and $c$ in Example 1, we have: $c_1 = \emptyset, c_2 = \{1, 2\}, c_3 = \{6\}, c_4 = c_5 = c_6 = c_7 = c_8 = \emptyset$.

††† The *c-welfare-kernel* of allocation $a$, denoted by $w(c, a)$, is the ordered list: $(|c_n|, |c_{n-1}|, \cdots, |c_1|)$. For instance, for $a$ and $c$ in Example 1, we have $w(c, a) = (0, 0, 0, 0, 0, 1, 2, 0)$. Looking at $w(c, a)$, we can say that under $a$, no member of coalition $c$ is assigned to a resource with cardinality 8, 7, 6, 5, 4, or 1. And one coalition member is assigned to a resource with cardinality 3, and two coalition members are assigned to resources with cardinality 2.

Below we present three lemmas. They are straightforward observations related to kernel values, and hence, we omit their proofs. The lemmas become instrumental in showing Theorem 1. Before presenting them, however, we need to introduce the notion of a "chain", which helps simplify our exposition.

Consider two allocations, say $a$ and $\overline{a}$. An *$a\overline{a}$-chain*, represented as $i_1 \rightarrow (j_1) \rightarrow i_2 \rightarrow (j_2) \rightarrow \cdots \rightarrow (j_{s-1}) \rightarrow i_s$ $(s \geq 2)$, refers to the situation such that:

- $j_1 \in a_{i_1}$, $j_1 \in \overline{a}_{i_2}$, and $i_1 \neq i_2$;

$\vdots$

- $j_{s-1} \in a_{i_{s-1}}$, $j_{s-1} \in \overline{a}_{i_s}$, and $i_{s-1} \neq i_s$.

In essence, an *$a\overline{a}$-chain* specifies how the allocation $a$ can be transformed into allocation $\overline{a}$: The transformation involves moving agent $j_1$ from resource $i_1$ to $i_2$, agent $j_2$ from resource $i_2$ to $i_3$, and so on. To transform $a$ into $\overline{a}$, it is necessary to "execute" this *$a\overline{a}$-chain*, and perhaps some others, too. Above, we refer to $s$ as the *length* of the $a\overline{a}$-chain. Note that the minimum length of an $a\overline{a}$-chain is 2. Also, note that there always exists an $a\overline{a}$-chain unless $a = \overline{a}$.

We also introduce the "chain addition operator," denoted by $\oplus$. We write: $\overline{a} = a \oplus i_1 \rightarrow (j_1) \rightarrow i_2 \rightarrow (j_2) \rightarrow \cdots \rightarrow (j_{s-1}) \rightarrow i_s$ if

- $i_1 \rightarrow (j_1) \rightarrow i_2 \rightarrow (j_2) \rightarrow \cdots \rightarrow (j_{s-1}) \rightarrow i_s$ is an $a\overline{a}$-chain,
- for each $j \in N \setminus \{j_1, j_2, \cdots, j_{s-1}\}$, the resource to which $j$ is assigned is the same under $a$ and $\overline{a}$.

The above statement basically states that the allocation $a$ can be transformed into allocation $\overline{a}$ by "executing" a single $a\overline{a}$-chain. Notice that if $\overline{a} = a \oplus i_1 \rightarrow (j_1) \rightarrow i_2 \rightarrow (j_2) \rightarrow i_3$, then we can also write $\overline{a} = (a \oplus i_1 \rightarrow (j_1) \rightarrow i_2) \oplus i_2 \rightarrow (j_2) \rightarrow i_3$.

Lemma 1 below states that given a partition coalition structure $C$, under some allocation, if two members of some coalition $c \in C$ switch their positions, then the kernel values remain the same as before.

**Lemma 1.** *Let $C$ be a partition coalition structure. Let $j_1, j_2 \in c \in C$. Let allocations $a, \overline{a} \in \mathcal{A}$ be such that $\overline{a} = a \oplus i_1 \rightarrow (j_1) \rightarrow i_2 \rightarrow (j_2) \rightarrow i_1$. Then, the kernel values of $a$ and $\overline{a}$ are the same. Also, for each $\widetilde{c} \in C$, the $\widetilde{c}$-kernel and $\widetilde{c}$-welfare-kernel values of $a$ and $\overline{a}$ are the same.*

We now introduce two asymmetric and transitive relations that become essential in showing Theorem 1. These relations involve lexicographical comparisons of kernel values defined above. We will use the notation $\prec$ to denote the "lexicographically smaller than" relation. For instance, $(1,4,5) \prec (2,3,4)$ because $1 < 2$ (by comparison of the first entries). And $(2,3,4) \prec (2,5,2)$ because $2 = 2, 3 < 5$ (by comparison of the second entries because the first entries are the same).

Our first asymmetric and transitive relation is defined for partition coalition structures. Given a partition coalition structure $C$, we say that allocation $a$ *$C$-balance dominates* $\overline{a}$,

- if $k(a) \prec k(\overline{a})$;
- or if $k(a) = k(\overline{a})$ and
  - * for each $c \in C$, $k(c, a) \prec k(c, \overline{a})$ or $k(c, a) = k(c, \overline{a})$,
  - * for some $c \in C$, $k(c, a) \prec k(c, \overline{a})$.

In loose terms: Allocation $a$ $C$-balance dominates $\overline{a}$ if under $a$, the distribution of agents to resources is more even. If there is a tie in this regard, then $a$ $C$-balance dominates $\overline{a}$ if under $a$, the distribution of coalition members to resources is more even. Lemma 2 below is a straightforward observation on the $C$-balance dominance relation.

**Lemma 2.** *Let $C$ be a partition coalition structure. Let $j_1 \in c \in C$. Let allocations $a$ and $\overline{a}$ be such that $\overline{a} = a \oplus i_1 \rightarrow (j_1) \rightarrow i_2$. Then, allocation $\overline{a}$ $C$-balance dominates $a$: **(a)** if $|a_{i_1}| \geq |a_{i_2}| + 2$, or **(b)** if $|a_{i_1}| = |a_{i_2}| + 1$ and $|c \cap a_{i_1}| \geq |c \cap a_{i_2}| + 2$.*

The second relation that we introduce compares $c$-welfare-kernel values of allocations, again, lexicographically. For a coalition $c$, we say that allocation $a$ *$c$-welfare-dominates* $\overline{a}$ if $w(c, a) \prec w(c, \overline{a})$. Lemma 3 below pertains to this relation.

**Lemma 3.** *Let $C$ be a partition coalition structure. Let $c \in C$ be a coalition. Let $a$ be an allocation.*
***(a)*** *Suppose that coalition $c$ blocks allocation $a$. Let $\widetilde{a}$ be the allocation induced when $c$ blocks $a$. Then, $\widetilde{a}$ $c$-welfare-dominates $a$.*
***(b)*** *Suppose that allocation $\overline{a}$ $c$-welfare-dominates $a$. Also, suppose that $\widetilde{M} \subset M$ is such that for each $s \in \{1, 2, \cdots, n\}$,*

$$\sum_{i \in \widetilde{M}, |a_i| = s} |c \cap a_i| = \sum_{i \in \widetilde{M}, |\overline{a}_i| = s} |c \cap \overline{a}_i|.$$

*Let $k_{max} = max_{i \in M \setminus \widetilde{M}} |a_i|$ and $\overline{k}_{max} = max_{i \in M \setminus \widetilde{M}} |\overline{a}_i|$. Then,*

$$\overline{k}_{max} \leq k_{max}, \text{ and } \sum_{i \in M \setminus \widetilde{M}, |a_i| = k_{max}} |c \cap a_i| \geq \sum_{i \in M \setminus \widetilde{M}, |\overline{a}_i| = k_{max}} |c \cap \overline{a}_i|.$$

## 3 The Main Result

This section is devoted to our main result: In Theorem 1, we show that in an ISCG, there always exists an allocation that satisfies the properties **P1**, **P2**, and **P3**, given in Sect. 1. The proof of the theorem is fairly involved. We prove the theorem by showing that every maximal outcome with respect to the $C$-balance dominance relation satisfies the three properties. At the end of the section, we also present an example and show that it is possible that an outcome satisfies the three properties and yet not be maximal with respect to the $C$-balance dominance relation.

**Theorem 1.** *In an ISCG, for any given partition coalition structure $C$, there always exists an allocation $a \in A^f$ such that $a$ is $\mathcal{P}_{=1}(N)$-stable, $\{N\}$-stable, and $C$-stable. That is, in an ISCG there always exists a Pareto efficient outcome that is a Nash equilibrium and a partition equilibrium.*

*Proof.* Let $C$ be a given partition coalition structure. Let $\kappa\left(C, A^f\right) \subseteq A^f$ be such that for each $a \in \kappa\left(C, A^f\right)$, there exists no $\overline{a} \in A^f$ such that $\overline{a}$ $C$-balance dominates $a$. That is, $\kappa\left(C, A^f\right)$ is the set of maximal allocations in $A^f$ with respect to the $C$-balance dominance relation. Note that $\kappa\left(C, A^f\right) \neq \emptyset$ since the $C$-balance dominance relation is transitive and asymmetric. Let $a \in \kappa\left(C, A^f\right)$. To show the theorem, we need to show that $a$ is $\mathcal{P}_{=1}(N)$-stable (**P1**), $\{N\}$-stable (**P2**), and $C$-stable (**P3**). The observation that $a$ is $\mathcal{P}_{=1}(N)$-stable and $\{N\}$-stable is fairly easy, and it is known due to [8]. Thus, below, we only show that $a$ is $C$-stable.

By way of contradiction, suppose that $a$ is not $C$-stable. Then, there exists $c \in C$ such that $c$ blocks $a$. Let $\overline{a} \in A^f$ be the allocation induced when $c$ blocks $a$. Then, by Lemma 3(a), we obtain that $\overline{a}$ $c$-welfare-dominates $a$. Thus, the following two statements are true:

**(S1)** $a \in \kappa\left(C, A^f\right)$,
**(S2)** $\overline{a} \in A^f$ and $\overline{a}$ $c$-welfare-dominates $a$.

We will prove that $a$ is $C$-stable by showing the supposition that **S1** and **S2** are true leads to a contradiction. The outline of the proof is as follows: Below, in Steps 1 and 2, we update $\overline{a}$ and $A^f$ iteratively. We show that after each update, **S1** and **S2** remain to be true. And in the end, once we are done with all our updates, we derive a contradiction.
**Step 1:** Update the feasibility constraints as follows:

- for each $i \in M$, let $f_i = a_i \cup \overline{a}_i$.

**Step 2:** As long as there exists an $a\overline{a}$-chain whose length is bigger than 2, represented as

$$i_1 \to (j_1) \to i_2 \to (j_2) \to i_3 \to \cdots (j_{s-1}) \to i_s,$$

update the feasibility constraint $f_{i_3}$ and the allocation $\overline{a}$ as follows:

- $f_{i_3} := f_{i_3} \setminus \{j_2\} \cup \{j_1\}$;
- $\overline{a} := \overline{a} \oplus i_3 \to (j_2) \to i_2 \to (j_1) \to i_3$.

Consider the update at Step 1: After the update, note that the set $A^f$ becomes smaller, but we still have $a, \overline{a} \in A^f$. Hence, **S1** and **S2** remain to be true.
Consider an update at Step 2: After the update, note that $\overline{a}$ and $A^f$ change, but we still have $a, \overline{a} \in A^f$. Also, by Lemma 1, for $\overline{a}$ the kernel values remain the same as before. Hence, after the update, **S2** remains to be true.

Let $\widetilde{a}$ be an allocation that was infeasible before the update but that becomes feasible after the update. It is clear that under $\widetilde{a}$, $j_1 \in \widetilde{a}_{i_3}$ and $j_2 \in \widetilde{a}_{i_2}$. To show that after the update **S1** remains to be true, we need to show that $\widetilde{a}$ does not $C$-balance dominate $a$. Let $\widehat{a} = \widetilde{a} \oplus i_3 \to (j_1) \to i_2 \to (j_2) \to i_3$. Note that before the update, $\widehat{a}$ was feasible. Thus, $\widehat{a}$ does not $C$-balance dominate $a$. But by Lemma 1, for $\widetilde{a}$ and $\widehat{a}$ the kernel values are the same. Then, $\widetilde{a}$ does not $C$-balance dominate $a$ either. Hence, after the update, **S1** remains to be true.

Also, notice that at Step 2 each time $\overline{a}$ and $A^f$ are updated, the length of an $a\overline{a}$-chain becomes smaller. Therefore, Step 2 terminates after a finite number of iterations. And when it terminates, we obtain that:

- for $a$, $\overline{a}$, $A^f$, the statements **S1** and **S2** are true,
- every $a\overline{a}$-chain that remains is of length 2.

We are now ready to derive a contradiction. Since every $a\overline{a}$-chain that remains is of length 2, we can divide the set $M$ into the following three subsets:

$$M^0 = \{i \in M : a_i = \overline{a}_i\}, \quad M^- = \{i \in M : \overline{a}_i \subset a_i\}, \quad M^+ = \{i \in M : \overline{a}_i \supset a_i\}.$$

Note that for $a$ and $\overline{a}$, each $a\overline{a}$-chain is of the form $i_1 \rightarrow (j) \rightarrow i_2$, where $i_1 \in M^-$ and $i_2 \in M^+$. We also divide the sets $M^-$ and $M^+$ into their partitions with respect to the cardinalities of resources under $a$: Let $M^-(k) = \{i \in M^- : |a_i| = k\}$ and $M^+(k) = \{i \in M^+ : |a_i| = k\}$.

We derive a contradiction using Lemma 3(b). We proceed as follows: First, we identify a set $\widetilde{M} \subset M$ for which the suppositions in Lemma 3(b) are satisfied. Then, using some arguments (presented below under the heading **Iteration**), we show that the set $\widetilde{M}$ can be updated iteratively such that after each iteration, its cardinality increases and yet the suppositions in Lemma 3(b) are still satisfied. Eventually, we obtain that $\widetilde{M} = M$ and for each $s \in \{1, 2, \cdots, n\}$,

$$\textstyle\sum_{i \in M, |a_i| = s} |c \cap a_i| = \sum_{i \in M, |\overline{a}_i| = s} |c \cap \overline{a}_i|.$$

This will imply that for $a$ and $\overline{a}$, the $c$-welfare-kernel values are indeed the same, which will contradict that $\overline{a}$ $c$-welfare-dominates $a$ (**S2**) and complete our proof.

Initially, we set $\widetilde{M} = M^0$.

For $\widetilde{M} = M^0$, it is easy to verify that the suppositions in Lemma 3(b) are satisfied. Let $k_{max} = max_{i \in M \setminus \widetilde{M}} |a_i|$ and $\overline{k}_{max} = max_{i \in M \setminus \widetilde{M}} |\overline{a}_i|$. Then, by Lemma 3(b), we have $\overline{k}_{max} \leq k_{max}$ and

$$\textstyle\sum_{i \in M \setminus \widetilde{M}, |a_i| = k_{max}} |c \cap a_i| \geq \sum_{i \in M \setminus \widetilde{M}, |\overline{a}_i| = k_{max}} |c \cap \overline{a}_i|. \qquad (*)$$

**Iteration:** Consider $M^+(k)$ where $k \geq k_{max}$. Suppose that $M^+(k) \neq \emptyset$. Let $i \in M^+(k)$. Since $|a_i| = k$ and $\overline{a}_i \supset a_i$, we obtain that $|\overline{a}_i| \geq k + 1 \geq k_{max} + 1$. But this contradicts that $\overline{k}_{max} \leq k_{max}$. Thus, for $k \geq k_{max}$, $M^+(k) = \emptyset$.

We now restrict our attention to the sets $M^-(k_{max})$ and $M^+(k_{max} - 1)$: Let $i_1 \in M^-(k_{max})$. Consider an $a\overline{a}$-chain of the form $i_1 \rightarrow (j_1) \rightarrow i_2$. Let $\widetilde{a} = a \oplus i_1 \rightarrow (j_1) \rightarrow i_2$. Note that $\widetilde{a} \in A^f$. By Lemma 2, if $|a_{i_2}| \leq k_{max} - 2$, or if $|a_{i_2}| = k_{max} - 1$ and $|c \cap a_{i_1}| \geq |c \cap a_{i_2}| + 2$, we obtain that $\widetilde{a}$ $C$-balance dominates $a$. But this contradicts that $a \in \kappa(C, A^f)$. Thus, $i_2 \in M^+(k_{max} - 1)$ and $|c \cap a_{i_1}| \leq |c \cap a_{i_2}| + 1$. Suppose that there exists an $a\overline{a}$-chain of the form $i \rightarrow (j) \rightarrow i_2$ where $j \neq j_1$. Then, $a_{i_2} \cup \{j, j_1\} \subseteq \overline{a}_{i_2}$. Then, $|\overline{a}_{i_2}| \geq |a_{i_2}| + 2 = k_{max} + 1$. But this contradicts that $\overline{k}_{max} \leq k_{max}$. Thus, there exists a one-to-one function $v : M^-(k_{max}) \rightarrow M^+(k_{max} - 1)$ such that:

– for each $i \in M^-(k_{max})$, there exists an $a\overline{a}$-chain of the form $i \to (j) \to \upsilon(i)$.

And for $\upsilon$, as argued above, the following holds:

– for each $i \in M^-(k_{max})$, $|c \cap \overline{a}_{\upsilon(i)}| = |c \cap a_{\upsilon(i)}| + 1 \geq |c \cap a_i|$.

But then the inequality above in $(*)$ holds only if $\upsilon$ is a bijection (i.e., $|M^-(k_{max})| = |M^+(k_{max}-1)|$) and for each $i \in M^-(k_{max})$, $|c \cap \overline{a}_{\upsilon(i)}| = |c \cap a_i|$. Therefore,

$$\sum_{i \in M \setminus \widetilde{M}, |a_i| = k_{max}} |c \cap a_i| = \sum_{i \in M \setminus \widetilde{M}, |\overline{a}_i| = k_{max}} |c \cap \overline{a}_i|.$$

But then if we update $\widetilde{M}$ and set $\widetilde{M} := \widetilde{M} \cup M^-(k_{max}) \cup M^+(k_{max}-1)$, the suppositions in Lemma 3(b) are still satisfied.

Therefore, as argued above, we can iterate these arguments and update $\widetilde{M}$ until we obtain that $\widetilde{M} = M$. And then we can conclude that for both $a$ and $\overline{a}$, the $c$-welfare-kernel values are actually the same. However, this contradicts that $\overline{a}$ $c$-welfare-dominates $a$ (**S2**). Therefore, our initial supposition must be wrong, i.e., allocation $a$ must be $C$-stable. This completes our proof. □

In the proof of Theorem 1, we showed that if $a \in \kappa(C, A^f)$, then $a$ satisfies the properties **P1**, **P2**, and **P3**, given in Sect. 1. The following example shows that the converse of this statement is not true.

*Example 2.* Consider an ISCG with two resources and fifteen agents such that both resources are accessible to every agent. Consider the following partition coalition structure $C = \{\{1,2,3\},\{4,5,6\},\{7,8,9\},\{10,11,12\},\{13,14,15\}\}$. Let allocation $a$ be as follows: $a_1 = \{1,2,4,5,7,8,10,11\}$, $a_2 = \{3,6,9,12,13,14,15\}$. Also, let allocation $\overline{a}$ be as follows: $\overline{a}_1 = \{1,2,4,7,8,10,13,14\}$, $\overline{a}_2 = \{3,5,6,9,11,12,15\}$. It is easy to verify that the allocation $a$ is $\mathcal{P}_{=1}(N)$-stable, $\{N\}$-stable, and $C$-stable. However, $a \notin \kappa(C, A^f)$ since $\overline{a}$ $C$-balance dominates $a$. ◊

As a technical note, in the example below, we show that the $C$-balance dominance relation is not a potential function. More specifically, we show that when a coalition $c \in C$ blocks an allocation $a$, the induced allocation, say $\bar{a}$, does not necessarily $C$-balance dominate $a$.

*Example 3.* Consider an ISCG with four resources and eighteen agents such that all resources are accessible to every agent. Consider the following partition coalition structure: $C = \{\{1,\ldots,13\},\{14,\ldots,18\}\}$. Let allocation $a$ be as follows: $a_1 = \{1,14,15,16\}$, $a_2 = \{2,3,4,5\}$, $a_3 = \{6,7,8,9,17\}$, $a_4 = \{10,11,12,13,18\}$. Notice that coalition $c = \{1,\ldots,13\}$ blocks allocation $a$ inducing the following allocation, $\bar{a}$: $\bar{a}_1 = \{6,10,14,15,16\}$, $\bar{a}_2 = \{7,8,9,11,12\}$, $\bar{a}_3 = \{1,2,3,17\}$, $\bar{a}_4 = \{4,5,13,18\}$. Note that under $\bar{a}$, no agent in coalition $c$ becomes worse off and agent 13 becomes better off. However, $\bar{a}$ does not $C$-balance dominate allocation $a$. This is because $k(a) = k(\bar{a})$ yet $k(c,a) \not\prec k(c,\bar{a})$ since $k(c,a) = (4,4,4,1)$ and $k(c,\bar{a}) = (5,3,3,2)$. ◊

It is also natural to ask whether Theorem 1 can be extended in a way that also incorporates overlapping coalitions, instead of a partition coalition structure as in property **P3**. The following example shows that this is not possible even in very restricted settings.

*Example 4.* Consider an ISCG with two resources and three agents such that agent 1 can only access resource 1, agent 2 can only access resource 2, and agent 3 can access both resources. Consider the following coalition structure $C = \{\{1,3\},\{2,3\}\}$. Notice that there are two possible allocations and neither of them is $C$-stable. $\diamondsuit$

## References

1. Altman, E., et al.: Blockchain competition between miners: a game theoretic perspective. Front. Blockchain **2** (2020)
2. Anshelevich, E., Caskurlu, B., Hate, A.: Partition equilibrium always exists in resource selection games. Theory Comput. Syst. **53**(1), 73–85 (2013)
3. van den Berg, J., Guy, S.J., Lin, M., Manocha, D.: Reciprocal n-body collision avoidance. In: Pradalier, C., Siegwart, R., Hirzinger, G. (eds.) Robotics Research, pp. 3–19. Springer, Heidelberg (2011). https://doi.org/10.1007/978-3-642-19457-3_1
4. Bilò, V., Gourvès, L., Monnot, J.: Project games. In: Heggernes, P. (ed.) CIAC 2019. LNCS, vol. 11485, pp. 75–86. Springer, Cham (2019). https://doi.org/10.1007/978-3-030-17402-6_7
5. Caskurlu, B., Ekici, O., Erdem Kizilkaya, F.: On existence of equilibrium under social coalition structures. In: Chen, J., Feng, Q., Xu, J. (eds.) TAMC 2020. LNCS, vol. 12337, pp. 263–274. Springer, Cham (2020). https://doi.org/10.1007/978-3-030-59267-7_23
6. Caskurlu, B., Ekici, Ö., Kizilkaya, F.E.: On efficient computation of equilibrium under social coalition structures. Turkish J. Electr. Eng. Comput. Sci. **28**(3), 1686–1698 (2020)
7. Feldman, M., Tennenholtz, M.: Structured coalitions in resource selection games. ACM Trans. Intell. Syst. Technol. **1**(1), 1–21 (2010)
8. Harks, T., Klimm, M., Möhring, R.H.: Strong Nash equilibria in games with the lexicographical improvement property. In: Leonardi, S. (ed.) WINE 2009. LNCS, vol. 5929, pp. 463–470. Springer, Heidelberg (2009). https://doi.org/10.1007/978-3-642-10841-9_43
9. Hoefer, M., Penn, M., Polukarov, M., Skopalik, A., Vöcking, B.: Considerate equilibrium. In: International Joint Conference on Artificial Intelligence (IJCAI). ACM Press (2011)
10. Holzman, R., Law-Yone, N.: Strong equilibrium in congestion games. Games Econ. Behav. **21**(1–2), 85–101 (1997)
11. Ieong, S., McGrew, R., Nudelman, E., Shoham, Y., Sun, Q.: Fast and compact: a simple class of congestion games. In: Proceedings of the 20th National Conference on Artificial Intelligence (AAAI), vol. 2, pp. 489–494. AAAI Press (2005)
12. Rahwan, T., Michalak, T.P., Wooldridge, M., Jennings, N.R.: Coalition structure generation: a survey. Artif. Intell. **229**, 139–174 (2015)
13. Rosenthal, R.W.: A class of games possessing pure-strategy Nash equilibria. Int. J. Game Theory **2**(1), 65–67 (1973)
14. Wardrop, J.G.: Some theoretical aspects of road traffic research. Proc. Inst. Civ. Eng. **1**(3), 325–362 (1952)

# A Pivot Gray Code Listing for the Spanning Trees of the Fan Graph

Ben Cameron, Aaron Grubb(✉), and Joe Sawada

University of Guelph, Guelph, Canada
{ben.cameron,agrubb,jsawada}@uoguelph.ca

**Abstract.** We use a greedy strategy to list the spanning trees of the fan graph, $F_n$, such that successive trees differ by pivoting a single edge around a vertex. It is the first greedy algorithm for exhaustively generating spanning trees using such a minimal change operation. The resulting listing is then studied to find a recursive algorithm that produces the same listing in $O(1)$-amortized time using $O(n)$ space. Additionally, we present $O(n)$-time algorithms for ranking and unranking the spanning trees for our listing; an improvement over the generic $O(n^3)$-time algorithm for ranking and unranking spanning trees of an arbitrary graph.

**Keywords:** Spanning tree · Greedy algorithm · Fan graph · Combinatorial generation

## 1 Introduction

This paper is concerned with the algorithmic problem of listing all spanning trees of the fan graph. Applications of efficiently listing all spanning trees of general graphs are ubiquitous in computer science and also appear in many other scientific disciplines [3]. In fact, one of the earliest known works on listing all spanning trees of a graph is due to the German physicist Wilhelm Feussner in 1902 who was motivated by an application to electrical networks [7]. In the 120 years since Feussner's work, many new algorithms have been developed, such as those in the following citations [1,4,6,8,9,12–17,19–22,24].

For any application, it is desirable for spanning tree listing algorithms to have the asymptotically best possible running time, that is, $O(1)$-amortized running time. The algorithms due to Kapoor and Ramesh [14], Matsui [16], Smith [22], Shioura and Tamura [20] and Shioura et al. [21] all run in $O(1)$-amortized time. Another desirable property of such listings is to have the *revolving-door* property, where successive spanning trees differ by the addition of one edge and the removal of another. Such listings where successive objects in a listing differ by a constant number of simple operations are more generally known as *Gray codes.* The algorithms due to Smith [22], Kamae [13], Kishi and Kajitani [15], Holzmann and Harary [12] and Cummins [6] all produce Gray code listings of spanning trees for an arbitrary graph. Of all of these algorithms, Smith's is the only one that


C.-Y. Chen et al. (Eds.): COCOON 2021, LNCS 13025, pp. 49–60, 2021.
https://doi.org/10.1007/978-3-030-89543-3_5

produces a Gray code listing in $O(1)$-amortized time. A stronger notion of a Gray code for spanning trees is where the revolving-door makes strictly local changes. More specifically, we would like the differing edges to share a common endpoint. Such a Gray code property, which we call a *pivot Gray code*, is not given by any previously known algorithm. This leads to our first research question.

**Research Question #1** Given a graph $G$ (perhaps from a specific class), does there exist a pivot Gray code listing of all spanning trees of $G$? Furthermore, can the listing be generated in polynomial (ideally constant) time per tree using polynomial space?

A related question that arises for any listing is how to *rank*, that is, find the position of the object in the listing, and *unrank*, that is, return the object at a specific rank. For spanning trees, an $O(n^3)$-time algorithm for ranking and unranking a spanning tree of a specific listing for an arbitrary graph is known [5].

**Research Question #2** Given a graph $G$ (perhaps from a specific class), does there exist a (pivot Gray code) listing of all spanning trees of $G$ that can be ranked and unranked in $O(n^2)$ time or better?

An algorithmic technique recently found to have success in the discovery of Gray codes is the greedy approach. An algorithm is said to be *greedy* if it can prioritize allowable actions according to some criteria, and then choose the highest priority action that results in a unique object to obtain the next object in the listing. When applying a greedy algorithm, there is no backtracking; once none of the valid actions lead to a new object in the set under consideration, the algorithm halts, even if the listing is not exhaustive. The work by Williams [23] notes that some very well-known combinatorial listings can be constructed greedily, including the binary reflected Gray code (BRGC) for binary strings, the plain change order for permutations, and the lexicographically smallest de Bruijn sequence. Recently, a very powerful greedy algorithm on permutations (known as Algorithm J, where J stands for "jump") generalizes many known combinatorial Gray code listings including many related to permutation patterns, rectangulations, and elimination trees [10,11,18]. However, no greedy algorithm was previously known to list the spanning trees of an arbitrary graph.

**Research Question #3** Given a graph $G$ (perhaps from a specific class), does there exist a greedy strategy to list all spanning trees of $G$? Moreover, does such a greedy strategy exist where the resulting listing is a pivot Gray code?

In most cases, a greedy algorithm requires exponential space to recall which objects have already been visited in a listing. Thus, answering this third question would satisfy only the first part of **Research Question #1**. However, in many cases, an underlying pattern can be found in a greedy listing which can result in space efficient algorithms [10,23].

To address these three research questions, we applied a variety of greedy approaches to structured classes of graphs including the fan, wheel, $n$-cube, and the compete graph. From this study, we were able to affirmatively answer each of the research questions for the fan graph. It remains an open question to find similar results for other classes of graphs.

### 1.1 New Results

The *fan graph* on $n$ vertices, denoted $F_n$, is obtained by joining a single vertex (which we label $v_\infty$) to the path on $n-1$ vertices (labeled $v_2, ..., v_n$) - see Fig. 1. Note that we label the smallest vertex $v_2$ so that the largest non-infinity labeled vertex equals the total number of vertices. Let $\mathbf{T}_n$ denote the set of all spanning trees of $F_n$. We discover a greedy strategy to generate $\mathbf{T}_n$ in a pivot Gray code order. We describe this greedy strategy in Sect. 2. The resulting listing is studied to find an $O(1)$-amortized time recursive algorithm that produces the same listing using only $O(n)$ space, which is presented in Sect. 3. We also show how to rank and unrank a spanning tree of the greedy listing in $O(n)$ time in Sect. 3, which is a significant improvement over the general $O(n^3)$-time ranking and unranking that is already known. We conclude with a summary in Sect. 4.

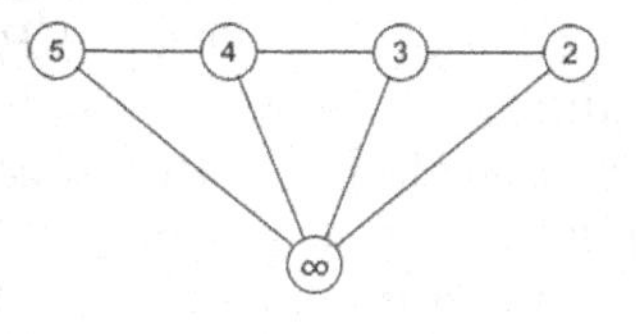


**Fig. 1.** $F_5$

## 2 A Greedy Generation for $\mathbf{T}_n$

With our goal to discover a pivot Gray code listing of $\mathbf{T}_n$, we tested a variety of greedy approaches. There are two important issues when considering a greedy approach to list spanning trees: (1) the labels on the vertices (or edges) and (2) the starting tree. For each of our approaches, we prioritized our operations by first considering which vertex $u$ to pivot on, followed by an ordering of the endpoints considered in the addition/removal. We call the vertex $u$ the *pivot*.

Our initial attempts focused only on pivots that were leaves. As a specific example, we ordered the leaves (pivots) from smallest to largest. Since each leaf $u$ is attached to a unique vertex $v$ in the current spanning tree, we then considered the neighbours $w$ of $u$ in increasing order of label. We restricted the labeling of the vertices to the most natural ones, such as the one presented in Sect. 1.1. For each strategy we tried all possible starting trees. Unfortunately, none of our attempts lead to exhaustive listings. Applying these strategies on the wheel, $n$-cube, and complete graph was also unsuccessful.

By allowing the pivot to be any arbitrary vertex, we experimentally discovered several exhaustive listings for $\mathbf{T}_n$ for $n$ up to 12 (testing every starting tree for $n = 12$ took about eight hours). One listing stood out as having an easily defined starting tree as well as a nice pattern which we could study to construct the listing more efficiently. It applied the labeling of the vertices as described in Sect. 1.1 with the following prioritization of pivots and their incident edges:

Prioritize the pivots $u$ from smallest to largest and then for each pivot, prioritize the edges $uv$ that can be removed from the current tree in increasing order of the label on $v$, and for each such $v$, prioritize the edges $uw$ that can be added to the current tree in increasing order of the label on $w$.

Since this is a greedy strategy, if an edge pivot results in a spanning tree that has already been generated or a graph that is not a spanning tree, then the next highest priority edge pivot is attempted. Let GREEDY($T$) denote the listing that results from applying this greedy approach starting with the spanning tree $T$. The starting tree that produced a nice exhaustive listing was the path $v_\infty, v_2, v_3, \ldots, v_n$, denoted $P_n$ throughout the paper. Figure 2 shows the listings GREEDY($P_n$) for $n = 2, 3, 4, 5$. The listing GREEDY($P_6$) is illustrated in Fig. 3. It is worth noting that starting with the path $v_\infty, v_n, v_{n-1}, \ldots, v_2$ or the star (all edges incident to $v_\infty$) did not lead to an exhaustive listing of $\mathbf{T}_n$.

As an example of how the greedy algorithm proceeds, consider the listing GREEDY($P_5$) in Fig. 2. When the current tree $T$ is the 16th one in the listing (the one with edges $\{v_2v_\infty, v_2v_3, v_3v_4, v_5v_\infty\}$), the first pivot considered is $v_2$. Since both $v_2v_3$ and $v_2v_\infty$ are present in the tree, no valid move is available by pivoting on $v_2$. The next pivot considered is $v_3$. Both edges $v_3v_2$ and $v_3v_4$ are incident with $v_3$. First, we attempt to remove $v_3v_2$ and add $v_3v_\infty$, which results in a tree previously generated. Next, we attempt to remove $v_3v_4$ and add $v_3v_\infty$, which results in a cycle. So, the next pivot, $v_4$, is considered. The only edge incident to $v_4$ is $v_4v_3$. By removing $v_4v_3$ and adding $v_4v_5$ we obtain a new spanning tree, the next tree in the greedy listing.

To prove that GREEDY($P_n$) does in fact contain all trees in $\mathbf{T}_n$, we demonstrate it is equivalent to a recursively constructed listing that we obtain by studying the greedy listings. Before we describe this recursive construction we mention one rather remarkable property of GREEDY($P_n$) that we will also prove in the next section: If $X_n$ is last tree in the listing GREEDY($P_n$), then GREEDY($X_n$) is precisely GREEDY($P_n$) in reverse order.

## 3 An $O(1)$-Amortized Time Pivot Gray Code Generation for $\mathbf{T}_n$

In this section we develop an efficient recursive algorithm to construct the listing GREEDY($P_n$). The construction generates some sub-lists in reverse order, similar to the recursive construction of the BRGC. The recursive properties allow us to provide efficient ranking and unranking algorithms for the listing based on counting the number of trees at each stage of the construction. Let $t_n$ denote the number of spanning trees of $F_n$. It is known that

$$t_n = f_{2(n-1)} = 2\frac{((3-\sqrt{5})/2)^n - ((3+\sqrt{5})/2)^{n-2}}{5-3\sqrt{5}},$$

where $f_n$ is the $n$th number of the Fibonacci sequence with $f_1 = f_2 = 1$ [2].

By studying the order of the spanning trees in GREEDY($P_n$), we identified four distinct stages S1, S2, S3, S4 that are highlighted for GREEDY($P_6$) in Fig. 3.

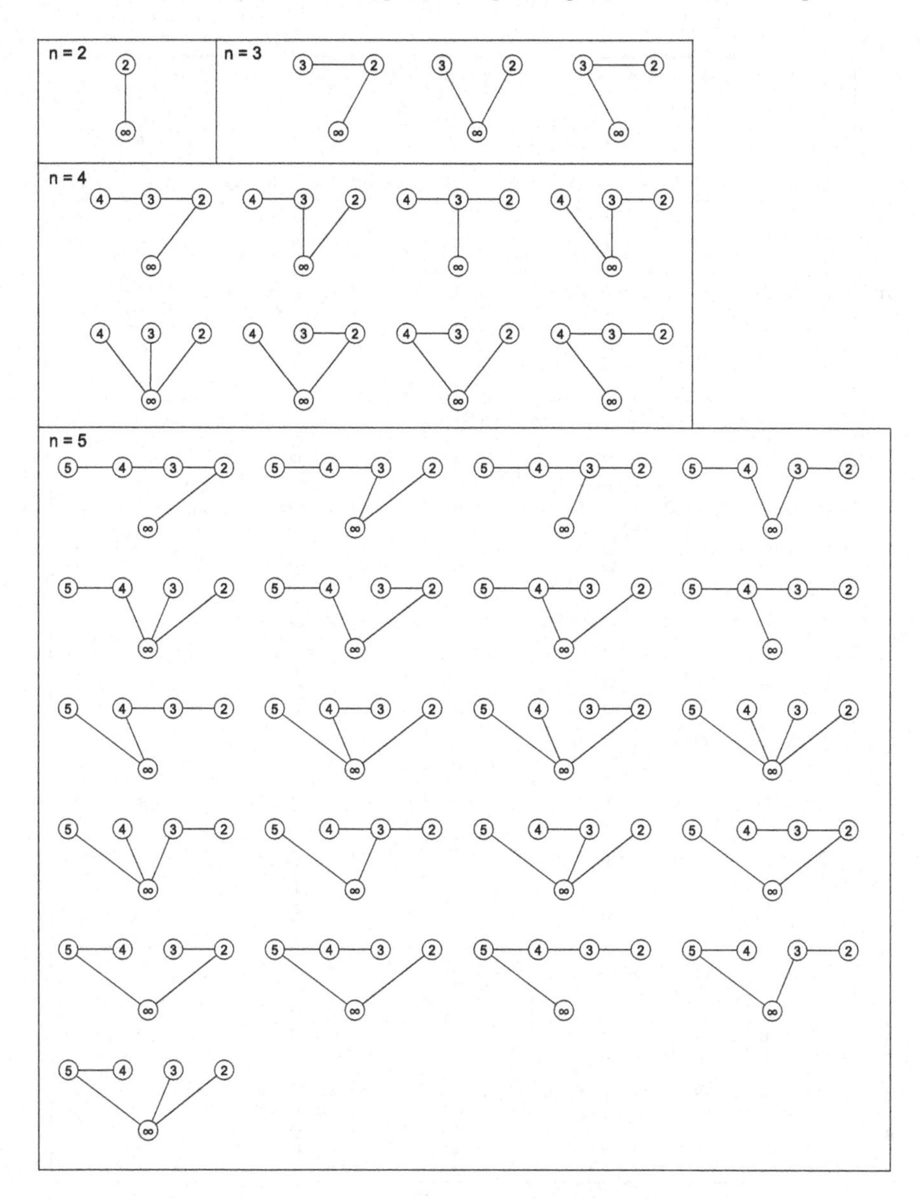


**Fig. 2.** GREEDY($P_n$) for $n = 2, 3, 4, 5$. Read left to right, top to bottom.

From this figure, and referring back to Fig. 2 to see the recursive properties, observe that:

- The trees in S1 are equivalent to GREEDY($P_5$) with the added edge $v_6v_5$.
- The trees in S2 are equivalent to the reversal of the trees in GREEDY($P_5$) with the added edge $v_6v_\infty$.

The trees in S3 and S4 have both edges $v_6v_5$ and $v_6v_\infty$ present.

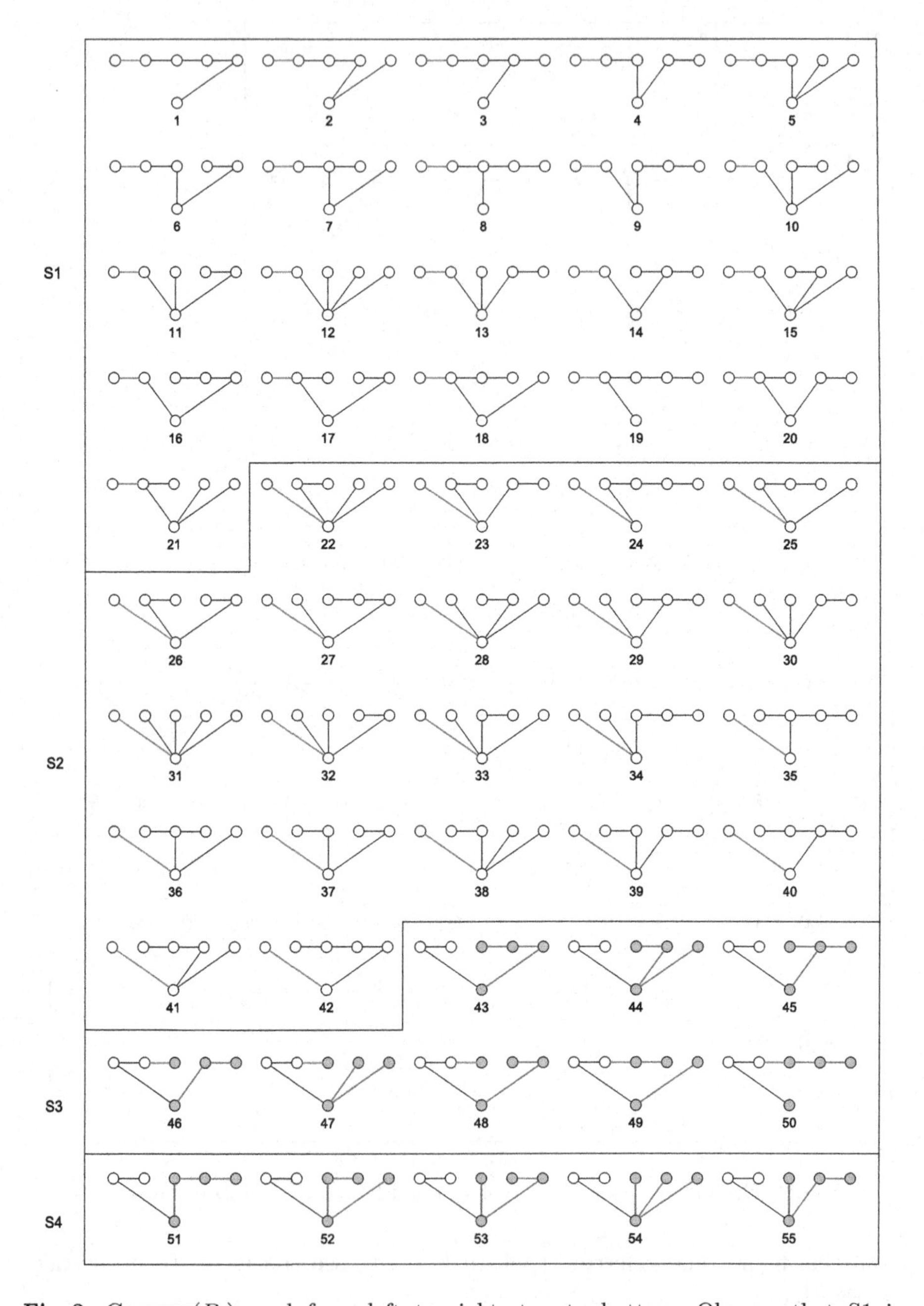


**Fig. 3.** GREEDY($P_6$) read from left to right, top to bottom. Observe that S1 is GREEDY($P_5$) with $v_6v_5$ added, S2 is the reverse of GREEDY($P_5$) with $v_6v_\infty$ added, S3 is GREEDY($P_4$) with $v_6v_5$ and $v_6v_\infty$ added, except the edge $v_4v_\infty$ is replaced by $v_4v_5$, and S4 is the last five trees of GREEDY($P_4$) in reverse order ($v_4v_\infty$ is now present) with $v_6v_5$ and $v_6v_\infty$ added.

- In S3, focusing only on the vertices $v_4, v_3, v_2, v_\infty$, the induced subgraphs correspond to GREEDY($P_4$), except whenever $v_4v_\infty$ is present, it is replaced with $v_4v_5$ (the last five trees).
- In S4, focusing only on the vertices $v_4, v_3, v_2, v_\infty$, the induced subgraphs correspond to the trees in GREEDY($P_4$) where $v_4v_\infty$ is present, in reverse order.

Generalizing these observations for all $n \geq 2$ leads to the recursive procedure GEN($k, s_1, varEdge$) given in Algorithm 1, which uses a global variable $T$ to store the current spanning tree with $n$ vertices. The parameter $k$ indicates the number of vertices under consideration; the parameter $s_1$ indicates whether or not to generate the trees in stage S1, as required by the trees for S4; and the parameter *varEdge* indicates whether or not a variable edge needs to be added as required by the trees for S3. The procedure REVGEN($k, s_1, varEdge$), which is left out due to space constraints, simply performs the operations from GEN($k, s_1, varEdge$) in reverse order. For each algorithm the base cases correspond to the edge moves in the listings GREEDY($P_2$) and GREEDY($P_3$).

Let $\mathcal{G}_n$ denote the listing obtained by initializing $T$ to $P_n$, printing $T$, and calling GEN($n, 1, 0$). Let $L_n$ denote the last tree in this listing. Let $\mathcal{R}_n$ denote the listing obtained by initializing $T$ to $L_n$, printing $T$, and calling REVGEN($n, 1, 0$). Thus, $\mathcal{R}_n$ is the listing $\mathcal{G}_n$ in reverse order.

**Algorithm 1**

```
 1: procedure GEN(k, s1, varEdge)
 2:     if k = 2 then                                            ▷ F2 base case
 3:         if varEdge then T ← T − v2v∞ + v2v3; PRINT(T)
 4:     else if k = 3 then                                       ▷ F3 base case
 5:         if s1 then
 6:             if varEdge then T ← T − v3v2 + v3v4; PRINT(T)
 7:             else T ← T − v3v2 + v3v∞; PRINT(T)
 8:         T ← T − v2v∞ + v2v3; PRINT(T)
 9:     else
10:         if s1 then
11:             GEN(k − 1, 1, 0)                                 ▷ S1
12:             if varEdge then T ← T − vkvk−1 + vkvk+1; PRINT(T)
13:             else T ← T − vkvk−1 + vkv∞; PRINT(T)
14:         REVGEN(k − 1, 1, 0)                                  ▷ S2
15:         T ← T − vk−1vk−2 + vk−1vk; PRINT(T)
16:         GEN(k − 2, 1, 1)                                     ▷ S3
17:         if k > 4 then T ← T − vk−2vk−1 + vk−2v∞; PRINT(T)
18:         REVGEN(k − 2, 0, 0)                                  ▷ S4
```

Our goal is to show that $\mathcal{G}_n$ exhaustively lists all trees in $\mathbf{T}_n$ and moreover, the listing is equivalent to GREEDY($P_n$). We accomplish this in two steps: first we show that $\mathcal{G}_n$ has the required size, then we show that $\mathcal{G}_n$ is equivalent to GREEDY($P_n$). Before doing this, we first comment on some notation. Let $T - v_i$

denote the tree obtained from $T$ by deleting the vertex $v_i$ along with all edges that have $v_i$ as an endpoint. Let $T + v_iv_j$ (resp. $T - v_iv_j$) denote the tree obtained from $T$ by adding (resp. deleting) the edge $v_iv_j$. For the remainder of this section, we will let $T_n$ denote the tree $T$ specified as a global variable for Gen and RevGen, and we let $T_{n-1} = T - v_n$ and $T_{n-2} = T - v_n - v_{n-1}$.

**Lemma 1.** *For* $n \geq 2$, $|\mathcal{G}_n| = |\mathcal{R}_n| = t_n$.

*Proof.* This result applies the Fibonacci recurrence and straightforward induction by counting the number of trees recursively generated in each stage S1, S2, S3, S4 as described earlier in this section. The base cases for $n = 2, 3, 4$ are easily verified by stepping through the algorithms. A formal proof is omitted due to space constraints. □

To prove the next result, we first detail some required terminology. If $T \in \mathbf{T}_n$, then we say that the operation of deleting an edge $v_iv_j$ and adding an edge $v_iv_k$ is a *valid* edge move of $T$ if the result is a tree in $\mathbf{T}_n$ that has not been generated yet. Conversely, if the result is not a tree in $\mathbf{T}_n$, or the result is a tree that has already been generated, then it is not a *valid* edge move of $T$. We say an edge $v_iv_j$ is *smaller* than edge $v_iv_k$ if $j < k$. An edge move $T_n - v_iv_j + v_iv_k$ is said to be *smaller* than another edge move $T_n - v_xv_y + v_xv_z$ if $i < x$, if $i = x$ and $j < y$, or if $i = x$, $j = y$, and $k < z$.

**Lemma 2.** *For* $n \geq 2$, $\mathcal{G}_n = \text{Greedy}(P_n)$ *and* $\mathcal{R}_n = \text{Greedy}(L_n)$.

*Proof.* By induction on $n$. It is straightforward to verify that the result holds for $n = 2, 3, 4$ by iterating through the algorithms. Assume $n > 4$, and that $\mathcal{G}_j = \text{Greedy}(P_j)$ and $\mathcal{R}_j = \text{Greedy}(L_j)$ for $2 \leq j < n$. We begin by showing $\mathcal{G}_n = \text{Greedy}(P_n)$, breaking the proof into each of the four stages for clarity.

S1: Since $n > 4$ and $s_1 = 1$, Gen$(n-1, 1, 0)$ is executed. By our inductive hypothesis, $\mathcal{G}_{n-1} = \text{Greedy}(P_{n-1})$. These must be the first trees for Greedy$(P_n)$, as any edge move involving $v_nv_{n-1}$ or $v_nv_\infty$ is larger than any edge move made by Greedy$(P_{n-1})$. Since Greedy$(P_{n-1})$ halts, it must be that no edge move of $T_{n-1}$ is possible. So Greedy$(P_n)$ must make the next smallest edge move, which is $T_n - v_nv_{n-1} + v_nv_\infty$. Since $T_n$ is a spanning tree, it follows that $T_n - v_nv_{n-1} + v_nv_\infty$ is also a spanning tree (and has not been generated yet), and therefore the edge move is valid. At this point, Gen$(n, 1, 0)$ also makes this edge move, by line 13.

S2: RevGen$(n-1, 1, 0)$ $(T_{n-1} = L_{n-1})$ is then executed. By our inductive hypothesis, $\mathcal{R}_n = \text{Greedy}(L_{n-1})$. Since Greedy$(L_{n-1})$ halts, it must be that no edge moves of $T_{n-1}$ are possible. At this point, $T_{n-1} = P_{n-1}$ because RevGen$(n-1, 1, 0)$ was executed. The smallest edge move now remaining is $T_n - v_{n-2}v_{n-1} + v_nv_{n-1}$. This results in $T_n = P_{n-2} + v_nv_{n-1} + v_nv_\infty$, which is a spanning tree that has not been generated. So, Greedy$(P_n)$ must make this move. Gen$(n, 1, 0)$ also makes this move, by line 15. So, $\mathcal{G}_n$ must equal Greedy$(P_n)$ up to the end of S2.

S3: Next, GEN$(n-2,1,1)$ starting with $T_{n-2} = P_{n-2}$ is executed. Since $varEdge = 1$, $v_{n-2}v_{n-1}$ is added instead of $v_{n-2}v_\infty$. GREEDY$(P_n)$ also adds $v_{n-2}v_{n-1}$ instead of $v_{n-2}v_\infty$ since $v_{n-2}v_{n-1}$ is smaller than $v_{n-2}v_\infty$ and this edge move results in a tree not yet generated. Other than the difference in this one edge move, which occurs outside the scope of $T_{n-2}$, GEN$(n-2,1,0)$ and GEN$(n-2,1,1)$ (both starting with $T_{n-2} = P_{n-2}$) make the same edge moves. Since we also know that $\mathcal{G}_{n-2} =$ GREEDY$(P_{n-2})$ by the inductive hypothesis, it follows that $\mathcal{G}_n$ continues to equal GREEDY$(P_n)$ after line 16 of GEN$(n,1,0)$ is executed. We know that $T_{n-2} = L_{n-2}$ after GEN$(n-2,1,0)$. However, $T_{n-2} = L_{n-2} - v_{n-2}v_\infty + v_{n-2}v_{n-1}$ instead because GEN$(n-2,1,1)$ was executed ($varEdge = 1$). It must be that no edge moves of $T_{n-2}$ are possible because GREEDY$(P_{n-2})$ (and GEN$(n-2,1,1)$) halted. The smallest edge move now remaining is $T_n - v_{n-2}v_{n-1} + v_{n-2}v_\infty$. This results in $T_{n-2} = L_{n-2}$. Also, $T_n = T_{n-2} + v_nv_{n-1} + v_nv_\infty$ is a spanning tree since $T_{n-2}$ is a spanning tree of $F_{n-2}$. So GREEDY$(P_n)$ makes this move. GEN$(n,1,0)$ also makes this move, by line 17, and thus $\mathcal{G}_n =$ GREEDY$(P_n)$ up to the end of S3.

S4: Finally, REVGEN$(n-2,0,0)$ starting with $T_{n-2} = L_{n-2}$ is executed. By our inductive hypothesis, $\mathcal{R}_{n-2} =$ GREEDY$(L_{n-2})$. From the recursive definition of REVGEN, it is clear that REVGEN$(n-2,0,0)$ and REVGEN$(n-2,1,0)$ make the same edge moves until REVGEN$(n-2,0,0)$ finishes executing. So, by the inductive hypothesis, the listings produced by REVGEN$(n-2,0,0)$ and GREEDY$(L_{n-2})$ are the same until this point, which is where GEN$(n,1,0)$ finishes execution. By Lemma 1 we have that $|\mathcal{G}_n| = t_n$. Therefore, GREEDY$(P_n)$ has also produced this many trees, and each tree is unique. Thus, it must be that all $t_n$ trees of $F_n$ have been generated. Thus, GREEDY$(P_n)$ also halts.

Since $\mathcal{G}_n$ and GREEDY$(P_n)$ start with the same tree, produce the same trees in the same order, and halt at the same place, it follows that $\mathcal{G}_n =$ GREEDY$(P_n)$. It is relatively straightforward to show that $\mathcal{R}_n =$ GREEDY$(L_n)$ by using similar arguments as above. This proof is omitted due to space constraints. □

Since $\mathcal{G}_n$ is the reversal of $\mathcal{R}_n$, we immediately obtain the following corollary.

**Corollary 1.** *For $n \geq 2$,* GREEDY$(P_n)$ *is equivalent to* GREEDY$(L_n)$ *in reverse order.*

Because GREEDY$(P_n)$ generates unique spanning trees of $F_n$, Lemma 1 together with Lemma 2 implies our first main result. This result answers **Research Question #3** and the first part of **Research Question #1** for fan graphs.

**Theorem 1.** *For $n \geq 2$, $\mathcal{G}_n =$* GREEDY$(P_n)$ *is a pivot Gray code listing of* $\mathbf{T}_n$.

To efficiently store the global tree $T$, the algorithms GEN and REVGEN can employ an adjacency list model where each edge $uv$ is associated only with the smallest labeled vertex $u$ or $v$. This means $v_\infty$ will never have any edges associated with it, and every other vertex will have at most 3 edges in its list. Thus the tree $T$ requires at most $O(n)$ space to store, and edge additions and deletions can be done in constant time. Our next result answers the second part of **Research Question #1** for fan graphs.

**Theorem 2.** *For $n \geq 2$, $\mathcal{G}_n$ and $\mathcal{R}_n$ can be generated in $O(1)$-amortized time using $O(n)$ space.*

*Proof.* For each call to $\textsc{Gen}(n, s_1, varEdge)$ where $n > 3$, there are at most four recursive function calls, and at least two new spanning trees generated. Thus, the total number of recursive calls made is at most twice the number of spanning trees generated. Each edge addition and deletion can be done in constant time as noted earlier. Thus each recursive call requires a constant amount of work, and hence the overall algorithm will run in $O(1)$-amortized time. There is a constant amount of memory used at each recursive call and the recursive stack goes at most $n-3$ levels deep; this requires $O(n)$ space. As mentioned earlier, the global variable $T$ stored as adjacency lists also requires $O(n)$ space. □

### 3.1 Ranking and Unranking

We now provide ranking and unranking algorithms for the listing $\mathcal{G}_n$ of all spanning trees for the fan graph $F_n$.

Given a tree $T$ in $\mathcal{G}_n$, we calculate its rank by recursively determining which stage (recursive call) $T$ is generated. We can determine the stage by focusing on the presence/absence of the edges $v_n v_{n-1}$, $v_n v_\infty$, $v_{n-2} v_\infty$, and $v_{n-2} v_{n-1}$. Based on the discussion of the recursive algorithm, there are $t_{n-1}$ trees generated in S1, $t_{n-1}$ trees generated in S2, $t_{n-2}$ trees generated in S3, and $t_{n-2} - t_{n-3}$ trees generated in S4. S3 is partitioned into two cases based on whether $v_{n-2} v_{n-1}$ ($varEdge$) is present. For the remainder of this section we will let $T_{n-1} = T - v_n$ and $T_{n-2} = T - v_n - v_{n-1}$.

For $n > 1$, let $R_n(T)$ denote the rank of $T$ in the listing $\mathcal{G}_n$. If $n = 2, 3, 4$, then $R_n(T)$ can easily be derived from Fig. 2. Based on the above discussion, for $n \geq 5$:

$$R_n(T) = \begin{cases} 2t_{n-1} + 2t_{n-2} - R_{n-2}(T_{n-2}) + 1 & \text{if } e_1, e_2, e_3 \in T \\ 2t_{n-1} + R_{n-2}(T_{n-2} + e_3) & \text{if } e_1, e_2, e_4 \in T, e_3 \notin T \\ 2t_{n-1} + R_{n-2}(T_{n-2}) & \text{if } e_1, e_2 \in T, e_3, e_4 \notin T \\ 2t_{n-1} - R_{n-1}(T_{n-1}) + 1 & \text{if } e_2 \in T, e_1 \notin T \\ R_{n-1}(T_{n-1}) & \text{if } e_1 \in T, e_2 \notin T \end{cases}$$

where $e_1 = v_n v_{n-1}$, $e_2 = v_n v_\infty$, $e_3 = v_{n-2} v_\infty$, and $e_4 = v_{n-2} v_{n-1}$.

Determining the tree $T$ at rank $r$ in the listing $\mathcal{G}_n$ follows similar ideas by constructing $T$ starting from a set of $n$ isolated vertices one edge at a time. Let $U_n(T, r, e)$ return the tree $T$ at rank $r$ for the listing $\mathcal{G}_n$. Initially, $T$ is the set of $n$ isolated vertices, $r$ is the specified rank, and $e = v_n v_\infty$. If $n = 2, 3, 4$, then $T$

is easily derived from Fig. 2. For these cases, if the edge $v_n v_\infty$ is present, then it is replaced by the edge $e$ that is passed in.

$$U_n(T,r,e) = \begin{cases} U_{n-1}(T{+}e_1, r, v_{n-1}v_\infty) & \text{if } 0 < r \leq t_{n-1}, \\ U_{n-1}(T{+}e, 2t_{n-1}{-}r{+}1, v_{n-1}v_\infty) & \text{if } t_{n-1} < r \leq 2t_{n-1}, \\ U_{n-2}(T{+}e_1{+}e, r{-}2t_{n-1}, e_4) & \text{if } 2t_{n-1} < r \leq 2t_{n-1}{+}t_{n-2}, \\ U_{n-2}(T{+}e_1{+}e, 2t_{n-1}{+}2t_{n-2}{-}r{+}1, e_3) & \text{otherwise.} \end{cases}$$

where $e_1 = v_n v_{n-1}$, $e_3 = v_{n-2}v_\infty$, and $e_4 = v_{n-2}v_{n-1}$.

Since the recursive formulae to perform the ranking and unranking operations each perform a constant number of operations and the recursion goes $O(n)$ levels deep, we arrive at the following result provided the first $2(n{-}2)$ Fibonacci numbers are precomputed. We note that the calculations are on numbers up to size $t_{n-1}$.

**Theorem 3.** *The listing $\mathcal{G}_n$ can be ranked and unranked in $O(n)$ time using $O(n)$ space under the unit cost RAM model.*

This answers **Research Question #2** for fan graphs.

## 4 Conclusion

We answer each of the three Research Questions outlined in Sect. 1 for the fan graph, $F_n$. First, we discovered a greedy algorithm that exhaustively listed all spanning trees of $F_n$ experimentally for small $n$ with an easy to define starting tree. We then studied this listings which led to a recursive construction producing the same listing that runs in $O(1)$-amortized time using $O(n)$ space. We also proved that the greedy algorithm does in fact exhaustively list all spanning trees of $F_n$ for all $n \geq 2$, by demonstrating the listing is equivalent to the aforementioned recursive algorithm. It is the first greedy algorithm known to exhaustively list all spanning trees for a non-trivial class of graphs. Finally, we provided an $O(n)$ time ranking and unranking algorithms for our listings, assuming the unit cost RAM model. It remains an interesting open problem to answer the research questions for other classes of graphs including the wheel, $n$-cube, and complete graph.

## References

1. Berger, I.: The enumeration of trees without duplication. IEEE Trans. Circuit Theory **14**(4), 417–418 (1967). https://doi.org/10.1109/TCT.1967.1082758
2. Bogdanowicz, Z.R.: Formulas for the number of spanning trees in a fan. Appl. Math. Sci. **2**(16), 781–786 (2008)
3. Chakraborty, M., Chowdhury, S., Chakraborty, J., Mehera, R., Pal, R.K.: Algorithms for generating all possible spanning trees of a simple undirected connected graph: an extensive review. Complex Intell. Syst. **5**(3), 265–281 (2018). https://doi.org/10.1007/s40747-018-0079-7

4. Char, J.: Generation of trees, two-trees, and storage of master forests. IEEE Trans. Circuit Theory **15**(3), 228–238 (1968)
5. Colbourn, C.J., Day, R.P., Nel, L.D.: Unranking and ranking spanning trees of a graph. J. Algorithms **10**(2), 271–286 (1989)
6. Cummins, R.: Hamilton circuits in tree graphs. IEEE Trans. Circuit Theory **13**(1), 82–90 (1966). https://doi.org/10.1109/TCT.1966.1082546
7. Feussner, W.: Ueber stromverzweigung in netzförmigen leitern. Ann. Phys. **314**(13), 1304–1329 (1902)
8. Gabow, H.N., Myers, E.W.: Finding all spanning trees of directed and undirected graphs. SIAM J. Comput. **7**(3), 280–287 (1978)
9. Hakimi, S.: On trees of a graph and their generation. J. Franklin Inst. **272**(5), 347–359 (1961)
10. Hartung, E., Hoang, H.P., Mütze, T., Williams, A.: Combinatorial generation via permutation languages. In: Proceedings of the Fourteenth Annual ACM-SIAM Symposium on Discrete Algorithms, pp. 1214–1225. SIAM (2020)
11. Hoang, H.P., Mütze, T.: Combinatorial generation via permutation languages. II. Lattice congruences. arXiv preprint arXiv:1911.12078 (2019)
12. Holzmann, C.A., Harary, F.: On the tree graph of a matroid. SIAM J. Appl. Math. **22**(2), 187–193 (1972). https://doi.org/10.1137/0122021
13. Kamae, T.: The existence of a Hamilton circuit in a tree graph. IEEE Trans. Circuit Theory **14**(3), 279–283 (1967)
14. Kapoor, S., Ramesh, H.: Algorithms for enumerating all spanning trees of undirected and weighted graphs. SIAM J. Comput. **24**(2), 247–265 (1995)
15. Kishi, G., Kajitani, Y.: On Hamilton circuits in tree graphs. IEEE Trans. Circuit Theory **15**(1), 42–50 (1968). https://doi.org/10.1109/TCT.1968.1082762
16. Matsui, T.: A flexible algorithm for generating all the spanning trees in undirected graphs. Algorithmica **18**, 530–543 (1997)
17. Mayeda, W., Seshu, S.: Generation of trees without duplications. IEEE Trans. Circuit Theory **12**(2), 181–185 (1965)
18. Merino, A., Mütze, T.: Efficient generation of rectangulations via permutation languages. In: 37th International Symposium on Computational Geometry (SoCG 2021). Schloss Dagstuhl-Leibniz-Zentrum für Informatik (2021)
19. Minty, G.: A simple algorithm for listing all the trees of a graph. IEEE Trans. Circuit Theory **12**(1), 120 (1965)
20. Shioura, A., Tamura, A.: Efficiently scanning all spanning trees of an undirected graph. J. Oper. Res. Soc. Jpn. **38**(3), 331–344 (1995)
21. Shioura, A., Tamura, A., Uno, T.: An optimal algorithm for scanning all spanning trees of undirected graphs. SIAM J. Comput. **26**(3), 678–692 (1997)
22. Smith, M.J.: Generating spanning trees. Master's thesis, University of Victoria (1997)
23. Williams, A.: The greedy gray code algorithm. In: Dehne, F., Solis-Oba, R., Sack, J.-R. (eds.) WADS 2013. LNCS, vol. 8037, pp. 525–536. Springer, Heidelberg (2013). https://doi.org/10.1007/978-3-642-40104-6_46
24. Winter, P.: An algorithm for the enumeration of spanning trees. BIT Numer. Math. **26**, 44–62 (1985)

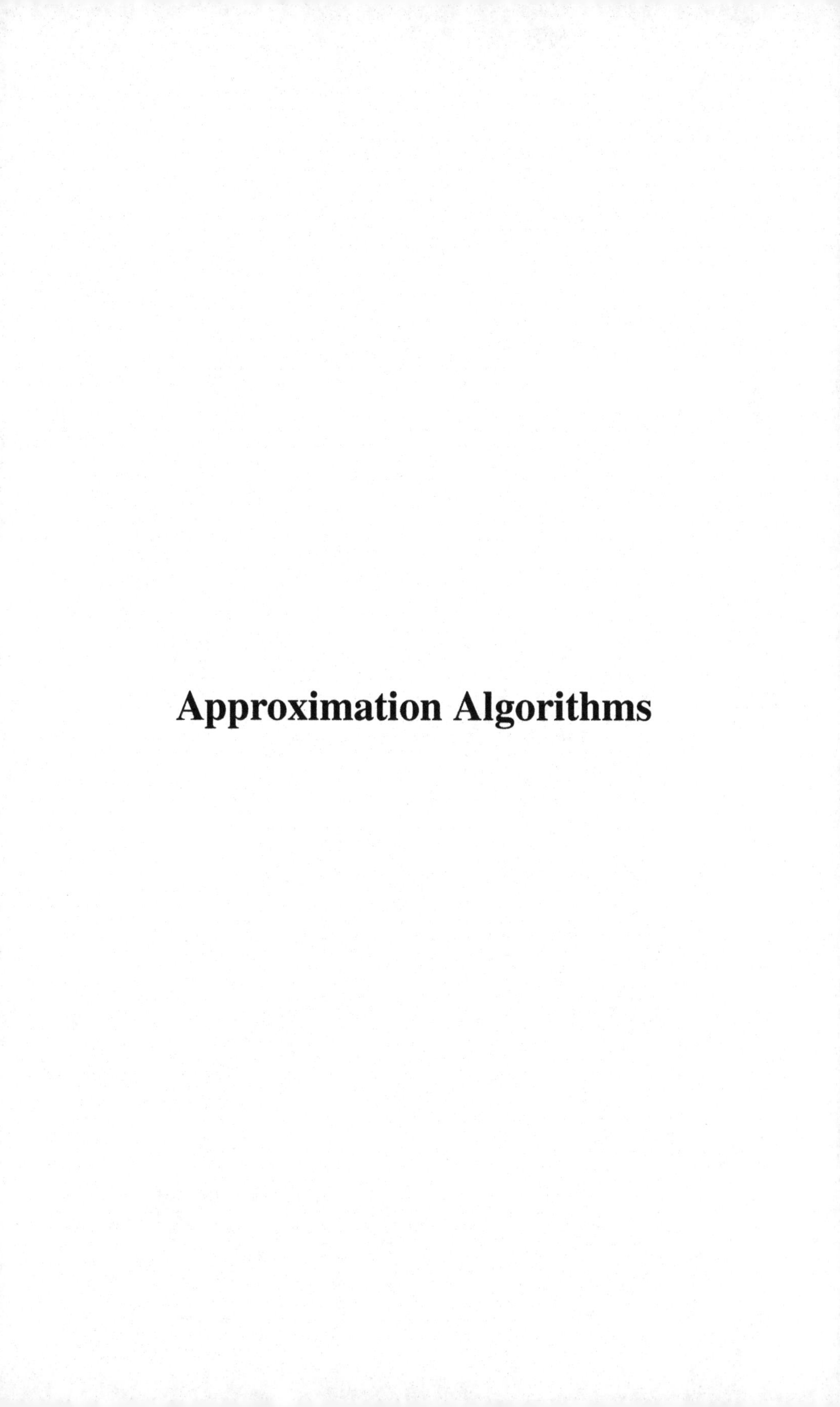

# Approximation Algorithms

# General Max-Min Fair Allocation

Sheng-Yen Ko[1], Ho-Lin Chen[2], Siu-Wing Cheng[3](✉), Wing-Kai Hon[4], and Chung-Shou Liao[1]

[1] Department of Industrial Engineering and Engineering Management, National Tsing Hua University, Hsinchu 30013, Taiwan
s107034524@m107.nthu.edu.tw, csliao@ie.nthu.edu.tw

[2] Department of Electrical Engineering, National Taiwan University, Taipei 106, Taiwan
holinchen@ntu.edu.tw

[3] Department of Computer Science and Engineering, HKUST, Hong Kong, China
scheng@cse.ust.hk

[4] Department of Computer Science, National Tsing Hua University, Hsinchu 30013, Taiwan
wkhon@cs.nthu.edu.tw

**Abstract.** In the general max-min fair allocation, also known as the Santa Claus problem, there are $m$ players and $n$ indivisible resources, each player has his/her own utilities for the resources, and the goal is to find an assignment that maximizes the minimum total utility of resources assigned to a player. We introduce an over-estimation strategy to help overcome the challenges of each resource having different utilities for different players. When all resource utilities are positive, we transform it to the machine covering problem and find a $\left(\frac{c}{1-\epsilon}\right)$-approximate allocation in polynomial running time for any fixed $\epsilon \in (0,1)$, where $c$ is the maximum ratio of the largest utility to the smallest utility of any resource. When some resource utilities are zero, we apply the approximation algorithm of Cheng and Mao [9] for the restricted max-min fair allocation problem. It gives a $\left(1 + 3\hat{c} + O(\delta\hat{c}^2)\right)$-approximate allocation in polynomial time for any fixed $\delta \in (0,1)$, where $\hat{c}$ is the maximum ratio of the largest utility to the smallest positive utility of any resource. The approximation ratios are reasonable if $c$ and $\hat{c}$ are small constants; for example, when the players rate the resources on a 5-point scale.

**Keywords:** Max-min allocation · Hypergraph matching · Approximation algorithms

## 1 Introduction

We consider the general max-min fair allocation problem. The input consists of a set $P$ of $m$ players and a set $R$ of $n$ indivisible resources. Each player $p \in P$

Cheng is supported by Research Grants Council, Hong Kong, China (project no. 16207419). Liao is supported by MOST Taiwan under Grant 109-2634-F-007-018. Chen is supported by MOST Taiwan under Grant 107-2221-E-002-031-MY3.


C.-Y. Chen et al. (Eds.): COCOON 2021, LNCS 13025, pp. 63–75, 2021.
https://doi.org/10.1007/978-3-030-89543-3_6

has his/her own non-negative utilities for the $n$ resources, and we denote the utility of the resource $r$ for $p$ by $v_{p,r}$. In other words, each resource $r$ has a set of non-negative utilities $\{v_{p,r} : p \in P\}$, one for each player. For every subset $S$ of resources, define the total utility of $S$ for $p$ to be $v_p(S) = \sum_{r \in S} v_{p,r}$. An allocation is a disjoint partition of $R$ into $\{S_p : p \in P\}$ such that $S_p \subseteq R$ for all $p \in P$, and $S_p \cap S_q = \emptyset$ for any $p \neq q$. That is, $p$ is assigned the resources in $S_p$. The *max-min fair allocation problem* is to find an allocation that maximizes $\min\{v_p(S_p) : p \in P\}$.

The problem has received considerable attention in recent decades. A related problem is a classic scheduling problem that minimizes the maximum makespan of scheduling on unrelated parallel machines. The problem has the same input as the max-min fair allocation problem. The only difference between them is that the goal of the scheduling problem is to minimize the maximum load over all machines. Lenstra et al. [16] proposed a 2-approximation algorithm by rounding the relaxation of the assignment linear programming model (LP). However, Bezáková and Dani [6] proved that the assignment LP cannot guarantee the same performance on the general max-min allocation problem.

*The machine covering problem* is a special case of the general max-min fair allocation problem where the objective is to assign $n$ jobs to $m$ parallel identical machines so that the minimum machine load is maximized. Every job (resource) has the same positive utility for every machine (player), i.e., every $r \in R$ has a positive value $v_r$ such that $v_{p,r} = v_r > 0$ for all $p \in P$. Deuermeyer et al. [13] proved that the heuristic LPT algorithm returns a $\frac{4}{3}$-approximation allocation. Csirik et al. [11] improved the approximation ratio to $\frac{4m-2}{3m-1}$. Later, Woeginger [18] presented a polynomial time approximation scheme to develop a $\frac{1}{1-\epsilon}$-approximation algorithm, where $\epsilon > 0$. For another machine covering problem that considers the machine speed $s_p$ and the processing time $v_{p,r} = v_r/s_p$, Azar et al. [4] also proposed a polynomial time approximation scheme. Furthermore, the online machine covering problem was studied in [14,15] for identical machines.

Bansal and Sviridenko [5] proposed a stronger LP relaxation, the configuration LP, for the general max-min fair allocation problem. They showed that the integrality gap of the configuration LP is $\Omega(\sqrt{m})$, where $m$ is the number of players. Based on the configuration LP, Asadpour and Saberi [3] developed an approximation algorithm that achieves an approximation ratio of $O(\sqrt{m}\log^3 m)$ by rounding. Later, Saha and Srinivasan [17] reduced the approximation ratio to $O(\sqrt{m \log m})$. Chakrabarty et al. [7] developed a method to provide a trade-off between the approximation ratio and the running time: for all $\delta \in (0, 1)$, an approximation ratio of $O(m^\delta)$ can be obtained in $O(m^{1/\delta})$ time.

Bansal and Sviridenko [5] also introduced an interesting restricted max-min fair allocation. In the restricted case, each resource has the same utility $v_r$ for all players who are interested in it, that is, $v_{p,r} \in \{0, v_r\}$ for all $p \in P$. They proposed an $O(\frac{\log\log m}{\log\log\log m})$-approximation algorithm by rounding the configuration LP. Later, Asadpour et al. [2] used the bipartite hypergraph matching technique to attack the restricted max-min fair allocation problem. They used local search to show that the integrality gap of the configuration LP is at most

4. However, it is not known whether the local search in [2] runs in polynomial time. Inspired by [2], Annamalai et al. [1] designed an approximation algorithm that enhances the local search with a *greedy player* strategy and a *lazy update* strategy. Their algorithm runs in polynomial time and achieves an approximation ratio of $12.325 + \delta$. Cheng and Mao [8] adjusted the greedy strategy in a more flexible and aggressive way, and they successfully lowered the approximation ratio to $6 + \delta$. Very recently, they introduced the *limited blocking* idea and improved the ratio to $4 + \delta$ [9,10]. This ratio was also obtained by Davies et al. [12]. Table 1 lists the related results.

**Table 1.** Results on the max-min allocation problem

| Problem | Approximation ratio | Running time | Ref. |
|---|---|---|---|
| Restricted | $O(\frac{\log\log m}{\log\log\log m})$ | $poly(m, n)$ | [5] |
| General | $O(\sqrt{m}\log^3 m)$ | $poly(m, n)$ | [3] |
| General | $O(\sqrt{m\log m})$ | $poly(m, n)$ | [17] |
| General | $O(m^\delta)$ | $n^{O(1/\delta)}$ | [7] |
| Restricted | $12.325 + \delta$ | $poly(m, n) \cdot m^{poly(1/\delta)}$ | [1] |
| Restricted | $6 + \delta$ | $poly(m, n) \cdot m^{poly(1/\delta)}$ | [8] |
| Restricted | $4 + \delta$ | $poly(m, n) \cdot m^{poly(1/\delta)}$ | [9,12] |
| General (positive utilities) | $c/(1 - \epsilon)$ | $O(n \log m)$ | This paper |
| General | $1 + 3\hat{c} + O(\delta\hat{c}^2)$ | $poly(m, n) \cdot m^{poly(1/\delta)}$ | This paper |

The major challenge for the general max-min fair allocation problem is that a resource may have different utilities for different players. Our key idea is to use a *player-independent* value to estimate the value of a particular set: for every resource $r \in R$, its over-estimated utility is $v_r^{max} = \max\{v_{p,r} : p \in P\}$. Consider an instance where every utility in the set $\{v_{p,r} : p \in P, r \in R\}$ is positive. Such a problem setting is closely related to the machine covering problem. The main difference is that players are not identical and players have their own preferences for the resources. Using the player-independent over-estimation, we can transform this case to the machine covering problem. Then, we can apply the currently best algorithm proposed by [18] to obtain an allocation in which every player gets at least $(1 - \epsilon)T_{oe}^*$ worth of resources, where $T_{oe}^*$ denotes the optimal solution for the transformed machine covering problem. Due to the over-estimation before the transformation, the allocation may be off by an additional factor of $c = \max\{v_{p,r}/v_{q,r} : r \in R \wedge p, q \in P\}$. This gives our first result which is summarized in Theorem 1 below. Further discussions are provided in Sect. 2.

**Theorem 1.** *For all $\epsilon \in (0, 1)$, there is a polynomial-time $(\frac{c}{1-\epsilon})$-approximation algorithm for the case that every resource has a positive utility for every player.*

The problem becomes much more complicated when some resource has zero utility for some players. Our strategy is to apply the algorithm of Cheng and Mao [9,10] for the restricted max-min fair allocation. To this end, we use the over-estimation strategy so that we can discuss active and inactive resources using maximum resource utilities when applying the technique of limited blocking as in [9,10]. Interestingly, we use each player's own utilities for the resources when running the algorithm although players may have different utilities for the same resource. Then, in the analysis, we use the over-estimation strategy again to reconcile the analysis in [9,10] for the restricted case with the general case that we consider. Since the approximation ratio given in [9] is $4 + \delta$, one may think that the approximation ratio is 4 times the over-estimation factor plus some low-order terms. We adapt the analysis in [9,10] to show a better approximation ratio of $1+3\hat{c}+O(\delta\hat{c}^2)$, where $\hat{c} = \max\{v_{p,r}/v_{q,r} : r \in R \wedge p,q \in P \wedge v_{q,r} > 0\}$.

**Theorem 2.** *For all $\delta \in (0,1)$, there is a $(1+3\hat{c}+O(\delta\hat{c}^2))$-approximation algorithm that runs in polynomial time for the case that some resources have zero utility for some players.*

The approximation ratios in Theorems 1 and 2 are reasonable if $c$ and $\hat{c}$ are small; for example, when the players rate the resources on a 5-point scale.

## 2 Every Resource Has a Positive Utility for Every Player

In this section, we consider the general max-min fair allocation problem in which all utilities are positive. This model is similar to the machine covering problem, but the major difference is that every player has his/her own preferences for the resources. There is a polynomial time approximation scheme for the machine covering problem, proposed by Woeginger [18], that achieves an approximation ratio of $1/(1-\epsilon)$ in polynomial time.

Fr each resource $r \in R$, we use $v_r^{\max} = \max\{v_{p,r} : p \in P\}$ as the player-independent utility for $r$. Hence all players become identical. This transformed problem is exactly the machine covering problem which we denote by $H'$. Let $H$ denote the original problem before the transformation. Let $T^*$ be the optimal target value for $H$, and let $T^*_{oe}$ be the optimal target value for $H'$. So $T^* \leq T^*_{oe}$. Using the PTAS algorithm [18], we can find an approximation allocation $\{S_p : p \in P\}$, where $\sum_{r\in S_p} v_r^{\max} \geq (1-\epsilon)T^*_{oe}$ for every $p \in P$. When we consider the actual value $v_p(S_p)$, we have to allow for the over-estimation factor $c = \max\{v_{p,r}/v_{q,r} : r \in R \wedge p,q \in P\}$.

The definition of $c$ implies that $\sum_{r\in S_p} v_r^{\max} \leq c\cdot v_p(S_p)$, which further implies that $(1-\epsilon)T^*_{oe} \leq \sum_{r\in S_p} v_r^{\max} \leq c\cdot v_p(S_p)$. That is, we guarantee that every player receives at least $(1-\epsilon)T^*_{oe}/c$ worth of resources. Since $T^*_{oe} \geq T^*$, the allocation is a $\left(\frac{c}{1-\epsilon}\right)$-approximation for the original problem $H$. This completes the discussion of Theorem 1.

# 3 Some Resources Have Zero Utility for Some Players

As mentioned previously, we will combine the over-estimation and the algorithm in [9,10] for the restricted max-min allocation problem. We will guess a target value $T$ of the general max-min allocation problem and then try to find an allocation in which the resources assigned to every player have a total utility of at least $\lambda T$. Depending on whether we succeed or not, we increase or decrease $T$ correspondingly in order to zoom into the value $T^*$ of the optimal max-min allocation. The initial range for $T$ for binary search is $(0, \frac{1}{m}\sum_{r\in R} v_r^{max})$.

In the rest of this section, we assume that $T = 1$, which can be enforced by scaling all resource utilities, and we describe how to find an allocation such that every player obtains resources with a total utility of at least $\lambda$.

## 3.1 Resources and Over-Estimation

We call a resource $r$ *fat* if $v_{p,r} \geq \lambda$ for all $p \in P$. Otherwise, there exists a player $p$ such that $v_{p,r} < \lambda$, and we call $r$ *thin* in this case. The input resources are thus divided into fat and thin resources.

Furthermore, we modify the resource utilities as follows: for every $r \in R$ and every $p \in P$, if $v_{p,r} > \lambda$, we reset $v_{p,r} := \lambda$. This modification does not affect our goal of finding an allocation in which the resources assigned to every player have a total utility of at least $\lambda$. Note that $v_{p,r}$ is left unchanged if it is at most $\lambda$. Therefore, fat resources remain fat, and thin resources remain thin. Note that $r$ still has zero utility for those players who are not interested in $r$, and different players may have different utilities for the same resource.

Since we have reset each $v_{p,r}$ so that it is at most $\lambda$, we have $v_r^{\max} \leq \lambda$. For any subset $D$ of thin resources, let $v_{\max}(D) = \sum_{r\in D} v_r^{\max}$. For every player $p$, $v_p(D)$ still denotes $\sum_{r\in D} v_{p,r}$.

## 3.2 Fat Edges and Thin Edges

For better resource utilization, it suffices to assign a player $p$ either a single fat resource (whose utilities are all equal to $\lambda$ after the above modification), or a subset $D$ of thin resources such that $v_p(D) \geq \lambda$. We model the above possible assignment of resources to players using a bipartite graph $G$ and a bipartite hypergraph $H$. The vertices of $G$ are the players and fat resources. For every player $p$ and every fat resource $r_f$, $G$ includes the edge $(p, r_f)$ which we call a *fat edge*. The vertices of $H$ are the players and thin resources. For every subset $D$ of thin resources and every player $p$, the hypergraph $H$ includes the edge $(p, D)$ if $v_p(D) \geq \lambda$.

## 3.3 Overview of the Algorithm

We focus on finding an allocation that corresponds to a maximum matching $M$ in $G$ and a subset $\mathcal{E}$ of hyperedges $H$ such that every player is incident to an edge in $M$ or $\mathcal{E}$, and no two edges in $M \cup \mathcal{E}$ share any resource.

To construct such an allocation, we start with an arbitrary maximum matching $M$ of $G$ and an empty $\mathcal{E}$, process unmatched players one by one in an arbitrary order, and update $M$ and $\mathcal{E}$ in order to match the next unmatched player. Once the algorithm matches a player, that player remains matched until the end of the algorithm. Also, although $M$ may be updated, it is always some maximum matching of $G$. We call any intermediate $M \cup \mathcal{E}$ a *partial allocation*.

Let $G_M$ be a directed graph obtained by orienting the edges of $G$ with respect to $M$ of $G$ as follows. If a fat edge $\{p, r_f\}$ belongs to the matching $M$, we orient $\{p, r_f\}$ from $r_f$ to $p$ in $G_M$. Conversely, if $\{p, r_f\}$ does not belong to the matching $M$, we orient $\{p, r_f\}$ from $p$ to $r_f$ in $G_M$.

Let $p_0$ be an arbitrary unmatched player with respect to the current partial allocation $M \cup \mathcal{E}$. We find a directed path $\pi$ from $p_0$ to a player $q_0$ in $G_M$. If $p_0 = q_0$, then $\pi$ a *trivial path*. In this case, if $q_0$ is covered by a thin edge $a$ that does not share any resource with the edges in $M \cup \mathcal{E}$, then we can update the partial allocation to be $M \cup (\mathcal{E} \cup \{a\})$ to match $p_0$. If $p_0 \neq q_0$, then $\pi$ a *non-trivial path*. Note that $\pi$ has an even number of edges because both $p_0$ and $q_0$ are players. For $i \geq 0$, every $(2i+1)$-th edge in $\pi$ does not belong to $M$, but every $(2i+2)$-th edge in $\pi$ does. It is an *alternating path* in the matching terminology. The last edge in $\pi$ is a matching edge $(r_f, q_0)$ in $M$ for some fat resource $r_f$. Suppose that $q_0$ is incident to a thin edge $a$ that does not share any resource with any edge in $M \cup \mathcal{E}$. Then, we can update $M$ to another maximum matching of $G$ by flipping the edge in $\pi$. That is, delete every $(2i+2)$-th edge in $\pi$ from $M$ and add every $(2i+1)$-th edge in $\pi$ to $M$. Denote this update of $M$ by flipping $\pi$ as $M \oplus \pi$. Consequently, $p_0$ is now matched by $M$. Although $q_0$ is no longer matched by $M$, we can regain $q_0$ by including the thin edge $a$. In all, the updated partial allocation is $(M \oplus \pi) \cup (\mathcal{E} \cup \{a\})$.

However, sometimes we cannot find a thin edge $a$ that is incident to $q_0$ and shares no resource with the edges in $M \cup \mathcal{E}$. Let $b$ be an edge in $M \cup \mathcal{E}$. If $a$ and $b$ share some resource, then $a$ is blocked by $b$. That is, if we want to add $a$ into $\mathcal{E}$, we must release the resources in $b$ first. Thus, $a$ is an *addable edge* and $b$ is a *blocking edge* that forbids the addition of $a$. We will provide the formal definitions of addable and blocking edges shortly. To release the resources covered by $b$, we need to reconsider how to match the player covered by $b$. This defines a similar intermediate subproblem that needs to be solved first, namely, finding a thin edge that is incident to the player covered by $b$ and shares no resource with the edges in $M \cup \mathcal{E}$. In general, the algorithm maintains a stack that consists of layers of addable and blocking edges; each layer correspond to some intermediate problems that need to be solved. Eventually, every blocking edge needs to be released in order that we can match $p_0$ in the end.

Annamalai et al. [1] introduced two ideas to enhance the above local search for the restricted max-min allocation problem. They are instrumental in obtaining a polynomial running time. First, when an *unblocked* addable edge is found, it is not used immediately to update the partial allocation. Instead, the algorithm waits until there are enough unblocked addable edges to reduce the number of blocking edges significantly. This ensures that the algorithm makes a substantial progress with each update of the partial allocation. This is called the *lazy update strategy*. Second, when the algorithm considers an addable edge $(p, D)$, it requires $v_p(D)$ to be a constant factor larger than $\lambda$. As a result, $(p, D)$ will induce more blocking edges, which will result in a *geometric growth* of the blocking edges in the layers from the bottom of the stack towards the top of the stack. This is called the *greedy player strategy*.

The greedy player strategy causes trouble sometimes, and a blocking edge may block too many addable edges. To this end, Cheng and Mao [9,10] introduced *limited blocking* which stops the resources in a blocking edge $b$ from being picked in an addable edge if $b$ shares too many resources with addable edges.

We provide more details of the algorithm in the remaining subsections.

### 3.4 Layers of Addable and Blocking Edges

For every thin edge $e$, we use $R_e$ to denote the resources covered by $e$. Given a set $\mathcal{X}$ of thin edges, we use $R(\mathcal{X})$ to denote the set of resources covered by the edges in $\mathcal{X}$.

Let $\Sigma = (L_0, L_1, \ldots, L_\ell)$ denote the current stack maintained by the algorithms, where each $L_i = (\mathcal{A}_i, \mathcal{B}_i)$ is a layer that consists of a set $\mathcal{A}_i$ of addable edges and a set $\mathcal{B}_i$ of blocking edges. That is, $\mathcal{B}_i = \{e \in \mathcal{E} : R_e \cap R(\mathcal{A}_i) \neq \emptyset\}$. The layer $L_{i+1}$ is on top of the layer $L_i$. The layer $L_0 = (\mathcal{A}_0, \mathcal{B}_0)$ at the stack bottom is initialized to be $(\emptyset, \{(p_0, \emptyset)\})$. It signifies that there is no addable edge initially, and replacing $(p_0, \emptyset)$ by some edge is equivalent to finding an edge that covers $p_0$ without causing any blocking. In general, when building a new layer $L_{\ell+1}$ in $\Sigma$, the algorithm starts with $\mathcal{A}_{\ell+1} = \emptyset$, $\mathcal{B}_{\ell+1} = \emptyset$, and addable and blocking edges will be added to $\mathcal{A}_{\ell+1}$ and $\mathcal{B}_{\ell+1}$.

We use $\mathcal{A}_{\leq i}$ to denote $\mathcal{A}_0 \cup \ldots \cup \mathcal{A}_i$. Similarly, $\mathcal{B}_{\leq i} = \mathcal{B}_0 \cup \ldots \cup \mathcal{B}_i$.

The current configuration of the algorithm can be specified by a tuple $(M, \mathcal{E}, \Sigma, \ell, \mathcal{I})$, where $M \cup \mathcal{E}$ is the current partial allocation, $\Sigma$ is the current stack of layers, $\ell$ is the index of the highest layer, and $\mathcal{I}$ is a set of thin edges in $H$ such that they cover the players of some edges in $\mathcal{B}_{\leq \ell}$ and each edge in $\mathcal{I}$ does not share any resource with any edge in $\mathcal{E}$. Although the edges in $\mathcal{I}$ can be added to $\mathcal{E}$ immediately to release some blocking edges in $\mathcal{B}_{\leq \ell}$, we do not do so right away in order to accumulate a larger $\mathcal{I}$ which will release more blocking edges in the future.

**Definition 1.** *Let $(M, \mathcal{E}, \Sigma, \ell, \mathcal{I})$ be the current configuration. A thin resource $r$ can be* **active** *or* **inactive**. *It is inactive if at least one of the following three conditions is satisfied: (a) $r \in R(\mathcal{A}_{\leq \ell} \cup \mathcal{B}_{\leq \ell})$, (b) $r \in R(\mathcal{A}_{\ell+1} \cup \mathcal{I})$, and (c) $r \in R_b$ for some $b \in \mathcal{B}_{\ell+1}$ and $v_{\max}(R_b \cap R(\mathcal{A}_{\ell+1})) > \beta\lambda$, where $\beta$ is a positive parameter to be specified later. If none is satisfied, then $r$ is active.*

Condition (c) is a modification of the *limited blocking strategy* introduced in [9,10] that fits with our over-estimation strategy. The utilities of inactive resources are disregarded in judging whether a thin edge contributes enough total utility to be considered an addable edge. Avoiding inactive resources, especially those in condition (c), helps to improve the approximation ratio.

Let $A_i$, $B_i$, and $I$ denote the sets of players covered by the edges in $\mathcal{A}_i$, $\mathcal{B}_i$, and $\mathcal{I}$, respectively. Let $A_{\leq i} = A_0 \cup \ldots A_i$, and let $B_{\leq i} = B_0 \cup \ldots \cup B_i$. Given two subsets of players $S$ and $T$, we use $f_M[S, T]$ to denote the maximum number of node-disjoint paths from $S$ to $T$ in $G_M$. The alternating paths in $G_M$ from $B_{\leq \ell}$ to $I$ and other players are relevant. If we flip the alternating paths to $I$, we can release some blocking edges in $\mathcal{B}_{\leq \ell}$ because they will be matched to fat resources instead. Also, if there is an alternating path from $B_{\leq \ell}$ to a player $p$, then we can look for a thin edge that covers $p$ to release a blocking edge.

**Definition 2.** *Let $(M, \mathcal{E}, \Sigma, \ell, \mathcal{I})$ be the current configuration. A player $p$ is* **addable** *if $f_M[B_{\leq \ell}, A_\ell \cup I \cup \{p\}] = f_M[B_{\leq \ell}, A_\ell \cup I] + 1$.*

**Definition 3.** *Let $(M, \mathcal{E}, \Sigma, \ell, \mathcal{I})$ be the current configuration. Given an addable player $p$, a thin edge $(p, D)$ in $H$ is* **addable** *if $D$ is a set of active thin resources and $v_p(D) \geq \lambda$.*

As mentioned before, the edges in $\mathcal{I}$ can be deployed any time to replace some blocking edges, but we only do so when we can release a significant number of blocking edges. When this is possible for a layer in $\Sigma$, we call that layer collapsible as defined below.

**Definition 4.** *Let $(M, \mathcal{E}, \Sigma, \ell, \mathcal{I})$ be the current configuration. Let $\mu \in (0, 1)$ be a parameter to be specified later. The layer $L_0$ in $\Sigma$ is collapsible if $f_M[B_0, I] = 1$ (note that $|B_0| = 1$), and for $i \in [1, \ell]$, the layer $L_i$ is* **collapsible** *if $f_M[B_{\leq i}, I] - f_M[B_{\leq i-1}, I] > \mu |B_i|$.*

### 3.5 The Local Search Step

We discuss how to match the next unmatched player $p_0$. Let $M \cup \mathcal{E}$ be the current partial allocation. Let $\Sigma = (L_0)$ be the initial stack. Let $\mathcal{I} = \emptyset$. We go into the Build phase to add a new layer to $\Sigma$. Afterwards, if some layer becomes collapsible, we go into the Collapse phase to prune $\Sigma$ and update the current partial allocation. Afterwards, we go back into the Build phase to add new layers to $\Sigma$ again. The above is repeated until $\Sigma$ becomes empty, which means that $p_0$ is matched eventually. We describe the Build and Collapse phases below.

**Build Phase.** We start with $\mathcal{A}_{\ell+1} = \mathcal{B}_{\ell+1} = \emptyset$. We grow $\mathcal{A}_{\ell+1}$ and $\mathcal{B}_{\ell+1}$ as long as we can find some appropriate thin edge $(p, D)$ in $H$:

- Suppose that there is an unblocked addable thin edge $(p, D)$. That is, $R(D) \cap R(\mathcal{E}) = \emptyset$. It is natural to add such an edge to $\mathcal{I}$, but for better resource utilization, there is no need to use the whole $D$ if $v_p(D)$ is way larger than $\lambda$. We greedily extract a *$\lambda$-minimal thin edge* $(p, D')$ from $(p, D)$: (i) $D'$ is a subset of $D$ such that $v_p(D') \geq \lambda$, and (ii) $v_p(D'') < \lambda$ for all subset $D'' \subset D'$. We add $(p, D')$ to $\mathcal{I}$.

- Assume that all addable thin edges are blocked. Suppose that there is blocked addable thin edge $(p, D)$ that is $(1+\gamma)\lambda$-minimal for an appropriate $\gamma \in (0,1)$ that will be specified later. That is, $v_p(D) \geq (1+\gamma)\lambda$ and for all $D' \subset D$, $v_p(D') < (1+\gamma)\lambda$. This is in accordance with the greedy player strategy. Let $E$ be the subset of thin edges in $\mathcal{E}$ that block $(p, D)$, i.e., $E = \{e \in \mathcal{E} : R_e \cap R(D) \neq \emptyset\}$. Add $(p, D)$ to $\mathcal{A}_{\ell+1}$ and update $\mathcal{B}_{\ell+1} := \mathcal{B}_{\ell+1} \cup E$.

Our definitions of $\lambda$-minimal and $(1+\gamma)\lambda$-minimal thin edges are player-dependent, in contrast to their player-independent counterparts in the restricted max-min case [9,10].

If no more edge can be added to $\mathcal{I}$, or $\mathcal{A}_{\ell+1}$ and $\mathcal{B}_{\ell+1}$, then we push $(\mathcal{A}_{\ell+1}, \mathcal{B}_{\ell+1})$ onto $\Sigma$ and increment $\ell$. If some layer becomes collapsible, we go into the Collapse phase; otherwise, we repeat the Build phase to construct another new layer.

**Collapse Phase.** Let $L_k$ be the lowest collapsible layer in $\Sigma$. We are going to prune $\mathcal{B}_k$ which will make all layers above $L_k$ invalid. Correspondingly, some of the unblocked addable edges in $\mathcal{I}$ also become invalid because they are generated using blocking edges in $\mathcal{B}_i$ for $i \in [k+1, \ell]$. So a key step is to decompose $\mathcal{I}$ into a disjoint partition $\bigcup_{i=0}^{\ell} \mathcal{I}_i$ such that, among the $f_M[B_{\leq \ell}, I]$ paths in $G_M$ from $B_{\leq \ell}$ to $I$, there are exactly $|I_i|$ paths from $B_i$ to $I_i$ for $i \in [0, \ell]$, where $I_i$ denotes the set of players covered by $\mathcal{I}_i$.

We remove $L_i$ for $i \geq k+1$ from $\Sigma$, and we also forget about $\mathcal{I}_i$ for $i \geq k+1$. We change $M$ by flipping the alternating paths from $B_k$ to $I_k$. The sources of these paths form a subset of $B_k$, which are covered by a subset $\mathcal{B}_k^* \subseteq \mathcal{B}_k$. The flipping of the alternating paths from $B_k$ to $I_k$ has the effect of replacing $\mathcal{B}_k^*$ by $\mathcal{I}_k$ in $\mathcal{E}$.

If $k = 0$, it means that the next unmatched player $p_0$ is now matched and the local search has succeeded. Otherwise, some of the addable edges in $\mathcal{A}_k$ may no longer be blocked due to the removal of $\mathcal{B}_k^*$ from $\mathcal{E}$. We reset $\mathcal{I} := \mathcal{I}_{\leq k-1}$. For each edge $(p, D) \in \mathcal{A}_k$ that becomes unblocked, we delete $(p, D)$ from $\mathcal{A}_k$, and if $f_M[B_{\leq k-1}, I \cup \{p\}] = f_M[B_{\leq k-1}, I] + 1$,[1] then we extract a $\lambda$-minimal thin edge $(p, D')$ from $(p, D)$ and add $(p, D')$ to $\mathcal{I}$. After pruning $\mathcal{A}_k$, we reset $\ell := k$.

We repeat the above as long as some layer in $\Sigma$ is collapsible. When this no longer the case and $p_0$ is not matched yet, we go back to the Build phase.

### 3.6 Analysis

For any $\delta \in (0,1)$, we show that we can set $\gamma = \Theta(\delta)$, $\beta = \gamma^2$, and $\mu = \gamma^3$ so that the local search runs in polynomial time, and if the target value 1 is feasible, the local search returns an allocation that achieves a value of $\lambda = 1/(1+3\hat{c}+O(\delta\hat{c}^2))$. Recall that $\hat{c} = \max\{v_{p,r}/v_{q,r} : r \in R \wedge p, q \in P \wedge v_{q,r} > 0\}$.

The key is to show that the stack $\Sigma$ has logarithmic depth for these choices of $\gamma$, $\beta$ and $\mu$. The numbers of blocking edges $|\mathcal{B}_i|$ for $i \in [0, \ell]$ induce a signature

[1] As proved in [9,10], this is equivalent to checking the addability of player $p$.

vector that increases lexicographically as the local search proceeds. Therefore, if $\Sigma$ has logarithmic depth, the local search must terminate in polynomial time before we run out of all possible signature vectors. To show that $\Sigma$ has logarithmic depth, we are to prove that the number of blocking edges increases geometrically from one layer to the next as we go up $\Sigma$.

First, we extract Lemmas 1–3 from [9,10]; either the original proofs still hold or minor adaptations work for our setting.

**Lemma 1.** *For $i \in [0, \ell]$, let $z_i = |\mathcal{A}_i|$ right after the creation of the layer $L_i$. Whenever no layer in collapsible, $|A_{i+1}| \geq z_{i+1} - \mu|B_{\leq i}|$ for $i \in [0, \ell - 1]$.*

**Lemma 2.** *For each blocking edge $b \in \mathcal{B}_i$, there exists an edge $a \in \mathcal{A}_i$ such that $v_{max}\big(R_b \cap R(\mathcal{A}_i \setminus \{a\})\big) \leq \beta\lambda$.*

**Lemma 3.** *For $i \in [0, \ell]$, $|\mathcal{A}_i| < \big(1 + \frac{\beta}{\gamma}\big)|\mathcal{B}_i|$.*

The analogous version of Lemma 4 below in [9,10] gives the inequality $|\mathcal{B}'_i| < \frac{(2+\gamma)}{\beta}|\mathcal{A}_i|$ in the restricted max-min case. We prove that a similar bound with the extra over-estimation factor $\hat{c}$ holds for the general max-min case.

**Lemma 4.** *Let $\mathcal{B}'_i$ be the subset of $\mathcal{B}_i$ such that all resources in $\mathcal{B}'_i$ are inactive, i.e., $\mathcal{B}'_i = \big\{b \in \mathcal{B}_i \mid v_{max}\big(R_b \cap R(\mathcal{A}_i)\big) > \beta\lambda\big\}$. Then, $|\mathcal{B}'_i| < \frac{(2+\gamma)\hat{c}}{\beta}|\mathcal{A}_i|$.*

*Proof.* Summing over the edges in $\mathcal{B}'_i$ gives $v_{max}\big(R(\mathcal{B}'_i) \cap R(\mathcal{A}_i)\big) > \beta\lambda|\mathcal{B}'_i|$. Every edge $(p, D) \in \mathcal{A}_i$ is $(1+\gamma)\lambda$-minimal by definition, so $v_p(D) \leq (2+\gamma)\lambda$. Summing over the edges in $\mathcal{A}_i$ gives $\sum_{(p,D)\in\mathcal{A}_i} v_p(D) \leq (2+\gamma)\lambda|\mathcal{A}_i|$. The definition of $\hat{c}$ implies that $v_r^{\max} \leq \hat{c}\, v_{p,r}$, which implies that $v_{max}(R(\mathcal{A}_i)) \leq \sum_{(p,D)\in\mathcal{A}_i} \hat{c}\, v_p(D) \leq (2+\gamma)\lambda\hat{c}\,|\mathcal{A}_i|$. Combining the inequalities above gives $\beta\lambda|\mathcal{B}'_i| < v_{max}(R(\mathcal{B}'_i) \cap R(\mathcal{A}_i)) \leq v_{max}(R(\mathcal{A}_i)) \leq (2+\gamma)\lambda\hat{c}\,|\mathcal{A}_i|$, which implies that $|\mathcal{B}'_i| < \frac{(2+\gamma)\hat{c}}{\beta}|\mathcal{A}_i|$. □

Lemma 4 is instrumental to proving Lemma 5 which is the key to showing a geometric growth in the numbers of blocking edges.

**Lemma 5.** *Assume that the target value 1 is feasible for the general max-min allocation problem. Then, immediately after the construction of a new layer $L_{\ell+1}$, if no layer is collapsible, then $|\mathcal{A}_{\ell+1}| > 2\mu|\mathcal{B}_{\leq \ell}|$.*

We show how to use Lemma 5 to obtain the geometric growth.

**Lemma 6.** *If no layer is collapsible, then $\big|\mathcal{B}_{i+1}\big| > \frac{\gamma^3}{1+\gamma}\big|\mathcal{B}_{\leq i}\big|$. Hence, $\big|\mathcal{B}_{\leq i+1}\big| > \big(1 + \frac{\gamma^3}{1+\gamma}\big|\mathcal{B}_{\leq i}\big|\big)$.*

*Proof.* Let $(L_0, L_1, \ldots, L_\ell)$ be the current stack $\Sigma$. Take any $i \in [0, \ell - 1]$. Since the most recent construction of $L_{i+1}$, $L_{i+1}$ and any layer below it is not collapsible. If not, $L_{i+1}$ would be deleted, which means that there would be another construction of it after the most recent construction, a contradiction. Therefore,

Lemma 5 implies that $z_{i+1} > 2\mu|B_{\leq i}|$, where $z_{i+1}$ is the value of $|A_{i+1}|$ right after the construction of $L_{i+1}$. By Lemma 1, $|A_{i+1}| \geq z_{i+1} - \mu|B_{\leq i}|$. Substituting $z_{i+1} > 2\mu|B_{\leq i}|$ into this inequality gives $|A_{i+1}| > 2\mu|B_{\leq i}| - \mu|B_{\leq i}| = \mu|B_{\leq i}|$. Lemma 3 implies that $|B_{i+1}| > \frac{\gamma}{\gamma+\beta}|A_{i+1}| > \frac{\gamma\mu}{\gamma+\beta}|B_{\leq i}|$. Plugging in $\beta = \gamma^2$ and $\mu = \gamma^3$ gives $|B_{i+1}| > \gamma^3|B_{\leq i}|/(1+\gamma)$. □

The next result shows that a polynomial running time follows from Lemma 6.

**Lemma 7.** *Suppose that the target value 1 is feasible for the general max-min allocation problem. Then, the local search matches a player in poly(m, n) ·* $m^{poly(1/\delta)}$.

*Proof.* The proof follows the argument in [1]. Let $h = \gamma^3/(1+\gamma)$. Define the signature vector $(s_1, s_2, ..., s_\ell, \infty)$, where $s_i = \lfloor \log_{1/(1-\mu)}(|\mathcal{B}_i| h^{-i-1}) \rfloor$. By Lemma 6, $|\mathcal{B}_\ell| \geq h|\mathcal{B}_{\leq \ell-1}| \geq h|\mathcal{B}_{\ell-1}|$. So $\infty > s_\ell \geq s_{\ell-1}$, which means that the coordinates of the signature vector are non-decreasing. When a (lowest) layer $L_t$ is collapsed in the Collapse phase, we update $\mathcal{B}_t$ to $\mathcal{B}'_t$ where $(1-\mu)|\mathcal{B}_t| > |\mathcal{B}'_t|$. The signature vector is updated to $(s_1, s_2, ..., s'_t)$ where $s'_t \leq s_t - 1$. So the signature vector decreases lexicographically. By Lemma 6, the number of layers in $\Sigma$ is at most $\log_{1+h} m$, where $m$ is the number of players. One can verify that the sum of coordinates in every signature vector is at most $U^2$ where $U = \log m \cdot O(\frac{1}{\mu h} \log \frac{1}{h})$. Every signature vector corresponds to a distinct partition of an integer that is no more than $U^2$. By summing up the number of distinct partitions of integers that are no more than $U^2$, we get that the upper bound of $m^{O(\frac{1}{\mu h} \log \frac{1}{h})}$ on the number of signature vectors. Since $\gamma = \Theta(\delta)$, $\mu = \gamma^3$, and $h = \gamma^3/(1+\gamma)$, this upper bound is $m^{poly(1/\delta)}$. This also bounds the number of calls on BUILD and COLLAPSE. It is not difficult to make the construction of a layer and the collapse of a layer run in polynomial time. □

We have not discussed how to handle the case that the target value 1 is infeasible for the general max-min allocation problem. In this case, the local search must fail at some point. From the previous proofs, we know that as long as the conclusion of Lemma 5 holds immediately after the construction of a new layer $L_{\ell+1}$, that is, if no layer is collapsible, then $|\mathcal{A}_{\ell+1}| > 2\mu|\mathcal{B}_{\leq \ell}|$, the local search must succeed and finish in polynomial time. As a result, we must encounter a situation that no layer is collapsible and yet $|\mathcal{A}_{\ell+1}| \leq 2\mu|\mathcal{B}_{\leq \ell}$ for the first time during the local search. This situation can be checked explicitly and we can abort and guess the next target value. Since the conclusion of Lemma 5 has held so far, the running time up to the point of abortion is polynomial.

## 4 Conclusion

We provide two solutions for the general max-min fair allocation problem. If every resource has a positive utility for every player, the problem can be transformed to the machine covering problem using our over-estimation strategy. By using an existing polynomial time approximation scheme for the machine covering problem, we obtain a $(\frac{c}{1-\epsilon})$-approximation algorithm which runs in polynomial time, where $\epsilon$ is any constant in the range $(0, 1)$, and $c = \max\{v_{p,r}/v_{q,r} : r \in R \wedge p, q \in P\}$. If some resource has zero utility for some players, we show how to combine the over-estimation strategy with the approximation algorithm in [9] for the restricted max-min allocation problem to obtain an approximation ratio of $1 + 3\hat{c} + O(\delta\hat{c}^2)$ for any $\delta \in (0, 1)$ in polynomial time, where $\hat{c} = \max\{v_{p,r}/v_{q,r} : r \in R \wedge p, q \in P \wedge v_{q,r} > 0\}$. We conclude with two research questions. The first question is whether the approximation ratios presented here can be improved further. Despite its theoretical guarantee, the local search step is still quite challenging to implement. So the second question is whether there is a simpler algorithm that can also achieve a good approximation ratio in polynomial time.

## References


1. Annamalai, C., Kalaitzis, C., Svensson, O.: Combinatorial algorithm for restricted max-min fair allocation. ACM Trans. Algorithms **13**(3), 1–28 (2017)
2. Asadpour, A., Feige, U., Saberi, A.: Santa Claus meets hypergraph matchings. ACM Trans. Algorithms **8**(3), 1–9 (2012)
3. Asadpour, A., Saberi, A.: An approximation algorithm for max-min fair allocation of indivisible goods. In: Proceedings of the 39th Annual ACM Symposium on Theory of Computing, pp. 114–121 (2007)
4. Azar, Y., Epstein, L.: Approximation schemes for covering and scheduling in related machines. In: Jansen, K., Rolim, J. (eds.) APPROX 1998. LNCS, vol. 1444, pp. 39–47. Springer, Heidelberg (1998). https://doi.org/10.1007/BFb0053962
5. Bansal, N., Sviridenko, M.: The Santa Claus problem. In: Proceedings of the 38th Annual ACM Symposium on Theory of Computing, pp. 31–40 (2006)
6. Bezáková, I., Dani, V.: Allocating indivisible goods. ACM SIGecom Exchanges **5**(3), 11–18 (2005)
7. Chakrabarty, D., Chuzhoy, J., Khanna, S.: On allocating goods to maximize fairness. In: Proceedings of the 50th Annual IEEE Symposium on Foundations of Computer Science, pp. 107–116 (2009)
8. Cheng, S.W., Mao, Y.: Restricted max-min fair allocation. In: Proceedings of the 45th International Colloquium on Automata, Languages, and Programming, pp. 37:1–37:13 (2018)
9. Cheng, S.W., Mao, Y.: Restricted max-min allocation: approximation and integrality gap. In: Proceedings of the 46th International Colloquium on Automata, Languages, and Programming, pp. 38:1–38:13 (2019)
10. Cheng, S.W., Mao, Y.: Restricted max-min allocation: approximation and integrality gap. arXiv:1905.06084 (2019)
11. Csirik, J., Kellerer, H., Woeginger, G.: The exact LPT-bound for maximizing the minimum completion time. Oper. Res. Lett. **11**(5), 281–287 (1992)
12. Davies, S., Tothvoss, T., Zhang, Y.: A tale of Santa Claus, hypergraphs and matroids. In: Proceedings of the 31st Annual ACM-SIAM Symposium on Discrete Algorithms, pp. 2748–2757 (2020)
13. Deuermeyer, B., Friesen, D., Langston, M.: Scheduling to maximize the minimum processor finish time in a multiprocessor system. SIAM J. Algebraic Discret. Methods **3**, 190–196 (1982)

14. He, Y., Jiang, Y.: Optimal semi-online preemptive algorithms for machine covering on two uniform machines. Theoret. Comput. Sci. **339**(2), 293–314 (2005)
15. He, Y., Tan, Z.: Ordinal on-line scheduling for maximizing the minimum machine completion time. J. Comb. Optim. **6**, 199–206 (2002). https://doi.org/10.1023/A:1013855712183
16. Lenstra, J.K., Shmoys, D.B., Tardos, E.: Approximation algorithms for scheduling unrelated parallel machines. In: Proceedings of the 28th Annual Symposium on Foundations of Computer Science, pp. 217–224 (1987)
17. Saha, B., Srinivasan, A.: A new approximation technique for resource-allocation problems. Random Struct. Algorithms **52**, 680–715 (2018)
18. Woeginger, G.J.: A polynomial-time approximation scheme for maximizing the minimum machine completion time. Oper. Res. Lett. **20**(4), 149–154 (1997)

# On the Approximation Hardness of Geodetic Set and Its Variants

Tom Davot[1(✉)], Lucas Isenmann[2], and Jocelyn Thiebaut[3]

[1] LIRMM, Univ Montpellier, CNRS, Montpellier, France
davot@lirmm.fr

[2] Aix-Marseille Université, Marseille, France
lucas.isenmann@laposte.net

[3] INSA Centre Val de Loire, Univ. Orléans, LIFO EA 4022, Orléans, France
jocelyn.thiebaut@univ-orleans.fr

**Abstract.** Given a graph, a geodetic set (resp. edge geodetic set) is a subset of its vertices such that every vertex (resp. edge) of the graph is on a shortest path between two vertices of the subset. A strong geodetic set is a subset S of vertices and a choice of a shortest path for every pair of vertices of S such that every vertex is on one of these shortest paths. The geodetic number (resp. edge geodetic number) of a graph is the minimum size of a geodetic set (resp. edge geodetic set) and the strong geodetic number is the minimum size of a strong geodetic set. We first prove that, given a subset of vertices, it is $\mathcal{NP}$-hard to determine whether it is a strong geodesic set. Therefore, it seems natural to study the problem of maximizing the number of covered vertices by a choice of a shortest path for every pair of a provided subset of vertices. We provide a tight 2-approximation algorithm to solve this problem. Then, we show that there is no $^{781}/_{780}$ polynomial-time approximation algorithm for edge geodetic number and strong geodetic number on subcubic bipartite graphs with arbitrarily high girth. We also prove that geodetic number and edge geodetic number are both LOG-$\mathcal{APX}$-hard, even on subcubic bipartite graphs with arbitrarily high girth. Finally, we disprove a conjecture of Iršič and Konvalinka by proving that the strong geodetic number can be computed in polynomial time in complete multipartite graphs.

## 1 Introduction

*Geodetic Number and Edge Geodetic Number.* A *geodesic* between two vertices of a graph $G$ is a path of minimum length between $x$ and $y$. The *geodetic number* of $G$ is the minimum size of a subset $X$ of the vertices such that, for every vertex $v$, there exists a geodesic between two vertices of $X$ containing $v$. The geodetic number of a graph has been introduced by Harary et al. in [13], where the authors showed that deciding whether a graph has a geodetic number less than an integer $k$ is $\mathcal{NP}$-complete. The complexity of this problem has also been investigated in several classes of graphs, such as bipartite graphs [11] and chordal graphs [10] where it remains $\mathcal{NP}$-complete. Recently, Chakraborty *et al.* proved that finding


C.-Y. Chen et al. (Eds.): COCOON 2021, LNCS 13025, pp. 76–88, 2021.
https://doi.org/10.1007/978-3-030-89543-3_7

the geodetic number of a graph is $\mathcal{NP}$-hard on planar graphs with maximum degree six and line graphs [6]. They also proved in [7] that, unless $\mathcal{P} = \mathcal{NP}$, there is no polynomial time $o(\log n)$-approximation algorithm for computing the geodetic number of a graph, even on graphs that have a universal vertex and where $n$ stands for the number of vertices in the input graph.

The edge version of the geodetic number has been introduced independently in [3] and in [22]. A subset $X$ of the vertices is an *edge geodetic set* if, for every edge $e$, there is a geodesic between two vertices of $X$ containing $e$. The *edge geodetic number* of $G$ is the size of the smallest geodetic set of $G$. Note that, given a graph $G$, the geodetic number of $G$ is smaller than its edge geodetic number. This edge version is also known to be $\mathcal{NP}$-hard [3]. This problem has been studied on several classes of graphs, such as Cartesian products [1,21] and fuzzy graphs [20]. From a structural point of view, Santhakumaran and Ullas Chandran characterized graphs with a prescribed edge geodetic number [23]. For more results and motivations about geodetic sets, see [5].

*Strong Geodetic Number and Strong Edge Geodetic Number.* A subset of vertices $X$ is a *strong geodetic set* if there exists a function $\tilde{I}$ that associates a unique geodesic to each pair of vertices of $X$ and such that every vertex $v$ is contained in a geodesic $\tilde{I}(a, b)$, where $\{a, b\} \subseteq X$. In the following, we call such a function a *geodesic assignation* for $X$. The *strong geodetic number* of a graph $G$ is the size of the smallest strong geodetic set. The strong geodetic number has been introduced recently by Arokiaraj *et al.* [2]. In their original paper, the authors motivate this variation by social network applications. Furthermore, they also prove that this problem is $\mathcal{NP}$-complete. Note that it remains $\mathcal{NP}$-hard even when restricted to bipartite graphs [15]. This problem has been studied on complete Apollonian networks [2], grids and cylinders [16], and on Cartesian product of graphs [12], on complete bipartite graphs [14], on complete multipartite graphs [15] and on outerplanar graphs [18]. Connections to the diameter of the graph were studied in [14] and to the isometric path problem [2].

Finally, the edge version of the strong geodetic problem, where we want to cover every edge of the graph, has been introduced by Manuel *et al.* [17] and were proved $\mathcal{NP}$-complete. Zmazek recently studied the values of the edge strong geodetic number on grids [25].

*Our Results.* The results of our paper are divided in four sections. In Sect. 3, we propose a variant of the strong geodetic problem where, given a subset $S$ of vertices, the question is to determine whether $S$ is a strong geodesic set of the graph. Using a reduction from MONOTONE BALANCED 3-SAT-(4), we prove that this problem is $\mathcal{NP}$-hard. Then, we consider in Sect. 4 the problem of maximizing the number of covered vertices by a choice of a geodesic for each pair of a provided subset of vertices, and provide a tight 2-approximation algorithm to solve it. In Sect. 5, we reduce the geodetic problems from SET COVER. We first give it in the general case, and then we adapt the previous construction on bipartite graphs with arbitrarily high girth. Using the previous reductions, we show in Sect. 6 that there is no approximation of EDGE GEODETIC NUMBER with an approximation factor better than 781/780. We also prove that geodetic number and edge geodetic number are both LOG-$\mathcal{APX}$-hard, even on subcubic bipartite graphs with arbitrarily high girth. Finally, in Sect. 7, we give a polynomial time algorithm which computes the STRONG GEODETIC NUMBER of complete multipartite graphs, disproving the conjecture of [15] which states that STRONG GEODETIC NUMBER is $\mathcal{NP}$-hard on complete multipartite graphs.

Due to space constraints, the proofs have been omitted. However, a full version can be found in https://hal-lirmm.ccsd.cnrs.fr/lirmm-03328636.

## 2 Notations

We first introduce some notations and formally define the problems. Given a set $X$, we denote by $\mathcal{P}_2(X)$ the set of its pairs. Given two sets $X$ and $Y$, we denote by $X \sqcup Y$ the union $X \cup Y$ when $X$ and $Y$ are disjoint. Let $G$ be a graph, we denote by $V(G)$ its set of vertices and by $E(G)$ its set of edges. We denote by $D_1(G)$ the set of vertices of degree one in $G$. Let $X$ be a subset of vertices of $G$ and $x$ be a vertex, we say that $x$ is *selected* by $X$ if $x \in X$ and that $x$ is *covered* by $X$ if $x$ is contained in a geodesic between two vertices of $X$ (or simply selected or covered if there is no ambiguity on $X$). Likewise, let $uv$ be an edge, we say that $uv$ is *covered* by $X$ (or simply covered) if $uv$ is contained in a geodesic between two vertices of $X$.

Let $g$ be a path that contains the vertices $u$ and $v$. We denote by $g[u, v]$ the subpath of $g$ with extremities $u$ and $v$. Furthermore, we denote by $V(g)$ the vertices of $g$. Similarly, given a geodesic assignation $\tilde{I}$ for a set of vertices $X$, we denote by $V(\tilde{I})$ the vertices covered by the geodesics of $\tilde{I}$.

We now introduce the problems studied in this work.

(STRONG) (EDGE) GEODETIC NUMBER
**Input:** a simple graph $G$ and an integer $k$.
**Question:** is there a (strong) (edge) geodetic set $X \subseteq V$ of size $k$?

The following already known property will be fundamental in the proofs of our reductions as it helps to force some vertices to be part of a (strong) (edge) geodetic set.

*Property 1.* If $G$ is a graph and $X$ is a solution of any geodesic problem, then we have $D_1(G) \subseteq X$.

## 3 Hardness to Find a Geodesic Assignation

In the proof that computing the strong geodesic number is $\mathcal{NP}$-complete, the geodesic assignation is rather trivial [2]. In this section, we show that determining if a set of vertices is a strong geodetic set (*i.e.* computing a geodesic assignation) is in itself $\mathcal{NP}$-complete. To do so, we reduce from a special case of 3-SAT called MONOTONE BALANCED 3-SAT-(4). In this variant, the boolean formula is composed of *monotone* clauses, that is, clauses that contains only positive literals or only negative literals. MONOTONE BALANCED 3-SAT-(4) is defined as follows.

MONOTONE BALANCED 3-SAT-(4)
**Input:** a monotone 3-SAT formula $\varphi$ where each variable occurs exactly two times positively and two times negatively.
**Question:** is $\varphi$ satisfiable?

Darman and Döcker showed that this problem is $\mathcal{NP}$-complete [8]. We introduce the following construction.

**Construction 1.** *Let $\varphi$ be a* MONOTONE BALANCED 3-SAT-(4) *formula, we construct the following graph $G$:*

- *For each clause $C_j$, introduce a vertex $q_j$.*
- *For each variable $x_i$, introduce two edges $v_i^0 v_i^1$ and $u_i^0 u_i^1$. Furthermore, let $C_j$ and $C_{j'}$, with $j < j'$ be the two clauses where $x_i$ occurs with the same polarity (i.e. it appears positively in both clauses, or negatively in both), construct a path $(v_i^1, q_j, q_{j'}, u_i^1)$.*
- *For each pair of vertices $v_i^1$ and $u_{i'}^1$ with $i \neq i'$, introduce a vertex $t_{i,i'}$ and construct the path $(v_i^1, t_{i,i'}, u_{i'}^1)$.*
- *Finally, construct two vertices $k_v$ and $k_u$, and for each variable $x_i$, introduce the edges $v_i^1 k_v$ and $u_i^1 k_u$.*

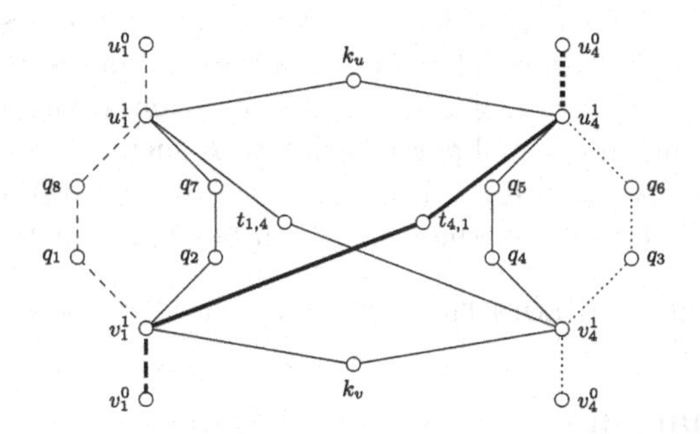


**Fig. 1.** Example of a subgraph induced by Construction 1. In the Boolean formula, the variable $x_1$ appears positively in $C_2$ and $C_7$ and negatively in $C_1$ and $C_8$. The variable $x_4$ appears positively in $C_3$ and $C_6$ and negatively in $C_4$ and $C_5$. The paths $p_1$, $\bar{p}_1$ and $p_{1,4}$ are depicted in dotted, dashed and bold, respectively.

For each variable $x_i$, let $C_j$ and $C_{j'}$ (resp. $C_k$ and $C_{k'}$) with $j < j'$ be the clauses where $x_i$ occurs positively (resp. negatively). We denote by $p_i$ the path $(v_i^0, v_i^1, q_j, q_{j'}, u_i^1, u_i^0)$, by $\bar{p}_i$ the path $(v_i^0, v_i^1, q_k, q_{k'}, u_i^1, u_i^0)$ and by $p_{i,i'}$ the path $(v_i^0, v_i^1, t_{i,i'}, u_{i'}^1, u_{i'}^0)$, for any $i \neq i'$. An example of a graph produced by Construction 1 is depicted in Fig. 1.

**Algorithm 1:** Greedy Algorithm

**Data:** A graph $G$ and a set of vertices $V' \subseteq V(G)$.
**Result:** A geodetic assignation $\tilde{I}$ for $V'$.

```
1 A ← P_2(V') ;
2 while A ≠ ∅ do
3     {u, v} ← first element of A;
4     g ← geodesic between u and v that maximizes |V(g) \ V(Ĩ)|;
5     Set Ĩ(u, v) := g; A ← A\{{u, v}};
6 end
7 return Ĩ;
```

**Lemma 1.** *Let $\varphi$ be a* MONOTONE BALANCED 3-SAT-(4) *formula and $G$ its graph resulting from Construction 1. Let $\tilde{I}$ be a geodesic assignation for $D_1(G)$. It is possible to construct a geodesic assignation $\tilde{I}'$ for $D_1(G)$ such that $|V(\tilde{I}')| \leq |V(\tilde{I})|$, and:*

*(1) for any $i \neq i'$, the geodesic between $v_i^0$ and $u_{i'}^0$ in $\tilde{I}'$ is $p_{i,i'}$, and*
*(2) for any $i$, the geodesic between $v_i^0$ and $u_i^0$ in $\tilde{I}'$ is either $p_i$ or $\bar{p}_i$.*

**Theorem 1.** *It is $\mathcal{NP}$-hard to determine if a set of vertices $V'$ is a strong geodetic set even if, for every strong geodetic set $V_{strong}$, we have $V' \subseteq V_{strong}$.*

From this theorem, we can derive a result about the residue variant of STRONG GEODETIC NUMBER. The *residue variant* of an optimisation problem has been defined recently in [24] and consists of, given a partial solution $P$ for an instance $I$, finding an optimal partial solution $R$ such that $P \cup R$ is a solution for $I$. The complexity class $\mathcal{RAPX}$ contains the residue variant optimisation problems such that the score of the residue can be approximated by a constant.

**Corollary 1.** STRONG GEODETIC NUMBER $\notin \mathcal{RAPX}$.

## 4 Approximation

Since it is hard to determine if a subset of vertices is a strong geodetic set, a natural question that arises is to find, given a subset of vertices, a geodetic assignation that maximizes the number of covered vertices. We call this problem MAX GEODESIC ASSIGNATION. By Theorem 1, this problem is also $\mathcal{NP}$-hard and we show that this problem belongs to $\mathcal{APX}$, *i.e.* approximable within a constant ratio. In this part, we show that this problem is 2-approximable using a simple greedy algorithm, defined in Algorithm 1.

**Theorem 2.** *Algorithm 1 computes in polynomial time a solution for* MAX GEODESIC ASSIGNATION *with an approximation ratio of* 2 *and this ratio is tight (Fig. 2).*

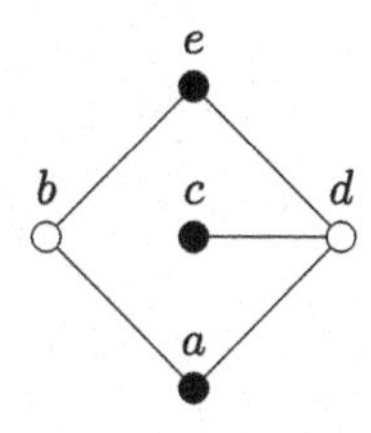


**Fig. 2.** Tightness of the approximation ratio of Algorithm 1. Consider $a, c$ and $e$ as selected (in black in the graph). The optimal solution consists in taking the geodesics $(a, b, e)$, $(c, d, e)$ and $(a, c, d)$ which cover the non-selected vertices $b$ and $d$. The greedy algorithm can start by taking the geodesic $(a, d, e)$ between $a$ and $e$. Then the algorithm will choose $(c, d, e)$ and $(a, c, d)$ for the last two pairs. This leads to a set of geodesics which only covers $d$.

# 5 Reduction from Set Cover

In this part, we prove preliminary results that will be used in the next section. More specifically, we reduce the geodetic problems from the classic $\mathcal{NP}$-complete problem SET COVER described as follows.

SET COVER (SC)
**Input:** A collection $C = \{S_1, \ldots, S_{m'}\}$ of finite sets over the universe $U = \{E_1, \ldots, E_{n'}\}$.
**Question:** Find a minimum $C' \subseteq C$ such that every element of $U$ is contained in a set of $C'$.

For the strong versions, we use a version of SET COVER, denoted $(k, k')$-SET COVER, where the size of the intersection between two sets is at most $k$ and the set sizes are bounded by $k'$. Notice that since VERTEX COVER is a particular case of $(1, k')$-SET COVER, then $(k, k')$-SET COVER is $\mathcal{NP}$-complete.

In the following, we first show how this reduction works in the general case and then, we adapt it in subcubic bipartite graphs with arbitrary high girth.

## 5.1 On General Case

**Construction 2.** *Let $(C, U)$ be an instance of* SET COVER. *We construct a graph $G$ as follows:*

- *For each set $S_i$, create a 3-path $sp_i = (v_i^0, v_i^1, v_i^2)$.*
- *For each element $E_j$, create a 4-path $ep_j = (u_j^0, u_j^1, u_j^2, u_j^3)$. We denote the edge $u_j^2u_j^3$ as $e_j$.*
- *For each set $S_i$ and each element $E_j \in S_i$, introduce a 3-path $cp_j^i$ between $v_i^1$ and $u_j^2$) and a 2-path $lp_j^i$ between $v_i^2$ and $u_j^3$.*
- *For each pair of elements $E_j$ and $E_{j'}$, introduce the edge $t_{j,j'} = u_j^1u_{j'}^1$.*

The paths $ep_j, sp_i, cp_j^i$ and $lp_k^i$ are called *element paths*, *set paths*, *cut paths* and *long paths*, respectively. An example of a graph produced by Construction 2 is depicted in Fig. 3.

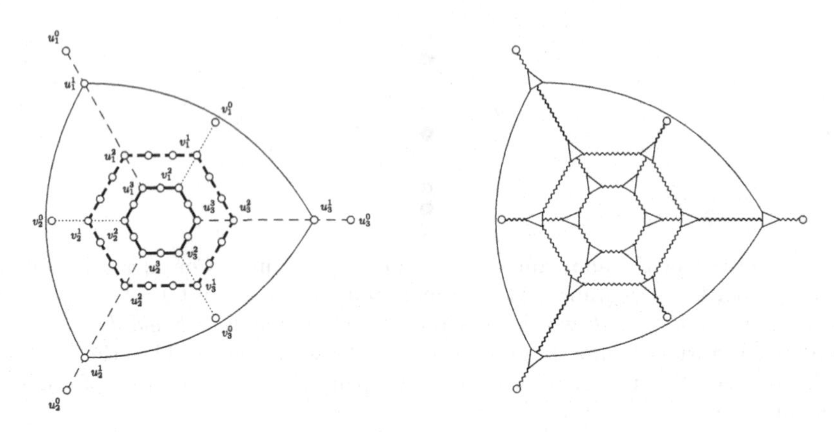

**Fig. 3.** Example of a graph produced by Construction 2 (left) and Construction 4 (right) on the collection containing $S_1 = \{E_1, E_3\}, S_2 = \{E_1, E_2\}$ and $S_3 = \{E_2, E_3\}$. Element paths, set paths, cut paths and long paths are depicted in dashed, dotted, bold dashed and bold, respectively. In the right graph, the squiggly edges represent paths of length $h, 2h$ or $3h$

Clearly, the construction can be carried in polynomial time. In order to show that Construction 2 constitutes a reduction, we introduce the following lemmas.

**Lemma 2.** *Let* $(C, U)$ *be an instance of* Set Cover *(resp.* $(1, k')$-Set Cover*) and let* $G$ *be its graph resulting from Construction 2. The set* $D_1(G)$ *covers (resp. strongly covers) every edge of* $G'$ *except the edges in* $\{e_j \mid E_j \in U\}$.

In the following, let $Y_i^S \subset V(G)$ denote the set containing $sp_i \backslash \{v_i^0\}$ and every long path $lp_j^i$ and cut path $cp_j^i$ incident to $cp_i$ minus vertices of every element path $ep_j$. Formally, $Y_i = (sp_i \backslash \{v_i^0\}) \cup \{(cp_j^i \cup lp_j^i) \backslash ep_j \mid \forall E_j \in S_i\}$. For each element $E_j \in U$, we also denote $Y_j^E = \{Y_i \mid E_j \in S_i\} \cup ep_j \backslash \{u_j^0\}$.

**Lemma 3.** *Let* $(C, U)$ *be an instance of* Set Cover *and let* $G$ *be its graph resulting from Construction 2. For each element* $E_j$*, every geodesic containing the edge* $e_j$ *has an extremity in* $Y_j^E$.

In order to easily produce a set cover in $G$ from a (strong) edge geodetic set $X$ of $G'$, we need $X$ to respect a certain property. Hence, we use the following lemma.

**Lemma 4.** *Let* $(C, U)$ *be an instance of* Set Cover *(resp.* $(1, k')$-Set Cover*) and* $G$ *its graph resulting from Construction 2. Let* $X \subseteq V(G)$ *be an edge geodetic set (resp. strong edge geodetic set) of* $G$*. It is possible to construct an edge geodetic set (resp. a strong edge geodetic set)* $X'$ *of* $G$ *such that* $|X'| \leq |X|$ *and*

$$X' \subseteq \{v_i^2 \mid S_i \in C\} \cup D_1(G)$$

.

**Lemma 5.** *Let $(C, U)$ be an instance of* SET COVER *(resp.* $(1, k')$-SET COVER*) and $G$ its graph resulting from Construction 2. Then the instance $(C, U)$ contains a set cover of size $k$ if and only if $G$ contains an edge geodetic set (resp. strong edge geodetic set) of size $k + |C| + |U|$.*

## 5.2 On Subcubic Bipartite Graphs

We now extend the previous result to subcubic and bipartite graphs. First, we show that the result holds in graph with maximum degree three. We introduce the following construction.

**Construction 3.** *Given a graph $G$, a vertex $u \in V(G)$, a set of non-adjacent neighbours $N^0 = \{v_0^0, \ldots, v_{k-1}^0\} \subseteq N(u)$ and an integer $h > \log k$,* emplace *a* $h$-pyramid *$Py(h, u, N^0)$ consists of removing all edges between $u$ and $N^0$ and replacing them with the following subgraph. For each $0 < i < h$, construct recursively the sets $N^i$:*

- *create $t = \lceil |N^{i-1}|/2 \rceil$ vertices $v_0^i, \ldots, v_t^i$, and*
- *introduce the edges $v_{t'}^i v_{2t'}^{i-1}$, and $v_{t'}^i v_{2t'+1}^{i-1}$ (if $v_{2t'+1}^{i-1}$ exists) for each $t' < t$.*

*Finally, introduce the edge $uv_0^{h-1}$ ($N^{h-1}$ consists of a single vertex since $h > \log k$) (Fig. 4).*

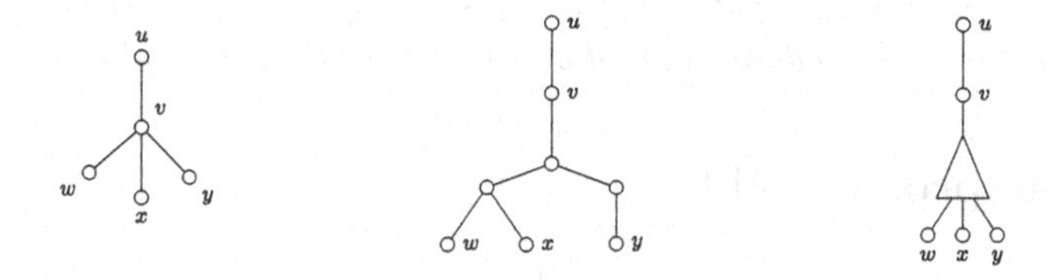


**Fig. 4.** Example of a 3-pyramid $Py(3, v, \{w, x, y\})$ produced by Construction 3. **Left:** $v$ and its neighbours in the original graph. **Center:** emplaced 3-pyramid. **Right:** Representation of the pyramid used in Fig. 3.

Let $Py(h, v, N)$ be a $h$-pyramid. We can make the following observations.

- The maximum degree of $Py_h(h, v, N)$ is three.
- Let $n_1, n_2 \in N$, the distance between $n_1$ and $n_2$ in $Py_h(h, v, N)$ is between 2 and $2h$ and the distance between $v$ and $n_1$ or $n_2$ is $h$.

We now use the previous structure to modify Construction 2 as follows.

**Construction 4.** *Let $(C, U)$ be an instance of* SET COVER *and $G$ be its graph resulting from Construction 2. Let $h > \log \Delta(G)$ be an integer. We modify $G$ as follows:*

- *for each set $S_i$ and each integer $k \in \{1, 2\}$, emplace a h-pyramid $Py(h, v_i^k, N(v_i^k)\backslash\{v_i^{k-1}\})$,*
- *for each element $E_j$ and each integer $k \in \{1, 2, 3\}$, emplace a h-pyramid $Py(h, u_j^k, N(u_j^k)\backslash\{u_j^{k-1}\})$, and*
- *replace each edge of $G$ that does not belong to a h-pyramid by a path of length $h$.*

Note that the resulting graph has maximum degree three. Moreover, if $k$ is odd then the resulting graph is bipartite. Finally, by taking an arbitrary high value of $k$, the resulting graph has an arbitrary high girth. We use a similar vocabulary than for Construction 2: an *element tree* $et_j$ is the tree induced by the vertices of $ep_j$ in the $h$-pyramids emplaced in it in the original graph. A *set tree* $st_i$ is defined the same way. For each element $E_j$, the $h$-path that replaces the edge $e_j$ is denoted $p_j$. An example of a graph produced by Construction 4 is depicted in Fig. 3. Since Construction 4 multiplies the length of every path of Construction 2 by $h$, we can adapt Lemmas 2 to 4 to it by replacing $ep_j$ by $et_j$ and $sp_i$ by $st_i$ in the geodesics descriptions. Using the same idea as for Lemma 5, we can now show that Construction 4 constitutes a reduction: if a path $p_j$ is (strongly) covered, then there is a vertex $v_i^2$, such that $E_j \in S_i$, is selected. Thus, given a solution for a geodetic problem $X$ the set $\{S_i \mid v_i^2 \in X\}$ is a set cover of $G$. Hence, we obtain the following result.

**Lemma 6.** *Let $(C, U)$ be an instance of* SET COVER *(resp.* $(1, k')$-SET COVER*) and $G$ its graph resulting from Construction 2. Then the instance $(C, U)$ contains a set cover of size $k$ if and only if $G$ contains an edge geodetic set and a geodetic set (resp. strong edge geodetic set and a strong geodetic set) of size $k + |C| + |U|$.*

# 6 Non-approximability

In this section, we use the results of the previous section to find hardness of approximation results for the geodetic problems.

## 6.1 Strong Geodetic Set and Strong Edge Geodetic Set

First, recall the definition of *L-reduction* between two hard problems $\Pi$ and $\Pi'$ (with respective cost functions $val_\Pi$ and $val_{\Pi'}$), as described by Papadimitriou and Yannakakis [19]. Let $OPT_\Pi(x)$ and $OPT_{\Pi'}(x)$ be the optimal value of $val_\Pi$ and $val_{\Pi'}$ on an instance $x$, respectively. An $L$-reduction consists of polynomial-time computable functions $f$ and $g$ such that, for each instance $x$ of $\Pi$, $f(x)$ is an instance of $\Pi'$ and for each feasible solution $y'$ for $f(x)$, $g(y')$ is a feasible solution for $x$. Moreover, there are constants $\alpha_1, \alpha_2 > 0$ such that:

1. $OPT_{\Pi'}(f(x)) \leq \alpha_1 \cdot OPT_\Pi(x)$ and
2. $|val_\Pi(g(y')) - OPT_\Pi(x)| \leq \alpha_2 \cdot |val_{\Pi'}(y') - OPT_{\Pi'}(f(x))|$.

Using Construction 4, we obtain an $L$-reduction with $\alpha_1 = (2k' + 2)$ and $\alpha_2 = 1$.

**Lemma 7.** *Let $\rho_{k'}$ be the best possible polynomial time approximation factor of* $(1, k')$-MINIMUM SET COVER. *Then* STRONG GEODETIC NUMBER *and* STRONG EDGE GEODETIC NUMBER *cannot be approximated with a factor better than*

$$1 + \frac{\rho_{k'} - 1}{2k' + 2},$$

*in subcubic bipartite graphs with arbitrary high girth.*

Since MINIMUM VERTEX COVER with bounded maximum degree $k'$ is a particular case of $(1, k')$-MINIMUM SET COVER, we can pick the value of $k'$ (and so the corresponding best-known value $\rho_{k'}$) that maximize the previous inapproximation ratio. Thus, since Berman and Karpinski showed that MINIMUM VERTEX COVER cannot be approximated with a factor better than $^{79}/_{78}$ in graphs with maximum degree four [4], we obtain the following result.

**Corollary 2.** STRONG GEODETIC NUMBER *and* STRONG EDGE GEODETIC NUMBER *cannot be approximated with a factor better than* $^{781}/_{780}$ *in subcubic bipartite graphs with arbitrary high girth.*

### 6.2 Geodetic Set and Edge Geodetic Set

Now, we provide approximation lower bounds for GEODETIC NUMBER and STRONG GEODETIC NUMBER. We apply the following modification to Construction 4.

**Construction 5.** *Let $(C, U)$ be an instance of* SET COVER, *$G$ be its graph produced by Construction 4 and $k > |V(G)|$ be an integer. We construct a graph $G'$ as follows:*

- *create $k$ disjoint copies $\{G_1, \ldots, G_k\}$ of $G$,*
- *for each vertex $x$ of $D_1(G)$,*
  - *create an edge $s_x^0 s_x^1$,*
  - *for each $G_\ell \in \{G_1, \ldots, G_k\}$ and for each vertex $x \in D_1(G_\ell)$, construct a $k$-path $p_x^\ell$ between $x$ and $s_x^1$, and*
  - *emplace a $k$-pyramid $Py(k, s_x^1, N(s_x^1) \backslash \{s_x^0\})$.*

Notice that the resulting graph has maximum degree three.

SET COVER is hard to approximate with a factor better than a logarithmic function [9]. Therefore, we can transfer the lower bounds of approximation of SET COVER to GEODETIC NUMBER and EDGE GEODETIC NUMBER. This result is in addition to the one proved by Chakraborty *et al.* [7].

**Theorem 3.** GEODETIC NUMBER *and* EDGE GEODETIC NUMBER *are LOG-$\mathcal{APX}$-hard, even in bipartite subcubic graphs with arbitrary high girth.*

## 7 Strong Geodetic Number on Complete Multipartite Graphs

First, remark that geodesics of complete multipartite graphs are easy to determine: for any pair of vertices which are not in the same part, the edge between them is the unique shortest path between them. For a pair of vertices which are in the same part, the shortest paths between them are all the paths of length two between them with all the vertices not in this part as middle vertices.

In this section, we develop a polynomial algorithm which computes the strong geodetic number of a complete multipartite graph. The algorithm is based on dynamic programming where we not only look after a minimum strong geodetic set of vertices covering all the graph, but we look after all sets of vertices maximizing the number of pairs not used to cover other vertices among sets with some fixed parameters.

Let $K_{n_1,\ldots,n_r}$ denotes a complete multipartite graph whose parts are noted $X_1,\ldots,X_r$ such that $|X_i| = n_i$ for every $i \in \{1,\ldots,r\}$. We denote $N_i = \sum_{j=1}^{i} n_j$ for every $i \in \{1,\ldots,r\}$ and $K_{n_1,\ldots,n_i}$ by $G_i$.

**Definition 1.** *A* selection *of $K_{n_1,\ldots,n_r}$ is a set of selected vertices $S$ in which we pick a set of pairs of non-adjacent vertices $C$ to cover some non-selected vertices. Formally, a selection is a triplet $(S, C, f)$, where*

- *$S \subseteq V$,*
- *$C \subseteq \bigcup_{j=1}^{r} \mathcal{P}_2(S \cap X_j)$ and,*
- *$f : C \to V \backslash S$ is an injective map such that $\forall c \in C \bigcap \mathcal{P}_2(S \bigcap X_i), f(c) \notin X_i$ (i.e. two vertices of $X_i$ can not cover another vertex of $X_i$).*

Given a selection $s(S, C, f)$, we denote by

- $s(S, C, f) = |S|$, the number of selected vertices,
- $r(S, C, f) = |V \backslash (S \sqcup f(C))| = n - s(S, C, f) - |C|$, the number of vertices that are neither selected nor covered, and
- $d(S, C, f) = |\cup_{j=1}^{n} \mathcal{P}_2(S \cap S_j) \backslash C| = \sum_{j=1}^{n} \binom{S \cap X_j}{2} - |C|$, the number of pairs of non-adjacent vertices that are not in $C$.

We denote by $d(i, j, r)$ the maximum of $d(S, C, f)$ for any selection $(S, C, f)$ of $G_i$ such that $s(S, C, f) = j$ and $r(S, C, f) = r$. This quantity is set to $-\infty$ if no such selection of $G_i$ exists.

**Lemma 8.** *For any integers $i, s, r$ and integers $k, u, q$ we define the following quantities:*

$$s' = s - k, \quad r' = r - n_i + u + q + k \quad \textit{and} \quad d' = d(i - 1, s', r')$$

*We deduce that:* $d(i, s, r) = \max \left\{ (d') + \binom{k}{2} - q - u \;\middle|\; \begin{array}{l} 0 \le k \le n_i \\ 0 \le u \le \min(n_i - k, d') \\ 0 \le q \le \min(\binom{k}{2}, r') \end{array} \right\}.$

From previous lemma we deduce the following theorem.

**Theorem 4.** *There exists an algorithm in $O(n^8)$ computing the geodetic number of a complete multipartite graph with $n$ vertices.*

Notice that a complete multipartite graph can be described with the list of integers $n_1, \ldots, n_k$. In that case, the dynamic programming that we described is not polynomial if the values of the $n_i$ are exponential. Thus, if we formulate EDGE GEODETIC NUMBER on complete multipartite graph as a specific problem on this class, the question whether such a problem is weak $\mathcal{NP}$-hard or not is open.

## 8 Conclusion

In this paper, we investigated the hardness of the approximation of the geodetic set problems. Given our approximation lower bound for GEODETIC NUMBER and EDGE GEODETIC NUMBER, the question of the existence of a $O(\log(n))$-approximation algorithm seems natural. We also proved that deciding whether a set admits a geodesic assignation NP-hard. Therefore, a second question arises: is it also hard to decide whether a set of vertices is a strong geodetic set. We also give a tight 2-approximation of this problem. Finding a lower bound for this problem is probably a good question for further work. Finally, for STRONG GEODETIC NUMBER, we proved that it was polynomial on complete multipartite graphs. What about other graph classes?

## References


1. Anand, B.S., Changat, M., Ullas Chandran, S.V.: The edge geodetic number of product graphs. In: Panda, B.S., Goswami, P.P. (eds.) CALDAM 2018. LNCS, vol. 10743, pp. 143–154. Springer, Cham (2018). https://doi.org/10.1007/978-3-319-74180-2_12
2. Arokiaraj, A., Klavzar, S., Manuel, P.D., Thomas, E., Xavier, A.: Strong geodetic problems in networks. Discuss. Math. Graph Theory **40**(1), 307–321 (2020)
3. Atici, M.: On the edge geodetic number of a graph. Int. J. Comput. Math. **80**(7), 853–861 (2003)
4. Berman, P., Karpinski, M.: On some tighter inapproximability results (extended abstract). In: Wiedermann, J., van Emde Boas, P., Nielsen, M. (eds.) ICALP 1999. LNCS, vol. 1644, pp. 200–209. Springer, Heidelberg (1999). https://doi.org/10.1007/3-540-48523-6_17
5. Bresar, B., Kovse, M., Tepeh, A.: Geodetic sets in graphs. In: Dehmer, M. (ed.) Structural Analysis of Complex Networks, pp. 197–218. Birkhäuser/Springer, Boston (2011). https://doi.org/10.1007/978-0-8176-4789-6_8
6. Chakraborty, D., Das, S., Foucaud, F., Gahlawat, H., Lajou, D., Roy, B.: Algorithms and complexity for geodetic sets on planar and chordal graphs. In: Cao, Y., Cheng, S.W., Li, M. (eds.) 31st International Symposium on Algorithms and Computation (ISAAC 2020). Leibniz International Proceedings in Informatics (LIPIcs), vol. 181, pp. 7:1–7:15. Schloss Dagstuhl-Leibniz-Zentrum für Informatik (2020)

7. Chakraborty, D., Foucaud, F., Gahlawat, H., Ghosh, S.K., Roy, B.: Hardness and approximation for the geodetic set problem in some graph classes. In: Changat, M., Das, S. (eds.) CALDAM 2020. LNCS, vol. 12016, pp. 102–115. Springer, Cham (2020). https://doi.org/10.1007/978-3-030-39219-2_9
8. Darmann, A., Döcker, J.: On simplified NP-complete variants of monotone 3-sat. Discret. Appl. Math. **292**, 45–58 (2021)
9. Dinur, I., Steurer, D.: Analytical approach to parallel repetition. In: Proceedings of the Forty-Sixth Annual ACM Symposium on Theory of Computing, STOC 2014, pp. 624–633. Association for Computing Machinery, New York (2014)
10. Doughat, A., Kong, M.: Computing geodetic bases of chordal and split graphs. J. Combin. Math. Combin. Comput. **22**, 67–78 (1996)
11. Douthat, A., Kong, M.: Computing the geodetic number of bipartite graphs. Congressus Numerantium 113–120 (1995)
12. Gledel, V., Irsic, V., Klavzar, S.: Strong geodetic cores and cartesian product graphs. Appl. Math. Comput. **363**, 124609 (2019)
13. Harary, F., Loukakis, E., Tsouros, C.: The geodetic number of a graph. Math. Comput. Model. **17**(11), 89–95 (1993)
14. Iršič, V.: Strong geodetic number of complete bipartite graphs and of graphs with specified diameter. Graphs Comb. **34**(3), 443–456 (2018)
15. Iršič, V., Konvalinka, M.: Strong geodetic problem on complete multipartite graphs. Ars Math. Contemp. **17**(2), 481–491 (2019)
16. Klavzar, S., Manuel, P.D.: Strong geodetic problem in grid-like architectures. Bull. Malays. Math. Sci. Soc. **41**(3), 1671–1680 (2018)
17. Manuel, P., Klavžar, S., Xavier, A., Arokiaraj, A., Thomas, E.: Strong edge geodetic problem in networks. Open Math. **15**(1), 1225–1235 (2017)
18. Mezzini, M.: Polynomial time algorithm for computing a minimum geodetic set in outerplanar graphs. Theor. Comput. Sci. **745**, 63–74 (2018). https://doi.org/10.1016/j.tcs.2018.05.032
19. Papadimitriou, C.H., Yannakakis, M.: Optimization, approximation, and complexity classes. J. Comput. Syst. Sci. **43**(3), 425–440 (1991)
20. Rehmani, S., Sunitha, M.: Edge geodesic number of a fuzzy graph. J. Intell. Fuzzy Syst. **37**(3), 4273–4286 (2019)
21. Santhakumaran, A., Ullas Chandran, S.: The edge geodetic number and cartesian product of graphs. Discuss. Math. Graph Theory **30**(1), 55–73 (2010)
22. Santhakumaran, A., John, J.: Edge geodetic number of a graph. J. Discrete Math. Sci. Cryptogr. **10**(3), 415–432 (2007)
23. Santhakumaran, A., Ullas Chandran, S.: Comment on" edge geodetic covers in graphs. Proyecciones (Antofagasta) **34**(4), 343–350 (2015)
24. Weller, M., Chateau, A., Giroudeau, R., König, J.-C., Pollet, V.: On residual approximation in solution extension problems. J. Comb. Optim. **36**(4), 1195–1220 (2017). https://doi.org/10.1007/s10878-017-0202-5
25. Zmazek, E.: Strong edge geodetic problem on grids. Bull. Malays. Math. Sci. Soc. **44**, 3705–3724 (2021). https://doi.org/10.1007/s40840-021-01137-4

# Approximate Distance Oracles with Improved Stretch for Sparse Graphs

Liam Roditty and Roei Tov(✉)

Bar Ilan University, Ramat Gan, Israel

**Abstract.** Thorup and Zwick [19] introduced the notion of approximate distance oracles, a data structure that produces for an $n$-vertices, $m$-edges weighted undirected graph $G = (V, E)$, distance estimations in *constant* query time. They presented a distance oracle of size $O(kn^{1+1/k})$ that given a pair of vertices $u, v \in V$ at distance $d(u, v)$ produces in $O(k)$ time an estimation that is bounded by $(2k - 1)d(u, v)$, i.e., a $(2k - 1)$-multiplicative approximation (stretch). Thorup and Zwick [19] presented also a lower bound based on the girth conjecture of Erdős.

For sparse unweighted graphs (i.e., $m = \tilde{O}(n)$) the lower bound does not apply. Pătraşcu and Roditty [10] used the sparsity of the graph and obtained a distance oracle that uses $\tilde{O}(n^{5/3})$ space, has $O(1)$ query time and a stretch of 2. Pătraşcu et al. [11] presented infinity many distance oracles with fractional stretch factors that for graphs with $m = \tilde{O}(n)$ converge exactly to the integral stretch factors and the corresponding space bound of Thorup and Zwick.

It is not known, however, whether graph sparsity can help to get a stretch which is better than $(2k - 1)$ using only $\tilde{O}(kn^{1+1/k})$ space. In this paper we answer this open question and prove a separation between sparse and dense graphs by showing that using sparsity it is possible to obtain better stretch/space tradeoffs than those of Thorup and Zwick. We show that for every $k \geq 2$ there is a distance oracle of size $O(knm^{1/k} \log n)$ that produces in $O(k)$ time an estimation $d^*(u, v)$ that satisfies $d(u, v) \leq d^*(u, v) \leq (2k - 1)d(u, v) - 4$, for $k > 2$, and $d(u, v) \leq d^*(u, v) \leq 3d(u, v) - 2$, for $k = 2$.

Another contribution of this paper is a refined stretch analysis of Thorup and Zwick distance oracles that allows us to obtain a better understanding of this important data structure. We present simple conditions for every $w \in V$ that characterizes the exact scenarios in which every query that involves $w$ produces an estimation of stretch strictly better than $2k - 1$, even in the case of dense graphs. We complement this contribution with an experiment on real world graphs. The main finding in the experiment is that different real world graphs are likely to satisfy the required conditions and hence the stretch of Thorup and Zwick distance oracles is much better than its worst case bound in these real world graphs.

**Keywords:** Graph algorithms · Approximate shortest paths · Approximate distance oracles


C.-Y. Chen et al. (Eds.): COCOON 2021, LNCS 13025, pp. 89–100, 2021.
https://doi.org/10.1007/978-3-030-89543-3_8

# 1 Introduction

An approximate distance oracle is a data structure that is required to produce distance estimations in *constant* query time. Thorup and Zwick [19] showed that given an undirected weighted graph $G = (V, E)$ with $m$ edges and $n$ vertices and an integer $k \geq 1$, there is a data structure of size $O(kn^{1+1/k})$ that for every pair of vertices $u, v \in V$ returns in $O(k)$ time an estimation $\hat{d}(u, v)$ which is a $(2k - 1)$ multiplicative approximation (stretch) of $d(u, v)$, that is, $d(u, v) \leq \hat{d}(u, v) \leq (2k-1)d(u, v)$, where $d(u, v)$ is the length of the shortest path between $u$ and $v$ in $G$.

Thorup and Zwick [19] presented also a lower bound based on the girth conjecture of Erdős[1]. More specifically, they proved that, for every $k \geq 1$, if there is a graph of $\Omega(n^{1+1/k})$ edges whose girth is $2k + 2$ then any distance oracle with stretch $t \leq 2k$, requires $\Omega(n^{1+1/k})$ bits on some input. A careful examination of their proof reveals that it relies on the stretch of the estimation for vertex pairs $u, v \in V$ for which $(u, v) \in E$, that is, $d(u, v) = 1$. Therefore, it still might be possible to obtain a data structure with constant query time and a stretch better than $2k-1$ using $O(kn^{1+1/k})$ space, for vertex pairs $u, v \in V$ that satisfy $d(u, v) \geq 2$, or for graphs with $m = o(n^{1+1/k})$, that is, sparse graphs[2].

We present a new distance oracle for unweighted undirected graphs, that uses $O(knm^{1/k} \log n)$ space and provides in $O(k)$ query time an estimation $d^*(u, v)$ that satisfies $d(u, v) \leq d^*(u, v) \leq (2k - 1)d(u, v) - 4$, for every $k > 2$, and $d(u, v) \leq d^*(u, v) \leq 3d(u, v) - 2$, for $k = 2$. This implies that for sparse graphs with $m = \tilde{O}(n)$[3] our new distance oracle uses the same space as Thorup and Zwick's distance oracle (up to poly-logarithmic factors) and produces in $O(k)$ time an estimation of strictly better stretch than the stretch of Thorup and Zwick's distance oracle. Sparse graphs with $m = \tilde{O}(n)$ edges are very interesting both from the practical perspective and the theoretical perspective.

From the practical perspective, it is important to note that many real world graphs are sparse and $m = \tilde{O}(n)$. This is usually the case in social networks and in many other types on networks[4].

From the theoretical perspective, Pătraşcu, Roditty and Thorup [11] proved a conditional lower bound for the case of sparse graphs with $m = \tilde{O}(n)$, based on a set intersection hardness conjecture. They showed that for any $\ell > 1$, a distance oracle that for every pair of vertices at distance $\ell+1$, provides in constant query time an estimation strictly smaller than $3(\ell+1) - 2$ requires $\tilde{\Omega}(n^{1+\frac{1}{2-1/\ell}})$ space. Notice that for $k = 2$ our distance oracle has an estimation that is at most $3d(u, v) - 2$, for every $u, v \in V$ and uses $\tilde{O}(n^{1.5})$ space for sparse graphs with

[1] The girth is the length of the shortest cycle in an unweighted graph.

[2] A trivial way to get a smaller space for sparse graphs is to simply save the graph and answer any query in $O(m)$ time by doing BFS, this however, violates the additional requirement for distance oracles of a constant or almost a constant query time.

[3] Throughout the paper we will use the $\tilde{O}(\cdot)$ notation to hide small poly-logarithmic factors.

[4] See for more examples https://snap.stanford.edu/index.html.

$m = \tilde{O}(n)$. It follows from [11] that bounding the estimation by a value strictly smaller than $3d(u,v) - 2$ requires $\tilde{\Omega}(n^{1.5+\varepsilon})$ space, where $\varepsilon > 0$.

Pătraşcu et al. [11] showed also that there are infinitely many distance oracles for sparse graphs with fractional stretch factors. Their distance oracles converge exactly to the integral stretch factors and the corresponding space bound of Thorup-Zwick distance oracles. Our new construction implies that for space $\tilde{O}(km^{1+1/k})$ a stretch that is strictly better than the corresponding integral stretch of $2k-1$ is possible.

The implications of our new distance oracles are not restricted only for sparse graphs with $m = \tilde{O}(n)$. Consider graphs with $m \in [n, o(n^{1+1/k})]$ edges. A natural question is whether a distance oracle for such graphs requires $\Omega(n^{1+1/k})$ for stretch $2k-1$. The girth based approach, as in the lower bound of Thorup and Zwick [19], is not possible here since we can store the entire graph. This implies that for vertex pairs $u, v \in V$ with $d(u,v) = 1$, we can store the exact distance. Our new distance oracle rules out also the option to use pairs of vertices $u, v \in V$ for which $d(u,v) = 2$, as a possible source of hardness for a possible lower bound. If we construct our new distance oracle with parameter $k+1$ then the space required is in the range $[n, o(n^{1+1/k})]$ and for every pair of vertices $u, v \in V$, for which $d(u,v) = 2$, the estimation is at most $(2(k+1)-1)2-4 = (2k-1)2$, and therefore, when $d(u,v) = 2$ the stretch is at most $2k-1$ .

The distance oracles of Thorup and Zwick, beside being an important data structure on their own, are also extremely useful as a tool in many applications. They were a crucial building block in several important dynamic graph algorithms along the last decade (e.g., [2,7,8,16]). They also play a pivotal role in designing distance labeling and compact routing schemes as was already shown by Thorup and Zwick [18] and in subsequent works (e.g., [1,3,13,14]). Distance oracles were also implemented and tested (e.g., [6,12]) and found useful on real world graphs. Therefore, any further understanding that we gain on the basic properties of distance oracles is of great interest.

We obtain our new distance oracle by a careful combination of a variant of Thorup and Zwick distance oracles with a new idea that interplays between a hitting set of vertices and a hitting set of edges to overcome a certain hard case that is relatively common in analysis of algorithms of shortest paths. Therefore, our new approach is of independent interest, as it might be found useful in other closely related problems.

Motivated by our theoretical finding, another contribution that we make in this paper is a refined analysis of the stretch of Thorup and Zwick distance oracles. At the base of the distance oracles there is an hierarchy of vertex sets $A_0, A_1, \ldots, A_k$, where $A_0 = V$, $A_k = \emptyset$ and $A_i$ is formed by picking each vertex of $A_{i-1}$, independently, with some probability $p$. For every $u \in V$ the distance $d(u, A_i)$ between $u$ and $A_i$ is computed and saved. We introduce a simple parameter, called the *average distance*, which is roughly defined[5] for every $i \in [1, k-1]$ as the distance between $u$ and $A_i$ divided by $i$, that is $d(u, A_i)/i$. Our refined analysis characterizes several cases in which the stretch is strictly better than

[5] In the formal definition we take the ceiling of the average distance.

$2k-1$ using only the average distance, which can be easily computed using the current information saved with the distance oracle. Roughly speaking, if there exist $i, j \in [1, k-1]$ such that $i \neq j$ and $d(u, A_i)/i \neq d(u, A_j)/j$, then the stretch is strictly better than $2k-1$ for every distance query that includes the vertex $u$.

Based on similar ideas we also show that if $D(u) = \{\Delta_1, \ldots, \Delta_\ell\}$ is the set of all possible distances of $u \in V$ with other vertices in the graph then there is at most one value $\Delta \in D(u)$ for which the stretch of the distance estimation is exactly $2k-1$, that is, only for vertices $v$ that satisfy $d(u, v) = \Delta$ it might be that $\hat{d}(u, v) = (2k-1)d(u, v)$.

We complement the refined stretch analysis by conducting a small experiment on real world graphs. In the experiment we check how frequent are the cases that allow for a better stretch in these real world graphs. Interestingly, these cases are quite frequent and thus in many cases the actual stretch is much better than the worst case stretch bound.

### 1.1 Related Work

Since their introduction by Thorup and Zwick [19] distance oracles were studied by many researchers. Chechik [4,5], presented a $(2k-1)$-stretch distance oracle with $O(1)$ query time and $O(n^{1+1/k})$ space. (See also [9,20].)

Pătraşcu and Roditty [10] showed a distance oracle for weighted undirected graphs with stretch 2 and size $O(n^{4/3}m^{1/3})$. For $m = o(n^2)$, this distance oracle has $o(n^2)$ size and stretch 2. Pătraşcu, Roditty and Thorup [11] showed for every integer $k \geq 0$ and $\ell > 0$ distance oracles, that use $\tilde{O}(m^{1+1/(k \pm 1/\ell)})$ space and answer distance query in $O(k+\ell)$ time with stretch $2k+1 \pm 2/\ell$. Sommer, Verbin, and Yu [17] provided a lower bound in the cell probe model. They showed that there are sparse graphs for which constant stretch and query time requires $m^{1+\Omega(1)}$ space[6].

Due to lack of space, we refer the reader to the full version of this paper [15] for the rest of the related work section.

### 1.2 Paper Organization

In the next section we present some necessary preliminaries, the distance oracles of Thorup-Zwick and a standard variant of it, that is required in order to obtain our new distance oracle. In Sect. 3 we present our new distance oracles. In Sect. 4 we present our refined stretch analysis for Thorup-Zwick distance oracles. In Sect. 5 we present some concluding remarks and open problems. Due to lack of space, we omit here some of the proofs of Sect. 2 and the technical part of Sect. 4. We refer the reader to [15] for the full version of this paper. Also, in [15] we present the experiment that we have conducted on real world graphs. In the experiment we examine how frequent are the cases that are characterized in our refined stretch analysis from Sect. 4.

[6] Using current techniques of cell probe lower bounds we cannot hope for more specific tradeoff since it is not possible to separate asymptotically the query times of data structures of size $m^{1.99}$ and $m^{1.01}$ for input size $m$.

## 2 Preliminaries and Previous Work

Let $G = (V, E)$ be an $n$-vertices $m$-edges undirected unweighted graph. For every $u, v \in V$, let $d(u, v)$ be the length of the shortest path between $u$ and $v$. Let $N(u)$ be the vertices that are neighbours of $u$ and let $deg(u) = |N(u)|$ be the degree of $u$.

For every set $A \subseteq V$, let $p_A(u)$ be the closest vertex to $u$ from $A$, that is $p_A(u) := \arg\min_{v \in A}(d(u, v))$, where ties are broken in favor of the vertex with a smaller identifier, and let $d(u, A) = d(u, p_A(u))$. Notice that it follows from this definition that if $v$ is on a shortest path between $u$ and $p_A(u)$, then $p_A(u) = p_A(v)$. For a set $E' \subseteq E$ let $V(E') = \{u \mid (u, v) \in E'\}$. Let $N(u, s, A)$ be the $s$ closest vertices to $u$ from the set $A$.

Let $B(u, r) = \{v \in V \mid d(u, v) < r\}$ and let $B(u, r, X) = \{v \in X \mid d(u, v) < r\}$, where $X \subseteq V$. Let $L(u, r) = \{v \in V \mid d(u, v) = r\}$.

The following Lemma is a standard tool in the area of approximate shortest paths and we provide it here for completeness.

**Lemma 1. (e.g. Lemma 3.6 in [19]).** *Let $U$ be a set of size $u$. Let $Q_1, \ldots, Q_n \subseteq U$. If $|Q_i| \geq s$, for every $1 \leq i \leq n$ then a hitting set $A$ of size $\tilde{O}(u/s)$ such that $Q_i \cap A \neq \emptyset$ can be found with a deterministic algorithm in $O(u + \sum_{i=1}^{n} |Q_i|)$ time.*

### 2.1 The Distance Oracle of Thorup and Zwick

In their seminal paper Thorup and Zwick [19] showed that there is a data structure of size $O(kn^{1+1/k})$ that returns a $(2k-1)$ multiplicative approximation (stretch) of the distances of an undirected weighted graph in $O(k)$ time. Let $k \geq 1$ and let $A_0, A_1, \ldots, A_k$ be sets of vertices, such that $A_0 = V$, $A_k = \emptyset$ and $A_i$ is a subset of $A_{i-1}$ of size at most $\tilde{O}(|A_{i-1}|/s)$ that hits for every $v \in V$ the set $N(v, s, A_{i-1})$, where $s$ is a parameter. The set $A_i$ is computed using Lemma 1. For every $u \in V$, let $p_i(u) = p_{A_i}(u)$ and $\ell_i(u) = d(u, A_i) = d(u, p_i(u))$. We set $p_0(u)$ to $u$, $p_k(u)$ to be null and $\ell_k(u)$ to $\infty$.

For every $0 \leq i \leq k-1$, let $B_i(u) = B(u, \ell_{i+1}(u), A_i)$. The *bunch* of $u \in V$ is $B(u) = \cup_{i=0}^{k-1} B_i(u)$.

The information saved in the distance oracle for every $u \in V$ is $B(u) = \cup_{i=0}^{k-1} B_i(u)$, the value of $d(u, v)$, for every $v \in B(u)$, in a 2-level hash table and the vertex $p_i(u)$, where $0 \leq i \leq k$.

Thorup and Zwick proved the following:

**Lemma 2.** *[Theorem 3.7 [19]]. For every $u \in V$ and $i \in [0, k-2]$, the size of $B_i(u)$ is at most $s$ and the size of $B_{k-1}(u)$ is $\tilde{O}(n/s^{k-1})$.*

Setting $s = n^{1/k} c \log n$ yields the desired size bound $O(kn^{1+1/k})$. The query algorithm $dist(u, v)$ of the distance oracle is presented in [15]. We look for the smallest even $i$ such that $p_i(u) \in B_i(v)$ or $p_{i+1}(v) \in B_{i+1}(u)$. Since both $p_{k-1}(u) \in B_{k-1}(v)$ and $p_{k-1}(v) \in B_{k-1}(u)$ the algorithm always stops. Let $f(u, v)$ be the largest value that $i$ reached to during the run of $dist(u, v)$. In other words, $f(u,v)$ is the largest value such that for every even $j < f(u,v)$, it holds that $p_j(u) \notin B_j(v)$ and for every odd $j < f(u,v)$ it holds that $p_j(v) \notin B_j(u)$. Since $dist(u,v)$ always stops it follows that $f(u,v) \leq k-1$.

To bound the stretch we first prove the following Lemma that is implicit in [19]. We prove it explicitly in [15] since we use it in our proofs

**Lemma 3.** *For every even $i \leq f(u,v)$ it holds that $\ell_i(u) \leq i \cdot d(u,v)$ and for every odd $i \leq f(u,v)$ it holds that $\ell_i(v) \leq i \cdot d(u,v)$.*

We proceed with the following useful observation on Thorup-Zwick distance oracle that we will use later on. Consider the set $A_{i-j}$, where $i$ and $j$ are even and $0 \leq j < i \leq f(u,v)$. From Lemma 3 it follows that $\ell_{i-j}(u) \leq (i-j) \cdot d(u,v)$ and $\ell_i(u) \leq i \cdot d(u,v)$. But what if we have a bound for $\ell_{i-j}(u)$ that is better than $(i-j) \cdot d(u,v)$, can we use it to obtain a better bound for $\ell_i(u)$? In the next Lemma we present a generalization of Lemma 3 and show that this is indeed possible. The proof is given in [15].

**Lemma 4.** *For every even $i \leq f(u,v)$: (i) $\ell_i(u) \leq \ell_{i-j}(u) + j \cdot d(u,v)$, for every even $j \leq i$, and (ii) $\ell_i(u) \leq \ell_{i-j}(v) + j \cdot d(u,v)$, for every odd $j \leq i$.*

*For every odd $i \leq f(u,v)$: (i) $\ell_i(v) \leq \ell_{i-j}(u) + j \cdot d(u,v)$, for every even $j \leq i$, and (ii) $\ell_i(v) \leq \ell_{i-j}(v) + j \cdot d(u,v)$, for every odd $j \leq i$.*

We finish the description of Thorup-Zwick distance oracle with a bound on $dist(u,v)$.

**Lemma 5.** *$dist(u,v)$ outputs an estimation that is bounded by $2\ell_{f(u,v)}(u) + d(u,v) \leq (2f(u,v)+1)d(u,v) \leq (2k-1)d(u,v)$, for even $f(u,v)$ and by $2\ell_{f(u,v)}(v) + d(u,v) \leq (2f(u,v)+1)d(u,v) \leq (2k-1)d(u,v)$, for odd $f(u,v)$.*

*Proof.* Let $i = f(u,v)$ be even. The algorithm returns $d(u,p_i(u)) + d(v,p_i(u))$. Using the triangle inequality we get $d(u,p_i(u)) + d(v,p_i(u)) \leq 2\ell_i(u) + d(u,v)$. From Lemma 3 we have $\ell_i(u) \leq i \cdot d(u,v)$ and since $i \leq k-1$ we get $d(u,p_i(u)) + d(v,p_i(u)) \leq (2i+1)d(u,v) \leq (2k-1)d(u,v)$. For the case that $f(u,v)$ is odd the proof is the same with $u$ and $v$ switching their roles.

## 2.2 A Standard Variant of the Distance Oracle of Thorup and Zwick

In order to obtain the new distance oracle we are using a slightly different but relatively standard variant of the distance oracle of Thorup and Zwick (e.g. [5]), which we present below.

In this variant we also save in the distance oracle the exact distance for every pair $\langle u,v \rangle \in A_{k/2} \times A_{k/2-1}$, when $k$ is even, and every pair $\langle u,v \rangle \in A_{(k-1)/2} \times A_{(k-1)/2}$ when $k$ is odd. In both cases the space remains $O(kn^{1+1/k} \log n)$, since $|A_{k/2}| \cdot |A_{k/2-1}| = O(kn^{1+1/k} \log n)$, when $k$ is even and $|A_{(k-1)/2}| \cdot |A_{(k-1)/2}| = O(kn^{1+1/k} \log n)$, when $k$ is odd.

The query will work as follows. Let $u, v \in V$. Let $f = \min(f(u,v), f(v,u))$. If $f \leq \lfloor k/2 \rfloor$ then we output $\min(dist(u,v), dist(v,u))$. If $f > \lfloor k/2 \rfloor$ then we output $\min\big(\ell_{k/2}(u) + d(p_{k/2}(u), p_{k/2-1}(v)) + \ell_{k/2-1}(v), \ell_{k/2}(v) + d(p_{k/2}(v), p_{k/2-1}(u)) + \ell_{k/2-1}(u)\big)$, for an even $k$, and $\ell_{(k-1)/2}(u) + d(p_{(k-1)/2}(u), p_{(k-1)/2}(v)) + \ell_{(k-1)/2}(v)$, for an odd $k$.

In the next Lemma we establish an upper bound on the query output when $f > \lfloor k/2 \rfloor$.

**Lemma 6.** *When* $f > \lfloor k/2 \rfloor$ *the query algorithm described above returns an estimation that is at most* $\min(2\ell_{k/2}(u) + 2\ell_{k/2-1}(v) + d(u,v), 2\ell_{k/2}(v) + 2\ell_{k/2-1}(u) + d(u,v))$, *when* $k$ *is even and at most* $2\ell_{(k-1)/2}(u) + 2\ell_{(k-1)/2}(v) + d(u,v)$, *when* $k$ *is odd.*

*Proof.* Let $a = \ell_{k/2}(u) + d(p_{k/2}(u), p_{k/2-1}(v)) + \ell_{k/2-1}(v)$. Let $b = \ell_{k/2}(v) + d(p_{k/2}(v), p_{k/2-1}(u)) + \ell_{k/2-1}(u)$. Let $A = 2\ell_{k/2}(u) + 2\ell_{k/2-1}(v) + d(u,v)$ and let $B = 2\ell_{k/2}(v) + 2\ell_{k/2-1}(u) + d(u,v)$. For even $k$, the query returns $\min\big(a, b\big)$. We show that this value is at most $\min(A, B)$.

Using the triangle inequality we get that $d(p_{k/2}(u), p_{k/2-1}(v)) \leq \ell_{k/2}(u) + d(u,v) + \ell_{k/2-1}(v)$. Therefore, $a \leq A$. Similarly, we get that $d(p_{k/2}(v), p_{k/2-1}(u)) \leq \ell_{k/2}(v) + d(u,v) + \ell_{k/2-1}(u)$. Therefore, $b \leq B$. Adding it all together we get that $\min(a,b) \leq \min(A,B)$, as required.

When $k$ is odd, the query returns $\ell_{(k-1)/2}(u) + d(p_{(k-1)/2}(u), p_{(k-1)/2}(v)) + \ell_{(k-1)/2}(v) \leq \ell_{(k-1)/2}(u) + (\ell_{(k-1)/2}(u) + d(u,v) + \ell_{(k-1)/2}(v)) + \ell_{(k-1)/2}(v) = 2\ell_{(k-1)/2}(u) + 2\ell_{(k-1)/2}(v) + d(u,v)$.

It is relatively straightforward to prove that the estimation produced by the updated query algorithm has $2k - 1$ stretch by combining Lemma 6 with Lemma 3.

Throughout the paper we will refer to this variant of Thorup-Zwick distance oracle as the standard variant of Thorup-Zwick distance oracle.

## 3 Distance Oracles with Improved Stretch

In this section we present our new distance oracle construction. We combine between two ideas. The first idea is to interplay between a hitting set of vertices and a hitting set of edges. This allows us to obtain, in some cases, a better bound on $\ell_1(u)$, for every $u \in V$. Consider a pair of vertices $u, v \in V$ such that $d(u,v) = \Delta$. In Thorup and Zwick distance oracles if $v \notin B_0(u)$ then it follows that $\ell_1(u) \leq \Delta$ and this bound is used, among other bounds, to bound the estimation. In our distance oracles we will have to use $\ell_1(u)$ to bound the estimation only in the case that $\ell_1(u) \leq \Delta - 1$. Our second idea is that in order to amplify the affect of this better bound we can use the standard variant of Thorup and Zwick distance oracles, presented in Sect. 2.2, since it allows to combine in the bound of the estimation both $\ell_1(u)$ and $\ell_1(v)$ in the case that both $\ell_1(u) \leq \Delta - 1$ and $\ell_1(v) \leq \Delta - 1$.

We now prove the following Theorem:

**Theorem 1.** *Let $G = (V, E)$ be an n-vertices m-edges undirected unweighted graph. For every $k > 2$ there is a distance oracle that uses $O(knm^{1/k} \log n)$ space and for every pair of vertices $u, v \in V$ returns in $O(k)$ time an estimation $d^*(u, v)$ such that:*

$$d(u,v) \leq d^*(u,v) \leq (2k-1)d(u,v) - 4.$$

*For $k = 2$, the estimation $d^*(u, v)$ satisfies: $d(u,v) \leq d^*(u,v) \leq 3d(u,v) - 2$.*

*Proof.* Our new distance oracle is constructed as follows. Let $s = m^{1/k} c \log n$. We start with the set $A_1$ that will be the union of two sets, $A_1^{\mathrm{v}}$ and $A_1^{\mathrm{e}}$. The set $A_1^{\mathrm{v}} \subseteq V$ is a hitting set of size $\tilde{O}(m/s)$ of the sets $N(v, s, V)$, for every $v \in V$, computed using Lemma 1.

The set $A_1^{\mathrm{e}}$ is computed as follows. We first compute for every $u \in V$ the set $L(u, d(u, A_1^{\mathrm{v}}))$. Let $V^H = \{u \mid |L(u, d(u, A_1^{\mathrm{v}}))| \geq s\}$. For every $u \in V^H$ let $E^H(u) = \{(x,y) \in E \mid x \in L(u, d(u, A_1^{\mathrm{v}}) - 1) \wedge y \in L(u, d(u, A_1^{\mathrm{v}}))\}$, that is, all the edges with one endpoint at distance $d(u, A_1^{\mathrm{v}}) - 1$ from $u$ and another endpoint at distance $d(u, A_1^{\mathrm{v}})$ from $u$. Consider now the sets $E^H(u)$, for every $u \in V^H$. Each such set contains at least $s$ edges and there are at most $n$ such sets. Thus, we can apply Lemma 1 to compute a hitting set $E^H \subseteq E$ of size $\tilde{O}(m/s)$. Let $A_1^{\mathrm{e}} = V(E^H)$. We set $A_1$ to $A_1^{\mathrm{v}} \cup A_1^{\mathrm{e}}$.

We now proceed with the sets $A_2, \ldots, A_{k-1}$ as in the distance oracle of Thorup and Zwick, that is, $A_i$ is a subset of $A_i$ of size at most $\tilde{O}(|A_{i-1}|/s)$ that hits for every $v \in V$ the set $N(v, s, A_{i-1})$. The set $A_k$ is empty.

We use the sets $V = A_0, A_1, \ldots, A_k$ to construct the standard variant of the distance oracle. The special way we used to compute the set $A_1$ allows us to prove the following crucial Lemma:

**Lemma 7.** $\sum_{u \in V} |L(u, \ell_1(u))| = \tilde{O}(nm^{1/k})$.

*Proof.* Assume, towards a contradiction, that there exists $u \in V$ such that $|L(u, \ell_1(u))| > s$. Since $A_1 = A_1^{\mathrm{v}} \cup A_1^{\mathrm{e}}$ we have $\ell_1(u) = \min(d(u, A_1^{\mathrm{v}}), d(u, A_1^{\mathrm{e}}))$. It cannot be that $\ell_1(u) = d(u, A_1^{\mathrm{v}})$ because this implies that $|L(u, d(u, A_1^{\mathrm{v}}))| > s$ and $u \in V^H$. In such a case, an edge $(x, y)$ from $E^H(u)$ is in $E^H$ and $x \in A_1^{\mathrm{e}}$ is added to $A_1$. Since $d(u, A_1^{\mathrm{e}}) \leq d(u, x) = d(u, A_1^{\mathrm{v}}) - 1$ and $\ell_1(u) = \min(d(u, A_1^{\mathrm{v}}), d(u, A_1^{\mathrm{e}}))$ we get that it must be that $\ell_1(u) < d(u, A_1^{\mathrm{v}})$.

So we have $|L(u, \ell_1(u))| > s$ and $\ell_1(u) = d(u, A_1^{\mathrm{e}}) < d(u, A_1^{\mathrm{v}})$. The set $A_1^{\mathrm{v}}$ is a hitting set for the sets $N(v, s, V)$, for every $v \in V$. From Lemma 2 it follows that $|B(u, d(u, A_1^{\mathrm{v}}))| \leq s$. Since $\ell_1(u) = d(u, A_1^{\mathrm{e}}) < d(u, A_1^{\mathrm{v}})$ we get that $L(u, \ell_1(u)) \subseteq B(u, d(u, A_1^{\mathrm{v}}))$, a contradiction to the fact that $|L(u, \ell_1(u))| > s$. Thus, we get that $\sum_{u \in V} |L(u, \ell_1(u))| = s \cdot n = \tilde{O}(nm^{1/k})$, as required.

It follows from the above Lemma that we can save also the set $L(u, \ell_1(u))$, for every $u \in V$, in a 2-level hash table, without increasing the total size of the distance oracle.

Given a pair $u, v \in V$ the query works as follows. First, we check if $(u, v) \in E$ and if so return 1 and stop. Otherwise, we check if either $v \in L(u, \ell_1(u))$ or $u \in L(v, \ell_1(v))$ and if so return the exact distance and stop. If this is not the case we use the query of the standard variant of Thorup-Zwick distance oracle on $u, v$ and on $v, u$ and report the minimum of these two estimations.

Next, we analyze the stretch of the distance oracle. Let $u, v \in V$ and let $\Delta = d(u, v)$. If $(u, v) \in E$ or $u \in B_0(v)$ or $v \in B_0(u)$ then the exact distance is returned. Therefore, we can assume that $(u, v) \notin E$, $u \notin B_0(v)$ and $v \notin B_0(u)$. Let $d(u', v) = d(u, v') = \Delta - 1$, where $u' \in N(u)$ and $v' \in N(v)$. If $u' \in B_0(v)$ (respectively, $v' \in B_0(u)$) then $u \in L(v, \ell_1(v))$ (respectively, $v \in L(u, \ell_1(u))$) and the exact distance is returned. Therefore, we can assume also that $u' \notin B_0(v)$ and $v' \notin B_0(u)$. This implies that $\ell_1(v) \leq \Delta - 1$ and $\ell_1(u) \leq \Delta - 1$.

For $k = 2$ the standard variant of Thorup-Zwick distance oracle degenerates to the regular one since the additional distances stored are for pairs from $A_1 \times A_0$. The query returns $\ell_1(u) + d(v, p_1(u))$ which is bounded by $2\ell_1(u) + \Delta$. Using the bound $\ell_1(u) \leq \Delta - 1$ we get that the estimation is bounded by $3\Delta - 2$, as required.

Consider now the case that $k \geq 3$. As we have checked whether $(u, v) \in E$, we can assume that $\Delta \geq 2$. Let $f = \min\big(f(u, v), f(v, u)\big)$. In the case that $f \leq \lfloor k/2 \rfloor$ the query returns $\min(dist(u, v), dist(v, u))$. From Lemma 5 it follows that this estimation is bounded by $(2(k/2)+1)d(u, v) = (k+1)\Delta \leq (2k-1)\Delta - 4$ for even $k \geq 4$ and $\Delta \geq 2$, and bounded by $(2((k-1)/2)+1)d(u, v) = k\Delta \leq (2k-1)\Delta - 4$ for odd $k \geq 3$ and $\Delta \geq 2$.

For $f > \lfloor k/2 \rfloor$ the query returns $\min\big(\ell_{k/2}(u) + d(p_{k/2}(u), p_{k/2-1}(v)) + \ell_{k/2-1}(v), \ell_{k/2}(v) + d(p_{k/2}(v), p_{k/2-1}(u)) + \ell_{k/2-1}(u)\big)$, for an even $k$, and $\ell_{(k-1)/2}(u) + d(p_{(k-1)/2}(u), p_{(k-1)/2}(v)) + \ell_{(k-1)/2}(v)$, for an odd $k$.

Consider the case of an even $k$. Let $i = k/2$ and assume that $i$ is even. It follows from Lemma 6 that $2\ell_i(u) + 2\ell_{i-1}(v) + d(u, v)$ is an upper bound for the estimation. From Lemma 4 we have $\ell_i(u) \leq \ell_1(v) + (i-1)\Delta$ and $\ell_{i-1}(v) \leq \ell_1(u) + (i-2)\Delta$. Thus, we get:

$$\begin{aligned} 2\ell_i(u) + 2\ell_{i-1}(v) + d(u, v) &\leq 2(\ell_1(v) + (i-1)\Delta) + 2((\ell_1(u) + (i-2)\Delta)) + \Delta \\ &\leq 2\ell_1(u) + 2\ell_1(v) + 4i\Delta - 5\Delta \\ &\leq 4(\Delta - 1) + 4(k/2)\Delta - 5\Delta \\ &\leq (2k-1)\Delta - 4 \end{aligned}$$

Assume now that $i$ is odd. It follows from Lemma 6 that $2\ell_i(v) + 2\ell_{i-1}(u) + d(u, v)$ is an upper bound for the estimation. From Lemma 4 we have $\ell_i(v) \leq \ell_1(v) + (i-1)\Delta$ and $\ell_{i-1}(u) \leq \ell_1(v) + (i-2)\Delta$. Thus, we get:

$$\begin{aligned} 2\ell_i(v) + 2\ell_{i-1}(u) + d(u, v) &\leq 4\ell_1(v) + 4i\Delta - 5\Delta \\ &\leq 4(\Delta - 1) + 4(k/2)\Delta - 5\Delta \\ &\leq (2k-1)\Delta - 4 \end{aligned}$$

Consider now the case that $k$ is odd. Let $i = (k-1)/2$. It follows from Lemma 6 that $2\ell_i(u) + 2\ell_i(v) + d(u, v)$ is an upper bound for the estimation.

From Lemma 4 we have $\ell_i(v) \leq \ell_1(u) + (i-1)\Delta$ and $\ell_i(u) \leq \ell_1(v) + (i-1)\Delta$ if $i$ is even or odd. Thus, we get:

$$\begin{aligned} 2\ell_i(u) + 2\ell_i(v) + d(u,v) &\leq 2(\ell_1(v) + (i-1)\Delta) + 2(\ell_1(u) + (i-1)\Delta) + \Delta \\ &\leq 4(\Delta - 1 + (i-1)\Delta) + \Delta \\ &\leq 4(i\Delta - 1) + \Delta \\ &\leq (2k-1)\Delta - 4 \end{aligned}$$

*Remark.* The hierarchal nature of the query algorithm that is based on the bunches induced by the sets $V = A_0, A_1, \ldots, A_k$ makes it tempting to try to apply the interplay between a hitting set of vertices and a hitting set of edges not only to $A_1$ but also to the sets $A_2, \ldots, A_k$. This however is not possible from the following reason. To obtain the improved bound on $\ell_1(u)$ we need that $p_{A_1}(u) \in A_1^{\mathrm{e}}$. Thus, in the next step of the query we need to check if $p_{A_1}(u) \in A_1^{\mathrm{e}}$ is in $B_2(v)$. To get a better bound now for $\ell_2(v)$ we need to be able to either save the vertices of $A_1$ that are at distance $\ell_2(v)$ from $v$, in case that there are at most $s$ such vertices or to improve the bound on $\ell_2(v)$ by a tighter hitting set of size $\tilde{O}(m/s^2)$, if there are strictly more than $s$ such vertices. However, in the later case, the fact that there are more than $s$ vertices of $A_1$, which all might be vertices of $A_1^{\mathrm{e}}$, at distance $\ell_2(v)$ does not imply that the number of edges with one endpoint at distance $\ell_2(v) - 1$ from $v$ and another endpoint at distance $\ell_2(v)$ from $v$ is more than $s^2$. It might be that there are many edges (strictly more than $s^2$) with both endpoints at distance $\ell_2(v)$ from $v$. These edges can cause to strictly more than $s$ vertices of $A_1^{\mathrm{e}}$ to be at distance $\ell_2(v)$ from $v$. On the other hand, hitting these set of edges might result with an edge whose both endpoints are at distance $\ell_2(v)$ and will not improve $\ell_2(v)$.

## 4 A Refined Stretch Analysis of Thorup-Zwick Distance Oracle

In this section we present several different conditions that can be easily checked and once fulfilled by the distance oracle of Thorup-Zwick guarantee that the estimation has a stretch which is strictly better than $2k-1$.

The main parameter that we use is the *average distance* between a vertex and the sets $A_1, \ldots, A_{k-1}$. We define the average distance between $u \in V$ and $A_i$ to be $\bar{\ell}_i(u) = \lceil \ell_i(u)/i \rceil$, where $i \in [1, k-1]$.

Let $\hat{d}(u,v) = \min(dist(u,v), dist(v,u))$. We prove the following properties:

*Property 1.* Let $u \in V$. If $\bar{\ell}_i(u) \neq \bar{\ell}_j(u)$ for some $i, j \in [1, k-1]$ then for every $v \in V$ the stretch of $\hat{d}(u,v)$ is strictly better than $(2k-1)$.

*Property 2.* Let $u, v \in V$. If $\bar{\ell}_i(u) \neq \bar{\ell}_i(v)$ for some $i \in [1, k-1]$ then the stretch of $\hat{d}(u,v)$ is strictly better than $(2k-1)$.

*Property 3.* Let $u, v \in V$. If $\bar{\ell}_i(u) = \bar{\ell}_i(v) = q$, for every $i \in [1, k-1]$ and $d(u,v) \neq q$ then the stretch of $\hat{d}(u,v)$ is strictly better than $(2k-1)$.

Before we turn into the technical part of this section we discuss these properties. First notice to the nice relation between these properties. If the conditions of Property 1 do not hold then the conditions of Property 2 can still hold, and if the conditions of both Properties 1 and 2 do not hold then the conditions of Property 3 can still hold.

From the implementation perspective we can verify whether Property 1 and Property 2 hold using a simple computation that does not require the actual computation of the distance oracle itself. Moreover, if Property 1 does not hold then we have $\bar{\ell}_i(u) = \ell_1(u)$, for every $i \in [1, k-1]$, since $\bar{\ell}_1(u) = \ell_1(u)$. Thus, $\ell_1(u) - 1 \leq \ell_i(u)/i \leq \ell_1(u)$ and we get that $\ell_i(u) \in [i\ell_1(u) - i, i\ell_1(u)]$. In such a scenario the shortest paths tree of $u$ has a relatively well defined structure in which $|B(u, \ell_1(u))| \leq n^{1/k}$ and for every $i \in [2, k-1]$ it holds that $|B(u, i\ell_1(u) - i)| \leq n^{i/k}$ and $n^{i/k} \leq |B(u, i\ell_1(u))|$. It is a plausible conjecture that such a well defined structure is not common. For the sake of completeness we do a small experiment on several different datasets of real world graphs to test how frequent these properties are. We elaborate more on this experiment in [15].

Due to lack of space, we omit the technical part of this section, which can be found in [15].

## 5 Concluding Remarks

In this paper we proved that for every $k \geq 2$ there is a distance oracle of size $O(knm^{1/k} \log n)$ that produces in $O(k)$ time an estimation $d^*(u,v)$ that satisfies $d(u,v) \leq d^*(u,v) \leq (2k-1)d(u,v) - 4$, for $k > 2$, and $d(u,v) \leq d^*(u,v) \leq 3d(u,v) - 2$, for $k = 2$.

An interesting open problem is whether it is possible to obtain a distance oracle with the same size and query time whose estimation $d^*(u,v)$ satisfies $d(u,v) \leq d^*(u,v) \leq (2k-1)d(u,v) - \Omega(k)$, for large enough $k$.

## References

1. Abraham, I., Gavoille, C.: On approximate distance labels and routing schemes with affine stretch. In: Peleg, D. (ed.) DISC 2011. LNCS, vol. 6950, pp. 404–415. Springer, Heidelberg (2011). https://doi.org/10.1007/978-3-642-24100-0_39
2. Bernstein, A.: Fully dynamic (2 + epsilon) approximate all-pairs shortest paths with fast query and close to linear update time. In: 50th Annual IEEE Symposium on Foundations of Computer Science, FOCS 2009, Atlanta, Georgia, USA, 25–27 October 2009, pp. 693–702 (2009)
3. Chechik, S.: Compact routing schemes with improved stretch. In: ACM Symposium on Principles of Distributed Computing, PODC 2013, Montreal, QC, Canada, 22–24 July 2013, pp. 33–41 (2013)
4. Chechik, S.: Approximate distance oracles with constant query time. In: STOC (2014)

5. Chechik, S.: Approximate distance oracles with improved bounds. In: STOC (2015)
6. Chen, W., Sommer, C., Teng, S.-H., Wang, Y.: Compact routing in power-law graphs. In: Keidar, I. (ed.) DISC 2009. LNCS, vol. 5805, pp. 379–391. Springer, Heidelberg (2009). https://doi.org/10.1007/978-3-642-04355-0_41
7. Henzinger, M., Krinninger, S., Nanongkai, D.: Dynamic approximate all-pairs shortest paths: breaking the O(mn) barrier and derandomization. SIAM J. Comput. **45**(3), 947–1006 (2016)
8. Lacki, J., Ocwieja, J., Pilipczuk, M., Sankowski, P., Zych, A.: The power of dynamic distance oracles: efficient dynamic algorithms for the Steiner tree. In: Proceedings of the Forty-Seventh Annual ACM on Symposium on Theory of Computing, STOC 2015, Portland, OR, USA, 14–17 June 2015, pp. 11–20 (2015)
9. Mendel, M., Naor, A.: Ramsey partitions and proximity data structures. In: 47th Annual IEEE Symposium on Foundations of Computer Science (FOCS 2006), 21–24 October 2006, Berkeley, California, USA, Proceedings, pp. 109–118 (2006)
10. Patrascu, M., Roditty, L.: Distance oracles beyond the Thorup-Zwick bound. SIAM J. Comput. **43**, 300–311 (2014)
11. Patrascu, M., Roditty, L., Thorup, M.: A new infinity of distance oracles for sparse graphs. In: FOCS (2012)
12. Qi, Z., Xiao, Y., Shao, B., Wang, H.: Toward a distance oracle for billion-node graphs. PVLDB **7**(1), 61–72 (2013)
13. Roditty, L., Tov, R.: New routing techniques and their applications. In: Proceedings of the 2015 ACM Symposium on Principles of Distributed Computing, PODC 2015, Donostia-San Sebastián, Spain, 21–23 July 2015, pp. 23–32 (2015)
14. Roditty, L., Tov, R.: Close to linear space routing schemes. Distrib. Comput. **29**(1), 65–74 (2015). https://doi.org/10.1007/s00446-015-0256-5
15. Roditty, L., Tov, R.: Approximate distance oracles with improved stretch for sparse graphs (2021). https://github.com/roei-tov/Approximate-Distance-Oracles-with-Improved-Stretch-for-Sparse-Graphs
16. Roditty, L., Zwick, U.: Dynamic approximate all-pairs shortest paths in undirected graphs. SIAM J. Comput. **41**(3), 670–683 (2012). https://doi.org/10.1137/090776573
17. Sommer, C., Verbin, E., Yu, W.: Distance oracles for sparse graphs. In: FOCS (2009)
18. Thorup, M., Zwick, U.: Compact routing schemes. In: SPAA, pp. 1–10 (2001)
19. Thorup, M., Zwick, U.: Approximate distance oracles. J. ACM **52**, 1–24 (2005)
20. Wulff-Nilsen, C.: Approximate distance oracles with improved query time. In: Proceedings of the Twenty-Fourth Annual ACM-SIAM Symposium on Discrete Algorithms, SODA 2013, New Orleans, Louisiana, USA, 6–8 January 2013, pp. 539–549 (2013)

# Hardness and Approximation Results of Roman {3}-Domination in Graphs

Pooja Goyal and B. S. Panda(✉)

Computer Science and Application Group, Department of Mathematics, Indian Institute of Technology Delhi, Hauz Khas, New Delhi 110016, India
{Pooja.Goyal,bspanda}@maths.iitd.ac.in

**Abstract.** A Roman {3}-dominating function (Double Italian dominating function) of a graph $G = (V, E)$ is a function $f : V \rightarrow \{0, 1, 2, 3\}$ having the property that for every vertex $v \in V$, if $f(v) = 0$, then $\sum_{u \in N(v)} f(u) \geq 3$ and if $f(v) = 1$, then $\sum_{u \in N(v)} f(u) \geq 2$. The weight, $f(V)$, of a Roman {3}-dominating function $f$ is $\Sigma_{u \in V} f(u)$. The minimum weight of a Roman {3}-dominating function in a graph $G$ is known as *Roman {3}-domination number* of $G$ and is denoted by $\gamma_{\{R3\}}(G)$. Minimum Roman {3}-Domination problem is to find a Roman {3}-dominating function of minimum weight and Decide Roman {3}-Domination is the decision version of Minimum Roman {3}-Domination problem. Decide Roman {3}-Domination is known to be NP-complete for bipartite graphs. In this paper, we show that Decide Roman {3}-Domination is NP-complete for chordal graphs. We show that Minimum Roman {3}-Domination problem is polynomial-time solvable for threshold graphs which is a subclass of chordal graphs. We propose an $O(\ln \Delta(G))$ approximation algorithm for Minimum Roman {3}-Domination problem for a graph $G$ with maximum degree $\Delta(G)$. Finally, we show that Minimum Roman {3}-Domination problem is APX-complete for bounded degree graphs.

**Keywords:** Roman {3}-Domination · NP-complete · Approximation algorithm

## 1 Introduction

Let $G = (V, E)$ be a finite, simple and undirected graph with vertex set $V$ and edge set $E$. A set $D \subseteq V$ is called a dominating set of $G$ if every vertex $v \in V \setminus D$ is adjacent to at least one vertex in $D$. The *domination number* of $G$ is the minimum cardinality among all dominating sets of $G$ and it is denoted by $\gamma(G)$. Minimum Domination problem is to find a dominating set of minimum cardinality and Decide Domination is the decision version of Minimum Domination problem. Domination in graphs has been studied extensively and has several applications (see [5,6]).

A Roman dominating function (RDF) of a graph $G$ is a function $f : V \rightarrow \{0, 1, 2\}$ such that any vertex $u$ with $f(u) = 0$ has at least one neighbor $v$


C.-Y. Chen et al. (Eds.): COCOON 2021, LNCS 13025, pp. 101–111, 2021.
https://doi.org/10.1007/978-3-030-89543-3_9

with $f(v) = 2$. The weight of $f$ is $f(V) = \Sigma_{u \in V} f(u)$. The minimum weight of a RDF in a graph $G$ is known as *Roman domination number* of $G$ and is denoted by $\gamma_R(G)$. MINIMUM ROMAN DOMINATION problem is to find a Roman dominating function of minimum weight and DECIDE ROMAN DOMINATION is the decision version of MINIMUM ROMAN DOMINATION problem. The concept of Roman domination was defined by Ian Stewart in an article entitled "Defend the Roman Empire!" [10].

A Roman {3}-dominating function (R3DF) of a graph $G = (V, E)$ is a function $f : V \rightarrow \{0, 1, 2, 3\}$ having the property that for every vertex $v \in V$, if $f(v) \in \{0, 1\}$, then $\sum_{u \in N[v]} f(u) \geq 3$. Formally, a Roman {3}-dominating function on a graph G is a function $f : V \rightarrow \{0, 1, 2, 3\}$ such that the following conditions are met:

(i) if $f(v) = 0$, then one of the following conditions must hold
  (a) there exist at least three vertices in $V_1 \cap N(v)$,
  (b) there exist one vertex in $V_1 \cap N(v)$ and one in $V_2 \cap N(v)$,
  (c) there exist two vertices in $V_2 \cap N(v)$,
  (d) there exist one vertex in $V_3 \cap N(v)$
(ii) if $f(v) = 1$, then one of the following conditions must hold
  (a) there exist at least two vertices in $V_1 \cap N(v)$,
  (b) there exist one vertex in $(V_2 \cup V_3) \cap N(v)$

The weight of $f$ is $f(V) = \Sigma_{u \in V} f(u)$. The minimum weight of a R3DF in a graph $G$ is known as *Roman {3}-domination number* of $G$ and is denoted by $\gamma_{R3}(G)$.

For a graph $G = (V, E)$, a Roman {3}-dominating function $f : V \rightarrow \{0, 1, 2, 3\}$ can be denoted by $(V_0, V_1, V_2, V_3)$, where $V_i = \{v \in V \mid f(v) = i\}$ for $i \in \{0, 1, 2, 3\}$. Note that there exists a one to one correspondence between the function $f : V \rightarrow \{0, 1, 2, 3\}$ and the ordered partition $(V_0, V_1, V_2, V_3)$ of $V$. Thus, we will write $f = (V_0, V_1, V_2, V_3)$.

Minimum Roman {3}-domination problem and its decision version are defined as follows:

MINIMUM ROMAN {3}-DOMINATION problem

**Instance:** A graph $G = (V, E)$.
**Solution:** A minimum Roman {3}-dominating function $f$ of $G$.

DECIDE ROMAN {3}-DOMINATION

**Instance:** A graph $G = (V, E)$ and a positive integer $r$.
**Question:** Deciding whether the Roman {3}-domination number of $G$ is equal to $r$?

Mojdeh and Volkmann [9] introduced the concept of Roman {3}-domination. Authors initiated the algorithmic study of Roman 3-domination and showed its relationship to domination, Roman domination, Roman {2}-domination (Italian domination) and double Roman domination. Further, Azvin and Jafari [2] obtained some bounds on Roman {3}-domination number.

In this paper, we extend the algorithmic study of MINIMUM ROMAN {3}-DOMINATION problem. The main contributions of the paper are as follows.

- We show that DECIDE ROMAN {3}-DOMINATION is NP-complete for chordal graphs.
- We propose a polynomial-time algorithm for MINIMUM ROMAN {3}-DOMINATION problem for threshold graphs, a subclass of chordal graphs.
- We propose an approximation algorithm with approximation ratio $O(\ln(\Delta(G)))$ for MINIMUM ROMAN {3}-DOMINATION problem for a graph $G$ with maximum degree $\Delta(G)$.
- We show that MINIMUM ROMAN {3}-DOMINATION problem for bounded degree graphs is APX-complete.

## 2 Preliminaries

Let $G = (V, E)$ be a finite, simple and undirected graph with no isolated vertex. The open neighborhood of a vertex $v$ in $G$ is $N_G(v) = \{u \in V \mid uv \in E\}$ and the closed neighborhood is $N_G[v] = \{v\} \cup N_G(v)$. The degree of a vertex $v$ is $|N_G(v)|$ and is denoted by $d_G(v)$. If $d_G(v) = 1$, then $v$ is called a *pendant* vertex and the neighbor of a pendant vertex is called a *support* vertex. The minimum and the maximum degree of $G$ will be denoted by $\delta(G)$ and $\Delta(G)$, respectively. For $D \subseteq V$, $G[D]$ denote the subgraph induced by $D$. For any $C \subseteq V$, if $G[C]$ is a complete subgraph of $G$, then $C$ is called a *clique* of $G$. For any $I \subseteq V$, if $G[I]$ has no edge, then $I$ is called an *independent set* of $G$. We use the standard notation $[k] = \{1, 2, \ldots, k\}$.

A *bipartite graph* is an undirected graph $G = (X, Y, E)$ whose vertices can be partitioned into two disjoint sets $X$ and $Y$ such that every edge has one end vertex in $X$ and the other in $Y$. A bipartite graph $G = (X, Y, E)$ is *complete bipartite* if for every $x \in X$ and $y \in Y$, there is an edge $xy \in E$. A complete bipartite graph with partitions of size $|X| = m$ and $|Y| = n$, is denoted $K_{m,n}$. In particular for $m = 1$, $K_{1,n}$ is known as *star* graph. A graph $G = (V, E)$ is said to be a *chordal graph* if every cycle of length at least four has a chord, i.e., an edge joining two non-consecutive vertices of the cycle.

**Observation 1.** *(see [9]) For any graph $G = (V, E)$, $\gamma(G) + 2 \leq \gamma_{R3}(G) \leq 3\gamma(G)$. These bounds are sharp.*

## 3 NP-completeness Result

Mojdeh and Volkmann [9] have shown that DECIDE ROMAN {3}-DOMINATION is NP-complete for bipartite graphs. In this section, we strengthen this NP-completeness result by showing that this problem remains NP-complete for chordal graphs. For this we recall the definition of EXACT-3-COVER.

**Exact-3-Cover (X3C)**
**Instance:** A finite set $X$ with $|X| = 3q$ and a collection $\mathcal{C}$ of 3-element subsets of $X$.
**Question:** Does $\mathcal{C}$ contain an exact cover for $X$, that is, a sub-collection $C' \subseteq \mathcal{C}$ such that every element in $X$ occurs in exactly one member of $C'$?

**Theorem 2.** Decide Roman {3}-Domination *is NP-complete for chordal graphs.*

*Proof.* Given a function $f : V \to \{0, 1, 2, 3\}$ of weight at most $r$ for a graph $G = (V, E)$, it can be checked in polynomial time whether $f$ is a Roman {3}-dominating function of $G$. Hence, Decide Roman {3}-Domination is in NP for split graphs. To show the hardness, we give a polynomial reduction from Exact-3-Cover, which is known to be NP-complete (see [7]). Given an arbitrary instance $(X, \mathcal{C})$ of $X3C$, $X = \{x_1, x_2, \ldots, x_{3q}\}$ and $\mathcal{C} = \{C_1, C_2, \ldots, C_t\}$. We construct a chordal graph $G = (V, E)$ from the system $(X, \mathcal{C})$ as follows:

- For each vertex $x_j \in X$, we build a graph $H_j$ obtained from a path $P_4$ : $x_j y_j z_j u_j$ by adding a star graph $K_{1,3}$ centered at $u_j$.
- For each $C_i \in \mathcal{C}$, we add a vertex $c_i$ and add a star graph $K_{1,3}$ centered at $d_i$. Further, add edges $\{d_i c_i, c_i c_k \mid i, k \in [t], i \neq k\}$.
- Finally add edges $x_j c_i$ if and only if $x_j \in C_i$.

We show an example in Fig. 1. A chordal graph $G$ is obtained from the system $(X, \mathcal{C})$, where $X = \{x_1, x_2, x_3, x_4, x_5, x_6\}$ and $\mathcal{C} = \{\{x_1, x_2, x_3\}, \{x_2, x_4, x_5\}, \{x_3, x_5, x_6\}, \{x_4, x_5, x_6\}\}$.

Now to complete the proof, it suffices to prove the following claim:

*Claim.* The system $(X, \mathcal{C})$ has an exact cover if and only if $G$ has a Roman {3}-dominating function with weight equal to $r = 16q + 3t$.

*Proof.* Suppose that $C'$ is a solution of $(X, \mathcal{C})$. We define a function $f : V \to \{0, 1, 2, 3\}$ as follows.

$$f(v) = \begin{cases} 3, & \text{if } v \in \{u_i \mid i \in [3q]\} \cup \{d_j \mid j \in [t]\} \\ 2, & \text{if } v \in \{y_i \mid i \in [3q]\} \\ 1, & \text{if } v \in \{u \mid u \in C'\} \\ 0, & \text{otherwise.} \end{cases}$$

We label all pendant vertices by 0. All $u_j$'s and $d_i$'s are labelled by 3. Also, every $y_j$ is labelled by 2 and every $x_j$ is labelled by 0. For any $i$, we label $c_i$ by 1 if $C_i \in C'$ and label $c_i$ by 0 if $C_i \notin C'$. Observe that since $C'$ exists, its cardinality is exactly $q$, and so the number of $c_i$'s with weight 1 is $q$. Since $C'$ is a solution for $X3C$, any vertex of $X$ has a neighbor $C_i$ labelled by 1. Every vertex $x_j$ has a neighbor $c_j$ labelled by 1 and a neighbor $y_j$ labelled by 2. Hence, $\sum_{v \in N_G(x_j)} f(v) \geq 3$. Every vertex $c_i$ has a neighbor $d_i$ labelled by 3. Hence, $f$ is a R3DF of $G$ with $f(V) = 3 \cdot 5q + 3t + q = r$.

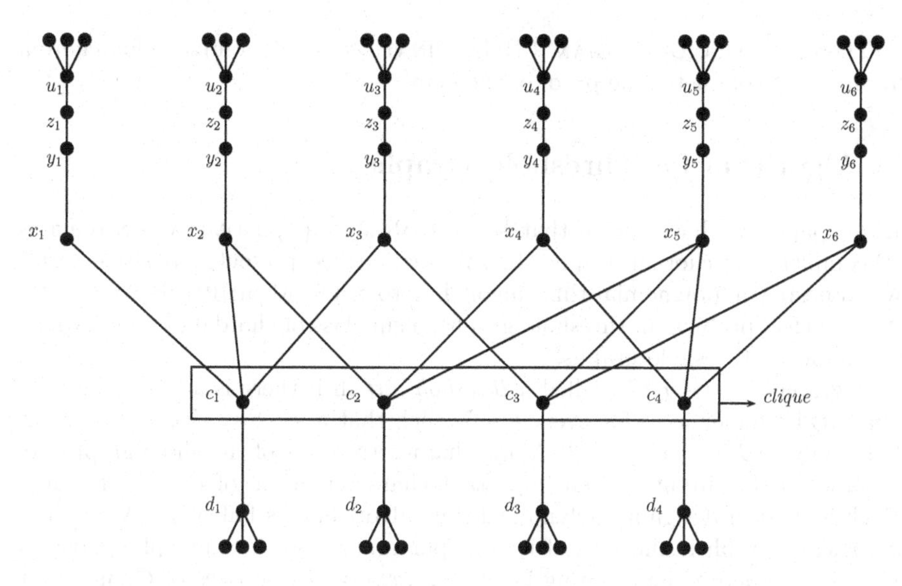


**Fig. 1.** An illustration of the construction of $G$ from system $(X, \mathcal{C})$ in the proof of Theorem 2.

Observe that, $H_j$ has a Roman {3}-domination number less than equal to 6, i.e. $f(V(H_j)) \leq 6$. More precisely, if $f(V(H_j)) = 6$, then we may assume, without loss of generality, that $f(u_j) = 3$, $f(z_j) = 0$, $f(y_j) = 2$ and $f(x_j) = 1$. Also, if $f(V(H_j)) = 5$, then clearly at least one vertex of $H_j$ (including $x_j$) is not Roman {3}-dominated. In this case, we may assume that vertices of $H_j$ are assigned as follows so that only $x_j$ is not Roman {3}-dominated: $f(u_j) = 3, f(y_j) = 2$ and $f(z_j) = f(x_j) = 0$.

Conversely assume that $G$ has a R3DF $f = (V_0, V_1, V_2, V_3)$ of weight $r = 16q + 3t$. Clearly, each star needs a weight of at least 3, and so we may assume, without loss of generality, that $f(u_j) = f(d_i) = 3$ and all its leaves are assigned 0. Since $d_ic_i \in E(G)$, it follows that each vertex $c_i$ may be assigned the value 0. Next we show that no $x_j$ needs to be assigned a positive value, where $f(x_j) \in \{0, 1\}$(as mentioned above). Assume $f(V) = r$ and there exist $l$ number of $x_j$'s such that $f(x_j) \neq 0$, i.e. $f(x_j) = 1$ (as discussed above). Then the number of $x_j$'s with $f(x_j) = 0$ is $3q - l$. In other words, there exist $l$ number of $H_j$'s such that $f(V(H_j)) = 6$ and $3q - l$ number of $H_j$'s such that $f(V(H_j) = 5$. Since $f$ is a R3DF, each $x_j$ with $f(x_j) = 0$ should have a neighbor $c_i$ with $f(c_i) = 1$(since $f(y_j) = 2$). Let $y$ be the number of $c_i$'s labelled with 1. Thus, $3y \geq 3q - l$. Hence $f(V) = r = 3t + 6 \cdot l + 5 \cdot (3q - l) + y \geq 3t + 15q + l + \lceil \frac{3q-l}{3} \rceil$. On solving this, we get $y = q$ and $l = 0$. Therefore for each $x_j \in X$, $f(x_j) = 0$. Clearly, there exist $q$ number of $c_i$'s with weight 1 such that each $x_j$ is adjacent to a $c_i$ of weight 1. Consequently, $C' = \{c_i \mid f(c_i) = 1\}$ is an exact cover for $\mathcal{C}$. This completes the proof of claim. □

Therefore, DECIDE ROMAN {3}-DOMINATION is NP-complete for chordal graphs. This completes the proof of theorem. □

# 4 Algorithm for Threshold Graphs

In this paper, we have shown that DECIDE ROMAN {3}-DOMINATION remains NP-complete for chordal graphs. In this section, we present a positive result by proposing a polynomial-time algorithm to solve MINIMUM ROMAN {3}-DOMINATION problem in threshold graphs, a subclass of chordal graphs. Firstly, we will define threshold graphs.

A graph $G = (V, E)$ is called a *threshold* graph if there is a real number $T$ and a real number $w(v)$ for every $v \in V$ such that a set $S \subseteq V$ is independent if and only if $\Sigma_{v \in S} w(v) \leq T$ [3]. Many characterizations of threshold graphs are available in the literature. An important characterization of threshold graph, which is used in designing polynomial-time algorithms is following: A graph $G$ is *threshold* graph if and only if it is a split graph and, for any split partition $(C, I)$ of $G$, there is an ordering $(x_1, x_2, \ldots, x_p)$ of the vertices of $C$ such that $N_G[x_1] \subseteq N_G[x_2] \subseteq \ldots \subseteq N_G[x_p]$, and there is an ordering $(y_1, y_2, \ldots, y_q)$ of the vertices of $I$ such that $N_G(y_1) \supseteq N_G(y_2) \supseteq \ldots \supseteq N_G(y_q)$ [8].

**Theorem 3.** *Let $G = (V, E)$ be a connected threshold graph with split partition $(C, I)$ as defined above, then $\gamma_{R3}(G) = 3$.*

*Proof.* Let $f : V \to \{0, 1, 2, 3\}$ be a function on $G$ defined as follows.

$$f(v) = \begin{cases} 3, & \text{if } v = x_p \\ 0, & \text{otherwise} \end{cases}$$

Since vertex $x_p$ of label 3 is adjacent to every other vertex of the graph of label 0. Thus, $f$ is Roman {3}-dominating function of $G$ of weight 3 and $\gamma_{R3}(G) \leq 3$. Since $\gamma(G) = 1$, from Observation 1 we have, $\gamma_{R3}(G) \geq 3$. Thus, $\gamma_{R3}(G) = 3$. □

# 5 Approximation Results

In this section, we propose an $O(\ln(\Delta(G)))$-approximation algorithm for MINIMUM ROMAN {3}-DOMINATION problem for a graph with maximum degree $\Delta(G)$. Next we show that MINIMUM ROMAN {3}-DOMINATION problem is APX-complete for bounded degree graphs.

## 5.1 Approximation Algorithm

In this subsection, we propose an approximation algorithm for MINIMUM ROMAN {3}-DOMINATION problem. To obtain the approximation ratio of MINIMUM ROMAN {3}-DOMINATION problem for any graph, we require approximation ratio of MINIMUM DOMINATION problem.

**Theorem 4.** [4] Minimum Domination *problem in graph $G$ with maximum degree $\Delta$ can be approximated with an approximation ratio of* $1 + \ln(\Delta + 1)$.

By Theorem 4, there exists a polynomial time algorithm, APPROX-DOM-SET algorithm, that outputs a dominating set $D$ of a graph $G$ and achieves the approximation ratio of $1 + \ln(\Delta + 1)$; that is, $|D| \leq (1 + \ln(\Delta + 1))\gamma(G)$.

Next, we propose an algorithm APPROX-R3D to compute an approximate solution of Minimum Roman {3}-Domination problem. In our algorithm, first we compute a dominating set $D$ of the input graph $G$ using the approximation algorithm APPROX-DOM-SET. Next, we construct an ordered partition $f = (V_0, V_1, V_2, V_3)$ of $V$ in which every vertex in $D$ will be labelled by 3 and the remaining vertices will be labelled by 0.

Now, let $f = (V \setminus D, \emptyset, \emptyset, D)$ be the ordered partition returned by APPROX-R3DF algorithm. It can be easily seen that every vertex $v \in V$ is assigned with weight either 0 or 3. Since $D$ is a dominating set of $G$, every vertex $v \in V \setminus D$ labelled by 0 is adjacent to a vertex $u \in D$ labelled by 3. Hence, $\sum_{u \in N[v]} f(u) \geq 3$, for every $v \in V$. Thus, $f$ gives a Roman {3}-dominating function of $G$. We note that the algorithm APPROX-R3DF computes a Roman {3}-dominating function of a given graph $G$ in polynomial time. Hence, we have the following algorithm.

---

**Algorithm 1.** APPROX-R3DF

---

**Input:** A graph $G = (V, E)$.
**Output:** A Roman {3}-dominating function $f = (V_0, V_1, V_2, V_3)$ of graph $G$.
**begin**
  Compute a dominating set $D$ of $G$ using algorithm APPROX-DOM-SET;
  $f = (V \setminus D, \emptyset, \emptyset, D)$;
  **return** $f$;

---

**Theorem 5.** Minimum Roman {3}-Domination *problem in a graph $G = (V, E)$ with maximum degree $\Delta$ can be approximated with an approximation ratio of* $3(1 + \ln(\Delta + 1))$.

*Proof.* Let $D$ be the dominating set returned by the algorithm APPROX-DOM-SET and $f$ be the Roman {3}-dominating function returned by the algorithm APPROX-R3DF. It can be observed that $f(V) = 3|D|$. It is known that $|D| \leq (1 + \ln(\Delta + 1))\gamma(G)$. Therefore, $f(V) \leq 3(1 + \ln(\Delta + 1))\gamma(G) \leq 3(1 + \ln(\Delta + 1))(\gamma(G)+2)$. Thus, Observation 1 leads to $f(V) \leq 3(1+\ln(\Delta+1))\gamma_{R3}(G)$. Thus, Minimum Roman {3}-Domination problem in a graph $G$ can be approximated with an approximation ratio of $3(1 + \ln(\Delta + 1))$. □

### 5.2 APX-completeness for bounded degree graphs

In this section, we show that Minimum Roman {3}-Domination problem is APX-complete for bounded degree graphs. Note that the class APX is the set of all optimization problems which admit a $c$-approximation algorithm, where $c$ is a constant.

By Theorem 5, Minimum Roman {3}-Domination problem for bounded degree graphs can be approximated within a constant ratio. Thus, Minimum Roman {3}-Domination problem for bounded degree graphs is in APX.

Next, we show that Minimum Roman {3}-Domination problem is APX-complete for graphs with maximum degree 7. For this purpose, we recall the concept of $L$-reduction.

**Definition 1.** *Given two NP optimization problems $\pi_1$ and $\pi_2$ and a polynomial time transformation $f$ from instances of $\pi_1$ to instances of $\pi_2$, we say that $f$ is an $L$-reduction if there are positive constants $\alpha$ and $\beta$ such that for every instance $x$ of $\pi_1$:*

1. *$opt_{\pi_2}(f(x)) \leq \alpha \cdot opt_{\pi_1}(x)$.*
2. *for every feasible solution $y$ of $f(x)$ with objective value $m_{\pi_2}(f(x), y) = c_2$, we can find a solution $y'$ of $x$ in polynomial time with $m_{\pi_1}(x, y') = c_1$ such that $|opt_{\pi_1}(x) - c_1| \leq \beta \cdot |opt_{\pi_2}(f(x)) - c_2|$.*

*To show the APX-completeness of a problem $\pi \in$ APX, it suffices to show that there is an $L$-reduction from some APX-complete problem to $\pi$.*

To show the APX-completeness of Minimum Roman {3}-Domination problem, we give an $L$-reduction from Minimum Vertex Cover problem. A set $S \subseteq V$ of a graph $G = (V, E)$ is called a vertex cover of $G$ if for every edge $uv \in E$, either $u \in S$ or $v \in S$. Minimum Vertex Cover problem for a graph $G$ is a problem to find a minimum cardinality vertex cover of $G$. Now, the following theorem is required.

**Theorem 6.** ([1]) Minimum Vertex Cover *problem is APX-complete for graphs with maximum degree* 3.

Now, we are ready to prove the following theorem.

**Theorem 7.** Minimum Roman {3}-Domination *problem is APX-complete for graphs with maximum degree* 7.

*Proof.* Since by Theorem 6, Minimum Vertex Cover problem is APX-complete for graphs with maximum degree 3. So, it is enough to construct an $L$-reduction $f$ from the instances of Minimum Vertex Cover problem for graphs with maximum degree 3 to the instances of Minimum Roman {3}-Domination problem. Given a graph $G = (V, E)$, where $V = \{v_1, v_2, \ldots, v_n\}$ and $E = \{e_1, e_2, \ldots, e_m\}$. We construct a graph $H = (V', E')$ from the graph $G$ in the following way.

(a) For every $v_i \in V$, add a vertex $v_i$ in $H$ and add a star graph $K_{1,3}$ centered at $p_i$ with pendant vertices $q_i, r_i$ and $s_i$. Add the edges $v_ip_i, p_iq_i, p_ir_i, p_is_i$.
(b) For every $e_j \in E$, we build a graph $H_j$ obtained from a path $P_5 : e_ja_jb_jc_jd_j$ by adding the edges $a_jc_j$ and $b_jd_j$.
(c) Finally add edges $e_jv_i$, $e_jv_k$ and $v_iv_k$ in $H$, where $e_j = v_iv_k \in E$.

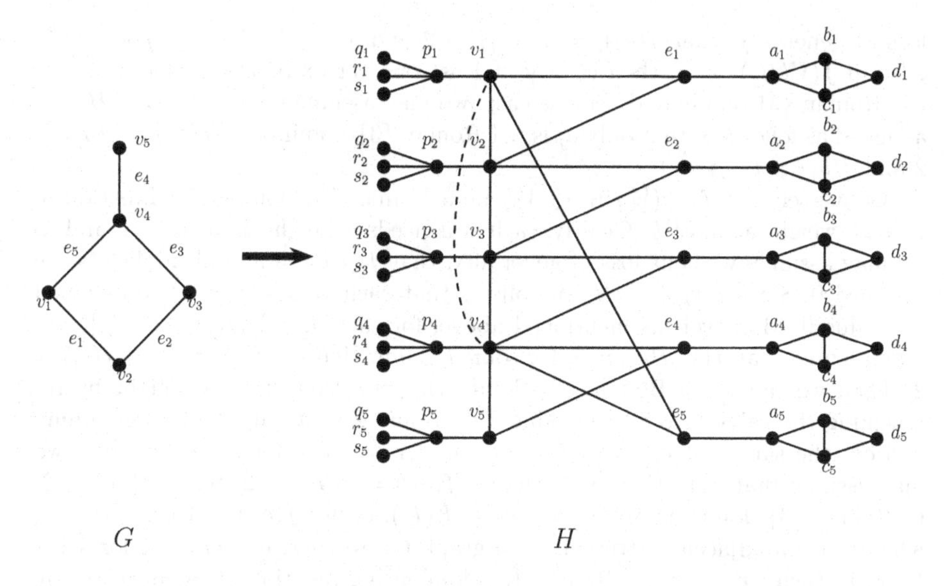


**Fig. 2.** An illustration of the construction of $H$ from $G$ in the proof of Theorem 7.

Formally, $V' = \{v_i, p_i, q_i, r_i, s_i \mid v_i \in V\} \cup \{e_j, a_j, b_j, c_j, d_j \mid e_j \in E\}$ and $E' = \{v_ip_i, p_iq_i, p_ir_i, p_is_i \mid v_i \in V\} \cup \{v_iv_k, e_jv_i, e_jv_k, e_ja_j, a_jb_j, b_jc_j, c_jd_j, c_ja_j, b_jd_j \mid e_j = v_iv_k \in E, j \in [m]\}$. Note that if maximum degree of $G$ is 3, then maximum degree of $H$ is 7. Now, we first prove the following claim (Fig. 2).

*Claim.* $\gamma_{R3}(H) = |VC^*(G)| + 4m + 3n$, where $VC^*(G)$ is the minimum cardinality vertex cover of $G$.

*Proof.* Let $C^*$ be a minimum vertex cover of $G$. We define a function $f : V' \rightarrow \{0, 1, 2, 3\}$ as follows.

$$f(v) = \begin{cases} 3, & \text{if } v \in \{p_i \mid i \in [n]\} \\ 2, & \text{if } v \in \{a_j, d_j \mid j \in [m]\} \\ 1, & \text{if } v \in \{u \mid u \in C^*\} \\ 0, & \text{otherwise} \end{cases}$$

Let $e_k = uv \in E(G)$. Since $C^*$ is a vertex cover for $G$, at least one of $u, v$ must be in $C^*$. Hence, $f(u) = 1$ or $f(v) = 1$. Without loss of generality, assume that $f(u) = 1$. Hence for every edge $e_k$ of weight 0 there exists an adjacent vertex $u$ of weight 1 and a neighbor $a_k$ of weight 2. Hence, for every edge $e_k$, $\sum_{w \in N[e_k]} f(w) \geq 3$. For every $j \in [m]$, vertex $b_j$ and $c_j$ have two neighbors $a_j$ and $d_j$ assigned 2. Also for every $i \in [n]$, vertex $v_i$ and all pendant vertices have a neighbor $p_i$ assigned weight 3. Hence, it is straightforward to see that $f$ is a Roman {3}-dominating function with weight $f(V') = 4m + 3n + |C^*|$. Thus, for a minimum weight R3DF $f^*$ of $H$, $f^*(V') \leq 4m + 3n + |C^*|$.

Observe that, $H_j$ has a Roman {3}-domination number less than equal to 5, i.e. $f(V(H_j)) \leq 5$. More precisely, if $f(V(H_j)) = 5$, then we may assume, without loss of generality, that $f(b_j) = 3, f(e_j) = 2$ and $f(a_j) = f(c_j) = f(d_j) = 0$. Also, if $f(V(H_j)) = 4$, then clearly at least one vertex of $H_j$ (including $e_j$) is not Roman $\{3\}$-dominated. In this case, we may assume that vertices of $H_j$ are assigned as follows so that only $e_j$ is not Roman $\{3\}$-dominated: $f(a_j) = f(d_j) = 2$ and $f(b_j) = f(c_j) = 0$.

Conversely, let $f = (V_0, V_1, V_2, V_3)$ be a Roman $\{3\}$-dominating function of $H$ with weight at most $k$. Clearly, each star needs a weight of at least 3, and so we may assume, without loss of generality, that $f(p_i) = 3$ and all its leaves are assigned 0. Since $p_i v_i \in E(H)$, it follows that each vertex $v_i$ may be assigned the value 0. Moreover, as mentioned above, for each $j$, $f(V(H_j)) \in \{4, 5\}$. We may assume that $f(V(H_l)) = 5$, for some $l \in [m]$. Hence, $f(b_l) = 3$ and $f(e_l) = 2$. Then we can get a Roman $\{3\}$-dominating function of same weight by re-assigning the value 1 to $e_l$, the value 2 to $a_l$ and 2 to $d_l$, and to the remaining vertices the same values as their values under $f$. Also, for some $l' \in [m]$, we may assume that $f(V(H_{l'})) = 4$. Hence, $f(a_{l'}) = f(d_{l'}) = 2$. Since $f(a_{l'}) = 2$, to Roman $\{3\}$-dominate the vertex $e_{l'} \in E(H)$, either $f(e_{l'}) = 1$ or $f(u) = 1$, where $u$ is an adjacent vertex of $e_{l'}$ in graph $G$. Further, if $f(e_k) = 1$ for some $k \in [m]$, then we can get a Roman $\{3\}$-dominating function of same weight by re-assigning the value 0 to $e_k$, the value 1 to $u$ where $u$ is adjacent vertex of $e_k$, and to the remaining vertices the same values as their values under $f$. Hence, there is a minimum weight R3DF, say $f^* = (V_0, V_1, V_2, V_3)$ of $H$ such that for every $i \in [n]$, $f^*(p_i) = 3$ and for every $j \in [m]$, $f^*(a_j) = f^*(d_j) = 2$ and $N_H(e_j) \cap V_1 \neq \emptyset$. Hence for each $e = uv \in E$, either $f^*(u) = 1$ or $f^*(v) = 1$. Thus, $C = V \cap V_1$ is a vertex cover of $G$. Hence, for a minimum cardinality vertex cover $C^*$ of $G$, $|C^*| \leq |C| \leq f^*(V') - 4m - 3n$.

Hence, $f^*(V') = |C^*| + 4m + 3n$. This completes the proof of claim. □

We now return to the proof of Theorem 7. Let $VC^*$ be a minimum vertex cover of $G$ and $f : V' \to \{0, 1, 2, 3\}$ be a minimum weight R3D function of $H$. From the above Claim it is evident that $f(V') = |VC^*(G)| + 4m + 3n$.

Since the maximum degree of $G$ is 3, so $m \leq \frac{3n}{2}$ and $|VC^*(G)| \geq \frac{n}{4}$. Thus, $\gamma_{R3}(H) = |VC^*(G)|+4m+3n \leq |VC^*(G)|+6n+3n \leq |VC^*(G)|+36|VC^*(G)| = 37|VC^*(G)|$.

Now consider a R3D function, say $f$, of $H$ of weight $k$, we can convert it into a R3D function, say $f^*$ with $f^*(V') \leq k$, such that $f^*(a_j) = f^*(d_j) = 2$ for every $j \in [m]$ and $f^*(p_i) = 3$ for every $i \in [n]$. Then $VC = \{v_i \in V \mid f^*(v_i) \neq 0\}$ is a vertex cover of $G$ (as explained in the proof of the above claim) of size less than or equal to $k - 4m - 3n$. Analogously as in the proof of above Claim, the set $VC$ is a vertex cover of $G$ and $|VC| \leq f^*(V') - 4m - 3n$. Hence, $||VC| - |VC^*(G)|| \leq |f^*(V') - 4m - 3n - (\gamma_{R3}(H) - 4m - 3n)| = |f^*(V') - \gamma_{R3}(H)|$.

From these two inequalities, it is clear that the above reduction is an $L$-reduction with $\alpha = 37$ and $\beta = 1$. Therefore, Minimum Roman $\{3\}$-Domination problem is APX-complete for graphs with maximum degree 7. □

## 6 Conclusion

In this paper, we have shown that DECIDE ROMAN {3}-DOMINATION is NP-complete for chordal graphs. On the positive side, we have shown that MINIMUM ROMAN {3}-DOMINATION problem can be solved in polynomial time for threshold graphs. We have then proposed an $O(\ln(\Delta(G)))$-approximation algorithm for finding minimum weight R3DF in any graph $G$ with maximum degree $\Delta(G)$. It would be interesting to give some inapproximability result of MINIMUM ROMAN {3}-DOMINATION problem. Finally, we have shown that MINIMUM ROMAN {3}-DOMINATION problem is APX-complete for bounded degree graphs. It would be interesting to study the complexity of this problem in other graph classes and also the relation between Roman {3}-domination number and other domination parameters.

## References

1. Alimonti, P., Kann, V.: Some APX-completeness results for cubic graphs. Theoret. Comput. Sci. **237**(1–2), 123–134 (2000)
2. Azvin, F., Rad, N.J., Volkmann, L.: Bounds on the outer-independent double Italian domination number. Commun. Comb. Optim. **6**(1), 123–136 (2021)
3. Chvátal, V., Hammer, P.L.: Aggregations of inequalities. Stud. Integer Program. Ann. Discret. Math. **1**, 145–162 (1977)
4. Cormen, T.H., Leiserson, C.E., Rivest, R.L., Stein, C.: Introduction to Algorithms. MIT Press, Cambridge (2009)
5. Haynes, T., Hedetniemi, S., Slater, P.: Domination in Graphs: Advanced Topics. Marcel Dekker Inc., New York (1998)
6. Haynes, T., Hedetniemi, S., Slater, P.: Fundamentals of Domination in Graphs. Marcel Dekker Inc., New York (1998)
7. Johnson, D.S., Garey, M.R.: Computers and Intractability: A Guide to the Theory of NP-Completeness. WH Freeman, New York (1979)
8. Mahadev, N.V.R., Peled, U.N.: Threshold Graphs and Related Topics. Elsevier, Amsterdam (1995)
9. Mojdeh, D.A., Volkmann, L.: Roman {3}-domination (double Italian domination). Discret. Appl. Math. **283**, 555–564 (2020)
10. Stewart, I.: Defend the Roman empire! Sci. Am. **281**(6), 136–138 (1999)

# Approximation Algorithms for Priority Steiner Tree Problems

Faryad Darabi Sahneh, Stephen Kobourov, and Richard Spence(✉)

University of Arizona, Tucson, AZ 85721, USA
rcspence@email.arizona.edu

**Abstract.** In the Priority Steiner Tree (PST) problem, we are given an undirected graph $G = (V, E)$ with a source $s \in V$ and terminals $T \subseteq V \setminus \{s\}$, where each terminal $v \in T$ requires a nonnegative priority $P(v)$. The goal is to compute a minimum weight Steiner tree containing edges of varying rates such that the path from $s$ to each terminal $v$ consists of edges of rate greater than or equal to $P(v)$. The PST problem with $k$ priorities admits a $\min\{2\ln|T| + 2, k\rho\}$-approximation [Charikar et al., 2004], and is hard to approximate with ratio $c \log\log n$ for some constant $c$ [Chuzhoy et al., 2008]. In this paper, we first strengthen the analysis provided by [Charikar et al., 2004] for the $(2\ln|T| + 2)$-approximation to show an approximation ratio of $\lceil \log_2 |T| \rceil + 1 \leq 1.443 \ln|T| + 2$, then provide a very simple, parallelizable algorithm which achieves the same approximation ratio. We then consider a more difficult node-weighted version of the PST problem, and provide a $(2\ln|T| + 2)$-approximation using extensions of the spider decomposition by [Klein & Ravi, 1995]. This is the first result for the PST problem in node-weighted graphs. Moreover, the approximation ratios for all above algorithms are tight.

**Keywords:** Priority steiner tree · Approximation algorithms · Network design

## 1 Introduction

We consider generalizations of the Steiner tree and node-weighted Steiner tree (NWST) problems in graphs where the terminals $T$ possess varying priority or quality of service (QoS) requirements, in which we seek to connect the terminals using edges of the appropriate rate or better. These problems have applications in multimedia and electric power distribution [4,21,27], multi-level graph visualization [1], and other network design problems where a source or root is to be connected to a set of heterogeneous receivers possessing different bandwidth or priority requests. We define a Priority Steiner Tree (PST) as follows:

**Definition 1 (Priority Steiner Tree (PST)).** *Given an undirected graph $G = (V, E)$, a source $s \in V$, and terminals $T \subseteq V \setminus \{s\}$, where each terminal*

Supported in part by NSF grants CCF-1740858, CCF-1712119, and DMS-1839274.


C.-Y. Chen et al. (Eds.): COCOON 2021, LNCS 13025, pp. 112–123, 2021.
https://doi.org/10.1007/978-3-030-89543-3_10

*$v \in T$ requires a nonnegative priority $P(v)$, a* PST *is a tree $\mathcal{T} \subseteq G$ rooted at $s$ containing edges of varying rates such that for all terminals $v \in T$, the $s$–$v$ path in $\mathcal{T}$ consists of edges of rate $P(v)$ or higher.*

We denote by $k$ the number of distinct priorities. Vertices in $V \setminus (T \cup \{s\})$ have zero priority but may be included in $\mathcal{T}$. Let $w(e,r)$ denote the weight of edge $e$ at rate $r$. We assume $w(e,0) = 0$ and $w(e,r_1) \leq w(e,r_2)$ for all $0 \leq r_1 \leq r_2$ and edges $e$ (i.e., higher-rate edges weigh at least as much as lower-rate edges). The weight of a PST $\mathcal{T}$ is the sum of the weights of the edges in $\mathcal{T}$ at their respective rates, namely $w(\mathcal{T}) := \sum_{e \in E(\mathcal{T})} w(e, R(e))$.

*Problem 1* (PRIORITY STEINER TREE *problem*). Given a graph $G = (V,E)$, source $s$, terminals $T \subseteq V$, priorities $P(\cdot)$, and edge weights $w : E \times \mathbb{R}_{\geq 0} \to \mathbb{R}_{\geq 0}$, compute a PST $\mathcal{T}$ with minimum weight.

While Problem 1 in the case where edge weights are proportional to rate (i.e., $w(e,r) = r \cdot w(e,1)$ for all $e \in E$ and $r \geq 0$) admits $O(1)$–approximations [1,6,18], the best known approximation ratio for PRIORITY STEINER TREE with arbitrary weights is $\min\{2\ln|T| + 2, k\rho\}$ by Charikar et al. [6] (see Sect. 2). On the other hand, Chuzhoy et al. [9] show that PRIORITY STEINER TREE cannot be approximated with ratio $c \log\log n$ for some constant $c$ unless NP $\subseteq$ DTIME($n^{O(\log\log\log n)}$), even with unit edge weights[1].

In Sect. 3, we introduce a node-weighted variant of PRIORITY STEINER TREE, called PRIORITY NWST (Definition 2). Here we assume edges have zero weight, as an instance with edge and vertex weights can be converted to an instance with only vertex weights by subdividing each edge $uv$ into two edges $uw$, $wv$ and assigning the weight of edge $uv$ to vertex $w$.

**Definition 2 (Priority Node-Weighted Steiner Tree (PNWST)).** *Given an undirected graph $G = (V,E)$, source $s$, and terminals $T \subseteq V \setminus \{s\}$, where each terminal $v \in T$ requires a nonnegative priority $P(v)$, a* priority node-weighted Steiner tree *(PNWST) is a tree $\mathcal{T}$ rooted at $s$ containing* vertices *of varying rates $R(v)$ such that for all terminals $v \in T$, the $s$–$v$ path in $\mathcal{T}$ consists of vertices of rate $P(v)$ or higher.*

In particular, we require $R(v) \geq P(v)$ for all $v \in T$. Further, we can assume w.l.o.g. that the path from $s$ to each terminal uses vertices of non-increasing rate (see Definition 3). As in the NWST problem, it is conventional to also assume terminals have zero weight, as they must be included in any feasible solution; thus, we assume $w(v,r) = 0$ for $0 \leq r \leq P(v)$ and $w(v,r_1) \leq w(v,r_2)$ for all $0 \leq r_1 \leq r_2$. The weight of a PNWST $\mathcal{T}$ with *vertex* rates $R(\cdot)$ is $w(\mathcal{T}) := \sum_{v \in V(\mathcal{T})} w(v, R(v))$.

[1] We remark that the formulation of PRIORITY STEINER TREE given in [9] is slightly more specific; each edge has a single weight $c_e$ as well as a quality of service (priority) $Q(e)$ on input, and the goal is to compute a Steiner tree such that the path from root to each terminal $v$ uses edges of quality of service greater than or equal to $P(v)$.

*Problem 2* (PRIORITY NWST *problem*). Given a graph $G = (V, E)$, source $s$, terminals $T \subseteq V \setminus \{s\}$, vertex priorities $P(\cdot)$, and vertex weights $w : V \times \mathbb{R}_{\geq 0} \to \mathbb{R}_{\geq 0}$, compute a PNWST $\mathcal{T}$ with minimum weight.

The PRIORITY NWST problem generalizes the NWST problem, and hence cannot be approximated with ratio $(1-o(1)) \ln |T|$ unless P = NP [12,13,19], via a reduction from the set cover problem. In Sect. 3, we show that the PRIORITY NWST problem admits a $2\ln(|T|+1)$–approximation (Theorem 2) using extensions of the spider decomposition given by Klein and Ravi [19] to accommodate the priority constraints of the PRIORITY NWST problem. The generalization is not immediately obvious; in particular it is not immediate whether an instance of PRIORITY NWST can be formulated as an instance of NWST. However, NWST and PRIORITY NWST can be easily reduced to Steiner arborescence (or directed Steiner tree), which admits a quasi-polynomial $O\left(\frac{\log^2 |T|}{\log\log |T|}\right)$-approximation [14].

*Notation.* A graph $G = (V, E)$ with $n = |V|$ and $m = |E|$ is undirected and connected, unless stated otherwise. Given terminals $u, v \in T$ for the PST problem, denote by $\sigma(u, v)$ the weight of a minimum weight $u$–$v$ path in $G$ using edges of rate $\min\{P(u), P(v)\}$, and let $p_{uv}$ denote such a path. For terminals $u, v \in T$ in the PRIORITY NWST problem, we define $\sigma(u, v)$ to be the weight of a minimum $u$–$v$ path using vertices of rate $\min\{P(u), P(v)\}$ not including the endpoints $u$ and $v$, and similarly define $\sigma_b(u, v)$ to be the weight of a minimum weight vertex-weighted path using vertices of rate $b$, so that $\sigma(u, v) = \sigma_{\min\{P(u),P(v)\}}(u, v)$. In particular, we have $\sigma_b(v, v) = 0$. Note that $\sigma$ is symmetric but does not satisfy the triangle inequality, and is not a metric. Let $\rho$ denote an approximation ratio for the (edge-weighted) Steiner tree problem, and let STEINER($n$) denote the running time of such an approximation algorithm on an $n$-vertex graph. We denote by OPT the weight of a min-weight PST or PNWST. Lastly, for $n \in \mathbb{Z}^+$, we denote by $[n]$ the set $\{1, 2, \ldots, n\}$.

### 1.1 Related Work

The Steiner tree problem in graphs has been studied in a wide variety of contexts; see the compendium [16]. The (edge-weighted) Steiner tree problem admits a folklore $2\left(1 - \frac{1}{|T|}\right)$–approximation, and is approximable with ratio $\rho = \ln 4 + \varepsilon \approx 1.387$ [5], but NP-hard to approximate with ratio $\frac{96}{95} \approx 1.01$ [8]. As stated previously, NWST cannot be approximated with ratio $(1 - o(1)) \ln |T|$ unless P = NP [12,13,19], but algorithms with logarithmic approximation ratio exist. Klein and Ravi [19] give a $2 \ln |T|$–approximation for NWST, which was improved to $1.61 \ln |T|$ and a less practical $(1.35 + \varepsilon) \ln |T|$ by Guha and Khuller [15]. Demaine et al. [11] give an $O(1)$–approximation for NWST when the input graph $G$ is $H$–minor free, and a 6-approximation when $G$ is planar. Naor et al. [23] give a randomized $O(\log n \log^2 |T|)$-approximation algorithm for the online version.

The (edge-weighted) PRIORITY STEINER TREE problem and variants thereof have been studied under various other names including Hierarchical Network Design [10], Multi-Level (or $k$-Level) Network Design [4], Multi-Tier Tree [22],

Grade of Service Steiner Tree [28], Quality of Service Multicast Tree [6,18], and Multi-Level Steiner Tree [1,2]. Earlier results on this problem typically consider a small number of priorities or restricted definition of weight [4,10]. In the special case where edge weights are proportional to rate, Charikar et al. [6] give the first $O(1)$–approximations with approximation ratios $4\rho$ and $e\rho \approx 4.214$ (with $\rho \approx 1.55$ [25]) independent of the number of priorities $k$. Karpinski et al. [18] give a slightly stronger variant of the $e\rho$-approximation [6] which achieves approximation ratio 3.802. Ahmed et al. [1] give an approximation ratio of $2.351\rho \approx 3.268$ for $k \leq 100$. Xue et al. [28] consider this problem where the terminals are embedded in the Euclidean plane, and give $\frac{4}{3}\rho$ (resp. $\frac{5+4\sqrt{2}}{7}\rho \approx 1.522\rho$)–approximations for two (resp. three) different priorities. Integer programming formulations have been proposed and evaluated over realistic problem instances [1,24].

If edge weights are not necessarily proportional to rate, Charikar et al. [6] give a simple $\min\{2\ln|T|+2, k\rho\}$-approximation (see Sect. 2), which remains the best known to date. Recently, Ahmed et al. [2] proposed an approximation based on Kruskal's MST algorithm which achieves the same approximation ratio, and provided an experimental study comparing the two methods. Chuzhoy et al. [9] show that PRIORITY STEINER TREE cannot be approximated with ratio $c\log\log n$ for some constant $c$ unless NP $\subseteq$ DTIME$(n^{O(\log\log\log n)})$. Angelopoulos [3] showed that every deterministic online algorithm for online PRIORITY STEINER TREE has ratio $\Omega(\min\{k\log\frac{|T|}{k}, |T|\})$. Interestingly, no node-weighted variant of PRIORITY STEINER TREE has been studied in existing literature. However, a related problem is the (single-source) node-weighted buy-at-bulk problem (NSS-BB) studied by Chekuri et al. [7], who show a $3H_{|T|} = O(\log|T|)$–approximation for NSS-BB by giving a randomized algorithm then derandomizing it using an LP relaxation, where $H_n = \frac{1}{1} + \frac{1}{2} + \ldots + \frac{1}{n}$ is the $n^{\text{th}}$ harmonic number.

### 1.2 Our Results

In Sect. 2, we strengthen the analysis of the simple $(2\ln|T|+2)$-approximation (Algorithm 1) by Charikar et al. [6] to show that it is a $\lceil\log_2|T|\rceil + 1 \leq (1.443\ln|T|+2)$-approximation. We then give a parallelizable algorithm (Algorithm 2) with the same approximation ratio that does not require that terminals be connected sequentially or in a particular order. This contrasts with the inherently serial Algorithm 1 [6], where the shortest path for each terminal depends on the partial PST computed at the previous iteration.

**Theorem 1.** *Algorithm 1* [6] *is a* $(\lceil\log_2|T|\rceil+1)$*-approximation for* PRIORITY STEINER TREE *with running time* $O(nm + n^2\log n)$*, and there is a parallelizable algorithm for* PRIORITY STEINER TREE *with the same approximation ratio.*

Moreover, the approximation ratio is tight up to a factor of 2, as there exists an input graph in which Algorithms 1–2 may output a PST with weight $\frac{1}{2}\log_2|T|+1$ times the optimum [17]. In Sect. 3, we show the following result for PRIORITY NWST:

**Theorem 2.** *There exists a* $2\ln(|T|+1)$*–approximation algorithm for* PRIORITY NWST *with running time* $O(n^4 k\log n)$*.*

This is the first known approximation algorithm for PRIORITY NWST, and is the main technical contribution of this paper. The analysis extends the spider decomposition of Klein and Ravi [19] in their greedy $(2\ln|T|)$–approximation for the NWST problem, to accommodate priority constraints in the PRIORITY NWST problem. Note the additional $+1$ arises as we do not consider the source $s$ a terminal. Proofs omitted for space are in the arXiv version [26].

## 2 Priority Steiner Tree: Two Logarithmic Approximations

We first review the greedy $\min\{2\ln|T|+2, k\rho\}$ approximation for PRIORITY STEINER TREE given by Charikar et al. [6]. This returns the better solution of two sub-algorithms; we focus primarily on the $(2\ln|T|+2)$-approximation (Algorithm 1). This algorithm sorts the terminals $T$ from highest to lowest priority. Then for $i = 1, \ldots, |T|$, the $i^{\text{th}}$ terminal $v_i$ in the sorted list is connected to the existing tree (containing the source $s$) using a minimum weight path of rate $P(v_i)$. The weight of this path is the *connection cost* of $v_i$. Cycles can be removed in the end by removing an edge from each cycle with the lowest rate.

---

**Algorithm 1** $R(\cdot)$ = QoSMT(graph $G$, priorities $P$, edge weights $w$, source $s$) [6]

---

1: Sort terminals $T$ by decreasing priority $P(\cdot)$
2: Initialize $V' = \{s\}$, $R(e) = 0$ for $e \in E$
3: **for** $i = 1, 2, \ldots, |T|$ **do**
4:     Connect $i^{\text{th}}$ terminal $v_i$ to $V'$ using minimum weight path $p_i$ of rate $P(v_i)$
5:     $R(e) = P(v_i)$ for $e \in p_i$
6:     $V' = V' \cup V(p_i)$
7: Remove lowest-rate edge from each cycle
8: **return** edge rates $R(\cdot)$

---

Algorithm 1 is based on a $(\log_2|T|)$-approximation for an online Steiner tree problem analyzed by Imase and Waxman [17]; however, Charikar et al. [6] give a simpler analysis which proves a weaker approximation ratio of $2\ln|T|+2$, based on the following lemma:

**Lemma 1** ([6]). *For $1 \le x \le |T|$, the $x^{th}$ most expensive connection cost incurred by Algorithm 1 is at most $\frac{2\text{OPT}}{x}$.*

Lemma 1 implies the weight of the PST is at most $2\text{OPT}\left(\frac{1}{1}+\frac{1}{2}+\ldots+\frac{1}{|T|}\right) = 2\text{OPT}H_{|T|} \le (2\ln|T|+2)\text{OPT}$. Line 4 can be executed by running Dijkstra's algorithm from $v_i$ with edge weights $w(\cdot, P(v_i))$ until reaching a vertex in $V'$; hence Algorithm 1 runs in $O(nm + n^2\log n)$ time.

We strengthen the analysis by Charikar et al. [6] to prove an approximation ratio of $\lceil\log_2|T|\rceil + 1$, thus matching the result for the online Steiner tree problem [17]. Instead of an upper bound on the $x^{\text{th}}$ most expensive connection cost, we establish a bound on the $\frac{|T|}{2}$ *least* expensive connection costs; a similar technique was used in [20] for a bicriteria diameter-constrained Steiner tree problem. For simplicity, we assume w.l.o.g. $|T|$ is a power of 2; this can be done by adding up to one dummy terminal of priority 1 to each terminal, connected with a zero-weight edge.

**Lemma 2.** *The sum of the $\frac{|T|}{2}$ least expensive connection costs incurred by Algorithm 1 is at most* OPT.

**Theorem 3.** *Algorithm 1 is a* $(\lceil \log_2 |T| \rceil + 1)$*-approximation for* PRIORITY STEINER TREE.

Lemma 2 is proved by considering pairs of consecutive terminals in a depth-first traversal of the optimum PST $\mathcal{T}^*$, and Theorem 3 is proved by applying Lemma 2 $\lceil \log_2 |T| \rceil + 1$ times.

In the following, we give a simpler, parallelizable algorithm for PRIORITY STEINER TREE which achieves the same approximation ratio of $\lceil \log_2 |T| \rceil + 1$. For simplicity we assume $P(s) = \infty$ and every (non-source) terminal has a different priority; ties between terminals of the same priority can be broken arbitrarily. The idea is to connect each terminal $v$ to the "closest" terminal or source with a greater priority than $v$. Specifically, for $v \in T$, find a vertex $u \in T \cup \{s\}$ with $P(u) > P(v)$ which minimizes $\sigma(u, v)$, and connect $v$ to $u$ with edges of rate $P(v)$. This can be done by executing Dijkstra's algorithm from $v$ using edge weights $w(\cdot, P(v))$ and stopping once we find a vertex with a greater priority than $v$. Moreover, this algorithm is parallelizable as the corresponding path for each terminal can be found in parallel. The weight of connecting $v$ to its parent $u$ is the *connection cost* of $v$. As before, cycles can be removed in the end by removing an edge from each cycle with the lowest rate.

**Algorithm 2** $R(\cdot) = \text{PST}(\text{graph } G,\ \text{priorities } P,\ \text{edge weights } w,\ \text{source } s)$

1: Initialize $R(e) = 0$ for $e \in E$
2: **for** $v \in T$ **do**
3: Find $u \in T \cup \{s\}$ with $P(u) > P(v)$ such that $\sigma(u, v)$ is minimized
4: $R(e) = \max\{R(e), P(v)\}$ for $e \in p_{vu}$
5: Remove lowest-rate edge from each cycle
6: **return** edge rates $R(\cdot)$

Algorithm 2 produces a valid PST which spans all terminals, since there is a path from each terminal $v$ to the source using edges of rate $P(v)$ or higher. Moreover, Lemma 1 and Theorem 3 extend easily:

**Lemma 3.** *The sum of the $\frac{|T|}{2}$ least expensive connection costs incurred by Algorithm 2 is at most* OPT.

**Theorem 4.** *Algorithm 2 is a* $(\lceil \log_2 |T| \rceil + 1)$*-approximation for* PRIORITY STEINER TREE.

One main difference compared to Algorithm 1 [6] is that Algorithm 2 is not required to connect the terminals sequentially, or even by order of priority. Further, unlike Algorithm 1, Algorithm 2 is not dependent on the solution computed at the previous iteration. If $k \ll |T|$, a simple $k\rho$-approximation given by Charikar et al. [6] is to compute a $\rho$-approximate Steiner tree over the terminals of each priority separately, taking $O(k \cdot \text{STEINER}(n))$ time. Executing both approximations and taking the better of the two solutions yields a $\min\{\lceil \log_2 |T| \rceil + 1, k\rho\}$-approximation as desired.

## 3 An $O(\log |T|)$-Approximation for Priority NWST

We remark that the analysis of Algorithms 1–2 does not extend to Priority NWST; one can construct an example input graph in which Algorithm 1 or 2 (considering minimum weight node-weighted paths) returns a poor NWST with weight $\Omega(|T|)$OPT. In this section, we extend the $(2 \ln |T|)$-approximation by Klein and Ravi [19] which maintains a collection of trees, and greedily merges a subset of these trees at each iteration to minimize a cost-to-connectivity ratio (Algorithm 3). For Priority NWST, we need to ensure that the priority constraint is always maintained throughout the construction process. To this end, we first define a *rate tree*:

**Definition 3 (Rate tree).** *Let $G = (V, E)$, and let $T_r$ be a subtree of $G$ (not necessarily a Steiner or spanning tree of $G$) which includes vertex $r$. Let $R : V \to \mathbb{R}_{\geq 0}$ be a function which assigns rates to the vertices in $G$. We say that $T_r$ is a* rate tree *rooted at $r$ if, for all $v \in V(T_r) \setminus \{r\}$, the path from $r$ to $v$ in $T_r$ consists of vertices of non-increasing rate.*

The main idea of Algorithm 3 is to maintain a set (not necessarily a forest) of rate trees. By simply connecting the *roots* of the rate trees with paths of appropriate vertex rates, we can satisfy the priority constraints.

Another challenge to tackle involves properly devising a definition of weight when greedily merging rate trees at each iteration. The greedy NWST algorithm by Klein and Ravi [19] simply sums the weights from a root vertex to each terminal. In our algorithm, we cannot simply connect the root of a rate tree to other roots of other rate trees of lower or equal priority and compute the weight similarly. This is due to a technical challenge needed for the analysis of the algorithm (see Sect. 3.2) that it is not possible, in general, to perform a spider decomposition (similar to [19]) on a rate tree such that paths from the center to leaves have non-increasing rates. To overcome this challenge, we introduce the notion of *rate spiders* and prove the existence of a *rate spider decomposition*, which further guides us to properly define weight computations at each iterative step.

### 3.1 Algorithm Description

In the following, let $p_1 < p_2 < \ldots < p_k$ denote the $k$ vertex priorities. Initialize a set $\mathcal{F}$ (not necessarily a forest) of $|T| + 1$ rate trees so that each terminal $v \in T$, including the source $s$, is a singleton rate tree whose root is itself. Initialize vertex rates $R(v) = P(v)$ for $v \in T$, $R(s) = p_k$, and $R(v) = 0$ for $v \notin T \cup \{s\}$. While $|\mathcal{F}| > 1$, the construction proceeds iteratively as follows. Each iteration consists of greedily selecting the following:

- a rate tree $\mathcal{T}_r \in \mathcal{F}$ rooted at $r$, called the *root tree*
- a special vertex $v \in V$ called the *center* (note $v$ could equal $r$)
- a real number $b \leq P(r)$ representing the rate which $v$ is "upgraded" to
- a nonempty subset $\mathcal{S} = \{\mathcal{T}_{r_1}, \ldots, \mathcal{T}_{r_{|\mathcal{S}|}}\} \subset \mathcal{F}$ of rate trees where $\mathcal{T}_r \notin \mathcal{S}$, and $P(r_j) \leq b$ for all roots $r_j$ associated with the rate trees in $\mathcal{S}$

By connecting $r$ to the center $v$ using vertices of rate $b$, upgrading $R(v)$ to $b$, then connecting $v$ to the root of each rate tree $\mathcal{T}_{r_j} \in \mathcal{S}$ using vertices of rate $P(r_j)$, we can replace the $|\mathcal{S}|+1$ rate trees in $\mathcal{F}$ with a new rate tree $\mathcal{T}_r^{\text{new}}$ rooted at $r$ (see Fig. 1).

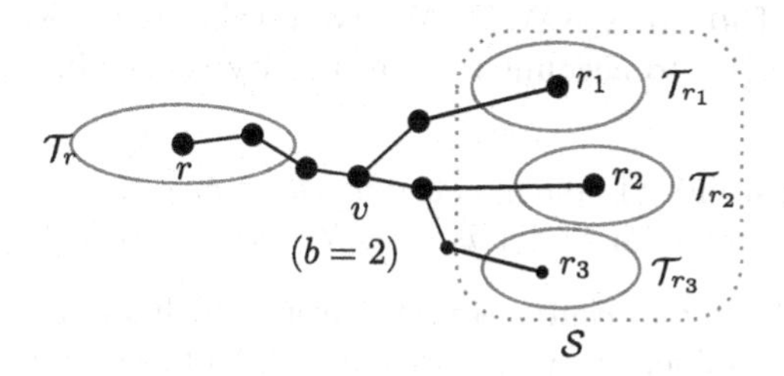


**Fig. 1.** Illustration of an iteration step in Algorithm 3 with $P(r) = 2$, $b = 2$, $P(r_1) = P(r_2) = 2$, and $P(r_3) = 1$. Vertices with larger circles (not necessarily terminals) have rate 2; vertices with smaller circles have rate 1.

The root tree, center, $b$, and $\mathcal{S}$ are greedily chosen to minimize a cost-to-connectivity ratio $\gamma$, defined as follows:

$$\gamma := \frac{1}{|\mathcal{S}|+1}\left(\sigma_b(r,v) + w(v,b) + \sum_{j=1}^{|\mathcal{S}|} \sigma_{P(r_j)}(v,r_j)\right) \tag{1}$$

where $r_j$ denotes the root of the $j^{\text{th}}$ rate tree $\mathcal{T}_{r_j}$ in $\mathcal{S}$. The second expression $\sigma_b(r,v) + w(v,b) + \sum_{j=1}^{|\mathcal{S}|} \sigma_{P(r_j)}(v,r_j)$ gives an upper bound on the weight of connecting $r$ to $v$, upgrading $R(v)$ to $b$, then connecting $v$ to $|\mathcal{S}|$ roots, and the denominator $|\mathcal{S}|+1$ represents the "connectivity", or the number of connected rate trees. Lemma 6 shows how to execute this iteration step in polynomial time.

Once $\mathcal{T}_r$, $v$, $b$, and $\mathcal{S}$ are chosen, we "upgrade" the vertex rates $R(\cdot)$ along a shortest $r$–$v$ path to $b$, then upgrade the vertex rates along each shortest $v$–$r_j$ path to $P(r_j)$. In the case that some vertex $u$ is on multiple $v$–$r_j$ paths, then $R(u)$ is upgraded to the maximum over all root priorities $P(r_j)$ for which $u$ appears on the corresponding path. Pseudocode is shown in Algorithm 3.

**Algorithm 3** $R(\cdot)$ = PNWST($G$, terminals $T$, priorities $P$, vertex weights $w$)

1: Initialize $\mathcal{F}$, $R(v) = P(v)$ if $v \in T \cup \{s\}$ and $R(v) = 0$ if $v \notin T \cup \{s\}$
2: **while** $|\mathcal{F}| > 1$ **do**
3: Find $\mathcal{T}_r$, $v$, $b$, $\mathcal{S}$ which minimize $\gamma$ (Lemma 6)
4: $R(u) = \max\{R(u), b\}$ for $u$ on $r$–$v$ path
5: $R(v) = \max\{R(v), b\}$
6: **for** $j = 1, \ldots, |\mathcal{S}|$ **do**
7: $R(u) = \max\{R(u), P(r_j)\}$ for $u$ on $v$–$r_j$ path
8: $\mathcal{F} = \mathcal{F} \setminus (\{\mathcal{T}_r\} \cup \mathcal{S})$
9: $\mathcal{F} = \mathcal{F} \cup \{\mathcal{T}_r^{\text{new}}\}$
10: **return** vertex rates $R(\cdot)$

### 3.2 Analysis of Algorithm 3

We show Theorem 2 by asserting that Algorithm 3 is a $2\ln(|T| + 1)$–approximation for PRIORITY NWST. We extend the spider decomposition given by Klein and Ravi [19] to account for the priority constraints in the PRIORITY NWST problem.

**Definition 4 (Spider).** *A* spider *is a tree where at most one vertex has degree greater than 2. A* nontrivial spider *is a spider with at least 2 leaves.*

A spider is identified by its *center*, a vertex from which all paths from the center to the leaves of the spider are vertex-disjoint. A foot of a spider is a leaf; if the spider has at least three leaves, then its center is unique and is also a foot. Klein and Ravi [19] show that given a graph $G$ and subset $M \subseteq V$ of vertices, $G$ can be decomposed into vertex-disjoint nontrivial spiders such that the union of the feet of the nontrivial spiders contains $M$. We extend the notions of spider and spider decomposition to the PRIORITY NWST problem.

**Definition 5 (Rate spider).** *A* rate spider *is a rate tree $\mathcal{X}$ which is also a nontrivial spider. It is identified by a* root *$r$ as well as a* center *$v$ such that:*

- *The root $r$ is either the center or a leaf of $\mathcal{X}$, and the path from $r$ to every vertex in $\mathcal{X}$ uses vertices of non-increasing rate $R(\cdot)$*
- *The paths from the center $v$ to each* non-root *leaf of $\mathcal{X}$ are vertex-disjoint and use vertices of non-increasing rate $R(\cdot)$.*

In Fig. 2, right, rate spiders $\mathcal{X}_2$ and $\mathcal{X}_3$ have centers distinct from their roots $r_2$, $r_3$ while $\mathcal{X}_1$ has center $v = r_1$. In Definition 6, we supply a notion of a "minimal" weight tree with respect to a subset $M$ of vertices.

**Definition 6 ($M$-optimized rate tree).** *Let $\mathcal{T}_r$ be a rate tree rooted at $r$ with vertex rates $R$. Let $M \subseteq V(\mathcal{T}_r)$ with $r \in M$. Then $\mathcal{T}_r$ is $M$–*optimized *if every leaf of $\mathcal{T}_r$ is in $M$, and if for every vertex $v \in V(\mathcal{T}_r) \setminus M$, we have $R(v) = \max R(w)$ over all vertices $w \in M$ in the subtree of $\mathcal{T}_r$ rooted at $v$.*

We show any $M$–optimized rate tree has a rate spider decomposition.

**Lemma 4 (Rate spider decomposition).** *Let $M \subseteq V(\mathcal{T}_r)$ with $|M| \geq 2$, and let $\mathcal{T}_r$ be an $M$–optimized rate tree where $r \in M$. Then $\mathcal{T}_r$ can be decomposed into vertex-disjoint rate spiders $\mathcal{X}_1, \ldots, \mathcal{X}_d$ rooted at $r_1, \ldots, r_d$ such that:*

- *the leaves and roots of the rate spiders are contained in $M$*
- *every vertex in $M$ is a either a leaf, root, or center of some rate spider*

Figure 2, right, shows an example of an $M$–optimized rate tree $\mathcal{T}_r$ for $|M| = 10$ and a rate spider decomposition $\mathcal{X}_1, \mathcal{X}_2, \mathcal{X}_3$ over $M$.

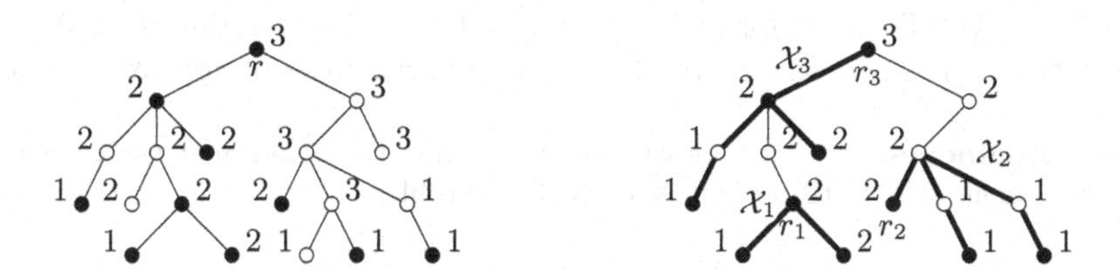

**Fig. 2.** *Left:* A rate tree rooted at $r$ with rates $R(\cdot)$ indicated and vertices in $M$ shown in black. *Right:* An $M$–optimized rate tree $\mathcal{T}_r$ and a rate spider decomposition $\mathcal{X}_1, \mathcal{X}_2, \mathcal{X}_3$ with roots $r_1, r_2, r_3$.

For $i \geq 1$, let $\mathcal{F}_i$ denote the set of rate trees at the beginning of iteration $i$ of Algorithm 3, and let $h_i \geq 2$ denote the number of rate trees in $\mathcal{F}_i$ which are connected on iteration $i$ (i.e., $h_i = |\mathcal{S}| + 1$). Let $\Delta C_i$ denote the actual weight incurred on iteration $i$ by upgrading vertex rates in line 7. Let $\gamma_i$ denote the minimum cost-to-connectivity ratio (Eq. (1)) computed by Algorithm 3 on iteration $i$. Lemma 4 (rate spider decomposition) yields the following lemma:

**Lemma 5.** *For each iteration $i$ of Algorithm 3, we have* $\dfrac{\Delta C_i}{h_i} \leq \dfrac{\text{OPT}}{|\mathcal{F}_i|}$.

Using Lemma 5, we can prove Theorem 2, by asserting that Algorithm 3 is a $2\ln(|T| + 1)$–approximation for Priority NWST. The remainder of the proof can be completed by following the analysis by Klein and Ravi [19].

It is worth noting that the extension of the $(2\ln|T|)$-approximation by Klein and Ravi [19] to the Priority NWST problem is not immediately obvious, as we must be careful when merging multiple rate trees while simultaneously satisfying the priority and rate requirements.

**Lemma 6.** *On iteration $i$ of Algorithm 3, a choice of $\mathcal{T}_r$, $v$, $b$, and $\mathcal{S}$ which minimizes $\gamma$ can be found in $O(n^3 k \log n)$ time.*

Algorithm 3 runs for $I \leq |T|$ iterations, as the size of $|\mathcal{F}|$ decreases by at least 1 at each iteration. By Lemma 6, the running time of Algorithm 3 is $O(n^4 k \log n)$. The approximation ratio is tight as is the case for the Ravi-Klein algorithm [19].

## 4 Conclusions and Future Work

By strengthening the analysis of [6], we showed that Priority Steiner tree is approximable with ratio $\min\{\lceil \log_2 |T| \rceil + 1, k\rho\} \leq \min\{1.443 \ln |T| + 2, k\rho\}$, then provided a simple, parallelizable algorithm with the same approximation ratio. Second, we showed that a natural node-weighted generalization of Priority Steiner tree admits a $O(\log |T|)$-approximation using a generalization of the Ravi-Klein algorithm [19] and spider decomposition. It remains open whether the approximability gap between $c \log \log n$ [9] and $O(\log n)$ for Priority Steiner tree can be tightened, or whether a more efficient approximation algorithm for Priority NWST can be formed. As both problems can be reduced to directed Steiner tree, this suggests a hierarchy in terms of hardness of approximation.

**Acknowledgments.** The authors wish to thank Alon Efrat and Spencer Krieger for their discussions related to the priority NWST problem.

## References

1. Ahmed, R., et al.: Multi-level Steiner trees. Proceedings of the 17th International Symposium on Experimental Algorithms (2018)
2. Ahmed, R., Sahneh, F.D., Kobourov, S., Spence, R.: Kruskal-based approximation algorithm for the multi-level Steiner tree problem. Proceedings of the 28th Annual European Symposium on Algorithms (2020)
3. Angelopoulos, S.: Online priority Steiner tree problems. In: Dehne, F., Gavrilova, M., Sack, J.-R., Tóth , C.D. (eds.) WADS 2009. LNCS, vol. 5664, pp. 37–48. Springer, Heidelberg (2009). https://doi.org/10.1007/978-3-642-03367-4_4
4. Balakrishnan, A., Magnanti, T.L., Mirchandani, P.: Modeling and heuristic worst-case performance analysis of the two-level network design problem. Manag. Sci. **40**(7), 846–867 (1994)
5. Byrka, J., Grandoni, F., Rothvoß, T., Sanità, L.: Steiner tree approximation via iterative randomized rounding. J. ACM **60**(1), 6:1–6:33 (2013)
6. Charikar, M., Naor, J.S., Schieber, B.: Resource optimization in QoS multicast routing of real-time multimedia. IEEE/ACM Trans. Netw. **12**(2), 340–348 (2004)
7. Chekuri, C., Hajiaghayi, M.T., Kortsarz, G., Salavatipour, M.R.: Approximation algorithms for nonuniform buy-at-bulk network design. SIAM J. Comput. **39**(5), 1772–1798 (2010)
8. Chlebík, M., Chlebíková, J.: The Steiner tree problem on graphs: inapproximability results. Theoret. Comput. Sci. **406**(3), 207–214 (2008)
9. Chuzhoy, J., Gupta, A., Naor, J.S., Sinha, A.: On the approximability of some network design problems. ACM Trans. Algorithms **4**(2), 23:1–23:17 (2008)
10. Current, J.R., ReVelle, C.S., Cohon, J.L.: The hierarchical network design problem. Eur. J. Oper. Res. **27**(1), 57–66 (1986)
11. Demaine, E.D., Hajiaghayi, M.T., Klein, P.N.: Node-weighted Steiner tree and group Steiner tree in planar graphs. In: Albers, S., Marchetti-Spaccamela, A., Matias, Y., Nikoletseas, S., Thomas, W. (eds.) ICALP 2009. LNCS, vol. 5555, pp. 328–340. Springer, Heidelberg (2009). https://doi.org/10.1007/978-3-642-02927-1_28

12. Dinur, I., Steurer, D.: Analytical approach to parallel repetition. In: Proceedings of the Annual ACM Symposium on Theory of Computing (2013). https://doi.org/10.1145/2591796.2591884
13. Feige, U.: A threshold of $\ln n$ for approximating set cover. J. ACM **45**(4), 634–652 (1998)
14. Grandoni, F., Laekhanukit, B., Li, S.: $O(\log^2 k/\log\log k)$-approximation algorithm for directed Steiner tree: a tight quasi-polynomial-time algorithm. In: Proceedings of the 51st Annual ACM SIGACT Symposium on Theory of Computing, STOC 2019, pp. 253–264. Association for Computing Machinery, New York (2019). https://doi.org/10.1145/3313276.3316349
15. Guha, S., Khuller, S.: Improved methods for approximating node weighted Steiner trees and connected dominating sets. J. Inform. Comput. **150**(1), 57–74 (1999)
16. Hauptmann, M., Karpinski, M.: A compendium on Steiner tree problems (2015). http://theory.cs.uni-bonn.de/info5/steinerkompendium/
17. Imase, M., Waxman, B.M.: Dynamic Steiner tree problem. SIAM J. Discret. Math. **4**(3), 369–384 (1991)
18. Karpinski, M., Măndoiu, I.I., Olshevsky, A., Zelikovsky, A.: Improved approximation algorithms for the quality of service multicast tree problem. Algorithmica **42**(2), 109–120 (2005). https://doi.org/10.1007/s00453-004-1133-y
19. Klein, P., Ravi, R.: A nearly best-possible approximation algorithm for node-weighted Steiner trees. J. Algorithms **19**(1), 104–115 (1995)
20. Marathe, M.V., Ravi, R., Sundaram, R., Ravi, S., Rosenkrantz, D.J., Hunt, H.B.: Bicriteria network design problems. J. Algorithms **28**(1), 142–171 (1998) https://doi.org/10.1006/jagm.1998.0930. https://www.sciencedirect.com/science/article/pii/S0196677498909300
21. Maxemchuk, N.F.: Video distribution on multicast networks. IEEE J. Sel. Areas Commun. **15**(3), 357–372 (1997)
22. Mirchandani, P.: The multi-tier tree problem. INFORMS J. Comput. **8**(3), 202–218 (1996)
23. Naor, J.S., Panigrahi, D., Singh, M.: Online node-weighted Steiner tree and related problems. In: 52nd Annual IEEE Symposium on Foundations of Computer Science FOCS 2011, pp. 210–219. IEEE (2011)
24. Risso, C., Robledo, F., Nesmachnow, S.: Mixed integer programming formulations for Steiner tree and quality of service multicast tree problems. Program. Comput. Softw. **46**(8), 661–678 (2020). https://doi.org/10.1134/S0361768820080174
25. Robins, G., Zelikovsky, A.: Improved Steiner tree approximation in graphs. In: Proceedings of the Eleventh Annual ACM-SIAM Symposium on Discrete Algorithms, SODA 2000, pp. 770–779. Society for Industrial and Applied Mathematics, USA (2000)
26. Sahneh, F.D., Kobourov, S., Spence, R.: Approximation algorithms for priority Steiner tree problems. arXiv preprint arXiv:2108.13544 (2021)
27. Turletti, T., Bolot, J.C.: Issues with multicast video distribution in heterogeneous packet networks. In: Proceedings of the Sixth International Workshop on Packet Video (1994)
28. Xue, G., Lin, G.H., Du, D.Z.: Grade of service Steiner minimum trees in the Euclidean plane. Algorithmica (New York) **31**(4), 479–500 (2001)

# Sublinear-Space Approximation Algorithms for Max $r$-SAT

Arindam Biswas(✉) and Venkatesh Raman

The Institute of Mathematical Sciences, HBNI, Chennai, India
{barindam,vraman}@imsc.res.in

**Abstract.** In the MAX $r$-SAT problem, the input is a CNF formula with $n$ variables where each clause is a disjunction of at most $r$ literals. The objective is to compute an assignment which satisfies as many of the clauses as possible. While there are many polynomial-time approximation algorithms for this problem, we take the viewpoint of space complexity following [Biswas et al., Algorithmica 2021] and design sublinear-space approximation algorithms for the problem.

We show that the classical algorithm of [Lieberherr and Specker, JACM 1981] can be implemented to run in $n^{O(1)}$ time while using $O(\log n)$ bits of space. The more advanced algorithms use linear or semi-definite programming, and seem harder to carry out in sublinear space. We show that a more recent algorithm with approximation ratio $\sqrt{2}/2$ [Chou et al., FOCS 2020], designed for the streaming model, can be implemented to run in time $n^{O(r)}$ using $O(r \log n)$ bits of space. While known streaming algorithms for the problem approximate optimum *values* and use randomization, our algorithms are deterministic and can output the approximately optimal assignments in sublinear space.

For instances of MAX $r$-SAT with planar incidence graphs, we devise a factor-$(1 - \epsilon)$ approximation scheme which computes assignments in time $n^{O(r/\epsilon)}$ and uses $\max\{\sqrt{n} \log n, (r/\epsilon) \log^2 n\}$ bits of space.

**Keywords:** Max SAT · Approximation · Sublinear space · Space-efficient · Memory-efficient · Planar incidence graph

## 1 Introduction

Starting in the 70's, there has been a long line of work on the approximation properties of NP-hard problems. The classical approach has been to obtain better-than-trivial approximations for such problems with polynomial-time algorithms. Later on, a number of such problems were also studied in the streaming model of computation, where an algorithm must read the input in a fixed (possibly adversarial) sequence. The goal is typically to compute an approximation by making a constant number of passes over the input using space sublinear in the input size. Recently, there has been some interest in studying approximation problems in the sublinear-space RAM model, a model halfway between the RAM and


C.-Y. Chen et al. (Eds.): COCOON 2021, LNCS 13025, pp. 124–136, 2021.
https://doi.org/10.1007/978-3-030-89543-3_11

streaming models of computation. In this paper, we continue the work initiated in [5] and devise sublinear-space approximation algorithms for MAX $r$-SAT.

An instance of MAX $r$-SAT is a CNF formula $F = C_1 \wedge \cdots \wedge C_m$, where each of the clauses $C_1, \ldots, C_m$ is a disjunction of at most $r$ literals over a variable set $\{x_1, ..., x_n\}$. The objective is to compute an assignment which satisfies as many of the clauses as possible. Viewing the variables and clauses as elements of an incidence structure yields an incidence graph where clauses and variables are vertices, and there is an edge between a variable $x$ and a clause $C$ whenever $x$ appears in $C$. We call the restriction of MAX $r$-SAT to instances with planar incidence graphs PLANAR MAX $r$-SAT.

The classical approximation algorithm [17] for MAX $r$-SAT achieves an approximation ratio of $1/2$ (shown to be $2/3$ in [7]). Later on, the ratio was improved to $(\sqrt{5} - 1)/2$ in [19]. Our first observation is that these ratios can be achieved using logarithmic space. Algorithms computing $(3/4)$-approximations are known [15], but they use linear or semi-definite programming. Under logarithmic-space reductions, it is P-complete to approximate LINEAR PROGRAMMING to any constant factor [27]. This makes it unlikely that such approaches will yield simultaneously polynomial-time and sublinear-space algorithms. In Sect. 3, we show that the previously mentioned factor-$((\sqrt{5}-1)/2)$ algorithm, and a more recent factor-$(\sqrt{2}/2)$ approximation algorithm [8], devised for the streaming model, can be implemented so as to run in polynomial time using logarithmic space.

For PLANAR MAX $r$-SAT, it is possible to compute factor-$(1 - \epsilon)$ approximations in polynomial time for any constant $\epsilon > 0$ [18]. In Sect. 4, we give a sublinear-space implementation of this scheme using recent results about computing tree decompositions [11] and BFS traversal sequences [1].

**The Model.** We use the standard RAM model and constrain the amount of space available to be sublinear in the input size. The input to an algorithm is provided using some canonical representation, which it can read but not modify, i.e. it has read-only access to the input. It also has read-write access to a certain amount of auxiliary space. Output is written to a stream: once something is output, the algorithm cannot read it back at a later point as it executes. We count the amount of auxiliary space in single-bit units, and the objective is to use as little auxiliary space as possible.

**Related Work.** In the RAM model, earlier works with an emphasis on space efficiency study problems such as reachability [4,24,26], sorting and selection [14,21,22] and graph recognition [2,12,23]. In recent years, new results on the computability of separators for planar graphs in sublinear space have been used to devise sublinear-space algorithms for BFS [1] and DFS [16] with better running times than algorithms for general graphs.

**Results.** Since our model is less restrictive than the streaming model, we are able to compute approximately optimal assignments for MAX $r$-SAT instead of approximating optimum values. On the other hand, our model is more restrictive than the RAM model of classical approximation algorithms where the amount of space used by an algorithm can potentially be polynomially large in the input size.

- For general MAX $r$-SAT (Sect. 3), we convert a classical algorithm of Lieberherr and Specker [19] to our model, obtaining a $((\sqrt{5}-1)/2)$-approximation algorithm which uses $O(\log n)$ bits of space. We also convert a more recent algorithm of Chou et al. [8] to obtain a $(\sqrt{2}/2)$-approximation algorithm which uses $O(r \log n)$ bits of space.
- For PLANAR MAX $r$-SAT (Sect. 4), we show how a $(1-\epsilon)$-approximation scheme of Khanna and Motwani [18] can be implemented to use $\max\{\sqrt{n}, (r/\epsilon)\log n\}$ bits of space.

## 2 Preliminaries

We use the following standard notation and concepts. The set $\{0, 1, \dots\}$ of natural numbers is denoted by $\mathbb{N}$ and the set $\{1, 2, \dots\}$ of positive integers is denoted by $\mathbb{Z}^+$. For $n \in \mathbb{Z}^+$, $[n]$ denotes the set $\{1, 2, \dots, n\}$. An $r$-CNF formula is a conjunction (OR) of disjunctions (AND) of at most $r$ literals (variables or their negations). The individual disjunctions are called clauses of the formula. A clause that consists of a single literal is called a unit clause. For $k \in [r]$, a $k$-clause is a clause which contains exactly $k$ literals.

### 2.1 Time and Space Overheads

In proofs, we measure resource costs in terms of overheads for individual steps. Since the space available to an algorithm is limited, objects created by processing the input are not stored, but recomputed on the fly. For example, consider a procedure (call it A) that reads an input formula $F$ and produces a subformula $F'$ consisting of the unit clauses of $F$. The procedure outputs $F'$ as a stream $S_{F'}$. Later on, when another procedure (call it B) reads a portion of $S_{F'}$, A recomputes the entire stream $S_{F'}$. Suppose the resource costs of A are $t_{\mathtt{A}}$ time and $s_{\mathtt{A}}$ space, and (assuming random, constant-time access to $F'$) suppose the resource costs of B are $t_{\mathtt{B}}$ time and $s_{\mathtt{B}}$ space.

In this scenario, we call $t_{\mathtt{B}}$ and $s_{\mathtt{B}}$ the resource overhead of B. Combining this overhead with resource costs of A, we obtain the actual resource costs of B: $t_{\mathtt{B}} \cdot t_{\mathtt{A}}$ time and $s_{\mathtt{B}} + s_{\mathtt{A}}$ space.

### 2.2 Universal Hash Families

Algorithms appearing later on use the trick of randomized sampling to show that certain good assignments exist, and then compute such assignments using *universal hashing*, a well-known derandomization technique. The following proposition arises from constructions of *universal hash families* described in [13] and the observation that the constructions can be carried out in logarithmic space.

**Proposition 1 (Fredman et al. [13]).** *Let $n, k, a, b \in \mathbb{Z}^+$ with $n \geq b \geq a$ and $n \geq k$. One can enumerate a family* $\mathrm{Univ}(n, k, a, b)$ *of functions from $[n]$ to $\{0, 1\}$ such that if $X_i = f(i)$ ($i \in [n]$) are random variables that arise when $f$ is sampled uniformly at random from* $\mathrm{Univ}(n, k, a, b)$, *then $X_1, \ldots, X_n$ are $k$-wise independent and for $i \in [n]$, $\mathrm{P}(X_i = 1) = a/b$. The procedure runs in time $n^{\mathrm{O}(k)}$ and uses $\mathrm{O}(k \log n)$ bits of space.*

## 3 Max $r$-SAT

In this section, we devise sublinear-space $(\sqrt{5}-1)/2$)- and $(\sqrt{2}/2)$-approximation algorithms for Max $r$-SAT.

### 3.1 Factor-$((\sqrt{5} - 1)/2)$ Approximation Algorithm

In what follows, we give a logarithmic-space implementation of the following result.

**Proposition 2 (Lieberherr and Specker [19], Theorem 1).** *Let $F$ be an $r$-CNF formula with $m$ clauses. There is an assignment for $F$ which satisfies at least $(\sqrt{5} - 1)m/2$ clauses.*

**Definition 1 (2-Satisfiability).** *An $r$-CNF formula $F$ is called 2-satisfiable if any two of its clauses can be simultaneously satisfied, i.e. $F$ does not contain a pair $(l, \neg l)$ of literals as clauses.*

The following proposition is based on arguments in [19] (see also [30]).

**Proposition 3.** *Let $F$ be a 2-satisfiable $r$-CNF formula with $m$ clauses in which all unit clauses are positive literals. For the pairwise-independent random assignment where each variable of $F$ is set to 1 with probability $p = 0.618 \approx (\sqrt{5}-1)/2$, the expected number of satisfied clauses is $0.618m$.*

We now show how the above proposition can be used to compute 0.618-approximate optimal Max $r$-SAT assignments for general $r$-CNF formulas in logarithmic space.

**Theorem 1.** *For any instance of* Max $r$-SAT *with $n$ variables, one can compute a 0.618-approximate optimal assignment in time $n^{\mathrm{O}(1)}$ using $\mathrm{O}(\log n)$ bits of space.*

*Proof.* Let $F$ be an $r$-CNF formula with variables $x_1, \ldots, x_n$. In what follows, we describe an algorithm which proves the claim.

**Computing an Equivalent 2-Satisfiable Formula $F'$.** For each clause $C$ in $F$ with at least two literals, check if any variables $x$ appearing in $C$ also appear as a negated clauses $\neg x$ in $F$. If they do, flip the $x$-literals (replace $x$ with $\neg x$ or $\neg x$ with $x$) in $C$ and output the resulting clause. Otherwise, output $C$. The clauses not output yet are unit clauses, i.e. they have exactly 1 literal. For each variable $x_i$, check if $x_i$ appears as a unit clause in $F$. If it does, output $x_i$. Then output the special flag #NEG, to indicate that clauses to follow appear negated in $F$. For each variable $x_i$, check if it appears as a unit clause $\neg x_i$ in $F$. If it does, check if the unit clause $x_i$ also appears in $F$. If both $\neg x_i$ and $x_i$ are clauses in $F$, output nothing. Otherwise, output $x_i$. Observe that the only clauses of $F$ not output are unit clauses that appear in pairs $(l, \neg l)$.

Let $F'$ be the conjunction of the clauses output and $S_{F'}$ be the stream output. With random access to $F$, $S_F$ is produced in time $n^{O(1)}$ using $O(\log n)$ bits of space. Clearly, $F'$ is 2-satisfiable. Let $\phi$ be an assignment for $F'$. Define $\phi'(x_i) = 1 - \phi(x_i)$ for every $x_i$ appearing after the #NEG flag in $S_{F'}$ and define $\phi'(x_i) = \phi(x_i)$ otherwise. It is easy to see that $\phi$ satisfies the same number of clauses in $F'$ as $\phi'$ does in $F$, and that given access to $S_{F'}$ and $\phi$, the overhead for computing $\phi'$ is $n^{O(1)}$ time and $O(\log n)$ space. We use this transformation later on to compute an assignment for $F$ from an assignment for $F'$.

**Computing an Assignment for $F'$.** Using the procedure of Proposition 1, compute the family $H = \mathrm{Univ}(n, 2, 618, 1000)$ and denote the stream of functions by $S_H$. Note that for $f$ sampled uniformly at random from $H$, the random variables $X_i = f(i)$ ($i \in [n]$) form a pairwise independent random assignment and for $i \in [n]$, $P(X_i = 1) = 0.618$. Thus, one of the assignments in $S_H$ achieves (for the 2-satisfiable formula $F'$) the expectation value in Proposition 3.

Let $m'$ be the number of clauses in $F'$. For each assignment $\phi$ in $S_H$, scan $S'_F$ to determine the number $c$ of clauses $\phi$ satisfies. If $c > 0.618m'$, output $\phi$ and skip to the next step. By Proposition 1, the family $H$ is computed in time $n^{O(1)}$ and $O(\log n)$ bits of space. The overhead of this step is therefore $n^{O(1)}$ time and $O(\log n)$ bits of space. Denote the output stream of this step by $S_\phi$.

**Computing an Assignment for $F$.** Now convert the assignment $\phi$ from the previous step to an assignment $\phi'$ (according to the transformation described earlier) as follows. For each $x_i$, scan $S_\phi$ to determine the value $v = \phi(x_i)$, and scan $S_{F'}$ to determine if $x_i$ appears after the #NEG (it was flipped). If it does, output the assignment $\phi'(x_i) = 1 - v$. Otherwise, output the assignment $\phi'(x_i) = v$. Since $\phi$ satisfies $c \geq 0.618m'$ clauses in $F'$, $\phi'$ satisfies the same number of clauses in $F$. In particular, it satisfies at least a 0.618-fraction of the non-unit clauses, and unit clauses that do not appear in $(l, \neg l)$ pairs.

Of the pairs $(l, \neg l)$ of unit clauses appearing in $F$, exactly half are satisfied by any assignment for the variables appearing in them. Now for each $x_i$, scan $S_\phi$, to determine if $\phi$ assigns it a value. If it does not, output the assignment $\phi'(x_i) = 1$. Clearly, $\phi'$ now also satisfies exactly half of the unit clauses in $F$ appearing in pairs $(l, \neg l)$, i.e. it is an optimal assignment for those clauses. Thus, $\phi'$ is 0.618-optimal assignment for all of $F$. The overhead of this conversion step is also $n^{O(1)}$ time and $O(\log n)$ bits of space.

Since the overheads for all steps are $n^{O(1)}$ time and $O(\log n)$ space, the overall running time is $(n^{O(1)})^3 = n^{O(1)}$ and the space used is $3 \cdot O(\log n) = O(\log n)$. □

### 3.2 Factor-($\sqrt{2}/2$) Approximation Algorithm

In the following, we adapt arguments in [8] to devise a ($\sqrt{2}/2$)-approximation algorithm which runs in time $n^{O(r)}$ and uses $O(r \log n)$ bits of space. Consider the following definitions.

**Definition 2 (Bias).** *Let $F$ be an $r$-CNF formula with variables $x_1, \ldots, x_n$. For $i \in [n]$, the bias of $x_i$ is* $\text{bias}(x_i) = \sum_{j\in[r]}(\#(j\text{-clauses containing } x_i) - \#(j\text{-clauses containing } \neg x_i))/2^j$.

*The bias of the entire formula is* $\text{bias}(F) = \sum_{i\in[n]}|\text{bias}(x_i)|$ *and the formula $F$ is called positively biased if* $\text{bias}(x_i) \geq 0$ *for each* $i \in [n]$.

The next proposition shows that depending on whether the bias of a formula is smaller than a certain value, one can satisfy a good proportion (in expectation) of the clauses in it by setting each variable to 1 with fixed (bias-dependent) probability.

**Proposition 4 (Chou et al. [8]).** *Let $F$ be a positively-biased $r$-CNF formula with $m$ clauses. For $i \in [r]$, let $m_i$ be the number of $i$-clauses in $F$. The following statements are true.*

- *The all-1's assignment satisfies at least* $\frac{\text{bias}(F)}{2} + \sum_{i\in[r]} \frac{im_i}{2^i}$ *clauses in $F$.*
- *When* $\text{bias}(F) \leq b^* = 4\sum_{i\in[r]}(1 - \frac{i+1}{2^i})m_i$, *an $r$-wise independent random assignment where variables are set to 1 with probability* $\frac{m-\text{bias}(F)}{2m-4\,\text{bias}(F)} \leq 1$ *satisfies, in expectation, at least* $\sum_{i\in[r]}(1 - \frac{1}{2^i})m_i + \frac{\text{bias}(F)^2}{4b^*}$ *clauses in $F$.*
- *The best of the two assignments above satisfies at least a ($\sqrt{2}/2$)-fraction of the maximum number of simultaneously-satisfiable clauses in $F$.*

We now show how the above proposition can be used to compute good approximations in sublinear space. For any $r$-CNF formula $F$, we first compute an equivalent positively-biased formula $F'$ and then using Proposition 1, compute an assignment for $F'$ which is a ($\sqrt{2}/2$)-approximation. We then convert this to an assignment for $F$ satisfying the same number of clauses.

**Theorem 2.** *For any instance of* Max $r$-SAT *with $n$ variables, one can compute a ($\sqrt{2}/2$)-approximate optimal assignment in time $n^{O(r)}$ using $O(r \log n)$ bits of space.*

*Proof.* Let $F$ be an $r$-CNF formula with variables $x_1, \ldots, x_n$ and for $i \in [n]$, let $m_i$ be the number of $i$-clauses in $F$. In what follows, we describe an algorithm which proves the claim.

**Computing** $\text{bias}(F)$ **and** $b^*$**.** Set $b_F, b^* \leftarrow 0$. For each $i \in [n]$, compute $b_i = \text{bias}(x_i)$ and $m_i$. It is easy to see that with random access to $F$, this can be done in logarithmic space. Set $b_F \leftarrow b_F + |b_i|$, $b^* \leftarrow b^* + (1 - (i+1)/2^i)m_i$, and if $b_i < 0$, output $x_i$ to indicate that $x_i$ has negative bias in $F$. Then discard $(b_i, m_i)$ and move to the next iteration. Finally, store $b_F$ and $b^* \leftarrow 4b^*$ for later steps using $O(\log n)$ bits of space. The entire loop takes time $n^{O(1)}$ and uses $O(\log n)$ bits of space. Let $S_B$ be the stream output.

**Computing an Equivalent Positively-Biased Formula $F'$.** For each clause $C$ in $F$, check if any variables $x$ appearing in $C$ also appear in the stream $S_B$. If they do, flip the $x$-literals (replace $x$ with $\neg x$ or $\neg x$ with $x$) in $C$ and output the resulting clause. Otherwise, output $C$. Observe that the variables $x$ flipped are precisely those for which $\text{bias}(x) < 0$ in the previous step. Thus, the clauses output form a positively-biased formula. Denote the output stream by $S_{F'}$. The overhead of this step is $n^{O(1)}$ time and $O(\log n)$ space.

**Computing an Assignment for $F'$.** If $b_F > b^*$, then output the all-1's assignment and skip to the next step. Otherwise, using the procedure of Proposition 1, compute the family $H = \text{Univ}(n, r, \lceil m - b_F \rceil, \lceil 2m - 4b_F \rceil)$ and denote the stream of functions by $S_H$. Similarly as in the proof of Theorem 1, one of the assignments in $S_H$ achieves the expectation value in Proposition 4.

For each assignment $\phi$ in $S_H$, scan $S'_F$ to determine the number $c$ of clauses $\phi$ satisfies. If $c \geq {b_F}^2/(16\sum_{i=2}^{k}(1-(i+1)/2^i)m_i)$, output $\phi$ and skip to the next step. The family of assignments is computed in time $n^{O(r)}$ and $O(r \log n)$ bits of space, so the overhead of this step is $n^{O(r)}$ time and $O(r \log n)$ bits of space. Denote the output stream of this step by $S_\phi$.

**Computing an Assignment for $F$.** Convert the assignment $\phi$ from the previous step to an assignment $\phi'$ for $F$ as follows. For each $x_i$, scan $S_\phi$ to determine the value $v = \phi(x_i)$, and scan $S_B$ to check if $x_i$ appears in it (it was flipped). If it does, output the assignment $\phi'(x_i) = 1 - v$. Otherwise, output the assignment $\phi'(x_i) = v$. Clearly, $\phi$ satisfies the same number of clauses in $F'$ as $\phi'$ does in $F$. By Proposition 4, this number is at least a $(\sqrt{2}/2)$-fraction of the maximum number of simultaneously-satisfiable clauses in $F$. With access to $S_\phi$ and $S_B$, the overhead of this step is $n^{O(1)}$ time and $O(\log n)$ space.

Thus, the algorithm outputs a $(\sqrt{2}/2)$-approximate optimal assignment as required. Observe that the maximum overhead of any of the steps is $n^{O(r)}$ time and $O(r \log n)$ space. Combining the (constantly many) overheads, the overall running time is $n^{O(r) \cdot O(1)} = n^{O(r)}$ and the space used is $O(r \log n) \cdot O(1) = O(r \log n)$. □

## 4 Planar Max $r$-SAT

In this section, we devise a sublinear-space PTAS for Planar Max $r$-SAT along the lines of [18] using the partitioning approach in [3] for planar graph problems. We use the following result to perform a BFS traversal of (the incidence graphs of) the input instances in sublinear space.

**Proposition 5. (Chakraborty and Tewari [9], Theorem 1).** *There is an algorithm which takes as input a planar graph on $n$ vertices and computes a BFS sequence for $G$ in time $n^{O(1)}$ using $O(\sqrt{n} \log n)$ bits of space.*

The next result shows how to use the BFS traversal procedure to partition—in sublinear space—the input formulas into subformulas of bounded diameter.

**Lemma 1.** *Let $F$ be an $r$-CNF formula with $n$ variables and $m$ clauses that has a planar incidence graph and let $k \in \mathbb{N}$. One can compute a sequence $F_1, \dots, F_l$ of subformulas of $F$ such that*

1. *the diameter of the incidence graph of each $F_i$ ($i \in [l]$) is at most $k$,*
2. *$F_i$ and $F_j$ have no variables in common for all $i, j \in [l]$ with $i \neq j$, and*
3. *$F_1, \dots, F_l$ together contain at least $(1 - 1/k)m$ clauses of $F$.*

*The procedure runs in time $n^{O(1)}$ and uses $O(\sqrt{n} \log n)$ bits of space.*

*Proof.* Let $x_1, \dots, x_n$ be the set of variables in $F$, $\{C_1, \dots, C_m\}$ be the set of clauses in $F$, $G_F$ be the incidence graph of $F$, and $V_F$ (resp. $C_F$) be the vertices of $G_F$ corresponding to the variables (resp. clauses) of $F$. In what follows, we describe a procedure which proves the claim.

**Adding a Dummy Vertex.** This step ensures that $G_F$ is connected. Determine the connected components of $G_F$ using the connectivity algorithm of Asano et al. [1]: for any two vertices, it runs in time $n^{O(1)}$ and uses $O(\sqrt{n} \log n)$ bits of space to check if the two vertices are connected. Then add a dummy variable vertex $x_{n+1}$ which has an edge to an arbitrary clause vertex in each connected component, making $G_F$ connected. Additionally, add the clause $\neg x_{n+1}$ (with an edge to $x_{n+1}$) to ensure that assignments for the formula $F'$ determined by the resulting graph $G_{F'}$ are in 1-1 correspondence with assignments for $F$. Now output $F'$ and $G_{F'}$, and denote this output stream by $S_{F'}$. With random access to $G_F$, it is not hard to see that this transformation runs in time $n^{O(1)}$ and uses $O(\sqrt{n} \log n)$ bits of space.

**Determining the BFS Levels of $G_{F'}$.** Consider a BFS traversal of $G_{F'}$ starting at (the variable vertex corresponding to) $x_{n+1}$. Suppose the depth of the traversal is $d_0$. Let $d = d_0$ if $d_0$ is even and $d = d_0 + 1$ otherwise. For $i \in [d]$, set $L_i = \{v \in V(G_{F'}) \mid \text{dist}(u, v) = i - 1\}$. Observe that $L_1, \dots, L_d$ are precisely the levels of the BFS tree, with $L_i \subseteq V_F$ for odd $i$ and $L_i \subseteq C_F$ for even $i$.

**Splitting $G_F$.** Consider the following subsets of $V(G_{F'})$.

- For $i \in [d/2 - 1]$, let $U_i = L_{2i} \cup L_{2i+1} \cup L_{2i+2}$. Observe that $U_i \cap U_j \neq \emptyset$ iff $|i - j| \leq 1$ and for $i \in [d/2 - 1]$, $U_i \cap U_{i+1} = L_{2i+2}$.
- For $i \in \{0, \dots, k-1\}$, let $W_i = \bigcup_{j \equiv i \pmod{k}} U_j$. Observe that $W_i \cap W_j \neq \emptyset$ iff $i - j \equiv \pm 1 \pmod{k}$ and for $i \in [d/2 - 1]$, $W_i \cap W_{i+1} = \bigcup_{j \equiv i \pmod{k}} L_{2j+2}$.
- For any $A \subseteq V_F \cup C_F$, let $C(A)$ be the clause vertices that appear in $A$, i.e. $C(A) = A \cap C_F$.

Clearly, for $i \in [d/2 - 1]$, $C(W_i) = \bigcup_{j \equiv i \pmod{k}} L_{2j} \cup L_{2j+2}$ and $C_F = \bigcup_{i \in 0, \dots, k-1} C(W_i)$. By the inclusion-exclusion principle, we have

$$\begin{aligned} |C(W_0)| + \dots + |C(W_{k-1})| &= |C_F| + |C(W_0) \cap C(W_1)| + \dots + |C(W_{k-1}) \cap C(W_0)| \\ &= |C_F| + \sum_{i \in \{0, \dots k-1\}} |L_{2j+1}| \leq |C_F| + |C_F| = 2|C_F|. \end{aligned}$$

Thus, for some $i \in \{0, \ldots, k-1\}$, we have $|C(W_i)| \leq 2|C_F|/k$, i.e. $W_i$ contains at most a $(2/k)$-fraction of the clauses in $F'$ (and $F$). Consider the graph $G_{F'} - W_i$. Observe that $W_i$ comprises groups of 3 consecutive layers of the BFS traversal, and consecutive groups are $k-2$ layers apart. Thus, removing $W_i$ from $G_{F'}$ disconnects $G_{F'}$ into connected components which contain at most $k-2$ layers of the BFS traversal each, i.e. their diameters are at most $k-2$. It follows that the formula $F^+$ corresponding to $G_{F'} - W_i$ satisfies the conditions of the claim.

To compute $F^+$, perform the following steps. Using the procedure of Proposition 5, perform a BFS traversal of the $G_{F'}$ portion of $S'_F$, starting at $x_{n+1}$. Let $S_B$ be the stream produced by this procedure. The overhead of the procedure is $n^{O(1)}$ time and $O(\sqrt{n} \log n)$ bits of space. For each $i \in [k]$, scan $S_B$ to determine the number $|C(W_i)|$ of clauses in $W_i$. For $i$ achieving the smallest $|C(W_i)|$ in the loop, scan $S_B$ and output only the levels (and edges between them) which do not appear in $W_i$. Let $S_{F^+}$ be this output stream. Now scan $S_{F^+}$, and for each sequence of consecutive (connected) levels, output the subformula of $F$ induced by those levels. Observe that $S_{F^+}$ is produced by scanning $S_{F'}$ and the final output is produced by scanning $S_{F^+}$. Each scan only involves counting elements in the stream and truncating parts of the stream to produce the output stream. Thus, the overhead of this entire step is $n^{O(1)}$ time and $O(\sqrt{n} \log n)$ bits of space.

For the various steps, the maximum overhead is $n^{O(1)}$ time and $O(\sqrt{n} \log n)$ bits of space. Thus, combining the overheads for the various steps, the resource costs of the entire algorithm are $n^{O(1)}$ time and $\sqrt{n} \log n$ bits of space. □

The next two results allow us to compute tree decompositions for incidence graphs of bounded diameter in sublinear space.

**Proposition 6 (Robertson and Seymour [25], Theorem 2.7).** *The treewidth of any planar graph with diameter $d$ is at most $3d + 1$.*

**Proposition 7 (Elberfeld et al. [11], Lemma III.1).** *Let $G$ be a graph on $n$ vertices with treewidth $k \in \mathbb{N}$. One can compute a tree decomposition of width $4k + 1$ for $G$ such that the decomposition tree is rooted, binary and has depth $O(\log n)$. The procedure runs in time $n^{O(k)}$ and uses $O(k \log n)$ bits of space.*

We now show how one can solve PLANAR MAX $r$-SAT exactly on formulas with incidence graphs of bounded diameter.

**Lemma 2.** *Let $F$ be an $r$-CNF formula with $n$ variables that has a planar incidence graph with diameter $k \in \mathbb{N}$. One can compute an assignment for $F$ satisfying the maximum number of clauses in time $n^{O(rk)}$ using $O(rk \log^2 n)$ bits of space.*

*Proof.* Let $G$ be the incidence graph of $F$. Since the diameter of $G$ is $k$, its treewidth is at most $3k + 1$ (Proposition 6). Consider a tree decomposition $(T, \mathcal{B})$ for $G$ computed by the procedure of Proposition 7. $T$ is the underlying tree (rooted at a vertex $v_r \in V(T)$) and $\mathcal{B} = \{B_v \mid v \in V(T)\}$ is the set of bags

**Procedure 1,** BdTWMaxSAT: find an optimal assignment

**Input**: $(T, v, \mathcal{B}, \psi)$
1 $max \leftarrow 0, \phi_{max} \leftarrow \emptyset$;
2 **if** *v has no children in T* **then**
3 store $V_v$, the set of variables in $B_v$;
4 let $\mathcal{A}_v$ be the set of assignments $V_v$ that extend $\psi$;
5 **foreach** $\phi \in \mathcal{A}_v$ **do**
6 determine $val$, the number of clauses appearing in $B_v$ that $\phi$ satisfies;
7 **if** $val > max$ **then** $max \leftarrow val,\ \phi_{max} \leftarrow \phi$
8 **return** $(max, \phi_{max})$
9 **else**
10 determine the left child $v_l$ and the right child $v_r$ of $v$ in $T$ if they exist;
11 store $V_v$, the set of variables in $B_v$, and those in $B_{v_l}$ and $B_{v_r}$ adjacent to clause variables in $B_v$;
12 let $\mathcal{A}_v$ be the set of assignments for $V_v$ that extend $\psi$;
13 **foreach** $\phi \in \mathcal{A}_v$ **do**
14 $(val_l, \phi_l) \leftarrow$ BdTWMaxSAT$(T, v_l, \mathcal{B}, \phi)$;
15 $(val_r, \phi_r) \leftarrow$ BdTWMaxSAT$(T, v_r, \mathcal{B}, \phi)$;
16 **if** $val_l + val_r > max$ **then** $max \leftarrow val_l + val_r,\ \phi_{max} \leftarrow \phi_l \cup \phi_r$
17 **return** $(max, \phi_{max})$

in the decomposition. By the proposition, the depth of $T$ is $O(\log n)$ and its width is at most $4 \cdot (3k+1)+1 = O(k)$, i.e. $|B_v| = O(k)$ for all $v \in V(T)$.

For each $v \in V(T)$, let $F_v$ be the subformula of $F$ consisting of all clauses appearing in bags of the subtree of $T$ rooted at $v$. Let $V_v$ be the set of variables in $B_v$, and those in the bags of $v$'s children (if they exist) that are adjacent to variables in $B_v$.

In what follows, we prove that BdTWMaxSAT $(T, v_r, \mathcal{B}, \emptyset)$ ($\emptyset$ denotes the empty assignment) computes an assignment for $F$ satisfying the maximum number of clauses. We momentarily assume constant-time access to $G$ and $(T, \mathcal{B})$.

Assume for induction that for any $v \in V(T)$, any assignment $\psi$ for $V_v$ and any child $v_c$ of $v$, that BdTWMaxSAT$(T, v_c, \mathcal{B}, \psi)$ returns an assignment for $F_{v_c}$ which extends $\psi$ and satisfies the maximum number of clauses in $F_{v_c}$ among all such assignments. Now consider a procedure call BdTWMaxSAT$(T, v, \mathcal{B}, \psi)$. The procedure first determines if $v$ has any children. If it does not, then the procedure iterates over all assignments for $V_v$ that extend $\psi$, finds one that satisfies the maximum number of clauses in $F_v$ and returns it. Thus, the procedure is correct in the base case. Since $|V_v| \le |B_v| = O(k)$, the number of such assignments is $2^{O(k)}$. The call stack stores $\psi$ and $V_v$, so the assignments can be enumerated in time $2^{O(k)} \cdot n^{O(1)}$ using $O(rk \log n)$ bits of extra space. Thus, this section of the procedure runs in time $2^{O(k)} \cdot n^{O(1)}$ and uses $O(k \log n)$ bits of space.

In the other case, i.e. $v$ has children, the procedure determines the left and right children of $v$ by scanning $(T, \mathcal{B})$ and stores $V_v$. Since each clause in $B_v$ has at most $r$ literals, we have $|V_v| \le r \cdot |B_v| = O(rk)$. The loop iterates over the set $\mathcal{A}_v$ of assignments $\phi$ for $V_v$ that extend $\psi$. The assignments can be enumerated (since $\psi$ and $V_v$ are stored on the call stack) in time $2^{O(rk)} \cdot n^{O(1)}$ using $O(rk \log n)$ bits of extra space. Next, the procedure calls itself recursively and stores the tuples returned. Because of the inductive assumption, $\phi_l$ (resp. $\phi_r$) extends $\phi$ and satisfies the maximum number of clauses in $F_{v_l}$ (resp. $F_{v_r}$) among all such assignments.

Observe that because $(T, \mathcal{B})$ is a tree decomposition, the variables outside of $V_v$ that $\phi_l$ sets are distinct from the variables outside of $V_v$ that $\phi_r$ sets. Thus, $\phi_l$ and $\phi_r$ do not conflict with each other. In the loop, the procedure finds an extension $\phi$ of $\psi$ such that its extensions $\phi_l$ and $\phi_r$, respectively, satisfy the maximum number of clauses in $F_l$ and $F_r$. Overall, $\phi$ is an extension of $\psi$ which satisfies the maximum possible number of clauses in $F_v$. This proves the inductive claim, and thus the procedure is correct.

We now prove the resource bounds of the procedure (assuming constant-time access to $G$ and $(T, \mathcal{B})$). Observe that in each recursive call, the individual steps use $O(rk \log n)$ bits of space and the loops also use $O(rk \log n)$ bits of space. Since $T$ has depth $O(\log n)$, the depth of the recursion tree is also $O(\log n)$, and therefore the call `BdTWMaxSAT`$(T, r, \mathcal{B}, \emptyset)$ uses a total of $O(rk \log^2 n)$ bits of space.

Outside of the recursive calls, the individual steps of the procedure are polynomial-time and the total running time for the other operations in the loops is $2^{O(rk)} \cdot n^{O(1)}$. Thus, if the recursive calls take time $T$, the overall running time of the procedure is $2^{O(rk)} \cdot 2T + 2^{O(rk)} \cdot n^{O(1)}$. Since the depth of the recursion tree is $O(\log n)$, this expression solves to $n^{O(rk)}$.

Now consider the overheads for computing $G$ and $(T, \mathcal{B})$. $G$ is clearly computable in polynomial time and logarithmic space and by Proposition 7, $(T, \mathcal{B})$ is computable in time $n^{O(k)}$ using $O(k \log n)$ bits of space. The real resource costs of `BdTWMaxSAT`$(T, v_r, \mathcal{B}, \emptyset)$ are therefore $n^{O(rk)} \cdot n^{O(k)} = n^{O(rk)}$ time and $O(rk \log^2 n) + O(k \log n) = O(rk \log^2 n)$ bits of space. □

The next theorem combines the previous results to devise a sublinear-space PTAS for Planar Max $r$-SAT.

**Theorem 3.** *For any* $0 < \epsilon < 1$, *one can compute* $(1 - \epsilon)$*-approximate optimal assignments for* Planar Max $r$-SAT *in time* $n^{O(r/\epsilon)}$ *using* $\max\{\sqrt{n} \log n, (r/\epsilon) \log^2 n\}$ *bits of space.*

*Proof.* Consider the following algorithm. Using the procedure of Lemma 1 with $k = \lceil 1/\epsilon \rceil$, partition $F$ into subformulas $F_1, \ldots, F_l$. Then for each $i \in [l]$, use the procedure of Lemma 2, compute an exact solution for $F_i$ and output an assignment. In the end, output assignments $x = 0$ for all variables $x$ not appearing in $F_1, \ldots, F_l$.

Observe that since the partitioning procedure outputs the subformulas as a stream $S_F = F_1, \ldots, F_l$, each access to $F_i$ costs a single pass over $S_F$, which adds only an $n^{O(1)}$-time, $O(\log n)$-space overhead. By Lemma 1, the partitioning procedure runs in time $n^{O(1)}$ and uses $O(\sqrt{n} \log n)$ bits of space. Combining the overhead for access to $F_i$ and the resource bounds from Lemma 2, solving $F_i$ exactly takes time $n^{O(1)} \cdot n^{O(rk)} = n^{O(r/\epsilon)}$ (since $k = \lceil 1/\epsilon \rceil$) and uses

$O(\log n)+O(\sqrt{n}\log n)+O(rk\log^2 n) = \max\{\sqrt{n}\log n, (r/\epsilon)\log^2 n\}$ bits of space. Finally, each $x = 0$ assignment for a variable not appearing in $F_1, \ldots, F_l$ costs a single pass over $S_F$. It follows that the total resource costs are $n^{O(r/\epsilon)}$ time and $\max\{\sqrt{n}\log n, (r/\epsilon)\log^2 n\}$ bits of space.

We now prove the approximation bound. Let $m$ be the number of clauses in $F$. Observe that Lemma 1 guarantees any two subformulas $F_i$ and $F_j$ $(i, j \in l$ with $i \neq j)$ have no variables in common, and the subformulas together contain at least $(1 - 1/k)m \geq (1 - \epsilon)m$ clauses of $F$. Thus, the assignment produced is valid and satisfies at least $(1 - \epsilon)m$ clauses, i.e. it is a $(1 - \epsilon)$-approximate optimal PLANAR MAX $r$-SAT assignment for $F$. □

## References


1. Asano, T., Kirkpatrick, D., Nakagawa, K., Watanabe, O.: $\widetilde{O}(\sqrt{n})$-space and polynomial-time algorithm for planar directed graph reachability. In: Csuhaj-Varjú, E., Dietzfelbinger, M., Ésik, Z. (eds.) MFCS 2014. LNCS, vol. 8635, pp. 45–56. Springer, Heidelberg (2014). https://doi.org/10.1007/978-3-662-44465-8_5
2. Allender, E., Mahajan, M.: The complexity of planarity testing. Inf. Comput. **189**(1), 117–134 (2004). ISSN 08905401
3. Baker, B.S.: Approximation algorithms for NP-complete problems on planar graphs. J. ACM **41**(1), 153–180 (1994). ISSN 0004-5411, 1557-735X
4. Barnes, G., Buss, J.F., Ruzzo, W.L., Schieber, B.: A sublinear space, polynomial time algorithm for directed s-t connectivity. SIAM J. Comput. **27**(5), 1273–1282 (1998). ISSN 0097-5397, 1095-7111
5. Biswas, A., Raman, V., Saurabh, S.: Approximation in (poly-) logarithmic space. Algorithmica **83**(7), 2303–2331 (2021). ISSN 0178-4617, 1432-0541
6. Bodlaender, H.L.: A partial k-arboretum of graphs with bounded treewidth. Theor. Comput. Sci. **209**(1–2), 1–45 (1998). ISSN 03043975
7. Chen, J., Friesen, D.K., Zheng, H.: Tight bound on Johnson's algorithm for maximum satisfiability. J. Comput. Syst. Sci. **58**(3), 622–640 (1999). ISSN 00220000
8. Chou, C.N., Golovnev, A., Velusamy, S.: Optimal streaming approximations for all Boolean max-2CSPs and max-kSAT. In: 61st Annual Symposium on Foundations of Computer Science, pp. 330–341 (2020). ISBN 978-1-72819-621-3
9. Chakraborty, D., Tewari, R.: Simultaneous Time-Space Upper Bounds for Certain Problems in Planar Graphs. arXiv Preprint arXiv: 1502.02135v1 (2015)
10. Crescenzi, P., Trevisan, L.: Max NP-completeness made easy. Theor. Comput. Sci. **225**(1–2), 65–79 (1999). ISSN 03043975
11. Elberfeld, M., Jakoby, A., Tantau, T.: Logspace versions of the theorems of bodlaender and courcelle. In: 51st Annual Symposium on Foundations of Computer Science, pp. 143–152 (2010). ISBN 978-1-4244-8525-3
12. Elberfeld, M., Kawarabayashi, K.i.: Embedding and canonizing graphs of bounded genus in logspace. In: 46th Annual Symposium on Theory of Computing, pp. 383–392 (2014). ISBN 978-1-4503-2710-7
13. Fredman, M.L., Komlós, J., Szemerédi, E.: Storing a sparse table with O(1) worst case access time. J. ACM **31**(3), 538–544 (1984). ISSN 00045411
14. Frederickson, G.N.: Upper bounds for time-space trade-offs in sorting and selection. J. Comput. Syst. Sci. **34**(1), 19–26 (1987). ISSN 00220000

15. Goemans, M.X., Williamson, D.P.: New (3/4)-approximation algorithms for the maximum satisfiability problem. SIAM J. Discret. Math. **7**(4), 656–666 (1994). ISSN 0895-4801, 1095-7146
16. Izumi, T., Otachi, Y.: Sublinear-space lexicographic depth-first search for bounded treewidth graphs and planar graphs. In: 47th International Colloquium on Automata, Languages, and Programming, pp. 67:1–67:17 (2020). ISBN 978-3-95977-138-2
17. Johnson, D.S.: Approximation algorithms for combinatorial problems. J. Comput. Syst. Sci. **9**(3), 256–278 (1974). ISSN 00220000
18. Khanna, S., Motwani, R.: Towards a syntactic characterization of PTAS. In: 28th Annual Symposium on Theory of Computing, pp. 329–337 (1996). ISBN 978-0-89791-785-8
19. Lieberherr, K.J., Specker, E.: Complexity of Partial Satisfaction. J. ACM **28**(2), 411–421 (1981). ISSN 00045411
20. Motwani, R., Raghavan, P.: Randomized Algorithms (1995). ISBN 978-0-511-81407-5
21. Munro, J.I., Paterson, M.S.: Selection and sorting with limited storage. Theor. Comput. Sci. **12**(3), 315–323 (1980). ISSN 03043975
22. Munro, J.I., Raman, V.: Selection from read-only memory and sorting with minimum data movement. Theor. Comput. Sci. **165**(2), 311–323 (1996). ISSN 03043975
23. Reif, J.H.: Symmetric complementation. J. ACM **31**(2), 401–421 (1984). ISSN 0004-5411, 1557-735X
24. Reingold, O.: Undirected connectivity in log-space. J. ACM **55**(4), 1–24 (2008). ISSN 00045411
25. Robertson, N., Seymour, P.D.: Graph minors. III. Planar tree-width. J. Combin. Theory Ser. B **36**(1), 49–64 (1984). ISSN 00958956
26. Savitch, W.J.: Relationships between nondeterministic and deterministic tape complexities. J. Comput. Syst. Sci. **4**(2), 177–192 (1970). ISSN 00220000
27. Serna, M.: Approximating linear programming is log-space complete for P. Inf. Process. Lett. **37**(4), 233–236 (1991). ISSN 00200190
28. Trevisan, L., Xhafa, F.: The parallel complexity of positive linear programming. Parallel Process. Lett. **08**(04), 527–533 (1998). ISSN 0129-6264, 1793-642X
29. Wegman, M.N., Carter, J.L.: New hash functions and their use in authentication and set equality. J. Comput. Syst. Sci. **22**(3), 265–279 (1981). ISSN 00220000
30. Williamson, D.P., Shmoys, D.B.: The Design of Approximation Algorithms (2011). ISBN 978-0-511-92173-5
31. Yannakakis, M.: On the approximation of maximum satisfiability. J. Algorithms **17**(3), 475–502 (1994). ISSN 01966774

# A Further Improvement on Approximating TTP-2

Jingyang Zhao(✉) and Mingyu Xiao(✉)

University of Electronic Science and Technology of China, Chengdu, China

**Abstract.** The Traveling Tournament Problem (TTP) is a hard but interesting sports scheduling problem inspired by Major League Baseball, which is to design a double round-robin schedule such that each pair of teams plays one game in each other's home venue, minimizing the total distance traveled by all $n$ teams ($n$ is even). In this paper, we consider TTP-2, i.e., TTP with one more constraint that each team can have at most two consecutive home games or away games. Due to the different structural properties, known algorithms for TTP-2 are different for $n/2$ being odd and even. For odd $n/2$, the best known approximation ratio is about $(1 + 12/n)$, and for even $n/2$, the best known approximation ratio is about $(1 + 4/n)$. In this paper, we further improve the approximation ratio from $(1+4/n)$ to $(1+3/n)$ for $n/2$ being even. Experimental results on benchmark sets show that our algorithm can improve previous results on all instances with even $n/2$ by 1% to 4%.

**Keywords:** Sports scheduling · Traveling tournament problem · Approximation algorithms · Timetabling combinatorial optimization

## 1 Introduction

The Traveling Tournament Problem (TTP), first systematically introduced in [5], is a hard but interesting sports scheduling problem inspired by Major League Baseball. This problem is to find a double round-robin tournament satisfying several constraints that minimizes the total distances traveled by all participant teams. There are $n$ participating teams in the tournament, where $n$ is always even. Each team should play $2(n-1)$ games in $2(n-1)$ consecutive days. Since each team can only play one game on each day, there are exact $n/2$ games scheduled on each day. There are exact two games between any pair of teams, where one game is held at the home venue of one team and the other one is held at the home venue of the other team. The two games between the same pair of teams could not be scheduled in two consecutive days. These are the constraints for TTP. We can see that it is not easy to construct a feasible schedule. Now we need to find an optimal schedule that minimizes the total traveling

The work is supported by the National Natural Science Foundation of China, under grant 61972070.



C.-Y. Chen et al. (Eds.): COCOON 2021, LNCS 13025, pp. 137–149, 2021.
https://doi.org/10.1007/978-3-030-89543-3_12

distances by all the $n$ teams. A well-known variant of TTP is TTP-$k$, which has one more constraint: each team is allowed to take at most $k$ consecutive home or away games. If $k$ is very large, say $k = n - 1$, then this constraint will lose its meaning and it becomes TTP again. For this case, a team can schedule its travel distance as short as the traveling salesmen problem. On the other hand, in a sports schedule, it is generally believed that home stands and road trips should alternate as regularly as possible for each team [3,15]. The smaller the value of $k$, the more frequently teams have to return their homes. TTP and its variants have been extensively studied in the literature [10,13,15,19].

### 1.1 Related Work

In this paper, we will focus on TTP-2. We mainly survey the results on TTP-$k$. For $k = 1$, TTP-1 is trivial and there is no feasible schedule [17]. But when $k \geq 2$, the problem suddenly becomes very hard. It is not easy to find a simple feasible schedule. Even no good brute force algorithm with a single exponential running time has been found yet. In the online benchmark [16], most instances with more than 10 teams are still unsolved completely even by using high-performance machines. The NP-hardness of TTP-$k$ with $k = 3$ or $k = n - 1$ has been proved [2,14]. Although the hardness of other cases has not been theoretically proved, most people believe TTP-$k$ with $k \geq 2$ is very hard. In the literature, there is a large number of contributions on approximation algorithms [8,9,12,15,18–20] and heuristic algorithms [1,4,6,7,11].

In terms of approximation algorithms, most results are based on the assumption that the distance holds the symmetry and triangle inequality properties. This is natural and practical in the sports schedule. For TTP or TTP-$k$ with $k \geq n - 1$, Westphal and Noparlik [18] proved an approximation ratio of 5.875 and Imahori *et al.* [9] proved an approximation ratio of 2.75 at the same time. For TTP-3, the current approximation ratio is $5/3+O(1/n)$ [20]. The first record of TTP-2 seems from the schedule of a basketball conference of ten teams in [3]. This paper did not discuss the approximation ratio. In fact, any feasible schedule for TTP-2 is a 2-approximation solution [15]. Although any feasible schedule will not have a very bad performance, no simple construction of feasible schedules is known now. In the literature, all known algorithms for TTP-2 are different for $n/2$ being even and odd. This may be caused by different structural properties. One significant contribution to TTP-2 was done by Thielen and Westphal [15]. They proposed a $(3/2+O(1/n))$-approximation algorithm for $n/2$ being odd and a $(1+16/n)$-approximation algorithm for $n/2$ being even. Now the approximation ratio was improved to $(1 + \frac{12}{n} + \frac{8}{n(n-2)})$ for odd $n/2$ [22] and to $(1 + \frac{4}{n} + \frac{4}{n(n-2)})$ for even $n/2$ [19].

### 1.2 Our Results

In this paper, we design an effective algorithm for TTP-2 with $n/2$ being even with an approximation ratio $(1 + \frac{3}{n} - \frac{6}{n(n-2)})$, improving the ratio from $(1 +$

$\frac{4}{n}+\Theta(\frac{1}{n(n-2)})$) to $(1+\frac{3}{n}-\Theta(\frac{1}{n(n-2)}))$. Now the ratio is small and improvement becomes harder and harder. Our major algorithm is based on packing minimum perfect matching. We first find a minimum perfect matching in the distance graph, then pair the teams according to the matching, and finally construct a feasible schedule based on the paired teams (called super-teams). Our algorithm is also easy to implement and runs fast. Experiments show that our results beat all previously-known solutions on the 17 tested instances in [19] with an average improvement of 2.10%. Due to limited space, the proofs of some lemmas and theorems are omitted, which can be found in the full version of this paper [21].

## 2 Preliminaries

We will always use $n$ to denote the number of teams and let $m = n/2$, where $n$ is an even number. We also use $\{t_1, t_2, \ldots, t_n\}$ to denote the set of the $n$ teams. A sports scheduling on $n$ teams is *feasible* if it holds the following properties.

- *Fixed-game-value*: Each team plays two games with each of the other $n-1$ teams, one at its home venue and one at its opponent's home venue.
- *Fixed-game-time*: All the games are scheduled in $2(n-1)$ consecutive days and each team plays exactly one game in each of the $2(n-1)$ days.
- *Direct-traveling*: All teams are initially at home before any game begins, all teams will come back home after all games, and a team travels directly from its game venue in the $i$th day to its game venue in the $(i+1)$th day.
- *No-repeat*: No two teams play against each other on two consecutive days.
- *Bounded-by-k*: The number of consecutive home/away games for any team is at most $k$.

The TTP-$k$ problem is to find a feasible schedule minimizing the total traveling distance of all the $n$ teams. The input of TTP-$k$ contains an $n \times n$ distance matrix $D$ that indicates the distance between each pair of teams. The distance from the home of team $i$ to the home of team $j$ is denoted by $D_{i,j}$. We also assume that $D$ satisfies the symmetry and triangle inequality properties, i.e., $D_{i,j} = D_{j,i}$ and $D_{i,j} \leq D_{i,h} + D_{h,j}$ for all $i, j, h$. We also let $D_{i,i} = 0$ for each $i$.

We will use $G$ to denote an edge-weighted complete graph on $n$ vertices representing the $n$ teams. The weight of the edge between two vertices $t_i$ and $t_j$ is $D_{i,j}$, the distance from the home of $t_i$ to the home of $t_j$. We also use $D_i$ to denote the weight sum of all edges incident on $t_i$ in $G$, i.e., $D_i = \sum_{j=1}^{n} D_{i,j}$. The sum of all edge weights of $G$ is denoted by $D_G$.

We let $M$ denote a minimum weight perfect matching in $G$. The weight sum of all edges in $M$ is denoted by $D_M$. We may consider the endpoint pair of each edge in $M$ as a *super-team*. We use $H$ to denote the complete graph on the $m$ vertices representing the $m$ super-teams. The weight of the edge between two super-teams $u_i$ and $u_j$, denoted by $D(u_i, u_j)$, is the sum of the weight of the four edges in $G$ between one team in $u_i$ and one team in $u_j$, i.e., $D(u_i, u_j) = \sum_{t_{i'} \in u_i \& t_{j'} \in u_j} D_{i',j'}$. We also let $D(u_i, u_i) = 0$ for any $i$. We give an illustration of the graphs $G$ and $H$ in Fig. 1.

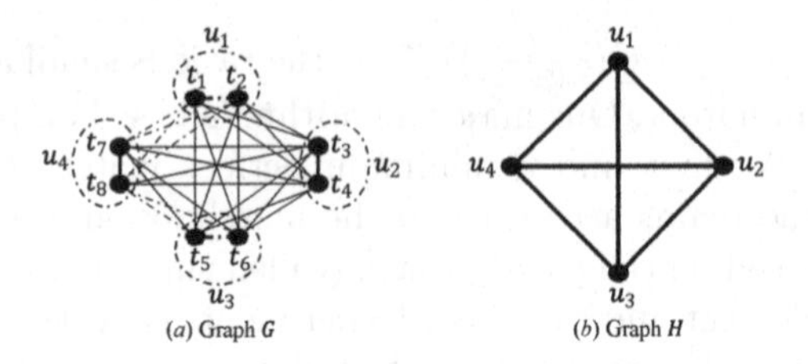


**Fig. 1.** An illustration of graphs $G$ and $H$, where there four dark lines form a minimum perfect matching $M$ in $G$

The sum of all edge weights of $H$ is denoted by $D_H$. It holds that

$$D_H = D_G - D_M. \tag{1}$$

## 2.1 Independent Lower Bound and Extra Cost

The *independent lower bound* for TTP-2 was firstly introduced by Campbell and Chen [3]. The basic idea of the independent lower bound is to obtain a lower bound $LB_i$ on the traveling distance of a single team $t_i$ independently without considering the feasibility of other teams.

The road of a team $t_i$ in TTP-2, starting at its home venue and coming back home after all games, is called an *itinerary* of the team. The itinerary of $t_i$ is also regarded as a graph on the $n$ teams, which is called the *itinerary graph* of $t_i$. In an itinerary graph of $t_i$, the degree of all vertices except $t_i$ is 2 and the degree of $t_i$ is greater than or equal to $n$ since team $t_i$ will visit each other team venue only once. Furthermore, for any other team $t_j$, there is at least one edge between $t_i$ and $t_j$, because $t_i$ can only visit at most 2 teams on each road trip and then team $t_i$ either comes from its home to team $t_j$ or goes back to its home after visiting team $t_j$. We decompose the itinerary graph of $t_i$ into two parts: one is a spanning star centered at $t_i$ and the forest of the remaining part. Note that in the forest, only $t_i$ may be a vertex of degree $\geq$2 and all other vertices are degree-1 vertices. See Fig. 2 for illustrations of the itinerary graphs.

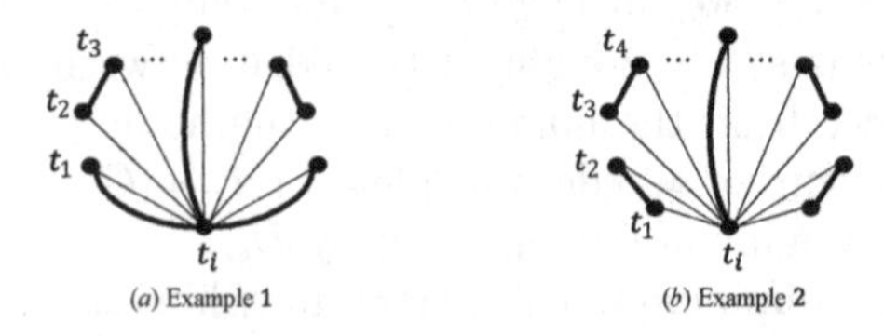


**Fig. 2.** The itinerary graph of $t_i$, where the light edges form a spanning star and the dark edges form the remaining forest. In the right example (b), the remaining forest is a perfect matching of $G$

For different itineraries of $t_i$, the spanning star is fixed and only the remaining forest may be different. The total distance of the spanning star is $\sum_{j \neq i} D_{i,j} = D_i$.

On the other hand, the distance of the remaining forest is at least as that of a minimum perfect matching of $G$ by the triangle inequality. Recall that we use $M$ to denote a minimum perfect matching of $G$. Thus, we have a lower bound $LB_i$ for each team $t_i$:

$$LB_i = D_i + D_M. \tag{2}$$

The itinerary of $t_i$ to achieve $LB_i$ is called the *optimal itinerary*. The *independent lower bound* for TTP-2 is the traveling distance such that all teams reach their optimal itineraries, which is denoted as

$$LB = \sum_{i=1}^{n} LB_i = \sum_{i=1}^{n} (D_i + D_M) = 2D_G + nD_M. \tag{3}$$

To analyze the quality of a schedule of the tournament, we will compare the itinerary of each team with the optimal itinerary. The different distance is called the *extra cost*. We may consider the extra cost for a subpart of the itinerary. A *road trip* in an itinerary of team $t_i$ is a simple cycle starting and ending at $t_i$. So an itinerary consists of several road trips. Let $L$ and $L'$ be two itineraries of team $t_i$, $L_s$ be a sub itinerary of $L$ consisting of several road trips in $L$, and $L'_s$ be a sub itinerary of $L'$ consisting of several road trips in $L'$. We say that the sub itineraries $L_s$ and $L'_s$ are *coincident* if they visit the same set of teams. We will only compare a sub itinerary of our schedule with a coincident sub itinerary of the optimal itinerary and consider the extra cost between them.

## 3 Constructing the Schedule

Our construction consists of two parts. First, we arrange *super-games* between *super-teams*, where each super-team contains a pair of normal teams. Then we extend super-games to normal games between normal teams. To make the itinerary as similar as the optimal itinerary, we take each team pair in the minimum perfect matching $M$ of $G$ as a *super-team*. There are $n$ normal teams and then there are $m = n/2$ super-teams. We denote the set of super-teams as $\{u_1, u_2, \ldots, u_m\}$ and relabel the $n$ teams such that $u_i = \{t_{2i-1}, t_{2i}\}$ for each $i$.

Each super-team will attend $m-1$ super-games in $m-1$ time slots. Each super-game on the first $m-2$ time slots will be extended to eight normal games between normal teams on four days, and each super-game on the last time slot will be extended to twelve normal games between normal teams on six days. So each normal team $t_i$ will attend $4 \times (m-2) + 6 = 4m - 2 = 2n - 2$ games. This is the number of games each team $t_i$ should attend in TTP-2. In our algorithm, the case of $n = 4$ is easy, and hence we assume here that $n \geq 8$.

We construct the schedule for super-teams from the first time slot to the last time slot $m-1$. In each of the $m-1$ time slots, we have $\frac{m}{2}$ super-games. In fact, our schedules in the first time slot and in the last time slot are different from the schedules in the middle time slots.

For the first time slot, the $\frac{m}{2}$ super-games are arranged as shown in Fig. 3. All of these super-games are called *normal super-games*. Each super-game is represented by a directed edge, the information of which will be used to extend super-games to normal games between normal teams.

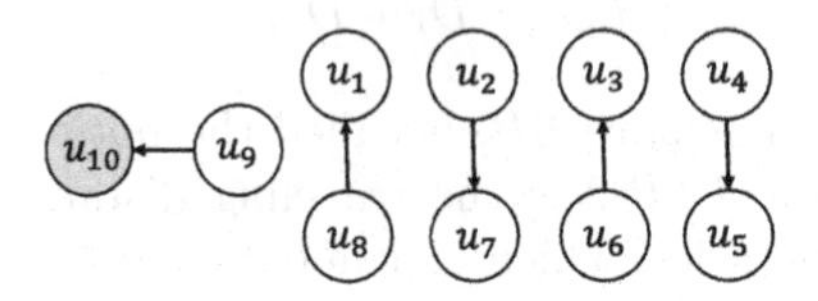


**Fig. 3.** The super-game schedule on the first time slot for an instance with $m = 10$

In Fig. 3, the last super-team $u_m$ is denoted as a dark node, and all other super-teams $u_1, \ldots, u_{m-1}$ are denoted as white nodes which form a cycle. In the second time slot, we keep the position of $u_m$ and change the positions of white super-teams in the cycle by moving one position in the clockwise direction, and also change the direction of each edge except for the most left edge incident on $u_m$. This edge will be replaced by a double arrow edge. The super-game including $u_m$ is also called a *left super-game* in the middle $m - 3$ time slots. So in the second time slot, there are $\frac{m}{2} - 1$ normal super-games and one left super-games. An illustration of the schedule in the second time slot is shown in Fig. 4.

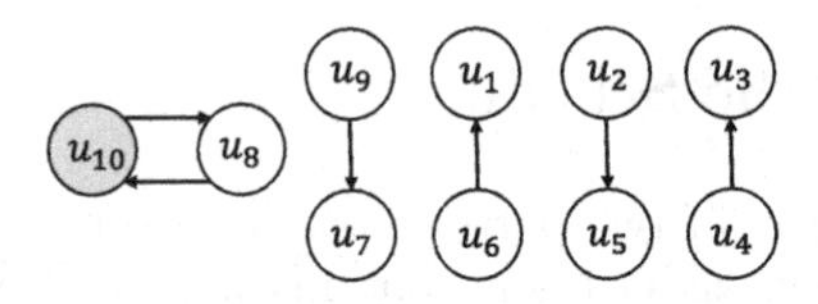


**Fig. 4.** The super-game schedule on the second time slot for an instance with $m = 10$

In the third time slot, there are also $\frac{m}{2} - 1$ normal super-games and one left super-games. We also change the positions of white super-teams in the cycle by moving one position in the clockwise direction while the direction of each edge is reversed. The position of the dark node will always keep the same. An illustration of the schedule in the third time slot is shown in Fig. 5.

The schedules for the other middle slots are derived analogously. Before we introduce the super-games in the last time slot $m - 1$, we first explain how to extend the super-games in the first $m - 2$ time slots to normal games. In these time slots, we have two different kinds of super-games: normal super-games and left super-games. We first consider normal super-games.

**Case 1. Normal Super-Games:** Each normal super-game will be extended to eight normal games on four days. Assume that in a normal super-game, super-team $u_i$ plays against the super-team $u_j$ on time slot $q$ ($1 \leq i, j \leq m$ and

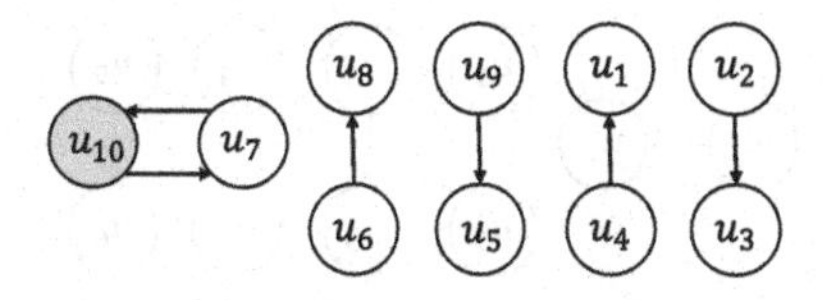


**Fig. 5.** The super-game schedule on the third time slot for an instance with $m = 10$

$1 \leq q \leq m-2$). Recall that $u_i$ represents normal teams $\{t_{2i-1}, t_{2i}\}$ and $u_j$ represents normal teams $\{t_{2j-1}, t_{2j}\}$. The super-game will be extended to eight normal games on four corresponding days from $4q-3$ to $4q$, as shown in Fig. 6. A directed edge from team $t_{i'}$ to team $t_{i''}$ means $t_{i'}$ plays against $t_{i''}$ at the home venue of $t_{i''}$. Note that if there is a directed edge from $u_j$ to $u_i$, then the direction of all the edges in Fig. 6 should be reversed.

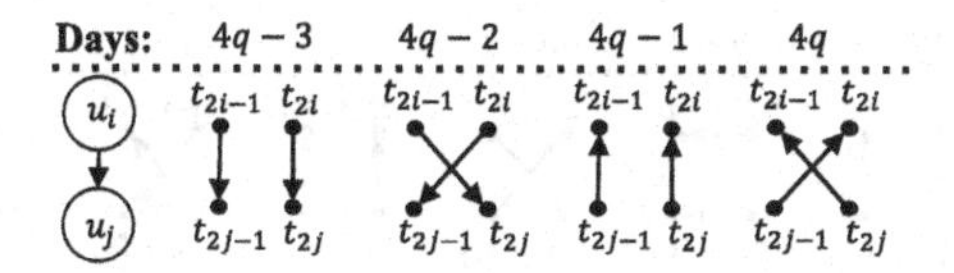


**Fig. 6.** Extending normal super-games

**Case 2. Left Super-Games:** Assume that in a left super-game, super-team $u_m$ plays against super-team $u_i$ on time slot $q$ ($2 \leq i \leq m-2$ and $2 \leq q \leq m-2$). Recall that $u_m$ represents normal teams $\{t_{2m-1}, t_{2m}\}$ and $u_i$ represents normal teams $\{t_{2i-1}, t_{2i}\}$. The super-game will be extended to eight normal games on four corresponding days from $4q-3$ to $4q$, as shown in Fig. 7 for even time slot $q$. For odd time slot $q$, the direction of edges in the figure will be reversed.

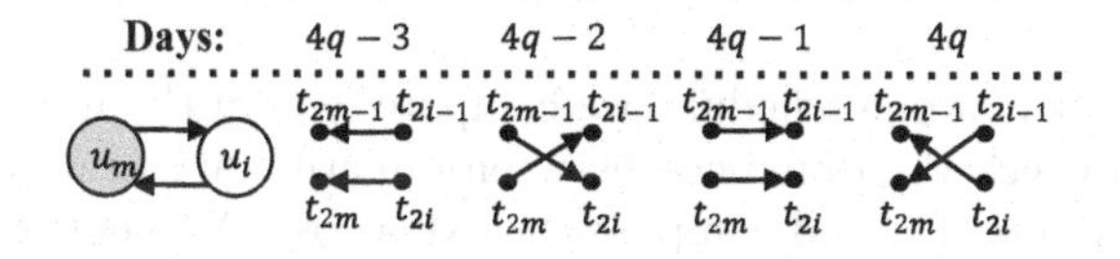


**Fig. 7.** Extending left super-games

The first $m-2$ time slots will be extended to $4(m-2) = 2n-8$ days according to the above rules. Each normal team will have six remaining games, which will be corresponding to the super-games on the last time slot. We will call a super-game on the last time slot a *last super-game*. Figure 8 shows an example of the schedule on the last time slot.

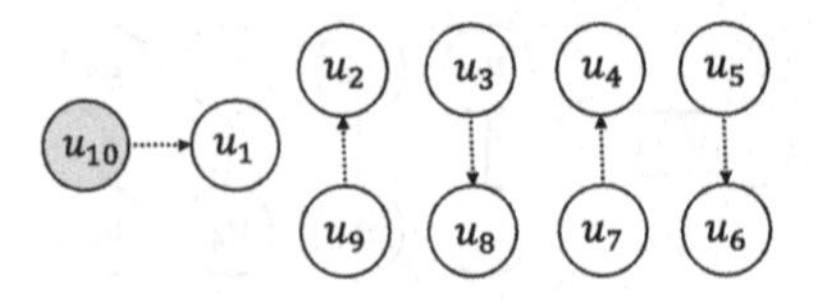


**Fig. 8.** The super-game schedule on the last time slot for an instance with $m = 10$

**Case 3. Last Super-Games:** Next, we extend a last super-game into twelve normal games on six days. Assume that on the last time slot $q = m - 1$, super-team $u_i$ plays against super-team $u_j$ $(1 \leq i, j \leq m)$. Recall that $u_i$ represents normal teams $\{t_{2i-1}, t_{2i}\}$ and $u_j$ represents normal teams $\{t_{2j-1}, t_{2j}\}$. The last super-game will be extended to twelve normal games on six corresponding days from $4q - 3$ to $4q + 2$, as shown in Fig. 9.

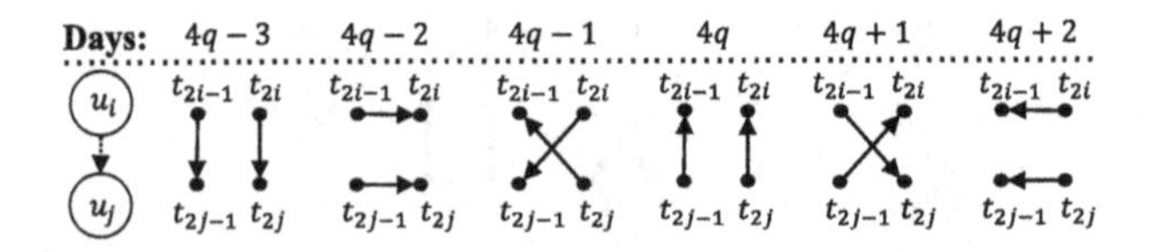


**Fig. 9.** Extending last super-games

**Theorem 1** *(*). For TTP-2 with $n$ teams such that $n \equiv 0 \pmod 4$, the above construction can generate a feasible schedule.*

We have introduced a method to construct a feasible schedule. Next, we will specify the order of some teams or super-teams to minimize the extra cost.

## 4 Approximation Quality of the Schedule

To show the quality of our schedule, we compare it with the independent lower bound. We will check the difference between our itinerary of each team $t_i$ and the optimal itinerary of $t_i$ and compute the extra cost. As mentioned in the last paragraph of Sect. 2, we will compare some sub itineraries of a team. We will look at the sub itinerary of a team on the four or six days in a super-game, which is coincident with a sub itinerary of the optimal itinerary: all teams stay at home before the first game in a super-game and return home after the last game in the super-game. In our algorithm, there are three types of super-games: normal super-games, left super-games, and last super-games. We analyze the total extra cost of all normal teams caused by each type of super-games.

**Lemma 1** *(*). Assume there is a super-game between super-teams $u_i$ and $u_j$ in our schedule.*

(a) *If the super-game is a normal super-game, then the extra cost of all normal teams in $u_i$ and $u_j$ is 0;*
(b) *If the super-game is a left or last super-game, then the extra cost of all normal teams in $u_i$ and $u_j$ is at most $D(u_i, u_j)$.*

In our schedule, there are $\frac{m}{2} + (m-3)(\frac{m}{2} - 1)$ normal super-games, which contribute 0 to the extra cost. There are $m-3$ left super-games on the $m-3$ middle time slots. By Lemma 1, we know that the total extra cost is $E_1 = \sum_{i=2}^{m-2} D(u_m, u_i)$. There are $\frac{m}{2}$ last super-games on the last time slot. By Lemma 1, we know that the total extra cost is $E_2 = \sum_{i=1}^{m/2} D(u_i, u_{m+1-i})$.

**Lemma 2.** *The total extra cost of our schedule is at most*

$$E_1 + E_2 = \sum_{i=2}^{m-2} D(u_m, u_i) + \sum_{i=1}^{m/2} D(u_i, u_{m+1-i}).$$

Next, we will make $E_1$ and $E_2$ as small as possible by reordering the teams.

First, we consider $E_2$. The extra cost is the sum of the weight of edges $\{u_i u_{m+1-i}\}_{i=1}^{m/2}$ in $H$, which form a matching in $H$. Our algorithm is to reorder $u_i$ such that $\{u_i u_{m+1-i}\}_{i=1}^{m/2}$ is a minimum perfect matching in $H$. Note that $H$ is a complete graph on $m$ (even) vertices and then we can use $O(m^3)$ time algorithm to find the minimum perfect matching $M_H$ in $H$. Our algorithm will reorder $u_i$ such that $\{u_i u_{m+1-i}\}_{i=1}^{m/2} = M_H$. For the cost of $M_H$, we have that

$$E_2 = D_{M_H} \leq \frac{1}{m-1} D_H. \tag{4}$$

Second, we consider $E_1$. Our idea is to choose $u_m$ such that $\sum_{i=2}^{m-2} D(u_m, u_i)$ is minimized. Note that once $u_m$ is determined, super-team $u_1$ is also determined by the matching $M_H$ ($u_m u_1$ should be an edge in $M_H$). After determining $u_m$ and $u_1$ together, we sill need to decider $u_{m-1}$. We first let $u_m$ be the super-team such that $\sum_{i=2}^{m-1} D(u_m, u_i)$ is minimized (There are $m$ possible candidates for $u_m$). Thus, we have that

$$\sum_{i=2}^{m-1} D(u_m, u_i) \leq \frac{2(D_H - D_{M_H})}{m}.$$

Then we let $u_{m-1}$ be the super-team such that $D(u_m, u_{m-1}) \geq D(u_m, u_i)$ for all $2 \leq i \leq m-2$. Thus, we have that

$$E_1 = \sum_{i=2}^{m-2} D(u_m, u_i) \leq \sum_{i=2}^{m-1} D(u_m, u_i) \frac{m-3}{m-2} \leq \frac{2(m-3)(D_H - D_{M_H})}{m(m-2)}. \tag{5}$$

By (1), (3), (4) and (5), we know that the total extra cost of our schedule is

$$\begin{aligned} E_1 + E_2 &\leq D_{M_H} + \frac{2(m-3)(D_H - D_{M_H})}{m(m-2)} \\ &= (1 - \frac{3}{m} + \frac{1}{m-2})D_{M_H} + (\frac{3}{m} - \frac{1}{m-2})D_H \\ &\leq (\frac{3}{m} - \frac{3}{m(m-1)})D_H \\ &\leq (\frac{3}{2m} - \frac{3}{2m(m-1)})LB = (\frac{3}{n} - \frac{6}{n(n-2)})LB. \end{aligned} \quad (6)$$

Next, we analyze the running-time bound of our algorithm. Our algorithm first uses $O(n^3)$ time to compute the minimum perfect matching $M$ and the minimum perfect matching $M_H$. It takes $O(n^2)$ time for us to determine $u_m$ and $u_{m-1}$ such that (4) and (5) hold and the remaining construction of the schedule also use $O(n^2)$ time. Thus, our algorithm runs in $O(n^3)$ time.

**Theorem 2.** *For TTP-2 on $n$ teams where $n \geq 8$ and $n \equiv 0 \pmod 4$, a feasible schedule can be computed in $O(n^3)$ time such that the total traveling distance is at most $(1 + \frac{3}{n} - \frac{6}{n(n-2)})$ times of the independent lower bound.*

## 5 Experimental Results

To test the performance of our schedule algorithm, we will implement it on well-known benchmark instances. For experiments, we will also use some simple heuristic methods to get further improvements.

### 5.1 Heuristics Based on Local Search

In the above analysis, we choose the permutation such that we can get a good approximation ratio. This is just for the purpose of the analysis. We do not guarantee this permutation is optimal. Other permutations may lead to better results on each concrete instance. However, the number of permutations is exponential and it is not effective to check all of them. Our idea is to only consider the permutations obtained by swapping the indexes of two super-teams and by swapping the indexes of the two teams in the same super-team. First, to check all possible swapping between two super-teams, we will have $O(m^2)$ loops, and the running-time bound will increase a factor of $m^2$. Second, for each last super-game between two super-teams, we consider the two orders of the two teams in each super-team and then we get four cases. We directly compute the extra cost for the four cases and select the best one. There are $m/2$ last super-games and then we only have $O(m)$ additional time. Note that we do not apply the second swapping for normal and left super-games since this operation will not get any improvement on them (this can be seen from the proof of Lemma 1).

### 5.2 Applications to Benchmark Sets

Our tested benchmark comes from [16], where introduces 62 instances and most of them are instances from the real world. There are 34 instances of $n$ teams

**Table 1.** Experimental results on the 17 instances with $n$ teams ($n$ is divisible by 4)

| Data set | ILB values | Previous results | Before swapping | After swapping | Our gap (%) | Improvement ratio (%) |
|---|---|---|---|---|---|---|
| Galaxy40 | 298484 | 307469 | 306230 | 305714 | 2.42 | 0.57 |
| Galaxy36 | 205280 | 212821 | 211382 | 210845 | 2.71 | 0.93 |
| Galaxy32 | 139922 | 145445 | 144173 | 144050 | 2.95 | 0.96 |
| Galaxy28 | 89242 | 93235 | 92408 | 92291 | 3.42 | 1.01 |
| Galaxy24 | 53282 | 55883 | 55486 | 55418 | 4.01 | 0.83 |
| Galaxy20 | 30508 | 32530 | 32082 | 32067 | 5.11 | 1.42 |
| Galaxy16 | 17562 | 19040 | 18614 | 18599 | 5.90 | 2.32 |
| Galaxy12 | 8374 | 9490 | 9108 | 9045 | 8.01 | 4.69 |
| NFL32 | 1162798 | 1211239 | 1199619 | 1198091 | 3.04 | 1.09 |
| NFL28 | 771442 | 810310 | 798208 | 798168 | 3.46 | 1.50 |
| NFL24 | 573618 | 611441 | 598437 | 596872 | 4.05 | 2.38 |
| NFL20 | 423958 | 456563 | 444426 | 442950 | 4.48 | 2.98 |
| NFL16 | 294866 | 321357 | 310416 | 309580 | 4.99 | 3.66 |
| NL16 | 334940 | 359720 | 351647 | 350727 | 4.71 | 2.50 |
| NL12 | 132720 | 144744 | 140686 | 140686 | 6.00 | 2.80 |
| Super12 | 551580 | 612583 | 590773 | 587387 | 6.49 | 4.11 |
| Brazil24 | 620574 | 655235 | 643783 | 642530 | 3.54 | 1.94 |

with $n \geq 4$ and $n \equiv 0 \pmod 4$. Half of them are very small ($n \leq 8$) or very special (all teams are in a cycle or the distance between any two teams is 1) and they were not tested in previous papers. So we only test the remaining 17 instances. The results are shown in Table 1, where the column '*ILB Values*' indicates the independent lower bounds, '*Previous Results*' lists previously known results in [19], '*Before Swapping*' is the results obtained by our schedule algorithm without using the local search method of swapping, '*After Swapping*' shows the results after swapping, '*Our Gap*' is defined to be $\frac{AfterSwapping\ -\ ILB\ Values}{ILB\ Values}$ and '*Improvement Ratio*' is defined as $\frac{Previous\ Results\ -\ AfterSwapping}{Previous\ Results}$.

From Table 1, we can see that our schedule algorithm can improve all the 17 instances with an average improvement of 2.10%. In these tested instances, the number of teams is at most 40. So our algorithm runs very fast. On a standard laptop with a 2.30 GHz Intel(R) Core(TM) i5-6200 CPU and 8 gigabytes of memory, all the 17 instances can be solved together within 0.1 s before applying the local search and within 8 s including local search.

## 6 Conclusion

In this paper, we introduce a new schedule for TTP-2 with $n \equiv 0 \pmod 4$ and prove an approximation ratio of $(1 + \frac{3}{n} - \frac{6}{n(n-2)})$, improving the previous ratio of $(1 + \frac{4}{n} + \frac{4}{n(n-2)})$ in [19]. The improvement looks small. However, the ratio is quite close to 1 now and further improvements become harder and harder. Furthermore, the new construction method is simpler and more intuitive, compared with the previous method in [19]. Experiments also show that the new schedule improves the results on all tested instances in the benchmark [16]. In the analysis, we can see that the extra cost of our schedule is contributed by left and last super-games. So we can decompose the analysis of the whole schedule into the analysis of left and last super-games. To get further improvements, we only need to reduce the number of left and last super-games.

## References


1. Anagnostopoulos, A., Michel, L., Van Hentenryck, P., Vergados, Y.: A simulated annealing approach to the traveling tournament problem. J. Sched. **9**(2), 177–193 (2006)
2. Bhattacharyya, R.: Complexity of the unconstrained traveling tournament problem. Oper. Res. Lett. **44**(5), 649–654 (2016)
3. Campbell, R.T., Chen, D.: A minimum distance basketball scheduling problem. Manage. Sci. Sports **4**, 15–26 (1976)
4. Di Gaspero, L., Schaerf, A.: A composite-neighborhood tabu search approach to the traveling tournament problem. J. Heurist. **13**(2), 189–207 (2007)
5. Easton, K., Nemhauser, G., Trick, M.: The traveling tournament problem description and benchmarks. In: Walsh, T. (ed.) CP 2001. LNCS, vol. 2239, pp. 580–584. Springer, Heidelberg (2001). https://doi.org/10.1007/3-540-45578-7_43
6. Easton, K., Nemhauser, G., Trick, M.: Solving the travelling tournament problem: a combined integer programming and constraint programming approach. In: Burke, E., De Causmaecker, P. (eds.) PATAT 2002. LNCS, vol. 2740, pp. 100–109. Springer, Heidelberg (2003). https://doi.org/10.1007/978-3-540-45157-0_6
7. Goerigk, M., Hoshino, R., Kawarabayashi, K., Westphal, S.: Solving the traveling tournament problem by packing three-vertex paths. In: AAAI 2014, pp. 2271–2277 (2014)
8. Hoshino, R., Kawarabayashi, K.I.: An approximation algorithm for the bipartite traveling tournament problem. Math. Oper. Res. **38**(4), 720–728 (2013)
9. Imahori, S., Matsui, T., Miyashiro, R.: A 2.75-approximation algorithm for the unconstrained traveling tournament problem. Ann. Oper. Res. **218**(1), 237–247 (2012). https://doi.org/10.1007/s10479-012-1161-y
10. Kendall, G., Knust, S., Ribeiro, C.C., Urrutia, S.: Scheduling in sports: an annotated bibliography. Comput. Oper. Res. **37**(1), 1–19 (2010)
11. Lim, A., Rodrigues, B., Zhang, X.: A simulated annealing and hill-climbing algorithm for the traveling tournament problem. Eur. J. Oper. Res. **174**(3), 1459–1478 (2006)
12. Miyashiro, R., Matsui, T., Imahori, S.: An approximation algorithm for the traveling tournament problem. Ann. Oper. Res. **194**(1), 317–324 (2012)
13. Rasmussen, R.V., Trick, M.A.: Round robin scheduling-a survey. Eur. J. Oper. Res. **188**(3), 617–636 (2008)
14. Thielen, C., Westphal, S.: Complexity of the traveling tournament problem. Theoret. Comput. Sci. **412**(4), 345–351 (2011)
15. Thielen, C., Westphal, S.: Approximation algorithms for TTP(2). Math. Methods Oper. Res. **76**(1), 1–20 (2012)
16. Trick, M.: Challenge traveling tournament instances (2013). http://mat.gsia.cmu.edu/TOURN/
17. de Werra, D.: Some models of graphs for scheduling sports competitions. Discret. Appl. Math. **21**(1), 47–65 (1988)

18. Westphal, S., Noparlik, K.: A 5.875-approximation for the traveling tournament problem. Ann. Oper. Res. **218**(1), 347–360 (2014)
19. Xiao, M., Kou, S.: An improved approximation algorithm for the traveling tournament problem with maximum trip length two. In: MFCS 2016, pp. 89:1–89:14 (2016)
20. Yamaguchi, D., Imahori, S., Miyashiro, R., Matsui, T.: An improved approximation algorithm for the traveling tournament problem. Algorithmica **61**(4), 1077–1091 (2011)
21. Zhao, J., Xiao, M.: A further improvement on approximating TTP-2. CoRR abs/2108.13060 (2021)
22. Zhao, J., Xiao, M.: The traveling tournament problem with maximum tour length two: a practical algorithm with an improved approximation bound. In: IJCAI 2021, pp. 4206–4212 (2021)

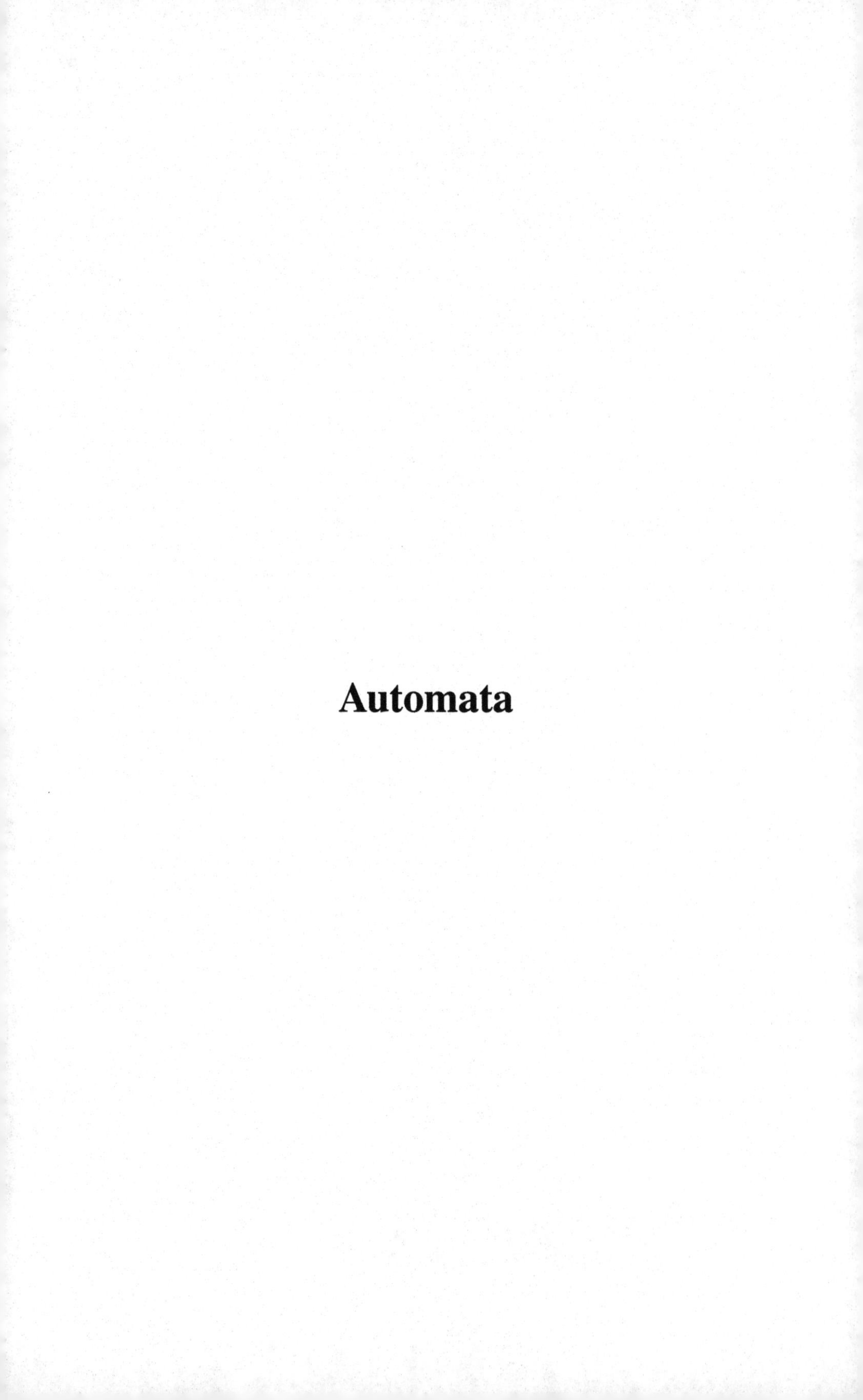

# Automata

# Sequence Graphs Realizations and Ambiguity in Language Models

Sammy Khalife[1(✉)], Yann Ponty[1], and Laurent Bulteau[2]

[1] LIX, CNRS, Ecole Polytechnique, Institut Polytechnique de Paris, 91128 Palaiseau, France
{khalife,yann.ponty}@lix.polytechnique.fr

[2] LIGM, CNRS, Université Gustave Eiffel, 77454 Marne-la-Vallée, France
laurent.bulteau@univ-eiffel.fr

**Abstract.** Several language models rely on an assumption modeling each local context as a (potentially oriented) bag of words, and have proven to be very efficient baselines. Sequence graphs are the natural structures encoding their information. However, a sequence graph may have several realizations as a sequence, leading to a degree of ambiguity. In this paper, we study such degree of ambiguity from a combinatorial and computational point of view. In particular, we present theoretical properties of sequence graphs. Several combinatorial problems are presented, depending on three levels of generalisation (window size, graph orientation, and weights), that we characterize with new complexity results. We establish different algorithms, including an integer program and a dynamic programming formulation to respectively recognize a sequence graph and to count the number of its distinct realizations.

**Keywords:** Graphs · Sequences · Combinatorics · Inverse problem · Complexity class

## 1 Introduction

The automated treatment of familiar objects, either natural or artifacts, always relies on a translation into entities manageable by computer programs. However, the correspondence between the object to be treated and "its" representation is not necessarily one-to-one. The representations used for learning algorithms are no exception to this rule. In particular, natural language words and textual documents representations are essential for several tasks, including document classification [12], role labelling [9], and named entity recognition [6]. The models based on pointwise mutual information, or graph-of-words (GOW), [3,7,10], supplement the content of bag-of-words (TF, TFIDF) with statistics of co-occurrences within a **window** of fixed size $w$, introduced to mitigate the degree of ambiguity. Several models [1,5,8,11] also use the same type of information and constitute strong baselines for natural language processing.

While these representations are more precise than the traditional bag-of-words (e.g. Parikh vectors), they still induce some level of ambiguity, *i.e.* a


C.-Y. Chen et al. (Eds.): COCOON 2021, LNCS 13025, pp. 153–163, 2021.
https://doi.org/10.1007/978-3-030-89543-3_13

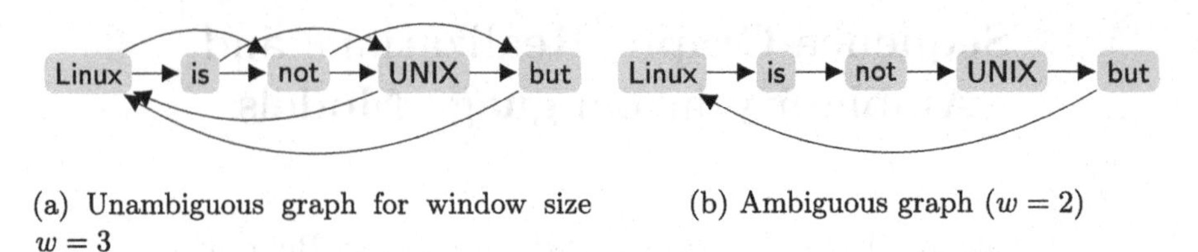


(a) Unambiguous graph for window size $w = 3$

(b) Ambiguous graph ($w = 2$)

**Fig. 1.** Sequence digraphs (or directed *graphs-of-words*) built for the sentence "Linux is not UNIX but Linux" using window sizes 3 (a) and 2 respectively (b). In the second case, the sequence graph is ambiguous, since any circular permutation of the words admits the same representation.

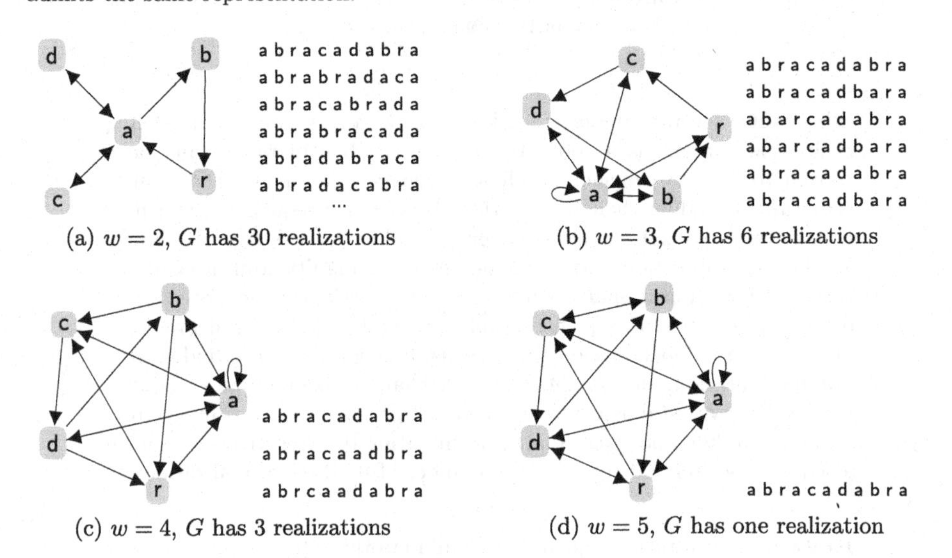


(a) $w = 2$, $G$ has 30 realizations

(b) $w = 3$, $G$ has 6 realizations

(c) $w = 4$, $G$ has 3 realizations

(d) $w = 5$, $G$ has one realization

**Fig. 2.** Sequence digraphs (or directed *graphs-of-words*) built for the sentence "a b r a c a d a b r a" using window sizes 2 (a), 3 (b), 4 (c) and 5 (d).

given graph can represent several sequences. Our study is thus motivated by a quantification of the level of ambiguity, seen as an algorithmic problem.

## Related Work

Sequence graphs encode the information of several co-occurences based models [1,8]. To the best of our knowledge, the ambiguity and realizability questions addressed in this work were never addressed by prior work in computational linguistics. Furthermore, we believe the problems studied in this paper are new and interesting from an algorithmic point of view, and appear to be devoid of reduction to other well-known problems.

The problem we consider in this paper can appear to be similar to the Universal Reconstruction of a String problem [2], which consists in determining the set of strings of a fixed length, with the most distinct letters satisfying substrings equations of the form: $s[q_1 \dots q_p] = s[q'_1 \dots q'_p], \dots, s[r_1 \dots r_m] = s[r'_1 \dots p'_m]$ where we denote the substring $s_{q_1} \dots s_{q_p}$ by $s[q_1 \dots q_p]$, and such that all the considered indexing sequences are strictly increasing. In this paper, we show that these problems are actually very different, and in particular, our complexity results imply the absence of reduction to the Universal Reconstruction of a String, which can be solved in linear time.

Furthermore, some similarities exist between our problem and others studied in the Distance Geometry (DG) literature. In distance geometry, the input consists of a set of pairwise distances between points, having unknown positions in a $d$-dimensional space. The problem then consists in determining a set of positions for the points (if they exist), satisfying the distance constraints. Since a position is fully characterized from $d+1$ constraining neighbors, the problem can be solved by finding a sequential order for processing points, such that the assignment of a point is always by at least $d+1$ among its neighbors [4]. This statement shares some level of similarity with our problem since a realization for a window $w = d+2$ also represents a linear ordering of its nodes, in which $w-1 = d+1$ of the neighbors have lower value with respect to the order.

The reasons for the insufficiency of linear ordering in DG to solve our realizability problem are threefold. First, each element of the sequence $x$ is associated a unique vertex. This is not the case we investigate here, since a symbol can be repeated several times, but only one vertex is created in the graph. This implies that the vertex associated to the $i^{\text{th}}$ element ($i \geq w$) of $x$ can have strictly less than $w-1$ distinct neighbors in its predecessors in $x$. Second, the absence of loops in distance geometry, because an element is at distance 0 from itself. Finally, the graphs are essentially undirected in distance geometry.

After introducing in Sect. 2 the formal definition of a sequence graph and the descriptions of our main problems, we establish in Sect. 3.1 complexity aspects of deciding the existence and counting sequences in GOWs associated with a window size $w = 2$. Then we consider in Sect. 3.2 the general case $w \geq 3$, present our theoretical results, and propose a integer program and a dynamic programming algorithm to respectively recognize a sequence graph and count its realizations.

## 2 Definitions and Problem Statement

Let $x = x_1, x_2, ..., x_p$ be a finite sequence of discrete elements among a finite vocabulary $X$. Without loss of generality, we can suppose that $X = \{1, ..., n\}$. In the following, let $X_p = \{1, ..., p\}$. This motivates the following definition:

**Definition 1.** *$G = (V, E)$ is the graph of the sequence $x$ with window size $w \in \mathbb{N}^*$ if and only if $V = \{x_i \mid i \in X_p\}$, and*

$$(i, j) \in E \iff \exists (k, k') \in X_p^2,\ |k - k'| \leq w - 1,\ x_k = i \text{ and } x_{k'} = j \quad (1)$$

*A sequence graph $G$ is endowed with a weights matrix $\Pi(G) = (\pi_{ij})$ such that*

$$\pi_{ij} = \text{Card}\ \{(k, k') \in X_p^2 \mid |k - k'| \leq w - 1,\ x_k = i \text{ and } x_{k'} = j\} \quad (2)$$

*Finally, for digraphs, the absolute value in the inequalities of Eq.* (1) *and Eq.* (2) *is replaced with* $k \leq k' \leq k + w - 1$. *We say that* $x$ *is a* $w$*-realization of* $G$ *(or a realization if there is no ambiguity), if* $G$ *is the graph of sequence* $x$ *with window size* $w$.

The natural integers $\pi_{ij}$ represent the number of co-occurrences of $i$ and $j$ in a window of size $w$. Hence, the graph of a sequence $x$ is unique given $w$. A linear time algorithm to construct a weighted sequence digraph is presented in a separate appendix. Other cases are obtained similarly. This procedure defines a correspondence between the sequence set $X^\star$ into the graph set $\mathcal{G}$: $\phi_w : X^\star \to \mathcal{G}, x \mapsto G_w(x)$. Based on these definitions, we consider the following problems:

*Problem 1 (Weighted-*REALIZABLE *(W-*REALIZABLE*)).*
**Input:** Possibly directed graph $G$, matrix weights $\Pi$, window size $w$
**Output:** True if $(G, \Pi)$ is the $w$-sequence graph of some sequence $x$, False otherwise.

*Problem 2 (Unweighted-*REALIZABLE *(U-*REALIZABLE*))).*
**Input:** Possibly directed graph $G$, window size $w$
**Output:** True if $G$ is the $w$-sequence graph of some sequence $x$, False otherwise.

We denote *D*-REALIZABLE (resp. *G*-) the restricted version of REALIZABLE where the input graph $G$ is directed (resp. undirected), and *W*-REALIZABLE (resp. *U*-) the restricted version of REALIZABLE where the input graph $G$ is weighted (resp. unweighted), possibly in combination with the D- or G- variants. We write REALIZABLE$_w$ for the case where $w$ is a fixed (given) constant. We also consider the variants of W-REALIZABLE, denoted GW-REALIZABLE and DW-REALIZABLE where the input graph is restricted to be respectively undirected and directed. We define GU-REALIZABLE and DU-REALIZABLE similarly. Finally, we write (GW-, DW-, ...) REALIZABLE$_w$ for the case where $w$ is a fixed positive integer.

*Problem 3 (Unweighted-*NUMREALIZATIONS *(U-*NUMREALIZATIONS*))).*
**Input:** Possibly directed graph $G$, window size $w$
**Output:** The number of **realizations** of $G$, *i.e.* preimages of $G$ through $\phi_w$ such that $|\{x \in X^\star \mid \phi_w(x) = G\}|$ if finite, or $+\infty$ otherwise.

*Problem 4 (Weighted-*NUMREALIZATIONS *(W-*NUMREALIZATIONS*)).*
**Input:** Possibly directed graph $G$, matrix weights $\Pi$, window size $w$
**Output:** The number of **realizations** of $G$ in the weighted sense.

Similarly, we use the same prefix for the directed or undirected versions of (D-, G-, i.e. DU- for directed and unweighted). We also denote NUMREALIZATIONS$_w$ for the case where $w$ is a fixed strictly positive integer. Note that NUMREALIZATIONS strictly generalizes the previous one, as REALIZABLE can be solved by testing the nullity of the number of suitable realization computed by NUMREALIZATIONS.

| | |
|---|---|
| **DW** Directed Weighted | **DU** Directed Unweighted |
| **GW** Undirected Weighted | **GU** Undirected Unweighted |

# 3 Theoretical Results

In this section, we present our main theoretical results. Due to length limitations, all of the proofs presented in this section are left in a separate appendix.

## 3.1 A Complete Characterization of 2-sequence Graphs

A graph has a sequential realization with $w = 2$ when there exists a path visiting every vertex and covering all of its edges (at least once for the unweighted case and exactly $\pi_e$ for the edge $e$ in the weighted case). This characterization enables relatively simple characterization and algorithmic treatment, leading to the results summarized in Table 1.

**Table 1.** Complexity for various instances of our problems ($w = 2$)

| Data instance | NUMREALIZATIONS$_2$ | | REALIZABLE$_2$ | |
|---|---|---|---|---|
| | Complexity | # Sequences | Complexity | Characterization |
| GU | P | $\{0, +\infty\}$ | P | $G$ connected |
| GW | #P-hard | $\{0, 1\} \cup 2\mathbb{N}^*$ | P | $\psi(G)$ (semi) Eulerian |
| DU | P | $\{0, 1, +\infty\}$ | P | Theorem 1 |
| DW | P | $\mathbb{N}$ (BEST Theorem) | P | $\psi(G)$ (semi) Eulerian |

**Theorem 1.** *Let $G = (V, E)$ be an unweighted digraph. Let $R^+(G)$ be the weighted DAG obtained by contracting the strongly connected components of $G$, such that the weight of an edge is attributed the number of distinct arcs from two strongly connected components in $G$. Then, $G$ is a 2-sequence graph if and only if $R^+(G)$ is a directed path and its weights are all equal to* 1.

## 3.2 General Sequence Graphs and REALIZABLE$_{w \geq 3}$

The characterization of more general sequence graphs, such as 3-graphs is not the same for 2-graphs, as shows the counterexample in Fig. 3a: the depicted graph has no self-edge so there must be at least one clique of size 3. Similarly, Fig. 3b depicts a counter example for directed graphs: $G$ does not have loops, so if it had a 3-realization, such sequence must be of the form $\{1\,2\,3\,1..., 1\,3\,2\,1..., 2\,3\,1\,2..., 3\,2\,1\,3..., 2\,1\,3\,2...\}$ but then $(2, 1)$ would form an edge.

**A Polynomial Time Algorithm for GU-REALIZABLE$_w$**

To construct our poly-time algorithm, we will use an auxiliary graph built on $G$. Let $H(G) = (E, E_H)$ be the new graph obtained with the following procedure. Two edges $e = (v_1, v_2)$, $f = (v_3, v_4)$ of $E$ are connected in $H(G)$ if and only if:

$$v_2 = v_3 \text{ and } (v_1, v_4) \in E \tag{3}$$

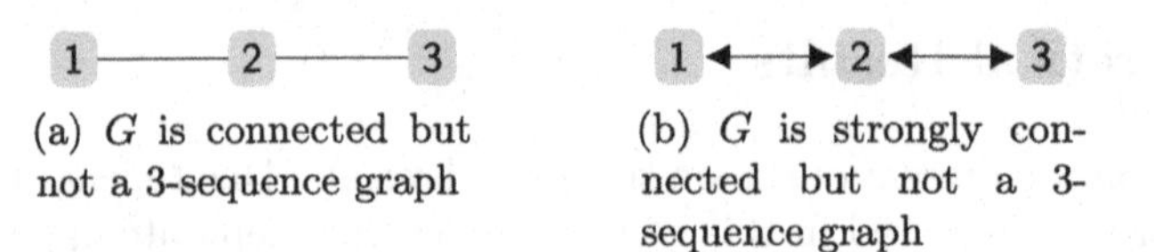

(a) $G$ is connected but not a 3-sequence graph

(b) $G$ is strongly connected but not a 3-sequence graph

**Fig. 3.** Counter examples for $w = 3$

An edge of $H(G)$ can be seen as an unique triplet $v_1, v_2, v_3$ where $(v_1, v_2), (v_1, v_3)$ and $(v_2, v_3) \in E$. Therefore, by definition, a walk $P$ in $H(G)$ is always of the form:

$$P = (t_1, t_2), ..., (t_{p-1}, t_p) \text{ s.t } \forall i \in \{1, ..., p-1\},\ (t_i, t_{i+1}) \in E \tag{4}$$

It is clear that if $H(G)$ is a 2-graph, then $G$ is a 3-graph since there is a walk going through all edges of $H(G)$ (so visiting every non isolated node and creating all edges of $G$). However, the converse is not true as depicted in Fig. 4. In order to determine if $G = (V, E)$ has a realization in the general case, a procedure is to recursively merge pairs of vertices, maintaining constraints depending on $E$. These constraints are similar to Eq. 3. We adopt the following notations, $u_{i,j} = (u_i, u_j)$ and $u_{1:k} = (u_1, ..., u_k)$. The iterative procedure for $w \geq 3$ is summed up in the following equation. Namely, $\forall k \in \{2, ..., w-2\}$, one has

$$E^{(k)} = \{u_{1:k+1} \in V^{k+1} \mid u_{1:k} \in E^{(k-1)}, u_{2:k+1} \in E^{(k-1)} \wedge (u_1, u_{k+1}) \in E\} \tag{5}$$

Let $H^{(k)} = (E^{(k)}, E^{(k+1)})$, it can be defined recursively through:

$$H^{(0)} = G \qquad \forall k \in \mathbb{N}^*,\ H^{(k)} = f(H^{(k-1)}) \tag{6}$$

where $f$ transforms edges into vertices and creates edges between new vertices that verify Eq. 5.

**Definition 2.** *Let $u$ be a vertex of $H^{(k)}$ for $k \in \mathbb{N}$, $u = (u_1, ..., u_k, u_{k+1})$. The sequence $u_1, ..., u_{k+1}$ is the* **authentic** *sequence of $u$. We also call an authentic sequence of a walk on $H^{(k)}$: $P = (x_1, ..., x_{k+1}), (x_2, ..., x_{k+2}), ..., (x_v, ..., x_{v+k})$ the sequence $x_1, x_2, ..., x_{v+k}$.*

In order to obtain realizations of length $p$, the computation of $H^{(p)}$ requires $p$ iterations, and the number of vertices and edges of $H^{(k)}$ can increase during iterations (the complete graph is an example for which these numbers increase exponentially). The next Proposition states a correspondence between realizations and authentic sequences.

**Proposition 1.** *Let $x = x_1, ..., x_p$ be a $w$-realization of a graph (or digraph) $G = (V, E)$. If $w \leq p$, then $x$ is an authentic sequence of a walk of length $p - w + 1$ on $H^{(w-2)}$.*

### Main Complexity Results

In this subsection we present the remaining complexity results, which are summarized in Theorem 2 and Table 2. We first show that GU-REALIZABLE$_w$ $\in P$, $\forall w \geq 3$. Besides, for GU, the number of realizations of a graph $G$ is either 0 (not realizable), $+\infty$ (realizable and there exists a cycle in a component of $H$ generating $G$), or 1 (realizable but no cycle in any component of $H$ generating $G$). These three cases can be tested in polynomial time using our algorithm, showing that GU-NUMREALIZATIONS$_w$ $\in P$, $\forall w \geq 3$.

**Theorem 2.** *All variations of* NUMREALIZATIONS$_w$ *and* REALIZABLE$_w$ *are NP-hard, except* ***GU****. Besides,* NUMREALIZATIONS, REALIZABLE *are para-NP-hard for all variations, except* ***GU****, in which case they are both W[1]-hard and XP.*

**Table 2.** Complexity for various instances of our problems ($w \geq 3$). We remind that a para-NP-hard problem does not admit any XP algorithm unless P = NP.

| Variation | Constant $w$, $w \geq 3$ | | Parameter $w$ | |
|---|---|---|---|---|
| | NUMREALIZATIONS$_w$ complexity | REALIZABLE$_w$ complexity | NUMREALIZATIONS complexity | REALIZABLE complexity |
| GU | P | P | W[1]-hard; XP | W[1]-hard; XP |
| GW | NP-hard | NP-hard | Para-NP-hard | Para-NP-hard |
| DU | NP-hard | NP-hard | Para-NP-hard | Para-NP-hard |
| DW | NP-hard | NP-hard | Para-NP-hard | Para-NP-hard |

In the remaining of this section, we present more details about these complexity results.

### A Special Case: GU

**Proposition 2.** *Let* $w \in \mathbb{N}^*$. GU-REALIZABLE$_w$ *is in* $P$.

For digraphs, the analogue of the procedure mentioned in the proof of Proposition 2 (left in the appendix) would consist in enumerating all paths in the DAG $R(H^{(w-2)})$, where $R(G)$ is the DAG obtained by contracting the strongly connected components of G. However, the number of paths can be exponential, even for a sequence graph. In the next subsection, we will prove that DU-REALIZABLE$_w$ is actually NP-hard. Finally, if $x_1, ..., x_c$ are vertices of a strongly component of $H^{(w-2)}$, which order should be considered to form a new vertex attribute $x_C$? The following lemma shows that this order is not important, as long as it represents a walk in the component. Moreover, it is possible to reconstruct all realizations from walks on $R(H^{w-2})$. With the same notations:

**Lemma 1.** *Let $x$ be a walk on $H^{(w-2)}$ whose authentic sequence is a $w$-realization for G. If $x$ goes through a strongly component $C$ of $H^{(w-2)}$, adding any supplementary path included in $C$ lets $x$ a $w$-realization. Any graph generated by a walk on $H^{(w-2)}$ can be generated by a walk on $R(H^{(w-2)})$.*

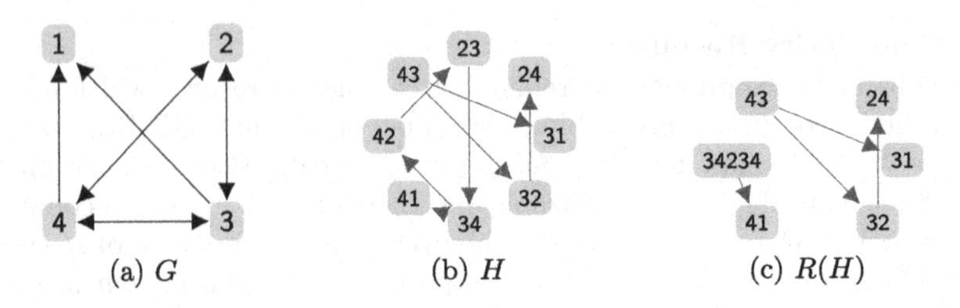


**Fig. 4.** Procedure to find a 3-realization. 34234, 41: is a 3-realization, with authentic sequence 3 4 2 3 4 1

### Other Variations

In the following, CLIQUE is the problem which takes as input an undirected graph $G$ and should return the maximal size of a clique in $G$.

**Proposition 3.** CLIQUE *admits a polynomial time parameterized reduction to* GU-REALIZABLE.

**Corollary 1.** GU-REALIZABLE *is* W[1]*-hard for parameter* $w$.

**Proposition 4.** *DU-Realizable*$_w$*, GW-Realizable*$_w$*, and DW-Realizable*$_w$ *are all NP-hard for any* $w \geq 3$.

All proofs of NP-hardness are left in a separate appendix.

## 4 Effective General Algorithms

### 4.1 REALIZABLE$_w$: Linear Integer Programming Formulation

Let $G = (V, E)$ be a graph with integer weights $\pi_{e \in E}$ . In this model, we represent a sequence $x$ over the alphabet $\{1, ...n\}$, as a $(0-1)$ matrix $X \in \mathbb{M}_{n,p}(\{0, 1\})$ encoding the sequence $x$:

$$X_{i,j} = \begin{cases} 1 & \text{if } x_j = i \\ 0 & \text{otherwise} \end{cases}$$

We represent the set of sequences over the alphabet $\{1, ...n\}$ by the $(0-1)$ matrices such that $\forall j \in \{1, ..., p\}, \sum_{i=1}^{n} X_{i,j} = 1$.

Given a window size $w$, a unit of $\pi_{e=(v_1,v_2)}$ corresponds to the appearance of two elements $v_1$, $v_2$ at a distance $i \in \{1, ..., w-1\}$ in the sequence. Now, let us consider a fixed distance $i$, and a starting index $j \in \{1, ..., p-i\}$, we use a intermediary slack variable $y_j^e(i) \in \{0, 1\}$ to model the presence of such appearance using the constraint:

$$X_{v_1,j} X_{v_2,j+i} = y_j^e(i) \tag{7}$$

Then, the Boolean variable $y_j^e(i)$ is equal to 1 when $v_1$ is located at position $j$ and $v_2$ at position $j+i$. We linearise Eq. 7 as:

$$\begin{aligned} -X_{v_1,j} \quad & \quad + y_j^e(i) \leq 0 \\ - X_{v_2,j+i} & \quad + y_j^e(i) \leq 0 \\ X_{v_1,1} + X_{v_2,j+i} & \quad - y_j^e(i) \leq 1 \end{aligned} \tag{8}$$

Each slack variable $y_k^e(i)$ is attributed to an edge $e$, a relative distance $i \in \{1, ..., w-1\}$ and a starting position $k \in \{1, ..., p-i\}$. Given our constraint formulation, every slack variable is attributed 3 constraints. For a digraph, the number of possible pair positions for a unit of $\pi_{e=(v_1,v_2)}$ is given by:

$$C = \sum_{i=1}^{w-1}(p-i) = p(w-1) - \frac{w(w-1)}{2} = (w-1)(p-\frac{w}{2})$$

Therefore, in our model, $C$ corresponds to the number of slack variables attributed to constraints for an edge of the graph.

On the contrary, the absence of an edge $e = (v_1, v_2)$, corresponding to $\pi_e = 0$, can be modeled for a distance $i \in \{1, ..., w-1\}$ and a starting position $j \in \{1, ..., p-i\}$ as:

$$X_{v1,j} + X_{v_2,j+i} \leq 1$$

Then, REALIZABLE$_w$ can be formulated as the following linear integer program:

$$\min_{X \in \{0,1\}^{p \times n}, y \in \{0,1\}^{|E| \times C}} \sum_{e \in E} \sum_{i \in \{1,...,w-1\}} y_1^e(i) + ... + y_{p-i}^e(i)$$

under the constraints

$$\forall j \in \{1, ..., p\} \qquad \sum_{i=1}^{n} X_{i,j} = 1$$

$$\begin{array}{l} \forall e = (v_1, v_2) \in E \\ \forall e' = (v_1', v_2') \notin E \\ \forall i \in \{1, ..., w-1\} \end{array} \left\{ \begin{array}{lll} -X_{v_1,1} & + y_1^e(i) \leq 0 & \\ - X_{v_2,1+i} & + y_1^e(i) \leq 0 & \\ X_{v_1,1} + X_{v_2,1+i} & - y_1^e(i) \leq 1 & X_{v_1',1} + X_{v_2',1+i} \leq 1 \\ \vdots & & \vdots \\ -X_{v_1,p-i} & + y_{p-i}^e(i) \leq 0 & X_{v_1',p-i} + X_{v_2',p} \leq 1 \\ - X_{v_2,p} & + y_{p-i}^e(i) \leq 0 & \\ X_{v_1,p-i} + X_{v_2,p} & - y_{p-i}^e(i) \leq 1 & \end{array} \right.$$

$$\text{and } \forall e \in E \qquad \sum_{i \in \{1,...,w-1\}} y_1^e(i) + ... + y_{p-i}^e(i) \geq \pi_e$$

If the objective function reaches $\sum_{e \in E} \pi_e$ at its minimum then the output of REALIZABLE$_w(G, \Pi)$ is True, and False otherwise.

### 4.2 NumRealizations$_w$: Dynamic Programming Formulation

We did not present a way to count realizations in the general case. We present in this subsection a method based on dynamic programming valid for all cases.

The recursion proceeds by extending a partial sequence, initially set to be empty, keeping track of for represented edges along the way. Namely, consider $N_w[\Pi, p, \mathbf{u}]$ to be the number of $w$-realizations of length $p$ for the graph $G = (V, E)$, respecting a weight matrix $\Pi = (\pi_{ij})_{i,j \in V^2}$, preceded by a sequence of nodes $\mathbf{u} := (u_1, \ldots, u_{|\mathbf{u}|}) \in V^\star$. It can be shown that, for all $\forall p \geq 1$, $\Pi \in \mathbb{N}^{|V^2|}$ and $\mathbf{u} \in V^{\leq w}$, $N_w[\Pi, p, \mathbf{u}]$ obeys the following formula, using the notations of Sect. 3.2:

$$N_w\left[\Pi, p, \mathbf{u}\right] = \sum_{v \in V} \begin{cases} N_w\left[\Pi'_{(\mathbf{u},v)}, p-1, (u_1, ..., u_{|\mathbf{u}|}, v)\right] & \text{if } |\mathbf{u}| < w-1 \\ N_w\left[\Pi'_{(\mathbf{u},v)}, p-1, (u_2, ..., u_{w-1}, v)\right] & \text{if } |\mathbf{u}| = w-1 \end{cases} \tag{9}$$

with $\Pi'_{(\mathbf{u},v)} := (\pi_{ij} - |\{k \in [1, |\mathbf{u}|] \mid (u_k, v) = (i,j)\}|)_{(i,j) \in V^2}$. The base case of this recurrence corresponds to $p = 0$, and is defined as

$$\forall\, \Pi,\ N_w[\Pi, 0, \mathbf{u}] = \begin{cases} 1 & \text{if } \Pi = (0)_{(i,j) \in V^2} \\ 0 & \text{otherwise.} \end{cases} \tag{10}$$

The total number of realizations is then found in $N_w[\Pi, p, \varepsilon]$, *i.e.* setting $\mathbf{u}$ to the empty prefix $\varepsilon$, allowing the sequence to start from any node.

The recurrence can be computed in $\mathcal{O}(|V|^w \times \prod_{i,j \in V^2}(\pi_{i,j} + 1))$ time using memoization, for $p$ the sequence length. The complexity can be refined by noting that:

$$\sum_{i,j \in V^2} \pi_{i,j} \leq w \times p$$

To investigate the worst case scenario, we can consider the optimisation problem:

$$\max_\Pi \prod_{i,j \in V^2}(\pi_{i,j} + 1) \text{ such that } \sum_{i,j} \pi_{i,j} = w\,p. \tag{11}$$

This problem is equivalent to maximise a product under a budget constraint. When $n^2 \geq w \times p$, which is the case in practice, the maximum is reached for a Boolean matrix $\Pi = (\pi_{i,j}) \in \{0,1\}^{|V|^2}$, verifying the constraint. This property can be deduced from the inequality:

$$\begin{aligned} 1 \leq a < b-1 \implies &\quad \log a + \log b < \log(a+1) + \log(b-1) \\ \implies &\quad ab < (a+1)(b-1) \end{aligned}$$

It follows that, in the worst-case scenario, $\prod_{i,j \in V^2}(\pi_{i,j} + 1) \in \mathcal{O}(2^{w\,p})$. Thus, despite the, apparently extreme complexity of our algorithm, it is still possible to compute $N_w[\Pi, p, u_{1:w}]$ for "reasonable" values of $p$ and $w$.

## 5 Conclusion

In this study, we visited a series of problems related to the ambiguity of sequence graphs representations, which are popular in the context of text mining and natural language processing. We derived theoretical properties and practical algorithms for the family of sequence graphs, which are suitable to estimate the ambiguity level of several pointwise mutual information models [1,5,8,11].

**Acknowledgments.** The authors wish to express their gratitude to Guillaume Fertin and an anonymous reviewer of an earlier version of this manuscript, for their valuable suggestions and constructive criticisms. Sammy Khalife acknowledges Agence Nationale de la Recherche for partially funding this paper.

## References

1. Arora, S., Li, Y., Liang, Y., Ma, T., Risteski, A.: A latent variable model approach to PMI-based word embeddings. Trans. Assoc. Comput. Linguist. **4**, 385–399 (2016)
2. Gawrychowski, P., Kociumaka, T., Radoszewski, J., Rytter, W., Waleń, T.: Universal reconstruction of a string. Theoret. Comput. Sci. **812**, 174–186 (2020)
3. Gibert, J., Valveny, E., Bunke, H.: Dimensionality reduction for graph of words embedding. In: Jiang, X., Ferrer, M., Torsello, A. (eds.) GbRPR 2011. LNCS, vol. 6658, pp. 22–31. Springer, Heidelberg (2011). https://doi.org/10.1007/978-3-642-20844-7_3
4. Liberti, L., Lavor, C., Maculan, N., Mucherino, A.: Euclidean distance geometry and applications. SIAM Rev. **56**(1), 3–69 (2014)
5. Mikolov, T., Chen, K., Corrado, G., Dean, J.: Efficient estimation of word representations in vector space. arXiv preprint arXiv:1301.3781 (2013)
6. Nadeau, D., Sekine, S.: A survey of named entity recognition and classification. Lingvisticae Invest. **30**(1), 3–26 (2007)
7. Peng, H., et al.: Large-scale hierarchical text classification with recursively regularized deep graph-CNN. In: Proceedings of the 2018 World Wide Web Conference, pp. 1063–1072 (2018)
8. Pennington, J., Socher, R., Manning, C.: GloVe: global vectors for word representation. In: Proceedings of the 2014 Conference on Empirical Methods in Natural Language Processing (EMNLP), pp. 1532–1543 (2014)
9. Roth, M., Woodsend, K.: Composition of word representations improves semantic role labelling. In: Proceedings of the 2014 Conference on Empirical Methods in Natural Language Processing (EMNLP), pp. 407–413 (2014)
10. Rousseau, F., Kiagias, E., Vazirgiannis, M.: Text categorization as a graph classification problem. In: Proceedings of the 53rd Annual Meeting of the ACL and the 7th IJCNLP (Volume 1: Long Papers), pp. 1702–1712 (2015)
11. Sanjeev, A., Yingyu, L., Tengyu, M.: A simple but tough-to-beat baseline for sentence embeddings. In: Proceedings of ICLR (2017)
12. Skianis, K., Malliaros, F., Vazirgiannis, M.: Fusing document, collection and label graph-based representations with word embeddings for text classification. In: Proceedings of the Twelfth Workshop on Graph-Based Methods for Natural Language Processing (TextGraphs-12), pp. 49–58 (2018)

# Between SC and LOGDCFL: Families of Languages Accepted by Polynomial-Time Logarithmic-Space Deterministic Auxiliary Depth-$k$ Storage Automata

Tomoyuki Yamakami(✉)

Faculty of Engineering, University of Fukui, 3-9-1 Bunkyo, Fukui 910-8507, Japan
TomoyukiYamakami@gmail.com

**Abstract.** The closure of deterministic context-free (dcf) languages under logarithmic-space many-one reductions (L-m-reductions), known as LOGDCFL, has been studied in depth from an aspect of parallel computability because it is nicely situated between L and $\mathrm{AC}^1 \cap \mathrm{SC}^2$. By changing a memory device from pushdown stacks to access-controlled storage tapes, we introduce a computational model of deterministic depth-$k$ storage automata ($k$-sda's) whose tape cells are freely modified during the first $k$ accesses and then erased and frozen forever. These $k$-sda's naturally induce the language family $k$SDA. Similarly to LOGDCFL, we study the closure LOG$k$SDA of all languages in $k$SDA under L-m-reductions. We demonstrate that DCFL $\subseteq$ $k$SDA $\subseteq$ $\mathrm{SC}^k$ by significantly extending Cook's early result (1979) of DCFL $\subseteq$ $\mathrm{SC}^2$. The entire hierarchy of LOG$k$SDA for all $k \geq 1$ therefore lies between LOGDCFL and SC. As an immediate consequence, we obtain the same simulation bounds for Hibbard's limited automata. We characterize LOG$k$SDA in terms of a new machine model, called logarithmic-space deterministic auxiliary depth-$k$ storage automata that run in polynomial time. These machine are also shown to be as powerful as a polynomial-time two-way multi-head deterministic depth-$k$ storage automata.

**Keywords:** Parallel computation · Deterministic context-free language · Logarithmic-space many-one reduction · LOGDCFL · SC · Depth-$k$ storage automata · Auxiliary storage automata · Multi-head storage automata

## 1 DCFL, LOGDCFL, and Beyond

In the literature, numerous computational models have been proposed to capture various aspects of *parallel computation*. Of those models, we wish to pay special attention to the model known as LOGDCFL, which is obtained from the family DCFL of all *deterministic context-free (dcf) languages* by taking the closure under logarithmic-space many-one reductions (or L-m-reductions,


C.-Y. Chen et al. (Eds.): COCOON 2021, LNCS 13025, pp. 164–175, 2021.
https://doi.org/10.1007/978-3-030-89543-3_14

for short) [2,12]. These dcf languages were first defined in 1966 by Ginsburg and Greibach [5] and their fundamental properties were studied extensively since then. Although dcf languages are accepted by *one-way deterministic pushdown automata* (or 1dpda's), these languages have a close connection to highly parallelized computation, and thus LOGDCFL has played a key role in discussing parallel complexity issues within P because of the nice inclusions $\mathrm{L} \subseteq \mathrm{LOGDCFL} \subseteq \mathrm{AC}^1 \cap \mathrm{SC}^2$.

It is known that LOGDCFL can be characterized without using L-m-reductions by several other intriguing machine models. Such a variety of characterizations prove LOGDCFL to be a robust and fully-applicable notion. A basis of LOGDCFL is of course 1dpda's, each of which is equipped with a read-once[1] input tape together with a storage device called a *stack*. Each stack allows two major operations. A pop operation is a deletion of a symbol and a push operation is an addition of extra symbols to the top of the stack. A rewriting of a topmost stack symbol is also allowed. The stack usage of pushdown storage seems too restrictive in practice and various extensions of such pushdown automata have been sought in the past literature. For instance, a *stack automaton* of Ginsburg, Greibach, and Harrison [6,7] is capable of freely traversing the inside of the stack to access each stored item but it is disallowed to modify them unless the scanning stack head eventually comes to the top of the stack. Thus, each cell of the stack could be accessed a number of times.

In real-life circumstances, it seems reasonable to limit the number of times to access data sets in the storage device. For instance, rewriting data items in blocks of a memory device, such as external hard drives or rewritable DVDs, is usually costly and it needs to be restricted during each execution of a computer program. Therefore, every memory cell on such a device must be permitted to be modified only during the first few accesses and, in case of exceeding the intended access limit, say, $k \geq 2$, the storage cell turns unusable and no more rewriting is possible. Such a storage device is referred to, in this exposition, as a *depth-k* storage tape and its scanning tape head is hereafter called a *depth-k storage-tape head* for convenience. The underlying machines are referred to as *deterministic depth-k storage automata* (or $k$-sda's, for short).

Our model of $k$-sda also expands the rewriting systems of Hibbard [9], known as *deterministic scan limited automata.*[2] Those rewriting system were lately remodeled in [10,11,13] as single input/storage-tape Turing machines that should modify the contents of tape cells whenever the associated tape heads access them; however, such modifications are limited to only the first $k$ accesses. A drawback of this model is that the use of a single tape prohibits us from accessing memory and input simultaneously.

We introduce the notation $k$SDA for each index $k \geq 2$ to express the family of all languages recognized by appropriately chosen $k$-sda's. It turns out that

[1] A read-only tape is called *read once* if, whenever it reads a tape symbol (except for $\varepsilon$-moves), it must move to the next unread cell.

[2] This claim comes from the fact that Hibbard's rewriting systems satisfy the so-called *blank-skipping property* [13], by which each tape cell becomes blank after the $k$ accesses and inner states cannot be changed while reading any blank symbol.

1SDA contains even non-context-free languages. With the use of L-m-reductions analogously to LOGDCFL, for any index $k \geq 2$, we consider the closure of $k$SDA under L-m-reductions, denoted LOG$k$SDA. It follows from the definition that LOGDCFL $\subseteq$ LOG$k$SDA $\subseteq$ LOG$(k+1)$SDA $\subseteq$ P. We raise two essential questions regarding this new language family LOG$k$SDA. (1) What is the computational complexity of language families $k$SDA as well as LOG$k$SDA? (2) Is there any natural machine model that can precisely characterize LOG$k$SDA in order to avoid the use of L-m-reductions? The sole purpose of this exposition is to answer these two questions.

All the omitted proofs and more elaborate explanations will appear in a forthcoming complete version of this exposition.

## 2 Storage Tapes and Storage Automata

The two notations $\mathbb{Z}$ and $\mathbb{N}$ represent the set of all *integers* and that of all *natural numbers* (i.e., nonnegative integers), respectively. Given two numbers $m, n \in \mathbb{Z}$ with $m \leq n$, $[m, n]_{\mathbb{Z}}$ denotes the *integer interval* $\{m, m+1, m+2, \ldots, n\}$. In particular, when $n \geq 1$, we abbreviate $[1, n]_{\mathbb{Z}}$ as $[n]$. Given a set $S$, $\mathcal{P}(S)$ denotes the *power set* of $S$, namely, the set of all subsets of $S$. An *alphabet* is a nonempty finite set of "symbols" or "letters". The *length* of a string $x$ is denoted by $|x|$. The special notation $\varepsilon$ is used to express the *empty string* of length 0.

Due to the page limit, we assume the reader's familiarity with multi-tape Turing machines and we abbreviate *deterministic Turing machines* as DTMs. An output tape is said to be *write-once* if its tape head never moves to the left and, whenever its tape head writes a nonempty symbol, it must move to the right. Given two languages $L_1$ over alphabet $\Sigma_1$ and $L_2$ over $\Sigma_2$, we say that $L_1$ is *L-m-reducible* to $L_2$ (denoted by $L_1 \leq_m^{\mathrm{L}} L_2$) if there exists a function $f$ computed by an appropriate polynomial-time DTM using only $O(\log n)$ space such that, for any $x \in \Sigma_1^*$, $x \in L_1$ iff $f(x) \in L_2$. Given an index $k \geq 1$, the $k$th Steve's class, $\mathrm{SC}^k$, is the family of all languages recognized by DTMs in polynomial time using $O(\log^k n)$ space.

We expand the standard model of pushdown automata by substituting its stack for a more flexible storage device, called a storage tape. A *storage tape* is a semi-infinite rewritable tape whose cells are initially blank (filled with distinguished initial symbols $\Box$) and are accessed sequentially by a storage-tape head that can move back and forth along the tape by changing tape symbols as it passes through.

Fix a constant $k \in \mathbb{N}^+$. A *(one-way) deterministic depth-k storage automaton* (or a $k$-sda, for short) $M$ is formally a 2-tape DTM of the form $(Q, \Sigma, \{\Gamma^{(e)}\}_{e \in [0,k]_{\mathbb{Z}}}, \{\vdash, \dashv\}, \delta, q_0, Q_{acc}, Q_{rej})$ with a finite set $Q$ of inner states, an input alphabet $\Sigma$, storage alphabets $\Gamma^{(e)}$ for indices $e \in [0,k]_{\mathbb{Z}}$ with $\Gamma = \bigcup_{e \in [0,k]_{\mathbb{Z}}} \Gamma^{(e)}$, a "deterministic" transition function $\delta$ from $(Q - Q_{halt}) \times \check{\Sigma}_{\varepsilon} \times \Gamma_{\varepsilon}$ to $\mathcal{P}(Q \times \Gamma_{\varepsilon} \times D_1 \times D_2)$ with $Q_{halt} = Q_{acc} \cup Q_{rej}$, $\check{\Sigma} = \Sigma \cup \{\vdash, \dashv\}$, $\check{\Sigma}_{\varepsilon} = \check{\Sigma} \cup \{\varepsilon\}$, $\Gamma_{\varepsilon} = \Gamma \cup \{\varepsilon\}$, $D_1 = \{0, +1\}$, and $D_2 = \{-1, 0, +1\}$, an initial state $q_0$ in $Q$, and sets $Q_{acc}$ and $Q_{rej}$ of accepting states and rejecting states, respectively, with

$Q_{acc} \cup Q_{rej} \subseteq Q$ and $Q_{acc} \cap Q_{rej} = \varnothing$, provided that $\Gamma^{(0)} = \{\Box\}$ (where $\Box$ is a distinguished symbol), $\Gamma^{(k)} = \{\vdash, B\}$ (where $B$ is a unique symbol indicating that a tape cell is completely erased and *frozen* forever) and $\Gamma^{(e_1)} \cap \Gamma^{(e_2)} = \varnothing$ for any distinct pair $e_1, e_2 \in [0,k]_{\mathbb{Z}}$. The two sets $D_1$ and $D_2$ indicate the direction of the input-tape head and that of the storage-tape head, respectively. A single move (or step) of $M$ is dictated by $\delta$. If $M$ is in inner state $q$, scanning $\sigma$ on the input tape and $\tau$ on the storage tape, a transition $\delta(q,\sigma,\tau) = (p,\xi,d_1,d_2)$ forces $M$ to change $q$ to $p$, overwrite $\tau$ by $\xi$, and move the input-tape head and the storage-tape head in directions $d_1$ and $d_2$, respectively.

Notice that a $k$-sda allows $\varepsilon$*-moves* (i.e., a tape head neither moves nor reads any tape symbol) on both the input tape and the storage tape so that the machine can continue working on one tape without scanning the content of the other tape. The tape head direction "0" indicates such an $\varepsilon$-move.

All tape cells are indexed by natural numbers from left to right. The leftmost tape cell is a *start cell* indexed 0. An input tape has endmarkers $\{\vdash, \dashv\}$ and a storage tape has only the left endmarker $\vdash$. When an input string $x$ is given to the input tape, it should be surrounded by the two endmarkers as $\vdash x \dashv$ so that $\vdash$ is located at the start cell and $\dashv$ is at the cell indexed $|x|+1$.

For any input string $x$ of length $n$ and any index $i \in [0, n+1]_{\mathbb{Z}}$, $x_{(i)}$ denotes the tape symbol written on the $i$th input-tape cell, provided that $x_{(0)} = \vdash$ (left endmarker) and $x_{(n+1)} = \dashv$ (right endmarker). Similarly, when $z$ represents the non-$\Box$ portion of the content of a storage tape, the notation $z_{(i)}$ expresses the symbol in the $i$th tape cell. Note that $z_{(0)} = \vdash$.

For the storage tape, we request the following rewriting restriction, called the *depth-k requirement*, to be satisfied. Whenever the storage-tape head passes through a tape cell containing a symbol in $\Gamma^{(e)}$ with $e < k$, the machine must replace it by another symbol in $\Gamma^{(e+1)}$ except for the case of the following "turns". We distinguish two types of turns. A *left turn at step t* refers to $M$'s step at which, after $M$'s tape head moves to the right at step $t-1$, it moves to the left at step $t$. Similarly, we say that $M$ makes a *right turn at step t* if $M$'s tape head moves from the left at step $t-1$ and changes its direction to the right at step $t$. Whenever a tape head makes a turn, we treat this case as "double accesses." More formally, at a turn, any symbol in $\Gamma^{(e)}$ with $e < k$ must be changed to another symbol in $\Gamma^{(\min\{k,e+2\})}$. No symbol in $\Gamma^{(k)}$ can be modified at any time. A storage tape that satisfies the depth-$k$ requirement is succinctly called a *depth-k storage tape*.

A *configuration* of $M$ on input $x$ is of the form $(x, q, l_1, l_2, z)$ with $q \in Q$, $l_1 \in [0, |x|+1]_{\mathbb{Z}}$, $l_2 \in \mathbb{N}$, and $z \in (\Gamma - \{\Box\})^*$, which indicates the situation where $M$ is in state $q$, the storage tape contains $z$ (except for the tape symbol $\Box$), and two tape heads scan the $l_1$th cell of the input tape containing $\vdash x \dashv$ and the $l_2$th cell of the storage tape. The *initial configuration* has the form $(x, q_0, \vdash, 0, 0)$ and $\delta$ describes how to reach the next configuration in a single step. For convenience, we define the *depth value* $dv(C)$ of a surface configuration $C = (q, l_1, l_2, z)$ to be the number $e$ satisfying $z_{(l_2)} \in \Gamma^{(e)}$. The $k$-sda $M$ *accepts* (resp., *rejects*) $x$ if $M$ starts with the initial configuration with the input $x$ and reaches an accepting configuration (resp., a rejecting configuration). Hereafter, we will pay attention only to $k$-sda's that always halt on any input. We say that

*M recognizes* a language $L$ over alphabet $\Sigma$ if, for any $x \in L$, $M$ accepts $x$ and, for any $x \in \Sigma^* - L$, $M$ rejects $x$. We write $k$SDA for the collection of all languages recognized by $k$-sda's and we set $\omega$SDA $= \bigcup_{k \in \mathbb{N}^+} k$SDA. Moreover, LOG$k$SDA and LOG$\omega$SDA denote the collections of all languages that are L-m-reducible to ceratin languages in $k$SDA and $\omega$SDA, respectively.

Recall that every tape cell of a $k$-sda is completely erased after the first $k$ accesses. If we allow such a tape cell to keep the last written symbol instead of erasing it, then the resulting machine gains enough power to solve even the circuit value problem (which is known to be P-complete). In this exposition, we do not further delve into this topic.

## 3 A Complexity Upper Bound of *k*SDA

A clear lower bound of the computational complexity of $k$SDA is DCFL. Hereafter, we discuss its non-trivial upper bound. Earlier, Cook [3] demonstrated that DCFL is included in SC$^2$. In the next theorem, we prove that $k$SDA is included in SC$^k$ for any index $k \geq 2$. Since DCFL $\subseteq k$SDA, our result significantly extends Cook's old result. Let SC $= \bigcup_{k \geq 1}$ SC$^k$.

**Theorem 1.** *For any integer $k \geq 2$, $k$SDA $\subseteq$ SC$^k$. Thus, $\omega$SDA $\subseteq$ SC holds.*

Since SC$^k$ is closed under L-m-reductions, Theorem 1 instantly implies that LOG$k$SDA $\subseteq$ SC$^k$ and LOG$\omega$SDA $\subseteq$ SC. However, whether the inclusion SC$^{k-1}$ $\subseteq$ LOG$k$SDA holds or not is unknown at this moment.

As another immediate consequence of Theorem 1, we obtain a complexity upper bound of Hibbard's *deterministic k-limited automata* ($k$-lda's, for short) because $k$-lda's (satisfying the blank-skipping property [13]) can be easily simulated by $k$-sda's simply by pretending that all input symbols are written on a storage tape. Notice that, in the past literature, no upper bound except for CFL has been shown for Hibbard's language families [9].

**Corollary 2.** *For any $k \geq 2$, all languages recognized by Hibbard's $k$-lda's are in* SC$^k$.

To prove Theorem 1, we attempt to employ a *divide-and-conquer argument*, which is based on [1] but expanded significantly to cope with more complex moves of $k$-sda's. Due to the page limit, we expect that the reader is familiar with [1]. Fix $k \geq 2$ and let $M = (Q, \Sigma, \{\Gamma^{(e)}\}_{e \in [0,k]_{\mathbb{Z}}}, \vdash, \dashv, \delta, q_0, Q_{acc}, Q_{rej})$ denote any $k$-sda. We further fix an arbitrary input $x \in \Sigma^*$ and set $n = |x|$.

To simulate the behavior of $M$ on $x$, we introduce an important notion of "marker". A *marker* $C$ is a sextuple $(q, l_1, l_2, \sigma, r, t)$ that indicates the following circumstance: at time $t$ with section time $r$ (which will be explained later), $M$ is in inner state $q$, its input-tape head is located at cell $l_1$, the storage-tape head is at cell $l_2$ containing symbol $\sigma$. To express each entry of $C$, we set *state*$(C) = q$, *in-loc*$(C) = l_1$, *st-loc*$(C) = l_2$, *symb*$(C) = \sigma$, *sectime*$(C) = r$, and *time*$(C) = t$. Let $\mathcal{M}_x$ denote the set of all markers of $M$ on $x$. To emphasize the value "$l_2$",

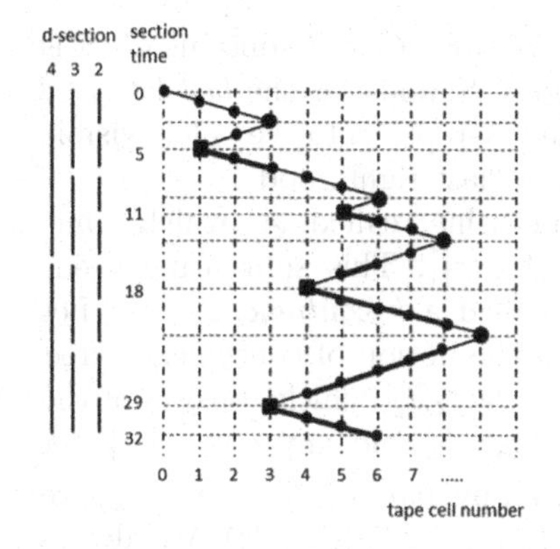


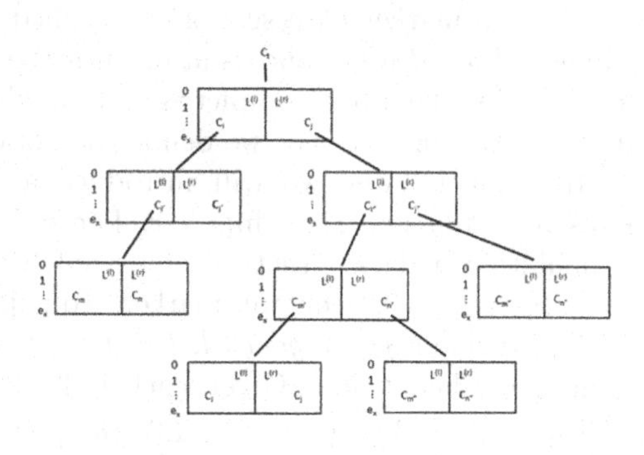

**Fig. 1.** [left] A history of consecutive moves of a storage-tape head for $k = 3$. In the leftmost vertical lines indicate $d$-sections for $d = 2, 3, 4$. The other vertical lines indicate the tape cell numbers from 0 to 10 and the horizontal lines show section time from 0 to 32. All storage-$\varepsilon$-moves are suppressed into filled circles and boxes. [right] A contingency tree, in which each node (except for the root) is a contingency list.

we occasionally call $C$ an $l_2$-*marker* if $st\text{-}loc(C) = l_2$. Similarly, we call $C$ a $t$-*marker* if $time(C) = t$.

A computation of a $k$-sda is characterized by a series of markers in the following way. Assume that $C_t$ has the form $(q, l_1, l_2, \sigma, r, t)$ and $\delta$ has a transition of the from $\delta(q, x_{(l_1)}, \sigma) = (p, \tau, d_1, d_2)$. We then set $next\text{-}state(C_t) = p$, $next\text{-}loc(C_t) = l_2 + d_2$, and $next\text{-}symb(C_t) = \tau$. To calculate the next marker $C_{t+1}$ (at time $t+1$) from $C_t$, we need the information on the most recent $(l_2+d_2)$-marker $C'$ with $time(C') < t$. Once $C'$ is obtained with $\xi' = next\text{-}symb(C')$, the sextuple $(p, l_1 + d_1, l_2 + d_2, \xi', r + |d_2|, t + 1)$ becomes the desired marker $C_{t+1}$. We remark that, when $\xi'$ is known, we do not need to use $C'$ to compute $C_{t+1}$.

Next, we introduce critical notions. Let $C = (q, l_1, l_2, \sigma, r, t)$ and $C' = (q', l'_1, l'_2, \sigma', r', t')$ be two arbitrary markers of $M$ on $x$. We say that $C'$ is *left-visible* from $C$ if (i) $C'$ is a marker with $l'_2 < l_2$ and $t' < t$ and (ii) there is no marker $\tilde{C}$ satisfying both $st\text{-}loc(\tilde{C}) \leq l'_2$ and $t' + 1 \leq time(\tilde{C}) < t$. Moreover, $C'$ is the *left-cut* of $C$ if $C'$ is left-visible from $C$ and $t'$ is the largest number with $t' < t$ (i.e., $C'$ is the most recent left-visible marker). In symmetry, we can define the notions of *right-visibility* and *right-cut*.

Recall that the storage-tape head may make $\varepsilon$-moves. To distinguish those from any other $\varepsilon$-move of the input-tape head, we call the former *storage-$\varepsilon$-moves*. A 0-*section* consists of either (a) a right/left turn (i.e., leftmost/rightmost point) or (b) a non-storage-$\varepsilon$-move to the right/left followed by a (possibly empty) series of consecutive storage-$\varepsilon$-moves. For any $d \geq 0$, a $(d+1)$-*section* is the union of two consecutive $d$-sections. The *section time* of $C$ is the total number of 0-sections before $C$ (thus, not including the 0-section containing $C$). The leftmost (resp., rightmost) marker in each $d$-section $S$ is called the *left-representative* (resp., right-representative) of $S$. Given a set of markers, $C$ is said to be the *leftmost marker* (resp., the rightmost marker) if $st\text{-}loc(C)$ is the smallest (resp., the largest) among the markers in the given set.

Given a marker $C$, a section $S$ is called *current* for $C$ if $S$ contains $C$, $S$ is *completed* for $C$ if $C$ appears at or after the end of $S$, and $S$ is the *last-left-good* for $C$ if $S$ is the latest completed section whose left-representative is left-visible from $C$. In a similar way, we define the notion of "last-right-good".

In what follows, we will introduce a subroutine, called $\mathcal{A}$, which simulates one step of $M$ on input $x$. Let $e_x = \lceil \log |x| \rceil$. This subroutine needs to maintain a data structure of nested lists, called a *k-contingency tree.* Let $C_t = (q, l_1, l_2, \sigma, r, t)$ be any marker. To explain the notion of contingency tree, we first define a *contingency list at time t*, $L(e_1, e_2, l_2, t)$, which consists of the items described below. Given each index $d \in [0, e_x]_{\mathbb{Z}}$ and any $a \in \{l, r\}$, let $L^{(a)}(d, l_2, t) = L^{(a)}_{last}(d, l_2, t) \cup L^{(a)}_{cur}(d, l_2, t)$. For any pair $e_1, e_2 \in [0, e_x]_{\mathbb{Z}}$, we set $L(e_1, e_2, l_2, t) = (\bigcup_{d\in[0,e_1]_{\mathbb{Z}}} L^{(l)}(d, l_2, t)) \cup (\bigcup_{d\in[0,e_2]_{\mathbb{Z}}} L^{(r)}(d, l_2, t))$. We always assume that all markers in $L(e_1, e_2, l_2, t)$ are enumerated according to their time.

(a) $L^{(l)}_{last}(d, l_2, t)$, which consists of all markers $C$ satisfying the following requirement: there exist a $d$-section $S$ and a $(d+1)$-section $S'$ such that (i) $S$ is enclosed in $S'$, (ii) $S'$ is the last-left-good section for $C_t$, and (iii) $C$ is the left-representative of $S$.
(b) $L^{(l)}_{cur}(d, l_2, t)$, which consists of all markers $C$ satisfying the following requirement: there exist a $d$-section $S$ and a $(d+1)$-section $S'$ such that (i) $S$ is enclosed in $S'$, (ii) $S'$ is current for $C_t$, (iii) $S$ is left-visible from $C_t$, and (iv) $C$ is the left-representative of $S$.
(c) Two more items, $L^{(r)}_{cur}(\cdot)$ and $L^{(r)}_{last}(\cdot)$, are defined similarly using right-visibility and right-representatives.

Notice that the left-cut (resp., the right-cut) for $C_t$ is the latest marker in $L^{(l)}(0, l_2, t)$ (resp., $L^{(r)}(0, , l_2, t)$). The contingency list $L(e_1, e_2, l_2, t)$ is said to be *linked to* $C_t$. A 2*-contingency tree at time t* is a two-node tree whose root is $C_t$ and its only child is a contingency list linked to $C_t$. A $(k+1)$*-contingency tree at time t* is a leveled, rooted tree whose top 2 levels form a 1-contingency tree at time $t$ and each $t'$-marker (except for $C_t$) appearing in this 2-contingency tree is a root of another $k$-contingency tree at time $t'$ with $t' < t$. See Fig. 1. In particular, $C_0$ links itself to a unique empty contingency list. To describe a $k$-contingency tree at time $t$, we use the notation $\mathcal{D}(k, e_1, e_2, l_2, t)$. The *principal contingency list* refers to the node attached directly to the root $C_t$.

Subroutine $\mathcal{A}$ takes an input of the form $(C_t, dv, l_2', e_1, e_2, , \mathcal{D}(dv, e, l_2, t))$ with $C_t = (q, l_1, l_2, \sigma, r, t) \in \mathcal{M}_t$, $l_2' = next\text{-}loc(C_t)$, $dv \in [0, k]_{\mathbb{Z}}$, and $e_1, e_2 \in [0, e_x]_{\mathbb{Z}}$. If $dv \geq dv(C_{t+1})$, then $\mathcal{A}$ returns $(C_{t+1}, \mathcal{D}(dv, e_1, e_2, l_2', t+1))$ with the $(t+1)$-marker $C_{t+1} = (q', l_1', l_2', \sigma', r', t+1)$ by making a series of recursive calls to itself. Otherwise, it may return anything.

**[Subroutine $\mathcal{A}$]**

(1) In the case where the storage-tape head stays still (i.e., makes a storage-$\varepsilon$-move), since $l_2' = l_2$, we define $r' = r$. Since we do not need to alter the contingency tree, in the principal contingency list, we automatically set $L^{(l)}(d, l_2', t+1) = L^{(l)}(d, l_2, t)$ and $L^{(r)}(d', l_2', t+1) = L^{(r)}(d', l_2, t)$ for any

$d \in [0, e_1]_{\mathbb{Z}}$ and $d' \in [0, e_2]_{\mathbb{Z}}$. We can compute $C_{t+1}$ from $C_t$ directly by applying $\delta$.

(2) We skip the case where the storage-tape head is moving to the right, that is, $l'_2 = l_2 + 1$.

(3) In the case where the storage-tape head is moving to the left, since $l'_2 = l_2 - 1$, we define $r' = r + 1$. We need to compute three sets: $L^{(l)}(\cdot)$ at location $< l'_2$, $L^{(r)}(\cdot)$ at location $> l'_2$, and $C_{t+1}$. Here, we focus only on $L^{(l)}(\cdot)$ and $C_{t+1}$. Consider the principal contingency list of $\mathcal{D}(dv, e, l_2, t)$. Define $d_0$ to be the maximum number in $[0, e_1]_{\mathbb{Z}}$ satisfying $r' \equiv 0 \pmod{2^{d_0}}$ and let $d_1$ denote the minimal number in $[0, e_1]_{\mathbb{Z}}$ satisfying $|L^{(l)}(d_1, l_2, t)| \geq 2$. If there is no such $d_1$, then we set $d_1 = e_1$. Initially, for each $d \in [0, d_1]_{\mathbb{Z}}$, we reset $L^{(l)}(d, l_2, t)$ to be $L^{(l)}(d, l_2, t) - \{C'\}$, where $C'$ is the latest marker in $L^{(l)}(d, l_2, t)$. (i) If $d_1 = 0$, then we define $L^{(l)}(d, l'_2, t+1) = L^{(l)}(d, l_2, t)$ for any $d \in [0, e_1]_{\mathbb{Z}}$. (ii) Assume otherwise. If there is no right-cut $C''$ for $C_t$ stored in $\mathcal{D}(dv, e_1, e_2, l_2, t)$, then we choose the latest marker $C'$ from $L^{(l)}(d_1, l_2, t)$. Let $l' = \textit{st-loc}(C')$ and $t' = \textit{time}(C')$. Since $C'$ is the left-representative of a completed $d_1$-section, we need to compute the new left-cut $C''$ for $C_{t+1}$, which appears in certain new last-left-good $d'$-sections with $d' < d_1$, where $\textit{st-loc}(C'') = l'_2$ and $t' < \textit{time}(C'') < t$. Starting from $C'$ with $\mathcal{D}(k, d_1 - 1, e_2, l', t')$, we inductively generate a pair of $\tilde{C}$ and $\mathcal{D}(k, d_1 - 1, e_2, \tilde{l}, \tilde{t})$ with $\tilde{t} = \textit{time}(\tilde{C})$ and $\tilde{l} = \textit{st-loc}(\tilde{C})$, and run $\mathcal{A}(\tilde{C}, k, \tilde{l}, d_1 - 1, e_2, \mathcal{D}(k, d_1 - 1, e_2, \tilde{l}, \tilde{t}))$ to obtain the next marker $\tilde{C}'$ and $\mathcal{D}(k, d_1 - 1, e_2, \tilde{l}', \tilde{t} + 1)$ with $\tilde{l}' = \textit{st-loc}(\tilde{C}')$ until we reach $C''$, except that the final recursive call is of the form $\mathcal{A}(\tilde{C}, dv - 1, l'_2, d_1 - 1, e_2, \mathcal{D}(dv - 1, d_1 - 1, e_2, l'_2 - 1, \tilde{t}))$ with $l'_2 - 1 = \textit{st-loc}(\tilde{C})$. Finally, we define $B_0 = L^{(l)}_{cur}(0, l'_2, t') \cup \{C''\}$ and $B_{i+1} = L^{(l)}_{cur}(i, l'_2, t') \cup \{\bar{C}_i\}$ for any $i \in [0, d_1 - 1]_{\mathbb{Z}}$, where $\bar{C}_i$ is the oldest marker in $B_i$. We then define $L^{(l)}_{last}(i, l'_2, t+1) = B_i$ and $L^{(l)}_{cur}(i, l'_2, t+1) = \varnothing$ for any $i \in [0, d_1 - 1]_{\mathbb{Z}}$, and $L^{(l)}_{cur}(d_1, l'_2, t+1) = \varnothing$ and $L^{(l)}_{last}(d_1, l'_2, t+1) = L^{(l)}_{last}(d_1, l_2, t) \cup B_{d_1}$. In the case of $d_0 > d_1$, we define $E_0 = \varnothing$ and $E_{i+1} = L^{(l)}_{cur}(i, l_2, t) \cup \{\bar{C}\}$ for every $i \in [d_1 + 1, d_0 - 1]_{\mathbb{Z}}$, where $\bar{C}$ is the oldest marker in $E_i$. We then define $L^{(l)}_{last}(i, l'_2, t+1) = E_i$ and $L^{(l)}_{cur}(i, l'_2, t+1) = \varnothing$ for any $i \in [d_1 + 1, d_0 - 1]_{\mathbb{Z}}$, and $L^{(l)}_{cur}(d_0, l'_2, t+1) = E_{d_0}$. We update all the others without changing the content of the old contingency lists. For all the remaining "nodes" in $\mathcal{D}(dv, e_1, e_2, l_2, t)$, we also update them according to the above changes.

To compute marker $C_{t+1}$, we first assume that $dv \geq 1$. If there is a right-cut $C''$ for $C_t$ at time $< t$ either stored inside $\mathcal{D}(dv, e, l_2, t)$ or obtained by the above procedure, then $C_{t+1}$ is directly calculated from $C_t$ and $\textit{symb}(C'')$ by applying $\delta$. Otherwise, if there is a left-representative for $C_t$ at time $t_0 < t$ and no left turn exists at location $\geq l'_2$ at time $< t_0$, then we start with the left-representative $C_{rep}$ and compute a series of markers $\tilde{C}$ one by one (by incrementing time) from $C_{rep}$ using $\Box$ as inputs *with no recursive call* until we obtain the $l'_2$-marker $C'$. We then compute $C_{t+1}$ from $C_t$ and $\textit{symb}(C')$ by applying $\delta$. In the case of $dv \leq 0$, by contrast, the cell $l'_2$ must be already blank $B$. Let $\sigma' = B$ and computer $C_{t+1}$ from $C_t$ alone by applying $\delta$.

**Lemma 3.** *Let $C_t$ and $C_{t+1}$ be two markers.*

1. *The total number of markers inside $\mathcal{D}(k, e_1, e_2, l_2, t)$ is $O(\log^{k-2} n)$.*
2. *If $dv \geq dv(C_{t+1})$, then Subroutine $\mathcal{A}(C_t, dv, l_2, e_1, e_2, \mathcal{D}(dv, e_1, e_2, l_2, t))$ correctly returns $(C_{t+1}, \mathcal{D}(dv, e_1, e_2, l_2', t+1))$ and the depth of recursion is $O(\log n)$.*

**Proof of Theorem 1.** Fix $k \geq 2$. Our goal is to prove that $k$SDA $\subseteq$ SC$^k$. Take an arbitrary language $L$ in $k$SDA and a $k$-sda $M$ that recognizes $L$. Consider the following procedure that simulates $M$ on input $x$ step by step. Initially, we prepare the marker $C_0 = (q_0, 0, 0, \vdash, 0)$ and the unique $k$-contingency list $\mathcal{D}(k, e_x, e_x, 0, 0)$ linked to $C_0$. Inductively, assuming that $C_t$ and $\mathcal{D}(k, e_x, e_x, l_2, t)$ have been already obtained, we run Subroutine $\mathcal{A}(C_t, k, e_x, e_x, l_2', \mathcal{D}(k, e_x, e_x, l_2, t))$ to compute $(C_{t+1}, \mathcal{D}(k, e_x, e_x, l_2', t+1))$ until $M$ enters a halting state (i.e., either an accepting or a rejecting states). We then decide whether $x$ is in $L$ by checking whether $M$ enters an accepting state. To store a contingency tree requires $O(\log^{k-1} n)$ bits by Lemma 3(1) since each contingency list in the tree needs $O(\log n)$ bits to express. Lemma 3(2) then concludes that we need only polynomial runtime and $O(\log^k n)$ memory bits to perform the entire simulation procedure. Therefore, $L$ belongs to SC$^k$.

## 4 Two Machine Models that Characterize LOG$k$SDA

In Sect. 2, LOG$k$SDA is defined to be the closure of $k$SDA under L-m-reductions. To remove the use of L-m-reductions from this definition, we wish to expand Cook's notion of deterministic auxiliary pushdown automata to *deterministic auxiliary depth-$k$ storage automata* (or aux-$k$-sda's, for short), each of which is equipped with a two-way read-only input tape, an auxiliary rewritable work tape, and a storage tape whose cells are rewritten only during the first $k$ accesses and then turns blank forever after $k$ accesses. We further introduce another machine model, called *(two-way) $\ell$-head deterministic depth-$k$ storage automata* (or $k$-sda$_2(\ell)$, for short), each of which is allowed to use $\ell$ two-way tape heads to access a single input tape.

### 4.1 Deterministic Auxiliary Depth-$k$ Storage Automata

To understand LOG$k$SDA better, we want to seek other characterizations of it *with no use of* L*-m-reduction*. For this purpose, we intend to expand Cook's auxiliary pushdown automata. For the description of the desired machine model, firstly we prepare a two-way read-only input tape and a depth-$k$ storage tape and secondly we supply a new space-bounded auxiliary rewritable work tape whose cells are freely modified by a two-way tape head. Notice that the storage-tape head is allowed to make $\varepsilon$-moves. A *deterministic auxiliary depth-$k$ storage automaton* (or an aux-$k$-sda, for short) $M$ is formally a 3-tape DTM $(Q, \Sigma, \Theta, \{\Gamma^{(e)}\}_{e\in[0,k]_{\mathbb{Z}}}, \vdash, \dashv, \delta, q_0, Q_{acc}, Q_{rej})$ with a read-only input tape, an auxiliary rewritable work tape with an alphabet $\Theta$, and a depth-$k$ storage tape. Initially, the input tape is filled with $\vdash x \dashv$, the auxiliary tape is blank, and the depth-$k$ storage tape has only designated blank symbols $\Box$ except for the left endmarker $\vdash$. The "deterministic" transition function $\delta$ maps $(Q - Q_{halt}) \times \check{\Sigma}_{\varepsilon} \times \Theta \times \Gamma_{\varepsilon}$ to $\mathcal{P}(Q \times \Theta \times \Gamma_{\varepsilon} \times D_1 \times D_2 \times D_3)$, where $D_1 = D_2 = D_3 = \{-1, 0, +1\}$. A transition $\delta(q, \sigma, \tau, \gamma) = (p, \theta, \xi, d_1, d_2, d_3)$ means that, on reading input symbol $\sigma$, $M$ changes inner state $q$ to $p$ by moving its tape head in direction $d_1$, changes auxiliary tape symbol $\tau$ to $\theta$ by moving its tape head in direction $d_2$, and changes storage tape symbol $\gamma$ to $\xi$ by moving its tape head in direction $d_3$. A string $x$ is *accepted* (resp., *rejected*) if $M$ enters an inner state in $Q_{acc}$ (resp., $Q_{rej}$). By excluding $(\Theta, D_3)$ from the definition of $M$, the resulting automaton must fulfill the depth-$k$ requirement of $k$-sda's.

### 4.2 Multi-head Deterministic Depth-*k* Storage Automata

We further argue another characterization of LOG$k$SDA using two-way multi-head machines. For each fixed number $\ell \geq 1$, we define an *$\ell$-head deterministic depth-$k$ storage automaton* as a 2-tape DTM with $\ell$ two-way read-only tape heads scanning over an input tape and a single read/write tape head over a depth-$k$ storage tape. For convenience, we call such a machine by a $k$-sda$_2(\ell)$, where the subscript "2" emphasizes that all input-tape heads move in both directions (except for $\varepsilon$-moves). Notice that each $k$-sda$_2(\ell)$ has actually $\ell + 1$ tape heads, including one tape head moving along the storage tape. For convenience, we call such a unique tape head the $(\ell + 1)$th *tape head*. More formally, a $k$-sda$_2(\ell)$ is a tuple $(Q, \Sigma, \{\Gamma^{(e)}\}_{e\in[0,k]_{\mathbb{Z}}}, \vdash, \dashv, \delta, q_0, Q_{acc}, Q_{rej})$ with a "deterministic" transition function $\delta$ mapping $(Q - Q_{halt}) \times \check{\Sigma}_{\varepsilon}^{\ell} \times \Gamma_{\varepsilon}$ to $\mathcal{P}(Q \times \Gamma_{\varepsilon} \times D^{\ell} \times D)$, where $D = \{-1, 0, +1\}$. A transition $\delta(q, \sigma_1, \ldots, \sigma_{\ell}, \gamma) = (p, \xi, d_1, \ldots, d_{\ell}, d_{\ell+1})$ means that $M$ is at present in state $q$, scanning $(\sigma_1, \ldots, \sigma_{\ell})$ on the input tape by the $\ell$ read-only tape heads and $\gamma$ on the rewritable depth-$k$ storage tape, and then $M$ enters state $p$ and writes $\xi$ on the depth-$k$ storage tape by moving the $i$th tape head in direction $d_i$ for every index $i \in [\ell + 1]$. The acceptance/rejection criteria is the same as $k$-sda's. A read-only tape head is called *sweeping* if it changes its direction (ignoring $\varepsilon$-moves) only at the two endmarkers.

### 4.3 Characterizations of LOG*k*SDA

We demonstrate that the two new machine models introduced in Sects. 4.1 and 4.2 precisely characterize LOG$k$SDA. This result naturally extends Sudborough's machine characterizations of LOGDCFL to LOG$k$SDA.

**Theorem 4.** *Let $k \geq 2$. Let $L$ be any language. The following three statements are logically equivalent.*

1. *$L$ is in* LOG$k$SDA.
2. *There exists an aux-$k$-sda that recognizes $L$ in polynomial time using logarithmic space.*

*3. There exist a number $\ell \geq 2$ and a k-sda$_2(\ell)$ that recognizes L in polynomial time.*

Sudborough's characterization of LOGDCFL [12] heavily relies on a simulation procedure [8, pp. 338–339] of Hartmanis and a proof argument [4, Lemma 4.3] of Galil; however, we cannot directly use them. This is because Sudborough's proof [12, Lemmas 3–6] is based on the use of pushdown-automata's stack operations, which are applied only to the topmost symbol of the stack but the other symbols in the stack are intact. In our case, on the contrary, we need to deal with the operations on a depth-$k$ storage tape whose head can move back and forth along the storage tape by modifying each cell's content during the first $k$ accesses. Thus, a new idea is definitely needed to establish Theorem 4. The proof of the theorem therefore requires a technically challenging simulation among three different computational models.

We begin with the following easy lemma.

**Lemma 5.** *For each fixed constant $s \in \mathbb{N}^+$, there exists a two-way 3-head deterministic finite automaton such that all input-tape heads are sweeping with making no $\varepsilon$-move and the automaton, on input of the form $(axb)^{|axb|}$ with $|x| > s$, moves one of the tape heads to cell $|x|^s$ in $O(|x|^s)$ steps, where a and b are designated tape symbols and x contains neither a nor b.*

For convenience, we call by the *3 marking heads* the 3 tape heads guaranteed to exist by Lemma 5. We then transform any given aux-$k$-sda to an equivalent $k$-sda$_2(5c+2)$ for a certain constant $c > 0$.

**Lemma 6.** *Let $k \geq 2$. Given a polynomial-time, log-space aux-k-sda M, there are a constant $c > 0$ and a k-sda$_2(5c+2)$ simulating M in polynomial time.*

To reduce the number of input-tape heads from $2\ell + 3$ to $\ell + 3$, we need to record the movement of multiple input-tape heads onto a depth-$k$ storage tape. For this purpose, we use the 3 marking heads of Lemma 5 to make enough blank space on the depth-$k$ storage tape.

**Lemma 7.** *Let $k \geq 2$, $s \geq 1$, and $\ell \geq 1$. Given a language L over alphabet $\Sigma$ and any k-sda$_2(2\ell+3)$ with the 3 marking heads recognizing L within $|x|^s$ steps (where x is an input), there exists a polynomial-time k-sda$_2(\ell+3)$ with the 3 marking heads that recognizes $L_{a,b} = \{(axb)^{|axb|} \mid x \in L\}$, where a and b are tape symbols not in $\Sigma$.*

Sudborough's proof also utilizes Galil's argument [4], which uses a stack for storing and removing specific symbols to remember the distance of a tape head from a particular input tape cell. However, since tape cells on a storage tape are not allowed to modify more than $k$ times, we need to develop a different strategy to prove Lemma 7. The last important lemma is stated below.

**Lemma 8.** *Let $s \in \mathbb{N}^+$. For any polynomial-time k-sda$_2(4)$ M with 3 marking heads running within $|x|^s$ steps, the language $L_{rev} = \{(\cent\tilde{x}\#\tilde{x}^R)^{|x|^s} \mid x \in L(M)\}$ is recognized by an appropriate polynomial-time k-sda, where $\cent$ and $\#$ are special separators and $\tilde{x} = 1^{|x|^s}x_1 1^{|x|^s}x_2 \cdots 1^{|x|^s}x_n$ for $x = x_1x_2\cdots x_n$.*

**Proof of Theorem 4.** The implication (1)⇒(2) is relatively easy. Lemma 6 obviously implies (2)⇒(3). Finally, we want to show that (3) implies (1). Given a language $L$, we assume that there is a polynomial-time $k$-sda$_2(\ell)$ $M$ recognizing $L$ for a certain number $\ell \geq 2$. Our goal is to define a polynomial-time, log-space computable function $f$ and a new $k$-sda $K$ such that $f$ reduces $L(M)$ to $L(K)$. This concludes that $L$ belongs to LOG$k$SDA. Take the smallest integer $\ell'$ satisfying $\ell \leq 2\ell' + 3$. We repeatedly apply Lemma 7 by adjusting the value of $\ell'$ (e.g., by setting $\ell' = 3$, we first reduce $2\ell' + 3 = 9$ to $\ell' + 3 = 6$ and, by resetting $\ell' = 2$, we then reduce $2\ell' + 3 = 7$ to $\ell' + 3 = 5$). Eventually, we obtain a polynomial-time $k$-sda$_2(4)$ $N$ with the 3 marking heads that can simulate $M$. By Lemma 8, there exists a $k$-sda $K$ that correctly recognizes the language $L_{rev}$. Next, we define $f(x) = (¢\tilde{x}\#\tilde{x}^R)^{|x|^s}$ for any $x$. We then obtain $L_{rev} = \{f(x) \mid x \in L(M)\}$. Since $L_{rev} = L(K)$, it follows that $x \in L(M)$ iff $f(x) \in L(K)$. Therefore, $L(M)$ is L-m-reducible to $L(K)$. □

## References


1. von Braunmühl, B., Cook, S., Mehlhorn, K., Verbeek, R.: The recognition of determinsitic CFLs in small time and space. Inf. Control **56**, 34–51 (1983)
2. Cook, S.A.: Characterizations of pushdown machines in terms of time-bounded computers. J. ACM **18**, 4–18 (1971)
3. Cook, S.A.: Determinsitic CFLs are accepted simultaneously in polynomial time and log squared space. In: Proceedings of STOC 1979, pp. 338–345 (1979)
4. Galil, Z.: Some open problems in the theory of computation as questions about two-way determinsitic pushdown automaton languages. Math. Syst. Theory **10**, 211–228 (1977)
5. Ginsburg, S., Greibach, S.: Deterministic context free languages. Inf. Control **9**, 620–648 (1966)
6. Ginsburg, S., Greibach, S.A., Harrison, M.A.: One-way stack automata. J. ACM **14**, 389–418 (1967)
7. Ginsburg, S., Greibach, S.A., Harrison, M.A.: Stack automata and compiling. J. ACM **14**, 172–201 (1967)
8. Hartmanis, J.: On non-determinacy in simple computing devices. Acta Inf. **1**, 336–344 (1972)
9. Hibbard, T.N.: A generalization of context-free determinism. Inf. Control **11**, 196–238 (1967)
10. Pighizzini, G., Pisoni, A.: Limited automata and regular languages. Int. J. Found. Comput. Sci. **25**, 897–916 (2014)
11. Pighizzini, G., Pisoni, A.: Limited automata and context-free languages. Fund. Inf. **136**, 157–176 (2015)
12. Sudborough, I.H.: On the tape complexity of deterministic context-free languages. J. ACM **25**, 405–414 (1978)
13. Yamakami, T.: Behavioral strengths and weaknesses of various models of limited automata. In: Catania, B., Královič, R., Nawrocki, J., Pighizzini, G. (eds.) SOFSEM 2019. LNCS, vol. 11376, pp. 519–530. Springer, Cham (2019). https://doi.org/10.1007/978-3-030-10801-4_40

# Ideal Separation and General Theorems for Constrained Synchronization and Their Application to Small Constraint Automata

Stefan Hoffmann(✉)

Informatikwissenschaften, FB IV, Universität Trier,
Universitätsring 15, 54296 Trier, Germany
hoffmanns@informatik.uni-trier.de

**Abstract.** In the constrained synchronization problem we ask if a given automaton admits a synchronizing word coming from a fixed regular constraint language. We show that intersecting a given constraint language with an ideal language does not increase the computational complexity. Additionally, we state a theorem giving PSPACE-hardness that broadly generalizes previously used constructions and a result on how to combine languages by concatenation to get polynomial time solvable constrained synchronization problems. We use these results to give a classification of the complexity landscape for small constraint automata of up to three states.

**Keywords:** Synchronization · Computational complexity · Automata theory · Finite automata · Constrained synchronization

## 1 Introduction

A deterministic semi-automaton is synchronizing if it admits a reset word, i.e., a word which leads to a definite state, regardless of the starting state. This notion has a wide range of applications, from software testing, circuit synthesis, communication engineering and the like, see [14,15]. The famous Černý conjecture [2] states that a minimal length synchronizing word, for an $n$-state automaton, has length at most $(n-1)^2$. We refer to the mentioned survey articles [14,15] for details.

Due to its importance, the notion of synchronization has undergone a range of generalizations and variations for other automata models. In some generalizations, related to partial automata [11], only certain paths, or input words, are allowed (namely those for which the input automaton is defined).

In [7] the notion of constrained synchronization was introduced in connection with a reduction procedure for synchronizing automata. The paper [5] introduced the computational problem of constrained synchronization. In this problem, we search for a synchronizing word coming from a specific subset of allowed input


C.-Y. Chen et al. (Eds.): COCOON 2021, LNCS 13025, pp. 176–188, 2021.
https://doi.org/10.1007/978-3-030-89543-3_15

sequences. For further motivation and applications we refer to the aforementioned paper [5]. In this paper, a complete analysis of the complexity landscape when the constraint language is given by small partial automata with up to two states and an at most ternary alphabet was done. It is natural to extend this result to other language classes, or even to give a complete classification of all the complexity classes that could arise. For commutative regular constraint languages, a full classification of the realizable complexities was given in [8]. In [9], it was shown that for polycyclic constraint languages, the problem is always in NP.

Let us mention that restricting the solution space by a regular language has also been applied in other areas, for example to topological sorting [1], solving word equations [3,4], constraint programming [12], or shortest path problems [13]. The road coloring problem asks for a labelling of a given graph such that a synchronizing automaton results. A closely related problem to our problem of constrained synchronization is to restrict the possible labeling(s), and this problem was investigated in [16].

**Contribution and Motivation:** In [5] a complete classification of the computational complexity for partial constraint automata with up to two states and an at most ternary alphabet was given. Additionally, an example of a a three-state automaton over a binary alphabet realizing an NP-complete constrained synchronization problem and a three-state automaton over a binary alphabet admitting a PSPACE-complete problem were given. The question was asked, if, and for what constraint automata, other complexity classes might arise. Here, we extend the classification by extending the two-state case to arbitrary alphabets and giving a complete classification for three-state automata. It turned out that only PSPACE-complete, or NP-complete, or polynomial time solvable constrained problems arise. In [5], the analysis for the small constraint automata were mainly carried out by case analysis. As for larger alphabets and automata this quickly becomes tedious, here we use, and present, new results to lift, extend and combine known results. Among these are three main theorems, which, when combined, allow many cases to be handled in an almost mechanical manner. More specifically, the motivation and application of these theorems is the following.

1. The *$UV^*W$-Theorem* describes how to combine languages with concatenation to get polynomial time solvable constrained problems.
2. The *uC-Theorem* gives a general condition on the form of a constraint language to yield a PSPACE-complete constrained synchronization problem.
3. The *Ideal Separation Theorem.* In general, if the constraint language could be written as the union of two languages, and for one of them the constrained problem is hard, we cannot deduce hardness for the original languages. However, under certain circumstances, namely if the hard language is contained in a unique regular ideal language, we can infer hardness for the original languages.

We apply these results to small constraint automata of up to three states.

## 2 General Notions and Definitions

By $\Sigma$ we will always denote a *finite alphabet*, i.e., a finite set of *symbols*, or *letters*. A *word* is an element of the free monoid $\Sigma^*$, i.e., the set of all finite sequences with concatenation as operation. For $u, v \in \Sigma^*$, we will denote their concatenation by $u \cdot v$, but often we will omit the concatenation symbol and simply write $uv$. The subsets of $\Sigma^*$ are also called *languages*. By $\Sigma^+$ we denote the set of all words of non-zero length. We write $\varepsilon$ for the empty word, and for $w \in \Sigma^*$ we denote by $|w|$ the length of $w$. Let $L \subseteq \Sigma^*$, then $L^* = \bigcup_{n \geq 0} L^n$, with $L^0 = \{\varepsilon\}$ and $L^n = \{u_1 \cdots u_n \mid u_1, \ldots, u_n \in L\}$ for $n > 0$, denotes the *Kleene star* of $L$. For some language $L \subseteq \Sigma^*$, we denote by $\mathrm{Pref}(L) = \{w \mid \exists u \in \Sigma^* : wu \in L\}$, $\mathrm{Suff}(L) = \{w \mid \exists u \in \Sigma^* : uw \in L\}$ and $\mathrm{Fact}(L) = \{w \mid \exists u, v \in \Sigma^* : uwv \in L\}$ the set of *prefixes*, *suffixes* and *factors* of words in $L$. The language $L$ is called *prefix-free* if for each $w \in L$ we have $\mathrm{Pref}(\{w\}) \cap L = \{w\}$. If $u, w \in \Sigma^*$, a prefix $u \in \mathrm{Pref}(\{w\})$ is called a *proper prefix* if $u \neq w$. A language $L \subseteq \Sigma^*$ is called a *right* (*left-*) *ideal* if $L = L \cdot \Sigma^*$ ($= \Sigma^* \cdot L$), or a *two-sided ideal* (or simply an *ideal* for short), if $L$ is both, a right and a left ideal. A language $L \subseteq \Sigma^*$ is called *bounded*, if there exist words $w_1, \ldots, w_n \in \Sigma^*$ such that $L \subseteq w_1^* \cdots w_n^*$.

Throughout the paper, we consider deterministic finite automata (DFAs). Recall that a DFA $\mathcal{A}$ is a tuple $\mathcal{A} = (\Sigma, Q, \delta, q_0, F)$, where the alphabet $\Sigma$ is a finite set of input symbols, $Q$ is the finite state set, with start state $q_0 \in Q$, and final state set $F \subseteq Q$. The transition function $\delta \colon Q \times \Sigma \to Q$ extends to words from $\Sigma^*$ in the usual way. The function $\delta$ can be further extended to sets of states in the following way. For every set $S \subseteq Q$ and $w \in \Sigma^*$, we set $\delta(S, w) := \{\, \delta(q, w) \mid q \in S \,\}$. We sometimes refer to the function $\delta$ as a relation and we identify a transition $\delta(q, \sigma) = q'$ with the tuple $(q, \sigma, q')$. We call $\mathcal{A}$ *complete* if $\delta$ is defined for every $(q, a) \in Q \times \Sigma$; if $\delta$ is undefined for some $(q, a)$, the automaton $\mathcal{A}$ is called *partial*. The set $L(\mathcal{A}) = \{w \in \Sigma^* \mid \delta(q_0, w) \in F\}$ denotes the language *recognized* by $\mathcal{A}$.

A *semi-automaton* is a finite automaton without a specified start state and with no specified set of final states. The properties of being *deterministic*, *partial*, and *complete* of semi-automata are defined as for DFA. When the context is clear, we call both deterministic finite automata and semi-automata simply *automata*. We call a deterministic complete semi-automaton a DCSA and a partial deterministic finite automaton a PDFA for short. If we want to add an explicit initial state $r$ and an explicit set of final states $S$ to a DCSA $\mathcal{A}$, which changes it to a DFA, we use the notation $\mathcal{A}_{r,S}$.

A complete automaton $\mathcal{A}$ is called *synchronizing* if there exists a word $w \in \Sigma^*$ with $|\delta(Q, w)| = 1$. In this case, we call $w$ a *synchronizing word* for $\mathcal{A}$. We call a state $q \in Q$ with $\delta(Q, w) = \{q\}$ for some $w \in \Sigma^*$ a *synchronizing state*.

For an automaton $\mathcal{A} = (\Sigma, Q, \delta, q_0, F)$, we say that two states $q, q' \in Q$ are *connected*, if one is reachable from the other, i.e., we have a word $u \in \Sigma^*$ such that $\delta(q, u) = q'$. A subset $S \subseteq Q$ of states is called *strongly connected*, if all pairs from $S$ are connected. A maximal strongly connected subset is called a *strongly connected component*. A state from which some final state is reachable is called *co-accessible*. An automaton $\mathcal{A}$ is called *returning*, if for every state $q \in Q$, there exists a word $w \in \Sigma^*$ such that $\delta(q, w) = q_0$, where $q_0$ is the start state of $\mathcal{A}$. A state $q \in Q$ such that for all $x \in \Sigma$ we have $\delta(q, x) = q$ is called a *sink state.*

The set of synchronizing words forms a two-sided ideal. We will use this fact frequently without further mentioning.

For a fixed PDFA $\mathcal{B} = (\Sigma, P, \mu, p_0, F)$, we define the *constrained synchronization problem*:

**Definition 2.1.** $L(\mathcal{B})$-CONSTR-SYNC
Input: DCSA $\mathcal{A} = (\Sigma, Q, \delta)$.
Question: *Is there a synchronizing word $w$ for $\mathcal{A}$ with $w \in L(\mathcal{B})$?*

The automaton $\mathcal{B}$ will be called the *constraint automaton.* If an automaton $\mathcal{A}$ is a yes-instance of $L(\mathcal{B})$-CONSTR-SYNC we call $\mathcal{A}$ *synchronizing with respect to* $\mathcal{B}$. Occasionally, we do not specify $\mathcal{B}$ and rather talk about $L$-CONSTR-SYNC. We are going to inspect the complexity of this problem for different (small) constraint automata. The unrestricted synchronization problem, i.e., $\Sigma^*$-CONSTR-SYNC in our notation, is in P [15].

We assume the reader to have some basic knowledge in computational complexity theory and formal language theory, as contained, e.g., in [10]. For instance, we make use of regular expressions to describe languages. We also identify singleton sets with its elements. And we make use of complexity classes like P, NP, or PSPACE. With $\leq_m^{\log}$ we denote a logspace many-one reduction. If for two problems $L_1, L_2$ it holds that $L_1 \leq_m^{\log} L_2$ and $L_2 \leq_m^{\log} L_1$, then we write $L_1 \equiv_m^{\log} L_2$.

## 3 Known Results on Constrained Synchronization

Here we collect results from [5,8,9], and some consequences, that will be used later.

**Lemma 3.1** ([8]). *Let $\mathcal{X}$ denote any of the complexity classes PSPACE, NP and P. If $L(\mathcal{B})$ is a finite union of languages $L(\mathcal{B}_1), L(\mathcal{B}_2), \ldots, L(\mathcal{B}_n)$ such that for each $1 \leq i \leq n$ we have $L(\mathcal{B}_i)$-CONSTR-SYNC $\in \mathcal{X}$, then $L$-CONSTR-SYNC $\in \mathcal{X}$.*

The next result from [5] states that the computational complexity is always in PSPACE.

**Theorem 3.2** ([5]). *For any constraint automaton $\mathcal{B} = (\Sigma, P, \mu, p_0, F)$ the problem $L(\mathcal{B})$-CONSTR-SYNC is in PSPACE.*

In [5, Theorems 24, 25 and 26], for a two-state partial constraint automaton with an at most ternary alphabet, the following complexity classification was proven. In Sect. 5.1, we will extend this result to arbitrary alphabets.

**Theorem 3.3** ([5]). *Let $\mathcal{B} = (\Sigma, P, \mu, p_0, F)$ be a PDFA. If $|P| \leq 1$ or $|P| = 2$ and $|\Sigma| \leq 2$, then $L(\mathcal{B})$-CONSTR-SYNC $\in$ P. For $|P| = 2$ with $|\Sigma| = 3$, up to* *symmetry by renaming of the letters, $L(\mathcal{B})$-*Constr-Sync *is PSPACE-complete precisely in the following cases for $L(\mathcal{B})$:*

| | | | |
|---|---|---|---|
| $a(b+c)^*$ | $(a+b+c)(a+b)^*$ | $(a+b)(a+c)^*$ | $(a+b)^*c$ |
| $(a+b)^*ca^*$ | $(a+b)^*c(a+b)^*$ | $(a+b)^*cc^*$ | $a^*b(a+c)^*$ |
| $a^*(b+c)(a+b)^*$ | $a^*b(b+c)^*$ | $(a+b)^*c(b+c)^*$ | $a^*(b+c)(b+c)^*$ |

*and polynomial time solvable in all other cases.*

The next result from [5, Theorem 17] will also be useful to single out certain polynomial time solvable cases.

**Theorem 3.4** ([5]). *If $\mathcal{B}$ is returning, then $L(\mathcal{B})$-*Constr-Sync $\in$ *P.*

The next result allows us to assume a standard form for two-state constraint automata. We will prove an analogous result for three-state constraint automata in Sect. 5.2.

**Lemma 3.5** ([5]). *Let $\mathcal{B} = (\Sigma, P, \mu)$ be a partial deterministic semi-automaton with two states, i.e., $P = \{1, 2\}$. Then, for each $p_0 \in P$ and each $F \subseteq P$, either $L(\mathcal{B}_{p_0,F})$-*Constr-Sync $\in$ *P, or $L(\mathcal{B}_{p_0,F})$-*Constr-Sync $\equiv_m^{\log}$ $L(\mathcal{B}')$-Constr-Sync *for a PDFA $\mathcal{B}' = (\Sigma, P, \mu', 1, \{2\})$.*

The next result combines results from [9] and [6] to show that for bounded constrained languages, the constrained synchronization problem is in NP.

**Theorem 3.6.** *For bounded constraint languages, the constrained synchronization problem is in NP.*

The following condition will be useful to single out, for bounded constraint languages, those problems that are NP-complete.

**Proposition 3.7** ([9]). *Suppose we find $u, v \in \Sigma^*$ such that we can write $L = uv^*U$ for some non-empty language $U \subseteq \Sigma^*$ with:*

$$u \notin \text{Fact}(v^*), \quad v \notin \text{Fact}(U), \quad \text{Pref}(v^*) \cap U = \emptyset.$$

*Then $L$-*Constr-Sync *is NP-hard.*

# 4 General Results

Here, we state various general results, among them our three main theorems: the Ideal Separartion Theorem, the $UV^*W$-Theorem and the $uC$-Theorem. The first result is a slight generalization of a Theorem from [5, Theorem 27].

**Theorem 4.1.** *Let $\varphi\colon \Sigma^* \to \Gamma^*$ be a homomorphism and $L \subseteq \Sigma^*$. Then $\varphi(L)$-*Constr-Sync $\le_m^{\log}$ *$L$-*Constr-Sync.

We will also need the next slight generalization of a Theorem from [5, Theorem 14].

**Theorem 4.2.** *Let $L, L' \subseteq \Sigma^*$. If $L \subseteq \mathrm{Fact}(L')$ and $L' \subseteq \mathrm{Fact}(L)$, then* $L$-CONSTR-SYNC $\equiv_m^{\log}$ $L'$-CONSTR-SYNC.

Next, we state a result on how we can combine languages using concatenation, while still getting polynomial time solvable problems. Another result, namely Theorem 4.5, is contrary in the sense that it states conditions for which the concatenation yields PSPACE-hard problems.

**Theorem 4.3 ($UV^*W$-Theorem).** *Let $U, V, W \subseteq \Sigma^*$ be regular and $\mathcal{B} = (\Sigma, P, \mu, p_0, \{p_0\})$ be a PDFA, whose initial state equals its single final state, such that (1) $V = L(\mathcal{B})$, (2) $U \subseteq \mathrm{Suff}(V)$ and (3) $W \subseteq \mathrm{Pref}(V)$. Then* $(UVW)$-CONSTR-SYNC $\in$ P.

*Proof (sketch).* Every synchronizing word from $UVW$ could be enlarged, by a suitable prefix and a suitable suffix, to a synchronizing word in $V$. Conversely, every synchronizing word in $V$ could be enlarged to a synchronizing word in $UVW$. So, searching for a synchronizing word in $UVW$ has the same complexity as searching in $V$, the latter being polynomial-time solvable by Theorem 3.4. □

*Remark 1.* Note that in Theorem 4.3, $U = \{\varepsilon\}$ or $W = \{\varepsilon\}$ is possible. In particular, $L(\mathcal{B})$-CONSTR-SYNC $\in$ P for every PDFA $\mathcal{B} = (\Sigma, P, \mu, p_0, \{p_0\})$.

The next theorem is useful, as it allows us to show PSPACE-hardness by reducing the problem, especially ones that are written as unions, to known PSPACE-hard problems. Please see Example 2, or the proof sketch of Theorem 5.6, for applications.

**Theorem 4.4 (Ideal Separation Theorem).** *Let $I \subseteq \Sigma^*$ be a fixed regular ideal language. Suppose $L \subseteq \Sigma^*$ is any regular language, then*

$$(I \cap L)\text{-CONSTR-SYNC} \leq_m^{\log} L\text{-CONSTR-SYNC}.$$

*In particular, let $u \in \Sigma^*$ and $L \subseteq \Sigma^*$. Then* $(L \cap \Sigma^* u \Sigma^*)$-CONSTR-SYNC $\leq_m^{\log}$ $L$-CONSTR-SYNC.

*Proof (sketch).* It could be shown that the minimal complete automaton $\mathcal{A}_I$ of $I$ has precisely $I$ as the set of synchronizing words. Then an input automaton has a synchronizing word in $I \cap L$ if and only if the product automaton of $\mathcal{A}_I$ and the input automaton has a synchronizing word in $L$. □

Mostly, we apply Theorem 4.4 with *principal ideals* $\Sigma^* u \Sigma^*$ for $u \in \Sigma^*$. The next proposition is a generalization of arguments previously used to establish PSPACE-hardness [5,8] for constraints as, for example, $a(b+c)^*$.

**Theorem 4.5 (uC-Theorem).** *Suppose $u \in \Sigma^+$ is a non-empty word.*

1. *Let $C \subseteq \Sigma^*$ be a finite prefix-free set of cardinality at least two with $C^* \cap \Sigma^* u \Sigma^* = \emptyset$.*
2. *Let $\Gamma \subseteq \Sigma$ be such that $u \notin \Gamma^*$, i.e., $u$ uses at least one symbol not in $\Gamma$.*

*Then, the problem* $(\Gamma^* u C^*)$-CONSTR-SYNC *is PSPACE-hard. If, additionally, we have* $\mathrm{Suff}(u) \cap \mathrm{Pref}(\{u\}) = \{\varepsilon, u\}$ *and the following is true:*

> *There exists* $x \in C$ *such that, for* $v, w \in \Sigma^*$, *if* $vxw \in (C \cup \{u\})^*$, *then* $vx \in (C \cup \{u\})^*$.

*Then,* $(C^* u \Gamma^*)$-CONSTR-SYNC *is PSPACE-hard.*

*Example 1.* Set $L = (a+b)^* ac(b+c)^*$. Using Theorem 4.5 with $\Gamma = \{a, b\}$, $u = ac$ and $C = \{b, c\}$ gives PSPACE-hardness. Hence, by Theorem 3.2, it is PSPACE-complete. Note that $(a+b)^* c(b+c)^* \cap \Sigma^* ac \Sigma^* = L$. Hence, together with Theorem 4.4, we get PSPACE-completeness for $(a+b)^* c(b+c)^*$. For the latter language, this was already shown in [5], as stated in Theorem 3.3, by more elementary means, i.e., by giving a reduction from a different problem.

*Example 2.* For the following $L \subseteq \{a, b\}^*$ we have that $L$-CONSTR-SYNC is PSPACE-hard. For the first two, this is implied by a straightforward application of Theorem 4.5, for the last one a more detailed proof is given.

1. $L = \Gamma^* aa(ba+bb)^*$ for $\Gamma \subseteq \{b\}$.
2. $L = \Gamma^* aba(a+bb)^*$ for $\Gamma \subseteq \{b\}$.
3. $L = b^* a(a+ba)^*$. Then $L = b^* bba(a+ba)^* \cup ba(a+ba)^* \cup a(a+ba)^*$. Set $U = L \cap \Sigma^* bba \Sigma^* = b^* bba(a+ba)^*$. By Theorem 4.4,

$$U\text{-CONSTR-SYNC} \le_m^{\log} L\text{-CONSTR-SYNC}.$$

   For $U$, with $\Gamma = \{b\}$, $u = bba$ and $C = \{a, ba\}$ and Theorem 4.5, we find that $U$-CONSTR-SYNC is PSPACE-hard. So, $L$-CONSTR-SYNC is also PSPACE-hard.

## 5 Application to Small Constraint Automata

Here, we apply the results obtained in Sect. 4. In Subsect. 5.1 we will give a complete overview of the complexity landscape for two-state constraint automata over an arbitrary alphabet, thus extending a result from [5], where it was only proven for an at most ternary alphabet. In Subsect. 5.2 we will give a complete overview of the complexity landscape for three-state constraint automata, the least number of states such that we get PSPACE-complete and NP-complete constrained synchronization problems [5].

<u>Notational Conventions in this Section:</u> Let $\mathcal{B} = (\Sigma, P, \mu, p_0, F)$ be a constraint PDFA with $|P| = n$. Here, we will denote the states by natural numbers $P = \{1, \ldots, n\}$, and we will assume that 1 always denotes the start state, i.e., $p_0 = 1$. In this section, $\mathcal{B}$ will always denote the fixed constraint PDFA. By Lemma 3.5, for $|P| = 2$, we can assume $F = \{2\}$. We will show in Sect. 5.2, stated in Lemma 5.5, that also for $|P| = 3$ we can assume $F = \{3\}$. So, if nothing else is said, by default we will assume $F = \{n\}$ in the rest of this paper.

Also, for a fixed constraint automaton[1], we set $\Sigma_{ij} := \{ a \in \Sigma \mid \mu(i,a) = j \}$ for $1 \le i,j \le n$. As $\mathcal{B}$ is deterministic, $\Sigma_{i1} \cap \Sigma_{i2} = \emptyset$.

## 5.1 Two States and Arbitrary Alphabet

Let $\mathcal{B} = (\Sigma, P, \mu, p_0, F)$ be a two-state constraint PDFA. Recall the definitions of the sets $\Sigma_{i,j}$, $1 \le i,j \le 2$ and that here, by our notational conventions, $P = \{1,2\}$, $p_0 = 1$ and $F = \{2\}$. In general, for two states, we have

$$L(\mathcal{B}) = (\Sigma_{1,1}^* \Sigma_{1,2} \Sigma_{2,2}^* \Sigma_{2,1})^* \Sigma_{1,1}^* \Sigma_{1,2} \Sigma_{2,2}^*.$$

First, as shown in [5], for two-state constraint automata, some easy cases could be excluded from further analysis by the next result, as they give polynomial time solvable instances.

**Proposition 5.1** ([5]). *If one of the following conditions hold, then $L(\mathcal{B}_{1,\{2\}})$--*CONSTR-SYNC $\in$ *P: (1) $\Sigma_{1,2} = \emptyset$, (2) $\Sigma_{2,1} \neq \emptyset$, (3) $\Sigma_{1,1} \cup \Sigma_{1,2} \subseteq \Sigma_{2,2}$, or (4) $\Sigma_{1,1} \cup \Sigma_{2,2} = \emptyset$.*

Next, we will single out those cases that give PSPACE-hard problem in Lemma 5.3 and Lemma 5.2. Finally, in Theorem 5.4 we will combine these results and show that the remaining cases all give polynomial time solvable instances.

**Lemma 5.2.** *Suppose $(\Sigma_{1,1} \cup \Sigma_{1,2}) \setminus \Sigma_{2,2} \neq \emptyset$, $\Sigma_{1,2} \neq \emptyset$, $\Sigma_{2,1} = \emptyset$ and $|\Sigma_{2,2}| \ge 2$. Then $L(\mathcal{B})$-*CONSTR-SYNC *is PSPACE-hard.*

*Proof.* Choose $a \in (\Sigma_{1,1} \cup \Sigma_{1,2}) \setminus \Sigma_{2,2}$. Then

$$L \cap \Sigma^* a \Sigma^* = \begin{cases} \Sigma_{1,1}^* a \Sigma_{2,2}^* & \text{if } a \in \Sigma_{1,2}; \\ \Sigma_{1,1}^* a \Sigma_{1,1}^* \Sigma_{1,2} \Sigma_{2,2}^* & \text{if } a \in \Sigma_{1,1}. \end{cases}$$

In the first case we can apply Theorem 4.5 with $\Gamma = \Sigma_{1,1}, u = a$ and $C = \Sigma_{2,2}$ to find that $(L \cap \Sigma^* a \Sigma^*)$-CONSTR-SYNC is PSPACE-hard. In the second case, choose some $x \in \Sigma_{1,2}$, then, as, by determinism of $\mathcal{B}$, $x \notin \Sigma_{1,1}$, we find $L \cap \Sigma^* ax \Sigma^* = \Sigma_{1,1}^* ax \Sigma_{2,2}^*$ and we can apply Theorem 4.5 with $\Gamma = \Sigma_{1,1}^*$, $u = ax$ and $C = \Sigma_{2,2}$ to find that $(L \cap \Sigma^* ax \Sigma^*)$-CONSTR-SYNC is PSPACE-hard. Finally, the claim follows by Theorem 4.4. □

The next lemma states a condition such that we get PSPACE-hardness if the set $\Sigma_{1,1}$ contains at least two distinct symbols.

**Lemma 5.3.** *Suppose $|\Sigma_{1,1}| \ge 2$, $\Sigma_{1,2} \neq \emptyset$, $\Sigma_{2,1} = \emptyset$ and $(\Sigma_{1,1} \cup \Sigma_{1,2}) \setminus \Sigma_{2,2} \neq \emptyset$. Then $L(\mathcal{B})$-*CONSTR-SYNC *is PSPACE-hard.*

[1] Note that this notation only makes sense with respect to a fixed alphabet and a fixed automaton, or said differently we have implicitly defined a function dependent on both of these parameters. But every more formal way of writing this might be cumbersome, and as the automaton used in this notation is always the (fixed) constraint automaton, in the following, usage of this notation should pose no problems. It is just a shorthand whose usage is restricted to the next two sections.

*Proof.* Set $C = \Sigma_{1,1}$ and $\Gamma = \Sigma_{2,2}$. By assumption, we find $a \in (\Sigma_{1,1} \cup \Sigma_{1,2}) \setminus \Sigma_{2,2}$. If $a \in \Sigma_{1,2}$, then set $u = a$. If $a \in \Sigma_{1,1} \setminus \Sigma_{1,2}$, then choose $b \in \Sigma_{1,2}$ and set $u = ab$. Note that, by determinism of the constraint automaton, we have $\Sigma_{1,1} \cap \Sigma_{1,2} = \emptyset$. Then, $L(\mathcal{B}) \cap \Sigma^* u \Sigma^* = C^* u \Gamma^*$. For this language, the conditions of Theorem 4.5 are fulfilled and hence, together with Theorem 4.4, the claim follows. □

Combining everything, we derive our main result of this section.

**Theorem 5.4.** *For a two-state constraint PDFA $\mathcal{B}$, $L(\mathcal{B})$-*CONSTR-SYNC *is PSPACE-complete precisely when $\Sigma_{1,2} \neq \emptyset$, $\Sigma_{2,1} = \emptyset$ and*

$$(\Sigma_{1,1} \cup \Sigma_{1,2}) \setminus \Sigma_{2,2} \neq \emptyset \text{ and } \max\{|\Sigma_{1,1}|, |\Sigma_{2,2}|\} \geq 2.$$

*Otherwise, $L(\mathcal{B})$-*CONSTR-SYNC $\in$ *P.*

*Proof.* We can assume $\Sigma_{1,2} \neq \emptyset$, $\Sigma_{2,1} = \emptyset$ and $(\Sigma_{1,1} \cup \Sigma_{1,2}) \setminus \Sigma_{2,2} \neq \emptyset$, for otherwise, by Proposition 5.1, we have $L(\mathcal{B})$-CONSTR-SYNC $\in$ P. If $|\Sigma_{1,1}| \geq 2$ or $|\Sigma_{2,2}| \geq 2$, by Lemma 5.3 or Lemma 5.2, we get PSPACE-hardness, and so, by Theorem 3.2, it is PSPACE-complete in these cases. Otherwise, assume $|\Sigma_{1,1}| \leq 1$ and $|\Sigma_{2,2}| \leq 1$. With the other assumptions,

$$L = \bigcup_{x \in \Sigma_{1,2}} \Sigma_{1,1}^* x \Sigma_{2,2}^*.$$

Each language of the form $\Sigma_{1,1}^* x \Sigma_{2,2}^*$ is over the at most ternary alphabet $\Sigma_{1,1} \cup \{x\} \cup \Sigma_{2,2}$. Hence, each such language has the form $y^* x z^*$, $x z^*$ or $y^* x$ with $|\{y, z, x\}| \leq 3$ and $\{x, y, z\} \subseteq \Sigma$. If a letter is not used in the constraint language, we can, obviously, assume the problem is over the smaller alphabet of all letters used in the constraint, as usage of letters not occurring in any accepting path in the constraint automaton is forbidden in any input semi-automaton. So, by Theorem 3.3, for the languages $y^* x z^*$ the constraint problem is polynomial time solvable, and by Lemma 3.1 we have $L$-CONSTR-SYNC $\in$ P. □

## 5.2 Three States and Arbitrary Alphabet

Let $\mathcal{B} = (\Sigma, P, \mu, p_0, F)$ be a three-state constraint PDFA. Recall the definitions of the sets $\Sigma_{i,j}$, $1 \leq i, j \leq 2$ and that here, by our notational conventions, $P = \{1, 2, 3\}$, $p_0 = 1$ and $F = \{3\}$. First, we will show an analogous result to Lemma 3.5 for the three-state case, which justifies the mentioned notational conventions.

**Lemma 5.5.** *Let $\mathcal{B} = (\Sigma, P, \mu, p_0, F)$ be a PDFA with three states. Then, either $L(\mathcal{B})$-*CONSTR-SYNC $\in$ *P, or $L(\mathcal{B})$-*CONSTR-SYNC $\equiv_m^{\log}$ *$L(\mathcal{B}')$-*CONSTR-SYNC *for a PDFA $\mathcal{B}' = (\Sigma, \{1, 2, 3\}, \mu', 1, \{3\})$.*

In the general theorem, stated next, the complexity classes we could realize depend on the number of strongly connected components in the constraint automaton.

**Table 1.** The constraint automata $\mathcal{B}_i$, $i \in \{1, \ldots, 6\}$, with the respective computational complexities of $L(\mathcal{B}_i)$-Constr-Sync, for which these complexities are proven in the proof sketch of Theorem 5.6. Please see the main text for more explanation.

| Type | Automaton | Complexity | Type | Automaton | Complexity |
|---|---|---|---|---|---|
| $\mathcal{B}_1$ | 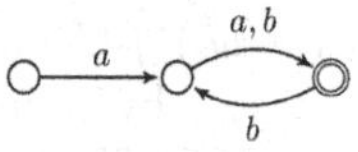 | PSPACE-c | $\mathcal{B}_2$ | 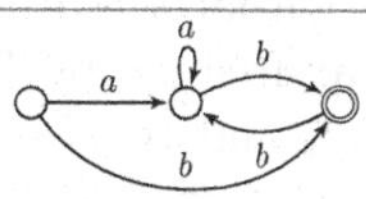 | P |
| $\mathcal{B}_3$ | 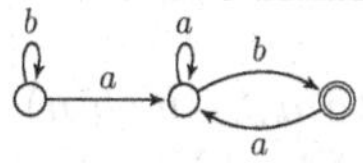 | PSPACE-c | $\mathcal{B}_4$ | 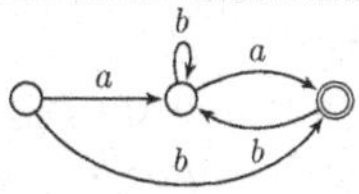 | PSPACE-c |
| $\mathcal{B}_5$ | 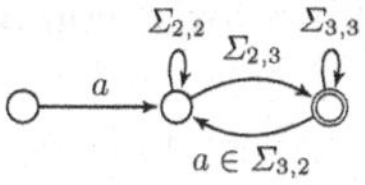 | P | $\mathcal{B}_6$ | 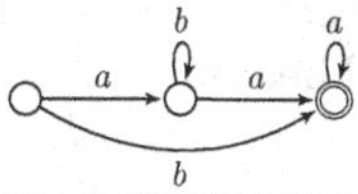 | NP-c |

**Theorem 5.6.** *For a constraint PDFA $\mathcal{B}$ with three states over an arbitrary alphabet $L(\mathcal{B})$-Constr-Sync is either in P, or NP-complete, or PSPACE-complete. More specifically,*

1. *if $\mathcal{B}$ is strongly connected the problem is always in P,*
2. *if the constraint automaton has two strongly connected components, the problem is in P or PSPACE-complete,*
3. *and if we have three strongly connected components, the problem is either in P or NP-complete.*

*Proof (sketch).* This is only a proof sketch for the binary alphabet $\Sigma = \{a, b\}$, as even up to symmetry, more than fifty cases have to be checked. We only show a few cases to illustrate how to apply the results from Sect. 4. We will handle the cases illustrated in Table 1, please see the table for the naming of the constraint automata. In all automata, the left state is the start state 1, the middle state is state 2 and the rightmost state is state 3. If not said otherwise, 3 will be the single final state, a convention in correspondence with Lemma 5.5. By Theorem 3.2, for PSPACE-completeness, it is enough to establish PSPACE-hardness.

1. The constraint automaton[2] $\mathcal{B}_1$: Here $L(\mathcal{B}_1) = a(a+b)(bb+ba)^*$. Set $U = L(\mathcal{B}_1) \cap \Sigma^* aa \Sigma^* = aa(bb+ba)^*$. By Theorem 4.4, $U$-Constr-Sync $\leq_m^{\log}$ $L(\mathcal{B}_1)$-Constr-Sync. As $(bb+ba) \cap \Sigma^* aa \Sigma^* = \emptyset$ and $\{bb, ba\}$ is prefix-free, by Theorem 4.5, $U$-Constr-Sync is PSPACE-hard, which gives PSPACE-hardness for $L(\mathcal{B}_1)$-Constr-Sync.
2. The constraint automaton $\mathcal{B}_2$: Here $L(\mathcal{B}_2) = aa^*b(ba^*b)^* \cup b(ba^*b)$. We have $aa^*b \subseteq \mathrm{Suff}((ba^*b)^*)$ and $b \subseteq \mathrm{Suff}((ba^*b)^*)$. By Theorem 4.3 and Lemma 3.1, $L(\mathcal{B}_2)$-Constr-Sync $\in$ P.

[2] This constraint automaton was already given in [5] as the single example of a three-state constraint automaton yielding a PSPACE-complete problem.

3. The constraint automaton $\mathcal{B}_3$: Here $L(\mathcal{B}_3) = b^*aa^*b(aa^*b)^*$. Set $U = L(\mathcal{B}_3) \cap \Sigma^*bbaba\Sigma^* = b^*bbaba(a+ba)^*b$. We have $b^*bbaba(a+ba)^*b \subseteq \text{Fact}(b^*bbaba(a+ba)^*)$ and $b^*bbaba(a+ba)^* \subseteq \text{Fact}(b^*bbaba(a+ba)^*b)$. Hence, by Theorem 4.2, $U$-Constr-Sync has the same computational complexity as synchronization for $b^*bbaba(a+ba)^*$. As $(a+ba)^* \cap \Sigma^*bbaba\Sigma^* = \emptyset$ and $\{a, ba\}$ is a prefix-free set, by Theorem 4.5, $(b^*bbaba(a+ba)^*)$-Constr-Sync is PSPACE-hard, and so also synchronization by $U$. As, by Theorem 4.4, $U$-Constr-Sync $\le_m^{\log}$ $L(\mathcal{B}_3)$-Constr-Sync, we get PSPACE-hardness for $L(\mathcal{B}_3)$-Constr-Sync.
4. The constraint automaton $\mathcal{B}_4$: Here $L(\mathcal{B}_4) = ab^*a(bb^*a)^* \cup b(bb^*a)^*$. Set $U = L(\mathcal{B}_4) \cap \Sigma^*aab\Sigma^* = aab(b+ab)^*a$. As $(b+ab)^* \cap \Sigma^*aab\Sigma^* = \emptyset$ and $\{b, ab\}$ is a prefix-free set, as above, PSPACE-hardness follows by a combination of Theorem 4.2, Theorem 4.4 and Theorem 4.5.
5. The constraint automaton $\mathcal{B}_5$: Here, $\mathcal{B}_5$ denotes an entire family of automata. In general, $L(\mathcal{B}_5) = a\Sigma_{2,2}^*\Sigma_{2,3}(\Sigma_{3,3}^*\Sigma_{3,2}\Sigma_{2,2}^*\Sigma_{2,3})^*$ with $a \in \Sigma_{3,2}$. As $a \in \Sigma_{3,2}$, we have $a\Sigma_{2,2}^*\Sigma_{2,3} \subseteq \Sigma_{3,2}\Sigma_{2,2}^*\Sigma_{2,3}$. So,

$$a\Sigma_{2,2}^*\Sigma_{2,3} \subseteq \text{Suff}((\Sigma_{3,3}^*\Sigma_{3,2}\Sigma_{2,2}^*\Sigma_{2,3})^*)$$

and by Theorem 4.3 we find $L(\mathcal{B}_5)$-Constr-Sync $\in$ P.
6. The constraint automaton $\mathcal{B}_6$: Here, $L(\mathcal{B}_6) = ab^*aa^* \cup ba^*$. As $L(\mathcal{B}_6) \subseteq a^*b^*a^*a^*$ the language $L(\mathcal{B}_6)$ is a bounded language, hence by Theorem 3.6 we have $L(\mathcal{B}_6)$-Constr-Sync $\in$ NP. Furthermore $L(\mathcal{B}_6) \cap \Sigma^*abb^*a\Sigma^* = abb^*aa^*$. So, by Theorem 4.4, the original problem is at least as hard as for the constraint language $abb^*aa^*$. As $ab \notin \text{Fact}(b^*)$, $b \notin \text{Fact}(aa^*)$ and $\text{Pref}(b^*) \cap aa^* = \emptyset$, by Proposition 3.7, for $abb^*aa^*$ the problem is NP-hard. So, by Theorem 4.4, $L(\mathcal{B}_6)$-Constr-Sync is NP-complete. □

## 6 Conclusion

We have presented general theorems to deduce, for a known constraint language, the computational complexity of the corresponding constrained synchronization problem. We applied these results to small constraint automata, generalizing the classification of two-state automata [5] from an at most ternary alphabet to an arbitrary alphabet. We also gave a full classification for three-state constraint automata. Hence, we were able, by using new tools, to strengthen the results from [5]. In light of the methods used and the results obtained so far, it seems probable that even for general constraint languages only the three complexity classes P, PSPACE-complete or NP-complete arise, hence giving a trichotomy result. However, we are still far from settling this issue, and much remains to be done to answer this question or maybe, surprisingly, present constraint languages giving complete problems for other complexity classes. Inspection of the results also shows that the NP-complete cases are all induced by bounded languages. Hence, the question arises if this is always the case, or if we can find non-bounded constraint languages giving NP-complete constrained problems.

**Acknowledgement.** I thank Prof. Dr. Mikhail V. Volkov for suggesting the problem of constrained synchronization during the workshop 'Modern Complexity Aspects of Formal Languages' that took place at Trier University 11.–15. February, 2019. I also thank anonymous referees of a very preliminary version of this work, whose detailed feedback directly led to this complete reworking of the three-state proof, and anonymous referees of another version and the present version.

## References

1. Amarilli, A., Paperman, C.: Topological sorting with regular constraints. In: Chatzigiannakis, I., Kaklamanis, C., Marx, D., Sannella, D. (eds.) ICALP 2018. LIPIcs, vol. 107, pp. 115:1–115:14. Leibniz-Zentrum für Informatik (2018)
2. Černý, J.: Poznámka k homogénnym experimentom s konečnými automatmi. Mat.-fyzikálny časopis **14**(3), 208–216 (1964)
3. Diekert, V.: Makanin's algorithm for solving word equations with regular constraints. Report, Fakultät Informatik, Universität Stuttgart, March 1998
4. Diekert, V., Gutiérrez, C., Hagenah, C.: The existential theory of equations with rational constraints in free groups is PSPACE-complete. Inf. Comput. **202**(2), 105–140 (2005)
5. Fernau, H., Gusev, V.V., Hoffmann, S., Holzer, M., Volkov, M.V., Wolf, P.: Computational complexity of synchronization under regular constraints. In: Rossmanith, P., Heggernes, P., Katoen, J. (eds.) MFCS 2019. LIPIcs, vol. 138, pp. 63:1–63:14. Schloss Dagstuhl - Leibniz-Zentrum für Informatik (2019)
6. Ginsburg, S., Spanier, E.H.: Bounded regular sets. Proc. Am. Math. Soc. **17**(5), 1043–1049 (1966)
7. Gusev, V.V.: Synchronizing automata of bounded rank. In: Moreira, N., Reis, R. (eds.) CIAA 2012. LNCS, vol. 7381, pp. 171–179. Springer, Heidelberg (2012). https://doi.org/10.1007/978-3-642-31606-7_15
8. Hoffmann, S.: Computational complexity of synchronization under regular commutative constraints. In: Kim, D., Uma, R.N., Cai, Z., Lee, D.H. (eds.) COCOON 2020. LNCS, vol. 12273, pp. 460–471. Springer, Cham (2020). https://doi.org/10.1007/978-3-030-58150-3_37
9. Hoffmann, S.: On a class of constrained synchronization problems in NP. In: Cordasco, G., Gargano, L., Rescigno, A. (eds.) Proceedings of the 21th Italian Conference on Theoretical Computer Science, ICTCS 2020, Ischia, Italy. CEUR Workshop Proceedings, CEUR-WS.org (2020)
10. Hopcroft, J.E., Motwani, R., Ullman, J.D.: Introduction to Automata Theory, Languages, and Computation, 2nd edn. Addison-Wesley, Boston (2001)
11. Martyugin, P.V.: Synchronization of automata with one undefined or ambiguous transition. In: Moreira, N., Reis, R. (eds.) CIAA 2012. LNCS, vol. 7381, pp. 278–288. Springer, Heidelberg (2012). https://doi.org/10.1007/978-3-642-31606-7_24
12. Pesant, G.: A regular language membership constraint for finite sequences of variables. In: Wallace, M. (ed.) CP 2004. LNCS, vol. 3258, pp. 482–495. Springer, Heidelberg (2004). https://doi.org/10.1007/978-3-540-30201-8_36
13. Romeuf, J.: Shortest path under rational constraint. Inf. Process. Lett. **28**(5), 245–248 (1988)
14. Sandberg, S.: 1 homing and synchronizing sequences. In: Broy, M., Jonsson, B., Katoen, J.-P., Leucker, M., Pretschner, A. (eds.) Model-Based Testing of Reactive Systems. LNCS, vol. 3472, pp. 5–33. Springer, Heidelberg (2005). https://doi.org/10.1007/11498490_2

15. Volkov, M.V.: Synchronizing automata and the Černý conjecture. In: Martín-Vide, C., Otto, F., Fernau, H. (eds.) LATA 2008. LNCS, vol. 5196, pp. 11–27. Springer, Heidelberg (2008). https://doi.org/10.1007/978-3-540-88282-4_4
16. Vorel, V., Roman, A.: Complexity of road coloring with prescribed reset words. J. Comput. Syst. Sci. **104**, 342–358 (2019)

# Most Pseudo-copy Languages Are Not Context-Free

Hyunjoon Cheon[1], Joonghyuk Hahn[1], Yo-Sub Han[1(✉)], and Sang-Ki Ko[2]

[1] Yonsei University, Seoul, Republic of Korea
{hyunjooncheon,greghahn,emmous}@yonsei.ac.kr
[2] Kangwon National University, Gangwon-Do, Republic of Korea
sangkiko@kangwon.ac.kr

**Abstract.** It is well known that the copy language $L = \{ww \mid w \in \Sigma^*\}$ is not context-free despite its simplicity. We study pseudo-copy languages that are defined to be sets of catenations of two similar strings, and prove non-context-freeness of these languages. We consider the Hamming distance and the edit-distance for the error measure of the two similar strings in pseudo-copy languages. When the error has an upper bound or a fixed value, we show that the pseudo-copy languages are not context-free. Similarly, if the error has a lower bound of at least four, then such languages are not context-free, either. Finally, we prove that all these pseudo-copy languages are context-sensitive.

**Keywords:** Context-freeness · Pseudo-copy languages · Hamming distance · Edit-distance

## 1 Introduction

For many years, people investigated the problems related to the repetition of strings from various perspectives such as bioinformatics [3,7,12], stringology [2, 5,16,20] and formal language theory [1,14]. For example, it was already proved in the early 80's that one can decide whether or not a given string contains a *square*—a string of the form $ww$ with $w$ nonempty—in $O(n \log n)$ time when $n$ is the length of an input string [2,5,16].

The problem of finding squares (also called *tandem repeat* or *contiguous repeat*) from biological sequences has been an intriguing topic in bioinformatics. Landau et al. [12] studied the problem of finding *approximate tandem repeats* from a given string, which can be described as $xy$, where $|x| = |y|$ and $d(x, y) \leq k$ for a given $k$ under the Hamming distance and the edit-distance metrics. They showed that all approximate tandem repeats can be found in $O(nk \log(n/k) + s)$ time, where $n$ is the length of the given string and $s$ is the number of repeats found. Later, Kolpakov and Kucherov [10] slightly improved the bound to $O(nk \log k + s)$ only in the case of the Hamming distance.

H. Cheon and J. Hahn—The first two authors contributed equally to this work.


C.-Y. Chen et al. (Eds.): COCOON 2021, LNCS 13025, pp. 189–200, 2021.
https://doi.org/10.1007/978-3-030-89543-3_16

We focus on the language-theoretic property related to the repetitions of strings. A string is *square-free* if none of its substrings is a square. It is easily seen that there are only finitely many square-free strings over one or two letters. Over a ternary alphabet, the set of square-free strings is infinite and, moreover, not context-free [15]. People also considered the complement of square-free languages—a language contains strings with at least one square as a substring. The language is also proved to be not context-free [6,18], and Ogden et al. [17] established a simpler proof using the interchange lemma.

The set of all squares, often called the *copy language* (denoted by COPY), is not context-free but can be recognized by realtime nondeterministic queue automata (NQAs) [11,21]. The class rtNQA of languages recognized by realtime NQAs is a proper subclass of context-sensitive languages (CS), and is incomparable to the class of context-free languages (CF). Therefore, it is immediate that the following relationship holds: COPY $\in \overline{\text{CF}} \cap \text{rtNQA} \subset \text{CS}$.

An interesting fact is that the complement of the copy language is context-free unlike COPY [9,19]. Since COPY $= \{xy \mid d_H(x, y) < 1\}$ and its complement of even-length[1] strings $\overline{\text{COPY}} = \{xy \mid d_H(x, y) > 0$, where $|x| = |y|\}$ can be defined using the Hamming distance $d_H$, one can consider the following question.

*Problem 1.* Consider the following language $L$:

$$L = \{xy \mid x, y \in \{0,1\}^*,\ |x| = |y|,\ d_H(x, y) < k\},$$

where $d_H(x, y)$ is the Hamming distance between $x$ and $y$.

**Q.** Is $L$ context-free?

We can think of the language $L$ in Problem 1 as a set of catenations of two similar strings—we call such $L$ a *pseudo-copy language.* In other words, the pseudo-copy language is a language with a bounded Hamming distance $k$ between two catenated strings.

Since one may consider different bound conditions such as threshold, inequality or equality relations, and error measures, a natural question that arises next is, whether or not such languages are context-free. In particular, many people conjecture that a complement of a pseudo-copy language with $k = 2$ would not be context-free, yet there is no formal proof and the problem is still open[2]. Even before, Bordihn [4] asked the following question, which has not been answered yet.

*Problem 2.* Consider the following language $L$:

$$L = \{xy \mid x, y \in \{0,1\}^*,\ |x| = |y|,\ |x| - d_H(x, y) \geq 2\}.$$

**Q.** Is $L$ context-free?

We consider several variants of pseudo-copy languages and their complements depending on the bound conditions, and demonstrate that most pseudo-copy languages and their complements are not context-free.

---

[1] We only consider even-length strings for the Hamming distance between two halves.
[2] https://cs.stackexchange.com/q/11585.

## 2 Preliminaries

Let $\Sigma$ denote a finite alphabet of symbols. Then a string $w$ is a finite sequence of symbols from $\Sigma$ and the length $|w|$ of $w$ is the number of symbols in $w$. The character $\lambda$ denotes an empty string.

For every string $w$ and every natural number $n$, we define the $n$-th power of the string $w$, denoted by $w^n$, by $w^0 = \lambda$ and $w^k = w^{k-1}w$ for $k = 1, 2, \ldots, n$. For a string $w$ of even length, we call two substrings $\alpha$ and $\beta$ of the same length, where $w = \alpha\beta$, *halves* of $w$.

A *context-free grammar* (CFG) $G$ is a tuple $G = (V, \Sigma, R, S)$, where $V$ is a set of nonterminals, $\Sigma$ is a set of terminals, $R \subseteq V \times (V \cup \Sigma)^*$ is a finite set of productions and $S \in V$ is the start symbol. Let $\alpha A \beta$ be a string over $V \cup \Sigma$, where $A \in V$ and $A \to \gamma \in R$. Then, we say that A can be rewritten as $\gamma$ and the corresponding *derivation step* is denoted $\alpha A \beta \Rightarrow \alpha\gamma\beta$. A production $A \to t \in R$ is a *terminating production* if $t \in \Sigma^*$. The reflexive, transitive closure of $\Rightarrow$ is denoted by $\stackrel{*}{\Rightarrow}$ and the context-free language generated by $G$ is $L(G) = \{w \in \Sigma^* \mid S \stackrel{*}{\Rightarrow} w\}$ [19].

The *Hamming distance* $d_H(x, y)$ measures the error between two strings $x$ and $y$ of the same length by counting the number of different symbols on the same position of each [8]. In other words, $d_H(x, y) = \sum_i d(x_i, y_i)$, where $d(a, b) = 0$ if $a = b$ and one otherwise. For example, $d_H(abca, acab) = 3$ since there are three positions with different symbols. $d_S(x, y) = |x| - d_H(x, y)$, on the other hand, can be seen as the *similarity* between $x$ and $y$, denoting the number of identical symbols at the same position of them.

An *alignment* of two strings $x$ and $y$ in $\Sigma^*$ is a sequence of $n$ pairs $(x_1, y_1)$, $(x_2, y_2), \ldots, (x_n, y_n)$ where $x_i, y_i \in \Sigma \cup \{\lambda\}$, $x_1 x_2 \cdots x_n = x$ and $y_1 y_2 \cdots y_n = y$. The *edit-distance* $d_E(x, y)$ of two strings $x$ and $y$ is the minimum number of pairs with different symbols in alignments of $x$ and $y$ [13]. For instance, strings $abca$ and $acab$ have two alignments $(a, a)$, $(b, c)$, $(c, a)$, $(a, b)$ and $(a, a)$, $(b, \lambda)$, $(c, c)$, $(a, a)$, $(\lambda, b)$. Although the first alignment is shorter, the number of different pairs is smaller for the second. Thus, the edit-distance of the two strings is two with the second alignment. Note that $(\lambda, \lambda)$, $(a, a)$, $(b, c)$, $(c, a)$, $(a, b)$ is also a valid alignment for the strings.

We generalize the pseudo-copy language in Problem 1 by allowing different conditions between the two catenated strings. First is to consider different error measures. While Problem 1 defines a language with the Hamming distance $d_H$ for the error measure. In Problem 2, we not only consider the conditions on the number for mismatches between two catenated strings but also matches by introducing the similarity measure $d_S$ as follows. For two equal-length strings $x, y$, we define $d_S(x, y) = |x| - d_H(x, y)$. Another measure is the edit-distance $d_E$ of $x$ and $y$, which does not require the two strings to be the same length. The edit-distance allows more operations than the Hamming distance. From the perspective of error correction, a symbol is not only tripped but added or removed in transmission, which resembles the edit operations: substitution, insertion and deletion, respectively.

Second is to consider the relations for error values. Similar to $\overline{\text{COPY}}$ where the Hamming distance is nonzero, we examine languages with different error bounds. Especially, these variants specify that the error (or similarity) of the two catenated strings should be bounded. For instance, one can think of a language with more than $k$ different symbol positions in its halves ($d_H > k$). Note that the languages with a lower bound is a natural extension of $\overline{\text{COPY}}$.

*Problem 3.* Given an integer $k \geq 0$ and an alphabet $\Sigma$, let $L = \{\alpha\beta \mid \alpha, \beta \in \Sigma^*, d(\alpha, \beta) \circ k\}$, where $d \in \{d_H, d_E, d_S\}$ and $\circ \in \{\leq, =, \geq\}$.

**Q.** Is such $L$ context-free?

Let $L_{X \circ k}$ denote the language under $d = d_X$. For example, $L_{H=k}$ is the language under $d_H$ and $\circ$ as $=$. For $d = d_E$, $L$ is the language with its minimum edit-distance considered. The languages with the same error measure define a class with bounded errors.

## 3 Pseudo-copy Languages

The first problem is for $L_{H=k}$, whose halves have exactly $k$ different symbols. Let us establish Lemma 4 for counting the Hamming distance on the specific form of strings for the problem.

**Lemma 4.** *For every string* $\alpha\beta = 0^a 1^b 0^c 1^d$, *where* $|\alpha| = |\beta|$, $d_H(\alpha, \beta) = \min(a + c, b + d, \max(|a - c|, |b - d|))$.

*Proof (Sketch).* If a 0-sequence occupies at least a half of $\alpha\beta$, then $d_H(\alpha, \beta)$ is the length $b + d$ of two 1-sequences. Otherwise, there is no sequence occupying a half. Without loss of generality, let us assume that a 0-sequence entirely aligns with the other 0-sequence. Then, $d_H(\alpha, \beta)$ is $|a - c|$, the number of 1's aligning with 0's. □

Based on the result of Lemma 4, we next show that $L_{H=k} = \{\alpha\beta \mid \alpha, \beta \in \Sigma^*, |\alpha| = |\beta|, d_H(\alpha, \beta) = k\}$ for every non-negative integer $k$ is not context-free.

**Theorem 5.** *For all* $k \geq 0$, $L_{H=k}$ *is not context-free.*

*Proof* (Proof by contradiction). Suppose that $L_{H=k}$ is context-free. Then $L' = L_{H=k} \cap \{0^a 1^b 0^c 1^d \mid a, b, c, d \geq k\}$ must be context-free and satisfies the pumping lemma. For an arbitrary pumping constant $p$, let $z = 0^l 1^{l+k} 0^{l+k} 1^l \in L'$ where $l = \max(p!, k)$. Then $z$ must have a decomposition of $uvwxy$ such that $|vx| > 0$, $|vwx| \leq p$ and $uv^n wx^n y \in L'$ for all $n \geq 0$. Note that $vx$ can only be a part of at most two consecutive sequences, each sequence of which consists of only 0's or only 1's. By pumping $v$ and $x$, we show that $d_H$ exceeds $k$, which contradicts the pumping lemma.

1. When $vx$ consists of only 0's or only 1's ($|vx|_0 = 0$ or $|vx|_1 = 0$)
   Without loss of generality, assume that $vx$ is in a sequence of 0's. We can pump $v$ and $x$ until the sequence of 0's that $vx$ is in occupies over half of the string. Let $z' = \alpha'\beta' = uv^{|z|}wx^{|z|}y$ and $|\alpha'| = |\beta'|$. Then, since the sequence containing $vx$ dominates $z'$, $d_H(\alpha', \beta') = 2l + k > k$. The same procedure can be applied when $vx$ is in the sequence of 1's.
2. When $vx$ consists of both 0's and 1's ($|vx|_0 \neq 0$ and $|vx|_1 \neq 0$)
   $vx$ is in consecutive sequences in forms such as $0^a1^b$ or $1^b0^a$. Apparently, when either $v$ or $x$ contains both 0 and 1, by pumping up $v$ and $x$, we obtain strings that are not in $L'$ which contradicts the pumping lemma. In the following, we assume that each of $v$ and $x$ contains only 0's or 1's. Without the loss of generality, let $|vx|_0 = a$ and $|vx|_1 = b$. Regarding which consecutive sequences $vx$ is placed in, one of the following holds:
   - $d_H(\alpha', \beta') = \min(2l + k + \min(a, b)i, \max(|ai - k|, |bi + k|))$,
   - $d_H(\alpha', \beta') = \min(2l + k + \min(a, b)i, \max(|ai + k|, |bi + k|))$ or
   - $d_H(\alpha', \beta') = \min(2l + k + \min(a, b)i, \max(|ai + k|, |bi - k|))$

   where $\alpha'\beta' = uv^{i+1}wx^{i+1}y$. For example, when $vx$ is in the first two sequences, applying Lemma 4 yields the first condition. Similarly, the other conditions can be computed from the remaining cases. All three cases show $d_H(\alpha', \beta') > k$ when $i = 2k+2$, contradicting the pumping lemma. Note that $l \geq k$ and the first part cannot be the minimum.

By the above, $L'$ is not context-free and, thus $L_{H=k}$ is not context-free. □

For different error bounds, we examine a language $L_{S=k} = \{\alpha\beta \mid \alpha, \beta \in \Sigma^*, |\alpha| = |\beta|, d_S(\alpha, \beta) = k\}$ that consists of strings whose halves have $k$ identical symbols.

**Theorem 6.** *For all $k \geq 0$, $L_{S=k}$ is not context-free.*

*Proof* (Proof by contradiction). Suppose that $L_{S=k}$ is context-free and let $L' = L_{S=k} \cap L(0^*1^*0^*1^*0^*1^*)$. Then $L'$ must satisfy the pumping lemma. For an arbitrary pumping constant $p$, choose $z = 0^P1^{P+k}0^P1^{P+k}0^P1^P$, where $P = 2(k+2)p$. Then $z$ must have a decomposition of $uvwxy$ that satisfies the pumping lemma. Let $t = |vx|/2$, and $\alpha$ and $\beta$ denote the first and the latter half of $z$. $z_i = uv^iwx^iy$ denotes the string after pumping $v$ and $x$ up $i - 1$ times, whose halves are $\alpha'$ and $\beta'$, respectively. Note that $|vx|$ must be even—otherwise $z_0 \notin L'$. The following case-by-case proof shows that $z_{k+3} \notin L'$.

1. $vx$ is in $\alpha$,
   When $vx$ is in the first half $\alpha$ of $z$, pumping sends latter part of $\alpha$ to $\beta$. This results in having identical substring in the head of $\alpha$ and $\beta$. By pumping $v$ and $x$ up $k+2$ times, the last $0^{t(k+2)}$ portion of $\alpha$ is pushed to the front of the latter half, thus $z_{k+3} = \alpha'0^{t(k+2)}1^{P+k}0^P1^P$, as illustrated in Fig. 1. Then $d_S(\alpha', \beta') \geq t(k+2) > k$.

2. $vx$ is in $\beta$,
   Similar to the case when $vx$ is in $\alpha$, we pump $vx$ to obtain identical substring in the tail of $\alpha$ and $\beta$. By pumping up $v$ and $x$ by $k+2$ times, the first $1^{t(k+2)}$ portion of $\beta$ is pushed to the first half, thus $z_{k+3} = 0^P 1^{P+k} 0^P 1^{t(k+2)} \beta'$. Then $d_S(\alpha', \beta') \geq t(k+2) > k$.
3. $vx$ is in both $\alpha$ and $\beta$, $vx = 0^a 1^b$.
   Contrary to the above, pumping $vx$ does not always result in having identical substring in the head or tail of $\alpha$ and $\beta$. We, therefore, examine the inner part of $\alpha$ and $\beta$, specifically, $1^{P+k}$.
   (a) When $a \leq b$, $\alpha' = \alpha 0^{a(k+2)} 1^{(b-a)(k+2)/2}$ and $\beta' = 1^{(a+b)(k+2)/2} \beta$. Since $\beta$ is pushed by $\frac{a+b}{2}(k+2)$ while $\alpha$ is not, the overlap in $1^{P+k}$ strictly increases. Thus $d_S(\alpha', \beta') \geq \frac{a+b}{2}(k+2) + k > k$.
   (b) When $a > b$, $\alpha' = \alpha 0^{(a+b)(k+2)/2}$ and $\beta' = 0^{(a-b)(k+2)/2} 1^{b(k+2)} \beta$. Thus $d_S(\alpha', \beta') \geq \frac{a+b}{2}(k+2) + k > k$.

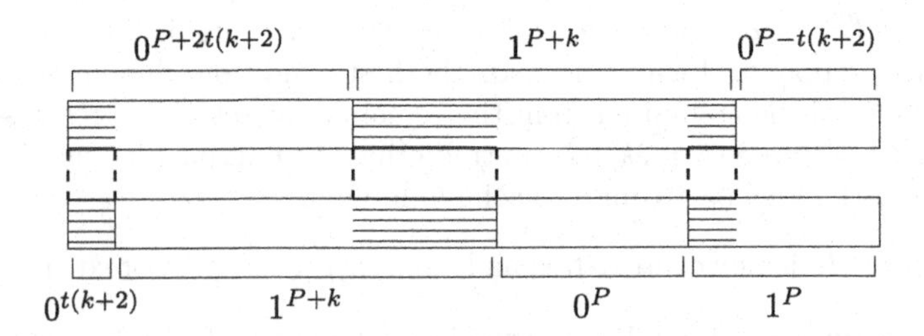


**Fig. 1.** Illustration of $z_{k+3}$ after pumping the first 0-sequence. The slanted lines denote the alignment pairs with the same symbols. Note that the second and the third overlaps already have $k$ symbols aligned.

Since every case contradicts the pumping lemma, $L'$ is not context-free, which leads to the fact that $L_{S=k}$ is not context-free. □

For the edit-distance case, we show that the Hamming distance and the edit-distance between the two catenated strings of a pseudo-copy language are the same. For a string $w = \alpha\beta$, we denote $\widehat{d_H}(w) = d_H(\alpha, \beta)$ and $\widehat{d_E}(w) = \min_{w=\alpha'\beta'} d_E(\alpha', \beta')$—the smallest edit-distance among all possible $\alpha', \beta'$ for $w$.

**Lemma 7.** *Let $w \in L(0^*1^*0^*1^*)$ be a string of even length. Then, $\widehat{d_E}(w) = \widehat{d_H}(w)$.*

*Proof* (Proof by induction). When $|w| = 0$, $\widehat{d_E}(w) = \widehat{d_H}(w) = 0$. Assume the claim holds for $|w| \leq n$. For $|w| = n+2$, suppose that the claim does not hold. Then, since $\widehat{d_E}(w) < \widehat{d_H}(w)$ is the case, there must be an optimal alignment with two symbols $u, v$ that matches to $\lambda$. Let $w'$ be the string without $u$ and $v$, then $\widehat{d_E}(w) = \widehat{d_E}(w') + 2 = \widehat{d_H}(w') + 2 < \widehat{d_H}(w)$. This cannot hold by case analysis on Lemma 4, contradicting the claim. □

**Theorem 8.** *For all $k \geq 0$, $L_{E=k}$ is not context-free.*

*Proof* (Proof by contradiction). Let $L' = L_{E=k} \cap \{0^a 1^b 0^c 1^d \mid a, b, c, d \geq k \text{ and } (a + b + c + d) \bmod 2 = 0\}$ and suppose $L_{E=k}$ is context-free. Then, $L'$ must be context-free and satisfies the pumping lemma. For an arbitrary pumping constant $p$, let $z = 0^l 1^{l+k} 0^{l+k} 1^l \in L'$, where $l = \max(p!, k)$. Then $z$ must have a factorization of $uvwxy$ such that $|vx| > 0$, $|vwx| \leq p$, and $uv^n wx^n y \in L'$ for all $n \geq 0$. By pumping $v$ and $x$, we show that $d_E$ exceeds $k$, which contradicts the pumping lemma. Referring to Lemma 7, $\widehat{d_E}(w) = \widehat{d_H}(w)$ for $w \in L'$. Instead of handling $d_E$, we can show that $d_H$ exceeds $k$ and this is already proven in Theorem 5. Therefore $L'$ is not context-free. By the above, $L_{E=k}$ is not context-free. □

One can define a hierarchy of pseudo-copy languages over exact error with these results. Theorem 5, 6 and 8 show that the class of languages with exact Hamming distance (exact similarities, edit-distance, resp.) is different from that of context-free languages.

From the proofs for the exact cases in Theorems 5, 6 and 8, one can observe that the error value of the chosen string strictly increases after pumping. These strings also apply to showing that the pseudo-copy languages are not context-free.

**Corollary 9.** *For all $k \geq 0$, $L_{H \leq k}$, $L_{S \leq k}$ and $L_{\leq k}$ are not context-free.*

*Proof.* In Theorem 5, we prove that $L_{H=k}$ is not context-free by showing the strings in $L_{H=k}$ have a larger error value when pumped, following Theorem 5. We can apply the exactly same procedure here. Instead of applying the pumping lemma directly to $L_{H \leq k}$, define $L' = L_{H=k} \cap \{0^a 1^b 0^c 1^d \mid a, b, c, d \geq k\}$. We know that $L'$ is not context-free as the pumped string has an error value larger than $k$. This is, in other words, the string which has an upper-bounded error value of $k$ can be pumped until the error value exceeds $k$. Therefore, the same string for $L_{H=k}$ contradicts the pumping lemma for $L_{H \leq k}$. Respectively on $L_{S \leq k}$ and $L_{E \leq k}$, we can use the proof procedure in each case similarly. □

## 4 Complements of Pseudo-copy Languages

The complements of pseudo-copy languages under the error measure $d_X \in \{d_H, d_S, d_E\}$ are defined as follows:

$$\overline{L_{X \leq k}} = L_{X \geq k+1}.$$

Therefore, only even-length strings exist in $\overline{L_{d_H \leq k}}$ and $\overline{L_{d_S \leq k}}$ for the Hamming distance and similarity, respectively. On the other hand, the complements of pseudo-copy languages under the edit-distance can have both odd-length and even-length strings.

**Theorem 10.** *For all $k \geq 4$, $L_{H \geq k}$ is not context-free.*

*Proof* (Proof by contradiction). The intuition is choosing a string of which a symbol, say 0, occupies the largest portion. Then we make an alignment of the sparse symbols, say 1's, to reduce the Hamming distance between its halves at least by one. For $n \geq 2$, suppose that $L_{H\geq 2n}$ is context-free and let $L' = L_{H\geq 2n} \cap \{w \mid |w|_1 = 2n+1\}$. Then, by the pumping lemma, there must be a pumping constant $p$ for $L'$.

Choose $z \in L'$ so that the position indices of 1's, $i_j \geq 1$ $(1 \leq j \leq 2n+1)$ are

$$i_j = \begin{cases} 2jP, & j \leq n, \\ (2j+1)P - 1, & j > n, \end{cases}$$

where $P = p!$. In other words, we place 1's in $z$ so that when we divide the string into halves, 1's from the first half alternate with 1's in the second half by $l = 2(n+1)P - 1$.

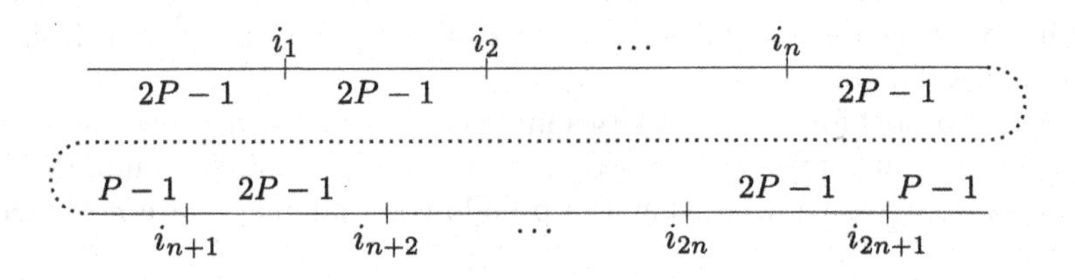


**Fig. 2.** An illustration of the chosen $z$

Let $i'_j$ denote the indices of 1's after pumping $v$ and $x$. It contradicts the pumping lemma if there exist $s$ and $t$ such that $i'_s \leq l + T < i'_t$ and $i'_t - i'_s = l + T$— two 1's in each half are aligned in the Hamming distance computation— where $2T$ is the length of the entire pumped string, and therefore, $v$ and $x$ duplicate $2T/|vx|$ times. The Hamming distance $d_H$ is at most $2n - 1$ in this case since two 1's do not contribute on the Hamming distance computation.

1. $|vx|$ is odd or $|vx|_1 = 1$, i.e., the pumping part contains 1.
   Since $uv^0xw^0y \notin L'$, it contradicts the pumping lemma.
2. $|vwx|_1 = 0$, i.e., the pumping occurs in a single 0-sequence. See Fig. 3 for an example.
   If $vwx$ is in the $h$-th 0-sequence, the indices $i'_j$ of 1's after pumping up $v$ and $x$ $2T/|vx|$ times is

$$i'_j = \begin{cases} i_j, & j < h, \\ i_j + 2T, & j \geq h, \end{cases}$$

   assuming $2T/|vx|$ is an integer.
   (a) If $h \leq n$, let $s = h$ and $t = h + n + 1$. $i'_t - i'_s = (2n+3)P - 1$. Figure 3(a) depicts how two 1's align. Note that the right-hand side of $h$-th 1 in the first half shortens by $T$ while that of $(h+n+1)$-th in the second half does not. These two 1's eventually meet after the pumping, when $l + T = i'_t - i'_s$, i.e., when $T = P$.

(b) If $n < h < 2n+1$, let $s = h - n$ and $t = h$. $i'_t - i'_s = (2n+1)P + 2T - 1$. Figure 3(b) depicts how two 1's align.
(c) If $h \in \{2n+1, 2n+2\}$, let $s = 1$ and $t = n+2$. $i'_t - i'_s = (2n+3)P - 1$.
Since, for all of three cases, $i'_t - i'_s = l + T$ holds if $T = P$, we pump up $v$ and $x$ $2P/|vx|$ times to contradict the pumping lemma for any positive integer $p$. Note that $P/|vx| = p!/|vx|$ is an integer.

3. $|vwx|_1 = 1, |vx|_1 = 0$, i.e., the pumping occurs in two 0-sequences.
For $h$ such that the $h$-th 1 is in $w$, the indices $i'_j$ after pumping up $v$ and $x$ $2T/|vx|$ is

$$i'_j = \begin{cases} i_j, & j < h, \\ i_j + a, & j = h, \\ i_j + 2T, & j > h, \end{cases}$$

where $a = |v| \cdot 2T/|vx|$.
(a) $h = 1$: Let $s = 2$ and $t = n+3$. $i'_t - i'_s = (2n+3)P - 1$.
(b) $2 \le h \le n$: Let $s = 1$ and $t = n+1$. $i'_t - i'_s = (2n+1)P + 2T - 1$.
(c) $n+1 \le h \le 2n-1$: Let $s = n$ and $t = 2n$. $i'_t - i'_s = (2n+1)P + 2T - 1$.
(d) $h \in \{2n, 2n+1\}$: Let $s = 1$ and $t = n+2$. $i'_t - i'_s = (2n+3)P - 1$.
For all cases, $i'_t - i'_s = l + T$ holds if $P = T$ and it contradicts the pumping lemma for CFLs.

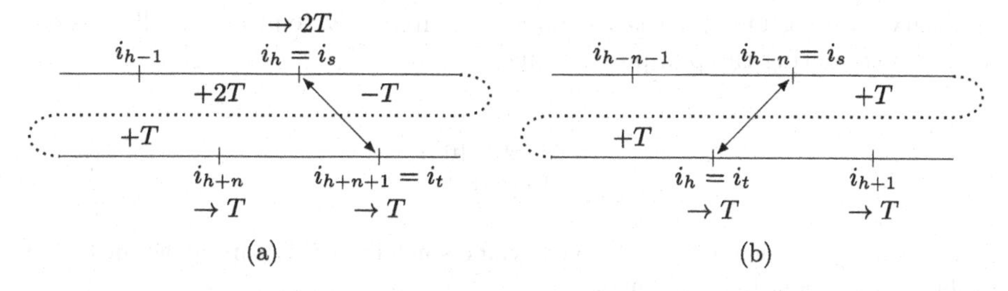


**Fig. 3.** Pumping a 0-block in (a): the first half and (b): the latter half. On each half, $\pm X$ denotes that the length of the sequence increases by $X$ and $\rightarrow X$ denotes that the specific point is pushed by $X$.

Because every case contradicts the pumping lemma, $L'$ is not context-free and neither is $L_{H \ge 2n}$. The case for $L_{H \ge 2n+1}$ is similar to the proof above. □

We have investigated languages with lower-bounded Hamming error values and in most cases, they are not context-free. However, it is still unknown whether or not $L_{H \ge 2}$ and $L_{H \ge 3}$ are context-free or non-context-free.

The case of $L_{S \ge k}$ starts with an obvious observation that $L_{S \ge 1}$ is context-free. $L_{S \ge 1}$ can be generated by the following CFG $G = (V, \Sigma, R, S)$:

$$S \rightarrow AA \mid BB,$$
$$A \rightarrow 0A0 \mid 0A1 \mid 1A0 \mid 1A1 \mid 0,$$
$$B \rightarrow 0B0 \mid 0B1 \mid 1B0 \mid 1B1 \mid 1.$$

Regarding $L_{S \geq k}$ with $k \geq 2$, we establish that the language over ternary alphabet is not context-free. Refer to the appendix for the full proof.

**Theorem 11.** *For all $k \geq 2$, $L_{S \geq k}$ over a ternary alphabet is not context-free.*

For binary case, some languages of the same type are not context-free.

**Theorem 12.** *For all $k \geq 5$, $L_{S \geq k}$ over a binary alphabet is not context-free.*

*Proof (Sketch).* This proof idea is similar to that of Theorem 10. Assume that $k$ is even and $L_{S \geq k}$ is context-free, then the following $L'$ is also context-free and should satisfy the pumping lemma.

$$L' = L_{S \geq k} \cap L([(01)^*00]^{k/2}[(01)^*11]^{k/2}(01)^*). \qquad (*)$$

For the illustration purpose, let $k = 6$. For a pumping constant $p$, let $P = (\max\{p, k\})!$ and choose

$$z = uvwxy = (01)^P 00(01)^{2P} 00(01)^{2P} 00(01)^{4P+1} 11(01)^{2P} 11(01)^{2P} 11(01)^P.$$

The 00's and 11's alternate like the 1's in Theorem 10. We can observe that $v$ and $x$ must be in $L((01)^* + (10)^*)$, otherwise, $uwy \notin L'$. It is also worth noting that Fig. 4 is the target alignment of 00 and 11 in each half, which reduces the similarity by two. Our goal is to show that similarity reduces for all possible cases, contradicting the pumping lemma.

$$\begin{array}{l} \ldots 01010\underline{00}01010\ldots \\ \ldots 10101\underline{11}0101\ldots \end{array}$$

**Fig. 4.** The target alignment for $z$. The symbols not from $(01)^*$ are underlined. The 11 shifts to the right by one symbol.

We make $z' = uv^iwx^iy$ to show such alignment by pumping up $v$ and $x$ sufficiently large. For example, when both $v$ and $x$ are in the first (01)-block, by shifting all 00's and 11's, $z'$ has similarity of $0 < k$ with $(k/2) = 3$ target alignments. The following is $z'$ after pumping up $v$ and $x$ sufficiently so that $(i-1) \cdot |vx| = 4P$, where $i$ is the number of duplications.

$$\begin{aligned} z' = &(01)^{P \underline{+2P}} \underline{00}(01)^{2P} \underline{00}(01)^{2P} \underline{00}(01)^{2P-P} 0 \\ &1(01)^{2P+P} \underline{11}(01)^{2P} \underline{11}(01)^{2P} \underline{11}(01)^P \end{aligned}$$

One can make at least one target alignment for every factorization of $uvwxy$ and it reduces the similarity at least by two. Thus, $z'$ has similarity of at most $k-2$, which contradicts the pumping lemma—$L_{S \geq k}$ is not context-free. This argument also holds for odd $k$, but with $k+1$ instead of $k$ for choosing a regular language to intersect with $L_{S \geq k}$ in $(*)$. □

We then provide Lemma 13 as a simple conversion scheme from a language with the edit-distance to a corresponding language with the Hamming distance.

**Lemma 13.** *Let $\Gamma = \{0, 1, \#\}$ be an alphabet and $h : \Gamma \to \Sigma$ be a homomorphism such that $h(0) = 0$, $h(1) = 1$ and $h(\#) = \lambda$. Then, $h^{-1}(L_{E \geq k}) \cap L((\Sigma^2)^*)$ is the language with $d_H \geq k$ over $\Gamma$.*

*Proof.* For $\alpha\beta \in L_{E \geq k}$, every alignment of $\alpha$ and $\beta$ has at least $k$ different pairs. Then, $h^{-1}$ replaces $\lambda$ in such alignment pairs in $L_{E \geq k}$ or inserts $(\#, \#)$ pairs. Thus, the strings with even length represent alignments of the strings in $L_{E \geq k}$, with at least $k$ differences.

On the other hand, let $L = L_{H \geq k}$ over $\Gamma$. Then, on its alignment of two halves, one can derive an alignment for strings with at least $k$ different pairs by replacing # with $\lambda$. □

Since context-free languages are closed under inverse homomorphism [9], if a language with the edit-distance is context-free, then the resulting language, which is one with the Hamming distance, must be context-free. We now show that such language is not context-free due to Theorem 10 and Lemma 13.

**Theorem 14.** *For $k \geq 4$, $L_{E \geq k}$ is not context-free.*

*Proof* (Proof by contradiction). Suppose that $L_{E \geq k}$ is context-free. Consider the alphabet $\Gamma$ and the homomorphism $h$ in Lemma 13. Since context-free languages are closed under these operations, $h^{-1}(L_{E \geq k}) \cap L((\Sigma^2)^*)$ must be context-free. However, this language is $L_{H \geq k}$ over $\Gamma$, which is proven to be non-context-free in Theorem 10 for $k \geq 4$. □

Finally, we can easily show that the pseudo-copy languages are strictly included in the class of context-sensitive languages by constructing realtime NQAs. Refer to the appendix for full proofs.

## 5 Conclusions

We have examined the problems of determining non-context-freeness of pseudo-copy languages and their complements defined under error measures such as the Hamming distance and the edit-distance. Unlike $\overline{\text{COPY}}$, the languages are proved to be non-context-free. Especially, our results show that most pseudo-copy languages as well as their complements are not context-free. It is interesting as the complements are not significantly different from $\overline{\text{COPY}}$ which is context-free.

There are, however, remaining problems that need further investigation to determine their context-freeness. Even though our results show that the answer for Problem 1 is not context-free, it still remains open for the complements of extended pseudo-copy languages. $L_{H \geq k}$, $L_{E \geq k}$ and $L_{S \geq k}$ regarding errors of small lower-bounds are to be examined in further study. We hope that our findings are helpful for answering these questions.

**Acknowledgments.** We wish to thank the referees for valuable suggestions that improve the presentation of the paper.

This research was supported by the NRF grant funded by MIST (NRF-2020R1A4A3079947, NRF-2018R1D1A1A09084107).

## References

1. Anderson, T., Rampersad, N., Santean, N., Shallit, J.: Finite automata, palindromes, powers, and patterns. In: Proceedings of the 2nd International Conference on Language and Automata Theory and Applications, pp. 52–63 (2008)
2. Apostolico, A., Preparata, F.: Optimal off-line detection of repetitions in a string. Theor. Comput. Sci. **22**(3), 297–315 (1983)
3. Bordihn, H., Mitrana, V., Păun, A., Păun, M.: Hairpin completions and reductions: semilinearity properties. Nat. Comput. **20**(2), 193–203 (2020). https://doi.org/10.1007/s11047-020-09797-0
4. Bordihn, H., Shallit, J.: Personal communication
5. Crochemore, M.: Recherche linéaire d'un carré dans un mot. Comptes rendus de l'Académie des sciences. Série I, Mathématique **296**(18), 781–784 (1983)
6. Ehrenfeucht, A., Rozenberg, G.: On the separating power of EOL systems. RAIRO Theor. Inform. Appl. **17**(1), 13–22 (1983)
7. Gusfield, D.: Algorithms on Strings, Trees, and Sequences: Computer Science and Computational Biology. Cambridge University Press, USA (1997)
8. Hamming, R.W.: Error detecting and error correcting codes. Bell Syst. Tech. J. **29**(2), 147–160 (1950)
9. Hopcroft, J.E., Motwani, R., Ullman, J.D.: Introduction to Automata Theory, Languages, and Computation, 3rd edn. Pearson Education Inc, Boston, MA, USA (2006)
10. Kolpakov, R., Kucherov, G.: Finding approximate repetitions under Hamming distance. Theor. Comput. Sci. **303**(1), 135–156 (2003)
11. Kutrib, M., Malcher, A., Wendlandt, M.: Queue automata: foundations and developments. In: Adamatzky, A. (ed.) Reversibility and Universality. ECC, vol. 30, pp. 385–431. Springer, Cham (2018). https://doi.org/10.1007/978-3-319-73216-9_19
12. Landau, G.M., Schmidt, J.P., Sokol, D.: An algorithm for approximate tandem repeats. J. Comput. Biol. **8**(1), 1–18 (2001)
13. Levenshtein, V.I.: Binary codes capable of correcting deletions, insertions, and reversals. Sov. Phys. Doklady **10**(8), 707–710 (1966)
14. Lischke, G.: Squares of regular languages. Math. Log. Q. **51**(3), 299–304 (2005)
15. Lothaire, M.: Combinatorics on Words, 2nd ed. Cambridge University Press, Cambridge (1997)
16. Main, M.G., Lorentz, R.J.: An $O(n \log n)$ algorithm for finding all repetitions in a string. J. Algorithms **5**(3), 422–432 (1984)
17. Ogden, W.F., Ross, R.J., Winklmann, K.: An "interchange lemma" for context-free languages. SIAM J. Comput. **14**(2), 410–415 (1985)
18. Ross, R.J., Winklmann, K.: Repetitive strings are not context-free. RAIRO Theor. Inform. Appl. **16**(3), 191–199 (1982)
19. Sipser, M.: Introduction to the Theory of Computation, 3rd edn. Cengage Learning, Boston, MA, USA (2013)
20. Stoye, J., Gusfield, D.: Simple and flexible detection of contiguous repeats using a suffix tree. Theor. Comput. Sci. **270**(1), 843–856 (2002)
21. Vollmar, R.: Über einen automaten mit pufferspeicherung. Computing **5**(1), 57–70 (1970)

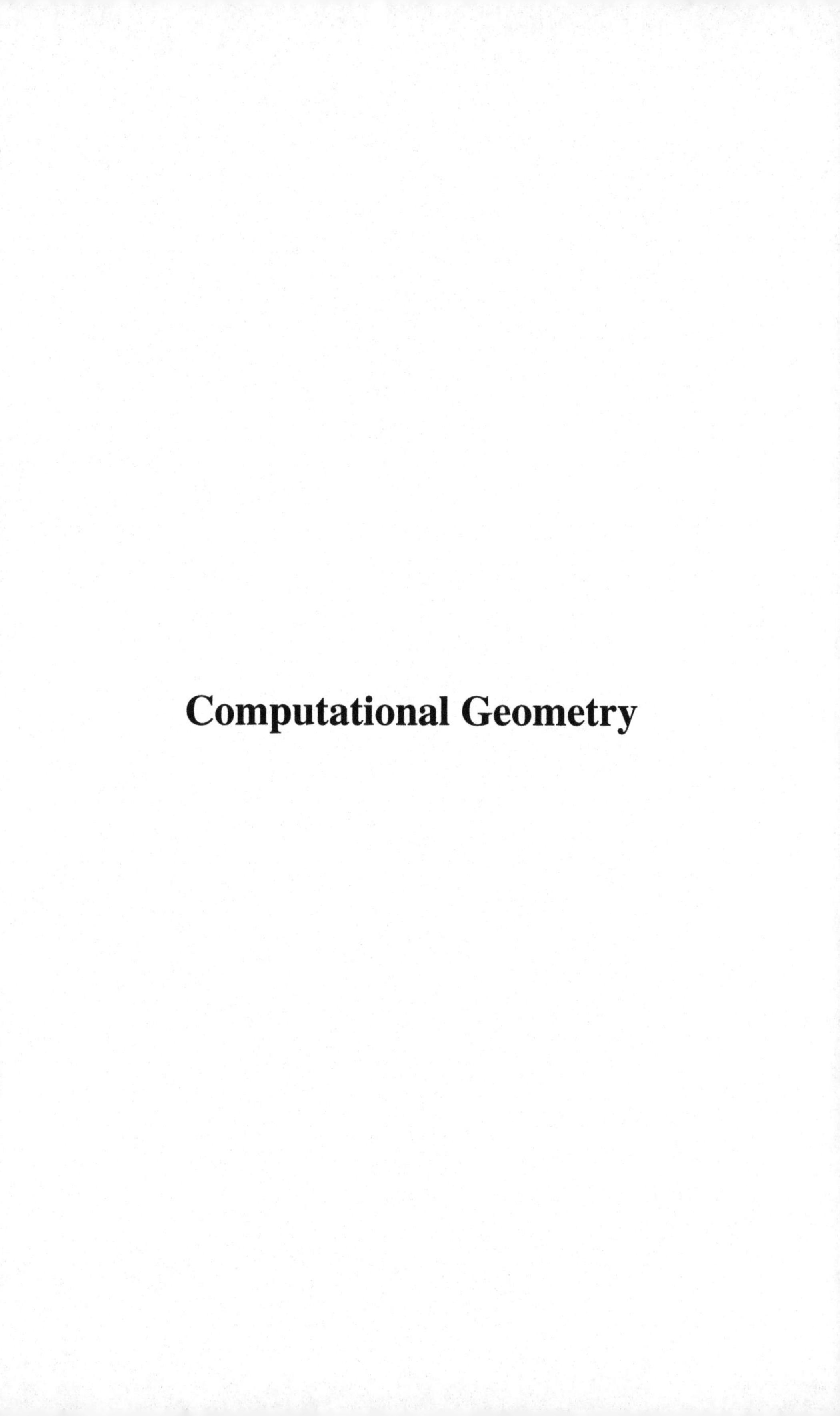

# Computational Geometry

# Bottleneck Convex Subsets: Finding $k$ Large Convex Sets in a Point Set

Stephane Durocher[1], J. Mark Keil[2(✉)], Saeed Mehrabi[3], and Debajyoti Mondal[2]

[1] University of Manitoba, Winnipeg, Canada
durocher@cs.umanitoba.ca
[2] University of Saskatchewan, Saskatoon, Canada
{keil,dmondal}@cs.usask.ca
[3] Carleton University, Ottawa, Canada

**Abstract.** Chvátal and Klincsek (1980) gave an $O(n^3)$-time algorithm for the problem of finding a maximum-cardinality convex subset of an arbitrary given set $P$ of $n$ points in the plane. This paper examines a generalization of the problem, the *Bottleneck Convex Subsets* problem: given a set $P$ of $n$ points in the plane and a positive integer $k$, select $k$ pairwise disjoint convex subsets of $P$ such that the cardinality of the smallest subset is maximized. Equivalently, a solution maximizes the cardinality of $k$ mutually disjoint convex subsets of $P$ of equal cardinality. We show the problem is NP-hard when $k$ is an arbitrary input parameter, we give an algorithm that solves the problem exactly, with running time polynomial in $n$ when $k$ is fixed, and we give a fixed-parameter tractable algorithm parameterized in terms of the number of points strictly interior to the convex hull.

**Keywords:** Computational geometry · Convex set · FPT-algorithm · NP-Hard

## 1 Introduction

A set $P$ of points in the plane is *convex* if for every $p \in P$ there exists a closed half-plane $H^+$ such that $H^+ \cap P = \{p\}$. Determining whether a given set $P$ of $n$ points in the plane is convex requires $\Theta(n \log n)$ time in the worst case, corresponding to the time required to determine whether the convex hull of $P$ has $n$ vertices on its boundary [19]. Chvátal and Klincsek [4] gave an $O(n^3)$-time and $O(n^2)$-space algorithm to find a maximum-cardinality convex subset of any given set $P$ of $n$ points in the plane. Later, Edelsbrunner and Guibas [8] improved the space complexity to $O(n)$. In this paper, we examine a generalization of the problems to multiple convex subsets of $P$. Given a set $P$ of points in the plane and

The work is supported in part by the Natural Sciences and Engineering Research Council of Canada (NSERC). The authors dedicate the work in memory of Saeed Mehrabi.


C.-Y. Chen et al. (Eds.): COCOON 2021, LNCS 13025, pp. 203–214, 2021.
https://doi.org/10.1007/978-3-030-89543-3_17

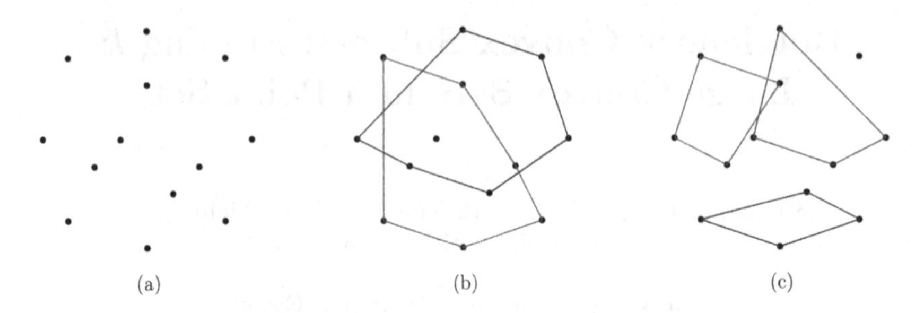


**Fig. 1.** (a) A point set $P$. (b) A solution to the Bottleneck Convex Subsets problem when $k = 2$. (c) A solution when $k = 3$.

a positive integer $k$, we examine the problem of finding $k$ convex and mutually disjoint subsets of $P$, such that the cardinality of the smallest set is maximized (e.g., see Fig. 1). We define the problem formally, as follows.

**BOTTLENECK CONVEX SUBSETS**
**Instance:** A set $P$ of $n$ points in $\mathbb{R}^2$, and a positive integer $k$.
**Problem:** Select $k$ sets $P_1, \ldots, P_k$ such that

- $\forall i \in \{1, \ldots k\}$, $P_i \subseteq P$,
- $\forall i \in \{1, \ldots k\}$, $P_i$ is convex,
- $\forall \{i, j\} \subseteq \{1, \ldots k\}$, $i \neq j \Rightarrow P_i \cap P_j = \varnothing$, and
- $\min_{i \in \{1, \ldots, k\}} |P_i|$ is maximized.

Since every subset of a convex set of points remains convex, any $k$ convex sets can be made to have equal cardinality by removing points from any set whose cardinality exceeds that of the smallest set. Therefore, an equivalent problem is to find $k$ mutually disjoint convex subsets of $P$ of equal cardinality, where the cardinality is maximized. The problem also relates to the problem of finding a convex point set embedding of a graph in a point set [7], where in this case the graph consists of $k$ cycles.

### 1.1 Our Contributions

In this paper we examine the problem of finding $k$ large convex subsets of a given point set with $n$ points. Our contributions are as follows:

1. We give a polynomial-time algorithm that solves Bottleneck Convex Subsets for any fixed $k$. The algorithm constructs a directed acyclic graph $G$ whose vertices correspond to distinct configurations of edges passing though vertical slabs between neighbouring points of $P$. A solution to the problem is found by identifying a node in $G$ associated with a maximum-cardinality set that is reachable from the source node.
2. Using a reduction from a restricted version of Numerical 3-Dimensional Matching, which is known to be NP-complete, we show that Bottleneck Convex Subsets is NP-hard when $k$ is an arbitrary input parameter.

3. We show that Bottleneck Convex Subsets is fixed-parameter tractable when parameterized by the number of points that are strictly interior to the convex hull of the given point set, i.e., the number of non-extreme points. Therefore, if the number of points interior to the convex hull is fixed, then for every $k$, Bottleneck Convex Subsets can be solved in polynomial time.

### 1.2 Related Work

A *convex k-gon* is a convex set with $k$ points. A convex *k-hole* within a set $P$ is a convex $k$-gon on a subset of $P$ whose convex hull is empty of any other points of $P$. A rich body of research examines convex $k$-holes in point sets [22]. By the Erdős-Szekeres theorem [12], every point set with $n$ points in the Euclidean plan contains a convex $k$-gon for some $k \in \Omega(\log n)$. Urabe [23] showed that by repeatedly extracting such a convex $\Omega(\log n)$-gon, one can partition a point set into $O(n/\log n)$ convex subsets, each of size $O(\log n)$.

Given a set $P$ of $n$ points in the plane, there exist $O(n^3)$-time algorithms to compute a largest convex subset of $P$ [4,8] and a largest empty convex subset of $P$ [2]. Both problems are NP-hard in $\mathbb{R}^3$ [15]. In fact, finding a largest empty convex subset is W[1]-hard in $\mathbb{R}^3$ [15]. González-Aguilar et al. [16] have recently examined the problem of finding a largest convex set in the rectilinear setting.

The *convex cover number* of a point set $P$ is the minimum number of disjoint convex sets that covers $P$. The *convex partition number* of a point set $P$ is the minimum number of convex sets with disjoint convex hulls (in addition to their vertex sets being pairwise vertex disjoint) that covers $P$. Urabe [23] examined lower and upper bounds on the convex cover number and the convex partition number. He showed that the convex cover number of a set of $n$ points in $\mathbb{R}^2$ is in $\Theta(n/\log n)$ and its convex partition number is bounded from above $\lceil \frac{2n}{7} \rceil$. Furthermore, there exist point sets with convex partition number at least $\lceil \frac{n-1}{4} \rceil$.

Arkin et al. [1] proved that both finding the convex cover number and the convex partition number of a point set are NP-hard problems, and gave a polynomial-time $O(\log n)$-approximation algorithm for both problems. Although the Bottleneck Convex Subsets problem appears to be similar to the convex cover number problem as both problems attempt to find disjoint convex sets, the objective functions are different. Neither the NP-hardness proof nor the approximation result for convex cover number [1] readily extends to the Bottleneck Convex Subsets problem. Previous work has also considered partitioning a point set into empty convex sets, where the convex hulls of the sets do not contain any interior point. For the number of empty convex point sets, an upper bound of $\lceil \frac{9n}{34} \rceil$ and a lower bound of $\lceil \frac{n+1}{4} \rceil$ is known [5]. We refer the readers to [10,11] for related problems on finding convex sets with various optimization criteria.

Another related problem in this context is to partition a given point set using a minimum number of lines (Point-Line-Cover), which Megiddo and Tamir [21] showed to be NP-hard, and was subsquently shown to be APX-hard [3,20]. Point-Line-Cover is known to be fixed-parameter tractable when parameterized on the number of lines. Whether the minimum convex cover problem is fixed-parameter tractable remains an open problem [9]. Note that for any fixed $k$, one can decide whether the minimum convex cover number of a point set is at most $k$ in polynomial time [1].

Previous work on the Ramsey-remainder problem provides insight into the Bottleneck Convex Subsets problem [13]. Given an integer $i$, the Ramsey-remainder is the smallest integer $rr(i)$ such that for every sufficiently large point set, all but $rr(i)$ points can be partitioned into convex sets of size at least $i$. Therefore, a Bottleneck Convex Subsets problem with sufficiently large $n$ and with $k \leq \lfloor \frac{n-rr(k)}{k} \rfloor$ must have a solution where the size of the smallest convex set is at least $k$. Note that the Bottleneck Convex Subsets problem is straightforward to solve for the case when $k \geq n/3$, i.e., one needs to compute a balanced partition without worrying about the convexity of the sets. However, the case when $k = n/4$ already becomes nontrivial. Károlyi [18] derived a necessary and sufficient condition for a set of $4n$ points in general position to admit a partition into $n$ convex quadrilaterals, and gave an $O(n \log n)$-time algorithm to decide whether such a partition exists.

## 2 A Polynomial-Time Algorithm for a Fixed $k$

Given a set $P$ of $n$ points in the plane and a fixed integer $k$, we describe an $O(kn^{5k+3})$-time algorithm that solves Bottleneck Convex Subsets for any fixed $k$. The idea is to construct a directed acyclic graph $G$ whose vertices each correspond to a vertical slab of the plane in a given state with respect to the selected subsets $P_1, \ldots, P_k$ of $P$, with an edge from one slab to the slab immediately to its right if the states of the two neighbouring slabs form a locally mutually compatible solution. A feasible solution ($P_1, \ldots, P_k$ are mutually disjoint convex subsets of $P$) corresponds to a directed path starting at the root node in $G$, i.e., a sequence of consecutive compatible slabs. Among the feasible solutions, an optimal solution ($\min_{i \in \{1,\ldots,k\}} |P_i|$ is maximized) corresponds to a path that ends at a node for which the cardinality of the smallest set is maximized.

Rotate $P$ such that no two of its points lie on a common vertical line. Partition the plane into $n-1$ vertical slabs, $S_1, \ldots, S_{n-1}$, determined by the $n$ vertical lines through points of $P$. Let $L$ be the set of $\binom{n}{2}$ line segments whose endpoints are pairs of points in $P$. Within each slab, $S_i$, consider the set of line segments $L_i = \{l \cap S_i \mid l \in L\}$. A convex point set corresponds to the vertices of a convex polygon; in a feasible solution, $j$ convex polygons intersect $S_i$ for some $j \in \{0, \ldots, k\}$. Each of these polygons has a top segment and a bottom segment in $L_i$. There are at most $\binom{|L_i|}{2}$ possible choices of segments in $L_i$ for the first polygon, $\binom{|L_i|-2}{2}$ for the second polygon, ..., and $\binom{|L_i|-2(j-1)}{2}$ for the $j$th polygon, giving $\prod_{x=0}^{j-1} \binom{|L_i|-2x}{2} \in O(|L_i|^{2j}) = O(n^{4j})$ possible combinations of edges in $S_i$ for a given $j \in \{0, \ldots, k\}$.

We construct an unweighted directed acyclic graph $G$. Each vertex in $V(G)$ corresponds to a slab $S_i$, a $j \in \{0, \ldots, k\}$, and a top edge and a bottom edge

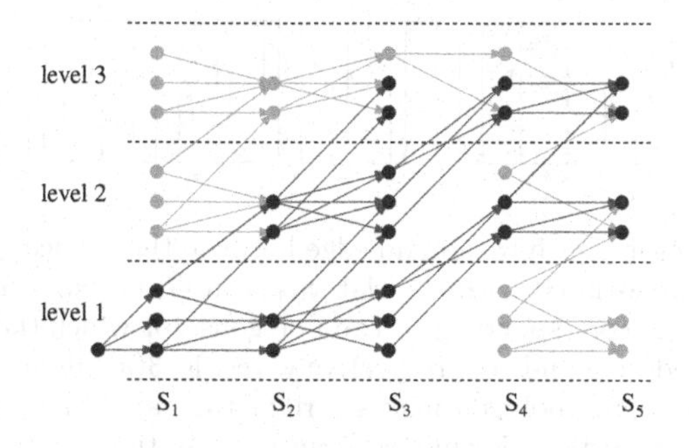


**Fig. 2.** Each slab $S_i$ has various combinations of pairs of edges possible, each of which corresponds to a vertex in $G$, which is copied at levels 1 through $n/k$. Directed edges are added from a vertex associated with slab $S_i$ to a vertex associated with a compatible slab $S_{i+1}$. The edge remains at the same level if the cardinality of the smallest set in $S_1 \cup \cdots \cup S_{i+1}$ remains unchanged; the level of $S_{i+1}$ is one greater than the level of $S_i$ if the cardinality of the smallest set in $S_1 \cup \cdots \cup S_{i+1}$ increases. Some vertices cannot be reached by any path from any source node at level 1 in slab $S_1$; these vertices and their out-edges are shaded gray. A feasible solution corresponds to a path rooted at a source node associated with the slab $S_1$ on level 1. An optimal solution ends at a sink node at the highest level among all feasible solutions.

for each of the $j$ convex polygons that intersect $S_i$. Consequently, the number of vertices in $G$ is $O(\sum_{i=1}^{n-1} \sum_{j=0}^{k} n^{4j}) = O(kn^{4k+1})$.

Furthermore, we create $(n/k)^k$ copies of each vertex associated with a slab $S_i$, each of which is assigned a distinct value $(\ell_1, \ldots, \ell_k) \in \mathbb{Z}^k$, where for each $j \in \{1, \ldots, k\}$, $\ell_j = |P_j \cap (S_1 \cup \cdots \cup S_i)|$, i.e., the number of points of $P_j$ that lie in the first $i$ slabs. We refer to $\ell = \min_{j \in \{1,\ldots,k\}} \ell_j$ as the vertex's *level*. Each vertex at level $\ell$ in $G$ corresponds to a slab $S_i$, such that the minimum cardinality of any polygon in $S_1 \cup \ldots \cup S_i$ (or partial polygon if it includes points to the right of $S_i$) is $\ell$. Therefore, the resulting graph $G$ has $O((n/k)^k kn^{4k+1}) \subseteq O(\frac{1}{k^{k-1}} \cdot n^{5k+1})$ vertices. See Fig. 2.

Every slab has exactly one point of $P$ on its left boundary and one on its right boundary. For each vertex $v$ in $G$, let $v_l$ and $v_r$ denote these two points of $P$ for the slab corresponding to $v$. We add an edge from vertex $u$ to vertex $v$ in $G$ if they are *compatible*. See Fig. 3. The vertices $u$ and $v$ are compatible if:

- $u$ and $v$ correspond to neighbouring slabs, $u$ to $S_i$ and $v$ to $S_{i+1}$, for some $i$, and
- all top and bottom segments associated with $u$ that do not pass through $p_i$ continue in $v$, where $p_i = u_r = v_l$ is the point of $P$ on the common boundary of $S_i$ and $S_{i+1}$, and
- one of the four following conditions is met:

Case 1. either (a) one top associated with $u$ ends at $p_i$ and one top associated with $v$ begins at $p_i$, forming a right turn at $p_i$, or (b) one bottom associated with $u$ ends at $p_i$ and one bottom associated with $v$ begins

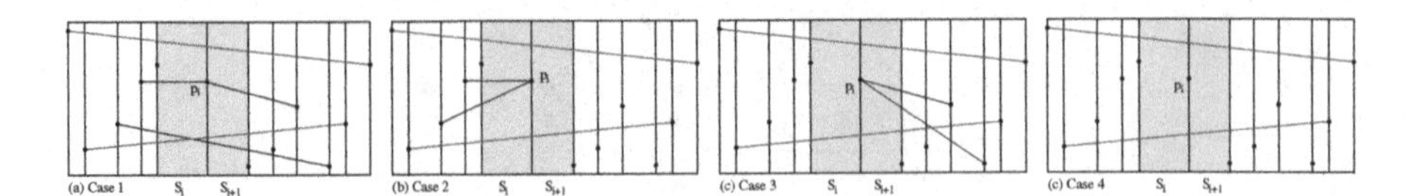


**Fig. 3.** The four cases in which we add an edge between the vertices $u$ (associated with the slab $S_i$) and $v$ (associated with the slab $S_{i+1}$) in $G$; i.e., $u$ and $v$ are *compatible.* In this example, $k = 2$, corresponding to two polygons, for which the edges through $S_i$ and $S_{i+1}$ are coloured blue and red, respectively. In Fig. 3(a), $p_i$ lies on the upper hull of the blue polygon, so the polygon makes a right turn at $p_i$, i.e., the angle below $p_i$ must be convex. Figure 3(d), $p_i$ is omitted from the selection.

at $p_i$, forming a left turn at $p_i$ (all polygons in $S_i$ continue in $S_{i+1}$; the number of edges in $S_i$ is equal to that in $S_{i+1}$);

Case 2. one top and one bottom associated with $u$ end at $p_i$, (one polygon ends in $S_i$ and all remaining polygons continue into $S_{i+1}$);

Case 3. no top or bottom associated with $u$ end at $p_i$, but one top and one bottom associated with $v$ start at $p_i$ (one polygon starts in $S_{i+1}$ and all remaining polygons continue from $S_i$ into $S_{i+1}$).

Case 4. all edges in $u$ continue into $v$ and no edge passes through $p_i = u_r = v_l$ (all polygons in $S_i$ continue into $S_{i+1}$; the number of edges in $S_i$ is equal to that in $S_{i+1}$).

For a given vertex $u$ at most $n - 2$ edges satisfy Case 1 (there are at most $n - 2$ possible edges that continue from $p_i$ to form a convex bend), at most one edge satisfies Case 2, at most $\binom{n-3}{2}$ edges satisfy Case 3, and at most one edge satisfies Case 4. Consequently, the number of edges in $G$ is $O(n^2|V(G)|) \subseteq O(\frac{1}{k^{k-1}} \cdot n^{5k+3})$.

Any path from a source on level 1 to a highest-level node corresponds to an optimal solution, and can be found using breadth-first search in time proportional to the number of edges in $G$. The resulting worst-case running time is proportional to the number of vertices and edges in $G$: $O(|V(G)| + |E(G)|) = O(\frac{1}{k^{k-1}} \cdot n^{5k+3})$. In addition to storing a single in-neighbour from which a longest path reaches each node $u$, we can maintain a list of all of its in-neighbours that give a longest path, allowing the algorithm to reconstruct all distinct optimal solutions with the running time increased only by the output size.

The time for constructing the graph $G$ is proportional to its number of edges. The combinations of $\binom{n}{2j}$ line segments in a slab $S_i$ on level $j$ can be enumerated and created in $O(1)$ time each, with $O(1)$ time per edge added if graph vertices are indexed according to their slab, their level, and the line segments they include. The level of each node in $G$ is determined in $O(1)$ time per node by examining the level of any of its in-neighbours; the level increases by one in Cases 1 and 2 if the point $p_i$ is added to the minimum-cardinality set and that set is the unique minimum.

**Theorem 1.** *Given a set $P$ of $n$ points in the plane, and a positive integer $k$, Bottleneck Convex Subsets can be solved exactly in $O(\frac{1}{k^{k-1}} \cdot n^{5k+3})$ time.*

# 3 NP-Hardness

In this section we show that Bottleneck Convex Subsets is NP-hard. We first introduce some notation. Let $x(p), y(p)$ be the $x$ and $y$-coordinates of a point $p$. An angle $\angle pqr$ determined by points $p, q$ and $r$ is called a *y-monotone angle* if $y(p) > y(q) > y(r)$. A $y$-monotone angle is *left-facing* (resp. *right-facing*) if the point $q$ lies interior to the left (resp., right) half-plane of the line through $pr$. If $q$ lies on the line through $pr$, then we refer to $\angle pqr$ as a *straight angle*.

The idea of the hardness proof is as follows. We first prove that given a set of $3n$ points in the Euclidean plane, it is NP-hard to determine whether the points can be partitioned into $n$ $y$-monotone angles, where none of them are right facing (Sect. 3.1). We then reduce this problem to Bottleneck Convex Subsets (Sect. 3.2).

## 3.1 Covering Points by Straight or Left-Facing Angles

In this section we show that given a set of $3n$ points in the Euclidean plane, it is NP-hard to determine whether the points can be partitioned into $n$ $y$-monotone angles, where none of them are right facing. In fact, we prove the problem to be NP-hard in a restricted setting, as follows:

**ANGLE PARTITION**
**Instance:** A set $P$ of $3n$ points lying on three parallel horizontal lines ($y = 0, y = 1$ and $y = 2$) in the plane, where each line contains exactly $n$ points.
**Problem:** Partition $P$ into at most $n$ $y$-monotone angles, where none of them are right facing.

We reduce Distinct 3-Numerical Matching with Target Sums (DNMTS), which is known to be strongly NP-complete [17, Corollary 8].

**DISTINCT NUMERICAL MATCHING WITH TARGET SUM**
**Instance:** Three sets $A = \{a_1, \ldots, a_n\}, B = \{b_1, \ldots, b_n\}, C = \{c_1, \ldots, c_n\}$, each with $n$ distinct positive integers, where $\sum_{i=1}^{n} a_i + \sum_{i=1}^{n} b_i = \sum_{i=1}^{n} c_i$.
**Problem:** Decide whether there exist $n$ triples $(a_i, b_j, c_k)$, where $1 \leq i, j, k \leq n$, such that $a_i + b_j = c_k$ and no two triples share an element.

**Theorem 2.** *Angle Partition is NP-hard.*

*Proof.* Let $M = (X, Y, Z)$ be an instance of DNMTS, where each set $A, B, C$ contains $n$ positive integers. We now construct an instance $Q$ of Angle Partition as follows: (I) For each $a \in A$, create a point at $(a, 0)$. (II) For each $b \in B$, create a point at $(b, 2)$. (III) For each $c \in C$, create a point at $(c/2, 1)$.

This completes the construction of the point set $P$ of the Angle Partition instance $Q$ (e.g., see Fig. 4(a)). Since the numbers in $A, B, C$ are distinct, no two points in $P$ will coincide. Note that by definition, a $y$-monotone angle must contain one point from each of the lines $y = 0, y = 1$ and $y = 2$. Furthermore, every straight angle $\angle pqr$ will satisfy the equation $\frac{x(p)+x(r)}{2} = x(q)$. This transformation is inspired by a 3-SUM hardness proof for 'GeomBase' [14].

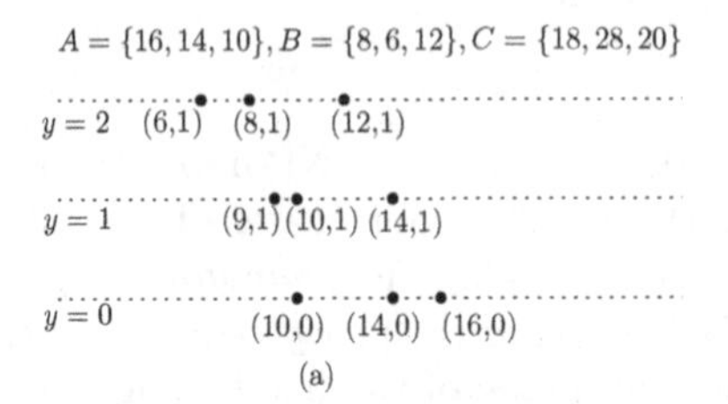


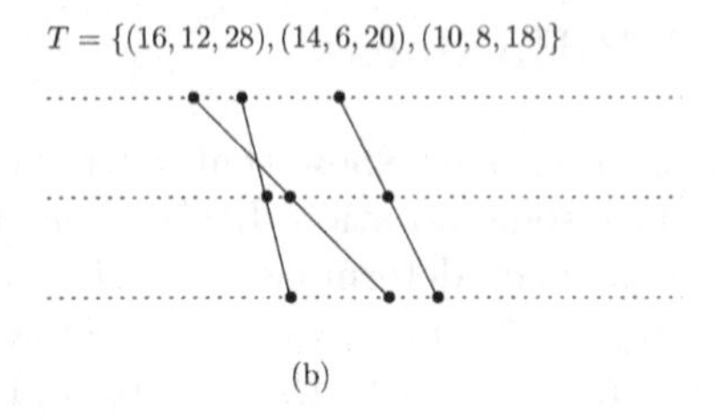


**Fig. 4.** (a) Construction of $Q$ from an instance $M$ of DNMTS. (b) A solution for $M$ and the corresponding angles of $Q$.

We now show that $M$ has an affirmative solution if and only if $P$ admits a partition into $n$ $y$-monotone angles where none of them are right facing.

First consider that $M$ has an affirmative answer, i.e., a set of $n$ triples $(a_i, b_j, c_k)$, where $1 \leq i, j, k \leq n$, such that $a_i + b_j = c_k$ and no two triples share an element. Therefore, we will have $\frac{(a_i+b_j)}{2} = \frac{c_k}{2}$. Hence we will find a straight line through $(a_i, 0), (b_k, 2), (c_j/2, 1)$. These lines will form $n$ $y$-monotone straight angles (e.g., see Fig. 4(b)). Since none of these angles are right facing, this provides an affirmative solution for the instance $Q$.

Consider now the case when $Q$ has an affirmative solution $T$, i.e., a partition of $P$ into $n$ $y$-monotone angles, where none of them are right facing. We first claim that (Step 1) all these $n$ $y$-monotone angles must be straight angles and then (Step 2) show how to construct an affirmative solution for $M$.

*Step 1:* Suppose for a contradiction that the solution $T$ contains one or more left-facing angles. For each left-facing angle $\angle rst$, where $r, s, t$ are on lines $y = 0, y = 1$ and $y = 2$, respectively, we have $x(s) < \frac{x(r)+x(t)}{2}$. For each straight angle $\angle rst$, we have $x(s) = \frac{x(r)+x(t)}{2}$. Since we do not have any right-facing angle, the following inequality holds: $\sum_{\angle rst \in T} x(s) < \sum_{\angle rst \in T} \frac{x(r)}{2} + \sum_{\angle rst \in T} \frac{x(t)}{2}$. Since no two angles share a point, we have $\sum_{i=1}^{n}(c_i/2) < \sum_{i=1}^{n}(a_i/2) + \sum_{i=1}^{n}(b_i/2)$, which contradicts that $M$ is an affirmative instance of DNMTS.

*Step 2:* We now transform the $y$-monotone straight angles of $T$ into $n$ triples for $M$. For each angle, $\angle rst$, where $r, s, t$ are on lines $y = 0, y = 1$ and $y = 2$, we construct a triple $(x(r), x(t), 2x(s))$. Since $\angle rst$ is a straight angle, $x(r) + x(t) = 2x(s)$. Since no two angles share a point, the triples will be disjoint. □

### 3.2 Bottleneck Convex Subsets Is NP-Hard

In this section we reduce Angle Partition to Bottleneck Convex Subsets. Let $P$ be an instance of Angle Partition, i.e., three lines $y = 0, y = 1$ and $y = 2$, each line containing $n$ points. We construct an instance $H$ of Bottleneck Convex Subsets with $k = n$.

**Construction of $H$:** We first take a copy $P'$ of the points of $P$ and include those in $H$. Let $\Delta$ be a sufficiently large number (to be determined later). We

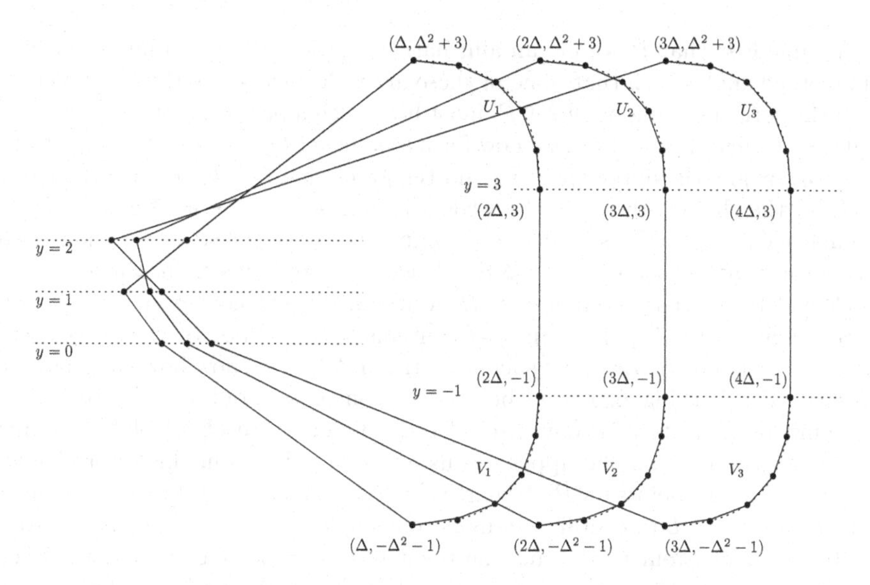


**Fig. 5.** Illustration for the construction of $H$. Note that this is only a schematic representation, which violates the property that all the chains are inside the wedge determined by the $y$-monotone angles.

now construct $n$ *upper chains*. The $i$th upper chain $U_i$, where $1 \leq i \leq n$, is constructed following the step below (see Fig. 5).

> *Construction of* $U_i$: Place two points at the coordinates $(i\Delta, \Delta^2 + 3)$ and $((i+1)\Delta, 3)$. Let $C$ be the curve determined by $y = \Delta^2 + 3 - (x - i\Delta)^2$, which passes through these two points. Place $2n$ points uniformly on $C$ between $(i\Delta, \Delta^2 + 3)$ and $((i+1)\Delta, 3)$.

Each upper chain contains $(2n+2)$ points. We define the $n$ lower chains symmetrically, where each lower chain $V_i$ starts at $(i\Delta, -\Delta^2 - 1)$ and ends at $((i+1)\Delta, -1)$.

We now choose the parameter $\Delta$. Let $t$ be the maximum $x$-coordinate of the points in $P$, and set $\Delta$ to be $t^4$. This ensures that for any line $\ell$ with non-zero slope passing through two points of $P$, the upper and lower chains lie on the right half-plane of $\ell$. This concludes the construction of the Bottleneck Convex Subsets instance $H$, where $k = n$. Note that $H$ has $3n + n(4n + 4) = n(4n + 7)$ points. In the best possible scenario, one may expect to cover all the points and have a partition into $n$ disjoint convex subsets, where each set contains $(4n + 7)$ points. The proof of Lemma 1 is in the full version [6].

**Lemma 1.** *Let $W$ be a partition of the upper and lower chains into a set $L$ of at most $n$ disjoint convex sets. Then each convex set in $L$ contains at least one point from an upper chain and one point from a lower chain.*

**Reduction:** We now show that the Angle Partition instance $P$ admits an affirmative solution if and only if the Bottleneck Convex Subsets instance $H$ admits $k(= n)$ disjoint convex sets with each set containing $(4n + 7)$ points.

Assume first that $P$ admits an affirmative solution, i.e., $P$ admits a set of $n$ $y$-monotone angles such that none of these are right facing. By the construction of $H$, the corresponding point set $P'$ must have such a partition into $y$-monotone angles. For each $i$ from 1 to $n$, we now form a point set $C_i$ that contains the $i$th $y$-monotone angle, the upper chain $U_i$ and the lower chain $V_i$. By the construction of $H$, all the chains are inside the wedge determined by the $y$-monotone angle and hence $C_i$ is a convex set with $(4n + 7)$ points. Since the sets are disjoint, we obtain the required solution to the Bottleneck Convex Subsets instance.

Consider now that the points of $H$ admits $n$ disjoint convex sets with each set containing $(4n+7)$ points. Since $H$ contains $n(4n+7)$ points, the convex sets form a partition of $H$. Let $L$ be such a partition. We now show how to construct a solution for $P$ using $L$. Let $L'$ be a set of convex sets obtained by removing the points of $P'$ from each convex set of $L$. By Lemma 1, each set of $L'$ contains at least one point from the upper chains and one point from the lower chains. Since there are $3n$ points on $P'$, to partition $P'$ into $n$ convex sets, we must need each convex set of $L$ to contain a $y$-monotone angle with exactly one point from $y = 0$, one point from $y = 1$ and one point from $y = 2$. Since each convex set contains one point from an upper chain and one point from a lower chain, none of these $y$-monotone angles can be right facing. Hence we obtain a partition of $P'$ into the required $y$-monotone angles, which implies a partition also for $P$. This completes the reduction. The following theorem summarizes the results.

**Theorem 3.** *The Bottleneck Convex Subsets problem is NP-hard.*

## 4 Point Sets with Few Points Inside the Convex Hull

In this section we show that the Bottleneck Convex Subsets problem is fixed-parameter tractable when parameterized by the number of points $r$ inside the convex hull, i.e., these points do not lie on the convex-hull boundary.

**Theorem 4.** *Let $P$ be a set of $n$ points and let $r$ be the number of points interior to the convex hull of $P$. Then one can solve the Bottleneck Convex Subsets problem on $P$ in $f(r) \cdot n^{O(1)}$ time, i.e., the Bottleneck Convex Subsets problem is fixed-parameter tractable when parameterized by $r$.*

*Proof. (Sketch: see the full version* [6] *for the complete proof)* Let $k$ be the number of disjoint convex sets that we need to construct. We guess the cardinality of the smallest convex set in an optimal solution and perform a binary search. For a guess $q$, we check whether there exists $k$ disjoint convex sets each with at least $q$ points. Assume that $j$ of the $k$ convex sets contain points from the interior. Since there are only $r$ interior points, we enumerate for each $j$ from 0 to $r$, all possible $j$ convex sets such that each set in these $j$ convex sets contains at most $q$ points from the interior of $P$. For each set of length $\ell \leq r$, we also consider all $\ell$ possible convex orderings of the points. Therefore, we have $\sum_{j=0}^{k} r\binom{2^r}{j}$ possibilities to consider. We need an additional consideration when all the points of a convex set lie on a straight line $L$. In that situation, we enumerate two further cases one that considers the left halfplane and the other that considers the right halfplane of $L$. Thus the number of elements in the enumeration is at most $\sum_{j=0}^{k} r\binom{2^r}{j}2^j \le \sum_{j=0}^{k} r2^{r^{j+1}} \le r2^{r^{k+2}}$. The idea now is to examine whether these $j$ sets can be extended to contain $q$ points each and to check whether the remaining points can be used to construct the remaining $(k-j)$ convex sets by modelling this with a maximum flow problem. □

## 5 Discussion

We examined the Bottleneck Convex Subsets problem of selecting $k$ mutually disjoint convex subsets of a given set of points $P$ such that the cardinality of the smallest set is maximized. We described an algorithm that solves Bottleneck Convex Subsets for small values of $k$, showed Bottleneck Convex Subsets is NP-hard for an arbitrary $k$, and proved Bottleneck Convex Subsets to be fixed parameter tractable when parameterized by the number of points interior to the convex hull. The problem is also solvable in polynomial time for specific large values of $k$. If $k > n/4$, then some subset has cardinality at most three; a solution is found trivially by arbitrarily partitioning $P$ into $k$ subsets of size $\lfloor n/k \rfloor$ or $\lceil n/k \rceil$. If $k \in \{\lfloor n/5 \rfloor + 1, \ldots, n/4\}$ then some subset has cardinality at most four. As discussed in Sect. 1.2, Károlyi [18] characterized necessary and sufficient conditions for a set of $n$ points in general position to admit a partition into $k = n/4$ convex quadrilaterals, and gave an $O(n \log n)$-time algorithm to decide whether such a partition exists; if no such partition exists, then some set must contain at most three points, which can be solved as described above. It remains open to determine whether Bottleneck Convex Subsets can be solved in polynomial time for all $k \in \Theta(n)$.

As a direction for future research, a natural question is to establish a good lower bound on the time required to solve these problems for small fixed values of $k$. In particular, is the $O(n^3)$-time algorithm of Chvátal and Klincsek [4] optimal for the case $k = 1$? Note that our algorithm has time $O(n^8)$ when $k = 1$. It would also be interesting to examine whether a fixed-parameter tractable algorithm exists for Bottleneck Convex Subsets when parameterized by $k$, and to find approximation algorithms for Bottleneck Convex Subsets when $k$ is an arbitrary input parameter, with running time polynomial in $n$ and $k$.

## References

1. Arkin, E.M., et al.: On the reflexivity of point sets. Discret. Comput. Geom. **25**, 139–156 (2003)
2. Avis, D., Rappaport, D.: Computing the largest empty convex subset of a set of points. In: Proceedings of the Symposium on Computational Geometry (SoCG), pp. 161–167 (1985)
3. Brodén, B., Hammar, M., Nilsson, B.J.: Guarding lines and 2-link polygons is APX-hard. In: Proceedings of the Canadian Conference on Computational Geometry (CCCG), pp. 45–48 (2001)

4. Chvátal, V., Klincsek, G.: Finding largest convex subsets. Congr. Numer. **29**, 453–460 (1980)
5. Ding, R., Hosono, K., Urabe, M., Xu, C.: Partitioning a planar point set into empty convex polygons. In: Akiyama, J., Kano, M. (eds.) JCDCG 2002. LNCS, vol. 2866, pp. 129–134. Springer, Heidelberg (2003). https://doi.org/10.1007/978-3-540-44400-8_13
6. Durocher, S., Keil, J.M., Mehrabi, S., Mondal, D.: Bottleneck convex subsets: Finding $k$ large convex sets in a point set. CoRR abs/2108.12464 (2021)
7. Durocher, S., Mondal, D.: On the hardness of point-set embeddability. In: Rahman, M.S., Nakano, S. (eds.) WALCOM 2012. LNCS, vol. 7157, pp. 148–159. Springer, Heidelberg (2012). https://doi.org/10.1007/978-3-642-28076-4_16
8. Edelsbrunner, H., Guibas, L.J.: Topologically sweeping an arrangement. J. Comput. Syst. Sci. **38**(1), 165–194 (1989)
9. Eppstein, D.: Forbidden Configurations in Discrete Geometry. Cambridge University Press, Cambridge (2018)
10. Eppstein, D., Erickson, J.: Iterated nearest neighbors and finding minimal polytopes. Discret. Comput. Geom. **11**(3), 321–350 (1994). https://doi.org/10.1007/BF02574012
11. Eppstein, D., Overmars, M.H., Rote, G., Woeginger, G.J.: Finding minimum area $k$-gons. Discret. Comput. Geom. **7**, 45–58 (1992)
12. Erdős, P., Szekeres, G.: A combinatorial problem in geometry. Compos. Math. **2**, 463–470 (1935)
13. Erdős, P., Tuza, Z., Valtr, P.: Ramsey-remainder. Eur. J. Comb. **17**(6), 519–532 (1996)
14. Gajentaan, A., Overmars, M.H.: On a class of $O(n^2)$ problems in computational geometry. Comput. Geom. **45**(4), 140–152 (2012)
15. Giannopoulos, P., Knauer, C., Werner, D.: On the computational complexity of Erdős-Szekeres and related problems in $R^3$. In: Proceedings of the European Symposium on Algorithms (ESA), pp. 541–552 (2013)
16. González-Aguilar, H., et al.: Maximum rectilinear convex subsets. In: Proceedings of the Symposium on Fundamentals of Computation Theory (FCT), pp. 274–291 (2019)
17. Hulett, H., Will, T.G., Woeginger, G.J.: Multigraph realizations of degree sequences: maximization is easy, minimization is hard. Oper. Res. Lett. **36**(5), 594–596 (2008)
18. Károlyi, G.: Ramsey-remainder for convex sets and the Erdős-szekeres theorem. Discret. Appl. Math. **109**(1–2), 163–175 (2001)
19. Kirkpatrick, D.G., Seidel, R.: The ultimate planar convex hull algorithm? SIAM J. Comput. **15**(1), 287–299 (1986)
20. Kumar, V.S.A., Arya, S., Ramesh, H.: Hardness of set cover with intersection 1. In: Proceedings of the International Colloquium on Automata, Languages and Programming (ICALP), pp. 624–635 (2000)
21. Megiddo, N., Tamir, A.: On the complexity of locating linear facilities in the plane. Oper. Res. Lett. **1**(5), 194–197 (1982)
22. Morris, W., Soltan, V.: The Erdős-Szekeres problem on points in convex position - a survey. Bull. Am. Math. Soc. **37**, 437–459 (2000)
23. Urabe, M.: On a partition into convex polygons. Discret. Appl. Math. **64**(2), 179–191 (1996)

# Deterministic Metric 1-median Selection with a $1 - o(1)$ Fraction of Points Ignored

Ching-Lueh Chang(✉)

Department of Computer Science and Engineering, Yuan Ze University, Taoyuan, Taiwan
clchang@saturn.yzu.edu.tw

**Abstract.** Given an $n$-point metric space $(M, d)$, METRIC 1-MEDIAN asks for a point $p \in M$ minimizing $\sum_{x \in M} d(p, x)$. We show that for each computable function $f\colon \mathbb{Z}^+ \to \mathbb{Z}^+$ satisfying $f(n) = \omega(1)$, METRIC 1-MEDIAN has a deterministic, $o(n)$-query, $o(f(n) \cdot \log n)$-approximation and nonadaptive algorithm. Previously, no deterministic $o(n)$-query $o(n)$-approximation algorithms are known for METRIC 1-MEDIAN.

**Keywords:** Median selection · 1-median problem · Metric space · Sublinear algorithm · Query complexity

## 1 Introduction

An $n$-point metric space $(M, d)$ is a size-$n$ set $M$ endowed with a distance function $d\colon M \times M \to [0, \infty)$ such that

- $d(x, y) = 0$ if and only if $x = y$,
- $d(x, y) = d(y, x)$, and
- $d(x, y) + d(y, z) \geq d(x, z)$ (triangle inequality)

for all $x$, $y$, $z \in M$ [13]. METRIC 1-MEDIAN asks for a point $p \in M$ minimizing $\sum_{x \in M} d(p, x)$. Clearly, it has a brute-force $O(n^2)$-time algorithm. Furthermore, it generalizes the classical median selection [5] and can be generalized further to metric $k$-median clustering. In social network analysis, METRIC 1-MEDIAN asks for an actor with the maximum closeness centrality [14]. For all $\beta \geq 1$, a $\beta$-approximate 1-median of $(M, d)$ is a point $p \in M$ satisfying $\sum_{y \in M} d(p, y) \leq \beta \cdot \min_{q \in M} \sum_{y \in M} d(q, y)$. By convention, a $\beta$-approximation algorithm for METRIC 1-MEDIAN must output a $\beta$-approximate 1-median of $(M, d)$. A query inspects $d(x, y)$ for some $x$, $y \in M$. An algorithm is nonadaptive if its $i$th query $(x_i, y_i) \in M^2$ is independent of the first $i - 1$ queries, for all $i > 1$.

Indyk [9,10] gives a Monte Carlo $O(n/\epsilon^2)$-time $(1 + \epsilon)$-approximation algorithm for METRIC 1-MEDIAN, where $\epsilon > 0$. His time complexity is optimal w.r.t. $n$. When restricted to $\mathbb{R}^D$, METRIC 1-MEDIAN has a Monte Carlo

Supported in part by the Ministry of Science and Technology of Taiwan under grant 109-2221-E-155-031.


C.-Y. Chen et al. (Eds.): COCOON 2021, LNCS 13025, pp. 215–220, 2021.
https://doi.org/10.1007/978-3-030-89543-3_18

$O(\exp(\text{poly}(1/\epsilon)))$-time $(1+\epsilon)$-approximation algorithm [12]. The more general $k$-median clustering in metric spaces has streaming approximation algorithms [8] and is inapproximable to within $(1 + 2/e - \Omega(1))$ unless NP $\subseteq$ DTIME$(n^{O(\log\log n)})$ [11]. For $\mathbb{R}^D$ and graph metrics, a well-studied problem is to find the average distance from a query point to a finite set of points [1,6,7].

Deterministic $o(n^2)$-query computation is almost completely understood for METRIC 1-MEDIAN: For all constants $\epsilon \in (0,1)$, the best approximation ratio achievable by deterministic $o(n^2)$-query and $O(n^{1+\epsilon})$-query algorithms is 4 and $2\lceil 1/\epsilon \rceil$, respectively [2,4,15]. The same holds with "query" replaced by "time" and regardless of whether the algorithms can be adaptive [2,4]. In contrast, we study the largely unknown deterministic $o(n)$-query computation. The query complexity of $o(n)$ has a special meaning: An $o(n)$-query algorithm must ignore a $1-o(1)$ fraction of all $n$ *points*. This is in contrast with $o(n^2)$-query algorithms, which ignore a $1-o(1)$ fraction of all $\binom{n}{2}$ distances.

It is folklore that every point is an $(n-1)$-approximate 1-median. Surprisingly, this is the current best upper bound for deterministic $o(n)$-query algorithms. In particular, no deterministic $o(n)$-query $o(n)$-approximation algorithms are known for METRIC 1-MEDIAN. Currently, the best lower bound against deterministic $o(n)$-query algorithms is that they cannot be $O(1)$-approximate; this remains true with "$o(n)$" replaced by "$O(n)$" [4].

We give a deterministic, $o(n)$-query, $o(f(n) \cdot \log n)$-approximation and non-adaptive algorithm for each computable function $f\colon \mathbb{Z}^+ \to \mathbb{Z}^+$ satisfying $f(n) = \omega(1)$. So, e.g., METRIC 1-MEDIAN has a deterministic $o(n)$-query $o(\alpha(n) \cdot \log n)$-approximation algorithm for the very slowly growing inverse Ackermann function $\alpha(\cdot)$. Previously, no deterministic $o(n)$-query $o(n)$-approximation algorithms are known. Our main technical discovery is that a $\beta$-approximate 1-median of $(S, d|_{S\times S})$ (where $d|_{S\times S}$ denotes $d$ restricted to $S \times S$) is an $O(\beta n/|S|)$-approximate 1-median of $(M,d)$, for all $\emptyset \subsetneq S \subseteq M$ and $\beta \geq 1$. We do not know whether METRIC 1-MEDIAN has a deterministic $o(n)$-query $O(\log n)$-approximation algorithm, though.

## 2 Main Result

Take an $n$-point metric space $(M,d)$ and $\emptyset \subsetneq S \subseteq M$. Define

$$x^* \equiv \operatorname*{argmin}_{x\in M} \sum_{y\in M} d(x,y),$$

$$x^*_S \equiv \operatorname*{argmin}_{x\in S} \sum_{y\in S} d(x,y)$$

to be a 1-median of $(M,d)$ and $(S, d|_{S\times S})$, respectively, breaking ties arbitrarily. Furthermore, pick $\boldsymbol{u}$ and $\boldsymbol{v}$ independently and uniformly at random from $S$. So

$$\bar{r}_S \equiv E\left[\, d\left(\boldsymbol{u}, \boldsymbol{v}\right)\right]$$

is the average distance in $(S, d|_{S\times S})$.

**Lemma 1.**

$$\sum_{y \in S} d(x^*, y) \geq \frac{|S| \, \bar{r}_S}{2}.$$

*Proof.* We have

$$\begin{aligned}
\sum_{y \in S} d(x^*, y) &= |S| \cdot E\left[ d(x^*, \boldsymbol{u}) \right] \\
&= \frac{1}{2} \cdot \left( |S| \cdot E\left[ d(x^*, \boldsymbol{u}) \right] + |S| \cdot E\left[ d(x^*, \boldsymbol{v}) \right] \right) \\
&\geq \frac{1}{2} \cdot |S| \cdot E\left[ d(\boldsymbol{u}, \boldsymbol{v}) \right].
\end{aligned}$$

□

**Lemma 2.**

$$\sum_{y \in S} d(x_S^*, y) \leq |S| \, \bar{r}_S.$$

*Proof.* By the optimality of $x_S^*$,

$$\sum_{y \in S} d(x_S^*, y) \leq E\left[ \sum_{y \in S} d(\boldsymbol{u}, y) \right].$$

Clearly,

$$E\left[ \sum_{y \in S} d(\boldsymbol{u}, y) \right] = |S| \cdot E\left[ d(\boldsymbol{u}, \boldsymbol{v}) \right].$$

□

For all $x'_S \in S$,

$$\sum_{y \in M} d(x'_S, y) \leq \sum_{y \in M} d(x'_S, x^*) + d(x^*, y) = n \cdot d(x'_S, x^*) + \sum_{y \in M} d(x^*, y). \quad (1)$$

The next two lemmas constitute our main discovery.

**Lemma 3.** *For all $x'_S \in S$ and $\beta \geq 1$ satisfying $\sum_{y \in S} d(x'_S, y) \leq \beta \cdot \sum_{y \in S} d(x_S^*, y)$ and $d(x'_S, x^*) \leq 2\beta \bar{r}_S$, $x'_S$ is an $O(\beta n / |S|)$-approximate 1-median of $(M, d)$.*

*Proof.* By Lemma 1,

$$n \cdot d(x'_S, x^*) \leq n \cdot d(x'_S, x^*) \cdot \frac{2}{|S| \, \bar{r}_S} \cdot \sum_{y \in S} d(x^*, y). \quad (2)$$

As $d(x'_S, x^*) \leq 2\beta\bar{r}_S$ and $S \subseteq M$,

$$\sum_{y\in M} d\left(x'_S, y\right) \leq O\left(\frac{\beta n}{|S|}\right) \cdot \sum_{y\in M} d\left(x^*, y\right)$$

by Eqs. (1)–(2). □

**Lemma 4.** *For all $x'_S \in S$ and $\beta \geq 1$ satisfying $\sum_{y\in S} d(x'_S, y) \leq \beta \cdot \sum_{y\in S} d(x^*_S, y)$ and $d(x'_S, x^*) > 2\beta\bar{r}_S$, $x'_S$ is an $O(n/|S|)$-approximate 1-median of $(M, d)$.*

*Proof.* By the triangle inequality,

$$\sum_{y\in S} d\left(x^*, y\right) \geq \sum_{y\in S} d\left(x'_S, x^*\right) - d\left(x'_S, y\right) = |S| \cdot d\left(x'_S, x^*\right) - \sum_{y\in S} d\left(x'_S, y\right). \quad (3)$$

Furthermore,

$$\sum_{y\in S} d\left(x'_S, y\right) \leq \beta \cdot \sum_{y\in S} d\left(x^*_S, y\right) \overset{\text{Lemma 2}}{\leq} \beta\,|S|\,\bar{r}_S. \quad (4)$$

As $d(x'_S, x^*) > 2\beta\bar{r}_S$,

$$\sum_{y\in S} d\left(x^*, y\right) \overset{(3)-(4)}{\geq} |S| \cdot d\left(x'_S, x^*\right) - \beta\,|S|\,\bar{r}_S > \frac{|S|}{2} \cdot d\left(x'_S, x^*\right).$$

So

$$n \cdot d\left(x'_S, x^*\right) = \frac{2n}{|S|} \cdot \frac{|S|}{2} \cdot d\left(x'_S, x^*\right) < \frac{2n}{|S|} \cdot \sum_{y\in S} d\left(x^*, y\right).$$

This and Eq. (1) imply

$$\sum_{y\in M} d\left(x'_S, y\right) \leq O\left(\frac{n}{|S|}\right) \cdot \sum_{y\in M} d\left(x^*, y\right).$$

□

Lemmas 3–4 imply the following.

**Lemma 5.** *For all $\beta \geq 1$, every $\beta$-approximate 1-median of $(S, d|_{S\times S})$ is an $O(\beta n/|S|)$-approximate 1-median of $(M, d)$.*

The following theorem is due to Chang [3].

**Theorem 1** ([3]). *For all constants $\epsilon > 0$,* METRIC 1-MEDIAN *has a deterministic, $O(\exp(O(1/\epsilon)) \cdot n \log n)$-time, $O(\exp(O(1/\epsilon)) \cdot n)$-query, $O(\epsilon \cdot \log n)$-approximation and nonadaptive algorithm.*

Below is our main theorem.

**Theorem 2.** *For each computable function $f\colon \mathbb{Z}^+ \to \mathbb{Z}^+$ satisfying $f(n) = \omega(1)$,* METRIC 1-MEDIAN *has a deterministic, $o(n)$-query, $o(f(n) \cdot \log n)$-approximation and nonadaptive algorithm.*

*Proof.* Take any $S \subseteq M$ of size $\Theta(n/\sqrt{f(n)})$. Applying Theorem 1 to $(S, d|_{S\times S})$, an $O(\log|S|)$-approximate 1-median $x'_S$ of $(S, d|_{S\times S})$ can be found deterministically and nonadaptively with $O(|S|)$ queries. By Lemma 5 (with $\beta = O(\log|S|)$), $x'_S$ is an $O((\log|S|) \cdot n/|S|)$-approximate 1-median of $(M, d)$. □

Taking a very slowly growing $f(\cdot)$ (e.g., the iterated logarithm or the inverse Ackermann function), Theorem 2 allows deterministic $o(n)$-query algorithms to be very close to being $O(\log n)$-approximate.

## 3 Conclusions

We give a deterministic $o(n)$-query $o(f(n) \cdot \log n)$-approximation algorithm for METRIC 1-MEDIAN, where $f(n) = \omega(1)$ is any computable function. This yields the first deterministic $o(n)$-query $o(n)$-approximation algorithm. Two questions remain:

- When we want an approximation ratio close to $O(\log n)$, our query complexity will be close to $\Theta(n)$ in the proof of Theorem 2. Could it be improved to be far below $\Theta(n)$ such as $O(\sqrt{n})$?
- Our approximation ratio, albeit close to $O(\log n)$, is still $\omega(\log n)$. Could it be improved to $O(\log n)$ or even to $o(\log n)$? Or is there a lower bound forbidding deterministic $o(n)$-query algorithms from being $o(\log n)$-approximate?

## References


1. Bose, P., Maheshwari, A., Morin, P.: Fast approximations for sums of distances, clustering and the Fermat-Weber problem. Comput. Geom. **24**(3), 135–146 (2003)
2. Chang, C.-L.: A lower bound for metric 1-median selection. J. Comput. Syst. Sci. **84**, 44–51 (2017)
3. Chang, C.-L.: Metric 1-median selection with fewer queries. In: Proceedings of the 2017 International Conference on Applied System Innovation, pp. 1056–1059 (2017)
4. Chang, C.-L.: Metric 1-median selection: query complexity vs. approximation ratio. ACM Trans. Comput. Theory **9**(4), 1–23 (2018). Article 20
5. Cormen, T.H., Leiserson, C.E., Rivest, R.L., Stein, C.: Introduction to Algorithms, 3rd edn. The MIT Press, Cambridge (2001)
6. Eppstein, D., Wang, J.: Fast approximation of centrality. J. Gr. Algorithms Appl. **8**(1), 39–45 (2004)
7. Goldreich, O., Ron, D.: Approximating average parameters of graphs. Random Struct. Algorithms **32**(4), 473–493 (2008)

8. Guha, S., Meyerson, A., Mishra, N., Motwani, R., O'Callaghan, L.: Clustering data streams: theory and practice. IEEE Trans. Knowl. Data Eng. **15**(3), 515–528 (2003)
9. Indyk, P.: Sublinear time algorithms for metric space problems. In: Proceedings of the 31st Annual ACM Symposium on Theory of Computing, pp. 428–434 (1999)
10. Indyk, P.: High-dimensional computational geometry. PhD thesis, Stanford University (2000)
11. Jain, K., Mahdian, M., Saberi, A.: A new greedy approach for facility location problems. In: Proceedings of the Thiry-Fourth Annual ACM Symposium on Theory of Computing, pp. 731–740 (2002)
12. Kumar, A., Sabharwal, Y., Sen, S.: Linear-time approximation schemes for clustering problems in any dimensions. J. ACM **57**(2), 5 (2010)
13. Rudin, W.: Principles of Mathematical Analysis, 3rd edn. McGraw-Hill, New York (1976)
14. Wasserman, S., Faust, K.: Social Network Analysis: Methods and Applications. Cambridge University Press, Cambridge (1994)
15. Wu, B.Y.: On approximating metric 1-median in sublinear time. Inf. Process. Lett. **114**(4), 163–166 (2014)

# The Coverage Problem by Aligned Disks

Shin-ichi Nakano(✉)

Gunma University, Maebashi, Japan
nakano@cs.gunma-u.ac.jp

**Abstract.** Given a set $C$ of points and a horizontal line $L$ in the plane and a set $F$ of points on $L$, we want to find a set of disks such that (1) each disk has the center at a point in $F$ (but with arbitrary radius), (2) each point in $C$ is covered by at least one disk, and (3) the cost of the set of disks is minimized. Here the (transmission) cost of a disk with radius $r$ is $r^\alpha$, where $\alpha$ is a constant depending on the power consumption model, and the cost of a set of disk is the sum of the cost of disks in the set.

In this paper we first give an algorithm based on dynamic programming method to solve the problem in $L_1$ metric. A naive dynamic programming algorithm runs in $O(|C|^3|F|^2)$ time. We design an algorithm which runs in $O(|C||F|^2)$ time.

Then we design another algorithm to solve the problem in $L_1$ metric based on a reduction to a shortest path problem in a directed acyclic graph. The running time of the algorithm is $O(|C|^2 + |C||F|)$.

**Keywords:** Algorithm · Disk coverage problem

## 1 Introduction

The traditional $k$-center problem finds a set of $k$ disks covering a given set of points on the plane minimizing the largest radius of the disks, and the traditional $k$-median problem finds a set of $k$ disks covering a given set of points on the plane minimizing the sum of the distances from the points to their nearest disk centers. Those problems are in general NP-hard [10]. However for some special cases polynomial time algorithms are known, for the 1D cases [2,4,5,7,11], and the 1.5 D cases (explained below) [1,3,8,9,12–14].

In this paper we consider a similar problem for the 1.5D case, called the aligned disk coverage problem.

It is known that, given a set $C$ of points and a line $L$ in the plane, one can find a set of disks such that (1) each disk has the center on the line $L$ (with arbitrary radius), (2) each point in $C$ is covered by at least one disk, and (3) the cost of the set of disks is minimized [12]. Here the (transmission) cost of a disk with radius $r$ is $r^\alpha$. We assume that we can compute $r^\alpha$ in a constant time, where $\alpha$ is a constant depending on the power consumption model. The cost of a set of disks is the sum of the cost of the disks in the set. The number of disks


C.-Y. Chen et al. (Eds.): COCOON 2021, LNCS 13025, pp. 221–230, 2021.
https://doi.org/10.1007/978-3-030-89543-3_19

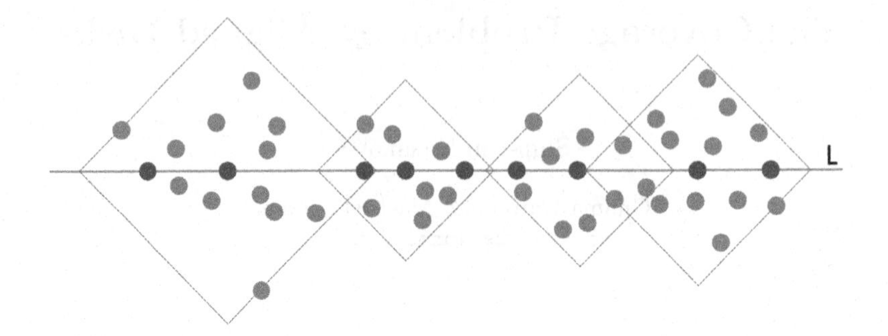


**Fig. 1.** An example of aligned disks covering a set of points in $L_1$ metric.

is not limited. The running time is $O(|C|^2)$ for any constant $\alpha$ in any fixed $L_p$ metric [12]. In this problem one can locate each disk center at any point on $L$.

In this paper we consider a "discrete" version of the problem. Given a set $C$ of points and a horizontal line $L$ in the plane and a (discrete) set $F$ of points on $L$, we want to find a set $S$ of disks such that (1) each disk has the center at a point in $F$ (with arbitrary radius), (2) each point in $C$ is covered by at least one disk, and (3) the cost of $S$ is minimized. See an example in Fig. 1. The cost of a disk and the cost of a set of disks are defined as above. The number of disks is not limited, but we want to minimize the cost of $S$. We call the problem *the discrete aligned disk coverage problem.*

Intuitively we are planning to install a set of base stations for a set of clients. We are allowed to locate each base station at a location among the candidate locations on the main (horizontal straight) road. For each base station at $f_i$ we can choose its radius $r_i$ (corresponding to the transmission distance from the base station), and we assume that if the distance to a client from $f_i$ is at most $r_i$, the client can receive a message from the base station. We want to choose a set $\{f_1, f_2, \cdots\}$ of base stations with their radii $\{r_1, r_2, \cdots\}$ so that the sum of $r_i^\alpha$ is minimized. Here $\alpha$ is a constant depending on the power consumption model, and the sum of $r_i^\alpha$ is the total power consumption of the base stations. Note that for the model in [12] one can locate each base station at any place on the main road. In contrast in the discrete model we can locate each base station only at a location among "the candidate locations" on the main road. So this model is more natural for some applications.

If all points in $C$ also lie on $L$ (the 1D problem) and $\alpha = 1$, one can solve the problem in $O((|C| + |F|)^3)$ time [9]. If the points in $F$ lie on anywhere in the plane (the 2D problem) the problem with any fixed $\alpha > 1$ is NP-hard [1], and if $\alpha = 1$ a PTAS is known [9].

In this paper we first give an algorithm to solve the discrete aligned disk coverage problem for any fixed $\alpha$. A naive algorithm runs in $O(|C|^3|F|^2)$ time. We design an algorithm which runs in $O(|C||F|^2)$ time. The algorithms are based on efficient dynamic programming method.

Then we design an $O(|C|^2 + |C||F|)$ time algorithm to solve the discrete aligned disk coverage problem for $L_1$ metric. If $|F|^2 > |C|$ then the algorithm runs faster than the first algorithm. The algorithm is based on a reduction to a shortest path problem in a directed acyclic graph.

The rest of the paper is organized as follows. In Sect. 2 we give some basic definitions and lemmas. In Sect. 3 we give an $O(|C||F|^2)$ time algorithm to solve the discrete aligned disk coverage problem in $L_1$ metric. In Sect. 4 we give an $O(|C|^2+|C||F|)$ time algorithm to solve the discrete aligned disk coverage problem in $L_1$ metric. Finally Sect. 5 is a conclusion.

## 2 Preliminary

In this section we give some definitions and lemmas.

Let $d(u, v)$ be the distance between two points $u$ and $v$. Let $C$ and $L$ be a set of points and a horizontal line in the plane, respectively, and $F$ a set of points on $L$. We assume $|C| > |F|$. *A disk* with center at $f \in F$ and radius $r$ is a set of points having distance at most $r$ from $f$. We denote such disk as $D(f, r)$. Note that a disk in $L_1$ metric is a region with a rotated square boundary. Without loss of generality we can assume that $L$ is a line on the $x$-axis. Let $x(p)$ and $y(p)$ be the $x-$ and $y-$coordinates of point $p$. We also assume that all points in $C$ are located above or on $L$, since if a point $p \in C$ is located below $L$ we can replace $p$ with its symmetrical point $p'$ with $L$ without affecting the (optimal) solution. Note that a disk $D$ centered at $f \in F$ contains $p$ iff $D$ contains $p'$. Also we can assume that no two point in $C$ have the same $x$-coordinate, since if a point $p \in C$ lies directly above another point $p' \in C$, we can remove $p'$ from $C$ without affecting the (optimal) solution. Note that any disk centered at $f \in F$ and containing $p$ also contains $p'$.

A set of disks are *aligned disks* if each disk in the set has the center at a point in $F$. We say that aligned disks $AD$ *cover* $C$ if each point in $C$ is contained in at least one disk in $AD$. The cost of aligned disks $AD$ is the sum of the costs of disks in $AD$, and the cost of a disk with radius $r$ is $r^\alpha$. We want to find aligned disks covering $C$ with the minimum cost. We call the problem the *discrete aligned disk coverage problem* in this paper.

Let $AD$ be aligned disks covering $C$ with the minimum cost. We have the following three lemmas. Those three lemmas hold for any constant $\alpha$.

**Lemma 1.** *One can assume that each disk in $AD$ does not contain another disk in $AD$.*

*Proof.* If $D \in AD$ contains $D' \in AD$ then by removing $D'$ from $AD$ one can have aligned disks covering $C$ with less cost. A contradiction. □

**Lemma 2.** *Let $D_\ell$ and $D_r$ be two consecutive disks in $AD$, $f_\ell$, $f_r \in F$ are their centers, and $x(f_\ell) < x(f_r)$. Every point $p \in C$ with $x(f_\ell) \leq x(p) \leq x(f_r)$ is covered by either $D_\ell$ or $D_r$.*

*Proof.* Assume otherwise. Then there are two consecutive disks $D_\ell$ and $D_r$ in $AD$, with centers at $f_\ell$ and $f_r$, and there exists a point $p \in C$ with $x(f_\ell) \leq x(p) \leq x(f_r)$ not covered by $D_\ell$ and $D_r$, but covered by some other disk, say $D'$ in $AD$. Then $D'$ contains either $D_\ell$ or $D_r$, which contradicts Lemma 1. □

We say a disk $D_\ell$ with the center at $f_\ell$ is *a possible left neighbour* of a disk $D_r$ with the center at $f_r$ if every point $p \in C$ with $x(f_\ell) \leq x(p) \leq x(f_r)$ is covered by either $D_\ell$ or $D_r$. Note that if $AD$ has a disk $D$ and it is not the leftmost one, then the disk located immediately left of $D$ in $AD$ is a possible left neighbour of $D$.

**Lemma 3.** *Each disk in $AD$ has some point $p \in C$ on its boundary.*

*Proof.* Assume otherwise. Now some disk $D(f, r)$ in $AD$ has no point in $C$ on its boundary. Then we can decrease the radius $r$ of $D(f, r)$ to $d(f, p')$, where $p'$ is the furthest point in $C$ from $f$ located in $D(f, r)$, so that the resulting aligned disks still cover $C$ with less cost. A contradiction. □

Thus the number of disks possibly appear in $AD$ is at most $|C||F|$.

## 3 Algorithm 1

A simple dynamic programming algorithm to solve the discrete aligned disk coverage problem in $L_1$ metric runs in $O(|C|^3|F|^2)$ time. (The number of possible disks in a solution is at most $|C||F|$ by Lemma 3. Then for each disk we compute if each other disk can appear as the preceding disk in an optimal solution in $O(|C|)$ time, so the running time is $O(|C||F|) \times O(|C||F|) \times O(|C|)$.) In this section we design a more efficient dynamic programming algorithm to solve the problem which runs in $O(|C||F|^2)$ time. We assume that we can compute $r^\alpha$ in a constant time.

Given a set $C$ of points and a horizontal line $L$ in the plane and a set $F$ of points on $L$, let $AD$ be aligned disks covering $C$ with the minimum cost. We append $f_\infty$ to $F$ as a hypothetical rightmost point. Let $C(f)$ be the subset of $C$ consisting of the points not located on the right of $f \in F$, that is $C(f) = \{p \in C | x(p) \leq x(f)\}$. Let $u_f$ be the farthest point in $C(f)$ from $f$, or equivalently the point $u_f \in C(f)$ attaining the maximum $y(u_f) - x(u_f)$.

Now we define the subproblems for our dynamic programming algorithm, as follows. The subproblem $P(f, r)$ is the problem of finding aligned disks $AD(f, r)$ including disk $D(f, r)$ (as the rightmost disk) covering $C(f)$ with the minimum cost. Let $C(f, r)$ be the cost of $AD(f, r)$. Note that $AD(f_\infty, 0) - \{D(f_\infty, 0)\}$ is a solution of our original problem. Note that disk $D(f_\infty, 0)$ has the radius 0, so it covers no point in $C$.

Fix $f \in F$. Let $p_i$ in $C$ be the $i$-th farthest point from $f$. Then set $r_i = d(p_i, f)$. We are going to compute $C(f, r_i)$ and $AD(f, r_i)$ for each $P(f, r_i)$. Fix $r_i$. We have the following two cases. Note that $AD(f, r_i)$ covers only $C(f)$ not entire $C$.

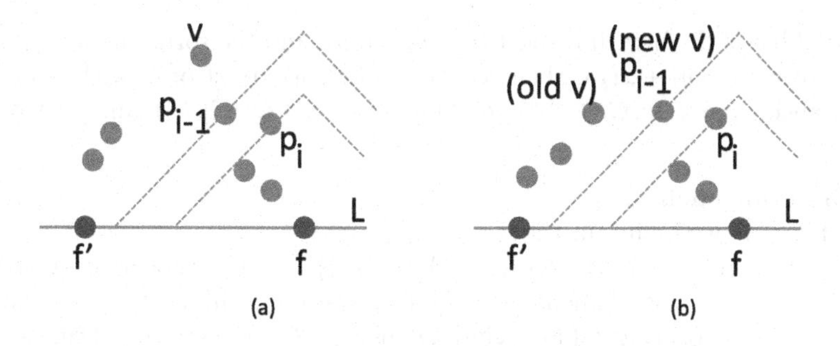


**Fig. 2.** An illustration for Case 2.

**Case 1**: $AD(f, r_i)$ consists of exactly one disk.

If $r_i \geq d(u_f, f)$ then the disk $D(f, r_i)$ covers $C(f)$, so $C(f, r_i) = r_i^\alpha$ and $AD(f, r_i) = \{D(f, r_i)\}$.

Otherwise, $r_i < d(u_f, f)$ holds, and then $D(f, r_i)$ cannot cover $u_f \in C(f)$, and to cover $C(f)$ we need more disks so this is not in Case 1.

**Case 2**: $AD(f, r_i)$ consists of two or more disks.

Now $f$ is not the leftmost point in $F$. For each $f' \in F$ with $x(f') < x(f)$, $D(f, r_i)$ has a possible left neighbor, say $D(f', r')$, for large enough $r'$.

If $AD(f, r_i)$ has the 2nd rightmost disk $D(f', r')$, then $P(f, r_i)$ has a solution $AD(f, r_i) = AD(f', r') \cup \{D(f, r_i)\}$ with cost $C(f, r_i) = C(f', r') + r_i^\alpha$. However we do not know whether $P(f, r_i)$ has a solution with two or more disks, and if it has, which one is the 2nd rightmost disk $D(f', r')$ in $AD(f, r_i)$.

For each $f'$ with $x(f') < x(f)$, we compute the following $r'$. Among all possible $r'$ we choose the $r'$ with (1) $D(f', r')$ is the possible left neighbor of $D(f, r_i)$ and (2) $C(f', r')$ is the minimum over the choice of $r'$.

Then, among all possible $f'$, we choose the 2nd rightmost disk $D(f', r')$ in $AD(f, r_i)$ which minimizes $C(f', r') + r_i^\alpha$. Now $AD(f, r_i) = AD(f', r') \cup \{D(f, r_i)\}$ and $C(f, r_i) = C(f', r') + r_i^\alpha$ hold.

Thus we can design a dynamic programming algorithm to solve the problem.

The running time of the algorithm is as follows.

**The search of $u_f$.**
In the preprocessing step, sort the points in $C$ with respect to their $x$-coordinates in the increasing order. This sort needs $O(|C| \log |C|)$ time. For each $f \in F$ in increasing order of $x(f)$, compute each $u_f$ above by scanning the prefix of the sorted list of $C$ upto $x(f)$ incrementally, corresponding to the current $C(f)$, then choose the point $u$ as $u_f$ having the maximum $y(u) - x(u)$. This needs $O(|C| + |F|)$ time in total for the whole algorithm. Since $|C| > |F|$ the running time is $O(|C|)$.

**The sorted list of $C$ with respect to the distance from $f$.**
In the preprocessing step, sort the points in $C$ with respect to their $y(p) - x(p)$'s in decreasing order, and sort the points in $C$ with respect to their $y(p) + x(p)$'s in decreasing order. By using the two sorted lists, for a fixed $f$, we can construct

the sorted list of $C(f)$ with respect to $y(p)-x(p)$, and the sorted list of $C-C(f)$ with respect to $y(p)+x(p)$, then construct the sorted list of $C$ with respect to the distance from $f$ in $O(|C|)$ time. Thus, we need $O(|C||F|)$ time in total for this part.

**The table for each $f$.**
For each $f \in F$ we maintain the list $((r_1, C(f, r_1), AD(f, r_1)), (r_2, C(f, r_2), AD(f, r_2)), \cdots)$ in the decreasing order of $r_i$. The list for $f$ has element $(r_i, C(f, r_i), AD(f, r_i))$ iff $P(f, r_i)$ has aligned disks $AD(f, r_i)$ covering $C(f)$ including disk $D(f, r_i)$ (as the rightmost disk) with the minimum cost $C(f, r_i)$. Each $AD(f, r_i)$ is stored as its rightmost disk $D(f, r_i)$ and a pointer to $AD(f', r')$, where $D(f', r')$ is the 2nd rightmost disk in $AD(f, r_i)$. So the size of space for each $AD(f, r_i)$ is a constant.

Each time if we find a solution of $P(f, r_i)$ we append $(r_i, C(f, r_i), AD(f, r_i))$ to the list of $f$.

**The computation of $r'$ for each $f'$.**
Fix $f$ and $f'$. For each $r_1, r_2, \cdots$ of $f$ if we compute every $r'$ of $f'$ independently, and choose the best $r'$ of $f'$, then we need $O(|C|)$ time for each $r_i$ of $f$. Our idea is, for each fixed $f'$, we can compute $r'$ for each $r_1, r_2, \cdots$ of $f$ incrementally in $O(|C|)$ time in total, as follows. Construct the list $S$ of the points in $C$ located between $x(f')$ and $x(f)$ in decreasing order of $y(p)-x(p)$. Since, in the preprocessing step, we have sorted $C$ with respect to $y(p)-x(p)$ we can construct $S$ in $O(|C|)$ time. Let $p_i \in C$ be the $i$-th farthest point from $f$, and set $r_i = d(p_i, f)$. We have two cases depending on whether $D(f, r_i)$ covers $S$ or not.

If $D(f, r_i)$ covers $S$, that is $d(u_f, f) \le r_i$, where $u_f$ is the farthest point from $f$ in $S$, then $r'$ is the one with the minimum $C(f', r')$.

If $D(f, r_i)$ does not cover $S$, then we first compute the point $v \in S$, (1) not covered by $D(f, r_i)$ and (2) having the maximum $d(v, f')$. (If we have computed $v$ for $r_{i-1}$ of $f$ then one can compute $v$ for $r_i$ by incrementally scanning $S$. The $v$ remains as it was or $p_{i-1}$ becomes new $v$. See Fig. 2.) Now $D(f', d(v, f'))$ is a possible left neighbor of $D(f, r_i)$. Then choose the minimum $r'$ such that (1) $r' \ge d(v, f')$ and (2) $C(f', r')$ is bounded. One can compute such $v$ incrementally for all $r_1, r_2, \cdots, r_{|C|}$ in $O(|C|)$ time in total. Thus, for $f$ and $f'$, we can choose each $r'$ for all $r_1, r_2, \cdots, r_{|C|}$ in $O(|C|)$ time in total.

Thus, for fixed $f$ and $f'$, our algorithm needs only $O(|C|)$ time. Therefore we need $O(|C||F|^2)$ time for this part.

**Theorem 1.** *One can solve the discrete aligned disk coverage problem in $L_1$ metric in $O(|C||F|^2 + |C|\log|C|)$ time.*

Our algorithm is shown in **Algorithm** Aligned-disk-coverage.

**Algorithm 1.** Aligned-disk-coverage

1: Append $f_\infty$ to $F$ as the rightmost point.
2: **for** each $f \in F\}$ /* in increasing order of $x(f)$ */ **do**
3: Let $C(f) = \{p \in C | x(p) \leq x(f)\}$
4: Let $u_f$ be the farthest point in $C(f)$ from $f$
5: **for** each $p_i \in C$ /* in decreasing order of $y(p_i) - x(p_i)$ */ **do**
6: set $r_i = d(p_i, f)$
7: /* Case 1: Aligned disk cover for $C(f)$ consists of exactly one disk */
8: **if** $r_i \geq d(u_s, f)$ **then**
9: $C(f, r_i) = {r_i}^\alpha$
10: $AD(f, r_i) = \{D(f, r_i)\}$
11: **else**
12: $C(f, r_i) = \infty$
13: **end if**
14: /* Case 2: Aligned disk cover for $C(f)$ consists of two or more disks */
15: **for** each $f' \in F$ with $x(f') < x(f)$ **do**
16: Let $r'$ be the one such that (1) $C(f', r')$ is a possible left neighbor of $D(f, r_i)$, and (2) $r'$ is the minimum one such that $C(f', r')$ is bounded
17: /* $AD(f', r') \cup \{D(f, r_i)\}$ covers $C(f)$ */
18: **if** $C(f', r') + {r_i}^\alpha < C(f, r_i)$ **then**
19: $C(f, r_i) = C(f', r') + r_i^\alpha$
20: $AD(f, r_i) = AD(f', r') \cup \{D(f, r_i)\}$
21: **end if**
22: **end for**
23: **end for**
24: **end for**
25: Output $AD(f_\infty, 0) - D(f_\infty, 0)$.

## 4 Algorithm 2

In this section we design an $O(|C|^2 + |C||F|)$ time algorithm to solve the discrete aligned disk coverage problem in $L_1$ metric.

Given a set $C$ of points and a horizontal line $L$ in the plane and a set $F$ of points on $L$, let $AD = \{D_1, D_2, \cdots, D_k\}$ be a set of aligned disks covering $C$ with the minimum cost. Note that we have assumed that all points in $C$ are located above or on $L$, and a disk in $L_1$ metric is a region with a rotated square boundary. So, to cover the points in $C$, we need only the upper half regions of disks in $AD$, each of which is a right isosceles triangle. See Fig. 3.

Let $\ell(D)$ be the line segments with slope 1 on the boundary of the right isosceles triangle corresponding to a disk $D$ in $AD$. Similarly, let $r(D)$ be the line segments with slope $-1$. For two consecutive disks $D_\ell, D_r$ in $AD$, let $c(D_\ell, D_r)$ be the intersection point of two segments $r(D_\ell)$ and $\ell(D_r)$ if $D_\ell$ and $D_r$ intersect, and the midpoint of two centers of $D_\ell$ and $D_r$ otherwise.

Fix $AD$. Now we define *the responsible point set* $C(D) \subset C$ for each $D \in AD$, as follows. For the leftmost disk $D_1$ in $AD$ we define $C(D_1) = \{p | x(p) < x(c(D_1, D_2))\}$. For the rightmost disk $D_k$ in $AD$ we define $C(D_k) = \{p | x(c(D_{k-1}, D_k)) \leq x(p)\}$. For other disk $D_i$ in $AD$, $C(D_i) =$

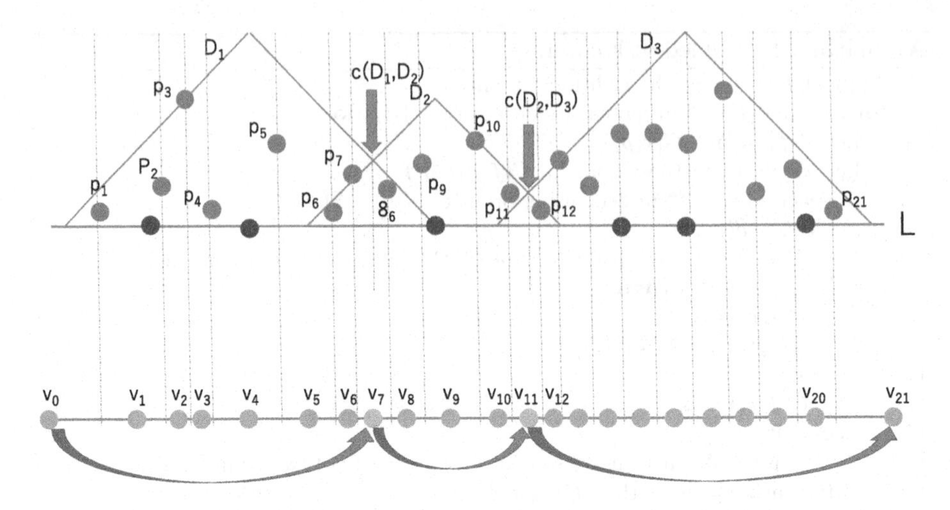


**Fig. 3.** An illustration for the graph.

$\{p|x(c(D_{i-1}, D_i)) \leq x(p) < x(c(D_i, D_{i+1}))\}$. Thus $C(D_1) \cup C(D_2) \cup \cdots \cup C(D_k)$ is a partition of $C$.

One can observe that each $D_i$ in $AD$ has at least one point $p \in C(D_i)$ with $x(c(D_{i-1}, D_i)) \leq x(p) < x(c(D_i, D_{i+1}))\}$ on its boundary, since otherwise we can shrink the disk so that it still covers $C(D_i)$ but with less radius. A contradiction. Thus each $D$ in $AD$ is a disk with the minimum radius covering some consecutive points $\{p_s, p_{s+1}, \cdots, p_t\}$, and the number of such disks, corresponding to some consecutive points, is at most $|C|^2$.

Therefore there is a natural correspondence between (1) aligned disks $AD = \{D_1, D_2, \cdots, D_k\}$ covering $C$ (with the minimum cost) and (2) a shortest path in the following graph. See Fig. 3.

The set of vertices of the graph is $\{v_0, v_1, \cdots, v_{|C|}\}$. (Intuitively $v_i$ corresponds to the midpoint of $p_i$ and $p_{i+1}$.) For each $i, j$ with $0 \leq i < j \leq n$ we append the directed edge from $v_i$ to $v_j$ with weight $w = r^\alpha$, where $r$ is the minimum radius of a disk covering $p_{i+1}, p_{i+2}, \cdots, p_j$ and having the center at some point in $F$.

Let $D$ be a disk centered at some point, say $f$, in $F$ covering $P = \{p_{i+1}, p_{i+2}, \cdots, p_j\} \subset C$ with the minimum radius, $p^\ell \in P$ the farthest point from $f$ among $P$ located on the left of $f$, and $p^r \in P$ the farthest point from $f$ among $P$ located on the right of $f$.

Given $P$, $D$, $p^\ell$ and $p^r$, one can compute the disk $D'$ centered at some point, say $f'$, in $F$ covering $P$ plus one more point $p_{j+1}$, that is, covering $P' = P \cup \{p_{j+1}\}$ with the minimum radius, the farthest point $q^\ell$ from $f'$ among $P'$ located on the left of $f'$, and the farthest point $q^r$ from $f'$ among $P'$ located on the right of $f'$, as follows.

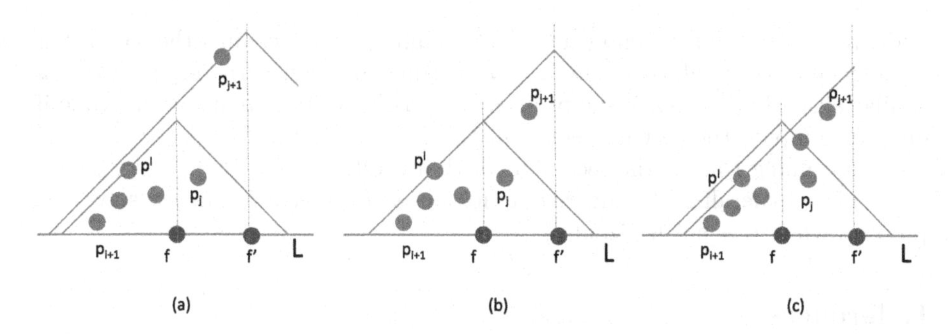


**Fig. 4.** An illustration for Case 2(b).

We have the following three cases.

**Case 1:** $D$ contains $p_{j+1}$.
Then $D' = D$ (so $f' = f$), $q^\ell = p^\ell$, and $q^r = p^r$ hold.

**Case 2:** Otherwise. ($D$ does not contain $p_{j+1}$.)
We have the following two subcases.

**Case 2(a):** $x(f') \le x(p_{j+1})$.
Now $f'$ is located on the left of $p_{j+1}$. The center $f'$ of $D'$ is the point $f' \in F$ with the minimum $max\{d(p^\ell, f'), d(p_{j+1}, f')\}$. One can find $f'$ by scanning $F$ from $f$ to right. The radius of $D'$ is $max\{d(p^\ell, f'), d(p_{j+1}, f')\}$. $q^\ell$ is $p^\ell$ and $q^r$ is $p_{j+1}$.

**Case 2(b):** $x(p_{j+1}) < x(f')$.
Now $f'$ is located on the right of $p_{j+1}$. The center $f'$ of $D'$ is the leftmost $f' \in F$ with $x(f') > x(p_{j+1})$. The radius of $D'$ is $\max\{d(p^\ell, f'), d(p_{j+1}, f')$. See Fig. 4. Now $q^\ell$ is $p^\ell$ or $p_{j+1}$, and $q^r$ is not defined.

Thus, for a fixed $i$, one can compute the weight of each possible edge in $(v_i, v_{i+1}), (v_i, v_{i+2}), \cdots, (v_i, v_|C|)$ in $O(|C| + |F|)$ time in total. So one can compute all weights of possible at most $|C|^2$ edges in $O(|C|^2 + |C||F|)$ time.

Since the graph is directed acyclic we can compute the shortest path from $v_0$ to $v_|C|$ in $O(|C|^2)$ time by a simple dynamic programming algorithm. See Chap. 24.2 of [6].

**Theorem 2.** *One can solve the discrete aligned disk coverage problem in $L_1$ metric in $O(|C|^2 + |C||F|)$ time.*

## 5 Conclusion

In this paper we have designed two algorithms to solve the discrete aligned disk coverage problem in $L_1$ metric. The running time of the first algorithm is $O(|C||F|^2)$. A similar algorithm can solve more generalized problem in which the cost of a disk $D_i$ centered at $f_i$ with radius $r_i$ is $r_i^\alpha + o_i$ where $o_i$ is an opening additive cost for a disk center at $f_i$. The running time remains the same. For more general cost, which is any non-decreasing function with respect to $r_i$, a similar algorithm can solve the problem. The running time remains the same if one can compute the cost in a constant time.

The running time of the second algorithm is $O(|C|^2 + |C||F|)$.

Can we generalize the algorithms for other $L_p$ metric? Can we solve the problem when $L$ is not horizontal?

## References


1. Alt, H., et al.: Minimum-cost coverage of point sets by disks. In: Proceedings of the Symposium on Computational Geometry 2006, pp. 449–458 (2005)
2. Auletta, V., Parente, D., Persiano, G.: Placing resources on a growing line. J. Algorithms **26**, 87–100 (1998)
3. Brass, P., et al.: The aligned k-center problem. Int. J. Comput. Geom. Appl. **21**, 157–178 (2011)
4. Chen, D., Li, J., Wang, H.: Efficient algorithms for the one-dimensional k-center problem. Theor. Comput. Sci. **592**, 135–142 (2015)
5. Chen, D., Wang, H.: New algorithms for facility location problems on the real line. Algorithmica **69**, 370–383 (2014)
6. Cormen, T.H., Leiserson, C.E., Rivest, R.L., Stein, C.: Introduction to Algorithms, Third Edition. The MIT Press, Cambridge (2009)
7. Hassin, R., Tamir, A.: Improved complexity bounds for location problems on the real line. Oper. Res. Lett. **10**, 395–402 (1991)
8. Karmakar, A., Das, S., Nandy, S.C., Bhattacharya, B.K.: Some variations on constrained minimum enclosing circle problem. J. Comb. Optim. **25**, 176–190 (2013)
9. Lev-Tov, N., Peleg, D.: Polynomial time approximation schemes for base station coverage with minimum total radii. Comput. Netw. **47**, 489–501 (2005)
10. Megiddo, N., Supowit, K.: On the complexity of some common geometric location problems. SIAM J. Comput. **13**, 182–196 (1984)
11. Megiddo, N., Tamir, A.: New results on the complexity of p-center problems. SIAM J. Comput. **12**, 751–758 (1983)
12. Pedersen, L., Wang, H.: On the coverage of points in the plane by disks centered at a line. In: Proceedings of the CCCG 2018, pp. 158–164 (2018)
13. Shin, C.: A note on minimum-sum coverage by aligned disks. Inf. Process. Lett. **113**, 871–875 (2013)
14. Wang, H., Zhang, J.: Line-constrained k-median, k-means, and k-center problems in the plane. Int. J. Comput. Geom. Appl. **26**, 185–210 (2016)

# Consistent Simplification of Polyline Tree Bundles

Yannick Bosch[1], Peter Schäfer[1], Joachim Spoerhase[2], Sabine Storandt[1], and Johannes Zink[2(✉)]

[1] University of Konstanz, Konstanz, Germany
[2] University of Würzburg, Würzburg, Germany
zink@informatik.uni-wuerzburg.de

**Abstract.** The POLYLINE BUNDLE SIMPLIFICATION (PBS) problem is a generalization of the classical polyline simplification problem. Given a set of polylines, which may share line segments and points, PBS asks for the smallest consistent simplification of these polylines with respect to a given distance threshold. Here, consistent means that each point is either kept in or discarded from all polylines containing it. In previous work, it was proven that PBS is NP-hard to approximate within a factor of $n^{\frac{1}{3}-\varepsilon}$ for any $\varepsilon > 0$ where $n$ denotes the number of points in the input. This hardness result holds even for two polylines. In this paper we first study the practically relevant setting of planar inputs. While for many combinatorial optimization problems the restriction to planar settings makes the problem substantially easier, we show that the inapproximability bound known for general inputs continues to hold even for planar inputs. We proceed with the interesting special case of PBS where the polylines form a rooted tree. Such tree bundles naturally arise in the context of movement data visualization. We prove that optimal simplifications of these tree bundles can be computed in $\mathcal{O}(n^3)$ for the Fréchet distance and in $\mathcal{O}(n^2)$ for the Hausdorff distance (which both match the computation time for single polylines). Furthermore, we present a greedy heuristic that allows to decompose polyline bundles into tree bundles in order to make our exact algorithm for trees useful on general inputs. The applicability of our approaches is demonstrated in an experimental evaluation on real-world data.

**Keywords:** Polyline simplification · Hardness of approximation · Tree graph · Dynamic program · Planarity

## 1 Introduction

Polyline simplification is a well-studied optimization problem [4,5,10,11,15] with a wide field of applications, e.g., in computer graphics, map visualization, or data

P. Schäfer—Funded by the Deutsche Forschungsgemeinschaft (DFG, German Research Foundation) – Project-ID 50974019 – TRR 161.


C.-Y. Chen et al. (Eds.): COCOON 2021, LNCS 13025, pp. 231–243, 2021.
https://doi.org/10.1007/978-3-030-89543-3_20

smoothing. In the classical sense, polyline simplification means removing some polyline bend points while keeping a small distance to the original polyline. Given a distance threshold, the optimal simplification of a single polyline, i.e., the simplification keeping as few polyline points as possible, respecting that threshold can be computed in polynomial time [11]. However, in case the input is a set of (partially) overlapping polylines, individual simplification of each polyline leads to visually unpleasing results as shared parts may be simplified in different ways. Moreover the visual complexity might even increase which opposes the simplification concept. Aiming at more appealing and sensible results, the problem of Polyline Bundle Simplification (PBS) was introduced in [14]. It adds as an additional constraint that shared parts must be simplified consistently (i.e. each point is either kept in or discarded from all polylines containing it).

**Definition 1 (Polyline Bundle Simplification** [14]**).** *An instance of PBS is a triple* $(P, \mathcal{L}, \delta)$ *where* $P = \{p_1, \ldots, p_n\}$ *is a set of* $n$ *points in the plane,* $\mathcal{L} = \{L_1, \ldots L_\ell\}$ *is a set of* $\ell$ *simple polylines, each represented as a list of points from* $P$ *(here, simple means that each point appears at most once in* $L_i$*), and* $\delta$ *is a distance parameter. The goal is to obtain a minimum size subset* $P^* \subseteq P$ *such that for each polyline* $L \in \mathcal{L}$ *its induced simplification* $L \cap P^*$ *contains the start and end point of* $L$ *and has a segment-wise distance of at most* $\delta$ *to* $L$.

PBS is a generalization of the classical polyline simplification problem but was proven to be NP-hard to approximate within a factor of $n^{\frac{1}{3}-\varepsilon}$ for any $\varepsilon > 0$ already for two polylines for the Hausdorff and the Fréchet distance [14]. Motivated by this strong hardness result, we investigate the complexity of practically interesting special cases of PBS, and we design and evaluate practical algorithms.

**Related Work.** Simultaneous simplification of multiple polylines was considered in previous work e.g. in the context of computational biology or for map generation. The so called *chain pair simplification problem* asks for two polylines for their simplifications such that for given $k \in \mathbb{N}$ and $\delta > 0$ each simplified chain contains at most $k$ segments, and the Fréchet distance between them is at most $\delta$ [1]. The problem arises in protein structure alignment or map matching tasks and was studied from a theoretical and practical perspective [6,7,16]. While the basic idea to preserve resemblance between polylines after simplification is similar to the motivation behind PBS, chain pair simplification only ever considers two polylines and does not put further restrictions on the simplification of shared parts. Analyzing bundles of (potentially overlapping and intersecting) movement trajectories is an important means to study group behavior and to generate maps. For example, the RoadRunner approach [9] infers high-precision maps from GPS trajectories. In [3], an approach was proposed that computes a concise graph that represents all trajectories in a given set sufficiently well. But these and similar methods do not produce valid simplifications of each input polyline, but allow to discard outliers or to let a polyline be represented by a completely disjoint polyline which is quite different from the PBS setting.

The PBS problem was introduced in [14]. In addition to the above mentioned inapproximability result, there were also two algorithms for PBS discussed in the paper. For PBS with the Fréchet distance, a bi-criteria $(O(\log(\ell + n)), 2)$-approximation algorithm was presented. This algorithm is allowed to return results within a distance threshold of $2\delta$, and based on this constraint relaxation achieves a logarithmic approximation factor (compared to the optimal solution for $\delta$) in polynomial time. Furthermore, it was shown that PBS is fixed-parameter tractable in the number $k$ of points that are shared by at least two polylines, based on a simple algorithm with a running time of $O(2^k \cdot \ell \cdot n^2 + \ell \cdot n^3)$.

**Contribution.** We present the following new theoretical and practical results:

- PBS remains NP-hard to approximate to within a factor $n^{\frac{1}{3}-\epsilon}$ for any $\epsilon > 0$ on *planar* inputs (Sect. 3).
- The special case of PBS where the polylines form a rooted tree can be solved optimally in polynomial time. Similar to the Imai-Iri algorithm for simplification of a single polyline [11], our algorithm precomputes the possible set of shortcuts for the given distance threshold and thereupon transforms the given geometric problem into a graph problem. But while in the Imai-Iri algorithm a simple search for the minimum link-path in the shortcut graph suffices, we need a more intricate dynamic programming approach (DP) to deal with the tree structure (Sect. 4).
- We devise a greedy heuristic that decomposes a general polyline bundle into tree bundles, which then can be simplified independently and optimally with our DP (Sect. 5).
- In the experimental evaluation, we use our new approach to simplify polyline bundles that model movement data or public transit maps. We compare our approach in terms of efficiency and quality to the bi-criteria approximation algorithm proposed in [14] (Sect. 6).

## 2 Preliminaries

The two most commonly used distance functions $d$ to govern polyline simplification are the Fréchet distance $d_F$ and the Hausdorff distance $d_H$. In the context of polyline simplification, the distance function $d$ is used to measure the distance of a line segment $(a, b)$ in the simplification to the corresponding sub-polyline of $L$, which we abbreviate by $L(a, b)$. For any line segment $(a, b)$ in a valid simplification, we require $d((a, b), L(a, b)) \leq \delta$. Given a single polyline $L$ of length $n$, a distance function $d$, and a distance threshold $\delta$, an optimal simplification of $L$ can be computed in time $\mathcal{O}(n^3)$ using the Imai-Iri algorithm [11]. The algorithm starts by constructing a so called shortcut graph, in which there is an edge between pairs of points $a, b$ in $L$ if $d((a, b), L(a, b)) \leq \delta$. Checking this property for each of the $\Theta(n^2)$ point pairs takes $\mathcal{O}(n^3)$ time when using the Fréchet or the Hausdorff distance. In the created shortcut graph, the best simplification can be identified by computing the minimum-link path between the start and the end node of $L$ with a BFS run. As this only takes time linear in the graph size, the shortcut graph computation dominates the overall running time.

An impoved method for shortcut graph construction presented by Chan and Chin [4] can reduce the running time to $\mathcal{O}(n^2)$ for $d = d_H$. The algorithm uses sweeps to first compute directed shortcuts from which then subsequently the valid undirected shortcuts can be deduced. More precisely, in each sweep, every point $p$ is considered as possible starting point of a directed shortcut in forward direction. To then efficiently decide whether a later point on the polyline is a valid endpoint of such a directed shortcut, a cone is maintained in which all valid endpoints have to lie. Updating that cone and making the containment check is possible in constant time, hence all directed shortcuts starting in $p$ can be computed in $\mathcal{O}(n)$. The respective total time for considering every point as starting point in both sweeps is $\mathcal{O}(n^2)$. A valid undirected shortcut exists if and only if both of its directed versions are constructed in the sweep phase. This obviously can be checked for each potential shortcut in constant time, leading to an overall shortcut graph construction time of $\mathcal{O}(n^2)$.

# 3 Hardness of Approximating Planar Polyline Bundle Simplification

In this section, we show that we cannot approximate polyline bundle simplification on planar inputs by the same polynomial factor that was previously shown for general, non-planar inputs [14]. Here, planar means that no two polyline segments touch or intersect each other unless they share a common endpoint.

**Theorem 1.** *PBS with a planar polyline bundle as input is NP-hard to approximate within a factor of $n^{\frac{1}{3}-\varepsilon}$ for any $\varepsilon > 0$, where $n$ is the number of points in the polyline bundle.*

We build upon the hardness reduction from minimum independent dominating set (MIDS) from [14] by modifying their gadgets and the arrangement of their gadgets such that the constructed polyline bundle is planar. In the MIDS problem, we are given a graph $G = (V, E)$ with $\hat{n}$ vertices and $c\hat{n}$ edges, and the task is to find a set $S$ of vertices that is independent (no two vertices in $S$ are adjacent) and dominating (each vertex not in $S$ has a neighbor in $S$). This problem has been shown to be NP-hard to approximate within a factor of $\hat{n}^{1-\varepsilon}$ for any $\varepsilon > 0$ [8], even for constant $c$, i.e., sparse graphs.

The construction uses vertex gadgets allowing exactly one shortcut; see Fig. 1a. Taking this shortcut represents that the corresponding vertex of the minimum independent dominating set instance is *not* included in $S$. Moreover it uses edge gadgets connecting for each edge its two corresponding vertex gadgets. They are comprised of long zizag pieces that can only be skipped if the independent set property for each edge is fulfilled; see Fig. 1c. Similarly, we have a neighborhood gadget connecting for each vertex the vertex gadgets of its neighborhood to ensure that the domination property is satisfied; see Fig. 1d.

On a high level, vertex gadgets are vertical pieces arranged horizontally next to each other and edge and neighborhood gadgets are horizontal pieces arranged vertically above each other and across the vertex gadgets. This yields a grid-like structure with many crossings between vertex gadgets and unrelated

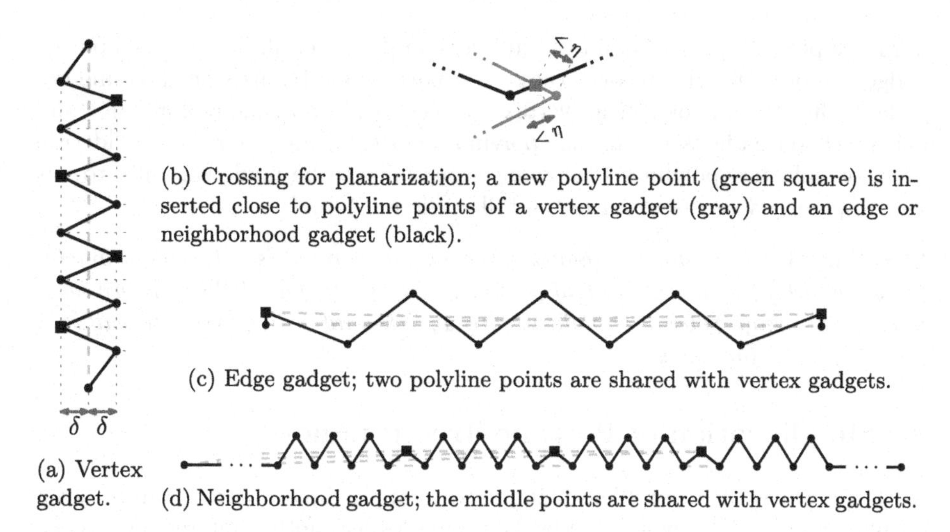


(a) Vertex gadget.

(b) Crossing for planarization; a new polyline point (green square) is inserted close to polyline points of a vertex gadget (gray) and an edge or neighborhood gadget (black).

(c) Edge gadget; two polyline points are shared with vertex gadgets.

(d) Neighborhood gadget; the middle points are shared with vertex gadgets.

**Fig. 1.** Schematization of the gadgets of the reduction from MIDS to PBS with planar instances. Shortcuts are indicated by dashed green line segments and points that are shared between two gadgets are drawn as squares. (Color figure online)

edge/neighborhood gadgets; see Fig. 2. The key idea is to planarize the non-planar construction by replacing crossings by new polyline points. However, we have to be careful where to insert these new points. Just inserting points wherever a crossing occurs would allow new shortcuts and hence destroy the mechanics of the gadgets. We can prevent this from happening by reshaping the construction so that crossings occur only close to existing polyline points. There, we can insert new polyline points onto the crossings sufficiently close to existing other polyline points. This ensures that for any shortcut starting or ending at a crossing point, this crossing point could either be replaced by an original polyline point or that the total saving incurred by the new crossing points does not severely impact the gap in the objective function between completeness and soundness in the

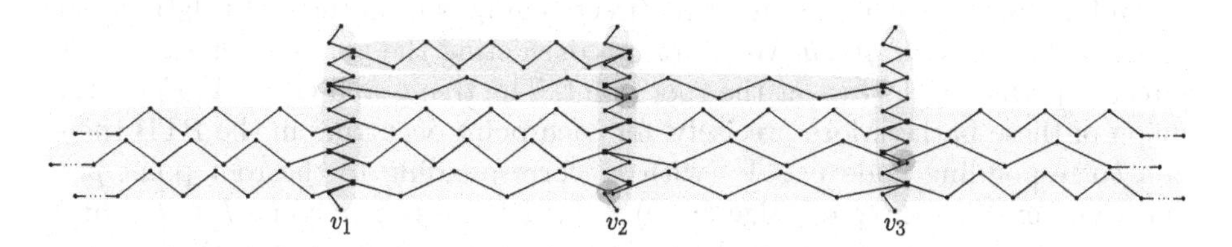


**Fig. 2.** Combination of three vertex gadgets (for the vertices $v_1, v_2, v_3$; blue background) with two edge gadgets (for the edges $v_1v_2$ and $v_1v_3$; red background) and three neighborhood gadgets (for the vertices $v_1, v_2, v_3$; green background). We use a crossing as in Fig. 1b between the vertex gadget of $v_2$ and the edge gadget of $v_1v_3$, between the vertex gadget of $v_3$ and the neighborhood gadget of $v_2$, and between the vertex gadget of $v_2$ and the neighborhood gadget of $v_3$. (Color figure online)

hardness proof. We describe in the full version [2] in detail, how to reshape the gadgets in order to ensure correctness. Our basic reduction uses one polyline per gadget. But we can connect all vertex gadgets to one polyline and all edge and neighborhood gadgets to another polyline, which means our results hold true even for only two polylines (the connections have to be made carefully to not violate planarity as discussed in the full version [2]).

**Corollary 1.** *PBS with a planar polyline bundle as input is not fixed-parameter tractable (FPT) in the number of polylines $\ell$. In particular, PBS with two polylines in a planar polyline bundle is already NP-hard to approximate within a factor of $n^{\frac{1}{3}-\varepsilon}$ for any $\varepsilon > 0$.*

## 4 Simplification of Polyline Tree Bundles

With the general PBS problem being hard to approximate better than $n^{\frac{1}{3}}$ even on planar inputs, we now consider tree bundles as another interesting special case of PBS. To form a tree bundle, the polylines have to start in a common root point and then branch out.

**Definition 2 (Polyline Tree Bundle (PTB)).** *An instance of PTB is a PBS instance $(P, \mathcal{L}, \delta)$ where we additionally require that $\mathcal{L}$ is a set of simple polylines such that all $L \in \mathcal{L}$ start at the root point $p_r$, and for any pair of polylines $L, L' \in \mathcal{L}$, the only intersection is a common prefix $L(p_r, p_i) = L'(p_r, p_i)$.*

We remark that this definition does not demand the tree bundle to be planar as the intersection constraint is only concerned with common points of the polylines. Moreover, we do not need to consider the case here where a polyline $L' \in \mathcal{L}$ is a sub-polyline of $L \in \mathcal{L}$. By definition, we will include the endpoints of all polylines in our simplification and hence if the endpoint of $L'$ lies on $L$, we could simply consider that point as the root of another PTB which can be simplified independently. We will show that tree bundles can be consistently simplified to optimality in polynomial time.

**Problem Transformation.** We transform the PTB simplification problem into a graph problem by constructing two directed graphs from the input data: a *tree graph* and a *shortcut graph*. We start by considering the polylines as embedded directed paths which start at the root point. The tree graph $G_t = (V, E_t)$ is the union of these paths. More precisely, for each point occurring in the PTB there is a corresponding node $v \in V$ (with $v_r$ corresponding to the root point $p_r$), and there exists a directed edge $(v, w) \in E_t$ if there is a polyline $L \in \mathcal{L}$ which contains the segment between the respective points (in that direction). For a given distance function $d$ and threshold $\delta > 0$, the shortcut graph $G_s = (V, E_s)$ is the union of all valid shortcut edges, i.e. edges $(v, w) \in \binom{V}{2}$ where for all polylines $L \in \mathcal{L}$ that contain $v$ and $w$ (in that order), we have $d((v, w), L(v, w)) \leq \delta$. Figure 3 shows $G_t$ and $G_s$ for an example PTB. Note that no matter the distance function and the value of $\delta$, we always have $E_t \subseteq E_s$, i.e. all tree graph edges are also contained in the shortcut graph.

**Lemma 1.** *The tree graph* $G_t = (V, E_t)$ *has size* $\mathcal{O}(n)$ *and can be constructed in time* $\mathcal{O}(n^2)$.

**Theorem 2.** *The shortcut graph* $G_s = (V, E_s)$ *has size* $\mathcal{O}(n^2)$ *and can be constructed for the Fréchet distance in time* $\mathcal{O}(n^3)$ *and for the Hausdorff distance in time* $\mathcal{O}(n^2)$.

The respective proofs are provided in the full version [2]. Based on the notion of the tree graph and the shortcut graph, we are now ready to restate the PTB simplification problem (PTBS) as a graph problem.

**Definition 3 (Polyline Tree Bundle Simplification (PTBS)).** *Given a tree graph* $G_t = (V, E_t)$ *and a shortcut graph* $G_s = (V, E_s)$, *the goal is to find a smallest node subset* $S \subseteq V$ *such that:*

- *The root node and all leaf nodes of the tree graph are contained in* $S$.
- *The induced subgraph* $G_s[S]$ *is connected.*

**Exact Polytime Algorithm.** Next, we describe a dynamic programming (DP) approach that only operates on $G_t$ and $G_s$, and returns an optimal PTBS solution in time $\mathcal{O}(n^2)$. Let $Sub(v) \subseteq G_t$ be the sub-tree rooted at node $v$ in the tree graph. Our main observation is that we can break down an optimal solution recursively. If a node $v$ is part of the solution, it's easy to see that there can't be shortcuts bypassing $v$. Thus, the solution $S$ can be split into two parts: an optimal solution for $Sub(v)$ and an optimal solution for $G_t \setminus Sub(v)$. We denote the size of an optimal solution for $Sub(v)$ by $s(v)$. As we don't know a priori which nodes will end up in the solution, we strive for computing $s(v)$ for each node $v \in V$ in an efficient manner. For leaf nodes $v$, we obviously get $s(v) = 1$. To compute $s(v)$ for an inner node $v$, we assume that $s(w)$ is already known for all nodes $w \in Sub(v) \setminus \{v\}$. Each path from a leaf $u$ to $v$ in $Sub(v)$ needs to contain a *cover node* $w$ such that $(v, w) \in E_s$ (that means there is a valid shortcut from $v$ to $w$). To identify the best selection of such cover nodes, we compute a helping function $h : V \to \mathbb{N}$ for each node $w \in Sub(v)$ as follows: Initially, $h(w) = s(w)$ if $(v, w) \in E_s$, and $h(w) = \infty$ otherwise. Then, in a post-order traversal of $Sub(v)$, for each non-leaf node $w$ we set $h(w) = \min\{h(w), \sum_{u \in N(w)} h(u)\}$ where $N(w)$ denotes the set of children (out-neighbors) of $w$ in $G_t$. In that way, $h(w)$ encodes

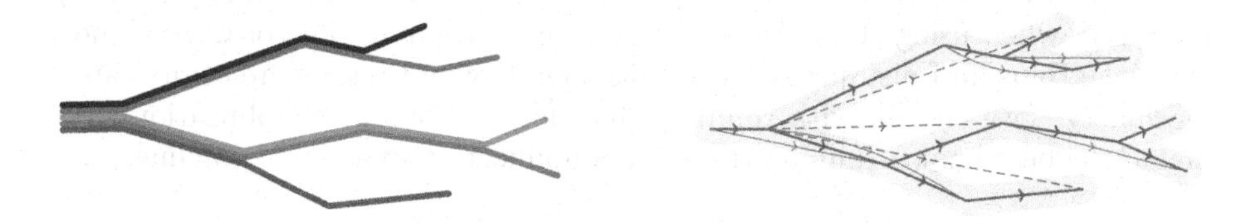

**Fig. 3.** Left: Example of a PTB instance. Right: Thick blue edges represent the tree graph $G_t$. The combination of the thick blue and the green edges build the shortcut graph $G_s$ for $d_H$ for the distance threshold $\delta$ (indicated via the light green tubes). Examples of invalid shortcuts are drawn dashed violet. (Color figure online)

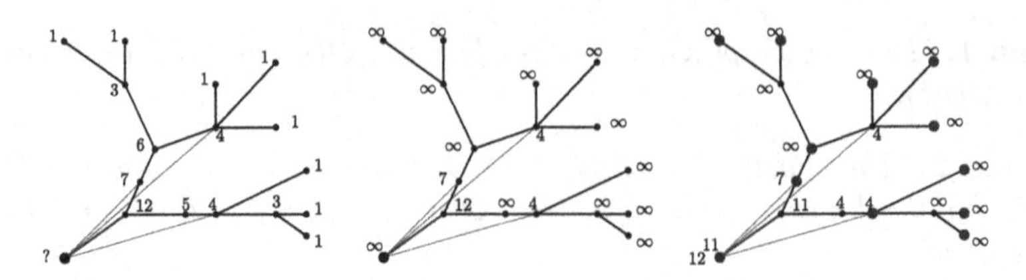

**Fig. 4.** The left image shows an example tree graph with optimum sub-tree simplification sizes (black) known for all nodes except the root node. The purple line segments indicate valid shortcuts from the root node. The middle image depicts the same tree after initial assignment of the helping values (red). Here, only end nodes of valid shortcuts have finite values assigned to them. The right image shows the final helping values (green) after propagation as well as the respective optimum simplification size of the tree assigned to the root node (black). The blue marked nodes are the ones that are contained in the optimal simplification. (Color figure online)

the smallest number of nodes that have to be kept in $Sub(w)$ if for all paths from $v$ to leaf nodes in $Sub(w)$ the respective cover node is contained in $Sub(w)$. The optimal solution size $s(v)$ for $Sub(v)$ is then $h(v)+1$ (as we have to additionally include $v$ itself). Note that $s(v)$ is always well-defined (i.e., finite) as the tree edges are all valid shortcuts in $G_s$. To make sure that at the time we compute $s(v)$ all values $s(w)$ for $w \in Sub(v) \setminus \{v\}$ are known, we also globally traverse the nodes in the tree graph in post-order. Figure 4 illustrates the computation of $s(v)$. The optimal set of simplification nodes can then be determined by backtracking.

For a faster running time of the DP in practice (used in our experiments), we only compute $h$-values for nodes in $Sub(v)$ which are on a path from $v$ to some node $w$ with $(v, w) \in E_s$. These nodes can easily be identified by computing the reverse path from each such node $w$ to $v$ and marking all nodes along the way (stopping as soon as a marked node is encountered to avoid redundancy). For marked nodes $w$ with an unmarked neighbor, we just set $\sum_{u \in N(w)} h(u)$ to $\infty$ to maintain correctness. Especially for small distance thresholds $\delta$ and large subtrees $Sub(v)$, this modification accelerates the computation of $s(v)$ significantly.

**Theorem 3.** *PTBS can be solved optimally in time $\mathcal{O}(n^2)$.*

The proof is given in the full version [2]. Combining the time for problem transformation with the time of the DP, we get an overall running time of $\mathcal{O}(n^3)$ for PTBS when using the Fréchet distance and a running time of $\mathcal{O}(n^2)$ when using the Hausdorff distance. Hence – although having to use more complicated machinery – we end up with running times for tree bundle simplification that match the best known running times for simplification of a single polyline.

## 5 Tree Bundle Decompositions

To leverage our algorithm for optimal tree bundle simplification for general bundles, we next consider the problem of decomposing a general bundle into (a small set of) tree bundles. To formalize the TREE BUNDLE DECOMPOSITION (TBD) problem we first introduce the notion of a $D$-decomposition of a polyline.

**Definition 4 ($D$-Decomposition).** *Let $L = (s, \dots, t)$ be a simple polyline (represented as a list of points) and let $D$ be a point set. Further let $d_1, d_2, \dots, d_k$ be the points in $L \cap D$ in the order in which they appear in $L$. The $D$-decomposition of $L$ denoted by $L(D)$ is the set of subpolylines $L(d_i, d_{i+1})$ for $i = 1, \dots, k-1$.*

We strive to find a sensible set $D$ that partitions a given bundle into tree bundles.

**Definition 5 (Tree Bundle Decomposition (TBD)).** *Given a PBS instance $(P, \mathcal{L}, \delta)$, we seek to find a point subset $D \subseteq P$ (the decomposition points) with the following requirements:*

- *Each polyline $L \in \mathcal{L}$ starts and ends in a point in $D$.*
- *Let $G_I$ be the intersection graph in which we have a node for each subpolyline in $\bigcup_{L \in \mathcal{L}} L(D)$ and an edge between two nodes if the subpolylines $\tilde{L}, \bar{L}$ share a point that is not in $D$, i.e. $(\tilde{L} \cap \bar{L}) \setminus D \neq \emptyset$. Then the subpolylines within a connected component in $G_I$ form a PTB.*

Based on a TBD, we can simplify the given bundle by simplifying each of the tree bundles induced by $D$ independently. The union of all tree simplifications then yields $S$. The goal, of course, is still to end up with a small set $S$. To achieve that, we aim at TBDs which induce few but large tree bundles with a small decomposition set $D$. In the following we assume that all polyline endpoints are already included in $D$ as they have to be part of $S$ by definition.

**A Simple Greedy Heuristic.** Nodes in the set $D$ might end up being the root node of a tree bundle or a leaf node (or both). One way to construct $D$ is hence to greedily select root nodes and grow trees from those (adding the respective leaf nodes to $D$ as well).

For root selection, we use the line degree of the point, that is, the number of polylines in $\mathcal{L}$ that contain the point. As only polylines that contain the root can be part of the respective PTB, we always choose the node with the highest line degree that is not already part of a tree bundle next. To compute the largest prossible tree bundle for a selected root node $r$, we first construct the union graph $G_U(V, E)$ of the polyline bundle. Here, each point in the bundle is represented by a node in $V$ and an edge exists between two nodes if there is a polyline segment between the respective points. Additionally, we assign to each edge in $G_U$ the set of polylines that traverse it. A tree bundle can then be computed in a BFS-like fashion in $G_U$ starting from $r$, always pushing edges instead of nodes in the queue. The edges incident to $r$ are always included in the PTB and are hence used for initialization of the queue (artificially directed away from $r$). In any later step, if an edge $(u, v)$ is extracted from the queue, we first check whether all other edges incident to $v$ are unvisited. If that is the case, we need to make sure that the polyline set assigned to each incident edge is a (not necessarily proper) subset of the polylines assigned to $(u, v)$. If and only if those conditions are met for all incident edges, these edges are included in the subtree, marked as visited, and inserted in the queue. Otherwise $v$ is added to $D$. The process takes $\mathcal{O}(\ell \cdot n)$ time.

## 6 Experimental Evaluation

We implemented the dynamic programming approach (DP) for exact tree bundle simplification as well as the greedy tree bundle decomposition algorithm (TBD) in C++. Furthermore, we also provide the first implementation of the bi-criteria approximation algorithm (BCA) from [14]. BCA demands to first compute a small *star cover* of the polyline bundle where a star is a point $p$ in the bundle together with selected shortcuts that end in $p$. A feasible star cover has to ensure that for each polyline $L \in \mathcal{L}$ and for each segment in $L$, there is a star in the cover with a shortcut that bridges said segment. The set of points of all stars in the star cover induces a simplified polyline bundle for a distance threshold of $2\delta$ (that is, twice the actual threshold). The number of retained points is at most a factor of $\mathcal{O}(\log(\ell + n))$ larger than the optimal solution for threshold $\delta$. The running time of BCA is in $\mathcal{O}(\ell \cdot n^3)$. As this result only holds when using the Fréchet distance, we will focus in the experiments on $d_F$. All experiments were run on a single core of an Intel Core i9 processor at 2.4 GHz.

**Benchmark Data.** We used two types of polyline bundle data to evaluate the algorithms: *(i) Path bundles from embedded road networks* (extracted from OpenStreetMap [13]). Such bundles are a good model for movement data. Bundles were constructed by first extracting a connected subgraph with a given number of nodes from the network. To obtain a tree bundle, we then performed a BFS run from a randomly selected root node in the subgraph and backtracked all paths from the leaves to the root of the BFS-tree. For general bundles, we select not one but several root nodes in the subgraph, construct a tree bundle for each and then combine those into a single bundle. *(ii) Public transit networks* (GTFS data provided by OpenMobilityData [12]). We used the data from Stuttgart, Freiburg, Manhattan and Chicago. Here each bus or train line forms a polyline in our bundle.

**Tree Bundle Simplification Results.** We compared the performance of DP and BCA on tree bundles of different sizes extracted from road networks. While it might seem to be an apples-to-oranges comparison, when we have an exact algorithm on the one side and a bicriteria approximation on the other, it is not a priori clear which algorithm would produce the smaller simplification when tested with the same $\delta$ (as BCA is allowed to exceed it by a factor of 2). We observe, however, that on all tested instances, the exact DP algorithm produces better simplification results than BCA, even though BCA is allowed to use a distance threshold of $2\delta$. If we call BCA with $\delta/2$ to then end up with a solution that obeys the $\delta$-constraint, the quality deteriorates significantly (with up to 50% larger outputs). Table 1 provides some selected results which reflect the general behavior. It is interesting that the BCA algorithm indeed produces solutions

**Table 1.** Comparison of DP and BCA on tree bundles (note that BCA is tested for the original $\delta$ and $\delta/2$). $\delta_F$ denotes the resulting Fréchet distance and $\delta_F/\delta$ the distance relative to the threshold $\delta$. $n$ is the input size, and $|S|$ the number of points in the computed solution for $\delta$ (in geo coordinates). Timings are in milliseconds.

| | $\delta \cdot 10^4$ | $\delta_F \cdot 10^4$ | $\delta_F/\delta$ | $n$ | $\|S\|$ | Time |
|---|---|---|---|---|---|---|
| DP | 5.00 | 4.99 | 0.99 | 500 | **204** | 3 |
| BCA | 5.00 | 9.81 | 1.96 | 500 | 216 | 3 |
| BCA | 2.50 | 3.17 | 1.27 | 500 | 251 | 6 |
| DP | 5.00 | 5.00 | 1.00 | 8,000 | **4009** | 21 |
| BCA | 5.00 | 8.77 | 1.75 | 8,000 | 4029 | 407 |
| BCA | 2.50 | 4.04 | 1.61 | 8,000 | 5276 | 350 |
| DP | 5.00 | 5.00 | 1.00 | 50,000 | **24,076** | 248 |
| BCA | 5.00 | 9.68 | 1.94 | 50,000 | 24,195 | 14,800 |
| BCA | 2.50 | 5.00 | 2.00 | 50,000 | 32,457 | 13,500 |

where the $\delta$ threshold is violated by a factor of 2, proving the theoretical analysis to be tight in this respect. We also observe that the DP approach scales much better, with running times up to a factor of 50 faster than BCA on our largest test instance.

**Results on General Bundles.** We used path bundles from road networks as well as public transit networks to evaluate the performance of TBD+DP and BCA. Again, BCA results are allowed to exceed the distance threshold $\delta$ by a factor of 2. This slack is indeed strongly exploited also on public transit networks as confirmed by our detailed BCA experiments reported in the full version [2]. We now focus on a comparative evaluation. We observe that our heuristic approach of first computing a tree decomposition and then simplifying the resulting trees individually is always faster than BCA, computing results within a second even for roadnetwork bundles with around 10,000 nodes while BCA takes 30 times longer. In terms of quality, TBD+DP produce comparable or even better results than BCA on the Stuttgart and Freiburg network, and clearly superior results on road network bundles. Detailed results and illustrations are provided in the full version [2]. The instances on which TBD+DP was outperformed by BCA in terms of simplification size are bundles with large grid-like structures as the Chicago and the Manhattan public transit network. Here, our tree decomposition results in a huge set of trees of which we need to keep all root and leaf nodes in the simplification. A post-processing step in which for each point in $S$, we test whether it could be removed without constraint violation could help to close that gap.

But especially for large instances, the simplicity and the fast computation time of TBD+DP is a great advantage over BCA; in particular as the TBD is independent of $\delta$ and individual tree simplification can be easily parallelized for further improvement.

## 7 Future Work

Based on our finding that the bi-criteria approximation algorithm indeed exceeds the distance threshold bound by a factor of 2 on practical instances but produces high-quality solutions, future work could investigate whether improved bi-criteria approximation factors can be proven. Furthermore, it might be interesting to investigate the existence of FPT algorithms for PBS for suitable parameters. Our results that tree bundles can be processed in polynomial time might hint at parameterizability by e.g. the treewidth of the union graph of the polylines. On the practical side, further development of heuristics for PBS or the consideration of non-simple polylines could be sensible avenues for future work.

## References

1. Bereg, S., Jiang, M., Wang, W., Yang, B., Zhu, B.: Simplifying 3D polygonal chains under the discrete Fréchet distance. In: Laber, E.S., Bornstein, C., Nogueira, L.T., Faria, L. (eds.) LATIN 2008. LNCS, vol. 4957, pp. 630–641. Springer, Heidelberg (2008). https://doi.org/10.1007/978-3-540-78773-0_54
2. Bosch, Y., Schäfer, P., Spoerhase, J., Storandt, S., Zink, J.: Consistent simplification of polyline tree bundles. CoRR abs/2008.10583 (2021). arXiv:2108.10790
3. Buchin, M., Kilgus, B., Kölzsch, A.: Group diagrams for representing trajectories. In: Proceedings of the 11th ACM SIGSPATIAL International Workshop on Computational Transportation Science, pp. 1–10 (2018). https://doi.org/10.1145/3283207.3283208
4. Chan, W.S., Chin, F.: Approximation of polygonal curves with minimum number of line segments or minimum error. Int. J. Comput. Geom. Appl. **6**(1), 59–77 (1996). https://doi.org/10.1142/s0218195996000058
5. Douglas, D.H., Peucker, T.K.: Algorithms for the reduction of the number of points required to represent a digitized line or its caricature. Cartographica **10**(2), 112–122 (1973). https://doi.org/10.3138/fm57-6770-u75u-7727
6. Fan, C., Filtser, O., Katz, M.J., Wylie, T., Zhu, B.: On the chain pair simplification problem. In: Dehne, F., Sack, J.-R., Stege, U. (eds.) WADS 2015. LNCS, vol. 9214, pp. 351–362. Springer, Cham (2015). https://doi.org/10.1007/978-3-319-21840-3_29
7. Fan, C., Filtser, O., Katz, M.J., Zhu, B.: On the general chain pair simplification problem. In: 41st International Symposium on Mathematical Foundations of Computer Science (MFCS 2016). Schloss Dagstuhl - Leibniz-Zentrum für Informatik (2016). https://doi.org/10.4230/LIPIcs.MFCS.2016.37
8. Halldórsson, M.M.: Approximating the minimum maximal independence number. Inf. Process. Lett. **46**(4), 169–172 (1993). https://doi.org/10.1016/0020-0190(93)90022-2
9. He, S., et al.: RoadRunner: improving the precision of road network inference from GPS trajectories. In: Proceedings of the 26th ACM SIGSPATIAL International Conference on Advances in Geographic Information Systems, pp. 3–12 (2018). https://doi.org/10.1145/3274895.3274974
10. Hershberger, J., Snoeyink, J.: Speeding up the Douglas-Peucker line-simplification algorithm. In: Proceedings of the 5th International Symposium on Spatial Data Handling (SDH 1992), pp. 134–143 (1992)

11. Imai, H., Iri, M.: Polygonal approximations of a curve-formulations and algorithms. In: Machine Intelligence and Pattern Recognition, vol. 6, pp. 71–86. Elsevier (1988). https://doi.org/10.1016/b978-0-444-70467-2.50011-4
12. MobilityData IO: OpenMobilityData. https://transitfeeds.com/
13. OpenStreetMap contributors: Planet dump retrieved from https://planet.osm.org (2017)
14. Spoerhase, J., Storandt, S., Zink, J.: Simplification of polyline bundles. In: Proceedings of the 17th Scandinavian Symposium and Workshops on Algorithm Theory (SWAT 2020) (2020). https://doi.org/10.4230/LIPIcs.SWAT.2020.35
15. Visvalingam, M., Whyatt, J.D.: Line generalisation by repeated elimination of points. Cartogr. J. **30**(1), 46–51 (1993). https://doi.org/10.1179/000870493786962263
16. Wylie, T., Zhu, B.: Protein chain pair simplification under the discrete Fréchet distance. IEEE/ACM Trans. Comput. Biol. Bioinfom. **10**(6), 1372–1383 (2013). https://doi.org/10.1109/tcbb.2013.17

# Improving Upper and Lower Bounds for the Total Number of Edge Crossings of Euclidean Minimum Weight Laman Graphs

Yuki Kobayashi[1(✉)], Yuya Higashikawa[2], and Naoki Katoh[2]

[1] Department of Engineering, Osaka City University, Osaka, Japan
kobayashi@osaka-cu.ac.jp
[2] Graduate School of Information Science, University of Hyogo, Kobe, Japan
{higashikawa,naoki.katoh}@gsis.u-hyogo.ac.jp

**Abstract.** We investigate the total number of edge crossings (i.e., the crossing number) of the Euclidean minimum weight Laman graph $\mathsf{MLG}(P)$ on a planar point set $P$. Bereg et al. (2016) showed that the upper and lower bounds for the crossing number of $\mathsf{MLG}(P)$ are $6|P|-9$ and $|P|-3$, respectively. In this paper, we improve these upper and lower bounds given by Bereg et al. (2016) to $2.5|P|-5$ and $(1.25-\varepsilon)|P|$ for any $\varepsilon > 0$, respectively. Especially, for improving the upper bound, we introduce a novel counting scheme based on some geometric observations.

**Keywords:** Laman graphs · Sparse and tight graphs · Plane graphs · Geometric graphs · Edge crossings

## 1 Introduction

A graph $G = (V, E)$ is called *Laman* if $|E| = 2|V| - 3$ and $|E(H)| \leq 2|V(H)| - 3$ for any subgraph $H$ of $G$ with $E(H) \neq \emptyset$. A Laman graph has a property of being *minimally rigid* in the plane if it is realized as a *generic bar-joint framework* [5,8]. A bar-joint framework is a straight-line realization of a graph in the plane, and by regarding each edge as a bar and each point as a joint the rigidity of such a graph can be defined in a natural way (see, e.g., [5]). One of the most fundamental results in combinatorial rigidity theory asserts that a graph $G$ realized on a generic point set (i.e., the set of the coordinates is algebraically independent over the rational field) is rigid if and only if $G$ contains a spanning Laman subgraph [8]. Laman graphs appear in a wide range of applications, not only statics but also mechanical design such as linkages, design of CAD systems, analysis of protein flexibility, and sensor network localization [9,10].

Given a set $P$ of $n$ points in the Euclidean plane, let $G(P)$ denote a *geometric graph* on $P$, i.e., $G(P) = (P, E)$ where $E$ is a set of edges each of which is drawn as a segment between two points in $P$. Throughout the paper, we assume that no

This work is supported by JST CREST Grant Number JPMJCR1402.


C.-Y. Chen et al. (Eds.): COCOON 2021, LNCS 13025, pp. 244–256, 2021.
https://doi.org/10.1007/978-3-030-89543-3_21

three points in $P$ are collinear and all interpoint distances are distinct. The point set satisfying these assumptions is called *semi-generic*. A two-dimensional bar-joint framework is considered as a geometric graph, thus in this paper, we deal with geometric graphs where the underlying graphs are Laman, called *Euclidean Laman graphs*. It is then natural to consider the Euclidean Laman graph on a planar point set $P$ with the minimum total edge-length over all Euclidean Laman graphs on $P$, i.e., the *Euclidean minimum weight Laman graph* on $P$ denoted by $\mathsf{MLG}(P)$.

In order to realize a geometric graph as a bar-joint framework in the real world, it is important to consider the *crossing property* of the geometric graph. A geometric graph is called *plane* (or *non-crossing*) if any two edges do not have a crossing except possibly at their endpoints. In fact, the Euclidean minimum spanning tree on a semi-generic planar point set $P$ ($\mathsf{MST}(P)$ for short) is plane. Observe that both Laman graphs and spanning trees are characterized by similar *sparsity conditions*: A graph $G$ is called $(k,l)$*-sparse* if $|E(H)| \leq k|V(H)| - l$ for any subgraph $H$ of $G$ with $E(H) \neq \emptyset$, and a $(k,l)$-sparse graph is called $(k,l)$*-tight* if it has exactly $k|V(H)| - l$ edges (see, e.g., [8]). A spanning tree is a $(1,1)$-tight graph while a Laman graph is a $(2,3)$-tight graph. Since $(k,l)$-sparse graphs have several common combinatorial properties such as being independent sets of a matroid, a natural question is whether the Euclidean minimum weight $(k,l)$-tight graph on a point set has a nice crossing property as does the Euclidean minimum weight $(1,1)$-tight graphs.

Bereg et al. [3] studied crossing properties of $\mathsf{MLG}(P)$. They proved as the main results that $\mathsf{MLG}(P)$ is 6-planar, i.e., each edge in $\mathsf{MLG}(P)$ has at most six crossings, and $\mathsf{MLG}(P)$ is also *quasi-planar*, i.e., no three edges in $\mathsf{MLG}(P)$ pairwise cross. In addition, they showed an instance $P$ for which there exists an edge that has six crossings in $\mathsf{MLG}(P)$.

In the following, we use the terminology *crossing number* to denote the total number of crossings. According to the results by Bereg et al. [3], it is easy to see that the crossing number of $\mathsf{MLG}(P)$ is at most $6 \times (2|P| - 3)/2 = 6|P| - 9$. Bereg et al. [3] also provided an instance $P$ for which the crossing number of $\mathsf{MLG}(P)$ is $|P| - 3$ (as shown in Fig. 3), therefore, there has been a gap between upper and lower bounds for the crossing number of $\mathsf{MLG}(P)$. In this paper, we improve these upper and lower bounds given by Bereg et al. [3] to $2.5|P| - 5$ and $(1.25 - \varepsilon)|P|$ for any $\varepsilon > 0$, respectively. Especially, for improving the upper bound, we introduce a novel counting scheme based on some geometric observations, which is the most important contribution presented in the paper.

As for the crossing number of geometric graphs, several classes of *proximity graphs* are studied by Ábrego et al. [1], e.g., *nearest neighbor graphs*, *relative neighborhood graphs*, *Gabriel graphs* and *Delaunay graphs*. In a *k-nearest neighbor graph* on a point set $P$ ($k - \mathsf{NNG}(P)$ for short), for $p, q \in P$, $pq$[1] is included if and only if $p$ is the $i$-th closest point among $p$ from $q$ for some $i \geq k$ or vice versa. In a *k-relative neighborhood graph* on a point set $P$ ($k - \mathsf{RNG}(P)$ for short), for $p, q \in P$, $pq$ is included if and only if $D_p(pq) \cap D_q(pq)$ (where $D_p(r)$

[1] Throughout the paper, for two points $p, q$, we abuse the notation $pq$ to denote the line segment between $p$ and $q$ or the length of itself, depending on the context.

denotes the closed disk with center $p$ and radius $r$) contains at most $k$ points among $P \setminus \{p, q\}$. In a *k-Gabriel graph* on a point set $P$ ($k - \mathsf{GG}(P)$ for short), for $p, q \in P$, $pq$ is included if and only the circle through $p$ and $q$ with diameter $pq$ contains at most $k$ points among $P \setminus \{p, q\}$. In a *k-Delaunay graphs* on a point set $P$ ($k - \mathsf{DG}(P)$ for short), for $p, q \in P$, $pq$ is included if and only if there is a circle through $p$ and $q$ that contains at most $k$ other points. Ábrego et al. [1] proved that for any set $P$ of $n$ points, $k - \mathsf{NNG}(P)$ has at most $k^3 n$ crossings, $k - \mathsf{RNG}(P)$ has at most $9k^3 n$ crossings, $k - \mathsf{GG}(P)$ has at most $3k^2 n^2$ crossings, and $k - \mathsf{DG}(P)$ has at most $3k^2 n^2$ crossings. Note that Bereg et al. [3] showed the relation among $k - \mathsf{NNG}(P)$, $k - \mathsf{RNG}(P)$, $k - \mathsf{GG}(P)$ and $\mathsf{MST}(P)$. See Lemma 1 for the details.

The rest of the paper is organized as follows. In Sect. 2, we introduce some notations and properties of $\mathsf{MLG}(P)$ given by Bereg et al. [3], that are used throughout the paper. In Sect. 3, we provide new geometric observations and give an efficient counting scheme for improving the upper bound based on the shown observations. In Sect. 4, we show how to construct an instance which achieves the improved lower bound. In Sect. 5, we discuss future works, which concludes the paper.

## 2 Preliminaries

First of all, we introduce some notations used throughout the paper. The closed disk (resp. circle) with center $p$ and radius $r$ is denoted $D_p(r)$ (resp. $C_p(r)$). Consider two points $p, q$ in the plane. Let $\mathtt{Lens}(pq) = D_p(pq) \cap D_q(pq)$. Let $\mathtt{bisect}(pq)$ denote the perpendicular bisector of segment $pq$. Let $\mathtt{Up\_Lens}(pq)$ (resp. $\mathtt{Low\_Lens}(pq)$) denote the intersection of $\mathtt{Lens}(pq)$ and the halfplane determined by $\mathtt{bisect}(pq)$ that contains $p$ (resp. $q$). Let $\mathtt{L\_Lens}(pq)$ (resp. $\mathtt{R\_Lens}(pq)$) denote the intersection of $\mathtt{Lens}(pq)$ and the halfplane determined by the supporting line of segment $pq$ that contains a point $p'$ such that $p, q$, and $p'$ are arranged on triangle $pqp'$ in clockwise (resp. counterclockwise) order. For a point $p$ and two half lines $\ell$ and $\ell'$ starting at $p$ in the plane, let $\mathtt{angle}_p(\ell, \ell')$ denote the smaller angle between $\ell$ and $\ell'$, and $\mathtt{Cone}_p(\ell, \ell')$ denote the cone with apex at $p$ delimited by $\ell$ and $\ell'$, which corresponds to $\mathtt{angle}_p(\ell, \ell')$.

In the rest of this section, we introduce several lemmas and theorems shown by Bereg et al. [3] since those are useful to prove our main lemmas provided in the next section. In the following, let $P$ be a set of semi-generic $n$ points in the Euclidean plane, and for a geometric graph $G(P)$, we abuse notation $G(P)$ to denote a set of edges in $G(P)$.

Let us start with a property based on which our counting scheme is.

**Lemma 1.** (Theorem 1.1 in [3]) *It holds*

$$\mathsf{MST}(P) \cup 2 - \mathsf{NNG}(P) \subseteq \mathsf{MLG}(P) \subseteq 1 - \mathsf{GG}(P) \cap 2 - \mathsf{RNG}(P).$$

Focusing on $\mathsf{MST}(P) \subseteq \mathsf{MLG}(P)$, we classify the edges in $\mathsf{MLG}(P)$ into ones in $\mathsf{MST}(P)$ and ones not in $\mathsf{MST}(P)$. The details will be given in the next section.

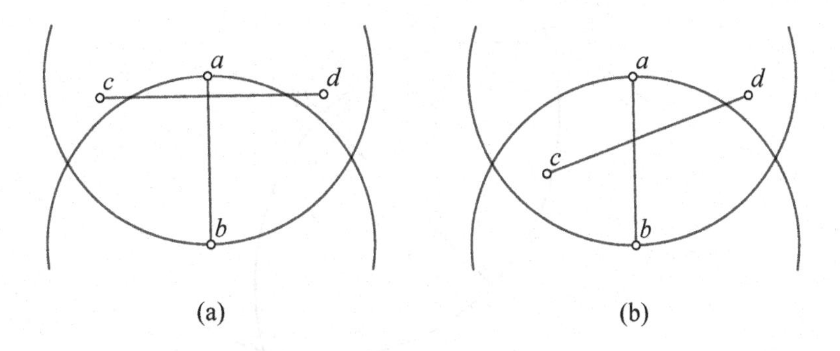


**Fig. 1.** (a) A lens-crossing edge $cd$ for $ab$. (b) A fan-crossing edge $cd$ for $ab$.

The following lemma shows properties which points of $P$ in $\texttt{Lens}(ab)$ for $ab \in \mathsf{MLG}(P)$ satisfy.

**Lemma 2.** (Lemma 2.3 in [3]) *Consider an edge* $ab \in \mathsf{MLG}(P)$.

*(i) There exists at most one point of* $P$ *in each of* $\texttt{L\_Lens}(ab)$ *and* $\texttt{R\_Lens}(ab)$.
*(ii) If there exists one point of* $P$ *in each of* $\texttt{L\_Lens}(ab)$ *and* $\texttt{R\_Lens}(ab)$, *i.e.,* $c \in \texttt{L\_Lens}(ab)$ *and* $d \in \texttt{R\_Lens}(ab)$, *it then holds that* $ab < cd$ *and* $cd \notin \mathsf{MLG}(P)$.

We introduce key concepts both in our paper and [3]. See also Fig. 1.

**Definition 1.** (lens-crossing edge) For four points $a, b, c, d \in P$, suppose that segments $ab$ and $cd$ cross each other, and $c, d \notin \texttt{Lens}(ab)$. Then, $cd$ is called a lens-crossing edge for $ab$.

**Definition 2.** (fan-crossing edge) For four points $a, b, c, d \in P$, suppose that segments $ab$ and $cd$ cross each other, and $c \in \texttt{Lens}(ab)$ and $d \notin \texttt{Lens}(ab)$. Then, $cd$ is called a fan-crossing edge for $ab$.

The following two lemmas show properties on lens-crossing edges for $ab \in \mathsf{MLG}(P)$.

**Lemma 3.** (Lemma 3.3 in [3]) *For four points* $a, b, c, d \in P$, *suppose that segment* $cd$ *is a lens-crossing edge for segment* $ab$, *and* $cd$ *cuts only* $\texttt{Up\_Lens}(ab)$ *(i.e., it does not cut* $\texttt{Low\_Lens}(ab)$*). Then, it holds* $a \in \texttt{Lens}(cd)$.

**Lemma 4.** (Lemma 4.1 in [3][2]) *Consider an edge* $ab \in \mathsf{MLG}(P)$. *Then,* $\mathsf{MLG}(P)$ *includes*

*(i) at most one lens-crossing edge for* $ab$ *that cuts only* $\texttt{Up\_Lens}(ab)$,
*(ii) at most one lens-crossing edge for* $ab$ *that cuts only* $\texttt{Low\_Lens}(ab)$, *and*
*(iii) no lens-crossing edge for* $ab$ *that cuts both* $\texttt{Up\_Lens}(ab)$ *and* $\texttt{Low\_Lens}(ab)$.

[2] Lemma 4.1 in [3] corresponds to Lemma 4(i)(ii).

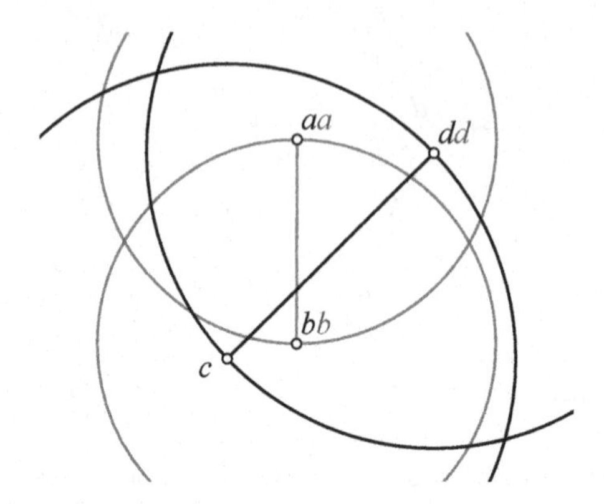


**Fig. 2.** Illustration of Lemma 4(iii).

Lemma 4 means that $\mathsf{MLG}(P)$ includes at most two lens-crossing edges for $ab \in \mathsf{MLG}(P)$. Especially, it is easy to see the proof of Lemma 4(iii) as follows: Suppose that a lens-crossing edge for $ab$, say $cd$, is also included in $\mathsf{MLG}(P)$, and $cd$ cuts both $\texttt{Up\_Lens}(ab)$ and $\texttt{Low\_Lens}(ab)$ as shown in Fig. 2. Then, it holds $ab < cd$ and $a, b \in \texttt{Lens}(cd)$, which contradicts Lemma 2(ii).

As for fan-crossing edges for $ab \in \mathsf{MLG}(P)$, we have the following lemma.

**Lemma 5.** (Lemma 4.3 in [3]) *Consider an edge* $ab \in \mathsf{MLG}(P)$*. Then,* $\mathsf{MLG}(P)$ *includes at most four fan-crossing edges for* $ab$*.*

Based on the proof of Lemma 5 written in [3], we analyze more details and obtain Lemmas 10 and 11 provided in the next section. Note that, indeed, the proof of Lemma 5 for the case where only one point exists in $\texttt{Lens}(ab)$ immediately follows from the proof of Lemma 10.

Let $\sigma(P)$ denote the crossing number of $\mathsf{MLG}(P)$. By Lemmas 4 and 5, we see that every edge in $\mathsf{MLG}(P)$ has at most six crossings. Therefore, $\sigma(P)$ is at most $6 \times (2n-3)/2$.

**Theorem 1** [3]**.** *For any set of semi-generic points* $P$*, it holds* $\sigma(P) \leq 6|P| - 9$*.*

Bereg et al. [3] also provide an instance $P$ whose crossing number is $|P| - 3$ as shown in Fig. 3.

**Theorem 2** [3]**.** *There exists a set of semi-generic points* $P$ *such that* $\sigma(P) \geq |P| - 3$*.*

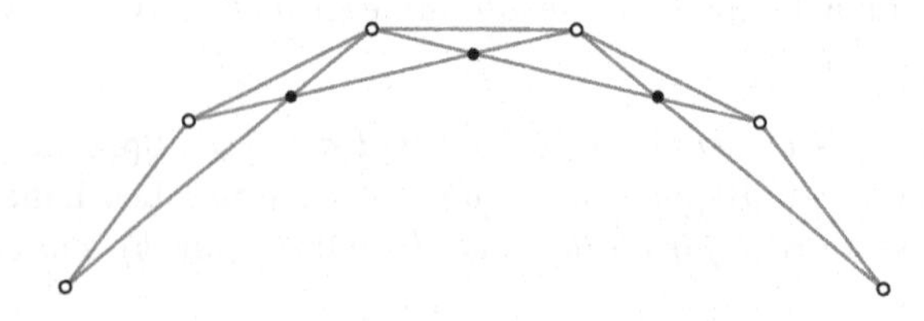

**Fig. 3.** $\mathsf{MLG}(P)$ that has $|P| - 3$ crossings (indicated by black-colored circles).

To conclude this section, we introduce the following useful lemmas implicitly shown in [11] and [3].

**Lemma 6** [11]. *For three points $a, b, c \in P$, suppose that $\texttt{angle}_a(\ell, \ell') < 60°$ and $b, c \in \texttt{Cone}_a(\ell, \ell')$. Then, the longer of $ab$ and $ac$ is not included in* $\mathsf{MST}(P)$.

**Lemma 7** [3]. *For four points $a, b, c, d \in P$, suppose that $\texttt{angle}_a(\ell, \ell') < 60°$ and $b, c, d \in \texttt{Cone}_a(\ell, \ell')$. Then, the longest of $ab$, $ac$, and $ad$ is not included in* $\mathsf{MST}(P)$.

## 3 Improved Upper Bound for $\sigma(P)$

In this section, we show a novel counting scheme based on some geometric observations, which improves the upper bound for $\sigma(P)$ shown in Theorem 1. Recall that $\mathsf{MST}(P) \subseteq \mathsf{MLG}(P)$ as shown in Lemma 1. In the following, let $\overline{\mathsf{MST}}(P) = \mathsf{MLG}(P) \setminus \mathsf{MST}(P)$. For counting $\sigma(P)$, we basically classify the edges in $\mathsf{MLG}(P)$ into ones in $\mathsf{MST}(P)$ and ones in $\overline{\mathsf{MST}}(P)$.

Let us first see the following lemma.

**Lemma 8.** *For an edge $ab \in \mathsf{MST}(P)$, there is no fan-crossing edge.*

*Proof.* We prove by contradiction that there exists no point of $P$ in $\texttt{Lens}(ab)$: Suppose that $c \in P$ lies in $\texttt{Lens}(ab)$. We then have $\max\{ab, bc, ca\} = ab$. Since for any triangle whose vertices are points in $P$ the longest edge is not in $\mathsf{MST}(P)$, it holds $ab \notin \mathsf{MST}(P)$, a contradiction. This completes the proof. □

At this point, it is easy to see $\sigma(P) \leq 4n - 7$ as follows: By Lemmas 4 and 8, an edge in $\mathsf{MST}(P)$ has at most two crossings in $\mathsf{MLG}(P)$. Therefore, we obtain

$$\sigma(P) \leq \frac{2|\mathsf{MST}(P)| + 6|\overline{\mathsf{MST}}(P)|}{2} = \frac{2(n-1) + 6(n-2)}{2} = 4n - 7.$$

In the rest of this section, we further improve this upper bound.

Next see the following lemma.

**Lemma 9.** *For four points $a, b, c, d \in P$, suppose that $cd$ is a lens-crossing edge for $ab$. Then, $ab$ is a fan-crossing edge for $cd$.*

*Proof.* By Lemma 4, without loss of generality, $cd$ cuts $\texttt{Up\_Lens}(ab)$ and does not cut $\texttt{Low\_Lens}(ab)$ (see Fig. 4). Then, by Lemma 3, we have $a \in \texttt{Lens}(cd)$. On the other hand, it holds $b \notin \texttt{Lens}(cd)$ by Lemma 2(ii). This completes the proof. □

We now consider classifying crossings in $\mathsf{MLG}(P)$ into two cases.

**Definition 3.** (f-f crossing/f-l crossing) For four points $a, b, c, d \in P$, suppose that two segments $ab$ and $cd$ intersect each other.

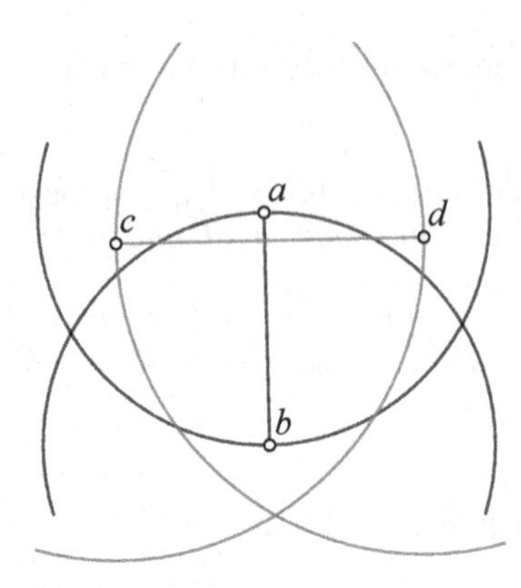


**Fig. 4.** Illustration of the proof of Lemma 9.

(i) If $cd$ is a fan-crossing edge for $ab$, and $ab$ is a fan-crossing edge for $cd$, we call the crossing between $ab$ and $cd$ an f-f crossing.
(ii) If $cd$ is a fan-crossing edge for $ab$, and $ab$ is a lens-crossing edge for $cd$, we call the crossing between $ab$ and $cd$ an f-l crossing.

Note that if $cd$ is a lens-crossing edge for $ab$, by Lemma 9, $ab$ must be a fan-crossing edge for $cd$. Therefore, every crossing in $\mathsf{MLG}(P)$ is an f-f crossing or an f-l crossing. Furthermore, since there is no fan-crossing edge for any edge in $\mathsf{MST}(P)$ by Lemma 8, every crossing in $\mathsf{MLG}(P)$ is a crossing between an edge $e \in \overline{\mathsf{MST}}(P)$ and a fan-crossing edge for $e$, which implies that $\sigma(P)$ can be counted only by checking fan-crossing edges for edges in $\overline{\mathsf{MST}}(P)$.

Prior to details of our counting scheme, we show the following two lemmas.

**Lemma 10.** *Consider an edge $ab \in \overline{\mathsf{MST}}(P)$. Among the crossings between $ab$ and fan-crossing edges for $ab$ in $\mathsf{MLG}(P)$, there exist at most two f-l crossings.*

*Proof.* First of all, if there exists no point of $P$ in $\texttt{Lens}(ab)$, the statement clearly holds. According to Lemma 2, we consider other two cases: [Case 1] There exists one point of $P$ in $\texttt{Lens}(ab)$. [Case 2] There exist two points of $P$ in $\texttt{Lens}(ab)$.

**Case 1:** Without loss of generality, $c \in P$ lies in $\texttt{L\_Lens}(ab)$. Let $\ell_a$ be a half line emanating from $c$ to $a$ and $\ell'_a$ be a half line emanating from $c$ such that $\texttt{angle}_c(\ell_a, \ell'_a) = 60°$ and $\ell'_a$ cuts $\texttt{R\_Lens}(ab)$. Similarly, let $\ell_b$ be a half line emanating from $c$ to $b$ and $\ell'_b$ be a half line emanating from $c$ such that $\texttt{angle}_c(\ell_b, \ell'_b) = 60°$ and $\ell'_b$ cuts $\texttt{R\_Lens}(ab)$.

Let $d$ be a point of $P$ such that $cd$ is a fan-crossing edge for $ab$ and $cd$ lies in $\texttt{Cone}_c(\ell_a, \ell'_a)$ (see Fig. 5(a)). Let $d'$ be the crossing between segment $cd$ and the boundary of $\texttt{Lens}(ab)$. Since $\texttt{bisect}(ad')$ passes through $b$ since $ad'$ is a chord of $C_b(ab)$, it is easy to see that $a$ and $c$ lie in the same side of $\texttt{bisect}(ad')$, which means $ca < cd'$. By $cd' \leq cd$, we obtain $ca < cd$. On the other hand, since $\angle dca \leq 60° < \angle dac$, we have $ad < cd$. Hence, it holds $a \in \texttt{Lens}(cd)$, i.e., a crossing between $ab$ and $cd$ is an f-f crossing. In a symmetric manner, we obtain

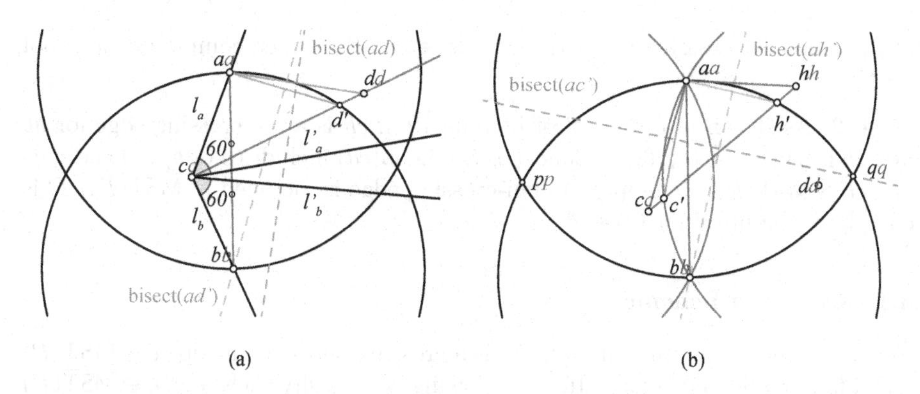


(a) (b)

**Fig. 5.** (a) Illustration of Case 1 in the proof of Lemma 10. (b) Illustration of Case 2 in the proof of Lemma 10.

the same conclusion even if $cd$ lies in $\texttt{Cone}_c(\ell_b, \ell'_b)$. Therefore, only if $cd$ lies in $\texttt{Cone}_c(\ell'_a, \ell'_b)$, a crossing between $ab$ and $cd$ can be an f-l crossing. By the fact of $\texttt{angle}_c(\ell'_a, \ell'_b) < 60°$ and Lemma 7, $\mathsf{MLG}(P)$ includes at most two fan-crossing edges lying in $\texttt{Cone}_c(\ell'_a, \ell'_b)$, which completes the proof for Case 1.

**Case 2:** According to Lemma 2, without loss of generality, $c, d \in P$ lie in $\texttt{L\_Lens}(ab)$ and $\texttt{R\_Lens}(ab)$, respectively. Let $p$ (resp. $q$) be the intersection point of two circles $C_a(ab)$ and $C_b(ab)$ in $\texttt{L\_Lens}(ab)$ (resp. $\texttt{R\_Lens}(ab)$) (see Fig. 5(b)). We can see $c \notin D_q(ab)$ and $d \notin D_p(ab)$ since otherwise $c \in D_q(ab)$ or $d \in D_p(ab)$ holds, and then $cd < ab$ holds, which contradicts Lemma 2.

We consider only fan-crossing edges for $ab$ emanating from $c$ since ones emanating from $d$ are symmetric. Let $\ell_a$, $\ell_b$, $\ell_q$ be a half line emanating from $c$ to $a$, $b$, $q$ respectively. Let $h$ be a point of $P$ such that $ch$ is a fan-crossing edge for $ab$ and $ch$ lies in $\texttt{Cone}_c(\ell_a, \ell_q)$. Let $h'$ be the crossing between segment $ch$ and the boundary of $\texttt{Lens}(ab)$, and $c'$ be the crossing (inside $\texttt{Lens}(ab)$) between segment $ch$ and $C_q(ab)$. Since $\texttt{bisect}(ah')$ passes through $b$, it is easy to see that $a$ and $c$ lie in the same side of $\texttt{bisect}(ah')$, which means $ca < ch'$. By $ch' \leq ch$, we obtain $ca < ch$. Similarly, by considering $\texttt{bisect}(ac')$, we obtain $ah < ch$. Hence, it holds $a \in \texttt{Lens}(ch)$, i.e., a crossing between $ab$ and $ch$ is an f-f crossing. In a symmetric manner, we obtain the same conclusion even if $ch$ lies in $\texttt{Cone}_c(\ell_b, \ell_q)$, which implies that there exists no f-l crossing between $ab$ and fan-crossing edges for $ab$. This completes the proof for Case 2. □

**Lemma 11.** *For an edge $ab \in \overline{\mathsf{MST}}(P)$, at most one fan-crossing edge is included in $\mathsf{MST}(P)$.*

*Proof.* Consider the same cases as in the proof of Lemma 10.

**Case 1:** As shown in the proof of Lemma 10, if $cd$ is a fan-crossing edge for $ab$ and $cd$ lies in $\texttt{Cone}_c(\ell_a, \ell'_a)$ or $\texttt{Cone}_c(\ell_b, \ell'_b)$, $\texttt{Lens}(cd)$ includes $a$ or $b$, respectively, i.e., $cd \notin \mathsf{MST}(P)$. Therefore, only if $cd$ lies in $\texttt{Cone}_c(\ell'_a, \ell'_b)$, $cd$ can be included in $\mathsf{MST}(P)$. By the fact of $\texttt{angle}_c(\ell'_a, \ell'_b) < 60°$ and Lemma 6, $\mathsf{MST}(P)$ includes at most one fan-crossing edges lying in $\texttt{Cone}_c(\ell'_a, \ell'_b)$, which completes the proof for Case 1.

**Case 2:** As shown in the proof of Lemma 10, if $ch$ is a fan-crossing edge for $ab$ and $ch$ lies in $\texttt{Cone}_c(\ell_a, \ell_q)$ or $\texttt{Cone}_c(\ell_b, \ell_q)$, $\texttt{Lens}(cd)$ includes $a$ or $b$, respectively, i.e., $ch \notin \mathsf{MST}(P)$. Therefore, no fan-crossing edge is included in $\mathsf{MST}(P)$. This completes the proof for Case 2. □

### 3.1 Counting Scheme

Recall that every crossing in $\mathsf{MLG}(P)$ is a crossing between an edge $e \in \overline{\mathsf{MST}}(P)$ and a fan-crossing edge for $e$. In the following, we classify each edge $e \in \overline{\mathsf{MST}}(P)$ into two types: [Type 1] There exists at least one f-l crossing among the crossings between $e$ and fan-crossing edges for $e$ in $\mathsf{MLG}(P)$. [Type 2] There exists no f-l crossing among the crossings between $e$ and fan-crossing edges for $e$ in $\mathsf{MLG}(P)$. Let $m_i$ be the number of edges of Type $i$ in $\overline{\mathsf{MST}}(P)$. Clearly, it holds

$$m_1 + m_2 = |\overline{\mathsf{MST}}(P)| = n - 2. \tag{1}$$

We then consider numbering edges in Type $i$ from 1 to $m_i$ in any order, and use $e_{ij}$ to denote the $j$-th edge in Type $i$. Recall that every crossing in $\mathsf{MLG}(P)$ is an f-f crossing or an f-l crossing. For every edge $e_{ij} \in \overline{\mathsf{MST}}(P)$, let $\sigma_{ij}^{\mathrm{f-f}}$ (resp. $\sigma_{ij}^{\mathrm{f-l}}$) be the number of f-f (resp. f-l) crossings between $e_{ij}$ and fan-crossing edges for $e_{ij}$ in $\mathsf{MLG}(P)$. By Lemma 10, it holds

$$\sigma_{1j}^{\mathrm{f-l}} \leq 2 \quad \text{for } j = 1, \ldots, m_1. \tag{2}$$

Also, we have by the definition

$$\sigma_{2j}^{\mathrm{f-l}} = 0 \quad \text{for } j = 1, \ldots, m_2. \tag{3}$$

Let $\sigma^{\mathrm{f-f}}$ (resp. $\sigma^{\mathrm{f-l}}$) denote the number of f-f (resp. f-l) crossings in $\mathsf{MLG}(P)$. Clearly, it holds

$$\sigma(P) = \sigma^{\mathrm{f-f}} + \sigma^{\mathrm{f-l}}. \tag{4}$$

First, we consider counting $\sigma^{\mathrm{f-f}}$ and $\sigma^{\mathrm{f-l}}$ by checking fan-crossing edges for every $e_{ij} \in \overline{\mathsf{MST}}(P)$. While counting, each f-f crossing is counted exactly twice. We thus have

$$\sigma^{\mathrm{f-f}} = \frac{1}{2}\left(\sum_{i=1}^{2}\sum_{j=1}^{m_i}\sigma_{ij}^{\mathrm{f-f}}\right). \tag{5}$$

On the other hand, each f-l crossing is counted exactly once. We thus have

$$\sigma^{\mathrm{f-l}} = \sum_{i=1}^{2}\sum_{j=1}^{m_i}\sigma_{ij}^{\mathrm{f-l}} = \sum_{j=1}^{m_1}\sigma_{1j}^{\mathrm{f-l}}. \tag{6}$$

Note that the second equality in Eq. (6) holds by applying Eq. (3). Summarizing Eq. (4), Eq. (5) and Eq. (6), we obtain

$$\sigma(P) = \frac{1}{2}\left(\sum_{i=1}^{2}\sum_{j=1}^{m_i}\sigma_{ij}^{\mathrm{f-f}}\right) + \sum_{j=1}^{m_1}\sigma_{1j}^{\mathrm{f-l}}$$
$$= \frac{1}{2}\left(\sum_{i=1}^{2}\sum_{j=1}^{m_i}\sigma_{ij}^{\mathrm{f-f}} + \sum_{j=1}^{m_1}\sigma_{1j}^{\mathrm{f-l}}\right) + \frac{1}{2}\sum_{j=1}^{m_1}\sigma_{1j}^{\mathrm{f-l}}. \qquad (7)$$

Next, consider counting the number of fan-crossing edges in $\overline{\mathsf{MST}}(P)$, say $\alpha$, for every $e_{ij} \in \overline{\mathsf{MST}}(P)$. Recall that the crossing between $e_{ij}$ and an edge in $\mathsf{MST}(P)$ is always an f-l crossing. Hence by Lemma 11, for an edge $e_{1j}$, the number of fan-crossing edges in $\overline{\mathsf{MST}}(P)$ is at least $\sigma_{1j}^{\mathrm{f-f}} + \sigma_{1j}^{\mathrm{f-l}} - 1$, and for an edge $e_{2j}$, it is exactly $\sigma_{2j}^{\mathrm{f-f}} + \sigma_{2j}^{\mathrm{f-l}} = \sigma_{2j}^{\mathrm{f-f}}$ (by Eq. (3)), i,e,. it holds

$$\alpha \geq \sum_{j=1}^{m_1}(\sigma_{1j}^{\mathrm{f-f}} + \sigma_{1j}^{\mathrm{f-l}} - 1) + \sum_{j=1}^{m_2}\sigma_{2j}^{\mathrm{f-f}} = \sum_{i=1}^{2}\sum_{j=1}^{m_i}\sigma_{ij}^{\mathrm{f-f}} + \sum_{j=1}^{m_1}\sigma_{1j}^{\mathrm{f-l}} - m_1. \qquad (8)$$

On the other hand, since each edge in $\overline{\mathsf{MST}}(P)$ is counted at most twice, and $|\overline{\mathsf{MST}}(P)| = n-2$, we have

$$\alpha \leq 2(n-2). \qquad (9)$$

Summarizing Eq. (8) and Eq. (9), we obtain

$$\sum_{i=1}^{2}\sum_{j=1}^{m_i}\sigma_{ij}^{\mathrm{f-f}} + \sum_{j=1}^{m_1}\sigma_{1j}^{\mathrm{f-l}} \leq 2(n-2) + m_1. \qquad (10)$$

Hence, we have an improved upper bound for $\sigma(P)$ as follows:

$$\sigma(P) \leq \frac{2(n-2)+m_1}{2} + \frac{1}{2}\sum_{j=1}^{m_1}\sigma_{1j}^{\mathrm{f-l}} \quad \text{(by substituting Eq. (10) into Eq. (7))}$$
$$\leq \frac{2(n-2)+m_1}{2} + \frac{1}{2}\cdot 2m_1 \quad \text{(by Eq. (2))}$$
$$\leq \frac{5}{2}(n-2) \quad \text{(since } m_1 \leq n-2 \text{ by Eq. (1))}.$$

**Theorem 3.** *For a set of any semi-generic points $P$, it holds $\sigma(P) \leq 2.5|P| - 5$.*

## 4 Improved Lower Bound for $\sigma(P)$

In this section, we show how to construct an instance $P$ for which there exist more crossings in $\mathsf{MLG}(P)$ than one shown by Bereg et al. [3].

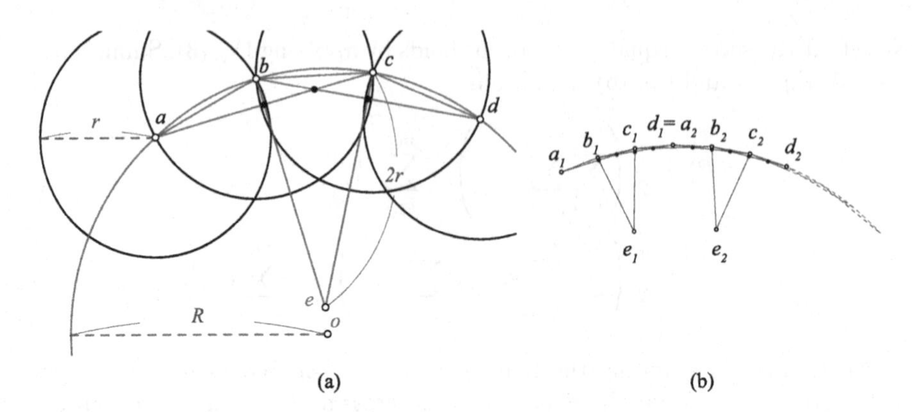


**Fig. 6.** (a) A unit w.r.t. $(o, R, r)$. (b) Illustration of how to connect two units.

Consider a set of five points $\{a, b, c, d, e\}$ as shown in Fig. 6(a). For a point $o$, and two real numbers $R$ and $r$ with $R > r > 0$, points $a, b, c, d$ are arranged on $C_o(R)$ in this order so that $ab = bc = cd = r$, and point $e$ is located in $D_o(R)$ such that $be = ce = 2r$. We call such a set of five points *unit* w.r.t. $(o, R, r)$ in the following.

For an integer $t > 0$, let us consider $t$ numbered units w.r.t. $(o, R, r)$. We identify points $a, b, c, d, e$ of $i$-th unit as $a_i, b_i, c_i, d_i, e_i$, respectively. We now put $t$ units so that $d_i = a_{i+1}$ (regarded as one point) for $i = 1, \ldots, t-1$ as shown in Fig. 6(b). Let $P(o, R, r, t)$ denote a set of points constructed in the above manner. It is then easy to see

$$|P(o, R, r, t)| = 4t + 1. \tag{11}$$

Let us consider $\mathsf{MLG}(P(o, R, r, t))$. In the following, we take values $R, r, t$ so that $R/t > 1 \gg r$. It is then easy to see that

$$\begin{aligned} a_ib_i = b_ic_i = c_id_i (= r) < a_ic_i = b_id_i = c_ib_{i+1} (\simeq 2r) \quad &< b_ie_i = c_ie_i (= 2r) \\ < a_ie_i = d_ie_i (\simeq \sqrt{6}r) \quad &< e_ie_{i+1} (\simeq 3r), \end{aligned}$$

thus $\mathsf{MLG}(P(o, R, r, t))$ consists of edges $a_ib_i$, $b_ic_i$, $c_id_i$, $a_ic_i$, $b_id_i$, $b_ie_i$, $c_ie_i$ for $i = 1, \ldots, t$, and edges $c_ib_{i+1}$ for $i = 1, \ldots, t-1$. Since there are three crossings in the $i$-th unit, and each edge $c_ib_{i+1}$ has two crossings, it holds

$$\sigma(P(o, R, r, t)) = 3t + 2(t-1) = 5t - 2. \tag{12}$$

By Eq. (11) and Eq.(12), we have

$$\frac{\sigma(P(o, R, r, t))}{|P(o, R, r, t)|} = \frac{5t-2}{4t+1} = \frac{5}{4} - \frac{13}{16t+4},$$

which can be larger than $5/4 - \varepsilon$ for any $\varepsilon > 0$ by taking $t$ as a sufficiently large integer. Notice that $P(o, R, r, t)$ is not semi-generic, however by moving each point in $P(o, R, r, t)$ infinitesimally, we can obtain a set of semi-generic points $P$ such that the topology of $\mathsf{MLG}(P)$ is the same as one of $\mathsf{MLG}(P(o, R, r, t))$ and $\sigma(P) = \sigma(P(o, R, r, t))$.

**Theorem 4.** *For any* $\varepsilon > 0$*, there exists a set of semi-generic points* $P$ *such that* $\sigma(P) \geq (1.25 - \varepsilon)|P|$.

## 5 Future Works

Several problems related to the crossing number of $\mathsf{MLG}(P)$ remain open.

One problem is to further improve upper or lower bounds for $\sigma(P)$. Although in this paper, we have improved upper and lower bounds for $\sigma(P)$ as shown in Theorems 3 and 4, respectively, there is still a gap.

Another interesting problem is to analyze the *thickness* of $\mathsf{MLG}(P)$. The thickness of a geometric graph $G(P)$[3] is the smallest number of layers necessary to partition the edges of $G(P)$ into layers in such a way that no two edges of the same layer cross. It is easy to see that the thickness of $\mathsf{MLG}(P)$ is at most 4 since it holds $\mathsf{MLG}(P) \subseteq 1 - \mathsf{GG}(P)$ by Lemma 1, and it is shown by Bose et al. [4] that the thickness of $1 - \mathsf{GG}(P)$ is at most 4. Therefore, a problem of whether the thickness of $\mathsf{MLG}(P)$ is at most 3 naturally arises. In order to prove this, one direction worth considering is as follows: Define a graph $H = (W, F)$ for $\mathsf{MLG}(P)$ such that each vertex $e \in W$ corresponds to each edge $e \in \mathsf{MLG}(P)$, and for two vertices $e, e' \in W$, edge $(e, e')$ is included in $F$ if and only if edges $e$ and $e'$ cross each other in $\mathsf{MLG}(P)$. We then notice that the thickness of $\mathsf{MLG}(P)$ is equal to the chromatic number of $H$. It is proved by Grötzsch [6] that a planar triangle-free graph is 3-colorable. On the other hand, Bereg et al. [3] show the quasi-planarity of $\mathsf{MLG}(P)$, i.e., no three edges in $\mathsf{MLG}(P)$ pairwise cross, which means that $H$ is triangle-free. Hence, once we prove the planarity of $H$, the claim immediately holds.

## References

1. Ábrego, B.M., Fabila-Monroy, R., Fernández-Merchant, S., Flores-Peñaloza, D., Hurtado, F., Sacristán, V., Saumell, M.: On crossing numbers of geometric proximity graphs. Comput. Geometry **44**(4), 216–233 (2011)
2. Avis, D., Katoh, N., Ohsaki, M., Streinu, I., Tanigawa, S.: Enumerating constrained non-crossing minimally rigid frameworks. Disc. Comput. Geometry **40**(1), 31–46 (2007). https://doi.org/10.1007/s00454-007-9026-x
3. Bereg, S., Hong, S.-H., Katoh, N., Poon, S.-H., Tanigawa, S.: On the edge crossing properties of Euclidean minimum weight Laman graphs. Comput. Geometry Theory Appl. **51**, 15–24 (2016)
4. Bose, P., et al.: Some properties of $k$-Delaunay and $k$-Gabriel graphs. Comput. Geometry **46**(2), 131–139 (2013)
5. Graver, J., Servatius, B., Servatius, H.: Combinatorial rigidity. graduate studies in mathematics, vol. 2. American Mathematical Society (1993)

[3] Another terminology *constrained geometric thickness of a graph* is used in [4].

6. Grötzsch, H.: Zur Theorie der diskreten Gebilde. VII. Ein Dreifarbensatz für dreikreisfreie Netze auf der Kugel. Wiss. Z. Martin-Luther-U., Halle-Wittenberg, Math.-Nat. Reihe, vol. 8, pp. 109–120 (1959)
7. Lee, A., Streinu, I.: Pebble game algorithms and sparse graphs. Disc. Math. **308**, 1425–1437 (2008)
8. Laman, G.: On graphs and rigidity of plane skeletal structures. J. Eng. Math. **4**, 331–340 (1970)
9. Servatius, B.: The geometry of frameworks: Rigidity, mechanisms and cad. In: Gorini, C.A. (Ed.), Geometry at Work: A Collection of Papers Showing Applications of Geometry. Cambridge University Press (2000)
10. Thorpe, M.F., Duxbury, P.M. (eds.): Rigidity Theory and Applications. Kluwer Academic/Plenum Publishers, New York (1999)
11. Yao, A.C.: On constructing minimum spanning trees in $k$-dimensional space and related Problems. SIAM J. Comput. **11**(4), 721–736 (1982)

# Minimum Color Spanning Circle in Imprecise Setup

Ankush Acharyya[1(✉)], Ramesh K. Jallu[1], Vahideh Keikha[1], Maarten Löffler[2], and Maria Saumell[1,3]

[1] The Czech Academy of Sciences, Institute of Computer Science, Prague, Czech Republic
{acharyya,jallu,keikha,saumell}@cs.cas.cz

[2] Department of Information and Computing Sciences, Utrecht University, Utrecht, The Netherlands
M.Loffler@uu.nl

[3] Department of Theoretical Computer Science, Faculty of Information Technology, Czech Technical University in Prague, Prague, Czech Republic

**Abstract.** Let $\mathcal{R}$ be a set of $n$ colored imprecise points, where each point is colored by one of $k$ colors. Each imprecise point is specified by a unit disk in which the point lies. We study the problem of computing the smallest and the largest possible minimum color spanning circle, among all possible choices of points inside their corresponding disks. We present an $O(nk \log n)$ time algorithm to compute a smallest minimum color spanning circle. Regarding the largest minimum color spanning circle, we show that the problem is NP-Hard and present a $\frac{1}{3}$-factor approximation algorithm. We improve the approximation factor to $\frac{1}{2}$ for the case where no two disks of distinct color intersect.

**Keywords:** Color spanning circle · Imprecise points · Algorithms · Computational complexity

## 1 Introduction

Recognition of color spanning objects of optimum size, in the classical (precise) setting, is a well-studied problem in the literature [2,3,7,14]. The motivation of color spanning problems stems from facility location problems. Here facilities of type $i \in \{1, 2, \ldots, k\}$ are modeled as points with color code $i$, and the objective is to identify the location of a desired geometric shape containing at least one facility of each type such that the desired measure parameter (width, perimeter, area, etc.) is optimized. Other applications of color spanning objects can be found in disk-storage management systems [5] and central-transportation systems [23].

The simplest type of two-dimensional problem considered in this setup is the *minimum color spanning circle* (MCSC) problem, defined as follows. Given a colored point set $P$ in the plane, such that each point in $P$ is colored with one of $k$ possible colors, compute a circle of minimum radius that contains at least one point of each color (see Fig. 1a). As observed in [1], the minimum color spanning


C.-Y. Chen et al. (Eds.): COCOON 2021, LNCS 13025, pp. 257–268, 2021.
https://doi.org/10.1007/978-3-030-89543-3_22

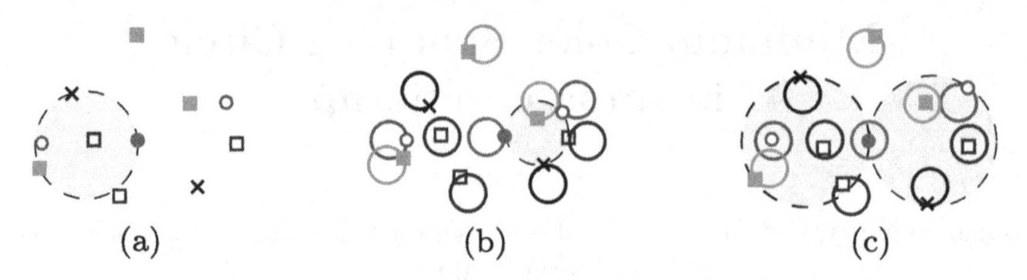


**Fig. 1.** (a) MCSC for a precise colored point set. (b, c) S-MCSC and L-MCSC for an imprecise colored point set. The representative for each disk is marked as a point of the corresponding color (for interpretation of the references to color in figure legends, the reader is referred to the web version).

circle can be computed in $O(nk \log n)$ time using results on the upper envelope of Voronoi surfaces obtained in [14].

In this work, the exact coordinates of the input points in $P$ are unknown. Instead, we are given a set $\mathcal{R} = \{R_1, R_2, \ldots, R_n\}$ of $n$ unit disks of diameter 1 in the plane, where each disk is colored with one of $k$ possible colors. A colored point set $P$ is a *realization* of $\mathcal{R}$ if there exists a color-preserving bijection between $P$ and $\mathcal{R}$ such that each point in $P$ is contained in the corresponding disk in $\mathcal{R}$. Each realization of $\mathcal{R}$ gives a MCSC of a certain radius. We are interested in finding realizations of $\mathcal{R}$ such that the corresponding MCSC has the smallest (S-MCSC) and largest (L-MCSC) possible radius (see Fig. 1b and c).

*Imprecise Points.* Uncertainty in data is paramount in contemporary geometric computations. The motivation of the studies on location-based data uncertainty stems from many real-life situations where the locations of the points are subject to errors and their exact coordinates are unknown. Such a set of points is known as an *imprecise* or *uncertain* point set and the set of all possible locations of a point is called its *region* [22,26]. In the literature, different variations have been considered where the regions are modelled as simple geometric objects such as line segments, disks or squares [22,26]. Computing the smallest circle intersecting a set of disks or convex regions of total complexity $n$ is called the *intersection radius problem*, and can be solved in $O(n)$ time [15]. Robert and Toussaint [25] studied the problem of computing the smallest width corridor intersecting a set of convex regions (disks and line segments) and proposed two $O(n \log n)$ time algorithms, where $n$ is the number of convex regions. Löffler and van Kreveld [22] considered the problem of computing the smallest and largest possible axis-parallel bounding box and circle of a set of regions modelled as circles or squares. Their proposed algorithms have running times ranging from $O(n)$ to $O(n \log n)$.

*Color Spanning Objects.* In the precise setting, an obvious variation of the smallest color spanning circle is the smallest color spanning square problem, which can also be computed using the upper envelope of Voronoi surfaces in $O(nk \log n)$ time [1]. Das et al. [7] showed that the smallest color spanning strip and axis-parallel rectangle can be computed in $O(n^2 \log n)$ and $O(n(n-k) \log k)$ time, respectively. They also proposed an algorithm to compute the arbitrarily oriented smallest color spanning rectangle, which runs in $O(n^3 \log k)$ time. The color spanning 2-interval [16], equilateral triangle of a fixed orientation [13], and axis-parallel square [18] can be computed in $O(n^2)$, $O(n \log n)$ and $O(n \log^2 n)$ time, respectively. Acharyya et al. [2] proposed several efficient algorithms to compute the narrowest color spanning annulus for circles, axis-parallel squares, rectangles, and equilateral triangles of a fixed orientation. The minimum diameter color spanning set (MDCS) problem has also been studied [11,17,27]. Its general version is known to be NP-Hard in $L_p$ metric, for $1 < p < \infty$, while in $L_1$ and $L_\infty$ metrics the problem can be solved in polynomial time [11].

Colored variations of other geometric problems have also been studied in the context of imprecise points [6,8,24]. Given a set of colored clusters, the problem of computing the minimum-weight color spanning tree (*generalized MST problem*) is APX-Hard [8]. Even when each cluster contains exactly 2 points the problem remains NP-Hard [12]. The problem admits a $2\delta$-approximation, where $\delta$ is the maximum size of the cluster for any imprecise vertex of the MST [24]. In the generalized TSP problem (GTSP), the imprecision is defined by neighborhoods (which are either continuous or discrete) and the goal is to find the shortest tour that visits all neighborhoods. It is known that GTSP with neighborhoods defined by subsets of cardinality two is inapproximable [8].

*Dispersion Problems.* The L-MCSC problem is closely related to the dispersion problem in unit disks, where for a given set of $n$ unit disks the goal is to select $n$ points, one from each disk, such that the minimum pairwise distance among the selected points is maximized. This problem was introduced by Fiala et al. [10] who proved that the problem is NP-Hard, unless $P = NP$. It is also known that the problem is APX-hard [9]. Constant factor approximation algorithms for this problem are given in [4,9].

### 1.1 Our Contribution

In this paper, we present the following results:

- The S-MCSC problem can be solved in $O(nk \log n)$ time.
- The L-MCSC problem is NP-Hard,
- $\frac{1}{2}$-factor and $\frac{1}{3}$-factor approximations to the L-MCSC problem can be computed in $O(nk \log n)$ time when no two distinct color disks intersect and after relaxing the disjointness criterion, respectively.

To the best of our knowledge there is no prior result on the minimum color spanning circle problem for imprecise point sets. Due to lack of space, some proofs are omitted; they will be provided in the full version of the paper.

## 2 The Smallest MCSC (S-MCSC) Problem

Given a set $\mathcal{R}$ of $n$ imprecise points modeled as unit disks, we present an algorithm that finds a S-MCSC, denoted by $C_{opt}$, and the realization of $\mathcal{R}$ achieving it. Let $r_{opt}$ be the radius of $C_{opt}$.

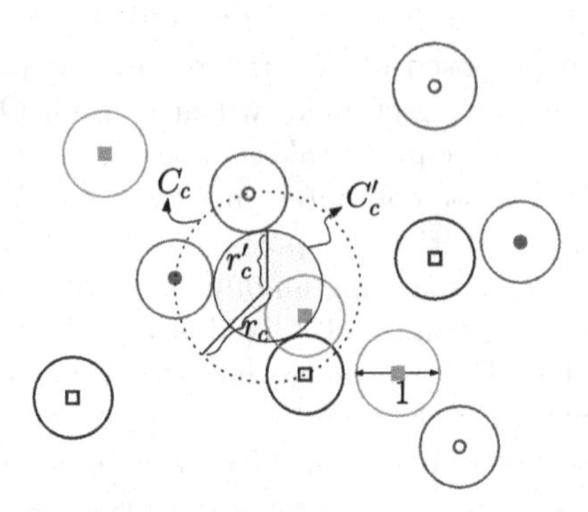


**Fig. 2.** Illustration of Lemma 1. The dotted circle $C_c$ is the MCSC of the disk centers and the circle $C'_c$ is obtained by decreasing $C_c$'s radius by $\frac{1}{2}$.

Let $\mathcal{C} = \{c_1, \ldots, c_n\}$ be the set of center points of the disks in $\mathcal{R}$. Let $C_c$ be a MCSC of the colored set $\mathcal{C}$, and let $r_c$ be its radius. The following relation holds:

**Lemma 1.** *If* $r_c > \frac{1}{2}$, *then* $r_c = r_{opt} + \frac{1}{2}$.

*Proof.* Consider a circle $C'_c$ concentric with $C_c$ with radius $r'_c = r_c - \frac{1}{2}$ (see Fig. 2). For every disk $R_i$ such that $c_i$ is contained in $C_c$, we have that $C'_c$ contains $c_i$ or the intersection between the boundary of $R_i$ and the segment connecting $c_i$ with the center of $C_c$. Thus, $C'_c$ contains at least one point of each color and $r_{opt} \leq r'_c = r_c - \frac{1}{2}$.

If $r_{opt} < r'_c$, we would get a feasible solution for the MCSC problem of $\mathcal{C}$ by increasing the radius of $C_{opt}$ by $\frac{1}{2}$. Since such a solution would have radius $r_{opt} + \frac{1}{2} < r'_c + \frac{1}{2} = r_c$, we would get a contradiction with the fact that $r_c$ is the radius of any MCSC of $\mathcal{C}$. □

Next, we compute $C_{opt}$ using Algorithm 1:

**Algorithm 1** Algorithm for the S-MCSC problem

---

**Input:** *A set* $\mathcal{R}$ *of* $n$ *unit disks*
**Output:** *A* S-MCSC *of* $\mathcal{R}$ *with radius* $r_{opt}$
1: compute $C_c$;
2: **if** $r_c > 1/2$ **then**
3: $C_{opt}$ is a circle concentric with $C_c$, $r_{opt} \leftarrow r_c - \frac{1}{2}$;
4: **else**
5: $C_{opt}$ is a circle concentric with $C_c$, $r_{opt} \leftarrow 0$; ▷ $C_{opt}$ is a point
6: **return** $C_{opt}$

**Theorem 1.** *A smallest minimum color spanning circle of* $\mathcal{R}$ *can be computed in* $O(nk \log n)$ *time.*

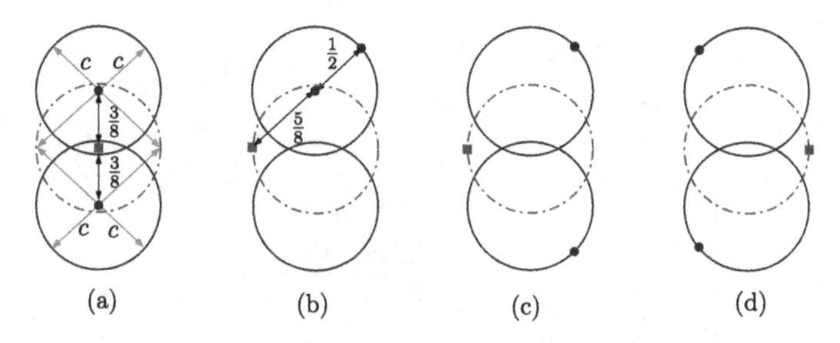


**Fig. 3.** (a) A stack of disks; (b) the distance from the left endpoint of the red disk to its farthest point in the top blue disk is $\frac{9}{8}$; (c, d) the two placements of points with red-blue distances equal to $c$.

## 3 The Largest MCSC (L-MCSC) Problem

In this section, we consider the L-MCSC problem, where for the given set $\mathcal{R}$ the goal is to find a realization such that any MCSC is as large as possible. We show that the problem is NP-Hard, already for $k = 2$, using a reduction from planar 3-SAT [21]. Our reduction is inspired by those described in [10,19].

Given a planar 3-SAT instance, we construct a set of colored unit disks with the following property: There exists a realization such that any MCSC has diameter $c$ if and only if the 3-SAT instance is satisfiable, where $c = \frac{9}{8}$. We use disks of colors red and blue. Thus, a point set having any MCSC of diameter at least $c$ is equivalent to saying that there is no red-blue pair of points at distance less than $c$. We denote the family of realizations with this property by $\mathcal{P}^c$.

A *stack of disks* is a set of three vertically aligned disks of alternating colors. As shown in Fig. 3a, for a blue-red-blue stack of disks, the distance between the centers of the blue disks and the center of the red disk is $\frac{3}{8}$.

**Lemma 2.** *There exist two realizations in $\mathcal{P}^c$ of a stack of disks.*

*Proof.* The red point can only be placed at the left or right extreme position of the red disk (see Fig. 3c and d), and such a placement forces the placement of points in the blue disks at distance $c$ (see Fig. 3b). Thus, there exist only two realizations in $\mathcal{P}^c$, shown in Fig. 3c and d. □

*Variable Gadget.* Our variable gadget (see Fig. 4) is an alternating chain of red and blue disks, whose centers lie on a hexagon, together with some stacks of disks. The distance between the centers of two consecutive red and blue disks along the same edge of the hexagon is $c$. Each edge of the hexagon contains two stacks of disks placed near the endpoints, and every pair of consecutive edges is joined by a blue disk. In the following description, we say that $p_i$ and $p_i'$ are the *leftmost* and *rightmost* points of disk $R_i$ if they are its leftmost and rightmost points after the hexagon has rotated so that the edge containing the center of $R_i$ is horizontal and the center of the hexagon is below the edge.

At the top-left corner of the variable gadgets, the disks are placed as follows (the other corners are constructed similarly). Let $R_i$ be the last disk in clockwise

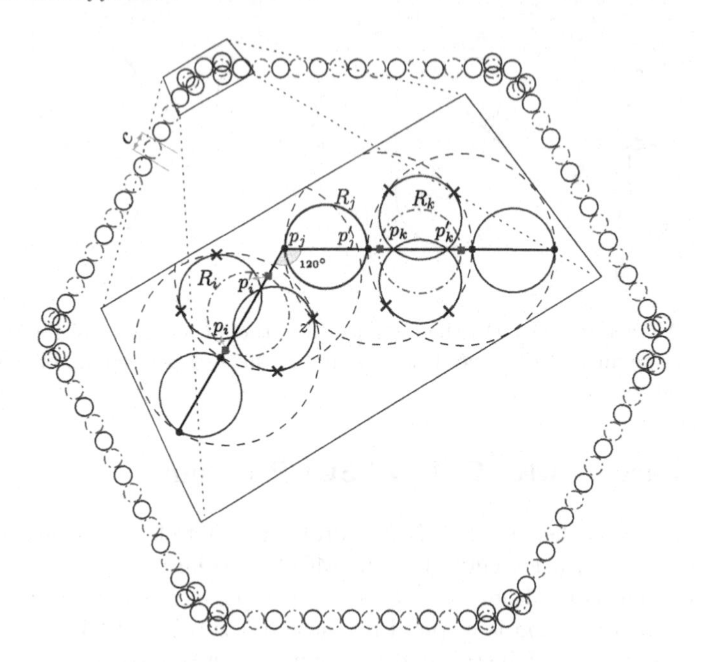


**Fig. 4.** A variable gadget with zoomed in view for the top-left corner. The dashed circles are centered at $p'_k, p_k, p'_i, p_i$ and have radius $c$.

order along the top-left edge of the hexagon, and let $R_j$ and $R_k$ be the first and second disks along the top edge (see Fig. 4). The point $p_j$ lies at the top left corner of the hexagon, and the centers of $R_j$ and $R_k$ are at distance $c$. Regarding $R_i$, it is placed in such a way that the lower blue disk of its stack contains a point $z$ which is at distance $c$ from both $p_k$ and $p_i$ (see Fig. 4). Notice that, if a realization in $\mathcal{P}^c$ chooses $p'_i$, the choice for $R_j$ is not unique; however, none of the points in $R_j$ at distance at least $c$ from $p'_i$ is compatible with the choice of $p_k$ for $R_k$. Therefore, the choice of $p'_i$ forces the choice of $p'_k$, and clearly the choice of $p_k$ forces the choice of $p_i$.

For a realization in $\mathcal{P}^c$ of a variable gadget, the following holds: By Lemma 2, the stack containing $R_k$ is constrained to choose either $p_k$ or $p'_k$. Let us assume that it chooses $p'_k$. This choice propagates to the right through the chain of disks in the top edge. The red disk of the stack on the right of the edge also chooses its rightmost point, and this forces the red disk of the first stack of the top right edge to choose its rightmost point too. Therefore, the choice of $p'_k$ propagates through the whole hexagon. If $R_k$ chooses $p_k$, the same phenomenon occurs. We conclude:

**Lemma 3.** *For any realization in $\mathcal{P}^c$ of a variable gadget, either all disks centered at the edges of the hexagon, except for the ones intersecting corners of the hexagon, choose their rightmost point, or they all choose their leftmost point.*

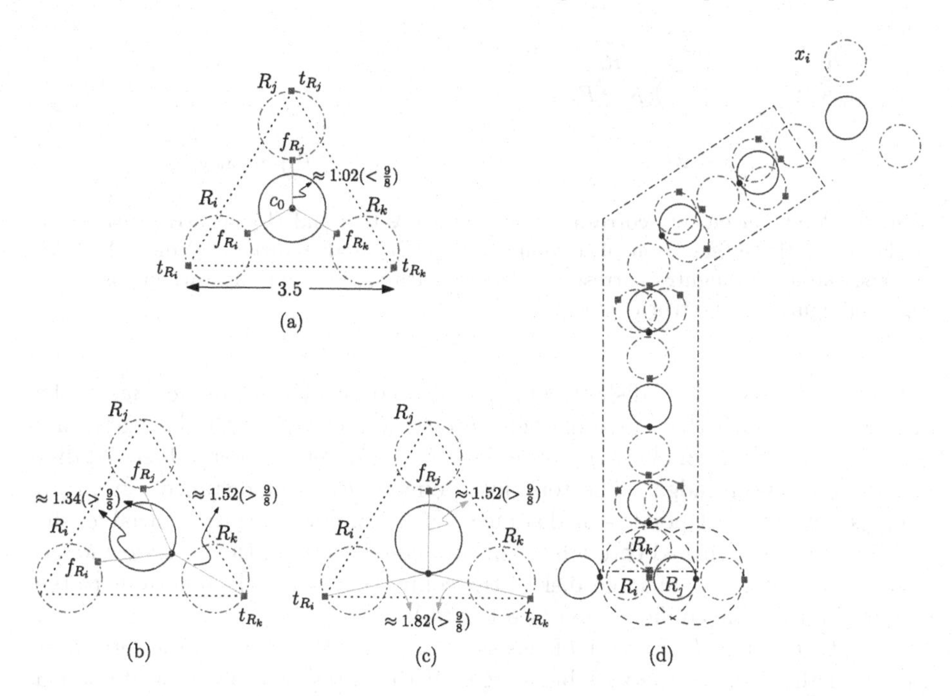


**Fig. 5.** (a-c) A clause gadget with distinct truth assignments. In the placement in (a), the red-blue pairs are at distance smaller than $c$, while in (b) and (c) they are at distance greater than $c$. (d) Connection gadget for a positive variable in a clause; the variable has truth value $T$.

*Clause Gadget.* Clause gadgets are illustrated in Fig. 5a-c. We consider an equilateral triangle of side length 3.5 and center $c_0$, and we place one red disk at every corner of the triangle in such a way that the center of the disk is aligned with $c_0$ and its nearest corner of the triangle. Then we place a blue disk centered at $c_0$. Each red disk of a clause gadget is associated to one of the literals occurring in the clause, and is connected to the corresponding variable gadget via a connection gadget. Intuitively, to decide if there exists any realization in $\mathcal{P}^c$, each red disk $R_\tau$ of the clause gadget has essentially two relevant placements, called $t_{R_\tau}$ and $f_{R_\tau}$ (see Fig. 5a). As we will see, when the associated literal is set to *true*, we can choose the placement $t_{R_\tau}$, and when it is set to *false*, we are forced to choose $f_{R_\tau}$. It is easy to see that any realization in $\mathcal{P}^c$ of the clause gadget does not choose $f_{R_\tau}$ for at least one of the disks $R_\tau$ (see Fig. 5b and c).

*Connection Gadget.* A variable gadget is connected to each of its corresponding clause gadgets with the help of a *connection* gadget. A connection gadget consists of an alternating chain of red and blue disks together with some stacks of disks (see Fig. 5d).

The precise connection of the variable gadget to the clause gadget by a connection gadget depends on whether the variable in the clause is positive or negative. For a positive variable, the connection is established through a pair of

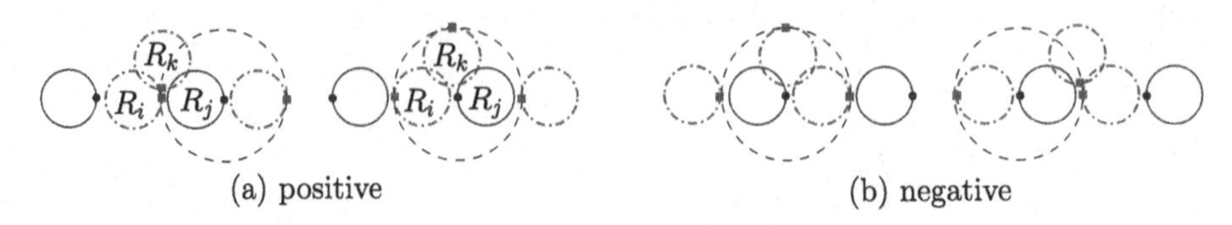


(a) positive (b) negative

**Fig. 6.** Point placements corresponding to (a) positive and (b) negative variable in a clause, at the intersection of a connection gadget and a variable gadget. In both figures, the left subfigures correspond to the truth value $T$ for the variable, and the right subfigures to the truth value $F$.

a red disk $R_i$ and a blue disk $R_j$ which appear consecutive along an edge of the hexagon, and such that none of them intersects a corner of the hexagon, and $R_i$ comes before $R_j$ in clockwise order (see Fig. 6a). Let $R_k$ be the first red disk of the connection gadget. The top-most point of $R_k$ is at distance $c$ from $p_j$, and its bottom-most point is at distance $c$ from $p'_j$. For a negative variable, the connection is established through a pair of blue-red disks in the variable gadget. The placement of the first red disk of the connection gadget is analogous to the one in the red-blue configuration (see Fig. 6b).

The truth value $T$ of a variable is associated with the choice of the rightmost points of the disks in the variable gadget. If the variable appears positive at a clause, this allows the choice of the bottom-most point of $R_k$, and this propagates through the connection gadget and eventually allows the choice of the associated point $t_{R_\tau}$ in the clause gadget (see Fig. 5d). If the truth value is $F$, $p_j$ is selected, which forces the choice of a point in a close vicinity of the top-most point of $R_k$, and eventually of $f_{R_\tau}$. The analysis of the other cases are similar.

The chain in a connection gadget might be bent by 120°, maintaining the planarity and the distance constraint $c$. The placement of disks at each bend of a connection gadget is similar to the placements at the corners of a variable gadget. Stacks of disks are used around the bends, next to the first disk $R_k$, and next to the red disk of the clause gadget associated to the literal.

Given a planar 3-SAT instance, we construct an instance of our problem as follows: Its dependency graph can be embedded so that all variables lie on a horizontal line, all clauses are on either side of them, and each edge connecting a clause to a variable is an orthogonal edge with at most one bend (see Fig. 7 for an example) [20]. Moreover, the construction can be easily modified so that edges only have bends of 120°. Then, we replace the vertices with variable and clause gadgets which are connected using the connection gadgets as described above. The number of disks in a variable gadget depends on the number of times the associated variable appears in the given 3-SAT formula. The whole construction has polynomial size. We have the following:

**Lemma 4.** *The planar* 3-SAT *formula has a satisfying assignment if and only if there exists a realization of the disks in* $\mathcal{P}^c$.

*Proof.* Let us consider the 3-SAT formula $\mathcal{F}$ with $n$ variables $x_1, x_2, \ldots, x_n$ and $m$ clauses $C_1, C_2, \ldots, C_m$.

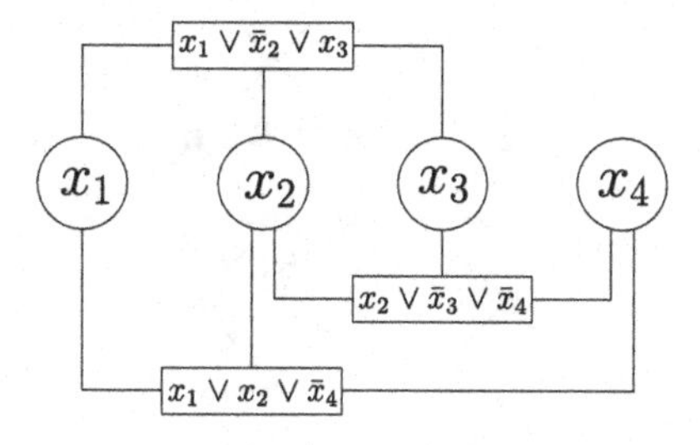


**Fig. 7.** Planar embedding of $\mathcal{F} = (x_1 \vee x_2 \vee \bar{x}_4) \wedge (x_1 \vee \bar{x}_2 \vee x_3) \wedge (x_2 \vee \bar{x}_3 \vee \bar{x}_4)$.

(Necessity) Suppose that $\mathcal{F}$ is satisfiable. Thus, every clause contains a literal whose truth value is $T$. We describe a realization of the disks in $\mathcal{P}^c$.

Let $C_j$ be a clause and let $\ell_i$ be the literal making the clause true. Then we pick $t_{R_i}$ from the associated disk $R_i$ in the clause. This choice propagates through the connection gadget and eventually forces one of the realizations for the variable gadget described in Lemma 3: the one choosing the rightmost points of the disks, if $\ell_i$'s truth value is $T$ and it appears in the clause positively; or the one choosing the leftmost points of the disks, if $\ell_i$'s truth value is $F$ and it appears in the clause negatively. Notice that a variable might appear in several clauses, but the obtained realizations of the variable gadget are consistent with each other.

For the cases where a variable's truth value is $T$ and it appears in a clause negatively, or a variable's truth value is $F$ and it appears in a clause positively, the placement described above forces the realizations of the corresponding connection gadget and associated disk in the clause gadget. Regarding the central blue disks in the clause gadgets, since at least one of the three surrounding red disks has its representative at the position $t_{R_r}$, it is possible to find a realization of the blue disk such that the realization of the clause gadget is in $\mathcal{P}^c$.

It is trivial to complete the realization to a realization of the whole construction so that the final realization is in $\mathcal{P}^c$.

(Sufficiency) Suppose that there exists a realization $R$ of the disks in $\mathcal{P}^c$. Let $C_j$ be a clause. Since the realization of the clause is in $\mathcal{P}^c$, at least one of the red disks $R_i$ of the clause does not have the representative at the position $f_{R_i}$. If the corresponding literal appears positive in the clause, we set the associated variable to $T$. Otherwise, we set it to $F$. In this way, we ensure that every clause is true, but we need to argue that we did not assign $T$ and $F$ simultaneously to the same variable due to two distinct clauses.

Since in the connection gadget there is a stack of disks next to $R_i$, the fact that $R_i$ does not have the representative at the position $f_{R_i}$ forces the choice of the representative of the blue disk in the stack. The choice propagates to the connection gadget and eventually to the variable gadget: If the literal associated to $R_i$ appears positive in $C_j$, the variable has been set to $T$ and the realization in the variable gadget is the one choosing the rightmost points of the disks. If the literal appears negative, the variable has been set to $F$ and the realization

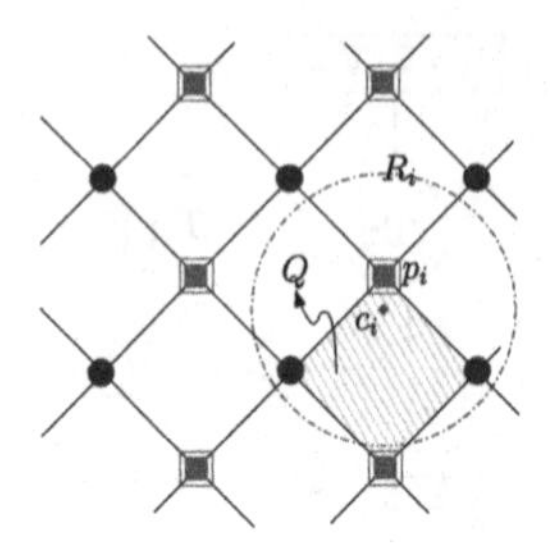


**Fig. 8.** The tilted grid. We choose $p_i$ as the red corner of $Q$ contained in $R_i$.

in the variable gadget chooses the leftmost points of the disks. Since in $R$ the realization of the variable gadget is either the rightmost or the leftmost, the variable has either been set to $T$ or $F$. □

**Theorem 2.** *The problem of finding the largest minimum color spanning circle of $\mathcal{R}$ is* NP-Hard.

*Remark 1.* Given a yes-instance, we can verify in $O(nk \log n)$ time whether the given realization is correct and the radius of its MCSC is at least $c$. Therefore, the decision version of the problem is NP-Complete.

## 4 Approximation Algorithms

In this section, we provide approximation algorithms for the L-MCSC problem. Let $\tilde{r}_{opt}$ denote the radius of a largest possible minimum color spanning circle of $\mathcal{R}$. We first prove bounds on $\tilde{r}_{opt}$.

**Lemma 5.** $\tilde{r}_{opt} \geq 1/4$.

*Proof.* It is easy to see that it is enough to prove the result for the case where $k = 2$. We show that the bound holds when $k = 2$ by providing a realization $P$ whose MCSC achieves the bound. Consider a regular square grid rotated by $\pi/4$ such that the side of every cell of the grid has length 1/2. We color the corners of the cells in red or blue in such a way that all corners lying in some vertical line are colored red, all corners lying in the next vertical line are colored blue, and so on (see Fig. 8). Now let $R_i \in \mathcal{R}$ have red color, and let $Q$ be the cell of the grid containing the center of $R_i$ (if the center of $R_i$ lies on an edge or vertex of the grid, we assign it to any of the adjacent cells). Notice that at least one of the two red corners of $Q$ lies inside $R_i$. We choose such a corner as $p_i \in P$. Similarly, for every $R_j \in \mathcal{R}$ of blue color, $P$ contains one of the blue corners of a cell containing the center of $R_j$. We obtain that $P$ is a subset of the grid corners. Since the distance between any pair of red and blue corners is at least 1/2, the radius of any MCSC is at least 1/4. □

**Lemma 6.** $\tilde{r}_{opt} \leq r_c + \frac{1}{2}$.

Let $P^g$ denote the realization of $\mathcal{R}$ described in the proof of Lemma 5. Our algorithm to compute the approximate L-MCSC is presented in Algorithm 2:

**Algorithm 2** $\frac{1}{3}$-factor approximation algorithm for the L-MCSC problem

**Input:** *A set* $\mathcal{R}$ *of* $n$ *unit disks*
**Output:** *A* MCSC *of a realization of* $\mathcal{R}$ *with radius at least* $\tilde{r}_{opt}/3$
1: compute $C_c$;
2: **if** $r_c \geq 1/4$ **then**
3:     **return** $C_c$;
4: **else**
5:     **return** a MCSC of $P^g$;

**Theorem 3.** *A* $\frac{1}{3}$*-factor approximation for the* L-MCSC *problem can be computed in* $O(nk \log n)$ *time. If no two distinct colored disks of* $\mathcal{R}$ *intersect, the approximation factor becomes* $\frac{1}{2}$.

**Acknowledgements.** The authors would like to thank Irina Kostitsyna for key discussions on the hardness reduction and Hans Raj Tiwary for the proof of Lemma 5. A.A., R.J., V.K., and M. S. were supported by the Czech Science Foundation, grant number GJ19-06792Y, and by institutional support RVO: 67985807. M.L. was partially supported by the Netherlands Organization for Scientific Research (NWO) under project no. 614.001.504. This project has received funding from the European Union's Horizon 2020 research and innovation programme under the Marie Skłodowska-Curie grant agreement No 734922.

## References

1. Abellanas, M., et al.: Smallest color-spanning objects. In: ESA, pp. 278–289 (2001)
2. Acharyya, A., Nandy, S.C., Roy, S.: Minimum width color spanning annulus. Theoret. Comput. Sci. **725**, 16–30 (2018)
3. de Berg, M., Gudmundsson, J., Katz, M.J., Levcopoulos, C., Overmars, M.H., van der Stappen, A.F.: TSP with neighborhoods of varying size. J. Algorithms **57**(1), 22–36 (2005)
4. Cabello, S.: Approximation algorithms for spreading points. J. Algorithms **62**(2), 49–73 (2007)
5. Consuegra, M.E., Narasimhan, G.: Geometric avatar problems. In: FSTTCS, pp. 389–400 (2013)
6. Daescu, O., Ju, W., Luo, J.: NP-completeness of spreading colored points. In: COCOA, pp. 41–50 (2010)
7. Das, S., Goswami, P.P., Nandy, S.C.: Smallest color-spanning object revisited. Internat. J. Comput. Geom. Appl. **19**(05), 457–478 (2009)
8. Dror, M., Orlin, J.B.: Combinatorial optimization with explicit delineation of the ground set by a collection of subsets. SIAM J. Discrete Math. **21**(4), 1019–1034 (2008)
9. Dumitrescu, A., Jiang, M.: Dispersion in disks. Theory Comput. Syst. **51**(2), 125–142 (2012)

10. Fiala, J., Kratochvíl, J., Proskurowski, A.: Systems of distant representatives. Discrete Appl. Math. **145**(2), 306–316 (2005)
11. Fleischer, R., Xu, X.: Computing minimum diameter color-spanning sets is hard. Inform. Process. Lett. **111**(21–22), 1054–1056 (2011)
12. Fraser, R.: Algorithms for geometric covering and piercing problems. Ph.D. thesis, University of Waterloo (2013)
13. Hasheminejad, J., Khanteimouri, P., Mohades, A.: Computing the smallest color spanning equilateral triangle. In: EuroCG, pp. 32–35 (2015)
14. Huttenlocher, D.P., Kedem, K., Sharir, M.: The upper envelope of Voronoi surfaces and its applications. Discrete Comput. Geom. **9**(3), 267–291 (1993)
15. Jadhav, S., Mukhopadhyay, A., Bhattacharya, B.: An optimal algorithm for the intersection radius of a set of convex polygons. J. Algorithms **20**(2), 244–267 (1996)
16. Jiang, M., Wang, H.: Shortest color-spanning intervals. In: COCOON, pp. 288–299 (2014)
17. Ju, W., Fan, C., Luo, J., Zhu, B., Daescu, O.: On some geometric problems of color-spanning sets. J. Comb. Optim. **26**(2), 266–283 (2013)
18. Khanteimouri, P., Mohades, A., Abam, M.A., Kazemi, M.R.: Computing the smallest color-spanning axis-parallel square. In: ISAAC, pp. 634–643 (2013)
19. Knauer, C., Löffler, M., Scherfenberg, M., Wolle, T.: The directed Hausdorff distance between imprecise point sets. Theoret. Comput. Sci. **412**(32), 4173–4186 (2011)
20. Knuth, D.E., Raghunathan, A.: The problem of compatible representatives. SIAM J. Discrete Math. **5**(3), 422–427 (1992)
21. Lichtenstein, D.: Planar formulae and their uses. SIAM J. Comput. **11**(2), 329–343 (1982)
22. Löffler, M., van Kreveld, M.: Largest bounding box, smallest diameter, and related problems on imprecise points. Comput. Geom. **43**(4), 419–433 (2010)
23. Manzini, R., Gamberini, R.: Design, management and control of logistic distribution systems. Int. J. Adv. Robot. Syst. 263–290 (2008)
24. Pop, P.C.: The generalized minimum spanning tree problem: an overview of formulations, solution procedures and latest advances. Eur. J. Oper. Res. (2019)
25. Robert, J.M., Toussaint, G.: Computational geometry and facility location. In: Operations Research and Management Science, pp. 11–15 (1990)
26. Salesin, D., Stolfi, J., Guibas, L.: Epsilon geometry: building robust algorithms from imprecise computations. In: SoCG, pp. 208–217 (1989)
27. Zhang, D., Chee, Y.M., Mondal, A., Tung, A.K.H., Kitsuregawa, M.: Keyword search in spatial databases: towards searching by document. In: ICDE, pp. 688–699 (2009)

# Fault Tolerant Computing and Fault Diagnosis

# Reliability Evaluation of Subsystem Based on Exchanged Hypercube

Yihong Wang, Shuming Zhou(✉), and Zhengqin Yu

School of Mathematics and Informatics, Fujian Normal University, Fuzhou, China
zhoushuming@fjnu.edu.cn

**Abstract.** As the scale of systems increased, the probability of a single node failure also increasing. Therefore, exploring the reliability of complex systems is crucial to the operation of the network. Few of Scholar focus on the reliability of subsystems of irregular networks. While exchanged hypercube $EH(s,t)$ was firstly proposed by Loh et al. [9] and it is a irregular network. In this paper, we evaluate the reliability of $EH(s,t)$ by studying the reliability of subsystems $EH(s-1,t-1)$. Specifically, we use the PIE (Principle of Inclusion-Exclusion) method to derive the approximate value of the reliability and the upper bound with the intersection of no more than 3 subgraphs $EH(s-1,t-1)$ under the probability fault model.

**Keywords:** $EH(s \cdot t)$ · Reliability · Probability fault model · Approximate value

## 1 Introduction

With the increase of the scale of multiprocessor systems, the probability of failure of subsystems also increases. We define the reliability of the system as the probability that the system is fault-free and it is a function of time. The longer the system runs, the reliability of the system deteriorates. There are many models for evaluating the reliability of the system, which can be referred to [1,3,7,14,16]. In recent years, many scholars have also devoted to studying the reliability of various network subsystems in [5,6,8,11,12,15,17,18].

As one of the variant of hypercube, exchanged hypercube was first introduced by Loh et al. [9], which remains some important performances, such as hierarchicality, strong connectivity, and hamiltonicity, while it is an irregular network. Recently, more properties of exchanged hypercube have been explored in Fan et al. [2] obtained an efficient algorithm for embedding exchanged hypercubes into grids. Ren and Wang [13] determined strong local diagnosability property of exchanged hypercubes under the comparison model, etc.

## 2 Preliminaries

### 2.1 Notations

To facilitate discussion, we usually use a undirected and simple graph $G = G(V(G), E(G))$ to simulate a network, where $V(G)$ represents the set of


C.-Y. Chen et al. (Eds.): COCOON 2021, LNCS 13025, pp. 271–282, 2021.
https://doi.org/10.1007/978-3-030-89543-3_23

processors in the network and $E(G)$ represents the set of communication links in the network. If two distinct nodes $u$ and $v$ are linked by an edge, then $uv \in E(G)$. We use $N_G(u) = \{v \in V \mid uv \in E\}$ to represent the neighborhood of any node $u$ in $G$, that is, the set of nodes are adjacent to $u$. We use $d_G(u)$ to denoted the degree of node $u$ in $G$, which represents the number of edges incident with $u$ in $G$. Then, we have $d_G(u) = |N_G(u)|$. We denote minimum degree $\delta(G) = \min\{d_G(u) \mid \forall u \in V(G)\}$ and maximum degree $\Delta(G) = \max\{d_G(u) \mid \forall u \in V(G)\}$. Moreover, $[n]$ represents the integers from 1 to $n$, where $n \geq 1$.

### 2.2 Exchanged Hypercube

**Definition 1.** *[10] For $s \geq 1$ and $t \geq 1$, the exchanged hypercube, $EH(s,t)$, has $2^{s+t+1}$ nodes. Its node set is $\{u_1 \cdots u_s u_{s+1} \cdots u_{s+t} u_{s+t+1} \mid u_i \in \{0,1\}\ \forall i \in [s+t+1]\}$. Two nodes $u = u_1 \cdots u_s u_{s+1} \cdots u_{s+t} u_{s+t+1}$ and $v = v_1 \cdots v_s v_{s+1} \cdots v_{s+t} v_{s+t+1}$ are linked by a $r$-dimensional edge for some $1 \leq r \leq s+t+1$, if and only if the following conditions are satisfied:*

*(a) $u$ and $v$ differ exactly in one bit on the $r$th bit;*
*(b) $u_{s+t+1} = v_{s+t+1} = 1$ if $u$ and $v$ are different in some bit $r \in [s+t] - [s]$;*
*(c) $u_{s+t+1} = v_{s+t+1} = 0$ if $u$ and $v$ are different in some bit $r \in [s]$.*

Obviously, $EH(s,t)$ is a bipartite from the construction, where $\delta(EH(s,t)) = \min\{s,t\} + 1$, and $\Delta(EH(s,t)) = \max\{s,t\} + 1$.

### 2.3 Subsystem Reliability Under the Probability Fault Model

If the probability for any node in exchanged hypercube to be fault-free is $p$, then we use $R_{s,t}^{s-1,t-1}(p)$ to denote the probability that at least one fault-free subgraph $EH(s-1,t-1)$ in $EH(s,t)$. Moreover, we use $R(i)(p)$ to represent the reliability of the $i$th subgraph; $R(i,j)(p)$ to represent the conjunctive reliability of the $i$th and $j$th subgraph, i.e., the probability that both $i$th subgraph and $j$th subgraph are fault-free; etc.

## 3 Approximate Value of $R_{s,t}^{s-1,t-1}(p)$

Under the probability fault model, we assume that each node in a graph such that the probability for a node to be fault-free is $p$, and the occurrence of failures among nodes are mutually independent. Thus, we can get $4st$ different subgraphs $EH(s-1,t-1)$ from $EH(s,t)$. While some subgraphs $EH(s-1,t-1)$ share common nodes, we use the PIE (Principle of Inclusion-Exclusion) method to calculate disjoint events under the probability fault model.

$$
\begin{aligned}
R_{s,t}^{s-1,t-1}(p) = &\sum_{i=1}^{4st} R(i)(p) + (-1)^1 \sum_{i<j} R(i,j)(p) \\
&+ (-1)^2 \sum_{i<j<k} R(i,j,k)(p) + (-1)^3 \sum_{i<j<k<l} R(i,j,k,l)(p) \\
&+ \cdots + (-1)^{4st-1} R(1,2,\ldots,4st)(p).
\end{aligned}
$$

In order to give the calculation of the approximate value of $R_{(s,t)}^{(s-1,t-1)}(p)$, we list the following $4st$ subgraphs $EH(s-1,t-1)$:

$$\begin{array}{cccc}
(0,0) & (0,1) & (1,0) & (1,1) \\
\underbrace{0Y\cdots Y}_{s}\underbrace{0Y\cdots Y}_{t} & \underbrace{0Y\cdots Y}_{s}\underbrace{1Y\cdots Y}_{t} & \underbrace{1Y\cdots Y}_{s}\underbrace{0Y\cdots Y}_{t} & \underbrace{1Y\cdots Y}_{s}\underbrace{1Y\cdots Y}_{t} \\
\cdots & \cdots & \cdots & \cdots \\
\underbrace{0Y\cdots Y}_{s}\underbrace{YY\cdots 0}_{t} & \underbrace{0Y\cdots Y}_{s}\underbrace{YY\cdots 1}_{t} & \underbrace{1Y\cdots Y}_{s}\underbrace{YY\cdots 0}_{t} & \underbrace{1Y\cdots Y}_{s}\underbrace{YY\cdots 1}_{t} \\
\underbrace{Y0\cdots Y}_{s}\underbrace{0Y\cdots Y}_{t} & \underbrace{Y0\cdots Y}_{s}\underbrace{1Y\cdots Y}_{t} & \underbrace{Y1\cdots Y}_{s}\underbrace{0Y\cdots Y}_{t} & \underbrace{Y1\cdots Y}_{s}\underbrace{1Y\cdots Y}_{t} \\
\cdots & \cdots & \cdots & \cdots \\
\underbrace{Y0\cdots Y}_{s}\underbrace{YY\cdots 0}_{t} & \underbrace{Y0\cdots Y}_{s}\underbrace{YY\cdots 1}_{t} & \underbrace{Y1\cdots Y}_{s}\underbrace{YY\cdots 0}_{t} & \underbrace{Y1\cdots Y}_{s}\underbrace{YY\cdots 1}_{t} \\
\cdots & \cdots & \cdots & \cdots \\
\underbrace{YY\cdots 0}_{s}\underbrace{0Y\cdots Y}_{t} & \underbrace{YY\cdots 0}_{s}\underbrace{1Y\cdots Y}_{t} & \underbrace{YY\cdots 1}_{s}\underbrace{0Y\cdots Y}_{t} & \underbrace{YY\cdots 1}_{s}\underbrace{1Y\cdots Y}_{t} \\
\cdots & \cdots & \cdots & \cdots \\
\underbrace{YY\cdots 0}_{s}\underbrace{YY\cdots 0}_{t} & \underbrace{YY\cdots 0}_{s}\underbrace{YY\cdots 1}_{t} & \underbrace{YY\cdots 1}_{s}\underbrace{YY\cdots 0}_{t} & \underbrace{YY\cdots 1}_{s}\underbrace{YY\cdots 1}_{t}
\end{array}$$

Note that each row above is divided by $i$th position and $j$th position, which can obtain four disjoint subgraphs, where $i \in \{1,\ldots,s\}$, $j \in \{s+1,\ldots,s+t\}$. We use $R_{s,t}^{s-1,t-1}(i,p)$ to denote the probability of the fault-free four disjoint subgraphs in the $i$th row mentioned above, where $1 \le i \le st$. Then

$$\begin{aligned}
R_{s,t}^{s-1,t-1}(i,p) &= 4p^{2^{s+t-1}} - \binom{4}{2}p^{2\cdot 2^{s+t-1}} + \binom{4}{3}p^{3\cdot 2^{s+t-1}} - \binom{4}{4}p^{4\cdot 2^{s+t-1}} \\
&= 1-\left(1-p^{2^{s+t-1}}\right)^4
\end{aligned}$$

Furthermore,

$$R_{s,t}^{s-1,t-1}(1,p) = R_{s,t}^{s-1,t-1}(2,p) = \cdots = R_{s,t}^{s-1,t-1}(st,p) = 1-(1-p^{2^{s+t-1}})^4.$$

By considering all the $R_{s,t}^{s-1,t-1}(i,p)$ and with the help of using PIE, we obtain the approximate value of $R_{s,t}^{s-1,t-1}(p)$:

$$\begin{aligned}
R_{s,t}^{s-1,t-1}(p) &\approx \sum_{i=1}^{st}(-1)^{i-1}\binom{st}{i}\left(R_{s,t}^{s-1,t-1}(i,p)\right)^i \\
&= 1-\left(1-p^{2^{s+t-1}}\right)^{4st}.
\end{aligned}$$

## 4 Upper Bound of $R_{s,t}^{s-1,t-1}(p)$

**Lemma 1.** *If the probability for any node in exchanged hypercube to be fault-free is $p$, then* $\sum_{j=1}^{4st} R(j)(p) = 4stp^{2^{s+t-1}}$.

*Proof.* Obviously, there are totally $4st$ distinct subgraphs $EH(s-1,t-1)$. Moreover, the size of each subgraph is $2^{s+t-1}$, the desired. □

**Lemma 2.** *If the probability for any node in exchanged hypercube to be fault-free is $p$, then*

$$\sum_{i<j} R(i,j)(p) = \left[6\binom{s}{1}\binom{t}{1} + 8\binom{s}{1}\binom{t}{2} + 8\binom{s}{2}\binom{t}{1}\right] \cdot p^{2^{s+t}}$$
$$+ \left[8\binom{s}{1}\binom{t}{2} + 8\binom{s}{2}\binom{t}{1}\right] \cdot p^{3\cdot 2^{s+t-2}}$$
$$+ 32\binom{s}{2}\binom{t}{2} \cdot p^{7\cdot 2^{s+t-3}}.$$

*Proof.* Without loss of generality, we assume that $EH[i_1, j_1 : u_1, v_1]$ and $EH[i_2, j_2 : u_2, v_2]$ are two distinct subgraphs. Let $S = \{i_1, i_2\}$, $T = \{j_1, j_2\}$, $|S| = \hat{s}$, $|T| = \hat{t}$. Obviously, $1 \le \hat{s} \le 2$, $1 \le \hat{t} \le 2$. Moreover, we denote $(\hat{s}, \hat{t})$ to select $\hat{s}$ positions from $\{1, \ldots, s\}$ and select $\hat{t}$ positions from $\{s+1, \ldots, s+t\}$. Thus, there are $\binom{s}{\hat{s}}$ ways to select $\hat{s}$ positions from $\{1, \ldots, s\}$, and there are $\binom{t}{\hat{t}}$ ways to select $\hat{t}$ positions from $\{s+1, \ldots, s+t\}$. We have the following cases.

**Case 1.** $(\hat{s}, \hat{t}) = (1, 1)$.
In this case, $i_1 = i_2$, $j_1 = j_2$. Therefore, there are totally $\binom{s}{1}\binom{t}{1}\binom{4}{2}$ 2-subgraph groups. Since $EH[i_1, j_1 : u_1, v_1]$ and $EH[i_2, j_2 : u_2, v_2]$ are disjoint, the size of the two subgraphs is $2 \cdot 2^{s+t-1} = 2^{s+t}$. Thus, the probability for $EH[i_1, j_1 : u_1, v_1]$ and $EH[i_2, j_2 : u_2, v_2]$ to be fault-free is $p^{2^{s+t}}$.

**Case 2.** $(\hat{s}, \hat{t}) = (1, 2)$.
Then, $i_1 = i_2$, $j_1 \ne j_2$. If $|\{u_1, u_2\}| = 1$, then $u_1 = u_2$. Thus, $EH[i_1, j_1 : u_1, v_1]$ and $EH[i_2, j_2 : u_2, v_2]$ are two intersecting subgraphs. Therefore, there are totally $2 \cdot 2^{s+t-1} - 2^{s+t-2} = 3 \cdot 2^{s+t-2}$ nodes in $EH[i_1, j_1 : u_1, v_1] \cup EH[i_2, j_2 : u_2, v_2]$. Moreover, each of position $i_1$, $j_1$, $j_2$ has two distinct codes (0 or 1) to select. Thus, there are totally $\binom{s}{1}\binom{t}{2} \cdot 2^3$ 2-subgraph groups. Otherwise, $|\{u_1, u_2\}| = 2$. That is, $u_1 \ne u_2$. Hence, $EH[i_1, j_1 : u_1, v_1]$ and $EH[i_2, j_2 : u_2, v_2]$ are two disjoint subgraphs. Therefore, there are totally $2 \cdot 2^{s+t-1} = 2^{s+t}$ nodes in $EH[i_1, j_1 : u_1, v_1] \cup EH[i_2, j_2 : u_2, v_2]$. Thus, there are totally $\binom{s}{1}\binom{t}{2} \cdot 2^3$ 2-subgraph groups.

**Case 3.** $(\hat{s}, \hat{t}) = (2, 1)$.
Similar to the Case 2, if $v_1 = v_2$, then there are totally $2 \cdot 2^{s+t-1} - 2^{s+t-2} = 3 \cdot 2^{s+t-2}$ nodes in $EH[i_1, j_1 : u_1, v_1] \cup EH[i_2, j_2 : u_2, v_2]$ and there are totally $\binom{s}{2}\binom{t}{1} \cdot 2^3$ 2-subgraph groups. Otherwise, if $v_1 \ne v_2$, then there are totally $2 \cdot 2^{s+t-1} = 2^{s+t}$ nodes in $EH[i_1, j_1 : u_1, v_1] \cup EH[i_2, j_2 : u_2, v_2]$ and there are totally $\binom{s}{2}\binom{t}{1} \cdot 2^3$ 2-subgraph groups.

**Case 4.** $(\hat{s}, \hat{t}) = (2, 2)$.
Obviously, $|\{i_1, i_2\}| = 2$, $|\{j_1, j_2\}| = 2$. Hence, there are totally $\binom{s}{2}\binom{t}{2} \cdot 2 \cdot 2^4$ 2-subgraph groups. □

**Lemma 3.** *If the probability for any node in exchanged hypercube to be fault-free is p, then we have*

$$\sum_{i<j<k} R(i,j,k)(p) = \left[8\binom{s}{1}\binom{t}{2} + 8\binom{s}{2}\binom{t}{1} + 64\binom{s}{2}\binom{t}{2}\right] \cdot p^{2^{s+t}} + \left[48\binom{s}{1}\binom{t}{3}\right.$$
$$+ 48\binom{s}{3}\binom{t}{1} + 32\binom{s}{1}\binom{t}{2} + 32\binom{s}{2}\binom{t}{1} + 96\binom{s}{2}\binom{t}{2}$$
$$\left. + 192\binom{s}{2}\binom{t}{3} + 192\binom{s}{3}\binom{t}{2}\right] \cdot p^{5\cdot 2^{s+t-2}} + \left[4\binom{s}{1}\binom{t}{1}\right.$$
$$\left. + 8\binom{s}{1}\binom{t}{2} + 8\binom{s}{2}\binom{t}{1}\right] \cdot p^{3\cdot 2^{s+t-1}} + \left[16\binom{s}{1}\binom{t}{3}\right.$$
$$\left. + 16\binom{s}{3}\binom{t}{1}\right] \cdot p^{7\cdot 2^{s+t-3}} + 128\binom{s}{2}\binom{t}{2} \cdot p^{9\cdot 2^{s+t-3}}$$
$$+ 64\binom{s}{2}\binom{t}{2} \cdot p^{11\cdot 2^{s+t-3}} + \left[192\binom{s}{2}\binom{t}{3}\right.$$
$$\left. + 192\binom{s}{3}\binom{t}{2}\right] \cdot p^{17\cdot 2^{s+t-4}} + 384\binom{s}{3}\binom{t}{3} \cdot p^{37\cdot 2^{s+t-5}}.$$

*Proof.* Without loss of generality, we assume that $EH[i_1, j_1 : u_1, v_1]$, $EH[i_2, j_2 : u_2, v_2]$, and $EH[i_3, j_3 : u_3, v_3]$ are three distinct subgraphs. Let $S = \{i_1, i_2, i_3\}$, $T = \{j_1, j_2, j_3\}$, $|S| = \hat{s}$, $|T| = \hat{t}$. Obviously, $1 \le \hat{s} \le 3$, $1 \le \hat{t} \le 3$. Moreover, we denote $(\hat{s}, \hat{t})$ to $\hat{s}$ positions from $\{1, \ldots, s\}$ and select $\hat{t}$ positions from $\{s+1, \ldots, s+t\}$. Thus, there are $\binom{s}{\hat{s}}$ ways to select $\hat{s}$ positions from $\{1, \ldots, s\}$, and there are $\binom{t}{\hat{t}}$ ways to select $\hat{t}$ positions from $\{s+1, \ldots, s+t\}$. We distinguish the following cases.

**Case 1.** $(\hat{s}, \hat{t}) = (1, 1)$.

Obviously, $|\{i_1, i_2, i_3\}| = 1$, $|\{j_1, j_2, j_3\}| = 1$, that is, $i_1 = i_2 = i_3$, $j_1 = j_2 = j_3$. According to the definition of $EH(s,t)$, as each position has two distinct codes 0 and 1 to select, we can obtain 4 disjoint subgraphs $EH(s-1, t-1)$ by fixing position $i$ and $j$, where $i \in \{1, \ldots, s\}$, $j \in \{s+1, \ldots, s+t\}$. Therefore, there are totally $\binom{s}{1}\binom{t}{1}\binom{4}{3}$ 3-subgraph groups. Since the three subgraphs are mutually disjoint (see Fig. 1(1)), there are $3 \cdot 2^{s+t-1}$ nodes in $EH[i_1, j_1 : u_1, v_1] \cup EH[i_2, j_2 : u_2, v_2] \cup EH[i_3, j_3 : u_3, v_3]$. Thus, the probability of these nodes are fault-free is $p^{3\cdot 2^{s+t-1}}$.

**Case 2.** $(\hat{s}, \hat{t}) = (1, 2)$.

Obviously, $|\{i_1, i_2, i_3\}| = 1$, $|\{j_1, j_2, j_3\}| = 2$, that is, $i_1 = i_2 = i_3$. Without loss of generality, we assume that $j_1 = j_2$ and $j_1 \ne j_3$. We distinguish the following cases.

**Subcase 2.1.** $|\{u_1, u_2\}| = 2$.

Thus, $u_1 \ne u_2$. $u_i \in \{0, 1\}$ for any $i \in \{1, 2\}$, then $u_3 \in \{u_1, u_2\}$. Without loss of generality, we assume that $u_3 = u_1$. It means that $EH[i_1, j_1 : u_1, v_1]$ and $EH[i_3, j_3 : u_3, v_3]$ are intersecting, while both of them are not intersecting with

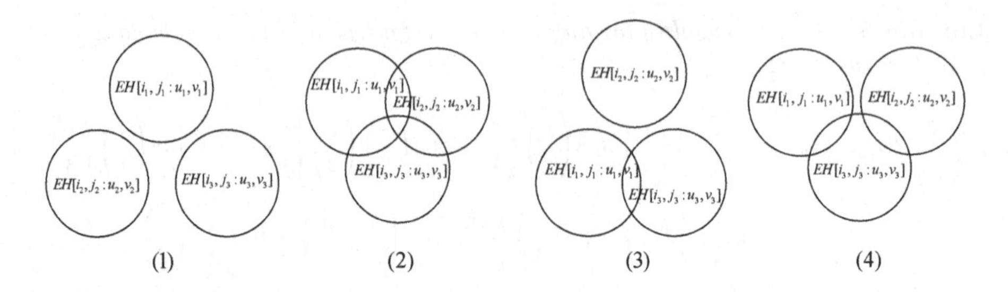


**Fig. 1.** Intersection of three subgraphs

$EH[i_2, j_2 : u_2, v_2]$ (see Fig. 1(3)). Therefore, there are totally $3 \cdot 2^{s+t-1} - 2^{s+t-2} = 5 \cdot 2^{s+t-2}$ nodes in $EH[i_1, j_1 : u_1, v_1] \cup EH[i_2, j_2 : u_2, v_2] \cup EH[i_3, j_3 : u_3, v_3]$. In view of $u_1 \neq u_2$, there are 2 ways to put codes in positions $i_1$ and $i_2$ ($u_1 = 0, u_2 = 1$ or $u_1 = 1, u_2 = 0$). In addition, since each of position $j_1$, $j_2$, $i_3$, $i_4$ has two distinct codes (0 or 1) to select, there $2^4$ ways to select codes to put the four positions. Hence, there are totally $\binom{s}{1}\binom{t}{2} \cdot 2 \cdot 2^4 = 32\binom{s}{1}\binom{t}{2}$ 3-subgraph groups.

**Subcase 2.2.** $|\{u_1, u_2\}| = 1$.

Obviously, $u_1 = u_2$. In order to facilitate the discussion, we distinguish the following situations. If $u_1 = u_2 = u_3$. Then, $EH[i_1, j_1 : u_1, v_1]$ and $EH[i_2, j_2 : u_2, v_2]$ are disjoint, while both of them are intersecting with $EH[i_3, j_3 : u_3, v_3]$ (see Fig. 1(4)). Therefore, there are totally $3 \cdot 2^{s+t-1} - 2 \cdot 2^{s+t-2} = 2^{s+t}$ nodes in $EH[i_1, j_1 : u_1, v_1] \cup EH[i_2, j_2 : u_2, v_2] \cup EH[i_3, j_3 : u_3, v_3]$. Discuss similarly, there are totally $\binom{s}{1}\binom{t}{2} \cdot 2 \cdot 2 \cdot 2 = 8\binom{s}{1}\binom{t}{2}$ 3-subgraph groups. Otherwise, $|\{u_1, u_2, u_3\}| = 2$, then $u_1 = u_2 \neq u_3$. Moreover, $EH[i_1, j_1 : u_1, v_1]$, $EH[i_2, j_2 : u_2, v_2]$, and $EH[i_3, j_3 : u_3, v_3]$ are mutually disjoint (see Fig. 1(1)). Therefore, there are totally $3 \cdot 2^{s+t-1}$ nodes in $EH[i_1, j_1 : u_1, v_1] \cup EH[i_2, j_2 : u_2, v_2] \cup EH[i_3, j_3 : u_3, v_3]$. Discuss similarly, there are totally $\binom{s}{1}\binom{t}{2} \cdot 2 \cdot 2 \cdot 2 = 8\binom{s}{1}\binom{t}{2}$ 3-subgraph groups.

**Case 3.** $(\hat{s}, \hat{t}) = (2, 1)$.

Similar to the Case 2, we have $|\{i_1, i_2, i_3\}| = 2$, $|\{j_1, j_2, j_3\}| = 1$, that is, $j_1 = j_2 = j_3$. Without loss of generality, we assume that $i_1 = i_2$ and $i_1 \neq i_3$. We distinguish the following cases.

**Subcase 3.1.** $|\{v_1, v_2\}| = 2$.

Obviously, $v_1 \neq v_2$. Therefore, there are totally $3 \cdot 2^{s+t-1} - 2^{s+t-2} = 5 \cdot 2^{s+t-2}$ nodes in $EH[i_1, j_1 : u_1, v_1] \cup EH[i_2, j_2 : u_2, v_2] \cup EH[i_3, j_3 : u_3, v_3]$. Hence, there are totally $\binom{s}{2}\binom{t}{1} \cdot 2 \cdot 2^4 = 32\binom{s}{2}\binom{t}{1}$ 3-subgraph groups.

**Subcase 3.2.** $|\{v_1, v_2\}| = 1$.

Obviously, $v_1 = v_2$. If $|\{v_1, v_2, v_3\}| = 1$, then $v_1 = v_2 = v_3$. Hence, there are totally $3 \cdot 2^{s+t-1} - 2 \cdot 2^{s+t-2} = 2^{s+t}$ nodes in $EH[i_1, j_1 : u_1, v_1] \cup EH[i_2, j_2 : u_2, v_2] \cup EH[i_3, j_3 : u_3, v_3]$. Hence, there are totally $\binom{s}{2}\binom{t}{1} \cdot 2 \cdot 2 \cdot 2 = 8\binom{s}{2}\binom{t}{1}$ 3-subgraph groups. Otherwise, $|\{v_1, v_2, v_3\}| = 2$. Then, $v_1 = v_2 \neq v_3$. Hence, there are totally $3 \cdot 2^{s+t-1}$ nodes in $EH[i_1, j_1 : u_1, v_1] \cup EH[i_2, j_2 : u_2, v_2] \cup EH[i_3, j_3 : u_3, v_3]$. Hence, there are totally $\binom{s}{2}\binom{t}{1} \cdot 2 \cdot 2 \cdot 2 = 8\binom{s}{2}\binom{t}{1}$ 3-subgraph groups.

**Case 4.** $(\hat{s}, \hat{t}) = (1, 3)$.

In this case, we have $i_1 = i_2 = i_3$, $|\{j_1, j_2, j_3\}| = 3$. We distinguish the following cases.

If $|\{u_1, u_2, u_3\}| = 1$, then $u_1 = u_2 = u_3$. Thus, $EH[i_1, j_1 : u_1, v_1]$, $EH[i_2, j_2 : u_2, v_2]$, and $EH[i_3, j_3 : u_3, v_3]$ are mutually intersecting (see Fig. 1(4)). Therefore, there are totally

$$3 \cdot 2^{s+t-1} - \binom{3}{2} \cdot 2^{s+t-2} + \binom{3}{3} \cdot 2^{s+t-3} = 7 \cdot 2^{s+t-3}$$

nodes in $EH[i_1, j_1 : u_1, v_1] \cup EH[i_2, j_2 : u_2, v_2] \cup EH[i_3, j_3 : u_3, v_3]$. Discuss similarly, there are totally $\binom{s}{1}\binom{t}{3} \cdot 2 \cdot 2^3 = 16\binom{s}{1}\binom{t}{3}$ 3-subgraph groups. Otherwise, $|\{u_1, u_2, u_3\}| = 2$. Then, $u_1 \neq u_2 = u_3$ or $u_2 \neq u_1 = u_3$ or $u_3 \neq u_1 = u_2$. Without loss of generality, we assume that $u_1 \neq u_2 = u_3$. Therefore, $EH[i_1, j_1 : u_1, v_1]$ is not intersecting with $EH[i_2, j_2 : u_2, v_2]$ and $EH[i_3, j_3 : u_3, v_3]$, while $EH[i_2, j_2 : u_2, v_2]$ and $EH[i_3, j_3 : u_3, v_3]$ are intersecting (see Fig. 1(3)). So, there are totally $3 \cdot 2^{s+t-1} - 2^{s+t-2} = 5 \cdot 2^{s+t-2}$ nodes in $EH[i_1, j_1 : u_1, v_1] \cup EH[i_2, j_2 : u_2, v_2] \cup EH[i_3, j_3 : u_3, v_3]$. Discuss similarly, there are totally

$$\binom{s}{1}\binom{t}{3} \cdot 2^3 \cdot 2 \cdot \binom{3}{2} = 48\binom{s}{1}\binom{t}{3}$$

3-subgraph groups.

**Case 5.** $(\hat{s}, \hat{t}) = (3, 1)$.

Similar to the Case 4, $|\{i_1, i_2, i_3\}| = 3$, $j_1 = j_2 = j_3$. We distinguish the following cases.

If $|\{v_1, v_2, v_3\}| = 1$, then, $v_1 = v_2 = v_3$. Hence, there are totally

$$3 \cdot 2^{s+t-1} - \binom{3}{2} \cdot 2^{s+t-2} + \binom{3}{3} \cdot 2^{s+t-3} = 7 \cdot 2^{s+t-3}$$

nodes in $EH[i_1, j_1 : u_1, v_1] \cup EH[i_2, j_2 : u_2, v_2] \cup EH[i_3, j_3 : u_3, v_3]$. Hence, there are totally $\binom{s}{3}\binom{t}{1} \cdot 2 \cdot 2^3 = 16\binom{s}{3}\binom{t}{1}$ 3-subgraph groups. Otherwise, $|\{v_1, v_2, v_3\}| = 2$. Thus, we have $v_1 \neq v_2 = v_3$ or $v_2 \neq v_1 = v_3$ or $v_3 \neq v_1 = v_2$. Then, there are totally $3 \cdot 2^{s+t-1} - 2 \cdot 2^{s+t-2} = 2^{s+t}$ nodes in $EH[i_1, j_1 : u_1, v_1] \cup EH[i_2, j_2 : u_2, v_2] \cup EH[i_3, j_3 : u_3, v_3]$. Hence, there are totally $\binom{s}{3}\binom{t}{1} \cdot 2^3 \cdot 2 \cdot \binom{3}{2} = 48\binom{s}{3}\binom{t}{1}$ 3-subgraph groups.

**Case 6.** $(\hat{s}, \hat{t}) = (2, 2)$.

Obviously, $|\{i_1, i_2, i_3\}| = 2$, $|\{j_1, j_2, j_3\}| = 2$. We distinguish the following cases.

**Subcase 6.1.** $i_1 = i_2$, $j_1 = j_2$.

Then, $EH[i_1, j_1 : u_1, v_1]$ and $EH[i_2, j_2 : u_2, v_2]$ are not intersecting, while they are not intersecting with $EH[i_3, j_3 : u_3, v_3]$ (see Fig. 1(4)). So, there are totally

$$3 \cdot 2^{s+t-1} - 2 \cdot 2^{s+t-3} = 5 \cdot 2^{s+t-2}$$

nodes in $EH[i_1, j_1 : u_1, v_1] \cup EH[i_2, j_2 : u_2, v_2] \cup EH[i_3, j_3 : u_3, v_3]$. Discuss similarly, there are totally

$$\binom{s}{2}\binom{t}{2}\binom{4}{2} \cdot 2^2 \cdot 2 \cdot 2 = 96\binom{s}{2}\binom{t}{2}$$

3-subgraph groups.

**Subcase 6.2.** $i_2 = i_3,\ j_2 \neq j_3$.

Since $\hat{t} = 2$, $j_1 \in \{j_2, j_3\}$. In order to facilitate discussion, we assume that $j_1 = j_3$. Then, $EH[i_1, j_1 : u_1, v_1]$ and $EH[i_2, j_2 : u_2, v_2]$ are intersecting.

**Subcase 6.2.1.** $|\{u_2, u_3\}| = |\{v_1, v_3\}| = 1$.

Obviously, $u_2 = u_3,\ v_1 = v_3$, three subgraphs $EH[i_1, j_1 : u_1, v_1]$, $EH[i_2, j_2 : u_2, v_2]$, and $EH[i_3, j_3 : u_3, v_3]$ are mutually intersecting. So, there are totally

$$3 \cdot 2^{s+t-1} - 2 \cdot 2^{s+t-2} - 2^{s+t-3} + 2^{s+t-3} = 2^{s+t}$$

nodes in $EH[i_1, j_1 : u_1, v_1] \cup EH[i_2, j_2 : u_2, v_2] \cup EH[i_3, j_3 : u_3, v_3]$. Discuss similarly, there are totally

$$\binom{s}{2}\binom{t}{2} \cdot 2 \cdot 2 \cdot 2 \cdot 2 \cdot 2 \cdot 2 = 64\binom{s}{2}\binom{t}{2}$$

3-subgraph groups.

**Subcase 6.2.2.** $|\{u_2, u_3\}| = 1$, $|\{v_1, v_3\}| = 2$ or $|\{u_2, u_3\}| = 2$, $|\{v_1, v_3\}| = 1$.

Then, $u_2 = u_3,\ v_1 \neq v_3$ or $u_2 \neq u_3,\ v_1 = v_3$. Without loss of generality, we assume that $u_2 = u_3,\ v_1 \neq v_3$. Therefore, $EH[i_1, j_1 : u_1, v_1]$ and $EH[i_3, j_3 : u_3, v_3]$ are disjoint, while they are not intersecting with $EH[i_2, j_2 : u_2, v_2]$ (see Fig. 1(4)). So, there are totally

$$3 \cdot 2^{s+t-1} - 2^{s+t-3} - 2^{s+t-2} = 9 \cdot 2^{s+t-3}$$

nodes in $EH[i_1, j_1 : u_1, v_1] \cup EH[i_2, j_2 : u_2, v_2] \cup EH[i_3, j_3 : u_3, v_3]$. Since $u_2 = u_3,\ v_1 \neq v_3$, , there are totally

$$\binom{s}{2}\binom{t}{2} \cdot 2 \cdot 2 \cdot 2 \cdot 2 \cdot 2 \cdot 2 = 64\binom{s}{2}\binom{t}{2}$$

3-subgraph groups. Since $u_2 \neq u_3, v_1 = v_3$ is similar to the situation $u_2 = u_3, v_1 \neq v_3$. Hence, there are totally

$$\binom{s}{2}\binom{t}{2} \cdot 2 \cdot 2 \cdot 2 \cdot 2 \cdot 2 \cdot 2 \cdot 2 = 128\binom{s}{2}\binom{t}{2}$$

3-subgraph groups.

**Subcase 6.2.3.** $|\{u_2, u_3\}| = |\{v_1, v_3\}| = 2$.

Obviously, $u_2 \neq u_3$ and $v_1 \neq v_3$. In this situation, $EH[i_1, j_1 : u_1, v_1]$ and $EH[i_2, j_2 : u_2, v_2]$ are intersecting, while they are not intersecting with $EH[i_3, j_3 : u_3, v_3]$ (see Fig. 1(4)). So, there are totally $3 \cdot 2^{s+t-1} - 2^{s+t-3} =$

$11 \cdot 2^{s+t-3}$ nodes in $EH[i_1, j_1 : u_1, v_1] \cup EH[i_2, j_2 : u_2, v_2] \cup EH[i_3, j_3 : u_3, v_3]$. Discuss similarly, there are totally

$$\binom{s}{2}\binom{t}{2} \cdot 2 \cdot 2 \cdot 2 \cdot 2 \cdot 2 \cdot 2 = 64\binom{s}{2}\binom{t}{2}$$

3-subgraph groups.

**Case 7.** $(\hat{s}, \hat{t}) = (2, 3)$.

Obviously, $|\{i_1, i_2, i_3\}| = 2$, $|\{j_1, j_2, j_3\}| = 3$. Then, $i_1 \neq i_2 = i_3$ or $i_2 \neq i_1 = i_3$ or $i_3 \neq i_1 = i_2$. Without loss of generality, we assume that $i_1 \neq i_2 = i_3$. In this situation, $EH[i_2, j_2 : u_2, v_2]$ and $EH[i_3, j_3 : u_3, v_3]$ are intersecting with $EH[i_1, j_1 : u_1, v_1]$. We distinguish the following cases. If $|\{u_2, u_3\}| = 1$, then $u_2 = u_3$, and $EH[i_1, j_1 : u_1, v_1]$, $EH[i_2, j_2 : u_2, v_2]$, and $EH[i_3, j_3 : u_3, v_3]$ are mutually intersecting (see Fig. 1(2)). So, there are totally

$$3 \cdot 2^{s+t-1} - 2 \cdot 2^{s+t-3} - 2^{s+t-2} + 2^{s+t-4} = 17 \cdot 2^{s+t-4}$$

nodes in $EH[i_1, j_1 : u_1, v_1] \cup EH[i_2, j_2 : u_2, v_2] \cup EH[i_3, j_3 : u_3, v_3]$. Discuss similarly, there are totally

$$\binom{s}{2}\binom{t}{3} \cdot 2 \cdot 2^3 \cdot 2 \cdot 2 \cdot \binom{3}{2} = 192\binom{s}{2}\binom{t}{3}$$

3-subgraph groups. Otherwise, $|\{u_2, u_3\}| = 2$. Obviously, $u_2 \neq u_3$. In this situation, $EH[i_2, j_2 : u_2, v_2]$ and $EH[i_3, j_3 : u_3, v_3]$ are disjoint, while they are intersecting with $EH[i_1, j_1 : u_1, v_1]$ (see Fig. 1(4)). So, there are totally $3 \cdot 2^{s+t-1} - 2 \cdot 2^{s+t-3} = 5 \cdot 2^{s+t-2}$ nodes in $EH[i_1, j_1 : u_1, v_1] \cup EH[i_2, j_2 : u_2, v_2] \cup EH[i_3, j_3 : u_3, v_3]$. Discuss similarly, there are totally

$$\binom{s}{2}\binom{t}{3} \cdot 2 \cdot 2^3 \cdot 2 \cdot 2 \cdot \binom{3}{2} = 192\binom{s}{2}\binom{t}{3}$$

3-subgraph groups.

**Case 8.** $(\hat{s}, \hat{t}) = (3, 2)$.

Similar to the proof of Case 7, $|\{i_1, i_2, i_3\}| = 3$, $|\{j_1, j_2, j_3\}| = 2$. Then, $j_1 \neq j_2 = j_3$ or $j_2 \neq j_1 = j_3$ or $j_3 \neq j_1 = j_2$. Without loss of generality, we assume that $j_1 \neq j_2 = j_3$. In this situation, $EH[i_1, j_1 : u_1, v_1]$ is intersecting with $EH[i_2, j_2 : u_2, v_2]$ and $EH[i_3, j_3 : u_3, v_3]$. We distinguish the following cases.

If $|\{v_2, v_3\}| = 1$, then $v_2 = v_3$. Then, there are totally

$$3 \cdot 2^{s+t-1} - 2 \cdot 2^{s+t-3} - 2^{s+t-2} + 2^{s+t-4} = 17 \cdot 2^{s+t-4}$$

nodes in $EH[i_1, j_1 : u_1, v_1] \cup EH[i_2, j_2 : u_2, v_2] \cup EH[i_3, j_3 : u_3, v_3]$. Hence, there are totally

$$\binom{s}{3}\binom{t}{2} \cdot 2 \cdot 2^3 \cdot 2 \cdot 2 \cdot \binom{3}{2} = 192\binom{s}{3}\binom{t}{2}$$

3-subgraph groups. Otherwise, $|\{v_2, v_3\}| = 2$. Then $v_2 \neq v_3$. Thus, there are totally

$$3 \cdot 2^{s+t-1} - 2 \cdot 2^{s+t-3} = 5 \cdot 2^{s+t-2}$$

nodes in $EH[i_1, j_1 : u_1, v_1] \cup EH[i_2, j_2 : u_2, v_2] \cup EH[i_3, j_3 : u_3, v_3]$. Hence, there are totally

$$\binom{s}{3}\binom{t}{2} \cdot 2 \cdot 2^3 \cdot 2 \cdot 2 \cdot \binom{3}{2} = 192\binom{s}{3}\binom{t}{2}$$

3-subgraph groups.

**Case 9.** $(\hat{s}, \hat{t}) = (3, 3)$.

Obviously, $|\{i_1, i_2, i_3\}| = |\{j_1, j_2, j_3\}| = 3$. $EH[i_1, j_1 : u_1, v_1]$, $EH[i_2, j_2 : u_2, v_2]$ and $EH[i_3, j_3 : u_3, v_3]$ are mutually intersecting (see Fig. 1(2)). So, there are totally

$$3 \cdot 2^{s+t-1} - \binom{s}{3} 2 \cdot 2^{s+t-3} + \binom{s}{3} 3 \cdot 2^{s+t-5} = 37 \cdot 2^{s+t-5}$$

nodes in $EH[i_1, j_1 : u_1, v_1] \cup EH[i_2, j_2 : u_2, v_2] \cup EH[i_3, j_3 : u_3, v_3]$. Discuss similarly, there are totally

$$\binom{s}{3}\binom{t}{3} \cdot 2^6 \cdot 3! = 384\binom{s}{3}\binom{t}{3}$$

3-subgraph groups.

By Lemmas 1, 2, and 3, we have the following result.

**Theorem 1.** *If the probability for any node in exchanged hypercube to be fault-free is p, then we have*

$$\begin{aligned}
R_{s,t}^{s-1,t-1}(p) \leq 4stp^{2^{s+t-1}} &- \left\{ \left[ 6\binom{s}{1}\binom{t}{1} + 8\binom{s}{1}\binom{t}{2} + 8\binom{s}{2}\binom{t}{1} \right] \cdot p^{2^{s+t}} \right. \\
&+ \left. \left[ 8\binom{s}{1}\binom{t}{2} + 8\binom{s}{2}\binom{t}{1} \right] \cdot p^{3 \cdot 2^{s+t-2}} + 32\binom{s}{2}\binom{t}{2} \cdot p^{7 \cdot 2^{s+t-3}} \right\} \\
&+ \left\{ \left[ 8\binom{s}{1}\binom{t}{2} + 8\binom{s}{2}\binom{t}{1} + 64\binom{s}{2}\binom{t}{2} \right] \cdot p^{2^{s+t}} + \left[ 48\binom{s}{1}\binom{t}{3} \right. \right. \\
&+ 48\binom{s}{3}\binom{t}{1} + 32\binom{s}{1}\binom{t}{2} + 32\binom{s}{2}\binom{t}{1} + 96\binom{s}{2}\binom{t}{2} \\
&+ \left. 192\binom{s}{2}\binom{t}{3} + 192\binom{s}{3}\binom{t}{2} \right] \cdot p^{5 \cdot 2^{s+t-2}} + \left[ 4\binom{s}{1}\binom{t}{1} + 8\binom{s}{1}\binom{t}{2} \right. \\
&+ \left. 8\binom{s}{2}\binom{t}{1} \right] \cdot p^{3 \cdot 2^{s+t-1}} + \left[ 16\binom{s}{1}\binom{t}{3} + 16\binom{s}{3}\binom{t}{1} \right] \cdot p^{7 \cdot 2^{s+t-3}} \\
&+ 128\binom{s}{2}\binom{t}{2} \cdot p^{9 \cdot 2^{s+t-3}} + 64\binom{s}{2}\binom{t}{2} \cdot p^{11 \cdot 2^{s+t-3}} \\
&+ \left[ 192\binom{s}{2}\binom{t}{3} + 192\binom{s}{3}\binom{t}{2} \right] \cdot p^{17 \cdot 2^{s+t-4}} \\
&+ \left. 384\binom{s}{3}\binom{t}{3} \cdot p^{37 \cdot 2^{s+t-5}} \right\}.
\end{aligned}$$

## 5 Comparative Analysis and Discussion

In this section, we compare the value $R_{s,t}^{s-1,t-1}(p)$ of the approximation in Sect. 3 and upper bound in Sect. 4 by drawing, respectively. Wu and Latifi [15] investigated the reliability $p$ of node as a function of time. And $f$ increases with time passes, where $f$ is the number of faulty nodes, with a constant failure rate $a$:

$$f = 2^{s+t+1}(1 - e^{-at}),$$

where $2^{s+t+1}$ is the order of $EH(s,t)$.

Then, we have

$$p = 1 - \frac{f}{2^{s+t+1}} = e^{-at}.$$

It is not difficult to observe that when the system do not operate ($t = 0$), each node is fault-free ($p = 1$). However, as the running time passes, each node is fault-free probability will decrease.

Figure 2 depicts the changes about reliability when $s$, $t$, and $a$ are different. It is easy to see that when $p$ is higher, the approximate value of reliability is obviously lower than the upper bound value. But when $p$ is small or even tends to 0, the approximate value of reliability and the upper bound value almost coincide, that is, the upper bound of reliability and the approximate value of reliability are close to the true reliability of the system.

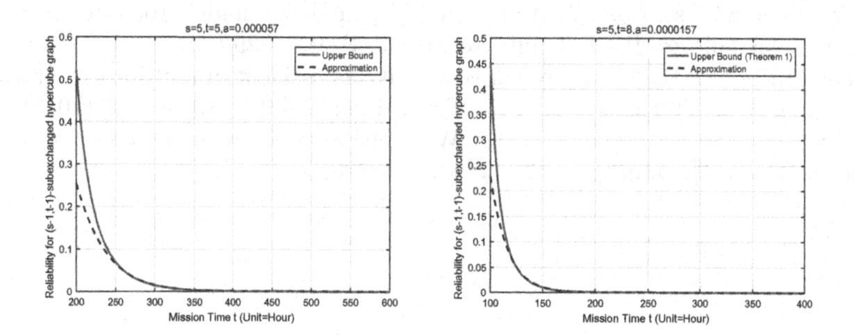


**Fig. 2.** The reliability of subgraph $EH(s-1,t-1)$.

## References


1. Eryilmaz, S.: Joint reliability importance in linear $m$-consecutive-$k$-out-of-$n$: $F$ systems. IEEE Trans. Reliab. **62**(4), 862–869 (2013)
2. Fan, W., Fan, J., Lin, C.-K., Wang, G., Cheng, B., Wang, R.: An efficient algorithm for embedding exchanged hypercubes into grids. J. Supercomput. **75**(2), 783–807 (2019). https://doi.org/10.1007/s11227-018-2612-2
3. Hebert, P.-R.: Sixty years of network reliability. Math. Comput. Sci. **12**(3), 275–293 (2018)

4. Hsu, L.-H., Lin, C.-K.: Graph Theory and Interconnection Networks. CRC Press, Boca Raton, FL (2009)
5. Huang, Y., Lin, L., Wang, D.: On the reliability of alternating group graph-based networks. Theor. Comput. Sci. **728**, 9–28 (2018)
6. Kung, T.-L., Hung, Y.-N.: Estimating the subsystem reliability of bubblesort networks. Theor. Comput. Sci. **670**, 45–55 (2017)
7. Kuo, S.-Y., Huang, C.-Y., Lyu, M.-R.: Framework for modeling software reliability, using various testing-efforts and fault-detection rates. IEEE Trans. Reliab. **50**(3), 310–320 (2001)
8. Kung, T.-L., Teng, Y.-H., Lin, C.-K., Hsu, Y.-L.: Combinatorial analysis of the subsystem reliability of the split-star network. Inf. Sci. **415–416**, 28–40 (2017)
9. Loh, P.K.K., Hsu, W.-J., Pan, Y.: The exchanged hypercube. IEEE Trans. Parallel Distrib. Syst. **16**(9), 866–874 (2005)
10. Li, X.-J., Xu, J.-M.: Generalized measures of fault tolerance in exchanged hypercubes. Inf. Process. Lett. **113**, 533–537 (2013)
11. Lin, L., Xu, L., Zhou, S., Wang, D.: The reliability of subgraph in the arrangement graph. IEEE Trans. Reliab. **62**, 807–818 (2015)
12. Li, X., Zhou, S., Xu, X., Lin, L., Wang, D.: The reliability analysis based on subsystems of $(n, k)$-Star Graph. IEEE Trans. Reliab. **65**, 1700–1709 (2016)
13. Ren, Y., Wang, S.: A short note on strong local diagnosability property of exchanged hypercubes under the comparison model. Int. J. Parallel, Emerg. Distrib. Syst. **35**(1), 9–15 (2020)
14. Soh, S., Rai, S., Trahan, J.L.: Improved lower bounds on the reliability of hypercube architectures. IEEE Trans. Parallel Distrib. Syst. **5**(4), 364–378 (1994)
15. Wu, X., Latifi, S.: Substar reliability analysis in star networks. Inf. Sci. **178**, 2337–2348 (2008)
16. Zarezadeh, S., Asadi, M.: Network reliability modeling under stochastic process of component failures. IEEE Trans. Reliab. **62**(4), 917–929 (2013)
17. Zhou, S., Li, X., Li, J., Wang, D.: Reliability assessment of multiprocessor system based on $(n, k)$-star network. IEEE Trans. Reliab. **66**(4), 1025–1035 (2017)
18. Zhang, Q., Xu, L., Zhou, S., Yang, W.: Reliability analysis of subsystem in dual cubes. Theor. Comput. Sci. **816**, 249–259 (2020)

# Fault Diagnosability of Regular Networks Under the Hybrid PMC Model

Jiafei Liu[1,2], Qianru Zhou[2], Zhengqin Yu[2], and Shuming Zhou[1,2(✉)]

[1] College of Computer and Cyber Security, Fujian Normal University, Fuzhou 350117, Fujian, People's Republic of China
zhoushuming@fjnu.edu.cn

[2] School of Mathematics and Statistics, Fujian Normal University, Fuzhou 350117, Fujian, People's Republic of China

**Abstract.** Large scale multiprocessor systems or multicomputer systems, taking interconnection networks as underlying topologies, have been widely used in the big data era. System level diagnosis is a primary strategy to identify the faulty processors in multiprocessor systems. To enhance the robustness of networks against processors and links fail simultaneously, Zhu et al. [21] proposed a novel fault diagnostic model, the hybrid PMC diagnostic model, which involves the failing of vertices and edges. In this paper, we determine the diagnosability of the triangle-free regular networks under the hybrid PMC model. As by-products, we apply the general results to the state-of-the-art regular networks, such as hypercube-like network as well as hypercube-based compound network, for example, DQcube, exchanged hypercube, dual cube, half-hypercube, hierarchical cubic network and so on.

**Keywords:** Multiprocessor systems · Fault tolerance · Diagnosability · Hybrid PMC model

## 1 Introduction

Multiprocessor system commonly use interconnection networks (or graphs) as the underlying topological model, in which the processors and communication links are portrayed as vertices and edges, respectively. Networks may fail due to different ways of attacks and different mechanisms of failure. The first type is physical attack via removal of some vertices or edges. It has been shown that in multiprocessor systems attacked by the malicious attackers, the overall network connectivity is measured by the sizes of the giant connected components and the diameters does not change significantly in response to random removal of a small fraction of vertices. The second type is the cascading failure of attacks,

Supported by the National Natural Science Foundation of China (Nos. 61977016 and 61572010), Natural Science Foundation of Fujian Province (Nos. 2020J01164, 2017J01738).


C.-Y. Chen et al. (Eds.): COCOON 2021, LNCS 13025, pp. 283–297, 2021.
https://doi.org/10.1007/978-3-030-89543-3_24

which naturally appears in rumour spreading, disease spreading, voting, and advertising. It has been shown that in social networks generated by the PA model even a weakly virulent virus can spread. This result explains a fundamental characteristic of the security of networks.

Fault tolerance is becoming an essential attribute in multiprocessor systems as the number of processors is getting larger. In order to ensure the stable running of the systems, we must find out the faulty processors and repair or replace them accurately. System-level diagnosis, as a powerful tool, has been widely deployed. The field of system-level diagnosis has evolved from the pioneering work of Preparata, Metze, and Chien [16], who established the first diagnostic model, namely, the PMC model. The PMC model assumes that each processor can test its neighbors, and the test results are either faulty or fault-free, which has been extensively applied in kinds of multiprocessor systems. The diagnosability of a system, is the maximum number of faulty nodes that the system is guaranteed to identify.

As a fundamental issue in the robustness analysis of multiprocessor systems, the topic on diagnosability has attracted great attention over the few decades, which has been regarded as a crucial technique to enhance the invulnerability of large scale networks. A great amount of effort has been devoted to the development to investigate diagnosability of networks under the PMC model. For example, Chang [1] investigated the diagnosabilities of regular networks. Chang and Hsieh [2,3] explored the conditional diagnosability of $(n, k)$-star graphs and alternating group networks under the PMC model. Guo et al. [6] studied the $g$-good neighbor conditional diagnosability of crossed cubes under the PMC model and MM* model. Liu et al. [12] studied the $g$-good neighbor conditional diagnosability of twisted hypercubes under the PMC model and MM* model. Lin et al. [11] explored the relationship between the $h$-extra vertex-connectivity and $h$-extra conditional diagnosability for regular networks under the PMC diagnostic model. Later, Cheng, Qiu and Shen [4,5] outlined the research development of diagnosability of interconnection networks. Zhu et al. [22] resolved hybrid fault identification capability of hypercubes under the PMC model and MM* model. Furthermore, studies on $h$-edge tolerable diagnosability for some interconnected networks under the PMC model are explored by many scholars [10,17,18,24].

However, the classic diagnostic model, the PMC model, assumes that only the processor can occur faulty. In practice, processor and communication link can be attacked by the adversary simultaneously in real system. To address this deficiency, Zhu et al. [21] introduced the Hybrid PMC model, marked as the HPMC model, accompanied by the emergence of vertex and link failure. Furthermore, they also determined the fault diagnosability of hypercube under the HPMC model. Later, Zhang et al. [23] established the hybrid diagnosability of exchanged hypercube under the HPMC model. Motivated by the idea above, we addressed fault diagnosability of a class of triangle-free regular networks under the HPMC model. As an empirical study, we apply the newly obtained results to a class of the state-of-the-art regular networks, including hypercube-like net-

**Table 1.** An illustration of some notions

| Notions | Interpretation and Significations |
|---|---|
| $N_G(u)$ | The open neighborhood set of $u$ in $G$ |
| $N_G[u]$ | The close neighborhood set $N_G(u) \cup \{u\}$ |
| $NE_G(u)$ | The edge set of all its incident edge in $G$ |
| $(u, v)$ | An edge between vertices $u$ and $v$ |
| $N_G(S)$ | $(\bigcup_{u \in S} N_G(u)) \setminus S$ |
| $G[S]$ | The subgraph induced by $S$ in $G$ |
| $G - F$ | The graph obtained by removing the vertex of $F$ and their incident edges from $G$ |
| $\langle n \rangle$ | $\{1, 2, \ldots, n\}$ |
| $\|S\|$ | The cardinality of $S$ |
| $F_1 \Delta F_2$ | $(F_1 \cup F_2) \setminus (F_2 \cap F_1)$ |

work as well as hypercube-based compound network, such as DQcube, exchanged hypercube, dual cube, half-hypercube, hierarchical cubic network, etc.

The rest of this paper is organized as follows. Section 2 introduces some preliminaries and some related definition and some key lemmas. Our main results as well as the detailed proofs are presented in Sect. 3. Empirical results are reported and analyzed in Sect. 4, where we apply the obtained results to a class of regular networks to directly determine their fault diagnosability under the HPMC model. Finally, we conclude the paper with a summary and future work in Sect. 5.

## 2 Preliminaries

Throughout this paper, a graph $G = (V(G), E(G))$ represents an interconnection network, where each vertex $u \in V(G)$ denotes a processor and each edge $(u, v) \in E(G)$ denotes a communication link between two processors $u$ and $v$. Table 1 demonstrates some basic notations in this context. One can review [19,20] to learn more concepts and notations of graph and network theory.

In fault diagnosis, the PMC model was initially introduced by Preparata, Metze, and Chien [16], the well-known diagnostic model, which is based on graph modelling and has been widely applied. Along with the model above, considering comprehensively the failure of processor and link within a network, the hybrid PMC model has been proposed by Zhu et al. [21] recently. A test of the HPMC model, denoted by $\delta(u, v; e)$, involves three elements, including the tester $u$, the testee $v$, the test link $e$. Figure 1 demonstrates distinct test results under the HPMC model (where X represents that the testee might be faulty or fault-free). It is widely recognized that, if the tester is fault-free, the test result is reliable; while the tester is faulty, the test result is unreliable. Furthermore, when a faulty processor is replaced or removed, its incident links have to be rechecked or removed. In the HPMC model, the vertices incident with any faulty link are fault-free, and the links incident with any faulty vertex are fault-free [21].

| tester ($u$) | testee ($v$) | test link ($e$) | $\delta(u, v; e)$ |
|---|---|---|---|
| Fault-free | Fault-free | Fault-free | 0 |
| Fault-free | Faulty | Fault-free | 1 |
| Fault-free | Fault-free | Faulty | 1 |
| Faulty | X | Fault-free | 0/1 |

**Fig. 1.** Illustration of test results in the HPMC model.

**Definition 1** [21]. For a multiprocessor system $G$, $(F, S)$ is a consistent faulty pair of $G$ if all vertices in $F$ cannot be incident to any edge in $S$.

**Definition 2** [21]. For a multiprocessor system $G$,

(1) given two positive integers $t, s$, $G$ is $(t, s)$-diagnosable if and only if for any two distinct faulty pairs $(F_1, S_1)$ and $(F_2, S_2)$ of $V(G)$ such that $|F_1|, |F_2| \leq t$ and $|S_1|, |S_2| \leq s$, $(F_1, S_1)$ and $(F_2, S_2)$ are distinguishable;
(2) For any two positive integers $h$ and $r$, the $h$-restricted vertex diagnosability of $G$, denoted by $t_h^e(G)$, is the maximum value of $t$ such that $G$ is $(t, h)$-diagnosable; the $r$-restricted edge diagnosability of $G$, denoted by $s_r^v(G)$, is the maximum value of $s$ such that $G$ is $(r, s)$-diagnosable.

Intuitively, when $h = 0$, the $h$-restricted vertex diagnosability of $G$ equals to its vertex diagnosability, i.e., $t_h^e(G) = t(G)$; when $r = 0$, the $r$-restricted edge diagnosability of $G$ equals to its edge diagnosability.

**Lemma 1** [21]. Let $G = (V(G), E(G))$ be a multiprocessor system, $F_1, F_2 \subseteq V(G)$ and $S_1, S_2 \subseteq E(G)$. Then, for any two distinct faulty pairs $(F_1, S_1)$ and $(F_2, S_2)$, they are distinguishable under the HPMC model if and only if one of the following conditions holds:

(1) there exists a vertex $u \in V(G) \setminus (F_1 \cup F_2)$, which is adjacent to a vertex $v \in F_1 \Delta F_2$ and $e = (u, v) \notin S_2$;
(2) there exists a vertex $u \in V(G) \setminus (F_1 \cup F_2)$, which is adjacent to a vertex $v \in F_2 \Delta F_1$ and $e = (u, v) \notin S_1$;
(3) there exists an edge $e = (u, v) \in S_1 \setminus S_2$ and $u, v \notin F_2$;
(4) there exists an edge $e = (u, v) \in S_2 \setminus S_1$ and $u, v \notin F_2$.

**Lemma 2** [21]. Let $G$ be a multiprocessor system. For any two positive integers $h, r$, if $t_h^e(G) \geq r$, then $s_r^v(G) \geq h$.

**Lemma 3** [23]. For the multiprocessor system $G = (V(G), E(G))$ and two distinct faulty pairs $(F_1, S_1)$ and $(F_2, S_2)$, if $F_1, F_2 \subseteq V(G)$ and $S_1, S_2 \subseteq E(G)$, $(F_1, S_1)$ and $(F_2, S_2)$ are distinguishable under the HPMC model, then $F_1 \Delta F_2 \neq \emptyset$.

# 3 Fault Diagnosability of Regular Networks Under the HPMC Model

In general, fault diagnosis has been viewed as an essential strategy for the robustness analysis of multiprocessor systems. In this section, a characterization of the vertex and edge restricted diagnosability of a class of regular networks under the HPMC model is presented.

## 3.1 *h*-Restricted Vertex Diagnosability of Networks

First, we establish the $h$-restricted vertex diagnosability of regular networks under the HPMC model.

**Theorem 1.** Given a $k$-regular connected network $G$ and an integer $h$ with $1 \leq h \leq k-2$. Let $t_h^e(G)$ be the $h$-restricted vertex diagnosability of $G$. If $G$ satisfies the following conditions:

**(1)** $|V(G)| \geq 2k - h$ for $t \geq 3$;
**(2)** $G$ is triangle-free,

then, the $h$-restricted vertex diagnosability of $G$ under the HPMC model is $t_h^e(G) = k - h$.

**Proof.** We first show that the upper bound of $h$-restricted vertex diagnosability of $G$ under the HPMC model is $t_h^e(G) \leq k - h$.

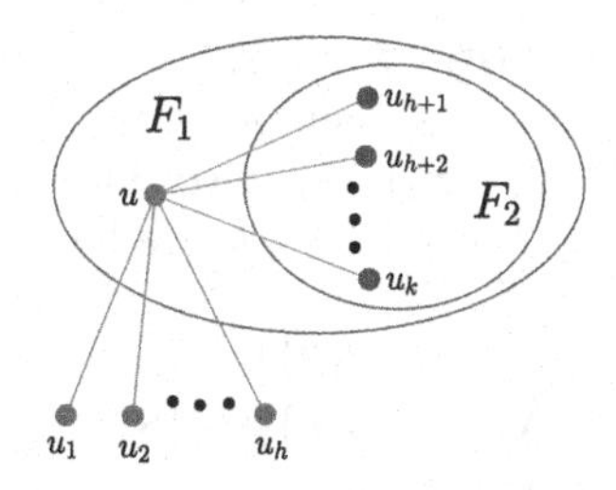


**Fig. 2.** Illustration of the upper bound of $t_h^e(G)$.

For any vertex $u$ in $V(G)$, we select a subgraph $A$ induced by the vertex set $V(A) = \{u, u_1, u_2, \ldots, u_k\}$ such that $A$ is isomorphic to $K_{1,k}$ with $(u, u_i) \in E(G)$ (see Fig. 2). Furthermore, we set $F_1 = \{u, u_{h+1}, u_{h+2}, \ldots, u_k\}$ and $F_2 = \{u_{h+1}, u_{h+2}, \ldots, u_k\}$. It follows that $|F_1| \leq k - h + 1$ and $|F_2| \leq k - h$. Take $S_1 = \emptyset$ and $S_2 = \{(u, u_1), (u, u_2), \ldots, (u, u_h)\}$. Clearly, $|S_2| \leq h$. In terms of Lemma 2, $(F_1, S_1)$ and $(F_2, S_2)$ are indistinguishable under the HPMC model. Therefore, $t_h^e(G) \leq k - h$ for $1 \leq h \leq k - 2$.

Next, it suffices to show that $t_h^e(G) \geq k - h$ under the HPMC model. To this aim, we shall adopt a method of contradiction and assume that $t_h^e(G) \leq k - h - 1$

for $1 \leq h \leq k-2$. Let $(F_1, S_1)$ and $(F_2, S_2)$ be two distinct indistinguishable faulty pairs of $G$ with $|F_1| \leq k-h$, $|F_2| \leq k-h$ as well as $|S_1| \leq h$, $|S_2| \leq h$. Because $(F_1, S_1)$ and $(F_2, S_2)$ are different, we need to distinguish two cases, i.e., $F_1 \Delta F_2 \neq \emptyset$ or $S_1 \Delta S_2 \neq \emptyset$.

**Case 1:** $F_1 \Delta F_2 \neq \emptyset$.

In this situation, by the symmetry of $F_1$ and $F_2$, three possibilities need to be explored.

**Subcase 1.1:** $|F_1 \setminus F_2| = 1$ and $|F_2 \setminus F_1| = 0$.

Since $|F_1 \setminus F_2| = 1$ and $|F_2 \setminus F_1| = 0$, we have $F_2 \subset F_1$. Denote $F_1 \setminus F_2 = \{u\}$. As $(F_1, S_1)$ and $(F_2, S_2)$ are indistinguishable under the HPMC model, $NE_{V(G)\setminus(F_1 \cup F_2)}(u) \subset S_2$. In terms of $|S_2| \leq h$, $u$ has at least $k-h$ neighbors in $F_2$. Thus, $|F_2| \geq k-h$. Due to the assumption of $|F_2| \leq k-h$, we have $|F_2| = k-h$. As a result, $|F_1| = |F_1 \setminus F_2| + |F_2| = k-h+1$ which contradicts with $|F_1| \leq k-h$. Therefore, $t_h^e(G) \geq k-h$.

**Subcase 1.2:** $|F_1 \setminus F_2| = 1$ and $|F_2 \setminus F_1| = 1$.

In this situation, we set $F_1 \setminus F_2 = \{u\}$ and $F_2 \setminus F_1 = \{v\}$. As $(F_1, S_1)$ and $(F_2, S_2)$ are indistinguishable, $NE_{V(G)\setminus(F_1 \cup F_2)}(u) \subset S_2$. In view of $|S_2| \leq h$, $u$ has at least $k-h$ neighbors in $F_2$. Thus, $|F_2| \geq k-h$. Due to the assumption of $|F_2| \leq k-h$, we have $|F_2| = k-h$. Analogously, $|F_1| = k-h$.

If $u$ is not adjacent to $v$, then $N_{F_1 \cup F_2}(u) \subset F_1 \cap F_2$ and $|F_1 \cap F_2| \geq k-h$. Therefore, $|F_1| \geq k-h+1$, a contradiction. If $u$ is adjacent to $v$, then there are $k-h-1$ common neighbors for $u$ and $v$. Because of $1 \leq h \leq k-2$, we have $k-h-1 \geq 1$. And so there exists at least one triangle in $G$ which yields a contradiction with the fact that $G$ is triangle-free. Therefore, $t_h^e(G) \geq k-h$.

**Subcase 1.3:** $|F_1 \setminus F_2| \geq 2$ or $|F_2 \setminus F_1| \geq 2$.

Without loss of generality, we suppose $|F_1 \setminus F_2| \geq 2$, and $u, v \in F_1 \setminus F_2$. As $(F_1, S_1)$ and $(F_2, S_2)$ are two indistinguishable faulty pairs, $NE_{V(G)\setminus(F_1 \cup F_2)}(\{u, v\}) \subset S_2$. Since $|S_2| \leq h$, we have

$$\begin{aligned}
|N_{F_1 \cup F_2}(\{u, v\})| &= |N_{V(G)}(\{u, v\})| - |N_{V(G)\setminus(F_1 \cup F_2)}(\{u, v\})| \\
&\geq |N_{V(G)}(\{u, v\})| - |S_2| \\
&\geq 2k-2-h.
\end{aligned}$$

Thus,

$$\begin{aligned}
|F_1 \cup F_2| &= |F_1 \setminus F_2| + |F_1 \cap F_2| + |F_2 \setminus F_1| \\
&\geq |F_1 \setminus F_2| + |N_{F_1 \cup F_2}(\{u, v\})| \\
&\geq 2 + (2k-2-h) \\
&\geq 2(k-h) \\
&\geq |F_1| + |F_2|,
\end{aligned}$$

which is a contradiction. Therefore, $t_h^e(G) \geq k-h$.

**Case 2:** $S_1 \Delta S_2 \neq \emptyset$.

Without loss of generality, we suppose $S_1 \setminus S_2 \neq \emptyset$, and $(u, v) \in S_1 \setminus S_2$.

We first consider the situation subject to $u \in \overline{F_1 \cup F_2}$, $v \in \overline{F_1 \cup F_2}$. When $(F_1, S_1)$ is faulty and $(F_2, S_2)$ is fault-free, $\delta(u, v; e) = 1$; while $(F_1, S_1)$ is fault-free and $(F_2, S_2)$ is faulty, $\delta(u, v; e) = 0$. Thus, $(F_1, S_1)$ and $(F_2, S_2)$ are distinguishable, a contradiction.

When both of $u$ and $v$ are in $F_1$, clearly, both of $u$ and $v$ are not in $F_2$. In terms of $(u, v) \in S_1 \setminus S_2$, by condition (3) of Lemma 1, $(F_1, S_1)$ and $(F_2, S_2)$ are distinguishable, a contradiction.

Hence, at least one of $u$ and $v$ in $F_2 \setminus F_1$, i.e., $F_1 \Delta F_2 \neq \emptyset$. However, by the proof of Case 1 in Theorem 1, when $F_1 \Delta F_2 \neq \emptyset$, it implies a contradiction. Therefore, $t_h^e(G) \geq k - h$, the desired.

From what has been analysed above, the $h$-restricted vertex diagnosability of $G$ under the HPMC model is $t_h^e(G) = k - h$. □

### 3.2 $r$-Restricted Edge Diagnosability of Networks

Here, the $r$-restricted edge diagnosability of $G$ under the HPMC model is characterized in the following theorem.

**Theorem 2.** Given a $k$-regular connected network $G$ and an integer $r$ with $1 \leq r \leq k - 1$. Let $s_r^v(G)$ be the $r$-restricted edge diagnosability of $G$. If $G$ satisfies the following conditions:

**(1)** $|V(G)| \geq 2k - r$ for $k \geq 3$;
**(2)** $G$ is triangle-free,

then, the $r$-restricted edge diagnosability of $G$ under the HPMC model is

$$s_r^v(G) = \begin{cases} k-2, \ r = 1 \text{ and } k \geq 3; \\ k - r, \ \ 2 \leq r \leq k-1. \end{cases}$$

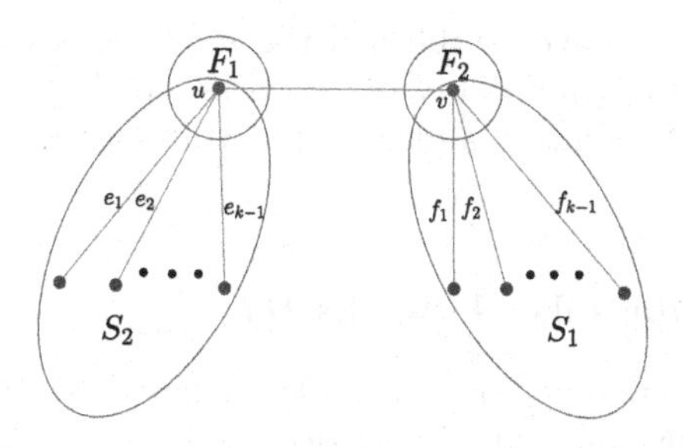


**Fig. 3.** Illustration of the upper bound of $s_1^v(G)$.

**Proof.** We first show that, under the HPMC model, $s_r^v(G) = k - 2$ when $r = 1$ and $k \geq 3$. For any edge $e = (u, v)$, we set $NE(u) = \{e_1, e_2, \ldots, e_{k-1}, (u, v)\}$ and $NE(v) = \{f_1, f_2, \cdots, f_{k-1}, (u, v)\}$. We set $F_1 = \{u\}$ and $F_2 = \{v\}$, $S_1 =$

$\{f_1, f_2, \ldots, f_{k-1}\}$ and $S_2 = \{e_1, e_2, \ldots, e_{k-1}\}$ (see Fig. 3). By Lemma 1, $(F_1, S_1)$ and $(F_2, S_2)$ are indistinguishable under the HPMC model. Therefore, $s_1^v(G) \leq k-2$ for $k \geq 3$.

Next, it suffices to show that $s_1^v(G) \geq k-2$ by contradiction. Suppose, to the contrary, that $s_1^v(G) < k-2$. Let $(F_1, S_1)$ and $(F_2, S_2)$ be two distinct indistinguishable faulty pairs of $G$ with $|F_1| \leq 1$, $|F_2| \leq 1$ as well as $|S_1| \leq k-2$, $|S_2| \leq k-2$. As $(F_1, S_1)$ and $(F_2, S_2)$ are different, we have $F_1 \Delta F_2 \neq \emptyset$ or $S_1 \Delta S_2 \neq \emptyset$. If $S_1 \Delta S_2 \neq \emptyset$, we assume $S_1 \setminus S_2 \neq \emptyset$ and denote $S_1 \setminus S_2 = \{(u, v)\}$. Because $(F_1, S_1)$ and $(F_2, S_2)$ are indistinguishable, by Lemma 1, at least one of $u$ and $v$ is in $F_2$. Thus, $F_1 \Delta F_2 \neq \emptyset$. Without loss of generality, set $F_1 = \{u\}$. By Lemma 1, all edges of $NE_{V(G)\setminus F_2}(u)$ are located in $S_2$. Since $G$ is $k$-regular and $|F_2| \leq 1$, we have $|S_2| \geq k-1$, which contradicts with $|S_2| \leq k-2$.

Furthermore, we prove that $s_r^v(G) = k - r$ for $2 \leq r \leq k-1$. Let $u$ be any vertex of $V(G)$ and $N_G(u) = \{u_1, u_2, \ldots, u_k\}$. Take $F_1 = \{u_1, u_2, \ldots, u_{r-1}\}$, $F_2 = \{u, u_1, u_2, \ldots, u_{r-1}\}$, and $S_1 = \{(u, u_r), (u, u_{r+1}), \ldots, (u, u_k)\}$, $S_2 = \emptyset$. Clearly, $|S_1| \leq k - r + 1$. By means of Lemma 1, $(F_1, S_1)$ and $(F_2, S_2)$ are indistinguishable under the HPMC model. Therefore, $s_r^v(G) \leq k - r$ for $2 \leq r \leq k-1$.

Now, it suffices to show the lower bound of $s_r^v(G)$ for $2 \leq r \leq k-1$. Set $h = k - r$. In terms of $2 \leq r \leq k-1$, we obtain $1 \leq h \leq k-2$. By Theorem 1, $t_h^e(G) \geq k - h$. Applying Lemma 2, we have $s_r^v(G) \geq h = k - r$.

Summing up above, we have

$$s_r^v(G) = \begin{cases} k-2, r = 1 \text{ and } k \geq 3; \\ k-r, \quad 2 \leq r \leq k-1. \end{cases}$$

as desired. □

## 4 Applications to Regular Networks

In previous section, we have established the $h$-restricted vertex diagnosability and $r$-restricted edge diagnosability of networks. In this section, we will apply the key results of Theorems 1 and 2 to the state-of-the-art regular networks without triangle.

### 4.1 The Hypercube-Like Networks $HL_n$

The $n$-dimensional hypercube-like networks, denoted by $HL_n$, is a class of $n$-regular $n$-connected graphs with $2^n$ vertices and $n2^{n-1}$ edges that are defined recursively as follows.

**Definition 3.** Given a positive integer $n(n \geq 1)$, the one-dimensional hypercube-like networks is $K_2$. A graph $G$ is called the $n$-dimensional hypercube-like networks, denoted by $HL_n$, if there exist two disjoint subsets of $V(HL_n)$, $V_0$ and $V_1$, subject to the following two conditions:

(1) $V(HL_n) = V_0 \cup V_1$, where $G[V_0], G[V_1] \in HL_{n-1}$;
(2) $E(HL_n) = E(G[V_0)] \cup E(G[V_1]) \cup M$, where $M$ is a perfect matching between $G[V_0]$ and $G[V_1]$.

Note that the $HL_n$ network is triangle-free. Every $HL_n$, by fixing the last dimensional ordinate of nodes, can be divided into two subgraphs $HL_n^0$ and $HL_n^1$, each of which is isomorphic to $HL_{n-1}$. Additionally, there exists a perfect matching between $HL_n^0$ and $HL_n^1$. Based on the properties above, we establish the $h$-restricted vertex diagnosability and $r$-restricted edge diagnosability of $HL_n$.

**Theorem 3.** For the hypercube-like network $HL_n$ with $n \geq 3$,

(1) the $h$-restricted vertex diagnosability of $HL_n$ is $t_h^e(HL_n) = n - h$ for $1 \leq h \leq n-2$;
(2) the $r$-restricted edge diagnosability of $HL_n$ under the HPMC model is

$$s_r^v(HL_n) = \begin{cases} n-2, r=1 \text{ and } n \geq 3; \\ n-r, \ \ 2 \leq r \leq n-1. \end{cases}$$

Zhu et al. [21] have determined the $h$-restricted vertex diagnosability and $r$-restricted edge diagnosability of $Q_n$ under the HPMC model, which is easily derived by our technical route.

**Corollary 1** [21]. For the hypercube $Q_n$ with $n \geq 3$,

(1) the $h$-restricted vertex diagnosability of $Q_n$ is $t_h^e(Q_n) = n - h$ for $1 \leq h \leq n-2$;
(2) the $r$-restricted edge diagnosability of $Q_n$ is

$$s_r^v(Q_n) = \begin{cases} n-2, r=1 \text{ and } n \geq 3; \\ n-r, \ \ 2 \leq r \leq n-1. \end{cases}$$

## 4.2 The DQcube $DQ(m, d, n)$

DQcube, introduced by Hung [7], is a new compound network based on hypercube network. Sort all vertices of hypercube cluster in the ascending order of the vertex labels. Then the smallest, second smallest, and the largest nodes will be $\alpha_1 = 00\cdots00$, $\alpha_2 = 00\cdots01$, and $\beta = 11\cdots11$, respectively. Let $\gamma_i$ be the $i$-th smallest node excluding any one of $\{\alpha_1, \alpha_2, \beta\}$ for $1 \leq i \leq 2^n - 3$. According to the construction methodology of $DQ(m, d, n)$ proposed by Hung [7], we review the definition of DQcube network, and some available properties of it as follows.

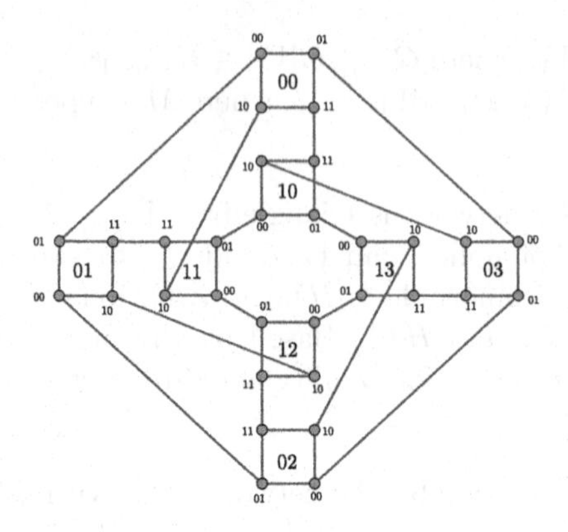


**Fig. 4.** Illustration of $DQ(4,2,2)$.

**Definition 4** [14]. The DQcube is characterized by $DQ(m,d,n)$ where $1 \leq d \leq m$ and $d+2=2^n$. The vertex-set $V$ is represented as $\{(z_1z_2, b_{n-1}b_{n-2}\cdots b_0)\}$ where $z_1z_2$ is the label of cluster in $D(m,d)$ and $b_{n-1}b_{n-2}\cdots b_0$ is the label of the node in $Q_n$. Two vertices $u=(z_1z_2, b_{n-1}b_{n-2}\cdots b_0)$ and $v=(z_1'z_2', b_{n-1}'b_{n-2}'\cdots b_0')$ are linked if and only if one of the following conditions is satisfied (see $DQ(4,2,2)$ in Fig. 4):

(1) $z_1z_2 = z_1'z_2'$ and $\sum_{i=0}^{n-1} \mid b_i - b_i' \mid = 1$;
(2) $z_1 = z_1'$, $z_2' = z_2 + 1$, $b_{n-1}b_{n-2}\cdots b_0 = \alpha_1$ and $b_{n-1}'b_{n-2}'\cdots b_0' = \alpha_2$;
(3) $z_1 - z_1' = 1$, $z_2 = z_2'$ and $b_{n-1}b_{n-2}\cdots b_0 = b_{n-1}'b_{n-2}'\cdots b_0' = \beta$;
(4) $z_1 = 0$, $z_1' = 1$, $z_2' = z_2 + i$ and $b_{n-1}b_{n-2}\cdots b_0 = b_{n-1}'b_{n-2}'\cdots b_0' = \gamma_i$.

**Lemma 4** [14]. $DQ(m,d,n)$ has the following properties:

(1) $2^n = d+2$;
(2) $DQ(m,d,n)$ is $(n+1)$-regular, and $DQ(m,d,n)$ has $m2^{n+1}$ vertices and $(n+1)m2^n$ edges;
(3) $DQ(m,d,n)$ is triangle-free and contains no 5-cycle.

**Theorem 4.** For the DQcube network $DQ(m,d,n)$ with $1 \leq d \leq m$ and $d+2 = 2^n$,

(1) the $h$-restricted vertex diagnosability of $DQ(m,d,n)$ is $t_h^e(DQ(m,d,n)) = n+1-h$ for $1 \leq h \leq n-1$ and $n \geq 3$;
(2) the $r$-restricted edge diagnosability of $DQ(m,d,n)$ is

$$s_r^v(DQ(m,d,n)) = \begin{cases} n-1, & r=1 \text{ and } n \geq 3; \\ n+1-r, & 2 \leq r \leq n-1. \end{cases}$$

**Proof.** (1) Since $|V(DQ(m,d,n))| = m2^{n+1}$ as well as $1 \leq h \leq n-1$, we have

$$\begin{aligned}|V(DQ(m,d,n))| - (2(n+1)-h) &\geq m2^{n+1} - 2n - 2 + h\\ &\geq m2^{n+1} - 2n - 2 + n - 1\\ &\geq m2^{n+1} - n - 3\\ &> 0,\end{aligned}$$

for $n \geq 3$. Moreover, By Lemma 4, $DQ(m,d,n)$ has no triangle. Applying Theorem 1, $t_h^e(DQ(m,d,n)) = n+1-h$, as required.

(2) Similar to the proof of Theorem 4(1), one can easily verify that these conditions that $DQ(m,d,n)$ is triangle-free and $|V(DQ(m,d,n))| = m2^{n+1}$ are in accord with conditions of Theorem 1. Thus, Theorem 4 holds. □

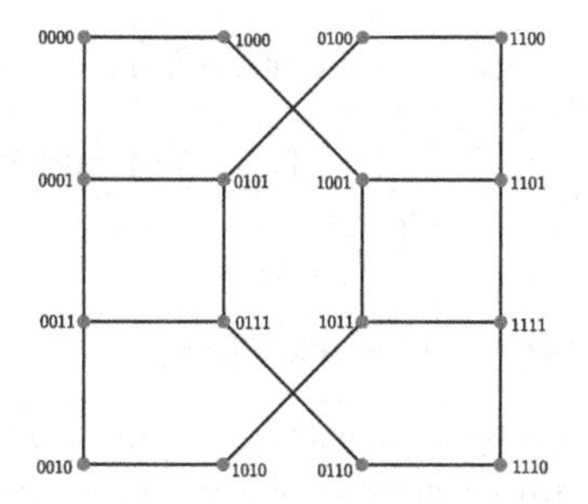


**Fig. 5.** Illustration of the exchanged hypercubes $EH(1,2)$.

### 4.3 The Exchanged Hypercube $EH(s,t)$

The exchange hypercube, was initially proposed by Loh et al. [15]. Next, we introduce the definition of the exchange hypercube.

**Definition 5** [9]**.** The $(s,t)$-dimensional exchanged cubes is defined as a graph $EH(s,t) = (V(EH(s,t)), E(EH(s,t)))$ for $s,t \geq 1$. $EH(s,t)$ consists of two disjoint subgraphs $L'$ and $R'$. And $L'$ contains $2^t$ subgraphs, denoted by $L'_i$, $i \in \langle 2^t \rangle$. Analogously, $R'$ contains $2^s$ subgraphs, denoted by $R'_j$, $j \in \langle 2^s \rangle$. Furthermore, $EH(s,t)$ satisfies the following conditions: (see Fig. 5)

(1) for any $i \in \langle 2^t \rangle$ and $j \in \langle 2^s \rangle$, $L'_i \cong Q_s$ and $R'_i \cong Q_t$;
(2) each vertex in $V(L'_i)$ has a unique neighbor in $V(R'_j)$ and vice versa. In addition, for distinct vertices in each $L'_i$, their neighbors locate in distinct $R'_j$s;
(3) For any two different subgraphs $L'_p$ and $L'_q$ with $p \neq q$, there exists no edge between them. Similar for $R'_j$ and $R'_k$ with $j \neq k$.

Zhang and Zhu [23] recently addressed the $h$-restricted vertex diagnosability and $r$-restricted edge diagnosability of $EH(s,t)$ under the HPMC model. The obtained results are presented as follows.

**Theorem 5** [23]. For the exchange hypercube $EH(s,t)$ with $1 \leq s \leq t$,

(1) the $h$-restricted vertex diagnosability of $EH(s,t)$ is $t_h^e(EH(s,t)) = s-h+1$ for $1 \leq h \leq s-1$;
(2) the $r$-restricted edge diagnosability of $EH(s,t)$ is

$$s_r^v(EH(s,t)) = \begin{cases} s-1, & r=1 \text{ and } s \geq 2; \\ s-r+1, & 2 \leq r \leq s. \end{cases}$$

Note that, when $s = t = n$, $EH(s,t)$ is isomorphic to the dual-cube $DC_n$. Hence, we have the following corollaries.

**Corollary 2.** For the dual-cube $DC_n$ with $n \geq 2$,

(1) the $h$-restricted vertex diagnosability of $DC_n$ is $t_h^e(DC_n) = n-h+1$ for $1 \leq h \leq n-1$;
(2) the $r$-restricted edge diagnosability of $DC_n$ is

$$s_r^v(DC_n) = \begin{cases} n-1, & r=1 \text{ and } n \geq 2; \\ n-r+1, & 2 \leq r \leq n. \end{cases}$$

### 4.4 The Half Hypercube Network $HH_n$

The $n$-dimensional half-hypercube network, denoted by $HH_n$, proposed by Kim et al. [8], which owns the same number of vertices as the hypercube but reduces the degree by approximately half, is constructed as follows.

**Definition 6** [8]. Every vertex of $HH_n$ is expressed by an $n$-bit binary strings, i.e., the vertex set $V(HH_n) = \{u_n u_{n-1} \cdots u_1 \mid u_i \in \{0,1\}, i = 1,2,\ldots,n\}$.

Suppose $u \in HH_n$, we denote $C(u)$ the leftmost $\lfloor \frac{n}{2} \rfloor$-bit binary string of $u$, and $P(u)$ the rest, i.e.,

$$C(u) = u_n \cdots u_{\lceil \frac{n}{2} \rceil + 1},\ P(u) = u_{\lceil \frac{n}{2} \rceil} \cdots u_2 u_1.$$

In fact

$$P(u) = \begin{cases} u_{\lceil \frac{n}{2} \rceil} \cdots u_2 u_1, & n \text{ is even}; \\ u_{\lceil \frac{n}{2} \rceil} \cdots u_2, & n \text{ is odd}. \end{cases}$$

Two vertices $u = u_n u_{n-1} \cdots u_1$ and $v = v_n v_{n-1} \cdots v_1$ in $HH_n$ have an edge if and only if $(u,v)$ satisfies one of the following two collections (Fig. 6):

$$E(HH_n) = \{(u,v) \mid C(u) = C(v), \sum_{i=1}^{\lceil \frac{n}{2} \rceil} |u_i - v_i| = 1\},$$

$$E\ \ (HH_n) = \{(u,v) \mid \text{if } C(u) \neq P(u), \text{ then } C(v) = P(u), P(v) = C(u);\ \text{ otherwise, } v = \overline{u}\}.$$

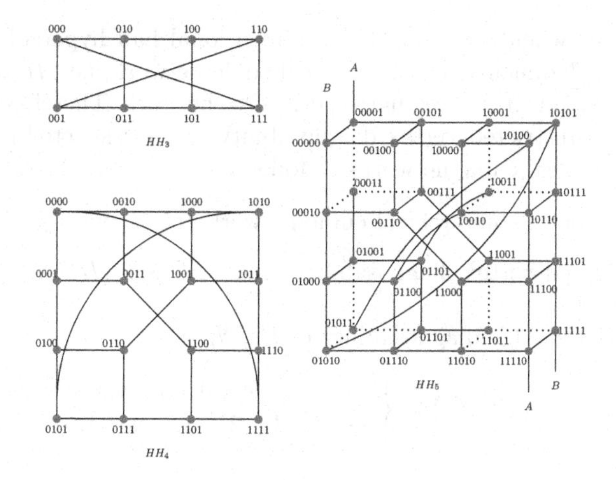


**Fig. 6.** Illustration of half hypercube $HH_n$ $(n = 3, 4, 5)$.

**Lemma 5** [13]. The half hypercube network $HH_n$ has the following properties:

(1) $HH_n$ is $(\lceil\frac{n}{2}\rceil + 1)$-regular, and has $2^n$ vertices and $2^{n-1}(\lceil\frac{n}{2}\rceil + 1)$ edges;
(2) $HH_n$ is triangle-free;
(3) Let $u$ and $v$ be any two vertices of $HH_n(n \geq 3)$. Then, $cn(u, v) \leq 2$.

**Theorem 6.** For the half hypercube $HH_n$ with $n \geq 3$,

(1) the $h$-restricted vertex diagnosability of $HH_n$ is $t_h^e(HH_n) = \lceil\frac{n}{2}\rceil - h + 1$ for $1 \leq h \leq \lceil\frac{n}{2}\rceil - 2$;
(2) the $r$-restricted edge diagnosability of $HH_n$ under the HPMC model is

$$s_r^v(HH_n) = \begin{cases} \lceil\frac{n}{2}\rceil - 1, & r = 1 \text{ and } n \geq 3; \\ \lceil\frac{n}{2}\rceil - r + 1, & 2 \leq r \leq \lceil\frac{n}{2}\rceil - 1. \end{cases}$$

**Proof.** (1) In terms of $|V(HH_n)| = 2^n$ and $1 \leq h \leq \lceil\frac{n}{2}\rceil - 2$, we have

$$\begin{aligned} |V(HH_n)| - (2(\lceil\frac{n}{2}\rceil + 1) - h) &\geq 2^n - 2\lceil\frac{n}{2}\rceil - 2 + h \\ &\geq 2^n - 2\lceil\frac{n}{2}\rceil - 2 + (\lceil\frac{n}{2}\rceil - 2) \\ &\geq 2^n - \lceil\frac{n}{2}\rceil - 4 \\ &\geq 2^n - \frac{n+1}{2} - 4 \\ &> 0, \end{aligned}$$

for $n \geq 3$. Moreover, By Lemma 5, $HH_n$ has no triangle. Applying Theorem 1, $t_h^e(HH_n) = \lceil\frac{n}{2}\rceil - h + 1$, as required.

(2) By Lemma 5, the half hypercube network $HH_n(n \geq 3)$ satisfies conditions of Theorem 2. Similar to verification of Theorem 6(1), we directly determine the $r$-restricted edge diagnosability of $HH_n$ under the HPMC model. □

In particular, when $n$ is even, the $n$-dimensional half hypercube network is isomorphic to a $\frac{n}{2}$-dimensional hierarchical cubic network, i.e., $HH_n \cong HCN_{\frac{n}{2}}$. Note that $HCN_n$ an $(n+1)$-regular triangle-free network. Therefore, we directly determine the $h$-restricted vertex diagnosability and $r$-restricted edge diagnosability of hierarchical cubic networks as follows.

**Corollary 3.** For the hierarchical cubic network $HCN_n$ with $n \geq 3$,

(1) the $h$-restricted vertex diagnosability of $HCN_n$ is $t_h^e(HCN_n) = n - h + 1$ for $1 \leq h \leq n$;
(2) the $r$-restricted edge diagnosability of $HCN_n$ is

$$s_r^v(HCN_n) = \begin{cases} n-1, & r = 1 \text{ and } n \geq 3; \\ n-r+1, & 2 \leq r \leq n-1. \end{cases}$$

## 5 Conclusions

Hybrid fault diagnosis against processors and links fault may be more useful for the robustness of a network. In this paper, we establish the $h$-restricted vertex diagnosability and $r$-restricted edge diagnosability of regular networks under the hybrid PMC model. As applications, we apply the general results obtained to the state-of-the-art regular networks, such as hypercube-like network as well as hypercube-based compound network, DQcube, exchanged hypercube, dual cube, half hypercube, hierarchical cubic network and so on. As for the future research, we will consider the novel fault diagnostic model, namely, the hybrid MM* model, and apply it to general networks.

## References

1. Chang, G.-Y., Chang, G.-J., Chen, G.-H.: Diagnosabilities of regular networks. IEEE Trans. Parallel Distrib. Syst. **16**, 314–323 (2005)
2. Chang, N.-W., Hsieh, S.-Y.: Conditional diagnosability of $(n, k)$-star graphs under the PMC model. IEEE Trans. Dependable Secure Comput. **15**(2), 207–216 (2018)
3. Chang, N.-W., Hsieh, S.-Y.: Conditional diagnosability of alternating group networks under the PMC model. IEEE/ACM Trans. Netw. **28**(5), 1968–1980 (2020)
4. Cheng, E., Qiu, K., Shen, Z.: On diagnosability of interconnection networks. Int. J. Unconv. Comput. **13**, 245–251 (2018)
5. Cheng, E., Qiu, K., Shen, Z.: Diagnosability of interconnection networks: past, present and future. Int. J. Parallel Emergent Distrib. Syst. **35**, 1–7 (2019)
6. Guo, J., Li, D., Lu, M.: The $g$-good-neighbor conditional diagnosability of the crossed cubes under the PMC and MM* model. Theor. Comput. Sci. **755**, 81–88 (2019)
7. Hung, R.-W.: DQcube: a novel compound architecture of disc-ring graph and hypercube-like graph. Theor. Comput. Sci. **498**, 28–45 (2013)
8. Kim, J.-S., Kim, M.-H., Lee, H.-O.: Analysis and Design of a Half Hypercube Interconnection Network. In: Park, J.J.J.H., Ng, J.K.-Y., Jeong, H.Y., Waluyo, B. (eds.) Multimedia and Ubiquitous Engineering. LNEE, vol. 240, pp. 537–543. Springer, Dordrecht (2013). https://doi.org/10.1007/978-94-007-6738-6_65

9. Li, X., Zhuang, H., Zhou, S., et al.: Reliability evaluation of generalized exchanged $X$-cubes based on the condition of $g$-good-neighbor. Wirel. Commun. Mob. Comput. **4**, 1–16 (2020)
10. Lian, G., Zhou, S., Hsieh, S.-Y., et al.: Performance evaluation on hybrid fault diagnosability of regular networks. Theor. Comput. Sci. **796**, 147–153 (2019)
11. Lin, L., Xu, L., Chen, R., Hsieh, S.-Y., Wang, D.: Relating extra connectivity and extra conditional diagnosability in regular networks. IEEE Trans. Dependable Secure Comput. **16**(6), 1086–1097 (2019)
12. Liu, H., Hu, X., Gao, S.: The $g$-good neighbor conditional diagnosability of twisted hypercubes under the PMC and MM* model. Appl. Math. Comput. **332**, 484–492 (2018)
13. Lv, M., Zhou, S., Sun, X., Lian, G., Chen, G.: The $g$-good-neighbor conditional diagnosability of multiprocessor system based on half hypercube. Int. J. Comput. Math. Comput. Syst. Theor. **3**(3), 1–15 (2018)
14. Lv, M., Zhou, S., Liu, J., Sun, X., Lian, G.: Fault diagnosability of DQcube under the PMC model. Discret. Appl. Math. **259**, 180–192 (2019)
15. Loh, P.K., Hsu, W.-J., Pan, Y.: The exchanged hypercube. IEEE Trans. Parallel Distrib. Syst. **16**, 866–874 (2005)
16. Preparata, F.P., Metze, G., Chien, R.T.: On the connection assignment problem of diagnosable systems. IEEE Trans. Electron. Comput. EC **16**(6), 848–854 (1967)
17. Wei, Y., Xu, M.: Hybrid fault diagnosis capability analysis of regular graphs. Theor. Comput. Sci. **760**, 1–14 (2019)
18. Wei, Y., Li, R.-H., Yang, W.: Hybrid fault diagnosis capability analysis of highly connected graphs. arXiv:2007.03455
19. Xu, J.-M.: Combinational Theory in Networks. Science Press, Beijing/ China (2013)
20. Zhou, S., Dong, Q.: Reliability analysis of multiprocessor systems. Science Press, Beijing (2020)
21. Zhu, Q., Thulasiraman, K., Xu, M., Radhakrishnan, S.: Hybrid PMC (HPMC) fault model and diagnosability of interconnection networks. AKCE Int. J. Graphs Comb. **17**(18), 755–760 (2020)
22. Zhu, Q., Li, L., Liu, S.Y., Zhang, X.: Hybrid fault diagnosis capability analysis of hypercubes under the PMC model and MM* model. Theor. Comput. Sci. **758**, 1–8 (2019)
23. Zhang, N.P., Zhu, Q.: Reliability and hybrid diagnosis of exchanged hypercube. Theor. Comput. Sci. **849**, 202–209 (2021)
24. Zhang, S., Liu, H., Hu, X.: Hybrid fault diagnosis capability analysis of triangle-free graphs. Theor. Comput. Sci. **799**, 59–70 (2019)

# A Study for Conditional Diagnosability of Pancake Graphs

Nai-Wen Chang(✉), Hsuan-Jung Wu, and Sun-Yuan Hsieh

Department of Computer Science and Information Engineering, National Cheng Kung University, No. 1, University Road, Tainan 70101, Taiwan
hsiehsy@mail.ncku.edu.tw

**Abstract.** Due to the increasing size of a multi-processor system, processor fault diagnosis has played an important role in measuring the reliability of the system. The diagnosability of many well-known multiprocessor systems has been widely investigated. The conditional diagnosability is a new measure of diagnosability by restricting an additional condition that any faulty set cannot contain all the neighbors of any node in a system. In this paper, we evaluate the conditional diagnosability for pancake graphs under the PMC model. We first derive several properties of pancake graphs, and then based on these properties, the conditional diagnosability of an $n$-dimensional pancake graph is shown to be 2 for $n = 3$ and $8n - 21$ for $n \geq 4$.

**Keywords:** Interconnection networks · Diagnosis model · Conditional diagnosability · Pancake graphs · Fault tolerance · Multiprocessor systems

## 1 Introduction

With the rapid development of multi-processor systems, a multi-processor system may consist of hundreds or even thousands of processors (nodes), and some of them may be failed when the system is put to use. As the number of processors in a system increases, faulty processors grows at the same time. In order to ensure the reliability of the systems, we need to find out the faulty processors to replace or repair them, this motivates us to the issue of reliability of multiprocessor systems. The system reliability means that it can distinguish between faulty processors and faulty-free processors, and the faulty processors can be replaced by a faulty-free one. The process of identifying the faulty processors is called *diagnosis* of the system, and the *diagnosability* of a system is the maximum number of faulty processors that the system can guarantee to identify.

There are several diagnosis models has been proposed in the diagnosability of multi-processors systems. In particular, there are two of well-known and widely used in the above models. The PMC model [27], named and proposed by

Supported by the Ministry of Science and Technology in Taiwan.


C.-Y. Chen et al. (Eds.): COCOON 2021, LNCS 13025, pp. 298–305, 2021.
https://doi.org/10.1007/978-3-030-89543-3_25

Preparata, Metze, and Chien's model, is a tested-based diagnosis model; and the MM model [24,25], proposed by Maeng and Malek's model, which is comparison-based diagnosis model. In the PMC model, a processor performs the diagnosis by testing on their neighboring processors via the communication links between them. It is assumed that a test result is reliable (resp. unreliable) if the processor evaluating the test is fault-free (resp. faulty). There are numerous studies on diagnoability of interconnection network that have been dedicated on the PMC model [2–6,11,22,30,34]. In the MM model, a comparison is performed by choosing a processor called comparator, which deals with the fault diagnosis by sending the same input or task to a pair of its neighboring processors and compare their response. As well as the PMC model, many researches studied the diagnosability of multi-processor systems under the MM model [8–10,12–14,20,21,23,29,31,32].

It is well known that the pancake graphs is one of the most popular multiprocessor systems for parallel computer/communication system. An $n$-dimensional Pancake graph $\mathcal{P}_n$ is an undirected regular graph with $n!$ nodes and with degree $n-1$. The pancake graph is a great substitute to the hypercube in a parallel system, which can interconnect processors with a lower degree, smaller diameter, and recursive structure. There are several properties have been investigated on pancake graphs [16–18,26].

In classical measures of system-level diagnosability for multi-processor systems, if all the neighbor of a processor are faulty nodes, the processor must can not be determined is faulty-free or faulty. Consequently, the diagnosability of a system is bounded from its minimum node degree. However, in some large-scale multi-processor systems, the probability that all the neighbors of a processor are failed concurrently is extremely small. Based on this assumption, Lai *et al.* [19] introduced the concept of *conditional diagnosability*, which assumed that all the neighbors of any node do not failed at the same time, and showed that the conditional diagnosability of an $n$-dimensional hypercube $Q_n$ is $4n-7$ for $n \geq 5$. Moreover, the study of conditional diagnosability for some interconnection networks has attracted the attention of research worker [5,12–14,19,23,30,33,34], In this paper, we evaluate the conditional diagnosability for $\mathcal{P}_n$ under the PMC model to be $8n-21$, for $n \geq 5$.

The rest of this paper is organized as follows: In Sect. 2, we provides the terminology and preliminaries for system-level diagnosis. In Sect. 3, some new properties of $\mathcal{P}_n$ are derived, which are used in Sect. 4 to evaluate the conditional diagnosability of $\mathcal{P}_n$. Finally, some concluding remarks are given in Sect. 5.

## 2 Preliminaries

An *undirected graph* (*graph* for short) $G$ consists of a *vertex* (*node*) *set* $V(G)$ and an *edge set* $E(G)$, where $V(G)$ is a finite set and $E(G)$ is a subset of $\{(u,v)|\ (u,v)$ is an unordered pair of $V(G)\}$. We usually represents a multiprocessor system or an interconnection network by an *undirected graph* $G$, where a node $u \in V(G)$ represents a processor in the system and an edge

$uv \in E(G)$ represents a communication link between the nodes $u$ and $v$. Given a vertex subset $U \subseteq V(G)$, the *subgraph of $G$ induced by $U$* is defined as $G[U] = (U, \{uv \in E(G) \mid u, v \in U\})$.

Let $G$ be a graph, and $v \in V(G)$ be a node. The *neighborhood* of a node $v$ in $G$, denoted by $N_G(v)$, is the set of all nodes adjacent to $v$ in $G$. The *degree* of a node $v$, denoted by $deg_G(v)$, is the number of edges incident to $v$, which equals to the cardinality $N_G(v)$ (i.e., $deg_G(v) = |N_G(v)|$). The *minimum degree* $\delta(G)$ equals $\min\{deg_G(v) | \ v \in V(G)\}$. Graph $G$ is said to be *$k$-regular* if all the nodes in $G$ has the same degree $k$. For a subset of nodes $V' \subseteq V(G)$, the *neighborhood set* of $V'$ in $G$ is defined as $N_G(V') = \left(\bigcup_{u \in V'} N_G(u)\right) \setminus V'$. Let $N_G[V'] = N_G(V') \cup V'$. For a node set $T \subseteq V(G)$, the notation $G \setminus T$ is used to denote the graph obtained by $G$ with removing all the nodes in $T$ from $G$. The *components* of a graph $G$ are its maximal connected subgraphs. A component called *trivial component* if it has no edges; otherwise, it called *non-trivial component.* A *path* $P[v_0, v_t] = \langle v_0, v_1, \ldots, v_t \rangle$ in a graph $G$ is a sequence of distinct vertices such that any two consecutive vertices are adjacent, and vertices $v_0$ and $v_t$ are called the *endpoints* of the path.

(Formal definitions for several terms about diagnosability are omitted due to space constraints.)

Since the test result performed by a faulty tester is unreliable, a given node set $F$ can produce different syndromes. Let $\sigma(F)$ represents the set of all syndromes that consistent with the node set $F$. For two distinct node sets $F_1$ and $F_2$ in a system $G$, $F_1$ and $F_2$ are said to be *distinguishable* if $\sigma(F_1) \cap \sigma(F_2) = \emptyset$; otherwise, $F_1$ and $F_2$ are said to be *indistinguishable.* Moreover, $(F_1, F_2)$ is an *distinguishable pair* if $\sigma(F_1) \cap \sigma(F_2) = \emptyset$; otherwise, $(F_1, F_2)$ is an *indistinguishable pair.*

Some known results about the definition of a $t$-diagnosable system and related concepts are described as follows.

**Definition 1.** [27] A system of $n$ nodes is *$t$-diagnosable* if all the faulty nodes can be identified without replacement, provided that the number of faulty nodes presented does not exceed $t$.

**Lemma 1.** [7] *A system $G$ is $t$-diagnosable if and only if for any pair $F_1, F_2 \subseteq V(G)$ with $|F_1|, |F_2| \leq t$ and $F_1 \neq F_2$, there exists at least one test from $V(G) \setminus (F_1 \cup F_2)$ to $F_1 \Delta F_2$.*

The following lemma follows directly from Lemma 1, which gives a necessary and sufficient condition for a pair of fault sets $F_1$ and $F_2$ to be distinguishable.

**Lemma 2.** [19] *Let $G$ be a system. Then, for any two sets $F_1, F_2 \subseteq V(G)$ with $F_1 \neq F_2$, $(F_1, F_2)$ is a distinguishable pair if and only if there exists a node $u \in V(G) \setminus (F_1 \cup F_2)$ which is adjacent to a node $v \in F_1 \Delta F_2$.*

## 3 Properties of Pancake Graphs

The $n$-dimensional pancake graph, denoted by $\mathcal{P}_n$, is a graph whose node set consists of all permutations on $\langle n \rangle$. Each node is assigned a unique label $\mathrm{x}_1\mathrm{x}_2 \ldots \mathrm{x}_n$, where $x_i \in \langle n \rangle$ for $1 \le i \le n$ and $x_i \ne x_j$ for $i \ne j$. Each node $x_1x_2 \ldots x_{i-1}x_ix_{i+1} \ldots x_n$ is adjacent to node $x_ix_{i-1} \ldots x_2x_1x_{i+1} \ldots x_n$, also referred as *i-neighbor*, by an edge in dimension $i$ for $2 \le i \le n$. In other words, an $i$-neighbor of $x_1x_2 \ldots x_{i-1}x_ix_{i+1} \ldots x_n$ is obtained by reversing from the first coordinate to the $i$th coordinate of the node. As a result, there exist $n!$ nodes in an $n$-dimensional pancake graph, and each node has degree $n-1$. Let $u = x_1x_2 \ldots x_n$ be a node in an $n$-dimensional pancake graph $\mathcal{P}_n$. We use $[u]_i$ to denote the $i$th coordinate of $u$ for $1 \le i \le n$ (i.e., $[u]_i = x_i$). Moreover, we use $(u)^i$ to denote the unique $i$-neighbor of $u$ for $2 \le i \le n$ (i.e., $(u)^i = x_ix_{i-1} \ldots x_2x_1x_{i+1} \ldots x_n$). Obviously, $((u)^i)^i = u$. In addition, $\mathcal{P}_n$ can be decomposed into $n$ copies of $(n-1)$-dimensional pancake graphs (*subpancakes* for short), denoted by $\mathcal{P}_n^i|_j$ over dimension $j$, that is, each subpancake $\mathcal{P}_n^i|_j$ is the subgraph induced by the nodes $u$ with $[u]_j = i$, where $1 \le i \le n$ and $2 \le j \le n$. The notation $\mathcal{P}_n^i|_j$ can be simplified as $\mathcal{P}_n^i$ when the index $j$ is not important. Clearly, each subpancake $\mathcal{P}_n^i$ is isomorphic to $\mathcal{P}_{n-1}$. For $1 \le i \ne j \le n$, we use $E^{i,j}$ to denote the set of edges between two subpancakes $\mathcal{P}_n^i$ and $\mathcal{P}_n^j$.

Some well-known properties about pancake graphs are described as follows.

*Property 1.* [1,15,26,28]

1. $\mathcal{P}_n$ is $(n-1)$-regular with $n!$ nodes for $n \ge 1$.
2. The connectivity of $\mathcal{P}_n$ is $n-1$ (i.e., $\kappa(\mathcal{P}_n) = n-1$), where $n \ge 1$.
3. The girth of $\mathcal{P}_n$, denoted by $girth(\mathcal{P}_n)$, equals 6 (i.e., $girth(\mathcal{P}_n) = 6$), where $n \ge 3$.
4. $|E^{i,j}| = (n-2)!$, where $1 \le i \ne j \le n$.

**Lemma 3.** [18] *The $n$-dimensional Pancake graph $\mathcal{P}_n$, where $n \ge 3$, has 6-cycle of canonical form $C_6 = r_3r_2r_3r_2r_3r_2$, where $r_i$ represents the reversion of the first coordinate to the $i$th coordinate of label, and each node of $\mathcal{P}_n$ belongs to exactly one 6-cycle.*

Let $T \subseteq V(\mathcal{P}_n)$ be a subset of nodes in $\mathcal{P}_n$. We need to prove some lemmas on the cardinalities of $T$ such that $\mathcal{P}_n \backslash T$ has a large component with some small components. The results are shown in the following lemmas.

**Lemma 4.** *Let $T$ be a node subset of $\mathcal{P}_n$ with $|T| \le 2n-5$, where $n \ge 4$. Then, $\mathcal{P}_n \setminus T$ has a large component and up to one trivial component.*

*Proof.* (The remaining proof is omitted due to space constraints.)

**Lemma 5.** *Let $T$ be a node subset of $\mathcal{P}_n$ with $|T| \le 3n-8$, where $n \ge 4$. Then, $\mathcal{P}_n \backslash T$ has a large component and at most two small components containing up to 2 nodes in total.*

*Proof.* (The detailed proof is omitted due to space constraints.)

Lemma 6 determines the cardinality of neighborhood of a path with length seven in $\mathcal{P}_4$, which is required to our method.

**Lemma 6.** *Let $S$ to be a path of length seven in $\mathcal{P}_4$, then $|N_{\mathcal{P}_4}(S)| \ge 8$ for $n = 4$.*

*Proof.* (The detailed proof is omitted due to space constraints.)

# 4 Conditional Diagnosability of Pancake Graphs

In an $n$-dimensional pancake graph $\mathcal{P}_n$, there are $\binom{n!}{n-1}$ node subsets of size $n-1$, among which there are only $n!$ node subsets containing all the neighbors of some nodes. Since the ratio $\frac{n!}{\binom{n!}{n-1}}$ is extremely small for large $n$, the probability that a fault set contains all the neighbors of some node is relatively low. For this reason, Lai *et al.* [19] proposed a new restricted diagnosis strategy called *conditional diagnosability* for multiprocessor systems. They considered the situation that any fault set cannot contain all the neighbors of any node in a system. A fault set $F \subseteq V(G)$ is called a *conditional fault set* if $N_G(u) \nsubseteq F$ for every node $u \in V(G)$. The definition of a *conditionally t-diagnosable* system is given as follows:

**Definition 2.** [19] *A system $G$ is* conditionally $t$-diagnosable *if and only if $F_1$ and $F_2$ are distinguishable for each pair of conditional fault sets $F_1, F_2 \subseteq V(G)$, where $F_1 \neq F_2$ and $|F_1|, |F_2| \leq t$.*

Let $F_1, F_2$ be two sets with $F_1 \neq F_2$. We say that $(F_1, F_2)$ is a distinguishable *conditional-pair* (resp. an *indistinguishable conditional-pair*) if $F_1$ and $F_2$ are both conditional fault sets and distinguishable (resp. indistinguishable). An equivalent way of representing the above definition is listed below, which will be used in our main theorem.

**Lemma 7.** [19] *A system $G$ is conditionally $t$-diagnosable if and only if for each indistinguishable conditional-pair $F_1, F_2 \subseteq V(G)$ with $F_1 \neq F_2$, it implies that $|F_1| > t$ or $|F_2| > t$.*

The *conditional diagnosability* of $G$, denoted by $t_c(G)$, is defined to be the maximum value of $t$ such that $G$ is conditionally $t$-diagnosable.

**Lemma 8.** *For a system $G$, $t_c(G) \geq t(G)$.*

Before discussing the conditional diagnosability, we have some observations about the neighborhood of a node in a system having an indistinguishable conditional-pair $(F_1, F_2)$. Lai *et al.* [19] state this phenomenon in the following lemma.

**Lemma 9.** [19] *Let $(F_1, F_2)$, where $F_1 \neq F_2$, be an indistinguishable conditional-pair in a system $G$. Denote $X = G \setminus (F_1 \cup F_2)$. Then, the following two conditions hold:*

1. *$|N_G(u) \cap (V(G) \setminus (F_1 \cup F_2))| \geq 1$ for $u \in V(X)$, and*
2. *$|N_G(v) \cap (F_1 \setminus F_2)| \geq 1$ and $|N_G(v) \cap (F_2 \setminus F_1)| \geq 1$ for $v \in F_1 \Delta F_2$.*

The following lemma exploits some properties of graphs with the girth at least six.

**Lemma 10.** [4] *Let $G$ be a system with $girth(G) \geq 6$. Given an indistinguishable conditional-pair $(F_1, F_2)$ in $G$ with $F_1 \neq F_2$, and let $T = F_1 \cap F_2$. Then $G \setminus T$ has a component $H$ with*

**1)** $V(H) \subseteq F_1 \Delta F_2$.
**2)** $\delta(H) \geq 2$.
**3)** *$H$ must contain a path of length seven $P$ as a subgraph.*
**4)** $|V(H)| \geq 8$.

Next, we demonstrate that the conditional diagnosability of a pancake graph $\mathcal{P}_n$ does not exceed $8n - 20$ for $n \geq 4$. We consider a subgraph of $\mathcal{P}_n$ that is an 8-cycle $C_8 = \langle v_1, v_2, v_3, v_4, v_5, v_6, v_7, v_8, v_1 \rangle$, where $v_2 = (v_1)^3$, $v_3 = (v_2)^2$, $v_4 = (v_3)^3$, $v_5 = (v_4)^4$, $v_6 = (v_5)^2$, $v_7 = (v_6)^3$, and $v_8 = (v_7)^2 = (v_1)^4$. Let $F_1 = N_{\mathcal{P}_n}(C_8) \cup \{v_1, v_2, v_5, v_6\}$ and $F_2 = N_{\mathcal{P}_n}(C_8) \cup \{v_3, v_4, v_7, v_8\}$. First, it is not difficult to verify that both $F_1$ and $F_2$ are conditional fault sets. Further, by the fact that $N_{\mathcal{P}_n}(C_8) = F_1 \cap F_2$, there exist no edges between $V(G) \setminus (F_1 \cup F_2)$ and $F_1 \Delta F_2$. According to Lemma 2, $(F_1, F_2)$ is an indistinguishable conditional pair. In addition, because $|F_1 \cap F_2| = |N_{\mathcal{P}_n}(C_8)| = 8(n-3) = 8n - 24$ and $|F_1 \setminus F_2| = |F_2 \setminus F_1| = 4$, we have $|F_1| = |F_2| = 8n - 20$. By Lemma 7, $\mathcal{P}_n$ is not conditionally $(8n-20)$-diagnosable. Hence, the following result can be obtained.

**Lemma 11.** *$t_c(S_n) \leq 8n - 21$ for $n \geq 4$.*

We are now ready to show that the conditional diagnosability of $\mathcal{P}_n$ is $8n-21$ for $n \geq 5$. Let $(F_1, F_2)$ be an indistinguishable conditional-pair in $\mathcal{P}_n$, where $n \geq 5$. The following shows that $|F_1| \geq 8n - 20$ or $|F_2| \geq 8n - 20$.

**Lemma 12.** *Let $(F_1, F_2)$, where $F_1 \neq F_2$, be an indistinguishable conditional-pair in $\mathcal{P}_n$ for $n \geq 5$. Then, we have $|F_1| \geq 8n - 20$ or $|F_2| \geq 8n - 20$.*

*Proof.* (The remaining proof is omitted due to space constraints.)

By Lemma 11, $t_c(\mathcal{P}_n) \leq 8n - 21$, and by Lemmas 7 and 12, we conclude that $\mathcal{P}_n$ is conditionally $(8n - 21)$–diagnosable for $n \geq 5$. Hence, $t_c(\mathcal{P}_n) = 8n - 21$, for $n \geq 5$. Finally, the conditional diagnosability of a pancake graph $\mathcal{P}_n$ is represented as follows:

**Theorem 1.** *$t_c(\mathcal{P}_n) = 8n - 21$ for $n \geq 5$.*

## 5 Conclusion

In a multiprocessor system, the processors in the system may fail independently. Thus, the diagnosis of system is an important aspect in system design. Because of the probability that an faulty set contains all the neighbors of some processor is extremely small, we are interested in conditional diagnosability, which restricts that each processor of a system is adjacent to at least one fault-free processor.

In this paper, we studied the conditional diagnosability of the $n$-dimensional pancake graph $\mathcal{P}_n$ under the PMC model. We discovered several new properties of $\mathcal{P}_n$ and then utilize these properties to prove that $t_c(\mathcal{P}_n) = 8n - 21$ for $n \geq 5$.

## References

1. Akers, S.B., Krishnamurthy, B.: A group-theoretic model for symmetric interconnection networks. IEEE Trans. Comput. **38**(4), 555–566 (1989)
2. Chang, G.Y., Chang, G.J., Chen, G.H.: Diagnosabilities of regular networks. IEEE Trans. Parallel Distrib. Syst. **16**(4), 314–323 (2005)
3. Chang, N.W., Hsieh, S.Y.: Conditional diagnosability of augmented cubes under the PMC model. IEEE Trans. Dependable Secur. Comput. **9**(1), 46–60 (2012)
4. Chang, N.W., Hsieh, S.Y.: Structure properties and conditional diagnosability of star graphs under the PMC model. IEEE Trans. Parallel Distrib. Syst. **25**, 1–11 (2014)
5. Chang, N.W., Lin, T.Y., Hsieh, S.Y.: Conditional diagnosability of $k$-ary $n$-cubes under the PMC model. ACM Trans. Des. Autom. Electron. Syst. **17**(4), 46 (2012)
6. Chen, C.A., Hsieh, S.Y.: t/t-Diagnosability of regular graphs under the PMC model. ACM Trans. Des. Autom. Electron. Syst. (TODAES) **18**(2), 20 (2013)
7. Dahbura, A.T., Masson, G.M.: An $O(n^{2.5})$ fault identification algorithm for diagnosable systems. IEEE Trans. Comput. **33**(6), 486–492 (1984)
8. Hong, W.S., Hsieh, S.Y.: Strong diagnosability and conditional diagnosability of augmented cubes under the comparison diagnosis model. IEEE Trans. Reliab. **61**(1), 140–148 (2012)
9. Hsieh, S.Y., Chen, Y.S.: Strongly diagnosable product networks under the comparison diagnosis model. IEEE Trans. Comput. **57**(6), 721–732 (2008)
10. Hsieh, S.Y., Chen, Y.S.: Strongly diagnosis systems under the comparison diagnosis model. IEEE Trans. Comput. **57**(12), 1720–1725 (2008)
11. Hsieh, S.Y., Chuang, T.Y.: The strong diagnosability of regular networks and product networks under the PMC model. IEEE Trans. Parallel Distrib. Syst. **20**(3), 367–378 (2009)
12. Hsieh, S.Y., Kao, C.Y.: The conditional diagnosability of $k$-ary $n$-cubes under the comparison diagnosis model. IEEE Trans. Comput. **62**(4), 839–843 (2013)
13. Hsu, G.H., Chiang, C.F., Shih, L.M., Hsu, L.H., Tan, J.J.M.: Conditional diagnosability of hypercubes under the comparison diagnosis model. J. Syst. Architect. **55**(2), 140–146 (2009)
14. Hsu, G.H., Tan, J.J.M.: Conditional diagnosability of the BC networks under the comparison diagnosis model. Int. Comput. Symp. **1**, 269–274 (2008)
15. Hung, C.N., Hsu, H.C., Liang, K.-Y., Hsu, L.-H.: Ring embedding in faulty pancake graphs. Inf. Process. Lett. **86**(5), 271–275 (2003)
16. Kaneko, K., Suzuki, Y.: Node-to-set disjoint paths problem in pancake graphs. IEICE Trans. Inf. Syst. **86**(9), 1628–1633 (2003)
17. Konstantinova, E.: On some structural properties of star and pancake graphs. In: Aydinian, H., Cicalese, F., Deppe, C. (eds.) Information Theory, Combinatorics, and Search Theory. LNCS, vol. 7777, pp. 472–487. Springer, Heidelberg (2013). https://doi.org/10.1007/978-3-642-36899-8_23
18. Konstantinova, E., Medvedev, A.: Small cycles in the pancake graph. ARS Math. Contemporanea **7**, 237–246 (2014)
19. Lai, P.L., Tan, J.J.M., Chang, C.P., Hsu, L.H.: Conditional diagnosability measures for large multiprocessor systems. IEEE Trans. Comput. **54**(2), 165–175 (2005)
20. Lai, P.L., Tan, J.J.M., Tsai, C.H., Hsu, L.H.: The diagnosability of the matching composition network under the comparison diagnosis model. IEEE Trans. Comput. **53**(8), 1064–1069 (2004)

21. Lee, C.W., Hsieh, S.Y.: Diagnosability of two-matching composition networks under the MM model. IEEE Trans. Dependable Secur. Comput. **8**(2), 246–255 (2011)
22. Lin, C.K., Kung, T.L., Tan, J.J.M.: Conditional-fault diagnosability of multiprocessor systems with an efficient local diagnosis algorithm under the PMC model. IEEE Trans. Parallel Distrib. Syst. **22**(10), 1669–1680 (2011)
23. Lin, C.K., Tan, J.J.M., Hsu, L.H., Cheng, E., Lipták, L.: Conditional diagnosability of Cayley graphs generated by transposition trees. J. Interconnection Netw. **9**(1–2), 83–97 (2008)
24. Maeng, J., Malek, M.: A comparison connection assignment for self-diagnosis of multiprocessors systems. In: Proceedings of the 11th International Symposium on Fault-Tolerant Computing, pp. 173–175 (1981)
25. Malek, M.: A comparison connection assignment for diagnosis of multiprocessors systems. In: Proceedings of the 7th International Symposium on Computer Architecture, pp. 31–36 (1980)
26. Nguyen, Q.T., Bettayeb, S.: On the genus of pancake network. Int. Arab. J. Inf. Technol. **8**(3), 289–292 (2011)
27. Preparata, F.P., Metze, G., Chien, R.T.: On the connection assignment problem of diagnosis systems. IEEE Trans. Electron. Comput. **16**(12), 848–854 (1967)
28. Suzuki, Y., Kaneko, K.: An algorithm for node-disjoint paths in pancake graphs. IEICE Trans. Inf. Syst. **E86–D**(3), 610–615 (2003)
29. Wang, D.: Diagnosability of hypercubes and enhanced hypercubes under the comparison diagnosis model. IEEE Trans. Comput. **48**, 1369–1374 (1999)
30. Xu, M., Thulasiraman, K., Hu, X.D.: Conditional diagnosability of matching composition networks under the PMC model. IEEE Trans. Circ. Syst. II **56**(11), 875–879 (2009)
31. Zheng, J., Latifi, S., Regentova, E., Luo, K., Wu, X.: Diagnosability of star graphs under the comparison diagnosis model. Inf. Process. Lett. **93**(1), 29–36 (2005)
32. Zhou, S.: The conditional diagnosability of Möbius cubes under the comparison model. In: Proceedings of the 2009 IEEE International Conference on Information and Automation, pp. 96–100 (2009)
33. Zhu, Q.: On conditional diagnosability and reliability of the BC networks. J. Supercomput. **45**(2), 173–184 (2008)
34. Zhu, Q., Liu, S.Y., Xu, M.: On conditional diagnosability of the folded hypercubes. Inf. Sci. **178**(4), 1069–1077 (2008)

# A New Measure for Locally $t$-Diagnosable Under PMC Model

Meirun Chen[1], D. Frank Hsu[2], and Cheng-Kuan Lin[3](✉)

[1] School of Applied Mathematics, Xiamen University of Technology, Xiamen, China
mrchen@xmut.edu.cn

[2] Department of Computer and Information Sciences, Fordham University, New York, NY 10023, USA
hsu@cis.fordham.edu

[3] Department of Computer Science, National Yang Ming Chiao Tung University, Hsinchu, Taiwan
cklin@nycu.edu.tw

**Abstract.** PMC model is the test-based diagnosis which a vertex performs the diagnosis by testing the neighbor vertices via the edges between them. Hsu and Tan proposed two structures to diagnose a vertex. But these structures don't always exist for any vertex. Here, we propose a new testing structure to diagnose a vertex under PMC model to solve the problem above. It can fit more general networks. Let $S$ be a set of faulty edges of the $n$-dimensional hypercube $Q_n$. Using this structure, we show that every vertex $u$ of $Q_n$ is $\deg_{Q_n-S}(u)$-diagnosable if $\delta(Q_n - S) \geq 2$, $\deg_{Q_n-S}(x) + \deg_{Q_n-S}(y) \geq 5$ for every two adjacent vertices $x$ and $y$ in $Q_n - S$, and $n \geq 5$.

## 1 Introduction

Due to the growth of network related applications such as cloud computing, Internet of things (IoT), and vehicle to everything (V2X), it is important to ensure the reliable operation of devices. As the number of processors increases, the communication link between processors becomes more and more complex. Therefore, one cannot avoid some processors failure. How to identify the faulty processors accurately is the key to ensure the normal operation of the network. System level diagnostics is to distinguish each processor failure or not, and then replace the faulty processors by fault-free ones to ensure the reliable operation of the system. A system is called $t$-diagnosable if all faulty processors can be identified without replacement as long as the number of faulty processors does not exceed $t$ [20]. The diagnosability of a system is the maximum value of $t$ such that it is $t$-diagnosable [5,17,20]. That is, the maximum number of faulty processors that can be identified in this system.

PMC model proposed by Preparata, Metze and Chien in 1967 [20] is the original diagnosis model. Lin and Teng [18] characterized the diagnosability for triangle-free graphs under PMC model. Lee and Hsieh [15,16] characterized the


C.-Y. Chen et al. (Eds.): COCOON 2021, LNCS 13025, pp. 306–316, 2021.
https://doi.org/10.1007/978-3-030-89543-3_26

(1,2)-composition networks as $t$-diagnosable under PMC model. The diagnosability of the hypercubes, the crossed cubes, the Möbius cubes, the twisted cubes, the enhanced hypercubes, the exchanged generalized hypercubes under the PMC model were studied in [1,3,8,9,14,17,22,23].

In order to diagnose who are the faulty nodes in the system, we need to do $2m$ times tests under PMC model, where $m$ stands for the number of the links in the system. If we want to know a specific node $u$ is faulty or not, instead of doing the global diagnosis, Hsu and Tan proposed the concept of local diagnosis [12]. They showed that if there exists $T(u;t)$ or $T(u;t-2,2)$ structure for the node $u$ (see Fig. 1) then $u$ is $t$-diagnosable.

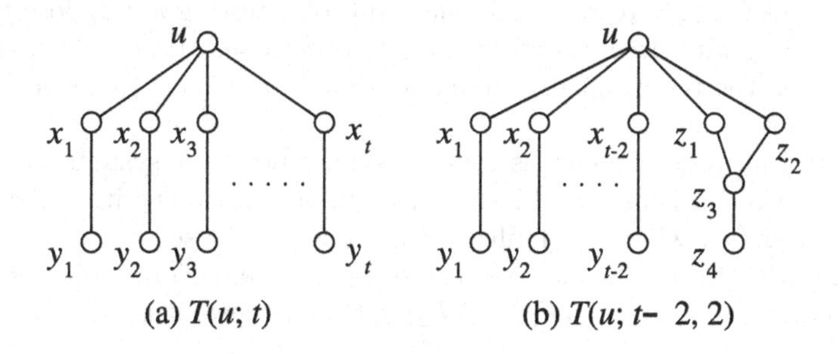


**Fig. 1.** Two diagnosis structures

However, such structures don't always exist for a node. To make up for that, we propose a new structure for $u$ and prove that if the number of faulty nodes in this structure doesn't exceed $t$ then $u$ can be correctly diagnosed, where $t$ is the degree of $u$. The rest of this paper is organized as follows. First, we give the necessary definitions and notations, the detail of PMC model, the definition of the local diagnosability in Sect. 2. In Sect. 3, we propose a new structure to diagnose a node, we also provide the corresponding algorithm to prove our theorem. In Sect. 4, we apply the new proposed structure to show that each node of an incomplete hypercube is locally $t$-diagnosable, where $t$ is the degree of the node in the incomplete hypercube.

## 2 Preliminaries

We model a network as a graph, the nodes (resp. links) of the network can be viewed as the vertices (resp. edges) of the graph. Let $G = (V, E)$, where $V$ represents the vertex set and $E$ represents the edge set. A matching $M$ of $G$ is a subset of $E$ such that any two distinct elements of $M$ are not incident to a common vertex. A graph $H$ is a subgraph of a graph $G$ if $V(H) \subseteq V(G)$ and $E(H) \subseteq E(G)$. Let $V'$ be a subset of $V(G)$. We say that $H$ is a subgraph of $G$ induced by $V'$ if $V(H) = V'$ and $E(H) = \{(u,v) \mid u, v \in V' \text{ and } (u,v) \in E(G)\}$. Let $u$ be any vertex in $G$. The neighborhood of $u$ in $G$, $N_G(u) = \{v \mid (u,v) \in$

$E(G)\}$, is the set of vertices adjacent to $u$. The neighbor of a vertex subset $A$ of graph $G$ is $N_G(A) = \bigcup_{u \in A} N_G(u) - A$. The degree of $u$ in $G$, $\deg_G(u) = |N_G(u)|$, is the number of edges incident with $u$ in $G$. We use $\delta(G) = \min_{v \in V(G)} \deg_G(v)$ to denote the minimum degree of the vertices of $G$. The distance of two vertices $u$ and $v$, $dist_G(u, v)$, is the number of the edges in the shortest path connecting $u$ and $v$ in $G$. The diameter of $G$, $d(G) = \max_{u,v \in V(G)} dist_G(u, v)$. A graph $G$ is bipartite if its vertices can be partitioned into two disjoint vertices subsets $V_1$ and $V_2$ such that for every edge in $G$, one of its end vertex is in $V_1$ and the other one is in $V_2$. For standard graph-theoretic terminology, we follow [2,13].

The diagnostic strategy of PMC model is proposed as follows. It assumes that two adjacent vertices can test each other. Let $x$ and $y$ be any two adjacent vertices. We use $\sigma(x, y)$ to represent the result of $x$ testing $y$. Suppose that $x$ is fault-free. If $y$ is fault-free, then $\sigma(x, y) = 0$; otherwise, $\sigma(x, y) = 1$. Suppose $x$ is faulty. Then the test result is unreliable, that is, $\sigma(x, y) \in \{0, 1\}$ no matter $y$ is faulty or not.

The set of all test outcomes is called a syndrome of the system. For a given syndrome $\sigma$, a vertex subset $F$ of $V(G)$ is said to be consistent with $\sigma$ if syndrome $\sigma$ can be produced when the faulty set of $G$ is $F$. We set $\sigma(F) = \{\sigma \mid F$ is consistent with $\sigma\}$. For any two distinct subsets $F_1$ and $F_2$ of $V(G)$, $(F_1, F_2)$ is an indistinguishable pair if $\sigma(F_1) \cap \sigma(F_2) \neq \emptyset$; otherwise, it is a distinguishable pair.

Let $A$ and $B$ be any two sets. The difference set for $A$ and $B$, $A - B$, is $\{x \mid x \in A$ and $x \notin B\}$, and the symmetric difference of $A$ and $B$ is $A \Delta B = (A - B) \cup (B - A)$.

**Theorem 1** [20]. *For any two distinct vertex subsets $F_1$ and $F_2$ of a graph $G$, $(F_1, F_2)$ is a distinguishable pair of $G$ under PMC model if and only if there is a vertex $u \in V(G) - (F_1 \cup F_2)$ and a vertex $v \in F_1 \Delta F_2$ such that $uv \in E(G)$.*

Lai et al. gave a necessary and sufficient condition of $t$-diagnosable under PMC model.

**Theorem 2** [17]. *A graph $G$ is $t$-diagnosable under PMC model if and only if, for each distinct pair of subsets $F_1$ and $F_2$ of $V(G)$ with $\max\{|F_1|, |F_2|\} \leq t$, $F_1$ and $F_2$ are distinguishable.*

If we are only interested in the status of some vertices, instead of doing the global diagnosis, Hsu and Tan proposed the concept of local diagnosis [12].

**Definition 3** [12]. *Let $G = (V, E)$ be a graph and $v \in V$ be a vertex. $G$ is locally $t$-diagnosable at vertex $v$ if, given a syndrome $\sigma_F$ produced by a set of faulty vertices $F \subseteq V$ containing vertex $v$ with $|F| \leq t$, every set of faulty vertices $F'$ compatible with $\sigma_F$ and $|F'| \leq t$ must also contain vertex $v$.*

**Definition 4** [12]. *Let $G = (V, E)$ be a graph and $v \in V$ be a vertex. The local diagnosability of vertex $v$, written as $t_l(v)$, is defined to be the maximum value of $t$ such that $G$ is locally $t$-diagnosable at vertex $v$.*

The following proposition related to locally $t$-diagnosable was proposed by Hsu and Tan.

**Proposition 5** [12]. *Let $G = (V, E)$ be a graph and $u \in V$ be a vertex. $G$ is locally $t$-diagnosable at vertex $u$ if and only if, for any two distinct sets of vertices $F_1, F_2 \subset V$, $|F_1| \leq t$, $|F_2| \leq t$, $u \in F_1 \Delta F_2$, and $(F_1, F_2)$ is a distinguishable pair.*

Hsu and Tan [12] showed that $t(G) = \min_{u \in V(G)} t_l(u)$. They gave two sufficient conditions for a vertex to be $t$-diagnosable. For a vertex $u$, if there exists Type I structure $T(u;t)$ or Type II structure $T(u;t-2,2)$ for $u$ then $u$ is $t$-diagnosable. However, it exists neither Type I nor Type II structure for some vertex. In this paper, we propose a new structure $T(u;a,2b)$ to diagnose a vertex $u$ under PMC model, where $a, b \geq 0$. Let $F$ be the faulty vertices in $G$ and $\deg_G(u) = t$, we prove that if there exists the structure $T(u;a,2b)$ for $u$ and $|F \cap V(T(u;a,2b))| \leq t$, then $u$ is $t$-diagnosable, where $t = a + 2b$. We also provide the corresponding algorithm. Let $S$ be a set of faulty edges of the $n$-dimensional hypercube $Q_n$. Using this structure, we show that every vertex $u$ of $Q_n$ is $\deg_{Q_n - S}(u)$-diagnosable if $\delta(Q_n - S) \geq 2$, $\deg_{Q_n-S}(x) + \deg_{Q_n-S}(y) \geq 5$ for every two adjacent vertices $x$ and $y$ in $Q_n - S$, and $n \geq 5$.

## 3 Local Diagnosis Algorithm

Let $u$ be any vertex of $G$. Suppose that $F$ is any set of faulty vertices of $G$. If there are two distinct vertices $p$ and $q$ of $G - \{u\}$ such that $\{(u,p),(p,q)\} \subset E(G)$, then Table 1 shows $\min|F \cap \{p,q\}|$. Moreover, Table 2 shows $\min|F \cap \{w,x,y,z\}|$ if $\{w,x,y,z\} \subset G - \{u\}$ and $\{(u,w),(u,x),(w,y),(x,y),(y,z)\} \subset E(G)$.

**Definition 6.** *Let $G = (V, E)$ be a graph and let $u$ be any vertex in $G$. The mix structure $T(u;a,2b)$ of order $a + 2b$ at vertex $u$ is a subgraph of $G$ where (1) $V(T(u;a,2b)) = \{u\} \cup \{p_i, q_i \mid 1 \leq i \leq a\} \cup \{w_j, x_j, y_j, z_j \mid 1 \leq j \leq b\}$ and (2) $E(T(u;a,2b)) = \{(u,p_i),(p_i,q_i) \mid 1 \leq i \leq a\} \cup \{(u,w_j),(u,x_j),(w_j,y_j),(x_j,y_j),(y_j,z_j) \mid 1 \leq j \leq b\}$. Figure 2 shows an illustration for $T(u;3,6)$.*

We set $B_0 = \{(0,0,0,0,0),(0,0,0,1,0),(0,0,1,0,0)\}$, $B_1 = \{(1,1,0,0,0),(1,1,0,1,0),(1,1,1,0,0)\}$, $B_2 = \{(0,0,i,j,1) \mid i,j \in \{0,1\}\} \cup \{(0,1,0,1,0),(1,0,1,0,0)\}$, $B_3 = \{(1,1,i,j,1) \mid i,j \in \{0,1\}\} \cup \{(0,1,1,0,0),(1,0,0,1,0)\}$, and $B_4 =$

**Table 1.** The minimum number of faulty vertices in the set $\{p,q\}$

| $(\sigma(p,u),\sigma(q,p))$ | $\min\|F \cap \{p,q\}\|$ | |
|---|---|---|
| | $u \in F$ | $u \notin F$ |
| $(0,0)$ | 2 | 0 |
| $(0,1)$ | 1 | 1 |
| $(1,0)$ | 0 | 2 |
| $(1,1)$ | 1 | 1 |

**Table 2.** The minimum number of faulty vertices in the set $\{w, x, y, z\}$

| $(\sigma(w,u), \sigma(x,u), \sigma(y,w), \sigma(y,x), \sigma(z,y))$ | $\min \lvert F \cap \{w,x,y,z\}\rvert$ | |
|---|---|---|
| | $u \in F$ | $u \notin F$ |
| $(0,0,0,0,0), (0,0,0,1,0), (0,0,1,0,0)$ | 4 | 0 |
| $(1,1,0,0,0), (1,1,0,1,0), (1,1,1,0,0)$ | 0 | 4 |
| $(0,0,0,0,1), (0,0,0,1,1), (0,0,1,0,1),$ $(0,0,1,1,1), (0,1,0,1,0), (1,0,1,0,0)$ | 3 | 1 |
| $(0,1,1,0,0), (1,0,0,1,0), (1,1,0,0,1),$ $(1,1,0,1,1), (1,1,1,0,1), (1,1,1,1,1)$ | 1 | 3 |
| $(0,0,1,1,0), (0,1,0,0,0), (1,0,0,0,0), (0,1,0,0,1),$ $(0,1,0,1,1), (0,1,1,0,1), (0,1,1,1,0), (0,1,1,1,1),$ $(1,0,0,0,1), (1,0,0,1,1), (1,0,1,0,1), (1,0,1,1,0),$ $(1,0,1,1,1), (1,1,1,1,0)$ | 2 | 2 |

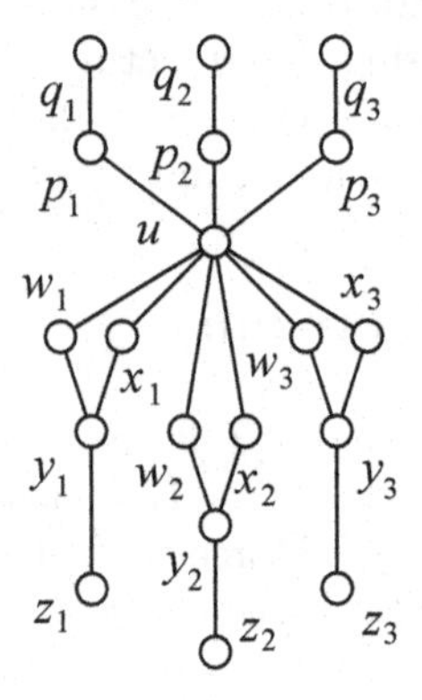


**Fig. 2.** Example of $T(u; 3, 6)$

$\{(i_1, i_2, i_3, i_4, i_5) \mid i_1, i_2, i_3, i_4, i_5 \in \{0,1\}\} - \bigcup_{j=0}^{3} B_j$. Let $T(u; a, 2b)$ be a mix structure of order $a + 2b$ at vertex $u$ of graph $G$. For any faulty vertices set $F$ of $G$, we set $\alpha_j(u, F) = \{i \mid i \in \{1, 2, \ldots, a\}$ and $(\sigma(p_i, u), \sigma(q_i, p_i)) = (j, 0)\}$ for $j \in \{0, 1\}$, $\alpha_2(u, F) = \{i \mid i \in \{1, 2, \ldots, a\}$ and $(\sigma(p_i, u), \sigma(q_i, p_i)) \in \{(0, 1), (1, 1)\}\}$, and $\beta_k(u, F) = \{i \mid i \in \{1, 2, \ldots, b\}$ and $(\sigma(w_i, u), \sigma(x_i, u), \sigma(y_i, w_i), \sigma(y_i, x_i), \sigma(z_i, y_i)) \in B_k\}$ for $k \in \{0, 1, 2, 3, 4\}$. Denote $|\alpha_i(u, F)|$ (resp. $|\beta_i(u, F)|$) by $\alpha_i(u)$ (resp. $\beta_i(u)$). If it is clear, we also use the symbol $\alpha_i$ (resp. $\beta_i$) for short. We have $\sum_{i=0}^{2} \alpha_i + \sum_{j=0}^{4} 2\beta_j = a + 2b$.

We propose the algorithm **LDAMIX** (see Algorithm 1) for the mix structure $T(u; a, 2b)$ and prove it can identify the state of $u$ correctly under PMC model if the number of faulty vertices do not exceed $O(a + b)$.

In **LDAMIX**, we initialize the variables of this algorithm from step 2 to step 6, and it takes $O(1)$ times. If $a > 0$, then the for loop from step 8 to step 11 executes $O(a)$ times to determine $\alpha_0$ and $\alpha_1$. Similarly, the for loop from step 14 to step 18 executes $O(4b)$ times to determine $\beta_0$, $\beta_1$, $\beta_2$ and $\beta_3$ if $b > 0$.

**Algorithm 1:** Local diagnosis algorithm for mix structure (**LDAMIX**)

```
  Input: A mix structure T(u; a, 2b) with min{a, b} ≥ 0.
  Output: The value is 0 or 1 if u is fault-free or faulty, respectively.
1 begin
2   B_0 ← {(0,0,0,0,0), (0,0,0,1,0), (0,0,1,0,0)};
3   B_1 ← {(1,1,0,1,0), (1,1,1,0,0), (1,1,0,0,0)};
4   B_2 ← {(0,0,i,j,1) | i,j ∈ {0,1}} ∪ {(0,1,0,1,0), (1,0,1,0,0)};
5   B_3 ← {(1,1,i,j,1) | i,j ∈ {0,1}} ∪ {(0,1,1,0,0), (1,0,0,1,0)};
6   (α_0, α_1, β_0, β_1, β_2, β_3) ← (0,0,0,0,0,0);
7   if a > 0 then
8     for (i = 1; i < a + 1; i = i + 1) do
9       if (σ(p_i, u), σ(q_i, p_i)) = (0,0) then α_0 ← α_0 + 1;
10      else if (σ(p_i, u), σ(q_i, p_i)) = (1,0) then α_1 ← α_1 + 1;
11    end
12  end
13  if b > 0 then
14    for (i = 1; i < b + 1; i = i + 1) do
15      for (j = 0; j < 4; j = j + 1) do
16        if (σ(w_i, u), σ(x_i, u), σ(w_i, y_i), σ(x_i, y_i), σ(y_i, z_i)) ∈ B_j then
          β_j ← β_j + 1;
17      end
18    end
19  end
20  if α_0 + 2β_0 + β_2 ≥ α_1 + 2β_1 + β_3 then return 0;
21  else return 1;
22 end
```

Finally, it takes $O(1)$ to determine its output at step 20 and step 21. So the time complexity of **LDAMIX** is $O(a + 4b) = O(a + b)$.

**Theorem 7.** *Let $T(u; a, 2b)$ be any mix structure of order $a + 2b$ at vertex $u$, and let $F$ be any faulty set of $G$. If $|F| \leq a + 2b$, then* **LDAMIX** *(Algorithm 1) can identify the state of $u$ correctly under PMC model. That is, $u$ is fault-free if $\alpha_0(u) + 2\beta_0(u) + \beta_2(u) \geq \alpha_1(u) + 2\beta_1(u) + \beta_3(u)$; otherwise, $u$ is faulty.*

*Proof.* We prove this Theorem by contradiction.

Suppose that $u$ is fault-free and $\alpha_1(u) + 2\beta_1(u) + \beta_3(u) > \alpha_0(u) + 2\beta_0(u) + \beta_2(u)$. According to Table 1 and Table 2, we have $|F| \geq \beta_2(u)+2\beta_4(u)+3\beta_3(u)+4\beta_1(u)+2\alpha_1(u)+\alpha_2(u) > \sum_{i=0}^{2} \alpha_i(u)+\sum_{i=0}^{4} 2\beta_i(u) = a+2b$ which contradicts to $|F| \leq a+2b$. Thus, $u$ is faulty if $\alpha_1(u)+2\beta_1(u)+\beta_3(u) > \alpha_0(u)+2\beta_0(u)+\beta_2(u)$.

Suppose that $u$ is faulty and $\alpha_0(u)+2\beta_0(u)+\beta_2(u) \geq \alpha_1(u)+2\beta_1(u)+\beta_3(u)$. According to Table 1 and Table 2, we have $|F| \geq 2\alpha_0(u) + \alpha_2(u) + 4\beta_0(u) + 3\beta_2(u) + 2\beta_4(u) + \beta_3(u) + 1 = (\alpha_0(u) + 2\beta_0(u) + 2\beta_2(u)) + (\alpha_0(u) + 2\beta_0(u) + \beta_2(u))+2\beta_4(u)+\beta_3(u)+\alpha_2(u)+1 \geq (\alpha_0(u)+2\beta_0(u)+2\beta_2(u))+(\alpha_1(u)+2\beta_1(u)+\beta_3(u))+2\beta_4(u)+\beta_3(u)+\alpha_2(u)+1 = a+2b+1$ which contradicts to $|F| \leq a+2b$. Thus, $u$ is fault-free if $\alpha_0(u) + 2\beta_0(u) + \beta_2(u) \geq \alpha_1(u) + 2\beta_1(u) + \beta_3(u)$. □

Next, we provide an example to show that by our new structure, the local diagnosability can be improved a lot for some vertices compared to the previous two structures.

**Example 8.** *Let* $G = (V, E)$, *where* $V = \{p\} \cup \{x_i | 1 \le i \le 4n\} \cup \{u_i, z_i | 1 \le i \le 2n\}$ *and* $E = \{(p, x_i) | 1 \le i \le 4n\} \cup \{(x_i, u_j), (x_{2n+i}, u_{n+j}) | 1 \le i \le 2n, 1 \le j \le n\} \cup \{(u_i, z_j) | 1 \le i, j \le 2n\}$. *For any* $1 \le i \le 4n$ *and* $1 \le j \le 2n$, *we know that* $\deg_G(p) = 4n$, $\deg_G(x_i) = n + 1$, $\deg_G(u_j) = 4n$, $\deg_G(z_j) = 2n$.

*By Type I and Type II structures, we have* $t_l(p) \ge 2n$, $t_l(x_i) \ge n+1$, $t_l(u_j) \ge 2n$, $t_l(z_j) \ge 2n$ *for any* $i \in \{1, 2, \ldots, 4n\}$, $j \in \{1, 2, \ldots, 2n\}$.

*By the new structure we propose in this paper, we get that* $t_l(p) \ge 4n$, $t_l(x_i) \ge n + 1$, $t_l(u_j) \ge 2n + 1$, $t_l(z_j) \ge 2n$ *for any* $i \in \{1, 2, \ldots, 4n\}$, $j \in \{1, 2, \ldots, 2n\}$.

*As we can see, the local diagnosability of the vertices* $p, u_j$ $(j \in \{1, 2, \ldots, 2n\})$ *has been improved by our new structure compared to the Type I and Type II structures.*

## 4 Local Diagnosis of Incomplete Hypercube

In this section, we show that every vertex $u$ of an incomplete hypercube is locally $t$-diagnosable, where $t$ is the degree of the vertex $u$ in this incomplete hypercube. The incomplete hypercube is obtained by removing some edges from hypercube.

Let $\mathbf{x} = x_{n-1}x_{n-2} \cdots x_0$ and $\mathbf{y} = y_{n-1}y_{n-2} \cdots y_0$ be two binary strings with length $n$. We set $\mathbf{x^i} = q_{n-1}q_{n-2} \cdots q_0$ being the binary string such that $q_i = 1 - x_i$ and $q_j = x_j$ for each $j \ne i$. The Hamming distance between $\mathbf{x}$ and $\mathbf{y}$, $H(\mathbf{x}, \mathbf{y})$, is the number of distinct positions between them, i.e., $H(\mathbf{x}, \mathbf{y}) = \sum_{i=0}^{n-1} |x_i - y_i|$.

The $n$-dimensional hypercube, $Q_n$ is one of the most popular interconnection networks [10,19,21]. It is high symmetrical such as edge-transitivity and vertex-transitivity. And there are several well-known graphs of its variants such as the crossed cube [7], Möbius cube [4], twisted cube [11], folded hypercube [6], etc. Its vertex set is $V(Q_n) = \{x_{n-1}x_{n-2} \cdots x_0 \mid x_i \in \{0, 1\}$ for each $i \in \{0, 1, \ldots, n-1\}\}$, and its edge set $E(Q_n) = \{(\mathbf{x}, \mathbf{y}) \mid \mathbf{x}, \mathbf{y} \in V(Q_n)$ with $H(\mathbf{x}, \mathbf{y}) = 1\}$. That is, two vertices in $Q_n$ are adjacent if they differ in exactly one coordinate. Figure 3 shows $Q_n$ for $n \in \{1, 2, 3, 4\}$.

In order to construct the $T(u; a, 2b)$ structure for every vertex $u \in V(Q_n)$, we need the following known results.

**Theorem 9** *(Hall Theorem). Suppose that* $G = (V_1 \cup V_2, E)$ *is a bipartite graph with* $|V_1| \le |V_2|$. *Then* $G$ *has a matching saturating every vertex of* $V_1$ *if and only if* $|N(A)| \ge |A|$ *for every subset* $A \subseteq V_1$.

Next is the weak version of Hall Theorem.

**Theorem 10.** *Suppose that* $G = (V_1 \cup V_2, E)$ *is a bipartite graph with* $|V_1| \le |V_2|$. *If every vertex from* $V_1$ *has at least* $t$ *neighbors in* $V_2$ *and every vertex from* $V_2$ *has at most* $t$ *neighbors in* $V_1$ *for some* $t \ge 1$. *Then* $G$ *has a matching saturating every vertex of* $V_1$.

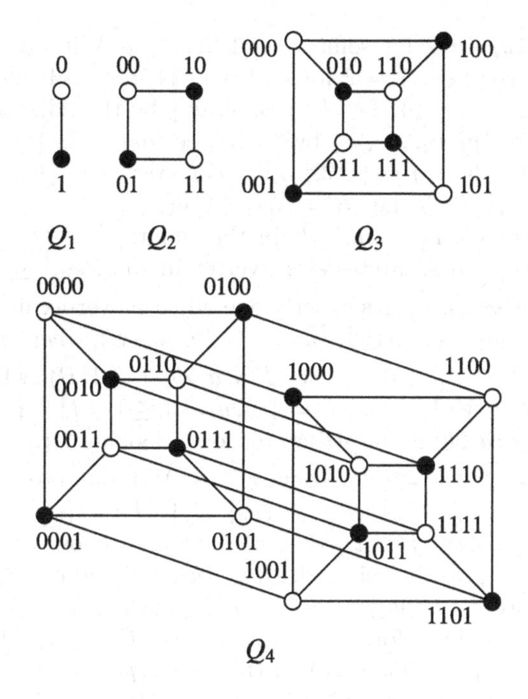


**Fig. 3.** $Q_n$ for $n \in \{1, 2, 3, 4\}$

Let $S$ be a set of faulty edges in $Q_n$. For any vertex $u \in V(Q_n)$ and a number $i \in \{1, 2, \ldots, d\}$ where $d$ stands for the diameter of $Q_n - S$, we define $\Gamma_i(u) = \{v \in V(Q_n) : dist_{Q_n - S}(u, v) = i\}$. We have the following observation.

**Lemma 11.** *For any vertex $u \in V(Q_n)$ and any two vertices $x, y \in \Gamma_2(u)$, if $u$ and $x$, $u$ and $y$ are in two edge-disjoint 4-cycles, then $x$ and $y$ are distance four.*

We set $\delta_{adj}(G) = \min\{\deg_G(x) + \deg_G(y) \mid (x, y) \in E(G)\}$.

**Theorem 12.** *For $n \geq 5$, let $S$ be any set of faulty edges in $Q_n$. Then the local diagnosability of any vertex $u$ in the incomplete hypercube $Q_n - S$ under PMC model is $\deg_{Q_n - S}(u)$ if $\delta(Q_n - S) \geq 2$ and $\delta_{adj}(Q_n - S) \geq 5$.*

*Proof.* Let $\Gamma_1(u) = \{u_1, u_2, \ldots, u_t\}$, where $t = \deg_{Q_n - S}(u)$. We want to show that there exists $T(u; a, 2b)$ in $Q_n - S$, where $t = a + 2b$. We classify into two cases.

*Case 1.* $\deg_{Q_n - S}(u_i) \geq 3$ for each $i \in \{1, 2, \ldots, t\}$. Since $Q_n$ is triangle-free, $u_i$ has at least two adjacent vertices in $\Gamma_2(u)$ for each $i \in \{1, 2, \ldots, t\}$. By the definition of $Q_n$, every vertex $v \in \Gamma_2(u)$ has at most two adjacent vertices in $\Gamma_1(u)$. By Theorem 10, there is a matching $M$ in the subgraph of $Q_n - S$ induced by $\Gamma_1(u) \cup \Gamma_2(u)$ saturates every vertex in $\Gamma_1(u)$. Thus, $T(u; a, 2b) = (A; B)$ is the structure we are looking for, where $A = \{u\} \cup V(M)$ and $B = \{(u, u_i) \mid 1 \leq i \leq t\} \cup M$.

*Case 2.* $\deg_{Q_n-S}(u_i) = 2$ for some $i \in \{1, 2, \ldots, t\}$. Without loss of generality, we assume $\deg_{Q_n-S}(u_i) = 2$ for each $i \in \{1, 2, \ldots, k\}$ where $1 \leq k \leq t$. We set $Z = \{u_1, u_2, \ldots, u_k\}$. Let $\{v_1, v_2, \ldots, v_l\}$ be the adjacent vertex set of $Z$ in $\Gamma_2(u)$. Denote $\{v_1, v_2, \ldots, v_l\}$ by $Y$. Notice that $v_j$ has at most two adjacent vertices in $Z$, where $j \in \{1, 2, \ldots, l\}$. For every $i \in \{k+1, k+2, \ldots, t\}$, we know $\deg_{Q_n-S}(u_i) \geq 3$. Let $R = N_{Q_n-S}(\{u_{k+1}, u_{k+2}, \ldots, u_t\}) \cap \Gamma_2(u)$. By Theorem 10, there is a matching $M'$ in the subgraph of $Q_n - S$ induced by $\{u_{k+1}, u_{k+2}, \ldots, u_t\} \cup R$ saturates every vertex in $\{u_{k+1}, u_{k+2}, \ldots, u_t\}$.

*Case 2.1.* $k = l$. That is, $v_j$ has exactly one adjacent vertex in $Z$ for every $j \in \{1, 2, \ldots, l\}$. Thus, we assume that $(u_j, v_j) \in E(Q_n - S)$, where $j \in \{1, 2, \ldots, k\}$.

*Case 2.1.1.* $|V(M') \cap Y| = 0$. We set $T(u; a, 2b) = (A; B)$, where $A = \{u\} \cup \{u_i, v_i \mid 1 \leq i \leq k\} \cup V(M')$ and $B = \{(u, u_i) \mid 1 \leq i \leq t\} \cup \{(u_i, v_i) \mid 1 \leq i \leq k\} \cup M'$. Then $T(u; a, 2b)$ is the structure we are looking for.

*Case 2.1.2.* $|V(M') \cap Y| = r$ for some $r \geq 1$. Without loss of generality, we assume that $V(M') \cap Y = \{v_1, v_2, \ldots, v_r\}$ and $M' = \{(u_{k+i}, v_i) \mid 1 \leq i \leq r\} \cup \{(u_{k+r+j}, p_j) \mid 1 \leq j \leq t-k-r\}$. Then we know that for every $i \in \{1, 2, \ldots, r\}$, $\deg_{Q_n-S}(v_i) \geq 3$ since $\deg_{Q_n-S}(u_i) = 2$ and $\delta_{adj}(Q_n - S) \geq 5$. By Lemma 11, there exist $\{w_1, w_2, \ldots, w_r\} \subseteq \Gamma_3(u)$ such that $(v_i, w_i) \in E(Q_n - S)$ for $i \in \{1, 2, \ldots, r\}$. Therefore, $T(u; a, 2b) = (A; B)$ is the structure we are looking, where $A = \{u\} \cup \Gamma_1(u) \cup Y \cup \{p_1, p_2, \ldots, p_{t-k-r}, w_1, w_2, \ldots, w_r\}$ and $B = \{(u, u_i) \mid 1 \leq i \leq t\} \cup \{(u_j, v_j) \mid 1 \leq i \leq k\} \cup \{(v_i, w_i) \mid 1 \leq i \leq r\} \cup M'$.

*Case 2.2.* $k > l$. That is to say, there are some index $i$ in $\{1, 2, \ldots, l\}$ such that $v_i$ has two adjacent vertices in $Z$. Without loss of generality, we assume that $v_i$ is adjacent to $u_{2i-1}$ and $u_{2i}$ for each $i \in \{1, 2, \ldots, s\}$, and $v_{s+j}$ is adjacent to $u_{2s+j}$ for each $j \in \{1, 2, \ldots, l-s\}$. Thus, $k = s+l$. Obviously, $V(M') \cap \{v_1, v_2, \ldots, v_s\} = \emptyset$.

*Case 2.2.1.* $|V(M') \cap \{v_{s+1}, v_{s+2}, \ldots, v_l\}| = 0$. Without loss of generality, we assume that $M' = \{(u_{k+j}, p_j) \mid 1 \leq j \leq t-k\}$. Similar to Case 2.1.2, there exist $s$ distinct vertices $w_1, w_2, \ldots, w_s$ in $\Gamma_3(u)$ such that $(v_i, w_i) \in E(Q_n - S)$ for each $i \in \{1, 2, \ldots, s\}$. Then $T(u; a, 2b) = (A; B)$ forms the structure we are looking for, where $A = \{u\} \cup Z \cup \{v_i \mid 1 \leq i \leq l\} \cup V(M') \cup \{w_i \mid 1 \leq i \leq s\}$ and $B = \{(u, u_i) \mid 1 \leq i \leq t\} \cup \{(u_{2i-1}, v_i), (u_{2i}, v_i), (v_i, w_i) \mid 1 \leq i \leq s\} \cup \{(u_{2s+j}, v_{s+j}) \mid 1 \leq j \leq l-s\} \cup M'$.

*Case 2.2.2.* $|V(M') \cap \{v_{s+1}, v_{s+2}, \ldots, v_l\}| \geq 1$. Without loss of generality, we assume that $V(M') \cap \{v_{s+1}, v_{s+2}, \ldots, v_l\} = \{v_{s+1}, v_{s+2}, \ldots, v_{s+q}\}$. Notice that $\deg_{Q_n-S}(v_i) \geq 3$ for $i \in \{1, 2, \ldots, l\}$ since $\deg_{Q_n-S}(u_i) = 2$ and $\delta_{adj}(Q_n - S) \geq 5$. We know that $u$ and every vertex from $\{v_1, v_2, \ldots, v_{s+q}\}$ are in edge-disjoint 4-cycles. By Lemma 11, any two vertices from $\{v_1, v_2, \ldots, v_{s+q}\}$ have no common neighbor. So there exist $\{w_1, w_2, \ldots, w_{s+q}\} \subseteq \Gamma_3(u)$ such that $(v_i, w_i) \in E(Q_n - S)$ for each $i \in \{1, 2, \ldots, s+q\}$. Therefore, $T(u; a, 2b) = (A; B)$ forms the structure we are looking for, where $a = \{u\} \cup \{u_i \mid 1 \leq i \leq k\} \cup \{v_j \mid 1 \leq j \leq l\} \cup \{w_i \mid 1 \leq i \leq s+q\} \cup V(M')$ and $B = \{(u, u_i) \mid 1 \leq i \leq t\} \cup \{(u_{2i-1}, v_i), (u_{2i}, v_i) \mid 1 \leq i \leq s\} \cup \{(u_{2s+j}, v_{s+j}) \mid 1 \leq i \leq l-s\} \cup M' \cup \{(v_i, w_i) \mid 1 \leq i \leq s+q\}$. □

Next, we provide two examples to show that our bound is optimal.

**Example 13.** *For $n \geq 5$, let $u = 000 \cdots 0, v = 100 \cdots 0, w = 110 \cdots 0, z = 010 \cdots 0 \in V(Q_n)$. Let $S$ be the set of edges incident with $v, w, z$ except the four edges in the four cycle $uvwzu$, then $\deg_{Q_n - S}(u) = n$, $\delta(Q_n - S) = 2$ and $\delta_{adj}(Q_n - S) = 4$. Let $F_1 = \{u\} \cup N(u) - \{v\}$, $F_2 = \{w\} \cup N(u) - \{z\}$, we have $|F_1| = |F_2| = n$ and $(F_1, F_2)$ is indistinguishable in $Q_n - S$. By Proposition 5, $u$ is not locally $n$-diagnosable.*

**Example 14.** *For $n \geq 5$, let $u = 000 \cdots 0, v = 100 \cdots 0 \in V(Q_n)$. Let $S$ be the set of edges incident with $v$ except the edge $(u, v)$, then $\deg_{Q_n - S}(u) = n$, $\delta(Q_n - S) = 1$ and $\delta_{adj}(Q_n - S) \geq 5$. Let $F_1 = \{u\} \cup N(u) - \{v\}$, $F_2 = N(u)$, we have $|F_1| = |F_2| = n$ and $(F_1, F_2)$ is indistinguishable in $Q_n - S$. By Proposition 5, $u$ is not locally $n$-diagnosable.*

**Acknowledgment.** This work was supported by Fujian Provincial Department of Science and Technology (2020J01268).

## References

1. Armstrong, J.R., Gray, F.G.: Fault diagnosis in a Boolean $n$-cube array of multiprocessors. IEEE Trans. Comput. **30**(8), 587–590 (1981)
2. Bondy, J.A., Murty, U.S.R.: Graph Theory. Springer, New York (2008)
3. Chang, G.-Y., Chang, G.J., Chen, G.-H.: Diagnosabilities of regular networks. IEEE Trans. Parallel Distrib. Syst. **16**(4), 314–323 (2005)
4. Cull, P., Larson, S.: The Möbius cubes. IEEE Trans. Comput. **44**(5), 647–659 (1995)
5. Dahbura, A.T., Masson, G.M.: An $O(n^{2.5})$ fault identification algorithm for diagnosable systems. IEEE Trans. Comput. **33**(6), 486–492 (1984)
6. EI-Awawy, A., Latifi, S.: Properties and performance of folded hypercubes. IEEE Trans. Parallel Distrib. Syst. **2**(1), 31–42 (1991)
7. Efe, K.: The crossed cube architecture for parallel computation. IEEE Trans. Parallel Distrib. Syst. **3**(5), 513–524 (1992)
8. Fan, J.: Diagnosability of crossed cubes under the two strategies. Chin. J. Comput. **21**(5), 456–462 (1998)
9. Fan, J.: Diagnosability of the Möbius cubes. IEEE Trans. Parallel Distrib. Syst. **9**(9), 923–928 (1998)
10. Harary, F., Hayes, J.P., Wu, H.-J.: A survey of the theory of hypercube graphs. Comput. Math. Appl. **15**(4), 277–289 (1988)
11. Hilbers, P.A.J., Koopman, M.R.J., van de Snepscheut, J.L.A.: The twisted cube. In: de Bakker, J.W., Nijman, A.J., Treleaven, P.C. (eds.) PARLE 1987. LNCS, vol. 258, pp. 152–159. Springer, Heidelberg (1987). https://doi.org/10.1007/3-540-17943-7_126
12. Hsu, G.H., Tan, J.J.M.: A local diagnosability measure for multiprocessor systems. IEEE Trans. Parallel Distrib. Syst. **18**, 598–607 (2007)
13. Hsu, L.-H., Lin, C.-K.: Graph Theory and Interconnection Networks. CRC Press, Boca Raton (2009)
14. Kavianpour, A., Kim, K.H.: Diagnosability of hypercube under the pessimistic one-step diagnosis strategy. IEEE Trans. Comput. **40**(2), 232–237 (1991)

15. Lee, C.-W., Hsieh, S.-Y.: Diagnosability of two-matching composition networks under the MM* model. IEEE Trans. Dependable Secure Comput. **8**(2), 246–255 (2011)
16. Lee, C.-W., Hsieh, S.-Y.: Determining the diagnosability of $(1, 2)$-matching composition networks and its applications. IEEE Trans. Dependable Secure Comput. **8**(3), 353–362 (2011)
17. Lai, P.-L., Tan, J.J.M., Chang, C.-P., Hsu, L.-H.: Conditional diagnosability measures for large multiprocessor systems. IEEE Trans. Comput. **54**(2), 165–175 (2005)
18. Lin, C.-K., Teng, Y.-H.: The diagnosability of triangle-free graphs. Theoret. Comput. Sci. **530**, 58–65 (2014)
19. Lin, C.-K., Zhang, L., Fan, J., Wang, D.: Structure connectivity and substructure connectivity of hypercubes. Theoret. Comput. Sci. **634**, 97–107 (2016)
20. Preparata, F.P., Metze, G., Chien, R.T.: On the connection assignment problem of diagnosable systems. IEEE Trans. Electron. Comput. **16**(12), 848–854 (1967)
21. Saad, Y., Schultz, M.H.: Topological properties of hypercubes. IEEE Trans. Comput. **37**(7), 867–872 (1988)
22. Wang, D.: Diagnosability of enhanced hypercubes. IEEE Trans. Comput. **43**(9), 1054–1061 (1994)
23. Wang, G., Lin, C.-K., Fan, J., Cheng, B., Jia, X.: A novel low cost interconnection architecture based on the generalized hypercube. IEEE Trans. Parallel Distrib. Syst. **31**(3), 647–662 (2020)

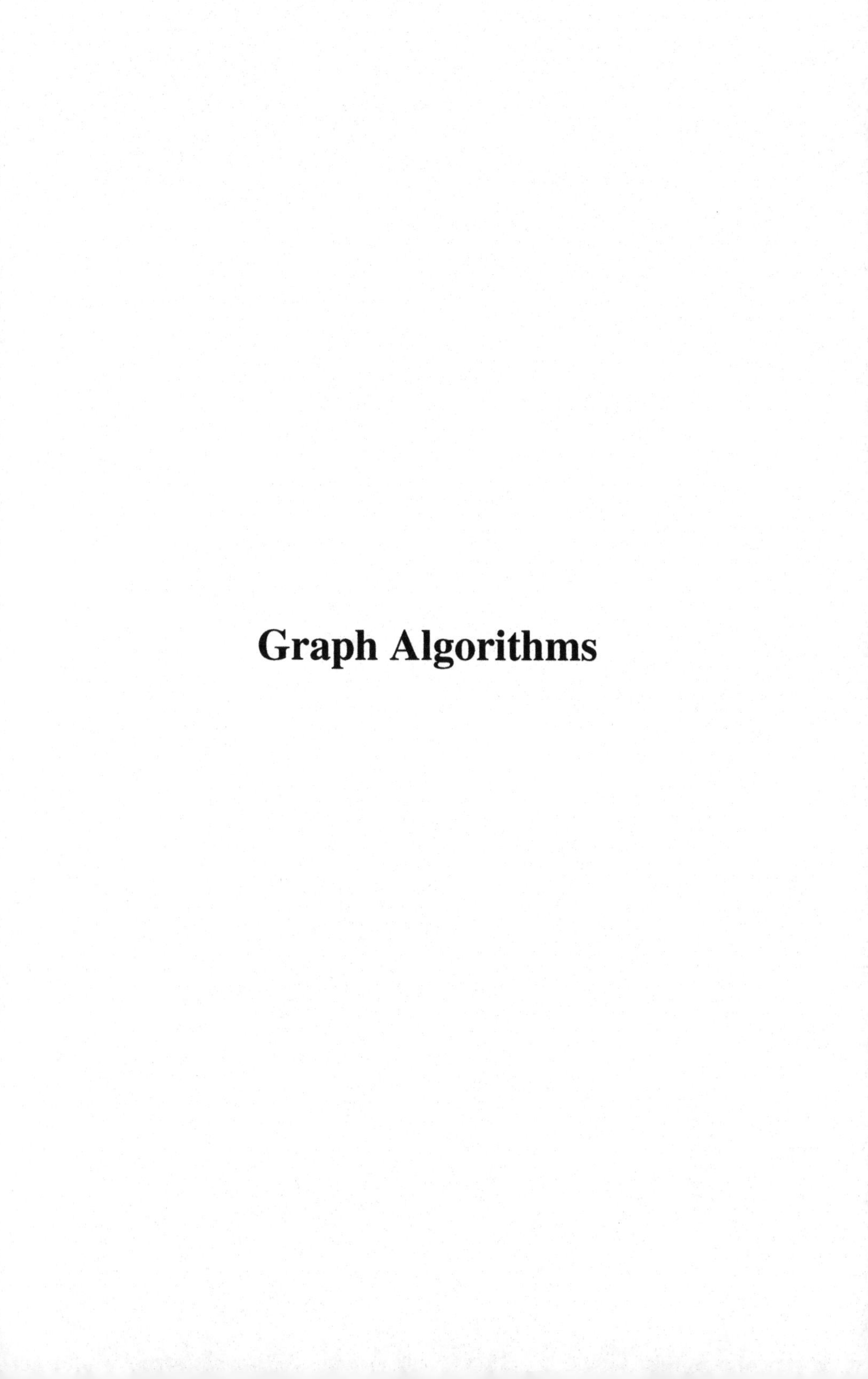

# Graph Algorithms

# Colouring Graphs with No Induced Six-Vertex Path or Diamond

Jan Goedgebeur[1,2], Shenwei Huang[3], Yiao Ju[3](✉), and Owen Merkel[4]

[1] Department of Applied Mathematics, Computer Science and Statistics, Ghent University, 9000 Ghent, Belgium
jan.goedgebeur@ugent.be
[2] Department of Computer Science, KU Leuven Kulak, 8500 Kortrijk, Belgium
[3] College of Computer Science, Nankai University, Tianjin 300350, China
shenweihuang@nankai.edu.cn, 2120190414@mail.nankai.edu.cn
[4] Department of Mathematics, Wilfrid Laurier University, Waterloo, Canada

**Abstract.** The diamond is the graph obtained by removing an edge from the complete graph on 4 vertices. A graph is ($P_6$, diamond)-free if it contains no induced subgraph isomorphic to a six-vertex path or a diamond. In this paper we show that the chromatic number of a ($P_6$, diamond)-free graph $G$ is no larger than the maximum of 6 and the clique number of $G$. We do this by reducing the problem to imperfect ($P_6$, diamond)-free graphs via the Strong Perfect Graph Theorem, dividing the imperfect graphs into several cases, and giving a proper colouring for each case. We also show that there is exactly one 6-vertex-critical ($P_6$, diamond, $K_6$)-free graph. Together with the Lovász theta function, this gives a polynomial time algorithm to compute the chromatic number of ($P_6$, diamond)-free graphs.

**Keywords:** Graph colouring · $k$-critical graph · $P_6$-free graph · Diamond-free graph

## 1 Introduction

All graphs in this paper are finite and simple. For general graph theory notation we follow [1]. A *q-colouring* of a graph $G$ assigns a colour from a colour set $\{1, \ldots, q\}$ to each vertex of $G$ such that adjacent vertices are assigned different colours. We say that a graph $G$ is *q-colourable* if $G$ admits a $q$-colouring. The problem of deciding if a graph is *q-colourable* is called the *q-colouring problem*. This decision problem is NP-complete for general graphs for every $q \geq 3$. However, there exist polynomial time algorithms when the input graphs are restricted to certain graph classes. For example, Chudnovsky, Spirkl, and Zhong [9] recently showed that the 4-colouring problem can be solved in polynomial time for the

S. Huang—Supported by the National Natural Science Foundation of China (11801284) and the Natural Science Foundation of Tianjin (20JCYBJC01190).


C.-Y. Chen et al. (Eds.): COCOON 2021, LNCS 13025, pp. 319–329, 2021.
https://doi.org/10.1007/978-3-030-89543-3_27

class of $P_6$-free graphs. The *chromatic number* of a graph $G$, denoted by $\chi(G)$, is the minimum number $q$ for which $G$ is $q$-colourable.

Let $P_n$, $C_n$ and $K_n$ denote the path, cycle and complete graph on $n$ vertices, respectively. The *diamond* is the graph obtained from $K_4$ by removing an edge. For two graphs $G$ and $H$, we use $G + H$ to denote the *disjoint union* of $G$ and $H$, and $G \vee H$ to denote the graph obtained from the disjoint union of $G$ and $H$ by adding an edge between every vertex in $G$ and every vertex in $H$. For a positive integer $r$, we use $rG$ to denote the disjoint union of $r$ copies of $G$. A *hole* is an induced cycle on 4 or more vertices. An *antihole* is the complement of a hole. A hole or antihole is *odd* or *even* if it has an odd or even number of vertices. We say that a graph $G$ *contains* a graph $H$ if an induced subgraph of $G$ is isomorphic to $H$. A graph $G$ is *$H$-free* if it does not contain $H$. For a family $\mathcal{H}$ of graphs, $G$ is *$\mathcal{H}$-free* if $G$ is $H$-free for every $H \in \mathcal{H}$. The graphs in $\mathcal{H}$ are the *forbidden induced subgraphs* of the family of $\mathcal{H}$-free graphs. We write $(H_1, \ldots, H_n)$-free instead of $\{H_1, \ldots, H_n\}$-free. A *clique* is a vertex set whose elements are pairwise adjacent. A vertex set is *stable* if its elements are pairwise nonadjacent, and *nonstable* otherwise. The *clique number* of $G$, denoted by $\omega(G)$, is the size of a largest clique in $G$. Obviously, $\chi(G) \geq \omega(G)$ for any graph $G$.

A graph family $\mathcal{G}$ is *hereditary* if $G \in \mathcal{G}$ implies that every induced subgraph of $G$ belongs to $\mathcal{G}$. Obviously, $\mathcal{G}$ is hereditary if and only if $\mathcal{G}$ is the class of $\mathcal{H}$-free graphs for some $\mathcal{H}$. A graph $G$ is *perfect* if $\chi(H) = \omega(H)$ for each induced subgraph $H$ of $G$, and *imperfect* otherwise. As a generalisation of perfect graphs, Gyárfás [16] introduced the $\chi$-bounded graph families. A hereditary graph family $\mathcal{G}$ is *$\chi$-bounded* if there is a function $f$ such that $\chi(G) \leq f(\omega(G))$ for every $G \in \mathcal{G}$. The function $f$ is called a *$\chi$-binding function*. Chudnovsky, Robertson, Seymour and Thomas [8] characterized the family of perfect graphs by forbidden induced subgraphs:

**Theorem 1** ([8]). *A graph is perfect if and only if it does not contain an odd hole or an odd antihole as an induced subgraph.*

In other words, the family of (odd hole, odd antihole)-free graphs is $\chi$-bounded, and its $\chi$-binding function is the identity function. Based on this theorem, researchers studied various graph families with two forbidden induced subgraphs and found several families that have a linear $\chi$-binding function. Note that every graph family mentioned in this paragraph forbids odd holes and odd antiholes on 7 or more vertices. It then follows from Theorem 1 that every graph in these graph families is either perfect or contains a $C_5$. In particular, Gaspers and Huang [11] showed that $\chi(G) \leq \frac{3}{2}\omega(G)$ for every ($P_6$, $C_4$)-free graph $G$. Karthick and Maffray [18] improved the $\chi$-binding function of ($P_6$, $C_4$)-free graphs to $\frac{5}{4}\omega(G)$. The graph $P_4 \vee K_1$ is called a *gem*, and a *co-gem* is the complement of a gem. Cameron, Huang and Merkel [4] showed that $\chi(G) \leq \lfloor \frac{3}{2}\omega(G) \rfloor$ for every ($P_5$, gem)-free graph $G$. Chudnovsky, Karthick, Maceli, and Maffray [7] improved the $\chi$-binding function of ($P_5$, gem)-free graphs to $\lceil \frac{5}{4}\omega(G) \rceil$. Karthick and Maffray [17] showed that $\chi(G) \leq \lceil \frac{5}{4}\omega(G) \rceil$ for every (gem, co-gem)-free graph $G$, and in [19] they showed that every ($P_5$, diamond)-free graph $G$ satisfies $\chi(G) \leq \omega(G) + 1$. For the family of ($P_6$, diamond)-free graphs, Karthick and Mishra [20] showed that every ($P_6$, diamond)-free graph $G$ satisfies $\chi(G) \leq 2\omega(G) + 5$. In the same paper, they proved that every ($P_6$, diamond, $K_4$)-free graph is 6-colourable. Finally, Cameron, Huang, and Merkel [5] improved the $\chi$-binding function of ($P_6$, diamond)-free graphs to $\omega(G) + 3$.

### Our Contributions

In this paper, we prove that every ($P_6$, diamond)-free graph $G$ satisfies $\chi(G) \leq \max\{6, \omega(G)\}$ (cf. Theorem 5 in Sect. 4). We do this by reducing the problem to imperfect ($P_6$, diamond)-free graphs via the Strong Perfect Graph Theorem, dividing the imperfect graphs into several cases, and giving a proper colouring for each case. Furthermore, we prove that the chromatic number of ($P_6$, diamond)-free graphs can be determined in polynomial time (cf. Theorem 7 in Sect. 5). In particular, we show that there is exactly one 6-vertex-critical ($P_6$, diamond, $K_6$)-free graph. Together with the Lovász theta function, this gives a polynomial time algorithm to compute the chromatic number of ($P_6$, diamond)-free graphs. Note that Dabrowski, Dross and Paulusma [10] proved the following dichotomy for computing the chromatic number of (diamond, $H$)-free graphs when $H$ has at most 5 vertices: they proved that this problem is NP-complete when $H$ contains a claw or a cycle, and polynomial time solvable otherwise. Our result thus generalises the polynomial time solvability of computing the chromatic number of ($P_5$, diamond)-free graphs.

Our results are an improvement of the result of Cameron, Huang and Merkel [5], answer an open question from [5], and are a natural next step of [10]. We believe that our proof technique for polynomial time solvability may also be useful for other graph families (see Sect. 6).

The remainder of the paper is organised as follows. We present some preliminaries in Sect. 2 and show some structural properties of imperfect ($P_6$, diamond)-free graphs in Sect. 3. We prove the $\chi$-bound in Sect. 4 and prove that the chromatic number can be determined in polynomial time in Sect. 5. We end with some open problems in Sect. 6.

Due to space constraints we had to omit several proofs. These proofs can be found in the full version of this paper, of which a preprint is already available on arXiv [13].

## 2 Preliminaries

Let $G = (V, E)$ be a graph. A *neighbour* of a vertex $v$ is a vertex adjacent to $v$. The *neighbourhood* of a vertex $v$, denoted by $N_G(v)$, is the set of neighbours of $v$. The *degree* of a vertex $v$, denoted by $d(v)$, is the number of neighbours of $v$. We denote the minimum degree of the vertices of $G$ by $\delta(G)$. For $X \subseteq V$, let $N_G(X) = \bigcup_{v \in X} N_G(v) \setminus X$. For $x \in V$ (or $X \subseteq V$) and $S \subseteq V$, let $N_S(x) = N_G(x) \cap S$ (or $N_S(X) = N_G(X) \cap S$). For $x \in V$ (or $X \subseteq V$) and $Y \subseteq V$, we say that $x$ (or $X$) is *complete* (resp. *anti-complete*) to $Y$ if $x$ (or every vertex in $X$) is adjacent (resp. nonadjacent) to every vertex in $Y$. We denote the complement of $G$ by $\overline{G}$. For $S \subseteq V$, let $G[S]$ denote the subgraph of $G$ induced by $S$. We often write $S$ for $G[S]$ if the context is clear. We say that $S$ induces an $H$ if $G[S]$ is isomorphic to $H$. A clique $K \subseteq V$ is a *clique cutset* if $G-K$ has more components than $G$. Two vertices $u, v \in V$ are *comparable* if they are nonadjacent, and either $N_G(u) \subseteq N_G(v)$ or $N_G(v) \subseteq N_G(u)$. A component of a graph is *trivial* if it has only one vertex, and *nontrivial* otherwise. We say that the edges between two vertex sets $X$ and $Y$ form a *matching* if every vertex in $X$ has at most one neighbour in $Y$, and every vertex in $Y$ has at most one neighbour in $X$.

Grötschel, Lovász and Schrijver [15] showed that the chromatic number of a perfect graph can be computed in polynomial time. In that paper, the authors used the Lovász theta function:

$$\vartheta(G) := max\{\sum_{i,j=1}^{n} b_{ij} :$$

$$B = (b_{ij}) \text{ is positive semidefinite with trace at most 1, and } b_{ij} = 0 \text{ if } ij \in E\}.$$

The Lovász theta function satisfies that $\omega(G) \leq \vartheta(\overline{G}) \leq \chi(G)$ for any graph $G$, and can be calculated in polynomial time [15].

A graph $G$ is *$k$-vertex-critical* if $\chi(G) = k$, and every proper induced subgraph of $G$ has chromatic number smaller than $k$. The following properties of $k$-vertex-critical graphs are well-known.

**Lemma 1** ([1]). *If $G$ is a $k$-vertex-critical graph, then $G$ is connected, has no clique cutsets or comparable vertices, and $\delta(G) \geq k-1$.*

It is easy to see that $G$ is not $(k-1)$-colourable if and only if $G$ contains a $k$-vertex-critical graph. This simple observation has an important algorithmic implication, the proof of which can be found in [3].

**Theorem 2 (Folklore).** *If a hereditary graph family $\mathcal{G}$ has a finite number of $k$-vertex-critical graphs, then the $(k-1)$-colouring problem can be solved in polynomial time for $\mathcal{G}$ by simply testing if the input graph contains any of these $k$-vertex-critical graphs as induced subgraph.*

The concept of $k$-vertex-critical graphs is also important in the context of certifying algorithms [21]. An algorithm is *certifying* if, along with the answer given by the algorithm, it also gives a certificate which allows to verify in polynomial time that the output of the algorithm is indeed correct. In case of the $k$-colouring problem, a canonical certificate for yes-instances would be a proper $k$-colouring of the graph while a canonical certificate for no-instances would be a $(k+1)$-vertex-critical graph.

We will also use the following two theorems in our proof:

**Theorem 3** ([9]). *The 4-colouring problem can be solved in polynomial time for the class of $P_6$-free graphs.*

**Theorem 4** ([22]). *Let $G$ be a $(P_6, K_3)$-free graph with no comparable vertices. Then $G$ is 4-colourable. Furthermore, $G$ is not 3-colourable if and only if it contains the Grötzsch graph as an induced subgraph and is an induced subgraph of the 16-vertex Clebsch graph. (See Fig. 1 for drawings of the Clebsch graph and the Grötzsch graph.)*

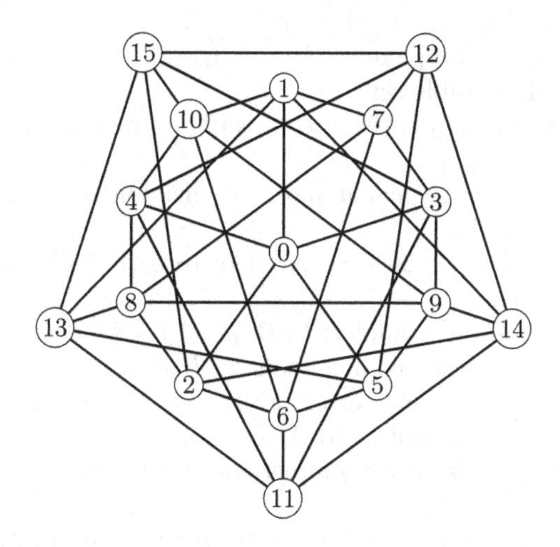


**Fig. 1.** The Clebsch graph. The 11 vertices with indices $0, \dots, 10$ induce the Grötzsch graph.

## 3 The Structure of Imperfect ($P_6$, diamond)-Free Graphs

In this section we study the structure of imperfect ($P_6$, diamond)-free graphs. By Theorem 1, every imperfect ($P_6$, diamond)-free graph contains a $C_5$. Let $G = (V, E)$ be an imperfect ($P_6$, diamond)-free graph. We follow the notation of [5] and partition $V$ into the following subsets:

Let $Q = \{v_1, v_2, v_3, v_4, v_5\}$ induce a $C_5$ in $G$ with edges $v_i v_{i+1}$ for $i = 1, \dots, 5$, with all indices modulo 5.

$$A_i = \{v \in V \backslash Q : N_Q(v) = \{v_i\}\},$$
$$B_{i,i+1} = \{v \in V \backslash Q : N_Q(v) = \{v_i, v_{i+1}\}\},$$
$$C_{i,i+2} = \{v \in V \backslash Q : N_Q(v) = \{v_i, v_{i+2}\}\},$$
$$F_i = \{v \in V \backslash Q : N_Q(v) = \{v_i, v_{i-2}, v_{i+2}\}\},$$
$$Z = \{v \in V \backslash Q : N_Q(v) = \emptyset\}.$$

Let $A = \bigcup_{i=1}^{5} A_i$, $B = \bigcup_{i=1}^{5} B_{i,i+1}$, $C = \bigcup_{i=1}^{5} C_{i,i+2}$, $F = \bigcup_{i=1}^{5} F_i$. Since $G$ is diamond-free, any vertex in $V \backslash Q$ cannot be adjacent to three sequential vertices $v_i, v_{i+1}, v_{i+2}$ in $Q$, then $V = Q \cup A \cup B \cup C \cup F \cup Z$.

In [5] the following 21 properties of these subsets are proved:

(P1) Each component of $A_i$ is a clique.
(P2) The sets $A_i$ and $A_{i+1}$ are anti-complete.
(P3) The sets $A_i$ and $A_{i+2}$ are complete.
(P4) Each $B_{i,i+1}$ is a clique.
(P5) The set $B = B_{i,i+1} \cup B_{i+2,i+3}$ for some $i$.
(P6) The set $B_{i,i+1}$ is anti-complete to $A_i \cup A_{i+1}$.
(P7) The set $B_{i,i+1}$ is complete to $A_{i-1} \cup A_{i+2}$.
(P8) Each $C_{i,i+2}$ is a stable set.
(P9) Each vertex in $C_{i,i+2}$ is either complete or anti-complete to each component of $A_i$ and $A_{i+2}$.
(P10) Each vertex in $C_{i,i+2}$ has at most one neighbour in each component of $A_{i+1}$, $A_{i+3}$ and $A_{i+4}$.
(P11) Each vertex in $C_{i,i+2}$ is anti-complete to each nontrivial component of $A_{i+1}$.
(P12) The set $C_{i,i+2}$ is anti-complete to $B_{j,j+1}$ if $j \neq i+3$. Moreover each vertex in $C_{i,i+2}$ has at most one neighbour in $B_{i+3,i+4}$.
(P13) Each $F_i$ has at most one vertex. Moreover, $F$ is a stable set.
(P14) The set $F_i$ is anti-complete to $A_{i+2} \cup A_{i+3}$.
(P15) Each vertex in $F_i$ is either complete or anti-complete to each component of $A_i$.
(P16) Each vertex in $F_i$ has at most one neighbour in each component of $A_{i+1}$ and $A_{i+4}$.
(P17) The set $F_i$ is anti-complete to $B_{j,j+1}$ if $j \neq i+2$ and complete to $B_{j,j+1}$ if $j = i+2$.
(P18) The set $F_i$ is anti-complete to $C_{j,j+2}$ if $j \neq i-1$.
(P19) If $A_i$ is not stable, then $A_{i+2} = A_{i+3} = B_{i+1,i+2} = B_{i-1,i-2} = \emptyset$.
(P20) If $A_i$ is not empty, then each of $B_{i+1,i+2}$ and $B_{i-1,i-2}$ contains at most one vertex.
(P21) The set $Z$ is anti-complete to $A \cup B$.

Now we prove some new properties which we will use in our proofs in Sect. 4 and Sect. 5:

(P22) If $A_i$, $B_{i+1,i+2}$ and $B_{i+3,i+4}$ are all nonempty, then $C_{i+1,i+3}$ (resp. $C_{i+2,i+4}$) is anti-complete to $A_{i+2}$ (resp. $A_{i+3}$).

*Proof.* Suppose that $a_1 \in A_i$, $a_2 \in A_{i+2}$, $b_1 \in B_{i+1,i+2}$, $b_2 \in B_{i+3,i+4}$, $c \in C_{i+1,i+3}$ such that $a_2c \in E$. By (P3), $a_1a_2 \in E$. By (P6), $b_1a_2 \notin E$. By (P7), $a_1b_1, a_1b_2, a_2b_2 \in E$. By (P12), $b_1c, b_2c \notin E$. Then either $\{a_1, a_2, b_2, c\}$ induces a diamond or $\{b_1, a_1, a_2, c, v_{i+3}, v_{i+4}\}$ induces a $P_6$, depending on whether $a_1$ and $c$ are adjacent. □

(P23) Suppose that $B_{i,i+1}$ is nonempty. Then $C_{i,i+2} \cup C_{i-2,i}$ (resp. $C_{i-1,i+1} \cup C_{i+1,i+3}$) is complete to $A_{i-1}$ (resp. $A_{i+2}$). Moreover if $A_{i-1}$ (resp. $A_{i+2}$) is nonempty, then $C_{i,i+2}$ and $C_{i-2,i}$ (resp. $C_{i-1,i+1}$ and $C_{i+1,i+3}$) are anti-complete.

*Proof.* Let $a \in A_{i-1}$, $b \in B_{i,i+1}$. Suppose there is a vertex $c_1 \in C_{i,i+2}$ such that $ac_1 \notin E$ or $c_2 \in C_{i-2,i}$ such that $ac_2 \notin E$. By (P12), $bc_1, bc_2 \notin E$. Then $\{c_1, v_{i+2}, v_{i-2}, v_{i-1}, a, b\}$ or $\{c_2, v_{i-2}, v_{i-1}, a, b, v_{i+1}\}$ induces a $P_6$. This proves the first part of the claim. Suppose that $c_1 \in C_{i,i+2}$ and $c_2 \in C_{i-2,i}$ are adjacent, then $\{a, c_1, c_2, v_i\}$ induces a diamond. □

(P24) Suppose that $A_i \cup A_{i+1}$ is nonempty. Then $B_{i,i+1}$ is complete to $A_{i+3}$. Moreover if $B_{i,i+1}$ contains two or more vertices, then $A_{i+3}$ is empty.

*Proof.* By symmetry suppose that $a_1 \in A_i$, $a_2 \in A_{i+3}$, $b \in B_{i,i+1}$ such that $ba_2 \notin E$. By (P3), $a_1a_2 \in E$. By (P6), $ba_1 \notin E$. Then $\{b, v_i, a_1, a_2, v_{i+3}, v_{i+2}\}$ induces a $P_6$. This proves the first part of the claim. Suppose that $b_1, b_2 \in B_{i,i+1}$, $a \in A_{i+3}$. By (P4), $b_1b_2 \in E$. Then $\{v_i, b_1, b_2, a\}$ induces a diamond. □

(P25) If $B_{i,i+1}$ is nonempty, then $C_{i-2,i}$ (resp. $C_{i+1,i+3}$) is anti-complete to $A_i$ (resp. $A_{i+1}$).

*Proof.* Suppose that $a \in A_i$, $b \in B_{i,i+1}$, $c \in C_{i-2,i}$ such that $ac \in E$. By (P6), $ab \notin E$. By (P12), $bc \notin E$. Then $\{a, c, v_{i-2}, v_{i+2}, v_{i+1}, b\}$ induces a $P_6$. □

(P26) If $B_{i,i+1}$ is nonempty, then $Z$ is anti-complete to $C_{i+1,i+3} \cup C_{i+3,i}$.

*Proof.* By symmetry suppose that $b \in B_{i,i+1}$, $c \in C_{i+1,i+3}$ and $z \in Z$ such that $cz \in E$. By (P12), $bc \notin E$. By (P21), $bz \notin E$. Then $\{z, c, v_{i+3}, v_{i+4}, v_i, b\}$ induces a $P_6$. □

## 4 The $\chi$-Bound of ($P_6$, diamond)-Free Graphs

In this section, we prove the following theorem.

**Theorem 5.** *Let $G$ be a ($P_6$, diamond)-free graph, then $\chi(G) \leq max\{6, \omega(G)\}$.*

This theorem is an improvement of the result of Cameron, Huang and Merkel [5] that $\chi(G) \leq \omega(G) + 3$ for a ($P_6$, diamond)-free graph $G$. To prove Theorem 5, we use Theorem 6 below.

**Theorem 6.** *Let $G$ be a connected imperfect ($P_6$, diamond)-free graph with no clique cutsets or comparable vertices. Then $\chi(G) \leq max\{6, \omega(G)\}$.*

*Proof (Proof of Theorem 5).* If $G$ is perfect, then $\chi(G) = \omega(G)$. If $G$ is disconnected and is the union of components $\{H_1, \ldots, H_n\}$, then $\chi(G) = max\{\chi(H_1), \ldots, \chi(H_n)\}$ and $\omega(G) = max\{\omega(H_1), \ldots, \omega(H_n)\}$. If $G$ is connected and contains a clique cutset $S$, $G - S$ is the disjoint union of subgraphs $\{H_1, \ldots, H_n\}$, then $\chi(G) = \max\{\chi(G[V(H_1) \cup S], \ldots, \chi(G[V(H_n) \cup S]\}$ and $\omega(G) = \max\{\omega(G[V(H_1) \cup S], \ldots, \omega(G[V(H_n) \cup S]\}$. If $u$ and $v$ are non-adjacent vertices in $G$ such that $N(u) \subseteq N(v)$, then $\chi(G) = \chi(G - u)$ and $\omega(G) = \omega(G - u)$. Then the theorem follows from Theorem 6. □

Due to page limits we omit the proof of Theorem 6.

## 5 Computing the Chromatic Number of ($P_6$, diamond)-Free Graphs in Polynomial Time

In this section we prove the following Theorem 7:

**Theorem 7.** *The chromatic number of ($P_6$, diamond)-free graphs can be computed in polynomial time.*

This answers an open question from [5]. To prove Theorem 7, we use Theorem 8 below.

**Theorem 8.** *The chromatic number of ($P_6$, diamond, $K_6$)-free graphs can be computed in polynomial time.*

To prove Theorem 8, we use the following three theorems.

**Theorem 9.** *There is one 6-vertex-critical ($P_6$, diamond)-free graph with clique number 3.*

**Theorem 10.** *There are no 6-vertex-critical ($P_6$, diamond)-free graphs with clique number 4.*

**Theorem 11.** *There are no 6-vertex-critical ($P_6$, diamond)-free graphs with clique number 5.*

**Fig. 2.** The complement of the 27-vertex Schläfli graph. The unique 6-vertex-critical ($P_6$, diamond)-free graph $\mathcal{G}$ with clique number 3 is obtained by removing the vertices with labels $x$ and $y$.

*Proof (Proof of Theorem 7).* Let $G$ be a ($P_6$, diamond)-free graph. We can calculate $\vartheta(\overline{G})$ in polynomial time [15]. When $\vartheta(\overline{G}) > 5$, since $\omega(G) \leq \vartheta(\overline{G}) \leq \chi(G)$ [15] and $\chi(G) \leq max\{\omega(G), 6\}$ (i.e. Theorem 5), we have $\chi(G) = \lceil \vartheta(\overline{G}) \rceil$. Then we just have to deal with the case that $\vartheta(\overline{G}) \leq 5$, which implies that $\omega(G) \leq 5$. Then the theorem follows from Theorem 8. □

**Remark.** As pointed out by one of the referees, we could also prove Theorem 7 using the fact that the clique number of a diamond-free graph can be computed in polynomial time: if $\omega(G) \geq 6$, then $\chi(G) = \omega(G)$; otherwise $\omega(G) \leq 5$ and then we use Theorem 8. However, we chose to use the Lovász theta function, since it is a more general strategy that could possibly also be useful for other graph classes.

*Proof (Proof of Theorem 8).* By Theorem 5 and Theorem 3, we only need to consider 5-colouring. By Theorem 2, we need to prove that there are a finite number of 6-vertex-critical ($P_6$, diamond, $K_6$)-free graphs. By Theorem 4, every 6-vertex-critical ($P_6$, diamond)-free graph has clique number at least 3. Then the theorem follows from Theorem 9, Theorem 10, and Theorem 11. □

The unique 6-vertex-critical ($P_6$, diamond)-free graph $\mathcal{G}$ with clique number 3 from Theorem 9 has 25 vertices and can be obtained from the complement of the Schläfli graph by deleting the vertices labelled $x$ and $y$ in Fig. 2. This graph can also be inspected at *the House of Graphs* [2] at: https://hog.grinvin.org/ViewGraphInfo.action?id=45613.

The proof of Theorem 9 uses computational methods. In the full version of this paper (cf. [13] for a preprint) we give a computer-free proof of a weaker version of Theorem 9: there we show that there are finitely many 6-vertex-critical ($P_6$, diamond)-free graphs with clique number 3. This weaker theorem still suffices to give a complete computer-free proof of Theorem 7 and Theorem 8.

*Proof (Proof of Theorem 9).* We used the generation algorithm for $k$-critical $H$-free graphs from [6,14] and extended it to generate 6-vertex-critical ($P_6$, diamond, $K_4$)-free graphs. The algorithm terminated in less than 5 min and yielded the graph $\mathcal{G}$ as the only 6-vertex-critical ($P_6$, diamond, $K_4$)-free graph. The source code of the program can be downloaded from [12]. We refer to [6,14] for more details on the algorithm and the proof of its correctness. □

We need the following lemmas for our proofs of Theorem 10 and Theorem 11.

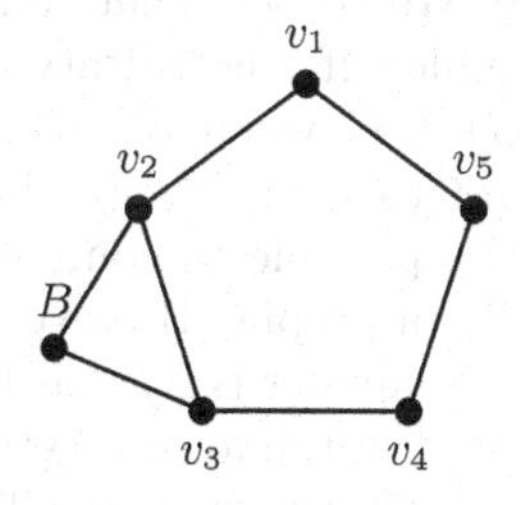


**Fig. 3.** The graph series $S_n$. $B$ is a clique of size $n$ such that $B$ is complete to $\{v_2, v_3\}$ and anti-complete to $\{v_1, v_4, v_5\}$.

**Lemma 2.** *Every 6-vertex-critical ($P_6$, diamond)-free graph with clique number $\omega$ ($\omega = 4, 5$) is $S_{\omega-2}$-free. (See Fig. 3 for $S_n$.)*

**Lemma 3.** *Every 6-vertex-critical ($P_6$, diamond)-free graph with clique number 3 is $D_1$-free. (See Fig. 4a for $D_1$.)*

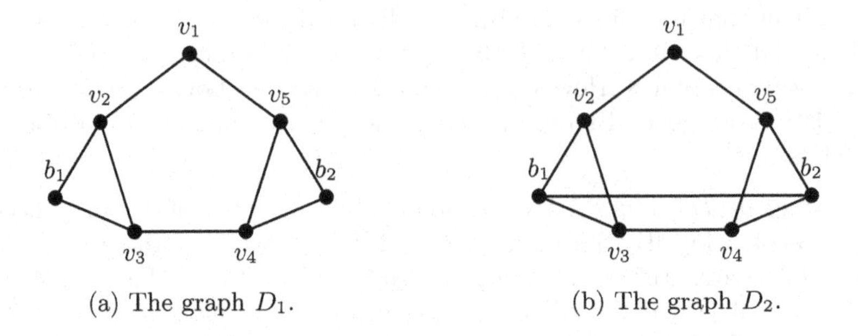


(a) The graph $D_1$.

(b) The graph $D_2$.

**Fig. 4.** The graphs $D_1$ and $D_2$.

**Lemma 4.** *Every 6-vertex-critical ($P_6$, diamond)-free graph with clique number 3 is $D_2$-free. (See Fig. 4b for $D_2$.)*

**Lemma 5.** *Every 6-vertex-critical ($P_6$, diamond)-free graph with clique number 3 is $S_1$-free. (See Fig. 3 for $S_1$.)*

**Lemma 6.** *Every 6-vertex-critical ($P_6$, diamond)-free graph with clique number $\omega$ ($\omega = 3, 4, 5$) is ($K_\omega + K_1$)-free.*

Due to page limits we omit the proof of Lemmas 2–6, Theorem 10 and Theorem 11.

## 6 Conclusion

In this paper, we improved the $\chi$-bound of ($P_6$, diamond)-free graphs from $\chi(G) \leq \omega(G) + 3$ [5] to $\chi(G) \leq \max\{6, \omega(G)\}$. Moreover, we proved that the chromatic number of graphs in the class of ($P_6$, diamond)-free graphs can be calculated in polynomial time. We suspect that similar results can be obtained for other hereditary graph families: if a hereditary graph family has a $\chi$-bound in the form of $\chi(G) \leq \omega(G) + C$ where $C$ is a constant, then it may be possible to improve the bound to $\max\{C', \omega(G)\}$ where $C'$ is a constant. If that is the case, it may also be possible to compute the chromatic number in polynomial time for this family of graphs. However, this is not always possible since determining the chromatic number is NP-hard for the hereditary family of line graphs, which has a $\chi$-bound in the form of $\chi(G) \leq \omega(G) + C$. So it would be interesting to find other hereditary graph families for which the chromatic number can be determined in polynomial time in this way.

## References


1. Bondy, J.A., Murty, U.S.R.: Graph Theory. Springer, London (2008)
2. Brinkmann, G., Coolsaet, K., Goedgebeur, J., Mélot, H.: House of graphs: a database of interesting graphs. Discret. Appl. Math. **161**(1–2), 311–314 (2013). https://hog.grinvin.org/
3. Cai, Q., Huang, S., Li, T., Shi, Y.: Vertex-critical ($P_5$, banner)-free graphs. In: Chen, Y., Deng, X., Lu, M. (eds.) FAW 2019. LNCS, vol. 11458, pp. 111–120. Springer, Cham (2019). https://doi.org/10.1007/978-3-030-18126-0_10
4. Cameron, K., Huang, S., Merkel, O.: A bound for the chromatic number of ($P_5$, gem)-free graphs. Bull. Aust. Math. Soc. **100**(2), 182–188 (2019)
5. Cameron, K., Huang, S., Merkel, O.: An optimal $\chi$-bound for ($P_6$, diamond)-free graphs. J. Graph Theory **97**(3), 451–465 (2021)
6. Chudnovsky, M., Goedgebeur, J., Schaudt, O., Zhong, M.: Obstructions for three-coloring graphs without induced paths on six vertices. J. Comb. Theory Ser. B **140**, 45–83 (2020)
7. Chudnovsky, M., Karthick, T., Maceli, P., Maffray, F.: Coloring graphs with no induced five-vertex path or gem. J. Graph Theory **95**(4), 527–542 (2020)
8. Chudnovsky, M., Robertson, N., Seymour, P., Thomas, R.: The strong perfect graph theorem. Ann. Math. **164**, 51–229 (2006)
9. Chudnovsky, M., Spirkl, S., Zhong, M.: Four-coloring $P_6$-free graphs. In: Proceedings of the Thirtieth Annual ACM-SIAM Symposium on Discrete Algorithms (SODA 2019), pp. 1239–1256 (2019)
10. Dabrowski, K.K., Dross, F., Paulusma, D.: Colouring diamond-free graphs. J. Comput. Syst. Sci. **89**, 410–431 (2017)
11. Gaspers, S., Huang, S.: Linearly $\chi$-bounding $(P_6, C_4)$-free graphs. In: Proceedings of the 43rd International Workshop on Graph-Theoretic Concepts in Computer Science (WG 2017), pp. 263–274 (2017)
12. Goedgebeur, J.: Homepage of a generator for $k$-critical $H$-free graphs. https://caagt.ugent.be/criticalpfree/
13. Goedgebeur, J., Huang, S., Ju, Y., Merkel, O.: Colouring graphs with no induced six-vertex path or diamond. arXiv preprint arXiv:2106.08602 (2021)
14. Goedgebeur, J., Schaudt, O.: Exhaustive generation of $k$-critical $\mathcal{H}$-free graphs. J. Graph Theory **87**, 188–207 (2018)
15. Grötschel, M., Lovász, L., Schrijver, A.: The ellipsoid method and its consequences in combinatorial optimization. Combinatorica **1**(2), 169–197 (1981)
16. Gyárfás, A.: Problems from the world surrounding perfect graphs. Appl. Math. **19**(3–4), 413–441 (1987)
17. Karthick, T., Maffray, F.: Coloring (gem, co-gem)-free graphs. J. Graph Theory **89**(3), 288–303 (2018)
18. Karthick, T., Maffray, F.: Square-free graphs with no six-vertex induced path. SIAM J. Discret. Math. **33**(2), 874–909 (2019)
19. Karthick, T., Maffray, F.: Vizing bound for the chromatic number on some graph classes. Graphs Comb. **32**, 1447–1460 (2016)
20. Karthick, T., Mishra, S.: On the chromatic number of ($P_6$, diamond)-free graphs. Graphs Comb. **34**, 677–692 (2018)
21. McConnell, R.M., Mehlhorn, K., Näher, S., Schweitzer, P.: Certifying algorithms. Comput. Sci. Rev. **5**(2), 119–161 (2011)
22. Randerath, B., Schiermeyer, I., Tewes, M.: Three-colourability and forbidden subgraphs II: polynomial algorithms. Discret. Math. **251**, 137–153 (2002)

# Constructing Tri-CISTs in Shuffle-Cubes

Yu-Han Chen[1], Kung-Jui Pai[2], Hsin-Jung Lin[1], and Jou-Ming Chang[1](✉)

[1] Institute of Information and Decision Sciences, National Taipei University of Business, Taipei, Taiwan
{10866005,10966013,spade}@ntub.edu.tw

[2] Department of Industrial Engineering and Management, Ming Chi University of Technology, New Taipei City, Taiwan
poter@mail.mcut.edu.tw

**Abstract.** A set of spanning trees $\mathcal{T} = \{T_1, T_2, \ldots, T_k\}$ with $k \geqslant 2$ in a graph $G$ is called completely independent spanning trees (CISTs for short) if the paths joining every pair of vertices $x$ and $y$ in any two trees have neither vertex nor edge in common except for $x$ and $y$. Particularly, $\mathcal{T}$ is called a dual-CIST (resp. tri-CIST) provided $k = 2$ (resp. $k = 3$). Recently, the construction of a dual-CIST has been proposed in a shuffle-cube $SQ_n$, which is an innovative hypercube-variant network that possesses both short diameter and connectivity advantages. This paper uses the CIST-partition technique to construct a tri-CIST of $SQ_6$, and shows that the diameters of three CISTs are 22, 22, and 13. Then, by the hierarchical structure of $SQ_n$, we propose a recursive algorithm for constructing a tri-CIST for high-dimensional shuffle-cubes. When $n \geqslant 10$, the diameters of $T_i$, $i = 1, 2, 3$, we constructed for $SQ_n$ are as follows: $2n + 11$, $2n + 9$, and $2n + 1$.

**Keywords:** Shuffle-cubes · Completely independent spanning trees · Interconnection networks · Diameter

## 1 Introduction

Let $k \geqslant 2$ be an integer and $T_1, T_2, \ldots, T_k$ be spanning trees of a graph $G = (V, E)$. A vertex $v$ is called an *inner-vertex* of $T_i$ if it has at least two neighbors in $T_i$, and a *leaf* otherwise. Two spanning trees $T_i$ and $T_j$, $1 \leqslant i, j \leqslant k$, are *edge-disjoint* if they share no common edge, and are *inner-vertex-disjoint* if the paths joining any two vertices $u, v \in V$ in both trees have no vertex in common except for $u$ and $v$. The spanning trees $T_1, T_2, \ldots, T_k$ are *completely independent spanning trees* (CISTs for short) if they are pairwise inner-vertex-disjoint (and thus are edge-disjoint, see Theorem 1). Particularly, two CISTs and three CISTs are call a *dual-CIST* and a *tri-CIST*, respectively, in this paper.

Hasunuma [7,8] first devoted to the theoretical study on CISTs and showed that the problem of determining whether if a graph $G$ admits $k$ CISTs is NP-complete, even for $k = 2$ (i.e., a dual-CIST). Moreover, Hasunuma [8] posted a conjecture which says that there exist $\lfloor k/2 \rfloor$ CISTs in a $k$-connected graph.


C.-Y. Chen et al. (Eds.): COCOON 2021, LNCS 13025, pp. 330–342, 2021.
https://doi.org/10.1007/978-3-030-89543-3_28

However, this conjecture has been proved to fail by counterexamples [18]. With the help of constructions, it has been confirmed that many classes of graphs possess a dual-CIST, e.g., 4-connected maximal planar graphs [8], Cartesian product of any 2-connected graphs [9], hypercube-variant networks [2,4,12,19], subclasses of Cayley graphs [13], and DCell data center networks [20]. Also, it was known that a few specific classes of graphs possess a tri-CIST, e.g., locally twisted cubes and crossed cubes [16,17], Möbius cubes [14], and alternating group graphs [15]. Moreover, researches on constructing multiple CISTs and other issues related to CISTs can be found in [3,5,11,21]. The following two characterizations are important for studying CISTs.

**Theorem 1 (Hasunuma [7]).** *$T_1, T_2, \ldots, T_k$ are CISTs in a graph $G$ if and only if they are edge-disjoint and for any vertex $v \in V(G)$, there is at most one tree $T_i$ for $i \in \{1, 2, \ldots, k\}$ such that $v$ is an inner-vertex of $T_i$.*

**Theorem 2 (Araki [1]).** *A graph $G = (V, E)$ admits $k$ CISTs if and only if there is a partition of $V$ into $V_1, V_2, \ldots, V_k$, which is called a $k$-CIST partition, such that the following conditions hold:*

(i) *For $i \in \{1, 2, \ldots, k\}$, the subgraph of $G$ induced by $V_i$, denoted by $G[V_i]$, is connected;*
(ii) *For distinct $i, j \in \{1, 2, \ldots, k\}$, the bipartite graph with bipartition $V_i \cup V_j$ and edge set $\{(x, y) \in E(G)\colon x \in V_i, y \in V_j\}$, denoted by $B(V_i, V_j, G)$, has no tree component.*

The $n$-dimensional shuffle-cube, denoted by $SQ_n$ with $n = 4k + 2$, is an interconnection network proposed by Li et al. [10], which is a variation of the hypercube $Q_n$ obtained by changing some edges. Then, $SQ_n$ still has connectivity $n$, while its diameter becomes smaller and approximately $\frac{n}{4}$. For the related research of shuffle-cubes, the reader can refer to [6,10,19,22]. Recently, Qin and Hao [19] provided a recursive algorithm to construct a dual-CIST of $SQ_n$ for $n \geqslant 6$.

In this paper, based on the major consideration of fault tolerance and confidentiality, we would like to solve the problem of constructing a tri-CIST of $SQ_n$ with $n = 4k + 2$ and $k \geqslant 1$. The remaining part of this paper is organized as follows. Section 2 introduces the necessary definitions and properties of $SQ_n$. Section 3 shows the constructing scheme of a tri-CIST in $SQ_n$ and its correctness. Section 4 makes our conclusions.

## 2 Preliminary

In this section, we introduce some terminologies, notions, and basic properties of shuffle-cubes. For a simple undirected graph $G$, the vertex set and edge set of $G$ are denoted by $V(G)$ and $E(G)$, respectively. The diameter of $G$, denoted by $\mathrm{diam}(G)$, is the greatest distance between any pair of vertices in $G$. A simple path with length equal to the diameter is called a *diametral path*. For $SQ_n$, we use $n$-bit binary strings to represent its vertices. For example, a vertex with label

$u$ is denoted by $u = u_{n-1}u_{n-2}\cdots u_0$, where $u_i \in \{0,1\}$ for $0 \leqslant i \leqslant n-1$. The complement of $u_i$ is denoted by $\bar{u}_i = 1 - u_i$. Thus, $\bar{u} = \bar{u}_{n-1}\bar{u}_{n-2}\cdots\bar{u}_0$. For two binary strings $a$ and $b$, we write $ab$ to mean the concatenation of $a$ and $b$. Let $p_j(u)$ denote the $j$-prefix of $u$, i.e., $p_j(u) = u_{n-1}u_{n-2}\cdots u_{n-j}$, and $s_i(u)$ the $i$-suffix of $u$, i.e., $s_i(u) = u_{i-1}u_{i-2}\cdots u_1u_0$. Hence, we use $SQ_{n-j}^{p_j(u)}$ to indicate the subcube obtained from $SQ_n$ in which every vertex has $p_j(u)$ as its $j$-prefix. To construct shuffle-cubes, four specific sets are defined below:

$$V_{00} = \{1111, 0001, 0010, 0011\},\ V_{01} = \{0100, 0101, 0110, 0111\},$$
$$V_{10} = \{1000, 1001, 1010, 1011\},\ V_{11} = \{1100, 1101, 1110, 1111\}.$$

Note that 1111 presents at two sets $V_{00}$ and $V_{11}$, and 0000 is absent in the above four sets.

**Definition 1 (Li et al. [10]).** The *n-dimensional shuffle-cube* with $n = 4k+2$ and $k \geqslant 0$, denoted by $SQ_n$, is recursively defined as follows: (i) for $k = 0$, $SQ_2$ is a 2-dimensional hypercube $Q_2$; (ii) for $k \geqslant 1$, $SQ_n$ consists of 16 disjoint subcubes $SQ_{n-4}^{i_1i_2i_3i_4}$, where $i_j \in \{0,1\}$ for $1 \leqslant j \leqslant 4$, and two vertices $u = u_{n-1}u_{n-2}\cdots u_0$ and $v = v_{n-1}v_{n-2}\cdots v_0$ in different $(n-4)$-dimensional subcubes are adjacent in $SQ_n$ if and only if $s_{n-4}(u) = s_{n-4}(v)$ and $p_4(u) \oplus p_4(v) \in V_{s_2(u)}$, where $\oplus$ denotes the addition with modulo 2. In this case, $(u, v)$ is called an *out-edge*.

Figure 1 demonstrates a partial view of $SQ_6$, where vertices in $SQ_6$ are labeled by binary strings and their decimals, and a binary string is divided into two parts (i.e., 4-prefix and 2-suffix) such that vertices with the same 4-prefix form a subcube $SQ_2$. In this figure, we only draw the out-edges incident with vertices in the subcube $SQ_2^{0000}$ and omits others. For example, if we consider $u = 001100$ (12), then $p_4(u) = 0011$ and $s_2(u) = 00$. We can easily check the adjacency of $u$ as follows:

$$0011 \oplus 1100 = 1111 \in V_{00},\quad 0011 \oplus 0010 = 0001 \in V_{00},$$
$$0011 \oplus 0001 = 0010 \in V_{00},\quad 0011 \oplus 0000 = 0011 \in V_{00}.$$

Thus, $u = 001100$ is connected to vertices 110000 (48), 001000 (8), 000100 (4), and 000000 (0) in $SQ_6$ by out-edges.

From Definition 1, it is obvious that $SQ_n$ is $n$-regular, and the numbers of vertices and edges of $SQ_n$ are the same as those of $Q_n$. Qin and Hao [19] showed that for two distinct subcubes $SQ_2^{i_1i_2i_3i_4}$ and $SQ_2^{j_1j_2j_3j_4}$ in $SQ_6$, if $j_1j_2j_3j_4 = \overline{i_1i_2i_3i_4}$, then there exist two out-edges between them; otherwise, there is only one out-edge between them. In general, Xu et al. [22] showed the existence of at least $2^{n-6}$ out-edges between any two distinct subcubes $SQ_{n-4}^{t}$ and $SQ_{n-4}^{t'}$ for $n = 4k+2$ and $k \geqslant 1$, where $t, t' \in \{0, 1, \ldots, 15\}$ are 4-bit binary strings with $t \neq t'$. Let $P_{ij}$ be the set of 4-bit binary strings with the 2-prefix $ij$, i.e.,

$$P_{00} = \{0000, 0001, 0010, 0011\},\ P_{01} = \{0100, 0101, 0110, 0111\},$$
$$P_{10} = \{1000, 1001, 1010, 1011\},\ P_{11} = \{1100, 1101, 1110, 1111\}.$$

For simplicity, we can also check an out-edge of $SQ_n$ by using the following property. Here we omit the proof because it is easy to obtain from Definition 1.

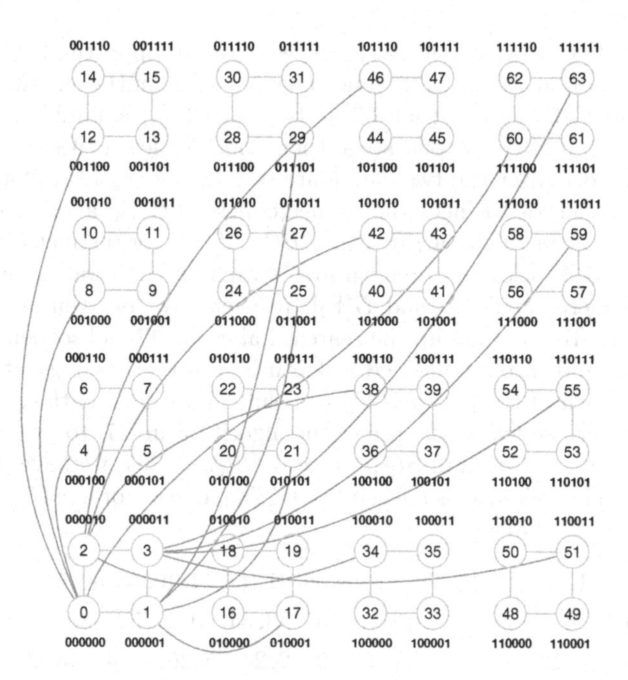


**Fig. 1.** The shuffle-cube $SQ_6$.

**Proposition 1.** *For every vertex* $u \in V(SQ_n)$ *with* $n \geqslant 6$, *let* $ij = p_2(u) \oplus s_2(u)$. *Then, the following conditions hold:*

(i) *If* $s_2(u) = 00$, *there exists a vertex* $v \in \{ps_{n-4}(u)\colon\, p \in (P_{ij} \setminus \{p_4(u)\}) \cup \{p_4(u) \oplus 1111\}\}$ *such that* $(u, v)$ *forms an out-edge.*

(ii) *Otherwise, there exists a vertex* $v \in \{ps_{n-4}(u)\colon\, p \in P_{ij}\}$ *such that* $(u, v)$ *forms an out-edge.*

# 3 Constructing Tri-CISTs in Shuffle-Cubes

In this section, we first find a 3-CIST-partition of $SQ_6$ by the tree searching algorithm. From this partition, we can easily construct a tri-CIST of $SQ_6$, which can be viewed as the induction base in our construction. Hence, for high-dimensional shuffle-cubes $SQ_n$ with $n = 4k + 2$ and $k \geqslant 1$, we can construct a tri-CIST by recursion.

## 3.1 A Tri-CIST of $SQ_6$

In [16], Pai et al. developed a two stages tree searching algorithm called $\mathbf{TS}^2$ to find a 3-CIST-partition $\{V_1, V_2, V_3\}$ of a 6-regular graph $G = (V, E)$ (or a graph with the minimum degree at least 6) if it exists. Note that this algorithm takes exponential complexity to find a nearly equalized 3-CIST-partition (i.e., $|V_i| \approx |V|/3$ for $i = 1, 2, 3$) to fulfill the conditions of Theorem 2. Therefore, if the number of vertices of $G$ is not too large, we have an opportunity to obtain the required partition in a reasonable time. The first stage of the algorithm attempts to find vertices of $V_1$ such that $G[V_1]$ is a tree (i.e., a minimally connected subgraph) in $G$, and the second stage then finds vertices of $V_2$ such that $G[V_2]$ also forms a tree. After the two candidate sets $V_1$ and $V_2$ being found out, the remaining work needs to check the connectedness of $G[V_3]$ and non-existence of a tree component in the bipartite graph $B(V_i, V_j, G)$ for distinct $i, j \in \{1, 2, 3\}$. In particular, $\mathbf{TS}^2$ uses the breadth-first search as the strategy in pursuit of a shorter diameter of $G[V_1]$ and $G[V_2]$, respectively, and thus it needs more searching time. To speeding up the search, Pai et al. [17] subsequently amended the algorithm and instead adopted a depth-first search strategy at each stage, which causes each of $G[V_1]$ and $G[V_2]$ to result in a simple path easily.

In this paper, we follow the use of the algorithm in [17] to produce a nearly equalized 3-CIST-partition of $SQ_6$. Since $|V(SQ_6)| = 64$, we have $|V_1| = |V_2| = |V_3| - 1 = 21$. The following is a feasible 3-CIST partition of $V(SQ_6)$ we obtained by the searching algorithm:

$$V_1 = \{0, 1, 6, 11, 23, 26, 28, 29, 32, 38, 39, 41, 43, 44, 46, 48, 50, 54, 55, 56, 57\}$$
$$V_2 = \{2, 5, 8, 10, 15, 20, 21, 25, 27, 30, 31, 33, 34, 40, 47, 52, 53, 58, 59, 61, 63\}$$
$$V_3 = \{3, 4, 7, 9, 12, 13, 14, 16, 17, 18, 19, 22, 24, 35, 36, 37, 42, 45, 49, 51, 60, 62\}$$

From the adjacency of $SQ_6$ (see Fig. 1) and by a lengthy checking, we can confirm that both $SQ_6[V_1]$ and $SQ_6[V_2]$ are paths, and $SQ_6[V_3]$ is a unicyclic graph (i.e., a connected graph with exactly one cycle), as shown in Fig. 2. This shows that each $SQ_6[V_i]$ for $i \in \{1, 2, 3\}$ is connected, which fulfills the condition of Theorem 2(i).

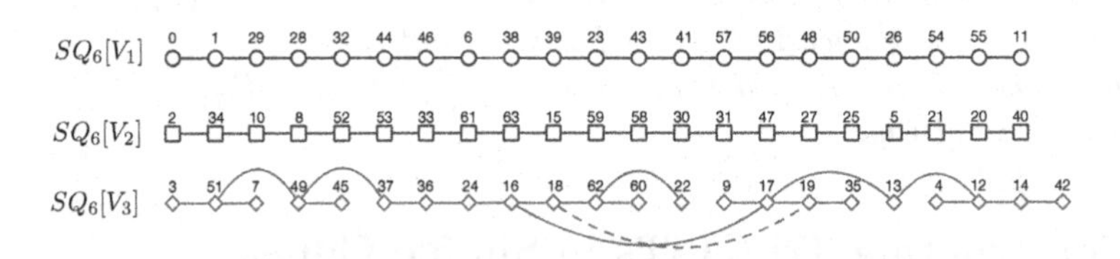


**Fig. 2.** Three connected graphs $SQ_6[V_i]$ for $i \in \{1, 2, 3\}$.

To confirm that the above partition $\{V_1, V_2, V_3\}$ is indeed a 3-CIST-partition of $SQ_6$, by Theorem 2(ii), we need to verify that all bipartite graphs $B(V_i, V_j, SQ_6)$ for $i, j \in \{1, 2, 3\}$ with $i \neq j$ have no tree component. Figure 3 shows the three bipartite graphs, and we can check as follows:

- $B(V_1, V_2, SQ_6)$ has two components, each of which contains a cycle $(46, 2, 38, 10, 46)$ and $(43, 27, 39, 31, 43)$, respectively.
- $B(V_1, V_3, SQ_6)$ has two components, each of which contains a cycle $(56, 60, 48, 12, 0, 4, 56)$ and $(50, 22, 54, 18, 50)$, respectively.
- $B(V_2, V_3, SQ_6)$ has one component with three cycles $(3, 63, 7, 59, 3)$, $(25, 13, 21, 9, 25)$, and $(53, 37, 61, 45, 53)$.

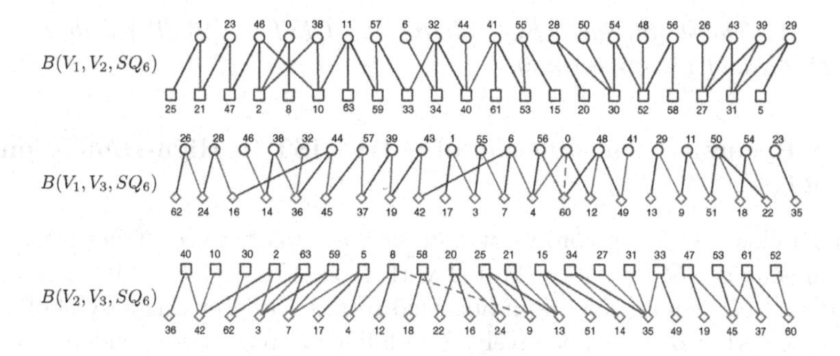


**Fig. 3.** Bipartite graphs $B(V_i, V_j, SQ_6)$ for $i, j \in \{1, 2, 3\}$ with $i \neq j$. (Color figure online)

Based on the above partition, we construct a tri-CIST of $SQ_6$ as follows. For each $i \in \{1, 2, 3\}$, the set of inner-vertices of $T_i$ is composed by $V_i$, and particularly, the edge $(18, 19)$ is removed from $SQ_6[V_3]$ (see the dashed line in Fig. 2). For each vertex $u \in V_i$, if $(u, v)$ is an edge of $B(V_i, V_j, SQ_6)$ such that its color is the same as $u$'s color in Fig. 3, then we take $v$ as a leaf to join $u$ in $T_i$. It is easy to examine that every vertex $v \in V_j$ for $j \in \{1, 2, 3\} \setminus \{i\}$ is exactly joined to an inner-vertex of $T_i$, and thus $T_i$ is a spanning tree of $SQ_6$. Since $|E(SQ_6)| = 192$ and each spanning tree requires 63 edges, except the edge $(18, 19)$, another two unused edges are $(0, 60)$, and $(8, 9)$ (see the two dashed lines in Fig. 3). Figure 4 shows the constructed tri-CIST of $SQ_6$, where vertices with a round-corner rectangle are called *port vertices* in this figure. For example, $T_1$ contains port vertices 38 and 39, $T_2$ contains port vertices 58 and 59, and $T_3$ contains port vertices 16 and 18. Note that a port vertex can be used as a connection for constructing CISTs in high-dimensional $SQ_n$ in Sect. 3.2. Since the diameters of the above-constructed CISTs are easy to determine from Fig. 4, we deduce the following lemma.

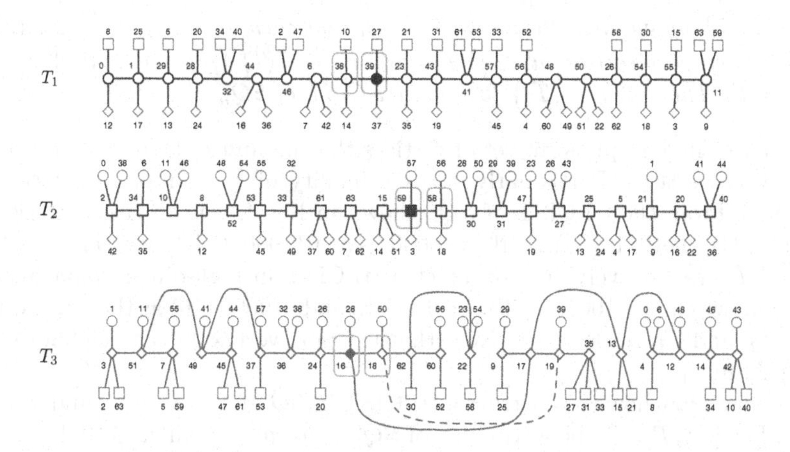


**Fig. 4.** A tri-CIST of $SQ_6$.

**Lemma 1.** *The shuffle-cubes $SQ_6$ admits a tri-CIST $\{T_1, T_2, T_3\}$ with the diameters* 22, 22, *and* 13, *respectively.*

### 3.2 A Recursive Construction of a Tri-CIST in High-Dimensional $SQ_n$

In what follows, by recursion, we will give an approach to construct a tri-CIST in shuffle-cubes $SQ_n$ for $n = 4k + 2$ with $k \geqslant 2$. Let $G$ be a labeled graph, and $G^0$ and $G^1$ denote the graph obtained from $G$ by prefixing a symbol 0 and 1 in every vertex of $G$, respectively. The following property is widely used for constructing CISTs in high-dimensional hypercube-variant networks.

**Theorem 3 (Pai and Chang [12]).** *Let $G_{n-1}$ be the $(n-1)$-dimensional variant hypercube for $n \geqslant 5$ and suppose that $\{\hat{T}_1, \hat{T}_2\}$ is a dual-CIST of $G_{n-1}$. For $i \in \{1, 2\}$, let $T_i$ be a spanning tree of $G_n$ constructed from $\hat{T}_i^0$ and $\hat{T}_i^1$ by adding an edge $(u_i, v_i) \in E(G_n)$ to connect two inner-vertices $u_i \in V(\hat{T}_i^0)$ and $v_i \in V(\hat{T}_i^1)$. Then, $\{T_1, T_2\}$ is a dual-CIST of $G_n$.*

In the above theorem, the two inner-vertices $u_i \in V(\hat{T}_i^0)$ and $v_i \in V(\hat{T}_i^1)$ for $i \in \{1, 2\}$ are called *port vertices* of $\hat{T}_i^0$ and $\hat{T}_i^1$, respectively. Also, the edge $(u_i, v_i) \in E(G_n)$ is called the *bridge* in the construction. Recall that we have mentioned port vertices in the construction of a tri-CIST of $SQ_6$ in the previous subsection (see the description of Fig. 4). By Definition 1, the $n$-dimensional shuffle-cube $SQ_n$ with $n = 4k+2$ and $k \geqslant 2$ consists of sixteen $(n-4)$-dimensional subcubes, denoted by $SQ_{n-4}^{i_1 i_2 i_3 i_4}$, where $i_j \in \{0, 1\}$ for $1 \leqslant j \leqslant 4$. For convenience, if $H$ is a subgraph of $SQ_{n-4}$ and $t \in \{0, 1, \ldots, 15\}$ is an integer represented by binary string, we denote by $H^t$ the graph obtained from $H$ by prefixing $t$. By Theorem 1 and using the same proof technique as Theorem 3, we can quickly obtain the following result for $SQ_n$.

**Corollary 1.** *For $n = 4k+2$ and $k \geqslant 2$, let $\{\hat{T}_1, \hat{T}_2, \hat{T}_3\}$ be a tri-CIST of $SQ_{n-4}$. For $i \in \{1, 2, 3\}$, let $T_i$ be a spanning tree of $SQ_n$ constructed from $\hat{T}_i^t$ for all $t \in \{0, 1, \ldots, 15\}$ by adding fifteen bridges such that each bridge $(u_i^t, v_i^{t'}) \in E(SQ_n)$ connects two port vertices $u_i^t \in V(\hat{T}_i^t)$ and $v_i^{t'} \in V(\hat{T}_i^{t'})$ for $t, t' \in \{0, 1, \ldots, 15\}$ with $t \neq t'$. Then, $\{T_1, T_2, T_3\}$ forms a tri-CIST of $SQ_n$.*

A *center* in a graph is the set of vertices that minimize the maximal distance from other vertices. Particularly, the cardinality of the center in a tree is at most two. From the tri-CIST of $SQ_6$ shown in Fig. 4, we can easily check that 010111 (23), 111011 (59), and {010000 (16), 011000 (24)} are centers of $T_1$, $T_2$, and $T_3$, respectively. To construct a tri-CIST in high-dimensional $SQ_n$ by recursion, we choose 100111 (39), 11011 (59), and 010000 (16) as the port vertices of $\hat{T}_1$, $\hat{T}_2$, and $\hat{T}_3$, respectively. Note that the port vertex 39 is a neighbor of the center 23 in $\hat{T}_1$.

We now show how to construct a tri-CIST in $SQ_n$ for $n = 4k+2$ and $k \geqslant 2$ in detail. Let $\{\hat{T}_1^t, \hat{T}_2^t, \hat{T}_3^t\}$ be a tri-CIST of $SQ_{n-4}^t$ for $n \geqslant 10$ and $t \in \{0, 1, \ldots, 15\}$.

According to Corollary 1, for $i \in \{1,2,3\}$, we need to connect all subtrees $\hat{T}_i^t$ for $t \in \{0,1,\ldots,15\}$ by using fifteen bridges to produce $T_i$ in $SQ_n$. Recall that we have defined $P_{ij}$ as the set of 4-bit binary string with the 2-prefix $ij$. For notational convenience, we write $P_{ij}s$ to mean all possible binary strings with 4-prefix in $P_{ij}$ and suffix $s$. Then the tri-CIST construction of $SQ_n$ can be described by Algorithm 1, where the constructing rules are depending on the parity of $k$. In each way, we accurately represent fifteen bridges by their two port vertices for every tree.

**Algorithm 1:** Constructing a Tri-CIST on Shuffle-Cubes

**Input:** An $n$-dimensional shuffle-cubes $SQ_n$, with $n = 4k+2$ and $k \geqslant 1$, which is composed of sixteen subcubes $SQ_{n-4}^t$ for $t \in \{0,1,\ldots,15\}$ represented by binary strings.
**Output:** A tri-CIST $\{T_1, T_2, T_3\}$ of $SQ_n$.

```
if k = 1 then
    return {T1, T2, T3} shown in Fig. 4;
for t ← 0 to 15 do
    Let {T̂1^t, T̂2^t, T̂3^t} be a tri-CIST of SQ^t_{n-4} constructed by Algorithm 1;
if k is even then
    E1 ← {(P00x11, 1111x11), (P11x11, 0011x11), (P01x11, 1011x11),
          (P10x11, 0111x11), (1111x10, 0111x10): x = 1^{n-10}1001};
    E2 ← {(P00x11, 1111x11), (P11x11, 0011x11), (P01x11, 1011x11),
          (P10x11, 0111x11), (1111x10, 0111x10): x = 1^{n-10}1110};
    E3 ← {(P00x00, 0000x00), (P11x00, 1111x00), (P01x00, 0100x00),
          (P10x00, 1011x00), (0000x00, 1111x00), (0100x00, 1011x00),
          (1111x10, 0100x10): x = 1^{n-10}0100};
else        // k is odd
    E1 ← {(P00x10, 1011x10), (P11x10, 0111x10), (P01x10, 1111x10),
          (P10x10, 0011x10), (1111x11, 0011x11): x = 1^{n-10}1001};
    E2 ← {(P00x10, 1011x10), (P11x10, 0111x10), (P01x10, 1111x10),
          (P10x10, 0011x10), (1111x11, 0011x11): x = 1^{n-10}1110};
    E3 ← {(P00x10, 1011x10), (P11x10, 0100x10), (P01x10, 1111x10),
          (P10x10, 0000x10), (1111x00, 0000x00): x = 1^{n-10}0100};
for i ← 1 to 3 do
    Construct Ti such that E(Ti) ← ⋃_{t=0}^{15} E(T̂i^t) ⋃ Ei;
return {T1, T2, T3}
```

For insight into Algorithm 1, we provide a schematic view to auxiliary illustrate (see Fig. 5). If $k = 1$ (i.e., $n = 6$), the algorithm output a tri-CIST produced in Sect. 3.1 (see Lines 1–2). Otherwise, the construction of CIST carries out by a recursive fashion. We first consider the case of even $k$. For $T_1$, we have four bridges with each of the form $(P_{00}x11, 1111x11)$, $(P_{11}x11, 0011x11)$, $(P_{01}x11, 1011x11)$, and $(P_{10}x11, 0111x11)$ in Lines 6–7. For the first form of bridges, it means that there is a bridge that connects a port vertex in $\hat{T}_1^t$ with $t \in P_{00}$ and a port vertex in $\hat{T}_1^{1111}$. We check the existence of such a bridge by Proposition 1. For instance, if we consider the port vertex $u = 1111x11$ in $\hat{T}_1^{1111}$, we have $s_2(u) = 11$ and $p_2(u) \oplus s_2(u) = 00$, and thus by condition (ii), it follows that $u$ connects to other port vertices $P_{00}s_{n-4}(u) = P_{00}x11$ by bridges (see Fig. 5(a)). Also, the other bridges of this form and bridges in other three forms can easily be verified by a similar way. Note that there is a duplicate bridge in the former two forms (resp. latter two forms). Except above, there is a remain-

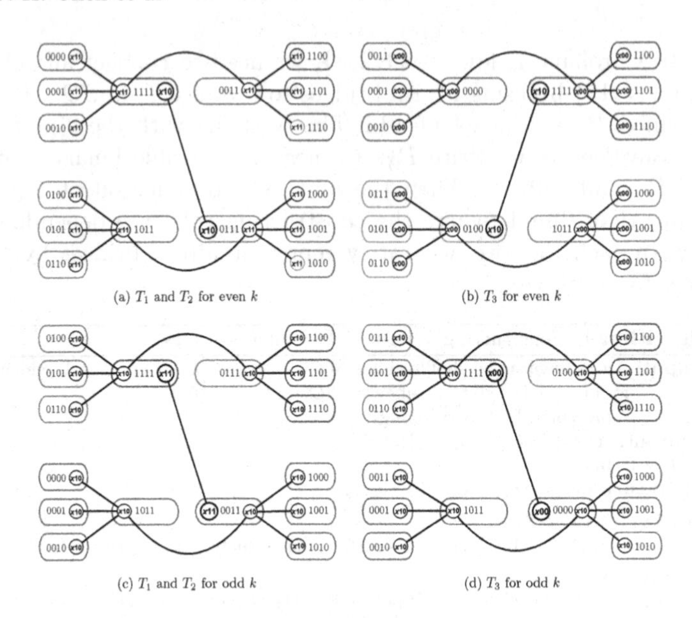


(a) $T_1$ and $T_2$ for even $k$

(b) $T_3$ for even $k$

(c) $T_1$ and $T_2$ for odd $k$

(d) $T_3$ for odd $k$

**Fig. 5.** The schematic concept of tree construction.

ing bridge with the form $(1111x\mathbf{10}, 0111x\mathbf{10})$ (see Line 7) that connects two specific vertices called *main port vertices*. Hereafter, to make the constructed tree with a short diameter, we always select a neighbor of the port vertex as the main port vertex in a tree. Since $s_2(1111x\mathbf{10}) = s_2(0111x\mathbf{10}) \neq 00$, we have $p_2(1111x\mathbf{10}) \oplus s_2(1111x\mathbf{10}) = 01$ and $p_2(0111x\mathbf{10}) \oplus s_2(0111x\mathbf{10}) = 11$. Again, by condition (ii) of Proposition 1, the bridge $(1111x\mathbf{10}, 0111x\mathbf{10})$ exists. In Fig. 5, we draw a tree $\hat{T}_1^t$ with $t$ belonging to distinct $P_{ij}$ by a round-corner rectangle with the different color and the label of 4-prefix $t$. Also, a port vertex $u$ connecting by a bridge is labeled by a binary string $xs_2(u)$ inside a circle, where $x$ is an $(n-6)$-bit string. In particular, a bridge connecting two main port vertices is drawn by a bold line. For instance, if $n = 10$, we have $x = 1001$ for $T_1$ (see Line 7). Let $P = P_{00} \cup P_{01} \cup P_{10} \cup P_{11}$, and let $C$ and $C'$ be the two main port vertices. Since we have chosen 100111 (39) as the port vertex in $\hat{T}_1$ for $SQ_6$ (see also $T_1$ in Fig. 4), the set of port vertices in $\hat{T}_1^t$ for $t \in P$ can be represented by $P100111$. In particular, two main port vertices are $C = 1111100110$ in $\hat{T}_1^{1111}$ and $C' = 0111100110$ in $\hat{T}_1^{0111}$.

Similarly, for constructing $T_2$ and $T_3$ under $k$ even, all bridges are described in Lines 8–9 and Lines 10–11 of Algorithm 1, respectively, and the corresponding schematics are shown in Fig. 5(a) and Fig. 5(b), respectively. Here, we only check the existence of the bridge $(0100x00, 1011x00)$ for $T_3$ and omit others (because they can be checked by a similar way). For instance, if we consider the port vertex $u = 0100x00$ in $\hat{T}_3^{0100}$, we have $s_2(u) = 00$ and $p_2(u) \oplus s_2(u) = 01$. By condition (i) of Proposition 1, it follows that $u$ connects to other port vertices $ps_{n-4}(u) = px00$ for $p \in P_{01} \setminus \{0100\} \cup \{1011\}$ (see Fig. 5(b)). For instance, if

$n = 10$, we have $x = 1110$ for $T_2$ (see Line 9) and $x = 0100$ for $T_3$ (see Line 11). Since we have respectively chosen 111011 (59) and 010000 (16) as port vertices in $\hat{T}_2$ and $\hat{T}_3$ for $SQ_6$ (see also $T_2$ and $T_3$ in Fig. 4), the sets of port vertices in $\hat{T}_2^t$ and $\hat{T}_3^t$ for $t \in P$ can be represented by $P111011$ and $P010011$, respectively. In particular, two main port vertices are $C = 1111111010$ in $\hat{T}_2^{1111}$ and $C' = 0111111010$ in $\hat{T}_2^{0111}$ for $T_2$, and $C = 1111010010$ in $\hat{T}_3^{1111}$ and $C' = 0100010010$ in $\hat{T}_3^{0111}$ for $T_3$.

For odd $k$, the bridges of $T_i$ for $i \in \{1, 2, 3\}$ are described in Lines 12–18 of Algorithm 1, and the corresponding schematics are shown in Fig. 5(c) for $T_1$ and $T_2$ and Fig. 5(d) for $T_3$, respectively. We omit to check the existence of bridges because it can be done similarly to the previous case. For example, if $n = 14$, we have $x = 1^{n-10}1001$ for $T_1$ (see Line 14), $x = 1^{n-10}1110$ for $T_2$ (see Line 16), and $x = 1^{n-10}0100$ for $T_3$ (see Line 18). Thus, port vertices are $P1^{n-10}100110$ for $T_1$, $P1^{n-10}111010$ for $T_2$, and $P1^{n-10}010010$ for $T_3$. Also, two main port vertices are $C = 11111^{n-10}100111$ and $C' = 00111^{n-10}100111$ for $T_1$, $C = 11111^{n-10}111011$ and $C' = 00111^{n-10}111011$ for $T_2$, and $C = 11111^{n-10}010000$ and $C' = 00111^{n-10}010000$ for $T_3$.

We summarize the binary representations of port vertices and main port vertices mentioned above, as shown in Table 1. Note that the guidance symbols (i.e., arrows) appearing in the table mean that each port vertex of a tree $\hat{T}_i^t$ in $SQ_n$ come from the port vertex or the main port vertex $C$ of the tree $\hat{T}_i$ in $SQ_{n-4}$ by adding a 4-prefix $t \in P$, where $P = P_{00} \cup P_{01} \cup P_{10} \cup P_{11}$.

**Table 1.** The binary strings of port vertices and main port vertices

| | $n$ | port vertices | main port vertices | |
|---|---|---|---|---|
| | | | $C$ | $C'$ |
| $T_1$ | 6 | 1001 11 | - | - |
| | 10 | $P$ 1001 11 | 1111 1001 10 | 0111 1001 10 |
| | 14 | $P$ $1^{n-10}$ 1001 10 | 1111 $1^{n-10}$ 1001 11 | 0011 $1^{n-10}$ 1001 11 |
| | 18 | $P$ $1^{n-10}$ 1001 11 | 1111 $1^{n-10}$ 1001 10 | 0111 $1^{n-10}$ 1001 10 |
| | 22 | $P$ $1^{n-10}$ 1001 10 | 1111 $1^{n-10}$ 1001 11 | 0011 $1^{n-10}$ 1001 11 |
| $T_2$ | 6 | 1110 11 | - | - |
| | 10 | $P$ 1110 11 | 1111 1110 10 | 0111 1110 10 |
| | 14 | $P$ $1^{n-10}$ 1110 10 | 1111 $1^{n-10}$ 1110 11 | 0011 $1^{n-10}$ 1110 11 |
| | 18 | $P$ $1^{n-10}$ 1110 11 | 1111 $1^{n-10}$ 1110 10 | 0111 $1^{n-10}$ 1110 10 |
| | 22 | $P$ $1^{n-10}$ 1110 10 | 1111 $1^{n-10}$ 1110 11 | 0011 $1^{n-10}$ 1110 11 |
| $T_3$ | 6 | 0100 00 | - | - |
| | 10 | $P$ 0100 00 | 1111 0100 10 | 0100 0100 10 |
| | 14 | $P$ $1^{n-10}$ 0100 10 | 1111 $1^{n-10}$ 0100 00 | 0000 $1^{n-10}$ 0100 00 |
| | 18 | $P$ $1^{n-10}$ 0100 00 | 1111 $1^{n-10}$ 0100 10 | 0100 $1^{n-10}$ 0100 10 |
| | 22 | $P$ $1^{n-10}$ 0100 10 | 1111 $1^{n-10}$ 0100 00 | 0000 $1^{n-10}$ 0100 00 |

**Lemma 2.** *For $n = 4k + 2$ and $k \geqslant 2$, Algorithm 1 produces three edge-disjoint spanning trees of $SQ_n$. Moreover, every main port vertex chosen in the algorithm is a center vertex in the corresponding tree.*

*Proof.* Let $\{T_1, T_2, T_3\}$ be the constructed tri-CIST of $SQ_n$. For $i \in \{1, 2, 3\}$, we have $E(T_i) = \bigcup_{t=0}^{15} E(\hat{T}_i^t) \bigcup E_i$ (see Line 20 of Algorithm 1). By Lemma 1, $SQ_6$

offers three edge-disjoint spanning trees, which are the base of the construction. Obviously, each tree $T_i$ mainly collects its own edges from distinct $\hat{T}_i^t$ for all $t \in \{0, 1, \ldots, 15\}$ by recursion, and then supplements fifteen bridges contained in the set $E_i$ according to Corollary 1. Since $E_i \cap E_j = \emptyset$ for distinct $i, j \in \{1, 2, 3\}$, this yields the desired three edge-disjoint spanning trees of $SQ_n$.

Next, we claim that both main port vertices $C$ and $C'$ are center vertices of each tree $T_i$ when $n \geqslant 10$. Recall that $C$ and $C'$ are neighbors of the port vertices in the two specific subtrees, and each subtree $\hat{T}_i^t$ for $t \in \{0, 1, \ldots, 15\}$ preserves the same structure of the tree $\hat{T}_i$ produced in $SQ_{n-4}$. Hence, if the port vertex in each subtree $\hat{T}_i^t$ is a center, the symmetric structure of $T_i$ ensures that the claim holds (see Fig. 5). From the recursive construction of CIST, each port vertex of a tree $\hat{T}_i^t$ in $SQ_n$ is obtained from the port vertex or the main port vertex $C$ of the tree $\hat{T}_i$ in $SQ_{n-4}$ when $n \geqslant 10$. The result is clear for $T_2$ and $T_3$ because we had initially chosen the centers 11011 (59) and 010000 (16), respectively, as the corresponding port vertices when we dealt with $SQ_6$. It remains to confirm the case of $T_1$. For $SQ_6$, although the port vertex 100111 (39) of $T_1$ is not a center vertex (indeed, it is a neighbor of a center), the set of main port vertices $\{C, C'\}$ will become the center in the newly constructed $T_1$ when we dealt with $SQ_{10}$. Therefore, all succeeding main port vertices $C$ and $C'$ will continue to be the center vertices during the recursive process. □

From Algorithm 1, it is evident that all port vertices and the main port vertices are inner-vertices in the corresponding trees. Thus, adding bridges in the recursive construction does not change a leaf to an inner-vertex. Together with Theorem 1 and Lemma 2, this result shows that the constructed trees form a tri-CIST. Moreover, since Lemma 2 showing that every main port vertex is a center and at most seven bridges are contained in the diametral path of $T_i$ (see Fig. 5), we can formulate the diameters of $T_i$ for $i \in \{2, 3\}$ as follows: for $n \geqslant 10$,

$$\mathrm{diam}(T_i) = 2 \cdot \lceil \mathrm{diam}(\hat{T}_i)/2 \rceil + 7 \quad \text{for } i \in \{2, 3\}. \tag{1}$$

As for $T_1$ of $SQ_6$, since the port vertex is a center's neighbor, we must increase the diameter by a constant two after solving Eq. (1). By Lemma 1, we have a tri-CIST $\{T_1, T_2, T_3\}$ in $SQ_6$ with diameters 22, 22, and 13. Then solving Eq. (1), we immediately obtain the following result.

**Theorem 4.** *For $n \geqslant 6$, $SQ_n$ admits a tri-CIST $\{T_1, T_2, T_3\}$. Particularly, if $n = 6$, the diameters of $T_i$ for $i = 1, 2, 3$ are 22, 22, and 13, respectively. Moreover, for $n \geqslant 10$, the diameter can be described below:*

$$diam(T_i) = \begin{cases} 2n + 13 - 2i, & \text{for } i \in \{1, 2\}; \\ 2n + 1, & i = 3. \end{cases}$$

## 4 Concluding Remarks

This article shows the existence of a tri-CIST in shuffle-cubes $SQ_n$ with $n = 4k+2$ and $k \geqslant 1$ by a recursive algorithm. Since a dual-CIST suffices to configure a protection routing with a single link or node failure, we intuitively think that a route can tolerate more links or nodes failure if we could provide a configuration using more CISTs. Moreover, by a secure-protection routing scheme [14,17], we can configure a protection routing using a tri-CIST such that data transmitted through the route are secure. Based on the consideration of the application in data transmission, future research can focus on how to shorten the diameter of CISTs.

**Acknowledgments.** This research was supported by the Ministry of Science and Technology of Taiwan under Grant MOST110-2221-E-141-004 (J.-M. Chang).

## References

1. Araki, T.: Dirac's condition for completely independent spanning trees. J. Graph Theory **77**(3), 171–179 (2014)
2. Chang, Y.-H., Pai, K.-J., Hsu, C.-C., Yang, J.-S., Chang, J.-M.: Constructing dual-CISTs of folded divide-and-swap cubes. Theor. Comput. Sci. **856**, 75–87 (2021)
3. Chen, G., Cheng, B., Wang, D.: Constructing completely independent spanning trees in data center network based on augmented cube. IEEE Trans. Parallel Distrib. Syst. **32**(3), 665–673 (2021)
4. Cheng, B., Wang, D., Fan, J.: Constructing completely independent spanning trees in crossed cubes. Discrete Appl. Math. **219**, 100–109 (2017)
5. Darties, B., Gastineau, N., Togni, O.: Completely independent spanning trees in some regular graphs. Discrete Appl. Math. **217**(2), 163–174 (2017)
6. Ding, T., Li, P., Xu, M.: The component (edge) connectivity of shuffle-cubes. Theor. Comput. Sci. **835**, 108–119 (2020)
7. Hasunuma, T.: Completely independent spanning trees in the underlying graph of a line digraph. Discrete Math. **234**(1–3), 149–157 (2001)
8. Hasunuma, T.: Completely independent spanning trees in maximal planar graphs. In: Goos, G., Hartmanis, J., van Leeuwen, J., Kučera, L. (eds.) WG 2002. LNCS, vol. 2573, pp. 235–245. Springer, Heidelberg (2002). https://doi.org/10.1007/3-540-36379-3_21
9. Hasunuma, T., Morisaka, C.: Completely independent spanning trees in torus networks. Networks **60**(1), 59–69 (2012)
10. Li, T.-K., Tan, J.M., Hsu, L.-H., Sung, T.Y.: The shuffle-cubes and their generalization. Inf. Process. Lett. **77**(1), 35–41 (2001)
11. Mane, S.A., Kandekar, S.A., Waphare, B.N.: Constructing spanning trees in augmented cubes. J. Parallel Distrib. Comput. **122**, 188–194 (2018)
12. Pai, K.-J., Chang, J.-M.: Constructing two completely independent spanning trees in hypercube-variant networks. Theor. Comput. Sci. **652**, 28–37 (2016)
13. Pai, K.-J., Chang, J.-M.: Dual-CISTs: configuring a protection routing on some Cayley networks. IEEE/ACM Trans. Netw. **27**(3), 1112–1123 (2019)
14. Pai, K.-J., Chang, R.-S., Chang, J.-M.: A protection routing with secure mechanism in Möbius cubes. J. Parallel Distrib. Comput. **140**, 1–12 (2020)
15. Pai, K.-J., Chang, R.-S., Chang, J.-M.: A well-equalized 3-CIST partition of alternating group graphs. Inf. Process. Lett. **155**, 105874 (2020)
16. Pai, K.-J., Chang, R.-S., Wu, R.-Y., Chang, J.-M.: A two-stages tree-searching algorithm for finding three completely independent spanning trees. Theor. Comput. Sci. **784**, 65–74 (2019)

17. Pai, K.-J., Chang, R.-S., Wu, R.-Y., Chang, J.-M.: Three completely independent spanning trees of crossed cubes with application to secure-protection routing. Inf. Sci. **541**, 516–530 (2020)
18. Péterfalvi, F.: Two counterexamples on completely independent spanning trees. Discrete Math. **312**(4), 808–810 (2012)
19. Qin, X.-W., Hao, R.-X.: Reliability analysis based on the dual-CIST in shuffle-cubes. Appl. Math. Comput. **397**, 125900 (2021)
20. Qin, X.-W., Hao, R.-X., Chang, J.-M.: Constructing dual-CISTs of DCell data center networks. Appl. Math. Comput. **362**, 124546 (2019)
21. Qin, X.-W., Hao, R.-X., Chang, J.-M.: The existence of completely independent spanning trees for some compound graphs. IEEE Trans. Parallel Distrib. Syst. **31**(1), 201–210 (2020)
22. Xu, J.-M., Xu, M., Zhu, Q.: The super connectivity of shuffle-cubes. Inf. Process. Lett. **96**(4), 123–127 (2005)

# Reconfiguring Directed Trees in a Digraph

Takehiro Ito[1], Yuni Iwamasa[2], Yasuaki Kobayashi[2], Yu Nakahata[2(✉)], Yota Otachi[3], and Kunihiro Wasa[4]

[1] Graduate School of Information Sciences, Tohoku University, Sendai, Japan
takehiro@tohoku.ac.jp
[2] Graduate School of Informatics, Kyoto University, Kyoto, Japan
{iwamasa,koba}@i.kyoto-u.ac.jp, nakahata.yu.27e@st.kyoto-u.ac.jp
[3] Graduate School of Informatics, Nagoya University, Nagoya, Japan
otachi@nagoya-u.jp
[4] Department of Computer Science and Engineering, Toyohashi University of Technology, Toyohashi, Japan
wasa@cs.tut.ac.jp

**Abstract.** In this paper, we investigate the computational complexity of subgraph reconfiguration problems in directed graphs. More specifically, we focus on the problem of reconfiguring directed trees in a digraph, where a directed tree is a directed graph such that its underlying undirected graph forms a tree and every vertex except for exactly one vertex has in-degree 1. Given two directed trees in a digraph, the goal of the problem is to determine whether there is a (reconfiguration) sequence of directed trees between two given ones such that each tree in the sequence can be obtained from the previous one by removing an arc and then adding another arc. We show that this problem can be solved in polynomial time, whereas the problem is PSPACE-complete when we restrict directed trees in a reconfiguration sequence to form directed paths. We also show that there is a polynomial-time algorithm for finding a shortest reconfiguration sequence between two spanning directed trees.

**Keywords:** Combinatorial reconfiguration · Graph algorithm

## 1 Introduction

Let $\Pi$ be a graph structure property. For a graph $G$, we denote by $\mathcal{S}_\Pi(G)$ the set of all subgraphs of $G$ that satisfy $\Pi$. In this paper, we study the reachability of the solution space formed by $\mathcal{S}_\Pi(G)$, where two subgraphs $H$ and $H'$ in $\mathcal{S}_\Pi(G)$ are *adjacent* in the solution space if and only if they can be obtained from each

This work is partially supported by JSPS KAKENHI Grant Numbers JP18H04091, JP18K11168, JP18K11169, JP19K11814, JP19K20350, JP19J21000, JP20H05793, JP20H05795, and JP21K11752, and JST, CREST Grant Number JPMJCR18K3, Japan.


C.-Y. Chen et al. (Eds.): COCOON 2021, LNCS 13025, pp. 343–354, 2021.
https://doi.org/10.1007/978-3-030-89543-3_29

**Fig. 1.** There is no reconfiguration sequence between the black and gray directed trees.

other by swapping a pair of edges, that is, $|E(H) \setminus E(H')| = |E(H') \setminus E(H)| = 1$. Our target is to decide whether there is a (reconfiguration) sequence of adjacent subgraphs in $\mathcal{S}_\Pi(G)$ between two given subgraphs $H^{\mathrm{s}}$ and $H^{\mathrm{t}}$ in $\mathcal{S}(G)$. To avoid a confusion, we sometimes call the problem the *reachability variant*, because we will study the shortest variant later.

The problem has been studied for several graph structure properties $\Pi$ (on undirected graphs), although most of the related results appear under the name of the property $\Pi$ under consideration. For example, Spanning Tree Reconfiguration, which is solvable in polynomial time [5], can be seen as the problem when the property $\Pi$ is a spanning tree. Ito et al. [5] also showed that the problem is solvable in polynomial time when $\Pi$ is a matching, and Mühlenthaler [6] extended the result to degree-constrained subgraphs. Hanaka et al. [4] introduced the framework of subgraph reconfiguration problems, and studied the problem for several properties $\Pi$, including trees and paths. In particular, they showed that when $\Pi$ is a tree, every instance of the problem is a yes-instance unless two input trees have different numbers of edges. Motivated by applications in motion planning, Biasi and Ophelders [1], Demaine et al. [2], and Gupta et al. [3] studied some variants of reconfiguring undirected paths. These variants are shown to be PSPACE-complete in general, while they are fixed-parameter tractable when parameterized by the length of input paths.

In contrast to various results for undirected graphs, the problem was not studied well for directed graphs. In this paper, we investigate the complexity of subgraph reconfiguration problems on directed graphs. We mainly study the problem when the property $\Pi$ is a directed tree, where a directed tree is a directed graph such that its underlying undirected graph forms a tree and every vertex except for exactly one vertex has in-degree 1. Note that two (directed) subgraphs in $\mathcal{S}_\Pi(G)$ are *adjacent* if and only if they can be obtained from each other by swapping a pair of arcs (instead of a pair of edges). We refer to this problem as Directed Tree Reconfiguration. (Formal definitions will be given in Sect. 2.) Interestingly, Directed Tree Reconfiguration has no-instances as shown in Fig. 1, in contrast to the fact that any two undirected trees are reconfigurable as long as they have the same number of edges [4]. Nonetheless we give the following theorem, as our main result.

**Theorem 1.** *Let $G = (V, A)$ be a directed graph.* Directed Tree Reconfiguration *can be solved in time $O(|V||A|)$. Moreover, if the answer is affirmative, we can construct a reconfiguration sequence between two given directed trees of length $O(|V|^2)$ within the same running time bound.*

**Table 1.** Summary of our results.

| Property $\Pi$ | Reachability variant | Shortest variant |
|---|---|---|
| Directed tree | P | Open |
| $r$-directed tree | Always yes | Open |
| Spanning directed tree | Always yes | P |
| Spanning $r$-directed tree | Always yes | P |
| Directed path | PSPACE-complete | – |
| Directed acyclic graph | PSPACE-complete | – |

We further investigate the problem for specific directed trees. By the definition, a directed tree has a unique vertex whose in-degree is 0. We call such a vertex the *root* of the directed tree, and call the tree an *$r$-directed tree.* We will show that any two $r$-directed trees are reconfigurable when the property $\Pi$ is an $r$-directed tree for a prescribed vertex $r$. This result gives an interesting contrast to Directed Tree Reconfiguration (recall the no-instance in Fig. 1), and will play an important role in our proof of Theorem 1. We also consider the cases where the property $\Pi$ is either a directed path or a directed acyclic graph (DAG). Formal definitions will be given in Sect. 5. We will prove that these cases are PSPACE-complete. Our results are summarized in Table 1.

In this paper, we also study the shortest variant, which computes the shortest length of a reconfiguration sequence between two given subgraphs in $\mathcal{S}_\Pi(G)$. In particular, Shortest Spanning Directed Tree Reconfiguration is the shortest variant when $\Pi$ is a spanning directed tree. We will prove the following theorem, by constructing a reconfiguration sequence between two spanning directed trees $T^{\mathrm{s}}$ and $T^{\mathrm{t}}$ of length $|A(T^{\mathrm{s}}) \setminus A(T^{\mathrm{t}})| = |A(T^{\mathrm{t}}) \setminus A(T^{\mathrm{s}})|$.

**Theorem 2.** Shortest Spanning Directed Tree Reconfiguration *can be solved in linear time.*

When $\Pi$ is a spanning directed trees, the reachability variant can be seen as a special case of Matroid Intersection Reconfiguration for a graphic matroid and (a truncation of) a partition matroid. Here, given two matroids and their two common bases $B^{\mathrm{s}}$ and $B^{\mathrm{t}}$, Matroid Intersection Reconfiguration asks to determine if there is a reconfiguration sequence of common bases between $B^{\mathrm{s}}$ and $B^{\mathrm{t}}$; see [7] for matroids. It is shown in [5] that Maximum Bipartite Matching Reconfiguration is solvable in polynomial time. While this problem can be seen as Matroid Intersection Reconfiguration for two (truncations of) partition matroids, the complexity of Matroid Intersection Reconfiguration remains open. Theorem 2 provides a new tractable class of Matroid Intersection Reconfiguration, particularly, its shortest version.

Due to the space limitation, several proofs (marked with $\star$) are omitted from this extended abstract.

## 2 Preliminaries

Let $G = (V, A)$ be a directed graph. We denote by $V(G)$ and $A(G)$ the vertex and arc sets of $G$, respectively. We may abuse $G$ to denote the arc set of $G$ when no confusions arise, and then we use $|G|$, called the *size* of $G$, to denote the number of arcs in $G$. Let $e = (u, v)$ be an arc of $G$. We say that $e$ is *directed from* $u$ or *directed to* $v$. The vertex $u$ (resp. $v$) is called the *tail* (resp. *head*) of $e$. For each $v \in V$, we denote by $N_G^+(v)$ the set of out-neighbors of $v$ in $G$, i.e., $N_G^+(v) = \{w \in V : (v, w) \in A\}$. The *in-degree* (resp. *out-degree*) of $v$ is the number of arcs directed to $v$ (resp. directed from $v$) in $G$. For a subset $X \subseteq V$, the subgraph of $G$ induced by $X$ is denoted by $G[X]$. For an arc $(u, v) \in G$ and a subgraph $H$ of $G$, we denote by $H + (u, v)$ and $H - (u, v)$ the directed graphs obtained from $H$ by adding $(u, v)$ and by removing $(u, v)$, respectively.

A *directed tree* $T$ is a directed graph such that its underlying undirected graph forms a tree and every vertex except for a vertex $r \in V(T)$ has in-degree exactly 1. The unique vertex $r$ of in-degree 0 is called the *root* of $T$, and $T$ is called an *$r$-directed tree.* A directed graph consisting of a disjoint union of directed trees is called a *directed forest* or an *$R$-directed forest*, where $R$ is the set of roots of its (weakly) connected components. An arc in a directed tree $T$ is called a *leaf arc* if the out-degree of its head is 0 in $T$. A *directed path* is a directed tree that has at most one leaf arc.

Let $\Pi$ be a graph structure property. For a graph $G$, we denote by $\mathcal{S}_\Pi(G)$ the set of all subgraphs of $G$ that satisfy $\Pi$. Let $H$ and $H'$ be two subgraphs in $\mathcal{S}_\Pi(G)$ that have the same size. A sequence $\langle H_0, H_1, \ldots, H_\ell \rangle$ of subgraphs in $\mathcal{S}_\Pi(G)$ is called a *reconfiguration sequence between $H$ and $H'$* if $H_0 = H$, $H_\ell = H'$, and $|A(H_i) \setminus A(H_{i+1})| = |A(H_{i+1}) \setminus A(H_i)| = 1$ for all $i$, $0 \le i < \ell$. In other words, $H_{i+1}$ can be obtained by removing an arc from $H_i$ and then adding another arc to it for each $i$, $0 \le i < \ell$. We call $\ell$ the *length* of the reconfiguration sequence. If there is a reconfiguration sequence between $H$ and $H'$, we say that $H$ is *reconfigurable* from $H'$. Note that any reconfiguration sequence is reversible: $H'$ is reconfigurable from $H$ if and only if $H$ is reconfigurable from $H'$. For simplicity, we assume without loss of generality that all subgraphs in $\mathcal{S}_\Pi(G)$ have the same size; otherwise they are not reconfigurable.

## 3 Always Reconfigurable Cases

In this section, we show that every instance of the reachability variant is a yes-instance for some graph properties $\Pi$. Our proof indeed implies that the shortest variant is solvable in linear time for the properties.

### 3.1 Directed Forests

Let $\mathcal{S} \subseteq 2^U$ be a collection of subsets of a finite set $U$. Suppose that every set in $\mathcal{S}$ has the same cardinality. We say that $\mathcal{S}$ satisfies the *weak exchange property* if for $S, S' \in \mathcal{S}$ with $S \neq S'$, there exist $e \in S \setminus S'$ and $e' \in S' \setminus S$ such that $S \setminus \{e\} \cup \{e'\} \in \mathcal{S}$. This property is closely related to the (simultaneous) exchange property of bases of matroids: Recall that if $\mathcal{B}$ is the collection of bases of a matroid, then for $B, B' \in \mathcal{B}$ with $B \neq B'$ and for $e \in B \setminus B'$, there is $e' \in B' \setminus B$ such that $B \setminus \{e\} \cup \{e'\} \in \mathcal{B}$. The weak exchange property is not only a weaker version of the exchange property but also gives an important consequence for reconfiguration problems.

In this subsection, we show that $\mathcal{S}_\Pi(G)$ satisfies the weak exchange property for some graph structure properties $\Pi$. Observe that if $\mathcal{S}_\Pi(G)$ satisfies the weak exchange property, then any two subgraphs $H$ and $H'$ in $\mathcal{S}_\Pi(G)$ admit a reconfiguration sequence of length $|A(H') \setminus A(H)| = |A(H) \setminus A(H')|$. Such a reconfiguration sequence is shortest, because we can exchange only one pair of arcs at a time. Therefore, if $\mathcal{S}_\Pi(G)$ satisfies the weak exchange property, then the shortest variant can be solved in linear time for the property $\Pi$.

Similar to the undirected case [5], we show that the weak exchange property holds when $\Pi$ is a spanning directed tree. Note that the following theorem proves Theorem 2.

**Theorem 3.** *$\mathcal{S}_\Pi(G)$ satisfies the weak exchange property when $\Pi$ is a spanning directed tree.*

*Proof.* Let $T$ and $T'$ be arbitrary spanning directed trees in $G$ with $T \neq T'$. Suppose first that $T$ and $T'$ have a common root $r$. Let $e' = (u, v)$ be an arc in $T' \setminus T$ such that the path from $r$ to $u$ in $T'$ is contained in $T$. Clearly, we have $v \neq r$. Let $e$ be the unique arc directed to $v$ in $T$. From the definition of $e$ and $e'$, we have $e \neq e'$. Let $R = T + e' - e$. Now in $T + e'$, the vertex $v$ is the only vertex that has two arcs ($e$ and $e'$) directed to it. Thus, in $R$, no vertex has in-degree 2 or more. Moreover, all vertices in $R$ are reachable from $r$: the paths in $T$ that use $e$ are rerouted to use $e'$ in $R$, and all other paths in $T$ still exist in $R$. Since $|T| = |R|$, $R$ is a spanning directed tree in $G$.

Suppose next that $T$ and $T'$ have different roots $r$ and $r'$, respectively. Let $e'$ be the unique arc in $T'$ directed to $r$, that is, $e' = (u, r)$ for some $u \in V$. Let $P$ be the path from $r$ to $u$ in $T$. Since $P + e'$ is a directed cycle, there is an arc $e = (v, w) \in P$ that does not belong to $T'$. Let $R = T + e' - e$. Observe that no vertex in $R$ has in-degree 2 or more since it holds already in $P + e'$. Observe also that all vertices in $R$ are reachable from $w$: for the descendants of $w$ in $T$, $R$ contains the same path from $w$; and for the other vertices, we first follow the path from $w$ to $u$ in $T$, use the arc $e' = (u, r)$, and then follow the path in $T$ from $r$. Since $|T| = |R|$, $R$ is a spanning directed tree (rooted at $w$) in $G$. □

From the proof of Theorem 3, we obtain the following corollary.

**Corollary 1.** *$\mathcal{S}_\Pi(G)$ satisfies the weak exchange property when $\Pi$ is a spanning $r$-directed tree.*

We then prove the following theorem, which implies that the shortest variant is solvable in linear time when $\Pi$ is a directed forest.

**Theorem 4.** *$S_\Pi(G)$ satisfies the weak exchange property when $\Pi$ is a directed forest.*

*Proof.* Let $F$ and $F'$ be distinct directed forests in $G$ with $|F| = |F'|$. We first consider the case where there is some arc $e' \in F' \setminus F$ such that the endpoints of $e'$ do not belong to the same (weakly) connected component of $F$, that is, either $e'$ connects two connected components of $F$ or at least one of the endpoints of $e'$ does not belong to $F$. Now, we show that there is an arc $e \in F \setminus F'$ such that $F + e' - e$ is a directed forest of $G$. If $F + e'$ is a directed forest, then we can select any arc in $F \setminus F'$ as $e$. Assume that $F + e'$ is not a directed forest. By the assumption in this case, the underlying undirected graph of $F + e'$ contains no (undirected) cycle. Thus there is a vertex of in-degree at least 2 in $F + e'$. Since $F$ is a directed forest, only the head of $e'$, say $v$, can be such a vertex, and its in-degree is exactly 2. As $e$, we select the other arc in $F + e'$ that has $v$ as its head. Since $e' \in F' \setminus F$, this arc $e$ does not belong to $F'$. Since $F + e' - e$ does not contain any cycle in the underlying graph nor any vertex of in-degree 2 or more, it is a directed forest in $G$.

Next we consider the case where every arc $e' \in F' \setminus F$ has both endpoints in the same connected component of $F$. Let $F_1, \ldots, F_c \subseteq F$ be the connected components of $F$, and let $F'_1, \ldots, F'_c \subseteq F'$ be the subsets of $F'$ such that $F'_i = \{e' \in F' \mid e' \text{ has both endpoints in } F_i\}$. We claim that $|F_i| = |F'_i|$. To see this, observe that if $|F_i| < |F'_i|$ for some $i$, then $F'_i$ is not a directed tree since $V(F'_i) \subseteq V(F_i)$ and $F_i$ is a spanning directed tree of the subgraph of $G$ induced by $V(F_i)$. This proves the claim as $|F| = |F'|$. Since both endpoints of every arc in $F'_i$ belong to $F_i$, we also have $V(F_i) = V(F'_i)$ for all $1 \leq i \leq c$. As $F \neq F_i$, there is a connected component $F_i$ in $F$ with $F_i \neq F'_i$ and by Theorem 3, the theorem follows. □

As mentioned in Introduction, when $\Pi$ is a spanning directed tree (or a directed forest), the reachability variant is a subclass of Matroid Intersection Reconfiguration. Theorems 3 and 4 give a new insight on matroid intersection in terms of the weak exchange property.

### 3.2 Directed Forests with Fixed Roots

In this subsection, we consider the case where the property $\Pi$ is an $r$-directed tree for a fixed vertex $r$. Then, every instance of the reachability variant is a yes-instance, and admits a reconfiguration sequence of linear length. More precisely, we prove the following theorem.

**Theorem 5.** *For every pair of $r$-directed trees $T$ and $T'$ in $G$ with $|T| = |T'| = k$, there is a reconfiguration sequence $\langle T = T_0, T_1, \ldots, T_\ell = T' \rangle$ such that all intermediate directed trees have the same root $r$. Moreover, the length $\ell$ of the reconfiguration sequence is at most $k$.*

*Proof.* We say that an arc $e$ in a directed tree $T''$ is *fixed* (with respect to $T'$) if the directed path from $r$ to the head of $e$ in $T''$ appears in $T'$. An arc is *unfixed* if it is not fixed. Let $h$ be the number of unfixed arcs in $T$. We prove that there is a reconfiguration sequence between $T$ and $T'$ of length at most $h$ by induction on $h$. If $h = 0$, then we have $T = T'$. In the following, we assume that $h \geq 1$ and that for every $r$-directed tree $T''$ that has $k$ arcs and contains fewer than $h$ unfixed arcs with respect to $T'$, there is a reconfiguration sequence from $T''$ to $T'$ of length at most $h - 1$.

Let $e = (u, v)$ be an arc in $T'$ such that $e$ is not included in $T$ but all other arcs in the path $P$ from $r$ to the tail of $e$ in $T'$ are included in $T$. Such an arc exists since $T \neq T'$ and they share the root $r$. Note that all arcs in $P$ are fixed.

Assume for now that there is an unfixed arc $f$ in $T$ such that $T'' := T + e - f$ is a directed tree in $G$. Note that $T''$ is still rooted at $r$ since $e$ is an arc of a directed tree rooted at $r$. Observe that arc $e$ is fixed in $T''$ as both $T''$ and $T'$ contain the path $P$ and that the fixed arcs of $T$ remain fixed in $T''$ since we only removed the unfixed arc $f$. Thus $T''$ has fewer than $h$ unfixed arcs. By the induction hypothesis, there is a reconfiguration sequence from $T''$ to $T'$ of length at most $h - 1$, and thus $T'$ is reconfigurable from $T$ as $|T \setminus T''| = |T'' \setminus T| = 1$. Therefore, it suffices to find such an arc $f$.

If the head $v$ of $e$ is included in $T$, then we set $f$ to the arc directed to $v$ in $T$. Then $f$ is unfixed since $T'$ cannot contain it and $T + e - f$ is a directed tree obtained from $T$ by changing the parent of $v$ to $u$. Otherwise, $v$ is not included in $T$, then we set $f$ to an unfixed leaf arc of $T$, which exists since $h \geq 1$. Since $T + e$ is a directed tree and $f$ is a leaf arc of $T + e$ as well, $T + e - f$ is a directed tree. □

This result can be extended to $R$-directed forests.

**Theorem 6 (⋆).** *For every pair of $R$-directed forests $F$ and $F'$ in $G$ with $|F| = |F'| = k$, there is a reconfiguration sequence $\langle F = F_0, F_1, \ldots, F_\ell = F' \rangle$ such that all intermediate forests are $R$-directed. Moreover, the length $\ell$ of the reconfiguration sequence is at most $k$.*

## 4 Algorithm for Directed Tree Reconfiguration

This section is devoted to proving our main result, Theorem 1, which is a polynomial-time algorithm for Directed Tree Reconfiguration. Recall that there are no-instances for the problem, as shown in Fig. 1.

The idea of our algorithm is as follows. Let $G = (V, A)$ be a directed graph, and let $k$ be a positive integer. For each $v \in V$, we denote by $\mathcal{T}(v)$ the collection of all $v$-directed trees $T$ in $G$ with $|T| = k$. By Theorem 5, there is a reconfiguration sequence between any pair of $v$-directed trees in $\mathcal{T}(v)$ such that all internal directed trees in the sequence belong to $\mathcal{T}(v)$. This enables us to "compress" all directed trees in $\mathcal{T}(v)$ into a single representative for each $v \in V$, and it suffices to seek the reachability in the "compressed" solution space. In the rest of this section, when we refer to reconfiguration sequences, every subgraph in these sequences are directed trees with $k$ arcs.

Let $u$ and $v$ be distinct vertices in $G$, and let $T \in \mathcal{T}(u)$ and $T' \in \mathcal{T}(v)$.

**Lemma 1.** *Suppose that $G$ has an arc $(u, v)$ or $(v, u)$. Then, there is a reconfiguration sequence between $T$ and $T'$.*

*Proof.* Assume without loss of generality that $G$ has an arc $(u, v)$. Since $v$ is the root of $T'$, we have $(u, v) \notin T'$. If $u \notin V(T')$, the subgraph $T''$ obtained from $T' + (u, v)$ by removing arbitrary one of the leaf arcs is a directed tree in $\mathcal{T}(u)$. Thus, by Theorem 5, there is a reconfiguration sequence between $T$ and $T''$ and then we are done in this case. Otherwise, $T + (u, v)$ has a directed cycle passing through $(u, v)$. Then, the graph obtained from $T + (u, v)$ by removing the arc directed to $u$ in the cycle is a directed tree in $\mathcal{T}(u)$. Again, by Theorem 5, the lemma follows. □

By inductively applying Theorem 5 and this lemma, we have the following corollary.

**Corollary 2.** *Suppose that $G$ has a directed path from $u$ to $v$ or from $v$ to $u$. Then, there is a reconfiguration sequence between $T$ and $T'$.*

**Lemma 2.** *If there is a vertex $w \in N^+_G(u) \cap N^+_G(v)$ such that $G[V \setminus \{u, v\}]$ has a $w$-directed tree of size $k - 1$, then there is a reconfiguration sequence between $T$ and $T'$.*

*Proof.* Let $T''$ be a directed tree in $G[V \setminus \{u, v\}]$ that has $k - 1$ arcs and root $w \in N^+_G(u) \cap N^+_G(v)$. Since $T'' + (u, w)$ and $T'' + (v, w)$ are directed trees that belong to $\mathcal{T}(u)$ and $\mathcal{T}(v)$, respectively, by Theorem 5, there are reconfiguration sequences between $T$ and $T'' + (u, w)$ and between $T'' + (v, w)$ and $T'$. As $T'' + (v, w)$ is reconfigurable from $T'' + (u, w)$, concatenating these sequences yields a reconfiguration sequence between $T$ and $T'$. □

The above corollary and lemma give sufficient conditions for finding a reconfiguration sequence between $T$ and $T'$. The following lemma ensures that these conditions are also necessary conditions for a "single step".

**Lemma 3.** *Suppose that $|T \setminus T'| = |T' \setminus T| = 1$. Then, at least one of the following conditions hold: (1) $G$ has a directed path from $u$ to $v$ or from $v$ to $u$ or (2) there is $w \in N^+_G(u) \cap N^+_G(v)$ such that $G[V \setminus \{u, v\}]$ has a $w$-directed tree of size $k - 1$.*

*Proof.* Suppose that $v \in V(T)$. Then, there is a directed path $P$ from $u$ to $v$ in $T$ and hence we are done. Symmetrically, the lemma follows when $u \in V(T')$. Thus, we assume that $v \notin V(T)$ and $u \notin V(T')$. This assumption implies that there is a unique arc $e$ directed from $u$ in $T$ as otherwise we have $|T \setminus T'| \geq 2$. Also, there is a unique arc $e'$ directed from $v$ in $T'$. By the fact that $|T \setminus T'| = |T' \setminus T| = 1$, $T - e$ $(= T' - e')$ must be a directed tree with root $w \in N^+_G(u) \cap N^+_G(v)$ that has $k - 1$ arcs in $G[V \setminus \{u, v\}]$. □

To find a reconfiguration sequence between $T^{\mathrm{s}}$ and $T^{\mathrm{t}}$, we construct an auxiliary graph $\mathcal{G}$ as follows. We assume that $G$ is (weakly) connected. For each

$v \in V$, $\mathcal{G}$ contains a vertex $v$ if $G$ has a $v$-directed tree of size $k$. For each pair of distinct $u$ and $v$ in $V(\mathcal{G})$, we add an (undirected) edge between them if (1) $G$ has a directed path from $u$ to $v$ or from $v$ to $u$; or (2) there is a vertex $w \in N_G^+(u) \cap N_G^+(v)$ such that $G[V \setminus \{u, v\}]$ has a $w$-directed tree of size $k-1$. The graph $\mathcal{G}$ can be constructed in $O(|V||A|)$ time. Our algorithm simply finds a path in $\mathcal{G}$ between the two roots of given directed trees $T^{\mathsf{s}}$ and $T^{\mathsf{t}}$. The correctness of the algorithm immediately follows from the following lemma, which also proves the first part of Theorem 1.

**Lemma 4.** *Let $T^{\mathsf{s}}$ and $T^{\mathsf{t}}$ be directed trees in $G$ with $|T^{\mathsf{s}}| = |T^{\mathsf{t}}| = k$ whose roots are $r^{\mathsf{s}}$ and $r^{\mathsf{t}}$, respectively. Then, there is a path between $r^{\mathsf{s}}$ and $r^{\mathsf{t}}$ in $\mathcal{G}$ if and only if there is a reconfiguration sequence between $T^{\mathsf{s}}$ and $T^{\mathsf{t}}$.*

*Proof.* We first show the forward implication. Suppose that there is a path $\mathcal{P}$ between $r^{\mathsf{s}}$ and $r^{\mathsf{t}}$ in $\mathcal{G}$. By Corollary 2 and Lemma 2 there is a reconfiguration sequence between $T^{\mathsf{s}}$ and $T^{\mathsf{t}}$ that can be constructed along the path $\mathcal{P}$.

For the converse implication, suppose that there is a reconfiguration sequence between $T^{\mathsf{s}}$ and $T^{\mathsf{t}}$. Let $T$ and $T'$ be two directed trees that appear consecutively in the sequence. We claim that either $T$ and $T'$ have a common root or the roots of $T$ and $T'$ are adjacent in $\mathcal{G}$. If $T$ and $T'$ have a common root, the claim obviously holds. Suppose otherwise. Let $u$ and $v$ be the roots of $T$ and $T'$, respectively. By Lemma 3, at least one of the conditions (1) and (2) holds, implying that $u$ and $v$ are adjacent in $\mathcal{G}$. □

It is easy to check that our algorithm turns into the one that finds an actual reconfiguration sequence of length $O(|V|^2)$ if the answer is affirmative, and hence Theorem 1 follows.

## 5 Intractable Cases

Directed Path Reconfiguration is a variant of Directed Tree Reconfiguration, where the two input trees $T^{\mathsf{s}}$, $T^{\mathsf{t}}$ and intermediate trees are all directed paths in $G$. Here, we use $\langle P_0, P_1, \ldots, P_\ell \rangle$ with $P_0 = P^{\mathsf{s}}$ and $P_\ell = P^{\mathsf{t}}$ to denote a reconfiguration sequence between two directed paths $P^{\mathsf{s}}$ and $P^{\mathsf{t}}$. Directed Path Sliding consists of the same instance of Directed Path Reconfiguration and we are allowed the following adjacency relation in a valid reconfiguration sequence: for every pair of consecutive directed paths $P = (v_1, v_2, \ldots, v_k)$ and $P' = (v'_1, v'_2, \ldots, v'_k)$, either $v_i = v'_{i+1}$ holds for all $1 \leq i < k$ or $v_i = v'_{i-1}$ holds for all $1 < i \leq k$. Since $P'$ is obtained by "sliding" in a forward or backward direction, we call the problem Directed Path Sliding. In this section, we show that Directed Path Reconfiguration and Directed Path Sliding are both PSPACE-complete.

To this end, we first show that both problems are equivalent with respect to polynomial-time many-one reductions. Let $G$ be a directed graph and let $P = (v_1, v_2, \ldots, v_k)$ be a directed path in $G$ with arc $e_i = (v_i, v_{i+1})$ for $1 \leq i < k$. We denote by $t(P)$ the tail $v_1$ of $P$ and by $h(P)$ the head $v_k$ of $P$. Observe that

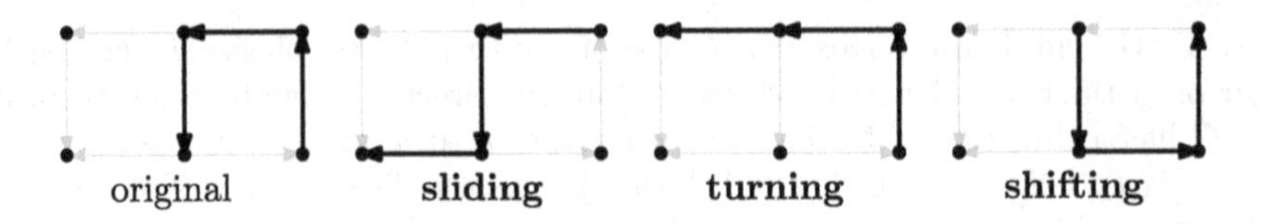


**Fig. 2.** An illustration of the three operations in DIRECTED PATH RECONFIGURATION.

for a directed path $P'$ in $G$ with $|P \setminus P'| = |P' \setminus P| = 1$, at least one of the following conditions hold:

- **sliding:** $P' = (v_2, v_3, \dots v_k, v)$ or $P' = (v, v_1, v_2, \dots, v_{k-1})$ for some $v \in V \setminus V(P)$;
- **turning:** $P' = (v_1, v_2, \dots, v_{k-1}, v)$ or $P' = (v, v_2, v_3, \dots, v_k)$ for some $v \in V \setminus V(P)$;
- **shifting:** $P' = (v_i, v_{i+1}, \dots, v_k, v_1, \dots, v_{i-1})$ for some $1 < i \leq k$. This can be done when $P + (v_k, v_1)$ forms a directed cycle.

See Fig. 2 for an illustration.

We can regard these conditions as operations to obtain $P'$ from $P$. Since **shifting** can be simulated by $i-1$ **sliding** operations along the directed cycle $P + (v_k, v_1)$, the essential difference between DIRECTED PATH RECONFIGURATION and DIRECTED PATH SLIDING is the **turning** operation in order to solve these problems. Now, we perform polynomial-time reductions between these problems in both directions.

Let $(G = (V, A), P^{\mathrm{s}}, P^{\mathrm{t}})$ be an instance of DIRECTED PATH RECONFIGURATION. For each vertex $v$ in $G$, we add two vertices $v^{\mathrm{in}}, v^{\mathrm{out}}$ and two arcs $(v^{\mathrm{in}}, v), (v, v^{\mathrm{out}})$. These two vertices are called *pendant vertices.* We let $G'$ be the graph obtained in this way. Then, we show the following lemma.

**Lemma 5.** *$(G, P^{\mathrm{s}}, P^{\mathrm{t}})$ is a yes-instance of* DIRECTED PATH RECONFIGURATION *if and only if $(G', P^{\mathrm{s}}, P^{\mathrm{t}})$ is a yes-instance of* DIRECTED PATH SLIDING.

*Proof.* Let $\langle P_0, P_1, \dots, P_\ell \rangle$ be a reconfiguration sequence between $P^{\mathrm{s}} = P_0$ and $P^{\mathrm{t}} = P_\ell$ of DIRECTED PATH RECONFIGURATION. By the above argument, we can assume that $P_{i+1}$ is obtained from $P_i$ by applying either **sliding** or **turning**. Let $P_i = (v_1, v_2, \dots, v_k)$. We replace the subsequence $\langle P_i, P_{i+1} \rangle$ with $\langle P_i, P', P_{i+1} \rangle$, where $P' = (v_1^{\mathrm{in}}, v_1, v_2, \dots, v_{k-1})$ if $t(P_i) = t(P_{i+1})$ and $P' = (v_2, v_3, \dots, v_k, v_k^{\mathrm{out}})$ otherwise. Clearly, $P'$ and $P_{i+1}$ are obtained from $P_i$ and $P'$ by applying **sliding** operations, respectively. By replacing each subsequence for $0 \leq i < \ell$, we have a reconfiguration sequence of DIRECTED PATH SLIDING in $G'$.

Conversely, let $\langle P_0, P_1, \dots, P_\ell \rangle$ be a reconfiguration sequence $P^{\mathrm{s}} = P_0$ and $P^{\mathrm{t}} = P_\ell$ of DIRECTED PATH SLIDING. Similarly to the other direction, we construct a reconfiguration sequence of DIRECTED PATH RECONFIGURATION. Assume that $P^{\mathrm{s}} \neq P^{\mathrm{t}}$ as otherwise we are done. Observe that each path $P_i = (v_1, v_2, \dots, v_k)$ contains at most one pendant vertex. This follows from the fact that if $P_i$ contains both $v^{\text{in}}$ and $w^{\text{out}}$ for some $v, w \in V$, then $P_i$ cannot move to a distinct position by **sliding** operations. Now, suppose $P_i$ is a directed path in $G$ with $P_i \neq P^{\text{t}}$, that is, it has no pendant vertices. As $P^{\text{t}}$ has no pendant vertices, we can find the smallest index $j > i$ such that $P_j$ has no pendant vertices. Since $P_j$ can be obtained from $P_i$ by **turning**, we can construct a reconfiguration sequence of DIRECTED PATH RECONFIGURATION by omitting paths having pendant vertices. □

For the converse direction, we let $(G, P^{\text{s}}, P^{\text{t}})$ be an instance of DIRECTED PATH SLIDING. Let $G'$ be the directed graph obtained from $G$ by subdividing each arc $e = (u, w)$ with a new vertex $v_e$, that is, we replace $e$ with $v_e$ and add two arcs $(u, v_e)$ and $(v_e, w)$. Let $Q^{\text{s}}$ and $Q^{\text{t}}$ be defined accordingly from $P^{\text{s}}$ and $P^{\text{t}}$, respectively. In $G'$, we say that a path $P'$ is a *standard path* if $h(P)$ and $t(P)$ belong to $V$ and it is a *nonstandard path* otherwise.

**Lemma 6** (⋆). *$(G, P^{\text{s}}, P^{\text{t}})$ is a yes-instance of* DIRECTED PATH SLIDING *if and only if $(G', Q^{\text{s}}, Q^{\text{t}})$ is a yes-instance of* DIRECTED PATH RECONFIGURATION.

Now, we show that the PSPACE-completeness of DIRECTED PATH SLIDING.

**Theorem 7** (⋆). DIRECTED PATH SLIDING *is PSPACE-complete.*

By Lemma 6, we immediately have the following corollary.

**Corollary 3.** DIRECTED PATH RECONFIGURATION *is PSPACE-complete.*

Suppose that subgraphs in a reconfiguration sequence are relaxed to be acyclic. Observe that the problem is equivalent to reconfiguring directed feedback arc sets in directed graphs. More specifically, given two directed acyclic subgraphs $H^{\text{s}}$ and $H^{\text{t}}$ in a directed graph $G = (V, A)$, the problem asks to determine whether there is a reconfiguration sequence of directed acyclic subgraphs $\langle H^{\text{s}} = H_0, H_1, \ldots, H_\ell = H^{\text{t}} \rangle$ such that $|A(H_i) \setminus A(H_{i+1})| = |A(H_{i+1}) \setminus A(H_i)| = 1$ for all $0 \leq i < \ell$. Seeing this problem from the complement, the problem is equivalent to finding a reconfiguration sequence $\langle A_1, A_2, \ldots, A_\ell \rangle$ of subsets of $A$ such that $H_i = G - A_i$ is acyclic for all $0 \leq i \leq \ell$. Since each $A_i$ is a feedback arc set of $G$, we call this problem DIRECTED FEEDBACK ARC SET RECONFIGURATION. There is another variant of this problem, called DIRECTED FEEDBACK VERTEX SET RECONFIGURATION, in which we are asked to determine given two subsets $V^s$ and $V^g$ of $V$, there is a sequence of vertex subsets $\langle V^s = V_0, V_1, \ldots, V_\ell = V^g \rangle$ of $V$ such that $G[V \setminus V_i]$ is acyclic and $|V_i \setminus V_{i+1}| = |V_{i+1} \setminus V_i| = 1$ for all $0 \leq i < \ell$.

**Theorem 8** (⋆). DIRECTED FEEDBACK ARC SET RECONFIGURATION *and* DIRECTED FEEDBACK VERTEX SET RECONFIGURATION *are PSPACE-complete.*

## 6 Concluding Remarks

There are several possible open questions related to our results. SHORTEST DIRECTED TREE RECONFIGURATION would be a notable open question arising in our work. Contrary to the cases of spanning directed trees and spanning $r$-directed trees, the sets of directed trees and $r$-directed trees with $k < |V| - 1$ arcs do not satisfy the weak exchange property, which makes Shortest Directed Tree Reconfiguration highly nontrivial. It would be also interesting to know whether Directed Path Reconfiguration and Directed Path Sliding are fixed-parameter tractable (FPT) when parameterized by the length of input paths. Although the undirected counterparts are known to be FPT [2,3], it would be difficult to apply their techniques directly to our cases.

## References

1. Biasi, M.D., Ophelders, T.: The complexity of snake and undirected NCL variants. Theor. Comput. Sci. **748**, 55–65 (2018). https://doi.org/10.1016/j.tcs.2017.10.031
2. Demaine, E.D., et al.: Reconfiguring undirected paths. In: Friggstad, Z., Sack, J.-R., Salavatipour, M.R. (eds.) WADS 2019. LNCS, vol. 11646, pp. 353–365. Springer, Cham (2019). https://doi.org/10.1007/978-3-030-24766-9_26
3. Gupta, S., Sa'ar, G., Zehavi, M.: The parameterized complexity of motion planning for snake-like robots. J. Artif. Intell. Res. **69**, 191–229 (2020). https://doi.org/10.1613/jair.1.11864
4. Hanaka, T., et al.: Reconfiguring spanning and induced subgraphs. Theor. Comput. Sci. **806**, 553–566 (2020). https://doi.org/10.1016/j.tcs.2019.09.018
5. Ito, T., et al.: On the complexity of reconfiguration problems. Theor. Comput. Sci. **412**(12–14), 1054–1065 (2011). https://doi.org/10.1016/j.tcs.2010.12.005
6. Mühlenthaler, M.: Degree-constrained subgraph reconfiguration is in P. In: Italiano, G.F., Pighizzini, G., Sannella, D.T. (eds.) MFCS 2015. LNCS, vol. 9235, pp. 505–516. Springer, Heidelberg (2015). https://doi.org/10.1007/978-3-662-48054-0_42
7. Oxley, J.: Matroid Theory, 2nd edn. Oxford University Press, Oxford (2011)

# Decremental Optimization of Vertex-Coloring Under the Reconfiguration Framework

Yusuke Yanagisawa(✉), Akira Suzuki, Yuma Tamura, and Xiao Zhou

Graduate School of Information Sciences, Tohoku University, Sendai, Japan
yusuke.yanagisawa.r7@dc.tohoku.ac.jp, {akira,tamura,zhou}@tohoku.ac.jp

**Abstract.** Suppose that we are given a positive integer $k$, and a $k$-(vertex-)coloring $f_0$ of a given graph $G$. Then we are asked to find a coloring of $G$ using the minimum number of colors among colorings that are reachable from $f_0$ by iteratively changing a color assignment of exactly one vertex while maintaining the property of $k$-colorings. In this paper, we give linear-time algorithms to solve the problem for graphs of degeneracy at most two and for the case where $k \leq 3$. These results imply linear-time algorithms for series-parallel graphs and grid graphs. In addition, we give linear-time algorithms for chordal graphs and cographs. On the other hand, we show that, for any $k \geq 4$, this problem remains NP-hard for planar graphs with degeneracy three and maximum degree four. Thus, we obtain a complexity dichotomy for this problem with respect to degeneracy of a graph and the number $k$ of colors.

## 1 Introduction

In *combinatorial reconfiguration*, we often consider the following problem: we are given two feasible solutions of a combinatorial search problem, then we are asked to determine whether one solution can be transformed into the other in a step-by-step fashion, such that each intermediate solution is also feasible. Such a problem is called *reconfiguration problem*. After Ito et al. proposed this framework [15], the reconfiguration problem has been extensively studied in the field of theoretical computer science. (See, e.g., the surveys of van den Heuvel [14] and Nishimura [22].)

Combinatorial reconfiguration models "dynamic" transformations of systems, where we wish to transform the current configuration of a system into a more desirable one by a step-by-step transformation. In the current framework of combinatorial reconfiguration, we need to have in advance a target (a more desirable) configuration. However, it is sometimes hard to decide a target configuration, because there may exist exponentially many desirable configurations.

A. Suzuki—Partially supported by JSPS KAKENHI Grant Numbers JP18H04091, JP20K11666 and JP20H05794, Japan.
X. Zhou—Partially supported by JSPS KAKENHI Grant Number JP19K11813, Japan.


C.-Y. Chen et al. (Eds.): COCOON 2021, LNCS 13025, pp. 355–366, 2021.
https://doi.org/10.1007/978-3-030-89543-3_30

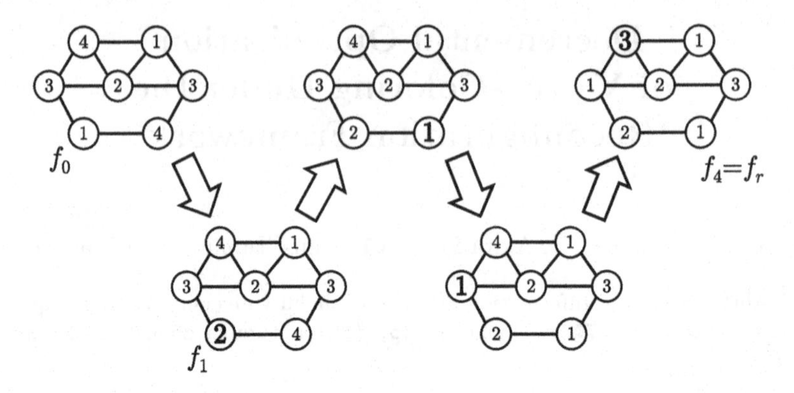


**Fig. 1.** A reconfiguration sequence between two colorings $f_0$ and $f_r$. A bold number implies a color that is changed from a previous one.

Based on this situation, Ito et al. introduced the new framework of reconfiguration problems, called the *optimization variant* [16].

In this variant, we are given a single solution as a current configuration, and asked for a more desirable solution reachable from the given one. This variant was introduced very recently, and hence it has only been applied to Independent Set Reconfiguration [16,17] and Dominating Set Reconfiguration [1] to the best of our knowledge. Therefore, since Coloring Reconfiguration is one of the most studied reconfiguration problems [2–6,8,11,13,18,24], we focus on this problem and study it under this framework.

### 1.1 Our Problem

For an integer $k \geq 1$, let $C$ be a *color set* consisting of $k$ colors $1, 2, \ldots, k$. Let $G$ be a graph with the vertex set $V(G)$ and the edge set $E(G)$. Recall that a *k-coloring* $f$ of $G$ is a mapping $f\colon V(G) \to C$ such that $f(v) \neq f(w)$ holds for each edge $vw \in E(G)$.

In the (Vertex-)Coloring Reconfiguration problem, we are given two $k$-colorings $f_0$ and $f_r$ of the same graph $G$. Then we are asked to determine whether there is a sequence $\langle f_0, f_1, \ldots, f_\ell \rangle$ of $k$-colorings of $G$ such that $f_\ell = f_r$ and $f_i$ can be obtained from $f_{i-1}$ by recoloring only a single vertex in $G$ for all $i$, $1 \leq i \leq \ell$. Such a sequence is called *reconfiguration sequence* from $f_0$ to $f_r$. See Fig. 1 as an example of reconfiguration sequence. The Coloring Reconfiguration is one of the most studied reconfiguration problems [2–6,8, 11,13,18,24]. See also the survey of Mynhardt and Nasserasr [21].

In this paper, we study the *optimization variant* of Coloring Reconfiguration. We denote this problem by Opt-Coloring Reconfiguration. In Opt-Coloring Reconfiguration, we are given only one $k$-coloring $f_0$ of the given graph $G$. Then we are asked to find a $k$-coloring $f_{\mathsf{sol}}$ of $G$ such that there exists a reconfiguration sequence of $k$-colorings from $f_0$ to $f_{\mathsf{sol}}$, and $f_{\mathsf{sol}}$ uses the minimum number of colors over all colorings which can be transformed from $f_0$

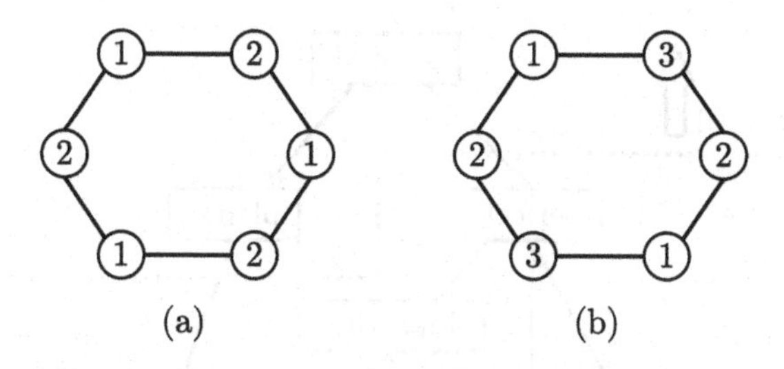


**Fig. 2.** (a) An optimal coloring and (b) a 3-coloring $f_0$ of a cycle of six vertices. When we use only three colors, $f_0$ cannot reach any optimal coloring of the graph.

through reconfiguration. We denote by $(G, k, f_0)$ an instance of OPT-COLORING RECONFIGURATION. Note that $f_{\mathsf{sol}}$ is not always a coloring of $G$ using the minimum number of colors among all colorings of $G$. For example, the graph of Fig. 2(a) has a 2-coloring, but the coloring $f_0$ depicted in Fig. 2(b) cannot be transformed into it when $k = 3$. Indeed, $f_{\mathsf{sol}} = f_0$ holds for this example.

### 1.2 Related Results

As we have mentioned above, COLORING RECONFIGURATION has been studied intensively.

For COLORING RECONFIGURATION, a sharp analysis under the number $k$ of colors has been obtained. It is known that COLORING RECONFIGURATION is PSPACE-complete for any fixed $k \geq 4$ [4]. On the other hand, it is known that COLORING RECONFIGURATION is solvable in linear time for any $k \leq 3$ [8,18]. In addition, given a yes-instance of COLORING RECONFIGURATION for any $k \leq 3$, a reconfiguration sequence with shortest length can be found in polynomial time [8].

COLORING RECONFIGURATION has also been studied from the viewpoint of graph classes. It is known that COLORING RECONFIGURATION is PSPACE-complete for bipartite planar graphs [4]. Since every bipartite planar graph is 3-degenerate, COLORING RECONFIGURATION is PSPACE-complete for 3-degenerate graphs. COLORING RECONFIGURATION is known to be PSPACE-complete also for graphs with bounded bandwidth [24] and chordal graphs [13]. On the other hand, COLORING RECONFIGURATION is solvable in polynomial time for split, trivially perfect, 2-degenerate, and $(k-2)$-connected chordal graphs for any number $k$ of colors [6,13].

The optimization variant of reconfiguration problems were recently proposed by Ito et al. [16]. To the best of our knowledge, it has only been applied to INDEPENDENT SET RECONFIGURATION [16,17] and DOMINATING SET RECONFIGURATION [1]. Therefore, in this paper, we apply this new framework to one of the most studied reconfiguration problems, namely COLORING RECONFIGURATION.

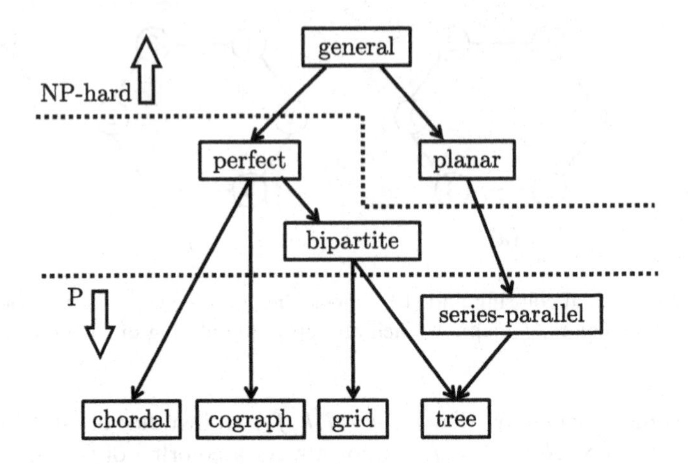


**Fig. 3.** Our results for Opt-Coloring Reconfiguration with respect to graphs classes. Each arrow represents the inclusion relationship between graph classes; $A \rightarrow B$ means that the graph class $B$ is a subclass of the graph class $A$.

**Table 1.** The complexity of Opt-Coloring Reconfiguration with respect to the number $k$ of colors and degeneracy $d$ of a graph.

| $d\backslash k$ | 1 | 2 | 3 | 4 | $\geq 5$ |
|---|---|---|---|---|---|
| 1 | P | P | P | P | P |
| 2 | P | P | P | P | P |
| 3 | P | P | P | NP-hard | NP-hard |
| 4 | P | P | P | NP-hard | NP-hard |
| $\geq 5$ | P | P | P | NP-hard | NP-hard |

### 1.3 Our Results

In this paper, we give linear-time algorithms to solve Opt-Coloring Reconfiguration for graphs of degeneracy two, and for any graph when $k \leq 3$. These results imply linear-time algorithms for series-parallel graphs and grid graphs. In addition, we give linear-time algorithms for chordal graphs and cographs for any $k$. Since Coloring Reconfiguration is PSPACE-hard for chordal graphs [13], we obtain a difference in complexity between Coloring Reconfiguration and Opt-Coloring Reconfiguration, that is, some difficulties disappear for the optimization variant, in a sense. On the other hand, we show that, for any $k \geq 4$, this problem remains NP-hard for planar graphs with degeneracy three and maximum degree four. Thus, we obtain a complexity dichotomy for this problem with respect to the number of colors and degeneracy of a graph. We summarize our results in Fig. 3 and Table 1.

## 2 Preliminaries

Let $G = (V, E)$ be a graph. We denote by $V(G)$ and $E(G)$ the vertex set and the edge set of $G$, respectively. We assume that all graphs in the remainder of this paper are simple, undirected, and have at least one edge. The *degeneracy* $d(G)$ of a graph $G$ is the minimum integer $d$ such that any subgraph $H$ of $G$ has a vertex of degree at most $d$. For a positive integer $k$, a graph $G$ is *$k$-colorable* if $G$ has a $k$-coloring. We say that a $k$-coloring $f$ of $G$ is *optimal* if $G$ has no $(k-1)$-coloring. We denote by $\chi(G)$, called the *chromatic number of* $G$, the integer $k$ such that $G$ has an optimal $k$-coloring.

A coloring $f$ of a graph $G$ is *$k$-reachable* from a coloring $f_0$ of $G$ if there is a sequence $\langle f_0, f_1, \ldots, f_\ell \rangle$ of $k$-colorings of $G$ such that $f_\ell = f$ and $f_i$ can be obtained from $f_{i-1}$ by recoloring only a single vertex of $G$ for every $i$, $1 \leq i \leq \ell$. For a coloring $f$ of $G$, let $\mathsf{col}(f)$ be the number of colors used in $f$. We define

$$\chi(G, k, f_0) = \min\{\mathsf{col}(f) \mid f \text{ is a coloring of } G \text{ and } f \text{ is } k\text{-reachable from } f_0\}$$

and $\chi(G, k, f_0) = +\infty$ if $k < \mathsf{col}(f_0)$. Note that $\chi(G, k, f_0)$ is at least $\chi(G)$. Opt-Coloring Reconfiguration is the problem of computing $\chi(G, k, f_0)$ for a given graph $G$, a positive integer $k$ and a coloring $f_0$ of $G$. We remark that, with minor adjustments, all algorithms in this paper can actually find a coloring $f_{\mathsf{sol}}$ of $G$ such that $\mathsf{col}(f_{\mathsf{sol}}) = \chi(G, k, f_0)$.

## 3 Linear-Time Algorithms

### 3.1 The Case Where the Number of Colors Is at Most Three

In this subsection, we show the following theorem:

**Theorem 1.** *Let $(G, k, f_0)$ be an instance of* Opt-Coloring Reconfiguration. *If $k \leq 3$, the problem can be solved in linear time.*

*Proof.* Recall that the input graph $G$ has at least one edge. This implies that $\chi(G) > 1$ and thus $\chi(G, k, f_0) > 1$. If $f_0$ is a 2-coloring of $G$, then we conclude that $\chi(G, k, f_0) = 2$. In the remainder of this proof, we assume that $k = 3$ and hence $f_0$ is a 3-coloring of $G$.

We give an algorithm for an instance $(G, 3, f_0)$. Our algorithm contains the following two steps. First, the algorithm checks in linear time whether $G$ is 2-colorable, that is, bipartite. Since $f_0$ is a 3-coloring, $\chi(G)$ is two or three. If $G$ is not 2-colorable, we have $\chi(G) = 3$. In this case, the algorithm concludes that $\chi(G, k, f_0) = 3$, otherwise we go to the next step.

In the next step, the algorithm finds an arbitrary 2-coloring $f_r$ of $G$ in linear time, and then checks whether $f_r$ is 3-reachable from $f_0$ or not. It is known that Coloring Reconfiguration is solvable in linear time if $k \leq 3$ [18]. If $f_r$ is 3-reachable from $f_0$, the algorithm concludes that $\chi(G, k, f_0) = 2$, otherwise $\chi(G, k, f_0) = 3$. This step correctly outputs a solution because one can see that any 2-coloring is 3-reachable from any other 2-coloring. The total running time of our algorithm is linear, completing the proof. □

### 3.2 The Graphs of Degeneracy at Most Two

In this subsection, we show the following theorem:

**Theorem 2.** *Let* $(G, k, f_0)$ *be an instance of* Opt-Coloring Reconfiguration. *If the degeneracy* $d(G)$ *is at most two, then the problem can be solved in linear time.*

*Proof.* For the case where $k \leq 3$, we use the algorithm given in Theorem 1. Suppose that $k \geq 4$. It is known that, if $k \geq d(G) + 2$, then any two $k$-colorings of $G$ are $k$-reachable from each other [7]. Thus, for the case where $d(G) \leq 2$ and $k \geq 4$, we have $\chi(G, k, f_0) = \chi(G)$, and hence it suffices to compute $\chi(G)$. One can easily check whether or not $G$ is 2-colorable, that is, $\chi(G) = 2$ in linear time. If $\chi(G) \neq 2$, then $\chi(G) = 3$ because $d(G) \leq 2$ and $\chi(G) \leq d(G)+1$. Thus, $\chi(G, k, f_0) = \chi(G)$ can be computed in linear time, completing the proof. □

Since both series-parallel and grid graphs have degeneracy at most two, we obtain the following corollary by Theorem 2:

**Corollary 1.** Opt-Coloring Reconfiguration *is solvable in linear time for series-parallel graphs and grid graphs.*

### 3.3 Chordal Graphs

In this subsection, we show the following theorem:

**Theorem 3.** Opt-Coloring Reconfiguration *is solvable in linear time for chordal graphs.*

*Proof.* Let $(G, k, f_0)$ be an instance of Opt-Coloring Reconfiguration, where $G$ is a chordal graph. Suppose that $k \geq \mathsf{col}(f_0)$ holds. Our algorithm computes $\chi(G)$ and concludes that $\chi(G, k, f_0) = \chi(G)$. Since we can compute $\chi(G)$ in linear time for any chordal graph $G$ [23], our algorithm takes linear time.

We give the correctness of the algorithm. Clearly, if $\chi(G) = k$, then $f_0$ itself is an optimal coloring of $G$ and hence $\chi(G, k, f_0) = \chi(G)$ holds. We show that $\chi(G, k, f_0) = \chi(G)$ holds also for $\chi(G) < k$. It suffices to prove that any optimal coloring of $G$ is $k$-reachable from $f_0$ if $\chi(G) < k$. For any chordal graph $G$, $\chi(G) = d(G) + 1$ holds [20]. Thus, we have $k \geq d(G) + 2$. It is known that, if $k \geq d(G) + 2$, then any two $k$-colorings of $G$ are $k$-reachable [7]. Therefore, any optimal coloring of $G$ is $k$-reachable from $f_0$ if $\chi(G) < k$, and hence $\chi(G, k, f_0) = \chi(G)$, completing the proof. □

### 3.4 Cographs

In this subsection, we give a linear-time algorithm for cographs. In fact, the algorithm is almost the same as the one for chordal graphs. For the correctness of the algorithm, we use the Grundy number. A $k$-coloring $f_g$ of a graph $G$ is called a *Grundy coloring* if each vertex $v \in V(G)$ such that $f_g(v) = i$ is adjacent to at least one vertex with color $j$ for each $j < i$. The *Grundy number* $\chi_g(G)$ of $G$ is the maximum integer $k$ such that $G$ has a Grundy coloring with $k$ colors.

**Theorem 4.** OPT-COLORING RECONFIGURATION *is solvable in linear time for cographs.*

*Proof.* Let $(G, k, f_0)$ be an instance of OPT-COLORING RECONFIGURATION, where $G$ is a cograph. Suppose that $k \geq \mathsf{col}(f_0)$ holds. Our algorithm computes $\chi(G)$ and concludes that $\chi(G, k, f_0) = \chi(G)$. Since we can compute $\chi(G)$ in linear time for any cograph $G$ [23], our algorithm takes linear time.

We give the correctness of the algorithm. As in the proof of Theorem 3, we show that any optimal coloring of $G$ is $k$-reachable from $f_0$ if $\chi(G) < k$. For any cograph $G$, $\chi(G) = \chi_g(G)$ holds [9]. Thus, we have $k \geq \chi_g(G) + 1$. It is known that, any two $k$-colorings of $G$ are $k$-reachable if $k \geq \chi_g(G) + 1$ [2]. Therefore, any optimal coloring of $G$ is $k$-reachable from $f_0$ if $\chi(G) < k$, and hence $\chi(G, k, f_0) = \chi(G)$, completing the proof. □

## 4 NP-Hardness

In this section, we show that OPT-COLORING RECONFIGURATION remains NP-hard even for any $k \geq 4$, planar graphs with degeneracy three and maximum degree four. We assume that $k = 4$ because our proof can easily be applicable to the case where $k > 4$. Our proof consists of the following three steps:

**Step 1** construct an instance $(G_\phi, 4, f_\phi)$ of OPT-COLORING RECONFIGURATION from an instance $\phi$ of 3-SAT so that $G_\phi$ has degeneracy three;

**Step 2** transform $(G_\phi, 4, f_\phi)$ into $(G_p, 4, f_p)$ where $G_p$ is a planar graph of degeneracy three; and

**Step 3** reduce the maximum degree of the graph $G_p$ and construct an instance $(G, 4, f_0)$.

In 3-SAT, we are given a CNF-formula $\phi$ with a collection $\{C_1, C_2, \ldots, C_m\}$ of $m$ clauses over $n$ variables $\{x_1, x_2, \ldots, x_n\}$, and each clause contains exactly three variables. Our task is to determine whether there exists a variable assignment which satisfies a given CNF-formula or not. 3-SAT is a well-known NP-complete problem [19].

In fact, our construction of $G$ follows the existing reduction which proves the NP-hardness of 3-COLORING problem for planar graphs with degeneracy three and maximum degree four [10,12]. Before we explain the construction of $G$ and $f_0$, we show that $\chi(G, 4, f_0) \geq 4$ if $\phi$ has no feasible variable assignment. In [10,12], the authors prove that $G$ has a 3-coloring if and only if $\phi$ has a feasible variable assignment. Therefore, if $\phi$ has no feasible variable assignment, any coloring $f_0$ cannot reach any 3-coloring of $G$, and hence $\chi(G, 4, f_0) \geq 4$. Thus, in the remainder of this section, it suffices to give a 4-coloring $f_0$ of $G$ so that $\chi(G, 4, f_0) \leq 3$ if $\phi$ has a feasible variable assignment.

### 4.1 Step 1: Constructing an Instance from a CNF-Formula

As the first step in our reduction, we explain how to construct an instance $(G_\phi, 4, f_\phi)$ of OPT-COLORING RECONFIGURATION from an instance $\phi$ of 3-SAT,

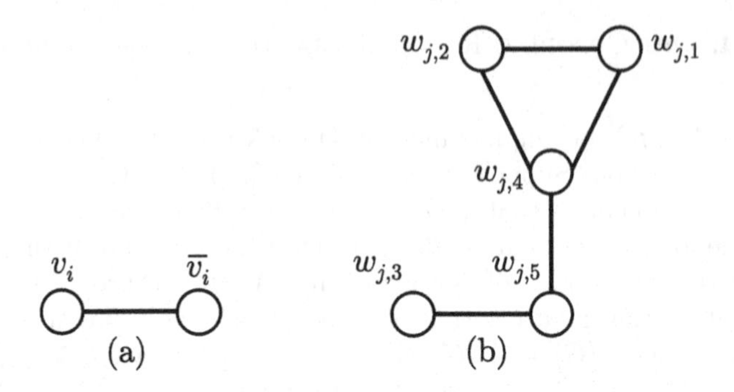


**Fig. 4.** (a) A variable gadget $X_i$ and (b) a clause gadget $Y_j$.

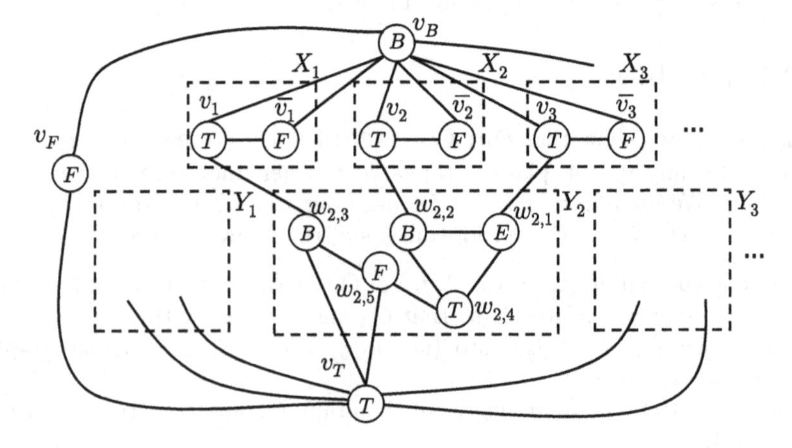


**Fig. 5.** An example of the construction of a graph $G_\phi$ and a 4-coloring $f_\phi$ of $G_\phi$, where $C_2 = x_3 \vee x_2 \vee x_1$.

where $G_\phi$ has degeneracy three. In the construction, we use a variable gadget and a clause gadget in Fig. 4, which appears in [10]. The variable gadget $X_i$, $1 \le i \le n$, consists of two vertices $v_i$ and $\bar{v}_i$. The clause gadget $Y_j$, $1 \le j \le m$, consists of five vertices $w_{j,1}, w_{j,2}, w_{j,3}, w_{j,4}, w_{j,5}$.

For a given CNF-formula $\phi$, we build a corresponding graph $G_\phi$. First, for each variable $x_i$, $1 \le i \le n$, and each clause $C_j$, $1 \le j \le m$, of $\phi$, we add one variable gadget $X_i$ and one clause gadget $Y_j$, respectively. We also add a cycle of three vertices $v_T, v_F$ and $v_B$. We connect $v_B$ to $v_i$ and $\bar{v}_i$ in each variable gadget $X_i$ by edges, and connect $v_T$ to $w_{j,3}$ and $w_{j,5}$ in each clause gadget $Y_j$ by edges. Then, if a variable $x_i$ (resp. $\bar{x}_i$) appears at the $\ell$-th position of a clause $C_j$ of $\phi$, we connect $v_i$ (resp. $\bar{v}_i$) of the variable gadget $X_i$ and $w_{j,\ell}$ of the clause gadget $Y_j$ by an edge, as illustrated in Fig. 5. This completes the corresponding graph $G_\phi$. Clearly, $G_\phi$ is constructed in polynomial time. From the construction of $G_\phi$, it is not hard to see that $G_\phi$ has degeneracy three.

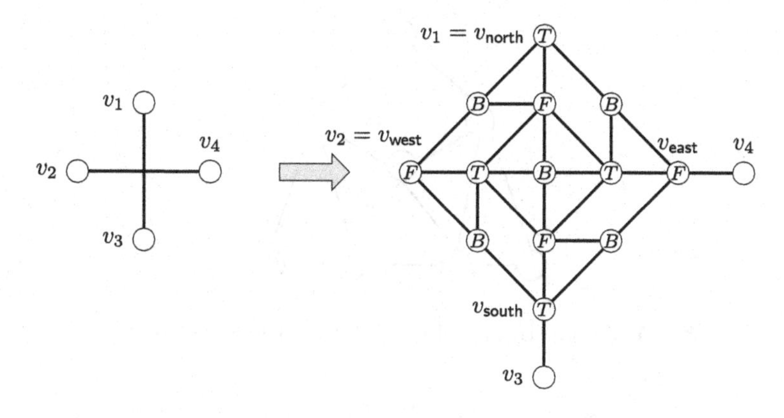


**Fig. 6.** A replacement of a crossing point of $G_\phi$ by a cross gadget $H_C$.

Next we explain the construction of $f_\phi$. Let $\{T, F, B, E\}$ be a color set. The vertices $v_T, v_F$ and $v_B$ are colored by $T, F$ and $B$, respectively. For each variable gadget $X_i$, $1 \leq i \leq n$, $v_i$ is colored by $T$ and $\bar{v}_i$ is colored by $F$. For each clause gadget $Y_j$, $1 \leq j \leq m$, $w_{j,1}$ is colored by $E$, $w_{j,2}$ and $w_{j,3}$ are colored by $B$, $w_{j,4}$ is colored by $T$, and $w_{j,5}$ is colored by $F$. Clearly, our construction of $f_\phi$ is done in polynomial time. Then, we have the following lemma.

**Lemma 1.** *$\chi(G_\phi, 4, f_\phi) \leq 3$ if $\phi$ has a feasible variable assignment.*

## 4.2 Step 2: Making a Graph Planar

In the second step of our reduction, we construct an instance $(G_p, 4, f_p)$ of Opt-Coloring Reconfiguration from the instance $(G_\phi, 4, f_\phi)$, where $G_p$ is a planar graph of degeneracy three.

In the construction of $G_p$, we use a cross gadget $H_C$ illustrated in Fig. 6, which appears in [12]. We assume that $G_\phi$ is embedded on a plane so that at most two edges of $G_\phi$ share the same coordinates on the plane. We replace crossing points as illustrated in Fig. 6. (For more details, refer to [12].) Repeating the replacement for all crossing points results in a planar graph $G_p$. Since $G_\phi$ and $H_C$ have degeneracy three, it is not hard to see that $G_p$ also has degeneracy three.

We now construct a 4-coloring $f_p$ of $G_p$ from the 4-coloring $f_\phi$ of $G_\phi$ by giving a 3-coloring of $H_C$. Figure 6 shows the 3-coloring of $H_C$ for the case where $f_\phi(v_1) = T$ and $f_\phi(v_2) = F$. We can give 3-colorings of $H_C$ for the other cases and we have the following lemma, although the colorings of $H_C$ and the proof of the lemma are omitted due to page limitation.

**Lemma 2.** *$\chi(G_p, 4, f_p) \leq 3$ if $\chi(G_\phi, 4, f_\phi) \leq 3$.*

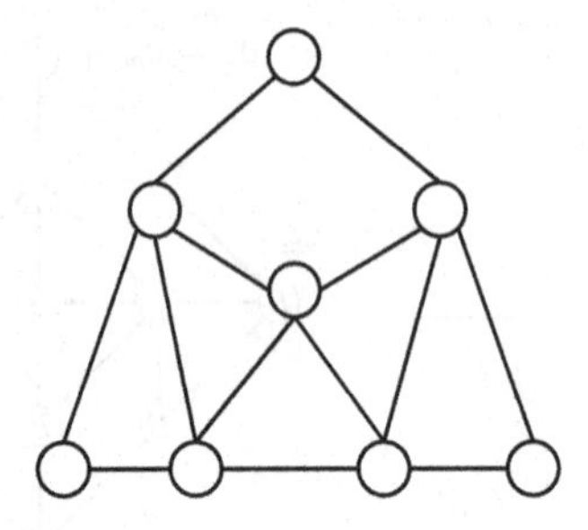

**Fig. 7.** An illustration of the reducing-degree gadget $H_3$.

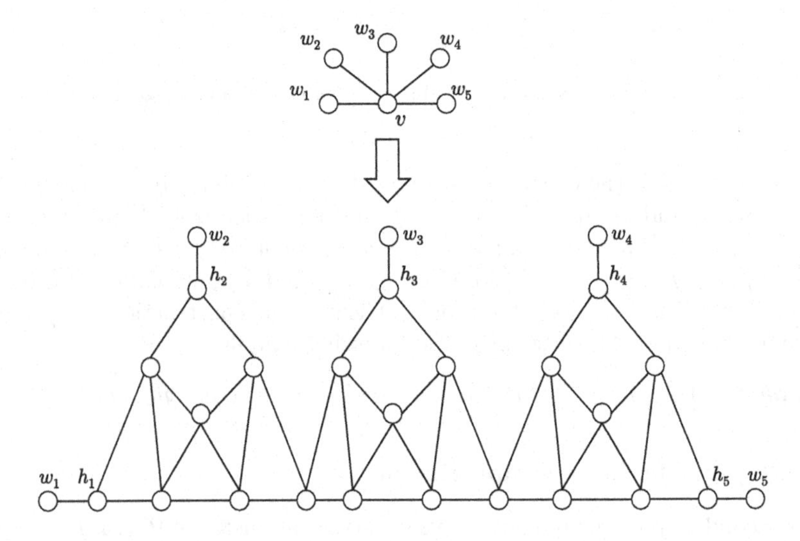


**Fig. 8.** A replacement of a vertex with degree five by the reducing-degree gadget $H_5$, where $H_5$ is constructed by merging three $H_3$'s.

### 4.3 Step 3: Reducing the Maximum Degree of a Graph

In the last step of our reduction, we construct an instance $(G, 4, f_0)$ of Opt-Coloring Reconfiguration from $(G_p, 4, f_p)$, where $G$ is a planar graph with degeneracy three and maximum degree four.

In the construction, we use a reducing-degree gadget $H_3$ illustrated in Fig. 7, which appears in [12]. Suppose that $G_p$ has a vertex $v$ of degree $\delta \geq 5$. We make a gadget $H_\delta$ by merging $\delta - 2$ reducing-degree gadgets, and replace $v$ with $H_\delta$ as illustrated in Fig. 8. If $v$ has a neighborhood $\{w_1, w_2, \ldots, w_\delta\}$, we connect $w_i$ to $h_i$ by an edge for each $i$, $1 \leq i \leq \delta$. By applying the above operation to all vertices of degree at least five, the construction of $G$ is completed. Since $H_\delta$ has no vertex of degree more than four, the maximum degree of $G$ is four. Since $H_\delta$ is planar and $h_i$ in $H_\delta$ has degree three, it is not hard to see that $G$ is planar and has degeneracy three.

We now give the 3-coloring of $H_\delta$ as depicted in Fig. 8, and we have the following lemma.

**Lemma 3.** $\chi(G, 4, f_0) \leq 3$ *if* $\chi(G_p, 4, f_p) \leq 3$.

It follows from Lemmas 1, 2 and 3 that $\chi(G, 4, f) \leq 3$ if $\phi$ has a feasible variable assignment. This completes the proof of NP-hardness for OPT-COLORING RECONFIGURATION on planar graphs with degeneracy three and maximum degree four if $k = 4$. Obviously, the above proof can be applicable to the case where $k > 4$, and hence we have the following theorem.

**Theorem 5.** *For any* $k \geq 4$, OPT-COLORING RECONFIGURATION *is NP-hard even for planar graphs with degeneracy three and maximum degree four.*

## 5 Conclusion

In this paper, we gave linear-time algorithms to solve the problem for graphs of degeneracy at most two and for the case where $k \leq 3$. These results imply linear-time algorithms for series-parallel graphs and grid graphs. In addition, we gave linear-time algorithms for chordal graphs and cographs. On the other hand, we showed that, for any $k \geq 4$, this problem remains NP-hard for planar graphs with degeneracy three and maximum degree four. In particular, our theorems give a sharp complexity dichotomies with respect to the degeneracy of the input graph and the number $k$ of colors.

It remains open to clarify the complexity status of perfect graphs, bipartite graphs, or graphs of maximum degree three.

**Acknowledgment.** We are grateful to Tatsuhiko Hatanaka, Takehiro Ito and Haruka Mizuta for valuable discussions with them. We thank the anonymous referees for their constructive suggestions and comments.

## References

1. Blanché, A., Mizuta, H., Ouvrard, P., Suzuki, A.: Decremental optimization of dominating sets under the reconfiguration framework. In: Gąsieniec, L., Klasing, R., Radzik, T. (eds.) IWOCA 2020. LNCS, vol. 12126, pp. 69–82. Springer, Cham (2020). https://doi.org/10.1007/978-3-030-48966-3_6
2. Bonamy, M., Bousquet, N.: Recoloring graphs via tree decompositions. Eur. J. Comb. **69**, 200–213 (2018)
3. Bonamy, M., Johnson, M., Lignos, I., Patel, V., Paulusma, D.: Reconfiguration graphs for vertex colourings of chordal and chordal bipartite graphs. J. Comb. Optim. **27**, 132–143 (2014)
4. Bonsma, P.S., Cereceda, L.: Finding paths between graph colourings: PSPACE-completeness and superpolynomial distances. Theoret. Comput. Sci. **410**(50), 5215–5226 (2009)
5. Bonsma, P., Mouawad, A.E., Nishimura, N., Raman, V.: The complexity of bounded length graph recoloring and CSP reconfiguration. In: Cygan, M., Heggernes, P. (eds.) IPEC 2014. LNCS, vol. 8894, pp. 110–121. Springer, Cham (2014). https://doi.org/10.1007/978-3-319-13524-3_10
6. Bonsma, P.S., Paulusma, D.: Using contracted solution graphs for solving reconfiguration problems. Acta Informatica **56**, 619–648 (2019)

7. Cereceda, L., van den Heuvel, J., Johnson, M.: Connectedness of the graph of vertex-colourings. Discret. Math. **308**(5), 913–919 (2008)
8. Cereceda, L., van den Heuvel, J., Johnson, M.: Finding paths between 3-colourings. J. Graph Theory **67**(1), 69–82 (2011)
9. Christen, C.A., Selkow, S.M.: Some perfect coloring properties of graphs. J. Comb. Theory Ser. B **27**(1), 49–59 (1979)
10. Erickson, J.: Algorithms (2019)
11. Feghali, C., Johnson, M., Paulusma, D.: A reconfigurations analogue of brooks' theorem and its consequences. J. Graph Theory **83**(4), 340–358 (2016)
12. Garey, M.R., Johnson, D.S.: Computers and Intractability: A Guide to the Theory of NP-Completeness. Freeman, San Francisco (1979)
13. Hatanaka, T., Ito, T., Zhou, X.: The coloring reconfiguration problem on specific graph classes. IEICE Trans. Fundam. Electron. Commun. Comput. Sci. **E102.D**(3), 423–429 (2019)
14. van den Heuvel, J.: The complexity of change. In: Blackburn, S.R., Gerke, S., Wildon, M. (eds.) Surveys in Combinatorics. London Mathematical Society Lecture Note Series, vol. 409, pp. 127–160. Cambridge University Press (2013)
15. Ito, T., et al.: On the complexity of reconfiguration problems. Theoret. Comput. Sci. **412**(12), 1054–1065 (2011)
16. Ito, T., Mizuta, H., Nishimura, N., Suzuki, A.: Incremental optimization of independent sets under the reconfiguration framework. In: Du, D.-Z., Duan, Z., Tian, C. (eds.) COCOON 2019. LNCS, vol. 11653, pp. 313–324. Springer, Cham (2019). https://doi.org/10.1007/978-3-030-26176-4_26
17. Ito, T., Mizuta, H., Nishimura, N., Suzuki, A.: Incremental optimization of independent sets under the reconfiguration framework. J. Comb. Optim. 1–16 (2020). https://doi.org/10.1007/s10878-020-00630-z
18. Johnson, M., Kratsch, D., Kratsch, S., Patel, V., Paulusma, D.: Finding shortest paths between graph colourings. Algorithmica **75**, 295–321 (2016)
19. Karp, R.M.: Reducibility among combinatorial problems. In: Miller, R.E., Thatcher, J.W., Bohlinger, J.D. (eds.) Complexity of Computer Computations. The IBM Research Symposia Series, pp. 85–103. Springer, Boston (1972). https://doi.org/10.1007/978-1-4684-2001-2_9
20. Markossian, S.E., Gasparian, G.S., Reed, B.A.: $\beta$-perfect graphs. J. Comb. Theory Ser. B **67**(1), 1–11 (1996)
21. Mynhardt, C.M., Nasserasr, S.: Reconfiguration of colourings and dominating sets in graphs. In: 50 Years of Combinatorics, Graph Theory, and Computing, pp. 171–191. Chapman and Hall/CRC (2019)
22. Nishimura, N.: Introduction to reconfiguration. Algorithms **11**(4), 52 (2018)
23. Rose, D.J., Tarjan, R.E., Lueker, G.S.: Algorithmic aspects of vertex elimination on graphs. SIAM J. Comput. **5**(2), 266–283 (1976)
24. Wrochna, M.: Reconfiguration in bounded bandwidth and tree-depth. J. Comput. Syst. Sci. **93**, 1–10 (2018)

# Embedding Three Edge-Disjoint Hamiltonian Cycles into Locally Twisted Cubes

Kung-jui Pai(✉)

Department of Industrial Engineering and Management, Ming Chi University of Technology, New Taipei City, Taiwan
poter@mail.mcut.edu.tw

**Abstract.** The $n$-dimensional locally twisted cube $LTQ_n$, a variation of the hypercube $Q_n$, has the same number of vertices and the same number of edges as $Q_n$, but it has only about half of the diameter of $Q_n$. The existence of the Hamiltonian cycle provides an advantage when implementing algorithms that require a ring structure. In addition, $k$ ($\geq 2$) edge-disjoint Hamiltonian cycles also provide the edge-fault tolerant Hamiltonicity for the interconnection network. Hung [Theoretical Computer Science 412, 4747–4753, 2011] proved that $LTQ_n$ with $n \geqslant 4$ contains two edge-disjoint Hamiltonian cycles, and posted an open problem what is the maximum number of edge-disjoint Hamiltonian cycles in $LTQ_n$ for n $\geqslant$ 6? In this paper, we show that there exist three edge-disjoint Hamiltonian cycles on $LTQ_n$ while $n \geqslant 6$.

**Keywords:** Edge-disjoint hamiltonian cycles · Locally twisted cubes · Interconnection networks

## 1 Introduction

The design of an interconnection network is one of the important issues for parallel computing systems and data centers. An interconnected network is usually modeled by a graph in which vertices representing processing units and edges representing communication links. We will use graphs and networks interchangeably in this paper.

Hypercubes [16, 17] have become one of the most popular interconnection networks due to their attractive features, including regularity, vertex symmetric, link symmetric, small diameter, strong connectivity, recursive construction, partition capability, and small link complexity. The locally twisted cube $LTQ_n$ [20] (defined later in Sect. 2) is one of the hypercube-variant networks. $LTQ_n$ is similar to $Q_n$ in that the vertices can be one-to-one labeled with 0–1 binary strings of length $n$, so the labels of any two adjacent vertices are different at most in two consecutive bits. One advantage of $LTQ_n$ is that the diameter is only about one-half of the diameter of an $n$-dimensional hypercube. Some properties of $LTQ_n$ have been obtained, such as a connectivity of $n$ [20], 4-pancyclic [20], edge-pancyclic [12], and $n$ edge-disjoint spanning trees [3]. For more study results about the locally twisted cube, please refer to [10, 19].


C.-Y. Chen et al. (Eds.): COCOON 2021, LNCS 13025, pp. 367–374, 2021.
https://doi.org/10.1007/978-3-030-89543-3_31

A Hamiltonian cycle in a graph is a cycle that visits through every vertex exactly once. Many communication algorithms, such as all-to-all broadcasting algorithms, are designed based on Hamilton cycles, and its benefits can be found in [11]. Hamiltonian cycles in the graph are said to be edge-disjoint if they do not share any common edges. Further, $k$ ($\geq 2$) edge-disjoint Hamiltonian cycles also provide the edge-fault tolerant hamiltonicity for the interconnection network. That is, when one edge in the Hamiltonian cycle fails, the other edge-disjoint Hamiltonian cycle can be used to replace it for transmission.

Previous related works on edge-disjoint Hamiltonian are described below. Rowley and Bose [15] show that a slightly modified degree $2r$ de Bruijn graph can be decomposed into $r$ Hamiltonian cycles when $r$ is a power of a prime. Barth and Raspaud [2] gave that there are two edge-disjoint Hamiltonian cycles on the butterfly networks. Lee and Shin [8] achieved reliable all-to-all broadcasting using edge-disjoint Hamiltonian cycles. Bae et al. [1] studied edge-disjoint Hamiltonian cycles in k-ary n-cubes and hypercubes. Petrovic and Thomassen [14] characterized the number of edge-disjoint Hamiltonian cycles in hypertournaments. Hung et al. presented how to construct two edge-disjoint Hamiltonian cycles in locally twisted cubes [4], crossed cubes [6], augmented cubes [5], twisted cubes [7], transposition networks, and hypercube-like networks [6], respectively. Wang et al. [18] showed that two edge-disjoint Hamiltonian cycles can be embedded into parity cubes. Recently, Pai [13] presented a parallel algorithm for constructing two edge-disjoint Hamiltonian cycles in crossed cubes. Li et al. [9] presented a parallel algorithm for constructing two edge-disjoint Hamiltonian cycles in locally twisted cubes.

Hung [7] proved that $LTQ_n$ with $n \geqslant 4$ contains two edge-disjoint Hamiltonian cycles, and posted an open problem what is the maximum number of edge-disjoint Hamiltonian cycles in $LTQ_n$ for n $\geqslant$ 6? Due to the constraint of edge-disjoint, the maximum number is either 2 or 3 for $LTQ_6$. In this paper, we show that there exist three edge-disjoint Hamiltonian cycles on $LTQ_n$ while $n \geqslant 6$.

The rest of the paper is organized as follows: In Sect. 2, the structure of locally twisted cubes is introduced and some notations are given. Section 3 presented three edge-disjoint Hamiltonian cycles in $LTQ_6$. Based on this result, we further show how to construct three edge-disjoint Hamiltonian cycles in $LTQ_n$ while $n \geq 6$. Finally, Sect. 4 is the concluding remarks of this paper.

## 2 Preliminaries

Interconnection networks are usually modeled as undirected simple graphs $G = (V, E)$, where the vertex set $V$ ($= V(G)$) and the edge set $E$ ($= E(G)$) represent the set of processing units and the set of communication links between vertices, respectively. The neighborhood of a vertex $v$ in a graph $G$, denoted by $N(v)$, is the set of vertices adjacent to $v$ in $G$. A cycle $C_m$ of length $m$ in $G$, denoted by $v_0$ - $v_1$ - $v_2$ - ... - $v_{m-2}$ - $v_{m-1}$ - $v_0$, is a sequence ($v_0$, $v_1$, $v_2$, ..., $v_{m-1}$, $v_0$) of vertices such that ($v_{m-1}$, $v_0$) $\in E$ and ($v_i$, $v_{i+1}$) $\in E$ for $0 \leq i \leq m-2$. Now, we introduce locally twisted cubes as follows:

**Definition 1:** (*Yang et al.* [20]) *The n-dimensional locally twisted cube* $LTQ_n$*is the labeled graph with the following recursive fashion*:

(1) *$LTQ_1$ is the complete graph on two vertices labeled by* 0 *and* 1. *$LTQ_2$ is a graph consisting of four vertices with labels* 00, 01, 10, 11 *together with four edges* (00, 01), (00, 10), (01, 11), *and* (10, 11).
(2) *For $n \geq 3$, $LTQ_n$ is composed of two subcubes $LTQ_{n-1}$, denoted as $LTQ_n^0$ and $LTQ_n^1$, such that each vertex $x = 0x_{n-2}x_{n-3}\ldots x_0 \in V(LTQ_n^0)$ is connected with the vertex $y = 1(x_{n-2} \oplus x_0)\, x_{n-3}\ldots x_0 \in V(LTQ_n^1)$ by an edge, where $x$ and $y$ are called the $(n-1)$-neighbors to each other.*

For example, Fig. 1 shows locally twisted cubes $LTQ_3$ and $LTQ_4$. In this paper, sometimes the labels of vertices are changed to their decimal.

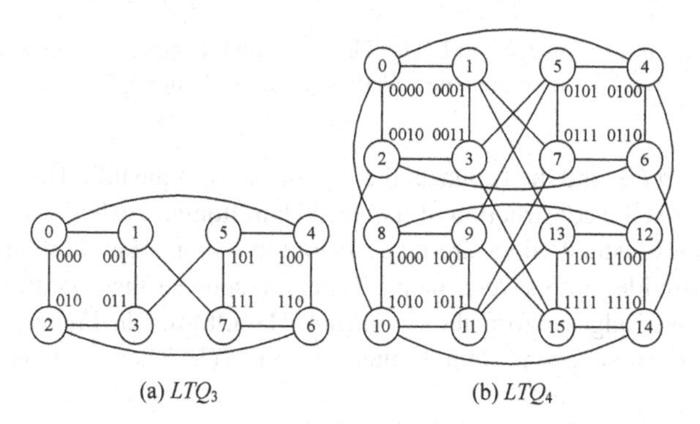


**Fig. 1.** Locally twisted cubes $LTQ_3$ and $LTQ_4$.

## 3 Main Results

In 2020, Li et al. [9] presented parallel algorithms for constructing two edge-disjoint Hamiltonian cycles in $LTQ_n$. We will use the above result to expand to the main contribution of this paper.

### 3.1 Three Edge-Disjoint Hamiltonian Cycles in $LTQ_6$

According to [9], there exist two edge-disjoint Hamiltonian cycles in $LTQ_6$. The first cycle $C_{64}$ is 0 - 1 - 13 - 15 - 9 - 11 - 7 - 6 - 38 - 39 - 43 - 41 - 47 - 45 - 33 - 32 - 36 - 37 - 35 - 34 - 42 - 40 - 44 - 46 - 62 - 60 - 56 - 58 - 50 - 51 - 53 - 52 - 48 - 49 - 61 - 63 - 57 - 59 - 55 - 54 - 22 - 23 - 27 - 25 - 31 - 29 - 17 - 16 - 20 - 21 - 19 - 18 - 26 - 24 - 28 - 30 - 14 - 12 - 8 - 10 - 2 - 3 - 5 - 4 - 0, and the second cycle $C_{64}$ is 0 - 2 - 6 - 4 - 12 - 13 - 11 - 10 - 14 - 15 - 3 - 1 - 7 - 5 - 9 - 8 - 24 - 25 - 21 - 23 - 17 - 19 - 31 - 30 - 26 - 27 - 29 - 28 - 20 - 22 - 18 - 16 - 48 - 50 - 54 - 52 - 60 - 61 - 59 - 58 - 62 - 63 - 51 - 49 - 55 - 53 - 57 - 56 - 40 - 41 - 37 - 39 - 33 - 35 - 47 - 46 - 42 - 43 - 45 - 44 - 36 - 38 - 34 - 32 - 0. Due to the constraint of edge-disjoint, there are two $C_8$s and twelve $C_4$s after $LTQ_6$ removes all edges of two Hamiltonian cycles. Our idea is to remove an edge in

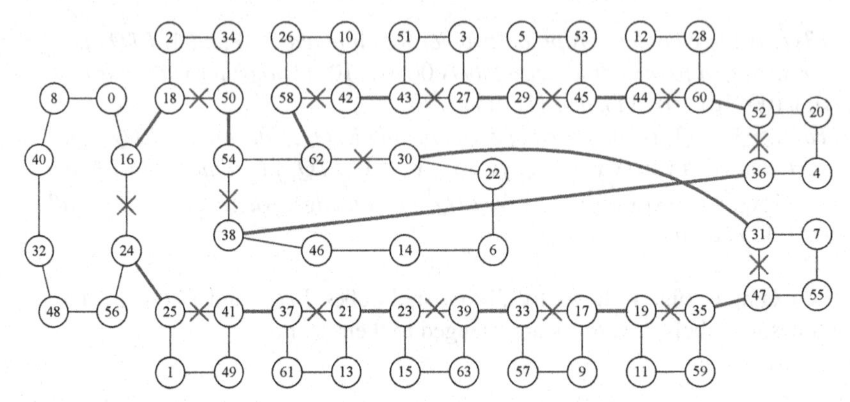


**Fig. 2.** The third Hamiltonian cycle in $LTQ_6$. The line with the red cross represents the removed edge, and the thick red line represents the added edge. (Color figure online)

each cycle $C_8$ and $C_4$, and then connect these paths to form the third Hamiltonian cycle. Figure 2 illustrates the construction of the third Hamiltonian cycle in $LTQ_6$.

However, these added edges are taken from the second Hamiltonian cycle, which causes it to be divided into several paths. Then, we connect these paths by adding the previously removed edges to form the second new Hamilton cycle. Figure 3 illustrates the construction of the new second Hamiltonian cycle in $LTQ_6$. Then, we have the following lemma.

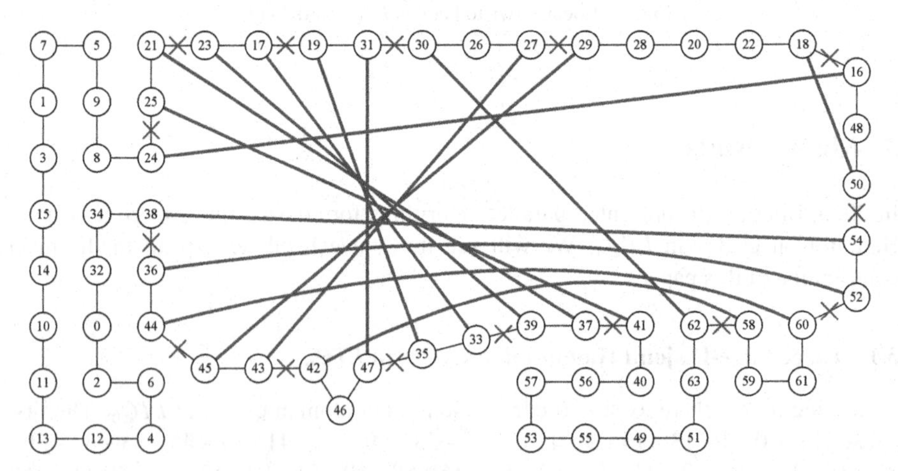


**Fig. 3.** The new second Hamiltonian cycle in $LTQ_6$. The line with the red cross represents the removed edge, and the thick red line represents the added edge. (Color figure online)

**Lemma 2.** *Let* $HC_1 = 0$ - 1 - 13 - 15 - 9 - 11 - 7 - 6 - 38 - 39 - 43 - 41 - 47 - 45 - 33 - 32 - 36 - 37 - 35 - 34 - 42 - 40 - 44 - 46 - 62 - 60 - 56 - 58 - 50 - 51 - 53 - 52 - 48 - 49 - 61 - 63 - 57 - 59 - 55 - 54 - 22 - 23 - 27 - 25 - 31 - 29 - 17 - 16 - 20 - 21 - 19 - 18 - 26 - 24 - 28 - 30 - 14 - 12 - 8 - 10 - 2 - 3 - 5 - 4 - 0, $HC_2 = 0$ - 2 - 6 - 4 - 12 - 13 - 11 - 10 - 14 - 15

- 3 - 1 - 7 - 5 - 9 - 8 - 24 - 16 - 48 - 50 - 18 - 22 - 20 - 28 - 29 - 45 - 43 - 27 - 26 - 30 - 62 - 63 - 51 - 49 - 55 - 53 - 57 - 56 - 40 - 41 - 25 - 21 - 37 - 39 - 23 - 17 - 33 - 35 - 19 - 31 - 47 - 46 - 42 - 58 - 59 - 61 - 60 - 44 - 36 - 52 - 54 - 38 - 34 - 32 - 0 *and* $HC_3 = 0$ - 16 - 18 - 2 - 34 - 50 - 54 - 62 - 58 - 26 - 10 - 42 - 43 - 51 - 3 - 27 - 29 - 5 - 53 - 45 - 44 - 12 - 28 - 60 - 52 - 20 - 4 - 36 - 38 - 46 - 14 - 6 - 22 - 30 - 31 - 7 - 55 - 47 - 35 - 59 - 11 - 19 - 17 - 9 - 57 - 33 - 39 - 63 - 15 - 23 - 21 - 13 - 61 - 37 - 41 - 49 - 1 - 25 - 24 - 56 - 48 - 32 - 40 - 8 - 0. *$HC_1$, $HC_2$ and $HC_3$ form three edge-disjoint Hamiltonian cycles in $LTQ_6$.*

**Proof.** Since each of $HC_1$, $HC_2$ and $HC_3$ is a $C_{64}$ through $LTQ_6$ that visits each vertex exactly once, they are all Hamiltonian cycles. As for edge-disjoint, it can be checked by Fig. 4. □

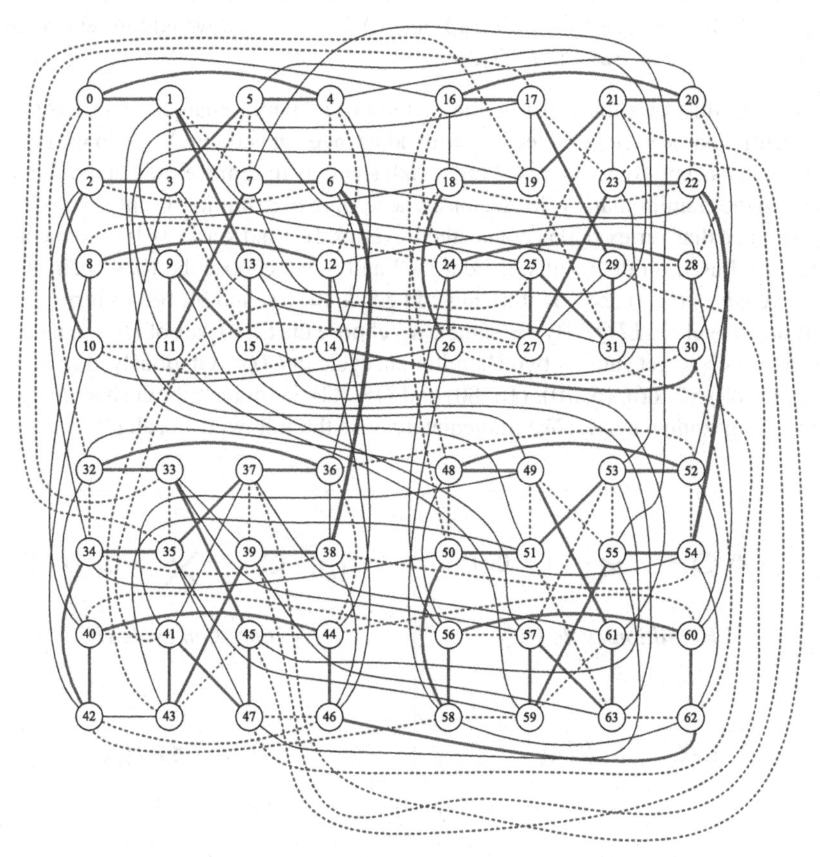


**Fig. 4.** Three edge-disjoint Hamiltonian cycles in $LTQ_6$. Red thick lines indicate edges of $HC_1$, green dashed lines indicate edges of $HC_2$, and blue thin lines indicate edges of $HC_3$. (Color figure online)

### 3.2 Three Edge-Disjoint Hamiltonian Cycles in $LTQ_n$ for $N \geq 7$

Based on the previous results, we provide a recursive construction method of the Hamiltonian cycle in $LTQ_n$ for $n \geq 7$.

**Lemma 3.** *If there exists a Hamiltonian cycle with a pair of adjacent even vertices, i.e. bit $b_0 = 0$, in $LTQ_n$, then there exists a Hamiltonian cycle in $LTQ_{n+1}$.*

**Proof.** By Definition 1, $LTQ_{n+1}$ is composed of two subcubes $LTQ_n$. We assume that there exists a Hamiltonian cycle with a pair of adjacent even vertices, called $u0$ and $v0$, in $LTQ_n$. We note that $u$ and $v$ represent the bit strings $b_{n-1}b_{n-2}\ldots b_1$ of these two vertices, respectively. Since both ${LTQ_{n+1}}^0$ and $LTQ^1_{n+1}$ are isomorphic with $LTQ_n$, we have the Hamiltonian cycle $HC_A$ in $LTQ^0_{n+1}$ and the Hamiltonian cycle $HC_B$ in $LTQ^1_{n+1}$. Then, we remove the edge ($0u0$, $0v0$) in $HC_A$ and remove the edge ($1u0$, $1v0$) in $HC_B$. By Definition 1, there exist edges ($0u0$, $1u0$) and ($0v0$, $1v0$), then we add these two edges to form a Hamiltonian cycle in $LTQ_{n+1}$. □

Figure 5 illustrates the construction method of a Hamiltonian cycle in $LTQ_{n+1}$. In this construction, we need to select a pair of adjacent even vertices. Why do we need the condition "adjacent even vertices"? First, "adjacent" is to remove the edge of adjacent vertices in the Hamiltonian cycle. Secondly, according to definition 1, only even vertices can guarantee that vertex 0u0 is connected to vertex 1u0 and vertex 0v0 is connected to vertex 1v0. Based on the results of Lemma 2 and 3, we choose the (0, 4) pair of $HC_1$ to remove edges (0, 4) and (64, 68) and add edges (0, 64) and (4, 68) to form the first Hamiltonian cycle in $LTQ_7$. By the same way, choose the (2, 6) pair of $HC_2$ and the (16, 18) pair of $HC_3$ to get another two Hamiltonian cycle in $LTQ_7$. Obviously, the six edges (0, 64), (4, 68), (2, 66), (6, 70), (16, 80) and (18, 82) are not common shared edges in any two Hamiltonian cycles. We immediately have the following corollary.

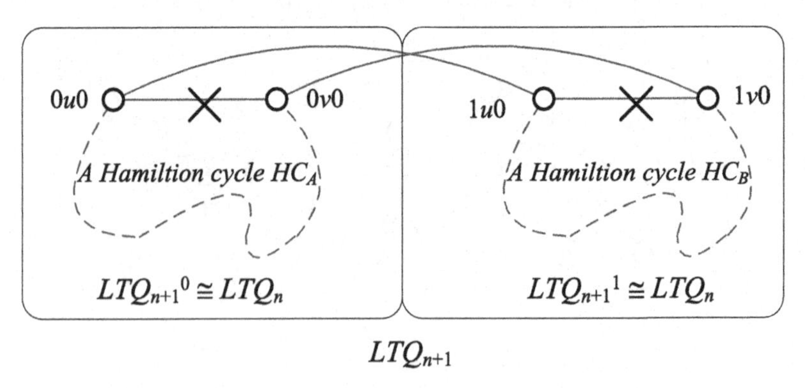


**Fig. 5.** The construction of a Hamiltonian cycle in $LTQ_{n+1}$.

**Corollary 4.** *There exist three edge-disjoint Hamiltonian cycles in $LTQ_7$.*

Based on the result of Lemma 2 and applying Lemma 3 recursively, we have the following theorem.

**Theorem 5.** *There exist three edge-disjoint Hamiltonian cycles in $LTQ_n$ for $n \geq 6$.*

# 4 Concluding Remarks

Hung [7] embedded two edge-disjoint Hamiltonian cycles into $LTQ_n$ while $n \geqslant 4$. In this paper, we show that $LTQ_n$ with $n \geqslant 6$ contains three edge-disjoint Hamiltonian cycles. It is interesting to see if there are four edge-disjoint Hamiltonian cycles in $LTQ_n$ for $n \geq 8$. So far, this is still an open problem.

**Acknowledgments.** This research was partially supported by MOST grants 107-2221-E-131-011 from the Ministry of Science and Technology, Taiwan.

# References

1. Bae, M.M., Bose, B.: Edge disjoint Hamiltonian cycles in k-ary n-cubes and hypercubes. IEEE Trans. Comput. **52**, 1271–1284 (2003)
2. Barth, D., Raspaud, A.: Two edge-disjoint Hamiltonian cycles in the butterfly graph. Inform. Process. Lett. **51**, 175–179 (1994)
3. Hsieh, S.Y., Tu, C.J.: Constructing edge-disjoint spanning trees in locally twisted cubes. Theor. Comput. Sci. **410**, 926–932 (2009)
4. Hung, R.W.: Embedding two edge-disjoint Hamiltonian cycles into locally twisted cubes. Theor. Comput. Sci. **412**, 4747–4753 (2011)
5. Hung, R.W.: Constructing two edge-disjoint Hamiltonian cycles and two-equal path cover in augmented cubes. J. Comput. Sci. **39**, 42–49 (2012)
6. Hung, R.W.: The property of edge-disjoint Hamiltonian cycles in transposition networks and hypercube-like networks. Discrete Appl. Math. **181**, 109–122 (2015)
7. Hung, R.W., Chan, S.J., Liao, C.C.: Embedding two edge-disjoint Hamiltonian cycles and two equal node-disjoint cycles into twisted cubes. Lecture Notes Eng. Comput. Sci. **2195**, 362–367 (2012)
8. Lee, S., Shin, K.: Interleaved all-to-all reliable broadcast on meshes and hypercubes. IEEE Trans. Parallel Distrib. Syst. **5**, 449–458 (1994)
9. Li, S.Y., Chang, J.M., Pai, K.J.: A parallel algorithm for constructing two edge-disjoint Hamiltonian cycles in locally twisted cubes. In: 2020 International Computer Symposium, pp. 116–119 (2020)
10. Lin, J.C., Yang, J.S., Hsu, C.C., Chang, J.M.: Independent spanning trees vs. edge-disjoint spanning trees in locally twisted cubes. Inf. Process. Lett. **110**, 414–419 (2010).
11. Lin, T.J., Hsieh, S.Y., Juan, J.S.-T.: Embedding cycles and paths in product networks and their applications to multiprocessor systems. IEEE Trans. Parallel Distrib. Syst. **23**, 1081–1089 (2012)
12. Ma, M., Xu, J.M.: Weakly edge-pancyclicity of locally twisted cubes. Ars Combin. **89**, 89–94 (2008)
13. Pai, K.-J.: A parallel algorithm for constructing two edge-disjoint hamiltonian cycles in crossed cubes. In: Zhang, Z., Li, W., Du, D.-Z. (eds.) AAIM 2020. LNCS, vol. 12290, pp. 448–455. Springer, Cham (2020). https://doi.org/10.1007/978-3-030-57602-8_40
14. Petrovic, V., Thomassen, C.: Edge-disjoint Hamiltonian cycles in hypertournaments. J. Graph Theory **51**, 49–52 (2006)

15. Rowley, R., Bose, B.: Edge-disjoint Hamiltonian cycles in de Bruijn networks. In: Proceedings of 6th Distributed Memory Computing Conference, pp. 707–709 (1991)
16. Saad, Y., Schultz, M.H.: Topological properties of hypercubes. IEEE Trans. Comput. **37**, 867–872 (1988)
17. Wang, D.: A low-cost fault-tolerant structure for the hypercube. J. Supercomput. **20**, 203–216 (2001)
18. Wang, Y., Fan, J., Liu, W., Wang, X.: Embedding two edge-disjoint Hamiltonian cycles into parity cubes. Appl. Mech. Mater. **336–338**, 2248–2251 (2013)
19. Yang, H., Yang, X.: A fast diagnosis algorithm for locally twisted cube multiprocessor systems under the MM∗ model. Comput. Math. **53**, 918–926 (2007)
20. Yang, X., Evans, D.J., Megson, G.M.: The locally twisted cubes. Int. J. Comput. Math. **82**, 401–413 (2005)

# On the Probe Problem for $(r, \ell)$-Well-Coveredness

Luerbio Faria[1(✉)] and Uéverton S. Souza[2]

[1] Universidade do Estado do Rio de Janeiro, Rio de Janeiro, Brazil
luerbio@ime.uerj.br
[2] Universidade Federal Fluminense, Niterói, Brazil
ueverton@ic.uff.br

**Abstract.** Let $\mathcal{C}$ be a class of graphs. A graph $G = (V, E)$ is $\mathcal{C}$–probe if $V(G)$ can be partitioned into two sets: *non-probes* $\mathcal{N}$ and *probes* $\mathcal{P}$, where $\mathcal{N}$ is an independent set and new edges may be added between some non-probe vertices such that the resulting graph is in the class $\mathcal{C}$. In this case, we say that $(\mathcal{N}, \mathcal{P})$ is a *$\mathcal{C}$–probe partition* of $G$. In the UNPARTITIONED PROBE problem for a graph class $\mathcal{C}$ we are given a graph $G$ and asked whether $G$ has a $\mathcal{C}$–probe partition, i.e., such a problem consist of recognizing the class of $\mathcal{C}$-probe graphs. A graph $G = (V, E)$ is an $(r, \ell)$-graph when $V$ can be partitioned into $(S^1, S^2, \ldots, S^r, K^1, K^2, \ldots, K^\ell)$ such that $S^1, S^2, \ldots, S^r$ are independent sets, and $K^1, K^2, \ldots, K^\ell$ are cliques. A graph $G$ is *well-covered* if every maximal independent set is also maximum, and it is $(r, \ell)$-well-covered if it is well-covered as well as an $(r, \ell)$-graph. In this paper, we study the complexity of the UNPARTITIONED PROBE problem for the class of $(r, \ell)$-well-covered graphs. We classify all but the $(2, 0)$ and $(1, 2)$ cases.

**Keywords:** Well-covered · $(r, \ell)$-graph · Probe · Recognition · Complexity

## 1 Introduction

Well-covered graphs were first introduced 50 years ago by Plummer [25] in 1970 as the class of graphs in which every maximal independent set has the same cardinality. In other words, every maximal independent set is maximum. The problem of recognizing a well-covered graph, which we denote by WELL-COVEREDNESS, was proved to be coNP-complete by Chvátal and Slater [12] and Sankaranarayana and Stewart [28]. In addition, fixed-parameter tractable algorithms for WELL-COVEREDNESS were presented in [2,3]. The importance of well-covered graphs is related to the fact that such a class is exactly the class of graphs in which polynomial-time greedy algorithms for maximal independent sets always returns maximum independent sets. Well-covered graphs were

Supported by CNPq, FAPERJ and CAPES. Brazilian research agencies.


C.-Y. Chen et al. (Eds.): COCOON 2021, LNCS 13025, pp. 375–386, 2021.
https://doi.org/10.1007/978-3-030-89543-3_32

studied on bipartite graphs [27], bounded degree graphs [6], graphs without large cycles [17], graphs with large girth [18], claw-free graphs [30], planar 3-connected, simplicial, chordal, circular arc graphs [26], cographs, and $P_4$-sparse graphs [22].

An $(r, \ell)$-*partition* of a graph $G = (V, E)$ is a partition of $V(G)$ into $r$ independent sets $S^1, \ldots, S^r$ and $\ell$ cliques $K^1, \ldots, K^\ell$. A graph is $(r, \ell)$ if it admits an $(r, \ell)$-partition. For convenience, some of these sets are allowed to be empty. A set containing a single vertex can be counted as either an independent set or a clique. The P versus NP-complete dichotomy of recognizing $(r, \ell)$-graphs is well-known [4]: the problem is in P if $\max\{r, \ell\} \leq 2$, and NP-complete otherwise. Subclasses of $(r, \ell)$-graphs have been extensively studied in the literature. For instance, the complexity of list partition problems on $(r, \ell)$-graphs was studied in [16], and polynomial-time algorithms to recognize cographs and chordal graphs that are $(r, \ell)$ were presented in [14,15].

Let $r, \ell \geq 0$ be two fixed integers not simultaneously zero. A graph is $(r, \ell)$-*well-covered* if it is both $(r, \ell)$ and well-covered. Alves et al. [2] established the complete classification of the complexity of recognizing $(r, \ell)$-well-covered graphs.

Let $\mathcal{C}$ be a class of graphs. A graph $G = (V, E)$ is a $\mathcal{C}$–probe graph if $V(G)$ can be partitioned into two sets: *non-probes* $\mathcal{N}$ and *probes* $\mathcal{P}$, where $\mathcal{N}$ is an independent set and new edges may be added between some non-probe vertices such that the resulting graph is in the class $\mathcal{C}$. Thus, the class of $\mathcal{C}$–probe graphs is the superclass of $\mathcal{C}$ where for each $G$ in $\mathcal{C}$–probe there is a graph $G' \in \mathcal{C}$ such that $V(G) = V(G')$ and the set of vertices incident to edges in $E(G') \setminus E(G)$ is an independent set of $G$ (denoted by $\mathcal{N}$). In this case, we say that $(\mathcal{N}, \mathcal{P})$ is a *$\mathcal{C}$–probe partition* of $G$. Note that it is allowed $\mathcal{N}$ or $\mathcal{P}$ to be empty. In the Unpartitioned Probe problem for a graph class $\mathcal{C}$ we are given a graph $G$ and asked whether $G$ has a $\mathcal{C}$–probe partition. In other words, Unpartitioned Probe is the problem of recognizing the class of $\mathcal{C}$–probe graphs. It is worth mentioning that depending on the structural properties of $\mathcal{C}$, $\mathcal{C}$-probe can preserve several interesting properties. For instance, when $\mathcal{C}$ is the class of chordal graphs, $\mathcal{C}$-probe graphs $G$ have neither odd chordless cycles nor a complement of a chordless cycle of length greater than 4 (see [20]), which implies, by the Strong Perfect Graph Theorem [11], that Chordal probe graphs are perfect. Therefore, studies regarding the recognition of $\mathcal{C}$-probe graphs can help us identify superclasses of $\mathcal{C}$ for which it may be possible to extend polynomial-time algorithms primarily designed to solve problems on $\mathcal{C}$, besides being a natural task for polynomial-time recognizable graph classes $\mathcal{C}$. In addition, in the Partitioned Probe problem for a graph class $\mathcal{C}$ we are given a graph $G$ and a partition $(\mathcal{N}, \mathcal{P})$ of $V(G)$ and asked whether $(\mathcal{N}, \mathcal{P})$ is a $\mathcal{C}$–probe partition of $G$.

In 1994, Peisen Zhang et al. [32] defined the partitioned interval probe graphs. One of the first papers to mention generalizing the concept of $\mathcal{C}$-probe graphs from interval to other graph classes is [21]. From then on, $\mathcal{C}$-probe graphs have been studied for several graph classes $\mathcal{C}$ such as interval [24], chain [21,23], chordal [20], Ptolomaic [7], permutation [9], self-complementary classes of perfect graphs [10], and cographs [8].

In 1995, Golumbic, Kaplan, and Shamir [19] stated the GRAPH SANDWICH problem for property $\Pi$. The input of such a problem is a pair of graphs $G^1 = (V, E^1)$ and $G^2 = (V, E^2)$ with $E^1 \subseteq E^2$, and the question is whether there is a graph $G = (V, E)$ with $E^1 \subseteq E \subseteq E^2$ such that $G$ satisfies the property $\Pi$.

GRAPH SANDWICH is a generalization of PARTITIONED PROBE, since PARTITIONED PROBE instances $(G = (V, E), \mathcal{N}, \mathcal{P})$ can be seen as GRAPH SANDWICH instances $(G^1, G^2)$, where $G^1 = G$, and $E(G^2) = E(G) \cup \{uv : u, v \in \mathcal{N}\}$. Hence, whenever GRAPH SANDWICH is polynomial-time solvable for a property $\Pi$, we have that PARTITIONED PROBE is polynomial-time solvable for the class of graphs satisfying the property $\Pi$, and whenever PARTITIONED PROBE is hard for a class of graphs satisfying a property $\Pi$, we have that GRAPH SANDWICH is hard for the property $\Pi$. On the other hand, nothing is known about the relationship between the time complexities for UNPARTITIONED PROBE, PARTITIONED PROBE, and GRAPH SANDWICH.

Dantas et al. [13] established the P versus NP-complete dichotomy for UNPARTITIONED PROBE for $(r, \ell)$-graphs, showing that UNPARTITIONED PROBE for $(r, \ell)$-graphs is polynomial-time solvable if $r + \ell \leq 2$, and NP-complete otherwise. Recently, Alves et al. [1] studied the complexity of GRAPH SANDWICH for the property of being $(r, \ell)$-well-covered.

In this paper, we are interested in the time complexity of UNPARTITIONED PROBE for the property of being $(r, \ell)$-WELL-COVERED. We focus on the following family of decision problems indexed with respect to the values of $r$ and $\ell$:

UNPARTITIONED PROBE FOR $(r, \ell)$-WELL-COVEREDNESS
*Input:* A graph $G = (V, E)$.
*Question:* Is there a partition $(\mathcal{N}, \mathcal{P})$ of $V(G)$ such that there is a set $E' \subseteq \{uv : u, v \in \mathcal{N}\}$, where the graph $H = (V, E \cup E')$ is $(r, \ell)$-well-covered?

We prove that UNPARTITIONED PROBE FOR $(k, \ell)$-WELL-COVEREDNESS is polynomial-time solvable when $(r, \ell) = (0, 1)$, $(1, 0)$, $(1, 1)$ or $(0, 2)$, the cases $(1, 2)$ and $(2, 0)$ are left as open, and the other cases we classified as either NP-hard or coNP-hard (see Table 1). Our polynomial-time algorithms generalize previous studies on recognizing $(r, \ell)$-well-covered graphs. In the present paper,

**Table 1.** Complexity of UNPARTITIONED PROBE FOR $(k, \ell)$-WELL-COVEREDNESS – P stands for polynomial problem, coNPh stands for coNP-hard, NPc stands for NP-complete, and NPh stands for NP-hard.

| $r \backslash \ell$ | 0 | 1 | 2 | $\geq 3$ |
|---|---|---|---|---|
| 0 | - | P (Prop. 4) | P (Alg. 2) | NPc (Thms. 4+6) |
| 1 | P (Prop. 2) | P (Alg. 1) | ? | NPh (Thms. 8+6) |
| 2 | ? | coNPh (Thm. 7) | coNPh (Thm. 6) | coNPh (Thm. 6) |
| $\geq 3$ | NPh (Thm. 5) | NPh (Thm. 6) | NPh (Thm. 6) | NPh (Thm. 6) |

we describe polynomial-time algorithms for UNPARTITIONED PROBE for $(0,1)$, $(1,0)$, $(1,1)$, and $(0,2)$-well-covered graphs (Theorems 2, and 4, and Algorithms 1 and 2); and we prove that: UNPARTITIONED PROBE for $(0,\ell)$, $\ell \geq 3$ (Theorem 4), $(3,0)$ (Theorem 5), $(2,1)$ (Theorem 7), and $(1,3)$-well-covered graphs (Theorem 8) are (co)NP-hard problems. In addition, we use the *monotonicity* (Theorem 6) in order to show that the other values are also (co)NP-hard.

**Notation.** This paper only deal with finite, simple and undirected graphs, for short we say graphs. In this context, a graph $G = (V,E)$ consist of a finite nonempty set $V$ of *vertices*, and a set $E$ of unordered pairs of distinct vertices of $V$, set of *edges*, with $n = |V|$, and $m = |E|$. An *independent set* $S \subseteq V$ of a graph $G = (V,E)$ satisfies that if $u,v \in S$, then $uv \notin E$. A *clique* $Q \subseteq V$ of a graph $G = (V,E)$ satisfies that if $u,v \in Q$, then $uv \in E$. A set $S$ is *maximal with respect to a property* $\Pi$ if $S$ satisfies $\Pi$, and every set $R$ containing and not being $S$, does not satisfies $\Pi$. Let $G = (V,E)$ be a graph and $v \in V$. The *open neighborhood* $N(v)$ of $v$, or *neighborhood* (for short) of $v$ is the set $N(v) = \{u : uv \in E\}$, and the *degree* $d(v)$ of $v$ in $G$ is $d(v) = |N(v)|$. The *closed neighborhood*, $N[v]$ of $v$ is the set $N[v] = N(v) \cup \{v\}$. If $B,S \subset V$, the *neighborhood of* $B$ *in* $S$ is $N_S(B) = \{x \in S : \exists y \in B, x \in N(y)\}$, and the *degree* $d_S(v)$ of $v$ in $S$ is $d_S(v) = |N_S(\{v\})|$. We say that $v$ is *universal* if $N[v] = V$, and *isolated* if $N[v] = \{v\}$. A graph $G$ is a *complete* graph $K_n$ if $V$ is a clique, and $G$ is *split* if it is $(1,1)$. Finally, a graph $G$ is a *complete split* graph if there is a partition $V = (S,Q)$, such that $S$ is an independent set and $Q$ is a clique and for each vertex $s \in S$, it holds that $N(s) = Q$.

## 2 Polynomial-Time Solvability

Since $(1,0)$-graphs have edge set empty, the following two propositions are clear.

**Proposition 1** [2]. *A graph is* $(1,0)$*-well-covered if and only if it is* $(1,0)$.

Note that when $E(G) = \emptyset$ the probe partition for $(1,0)$-well-coveredness is realized with $(\mathcal{N},\mathcal{P}) = (\emptyset, V)$.

**Proposition 2.** *There is a probe partition* $(\mathcal{N},\mathcal{P})$ *for an instance* $G = (V,E)$ *of* UNPARTITIONED PROBE FOR $(1,0)$-WELL-COVEREDNESS *if and only if* $E = \emptyset$.

Next, we discuss the $(0,1)$ case. Recall that $(0,1)$-graphs are complete graphs.

**Proposition 3** [2]. *A graph is* $(0,1)$*-well-covered if and only if it is* $(0,1)$.

**Proposition 4.** *There is a probe partition* $(\mathcal{N},\mathcal{P})$ *for an instance* $G = (V,E)$ *of* UNPARTITIONED PROBE FOR $(0,1)$-WELL-COVEREDNESS *if and only if there is an* $(1,1)$*-partition* $(S,K)$ *of* $V(G)$*, where* $S$ *is an independent set and* $K$ *is a clique, such that each vertex* $u$ *of* $S$ *is adjacent to each vertex* $v$ *of* $K$ *(*$G$ *is a complete split graph).*

*Proof.* Suppose that there is a probe partition $(\mathcal{N}, \mathcal{P})$ for an instance $G = (V, E)$ of UNPARTITIONED PROBE FOR $(0, 1)$-WELL-COVEREDNESS. Hence, $H = (V, E \cup \{uv : u, v \in \mathcal{N}\})$ is a complete graph. Then, the induced subgraph $G[\mathcal{P}]$ of $G$ must be a clique, and for each pair $u, v$ with $u \in \mathcal{N}$ and $v \in \mathcal{P}$, $uv \in E$. Thus, there is a partition $(S, K) = (\mathcal{N}, \mathcal{P})$ of $V$, where $S$ is an independent set and $K$ is a clique, such that each vertex $u$ of $S$ is adjacent to each vertex $v$ of $K$. Now, suppose that there is a partition $(S, K)$ of $V$, where $S$ is an independent set and $K$ is a clique, such that each vertex $u$ of $S$ is adjacent to each vertex $v$ of $K$. We can set $\mathcal{N} = S$, $\mathcal{P} = K$ and $H = (V, E \cup \{uv : u, v \in \mathcal{N}\})$ is a complete graph. □

Since the recognition of (complete) split graphs can be done in polynomial time, the problem for $(0, 1)$-well-coveredness is in P. Alves et al. [2] showed the following characterization for $(1, 1)$-well-coveredness.

**Proposition 5** [2]. *A graph $G = (V, E)$ is $(1, 1)$-well-covered if and only if there is a partition (called $(1, 1)$-well-covered partition) of $V$ into $(S, K)$ where $S$ is a independent set, $K$ is a clique, and either $d_S(v) = 1$ for each vertex $v \in K$, or $d_S(v) = 0$ for each vertex $v \in K$.*

**Lemma 1.** *Let $G = (V, E)$ be an instance of* UNPARTITIONED PROBE FOR $(1, 1)$-WELL-COVEREDNESS. *There is a probe partition $(\mathcal{N}, \mathcal{P})$ of $G$ for which there is a $(1, 1)$-well-covered graph $H = (V, E \cup E')$ with $E' \subseteq \{uv : u, v \in \mathcal{N}\}$ and having a $(1, 1)$-well-covered partition $V = (S, K)$ such that $K \subseteq \mathcal{N}$, if and only if each edge $uv$ of $G$ has a vertex $v$ of degree $d(v) = 1$ in $G$.*

**Lemma 2.** *Let $G = (V, E)$ be an instance of* UNPARTITIONED PROBE FOR $(1, 1)$-WELL-COVEREDNESS. *There is a probe partition $(\mathcal{N}, \mathcal{P})$ of $G$ for which there is a $(1, 1)$-well-covered graph $H = (V, E \cup E')$ with $E' \subseteq \{uv : u, v \in \mathcal{N}\}$ and having a $(1, 1)$-well-covered partition $V = (S, K)$ such that $K \cap \mathcal{P} \neq \emptyset$, if and only if there is a partition $(S^1, S^2, Q)$ of $V$, where $S^1$ and $S^2$ are independent sets and $Q$ is a non-empty clique, such that each vertex $u$ of $S^2$ is adjacent to each vertex $v$ of $Q$, and either $d_{S^1}(v) = 1$ for each vertex $v \in (S^2 \cup Q)$, or $d_{S^1}(v) = 0$ for each vertex $v \in (S^2 \cup Q)$.*

Our polynomial-time algorithm for UNPARTITIONED PROBE FOR $(1, 1)$-WELL-COVEREDNESS relies on the argument that a $(2, 1)$-partition of a graph $G$, if any, can be found in polynomial time [4,5], and once a $(2, 1)$-partition $V = (S^1, S^2, Q)$ is given, where $S^1$ and $S^2$ are independent sets and $Q$ is a clique, one can enumerate in polynomial-time all sparse-dense partitions of $G$ into a clique $Q_i$ and a bipartite graph $G[S_i^1 \cup S_i^2]$ (see [16]).

**Theorem 1.** *Algorithm 1 correctly asserts in polynomial time whether $G$ has a probe partition for $(1, 1)$-well-coveredness.*

**Algorithm 1.** Unpartitioned probe for $(1, 1)$-well-coveredness algorithm
**Output**: A probe partition $(\mathcal{N}, \mathcal{P})$ of $G = (V, E)$ and a set $E' \subseteq \{uv : u, v \in \mathcal{N}\}$, such that $H = (V, E \cup E')$ is $(1, 1)$-well-covered, or *No* when there is no such a partition.

1. **If** ($G$ is $(2, 0)$) and (each edge $uv$ of $E$ has a vertex $v$ of degree $d(v) = 1$) **then**

$\mathcal{N} \leftarrow \emptyset$;
**For each** edge $e \in E$ **do**
$\mathcal{N} \leftarrow \mathcal{N} \cup \{v\}$, where $v$ is an endpoint of $e$ and $d(v) = 1$;
**Return** $\Big((\mathcal{N}, P = V \setminus \mathcal{N}),\ E' = \{uv : u, v \in \mathcal{N}\}\Big)$;

2. **If** $G$ is not a $(2,1)$-graph **then Return** (No);
3. **For each** $Q_i$ in a sparse-dense partition of $G$ in a clique $Q_i$ and a bipartite $G[S_i^1 \cup S_i^2]$ **do**
   Let $(A, B)$ be a partition for $S^1 \cup S^2$, such that every vertex of $A$ is completely adjacent to any vertex of $Q_i$, and $A$ is maximal;
4. **If** ($B$ is an independent set) and ($G[A]$ has vertex cover number at most one) **then**
5. **If** ($B$ is a set of isolated vertices in $G$) **then**
   Let $\{a\}$ be a non-empty vertex cover[3] of $G[A]$ and let $b \in B$;
   $E' = \{uv : u, v \in A \setminus \{a\}\} \cup \{bx : x \in A \setminus N[a]\}$;
   **Return** $\Big((\mathcal{N} = (A \setminus \{a\}) \cup \{b\}, P = V \setminus \mathcal{N}), E'\Big)$;
6. **Else**
   **If** ($A$ is an independent set) and ($\exists\, uv \in E$ such that $u \in Q_i$ and $v \in B$) **then**
   **If** ($|N_B(q)| = 1,\ \forall\, q \in Q_i$) and ($|N_B(a)| \leq 1,\ \forall\, a \in A$) **then**
   **If** there is a vertex $a \in A$ such that $|N_B(a)| = 0$ **then**
   **If** there is a vertex $b \in B$ such that $|N_A(b)| = 0$ **then**
   $\mathcal{N} \leftarrow A \cup \{b\}$;
   $E' \leftarrow \{uv : u, v \in A\} \cup \{ba : a \in A \text{ and } |N_B(a)| = 0\}$;
   **Return** $\Big((\mathcal{N}, P = V \setminus \mathcal{N}), E'\}\Big)$;
   **Else Return** $\Big((\mathcal{N} = A, P = B \cup Q_i), \{uv : u, v \in A\}\Big)$;
7. **Else If** ($N(B) \subseteq A$) **then**
   **For each** non-empty vertex cover[1] $\{a\}$ of $G[A]$ **do**
   **If** ($B \cup \{a\}$) is an independent set) **then**
   **If** (for every $x \in A \setminus \{a\}$ it holds that $|N_{(B \cup \{a\})}(x)| \leq 1$) **then**
   **If** $A$ is an independent set **then**
   $E' \leftarrow \{uv : u, v \in A \setminus \{a\}\} \cup \{ax : x \in A \setminus (N_A(B) \cup \{a\})\}$
   **Return** $\Big((\mathcal{N} = A, P = B \cup Q), E'\Big)$;
   **Else**
   **If** there is a vertex $x \in A \setminus \{a\}$ such that $|N_{B \cup \{a\}}(x)| = 0$ **then**
   **If** there is a vertex $b \in B$ such that $|N_A(b)| = 0$ **then**
   $\mathcal{N} \leftarrow (A \setminus \{a\}) \cup \{b\}$;
   $E' \leftarrow \{uv : u, v \in (A \setminus \{a\})\} \cup \{bx : x \in (A \setminus \{a\}), |N_{B \cup \{a\}}(x)| = 0\}$;
   **Return** $\Big((\mathcal{N}, P = V \setminus \mathcal{N}), E'\}\Big)$;
   **Else**
   $E' \leftarrow \{uv : u, v \in A \setminus \{a\}\}$
   **Return** $\Big((\mathcal{N} = A \setminus \{a\}, P = B \cup Q \cup \{a\}), E'\Big)$;
8. **Return** $\Big(\text{No}\Big)$.

For the $(0,2)$ case we recall that $(0,2)$-graphs are co-bipartite graphs.

**Proposition 6** [2]. *A graph $G = (V, E)$ is $(0,2)$-well-covered if and only if $G$ is $(0,2)$ and either $G$ is a complete graph, or $G$ has no universal vertex.*

**Lemma 3.** *Let $G = (V, E)$ be an instance of* Unpartitioned Probe for $(0,2)$-Well-Coveredness. *There is a probe partition $(\mathcal{N}, \mathcal{P})$ of $V(G)$ for $(0,2)$-well-coveredness if and only if either $G$ is a complete split graph, or $G$ has no universal vertex and there is a partition $(S, Q^1, Q^2)$ of $V$, where $S$ is an independent set, and $Q^1$ and $Q^2$ are cliques, such that for each vertex $u$ of $S$, either $u$ is adjacent to each vertex $v$ of $Q^1$, or $u$ is adjacent to each vertex $v$ of $Q^2$.*

Again, our polynomial-time algorithm for Unpartitioned Probe for $(0,2)$-Well-Coveredness is based on the fact that a $(1,2)$-partition of a graph $G$, if any, can be found in polynomial time [4,5], and once a $(1,2)$-partition

---
[1] When $G[A]$ has no edges any vertex form a non-empty vertex cover.

$V = (S, Q^1, Q^2)$ is given, where $S$ is an independent sets and $Q^1$ and $Q^2$ are cliques, one can enumerate in polynomial-time all sparse-dense partitions of $G$ into an independent set $S_i$ and a co-bipartite graph (see [16]). Thus, we can "guess" the independent set $S$ of the partition $(S, Q^1, Q^2)$, i.e., we enumerate all the partitions and check each one. Finally, given a sparse-dense partition, we use 2-SAT to decide which clique a vertex belongs to, i.e., whether from such a partition there can arise a $(0, 2)$-well-covered partition.

**Theorem 2.** *Algorithm 2 correctly asserts in polynomial time whether G has a probe partition for* $(0, 2)$*-well-covered graphs.*

**Algorithm 2.** Unpartitioned probe for $(0, 2)$-well-coveredness algorithm.
**Output**: A probe partition $(\mathcal{N}, \mathcal{P})$ of $G = (V, E)$ and a set $E' \subseteq \{uv : u, v \in \mathcal{N}\}$, such that $H = (V, E \cup E')$ is a $(0, 2)$-well-covered graph; or *No* whether there is no such a partition.

1. **If** $G = (V, E)$ is a complete split graph with partition $(S, K)$ **then**
   **Return** $\Big((\mathcal{N} = S, \mathcal{P} = K), E' = \{uv : u, v \in \mathcal{N}\}\Big)$;
2. **If** ($G$ has a universal vertex) or ($G$ is not a $(1, 2)$-graph) **then**
   **Return** (No);
3. **For each** $S$ in a sparse-dense partition of $G$ into an independent set $S$ and a co-bipartite graph $G[Q^1 \cup Q^2]$ **do**
   $\mathcal{N} \leftarrow S$;
   $\mathcal{P} \leftarrow Q^1 \cup Q^2$;
   (to guess a $(0, 2)$-well-covered partition as well as the edges to be added between vertices of $\mathcal{N}$ we construct a 2-SAT instance, where $u_i = True$ means that vertex $u$ must belong to the clique $K^i$).
4. $I = (U, C) \leftarrow \Big(\{p_1, p_2, n_1, n_2 : p \in \mathcal{P}, n \in \mathcal{N}\},$
   $\{(p_1 \vee p_2), (\overline{p_1} \vee \overline{p_2}), (n_1 \vee n_2), (\overline{n_1} \vee \overline{n_2}) : p \in \mathcal{P}, n \in \mathcal{N}\}\Big)$;
   (the current clauses ensure that each vertex of $G$ belong to either $K^1$ or $K^2$).
5. **For each** pair $u, v$ such that $u \in \mathcal{P}$ and $v \in V$ with $u \neq v$, and $uv \notin E$ **do**
   $C \leftarrow C \cup \Big\{(u_1 \vee v_1), (\overline{u_1} \vee \overline{v_1}), \{u_2 \vee v_2), (\overline{u_2} \vee \overline{v_2})\Big\}$;
   (pairs $u, v$ of non-adjacent vertices with $u \in \mathcal{P}$ belong to distinct cliques).
6. **Run** a 2-SAT polynomial-time algorithm on $I$;
7. **If** $I$ has a satisfying truth assignment $\eta : V \rightarrow \{T, F\}$ **then**
   $K^1 \leftarrow K^2 \leftarrow \emptyset$;
   **For** $i \leftarrow 1$ **to** $n = |U|$ **do**
   **If** $\eta(v_1) = T$, **then** $K^1 \leftarrow K^1 \cup \{v\}$ **else** $K^2 \leftarrow K^2 \cup \{v\}$;
8. **For each** vertex $v \in V$ **do**
   **If** $v \in K^1$ and $v$ has edges for each vertex of $K^2$ **then**
   $K^1 \leftarrow K^1 \setminus \{v\}$; $K^2 \leftarrow K^2 \cup \{v\}$;
   **If** $v \in K^2$ and $v$ has edges for each vertex of $K^1$ **then**
   $K^2 \leftarrow K^2 \setminus \{v\}$; $K^1 \leftarrow K^1 \cup \{v\}$;
9. **Return** $\Big(\mathcal{N}, \mathcal{P}, E' = \{uv : u, v \in (\mathcal{N} \cap K^i), i \in \{1, 2\}\}\Big)$;
10. **Return** $\Big(No\Big)$.

## 3 Hardness Results

We use the NP-complete problem $(0, \ell)$-Well-Coveredness with $\ell \geq 3$ in [2], to prove that Unpartitioned Probe for $(0, \ell)$-Well-Coveredness, for each $\ell \geq 3$, is a NP-complete problem.

**Theorem 3** [2]**.** *If* $\ell \geq 3$ *then* $(0, \ell)$*-*Well-Coveredness *is NP-complete.*

**Theorem 4.** *If* $\ell \geq 3$ *then* Unpartitioned Probe for $(0, \ell)$-Well-Coveredness *is NP-complete.*

*Proof.* A positive certificate $C$ to a Yes-instance $G = (V, E)$ of UNPARTITIONED PROBE FOR $(0, \ell)$-WELL-COVEREDNESS is a partition $(K^1, K^2, \ldots, K^\ell)$ of $V$ where $K^1, K^2, \ldots, K^\ell$ are cliques of $G$. Recall that $\ell$ is fixed, so one can check whether $G$ is well-covered in $O(n^\ell)$ time. Therefore, the problem is in NP.

Now, let $G$ be an instance of $(0, \ell)$-WELL-COVEREDNESS, $\ell \geq 3$. In polynomial time with respect to the size of $G$ we construct an instance $H$ of UNPARTITIONED PROBE FOR $(0, \ell)$-WELL-COVEREDNESS, such that $G$ is $(0, \ell)$-well-covered if and only if $H$ is a $(0, \ell)$-well-covered probe graph. The construction of $H$ is as follows: The vertex set of the graph $H$ is $V(H) = \{v^1, v^2 : v \in V(G)\}$, and the edge set is $E(H) = \{u^1v^1, u^2v^2 : uv \in E(G)\} \cup \{u^1v^2 : u, v \in V(G)\}$. Note that the graph $H$ consists of the join of two copies $G^1$, $G^2$ of the graph $G$. Notice that $G$ is $(0, \ell)$-well-covered if and only if $H$ is a $(0, \ell)$-well-covered probe graph. Therefore, for each $\ell \geq 3$, UNPARTITIONED PROBE FOR $(0, \ell)$-WELL-COVEREDNESS is NP-complete. □

Next, we prove that for each $r \geq 3$ the UNPARTITIONED PROBE FOR $(r, 0)$-WELL-COVEREDNESS problem is NP-hard. We use the well-known result due to Stockmeyer [29] that $r$-COLORING for $r \geq 3$ is NP-complete, together with the following theorem.

**Proposition 7 (Topp and Volkmann [31]).** *Let $G = (V, E)$ be an $n$-vertex graph, $V = \{v_1, v_2, v_3, \ldots, v_n\}$, and let $H$ be obtained from $G$ such that $V(H) = V \cup \{u_1, u_2, u_3, \ldots, u_n\}$ and $E(H) = E \cup \{v_iu_i : i \in \{1, 2, 3, \ldots, n\}\}$. Then $H$ is a well-covered graph where every maximal independent set has size $n$.*

Observe that every maximal independent set $I$ of $H$ has a subset $I_G = I \cap V$. Let $\mathcal{U} \subseteq \{1, 2, 3, \ldots, n\}$ be the set of indices $i$ such that $v_i \in I$. Since $I$ is maximal, the set $\{u_i : i \in \{1, 2, 3, \ldots, n\} \setminus \mathcal{U}\}$ must be contained in $I$, so $|I| = n$.

**Theorem 5.** UNPARTITIONED PROBE FOR $(r, 0)$-WELL-COVEREDNESS *is NP-hard, for each $r \geq 3$.*

*Proof.* Let $G = (V, E)$ be an instance of $r$-COLORING ($r \geq 3$), and let $H$ be the graph obtained from $G$ by the transformation described in Proposition 7. Notice that $H$ is well-covered. Since adding pendant vertices does not increase the chromatic number of non-empty graphs, it holds that if $G$ is $r$-colorable then $H$ is a $(r, 0)$-well-covered graphs. Hence it is enough to consider $(\mathcal{N}, \mathcal{P}) = (\emptyset, V(H))$. Conversely, if $G$ is not $r$-colorable then the chromatic number of $G$ is at least $r + 1$. Since the addition of vertices and edges does not decrease the chromatic number, it holds that $H$ is not a $(r, 0)$-well-covered probe graph. □

**Theorem 6** *(Monotonicity). If* UNPARTITIONED PROBE FOR $(r, \ell)$-WELL-COVEREDNESS *is (co)NP-hard, then* UNPARTITIONED PROBE FOR $(r, \ell + 1)$-WELL-COVEREDNESS *is (co)NP-hard.*

To prove that UNPARTITIONED PROBE FOR $(2, 1)$-WELL-COVEREDNESS is coNP-hard we consider a special version of 3-SAT, where every satisfying truth assignment has at least one true variable and at least one false variable. This problem can be reduced from 3-SAT by adding the 3 new variables $x, y, z$ and the clauses $(x, y, z), (\overline{x}, \overline{y}, \overline{z})$.

**Theorem 7.** Unpartitioned Probe for $(2,1)$-Well-Coveredness *is coNP-hard.*

*Proof.* Let $I = (U, C)$ be a 3-SAT instance with $n = |U|$ boolean variables and $m = |C|$ clauses over $U$.

In polynomial time with respect to the size of $I$, we construct an instance $G$ of Unpartitioned Probe for $(2,1)$-well-coveredness, such that $I$ is *not* satisfiable if and only if $G$ is a $(2,1)$-well-covered probe graph. The construction of $G$ is as follows: The vertex set of $G$ is $V(G) = A \cup B \cup U \cup \overline{U} \cup C$, where $U = \{u_1, u_2, \ldots, u_n\}$, $\overline{U} = \{\overline{u}_1, \overline{u}_2, \ldots, \overline{u}_n\}$, $C = \{c_1, c_2, \ldots, c_m\}$, $A = \{a_1, a_2, \ldots, a_n, a_{n+1}, a_{n+2}\}$, and $B = \{b_1, b_2, \ldots, b_n, b_{n+1}, b_{n+2}\}$. The $G$ edge set is $E(G) = \{a_i u_j : i \in \{1, \ldots, n+2\}, j \in \{1, \ldots n\}\} \cup \{b_i \overline{u}_j : i \in \{1, \ldots, n+2\}, j \in \{1, \ldots, n\}\} \cup \{a_i c_j, b_i c_j : i \in \{1, \ldots, n+2\}, j \in \{1, \ldots m\}\} \cup \{u_i \overline{u}_i : u_i \in U\} \cup \{c_i c_j : c_i, c_j \in C, i \neq j\} \cup \{xc : x \in U \cup \overline{U}, c \in C$ and $x$ occurs in $c\}$. We observe that $I$ is not satisfiable if and only if $G$ is a $(2,1)$-well-covered probe graph. This completes the proof of Theorem 7. $\square$

To prove the NP-completeness of Unpartitioned Probe for $(1,3)$-Well-Coveredness, we present a reduction from $(0,3)$-Well-Coveredness [2]. Also, we remark that $(0,3)$-graphs have only independent sets of size at most three. Thus, $(0,3)$-well-covered graphs that are not complete are either 2-well-covered or 3-well-covered (both are recognizable in polynomial time). In particular, in the NP-completeness proof of $(0,3)$-Well-Coveredness, Alves et al. [2] constructed instances that are 3-well-covered (every maximal independent set has size three). Therefore, the problem of recognizing $(0,3)$-well-covered graphs still NP-complete even when it is known that the input graph is 3-well-covered. We assume that this is the case.

**Theorem 8.** Unpartitioned Probe for $(1,3)$-Well-Coveredness *is NP-hard.*

*Proof.* Let $G = (V, E)$ be an instance of $(0,3)$-Well-Coveredness such that every maximal independent set of $G$ has size three (3-well-covered). Recall that from [2] it holds that recognizing $(0,3)$-well-covered graphs $G$ is NP-complete even when it is known that $G$ is 3-well-covered. At this point, in polynomial time with respect to the size of $G$, we construct an instance $H$ of Unpartitioned Probe for $(1,3)$-Well-Coveredness, such that $G$ is a $(0,3)$-well-covered graph if and only if $H$ is a $(1,3)$-well-covered probe graph. The construction is as follows: First, we define graphs $A$ and $B$, which are used to define $H$. $A = (V(A), E(A))$, where $V(A) = (\{a_1, a_2, a_3, a_4, a_5, a_6, a_7\}$, $E(A) = \{a_1a_2, a_1a_3, a_2a_3, a_3a_4, a_2a_4, a_4a_5, a_5a_6, a_6a_1\})$. and $B = (V(B), E(B))$, where $V(B) = \{b_1, b_2, b_3, b_4\}$, $E(B) = \emptyset$. The vertex set of the graph $H$ is $V(H) = V(G) \cup V(A) \cup V(B)$, and the edge set of $H$ is the union of the following sets: $E(G)$, $E(A)$, $\{a_i b_j : i \in \{1,2,3,4,5,6,7\}, j \in \{1,2,3,4\}\}$, $\{a_i v : i \in \{1,2,3,4,5,6,7\}, v \in V(G)\}$, $\{b_i v : i \in \{1,2,3,4\}, v \in V(G)\}$.

This completes the construction of $H$. For the sake of the reader, with Theorem 8, we offer in Fig. 1, three examples: one positive and two negatives for Unpartitioned Probe for $(1,3)$-Well-Coveredness.

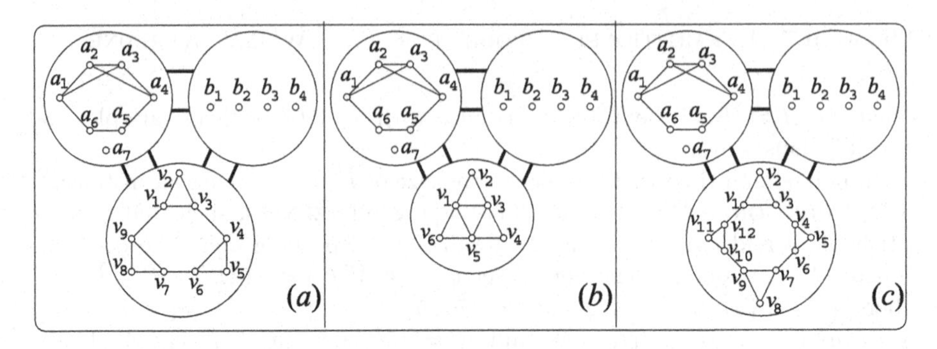


**Fig. 1.** Three instances of UNPARTITIONED PROBE FOR (1, 3)-WELL-COVEREDNESS obtained, according to Theorem 8, from graphs $G_1$, $G_2$, and $G_3$, respectively, depicted in Fig. 1 (a), (b), and (c). Graph $G_1 = (V_1, E_1)$ is a positive instance, while graphs $G_2 = (V_2, E_2)$ and $G_3 = (V_3, E_3)$ are negative instances of (0, 3)-WELL-COVEREDNESS. We observe that graph $G_2$ is not (0, 3)-well-covered because it is not well-covered, and graph $G_3$ is not (0, 3)-well-covered because it is not a (0, 3)-graph.

The key property of graphs $A$ and $B$ is that the minimum number of cliques to cover the vertices of $A$ or $B$ is 4. In particular, $A$ is a well-covered graph with a maximum independent set of size three and clique cover number of size four. In addition, among the graphs $A$, $B$, and $G$, the graph $B$ is the unique one with an independent set of size greater than three.

Now, suppose that $G$ is a (0, 3)-well-covered graph. We define $\mathcal{N} = V(B)$, and set $E' = \{b_1b_2\}$. Let $F = (V(H), E(H) \cup E')$. First, note that every maximal independent set $S$ of $F$ is entirely contained in $V(A)$, or in $V(G)$, or in $V(B)$, since each vertex in one of these sets dominates all the vertices in the other sets. In addition, it is easy to see that $F[V(A)]$ and $F[V(B)]$ are well-covered, and their maximum independent sets have size three, and as $G$ is 3-well-covered it holds that $F$ is well-covered. To see that $F$ is a (1, 3)-graph, consider $(Q^1, Q^2, Q^3)$ as a (0, 3)-partition of the graph $G$. Hence, $(S, K^1, K^2, K^3)$ where $S = \{a_7\}$, $K^1 = Q^1 \cup \{a_1, a_2, a_3\} \cup \{b_1, b_2\}$, $K^2 = Q^2 \cup \{a_4, a_5\} \cup \{b_3\}$, $K^3 = Q^3 \cup \{a_6\} \cup \{b_4\}$), is a (1, 3)-partition of $F$.

Conversely, suppose that $H$ is a (1, 3)-well-covered probe graph. Let $(\mathcal{N}, \mathcal{P})$ be a (1, 3)-well-covered probe partition of $H$. Let $F$ be a (1, 3)-well-covered graph, with (1, 3)-partition $(S, K^1, K^2, K^3) = V$ such that $E(H) = E(G) \cup E'$, with $E' \subseteq \{uv : u, v \in \mathcal{N}\}$. By construction, each vertex in one of the graphs $A, B$, or $G$ dominates the vertices in the other two graphs. Therefore, the independent set $\mathcal{N}$ as well as any other independent set of $H$ must be entirely contained in just one of these three subgraphs of $H$.

Since $G$ and $A$ are 3-well-covered then $\mathcal{N} \subseteq V(B)$ because $B$ is an independent set of size four. Now, since $\mathcal{N} \subseteq V(B)$, and the clique cover number of $A$ is 4, we have that $S \subset V(A)$. Hence, $(K^1 \cap V, K^2 \cap V, K^3 \cap V)$ is a (0, 3) partition for $G$, and $G$ is a (0, 3)-well-covered graph. □

## References

1. Alves, S.R., Couto, F., Faria, L., Gravier, S., Klein, S., Souza, U.S.: Graph sandwich problem for the property of being well-covered and partitionable into $k$ independent sets and $\ell$ cliques. In: Kohayakawa, Y., Miyazawa, F.K. (eds.) LATIN 2021. LNCS, vol. 12118, pp. 587–599. Springer, Cham (2020). https://doi.org/10.1007/978-3-030-61792-9_46
2. Alves, S.R., Dabrowski, K.K., Faria, L., Klein, S., Sau, I., Souza, U.S.: On the (parameterized) complexity of recognizing well-covered $(r, \ell)$-graph. Theor. Comput. Sci. **746**, 36–48 (2018). https://doi.org/10.1016/j.tcs.2018.06.024
3. Araújo, R.T., Costa, E.R., Klein, S., Sampaio, R.M., Souza, U.S.: FPT algorithms to recognize well covered graphs. Discret. Math. Theor. Comput. Sci. **21**(1) (2019). https://doi.org/10.23638/DMTCS-21-1-3
4. Brandstädt, A.: Partitions of graphs into one or two independent sets and cliques. Discret. Math. **152**(1–3), 47–54 (1996). https://doi.org/10.1016/0012-365X(94)00296-U
5. Brandstädt, A.: Corrigendum. Discret. Math. **186**(1–3), 295 (1998). https://doi.org/10.1016/S0012-365X(98)00014-4
6. Campbell, S., Ellingham, M., Royle, G.: A characterization of well-covered cubic graphs. J. Comb. Comput. **13**, 193–212 (1993)
7. Chandler, D.B., Chang, M.-S., Kloks, T., Le, V.B., Peng, S.-L.: Probe Ptolemaic graphs. In: Hu, X., Wang, J. (eds.) COCOON 2008. LNCS, vol. 5092, pp. 468–477. Springer, Heidelberg (2008). https://doi.org/10.1007/978-3-540-69733-6_46
8. Chandler, D.B., Chang, M.-S., Kloks, T., Liu, J., Peng, S.-L.: Recognition of probe cographs and partitioned probe distance hereditary graphs. In: Cheng, S.-W., Poon, C.K. (eds.) AAIM 2006. LNCS, vol. 4041, pp. 267–278. Springer, Heidelberg (2006). https://doi.org/10.1007/11775096_25
9. Chandler, D.B., Chang, M., Kloks, T., Liu, J., Peng, S.: On probe permutation graphs. Discret. Appl. Math. **157**(12), 2611–2619 (2009). https://doi.org/10.1016/j.dam.2008.08.017
10. Chang, M.-S., Kloks, T., Kratsch, D., Liu, J., Peng, S.-L.: On the recognition of probe graphs of some self-complementary classes of perfect graphs. In: Wang, L. (ed.) COCOON 2005. LNCS, vol. 3595, pp. 808–817. Springer, Heidelberg (2005). https://doi.org/10.1007/11533719_82
11. Chudnovsky, M., Robertson, N., Seymour, P., Thomas, R.: The strong perfect graph theorem. Ann. Math. **164**, 51–229 (2006)
12. Chvátal, V., Slater, P.J.: A note on well-covered graphs. Ann. Discret. Math. **55**, 179–181 (1993)
13. Dantas, S., Faria, L., de Figueiredo, C.M.H., Teixeira, R.B.: The $(k, \ell)$ unpartitioned probe problem NP-complete versus polynomial dichotomy. Inf. Process. Lett. **116**(4), 294–298 (2016). https://doi.org/10.1016/j.ipl.2015.11.004
14. Demange, M., Ekim, T., de Werra, D.: Partitioning cographs into cliques and stable sets. Discret. Optim. **2**(2), 145–153 (2005). https://doi.org/10.1016/j.disopt.2005.03.003
15. Feder, T., Hell, P., Klein, S., Nogueira, L.T., Protti, F.: List matrix partitions of chordal graphs. Theor. Comput. Sci. **349**(1), 52–66 (2005). https://doi.org/10.1016/j.tcs.2005.09.030
16. Feder, T., Hell, P., Klein, S., Motwani, R.: List partitions. SIAM J. Discret. Math. **16**(3), 449–478 (2003). https://doi.org/10.1137/S0895480100384055

17. Finbow, A., Hartnell, B., Nowakowski, R.J.: A characterization of well-covered graphs that contain neither 4- nor 5-cycles. J. Graph Theory **18**(7), 713–721 (1994). https://doi.org/10.1002/jgt.3190180707
18. Finbow, A., Hartnell, B., Nowakowski, R.: A characterization of well covered graphs of girth 5 or greater. J. Comb. Theory Ser. B **57**(1), 44–68 (1993). https://doi.org/10.1006/jctb.1993.1005
19. Golumbic, M.C., Kaplan, H., Shamir, R.: Graph sandwich problems. J. Algorithms **19**(3), 449–473 (1995). https://doi.org/10.1006/jagm.1995.1047
20. Golumbic, M.C., Lipshteyn, M.: Chordal probe graphs. Discret. Appl. Math. **143**(1–3), 221–237 (2004)
21. Golumbic, M.C., Maffray, F., Morel, G.: A characterization of chain probe graphs. Ann. Oper. Res. **188**(1), 175–183 (2011). https://doi.org/10.1007/s10479-009-0584-6
22. Klein, S., de Mello, C.P., Morgana, A.: Recognizing well covered graphs of families with special P 4-components. Graphs Comb. **29**(3), 553–567 (2013). https://doi.org/10.1007/s00373-011-1123-1
23. Le, V.B.: Two characterizations of chain partitioned probe graphs. Ann. Oper. Res. **188**(1), 279–283 (2011)
24. McMorris, F., Wang, C., Zhang, P.: On probe interval graphs. Discret. Appl. Math. **88**(1), 315–324 (1998). https://doi.org/10.1016/S0166-218X(98)00077-8
25. Plummer, M.D.: Some covering concepts in graphs. J. Comb. Theory **8**(1), 91–98 (1970). https://doi.org/10.1016/S0021-9800(70)80011-4
26. Prisner, E., Topp, J., Vestergaard, P.D.: Well covered simplicial, chordal, and circular arc graphs. J. Graph Theory **21**(2), 113–119 (1996). https://doi.org/10.1002/(SICI)1097-0118(199602)21:2<113::AID-JGT1>3.0.CO;2-U
27. Ravindra, G.: Well-covered graphs. J. Combin. Inform. Syst. Sci. **2**(1), 20–21 (1977)
28. Sankaranarayana, R.S., Stewart, L.K.: Complexity results for well-covered graphs. Networks **22**(3), 247–262 (1992)
29. Stockmeyer, L.: Planar 3-colorability is polynomial complete. ACM SIGACT News **5**(3), 19–25 (1973). https://doi.org/10.1145/1008293.1008294
30. Tankus, D., Tarsi, M.: Well-covered claw-free graphs. J. Comb. Theory Ser. B **66**(2), 293–302 (1996). https://doi.org/10.1006/jctb.1996.0022
31. Topp, J., Volkmann, L.: Well covered and well dominated block graphs and unicyclic graphs. Math. Pannon. **1**(2), 55–66 (1990)
32. Zhang, P., et al.: An algorithm based on graph theory for the assembly of contigs in physical mapping of DNA. Bioinformatics **10**(3), 309–317 (1994). https://doi.org/10.1093/bioinformatics/10.3.309

# Distinguishing Graphs via Cycles

Nina Klobas[1(✉)] and Matjaž Krnc[2]

[1] Durham University, Durham, UK
nina.klobas@durham.ac.uk
[2] The Faculty of Mathematics, Natural Sciences and Information Technologies, University of Primorska, Koper, Slovenia
matjaz.krnc@upr.si

**Abstract.** Recognizing graphs with high level of symmetries is hard in general, and usually requires additional structural understanding. In this paper we study a particular graph parameter and motivate its usage by devising efficient recognition algorithms for three highly symmetric graph families: folded cubes, $I$-graphs and double generalized Petersen graphs.

For integers $\ell, \lambda, m$ a simple graph is $[\ell, \lambda, m]$-cycle regular if every path of length $\ell$ belongs to exactly $\lambda$ different cycles of length $m$. We identify all $[1, \lambda, 8]$-cycle regular $I$-graphs and all $[1, \lambda, 8]$-cycle regular double generalized Petersen graphs. For $n \geq 7$ we show that for a folded cube $FQ_n$ is $[1, n-1, 4]$, $[1, 4n^2 - 12n + 8, 6]$ and $[2, 4n - 8, 6]$-cycle regular, and identify the corresponding exceptional values of cycle regularity for $n < 7$. As a consequence we describe linear recognition algorithms for $I$-graphs and double generalized Petersen graphs, and an $O(|E| \log |V|)$ recognition algorithm for the family of folded cubes.

We believe that the structural observations and methods used in the paper are of independent interest and could be used for solving other algorithmic problems.

**Keywords:** Recognition algorithm · Generalized Petersen graphs · Double generalized Petersen graph · Folded cubes

## 1 Introduction

Important graph classes such as bipartite graphs, (weakly) chordal graphs, perfect graphs and forests are defined or characterized by their cycle structure. A particularly strong description of a cyclic structure is the notion of *cycle-regularity*, introduced by Mollard [23]:

*For integers $l, \lambda, m$ a simple graph is $[l, \lambda, m]$-cycle regular if every path on $l + 1$ vertices belongs to exactly $\lambda$ different cycles of length $m$.*

It is perhaps natural that cycle-regularity mostly appears in the literature in the context of symmetric graph families such as hypercubes, Cayley graphs or circulants. Indeed Mollard showed that certain extremal $[3, 1, 6]$-cycle regular


C.-Y. Chen et al. (Eds.): COCOON 2021, LNCS 13025, pp. 387–398, 2021.
https://doi.org/10.1007/978-3-030-89543-3_33

graphs correspond exactly to the graphs induced by the two middle-layers of odd-dimensional hypercubes. Also, for $[2,1,4]$-cycle regular graphs Mulder [24] showed that their degree is minimized in the case of Hadamard graphs, or in the case of hypercubes. In relation with other graph families, Fouquet and Hahn [13] described the symmetric aspect of certain cycle-regular classes, while in [19] authors describe all $[1,\lambda,8]$-cycle regular members of generalized Petersen graphs, and use this result to obtain linear recognition algorithm for generalized Petersen graphs. Understanding the structure of subgraphs of hypercubes which avoid all 4-cycles does not seem to be easy. Indeed, a question of Erdős regarding how many edges can such a graph contain remains open after more than 30 years [11].

In this paper we study cycle-regularity and more general cyclic aspects of three graph families, namely $I$-graphs, double generalized Petersen graphs and folded cubes, with the focus of devising efficient recognition algorithms. In all three cases, if the input graph is a member of the observed family, we not only provide its parameters but also give a certificate of correctness, i.e. we give an exact isomorphism. In general the graph recognition problem can be difficult to solve. For instance it is $\mathcal{NP}$-hard to recognise unit disk graphs [5], coordinate graphs [29], string graphs [18], clique graphs [1] etc.

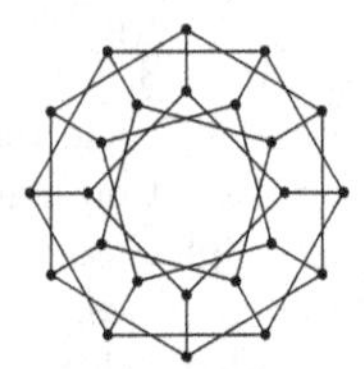
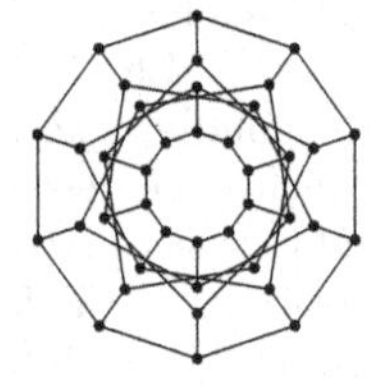
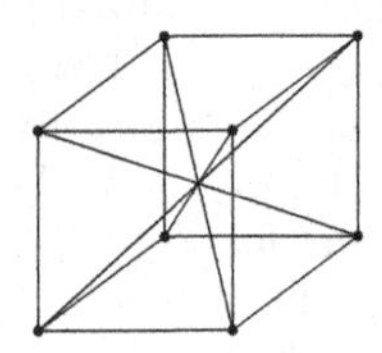

**Fig. 1.** $I$-graph $I(12,2,3)$, double generalized Petersen graph $\mathrm{DP}(10,2)$, and folded cube $\mathrm{FQ}_4$.

This paper is structured as follows. In Sect. 1.1 we provide basic definitions and notations that are used throughout the paper, in Sect. 2 we present two of above mentioned graph families, namely $I$-graphs and double generalized Petersen graphs, observe their cyclic structure and provide their linear recognition algorithm, in Sect. 3 we do the similar for the folded cubes. Due to the space constraints, some proofs are deferred to the full version of this paper [17].

### 1.1 Preliminaries

Unless specified otherwise, all graphs in this paper are finite, simple, undirected and connected. For a given graph $G$ we use a standard notation for a set of vertices $V(G)$ and a set of edges $E(G)$. A $k$-cycle $C$ in $G$, on vertices $v_1, v_2, \ldots, v_k$ from $V(G)$ using edges $e_1, e_2, \ldots, e_k$ from $E(G)$, will be denoted in two ways: as $(v_1, \ldots, v_k)$, or as $(e_1, \ldots, e_k)$. For integers $a$ and $b$ we denote with $\gcd(a, b)$ the greatest common divisor of $a$ and $b$ respectively. For a binary value $x$ we use $\overline{x}$ to denote $1 - x$.

**Definition 1.** *Let $l, \lambda, m$ be positive integers. A simple graph $G$ is $[l, \lambda, m]$-cycle regular if every path on $l+1$ vertices of $G$ belongs to exactly $\lambda$ different $m$-cycles of $G$.*

It is easy to see that $[1, \lambda, 8]$-cycle regular cubic graphs are also $[0, 3\lambda/2, 8]$-cycle regular, but the converse does not hold (for example in a 5-prism every vertex lies in four 8-cycles, whereas every edge lies in three or two different 8-cycles). Related to this we define a function $\sigma : E(G) \mapsto \mathbb{N}$, where $\sigma(e)$ corresponds to the number of distinct 8-cycles an edge $e$ belongs to. We call $\sigma(e)$ an *octagon value* of an edge $e$, and we say that a graph $G$ has a *constant octagon value* if $\sigma$ is a constant function.

## 2 *I*-Graphs and Double Generalized Petersen Graphs

$I$-graphs were introduced in the Foster census [12], and are trivalent or cubic graphs with edges connecting vertices of two star polygons. They form a natural generalization of the well-known *generalized Petersen graphs* introduced in 1950 by Coxeter [8] and later named by Watkins in 1969 [32].

***I*-graphs** are denoted by $I(n, j, k)$ for integers $n, j, k$, where $n \geq 3$ and $n \geq j, k \geq 1$. The vertex set of $I(n, j, k)$ is defined as $\{u_0, u_1, \ldots, u_{n-1}, w_0, w_1, \ldots, w_{n-1}\}$ while the edge set consists of outer edges $u_i u_{i+j}$, inner edges $w_i w_{i+k}$ and spoke edges $u_i w_i$, where the subscripts are taken modulo $n$.

**Generalized Petersen graphs** are denoted by $G(n, k)$ and form a subclass of $I$-graphs, where parameter $j$ has value 1. In other words $G(n, k) \simeq I(n, 1, k)$ (Fig. 2).

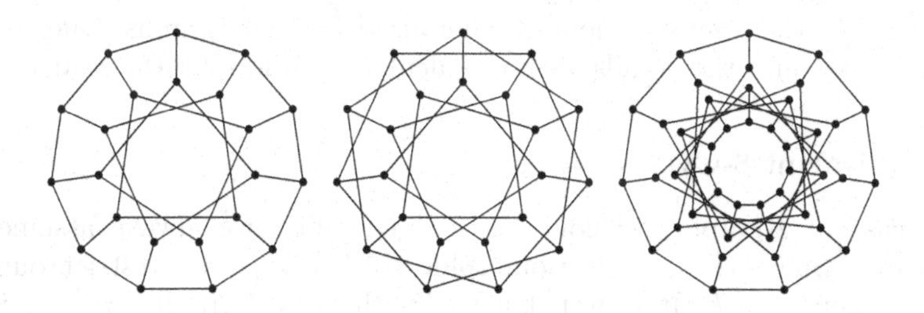

**Fig. 2.** Example of a $G(11, 3), I(11, 2, 3)$ and $\mathrm{DP}(11, 3)$.

The family of $I$-graphs has been studied extensively with respect to their automorphism group and isomorphisms [2,14,26], Hamiltonicity [4], spectrum [25], and independence number [10,16].

Our first result identifies all $[1, \lambda, 8]$-cycle regular members and determines the corresponding values of $\lambda$.

**Theorem 1.** *An arbitrary I-graph is never* $[1, \lambda, 8]$*-cycle regular, except when isomorphic to* $I(n, j, k)$ *where* $j = 1$ *and*

$$(n, k) \in \{(3,1), (4,1), (5,2), (8,3), (10,2), (10,3), (12,5), (13,5), (24,5), (26,5)\}.$$

Double generalized Petersen graphs consist of two identical copies of generalized Petersen graphs, where instead of connecting the vertices inside a star polygon, we connect the vertices from two different star polygons accordingly.

**Double generalized Petersen graphs** are denoted by $\mathrm{DP}(n, k)$, where $n \geq 3$ and $k < n/2$. The vertex set of $\mathrm{DP}(n, k)$ consists of $\{u_0, u_1, \ldots, u_{n-1}, w_0, w_1, \ldots, w_{n-1}, x_0, x_1, \ldots, x_{n-1}, y_0, y_1, \ldots, y_{n-1}\}$. The edge-set contains outer edges $u_i u_{i+1}, x_i x_{i+1}$, inner edges $w_i y_{i+k}$, $y_i w_{i+k}$ and spoke edges $u_i w_i, x_i y_i$, where the subscripts are taken modulo $n$.

These graphs were first introduced in 2012 by Zhou and Feng [35] and have been studied with respect to their Hamiltonicity [28], automorphisms [20], vertex-transitivity [36], determining number [9] and canonical double covers [27].

Similarly as in the case of $I$-graphs we identify all $[1, \lambda, 8]$-cycle regular members and determine corresponding values of $\lambda$.

**Theorem 2.** *A double generalized Petersen graph is never* $[1, \lambda, 8]$*-cycle regular, except when isomorphic to* $\mathrm{DP}(n, k)$ *where* $(n, k) \in \{(5,2), (10,2)\}$.

Using the above mentioned structural properties we get the following result.

**Theorem 3.** *I-graphs and double generalized Petersen graphs can be recognized in linear time.*

Let us now start with the proof of these structural results and the recognition algorithm. We start with the study of $I$-graphs and their properties. For a slightly changed definition of *equivalent cycles* the same approach can be used to obtain the results for the family of double generalized Petersen graphs. Due to space constraints we omit the details of those calculations and refer the reader to [17].

### 2.1 Equivalent 8-Cycles

In the case of $I$-graphs, without loss of generality, we always assume that $j, k < n/2$. Since $I(n, j, k)$ is isomorphic to $I(n, k, j)$, we restrict ourselves to cases when $j \leq k$. It is well known [2] that an $I$-graph $I(n, j, k)$ is disconnected whenever $d = \gcd(n, j, k) > 1$. In this case it consists of $d$ copies of $I(n/d, j/d, k/d)$. Therefore, throughout the paper we consider only graphs $I(n, j, k)$ where $\gcd(n, j, k) = 1$. We also know [14] that two $I$-graphs $I(n, j, k)$ and $I(n, j', k')$ are isomorphic if and only if there exists an integer $a$, which is relatively prime to $n$, for which either $\{j', k'\} = \{aj \pmod{n}, ak \pmod{n}\}$ or $\{j', k'\} = \{aj \pmod{n}, -ak \pmod{n}\}$. Throughout the paper, whenever we discuss $I$-graphs with certain parameters, we consider only the lexicographically smallest possible parameters by which the graph is uniquely determined.

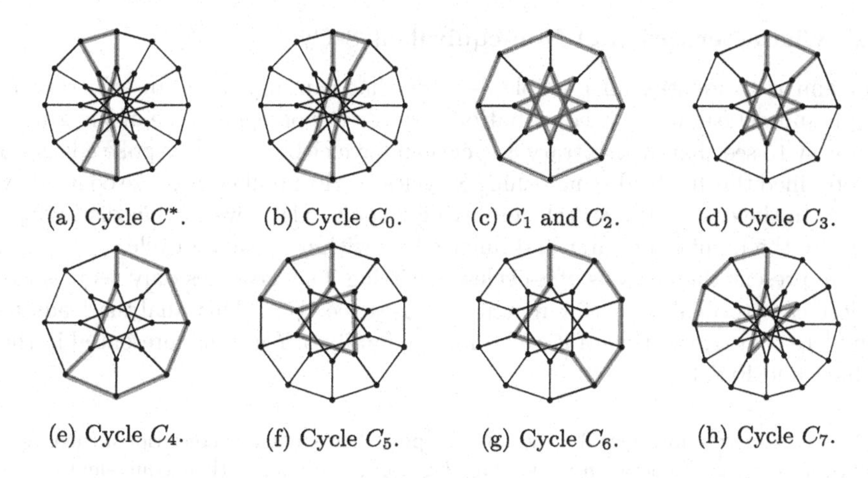

(a) Cycle $C^*$. (b) Cycle $C_0$. (c) $C_1$ and $C_2$. (d) Cycle $C_3$.

(e) Cycle $C_4$. (f) Cycle $C_5$. (g) Cycle $C_6$. (h) Cycle $C_7$.

**Fig. 3.** Examples of non-equivalent 8-cycles in $I$-graphs.

A particular member of automorphism group of every $I$-graph is a *rotation* defined as: $\rho(u_i) = u_{i+1}$, $\rho(w_i) = w_{i+1}$. Clearly, applying $n$ times the rotation $\rho$ yields an identity automorphism. When acting on $I$-graphs with $\rho$ we get 3 edge orbits: orbit of outer edges $E_J$, orbit of spoke edges $E_S$ and orbit of inner edges $E_I$. Edges from the same orbit $E_J, E_S$, or $E_I$ have the same octagon value, which we denote by $\sigma_J, \sigma_S$ and $\sigma_I$, respectively. Therefore the *octagon value of an I-graph* is said to be a triple $(\sigma_J, \sigma_S, \sigma_I)$.

We say that two 8-cycles of an $I$-graph are *equivalent* iff we can map one into the other using rotation $\rho$. Let $G \simeq I(n, j, k)$ be an arbitrary $I$-graph and let $C$ be one of its 8-cycles. With $\gamma(C)$ we denote the number of equivalent 8-cycles to $C$ in $G$. Each 8-cycle contributes to the octagon value of an $I$-graph. We denote the contributed amount with $\tau(C)$, defined as the triple $(\delta_j, \delta_s, \delta_i)$, where we calculate $\delta_j, \delta_s, \delta_i$ by counting the number of outer, spoke and inner edges of a cycle and multiply these numbers with $\gamma(C)/n$. If a graph $G$ admits $m$ non-equivalent 8-cycles $C_1, C_2, \ldots, C_m$, one may calculate its octagon value $(\sigma_J, \sigma_S, \sigma_I)$ as $\sum_{i=1}^{m} \tau(C_i)$.

The following claim serves also as an example of the above-mentioned definitions.

*Claim.* For $I(n, j, k)$ where $n > 3$ and integers $k, j < n/2$ there always exists an 8-cycle.

Indeed, if $k \neq j$ it is of the form

$$C^* = (w_0, w_{\pm k}, u_{\pm k}, u_{\pm k \pm j}, w_{\pm k \pm j}, w_{\pm j}, u_{\pm j}, u_0).$$

If $k = j$ it is of the form

$$C_7 = (u_0, u_k, u_{2k}, u_{3k}, w_{3k}, w_{2k}, w_k, w_0).$$

## 2.2 Characterization of Non-equivalent 8-Cycles

Our aim is to identify all possible 8-cycles that can appear in an arbitrary $I$-graph and determine their contribution towards the octagon value of the graph. It is easy to see that an arbitrary 8-cycle can have either 4, 0 or 2 spoke edges, so we obtained this list by distinguishing 8-cycles by the number of spoke edges they admit. In the case of the 8-cycle admitting 2 spoke edges, we further distinguish cases by the number of outer and inner edges within a given 8-cycle.

We present the analysis of 8-cycles admitting 4 spoke edges only, as it is the easiest case to deal with (for remaining cases see [17]). This analysis leads to complete characterisation of 8-cycles for the family of $I$-graphs, presented in the Table 1 and Fig. 3.

**Table 1.** All non-equivalent 8-cycles of $I$-graphs, their existence conditions, their contribution towards the octagon value of an $I$-graph $\tau$, number of their equivalent cycles in an $I$-graph $\gamma$.

| Label | A representative of an 8-cycle | Existence conditions | $\tau(C)$ | $\gamma(C)$ |
|---|---|---|---|---|
| $C^*$ | $(w_0, w_{\pm k}, u_{\pm k}, u_{\pm k\pm j}, w_{\pm k\pm j}, w_{\pm j}, u_{\pm j}, u_0)$ | $k \neq j$ and $n > 4$ | $(2,4,2)$ | $n$ |
| $C_0$ | $(w_0, w_{\pm k}, u_{\pm k}, u_{\pm k\pm j}, w_{\pm k\pm j}, w_{\pm 2k\pm j}, u_{\pm 2k\pm j}, u_{\pm 2k\pm 2j})$ | $2k + 2j = n$ | $(1,2,1)$ | $n/2$ |
| $C_1$ | $(u_0, u_j, u_{2j}, u_{3j}, u_{4j}, u_{5j}, u_{6j}, u_{7j})$ | $8j = n$ or $3n$ | $(0,0,1)$ | $n/8$ |
| $C_2$ | $(w_0, w_k, w_{2k}, w_{3k}, w_{4k}, w_{5k}, w_{6k}, w_{7k})$ | $8k = n$ or $3n$ | $(1,0,0)$ | $n/8$ |
| $C_3$ | $(w_0, w_k, w_{2k}, w_{3k}, w_{4k}, w_{5k}, u_{5k}, u_{5k+j})$ | $5k + j = n$ or $2n$ | $(5,2,1)$ | $n$ |
| | $(w_0, w_k, w_{2k}, w_{3k}, w_{4k}, w_{5k}, u_{5k}, u_{5k-j})$ | $5k - j = n$ or $2n$ | | |
| $C_4$ | $(u_0, u_j, u_{2j}, u_{3j}, u_{4j}, u_{5j}, w_{5j}, w_{5j+k})$ | $k + 5j = n$ or $2n$ | $(1,2,5)$ | $n$ |
| | $(u_0, u_j, u_{2j}, u_{3j}, u_{4j}, u_{5j}, w_{5j}, w_{5j-k})$ | $5j - k = 2n$ or $n$ or $0$ | | |
| $C_5$ | $(w_0, w_k, w_{2k}, w_{3k}, w_{4k}, u_{4k}, u_{4k+j}, u_{4k+2j})$ | $4k + 2j = n$ or $2k + j = n$ | $(4,2,2)$ | $n$ |
| | $(w_0, w_k, w_{2k}, w_{3k}, w_{4k}, u_{4k}, u_{4k-j}, u_{4k-2j})$ | $4k - 2j = n$ | | |
| $C_6$ | $(u_0, u_j, u_{2j}, u_{3j}, u_{4j}, w_{4j}, w_{4j+k}, w_{4j+2k})$ | $2k + 4j = n$ or $k + 2j = n$ | $(2,2,4)$ | $n$ |
| | $(u_0, u_j, u_{2j}, u_{3j}, u_{4j}, w_{4j}, w_{4j-k}, w_{4j-2k})$ | $4j - 2k = n$ or $0$ | | |
| $C_7$ | $(w_0, w_k, w_{2k}, w_{3k}, u_{3k}, u_{3k+j}, u_{3k+2j}, u_{3k+3j})$ | $3k + 3j = n$ or $2n$ | $(3,2,3)$ | $n$ |
| | $(w_0, w_k, w_{2k}, w_{3k}, u_{3k}, u_{3k-j}, u_{3k-2j}, u_{3k-3j})$ | $3k - 3j = n$ or $0$ | | |

*8-Cycles with 4 Spoke Edges.* In addition to 4 spoke edges the 8-cycle must also have two inner and two outer edges. When using the spoke edge there are two options for choosing an inner (outer) edge. After considering all cases it is easy to see that there can be just two such 8-cycles, $C^*$ (see Sect. 2.1), which exists whenever $j \neq k$, and $C_0$, which is of the following form:

$$C_0 = (w_0, w_{\pm k}, u_{\pm k}, u_{\pm k\pm j}, w_{\pm k\pm j}, w_{\pm 2k\pm j}, u_{\pm 2k\pm j}, u_{\pm 2k\pm 2j}).$$

Cycle $C_0$ exists whenever $2k + 2j = n$. One can verify easily, that $n$ applications of the rotation $\rho$ to $C^*$ and $n/2$ applications of the rotation $\rho$ to cycle $C_0$ maps the cycle back to itself. Therefore there are $n$ equivalent cycles to $C^*$ and $n/2$ equivalent cycles to $C_0$ in an $I$-graph $I(n, j, k)$ and they contribute $(2, 4, 2)$ and $(1, 2, 1)$, respectively, to the graph octagon value.

## 2.3 Obtaining Constant Octagon Value

Every 8-cycle of an $I$-graph contributes to the octagon value of each edge partition. It turns out that if we can identify at least one edge partition of a graph, we can easily determine its parameters (see Algorithm 1). Therefore, we want to find graphs with constant octagon value. These are graphs for which all edges touch the same number of 8-cycles. They are called $[1, \lambda, 8]$-cycle regular graphs. We consider all possible collections of 8-cycles and determine octagon values of $I$-graphs admitting those 8-cycles. Since $I$-graphs are defined with 3 parameters and all 8-cycles give constraints for these parameters, it is enough to consider collections of at most 4 cycles, to uniquely determine all $[1, \lambda, 8]$-cycle regular graphs. After a thorough analysis (see [17]) we obtain the result in Fig. 4. Surprisingly, it turns out that all $[1, \lambda, 8]$-cycle regular $I$-graphs are in the family of generalized Petersen graphs.

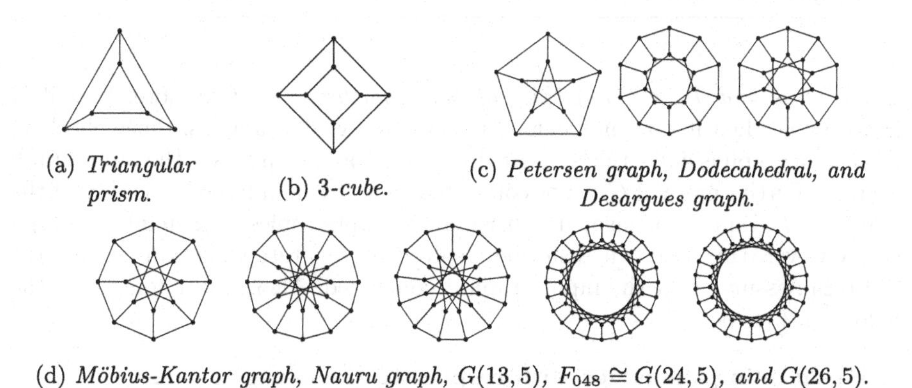

(a) *Triangular prism.* (b) *3-cube.* (c) *Petersen graph, Dodecahedral, and Desargues graph.*

(d) *Möbius-Kantor graph, Nauru graph,* $G(13, 5)$, $F_{048} \cong G(24, 5)$, *and* $G(26, 5)$.

**Fig. 4.** All $[1, \lambda, 8]$-cycle regular $I$-graphs. That is, (a) the graph on no 8-cycles; (b) the graph containing $C_0$ and $C_7$; (c) graphs containing $C_5, C_6$ and $C^*$; and (d) graphs containing $C_3, C_4$ and $C^*$.

In the case of double generalized Petersen graphs, the list of $[1, \lambda, 8]$-cycle regular graphs is even shorter; it consist of Dodecahedral graph, and of DP(10, 2). These two graphs are depicted on Fig. 4c and Fig. 1. The above calculation proves Theorems 1 and 2.

## 2.4 Recognition Algorithm

The recognition algorithm for both graph families (see Algorithm 1) relies on the fact that there is just a small number of $I$-graphs (ten) and double generalized Petersen graphs (two) with the constant octagon value (see previous section). In particular, whenever the input graph $G$ of the Algorithm 1 is a member of one of the observed graph families and is not $[1, \lambda, 8]$-cycle regular, we can immediately identify one of its edge orbits ($E_I, E_J$, or $E_S$), of size $|V(G)|/2$. Since the octagon value of each edge is computed in constant time and there is a finite number of the $[1, \lambda, 8]$-cycle regular $I$-graphs and double generalized Petersen graphs, the first part of Theorem 3 holds.

**Algorithm 1.** Recognition procedure for $I$-graphs or for DP graphs, depending on the subprocedure EXTEND$(G, U)$.

**Require:** connected cubic graph $G$
1: $\mathcal{P} \leftarrow$ an empty dictionary
2: **for** $e \in E(G)$ **do**
3: $\quad s = \text{OCTAGONVALUE}(e)$ ▷ calculate $\sigma(e)$
4: $\quad \mathcal{P}[s]$.append$(e)$
5: $U \leftarrow$ an item of $\mathcal{P}$ with minimum positive cardinality
6: **if** $G[U]$ is a 2-factor **then** ▷ 2-factor is a 2-regular graph
7: $\quad U \leftarrow \{e \mid e \in E(G), e$ is adjacent to an edge of $U\}$ ▷ $U$ is a perfect matching in $G$
8: **return** EXTEND$(G, U)$

*Correctness and Time Complexity of the Algorithm.* We first note, that if $G$ is not cubic then it does not belong to observed graph families. Since checking whether a graph is cubic takes linear time we simply assume that the input graph is cubic. Furthermore, if $G$ is not connected then it can only be a member of the family of $I$-graphs whenever it consists of multiple copies of a smaller $I$-graph $G'$. However, this case can easily be resolved by separately checking each part, so we can assume that the input graph is connected. Algorithm 1 consists of the following 3 parts.

1. **Partitioning the edges with respect to the octagon value**
   The algorithm determines the octagon value of each edge $e \in E(G)$ and builds a partition set $\mathcal{P}$ of graph edges (see lines 1–4). Since $G$ is cubic and all 8-cycles containing edge $e$ consist of edges which are at distance at most 4 from $e$, it is enough to check a subgraph $H$ of $G$ of order at most 62, to calculate the octagon value of an edge $e$. Therefore, calculation of OCTAGONVALUE$(e)$ takes $O(1)$ time for each edge $e$ and this whole part is performed in $\Theta(|E(G)|)$ time.
2. **Identifying the edge-orbit which corresponds to the set of spokes**
   Throughout lines 5–7 we determine the edge-orbit which corresponds to the set of spokes. It is easy to see that this requires additional $O(|E(G)|/3)$ time.
3. **Using set $U$ for determining parameters of a given graph**
   The algorithm uses computed set $U$ to determine exact isomorphism between $G$ and an $I$-graph or a double generalized Petersen graph, if it exists. This procedure differentiates regarding the graph family we are considering. The related procedure EXTEND$(G, U)$ is performed in $\Theta(|E(G)|)$ time (see [17] for details).

## 3 Folded Cubes

Folded cubes were studied already by Brouwer in 1983 [6] and are formed by identifying antipodal vertices of the hypercube graph.

**$n$-dimensional hypercubes** are denoted by $Q_n$ for a positive integer $n$. A graph $Q_n$ contains $2^n$ vertices represented by binary strings $(x_1x_2\dots x_n)$. Two vertices of $Q_n$ are adjacent whenever they differ in exactly one bit.

**Folded $n$-cubes** are denoted by $\mathrm{FQ}_n$ for a positive integer $n$. A graph $\mathrm{FQ}_n$ is a graph on $2^{n-1}$ vertices and is constructed from $Q_n$, by identifying pairs of antipodal vertices, these are vertices which are exactly $n$ apart. Alternatively it can be defined by adding *complementary edges* to the hypercube $Q_{n-1}$, i.e. edges between $(x_1x_2\dots x_{n-1})$ and $(\overline{x}_1\overline{x}_2\dots\overline{x}_{n-1})$, for example see (Fig. 5).

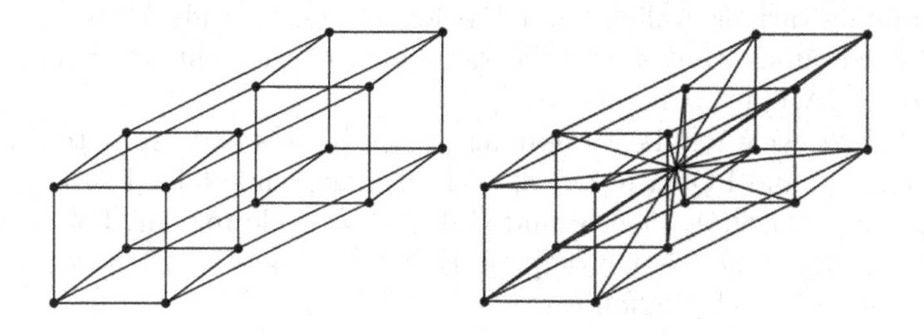

**Fig. 5.** Example of a 4-dimensional hypercube and a folded cube $\mathrm{FQ}_5$.

Some known results of these graphs include characterizing their cyclic structure [33], edge-fault-tolerant properties [34], their automorphism group [7,22, pg. 265] and Hamiltonian-connectivity and strongly Hamiltonian-laceability [15]. They were studied also in the context of efficient routing algorithms and were used in many other applications, see [6,21,30,31].

Since folded cubes are distance-transitive (i.e. the distance between any pair of vertices is the same), see for example [3], they are arc-transitive, so they are $[1, \lambda, m]$-cycle regular for all $m$. For this family we fully determine the values of cycle-regular parameters in the case of $(\ell, \lambda, m) \in \{(1,0,4),(1,0,6),(2,9,6)\}$.

**Theorem 4.** *Folded cubes* $\mathrm{FQ}_1$ *and* $\mathrm{FQ}_2$ *are* $[1,0,4]$*-cycle regular. Folded cube* $\mathrm{FQ}_4$ *is* $[1,9,4]$*-cycle regular. Any other folded cube* $\mathrm{FQ}_n$ *is* $[1,n-1,4]$*-cycle regular.*

**Theorem 5.** *Folded cubes* $\mathrm{FQ}_1, FQ_2$ *and* $\mathrm{FQ}_3$ *are* $[1,0,6]$*-cycle regular. Folded cubes* $\mathrm{FQ}_4$ *and* $\mathrm{FQ}_6$ *are* $[1,36,6]$ *and* $[1,200,6]$*-cycle regular. Any other folded cube* $\mathrm{FQ}_n$ *is* $[1,4(n-2)(n-1),6]$*-cycle regular.*

**Theorem 6.** *Folded cubes* $\mathrm{FQ}_1, \mathrm{FQ}_2$ *and* $\mathrm{FQ}_3$ *are* $[2,0,6]$*-cycle regular. Folded cubes* $\mathrm{FQ}_4$ *and* $\mathrm{FQ}_6$ *are* $[2,12,6]$ *and* $[2,40,6]$*-cycle regular. Any other folded cube* $\mathrm{FQ}_n$ *is* $[2,4(n-2),6]$*-cycle regular.*

Besides studying their cyclic structure we devise also the recognition algorithm.

**Theorem 7.** *Folded cubes can be recognized in* $O(N \log N)$ *time, for* $N = |V(G)| + |E(G)|$.

As the proofs of the above results are a bit technical, we skip them and invite the interested reader to check [17] for details.

# 4 Conclusion

It is easy to observe that the gap between linear lower bound and our upper bound for recognizing folded cubes can be expressed in a sub-logarithmic multiplicative factor. It is hence natural to ask whether the recognition can be done faster, i.e. in linear time. From the structural point of view, this paper focuses on determining various cyclic parameters for some well-studied families. Namely, for the folded cubes we describe the value of $\lambda$ in the context of their $[1, \lambda, 4]$, $[1, \lambda, 6]$ and $[2, \lambda, 6]$-cycle regularity.

The $[1, \lambda, 8]$-cycle regularity remains open. To this end, we note that folded cubes $FQ_1, FQ_2$ and $FQ_3$ are $[1, 0, 8]$-cycle regular, while folded cubes $FQ_4, FQ_6$ and $FQ_8$ are $[1, 36, 8], [1, 3580, 8]$ and $[1, 10794, 8]$-cycle regular. For $n \in \{5, 7, 9, 10, 11, 12, 13, 14\}$ it holds that $FQ_n$ is $[1, 27n^3 - 160n^2 + 291n - 158, 8]$-cycle regular.[1] Hence we ask the following:

*Conjecture 1.* For $n \geq 9$, folded cube $FQ_n$ is $[1, 27n^3 - 160n^2 + 291n - 158, 8]$-cycle regular.

For the families of $I$-graphs and double generalized Petersen graphs, we also settled $[1, \lambda, 8]$-cycle regularity. For the latter family we observe the following isomorphism property:

> Let $n, k$ be positive integers, where $n$ is even and $k < n/2$. Then the graph $DP(n, k)$ is isomorphic to $DP(n, n/2 - k)$.

This observation is a step towards the characterization of isomorphisms of double generalized Petersen graphs, which is an open question.

Studying the cyclic structure as described in this paper led to the construction of fast recognition algorithms for three parametric families. To the best of our knowledge, in addition to this work, such a procedure was so far only used in [19] for the family of generalized Petersen graphs. We believe that a similar approach should give interesting results for other parametric graph families of bounded degree, such as Johnson graphs, rose window graphs, Tabačjn graphs, $Y$-graphs, or $H$-graphs.

**Acknowledgements.** We would like to thank Assist. Prof. Nino Bašić, Štefko Miklavič, Martin Milanič, Tomaž Pisanski, and also to anonymous reviewers, for providing fruitful comments which led to the improvement of this work.

The second author acknowledges partial support of the Slovenian Research Agency (research programs P1-0383, P1-0297 and research projects J1-1692, J1-9187).

[1] The correctness was verified by computer.

## References

1. Alcón, L., Faria, L., de Figueiredo, C.M.H., Gutierrez, M.: Clique graph recognition is NP-complete. In: Fomin, F.V. (ed.) WG 2006. LNCS, vol. 4271, pp. 269–277. Springer, Heidelberg (2006). https://doi.org/10.1007/11917496_24
2. Boben, M., Pisanski, T., Žitnik, A.: $I$-graphs and the corresponding configurations. J. Comb. Des. **13**(6), 406–424 (2005)
3. van Bon, J.: Finite primitive distance-transitive graphs. Eur. J. Combin. **28**(2), 517–532 (2007). https://doi.org/10.1016/j.ejc.2005.04.014
4. Bonvicini, S., Pisanski, T.: Hamiltonian cycles in $I$-graphs. Electron. Notes Discret. Math. **40**, 43–47 (2013). https://doi.org/10.1016/j.endm.2013.05.009. Combinatorics 2012
5. Breu, H., Kirkpatrick, D.G.: Unit disk graph recognition is NP-hard. Comput. Geom. **9**(1–2), 3–24 (1998). https://doi.org/10.1016/S0925-7721(97)00014-X
6. Brouwer, A.E.: On the uniqueness of a certain thin near octagon (or partial 2-geometry, or parallelism) derived from the binary Golay code. IEEE Trans. Inform. Theory **29**(3), 370–371 (1983). https://doi.org/10.1109/TIT.1983.1056664
7. Brouwer, A.E., Cohen, A.M., Neumaier, A.: Distance-regular graphs, Ergebnisse der Mathematik und ihrer Grenzgebiete (3) [Results in Mathematics and Related Areas (3)], vol. 18. Springer, Berlin (1989). https://doi.org/10.1007/978-3-642-74341-2
8. Coxeter, H.S.M.: Self-dual configurations and regular graphs. Bull. Am. Math. Soc. **56**, 413–455 (1950). https://doi.org/10.1090/S0002-9904-1950-09407-5
9. Das, A.: Determining number of generalized and double generalized Petersen graph. In: Changat, M., Das, S. (eds.) CALDAM 2020. LNCS, vol. 12016, pp. 131–140. Springer, Cham (2020). https://doi.org/10.1007/978-3-030-39219-2_11
10. Dods, M.S.: Independence number of specified $I$-graphs. Ph.D. thesis, Naval Postgraduate School, Monterey, CA (2020)
11. Erdős, P.: Some of my favorite solved and unsolved problems in graph theory. Quaestiones Math. **16**(3), 333–350 (1993)
12. Foster, R.M.: The Foster census. Charles Babbage Research Centre, Winnipeg, MB (1988). R. M. Foster's census of connected symmetric trivalent graphs, With a foreword by H. S. M. Coxeter, With a biographical preface by Seymour Schuster, With an introduction by I. Z. Bouwer, W. W. Chernoff, B. Monson and Z. Star, Edited and with a note by Bouwer
13. Fouquet, J.L., Hahn, G.: Cycle regular graphs need not be transitive. Discret. Appl. Math. **113**(2–3), 261–264 (2001). https://doi.org/10.1016/S0166-218X(00)00292-4
14. Horvat, B., Pisanski, T., Žitnik, A.: Isomorphism checking of $I$-graphs. Graphs Combin. **28**(6), 823–830 (2012). https://doi.org/10.1007/s00373-011-1086-2
15. Hsieh, S.Y., Kuo, C.N.: Hamiltonian-connectivity and strongly Hamiltonian-laceability of folded hypercubes. Comput. Math. Appl. **53**(7), 1040–1044 (2007). https://doi.org/10.1016/j.camwa.2006.10.033
16. Klein, Z.J.: Structural properties of $I$-graphs: their independence numbers and Cayley graphs. Ph.D. thesis, Naval Postgraduate School, Monterey, CA (2020)
17. Klobas, N., Krnc, M.: Fast recognition of some parametric graph families. arXiv e-prints arXiv:2008.08856, August 2020
18. Kratochvíl, J.: String graphs. II. Recognizing string graphs is NP-hard. J. Combin. Theory Ser. B **52**(1), 67–78 (1991). https://doi.org/10.1016/0095-8956(91)90091-W

19. Krnc, M., Wilson, R.J.: Recognizing generalized Petersen graphs in linear time. Discret. Appl. Math. **283**, 756–761 (2020). https://doi.org/10.1016/j.dam.2020.03.007
20. Kutnar, K., Petecki, P.: On automorphisms and structural properties of double generalized Petersen graphs. Discret. Math. **339**(12), 2861–2870 (2016). https://doi.org/10.1016/j.disc.2016.05.032
21. Latifi, S., El-Amawy, A.: Properties and performance of folded hypercubes. IEEE Trans. Parallel Distrib. Syst. **2**(01), 31–42 (1991). https://doi.org/10.1109/71.80187
22. Mirafzal, S.M.: Some other algebraic properties of folded hypercubes. Ars Combin. **124**, 153–159 (2016)
23. Mollard, M.: Cycle-regular graphs. Discret. Math. **89**(1), 29–41 (1991)
24. Mulder, M.: (0, $\lambda$)-graphs and $n$-cubes. Discret. Math. **28**(2), 179–188 (1979). https://doi.org/10.1016/0012-365X(79)90095-5
25. Oliveira, A.S.S., Vinagre, C.T.M.: The spectrum of an $I$-graph (2015)
26. Petkovšek, M., Zakrajšek, H.: Enumeration of $I$-graphs: burnside does it again. Ars Math. Contemp. **2**(2), 241–262 (2009). https://doi.org/10.26493/1855-3974.113.3dc
27. Qin, Y., Xia, B., Zhou, S.: Canonical double covers of generalized Petersen graphs, and double generalized Petersen graphs. J. Graph Theory (2020). https://doi.org/10.1002/jgt.22642
28. Sakamoto, Y.: Hamilton cycles in double generalized Petersen graphs. Discuss. Math. Graph Theory **39**(1), 117–123 (2019). https://doi.org/10.7151/dmgt.2062
29. Soulignac, F.J., Sueiro, G.: NP-hardness of the recognition of coordinated graphs. Ann. Oper. Res. **169**, 17–34 (2009). https://doi.org/10.1007/s10479-008-0392-4
30. Terwilliger, P.: The classification of distance-regular graphs of type IIB. Combinatorica **8**(1), 125–132 (1988). https://doi.org/10.1007/BF02122560
31. Varvarigos, E.: Efficient routing algorithms for folded-cube networks. In: Proceedings International Phoenix Conference on Computers and Communications, pp. 143–151 (1995). https://doi.org/10.1109/PCCC.1995.472498
32. Watkins, M.E.: A theorem on Tait colorings with an application to the generalized Petersen graphs. J. Comb. Theory **6**, 152–164 (1969)
33. Xu, J.M., Ma, M.: Cycles in folded hypercubes. Appl. Math. Lett. **19**(2), 140–145 (2006). https://doi.org/10.1016/j.aml.2005.04.002
34. Xu, J.M., Ma, M., Du, Z.: Edge-fault-tolerant properties of hypercubes and folded hypercubes. Australas. J. Combin. **35**, 7–16 (2006)
35. Zhou, J.X., Feng, Y.Q.: Cubic vertex-transitive non-Cayley graphs of order $8p$. Electron. J. Combin. **19**(1), Paper 53, 13 (2012)
36. Zhou, J.X., Feng, Y.Q.: Cubic bi-Cayley graphs over abelian groups. Eur. J. Combin. **36**, 679–693 (2014). https://doi.org/10.1016/j.ejc.2013.10.005

# Graph Theory and Applications

# The Restrained Domination and Independent Restrained Domination in Extending Supergrid Graphs

Ruo-Wei Hung(✉) and Ming-Jung Chiu

Department of Computer Science and Information Engineering, Chaoyang University of Technology, Wufeng, Taichung 413310, Taiwan
rwhung@cyut.edu.tw

**Abstract.** Let $G$ be a graph with vertex set $V(G)$ and edge set $E(G)$. A set $D \subseteq V(G)$ is a dominating set of $G$ if every vertex not in $D$ is adjacent to at least one vertex of $D$. A restrained dominating set of $G$ is a dominating set $S$ such that every vertex not in $S$ is adjacent to another vertex in $V(G) - S$. An independent restrained dominating set of $G$ is a restrained dominating set such that it is also an independent set. The domination (resp., restrained domination, and independent restrained domination) number of $G$, denoted by $\gamma(G)$ (resp., $\gamma_{\mathrm{r}}(G)$, and $\gamma_{\mathrm{ir}}(G)$) is the minimum cardinality of a dominating (resp., a restrained dominating, and an independent restrained dominating) set of $G$. The domination (resp., restrained domination, and independent restrained domination) problem on a graph $G$ is to compute a dominating (resp., a restrained dominating, and an independent restrained dominating) set of $G$ with size $\gamma(G)$ (resp., $\gamma_{\mathrm{r}}(G)$, and $\gamma_{\mathrm{ir}}(G)$). Extending supergrid graphs are a natural extension of grid graphs. They are first appeared here and contain grid, supergrid, triangular supergrid, and diagonal supergrid graphs as subclasses. The domination problem on grid graphs was known to be NP-complete, and hence it is NP-complete for extending supergrid graphs. However, the complexities of the restrained and independent restrained domination problems on (extending) supergrid graphs are still unknown. In this paper, we will prove these two problems to be NP-complete for diagonal supergrid graphs, and hence they are NP-complete for extending supergrid graphs. These results can be easily applied to supergrid graphs. Then, we compute $\gamma_{\mathrm{r}}(R_{m\times n})$ and $\gamma_{\mathrm{ir}}(R_{m\times n})$, and verify that $\gamma_{\mathrm{r}}(R_{m\times n}) = \gamma_{\mathrm{ir}}(R_{m\times n}) = \gamma(R_{m\times n})$ for rectangular supergrid graph $R_{m\times n}$ which form a subclass of diagonal supergrid graphs excluding paths.

**Keywords:** Restrained domination · Independent restrained domination · Domination · Supergrid graph · Extending supergrid graph · Diagonal supergrid graph · Rectangular supergrid graph

Supported by Ministry of Science and Technology, Taiwan under grant no. MOST 110-2221-E-324-007-MY3.


C.-Y. Chen et al. (Eds.): COCOON 2021, LNCS 13025, pp. 401–412, 2021.
https://doi.org/10.1007/978-3-030-89543-3_34

## 1 Introduction

For two sets $A$ and $B$, let $A - B$ denote the set of elements in $A$ that are not in $B$. Let $G$ be a graph. We will use $V(G)$ and $E(G)$ to denote the vertex set and edge set of $G$, respectively. Let $v \in V(G)$ and let $S \subseteq V(G)$. The subgraph induced by $S$ is represented as $G[S]$. The *open neighborhood* of vertex $v$ is denoted by $N_G(v) = \{u \in V(G) | (u, v) \in E(G)\}$, while its *closed neighborhood* is given by $N_G[v] = N_G(v) \cup \{v\}$. In general, let $N_G(S)$ and $N_G[S]$ denote $\cup_{v \in S} N_G(v)$ and $\cup_{v \in S} N_G[v]$, respectively. The *degree* of vertex $v$ in $G$, denoted by $deg_G(v)$, is the number of vertices adjacent to $v$. A vertex $v$ of $G$ is called a *leaf* if $deg_G(v) = 1$. A set $S \subseteq V(G)$ is called an *independent set* if any two distinct vertices in $S$ are not adjacent. In addition, if two edges contain no common vertex, then they are called *independent edges.* Let $D \subseteq V(G)$. Set $D$ *dominates* vertex $v$ if $N_G[v] \cap D \neq \emptyset$. If $D$ dominates all vertices of a subset $S \subseteq V(G)$, then we say that $D$ *dominates* $S$. Set $D$ is called a *dominating set* of $G$ if and only if $D$ dominates $V(G)$; that is, every vertex not in $D$ is adjacent to one vertex in $D$. The *domination number* $\gamma(G)$ is the minimum cardinality of a dominating set of $G$. A *minimum dominating set* of $G$ is a dominating set with size $\gamma(G)$. The *domination problem* is to find a minimum dominating set of $G$, and it is well-known to be NP-complete for general graphs [4].

Variations of the domination problem seek to find a minimum dominating set with some additional properties, e.g., to be independent or to induce a connected graph. These problems arise in a number of distributed network applications, where the problem is to locate the smallest number of centers in networks such that every vertex is nearby at least one center. The concepts of domination and its variations have many applications and have been widely studied in literature (see [7,8]); a rough estimate says that it occurs in more than 6000 papers to date. In this paper, we will study two variants of the domination problem, namely *restrained domination* and *independent restrained domination problems.*

A set $S \subseteq V(G)$ is a *restrained dominating set* if every vertex in $V(G) - S$ is adjacent to a vertex in $S$ and another vertex in $V(G) - S$, i.e., $S$ is a dominating set and $G[V(G) - S]$ contains no isolated vertex. The concept of restrained domination was introduced by Telle and Proskurowski [17] in 1997. Note that every graph has a restrained dominating set since $S = V(G)$ is such a set. Let $\gamma_r(G)$ denote the size of a smallest restrained dominating set of $G$. A restrained dominating set of $G$ is called *minimum* if its size equals to $\gamma_r(G)$. The *restrained domination problem* is to compute a minimum restrained dominating set of a graph. This problem was known to be NP-complete for bipartite graphs, chordal graphs [3], and so on. In this paper, we will study the restrained domination problem and its one variant, namely independent restrained domination. The *independent restrained domination problem* is initially studied in [16]. A restrained dominating set of $G$ is called *independent* if it is an independent set. The *independent restrained domination number* of $G$, denoted by $\gamma_{ir}(G)$, is the minimum cardinality of an independent restrained dominating set of $G$. The *independent restrained domination problem* is to find an independent restrained dominating set of a graph $G$ with size $\gamma_{ir}(G)$.

The concept of *restrained domination* can be applied to the prisoners and guards system [17]. In this system, guards are to monitor the prisoners. Here, every vertex in the restrained dominating set corresponds to a position of a guard, and each vertex not in the restrained dominating set corresponds to a position of a prisoner. Each prisoner is observed by a guard for security while each prisoner is seen by at least one other prisoner for protecting the rights of prisoners. To minimize the cost, it is desirable to place as few guards as possible. Of course, it is possible to probe other applications of the restrained domination concept on the other fields, such as the sensor network monitoring system. A possible application for *independent restrained dominating set* $S$ is the location of product distribution centers or hospitals where a certain level of redundancy is desired. In this case, vertices in $S$ represent cities with a distribution center and edges represent transportation routes between cities. Selecting the cities in which to place centers using an independent restrained dominating set guarantees that every city without a distribution center is at least next to a city with one. It also guarantees that every city without a distribution center has a neighbor that also lacks a distribution center. In case of shortages at one distribution center, every city has access to a different center by going through one of its neighbors.

The *two-dimensional integer grid* $G^\infty$ is an infinite graph whose vertex set consists of all points of the Euclidean plane with integer coordinates and in which two vertices are adjacent if and only if the (Euclidean) distance between them is equal to 1. A *grid graph* is a finite, vertex-induced subgraph of $G^\infty$. For a node $v$ in the plane with integer coordinates, let $v_x$ and $v_y$ represent the $x$ and $y$ coordinates of node $v$, respectively, denoted by $v = (v_x, v_y)$. If $v$ is a vertex in a grid graph, then its possible adjacent vertices include $(v_x, v_y - 1)$, $(v_x - 1, v_y)$, $(v_x + 1, v_y)$, and $(v_x, v_y + 1)$. The *two-dimensional integer supergrid* $S^\infty$ is an infinite graph whose vertex set consists of all points of the plane with integer coordinates and in which two vertices are adjacent if and only if the difference of their $x$ or $y$ coordinates is not larger than 1. A *supergrid graph* is a finite, vertex-induced subgraph of $S^\infty$, and an *extending supergrid graph* is a finite and connected subgraph of $S^\infty$. Notice that a supergrid graph is also called *original* supergrid graph, and extending supergrid graphs, which are first appeared here, are not necessary to be vertex-induced subgraphs of $S^\infty$. Thus, supergrid graphs form a subclass of extending supergrid graphs, i.e., any supergrid graph is an extending supergrid graph but the reverse is not true. For a vertex $v$ in (extending) supergrid graph $G_s$, $N_{G_s}(v) \subseteq \{(v_x, v_y - 1), (v_x - 1, v_y), (v_x + 1, v_y), (v_x, v_y + 1), (v_x - 1, v_y - 1), (v_x + 1, v_y + 1), (v_x + 1, v_y - 1), (v_x - 1, v_y + 1)\}$ In the figures, we will assume that $(1, 1)$ is the coordinates of the most upper-left vertex of a grid or (extending) supergrid graph. Let $(u, v)$ be an edge of an extending supergrid graph with $u_x \leqslant v_x$. The edge $(u, v)$ is called *horizontal* (resp., *vertical*) if $u_y = v_y$ (resp., $u_x = v_x$), and is said to be *diagonal* if it is neither a horizontal nor a vertical edge. A diagonal edge $(u, v)$ is called $l$-type if $u_x = v_x - 1$ and $u_y = v_y - 1$, and is called $r$-type otherwise. A grid graph is an extending supergrid graph without any diagonal edge. Thus, grid graphs form a subclass of extending supergrid graphs. A *diagonal supergrid graph* is an

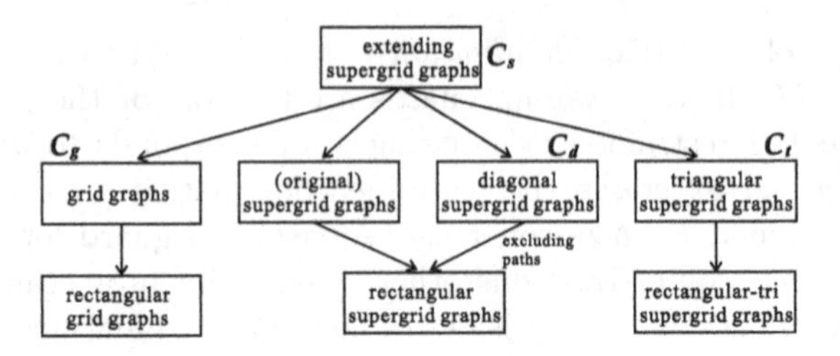


**Fig. 1.** The containment relations among the classes of grid, diagonal supergrid, triangular supergrid, and extending supergrid graphs, where $\mathcal{C} \rightarrow \mathcal{C}'$ indicates $\mathcal{C}'$ is a subclass of $\mathcal{C}$.

extending supergrid graph whose edge set contains at least one $l$-type and one $r$-type diagonal edges, and a *triangular supergrid graph* is an extending supergrid graph whose edge set contains at least one $l$-type diagonal edge and contains no $r$-type diagonal edge. Let $\mathcal{C}_s$ be the class of extending supergrid graphs, $\mathcal{C}_g$ be the class of grid graphs, $\mathcal{C}_d$ be the class of diagonal supergrid graphs, and let $\mathcal{C}_t$ be the class of triangular supergrid graphs. Then, $\mathcal{C}_g \subset \mathcal{C}_s$, $\mathcal{C}_d \subset \mathcal{C}_s$, $\mathcal{C}_t \subset \mathcal{C}_s$, and $\mathcal{C}_g \cap \mathcal{C}_d = \mathcal{C}_g \cap \mathcal{C}_t = \mathcal{C}_d \cap \mathcal{C}_t = \emptyset$. Figure 1 shows the relationship among these four graph classes. Obviously, all grid graphs are bipartite [15] but (extending, diagonal, triangular) supergrid graphs may not be bipartite. In general, a supergrid graph may not be a diagonal or triangular supergrid graph, and the reverse is also true.

A *rectangular grid* (or called *complete grid*) graph $G_{m\times n}$ has $mn$ nodes with vertex $u = (u_x, u_y)$ adjacent to $v = (v_x, v_y)$ if and only if $|u_x - v_x| + |u_y - v_y| = 1$. A *rectangular supergrid* graph $R_{m\times n}$ is a special supergrid graph with vertex set $\{v = (v_x, v_y) | 1 \leqslant v_x \leqslant n \text{ and } 1 \leqslant v_y \leqslant m\}$ and edge set $\{(u, v) | 0 \leqslant |u_x - v_x| \leqslant 1 \text{ and } 0 \leqslant |u_y - v_y| \leqslant 1\}$. Thus, for $\mu \in V(G_{m\times n})$ and $v \in V(R_{m\times n})$, $1 \leqslant deg_{G_{m\times n}}(\mu) \leqslant 4$ and $1 \leqslant deg_{R_{m\times n}}(v) \leqslant 8$.

Previous related works are summarized below. The domination problem on grid graphs has been shown to be NP-complete [2]. Many authors studied the domination numbers of rectangular grid graphs [1,5,6]. Gonçalves *et al.* computed $\gamma(G_{m\times n}) = \lfloor \frac{(m+2)(n+2)}{5} \rfloor - 4$ for $n \geqslant m \geqslant 16$ [5]. In [3], Domke *et al.* computed $\gamma_{\mathrm{r}}(P_n) = n - 2 \cdot \lfloor \frac{n-1}{3} \rfloor$ and $\gamma_{\mathrm{r}}(C_n) = n - 2 \cdot \lfloor \frac{n}{3} \rfloor$, where $P_n$ and $C_n$ are a path and a cycle with $n$ vertices, respectively. Supergrid graphs were first appeared in [9], in which the Hamiltonian problems on supergrid graphs were proved to be NP-complete, and every rectangular supergrid graph contains a Hamiltonian cycle. In [10], Hung proved that linear-convex supergrid graphs always contain Hamiltonian cycles. In 2017, Hung *et al.* proved that rectangular supergrid graphs (with one trivial exception) are always Hamiltonian connected [11]. Recently, we verified the Hamiltonicity and Hamiltonian connectivity of some shaped supergrid graphs, including triangular, parallelogram, and trapezoid, and alphabet [12,13]. In [14], we proved the domination and independent domination problems on supergrid graphs to be NP-complete. In this paper, we will prove the restrained domination problem on diagonal supergrid graphs to be

NP-complete. This result can be easily extended to the independent restrained domination problem. Thus, these two problems on extending supergrid graphs are also NP-complete. By a simple modification, these two problems can be easily verified to be NP-complete for supergrid graphs. Then, we compute $\gamma_{\rm r}(R_{m\times n})$ and $\gamma_{\rm ir}(R_{m\times n})$ for rectangular supergrid graph $R_{m\times n}$.

The paper is structured as follows. In Sect. 2, some notations and definitions are introduced. In Sect. 3, we prove that the restrained domination and independent restrained domination problems on diagonal supergrid graphs are NP-complete, and hence they are NP-complete for extending supergrid graphs. These NP-complete results can be easily applied to supergrid graphs. Section 4 computes $\gamma_{\rm r}(R_{m\times n})$ and $\gamma_{\rm ir}(R_{m\times n})$. Finally, we make some concluding remarks in Sect. 5.

## 2 Notations

In this section, we will introduce some notations used in this paper. A path in a graph $G$ is a sequence $(v_1, v_2, \ldots, v_{k-1}, v_k)$ of adjacent vertices starting from $v_1$ and ending at $v_k$ and is denoted by $(v_1, v_k)$-path, where all the vertices $v_1, v_2, \ldots, v_k$ are distinct except that possibly the path is a cycle when $v_1 = v_k$. In general, a path with $n$ vertices is denoted by $P_n$ if no ambiguity appears.

The *two-dimensional supergrid graph* $S^\infty$ is the infinite graph whose vertex set consists of all points of the plane with integer coordinates and in which two vertices are adjacent if the difference of their $x$ or $y$ coordinates is not larger than 1. A supergid graph is a finite, vertex-induced subgraph of $S^\infty$, and an extending supergrid graph $G_s$ is a finite and connected graph such that $V(G_s) \subset V(S^\infty)$ and $E(G_s) \subset E(S^\infty)$. For a vertex $v \in V(G_s)$, it is represented as $(v_x, v_y)$, where $v_x$ and $v_y$ are the $x$ and $y$ coordinates of $v$ respectively. Then, $1 \leqslant deg_{G_s}(v) \leqslant 8$. A diagonal supergrid graph is an extending supergrid graph such that it contains at least one $l$-type and one $r$-type diagonal edges. Then, diagonal supergrid graphs form a subcalss of extending supergrid graphs. However, a diagonal supergrid graph is not necessary to be a supergrid graph, and the reverse is also true.

Rectangular supergrid graphs first appeared in [9], in which the Hamiltonian cycle problem is solved in linear time. A rectangular supergrid graph $R_{m\times n}$ is a supergrid graph with vertex set $V(R_{m\times n}) = \{v = (v_x, v_y) | 1 \leqslant v_x \leqslant n$ and $1 \leqslant v_y \leqslant m\}$ and edge set $E(R_{m\times n}) = \{(u, v) | 0 \leqslant |u_x - v_x| \leqslant 1$ and $0 \leqslant |u_y - v_y| \leqslant 1\}$. In this paper, without loss of generality we will assume that $m \leqslant n$ for $R_{m\times n}$. Let $v$ be a vertex in $R_{m\times n}$ with $m \geqslant 2$. The vertex $v$ is called a *corner* of $R_{m\times n}$ if $deg_{R_{m\times n}}(v) = 3$. There are four corners of $R(m, n)$ including *upper-left*, *upper-right*, *down-left*, and *down-right* corners coordinated as $(1,1)$, $(n,1)$, $(1,m)$, and $(n,m)$, respectively (see Fig. 2). Note that we will assume that $(1,1)$ are coordinates of the upper-left corner of $R(m,n)$, except we explicitly change this assumption. For example, Fig. 2 shows a rectangular supergrid graph $R_{8\times 10}$ and it also indicates the types of edges and corners. Note that a grid graph contains horizontal and vertical edges, but it contains

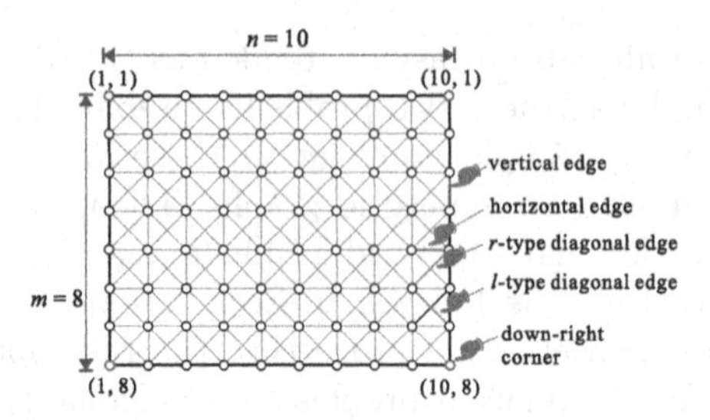


**Fig. 2.** A rectangular supergrid graph $R_{8\times10}$, where bold dashed lines indicate vertical and horizontal separations.

no diagonal edge. In addition, rectangular supergrid graphs form a subclass of diagonal supergrid graphs except paths.

## 3 NP-Completeness Results

In this section, we will prove that the restrained and independent restrained domination problems for diagonal supergrid graphs are NP-complete. In [14], the domination and independent domination problems on (original) supergrid graphs have been proved to be NP-complete. We will apply the similar concept to show that the restrained domination problem is NP-complete for diagonal supergrid graphs. By a simple extension, the independent restrained domination problem on diagonal supergrid graphs can be easily verified to be NP-complete. Then, these two problems are also NP-complete for extending supergrid graphs. To prove it, we establish a polynomial-time reduction from the domination problem on grid graphs. In [2], Clark *et al.* proved that the domination problem on grid graphs is NP-complete.

**Theorem 1** *(See* [2]*). The domination problem on grid graphs is NP-complete.*

Given a grid graph $G_g$, we will construct a diagonal supergrid graph $G_d$ such that $G_g$ has a dominating set $D$ with size $|D| \leqslant k$ if and only if $G_d$ contains a restrained dominating set $D'$ with size $|D'| \leqslant k + 2|E(G_g)|$. The construction of $G_d$ from $G_g$ is sketched as follows. First, we enlarge the input grid graph $G_g$ such that each edge of $G_g$ is transformed into a path $P_8$ with 7 edges; i.e., enlarge each edge of $G_g$ by 7 times. Let the enlarged grid graph be $G'_g$. For example, Fig. 3(b) shows grid graph $G'_g$ enlarged from grid graph $G_g$ in Fig. 3(a). In the second step, each enlarged path of graph $G'_g$ is replaced by a connected component which is a small diagonal supergrid graph. The connected component connecting $u$ and $v$ is called a *triangle* $(u, v)$*-tentacle*, denoted by $T(u, v)$, where $u, v \in V(G_g)$ and are called *connectors* of $T(u, v)$. Figure 3(c) depicts a triangle $(u, v)$-tentacle. Then, for each leaf $v$ of $G_g$ we replace it with a triangle containing $v$, as shown in Fig. 3(d), where $deg_{G_g}(v) = 1$. Finally, the constructed graph is a diagonal supergrid graph $G_d$ whose edge set contains at least one $l$-diagonal

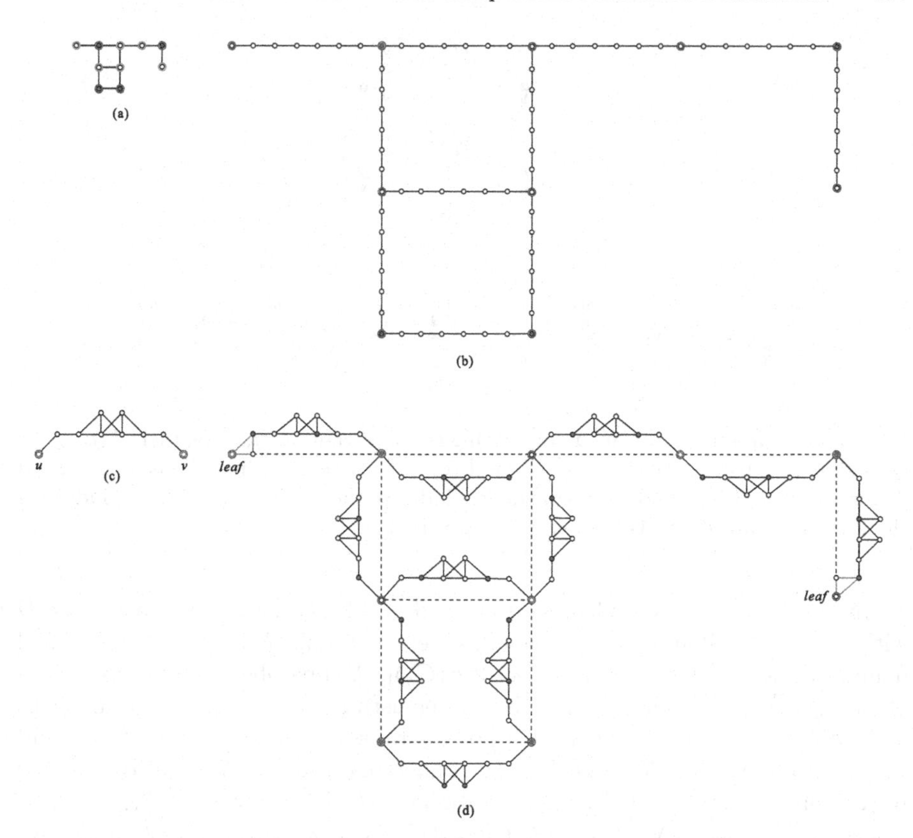


**Fig. 3.** (a) A grid graph $G_g$, (b) a grid graph $G'_g$ by enlarging each edge of $G_g$ 7 times, (c) a triangle $(u, v)$-tentacle $T(u, v)$ to replace the enlarged $(u, v)$-path of $G'_g$, and (d) a constructed diagonal supergrid graph $G_d$ obtained from $G'_g$ by replacing each enlarged path with the triangle tentacle, and by replacing each leaf with a triangle, where solid lines indicate the edges of $G_g$ and $G_d$, double circles represent the vertices of $G_g$, and solid circles indicate the vertices in a dominating set (resp., restrained dominating set) of $G_g$ (resp., $G_d$).

and $r$-diagonal edges. For example, Fig. 3(d) shows a diagonal supergrid graph $G_d$ constructed from grid graph $G_g$ in Fig. 3(a). The above construction is called Algorithm Construct-DiSupergrid and can be easily done in polynomial time.

In [14], we have provided a rule to arrange these triangle tentacles of $G_d$ such that they are disjoint except their connectors. Due to space limitation, we omit the arrangement of triangle tentacles. The arrangement rule is called *Rule AT*. Clearly, Algorithm Construct-DiSupergrid, together with Rule AT, can be done in polynomial time. Thus, the following lemma holds true.

**Lemma 1.** *Given a grid graph $G_g$, Algorithm Construct-DiSupergrid constructs a diagonal supergrid graph $G_d$ in polynomial time.*

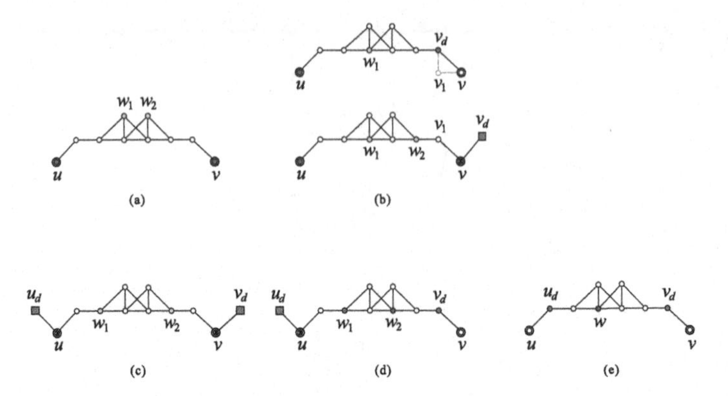


**Fig. 4.** The minimum restrained dominating set of a triangle $(u, v)$-tentacle $T(u, v)$ for (a) $u, v \in D'$, (b) $u \in D'$ and $v \notin D'$, and (c)–(e) $u, v \notin D'$, where $D'$ is a restrained dominating set of $G_d$, solid circles or squares indicate the vertices in $D'$, and $\otimes$ indicates the vertex not in $D'$ and is dominated by a vertex in $D' - T(u, v)$.

In the following, we will prove that grid graph $G_g$ has a dominating set $D$ with size $|D| \leqslant k$ if and only if diagonal supergrid graph $G_d$ contains a restrained dominating set $D'$ with size $|D'| \leqslant k+2|E(G_g)|$. We first observe some properties of triangle tentacles. These properties will be used in proving the above sufficient and necessary conditions. Let $D'$ be a restrained dominating set of $G_d$, and let $T(u, v)$ be a triangle tentacle with connectors $u$ and $v$. We will denote the restriction of $D'$ to $H$ by $D'_{\|H}$ for a subgraph $H$ of $G_d$. Let $u_d \in N_{G_d}(u) \cap D'$ if $u \notin D'$, and $v_d \in N_{G_d}(v) \cap D'$ if $v \notin D'$. Then, $u_d$ and $v_d$ dominate $u$ and $v$, respectively, if $u, v \notin D'$. Depending on whether $u$, $v$ are in $D'$, we consider three cases: (1) $u, v \in D'$, (2) $u \in D'$ and $v \notin D'$, and (3) $u, v \notin D'$. Suppose that $u, v \in D'$. Since $D'$ is a restrained dominating set of $G_d$, $G_d[T(u, v) - D']$ contains no isolated vertex. To dominate $T(u, v) - (N_{G_d}[u] \cup N_{G_d}[v])$, it needs at least two vertices. Thus, $|D'_{\|T(u,v)}| \geqslant 4$ (see Fig. 4(a)). Let $W = \{w_1, w_2\}$ in $T(u, v) - \{u, v\}$, as depicted in Fig. 4(a). Then, $W \cup \{u, v\}$ is a restrained dominating set of $T(u, v)$. Thus, $|D'_{\|T(u,v)-\{u,v\}}| \geqslant 2$. Due to space limitation, we omit the proofs of the other cases and they can be observed in Figs. 4(b)–(e). The following lemma summarizes these properties.

**Lemma 2.** *Let $D'$ be a restrained dominating set of $G_d$, and let $T(u, v)$ be a triangle tentacle with connectors $u$ and $v$. Then, the following statements hold true:*

(1) *If $u, v \in D'$, then $|D'_{\|T(u,v)-\{u,v\}}| \geqslant 2$ (see Fig. 4(a)).*
(2) *If $u \in D'$ and $v \notin D'$, then $|D'_{\|T(u,v)-\{u\}}| \geqslant 2$ and there exist vertices $v_d \in D'$ and $v_1 \notin D'$ such that $v_d$ dominates $v$ and $v$ is adjacent to $v_1$ (see Fig. 4(b)).*
(3) *If $u, v \notin D'$, then*
(3-1) *if $(N_{G_d}(u) - T(u, v)) \cap D' \neq \emptyset$ and $(N_{G_d}(v) - T(u, v)) \cap D' \neq \emptyset$, then $|D'_{\|T(u,v)}| \geqslant 2$ (see Fig. 4(c));*

(3-2) *if* $(N_{G_d}(u) - T(u,v)) \cap D' = \emptyset$ *or* $(N_{G_d}(v) - T(u,v)) \cap D' = \emptyset$, *then* $|D'_{\|T(u,v)}| \geqslant 3$ *(see Figs. 4(d)–(e)).*

In the following, we will prove that grid graph $G_g$ has a dominating set $D$ with size $|D| \leqslant k$ if and only if diagonal supergrid graph $G_d$ contains a restrained dominating set $D'$ with size $|D'| \leqslant k + 2|E(G_g)|$. We first prove the only if part in Lemma 3. Due to space limitation, we omit its proof.

**Lemma 3.** *Assume that grid graph $G_g$ has a dominating set $D$ with size $|D| \leqslant k$. Then, diagonal supergrid graph $G_d$ contains a restrained dominating set $D'$ with size $|D'| \leqslant k + 2|E(G_g)|$.*

In our proof of the above lemma, we will compute two vertices, not connectors, of each triangle tentacle in $G_d$ by using Lemma 2 (see Fig. 4) such that they are in $D'$. For example, Fig. 3(a) shows a dominating set $D$ of $G_g$ with size 4. The constructed restrained dominating set $D'$ of $G_d$ with size $4 + 2 \times 11 = 26$ is depicted in Fig. 3(d).

Next, we will prove the if part in Lemma 4. Due to space constraints, the details of its proof are omitted.

**Lemma 4.** *Assume that diagonal supergrid graph $G_d$ has a restrained dominating set $D'$ with size $|D'| \leqslant k + 2|E(G_g)|$. Then, grid graph $G_g$ contains a dominating set $D$ with size $|D| \leqslant k$.*

In our proof of the above lemma, we first construct a restrained dominating set $\hat{D}$ of $G_d$ obtained from $D'$ to satisfy the following properties:

(p1) $|\hat{D}| \leqslant |D'|$,
(p2) for each triangle $(u,v)$-tentacle $T(u,v)$, $|\hat{D} \cap (T(u,v) - \{u,v\})| = 2$,
(p3) for each triangle $(u,v)$-tentacle $T(u,v)$ with $u, v \notin \hat{D}$, there exist $z_1 \in N_{G_g}(u)$ and $z_2 \in N_{G_g}(v)$ such that $z_1, z_2 \in \hat{D}$ while $deg_{G_g}(u) \neq 1$ and $deg_{G_g}(v) \neq 1$, and
(p4) for each triangle $(u,v)$-tentacle $T(u,v)$ with $deg_{G_g}(v) = 1$ and $v \notin \hat{D}$, $u \notin \hat{D}$.

A dominating set $D$ of $G_g$ is then constructed from $\hat{D}$ as follows: (1) initially, let $D = \hat{D}$; (2) remove all vertices of $D$ not in $D_g$ from $D$; and (3) the resultant set $D$ will be a dominating set of $G_g$.

Since $\hat{D} \cap T(u,v)$ contains exactly two vertices not in $G_g$ for each triangle tentacle $T(u,v)$, we get that $|D| = |\hat{D}| - 2|E(G_g)|$. Then, $|D| = |\hat{D}| - 2|E(G_g)| \leqslant |D'| - 2|E(G_g)| \leqslant (k + 2|E(G_g)|) - 2|E(G_g)| = k$. Thus, we construct a dominating set $D$ of $G_g$ with size $|D| \leqslant k$.

Combining Lemmas 3 and 4, we summarize the following lemma:

**Lemma 5.** *Let $G_g$ be a grid graph and let $G_d$ be the diagonal supergrid graph constructed from $G_g$ by Algorithm Construct-DiSupergrid and Rule AT. Then, $G_g$ has a dominating set $D$ with size $|D| \leqslant k$ if and only if $G_d$ contains a restrained dominating set $D'$ with $|D'| \leqslant k + 2|E(G_g)|$.*

Obviously, the restrained domination problem for diagonal supergrid graphs is in NP. By Theorem 1, Lemma 1, and Lemma 5, we conclude the following theorem:

**Theorem 2.** *The restrained domination problem on diagonal supergrid graphs is NP-complete.*

A restrained dominating set of the supergrid graphs constructed in the proofs of Lemmas 3 and 4 is also an independent set (see Fig. 3(d) and Fig. 4). Thus, the independent restrained domination problem on diagonal supergrid graphs is also NP-complete and hence the following theorem holds.

**Theorem 3.** *The independent restrained domination problem on diagonal supergrid graphs is NP-complete.*

By Fig. 1 and Theorems 2–3, the restrained domination and independent restrained domination problems on extending supergrid graphs are NP-complete. By a simple modification of triangle tentacle, we can obtain a vertex-induced subgraph $G_d$ of $S^\infty$. The related properties of the modified tentacle are the same as those of original triangle tentacle (ref. Lemma 2). We then have the following corollary:

**Corollary 1.** *The restrained domination and independent restrained domination problems on (original) supergrid graphs are NP-complete.*

## 4 The Restrained and Independent Restrained Domination Number of Rectangular Supergrid Graphs

In this section, we will compute $\gamma_r(R_{m\times n})$, the restrained domination number, and $\gamma_{ir}(R_{m\times n})$, the independent restrained domination number, for rectangular supergrid graph $R_{m\times n}$. By symmetry, without loss of generality we will assume that $n \geqslant m$ for $R_{m\times n}$. We first compute the restrained domination number $\gamma_r(R_{m\times n})$ of $R_{m\times n}$. In [3], Domke *et al.* computed $\gamma_r(R_{1\times n}) = \gamma_r(P_n)$, where $P_n$ is a path with $n$ vertices, as follows.

**Lemma 6** *(see [3]).* $\gamma_r(R_{1\times n}) = \gamma_r(P_n) = n - 2 \cdot \lfloor \frac{n-1}{3} \rfloor$.

Next, we consider $R_{m\times n}$ with $m \geqslant 2$. Since a restrained dominating set of a graph $G$ is also a dominating set of $G$, $\gamma(G) \leqslant \gamma_r(G)$ for any graph $G$. In [14], we computed $\gamma(R_{m\times n}) = \lceil \frac{m}{3} \rceil \lceil \frac{n}{3} \rceil$. Since $\gamma(R_{m\times n}) \leqslant \gamma_r(R_{m\times n})$, $\lceil \frac{m}{3} \rceil \lceil \frac{n}{3} \rceil$ is a lower bound of $\gamma_r(R_{m\times n})$ for $m \geqslant 2$. That is, $\lceil \frac{m}{3} \rceil \lceil \frac{n}{3} \rceil \leqslant \gamma_r(R_{m\times n})$ for $n \geqslant m \geqslant 2$. The following lemma shows that $\gamma_r(R_{2\times n}) = \gamma_r(R_{3\times n}) = \lceil \frac{n}{3} \rceil$ by constructing its upper bound. Due to space limitation, we omit its proof.

**Lemma 7.** $\gamma_r(R_{2\times n}) = \gamma_r(R_{3\times n}) = \lceil \frac{n}{3} \rceil$.

Now, we consider $R_{m\times n}$ with $n \geqslant m \geqslant 3$ in the following lemma. Due to space constraints, we omit its proof.

**Lemma 8.** $\gamma_{\rm r}(R_{m\times n}) = \lceil \frac{m}{3} \rceil \lceil \frac{n}{3} \rceil$ *for* $n \geqslant m \geqslant 3$.

It immediately follows from Lemmas 6–8 that the following theorem holds true:

**Theorem 4.** *Let* $R_{m\times n}$ *be a rectangular supergrid graph with* $n \geqslant m \geqslant 1$. *Then,*

$$\gamma_{\rm r}(R_{m\times n}) = \begin{cases} n - 2 \cdot \lfloor \frac{n-1}{3} \rfloor, & \text{if } m = 1; \\ \lceil \frac{m}{3} \rceil \cdot \lceil \frac{n}{3} \rceil, & \text{otherwise.} \end{cases}$$

For the independent restrained domination number of $R_{1\times n}$, we can easily see that $\gamma_{\rm ir}(R_{1\times n})$ may not exist. For example, there exists no independent restrained dominating set in $R_{1\times 5}$. Consider that $R_{m\times n}$ with $n \geqslant m \geqslant 2$. For the restrained dominating set $D_{\rm r}$ of $R_{m\times n}$ constructed in our proofs, $D_{\rm r}$ is also an independent set. Thus, $D_{\rm r}$ is also an independent restrained dominating set of $R_{m\times n}$. Then, $\gamma_{\rm ir}(R_{m\times n}) = \lceil \frac{m}{3} \rceil \cdot \lceil \frac{n}{3} \rceil$, and hence the following theorem holds.

**Theorem 5.** *Let* $R_{m\times n}$ *be a rectangular supergrid graph with* $n \geqslant m \geqslant 2$. *Then,* $\gamma_{\rm ir}(R_{m\times n}) = \lceil \frac{m}{3} \rceil \lceil \frac{n}{3} \rceil$.

Let $\gamma_{\rm ind}(R_{m\times n})$ be the independent domination number of $R_{m\times n}$, where an independent dominating set of a graph $G$ is a dominating set and an independent set, and the size of a minimum independent dominating set of $G$ is called the independent domination number of $G$. In [14], we proved that $\gamma_{\rm ind}(R_{m\times n}) = \gamma(R_{m\times n})$. Based on the result in [14], together with Theorems 4 and 5, we conclude the following theorem:

**Theorem 6.** *Let* $R_{m\times n}$ *be a rectangular supergrid graph with* $n \geqslant m \geqslant 2$. *Then,* $\gamma(R_{m\times n}) = \gamma_{\rm ind}(R_{m\times n}) = \gamma_{\rm r}(R_{m\times n}) = \gamma_{\rm ir}(R_{m\times n}) = \lceil \frac{m}{3} \rceil \lceil \frac{n}{3} \rceil$.

## 5 Concluding Remarks

In this paper, we first prove that the restrained and independent restrained domination problems on diagonal supergrid graphs are NP-complete. These results can be applied to supergrid graphs, and hence they are NP-complete for supergrid graphs. Then, we solve these two problems on rectangular supergrid graphs in linear time. The restrained step domination problem is a variant of restrained domination problem. A restrained dominating set $S$ of a graph $G$ is called restrained step dominating set of $G$ if the induced subgraph by $V(G) - S$ has a perfect matching. The restrained step domination problem on $G$ is to compute a restrained step dominating set with minimum cardinality. For the restrained step domination problem on (diagonal) supergrid graphs, we conjecture that it is NP-complete. However, we can not verify it. We would like to post it as an open problem to interested readers.

## References

1. Chang, T.Y., Clark, W.E., Hare, E.O.: Domination numbers of complete grids I. Ars Comb. **38**, 97–112 (1994)
2. Clark, B.N., Colbourn, C.J., Johnson, D.S.: Unit disk graphs. Discret. Math. **86**, 165–177 (1990)
3. Domke, G.S., Hattingh, J.H., Hedetniemi, S.T., Laskar, R.C., Markus, L.R.: Restrained domination in graphs. Discret. Math. **203**, 61–69 (1999)
4. Garey, M.R., Johnson, D.S.: Computers and Intractability: A Guide to the Theory of NP-Completeness. Freeman, San Francisco (1979)
5. Gonçalves, D., Pinlou, A., Rao, M., Thomassé, A.: The domination number of grids. SIAM J. Discret. Math. **25**, 1443–1453 (2011)
6. Guichard, D.R.: A lower bound for the domination number of complete grid graphs. J. Comb. Math. Comb. Comput. **49**, 215–220 (2004)
7. Haynes, T.W., Hedetniemi, S.T., Slater, P.J.: Fundamentals of Domination in Graphs. Marcel Dekker, New York (1998)
8. Haynes, T.W., Hedetniemi, S.T., Slater, P.J.: Domination in Graphs: Advanced Topics. Marcel Dekker, New York (1998)
9. Hung, R.W., Yao, C.C., Chan, S.J.: The Hamiltonian properties of supergrid graphs. Theor. Comput. Sci. **602**, 132–148 (2015)
10. Hung, R.W.: Hamiltonian cycles in linear-convex supergrid graphs. Discret. Appl. Math. **211**, 99–112 (2016)
11. Hung, R.W., Li, C.F., Chen, J.S., Su, Q.S.: The Hamiltonian connectivity of rectangular supergrid graphs. Discret. Optim. **26**, 41–65 (2017)
12. Hung, R.W., Chen, H.D., Zeng, S.C.: The Hamiltonicity and Hamiltonian connectivity of some shaped supergrid graphs. IAENG Int. J. Comput. Sci. **44**, 432–444 (2017)
13. Hung, R.W., Keshavarz-Kohjerdi, F., Lin, C.B., Chen, J.S.: The Hamiltonian connectivity of alphabet supergrid graphs. IAENG Int. J. Appl. Math. **49**, 69–85 (2019)
14. Hung, R.W., Chiu, M.J., Chen, J.S.: The domination and independent domination problems in supergrid graphs. In: The 21st International Conference on Computational Science and its Applications (ICCSA 2021), 13–16 September, University of Cagliari, Italy (2021, to appear)
15. Itai, A., Papadimitriou, C.H., Szwarcfiter, J.L.: Hamiltonian paths in grid graphs. SIAM J. Comput. **11**(4), 676–686 (1982)
16. Sivagnanam, C., Kulandaivel, M.P.: Extended results on restrained domination number and connectivity of a graph. Int. J. Math. Soft Comput. **5**, 183–187 (2015)
17. Telle, J.A., Proskurowski, A.: Algorithms for vertex partitioning problems on partial $k$-trees. SIAM J. Discret. Math. **10**, 529–550 (1997)

# The Concentration of the Maximum Degree in the Duplication-Divergence Models

Alan Frieze[1], Krzysztof Turowski[2(✉)], and Wojciech Szpankowski[2,3]

[1] Department of Mathematical Sciences, Carnegie Mellon University, Pittsburgh, PA, USA
alan@random.math.cmu.edu

[2] Department of Theoretical Computer Science, Jagiellonian University, Kraków, Poland
{krzysztof.turowski,wojciech.szpankowski}@uj.edu.pl

[3] Center for Science of Information, Department of Computer Science, Purdue University, West Lafayette, IN, USA
szpan@purdue.edu

**Abstract.** We pursue the analysis of the maximum degree in a dynamic duplication-divergence graph model defined by Solé et al. in which a new node arriving at time $t$ first randomly selects an existing node and connects to its neighbors with probability $p$, and then connects to the other nodes with probability $r/t$. This model is often said to capture the growth of some real-world processes e.g. biological or social networks. However, there are only a handful of rigorous results concerning this model. Here we study the distribution of the maximum degree of a vertex in graphs generated by this model.

In this paper we prove that for $\frac{1}{2} < p < 1$ with high probability the maximum degree is asymptotically quite surely concentrated around $t^p$, i.e. it deviates from this value by at most a polylogarithmic factor. Our findings are a step towards a better understanding of the overall structure of graphs generated by this model, especially the degree distribution, compression, and symmetry.

**Keywords:** Random graphs · Duplication-divergence model · Degree distribution · Maximum degree · Large deviation

## 1 Introduction

Studying structural properties of graphs (e.g., symmetry, compressibility, vertex degree) is a popular topic of research in computer science and discrete

This work was supported by NSF Center for Science of Information (CSoI) Grant CCF-0939370, by NSF Grants CCF-1524312, CCF-2006440, CCF-2007238, DMS1952285 and, in addition, by the National Science Center, Poland, Grant 2020/39/D/ST6/00419.


C.-Y. Chen et al. (Eds.): COCOON 2021, LNCS 13025, pp. 413–424, 2021.
https://doi.org/10.1007/978-3-030-89543-3_35

mathematics ever since the seminal work of Paul Erdős and Alfréd Rényi [8]. Recently attention has turned to dynamic graphs such as preferential attachment (Barabási-Albert) graphs [1], Watts-Strogatz small world graphs [25] or duplication-divergence graphs. Dynamic graphs, in which the edge- and/or vertex-sets are functions of time, are ubiquitous in diverse application domains ranging from biology to finance to social science. Deriving novel insights and knowledge from dynamic structures is a key challenge and understanding the structural properties of such dynamic graphs is critical for new characterizations and insights of the underlying dynamic processes.

Numerous networks in the real world change over time, in the sense that nodes and edges enter and leave the networks. To explain their macroscopic properties (e.g., subgraph frequencies, diameter, degree distribution, symmetry) and to make predictions and other inferences (such as community detection, graph compression, order of node arrivals), several generative models have been proposed [19,24]. Typically, one tries to capture the behavior of well-known graph parameters under probability distributions induced by the models, e.g. the distribution of the number of vertices with a given degree, the number of connected components, the existence of Hamiltonian paths or other parameters like clique number and chromatic number (see [3,9,13] for overviews of the main results in the area).

In this paper we make further progress on structural properties of the *duplication-divergence* graph models, in which vertices arrive one by one, select an existing node as a parent, connect to the some neighbors of its parent and other vertices according to some pre-defined rule. More precisely, a newly arriving node at time $t$ first selects randomly an existing node and connects to its neighbors with probability $p$; and then connects to other nodes with probability $r/t$. The particular model which we bring under consideration is a duplication-divergence model, first defined by Solé, Pastor-Satorras et al. [21]. It has been a popular object of study because it has been shown empirically that its degree distribution, small subgraph (graphlets) counts and number of symmetries fit very well with the structure of some real-world biological and social networks, e.g. protein-protein and citation networks [5,20,22]. This suggests a possible real-world significance for the duplication-divergence model, which further motivates the studies of its structural properties. However, it is also one of the least understood models, much less so than the Erdős-Rényi or preferential attachment models. At the moment there exist only a handful of results related to the behavior of the degree distribution of the graphs generated by this model. Unlike other dynamic graphs such as the preferential attachment model, the graphs generated by the duplication-divergence model can be very symmetric or quite asymmetric. In Fig. 1 from [22] it is shown that there exist certain ranges of the model parameters $p$ and $r$ such that the graphs generated from the model are highly symmetric, and certain ranges such that the graphs are asymmetric. Here the symmetry is measured by the size of the automorphism group $|\mathrm{Aut}(G)|$, i.e. the number of distinct mapping of vertices onto themselves preserving the adjacency matrix. Still the basic question about the conditions under which the

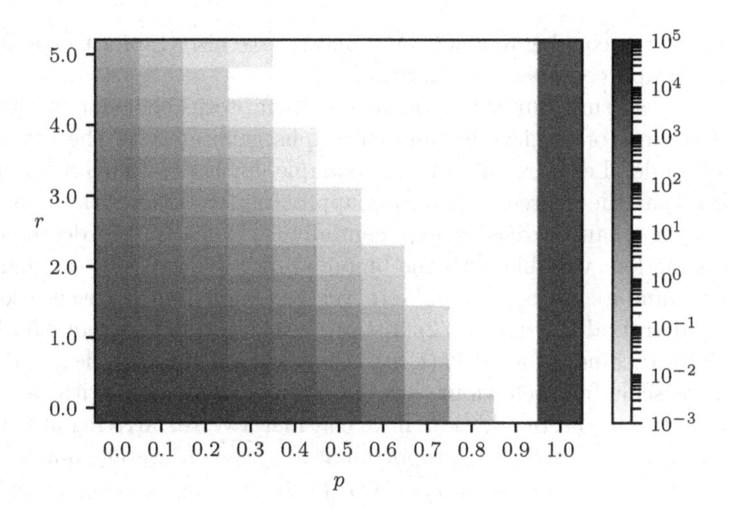


**Fig. 1.** Symmetry of graphs ($\log|\mathrm{Aut}(G)|$) generated by the Solé-Pastor-Satorras duplication-divergence model, based on simulations from [22].

generated graph is symmetric or not remains unanswered. We believe that proving results about the range of the maximum degree can be a stepping stone for rigorous general results regarding symmetries and compression, just as it has been in the case for other random graph models.

In particular, the parameters such as the maximum degree of a random graph and the degree of a given vertex are parameters that are studied not only for their own sake, but it turns out that their analysis opens the way to further results. Let us recall here two examples of these insights related to the questions of graph asymmetry and incompressibility.

First, Łuczak et al. [17] used the estimation of these parameters to prove that the preferential attachment model with $m \geq 3$ (where $m$ is the number of edges added when a new node arrives) generates asymmetric graphs (i.e. graphs with only one automorphism) with high probability. This was achieved by proving two properties: (A) for any pair of early vertices $t_1$ and $t_2$ the degrees of both nodes $t_1$ and $t_2$ are distinct, and (B) for any pair of late vertices their corresponding neighbors are not the same, in particular, they have different sets of early neighbors (and therefore, a permutation of $t_1$ and $t_2$ does not produce symmetry). We believe that this approach to asymmetry analysis can be extended to the duplication-divergence model and it requires knowledge of the maximum degree which is exactly the topic of this paper.

A second usage of these parameters was presented by Chierichetti et al. in [4]. For example, for the preferential attachment model they used an upper bound on the maximum degree and the degree of a vertex arriving at time $s$ to show that the entropy over all graphs on $t$ vertices generated by this model is bounded by $\Omega(t \log t)$. They also used their bound on vertex degrees to provide lower bounds on graph entropy for several other random graph models known in the literature, e.g. the copying model or ACL model (see also [18] for the preferential attachment graph compression algorithm).

Therefore, we turn our attention to the asymptotic behavior of the distribution of degrees of vertices in random graphs generated by the duplication-divergence model. Let us recall that, for example, for Erdős-Rényi model $\mathtt{ER}(t, p)$ it is known that the degree distribution approximately follows the Poisson distribution with a tail decreasing exponentially [2]. Clearly, the degree of each vertex is a random variable with the binomial distribution, so it is highly concentrated around its mean $(t-1)p$. Moreover, the maximum degree is also highly concentrated around $(t-1)p + \sqrt{2p(1-p)(t-1)\log t}$ [9, Theorem 3.5]. For the preferential attachment model $\mathtt{PA}(t, m)$ it was proved that the degree distribution exhibits scale-free behaviour, i.e. the number of vertices with degree $k$ is proportional to $k^{-3}$ [3]. In addition, if we consider a vertex arriving at time $s$, its degree in graph on $t$ vertices is proportional to $\sqrt{t/s}$ on average and with high probability it does not exceed $\sqrt{t/s}\log^3 t$ [6]. In the next section we discuss in some details recent results regarding the degree distribution of the duplication-divergence graph model.

Here we provide analogous results for the duplication-divergence model. The paper is organized as follows: in Sect. 2 we present a formal definition of the duplication-divergence model, recall previous results related to the properties of the degree distribution and introduce our main results. In Sect. 3.1 and Sect. 3.2 we prove upper bounds for the degrees for earlier and later vertices arriving in the graph, respectively. Finally, in Sect. 3.3 we give a proof of the lower bound for the maximum degree in the graph.

## 2 Model Definition and Main Results

We formally define the duplication-divergence model $\mathtt{DD}(t, p, r)$, introduced by Solé et al. [21]. Then we summarize our main results about high-probability bounds on the maximum degree.

Throughout the paper we use standard graph notation from [7], e.g. $V(G)$ denotes the vertex set of a graph $G$, $\deg_G(s)$ – the degree of node $s$ in $G$ and $\Delta(G)$ – the maximum degree of a vertex in $G$. All graphs considered in the paper are simple.

$G_t$ denotes a graph on $t$ vertices. Because in the paper we deal with graphs that are dynamically generated, we assume that the vertices are identified with the natural numbers according to their arrival time. We use the notation $\deg_t(s)$ for the random variable denoting the degree of vertex $s$ at time $t$ i.e. after $t$ vertices have been added in total.

Let us now formally define the model $\mathtt{DD}(t, p, r)$ as follows: let $G_T$ be a fixed graph on $T \leq t$ vertices, with vertices having distinct labels from 1 to $T$. Let also $0 \leq p \leq 1$ and $0 \leq r \leq T$ be the parameters of the model. Now, for every $n = T, T+1, \ldots, t-1$ we create $G_{n+1}$ from $G_n$ according to the following rules:

1. we add a new vertex with label $n+1$ to the graph,

2. we choose a vertex $u$ from $G_n$ uniformly at random – and we denote $u$ as parent$(n+1)$,
3. for every vertex $v$:
   (a) if $v$ is adjacent to $u$ in $G_n$, then add an edge between $v$ and $n+1$ with probability $p$,
   (b) if $v$ is not adjacent to $u$ in $G_n$, then add an edge between $v$ and $n+1$ with probability $\frac{r}{n}$.

All edge additions are independent random Bernoulli variables.

We now review in some detail, recent results on the degree distribution. For example, for $p < 1$ and $r = 0$, it is shown in [11] that even for large $p$ the limiting distribution of degree frequencies indicates that almost all vertices are isolated as $t \to \infty$. Moreover, from [16] we know that the number of vertices of degree one is $\Omega(\log t)$ but again the precise rate of growth of the number of vertices with any fixed degree $k > 0$ is currently unknown. Recently, also for $r = 0$, in [12,14] the authors showed that for $0 < p < e^{-1}$ the non-trivial connected component has a degree distribution that has a power-law behavior with the exponent is equal to $\gamma$ satisfying $3 = \gamma + p^{\gamma-2}$.

Now let us turn to results directly related to the question of maximum degree. For example, in [23] it was shown that for any fixed $s$ asymptotically as $t \to \infty$ it holds that

$$\mathbb{E}[\deg_t(s)] = \begin{cases} \Theta(\ln t) & \text{if } p = 0 \text{ and } r > 0, \\ \Theta(t^p) & \text{otherwise.} \end{cases}$$

Note that by the close relation between parameters $\Delta(G_t)$ and $\deg_t(s)$ we can establish easily that $\mathbb{E}[\Delta(G_t)] = \Omega(t^p)$ when $p > 0$ or $r = 0$, and $\mathbb{E}[\Delta(G_t)] = \Omega(\ln t)$ otherwise.

It turns out that a lower bound on maximum degree is easily established as a byproduct of existing results by Frieze et al. [10]: for $\frac{1}{2} < p < 1$ and $G_t \sim \mathtt{DD}(t,p,r)$ with $p > 0$ and $s = O(1)$ it holds that

$$\Pr\left[\deg_t(s) \le \frac{C}{A} t^p \log^{-3-\varepsilon}(t)\right] = O(t^{-A})$$

for some fixed constant $C > 0$ and any $A > 0$. This lower bound holds for the maximum degree because for any $s$ it holds that $\deg_t(s) \le \Delta(G_t)$. In the same paper, Frieze et al. also proved that for $\frac{1}{2} < p < 1$, $G_t \sim \mathtt{DD}(t,p,r)$ and $s = O(1)$ it holds that

$$\Pr[\deg_t(s) \ge A\,C\,t^p \log^2(t)] = O(t^{-A})$$

for some fixed constant $C > 0$ and any $A > 0$. They also left as an open problem the question of the behavior of the right tail of the maximum degree distribution or, equivalently, of the upper bound on $\deg_t(s)$ for larger $s$ that holds with high probability.

In this paper, we solve this problem. More precisely, we obtain two major results: first, we provide a bound $\deg_t(s) \le t^p \mathrm{polylog}(t)$ which holds quite surely (i.e. at least $1 - O(t^{-A})$ for any given $A > 0$ [15]). We prove that this bound is valid for all vertices in $G_t$, not only for $s = O(1)$ as before, leading to the estimate $\Delta(G_t) \leq t^p \text{polylog}(t)$ for any $\varepsilon > 0$ with high probability. Next, we provide a precise lower bound and we show that there exists an early vertex $s$ such that $\deg_t(s) \geq (1 - \varepsilon)t^p$ for any $\varepsilon > 0$ quite surely. Putting everything together we obtain the main result of this paper, that is:

**Theorem 1.** *Let $\frac{1}{2} < p < 1$. Asymptotically for $G_t \sim DD(t, p, r)$*

$$\Pr[(1 - \varepsilon)t^p \leq \Delta(G_t) \leq (1 + \varepsilon)t^p \log^{5-4p}(t)] = O(t^{-A})$$

*for any constants $\varepsilon > 0$ and $A > 0$,*

In other words, we are now certain that the maximum degree of the graph is concentrated in the sense that by moving only by some polylogarithmic factor from the mean to both left and right we observe the tail decay which is greater than any polynomial.

# 3 Analysis and Proofs

## 3.1 Upper Bound, Early Vertices

The main idea of the proof of the upper bound of the maximum degree is as follows: we first find for small $s$ (i.e. $s \leq t_0$) a Chernoff-type bound on the growth of $\deg_\tau(s)$ over an interval of certain length $h$.

Then, we introduce auxiliary deterministic sequences $t_i$ and $X_{t_i}$ such that $t_0 < \ldots < t_{k-1} < t \leq t_k$. The definition of these sequences stems from the bound mentioned above, in particular from the relation between $h$ and the growth of the degree, guaranteed with high probability. Ultimately, we prove $\deg_\tau(s) \leq X_\tau$ with high probability for all $s \leq t_0$.

Let us start with providing a Chernoff-type bound on the growth of the degree of a given early vertex (with proof in the appendix):

**Lemma 1.** *Let $1 \leq s \leq \tau \leq t$. Let $X_\tau$ be any value such that $\deg_\tau(s) \leq X_\tau$. Then for any $h \leq \varepsilon X_\tau$ with $\varepsilon \in (0, 1)$ it is true that*

$$\Pr\left[\deg_{\tau+h}(s) \geq \deg_\tau(s) + (1 + 3\varepsilon)\,\frac{h(pX_\tau+r)}{\tau}\right] \leq \exp\left(-\frac{h\varepsilon^2(1+\varepsilon)(pX_\tau+r)}{3\tau}\right).$$

We can immediately deduce how large $h$ has to be to get a polynomial tail:

**Corollary 1.** *Let $1 \leq s \leq \tau \leq t$. Let $X_\tau \geq 0$, $\varepsilon \in (0, 1)$ be values such that asymptotically for any $A > 0$, it holds that $\deg_\tau(s) \leq X_\tau$ and $3A\tau \log t \leq \varepsilon^3 X_\tau(pX_\tau + r)$. Then for any $h \in \left[\frac{3A\tau \log t}{\varepsilon^2(pX_\tau+r)}, \varepsilon X_\tau\right]$ it is true that*

$$\Pr\left[\deg_{\tau+h}(s) > \deg_\tau(s) + (1 + 3\varepsilon)\,\frac{h(pX_\tau+r)}{\tau}\right] = O(t^{-A}).$$

Now we provide the definitions for two auxiliary sequences that we mentioned earlier:

**Definition 1.** *Let $0 < p < 1$ be fixed with certain $\alpha$, $\beta_i$ and $\phi$. We define the increasing sequences $(t_i)_{i=0}^k$ and $(X_{t_i})_{i=0}^k$ and an integer $k$ in the following way:*

$$t_0 = \phi, \qquad t_{i+1} = t_i + \frac{\alpha\, t_i \log t_i}{X_{t_i}}, \qquad t_{k-1} < t \le t_k,$$
$$X_{t_0} = t_0, \qquad X_{t_{i+1}} = X_{t_i} + \beta_i \log t_i.$$

Note that $\alpha$, $\beta_i$ and $\phi$ can be, and indeed we will specify them, as dependent on $t$. However, for brevity, we assume the possible dependency on $t$ as implicit.

Observe that inductively from the definition it follows that if $\alpha \ge \beta_i$, then $X_{t_i} \le t_i$ for all $i = 0, 1, \ldots, k$.

Moreover, note that we do not specify the values of $X_\tau$ for $\tau$ other than $\{t_0, t_1, \ldots, t_k, \ldots\}$. In the rest of the paper we will be using precisely these values in the proofs, so such a definition is sufficient for our purposes. For convenience, we only assume that for any $\tau \in (t_l, t_{l+1})$ for some $l = 0, 1, \ldots, k-1$ the sequence is completed in any way such that $X_{t_l} \le X_\tau \le X_{t_{l+1}}$.

Now we analyze the asymptotic properties of these sequences. We start with a simple lower bound (see the respective appendix for proof):

**Lemma 2.** *Assume that $\phi \ge \log^2 t$, $\alpha \le \sqrt{\phi}$ and $\beta_i \ge \alpha(p - \delta)$ for some $\delta \in [0, p)$. Asymptotically as $t \to \infty$ for any $i = 0, 1, \ldots, k$ we have $X_{t_i} \ge t_i^{p-\delta}$.*

It enables us to we prove (in the appendix) the upper bound:

**Lemma 3.** *Assume that $\phi \ge \log^3 t$, $\alpha(p - \delta) \le \beta_i \le \alpha p + \frac{\alpha}{2 \log t_i}$ for some $\delta \in [0, p)$. It holds asymptotically as $t \to \infty$ that $X_{t_i} \le \phi^{1-p} t_i^p \log t_i$ for all $i = 0, 1, \ldots, k$.*

**Corollary 2.** *If $\alpha \le \phi$, then for the value of $k$ such that $t_{k-1} < t \le t_k$ it is true that $\alpha k < t$.*

*Proof.* We know from the definition of $t_i$ and Lemma 3 that

$$t > t_{k-1} \ge t_0 + \sum_{i=0}^{k-2} \frac{\alpha t_i \log t_i}{\phi^{1-p} t_i^p \log t_i} \ge t_0 + \sum_{i=0}^{k-2} \alpha \ge \phi + (k-1)\alpha > \alpha k$$

as needed.

Here let us note (and prove in one of the appendices) the relation between the last elements of the sequences $(t_i)_{i=0}^k$, $(X_{t_i})_{i=0}^k$ and the final values themselves:

**Lemma 4.** *Let $\varepsilon$ be any positive constant. Assume that $\phi \ge \log^3 t$, $\alpha \le \sqrt{\phi}$, $\alpha(p-\delta) < \beta_i \le \alpha p + \frac{\alpha}{2 \log t_i}$ for some $\delta \in [0, p)$. It holds asymptotically as $t \to \infty$ that $(1 - \varepsilon) t_k \le t \le (1 + \varepsilon) t_{k-1}$ and $(1 - \varepsilon) X_{t_k} \le X_{t_k} \le (1 + \varepsilon) X_{t_{k-1}}$.*

Observe that since we will use $\phi < t$, it holds that $k \ge 1$.

Let us denote by $\mathcal{A}_i(s)$ the event that $\deg_{t_i}(s) \le X_{t_i}$ for a fixed $s \le t_i$. Now we proceed with the main theorem:

**Theorem 2.** *For $G_t \sim DD(t, p, r)$ with $\frac{1}{2} < p < 1$ and $s \in [1, 74529(A+1)^2 \log^4 t]$ it holds asymptotically that*

$$\Pr\left[\deg_t(s) > (1+\varepsilon)t^p \log^{5-4p} t\right] = O(t^{-A})$$

*for any constants $\varepsilon > 0$ and $A > 0$.*

*Proof.* Throughout the proof we will use sequences $(t_i)_{i=0}^k$ and $(X_{t_i})_{i=0}^k$ with $\alpha = 273p^3(A+1)\log^2 t$, $\beta_i = \alpha p + \frac{\alpha}{2\log t_i}$ and $\phi = 74529(A+1)^2 \log^4 t$ and $t_{k-1} < t \le t_k$.

Observe that all the assumptions of Lemma 2, Lemma 3 and Corollary 2 are met so we know that $\max\{74529(A+1)^2\log^4 t, t_i^p\} \le X_{t_i} \le t_i^p \log^{5-4p} t$ for all $i = 0, 1, \ldots, k$ and also $k < \frac{t}{\log^2 t}$. Moreover, if $\mathcal{A}_i(s)$ holds, then the assumptions of Corollary 1 also are true for $\tau = t_i$ and $h = \frac{\alpha t_i \log t_i}{X_{t_i}}$ as $t_i \to \infty$ since for any constant $A > 0$ and $\varepsilon = \frac{1}{9p\log t_i}$ it holds that

$$\frac{3At_i \log t}{\varepsilon^2(pX_{t_i} + r)} < h = \frac{\alpha t_i \log t_i}{X_{t_i}} < \varepsilon X_{t_i}.$$

The left inequality is easy to verify as the left element is $\Theta\left(\frac{t_i \log^2 t_i \log t}{X_{t_i}}\right)$ and $h$ grows like $\Theta\left(\frac{t_i \log t_i \log^2 t}{X_{t_i}}\right)$. The right inequality follows directly from Lemma 2, provided we choose some $\delta \in \left[0, p - \frac{1}{2}\right)$ so that $X_{t_i}$ grows sufficiently fast.

Moreover, since $\beta_i > \alpha p$, we know that for $\varepsilon = \frac{1}{9p\log t_i}$ asymptotically

$$X_{t_{i+1}} - X_{t_i} = \beta_i \log t_i \ge (1+3\varepsilon)\frac{h(pX_{t_i} + r)}{t_i}.$$

where $h = \frac{1}{1+\frac{1}{2p\log t_i}} \frac{\beta_i t_i \log t_i}{pX_{t_i}} \le \varepsilon X_{t_i}$.

Therefore, Corollary 1 implies that for any constant $A > 0$ and $\varepsilon = \frac{1}{9p\log t_i}$ it is true that $\Pr[\neg\mathcal{A}_{i+1}(s)|\mathcal{A}_i(s)] = O(t^{-A})$.

Clearly, for any $1 \le s \le t_0$ we know that $\mathcal{A}_0(s)$ always holds so $\Pr[\neg\mathcal{A}_0(s)] = 0$. Finally, we obtain using Lemma 4 and Corollary 1 that

$$\begin{aligned}\Pr[\deg_t(s) > X_{t_k}] &\le \Pr[\deg_{t_k}(s) > X_{t_k}] = \Pr[\neg\mathcal{A}_k(s)] \\ &\le \sum_{i=0}^{k-1} \Pr[\neg\mathcal{A}_{i+1}(s)|\mathcal{A}_i(s)] + \Pr[\neg\mathcal{A}_0(s)] = \sum_{i=0}^{k-1} O(t^{-A}) = O(t^{-A+1}).\end{aligned}$$

### 3.2 Upper Bound, Late Vertices

In the second part of the proof we also use the sequences $(t_i)_{i=0}^k$ and $(X_{t_i})_{i=0}^k$ as defined in Definition 1. Moreover, in their definition throughout this section we use the same constants as in the proof of Theorem 2: $\alpha = 273p^3(A+1)\log^2 t$, $\beta_i = \alpha p + \frac{\alpha}{2\log t_i}$ and $\phi = 74529(A+1)^2 \log^4 t$.

The proof consists of showing that for $s \in [t_i, t_{i+1})$ for some $i = 0, 1, \ldots, k-1$ the degree of the vertex when it appears in the graph (i.e. $\deg_s(s)$) is with high probability significantly smaller than its respective $X_{t_{i+1}}$. Furthermore, we show that the increase in the degree between $\deg_s(s)$ and $\deg_{t_{i+1}}(s)$ with high probability cannot compensate for this difference. Thus, $X_t$ (or, to be more precise, $X_{t_k}$) gives us a good upper bound on $\deg_t(s)$ for all $s$ – and therefore also we obtain an upper bound for $\Delta(G_t)$.

Let us introduce an auxiliary event $\mathcal{B}_l(s) = \bigcup_{\tau=1}^{s} \mathcal{A}_l(\tau) = [\deg_{t_l}(s) \leq X_{t_l}$ for any $s$ and $l$ such that $s \leq t_l]$.

**Lemma 5.** *Let $s \in (t_l, t_{l+1}]$ for some $l = 0, 1, \ldots, k-1$. Then, for any $\varepsilon \in (0,1)$*

$$\Pr\left[\deg_s(s) \geq (1+\varepsilon)(pX_{t_{l+1}} + r) | \mathcal{B}_l(t_l) \wedge \mathcal{B}_{l+1}(s-1)\right] \leq \exp\left(-\frac{\varepsilon^2(pX_{t_{l+1}}+r)}{3}\right).$$

*Proof.* First, we notice the fact that $\max\{\deg_{t_{l+1}}(\tau) \colon 1 \leq \tau \leq s-1\} \leq X_{t_{l+1}}$ guarantees that $\max\{\deg_s(\tau) \colon 1 \leq \tau \leq s-1\} \leq X_{t_{l+1}}$. Therefore, $\deg_s(s)$ is stochastically dominated by $A_s \sim Bin\left(s, \frac{pX_{t_{l+1}}+r}{s}\right)$ so for any $\varepsilon \in (0,1)$ we obtain the result directly using the Chernoff bound with $\mathbb{E}[A_s] = pX_{t_{l+1}} + r$.

Note that the result implies that with high probability at most slightly more than a $p$ fraction of the maximum allowed degree is already used at time $s$. Therefore, we are interested in bounding the remaining part of the degree, i.e. $\deg_{t_{l+1}}(s) - \deg_s(s)$, by something smaller than the remaining $(1-p)$ fraction of the maximum allowed degree.

**Lemma 6.** *Let $\frac{1}{2} < p < 1$ and $s \in (t_l, t_{l+1}]$ for some $l = 0, 1, \ldots, k-1$. Then asymptotically as $t \to \infty$, for any constant $A > 0$ it holds that*

$$\Pr\left[\deg_{t_{l+1}}(s) \geq X_{t_{l+1}} | \mathcal{B}_l(t_l) \wedge \mathcal{B}_{l+1}(s)\right] = O(t^{-A}).$$

**Lemma 7.** *Let $\frac{1}{2} < p < 1$ and $s \in (t_l, t_{l+1}]$ for some $l = 0, 1, \ldots, k-1$. Then asymptotically as $t \to \infty$, for any constant $A > 0$ it holds that*

$$\Pr\left[\neg\mathcal{B}_{l+1}(t_{l+1}) | \mathcal{B}_l(t_l)\right] = O(t^{-A}).$$

The proofs of both lemmas above are presented in the respective appendices.

**Theorem 3.** *Let $\frac{1}{2} < p < 1$. Then asymptotically as $t \to \infty$, for any constant $A > 0$ it holds that*

$$\Pr\left[\Delta(G_t) \geq (1+\varepsilon)t^p \log^{5-4p} t\right] = O(t^{-A}).$$

*Proof.* From Lemma 3 we know that $X_{t_k} \leq (1+\varepsilon)t^p \log^{5-4p} t$ holds quite surely. It follows that

$$\begin{aligned}\Pr\left[\Delta(G_t) \geq (1+\varepsilon)t^p \log^{5-4p} t\right] &\leq \Pr\left[\Delta(G_t) \geq X_{t_k}\right] \leq \Pr\left[\neg\mathcal{B}_k(t_k)\right] \\ &\leq \sum_{l=0}^{k-1} \Pr\left[\neg\mathcal{B}_{l+1}(t_{l+1}) | \mathcal{B}_l(t_l)\right] + \Pr\left[\neg\mathcal{B}_0(t_0)\right].\end{aligned}$$

Now, from Theorem 2 and Lemma 7 we know that both $\Pr[\neg\mathcal{B}_0(t_0)] = O(t^{-A})$ and $\Pr[\neg\mathcal{B}_{l+1}(t_l)|\mathcal{B}_l(t_l)] = O(t^{-A})$ for any $A > 0$, respectively. Putting this all together with Lemma 4 we obtain the result.

### 3.3 Lower Bound

Here we proceed analogously to the case of the upper bound for early vertices. First, we provide an appropriate Chernoff-type bound for the degree of a given vertex with respect to some deterministic sequence. Then we again use a special sequence, which has the desired rate of growth and serves as a lower bound on $\deg_t(s)$. Note that we don't need to extend our analysis for the late vertices since a lower bound for the degree of any vertex $s$ at time $t$ is also a lower bound for the minimum degree of $G_t$.

First, we note that if we start the whole process from a non-empty graph, then there exists $s \in [1, t_0]$ such that $\deg_{t_0}(s) \geq 1$. Moreover, even if the starting graph is empty, but $r > 0$, then with high probability there exists a vertex with positive degree, as the probability of adding another isolated vertex to an empty graph on $t$ vertices is at most $(1 - \frac{r}{t})^t \leq \exp(-r)$, so within first $\frac{A}{r}\log t$ vertices for any $A > 0$ we have a non-isolated vertex with probability at least $1 - O(t^{-A})$. Of course, if we start from an empty graph and $r = 0$, then for any $p$ there cannot arise any edge in the duplication process. However, in this case it trivially follows that $\Delta(G_t) = 0$, so we omit this case in further analysis.

Let us now return to the aforementioned Chernoff-type lower bound:

**Lemma 8.** *Let $1 \leq s \leq \tau \leq t$. Let $X_\tau$ be any value such that $\deg_\tau(s) \geq X_\tau$. Then for any $h \leq \varepsilon\tau$ with $\varepsilon \in \left(0, \frac{1}{3}\right)$ it is true that*

$$\Pr\left[\deg_{\tau+h}(s) \leq \deg_\tau(s) + (1 - 2\varepsilon)\,\frac{hpX_\tau}{\tau}\right] \leq \exp\left(-\frac{h\varepsilon^2 pX_\tau}{3\tau}\right).$$

**Corollary 3.** *Let $1 \leq s \leq \tau \leq t$. Let $X_\tau \geq 0$, $A > 0$, $\varepsilon \in \left(0, \frac{1}{3}\right)$ be values such that $\deg_\tau(s) \leq \tau$ and $3A\log t \leq \varepsilon^3 pX_\tau$. Then for any $h \in \left[\frac{3A\log t}{\varepsilon^2 pX_\tau}, \varepsilon\tau\right]$ it is true that*

$$\Pr\left[\deg_{\tau+h}(s) \leq \deg_\tau(s) + (1 - 2\varepsilon)\,\frac{hpX_\tau}{\tau}\right] = O(t^{-A}).$$

In the following, we again use sequences $(t_i)_{i=1}^k$ and $(X_{t_i})_{i=1}^k$ from Definition 1. Let us also define $\mathcal{C}_i(s)$ as the event that $\deg_{t_i}(s) \geq X_{t_i} - \phi + 1$ for a fixed $s \leq t_i$. This allows us to proceed with the main theorem of this section:

**Theorem 4.** *For $G_t \sim DD(t, p, r)$ with $\frac{1}{2} < p < 1$ there exists $s$ such that for any constants $\varepsilon > 0$ and $A > 0$ it holds asymptotically that*

$$\Pr\left[\deg_t(s) < (1 - \varepsilon)t^p\right] = O(t^{-A}).$$

*Proof.* Again let us use sequences $(t_i)_{i=0}^k$ and $(X_{t_i})_{i=0}^k$ with $\alpha = 12p^3(A+1)\log^2 t$, $\beta_i = \alpha p - \frac{\alpha}{\log t_i}$ and $\phi = 144(A+1)^2 \log^4 t$. These parameters satisfy the assumptions of Lemma 3 and Corollary 2.

Moreover, if $\mathcal{C}_i(s)$ holds, then the assumptions of Corollary 3 are also true for $\tau = t_i$ and $h = \frac{\alpha t_i \log t_i}{X_{t_i}}$ as $t_i \to \infty$, since for any constant $A > 0$ and $\varepsilon = \frac{1}{2p \log t_i}$

$$\frac{3A\tau \log t}{\varepsilon^2 p X_{t_i}} < h = \frac{\alpha t_i \log t_i}{X_{t_i}} < \varepsilon t_i.$$

The left inequality is easy to verify as the left hand side is $\Theta\left(\frac{t_i \log^2 t_i \log t}{X_{t_i}}\right)$ and $h$ grows like $\Theta\left(\frac{t_i \log t_i \log^2 t}{X_{t_i}}\right)$. The right inequality follows directly from Lemma 2.

Next, $X_{t_{i+1}} - X_{t_i} = \beta_i \log t_i = (1-2\varepsilon)\frac{hpX_{t_i}}{t_i}$, where $h = \frac{1}{1-\frac{1}{p \log t_i}} \frac{\beta_i t_i \log t_i}{p X_{t_i}}$. Therefore, Corollary 3 implies that for any constant $A > 0$ and $\varepsilon = \frac{1}{2p \log t_i}$ it is true that $\Pr[\neg \mathcal{C}_{i+1}(s) | \mathcal{C}_i(s)] = O(t^{-A})$. Note that we apply this with a sequence $X_{t_i} - \phi + 1$, not with $X_{t_i}$ itself this time. This is so because to use Corollary 3 we need $\deg_{t_0}(s) \geq X_{t_0} - \phi + 1 = 1$, which holds with high probability – as e.g. $\deg_{t_0}(s) \geq X_{t_0}$ is false with high probability.

Since $X_{t_0} = 144(A+1)^2 \log^4 t$ we know that $\mathcal{C}_0(s)$ holds with high probability: either the starting graph is nonempty, or $r > 0$ and some edges appear before $t_0$. Using Lemma 4 and Corollary 3 for any $\varepsilon > 0$ and $A > 0$ we get

$$\begin{aligned}\Pr[\deg_t(s) <& (1-\varepsilon)t^p] \leq \Pr[\deg_t(s) < X_{t_{k-1}} - \phi + 1] \leq \Pr[\neg \mathcal{C}_{k-1}(s)] \\ &\leq \sum_{i=0}^{k-2} \Pr[\neg \mathcal{C}_{i+1}(s) | \mathcal{C}_i(s)] + \Pr[\neg \mathcal{C}_0(s)] = \sum_{i=0}^{k-1} O(t^{-A}) = O(t^{-A+1}).\end{aligned}$$

We conclude our analysis with the following corollary.

**Corollary 4.** *For $G_t \sim DD(t,p,r)$ with $\frac{1}{2} < p < 1$ for any constants $\varepsilon > 0$ and $A > 0$ it holds asymptotically that*

$$\Pr\left[\Delta(G_t) \leq (1-\varepsilon)t^p\right] = O(t^{-A}).$$

## References


1. Barabási, A.L., Albert, R.: Emergence of scaling in random networks. Science **286**(5439), 509–512 (1999)
2. Bollobás, B.: Random Graphs. Cambridge University Press, Cambridge (2001)
3. Bollobás, B., Riordan, O., Spencer, J., Tusnády, G.: The degree sequence of a scale-free random graph process. In: The Structure and Dynamics of Networks, pp. 384–395. Princeton University Press (2011)
4. Chierichetti, F., Kumar, R., Lattanzi, S., Panconesi, A., Raghavan, P.: Models for the compressible web. SIAM J. Comput. **42**(5), 1777–1802 (2013)

5. Colak, R., et al.: Dense graphlet statistics of protein interaction and random networks. In: Biocomputing 2009, pp. 178–189. World Scientific Publishing, Singapore (2009)
6. Cooper, C., Frieze, A.: The cover time of the preferential attachment graph. J. Comb. Theory Ser. B **97**(2), 269–290 (2007)
7. Diestel, R.: Graph Theory. Springer, Heidelberg (2005)
8. Erdős, P., Rényi, A.: On random graphs I. Publicationes Mathematicae **6**, 290–297 (1959)
9. Frieze, A., Karoński, M.: Introduction to Random Graphs. Cambridge University Press, Cambridge (2016)
10. Frieze, A., Turowski, K., Szpankowski, W.: Degree distribution for duplication-divergence graphs: large deviations. In: Adler, I., Müller, H. (eds.) WG 2020. LNCS, vol. 12301, pp. 226–237. Springer, Cham (2020). https://doi.org/10.1007/978-3-030-60440-0_18
11. Hermann, F., Pfaffelhuber, P.: Large-scale behavior of the partial duplication random graph. ALEA **13**, 687–710 (2016)
12. Jacquet, P., Turowski, K., Szpankowski, W.: Power-law degree distribution in the connected component of a duplication graph. In: Drmota, M., Heuberger, C. (eds.) 31st International Conference on Probabilistic, Combinatorial and Asymptotic Methods for the Analysis of Algorithms, AofA 2020, Klagenfurt, Austria, 15–19 June 2020 (Virtual Conference). LIPIcs, vol. 159, pp. 16:1–16:14. Schloss Dagstuhl - Leibniz-Zentrum für Informatik (2020)
13. Janson, S., Łuczak, T., Ruciński, A.: Random Graphs. Wiley, Hoboken (2011)
14. Jordan, J.: The connected component of the partial duplication graph. ALEA - Lat. Am. J. Probab. Math. Stat. **15**, 1431–1445 (2018)
15. Knuth, D., Motwani, R., Pittel, B.: Stable husbands. Random Struct. Algorithms **1**(1), 1–14 (1990)
16. Li, S., Choi, K.P., Wu, T.: Degree distribution of large networks generated by the partial duplication model. Theor. Comput. Sci. **476**, 94–108 (2013)
17. Łuczak, T., Magner, A., Szpankowski, W.: Asymmetry and structural information in preferential attachment graphs. Random Struct. Algorithms **55**(3), 696–718 (2019)
18. Łuczak, T., Magner, A., Szpankowski, W.: Compression of preferential attachment graphs. In: IEEE International Symposium on Information Theory, ISIT 2019, Paris, France, 7–12 July 2019, pp. 1697–1701. IEEE (2019)
19. Newman, M.: Networks: An Introduction. Oxford University Press, Oxford (2010)
20. Shao, M., Yang, Y., Guan, J., Zhou, S.: Choosing appropriate models for protein-protein interaction networks: a comparison study. Brief. Bioinform. **15**(5), 823–838 (2013)
21. Solé, R., Pastor-Satorras, R., Smith, E., Kepler, T.: A model of large-scale proteome evolution. Adv. Complex Syst. **5**(01), 43–54 (2002)
22. Sreedharan, J., Turowski, K., Szpankowski, W.: Revisiting parameter estimation in biological networks: influence of symmetries. IEEE/ACM Transactions on Computational Biology and Bioinformatics (2020)
23. Turowski, K., Szpankowski, W.: Towards degree distribution of a duplication-divergence graph model. Electron. J. Comb. **28**(1), P1.18 (2021)
24. Van Der Hofstad, R.: Random Graphs and Complex Networks. Cambridge University Press, Cambridge (2016)
25. Watts, D., Strogatz, S.: Collective dynamics of "small-world" networks. Nature **393**(6684), 440–442 (1998)

# Conditional Fractional Matching Preclusion for Burnt Pancake Graphs and Pancake-Like Graphs (Extended Abstract)

Sambhav Gupta[1], Eddie Cheng[2(✉)], and László Lipták[2]

[1] Northville High School, Northville, MI 48168, USA
[2] Department of Mathematics and Statistics, Oakland University, Rochester, MI 48309, USA
{echeng,liptak}@oakland.edu

**Abstract.** The *conditional fractional strong matching preclusion number* of a graph $G$ is the minimum size of $F$ such that $F \subset V(G) \cup E(G)$ and $G - F$ has neither a fractional perfect matching nor an isolated vertex. In this paper, we obtain the conditional fractional strong matching preclusion number for burnt pancake graphs and a subset of the class of pancake-like graphs.

**Keywords:** Fractional matching preclusion · Conditional fractional strong matching preclusion · Pancake graph

## 1 Introduction

Parallel computing is an important area of computer science and engineering. The underlying topology of such a parallel machine or a computer network is the interconnection network. Computing nodes are processors where the resulting system is a multiprocessor supercomputer, or they can be computers in which the resulting system is a computer network. It is unclear where the computing future is headed. It may lead to more research in multiprocessor supercomputers, physical networks or networks in the cloud. Nevertheless, the analysis of such networks will always be important. One important aspect of network analysis is fault analysis, that is, the study of how faulty processors/links will affect the structural properties of the underlying interconnection networks, or simply graphs.

All graphs considered in this paper are undirected, finite and simple. We refer to the book [3] for graph theoretical notation and terminology not described here. For a graph $G$, let $V(G)$, $E(G)$, and $(u, v)$ ($uv$ for short) denote the set of vertices, the set of edges, and the edge whose end vertices are $u$ and $v$, respectively. For any subset $X$ of $V(G)$ or $E(G)$, let $G[X]$ denote the subgraph induced by $X$. We use $G - F$ to denote the subgraph of $G$ obtained by removing all the vertices and (or) the edges of $F$. Some portions of this paper containing definitions are reused unchanged from [19].


C.-Y. Chen et al. (Eds.): COCOON 2021, LNCS 13025, pp. 425–435, 2021.
https://doi.org/10.1007/978-3-030-89543-3_36

### 1.1 Matchings

A *perfect matching* in a graph is a set of edges such that each vertex is incident to exactly one of them, and an *almost perfect matching* is a set of edges such that each vertex but one is incident to exactly one edge in the set, and the remaining vertex is incident to none. We can define a perfect matching as an indicator function as follows: let S be a set of edges in G. Then $f^S$ is the indicator function of $S$, with $f^S : E(G) \to \{0,1\}$ such that $f^S(e) = 1$ iff $e \in S$. Let $\delta'(v)$ be the set of edges for which one end is $v$. Then $M$ is a perfect matching of $G$ if $\sum_{e\in\delta'(v)} f^M(e) = 1$ for each vertex $v \in G$

A standard relaxation from an integer setting to a continuous setting is to extend the codomain of the indicator function from $\{0,1\}$ to $[0,1]$. Let $f : E(G) \to [0,1]$. Then $f$ is a *fractional perfect matching* if $\sum_{e\in\delta'(v)} f(e) = 1$ for each vertex $v \in G$. We note that the specification that such a matching be "perfect" is somewhat redundant, as unlike a perfect matching, a fractional perfect matching can exist on odd graphs; the concept of fractional almost perfect matchings is not really necessary to consider or study.

**Proposition 1.** [23] *The graph $G$ has a fractional perfect matching if and only if there is a partition $\{V_1, V_2, \ldots, V_n\}$ of the vertex set of $V(G)$ such that, for each $i$, the graph $G[V_i]$ is either $K_2$ or a Hamiltonian graph on odd number of vertices.*

Any graph with such a decomposition can be trivially assigned a fractional perfect matching by assigning each $K_2$ of the decomposition a weight of 1 and each edge in an odd cycle a weight of $\frac{1}{2}$, then replacing all removed edges. We call such a fractional perfect matching *nice*. For notational convenience, we assume that if a graph $G$ has a fractional perfect matching $f$, then $f$ is nice. Furthermore, if a nice fractional perfect matching contains an edge with weight 1, we refer to it as a *complete edge*, and if it contains an edge with weight $\frac{1}{2}$, we refer to it as a *half edge*. Finally, if we claim that any vertex $u$ or edge $vw$ is in an odd cycle in a fractional perfect matching, then we mean $u$ is in a half edge with each of two other vertices, and that $vw$ is a half edge.

### 1.2 Matching Preclusion

In [1], Brigham et al. first introduced the concept of matching preclusion. A set of edges $F$ of $G$ is called a *matching preclusion set* if $G - F$ has neither perfect matchings nor almost-perfect matchings, and it is called an *optimal matching preclusion set* if $|F|$ is minimal. Then if $F_1$ is an optimal matching preclusion set, any set $F_2$ for which $|F_2| < |F_1|$ is not a matching preclusion set. The matching preclusion number of $G$, denoted by $mp(G)$, is the cardinality of an optimal matching preclusion set. A set $F$ of edges and vertices of $G$ is a *strong matching preclusion set* (SMP set for short) if $G - F$ has neither perfect matchings nor almost-perfect matchings, and it is called an optimal strong matching preclusion set if $F$ is one with the smallest size. The *strong matching preclusion number* (SMP number for short) of $G$, denoted by $smp(G)$, is the cardinality of an optimal SMP set. An optimal SMP set is trivial if $G - F$ is even and there is a vertex $v$ such that every vertex in $F$ is a neighbour of $v$ and every edge in $F$ is incident to $v$. The concept of strong matching preclusion was proposed by Park and Ihm in 2011. We refer the readers to [4,7–9,16,20,22] for further details and additional references.

Recently, Liu and Liu in [18] introduced generalizations of the above concepts. An edge subset $F$ of $G$ is a *fractional matching preclusion set* (FMP set for short) if $G - F$ has no fractional perfect matchings. The fractional matching preclusion number (FMP number for short) of $G$, denoted by $fmp(G)$, is the minimum size of FMP sets of $G$, that is, $fmp(G) = \min\{|F| : F \text{ is an FMP set}\}$. A set $F$ of edges and vertices of $G$ is a *fractional strong matching preclusion set* (FSMP set for short) if $G - F$ has no fractional perfect matchings. The *fractional strong matching preclusion number* (FSMP number for short) of $G$, denoted by $fsmp(G)$, is the minimum size of FSMP sets of $G$, that is, $fsmp(G) = \min\{|F| : F \text{ is an FSMP set}\}$. A FMP (FSMP) set of minimal cardinality is called optimal, and an optimal FMP (FSMP) set $F$ is trivial if $G - F$ contains an isolated vertex $v$ ($\delta(v) = 0$). If $F$ is a trivial FMP (FSMP) set, every element in $F$ is adjacent or incident to some $v$ which is isolated in $G - F$. A graph $G$ is *fractional strongly super matched* if every optimal FSMP set is trivial.

We can further constrain the conditions for FSMP by requiring that $G - F$ has no isolated vertices. An edge and vertex subset $F$ of $G$ is a *conditional fractional strong matching preclusion set* (CFSMP set for short) if $G - F$ has neither a fractional perfect matching, nor an isolated vertex. The *conditional fractional strong matching preclusion number* (CFSMP number for short) of $G$ is the minimum size of a CFSMP set of $G$, that is, $cfsmp(G) = \min\{|F| : F \text{ is a CFSMP set}\}$. A CFSMP set of minimal cardinality is called optimal, and a CFSMP set $F$ is trivial if the graph $G - F$ contains some vertices $u, v, w$ for which $\delta(u) = 1, \delta(v) = 1$, and $uw, vw \in E(G - F)$. A graph $G$ is *conditionally fractional strongly super matched* if every optimal CFSMP set is trivial.

The pancake graphs and burnt pancake graphs, introduced in [11], are two well-studied interconnection networks. Although these graphs have nice structures, it seems that some problems in them are difficult and are still open, such as optimal routing problem. However, researchers have found that they are excellent candidates as interconnection networks. Some papers on the pancake graphs include [2,6,10,14,17,21,24] and papers on the burnt pancake graphs include [5], [6,12,13,15]. In particular, in [12], the burnt pancake graphs are used for genome analysis. In [19], the FSMP number of general pancake and burnt pancake graphs was found (using the same inductive proof strategy we use here).

The *pancake-like* graphs are a broad class of graphs which contain the pancake graphs and burnt pancake graphs. Although the main result of this paper applies to burnt pancake graphs, the result also extends to a subset of pancake-like graphs.

In this paper, we study the conditional fractional strong matching preclusion problems for the pancake graphs and obtain the following main result.

**Theorem 1.** *Let $n \geq 3$ be an integer, and let $B_n$ be the burnt pancake graph of dimension $n$. Then $cfsmp(B_n) = 2n - 2$, and $B_n$ is conditionally fractional strongly super matched.*

**Theorem 2.** *Let $L$ be an arbitrary pancake-like graph composed of $n \geq 3$ subgraphs, such that every subgraph is $k-1$-regular with $k \geq 4$ and has girth at least 5. If every subgraph is fractional strongly super matched and conditionally fractional strongly super matched, then $L$ must be fractional strongly super matched and conditionally fractional strongly super matched.*

The rest of this paper is organized as follows. In Sect. 2, we provide some definitions and known results regarding the pancake graphs and burnt pancake graphs, followed by the pancake-like graph class. In Sect. 3, we note the conditional fractional strong matching preclusion of $B_3$, to serve as a base case. In Sect. 4, we discuss the proof technique.

## 2 Preliminaries

We first review the construction of pancake and burnt pancake graphs, as stated in [19], and present some related results.

The pancake graph of dimension $n$, denoted by $P_n$, has as its vertex set the set of all $n!$ permutations on $\{1, 2, 3, \ldots, n\}$. Two vertices $[a_1, a_2, a_3, \ldots, a_n]$ and $[b_1, b_2, b_3, \ldots, b_n]$ are adjacent if there exists an integer $k$ with $2 \leq k \leq n$ such that $a_i = b_{k+1-i}$ for every $i$ with $1 \leq i \leq n$, and $a_i = b_i$ for $k+1 \leq i \leq n$; in other words, we take the "substring" $a_1, a_2, a_3, \ldots, a_n$ and reverse the order. The edge generated by such an adjacency is called a $k$-edge. It follows directly from the definition that $P_n$ is $(n-1)$-regular. Although $P_n$ is vertex-transitive, it is not edge-transitive except for $n = 3$. (However, it is not difficult to determine the edge-transitive classes.) Indeed, let $H_i$ be the subgraph of $P_n$ induced by the vertices with $i$ in the $n$th position, which is isomorphic to $P_{n-1}$. We call $H_i$ to be a copy of $P_n$. We remark that $P_n$ can be decomposed into $n$ copies, i.e., $P^1_{n-1}, P^2_{n-1}, \ldots, P^n_{n-1}$. We note that $P_2$ is a complete graph with two vertices, $P_3$ is the cycle with six vertices and $P_4$ is given in Fig. 1. The edges between different copies are $n$-edges, which form a perfect matching in $P_n$. To highlight this property, we will refer to these edges as cross edges, and if $uv$ is a cross edge, we call it the cross edge of $u$, and call $v$ the cross neighbour of $u$. It is easy to see from the definition that if $u$ is a vertex in $H_i$, then it has $n-2$ neighbours in $H_i$ (the set of these neighbours is denoted by $N_{H_i}(u)$), and the $n-1$ cross neighbours of the vertices in $\{u\} \cup N_{H_i}(u)$ are in different $H_j$ 's (one in each). It is also easy to see that there are exactly $(n-2)!$ independent cross edges between two different $H_i$'s. Hence by an inductive argument we get that for each $k$, the set of $k$-edges forms a perfect matching in $P_n$. So the edges of $P_n$ can be partitioned into $n-1$ edge-disjoint perfect matchings. See Fig. 1 for $P_4$, the pancake graph of dimension 4.

The definition of the burnt pancake graphs is related to the definition of the pancake graphs. Let $n \geq 3$. The burnt pancake graph of dimension $n$, denoted by $B_n$, is defined similarly to the pancake graphs. We say the list $[a_1, a_2, \ldots, a_n]$ is a signed permutation on $\{1, 2, 3, \ldots, n\}$ if $[|a_1|, |a_2|, \ldots, |a_n|]$ is a permutation on $\{1, 2, 3, \ldots, n\}$. For notational simplicity, which is customary for this class of graphs, we use the notation $\bar{a}$ instead of $-a$. The burnt pancake graph $B_n$ has the set of signed permutations on $\{1, 2, 3, \ldots, n\}$ as its vertex set. Two vertices $[a_1, a_2, \ldots, a_n]$ and $[b_1, b_2, \ldots, b_n]$ are adjacent if there exists a $k$ with $1 \leq k \leq n$ such that $a_i = b_{k+1-i}$ for every $i$ with $1 \leq i \leq k$, and $a_i = b_i$ for $k+1 \leq i \leq n$. The edge generated by such an adjacency is again called a $k$-edge. It follows directly from the definition that $B_n$ is $n$-regular with $n!2^n$ vertices. We remark that $B_n$ is vertex transitive but not edge transitive. Indeed, let $H_a$ be the subgraph of Bn induced by the vertices with $a$ in the $n$th position where $a \in \{1, 2, 3, \ldots, n\} \cup \{\bar{1}, \bar{2}, \bar{3}, \ldots, \bar{n}\}$, which is isomorphic to $B_{n-1}$. We call $H_a$ to be a copy of $B_n$. Like $P_n$, $B_n$ is recursive in structure and can be decomposed into $2n$ copies, i.e., $B^1_{n-1}, B^2_{n-1}, \ldots, B^n_{n-1}, B^{\bar{1}}_{n-1}, B^{\bar{2}}_{n-1}, \ldots, B^{\bar{n}}_{n-1}$ We note that $B_1$ is a complete graph with two vertices, $B_2$ is the cycle with eight vertices and $B_3$ is given in Fig. 2. The $n$-edges are the edges between different $H_a$'s, and they form a perfect matching in $B_n$. Again we will refer to the $n$-edges as cross edges, and if $uv$ is a cross edge, we call it the cross edge of $u$, and call $v$ the cross neighbour of $u$. It is easy to see from the definition that if $u = [a_1, a_2, \ldots, a_n]$ is a vertex in $H_{a_n}$, then it has $n-1$ neighbours in $H_{a_n}$ , and the cross neighbours of the vertices in $\{u\} \cup N_{a_n}(u)$ are in different $H_j$ 's. Indeed, the cross neighbour of $u$ is in $H_{a_1}$, and the cross neighbours of the neighbours of $u$ are in $H_{a_1}, H_{a_2}, \ldots, H_{a_{n-1}}$. It is easy to see that there are exactly $(n-2)!2^{n-2}$ independent cross edges between $H_a$ and $H_b$ if $a = b$, and there are no edges between $H_a$ and $H_{\bar{a}}$. We note that for each $k$, the set of $k$- edges forms a perfect matching in $B_n$. So the edges of $B_n$ can be partitioned into $n$-edge-disjoint perfect matchings. See Fig. 2 for $B_3$, the burnt pancake graph of dimension 3.

From the definition of $P_n$ and $B_n$, the following observations are immediate.

**Proposition 2.** *For $n \geq 4$, the cross neighbours of two adjacent vertices in a copy of $P_n$ are in different copies.*

**Proposition 3.** *For $n \geq 3$, the cross neighbours of two adjacent vertices in a copy of $B_n$ are in different copies.*

From [19], we have the following results on the FSMP numbers of pancake and burnt pancake graphs:

**Theorem 3.** [19] *Let $n \geq 4$ be an integer, and let $P_n$ be the pancake graph of dimension $n$. Then $fsmp(P_n) = n - 1$, and every optimal FSMP set of $P_n$ is trivial when $n \geq 5$.*

Directly, for integer $n \geq 5$, $P_n$ is fractional strongly super matched.

**Theorem 4.** [19] *Let $n \geq 3$ be an integer, and let $B_n$ be the burnt pancake graph of dimension $n$. Then $fsmp(B_n) = n$, and every optimal FSMP set of $B_n$ is trivial.*

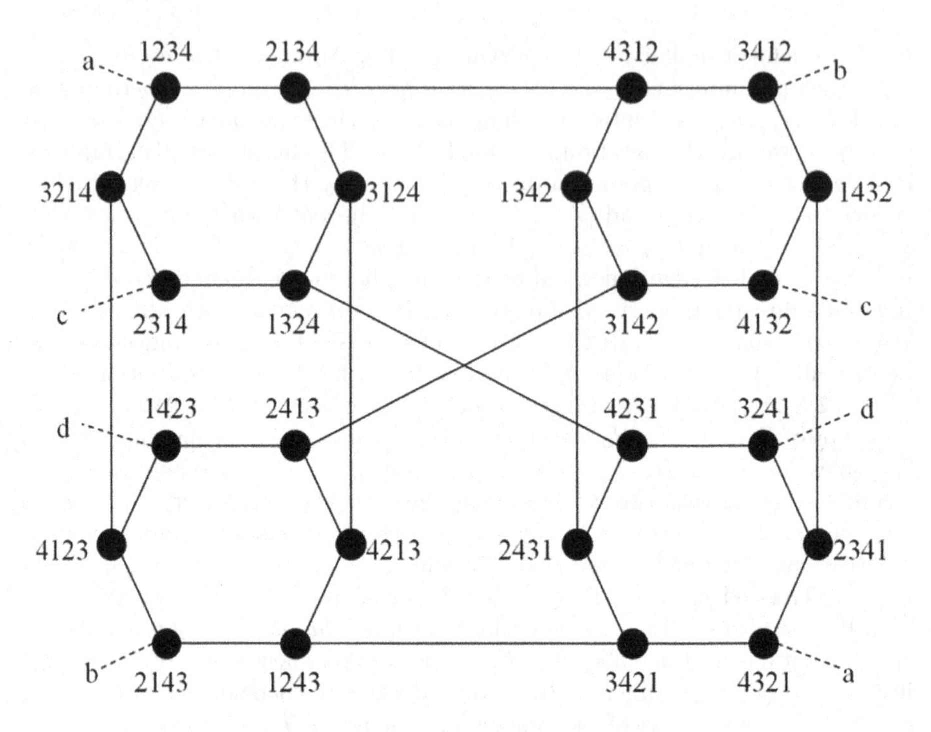


**Fig. 1.** The pancake graph of dimension 4.

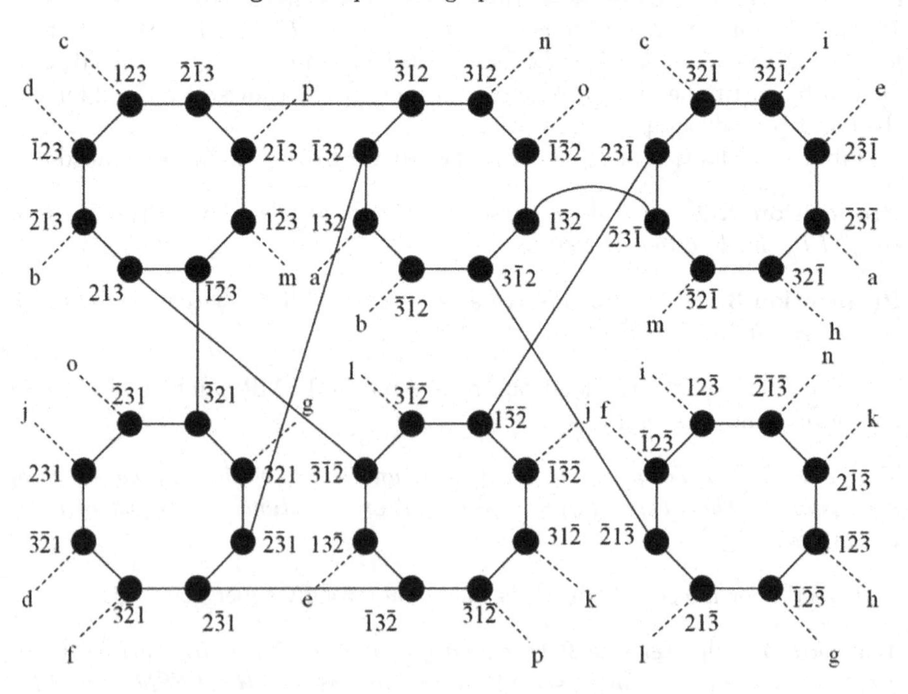

**Fig. 2.** The burnt pancake graph of dimension 3.

Directly, for integer $n \geq 3$, $B_n$ is fractional strongly super matched.

A *pancake-like* graph is any graph $G$ which exhibits the property that it has a partition $\{V_1, V_2, \ldots, V_n\}$ of the vertex set of $V(G)$ such that, for each $i$, the graph $G[V_i]$ satisfies the following two properties: (i) each vertex is incident to exactly one cross edge, and (ii) the cross neighbours of two adjacent vertices are not in the same vertex subset in the partition, where a cross edge is any edge in the set $E(G) - E(G[V_1]) - E(G[V_2]) - \ldots - E(G[V_n])$ and cross neighbors are any two vertices $a$, $b$ for which $ab$ is a cross edge. We further define each $G[V_i]$ as a subgraph of $G$. Clearly, this definition aims to replicate the top-level structure of pancake and burnt pancake graphs.

## 3 Results for $B_3$

The following results are important in our analysis.

**Lemma 1.** [19] *Let $G$ be a fractional strongly super matched graph with $\delta(G) \geq 2$. If $F$ is a trivial FSMP set of $G$ and $G - F$ has an isolated vertex $v$, then $G - F - v$ has a fractional perfect matching.*

We also prove an analogous result for conditionally fractional strongly super matched graphs.

**Lemma 2.** *Let a conditionally fractional strongly super matched and fractional strongly super matched graph $G$ with girth at least 5 and $\delta(G) = k$ have a trivial CFSMP set $F$ such that $G - F$ contains some vertices $u, v, w$ for which $\delta(u) = 1, \delta(v) = 1$, and $uw, vw \in E(G - F)$. Then $G - F - u$ and $G - F - v$ have fractional perfect matchings.*

*Proof.* Since $u$ and $v$ are transitive, we prove for $G - F - u$. Let $F$ contain at least one vertex $x$ adjacent to $u$. The graph $G - (F - x)$ must either contain an isolated vertex or contain a fractional perfect matching. If $G - (F - x)$ contains an isolated vertex, since $G - F$ does not contain an isolated vertex, the isolated vertex must be $x$, which is adjacent to $u$. Then $G - (F - x)$ must contain a fractional perfect matching $f$. Since $\delta(v) = 1$, then $f(vw) = 1$, so $f(ux) = 1$. Then $G - (F - x) - u - x = G - F - u$ has a fractional perfect matching.

If $F$ does not contain at least one vertex adjacent to $u$, then we let $F_u$ be the set of edges in $F$ adjacent to $u$. Then $G - F_u - u = G - u$ which must have a fractional perfect matching, so we consider $G - u - (F - F_u)$. Because $|u \cup (F - F_u)| = k - 1$, we have that $G - u - (F - F_u)$ must have either a fractional perfect matching or an isolated vertex, but since $v$ is adjacent to $w$, there are no isolated vertices. Then $G - u - (F - F_u) = G - F - u$ must have a fractional perfect matching.

We use the following result to find that fractionally strongly conditionally super matched graphs with isolated vertices removed must have a fractional perfect matching.

**Lemma 3.** *Let a fractionally strongly conditionally super matched graph $G$ of degree $k \geq 3$ have a fault set $F \subset V(G) \cup E(G)$ with $|F| \leq 2k-2$. If $G-F$ does not have an isolated vertex or three vertices $u, v, w$ for which $\delta(u) = 1, \delta(v) = 1$, and $uw, vw \in E(G-F)$, then it must have a fractional perfect matching.*

*Proof.* The proof follows from the definition of fractionally strongly conditionally super matched. For such a graph, every optimal CFSMP set of size $2k-2$ must be trivial, so the set $F$ must either be a trivial CFSMP set, a non-CFSMP set which fails to be a CFSMP set by isolating a vertex, or a non-CFSMP set which fails to be a CFSMP set by leaving a fractional perfect matching.

We use the next result to show that graphs must effectively "concentrate" faults towards one of two categories in order to preclude a fractional perfect matching; if faults are allocated towards isolating a vertex, there are not enough faults left over to create the 3-vertex precluding structure.

**Lemma 4.** *Let a conditionally fractional strongly super matched $k$-regular graph $G$ of girth at least 5 have a fault set $F \subseteq V(G) \cup E(G)$ with $|F| = 2k-2$. Then if $G-F$ contains an isolated vertex $e$, the graph $G-(F-\{e\})$ must contain a fractional perfect matching.*

*Proof.* There are two cases.

**Case 1.** $G-F$ contains two isolated vertices, $e$ and $g$. Then $e$ and $g$ are either adjacent or they share at most one common neighbor. If they are adjacent, then there are $k$ faults adjacent or incident to each of $e$ and $g$ with at most one of those faults shared (the edge $eg$). Since $2k-1 > 2k-2$, this can not occur. If they share a common neighbor, the minimum fault set size required is the same. If they do not share a common neighbor, then $|F| \geq 2k$. Thus $G-F$ contains at most one isolated vertex.

**Case 2.** $G-F-\{e\}$ contains three vertices $u, v, w$ for which $\delta(u) = 1, \delta(v) = 1$, and $uw, vw \in E(G-F-\{e\})$. There are $k-1$ faults adjacent to $u$ and another disjoint $k-1$ faults adjacent to $v$ in $G$. The vertex $e$ can be adjacent or incident to at most two of the faults adjacent or incident to either $u$ or $v$, so an additional $k-2$ faults are adjacent or incident to $e$. Then $|F| \geq 3k-4$, and $3k-4 > 2k-2$ for $k \geq 3$, so this construction can not be made.

Finally, we prove the base case result on $B_3$ using a computer check.

**Lemma 5.** $cfsmp(B_3) = 4$. *Moreover, every optimal CFSMP set of $B_3$ is trivial, that is, $B_3$ is fractional strongly super matched.*

*Proof.* This was verified by computer check. Due to the relatively small size of the minimum CFSMP set and the graph $B_3$, this was checked in about 1 day on a conventional computer. The program was written in Python and used the NetworkX package for graph analysis and the SciPy package for a linear program solver to check for fractional perfect matchings. The verification was completed by looping through all possible fault sets on $B_3$ of size 4. After optimizing for vertex transitivity, there are $1,302,609$ such fault sets.

We remark that a similar result for $P_5$ likely exists; however, due to the size of the graph, the linear optimization problem requires much longer to compute for each fault set case, and there are more cases for fault sets. This may be feasible by using a computer cluster.

## 4 The Main Results

First we note the following result.

**Theorem 5.** [19] *Let $H$ be a $(k-1)$-regular graph with $k \geq 4$. Let $G$ be a $k$-regular graph constructed from at least three copies of $H$ by adding edges between the copies, such that $G$ satisfies the following properties: (i) each vertex is incident to exactly one cross edge, and (ii) the cross neighbours of two adjacent vertices in a copy of $H$ are in different copies. Then if $H$ is fractional strongly super matched, $G$ is fractional strongly super matched.*

Although this theorem states that $G$ must be composed of copies of $H$, the proof of this theorem does not actually require every subgraph to be isomorphic. Thus, we restate the theorem as follows.

**Theorem 6.** *Let $H_1, H_2, \ldots, H_n$ be a set of $(k-1)$-regular graphs with $k \geq 4$ and $n \geq 3$. Let $G$ be a $k$-regular graph constructed from all $H_i$ with $1 \leq i \leq n$ by adding edges between the subgraphs, such that $G$ satisfies the following properties: (i) each vertex is incident to exactly one cross edge, and (ii) the cross neighbours of two adjacent vertices in any $H_i$ are in different copies for $1 \leq i \leq n$. Then if every $H_1, H_2, \ldots, H_n$ is fractional strongly super matched, $G$ is fractional strongly super matched.*

This modified theorem allows us to extend our main result from burnt pancake graphs to pancake-like graphs.

In this section, we use the same notation as in the previous sections, but in a more general context. Suppose $G$ is a graph constructed from disjoint copies of a graph $H$ by adding edges between the copies. If $uv$ is an edge joining two vertices from different copies of $H$, we call it a cross edge, and say $u$ and $v$ are cross neighbours.

**Theorem 7.** *Let $H_1, H_2, \ldots, H_n$ be a set of $(k-1)$-regular graphs with $k \geq 4$ and $n \geq 3$, and girth at least 5. Let $G$ be a $k$-regular graph constructed from all $H_i$ with $1 \leq i \leq n$ by adding edges between the subgraphs, such that $G$ satisfies the following properties: (i) each vertex is incident to exactly one cross edge, and (ii) the cross neighbours of two adjacent vertices in a copy of $H$ are in different copies. Then if every $H_1, H_2, \ldots, H_n$ is fractional strongly super matched and conditionally fractional strongly super matched, $G$ is fractional strongly super matched and conditionally fractional strongly super matched.*

We omit the proof in this extended abstract due to a space constraint. The proof is long and technical, involving a careful case analysis based on the distribution of faults in the substructures based on the recursive properties and other properties that we have developed here.

## 5 Conclusion

We apply Theorem 7 inductively to $B_n$, given Proposition 3 and Theorem 4 along with the base case Lemma 5 to complete the proof of Theorem 1.

We also note that Theorem 2 is a direct restatement of Theorem 7, and is thus proven.

Because the conditional fractional matching preclusion number must be greater than or equal to the conditional fractional strong matching preclusion number, it is not necessary to separately consider this problem for the burnt pancake graph or the subset of pancake-like graphs we described.

## References

1. Brigham, R.C., Harary, F., Violin, E.C., Yellen, J.: Perfect-matching preclusion. Congr. Numer. **174**, 185–192 (2005)
2. Bennes, R., Latifi, S., Kiruma, N.: A comparative study of job allocation and migration in the pancake network. Inform. Sci. **177**, 2327–2335 (2007)
3. Bondy, J.A., Murty, U.S.R.: Graph Theory. GTM, vol. 244. Springer, Heidelberg (2008)
4. Cheng, E., Jia, R., Lu, D.: Matching preclusion and conditional matching preclusion for augmented cubes. JOIN **11**, 35–60 (2010)
5. Cheng, E., Kelm, J.-T., Orzach, R., Xu, B.: Strong matching preclusion of burnt pancake graphs. Int. J. Parallel Emergent Distrib. Syst. **31**, 220–232 (2016)
6. Cheng, E., Hu, P., Jia, R., Liptak, L., Scholten, B., Voss, J.: Matching preclusion and conditional matching preclusion for pancake and burnt pancake graphs. Int. J. Parallel Emergent Distrib. Syst. **29**(5), 499–512 (2014)
7. Cheng, E., Liptak, L.: Matching preclusion for some interconnection networks. Networks **50**, 173–180 (2007)
8. Cheng, E., Lesniak, L., Lipman, M.J., Liptak, L.: Conditional matching preclusion sets. Inform. Sci. **179**, 1092–1101 (2009)
9. Cheng, E., Shah, S., Shah, V., Steffy, D.-E.: Strong matching preclusion for augmented cubes. Theor. Comput. Sci. **491**, 71–77 (2013)
10. Compeau, P.E.C.: Girth of pancake graphs. Discret. Appl. Math. **159**, 1641–1645 (2011)
11. Gates, W.H., Papadimitriou, C.H.: Bounds for sorting by prefix reversal. Discret. Math. **27**, 47–57 (1979)
12. Gu, Q.-P., Peng, S., Sudborough, I.H.: A 2-approximation algorithm for genome rearrangements by reversals and transpositions. Theor. Comput. Sci. **210**, 327–339 (1999)
13. Hu, X.L., Liu, H.Q.: The (conditional) matching preclusion for burnt pancake graph. Discret. Appl. Math. **161**, 1481–1489 (2013)
14. Hung, C.-N., Hsu, H.-C., Liang, K.-Y., Hsu, L.H.: Ring embedding in faulty pancake graphs. Inform. Proc. Lett. **86**, 271–275 (2003)
15. Kaneko, K.: Hamiltonian cycles and Hamiltonian paths in faulty burnt pancake graphs. IEICE-Trans. Inform. Syst. **E90–D**, 716–721 (2007)
16. Li, Q.L., He, J.H., Zhang, H.P.: Matching preclusion for vertex-transitive networks. Discret. Appl. Math. **207**, 90–98 (2016)

17. Lin, C.-K., Huang, H.-M., Hsu, L.-H.: The super connectivity of the pancake graph and the super laceability of the star graph. Theor. Comput. Sci. **339**, 257–271 (2005)
18. Liu, Y., Liu, W.: Fractional matching preclusion of graphs. J. Comb. Optim. **34**(2), 522–533 (2016). https://doi.org/10.1007/s10878-016-0077-x
19. Ma, T., Mao, Y., Cheng, E., Melekian, C.: Fractional matching preclusion for (burnt) pancake graphs. In: Proceedings to the Fifteen International Symposium on Pervasive, Algorithms, and Networks, pp. 133–141 (2018)
20. Mao, Y., Wang, Z., Cheng, E., Melekian, C.: Strong matching preclusion number of graphs. Theor. Comput. Sci. **713**, 11–20 (2018)
21. Mohamed, A., Ramakrishna, R.S.: Linear election in pancake graphs Inform. Proc. Lett. **106**, 127–131 (2008)
22. Park, J.-H., Ihm, I.: Strong matching preclusion. Theor. Comput. Sci. **412**, 6409–6419 (2011)
23. Scheinerman, E.R., Ullman, D.H.: Fractional Graph Theory: A Rational Approach to the Theory of Graphs. Wiley, New York (1997)
24. Tsai, P.-Y., Fu, J.-S., Chen, G.-H.: Edge-fault-tolerant Hamiltonicity of pancake graphs under the conditional fault model. Theor. Comput. Sci. **409**, 450–460 (2008)

# The Weakly Dimension-Balanced Pancyclicity on Toroidal Mesh Graph $T_{m,n}$ When Both $m$ and $n$ Are Odd

Justie Su-Tzu Juan(✉) and Zong-You Lai

Department of Computer Science and Information Engineering, National Chi Nan University, No. 1, University Road, Puli 545, Nantou, Taiwan
jsjuan@ncnu.edu.tw, s107321521@mail1.ncnu.edu.tw

**Abstract.** In a graph $G = (V, E)$, the edge set $E$ be partitioned into $k$ dimensions $E_1, E_2, \ldots, E_k$ for a positive integer $k$. For any cycle $C$ on $G$, the set of all $i$-dimensional edge of $C$, a subset of $E(C)$, is denoted as $E_i(C)$ for $1 \leq i \leq k$. If $||E_i(C)| - |E_j(C)|| \leq 1$ for $1 \leq i < j \leq k$, $C$ is called a dimension-balanced cycle (DBC for short). If $G$ contains a DBC of every (even, resp.) length between 3 (4, resp.) to $|V(G)|$, $G$ is called DBP (DBBP, resp.). Furthermore, if $||E_i(C)| - |E_j(C)||$ $\leq 3$ for $1 \leq i < j \leq k$, $C$ is called a weakly dimension-balanced cycle (WDBC for short). If $G$ contains a WDBC of every (even, resp.) length between 3 (4, resp.) to $|V(G)|$, $G$ is called WDBP (WDBBP, resp.). The weakly dimension-balanced pancyclicity on graph $G$ is to study whether $G$ is WDBP or WDBBP. For the toroidal mesh graph $T_{m,n}$, the weakly dimension-balanced pancyclicity has been discussed in 2019 when at least one of $m$, $n$ is even; 2021 when both $m$ and $n$ are odd. In this paper, we discuss the weakly dimension-balanced pancyclicity on $T_{m,n}$ when both $m$ and $n$ are odd.

**Keywords:** Toroidal mesh graph · Hamiltonian cycle · Weakly dimension-balanced cycle · Pancyclicity · Bipancyclicity

## 1 Introduction

Interconnection network is a popular issue in recent years. A topological structure of an interconnection network is usually modeled by a graph whose vertices represent processors/cores and edges represent communication links between processors. By this transformation, an interconnection network can be transformed to a graph. For a graph $G = (V, E)$, where $V(G)$ is the *vertex set* and $E(G)$ is the *edge set*, $|V(G)|$ denotes the number of vertices and $|E(G)|$ denotes the number of edges. A *Hamiltonian cycle* of $G$ is a cycle that contains every vertex of $G$. We say a graph $G$ is *pancyclic* (*p-pancyclic*, resp.) if it embeds cycles of every length ranging from 3 to $|V(G)|$ ($p$ to $|V(G)|$, resp.). Since there exist no odd cycles in bipartite graph, a bipartite graph $G$ is called *bipancyclic* (*p-bipancyclic*, resp.) if $G$ contains cycles of every even length between 4 to $|V(G)|$ ($p$ to $|V(G)|$, resp.). Note that the definition of bipancyclicity is intended for bipartite graphs, but can be applied to any graph. In interconnection graph, Hamiltonicity and pancyclicity


C.-Y. Chen et al. (Eds.): COCOON 2021, LNCS 13025, pp. 436–448, 2021.
https://doi.org/10.1007/978-3-030-89543-3_37

are two important properties to design the graph and have been widely discussed in the previous literature (see [1, 3–5, 7, 13, 17, 18, 25]).

In the 3D stereogram reconstruction problem, the research of optimal encode will use gray-code encode to signify the information of $n$-bits, which has been mentioned in the references [2, 16, 20]. The utility of gray-code will decrease the consumption of resource and increase the precision. However, there will be some problems when deal with too many data transformation between 1 and 0 of the same dimension, such as it will cost more or produce poor results. How to reduce the cost of dealing with such problems is very important. Hence, Wang [19] discusses a problem to decrease the efficiency of the information transmission on hypercube. It is called the *dimension-balanced Hamiltonicity* on hypercube. The design for network-on-chip (NoC) problem is an important issue recently. When congestion occurs, the NoC network performance will descend. Thus, how to select efficiently path and provide a strategy to solve this problem are critical factors for the performance of NoC. It also received a great attention in recent years [10–12]. One popular structure in NoC can be transformed into a toroidal mesh graph. We believe that a DBC of given length is helpful for designing simple algorithms with low communication cost and avoid congestion.

Given a graph $G = (V, E)$, the edge set $E$ be partitioned into $k$ dimensions $E_1$, $E_2$, ..., $E_k$ for a positive integer $k$. For any cycle $C$ on $G$, the set of all $i$-dimensional edge of $C$, which is a subset of $E(C)$, is denoted as $E_i(C) = E(C) \cap E_i$. If $||E_i(C)| - |E_j(C)|| \leq 1$ for any $1 \leq i < j \leq k$, $C$ is called a *dimension-balanced cycle* (*DBC*, for short). If the length of $C$ is $p$, we call $C$ is a $p$-DBC. Combining the concept of DBC, and Hamiltonian cycle, pancyclic, $p$-pancyclic, bipancyclic and $p$-bipancyclic, the following definitions are given. Let $C$ be a DBC on $G$, if $C$ is also a Hamiltonian cycle on $G$, $C$ is called a *DBH* and $G$ is called *DB-Hamiltonian.* If $G$ contains a DBC of every length between 3 to $|V(G)|$, $G$ is called *DBP*. Similarly, if $G$ contains a DBC of every length between $p$ to $|V(G)|$, $G$ is called *p-DBP*. Besides, if $G$ is called *DBBP*, $G$ may embed a DBC of every even length between 4 to $|V(G)|$. Also, if $G$ is called *p-DBBP*, $G$ may embed a DBC of every even length between $p$ to $|V(G)|$.

The dimension-balanced cycle problem is a quite new topic of graph theory. The first research about dimension-balanced cycle has been proposed by the ref. [19]. In 2011, Wang presented that $G$ contains a DBH when $G$ is the hypercube $Q_k$ for $k = 2$, 3 and 4; $T_{n,n}$, $C_n \times K_{m,n}$, $C_3 \times K_m$ and $C_4 \times C_m$ for $m, n \geq 3$ [19]. In 2012, Peng and Juan proposed a method for finding a DBH on $T_{m,n}$ if it exist, where $m, n \geq 3$ [14]. They proved that, there is no DBH on $T_{m,n}$ for $mn$ mod $4 = 2$. [15, 9], and [6] discussed the dimension-balanced pancyclicity on $T_{m,n}$ in 2014, 2017, and 2021, respectively. But in practical, although an interconnection network contain no DBC for some special length, it still need a suitable routing cycle for those interconnection network, to give some specific methods for solving the original problem. Hence, the *weakly dimension-balanced cycle* had been discussed. [23] defined a weaker version of DBC as follows. Given a graph $G = (V, E)$, and $\{E_1, E_2, ..., E_k\}$ is a partition of $E$ with dimension $k$. Let $C$ is a cycle on $G$ and $E_i(C) = E(C) \cap E_i$. If $||E_i(C)| - |E_j(C)|| \leq 3$ for all $1 \leq i < j \leq k$, $C$ is called a *weakly dimension-balanced cycle* (WDBC, for short). If the length of $C$ is $p$, we call $C$ is a $p$-WDBC. If $C$ is also a Hamiltonian cycle on $G$, $C$ is called a *WDBH*. [8] defined some similar definitions for pancyclicity and bipancyclicity as follows. For some integer $p \geq 3$, if $G$ contains a WDBC of every length from 3 ($p$, resp.) to $|V(G)|$, $G$ is called *WDBP* (*p-WDBP*, resp.). If $G$ contains a WDBC of every even length from 4 ($p$, resp.) to $|V(G)|$, $G$ is called *WDBBP* (*p-WDBBP*, resp.).

$T_{m,n}$ is called the *toroidal mesh graph* whose vertex set $V(T_{m,n}) = \{(x, y)|0 \leq x \leq m-1, 0 \leq y \leq n-1\}$, and edge set $E(T_{m,n}) = \{(x_1, y_1)(x_2, y_2)| x_1 = x_2 \text{ and } y_1 - y_2 \equiv \pm 1 \pmod n, \text{ or } y_1 = y_2 \text{ and } x_1 - x_2 \equiv \pm 1 \pmod m\}$. Figure 1 shows an example of $T_{4,3}$. For convenience, we define the set $cross^1 = \{(0, i)(m-1, i) \mid 0 \leq i \leq n-1\}$ and the set $cross^2 = \{(j, 0)(j, n-1) \mid 0 \leq j \leq m-1\}$. The toroidal mesh graph is a famous interconnection network which have been pay more attention recently [6, 8, 9, 14, 15, 21–24]. For the toroidal mesh graph $T_{m,n}$, we set $E_1 = \{(x_1, y_1)(x_2, y_2)| y_1 = y_2 \text{ and } x_1 - x_2 \equiv \pm 1 \pmod m\}$, and $E_2 = \{(x_1, y_1)(x_2, y_2)| x_1 = x_2 \text{ and } y_1 - y_2 \equiv \pm 1 \pmod n\}$ intuitively.

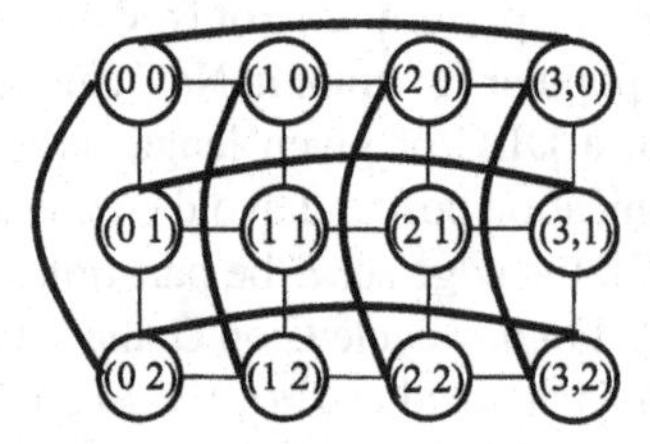


**Fig. 1.** The structure of $T_{4,3}$.

For toroidal mesh graph $T_{m,n}$, the dimension-balanced Hamiltonicity had been discussed in 2012 [14], and the $T_{m,n}$ contains a WDBC for any $m, n \geq 3$ are proved in 2018 [23]. The dimension-balanced pancyclicity on $T_{m,n}$ had been discussed separately in 2014 [15] (for both $m$ and $n$ are even), 2017 [9] (for one of $m$ and $n$ is even, the other is odd), and 2021 [6] (for both $m$ and $n$ are odd). Furthermore, the weakly dimension-balanced pancyclicity already been further discussed in 2019 [8] (for both $m$ and $n$ are even), [24] (for one of $m$ and $n$ is even, the other is odd). Therefore, the weakly dimension-balanced pancyclicity on $T_{m,n}$ for both $m$ and $n$ are odd discussed in this paper is a natural continuation of research, which will give a complete conclusion to the weakly dimension-balanced pancyclicity on $T_{m,n}$.

The remainder of this paper is organized as follows: Some related work and preliminary are presented in Sect. 2. Section 3 presents the main result. Finally, the conclusion and discussion are presented in Sect. 4.

## 2 Preliminary

By definition, a DBC is also a WDBC. The dimension-balanced Hamiltonicity and the weakly dimension-balanced Hamiltonicity on a toroidal mesh graph $T_{m,n}$ are concluded in the following two theorems.

**Theorem 1.** [14] For $m, n \geq 3$ and $mn \bmod 4 \neq 2$, there is a DBH on $T_{m,n}$.

**Theorem 2.** [23] For $m, n \geq 3$ and $mn \bmod 4 = 2$, there is a WDBH on $T_{m,n}$.

So, there is a WDBH on $T_{m,n}$ for any integers $m, n \geq 3$. That is, there is a $mn$-WDBC on $T_{m,n}$ for any integers $m, n \geq 3$. When we consider the weakly dimension-balanced pancyclicity on a toroidal mesh graph $T_{m,n}$, we only need to study whether there is a $l$-WDBC on $T_{m,n}$ for $3 \leq l < mn$. Table 1 and 2 summarized the current known results of the weakly dimension-balanced pancyclicity on a toroidal mesh graph $T_{m,n}$.

**Table 1.** The existence of WDBP on $T_{m,n}$ for both $m$ and $n$ are even [8].

| $T_{m,n}$ exist WDBC or not for even $m, n \geq 4$ | | |
|---|---|---|
| $4k$-WDBC | $(4k+2)$-WDBC | $(2k+1)$-WDBC |
| **Yes,** $1 \leq k \leq mn/4$ | **Yes,** $1 \leq k \leq mn/4 - 1$ | **No** |

**Table 2.** The existence of WDBP on $T_{m,n}$ for one of $m$ and $n$ is even, the other is odd [24].

| $T_{m,n}$ exist WDBC or not for one of $m$ and $n$ is even, the other is odd | | | |
|---|---|---|---|
| | $4k$-WDBC | $(4k+2)$-WDBC | $(2k+1)$-WDBC |
| $m \geq 4$ is even, $n \geq 4$ is odd | **Yes,** $1 \leq k \leq \lfloor mn/4 \rfloor$ | **Yes,** $1 \leq k \leq \lfloor (mn-2)/4 \rfloor$ | **Yes,** $n-2 \leq k \leq mn/2 - 1$ **No,** $1 < k < n-2$ |
| $m \geq 4$ is even, $n = 3$ | **Yes,** $k = 1, 2, 3$, or $m/2 \leq k \leq \lfloor mn/4 \rfloor$; **No,** $3 < k < m/2$ | **Yes,** $k = 1, 2, 3, 4$ or $m/2 - 1 \leq k \leq \lfloor (3m - 2)/4 \rfloor$; **No,** $4 < k < m/2 - 1$ | **Yes,** $k = 1 \sim 8, 10$ or $m - 2 \leq k \leq mn/2 - 1$ **No,** $k = 9$, or $10 < k < m - 2$ |

For both $m$ ($\geq 3$) and $n$ ($\geq 3$) are odd, Juan et al. studied the dimension-balanced pancyclicity on $T_{m,n}$ in 2021 [6]. The results of [6] shows in Table 3.

In next section, we will consider the "No" parts of Table 3 step by step, and try to find whether there exist $l$-WDBC for any $3 \leq l \leq mn$ in $T_{m,n}$ for both $m, n$ are odd. The following two properties stated in [6] are still useful in this article.

**Property 1.** [6] Let $C$ be a cycle on $T_{m,n}$. If $m$ ($n$, resp.) is even, then $|E_1(C)|$ ($|E_2(C)|$, resp.) is even.

**Property 2.** [6] If $m$ ($n$, resp.) is odd and there exist a cycle $C$ on $T_{m,n}$ such that $|E_1(C)|$ ($|E_2(C)|$, resp.) is odd, then $|E_1(C)| \geq m$ ($|E_2(C)| \geq n$, resp.) and with odd number of edges in cross$^1$ (cross$^2$, resp.).

**Table 3.** Summary of ref. [6]: $T_{m,n}$ has DBC or not for both $m$, $n$ are odd.

| $T_{m,n}$ exist DBC or not for odd $m, n \geq 3$ | | | |
|---|---|---|---|
| | $4k$-DBC | $(4k+2)$-DBC | $(2k+1)$-DBC |
| $m \geq n \geq 5$ | **Yes,** $1 \leq k \leq \lfloor (mn-3)/4 \rfloor$ | **Yes,** $(m-1)/2 \leq k \leq (mn-3)/4$ | **Yes,** $n-1 \leq k \leq (mn-1)/2$ **No,** $1 \leq k < n-1$ |
| $n = 3.$ | **Yes,** $k = 1, 2$ or $3$; **No,** $3 < k \leq \lfloor 3m/4 \rfloor$ | **No,** $1 \leq k < (m-1)/2$ | **Yes,** $k = 2, 3, 4, 5, 7$, or $m-1 \leq k \leq (3m-1)/2$ **No,** $k = 1, 6$ or $7 < k < m-1$ |

For convenience, we give the following definitions on $T_{m,n}$: for any $0 \leq y \leq n-1$, a cycle $R^{1,y}$ is an induced subgraph of $T_{m,n}$ by the vertex set $V(R^{1,y}) = \{(i, y)|0 \leq i \leq m-1\}$. Similarly, for any $0 \leq x \leq m-1$, a cycle $R^{2,x}$ is an induced subgraph of $T_{m,n}$ by the vertex set $V(R^{2,x}) = \{(x, j)|0 \leq j \leq n-1\}$. Let $R_{i,j}^{1,y,+} = \langle (i, y), (i+1, y), \ldots, (j, y) \rangle$ be a path for any $0 \leq i \leq j < m$; $R_{i,j}^{2,x,+} = \langle (x, i), (x, i+1), \ldots, (x, j) \rangle$ be a path for any $0 \leq i \leq j < n$. $R_{j,i}^{1,y,-} = \langle (j, y), (j-1, y), \ldots, (i, y) \rangle$ be a path for any $0 \leq i \leq j < m$; $R_{j,i}^{2,x,-} = \langle (x, j), (x, j-1), \ldots, (x, i) \rangle$ be a path for any $0 \leq i \leq j < n$.

## 3 The Main Discussion

In this section, the following lemma, theorem and corollary are the study result of the weakly dimension-balanced pancyclicity on toroidal mesh graph $T_{m,n}$ for both $m$ and $n$ are odd. According to Table 3, there are four "NO"s in three cases. Hence, we divide this section into three subsections based on the length of the WDBC we seek.

### 3.1 4*k*-WDBC

In this subsection, we prove that when $k$ is small $(3 < k < (m-1)/2)$, there is no $4k$-WDBC on $T_{m,3}$; otherwise, there exist a $4k$-WDBC on $T_{m,3}$. For example, there is no 16-WDBC on $T_{11,3}$, and there is a 20-WDBC on $T_{11,3}$.

**Lemma 1.** For $m$ is odd, $T_{m,3}$ does not contain every $4k$-WDBC for $3 < k < (m-1)/2$.

**Proof.** Assume that $C$ is a $4k$-WDBC on $T_{m,3}$. By Properties 1 and 2, we know that $|E_1(C)|$ is odd and $C$ has odd number of edges in cross[1]. So, $|E_1(C)| \geq m$ and $|E_2(C)| = 4k - |E_1(C)| \leq 4k - m \leq 4((m-1)/2 - 1) - m = m - 6$, and $||E_1(C)| - |E_2(C)|| \geq 6 > 3$ is a contradiction. Therefore, $T_{m,3}$ does not have $4k$-WDBC where $3 < k < (m-1)/2$. □

**Theorem 3.** If $m \geq 5$ is odd, $T_{m,3}$ contains $4k$-WDBC for $3 < (m-1)/2 \leq k < \lfloor 3m/4 \rfloor$.

**Proof.** Because $(m-1)/2 > 3$ and $m$ is odd, that means $m \geq 9$. Since there is no 4$k$-DBC in this case, we know that $\{|E_1(C)|, |E_2(C)|\} = \{2k+1, 2k-1\}$ if $C$ is a 4$k$-WDBC on $T_{m,3}$. In the following, we will construct a WDBC $C$ on $T_{m,3}$ with $|E_1(C)| = 2k+1$ and $|E_2(C)| = 2k-1$. Let $\rho = (2k+1-m)/2$. If $\rho$ is even, let $\beta_1 = (2k-1-3) \bmod (m-\rho-1)$ and $n_1 = (2k-1-3-\beta_1)/(m-\rho-1) = \lfloor(2k-1-3)/(m-\rho-1)\rfloor$; if $\rho$ is odd, let $\beta_2 = (2k-1-3) \bmod (m-\rho-2)$ and $n_2 = (2k-1-3-\beta_2)/(m-\rho-2) = \lfloor(2k-1-3)/(m-\rho-2)\rfloor$.

When $\rho$ is even, we construct a 4$k$-cycle $C_1 = \langle(0, 0), (m-1, 0), R^{1,0,-}_{m-1,m-\rho-1}, (m-\rho-1, 0), (m-\rho-1, 1), R^{1,1,+}_{m-\rho-1,m-1}, (m-1, 1), (m-1, 2), R^{1,2,-}_{m-1,m-\rho-2}, (m-\rho-2, 2), R^{2,m-\rho-2,-}_{2,1-n_1}, (m-\rho-2, 1-n_1), (m-\rho-3, 1-n_1), R^{2,m-\rho-3,+}_{1-n_1,2}, (m-\rho-3, 2), \ldots, (m-\rho-\beta_1-1, 2), (m-\rho-\beta_1-2, 2), (m-\rho-\beta_1-2, 2-n_1), (m-\rho-\beta_1-3, 2-n_1), (m-\rho-\beta_1-3, 2), \ldots, (0, 2), (0, 0)\rangle$.

The structure of $C_1$ is shown in Fig. 2. Obviously, $|E_1(C_1)| = m + 2\rho = 2k+1$ and $|E_2(C_1)| = 1 + (2-(2-n_1))(m-\rho-\beta_1-1) + (2-(1-n_1))((m-\rho-2)-(m-\rho-\beta_1-1)+1) + 2 = 3 + n_1(m-\rho-1) + \beta_1 = 2k-1$. So $C_1$ is a 4$k$-WDBC.

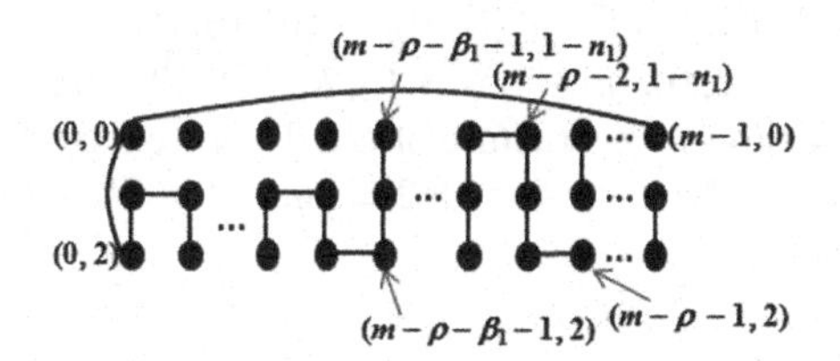


**Fig. 2.** The constructed WDBC $C_1$ on $T_{m,3}$ when $\rho$ is even of Theorem 3.

Note that 1. Since $k \leq \lfloor 3m/4\rfloor \leq 3m/4$, so $m-\rho-1 = m-(2k+1-m)/2-1 = 3m/2-k-3/2 \geq (3m-3)/2-3m/4 = 3m/4-3/2 \geq 0$ when $m \geq 2$. And 2. $k \geq (m-1)/2$ and $m \geq 9$ is odd, so $2k-1-3 \geq 2(m-1)/2-1-3 = m-5 \geq 0$, then $n_1 = \lfloor(2k-1-3)/(m-\rho-1)\rfloor \geq 0$. In the other hand, $n_1 = \lfloor(2k-1-3)/(m-\rho-1)\rfloor \leq (2k-4)/(m-\rho-1) \leq (2(3m)/4-4)/(m-\rho-1) = (3m-8)/(3m-2k-3) \leq (3m-8)/(3m-2(3m)/4-3) = (6m-16)/(3m-6) < 2$. Since $m-\rho-\beta_1-1 > m-\rho-(m-\rho-1)-1 = 0$, so $0 \leq n_1 \leq 1$ and $m-\rho-\beta_1-1 > 0$. That is, $C_1$ is a well-defined cycle.

When $\rho$ is odd and $n_2 < 2$, we construct a 4$k$-cycle $C_2 = \langle(0, 0), C_1((0,0), (m-\rho-2, 2)), (m-\rho-2, 2), R^{2,m-\rho-2,-}_{2,1-n_2}, (m-\rho-2, 1-n_2), (m-\rho-3, 1-n_2), R^{2,m-\rho-3,+}_{1-n_2,2}, (m-\rho-3, 2), \ldots, (m-\rho-\beta_2-1, 2), (m-\rho-\beta_2-2, 2), (m-\rho-\beta_2-2, 2-n_2), (m-\rho-\beta_2-3, 2-n_2), (m-\rho-\beta_2-3, 2), \ldots, (1, 2), (0, 2), (0, 0)\rangle$.

According to Fig. 3, we know that $|E_1(C_2)| = m + 2\rho = 2k+1$ and $|E_2(C_2)| = 1 + (2-(2-n_1))(m-\rho-\beta_2-2) + (2-(1-n_1))((m-\rho-2)-(m-\rho-\beta_2-1)+1) + 2 = 3 + n_2(m-\rho-2) + \beta_2 = 2k-1$. Hence $C_2$ is a 4$k$-WDBC.

Also note that 1. Since $k \leq\lfloor 3m/4\rfloor \leq 3m/4$, so $m-\rho-2 = m-(2k+1-m)/2-2 = 3m/2-k-5/2 \geq 3m/2-3m/4-5/2 = 3m/4-5/2 \geq 0$ when $m \geq 4$. And 2. $k \geq (m-1)/2$ and $m \geq 9$, hence $2k-1-3 \geq 2(m-1)/2-1-3 \geq m-5 \geq 0$, $n_2 =$

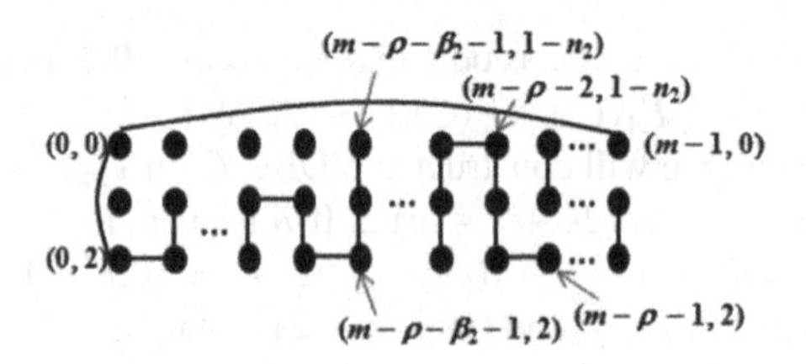


**Fig. 3.** The constructed WDBC $C_2$ on $T_{m,3}$ when $\rho$ is odd of Theorem 3.

$\lfloor(2k-1-3)/(m-\rho-2)\rfloor \geq 0$. Besides, $m-\rho-\beta_2-1 > m-\rho-(m-\rho-2)-1 = 1 > 0$. That is, $m-\rho-\beta_2-1 \geq 2$. Since $n_2 = \lfloor(2k-1-3)/(m-\rho-2)\rfloor \leq (2k-4)/(m-\rho-2) \leq (2(3m)/4-4)/(m-\rho-2) = (3m-8)/(3m-2k-5) \leq (3m-8)/(3m-2(3m/4)-5) = (6m-16)/(3m-10)$. Since $m \geq 9 > 14/3$, $14 < 3m$ and $6m-16 < 9m-30$, so $(6m-16)/(3m-10) < 3$. That is, $0 \leq n_2 \leq 2$. When $n_2 = 2$, this only occur in $k = (3m-1)/4$, because if $k < (3m-1)/4$, $n_2 = \lfloor(2k-1-3)/(m-\rho-2)\rfloor \leq (2k-4)/(m-\rho-2) < (2(3m-1)/4-4)/(m-\rho-2) = (3m-9)/(3m-2k-5) < (3m-9)/(3m-2(3m-1)/4-5) = (6m-18)/(3m-9) = 2$, $n_2 < 2$. So we construct a cycle $C_3$ for $k = (3m-1)/4$. Let $C_3 = \langle(0,0), C_1((0,0),(m-\rho-2,2)), (m-\rho-2,2), R_{2,0}^{2,m-\rho-2,-}, (m-\rho-2,0), (m-\rho-3,0), R_{0,2}^{2,m-\rho-3,+}, (m-\rho-3,2),\ldots,(1,2),(0,2),(0,0)\rangle$. Figure 4 shows the structure of $C_3$. Then we have $|E_1(C_3)| = m+2\rho = 2k+1$ and $|E_2(C_3)| = 1+2(m-\rho-2)+2 = 3m-1-2k-1 = 2k-1$. Hence $C_2$ and $C_3$ are well-defined. □

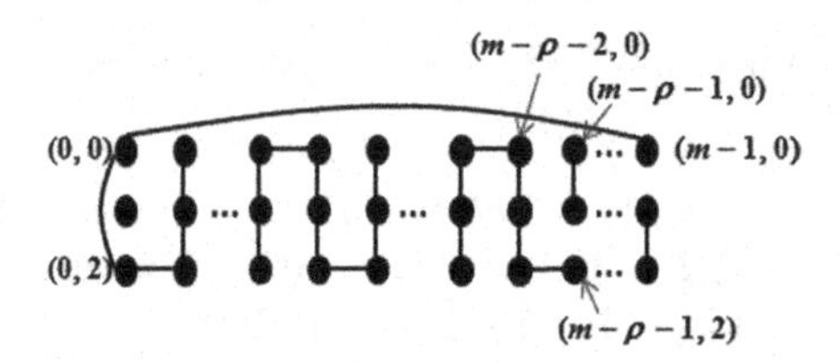


**Fig. 4.** The constructed WDBC $C_3$ on $T_{m,3}$ when $\rho$ is odd and $n_2 = 2$ of Theorem 3.

### 3.2 $(4k+2)$-WDBC

In this subsection, we prove that when $k$ is small (1, 2, 3 or 4), there is a $(4k+2)$-WDBC on $T_{m,n}$ for any odd $m, n \geq 3$; when $k$ is larger $(4 < k < (m-1)/2)$, although there is still no $(4k+2)$-WDBC on $T_{m,3}$, we can find a $(4k+2)$-WDBC on $T_{m,n}$ if $m \geq n \geq 5$.

**Lemma 2.** If $m$ is odd, $T_{m,3}$ contains no $(4k+2)$-WDBC for any $4 < k < (m-1)/2$.

**Proof.** Assume that $C$ is a $(4k+2)$-WDBC on $T_{m,3}$. Since there is no $(4k+2)$-DBC for $1 \leq k < (m-1)/2$, the only possible situation are $|E_1(C)| = 2k+2$ $(< m)$ and $|E_2(C)| = 2k$ ($|E_1(C)| = 2k$ $(< m)$ and $|E_2(C)| = 2k+2$, resp.). Therefore, $|E_2(C)|$ is even and $C$ has even number of edges in $cross^2$. Without loss of generality, say no edge in $cross^1$

be used in $C$. Let $x \in V(C)$ and $d_C(x)$ denote the degree of $x$ in $C$, be the number of edges incident with $x$ in $C$. Then let $Z_m = \{0,1,\ldots, m-1\}$ and $E_1^i(C) = \{(i, y)(x, y) \in E(C) | x = (i+1) \bmod m, 0 \le y \le 2\}$, $E_2^i(C) = \{(i, y_1)(i, y_2) \in E(C) | 0 \le y_1 \le 2$ and $y_2 = y_1 + 1 \bmod 3\}$ for $i \in Z_m$. According to the structure of $C$, $|E_1^i(C)| = 0$ or $2$ for every $i \in Z_m$. That is, $|\{i \in Z_m | |E_1^i(C)| = 2\}| = k+1$ ($k$, resp.) and those integers in the set $\{i \in Z_m | |E_1^i(C)| = 0\}$ must be consecutive. Without loss of generality, let $|E_1^i(C)| = 2$ for $0 \le i \le k$ ($k-1$, resp.) and $|E_1^j(C))| = 0$ for $k+1$ ($k$, resp.) $\le j \le m-1$. Since cycle is a 2-regular graph, we know that for $i \in Z_m$ and $j = (i+1) \bmod m$, $6 \ge d_C((j, 0)) + d_C((j, 1)) + d_C((j, 2)) = |E_1^i(C)| + |E_1^j(C)| + 2|E_2^j(C)|$ is even. If $|E_1^i(C)| + |E_1^j(C)| = 2$ then $|E_2^j(C)| \le 2$, this case only occur when $i = m-1$ or $i = k$ ($k-1$, resp.). If $|E_1^i(C)| + |E_1^j(C)| = 4$ then $|E_2^j(C)| \le 1$, this case only occur when $0 \le i \le k-1$ ($k-2$, resp.). Thus $|E_2(C)| \le 4 + k < 2k$ ($|E_2(C)| \le 4 + k - 1 < 2k + 2$, resp.) when $k > 4$. That is a contradiction. Therefore, $T_{m,3}$ does not have every $(4k+2)$-WDBC for $4 < k < (m-1)/2$. □

When $k = 1, 2, 3$ or $4$, we find a $(4k+2)$-WDBC on $T_{m,3}$ for $m$ is odd as Fig. 5 shows.

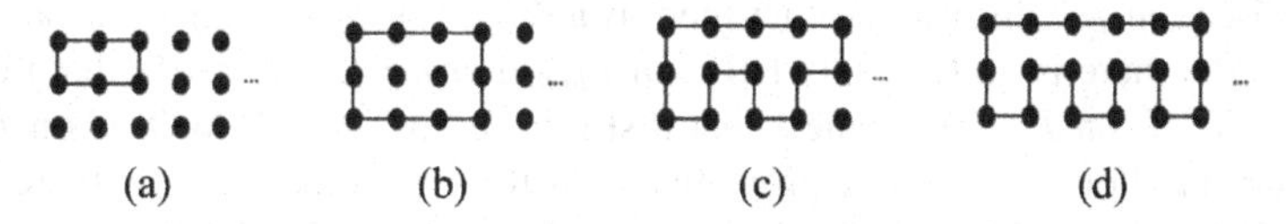


**Fig. 5.** When $m$ is odd, $T_{m,3}$ have a $(4k+2)$-WDBC for $k =$ (a) 1, (b) 2, (c) 3, (d) 4.

**Theorem 4.** For $m, n \ge 5$ are odd integers, $T_{m,n}$ contains every $(4k+2)$-WDBC for $1 \le k < (\max\{m, n\} - 1)/2$.

**Proof.** Without loss of generality, say $m \ge n$. According to the range of $k$, we divided this proof into two cases.

**Case 1:** $1 \le k \le n-2$.

Construct a $(4k+2)$-cycle $C_1 = \langle (0,0), R_{0,k}^{2,0,+}, (0, k), R_{0,k+1}^{1,k,+}, (k+1, k), R_{k,0}^{2,k+1,-}, (k+1,0), R_{k+1,0}^{1,0,-}, (0, 0)\rangle$. Since $k+1 \le n-1 \le m-1$, $C_1$ is a well-defined cycle. Obviously, $|E_1(C_1)| = 2k+2$ and $|E_2(C_1)| = k+k = 2k$. Hence $C_1$ is a $(4k+2)$-WDBC.

**Case 2:** $n-1 \le k < (m-1)/2$.

Let $\beta_1 = ((2k+2-2(n-1)) \bmod 2(n-2))/2 = (k-n+2) \bmod (n-2) = k \bmod (n-2)$ and $n_1 = ((2k+2) - (2n-2) - 2\beta_1)/2(n-2) = \lfloor (k-n+2)/(n-2) \rfloor$. Then construct a $(4k+2)$-cycle $C_2 = \langle (0, 0), (1, 0), R_{0,n-2}^{2,1,+}, (1, n-2), (2, n-2), R_{n-2,0}^{2,2,-}, (2, 0), (3, 0), R_{0,n-2}^{2,3,+}, (3, n-2), (4, n-2), \ldots, (2n_1, 0), (2n_1+1, 0), R_{0,\beta_1}^{2,2n_1+1,+}, (2n_1+1, \beta_1), (2n_1+2, \beta_1), R_{\beta_1,0}^{2,2n_1+2,-}, (2n_1+2, 0), R_{2n_1+2,k}^{1,0,+}, (k, 0), R_{0,n-1}^{2,k,+}, (k, n-1), R_{k,0}^{1,n-1,-}, (0, n-1), R_{n-1,0}^{2,0,-}, (0, 0)\rangle$.

Figure 6 shows the structure of $C_2$. Note that because $m \geq n \geq 5$ are odd integers, $2n_1 = 2\lfloor(k - n + 2)/(n - 2)\rfloor \leq 2(k/(n-2) - 1) \leq 2(k/3 - 1) < k - 3$ since $k \geq n - 1 \geq 4$, the structure of $C_2$ is well-defined. Besides, $|E_1(C_2)| = k + k = 2k$ and $|E_2(C_2)|$ $2n - 2 + 2n_1(n - 2) + 2\beta_1 + 2k + 2$. That is, $C_2$ is a $(4k + 2)$-WDBC.□

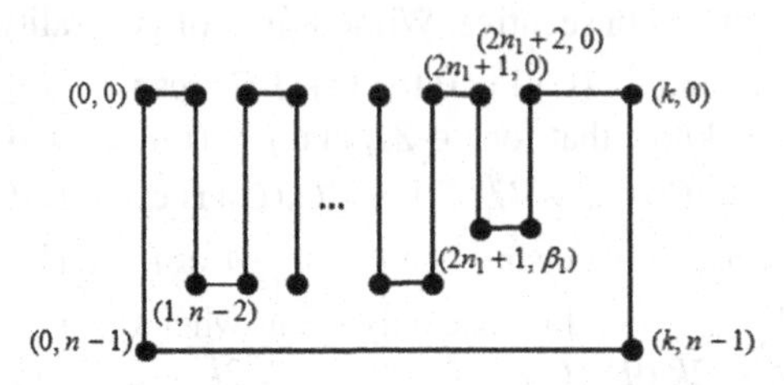


**Fig. 6.** The constructed WDBC $C_2$ on $T_{m,n}$ of Theorem 4.

### 3.3 (2k + 1)-WDBC

In this subsection, we discuss whether there is a $(2k + 1)$-WDBC on $T_{m,n}$ for some $m$, $n$, and $k$ while there is no $(2k + 1)$-DBC on $T_{m,n}$. For $n = 3$, we prove that there is a $(2k + 1)$-WDBC on $T_{m,3}$ for some $k$, but it still have no $(2k + 1)$-WDBC on $T_{m,3}$ for most cases of $k$. For $m, n \geq 5$, we prove that only for $k = \min\{m, n\} - 2$, there is a $(2k + 1)$-WDBC on $T_{m,n}$; for any other $k$, there is still no $(2k + 1)$-WDBC on $T_{m,n}$.

**Lemma 3.** If $m$ is odd, $T_{m,3}$ contains no $(2k + 1)$-WDBC for $k = 9$ or $10 < k < m - 2$.

**Proof.** Since we already know $T_{m,3}$ does not have any $(2k + 1)$-DBC for $k = 6$ or $7 < k < m - 1$ by [6]. Assume that $C$ is a $(2k + 1)$-WDBC on $T_{m,3}$. Let $k_1 + k_2 = 2k + 1$, $k_1$ be even, $k_2$ be odd and $|k_1 - k_2| = 3$. If $|E_1(C)| = k_2$ is odd number, $|E_1(C)| = k_2 \geq m$ by Property 2. Since $k < m - 2$, $k_2 \leq k + 2 < m$, a contradiction. Therefore, $|E_1(C)| = k_1$ is even and $|E_2(C)| = k_2$ is odd. Again, since $k < m - 2$, $k_1 \leq k + 2 < m$, so we can say there is no edge of $C$ in cross[1] without loss of generality. Let $x \in V(C)$ and $d_C(x)$ denote the degree of $x$ on $C$, be the number of edges incident with $x$ in $C$. Then let $Z_m = \{0, 1,\ldots, m - 1\}$ and $E_1^i(C) = \{(i, y)(x, y) \in E(C) | x = (i + 1) \bmod m, 0 \leq y \leq 2\}$, $E_2^i(C) = \{(i, y_1)(i, y_2) \in E(C) | 0 \leq y_1 \leq 2$ and $y_2 = y_1 + 1 \bmod 3\}$ for $i \in Z_m$. According to the structure of $C$, $|E_1^i(C)| = 0$ or $2$ for every $i \in Z_m$. That is, $|\{i \in Z_m | |E_1^i(C)| = 2\}| = k_1/2$ and those integers in the set $\{i \in Z_m | |E_1^i(C)| = 0\}$ must be consecutive. Without loss of generality, let $|E_1^i(C)| = 2$ for $0 \leq i \leq k_1/2 - 1$ and $|E_1^j(C)| = 0$ for $k_1/2 \leq j \leq m - 1$. Since cycle is a 2-regular graph, we know that for $i \in Z_m$ and $j = (i + 1) \bmod m$, $6 \geq d_C((j, 0)) + d_C((j, 1)) + d_C((j, 2)) = |E_1^i(C)|+|E_1^j(C)|+2|E_2^j(C)|$ is even. If $|E_1^i(C)|+|E_1^j(C)| = 2$ then $|E_2^j(C)| \leq 2$, this case only occur when $i = m - 1$ or $i = k_1/2 - 1$. If $|E_1^i(C)|+|E_1^j(C)| = 4$ then $|E_2^j(C)| \leq 1$, this case only occur when $0 \leq i \leq k_1 / 2 - 2$. Thus $k_2 = |E_2(C)| \leq 4 + k_1 / 2 - 1$. Note that $k_1 + k_2 = 2k + 1$, $k_1$ be even, $k_2$ be odd and $|k_1 - k_2| = 3$. If $k_1 = k_2 + 3$, then $k_2 \leq 9$ and $k_1 \leq 12$. If $k_1 = k_2 - 3$, then $k_2 \leq 3$ and $k_1 \leq 0$.

When $k = 9$, $2k + 1 = 19$, so that $k_1 = 8$ and $k_2 = 11$, is a contradiction. When $10 < k < m - 2$, $21 < 2k + 1 = k_1 + k_2 \le 21$, is a contradiction, too. Therefore, $T_{m,3}$ does not have every $(2k + 1)$-WDBC for $k = 9$ or $10 < k < m - 2$. □

Note that a DBC is a WDBC by the definition. According to Lemma 3 and Table 3, we still need to discuss when $k = 1, 6, 8, 10$ or $m - 2$, whether there is a $(2k + 1)$-WDBC on $T_{m,3}$ for $m$ is odd or not. Figure 7 give the answer, which shows the structure of $(2k + 1)$-WDBC on $T_{m,3}$ for $k = 1, 6, 8, 10$, respectively. And next lemma will give a $(2k + 1)$-WDBC on $T_{m,3}$ for $k = m - 2$.

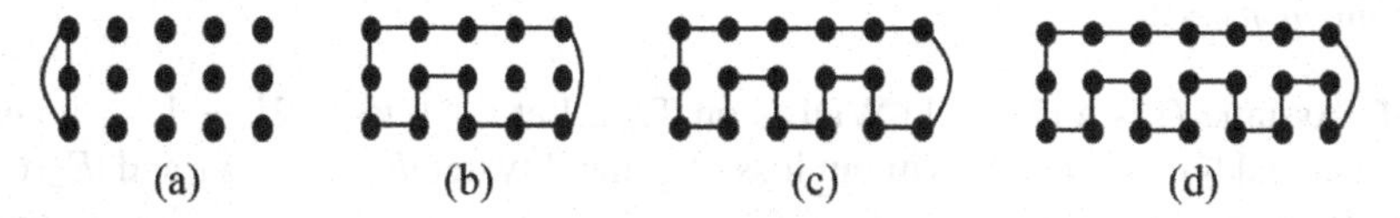


**Fig. 7.** For odd $m$, $T_{m,3}$ have a $(2k + 1)$-WDBC for $k =$ (a) 1, (b) 6, (c) 8, (d) 10.

**Lemma 4.** If $m$ is odd, $T_{m,3}$ contains $(2k + 1)$-WDBC for $k = m - 2$.

**Proof.** Note that we only need to consider $k \ge 11$ according to Table 3 and Fig. 7. For $k = m - 2$, let $|E_1(C)| = m = k + 2$, $|E_2(C)| = m - 3 = k - 1$ and $\beta_1 = (k - 1) - 4 = k - 5 = m - 7 > 0$. We can construct a $(2k + 1)$-cycle $C = \langle(0, 0), (0, 1), (0, 2), (1, 2), (1, 1), (2, 1), (2, 2),\ldots, (\beta_1, 2), R^{1,2,+}_{\beta_1,m-1}, (m - 1, 2), (m - 1, 1), (m - 1, 0), (0, 0)\rangle$. Figure 8 shows the structure of $C$. We have $|E_1(C)| = m = k + 2$ and $|E_2(C)| = 4 + \beta_1 = k - 1$. Then $||E_1(C)| - |E_2(C)|| = 3$. That is, $C$ is a $(2k + 1)$-WDBC for $k = m - 2$. □

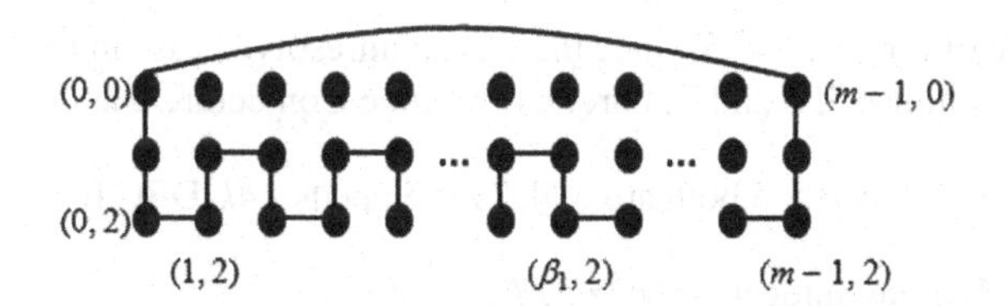


**Fig. 8.** The constructed WDBC $C$ on $T_{m,3}$ of Lemma 4.

**Lemma 5.** For both $m, n \ge 5$ are odd, $T_{m,n}$ have a $(2k + 1)$-WDBC where $k = \min\{m, n\} - 2$.

**Proof.** Without loss of generality, say $m \ge n$. So $k = n - 2$ and $2k + 1 = 2(n - 2) + 1 = 2n - 3$. Let $\gamma_1 = (k - 1) / 2$, we construct a $(2k + 1)$-cycle $C = \langle(0, 0), (0, n - 1), R^{1,n-1,+}_{0,\gamma_1}, (\gamma_1, n - 1), R^{2,\gamma_1,-}_{n-1,0}, (\gamma_1, 0), R^{1,0,-}_{\gamma_1,0}, (0, 0)\rangle$ shows in Fig. 9. In $C$, $|E_1(C)| = 2\gamma_1 = k - 1$ and $|E_2(C)| = n = k + 2$. Then $||E_1(C)|-|E_2(C)|| = 3$. That is, $C$ is a $(2k + 1)$-WDBC □

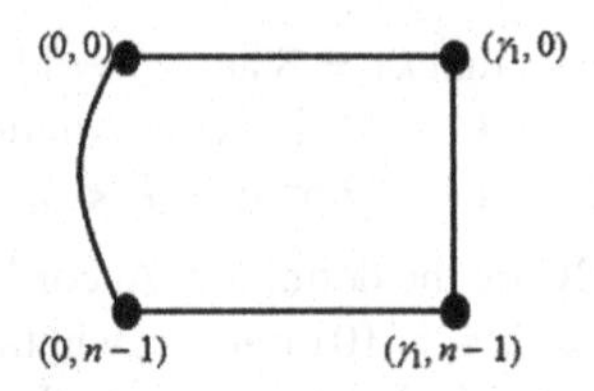


**Fig. 9.** The constructed WDBC $C$ on $T_{m,n}$ of Lemma 5.

**Lemma 6.** For both $m, n \geq 5$ are odd, $T_{m,n}$ does not have $(2k + 1)$-WDBC for any $1 \leq k < \min\{m, n\} - 2$.

**Proof.** Assume $C$ is a $(2k + 1)$-WDBC on $T_{m,n}$. Let $k_1 + k_2 = 2k + 1$, $k_1$ be even, $k_2$ be odd and $|k_1 - k_2| = 3$. Without loss of generality, let $|E_1(C)| = k_1$ and $|E_2(C)| = k_2$. By Property 2, we know that $|E_2(C)| \geq n$ and $C$ has odd number of edges in cross$^2$. Then $|E_1(C)| = 2k + 1 - |E_2(C)| \leq 2k + 1 - n < 2(\min\{m, n\} - 2) + 1 - n \leq 2(n - 2) + 1 - n = n - 3$, and $||E_1(C)| - |E_2(C)|| > 3$ is a contradiction. Therefore, $T_{m,n}$ does not have $(2k + 1)$-WDBC where $1 \leq k < \min\{m, n\} - 2$. □

## 4 Conclusion

In this paper, we study the weakly dimension-balanced pancyclicity on the toroidal mesh graph $T_{m,n}$ for both $m$ and $n$ are odd. Referring to previous research work, Table 4 lists the summary of this article. As a conclusion, according to [6], Lemmas 1, 2, 3, 4, Theorem 3, we get Corollary 1.

**Corollary 1.** For odd integer $m$, $T_{m,3}$ are $(2m - 3)$-WDBP.

For the case of both $m, n \geq 5$, since there is an unresolved case in [6] (as Conjecture 1), combining Lemmas 5, 6 and Theorem 4, we give Conjecture 2 as follows.

**Conjecture 1.** [6] If $m, n \geq 5$ both are odd, $T_{m,n}$ embeds a $4k$-DBC for $k = (mn - 1)/4$.

**Conjecture 2.** For any integers $m, n \geq 5$, $T_{m,n}$ are

(a) WDBBP when both $m$, $n$ are odd and;
(b) $(2n - 3)$-WDBP when $m \geq n$; $(2m - 3)$-WDBP when $m \leq n$.

**Table 4:** The existence of WDBP on $T_{m,n}$ for both $m$ and $n$ are odd.

| $T_{m,n}$ exist WDBC or not for odd $m, n \geq 3$ | | | |
|---|---|---|---|
| | $4k$-WDBC | $(4k+2)$-WDBC | $(2k+1)$-WDBC |
| $m \geq n \geq 5$ | **Yes,** $1 \leq k \leq \lfloor (mn-3)/4 \rfloor$ [6] | **Yes,** $1 \leq k \leq (mn-3)/4$ [6], Thm. 4 | **Yes,** $n-2 \leq k \leq (mn-1)/2$ [6], Lemma 5; **No,** $1 \leq k < n-2$ Lemma 6 |
| $n = 3.$ | **Yes,** $k = 1, 2, 3$ or $(m-1)/2 \leq k < \lfloor 3m/4 \rfloor$ Thm. 3; **No,** $3 < k < (m-1)/2$ Lemma 1 | **Yes,** $k = 1, 2, 3, 4$ and $(m-1)/2 \leq k \leq \lfloor (3m-2)/4 \rfloor$ [6], Fig. 5; **No,** $4 < k < (m-1)/2$ Lemma 2 | **Yes,** $k = 1 \sim 8, 10$ or $m-2 \leq k \leq (mn-1)/2$ [6], Fig. 7, Lemma 4; **No,** $k = 9$, or $10 < k < m-2$ Lemma 3 |

## References


1. Chen, H.C., Kung, T.L., Hsu, L.H.: An augmented pancyclicity problem of crossed cubes. Comput. J. **61**(1), 54–62 (2018)
2. Chin, Y.Y.: Binary and $M$-ary structure light patterns with Gray coding rule in active non-contact 3D surface scanning. Master's thesis of Department of Computer Science and Information Engineering, National Chi Nan University (2009)
3. Hao, R.X., Tian, Z.: The vertex-pancyclicity of data center networks. Theor. Comput. Sci. **855**, 74–89 (2021)
4. Heieh, S.Y., Lin, T.J.: Panconnectivity and edge-pancyclicity of $k$-Ary $n$-cubes. Networks **54**(1), 1–11 (2009)
5. Huang, C.H., Fang, J.F.: The pancyclicity and Hamiltonian-connectivity of the generalized base-b hypercube. Comput. Electr. Eng. **34**, 263–269 (2008)
6. Juan J.S.T., Peng, W.F., Lai, Z.Y.: Dimension-balanced pancyclicity on $T_{m,n}$ when both $m$ and $n$ are odd. In: International Conference on Parallel and Distributed Processing Techniques and Applications (PDPTA) (2021, accepted)
7. Kung, T.L., Chen, H.C., Lin, C.H., Hsu, L.H.: Three types of two-disjoint-cycle-cover pancyclicity and their applications to cycle embedding in locally twisted cubes. Comput. J. **64**(1), 27–37 (2021)
8. Lai, Z.Y., Juan, J.S.T.: The weakly dimension-balanced pancyclic on $T_{m,n}$ for both $m$ and $n$ are even. In: International Congress on Engineering and Information (ICEAI), pp. 35–44 (2019)
9. Lai, Z.Y., Peng, W.F., Juan, J.S.T.: The dimension-balanced pancyclicity on $T_{m,n}$ for one of $m$ and $n$ is even, the other is odd. In: Global Conference on Engineering and Applied Science (GCEAS), pp. 295–308 (2017)
10. Li, Z., Jiang, L.: A congestion avoidance communication protocol for network on chip. In: 2010 IEEE International Conference on Intelligent Computing and Intelligent Systems, pp. 88–92 (2010)

11. Lin, C.A.: A congestion-avoidance arbitration mechanism for adaptive traffic shaping in network on chip. Master's thesis of Department of Computer Science and Information Engineering, National Yunlin University of Science and Technology (2011)
12. Lin, Y.H.: A statistical traffic analysis for path selection in networks on chip. Master's Thesis of Department of Computer Science and Information Engineering, National Chung Hsing University (2013)
13. Lu, Y., Xu, J.M.: Bipanconnectivity of Cartesian product networks. Australas. J. Combin. **46**, 297–306 (2010)
14. Peng, W.F., Juan, J.S.T.: The balanced Hamiltonian cycle on toroidal mesh graphs. In: World Academy of Science, Engineering and Technology (WASET), 65, pp. 1096–1103 (2012)
15. Peng, W.F., Juan, J.S.T.: The dimension-balanced pancyclicity on $T_{m,n}$ for even $m$ and $n$. In: International Conference on Engineering and Applied Science (ICEAS), pp. 491–499 (2014)
16. Salvi, J., Pagès, J., Batlle, J.: Pattern codification strategies in structure light systems. Pattern Recogn. **37**, 827–849 (2004)
17. Stewart, I.A., Xiang, Y.: Bipanconnectivity and bipancyclicity in $k$-ary $n$-cubes. IEEE Trans. Parallel Distrib. Syst. **20**(1), 25–33 (2009)
18. Tsai, C.H., Jiang, S.Y.: Path bipancyclicity of hypercubes. Inf. Process. Lett. **101**, 93–97 (2007)
19. Wang, H.R.: The balanced Hamiltonian cycle problem. Master's thesis of Department of Computer Science and Information Engineering, Nation Chi Nan University (2011)
20. Wang, S.H.: 3-D surface acquisition with non-uniform Albedo using structured light range sensor. Master's thesis of Department of Computer Science and Information Engineering, National Chi Nan University (2007)
21. Wei, Y., Xu, M., Wang, K.: Strong rainbow connection numbers of toroidal meshes. Discrete Math. Algorithms Appl. **10**(03), 1850039 (2018)
22. Wibbels, M.J., Das, S., Takur, D.S., Nori, V., Stevens, K. S.: A transmission line enabled deadlock free toroidal network-on-chip using asynchronous handshake protocols. In: 2019 25th IEEE International Symposium on Asynchronous Circuits and Systems (ASYNC), pp. 36–45 (2019)
23. Wu, R.Y., Lai, Z.Y., Juan, J.S.T.: The weakly dimension-balanced Hamiltonian cycle on $T_{m,n}$. In: International Conference on Parallel and Distributed Processing Techniques and Applications (PDPTA), pp. 178–182 (2018)
24. Wu, R.Y., Lai, Z.Y., Juan, J.S.T.: The weakly dimension-balanced pancyclic on $T_{m,n}$ for $m$ or $n$ being even and the other being odd. In: International Conference on Parallel and Distributed Processing Techniques and Applications (PDPTA), pp. 59–65 (2019)
25. Zhang, Z.B., Zhang, X., Gutin, G., Lou, D.: Hamiltonicity, pancyclicity, and full cycle extendability in multipartite tournaments. J. Graph Theory **96**(2), 171–191 (2021)

# Hypercontractivity via Tensor Calculus

Maciej Skorski(✉)

University of Luxembourg, Luxembourg, Luxembourg

**Abstract.** This work improves best known numeric constants in the seminal hypercontractive inequality of Bonami, specialized to boolean polynomials of degree two. The novel approach builds on tensor calculus and a new tensor contraction inequality which is of independent interest. As an interesting byproduct, a closed-form formula for the number of perfect matchings in the cocktail-party graph is obtained.

**Keywords:** Hypercontractive inequality · Moment inequalities · Boolean polynomials · Rademacher chaos · Tensor calculus

## 1 Introduction

### 1.1 Motivation and Background

The seminal hypercontractive inequality, discovered independently first by Aline Bonami [8,9] and later by Leonard Gross [16], estimates the moments of sums known as Rademacher chaoses or boolean polynomials. Of particular importance are quadratic chaoses, which appear in high-dimensional statistics when studying the quality of random projections [1,22] or matrix trace estimators [5,28,31]. A quadratic chaos is formally defined as an expression of the form

$$F = X^T AX, \quad X \sim \{-1,1\}^n$$

where $A$ is an $n \times n$ off-diagonal matrix and the vector of random $\pm 1$ numbers $X$ is referred to as a Rademacher vector. The precise statement of the hypercontractive inequality relates the $q$-th central moment of $F$ (usually hard to determine exactly) to the variance (easy to compute explicitly). Namely, for every $q \geqslant 2$ and some constant $C(q)$ we have the following bound:

$$\mathbf{E}\left[F^q\right] \leqslant C(q) \cdot \mathbf{Var}[F]^{q/2}.$$

This goal of this work is to improve upon the best known constant $C(q)$.

The work supported by the FNR grant C17/IS/11613923.



C.-Y. Chen et al. (Eds.): COCOON 2021, LNCS 13025, pp. 449–459, 2021.
https://doi.org/10.1007/978-3-030-89543-3_38

### 1.2 Related Work

Probably the earliest proof of the hypercontractive inequality appears in Paley's work [27], albeit with no clear constant. Explicit, and best up to date, are combinatorial bounds given by Bonami [8,9]; another combinatorial albeit weaker bounds were given in parallel by Kiener [21]. In his monograph [25] O'Donnel gives a modernized and cleaned combinatorial argument to reproduce the best known bound. The hypercontractive inequality is so popular that alternative proofs, quantitatively weaker but conceptually simpler, were proposed; see for example [7,14] for an approach using entropy, [24] for a relatively short inductive argument and [6] for "sum-of-squares" proofs. Also survey articles often refer only simpler bounds, ignoring refined but more complicated versions [11].

In the modern literature, the general hypercontractive inequality has surprisingly many applications; these include information theory [3,23], analysis of Boolean functions [19], learning and approximations [12,20,26], SAT problems [13] and others. There are also extensions beyond the real domain, for example to complex numbers [18] and general Banach spaces [29].

### 1.3 Contribution

This work improves upon the best known constant $C(q)$ when $q$ is an even integer (improvements for other values will follow by a standard application of Riesz-Thorin interpolation). By doing so, it offers a novel perspective: the argument is built on tensor calculus and a novel contraction inequality, of independent interest, is obtained. In addition, a connection and application to the problem of counting perfect matching on certain graphs is discussed. The improvement is also demonstrated numerically, with the code available at GitHub [30].

### 1.4 Organization

The concepts and notation used throughout the paper are clarified in Sect. 2. Our results are stated precisely in Sect. 3, along with a comparison to prior work, and followed by the proofs in Sect. 4. The work is concluded in Sect. 5. The Python code for the empirical evaluation is published on GitHub [30].

## 2 Preliminaries

### 2.1 Basics

*Notation.* We adopt the standard notation $[n] = \{1 \ldots n\}$ and $(x)_n = \prod_{k=0}^{n-1}(x-k)$ (the falling factorial). The double-factorial is $n!! = n(n-2)(n-4)\cdots 1$ for odd $n$ and $n!! = n(n-2)(n-4)\cdots 2$ for even $n$. The gamma function $\Gamma$ extends the factorial to the complex domain via $\Gamma(z) = (z-1)!$ when $z$ is an integer.

*Partitions.* A partition $\Pi$ of a set $V$ is a family of disjoint subsets whose union is $V$. For convenience, we will refer to $k$-element subsets as simply $k$-subsets.

*Perfect Matchings.* For a graph with nodes $V$ and edges $E$ we define the subset $M$ of edges to be a perfect match if the edges in $M$ are non-overlapping (do not share a node) and cover all the nodes (the union of $M$ equals $V$).

## 2.2 Matrices

*Trace as Inner Product.* Recall (cf. [2,10]) that the mapping

$$\langle A, B\rangle \triangleq \operatorname{tr}\left(AB^T\right) = \sum_{i,j} A_{i,j}B_{i,j}$$

defines the inner product on the space of real matrices $A, B$ of same fixed shape.

*Frobenius Norm.* The trace inner product induces the Frobenius norm:

$$\|A\|_F^2 = \langle A, A\rangle = \sum_{i,j} A_{i,j}^2$$

Furthermore, this norm is sub-multiplicative (although not induced!) [17]:

$$\|AB\|_F \leqslant \|A\|_F\|B\|_F$$

## 2.3 Tensor Calculus

*Tensors.* For the purpose of this discussion, we define a tensor of rank $d$ as a $d$-dimensional array $u\left[k_1,\ldots,k_d\right]$ where dimension indices $k_1,\ldots,k_d$ take their values in, possibly different, discrete sets.

*Kroenecker Product.* Given two tensors $u\left[i_1,\ldots,i_p\right]$ and $v\left[j_1,\ldots,j_q\right]$ of rank $p$ and $q$ respectively, we define the Kronecker Product of $u$ and $v$ to be

$$(u\otimes v)\left[i_1,\ldots,i_p,j_1,\ldots,j_q\right] = u\left[i_1,\ldots,i_p\right]\cdot v\left[j_1,\ldots,j_q\right]$$

In other words, this is the tensor of rank $p+q$ storing all possible cross-products.

*Tensor Contraction.* For a tensor $u\left[k_1,k_2,\ldots,k_d\right]$ and a set $\mathcal{I}$ of non-overlapping pairs formed from $\{1,\ldots,d\}$ we define the contraction of $u$ on pairing $\mathcal{I}$ as

$$\operatorname{contr}(u;\mathcal{I}) = \sum_{k_1,\ldots,k_d}\prod_{\{i,j\}\in\mathcal{I}}\delta_{k_i=k_j}\cdot u\left[k_1,\ldots,k_d\right]$$

In other words, we sum over all the tensor entries, with the restriction that the pair of indices that are found in the paired axes set $\mathcal{I}$ are set equal.

*Examples.* We can think of a square matrix as a tensor $A[i,j]$ of rank 2. Then we can represent the matrix trace as the contraction

$$\mathrm{tr}(A) = \sum_{i=j} A[i,j] = \mathrm{contr}(A; \{\{1,2\}\})$$

that is the axes paired are $\{1,2\}$. In turn, the multiplication of matrices $A[i,j]$ and $B[j,k]$ can be represented as the contraction of their Kronecker product

$$A \cdot B = \mathrm{contr}(A\otimes; \{\{2,3\}\})$$

on axes 2 and 3 in $A \otimes B$, which correspond to the index $j$.

### 2.4 Hypergeometric Functions

Recall that the hypergeometric function is defined as (cf. [15]):

$${}_pF_q\left(\begin{matrix} a_1, \cdots, a_p \\ b_1, \cdots, b_q \end{matrix} \mid z\right) = \sum_{n=0}^{\infty} \frac{(a_1)_n \cdots (a_p)_n}{(b_1)_n \cdots (b_q)_n} \frac{z^n}{n!}$$

where the case $p = q = 1$ is referred to as confluential hypergeometric function.

## 3 Results

### 3.1 Product-Contraction Tensor Inequality

The main technical ingredient of our approach is the following inequality for contractions of the folded Kronecker product.

**Theorem 1.** *Let $a$ be a tensor of rank 2 and let $II$ be a partition of $[2q]$ into 2-subsets different from $\{1,2\},\ldots,\{2q-1,2q\}$. Then we have:*

$$\mathrm{contr}(\underbrace{a \otimes \cdots \otimes a}_{4} \pi) \leqslant \|a\|_P^q.$$

### 3.2 Improved Hypercontractivity

The main technical contribution of this work is stated below. It improves the hypercontractive inequality in the regime of even moments.

**Theorem 2.** *Let $X \sim \{-1,1\}^n$ be the Rademacher vector. Then for any off. diagonal matrix $A \in \mathbb{R}^{n\times n}$ and $F = X^T AX$ the following holds for even $q > 0$*

$$\mathrm{E}\left[F^q\right| \leqslant C(q) \cdot \mathrm{Var}\left|F\right]^{q/2}$$

*where the constant $C(q)$ is given by*

$$C(q) = 2^{-q/2} \sum_{k=0}^{q} (-1)^k \binom{q}{k} (2q-2k-1)!$$

**Table 1.** Best constants in the hypercontractive inequality (polynomials of degree 2).

| $C(q)$ | Reference | Technique |
|---|---|---|
| $(q-1)^q$ | Most popular version [11] | Noise operator, induction, others |
| $2^{-q/2}q!/(q/2)^q \cdot (q-1)^q$ | Bonami and O'Donnel [8,25] | Combinatorics, graph enumerating |
| $2^{-q/2}\sum_{k=0}^{q}(-1)^k \binom{q}{k} (2q-2k-1)!!$ | **This work** | **Tensor calculus** |

*Remark 1.* It is possible to derive a closed-form expression for the constant $C(q)$, with the help of hypergeometric summation techniques.

In order to accurately compare our bound with the results form prior works, we precisely state the relevant constants and highlight proof techniques in Table 1.

Consequently, Fig. 1 illustrates the gain of Theorem 2 with respect to prior works. For details on numerical experiments, we refer to the code repository [30].

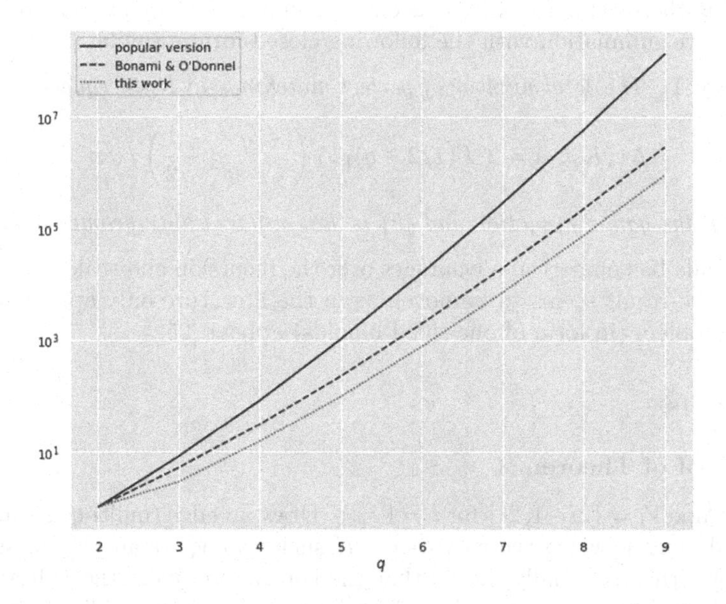


**Fig. 1.** The constant $C(q)$ in the hypercontractive inequality $\mathbf{E}[F^q] \leqslant C(q)\mathbf{Var}[F]^{q/2}$ for $F$ of degree 2. Theorem 2 is compared with prior works (note the logarithmic scale).

### 3.3 Application: Counting Perfect Matchings

The cocktail-party graph $K_{q\times 2}$ is the complete $q$-partite graph, where nodes are partitioned in 2-subsets $V_1,\ldots,V_q$ and edges appear only if nodes does not

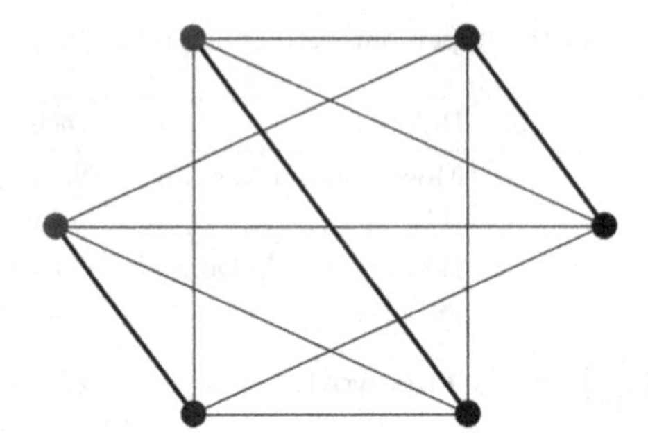

**Fig. 2.** The cocktail-party graph $K_{3\times 2}$ and its perfect matching (bold edges).

belong to the same component $V_k$ (intuitively: node subsets represent couples at the party and edges represent handshakes to be done - excluding those between partners); see Fig. 2 below for an illustrative example.

Looking into the task of enumerating tensor contractions in Theorem 1, we observe that contractions of interest correspond exactly to perfect matchings on the $q$-partite graph with nodes partitioned into $\{1,2\},\{3,4\},\ldots,\{2q-1,2q\}$. Leveraging the machinery of hypergeometric summation, we are able to replace the recursive summation with the following closed-form formula:

**Corollary 1.** *The total number of perfect matchings in $K_{q\times 2}$ equals*

$$\mathcal{M}\left(K_{q\times 2}\right) = 2^q\Gamma(1/2+q)_1F_1\left(\begin{matrix}-q\\ \frac{1}{2}-q\end{matrix} \mid -\frac{1}{2}\right)/\sqrt{\pi}$$

*where $\Gamma$ is the gamma function and ${}_1F_1$ is the confluent hypergeometric function.*

This formula has numerical advantages over the recursion and scales better with large $q$. The result seems to be novel, as in the literature only approximations for this problem (in form of one-sided bounds) appear [25].

## 4 Proofs

### 4.1 Proof of Theorem 1

Proof. Define $V_i = \{2i-1, 2i\}$ for $i = 1\ldots q$. Draw an edge (multi-edges allowed) between $V_i$ and $V_j$ when there exists $e\in\Pi$ such that $V_i\cap e$ and $V_j\cap e$ are both non-empty; this essentially means that the contraction pairs the $i$-th and $j$-th copy of the tensor $a$ in the $q$-fold product, on the dimensions explicitly indicated by e. By the properties of $\Pi$, every node is of degree 2 and thus connected components must be cycles. Thus, the graph can be seen as a union of disjoint cycles of length $m_1,\ldots,m_s$. Then, by the definition of the tensor product and contraction, we can write:

$$\operatorname{contr}\left(a^{\otimes q};\Pi\right) = \prod_s \sum_{i_1,i_2,i_{m_s},i_{m_s}+1=i_1} \prod_{k=1}^{m_s} a^{s,k}\left[i_k,i_{k+1}\right]$$

where for each $s, k$ either $a^{s,k} = a$ or $a^{k,s} = a^T$; we transpose accordingly so that the outer dimension of the left tensor same as the inner dimension of the right tensor. Now by the definition of the matrix trace we obtain:

$$\sum_{i_1, i_2, i_{m_x}, i_{m_*+1} = i_1} \prod_{k=1}^{m_*} a^{s,k}\left[i_k, i_{k+1}\right] = \operatorname{tr}\left(\prod_{k=1}^{m_s} a^{s,k}\right)$$

where the product on the right is the standard matrix dot product. We now invoke the inner-product properties of the matrix trace and the sub-multiplicativity of the Frobenius norm. Along with the symmetry $\left\|A^T\right\|_F = \|A\|_F$ this gives:

$$\begin{aligned} \left|\operatorname{tr}\left(\prod_{k=1}^{m_s} a^{s,k}\right)\right| &\leqslant \left\|a^{s,1}\right\|_F \cdot \left\|\prod_{k=2}^{m_s} a^{s,k}\right\|_F \\ &\leqslant \prod_{k=2}^{m_*} \left\|a^{s,k}\right\|_F \\ &= \|a\|_F^{m_s}. \end{aligned}$$

Using this estimate we finally obtain the inequality

$$\operatorname{contr}\left(a^{\otimes q}; \Pi\right) \leqslant \prod_s \|a\|_F^{m_s} = \|a\|_F^q,$$

which completes the proof.

## 4.2 Proof of Theorem 2

*Step 1: Expanding Product.* Recall that $A$ is off-diagonal. We have

$$\mathbf{E}\left[\left(X^T A X\right)^q\right] = \sum_{i_1, j_1, i_2, j_2, \ldots, i_q, j_q} \mathbf{E}\left[\prod_{k=1}^q a_{i_k, j_k} X_{i_k} X_{j_k}\right].$$

The expectation of each product is non-zero only when in the sequence of indices $i_1, j_1, i_2, j_2, \ldots, i_q, j_q$ every value occurs an even number of times, and when $i_k \neq j_k$ for $k = 1 \ldots q$. For convenience, we will call such a sequence *valid*.

*Step 2: Partition-Labeling Encoding.* Observe that valid indices $k_1, k_2, \ldots, k_{2q}$ are generated using labeled partitions of $[2q]$ into 2-subsets, as shown by Algorithm 1.

To prove this claim, consider the following Algorithm 2.

Observe that in each encoder's round one can find a novel pair $i \neq j$ such that $k_i = k_j$; this follows by the fact that each value appears in the sequence $(k_i)$ with even multiplicity. Thus, in the end we obtain a correct partition $\Pi$; the fact that $\Pi$ does not contain a set of form $\{2k-1, 2k\}$ follows by the assumption

**Algorithm 1:** Decoder

**Input**:
- partition $\Pi$ of $[2q]$ into 2 -subsets different from $\{1,2\},\ldots,\{2q-1,2q\}$
- labeling $\ell : \Pi \rightarrow [n]$

**Output**: sequence of valid indices $k_1,\ldots,k_{2q} \in [n]^{2q}$

1 **for** $\{i,j\} \in \Pi$ **do**
2 $\quad k_i, k_j \triangleq \ell(\{i,j\})$

**Algorithm 2:** Encoder

**Input**:
- sequence of valid indices $k_1,\ldots,k_{2q} \in [n]^{2q}$

**Output**:

- partition $\Pi$ of $[2q]$ into 2 -subsets different from $\{1,2\},\ldots,\{2q-1,2q\}$
- labeling $\ell : \Pi \rightarrow [n]$

1 $\Pi \leftarrow \emptyset$
2 **while** $\#\Pi < q$ **do**
3 $\quad$ find $i \neq j$ s.t. $k_i = k_j$ and $\{i,j\} \in \Pi$
4 $\quad \ell(i), \ell(j) \triangleq k_i$
5 $\quad \Pi \leftarrow \Pi \cup \{i,j\}$

that $k_{2i-1} \neq k_{2i}$. Thus, the decoder works as claimed. Furthermore, for every valid sequence $k_1,\ldots,k_{2q}$ we have that

$$\text{Decoder}\left(\text{Encoder}\left(k_1,\ldots,k_{2q}\right)\right) = k_1,\ldots,k_{2q}$$

which shows that the decoder is an "onto" mapping, and proves the claim. From this discussion it follows that the following inequality holds:

$$\mathbf{E}\left[\left(X^T A X\right)^q\right] \leqslant \sum_{\Pi,\ell,(i_1,j_1,\ldots,i_q,j_q)=\text{Decoder}(\Pi,\ell)} \prod_{k=1}^{q} |a_{i_k,j_k}|,$$

where the sum is over all valid inputs $\Pi, \ell$ to the decoder, and $i_1, j_1, \ldots, i_q, j_q$ denotes the decoded sequence.

*Step 3: Tensor Calculus.* Grouping the terms produced from the fixed partition and recalling the definition of tensor contraction we obtain the identity

$$\sum_{i_1,j_1,\ldots,i_q,j_q=\text{Decoder}(\Pi,\ell)} \prod_{k=1}^{q} |a_{i_k,j_k}| = \sum_{\Pi} \text{contr}(|a| \otimes \cdots \otimes |a|; \Pi),$$

where the tensor $|a|$ is understood as $|a|[i,j] \triangleq |a_{i,j}|$ and $\Pi$ runs over all partitions that can be inputs to the decoder. Therefore:

$$\mathbf{E}\left[(X^T A X)^q\right] \leqslant \sum_{\Pi} \mathrm{contr}(|a| \otimes \cdots \otimes |a|; \Pi).$$

By Theorem 1 we obtain now the following estimate

$$\mathbf{E}\left[(X^T A X)^q\right] \leqslant \#\{\Pi : \text{ as in Algorithm 1}\} \cdot \|a\|_F^q.$$

*Step 4: Counting Partitions.* How many partitions are valid for Algorithm 1? To this end, we will use the inclusion-exclusion principle. The number of partitions $[2q]$ into 2 -subsets for which $k$ given parts are on the list $\{1,2\},\ldots,\{2q-1,2q\}$ equals the number of partitions of $[2q-2k]$ into 2 -subsets. This number equals $(2q-2k-1)!!$. Thus, the total number of partitions equals

$$\#\{\Pi : \text{ as in Algorithm 1 }\} = \sum_{k=0}^{q}(-1)^k \binom{q}{k} (2q-2k-1)!!.$$

Define the constant

$$C(q) \triangleq \sum_{k=0}^{q}(-1)^k \binom{q}{k} (2q-2k-1)!!/2^{q/2}.$$

Since $\mathbf{Var}\left[X^T A X\right] = 2\|a\|_F^2$ (mind the coefficient 2!), we finally obtain

$$\mathbf{E}\left[(X^T A X)^q\right] \leqslant C(q) \cdot \mathbf{Var}\left[X^T A X\right]^{q/2},$$

which finishes the proof.

### 4.3 Proof of Corollary 1

We first note that the every partition $\Pi$ of the set $\{\Pi : \text{as in Algorithm 1}\}$ is a perfect matching on the graph $K_{q\times 2}$, with node subsets $\{1,2\},\ldots,\{2q-1,2q\}$. Thus, the following enumeration formula is valid:

$$\mathcal{M}\left(K_{q\times 2}\right) = \sum_{k=0}^{q}(-1)^k \binom{q}{k} (2q-2k-1)!!.$$

With the help of falling factorial notation we can write

$$(2q-2k-1)!! = \frac{(2q-1)!!}{(-2)^k(1/2-q)_k},$$

and also

$$\binom{q}{k} = \frac{(-1)^k(-q)_k}{k!}.$$

Thus, invoking the definition of the hypergeometric series, we finally get

$$\mathcal{M}\left(K_{q\times 2}\right) = (2q-1)!! \cdot \sum_{k=0}^{q} \frac{(-q)_k}{(1/2-q)_k} \cdot \frac{(-1/2)^k}{k!}.$$

Since the summation can be extended to $k > q$ without changing the value (yielding zero terms), we obtain

$$\mathcal{M}\left(K_{q\times 2}\right) = {}_1F_1\left( {-q \atop \frac{1}{2}-q} \mid -\frac{1}{2} \right).$$

Since $(2q-1)!! = 2^q \Gamma(1/2+q)/\sqrt{\pi}$ (see [4]), the result follows.

## 5 Conclusion

Although hypercontractivity is a well-researched topic, this work managed to improve upon the best numeric results in the interesting case of quadratic polynomials. The novel method which reduces the problem to tensor inequalities is of independent interest. The result is applicable, in particular, to determining the quality of random projections and probabilistic trace estimators.

## References


1. Achlioptas, D.: Database-friendly random projections: Johnson-lindenstrauss with binary coins. J. Comput. Syst. Sci. **66**(4), 671–687 (2003)
2. Amir-Moéz, A.R., Davis, C.: Generalized Frobenius inner products. Math. Ann. **141**(2), 107–112 (1960). https://doi.org/10.1007/BF01360166
3. Anantharam, V., Gohari, A., Kamath, S., Nair, C.: On hypercontractivity and a data processing inequality. In: 2014 IEEE International Symposium on Information Theory, pp. 3022–3026. IEEE (2014)
4. Arfken, G.B., Weber, H.J.: Mathematical methods for physicists (1999)
5. Avron, H., Toledo, S.: Randomized algorithms for estimating the trace of an implicit symmetric positive semi-definite matrix. J. ACM (JACM) **58**(2), 1–34 (2011)
6. Barak, B., Brandao, F.G., Harrow, A.W., Kelner, J., Steurer, D., Zhou, Y.: Hypercontractivity, sum-of-squares proofs, and their applications. In: Proceedings of the Forty-Fourth Annual ACM Symposium on Theory of Computing, pp. 307–326 (2012)
7. Blais, E., Tan, L.Y.: Hypercontractivity via the entropy method. Theory Comput. **9**(1), 889–896 (2013)
8. Bonami, A.: Ensembles $\lambda(p)$ dans le dual de $d^{\infty}$. In: Annales de l'institut Fourier, vol. 18, pp. 193–204 (1968)
9. Bonami, A.: Étude des coefficients de fourier des fonctions de $l^p(g)$. In: Annales de l'institut Fourier, vol. 20, pp. 335–402 (1970)
10. Böttcher, A., Wenzel, D.: The Frobenius norm and the commutator. Linear Algebra Appl. **429**(8–9), 1864–1885 (2008)
11. De Wolf, R.: A brief introduction to Fourier analysis on the Boolean cube. Theory Comput. 1–20 (2008)

12. Feige, U., Kilian, J.: Zero knowledge and the chromatic number. J. Comput. Syst. Sci. **57**(2), 187–199 (1998)
13. Friedgut, E., Bourgain, J., et al.: Sharp thresholds of graph properties, and the $k$-SAT problem. J. Am. Math. Soc. **12**(4), 1017–1054 (1999)
14. Friedgut, E., Rödl, V.: Proof of a hypercontractive estimate via entropy. Isr. J. Math. **125**(1), 369–380 (2001). https://doi.org/10.1007/BF02773387
15. Gasper, G., Rahman, M., George, G.: Basic Hypergeometric Series, vol. 96. Cambridge University Press, Cambridge (2004)
16. Gross, L.: Logarithmic sobolev inequalities. Am. J. Math. **97**(4), 1061–1083 (1975)
17. Hackbusch, W.: Tensor Spaces and Numerical Tensor Calculus, vol. 42. Springer, Heidelberg (2012). https://doi.org/10.1007/978-3-642-28027-6
18. Ivanisvili, P., Tkocz, T., et al.: Comparison of moments of Rademacher chaoses. Ark. Mat. **57**(1), 121–128 (2019)
19. Kahn, J.: The influence of variables on Boolean functions. In: 1988 Proceedings of the 29th Annual IEEE Symposium on Foundations of Computer Science (1988)
20. Khot, S., Kindler, G., Mossel, E., O'Donnell, R.: Optimal inapproximability results for MAX-CUT and other 2-variable CSPs? SIAM J. Comput. **37**(1), 319–357 (2007)
21. Kiener, K.: Uber Produkte von quadratisch integrierbaren Funktionen endlicher Vielfalt. Ph.D. thesis, PhD thesis, PhD thesis, Dissertation, Universität Innsbruck (1969)
22. Matoušek, J.: On variants of the Johnson-Lindenstrauss lemma. Random Struct. Algorithms **33**(2), 142–156 (2008)
23. Montanaro, A.: Some applications of hypercontractive inequalities in quantum information theory. J. Math. Phys. **53**(12), 122206 (2012)
24. Mossel, E., O'Donnell, R., Oleszkiewicz, K.: Noise stability of functions with low influences: invariance and optimality. In: 46th Annual IEEE Symposium on Foundations of Computer Science (FOCS 2005), pp. 21–30. IEEE (2005)
25. O'Donnell, R.: Analysis of Boolean Functions. Cambridge University Press, Cambridge (2014)
26. O'Donnell, R., Servedio, R.A.: Learning monotone decision trees in polynomial time. SIAM J. Comput. **37**(3), 827–844 (2007)
27. Paley, R.: A remarkable series of orthogonal functions (i). Proc. Lond. Math. Soc. **2**(1), 241–264 (1932)
28. Roosta-Khorasani, F., Ascher, U.: Improved bounds on sample size for implicit matrix trace estimators. Found. Comput. Math. **15**(5), 1187–1212 (2015). https://doi.org/10.1007/s10208-014-9220-1
29. Rosenthal, H.P.: Convolution by a biased coin. The Altgeld Book 76 (1975)
30. Skorski, M.: Improved hypercontractivity: Github repo (2021). https://github.com/maciejskorski/improved_hypercontractivity
31. Skorski, M.: Modern analysis of Hutchinson's trace estimator. In: 2021 55th Annual Conference on Information Sciences and Systems (CISS), pp. 1–5. IEEE (2021)

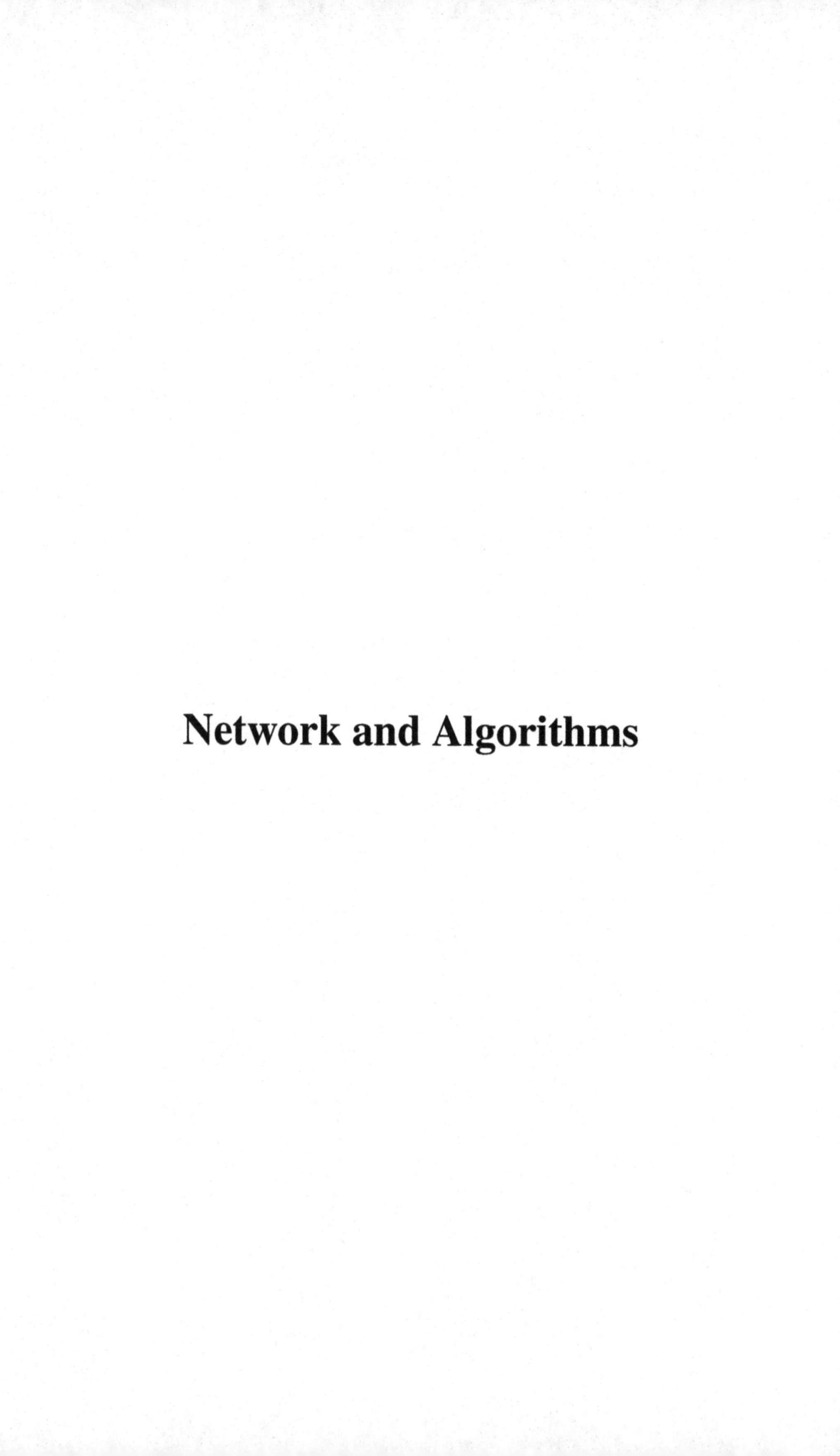

# Network and Algorithms

# Respecting Lower Bounds in Uniform Lower and Upper Bounded Facility Location Problem

Neelima Gupta, Sapna Grover(✉), and Rajni Dabas

Department of Computer Science, University of Delhi, Delhi, India
{ngupta,sgrover,rajni}@cs.du.ac.in

**Abstract.** With growing emphasis on e-commerce marketplace platforms where we have a central platform mediating between the seller and the buyer, it becomes important to keep a check on the availability and profitability of the central store. A store serving too less clients can be non-profitable and a store getting too many orders can lead to bad service to the customers which can be detrimental for the business. In this paper, we study the facility location problem (FL) with upper and lower bounds on the number of clients an open facility serves. Constant factor approximations are known for the restricted variants of the problem with only the upper bounds or only the lower bounds. The only work that deals with bounds on both sides violates both the bounds [7]. In this paper, we present the first (constant factor) approximation for the problem violating the upper bound by a factor of (5/2) without violating the lower bounds when both the lower and the upper bounds are uniform. We first give a tri-criteria (constant factor) approximation violating both the upper and the lower bounds and then get rid of violation in lower bounds by transforming the problem instance to an instance of capacitated facility location problem.

**Keywords:** Facility location · Lower bounds · Upper bounds · Approximation

## 1 Introduction

Facility location problem (FL) is a well motivated and extensively studied problem. Given a set of facilities with facility opening costs and a set of clients with a metric specifying the connection costs between facilities and clients, the goal is to select a subset of facilities such that the total cost of opening the selected facilities and connecting clients to the opened facilities is minimized.

With growing emphasis on e-commerce marketplace platforms where we have a central platform mediating between the seller and the buyer, it becomes important to keep a check on the availability and profitability of the central store. A store serving too less clients can be non-profitable and a store getting too many orders can lead to bad service to the customers. This scenario leads to what


C.-Y. Chen et al. (Eds.): COCOON 2021, LNCS 13025, pp. 463–475, 2021.
https://doi.org/10.1007/978-3-030-89543-3_39

we call the lower- and upper- bounded facility location (LBUBFL) problem. Another application of the problem is a real world transportation problem presented by Lim et al. [15] where-in customers wish to ship a certain number of cargoes through some carriers at minimum transportation cost. Each carrier has a minimum (to ensure profit) and a maximum number of cargoes it can carry.

In this paper, we study the facility location problem with lower and upper bounds. We are given a set $\mathcal{C}$ of clients and a set $\mathcal{F}$ of facilities with lower bounds $\mathcal{L}_i$ and upper bounds $\mathcal{U}_i$ on the minimum and the maximum number of clients a facility $i$ can serve, respectively. Setting up a facility at location $i$ incurs cost $f_i$(called the *facility opening cost*) and servicing a client $j$ by a facility $i$ incurs cost $c(i, j)$ (called the *service cost*). We assume that the costs are metric, i.e., they satisfy the triangle inequality. Our goal is to open a subset $\mathcal{F}' \subseteq \mathcal{F}$ and compute an assignment function $\sigma : \mathcal{C} \rightarrow \mathcal{F}'$ (where $\sigma(j)$ denotes the facility that serves $j$ in the solution) such that $\mathcal{L}_i \leq |\sigma^{-1}(i)| \leq \mathcal{U}_i \; \forall i \in \mathcal{F}'$ and, the total cost of setting up the facilities and servicing the clients is minimised. The problem is known to be NP-Hard. We present the first (constant factor) approximation for the problem with uniform lower and uniform upper bounds, i.e., $\mathcal{L}_i = \mathcal{L}$ and $\mathcal{U}_i = \mathcal{U} \; \forall i \in \mathcal{F}$ without violating the lower bounds, as stated in Theorem 1.

**Definition 1.** *A tri-criteria* $(\alpha, \beta, \gamma)$- *approximation for LBUBFL problem is a solution* $S = (\mathcal{F}', \sigma)$ *satisfying* $\alpha\mathcal{L} \leq |\sigma^{-1}(i)| \leq \beta\mathcal{U} \; \forall i \in \mathcal{F}', \alpha \leq 1, \beta \geq 1$, *with cost no more than* $\gamma OPT$, *where* $OPT$ *denotes the cost of an optimal solution of the problem.*

**Theorem 1.** *A* $(1, 5/2, O(1))$- *approximation can be obtained for LBUBFL in polynomial time.*

Constant factor approximations are known for the problem with upper bounds only (popularly known as Capacitated Facility Location (CFL)) with [5, 9,18] and without [1,3,4,6,13,16,17,20] violating the capacities using local search/LP rounding techniques. Constant factor approximations are also known for the problem with lower bounds only with [10–12] and without [2,14,19] violating the lower bounds. The only work that deals with the bounds on both the sides is due to Friggstad et al. [7], which deals with the problem with non-uniform lower bounds and uniform upper bounds. They gave a constant factor approximation for the problem using LP-rounding, violating both the upper and the lower bounds by a constant factor. The technique cannot be used to get rid of the violation in the lower bounds even if they are uniform as the authors show an unbounded integrality gap for the problem. Thus, our result is an improvement over them when the lower bounds are uniform in the sense that they violate both the bounds whereas we do not violate the lower bounds.

### 1.1 Related Work

For capacitated facility location with uniform capacities, Shmoys et al. [18] gave the first constant factor(7) algorithm with a capacity blow-up of 7/2 using LP rounding techniques. An $O(1/\epsilon^2)$ factor approximation, with $(2 + \epsilon)$ violation in capacities, follows as a special case of C$k$FLP by Byrka et al. [5]. Grover et al. [9] reduced the capacity violation to $(1+\epsilon)$. For non-uniform capacities, An et al. [3] gave the first LP-based constant factor approximation by strengthening the natural LP. The local search technique has been particularly useful to deal with capacities, with the current best being 3 by Aggarwal et al. [1] for uniform capacities and $(5+\epsilon)$ due to Bansal et al. [4] for non-uniform capacities.

Lower-Bounded Facility Location (LBFL) problem was introduced by Karger and Minkoff [12] and Guha et al. [10] independently in 2000. Both the works violate the lower bounds. Zoya Svitkina [19] presented the first true constant factor(448) approximation for uniform lower bounds by reducing the problem to CFL. This was later improved to 82.6 by Ahmadian and Swamy [2] using reduction to a special case of CFL called Capacitated Discounted Facility Location. Later, Shi Li [14] gave the first true constant (4000) factor approximation for the problem with non uniform lower bounds. Li obtained the result by reducing the problem to CFL via two intermediate problems called, LBFL-P(Lower bounded facility location with penalties) and TCSD(Transportation with configurable supplies and demands). As a particular case of lower bounded $k$-FL problem, Han *et al.* [11] gave a bi-criteria solution for the non-uniform version of the problem violating the lower bounds by a constant factor.

Friggstad et al. [7] is the only work that deals with the problem with bounds on both sides. They considered non-uniform lower bounds and uniform upper bounds and, gave a constant factor approximation violating both the bounds.

### 1.2 High Level Idea

Let $I$ be an input instance of the LBUBFL problem. We first present a tri-criteria solution $(> 1/2, 3/2, O(1))$ violating both the lower as well as the upper bound and then get rid of the violation in the lower bound by reducing the problem instance $I$ to an instance $I_{cap}$ of CFL via a series of reductions ($I \rightarrow I_1 \rightarrow I_2 \rightarrow I_{cap}$). We will see that maintaining $\alpha > 1/2$ is crucial in getting rid of the violation in the lower bound. Thus, the tri-criteria solution of Friggstad et al. [7] cannot be used here as it has $\alpha < 1/2$. Also, when the lower bounds are uniform, our approach is comparatively simpler and straightforward. This is one of the major contributions of our work and it might also be of independent interest for its simplicity.

Using the tri-criteria solution $S^t = (\mathcal{F}^t, \sigma^t)$, instance $I_1$ of LBUBFL is obtained by moving the clients assigned to a facility $i$ by $\sigma^t$ to $i$ and making the facilities opened by $S^t$ free. An instance $I_2$ of LBFL is obtained from $I_1$ by ignoring the upper bounds and removing the facilities not opened by $S^t$. Finally, $I_2$ is transformed into an instance $I_{cap}$ of CFL. The key idea in the reduction is: let $n_i$ be the number of clients assigned to a facility $i$ in $\mathcal{F}^t$. If $i$ violates the lower bound in $S^t$, we create a demand of $\mathcal{L} - n_i$ at $i$ in the CFL instance otherwise we create a supply of $n_i - \mathcal{L}$ at $i$.

An approximate solution [4] to the CFL instance is computed and used to obtain an approximate solution to $I_1$ getting rid of the violation in the lower bound, by increasing the assignments at some of the violating facilities to respect the lower bound and deciding to shut others. Facility trees are constructed and processed bottom-up for the purpose. The clients of a facility are either moved up in the tree to the parent or to a sibling (shutting down the facility) until we collect at least $\mathcal{L}$ clients at a facility. Whenever $\mathcal{L}$ clients are assigned to a facility, it is opened and the subtree rooted at it is chopped off the tree and the process is repeated with the remaining tree. The process results in an increased violation in the upper bounds by plus 1. This step is the main contribution of our work. Finally, an approximate solution to $I$ is obtained by moving the clients back to their original location.

### 1.3 Organisation of the Paper

In Sect. 2, we present a tri-criteria algorithm for LBUBFL using LP rounding techniques. We reduce instance $I$ to $I_{cap}$ via instances $I_1$ and $I_2$ in Sect. 3. Finally, an approximate solution to $I_1$, that does not violate the lower bounds, is presented in Sect. 4. The full version of the paper, with detailed proofs, is presented in [8].

## 2 Computing the Tri-Criteria Solution

In this section, we first give a tri-criteria solution that violates the lower bound by a factor of $\alpha = (1 - 1/\ell)$ and the upper bound by a factor of $\beta = (2 - 1/\ell)$, where $\ell \geq 2$ is a tunable parameter. We also give a slight modification to the solution to obtain $\alpha > 1/2$ and $\beta = 3/2$.[1] This section is one of the two major contributions of our work. Instance $I$ of LBUBFL can be formulated as the following integer program (IP):

$$Minimize\ CostLBUBFL(x, y) = \sum_{j \in \mathcal{C}} \sum_{i \in \mathcal{F}} c(i,\ j) x_{ij} + \sum_{i \in \mathcal{F}} f_i y_i$$

$$\begin{aligned} subject\ to \quad & \sum_{i \in \mathcal{F}} x_{ij} \geq 1 && \forall j \in \mathcal{C} && (1) \\ \mathcal{U} y_i \geq & \sum_{j \in \mathcal{C}} x_{ij} \geq \mathcal{L} y_i && \forall i \in \mathcal{F} && (2) \\ & x_{ij} \leq y_i && \forall i \in \mathcal{F},\ j \in \mathcal{C} && (3) \\ & y_i, x_{ij} \in \{0, 1\} && && (4) \end{aligned}$$

where $y_i$ is an indicator variable which is equal to 1 if facility $i$ is open and 0 otherwise. $x_{ij}$ is an indicator variable which is equal to 1 if client $j$ is served by facility $i$ and 0 otherwise. Constraints 1 ensure that every client is served. Constraints 2 make sure that the total demand assigned to an open facility is at least $\mathcal{L}$ and at most $\mathcal{U}$. Constraints 3 ensure that a client is assigned to an open facility. LP-relaxation is obtained by allowing the variables to be non-integral. Let $\zeta^* = <x^*, y^*>$ be an optimal solution to the LP and $LP_{opt}$ be its cost.

[1] This doesn't follow from the general values for any value of $\ell$.

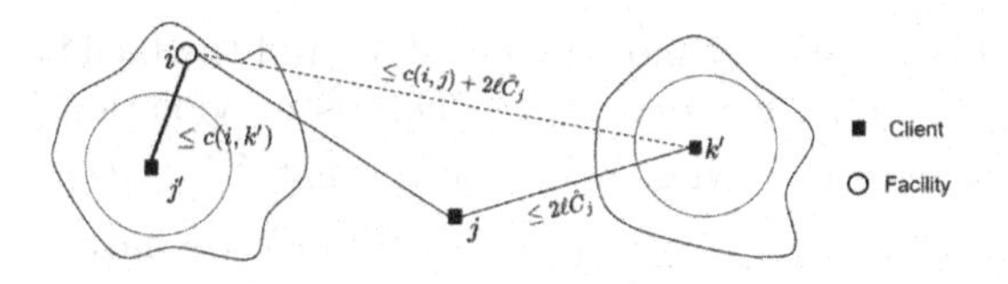


**Fig. 1.** $c(i,\ j') \leq c(i,\ j) + 2\ell\hat{C}_j$

We start by sparsifying the problem instance by removing some clients. For $j \in \mathcal{C}$, let $\hat{C}_j = \sum_{i \in \mathcal{F}} x^*_{ij} c(i,\ j)$ denote the average connection cost paid by $j$ in $\zeta^*$. Further, let $\ell \geq 2$ be a tunable parameter, $\mathcal{B}(j)$ be the ball of facilities within a radius of $\ell\hat{C}_j$ of $j$ and $Y^*(\mathcal{B}(j))$ be the total extent up to which facilities are opened in $\mathcal{B}(j)$ under solution $\zeta^*$, i.e., $Y^*(\mathcal{B}(j)) = \sum_{i \in \mathcal{B}(j)} y^*_i$. Then, $Y^*(\mathcal{B}(j)) \geq (1 - 1/\ell) \geq 1/2$. The clients are processed in the non-decreasing order of the radii of their balls, removing the close-by clients with balls of larger radii and dissolving their balls: let $\bar{\mathcal{C}} = \mathcal{C}$ and $\mathcal{C}'$ denote the sparsified set of clients. Initially $\mathcal{C}' = \phi$. Let $j'$ be a client in $\bar{\mathcal{C}}$ with a ball of the smallest radius (breaking the ties arbitrarily). Remove $j'$ from $\bar{\mathcal{C}}$ and add it to $\mathcal{C}'$. For all $j(\neq j') \in \bar{\mathcal{C}}$ with $c(j',\ j) \leq 2\ell\hat{C}_j$, remove $j$ from $\bar{\mathcal{C}}$. Repeat the process until $\bar{\mathcal{C}} = \phi$. Cluster of facilities are formed around the clients in $\mathcal{C}'$ by assigning a facility to the cluster of $j' \in \mathcal{C}'$ if and only if $j'$ is nearest to the facility amongst all $k' \in \mathcal{C}'$, i.e., if $\mathcal{N}_{j'}$ denotes the cluster centered at $j'$ then, $i \in \mathcal{N}_{j'}$ iff $c(i,\ j') < c(i,\ k')$ for all $k'(\neq j') \in \mathcal{C}'$ (assuming that the distances are distinct). The clients in $\mathcal{C}'$ are then called the cluster centers. ***Separation property:*** $c(j',\ k') > 2\ell\ max\{\hat{C}_{j'}, \hat{C}_{k'}\}$ for all $j' \neq k'$ in $\mathcal{C}'$.

**Lemma 1.** *Let* $j' \in \mathcal{C}',\ i \in \mathcal{N}_{j'},\ j \in \mathcal{C}$. *Then,* 1. $c(i,\ j') \leq c(i,\ j) + 2\ell\hat{C}_j$, 2. $c(j,\ j') \leq 2c(i,\ j) + 2\ell\hat{C}_j$ *and,* 3. *If* $c(j,\ j') \leq \ell\hat{C}_{j'}$, *then* $\hat{C}_{j'} \leq 2\hat{C}_j$.

*Proof.* Let $j' \in \mathcal{C}'$, $i \in \mathcal{N}_{j'}$, $j \in \mathcal{C}$. 1. Note that, $c(j,\ k') \leq 2\ell\hat{C}_j$ for some $k' \in \mathcal{C}'$. Then we have $c(i,\ j') \leq c(i,\ k') \leq c(i,\ j) + c(j,\ k') \leq c(i,\ j) + 2\ell\hat{C}_j$, where the first inequality follows because $i \in \mathcal{N}_{j'}$ and not $\mathcal{N}_{k'}$ whenever $k' \neq j'$. See Fig. 1. 2. Using triangle inequality, we have $c(j,\ j') \leq c(i,\ j) + c(i,\ j') \leq 2c(i,\ j) + 2\ell\hat{C}_j$. 3. Let $j \neq j'$. Note that $c(j,\ j') \leq \ell\hat{C}_{j'} \Rightarrow j \notin \mathcal{C}'$. Suppose if possible, $\hat{C}_{j'} > 2\hat{C}_j$. Since $j \notin \mathcal{C}', \exists$ some $k' \in \mathcal{C}' : c(j,\ k') \leq 2\ell\hat{C}_j$. Then, $c(k',\ j') \leq c(k',\ j) + c(j,\ j')$ $\leq 2\ell\hat{C}_j + \ell\hat{C}_{j'} < 2\ell\hat{C}_{j'}$. Thus, we arrive at a contradiction to the separation property. Hence, $\hat{C}_{j'} \leq 2\hat{C}_j$.

For $j' \in \mathcal{C}', j \in \mathcal{C}$, let $\phi(j,\ j')$ be the extent up to which $j$ is served by the facilities in the cluster of $j'$ under solution $\zeta^*$: $\phi(j,\ j') = \sum_{i \in \mathcal{N}_{j'}} x^*_{ij}$ and $d_{j'} = \sum_{j \in \mathcal{C}} \phi(j,\ j')$. We call a cluster to be *sparse* if $d_{j'} \leq \mathcal{U}$ and *dense* otherwise. Let $\mathcal{C}_S$ and $\mathcal{C}_D$ be the set of cluster centers of sparse and dense clusters respectively.

Let $j' \in \mathcal{C}_S$ and $i(j')$ be the cheapest (lowest facility opening cost) facility in $\mathcal{B}(j')$. We open $i(j')$ and transfer all the assignments coming into the cluster onto it. Since $Y^*(\mathcal{B}(j')) \geq (1 - 1/\ell)$, we have $\sum_{j \in \mathcal{C}} \hat{x}_{i(j')j} = d_{j'} \geq (1 - \frac{1}{\ell})\mathcal{L}$. Thus, the lower bound is violated at most by $(1 - 1/\ell)$ and the facility cost is bounded by $\ell/(\ell-1)\sum_{i \in \mathcal{N}_{j'}} f_i y_i^*$. Since $d_{j'} \leq \mathcal{U}$, there is no violation in upper bounds. To bound the service cost, it can be shown that $\sum_{j \in \mathcal{C}} \phi(j, j')c(i(j'), j) \leq 4\sum_{i \in \mathcal{N}_{j'}} \sum_{j \in \mathcal{C}} x_{ij}^*(c(i, j) + \ell\hat{C}_j)$: since $i(j') \in \mathcal{B}(j')$, we have $c(i(j'), j') \leq \ell\hat{C}_{j'}$. Thus, for $j \in \mathcal{C}$, we have $c(i(j'), j) \leq c(j', j) + c(i(j'), j') \leq c(j', j) + \ell\hat{C}_{j'}$. If $\ell\hat{C}_{j'} \leq c(j', j)$ then $c(i(j'), j) \leq 2c(j', j) \leq 4(c(i, j) + \ell\hat{C}_j)$ ($\forall i \in \mathcal{N}_{j'}$ by Lemma 1), else $c(i(j'), j) \leq 2\ell\hat{C}_{j'} \leq 4\ell\hat{C}_j$, where the second inequality in the else part follows by Lemma 1, Thus, in either case $c(i(j'), j) \leq 4(c(i, j) + \ell\hat{C}_j)$ for all $i \in \mathcal{N}_{j'}$. Substituting $\phi(j, j') = \sum_{i \in \mathcal{N}_{j'}} x_{ij}^*$ and summing over all $j \in \mathcal{C}$ we get the desired bound.

Next, let $j' \in \mathcal{C}_D$. To open the facilities integrally in $\mathcal{N}_{j'}$, we consider the following LP ($LP_1$): Min. $\sum_{i \in \mathcal{N}_{j'}} (f_i + \mathcal{U}c(i, j'))z_i$ s.t. $\mathcal{U}\sum_{i \in \mathcal{N}_{j'}} z_i \geq d_{j'}$, $\mathcal{L}\sum_{i \in \mathcal{N}_{j'}} z_i \leq d_{j'}$ and $0 \leq z_i \leq 1$ . Note that $z_i = \sum_{j \in \mathcal{C}} x_{ij}^*/\mathcal{U}$ is a feasible solution with the cost at most $\sum_{i \in \mathcal{N}_{j'}} (f_i y_i^* + \sum_{j \in \mathcal{C}} x_{ij}^*(c(i, j) + 2\ell\hat{C}_j))$ by Lemma 1.

We say that a solution to $LP_1$ is almost integral if it has at most one fractionally opened facility in $\mathcal{N}_{j'}$. We obtain an almost integral solution $z'$ by arranging the facilities opened in $z$, in non-decreasing order of $f_i + c(i, j')\mathcal{U}$ and greedily transferring the openings $z$ onto them. Note that $\sum_{i \in \mathcal{N}_{j'}} z'_i = \sum_{i \in \mathcal{N}_{j'}} z_i$ and hence $z'_i$ is a feasible solution to $LP_1$. Also, the cost of solution $z'$ is no more than that of solution $z$. We next convert $z'$ to an integral solution $\hat{z}$. Let $\hat{z} = z'$ initially. Let $i'$ be the fractionally opened facility, if any, in $\mathcal{N}_{j'}$. If $z'_{i'} \leq 1 - 1/\ell$: close $i'$ in $\hat{z}$. There must be at least one integrally opened facility, say $i(\neq i') \in \mathcal{N}_{j'}$ in $z'$, as $d_{j'} \geq \mathcal{U}$. Then, $d_{j'} \leq (2 - 1/\ell)\mathcal{U}\sum_{k \in \mathcal{N}_{j'}} \hat{z}_k$. There is no increase in cost as we have only (possibly) shut down one of the facilities. Else (i.e., $z'_{i'} > 1 - 1/\ell$): open $i'$ integrally at a loss of factor $\ell/(\ell-1)$ in cost and $d_{j'} \geq \mathcal{L}(\ell-1)/\ell\sum_{k \in \mathcal{N}_{j'}} \hat{z}_k$.

Next, we define our assignments, possibly fractional, in the dense clusters. For $j' \in \mathcal{C}_D$, we distribute $d_{j'}$ equally to the facilities opened in $\hat{z}$. Let $l_i$ be the amount assigned to facility $i$ under this distribution, i.e., $l_i = d_{j'}\hat{z}_i/\sum_{i \in \mathcal{N}_{j'}} \hat{z}_i$. Then, $((\ell-1)/\ell)\mathcal{L}\hat{z}_i \leq l_i \leq (2 - 1/\ell)\mathcal{U}\hat{z}_i$. A min-cost flow problem with relaxed lower and upper bounds is solved to obtain the integral assignments.

For $\ell = 2.01$, we get $\alpha > 1/2$, $\beta$ slightly more than $3/2$ and the approximation ratio slightly more than 24. $\beta$ can be reduced to $3/2$ by a slight modification in obtaining the integral solution $\hat{z}$: instead of comparing $z'_{i'}$ with $(1 - 1/\ell)$, we compare it with $1/2$.

## 3 Instance $I_{cap}$ of Capacitated Facility Location Problem

In this section, we create an instance $I_{cap}$ of capacitated facility location problem via a series of transformations: instance $I_1$ of LBUBFL is obtained by moving the clients assigned to a facility $i$ in the tri-criteria solution $S^t = (\mathcal{F}^t, \sigma^t)$ to $i$. Facilities opened by $S^t$ are free. Let $O$ be the optimal solution to $I$ and $\sigma^*(j)$ denote the facility serving $j$ in $O$. A solution to $I_1$ can be obtained by

**Table 1.** Instance $I_{cap}$: $d_i, u_i, f_i^t$ are demands, capacities and facility costs resp.

| Type | $n_i$ | $u_i$ | $d_i$ | $f_i^t$ |
|---|---|---|---|---|
| *small* | $n_i \leq \mathcal{L}$ | $\mathcal{L}$ | $\mathcal{L} - n_i$ | $\delta n_i l(i)$ |
| *big* | $n_i > \mathcal{L}$ | $(i_1)\mathcal{L}$ | 0 | $\delta \mathcal{L} l(i)$ |
| | | $(i_2)\ n_i - \mathcal{L}$ | 0 | 0 |

assigning a client $j$ to a facility $i$ if it was assigned to $i$ in $O$. The connection cost is bounded by the sum of costs $j$ pays in $S^t$ and in $O$: $(c(\sigma^t(j),\ \sigma^*(j))) \leq c(j,\ \sigma^t(j))+c(j,\ \sigma^*(j)))$. $I_2$, an instance of LBFL, is obtained from $I_1$ by ignoring the upper bounds and removing the facilities not opened in $S^t$. Let $O_1$ be the optimal solution to $I_1$ and $\sigma^1(j)$ denote the facility serving $j$ in $O_1$. If a client $j$ is assigned, by the optimal solution $O_1$, to a facility $i$ not in $\mathcal{F}^t$ then it is assigned to the facility $i' \in \mathcal{F}^t$ nearest to $i$ and we open $i'$. The connection cost is bounded by twice the cost $j$ pays in $O_1$ by a simple triangle inequality: $(c(\sigma^t(j),\ i') \leq c(\sigma^t(j),\ \sigma^1(j))+c(\sigma^1(j),\ i') \leq 2c(\sigma^t(j),\ \sigma^1(j))$ (since $i'$ is closest in $\mathcal{F}^t$ to $i = \sigma^1(j)$).

Finally, we transform $I_2$ to $I_{cap}$. Let $n_i$ be the number of clients co-located at facility $i$. The main idea is to create a demand of $\mathcal{L} - n_i$ units at locations where the number of clients served by the facility is less than $\mathcal{L}$ and a supply of $n_i - \mathcal{L}$ units at locations with surplus clients. For each facility $i \in \mathcal{F}^t$, let $l(i)$ be the distance of $i$ from the nearest facility $i' \in \mathcal{F}^t$, $i' \neq i$ and let $\delta$ be a constant to be chosen appropriately. A facility $i$ is called *small* if $0 < n_i \leq \mathcal{L}$ and *big* otherwise. A big facility $i$ is split into two co-located facilities $i_1$ and $i_2$. We also split the set of clients at $i$ into two sets: arbitrarily, $\mathcal{L}$ of these clients are placed at $i_1$ and the remaining $n_i - \mathcal{L}$ clients at $i_2$. Instance $I_{cap}$ is then defined as follows: A demand of $\mathcal{L} - n_i$ is created at a *small* facility $i$. A facility with capacity $\mathcal{L}$ and facility opening cost $\delta n_i l(i)$ is also created at $i$. For a *big* facility $i$, correspondingly two co-located facilities $i_1$ and $i_2$ are created with capacities $\mathcal{L}$ and $n_i - \mathcal{L}$ respectively. The facility opening cost of $i_1$ is $\delta \mathcal{L} l(i)$ whereas $i_2$ is free. The second type of big facilities are called *free*. Let $\bar{\mathcal{F}}^t$ be the set of locations so obtained. We also use $i \in \bar{\mathcal{F}}^t$ to refer to both the client (with demand) as well as the facility located at $i$. Table 1 summarizes the instance.

We will construct a feasible solution $S_{cap}$ to $I_{cap}$ of bounded cost from $O_2$, where $O_2$ is an optimal solution to $I_2$. As $O_2$ satisfies the lower bound, we can assume wlog that if $i$ is opened in $O_2$ then it serves all of its clients if $n_i \leq \mathcal{L}$ (before taking more clients from outside) and it serves at least $\mathcal{L}$ of its clients before sending out its clients to other (small) facilities otherwise. Also, if two big facilities are opened in $O_2$, one does not serve the clients of the other. Let $\rho^2(j_i, i')$ denote the number of clients co-located at $i$ and assigned to $i'$ in $O_2$ and $\rho^c(j_{i'}, i)$ denotes the amount of demand of $i'$ assigned to $i$ in $S_{cap}$.

1. If $i$ is closed in $O_2$ then open $i$ if $i$ is small and open $(i_1 \& i_2)$ if it is big. Assignments are defined as follows:

(a) If $i$ is small: Let $i$ serve its own demand. In addition, assign $\rho^2(j_i, i')$ demand of $i'$ to $i$ in our solution for small $i' \neq i$. Note that the same cannot be done when $i'$ is big as there is no demand at $i'_1$ and $i'_2$. Thus, $\rho^c(j_{i'}, i) = \rho^2(j_i, i')$, for small $i' \neq i$ and $\rho^c(j_i, i) = d_i$. Also, $\sum_{small\ i' \neq i} \rho^c(j_{i'}, i) = \sum_{small\ i' \neq i} \rho^2(j_i, i') \leq n_i$. Hence, $\sum_{i'} \rho^c(j_{i'}, i) \leq n_i + d_i = \mathcal{L} = u_i$.

(b) If $i$ is big: assign $\rho^2(j_{i_1}, i')$ $(/\rho^2(j_{i_2}, i'))$ demand of $i'$ to $i_1(/i_2)$ in our solution for small $i' \neq i$. Thus, $\sum_{small\ i' \neq i} \rho^c(j_{i'}, i_1) = \sum_{small\ i' \neq i} \rho^2(j_{i_1}, i') \leq \mathcal{L} = u_{i_1}$. Also, $\sum_{small\ i' \neq i} \rho^c(j_{i'}, i_2) = \sum_{small\ i' \neq i} \rho^2(j_{i_2}, i') \leq n_i - \mathcal{L} = u_{i_2}$.

2. If $i$ is opened in $O_2$ and is big, open the free facility $i_2$: assign $\rho^2(j_i, i')$ demand of $i'$ to $i_2$ in our solution for small $i' \neq i$. Thus, $\sum_{i' \neq i} \rho^c(j_{i'}, i_2) = \sum_{i' \neq i} \rho^2(j_i, i') \leq n_i - \mathcal{L}$ (the inequality holds by assumption on $O_2$) $= u_{i_2}$.

Next, we show that all the demands are satisfied. If a facility is opened in $S_{cap}$, it satisfies its own demand. Let $i$ be closed in $S_{cap}$. If $i$ is big, we need not worry as $i_1$ and $i_2$ have no demand. So, let $i$ be small. Then, it must be opened in $O_2$. Then, $\sum_{i' \neq i} \rho^c(j_i, i') = \sum_{small\ i' \neq i,} \rho^c(j_i, i') + \sum_{i'_1 : i'\ is\ big} \rho^c(j_i, i'_1) + \sum_{i'_2 : i'\ is\ big} \rho^c(j_i, i'_2) = \sum_{small\ i' \neq i,} \rho^2(j_{i'}, i) + \sum_{i'_1 : i'\ is\ big} \rho^2(j_{i'_1}, i) + \sum_{i'_2 : i'\ is\ big} \rho^2(j_{i'_2}, i) = \sum_{small\ i' \neq i,} \rho^2(j_{i'}, i) + \sum_{i'\ is\ big} \rho^2(j_{i'}, i) = \sum_{i' \neq i} \rho^2(j_{i'}, i) \geq \mathcal{L} - n_i$ (since $O_2$ is a feasible solution of $I_2$) $= d_i$.

Next, we bound the cost of the solution. The connection cost is at most that of $O_2$. For facility costs, consider a facility $i$ that is opened in our solution and closed in $O_2$. Such a facility must have paid a cost of at least $n_i l(i)$ to get its clients served by other (opened) facilities in $O_2$. The facility cost paid by our solution is $\delta min\{n_i, \mathcal{L}\} l(i) \leq \delta n_i l(i)$. If $i$ is opened in $O_2$, then it must be serving at least $\mathcal{L}$ clients and hence paying a cost of at least $\mathcal{L} l(i)$ in $O_2$. In this case also, the facility cost paid by our solution is $\delta min\{n_i, \mathcal{L}\} l(i) \leq \delta \mathcal{L} l(i)$. Summing over all $i$'s, we get that the facility cost is bounded by $2\delta Cost_{I_2}(O_2)$ and the total cost is bounded by $(1 + 2\delta) Cost_{I_2}(O_2)$. Factor 2 comes because we may have counted an edge $(i, i')$ twice, once as a client when $i$ was closed and once as a facility when $i'$ was opened in $O_2$.

## 4 Approximate Solution $AS_1$ to $I_1$

In this section, we obtain an approximate solution $AS_1$ to $I_1$ that violates the upper bounds by a factor of $(\beta + 1)$ without violating the lower bounds. This is the main contribution of our work. We first obtain a $(5 + \epsilon)$-approximate solution $AS_{cap}$ to $I_{cap}$ using approximation algorithm of Bansal *et al.* [4]. $AS_{cap}$ is then used to construct $AS_1$. Wlog assume that if a facility $i$ is opened in $AS_{cap}$ then it serves all its demand. (This is always feasible as $d_i \leq u_i$.) If this is not true, we can modify $AS_{cap}$ and obtain another solution, that satisfies the condition, of cost no more than that of $AS_{cap}$. $AS_1$ is obtained from $AS_{cap}$ by first defining the assignment of the clients and then opening the facilities

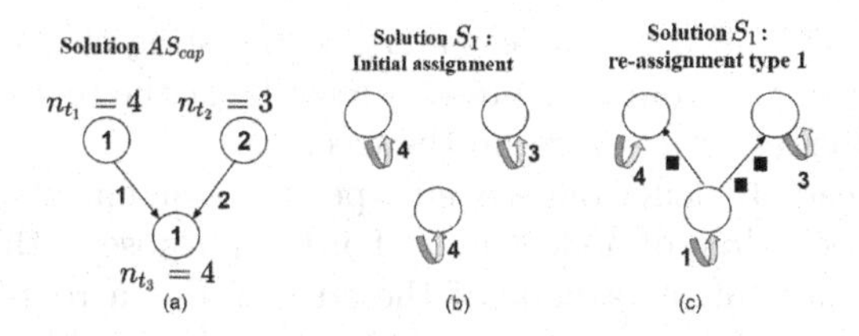


**Fig. 2.** $(a)$ $\mathcal{L} = 5$, $t_1, t_2, t_3$ have demands 1, 2 and 1 unit each respectively. In $AS_{cap}$, 1 unit of demand of $t_1$ and 2 units of demand of $t_2$ are assigned to $t_3$. (b) Solution $S_1$: $n_i$ clients are initially assigned to facility $i$. (c) Type-1 reassignments: 1 and 2 clients of $t_3$ reassigned to $t_1$ and $t_2$ respectively. After this reassignment, $t_3$ has only 1 client.

that get at least $\mathcal{L}$ clients. Let $\bar{\rho}^c(j_{i'}, i)$ denotes the amount of demand of $i'$ assigned to $i$ in $AS_{cap}$ and $\bar{\rho}^1(j_i, i')$ denotes the number of clients co-located at $i$ and assigned to $i'$ in $AS_1$. Clients are assigned in three steps. In the first step, assign the clients co-located at a facility to itself. For small $i'$, additionally, we do the following (type-1) re-assignments (see Fig. 2): $(i)$ For small $i \neq i'$, assign $\bar{\rho}^c(j_{i'}, i)$ clients co-located at $i$ to $i'$. Thus, $\bar{\rho}^1(j_i, i') = \bar{\rho}^c(j_{i'}, i)$ for small $i$. $(ii)$ For big $i$, assign $\bar{\rho}^c(j_{i'}, i_1) + \bar{\rho}^c(j_{i'}, i_2)$ clients co-located at $i$ to $i'$. Thus, $\bar{\rho}^1(j_i, i') = \bar{\rho}^c(j_{i'}, i_1) + \bar{\rho}^c(j_{i'}, i_2)$ for big $i$. The assignments are feasible i.e., $\sum_{i' \neq i} \bar{\rho}^1(j_i, i') \leq n_i$, $\forall\ i \in \bar{\mathcal{F}}^t$: since $\bar{\rho}^c(j_i, i) = d_i$ therefore $\sum_{i' \neq i} \bar{\rho}^1(j_i, i') = \sum_{i' \neq i} \bar{\rho}^c(j_{i'}, i) \leq u_i - d_i \leq n_i$ in all the cases. Also, upper bounds are violated only upto the extent to which they were violated in the tri-criteria solution $S^t$ i.e., $\sum_i \bar{\rho}^1(j_i, i') \leq max\{\mathcal{L}, n_{i'}\} \leq \beta\mathcal{U}$, $\forall\ i' \in \bar{\mathcal{F}}^t$: $\bar{\rho}^1(j_{i'}, i') = n_{i'}$. For $i \neq i'$, $\bar{\rho}^1(j_i, i') = \bar{\rho}^c(j_{i'}, i)$. Thus, $\sum_i \bar{\rho}^1(j_i, i') = n_{i'} + \sum_{i \neq i'} \bar{\rho}^c(j_{i'}, i) \leq n_{i'} + d_{i'} \leq max\{\mathcal{L}, n_{i'}\} \leq \beta\mathcal{U}$.

Note that $AS_1$ so obtained may still not satisfy the lower bound requirement. In fact, although each facility was assigned $n_i \geq \alpha\mathcal{L}$ clients initially, they may be serving less clients now after type-1 re-assignments. For example, in Fig. 2 $t_3$ had 4 clients initially which was reduced to 1 after type-1 reassignments. Let $P \subseteq \mathcal{F}^t$ be the set of facilities each of which is serving at least $\mathcal{L}$ clients after type-1 reassignments, and $\bar{P} = \mathcal{F}^t \setminus P$ be the set of remaining facilities. We open all the facilities in $P$ and let them serve the clients assigned to them after type-1 reassignments.

We observe that a small facility is in $\bar{P}$ only if it was open in $AS_{cap}$[2] and a big facility $i$ is in $\bar{P}$ only if $i_1$ was open in $AS_{cap}$. We now group these facilities so that each group serves at least $\mathcal{L}$ clients and open one facility in each group that serves all the clients in the group. For this we construct what we call *the facility trees*. We construct a graph $G$ with nodes corresponding to the facilities in $P \cup \bar{P}$. For $i \in \bar{P}$, let $\eta(i)$ be the nearest other facility to $i$ in $\mathcal{F}^t$. Then, $G$ consists of edges $(i, \eta(i))$ with edge costs $l(i) = c(i,\ \eta(i))$. Each component of $G$ is a tree except possibly a double-edge cycle at the root. In this case, we say

[2] If a small facility $i'$ was closed in $AS_{cap}$ then $\bar{\rho}^1(j_{i'}, i') + \sum_{i \neq i'} \bar{\rho}^1(j_i, i') = n_{i'} + \sum_{i \neq i'} \bar{\rho}^c(j_{i'}, i) = n_{i'} + d_{i'} \geq \mathcal{L}$.

that we have a root, called *root-pair* $< r_1, r_2 >$ consisting of a pair of facilities $r_1$ and $r_2$. Also, a facility $i$ from $P$, if present, must be at the root of a tree. Clearly, edge costs are decreasing as we go up the tree.

Now we are ready to define our second type of re-assignments. Let $children(x)$ denote the set of children of node $x$ in $\mathcal{T}$. For the purpose of this part, we treat the root-pair, if present, as a node of the tree. If $x$ is a root-pair $< r_1, r_2 >$, then $children(x) = children(r_1) \cup children(r_2)$. Process the tree bottom-up (level by level). While processing a node $x$ as explained in Algorithm 1, all its children with at least $\mathcal{L}$ clients are opened and removed along with the subtrees rooted at them; update $children(x)$; remaining children of $x$ are arranged and considered (left to right) in decreasing order of distance from $x$. For any child $y \in children(x)$, let $right - sibling(y)$ denote the adjacent right sibling of $y$ in the arrangement. If $y$ has at least $\mathcal{L}$ clients, it is opened else it sends its clients to its adjacent right sibling, if it has one, or to its parent ($x$) otherwise. The subtree rooted at $y$ is removed.

**Algorithm 1:** Process($x$)

```
   Input  : x(x can be a root-pair node)
 1 for y ∈ children(x) do
 2   | if n_y ≥ L then
 3   |   | Open facility y
 4   |   | Remove edge (y, η(y))          // Remove the connection of y from its parent
 5   |   | Delete y from children(x);
 6   | end
 7 end
 8 if children(x) = φ then
 9   | return;
10 end
11 Arrange children(x) in the sequence < y_1, ... y_k > such that
     c(y_i, η(y_i)) ≥ c(y_{i+1}, η(y_{i+1})) ∀ i = 1...k − 1
12 for i = 1 to k − 1 do
13   | if n_{y_i} ≥ L then
14   |   | Open facility y_i
15   | else
16   |   | n_{y_{i+1}} = n_{y_{i+1}} + n_{y_i}      // Send the clients of y_i to y_{i+1}
17   | end
18   | Remove edge (y_i, η(y_i))          // Remove the connection of y_i from its parent
19   | Delete y_i from children(x)
20 end
21 if n_{y_k} ≥ L then
22   | Open facility y_k
23   | Delete y_k from children(x)
24 else
25   | n_{η(y_k)} = n_{η(y_k)} + n_{y_k}      // Send the clients of y_k to η(y_k)
26   | Delete y_k from children(x)
27 end
```

There are two possibilities at the root: either we have a facility $i$ from $P$ or we have a root-pair $< r_1, r_2 >$. In the first case, we are done as $i$ is already open and has at least $\mathcal{L}$ clients. To handle the second case, we need to do a little more work. So, we define our assignments of third type as follows: $(i)$ If the total number of clients collected at the root-pair node is at least $\mathcal{L}$ and at most $2\mathcal{L}$ then open any one of the two facilities in the root-pair and assign all the clients to it. $(ii)$ If the total number of clients collected at the root-pair node is more than $2\mathcal{L}$ then open both the facilities at the root node and distribute the clients so that each one of them gets at least $\mathcal{L}$ clients. $(iii)$ If the total number of clients collected at the root-pair node is less than $\mathcal{L}$, then let $i$ be the node in $P$ nearest to the root-pair i.e. $i = argmin_{i' \in P} min\{c(i',\ r_1), c(i',\ r_2)\}$, then $i$ is already open and has at least $\mathcal{L}$ clients. Send the clients collected at the root-pair to $i$. Clearly, the opened facilities satisfy the lower bounds.

Next, we bound the violation in the upper bound. It is easy to see that the number of clients collected in the second type of re-assignments, at any non-root node is at most $2\mathcal{L}$. Next, let $r$ be a root/root-pair node and $i_c$ be the child of $r$, if any, that sends its clients to $r$. Let $i$ be the parent of $i_c$ ($i$ is either $r$ and is in $P$ or it is one of $r_1$ and $r_2$ in case $r = <r_1, r_2>$ is a root-pair). Let $\pi_i$ be the number of clients assigned to facility $i$ after type-1 re-assignments. Then $\pi_i \leq \beta\mathcal{U}$. We have the following cases: $(i)$ the root node $i$ is in $P$ and it gets additional $< \mathcal{L}$ clients from $i_c$ (in second type of reassignment) making a total of at most $\pi_i + \mathcal{L} \leq \beta\mathcal{U} + \mathcal{L} \leq (\beta+1)\mathcal{U}$ clients at $i$. $(ii)$ the total number of clients collected at the root-pair node $(n_{i_c} + n_{r_1} + n_{r_2})$ is at least $\mathcal{L}$ and at most $2\mathcal{L}$. The number of clients the opened facility gets is at most $2\mathcal{L} \leq 2\mathcal{U} \leq (\beta+1)\mathcal{U}$. $(iii)$ the total number of clients collected at the root-pair node is more than $2\mathcal{L}$. Note that the total number of clients collected at these facilities is at most $3\mathcal{L}$ and hence ensuring that each of them gets at least $\mathcal{L}$ clients also ensures that none of them gets more than $2\mathcal{L}$ clients. Thus, none of them gets more than $2\mathcal{U} \leq (\beta+1)\mathcal{U}$ clients. $(iv)$ the total number of clients collected at the root-pair node is less than $\mathcal{L}$. Thus, the number of clients the opened facility in $P$ gets is at most $\beta\mathcal{U} + \mathcal{L} \leq (\beta+1)\mathcal{U}$. Note that if $\mathcal{L} \leq \tau\mathcal{U}$ for a $\tau < 1$, the violation in upper bound is smaller; in particular, it is bounded by $(\beta+\tau)$.

We next bound the connection cost. The cost of first type of re-assignments is at most the connection cost of $AS_{cap}$. To bound the connection cost of second type, observe that we never send more than $\mathcal{L}$ clients on any edge. Thus, a facility $i$ that sends its (at most $\mathcal{L}$) clients to its parent or to its right sibling incurs a cost of at most $3\mathcal{L}l(i)$ (it is easy to show that $c(i,\ right-sibling(i)) \leq 3c(i,\ \eta(i)) = 3l(i)$: 3 factor loss can happen when $\eta(y) \neq \eta(right-sibling(y))$ (which can be the case when root is a root-pair. In that case, $c(y,\ right-sibling(y)) \leq c(y,\ \eta(y)) + c(\eta(y),\ \eta(right-sibling(y))) + c(\eta(right-sibling(y)),\ right-sibling(y)) \leq 3c(y,\ \eta(y))$.) Since the facility $i$ was opened in $AS_{cap}$, it pays facility opening cost of $\delta l(i) min\{n_i, \mathcal{L}\} \geq \delta l(i)\alpha\mathcal{L}$ in $AS_{cap}$.

For the third type of re-assignments, we bound the cost as follows: note that the total number of clients, initially at $r_1$ and $r_2$ was $\geq 2\alpha\mathcal{L}$. However, during type-1 re-assignments, some of them got reassigned to other facilities. Note that these (other) facilities must have been closed in $AS_{cap}$ and thus do not belong to $\bar{P}$. Hence, each of these reassignments correspond to an assignment in $AS_{cap}$ whose cost was $\geq l(i)$ and a total of $\geq (2\alpha - 1)\mathcal{L}l(i)$. Thus, the cost of type-3 re-assignments is bounded by $3/(2\alpha - 1)$ times the connection cost of $AS_{cap}$.

## 5 Conclusion and Future Work

In this paper, we presented the first (constant factor) approximation algorithm for facility location problem with uniform lower and upper bounds without violating the lower bounds. Upper bounds are violated by (5/2)-factor. In the future, if one can obtain a tri-criteria solution (with $\alpha > 1/2$) with uniform lower bounds and non-uniform upper bounds, then it can be simply plugged into our technique to obtain a similar result for the problem. It will be interesting to see if the technique can be modified to obtain a solution with non-uniform lower bounds and uniform upper bounds without violating the lower bounds.

## References

1. Aggarwal, A., et al.: A 3-approximation algorithm for the facility location problem with uniform capacities. Math. Program. **141**(1), 527–547 (2012). https://doi.org/10.1007/s10107-012-0565-4
2. Ahmadian, S., Swamy, C.: Improved approximation guarantees for lower-bounded facility location. In: Erlebach, T., Persiano, G. (eds.) WAOA 2012. LNCS, vol. 7846, pp. 257–271. Springer, Heidelberg (2013). https://doi.org/10.1007/978-3-642-38016-7_21
3. An, H.C., Singh, M., Svensson, O.: LP-based algorithms for capacitated facility location. In: FOCS, pp. 256–265, October 2014
4. Bansal, M., Garg, N., Gupta, N.: A 5-approximation for capacitated facility location. In: Epstein, L., Ferragina, P. (eds.) ESA 2012. LNCS, vol. 7501, pp. 133–144. Springer, Heidelberg (2012). https://doi.org/10.1007/978-3-642-33090-2_13
5. Byrka, J., Fleszar, K., Rybicki, B., Spoerhase, J.: Bi-factor approximation algorithms for hard capacitated k-median problems. In: SODA 2015, pp. 722–736 (2015)
6. Chudak, F.A., Williamson, D.P.: Improved approximation algorithms for capacitated facility location problems. In: Cornuéjols, G., Burkard, R.E., Woeginger, G.J. (eds.) IPCO 1999. LNCS, vol. 1610, pp. 99–113. Springer, Heidelberg (1999). https://doi.org/10.1007/3-540-48777-8_8
7. Friggstad, Z., Rezapour, M., Salavatipour, M.R.: Approximating connected facility location with lower and upper bounds via LP rounding. In: SWAT 2016, pp. 1:1–1:14 (2016)
8. Grover, S., Gupta, N., Dabas, R.: First approximation for uniform lower and upper bounded facility location problem avoiding violation in lower bounds. CoRR abs/2106.11372 (2021)
9. Grover, S., Gupta, N., Khuller, S., Pancholi, A.: Constant factor approximation algorithm for uniform hard capacitated knapsack median problem. In: FSTTCS, pp. 23:1–23:22 (2018)
10. Guha, S., Meyerson, A., Munagala, K.: Hierarchical placement and network design problems. In: FOCS, p. 603 (2000)
11. Han, L., Hao, C., Wu, C., Zhang, Z.: Approximation algorithms for the lower-bounded $k$-median and its generalizations. In: Kim, D., Uma, R.N., Cai, Z., Lee, D.H. (eds.) COCOON 2020. LNCS, vol. 12273, pp. 627–639. Springer, Cham (2020). https://doi.org/10.1007/978-3-030-58150-3_51
12. Karger, D.R., Minkoff, M.: Building Steiner trees with incomplete global knowledge. In: FOCS, pp. 613–623 (2000)

13. Korupolu, M.R., Plaxton, C.G., Rajaraman, R.: Analysis of a local search heuristic for facility location problems. J. Algorithms **37**(1), 146–188 (2000)
14. Li, S.: On facility location with general lower bounds. In: SODA, pp. 2279–2290 (2019)
15. Lim, A., Wang, F., Xu, Z.: A transportation problem with minimum quantity commitment. Transp. Sci. **40**, 117–129 (2006)
16. Mahdian, M., Pál, M.: Universal facility location. In: Di Battista, G., Zwick, U. (eds.) ESA 2003. LNCS, vol. 2832, pp. 409–421. Springer, Heidelberg (2003). https://doi.org/10.1007/978-3-540-39658-1_38
17. Pál, M., Tardos, E., Wexler, T.: Facility location with nonuniform hard capacities. In: FOCS, pp. 329–338 (2001)
18. Shmoys, D.B., Tardos, É., Aardal, K.: Approximation algorithms for facility location problems (extended abstract). In: STOC, pp. 265–274 (1997)
19. Svitkina, Z.: Lower-bounded facility location. ACM Trans. Algorithms **6**, 1–16 (2010)
20. Zhang, J., Chen, B., Ye, Y.: A multiexchange local search algorithm for the capacitated facility location problem. J. Math. Oper. Res. **30**, 389–403 (2005)

# Finding Cheapest Deadline Paths

Wei Ding(✉)

Zhejiang University of Water Resources and Electric Power,
Hangzhou 310018, Zhejiang, China

**Abstract.** This paper considers a novel quality-of-service (QoS) routing problem from a *source* to a *destination*, named the **Cheapest Deadline Path Problem (CDPP)**, which arises from the real-world scenario. Let $D = (V, A, c, d, s, t)$ be a double-weighted strongly connected digraph, where each arc $a \in A$ is associated with a *cost*, $c(a) \in \mathbf{Z}^+$, and a *delay*, $d(a) \in \mathbf{Z}^+$, and $s$ and $t$ are the indices of the designated source and destination, respectively, and let $\mathcal{B} = \{\mathbb{B}_1, \mathbb{B}_2, \ldots, \mathbb{B}_n\}$ be a set of positive constants, where $\mathbb{B}_i, 1 \leq i \leq n$ represents the upper bound on delay at $v_i \in V$. The objective of CDPP is to find a $v_s$-to-$v_t$ path of the minimum cost in $D$ such that the $v_s$-to-$v_k$ delay along the path is at most $\mathbb{B}_k$, for each vertex, $v_k$, appearing in the path. This paper presents a fully polynomial time approximation scheme (FPTAS) for CDPP in $D = (V, A, c, d, s, t)$ using a graph traverse based dynamic programming algorithm as a sub-procedure.

**Keywords:** Deadline path · Graph traverse · FPTAS

## 1 Introduction

The **Restricted Shortest Path Problem (RSPP)**, also called the **Constrained Shortest Path Problem (CSPP)**, asks for a cheapest source-to-destination path in the given graph such that the delay of it is bounded by a constant. This problem is one of the well-known combinatorial optimization problems and known to be NP-hard [7]. In past decades, it has been widely studied [1,4,8–10,14–17,19–22].

### 1.1 Overview of Related Works

There have been many FPTAS's and related algorithms for RSPP. In [19], Warburton gave the first FPTAS for acyclic digraphs, with a time cost of $O(\frac{n^3 \log n}{\epsilon} \cdot \log_2 \mathrm{UB})$, where $n$ is the number of vertices, UB is an upper bound of the optimal value and $\epsilon$ is a given positive constant. In [10], Hassin presented a significantly improved FPTAS for acyclic digraphs, with a time cost of $O(\frac{mn}{\epsilon} \cdot \log_2 \log_2 \frac{\mathrm{UB}}{\mathrm{LB}})$, where $m$ is the number of edges and LB is a lower bound of the optimal value. Hassin's FPTAS can be extended to general digraphs. The above two FPTAS's are both not strongly polynomial since their running times


C.-Y. Chen et al. (Eds.): COCOON 2021, LNCS 13025, pp. 476–486, 2021.
https://doi.org/10.1007/978-3-030-89543-3_40

depend on the arc costs of the given digraph. Later, a few strongly polynomial time algorithms were obtained. In [14], Lorenz and Raz obtained the best known algorithm for general digraphs, an FPTAS with a time cost of $O(mn(\frac{1}{\epsilon}+\log n))$. In [4], Ergun et al. obtained a slightly faster FPTAS with a time cost of $O(\frac{mn}{\epsilon})$, but it is only applicable to acyclic digraphs. In [8], Goel et al. obtained an asymptotically faster algorithm with a time cost of $O((m+n\log n)\frac{H}{\epsilon})$, where $H$ is the hop number of the longest path found by their algorithm. This algorithm can find a path of cost at most the optimal solution and of delay at most $(1+\epsilon)$ times the upper bound on delay. In [1], Bernstein obtained a so-called near linear time algorithm for undirected graphs, with a time cost of $\widetilde{O}(m(\frac{2}{\epsilon})^{O(\sqrt{\log n}\cdot\log\log n)})$, which can find $(1+\epsilon)$-approximation of delay at most $(1+\epsilon)$ times the upper bound.

Also, there have been a variety of heuristics for RSPP. In [9], Handler and Zang designed a dual algorithm, which established the basis of the Lagrangian relaxation based algorithms for RSPP. In [17], Siachalou and Georgiadis obtained two pseudopolynomial time exact algorithms. In [16], Orda and Sprintson used the precomputation methods to reduce the computational complexity. In [22], Xue considered a bicriteria QoS routing and designed a primal-dual heuristic to find a path to balance the cost and delay of path. In [21], Xiao et al. used the primal simplex method of linear programming to design two pseudopolynomial time algorithms.

The other similar QoS routing problems were also studied. In [15], Misra et al. studied the problem of finding a set of source-to-destination paths to minimize the delay of the longest path and meanwhile to make the aggregated bandwidth of the set of paths at least a given lower bound, and designed an FPTAS. In [20], Xiao et al. studied the problem of computing a most probable path with a delay mean at most the given upper bound of delay, and designed an FPTAS. Moreover, the extended version of RSPP, i.e., the problem of finding a path (or multiple paths) of minimum cost subject to multiple constraints, has been widely studied, see [5,6,11–13,23,24,26–28].

### 1.2 Motivation

In this paper, we consider such a real-world scenario where each vertex of the given digraph might be blocked (or polluted). It is of great value to study the source-to-destination *routing* problems in such a scenario (e.g., in a delay-sensitive network). Obviously, all the source-to-destination successful paths satisfy a common property that the arriving time at each vertex appearing in the path must be earlier than the time of the vertex to be blocked. A path satisfying this property is called a *deadline path*. This paper deals with the *offline* model where the time of each vertex in the digraph to be blocked is known.

In fact, a source-to-destination deadline path can be stated as a path satisfying that the delay from the source to each vertex appearing in the path is at most the upper bound on the delay at this vertex. When each arc of the digraph has another weight representing the cost of it, we have a big concern with the cost of deadline path and aim to find a cheapest deadline path.

### 1.3 Our Contribution

This paper proposes the **Cheapest Deadline Path Problem (CDPP)**. Let $D = (V, A, c, d, s, t)$ be a double-weighted strongly connected digraph, where each $a \in A$ has a *cost*, $c(a) \in \mathbf{Z}^+$, and a *delay*, $d(a) \in \mathbf{Z}^+$, and $s$ and $t$ are the indices of the designated *source* and *destination*, respectively, and let $\mathcal{B} = \{\mathbb{B}_1, \mathbb{B}_2, \ldots, \mathbb{B}_n\}$ be a set of positive constants, where $\mathbb{B}_i, 1 \leq i \leq n$ represents the upper bound on delay at $v_i \in V$. The objective of CDPP is to compute a $v_s$-to-$v_t$ path of minimum cost in $D = (V, A, c, d, s, t)$ such that the $v_s$-to-$v_k$ delay along the path is at most $\mathbb{B}_k$, for each vertex, $v_k$, appearing in the path. The classic RSPP is the special case of CDPP with $\mathbb{B}_t < \infty$ and $\mathbb{B}_k = \infty, k \neq t$.

First, we design an FPTAS for CDPP with a time cost of $O(\frac{mn \cdot \text{UB}}{\epsilon \cdot \text{LB}})$, using a graph traverse based dynamic programming algorithm as a sub-procedure, where LB and UB are two given positive constants and $\epsilon$ is an arbitrarily small positive real number. Next, we develop a faster FPTAS with a time cost of $O(\frac{mn}{\epsilon} \cdot \log_2 \log_2 \frac{\text{UB}}{\text{LB}})$, using a binary search technique. Our FPTAS's are both applicable to RSPP. Surprisingly, our FPTAS has the same time cost as the well-known Hassin's algorithm [10] although CDPP is a more general problem than RSPP, and even close to the best known algorithm by Lorenz and Raz [14].

## 2 Preliminaries

### 2.1 Notations

Let $D = (V, A, c, d, s, t)$ be a strongly connected digraph, where each arc $a \in A$ has a *cost*, $c(a) \in \mathbf{Z}^+$, and a *delay*, $d(a) \in \mathbf{Z}^+$, and $s$ and $t$ are the indices of the designated *source* and *destination*, respectively. Here, $|V| = n$ and $|A| = m$. All the vertices are numbered by $1, 2, \ldots, n$ in turn, and so $v_i$ denotes the vertex with number $i$, for $1 \leq i \leq n$. We use $\langle v_i, v_j \rangle \in V^2$ to denote a vertex pair, for $1 \leq i, j \leq n$. Note that $\langle v_i, v_j \rangle$ can be abbreviated as $\langle i, j \rangle$. For any $a = (v_i, v_j) \in A$, we also use $d(i, j)$ and $c(i, j)$ to denote the delay and cost of $(v_i, v_i)$, respectively, i.e., $d(a) = d(i, j)$ and $c(a) = c(i, j)$. In general, $d(i, j) \neq d(j, i)$ and $c(i, j) \neq c(j, i)$.

For $v_i, 1 \leq i \leq n$, we let $\mathbb{B}_i > 0$ be the *deadline* at $v_i$, representing the upper bound on delay at $v_i$. For $\langle i, j \rangle \in V^2$, a $v_i$-to-$v_j$ simple (acyclic) path is called a $v_i$-to-$v_j$ *deadline path*, denoted by $\pi_{ij}$, if it satisfies that the $v_i$-to-$v_k$ delay on the path is at most $\mathbb{B}_k$, for each $v_k$ appearing in the path. We use $d(\pi_{ij})$ to denote the delay of $\pi_{ij}$, which is defined as $d(\pi_{ij}) = \sum_{a \in \pi_{ij}} d(a)$, and use $V(\pi_{ij})$ to denote the vertex set of $\pi_{ij}$. For $v_k \in V(\pi_{ij})$, we use $d_{\pi_{ij}}(i, k)$ to denote the $v_i$-to-$v_k$ delay on $\pi_{ij}$, i.e., the delay of subpath $\pi_{ik} \subseteq \pi_{ij}$. So, $d_{\pi_{ij}}(i, k) \leq \mathbb{B}_k$ and

$$d_{\pi_{ij}}(i, k) = \sum_{a \in \pi_{ik} \subseteq \pi_{ij}} d(a), \quad \forall v_k \in V(\pi_{ij}). \tag{1}$$

We use $c(\pi_{ij})$ to denote the cost of $\pi_{ij}$, defined as $c(\pi_{ij}) = \sum_{a \in \pi_{ij}} c(a)$. A $v_i$-to-$v_j$ deadline path of the minimum cost is named a $v_i$-to-$v_j$ *cheapest deadline path*

*(CDP)* and denoted by $\pi^*_{ij}$. This minimum cost is named the $v_i$-to-$v_j$ *cheapest deadline path cost (CDPC)* and denoted by $c_i(j)$.

For $(v_i, v_j) \in A$, we call $v_i$ the *tail* of $(v_i, v_j)$ and $v_j$ the *head* of it, and call $(v_i, v_j)$ an *outgoing* arc from $v_i$ and an *incoming* arc to $v_j$. Let $\mathcal{O}_k$ be the head set of all outgoing arcs from $v_k$, i.e. $\mathcal{O}_k = \{k'|(v_k, v_{k'}) \in A\}$, and $\mathcal{I}_k$ be the tail set of all incoming arcs to $v_k$, i.e., $\mathcal{I}_k = \{k'|(v_{k'}, v_k) \in A\}$. For a directed path, $\pi = v_{i_1}v_{i_2}\dots v_{i_r}, r \geq 2$, we call $v_{i_j}$ the *precedent* vertex of $v_{i_{j+1}}$ in $\pi$, for $1 \leq j \leq r-1$.

### 2.2 Problem Statement

Let $\mathcal{B} = \{\mathbb{B}_1, \mathbb{B}_2, \dots, \mathbb{B}_n\}$ be a set of given positive constants (deadlines), called a *deadline set*. Below is the problem we focus on in this paper.

**Definition 1.** *Given $D = (V, A, c, d, s, t)$ and $\mathcal{B} = \{\mathbb{B}_1, \mathbb{B}_2, \dots, \mathbb{B}_n\}$, the* ***Cheapest Deadline Path Problem (CDPP)*** *asks for a $v_s$-to-$v_t$ deadline path, $\pi^*_{st}$, of minimum cost satisfying that $d_{\pi^*_{st}}(s, k) \leq \mathbb{B}_k$, for any $k \in V(\pi^*_{st})$.*

The classic RSPP is the special case of CDPP with $\mathbb{B}_t < \infty$ and $\mathbb{B}_k = \infty, k \neq t$. Below is the decision version of CDPP.

**Definition 2.** *Given $D = (V, A, c, d, s, t)$, $\mathcal{B} = \{\mathbb{B}_1, \mathbb{B}_2, \dots, \mathbb{B}_n\}$ and a constant $\mathbb{C}_0 > 0$, does there exist a $v_s$-to-$v_t$ deadline path, $\pi_{st}$, such that $d_{\pi_{st}}(s, k) \leq \mathbb{B}_k$, for any $k \in V(\pi_{st})$, and $c(\pi_{st}) \leq \mathbb{C}_0$.*

**Lemma 1.** *CDPP is NP-hard.*

## 3 Approximation for CDPP

In this section, we will present FPTAS's for CDPP in $D = (V, A, c, d, s, t)$. First, we define another problem that is closely related to CDPP. A $v_i$-to-$v_j$ deadline path of the shortest delay is named a $v_i$-to-$v_j$ *quickest deadline path (QDP)*.

**Definition 3.** *Given $D = (V, A, c, d, s, t)$, $\mathcal{B} = \{\mathbb{B}_1, \mathbb{B}_2, \dots, \mathbb{B}_n\}$, and a constant $\mathbb{C} > 0$, the* ***Cost-Constrained Quickest Deadline Path Problem (CCQDPP)*** *asks for a $v_s$-to-$v_t$ deadline path, $\pi^\star_{st}$, of shortest delay satisfying that $d_{\pi^\star_{st}}(s, k) \leq \mathbb{B}_k$, for any $k \in V(\pi^\star_{st})$, and $c(\pi^\star_{st}) \leq \mathbb{C}$.*

### 3.1 An Exact Algorithm for CCQDPP

For any $\langle i, j\rangle \in V^2$ and $C \geq 0$, we let $\pi^\star_{ij}(C)$ denote a $v_i$-to-$v_j$ quickest deadline path with cost at most $C$, abbreviated as a $v_i$-to-$v_j$ *$C$-constrained QDP*. The delay of $\pi^\star_{ij}(C)$ is named the $v_i$-to-$v_j$ *$C$-constrained QDP delay (QDPD)*, denoted by $\delta_i(j; C)$. As $c(a) \in \mathbf{Z}^+, \forall a \in A$, we let

$$\delta_i(j; C) = \infty, \quad \forall C < 0. \tag{2}$$

Below are several useful lemmas, which inspire us to design a graph traverse based dynamic programming algorithm for CCQDPP.

**Lemma 2.** *Given* $D = (V, A, c, d, s, t)$ *and* $1 \leq i \leq n, i \neq s$, *we have*

$$\delta_s(i; C) \leq \delta_s(i; C'), \quad \forall C' \leq C. \tag{3}$$

**Lemma 3.** *For* $1 \leq i \leq n, i \neq s$ *and* $C \geq 1$, *let* $\pi^\star_{si}(C) = v_s \cdots v_r v_i$ *be a* $v_s$*-to-*$v_i$ $C$*-constrained QDP in* $D = (V, A, c, d, s, t)$. *Then, the subpath* $\pi' = v_s \cdots v_r$ *of* $\pi^\star_{si}(C)$ *must be a* $v_s$*-to-*$v_r$ $C'$*-constrained QDP satisfying that* $d(\pi') + d(r, i) \leq \mathbb{B}_i$, *where* $C' = C - c(r, i)$.

Next, we consider the special $v_s$-to-$v_i$ deadline path of passing through $v_k$, for each $k \in \mathcal{I}_i$. Let $\pi^\star_{si}(k; C)$ and $\overline{\delta}_s(k, i; C)$ denote the $v_s$-to-$v_i$ $C$-constrained QDP and QDPD through $v_k$, respectively. Let

$$\overline{\delta}_s(k, i; C) = \infty, \quad \forall C < 0. \tag{4}$$

**Lemma 4.** *Given* $D = (V, A, c, d, s, t)$ *and* $1 \leq i \leq n, i \neq s$, *and for any* $k \in \mathcal{I}_i$ *and integer* $C \geq 1$, *we can compute* $\overline{\delta}_s(k, i; C)$ *by*

$$\overline{\delta}_s(k, i; C) = \delta_s(k; C - c(k, i)) + d(k, i). \tag{5}$$

*Proof.* By Lemma 3, it follows that each $v_s$-to-$v_i$ $C$-constrained QDP through $v_k$ must be the combination of a $v_s$-to-$v_k$ $(C - c(k, i))$-constrained QDP and arc $(v_k, v_i)$, for each $v_k, k \in \mathcal{I}_i$. □

**Lemma 5.** *Given* $D = (V, A, c, d, s, t)$ *and* $1 \leq i \leq n, i \neq s$, *and for any positive integer* $C \geq 1$, *we can compute* $\delta_s(i; C)$ *by*

$$\delta_s(i; C) = \min_{k \in \mathcal{I}_i} \overline{\delta}_s(k, i; C). \tag{6}$$

*Proof.* Each $v_s$-to-$v_i$ $C$-constrained QDP must reach $v_i$ via some vertex in $\mathcal{I}_i$. For each $k \in \mathcal{I}_i$, we only consider the $v_s$-to-$v_i$ $C$-constrained QDP through $v_k$. So, the $v_s$-to-$v_i$ $C$-constrained QDPD is equal to the minimum of all the $v_s$-to-$v_i$ $C$-constrained QDPD's through a vertex in $\mathcal{I}_i$. □

For any $1 \leq i \leq n, i \neq s$ and integer $C \geq 1$, Eq. (6) implies that we need to obtain all $\overline{\delta}_s(k, i; C), \forall k \in \mathcal{I}_i$ in advance so as to compute $\delta_s(i; C)$, and Eq. (5) implies that we need to obtain $\delta_s(k; C - c(k, i))$ in advance so as to compute $\overline{\delta}_s(k, i; C)$, for each $k \in \mathcal{I}_i$. Therefore, we can compute $\{\delta_s(i; \widehat{C}) : 1 \leq i \leq n\}, \widehat{C} = 1, 2, \ldots, C$, in turn.

For each $v_i \in V(\pi^\star_{st}(C))$, we let $f_{st}(i; C)$ be the precedent vertex of $v_i$ in $\pi^\star_{st}(C)$. All such precedent vertices in $\pi^\star_{st}(C)$ form a set, denoted by $\mathcal{F}_{st}(C)$. In other words, $\mathcal{F}_{st}(C)$ stores an optimum to CCQDPP with cost at most $C$. We have

$$\mathcal{F}_{st}(C) = \{f_{st}(i; C) : v_i \in V(\pi^\star_{st}(C))\}. \tag{7}$$

As for CCQDPP with cost at most $\zeta \geq 1$, we let $\zeta^\circ$ be the minimum in $C = 1, 2, \ldots, \zeta$ such that $\delta_s(t; C)$ does not exceed $\mathbb{B}_t$. If all $\delta_s(t; C), 1 \leq C \leq \zeta$ exceed $\mathbb{B}_t$, then $\zeta^\circ =$ null. If such $\zeta^\circ$ exists, then $\pi^\star_{st}(\zeta^\circ)$ is just the corresponding

| ALG($D, \mathcal{B}, \zeta$): PPT Algorithm for CCQDPP. |
|---|
| **Input:** $D = (V, A, c, d, s, t)$, $\mathcal{B} = \{\mathbb{B}_1, \mathbb{B}_2, \ldots, \mathbb{B}_n\}$, and $\zeta \in \mathbf{Z}^+$; |
| **Output:** YES with $\zeta^\circ$ and $\pi^\star_{st}(\zeta^\circ)$, or NO with $\zeta^\circ$ = null. |
| $\delta_s(s;0) \leftarrow 0; \delta_s(i;0) \leftarrow \infty, \forall 1 \le i \le n, i \ne s;\ f_{st}(i;0) \leftarrow$ null, $\forall 1 \le i \le n$;<br>**for** $C := 1, 2, \ldots, \zeta$ **do**<br>    $\delta_s(i;C) \leftarrow \delta_s(i;C-1), f_{st}(i;C) \leftarrow f_{st}(i;C-1), \forall 1 \le i \le n$;<br>    Use BFS to traverse $D$ with $v_s$ as the origin; Let $v_{\widetilde{k}}$ be the<br>    current vertex; When $v_{\widetilde{k}}$ is visited, the steps below are done:<br>    **for** each $j \in \mathcal{O}_{\widetilde{k}}$ **do**<br>        $\overline{\delta}_s(\widetilde{k}, j; C) \leftarrow \delta_s(\widetilde{k}; C - c(\widetilde{k}, j)) + d(\widetilde{k}, j)$;<br>        **if** $\overline{\delta}_s(\widetilde{k}, j; C) \le \mathbb{B}_j$ **then**<br>            **if** $\overline{\delta}_s(\widetilde{k}, j; C) < \delta_s(j;C)$ **then**<br>                $\delta_s(j;C) \leftarrow \overline{\delta}_s(\widetilde{k}, j; C);\ f_{st}(j;C) \leftarrow \widetilde{k}$;<br>            **endif**<br>        **endif**<br>    **endfor**<br>    **if** $\delta_s(t;C) \le \mathbb{B}_t$ **then**<br>        $\zeta^\circ \leftarrow C$; Call SOLT$(s, t, \zeta^\circ)$ to compute $\pi^\star_{st}(\zeta^\circ)$;<br>        Return YES with $\zeta^\circ$ and $\pi^\star_{st}(\zeta^\circ)$;<br>    **endif**<br>**endfor**<br>Return NO with $\zeta^\circ$ = null; |

$v_s$-to-$v_t$ CDP, $\pi^\star_{st}$, with cost equal to $\zeta^\circ$. This leads to our algorithm ALG for CCQDPP, which is a graph traverse based dynamic programming algorithm. Theorem 1 shows that ALG runs in a pseudo polynomial time. The framework of ALG is described as follows.

First of all, we do initialization: $\delta_s(s;0) = 0$ and $\delta_s(i;0) = \infty$, for $1 \le i \le n, i \ne s$, and $f_{st}(i;0) =$ null, for $1 \le i \le n$.

Next, for each $C = 1, 2, \ldots, \zeta$, we compute $\delta_s(i;C)$ and record $f_{st}(i;C)$. Initially, by Eq. (3), we let

$$\delta_s(i;C) = \delta_s(i;C-1), \quad f_{st}(i;C) = f_{st}(i;C-1).$$

Note that all $\delta_s(i;\widehat{C})$ and $f_{st}(i;\widehat{C}), \forall \widehat{C} = 1, 2, \ldots, C-1$ have been obtained. Then, we use the *breadth-first search* (BFS) to traverse $D$ with $v_s$ as the origin. When the current vertex, $v_{\widetilde{k}}$, is visited, the following steps are done. For each $j \in \mathcal{O}_{\widetilde{k}}$, we compute $\overline{\delta}_s(\widetilde{k}, j; C)$ by Eq. (5). If $\overline{\delta}_s(\widetilde{k}, j; C)$ is at most $\mathbb{B}_j$ and strictly smaller than the current value of $\delta_s(j;C)$, then this current value is updated as $\overline{\delta}_s(\widetilde{k}, j; C)$ and $f_{st}(j;C)$ is updated as $\widetilde{k}$. When BFS ends, all $\overline{\delta}_s(k, j; C), k \in \mathcal{I}_j$ are obtained and then $\delta_s(j;C)$ can be obtained by Eq. (6). Consequently, we obtain $\delta_s(1;C), \delta_s(2;C), \ldots, \delta_s(n;C)$. If $\delta_s(t;C) \le \mathbb{B}_t$ then we determine $\zeta^\circ = C$ and use a sub-procedure SOLT to compute $\pi^\star_{st}(\zeta^\circ)$, and return YES with $\zeta^\circ$ and $\pi^\star_{st}(\zeta^\circ)$. If $\delta_s(t;C) > \mathbb{B}_t$, for all $1 \le C \le \zeta$, then NO and $\zeta^\circ$ = null are returned.

The main idea of SOLT is to trace out $\pi^\star_{st}(\zeta)$ with $v_t$ as the starting point and $v_s$ as the ending point, according to Eq. (7). Let $v_\alpha$ be the current vertex and $C_0$

be the current cost bound of $v_s$-to-$v_\alpha$ subpath of $\pi^\star_{st}(\zeta)$. Initially, $\alpha = t, C_0 = \zeta$ and $\mathcal{F}_{st}(\zeta) = \{t\}$. The precedent vertex, $f_{st}(\alpha; C_0)$, of $v_\alpha$ in $\pi^\star_{st}(\zeta)$ is selected into $\mathcal{F}_{st}(\zeta)$, and the cost bound of subpath is updated as $C_0 - c(f_{st}(\alpha; C_0), \alpha)$ and the current vertex is updated as $f_{st}(\alpha; C_0)$. When $\alpha = s$, SOLT ends and returns $\mathcal{F}_{st}(\zeta)$.

| SOLT$(s, t, \zeta)$: |
|---|
| $\mathcal{F}_{st}(\zeta) \leftarrow \{t\}$; $\alpha \leftarrow t, C_0 \leftarrow \zeta$;<br>**while** $\alpha \neq s$ **do**<br>    $\mathcal{F}_{st}(\zeta) \leftarrow \mathcal{F}_{st}(\zeta) \cup \{f_{st}(\alpha; C_0)\}$;<br>    $C_0 \leftarrow C_0 - c(f_{st}(\alpha; C_0), \alpha)$;<br>    $\alpha \leftarrow f_{st}(\alpha; C_0)$;<br>**endwhile** |

**Theorem 1.** *Given $D = (V, A, c, d, s, t)$ with $|V| = n$ and $|A| = m$, a deadline set, $\mathcal{B} = \{\mathbb{B}_1, \mathbb{B}_2, \ldots, \mathbb{B}_n\}$, and a constant $\zeta > 0$, the worst-case time cost of* ALG$(D, \mathcal{B}, \zeta)$ *is $O(m\zeta)$. Furthermore, we claim that $c_s(t) \leq \zeta$ if the output is YES and $c_s(t) > \zeta$ if the output is NO.*

### 3.2 Test Procedure

Given a real number $C > 0$, it is NP-hard to decide whether $c_s(t) > C$ or $c_s(t) < C$ for CDPP in $D = (V, A, c, d, s, t)$. However, by using the standard technique of *scaling* and *rounding* [3,10,15,19,20,24,25], we can decide whether $c_s(t) > C$ or $c_s(t) < (1+\epsilon) \cdot C$ in a fully polynomial time, for any given constant $\epsilon > 0$. The above decision process can be described as an approximate testing procedure, named TEST. First of all, we need to introduce two auxiliary graphs and the related definitions.

| TEST$(D, \mathcal{B}, C, \epsilon)$: Approximate Test Procedure. |
|---|
| **Input:** $D = (V, A, c, d, s, t)$, $\mathcal{B} = \{\mathbb{B}_1, \mathbb{B}_2, \ldots, \mathbb{B}_n\}$, $C \in \mathbf{Z}^+$, and $\epsilon \in (0, 2(n-1)]$.<br>**Output:** YES or NO. |
| $\theta \leftarrow \frac{2(n-1)}{C \cdot \epsilon}$; $c^{\lfloor\theta\rfloor}(a) \leftarrow \lfloor c(a) \cdot \theta \rfloor, \forall a \in A$; $\zeta \leftarrow \lfloor C \cdot \theta \rfloor$;<br>Use ALG$(D^{\lfloor\theta\rfloor}, \mathcal{B}, \zeta)$ to obtain $\zeta^{\lfloor\theta\rfloor}$;<br>**if** $\zeta^{\lfloor\theta\rfloor} \leq \zeta$ **then** Return YES;<br>**else** Return NO; **endif** |

Let $c^\theta(\cdot)$ be such a *scaling-arc-cost function*, where $c^\theta(a) = c(a) \cdot \theta$, for each $a \in A$. An auxiliary graph, $D^\theta = (V, A, c^\theta, d, s, t)$, can be obtained by replacing $c(\cdot)$ with $c^\theta(\cdot)$ in $D = (V, A, c, d, s, t)$. Furthermore, we let $c^{\lfloor\theta\rfloor}(\cdot)$ be such a *scaling-rounding-arc-cost function*, where $c^{\lfloor\theta\rfloor}(a) = \lfloor c(a) \cdot \theta \rfloor$, for each $a \in A$.

Another auxiliary graph, $D^{\lfloor\theta\rfloor} = (V, A, c^{\lfloor\theta\rfloor}, d, s, t)$, can be obtained by replacing $c(\cdot)$ with $c^{\lfloor\theta\rfloor}(\cdot)$. Accordingly, we let $\pi_{st}^{\theta}$ and $\pi_{st}^{\lfloor\theta\rfloor}$ denote a $v_s$-to-$v_t$ CDP in $D^{\theta}$ and $D^{\lfloor\theta\rfloor}$, respectively.

**Theorem 2.** *Given $D = (V, A, c, d, s, t)$ with $|V| = n$ and $|A| = m$, a deadline set, $\mathcal{B} = \{\mathbb{B}_1, \mathbb{B}_2, \ldots, \mathbb{B}_n\}$, and two constants, $C \in \mathbf{Z}^+$ and $0 < \epsilon \leq 2(n-1)$, the worst-case time cost of* $\mathsf{TEST}(D, \mathcal{B}, C, \epsilon)$ *is $O(\frac{mn}{\epsilon})$. Furthermore, we claim that $c_s(t) < (1+\epsilon)\cdot C$ if the output is* YES *and $c_s(t) > C$ if the output is* NO.

*Proof.* From $\theta = \frac{2(n-1)}{C\cdot\epsilon}$, it follows that $\zeta = \lfloor C\cdot\theta\rfloor = \lfloor C\cdot\frac{2(n-1)}{C\cdot\epsilon}\rfloor = \lfloor\frac{2(n-1)}{\epsilon}\rfloor$. By $\epsilon \leq 2(n-1)$, we have $\zeta \geq 1$. The time cost of $\mathsf{TEST}(D, \mathcal{B}, C, \epsilon)$ is dominated by that of $\mathsf{ALG}(D^{\lfloor\theta\rfloor}, \mathcal{B}, \zeta)$. By substituting $\zeta = \lfloor\frac{2(n-1)}{\epsilon}\rfloor$ into the latter, see Theorem 1, we obtain that the time complexity of $\mathsf{TEST}(D, \mathcal{B}, C, \epsilon)$ is $O(\frac{mn}{\epsilon})$.

First, if the output is YES, then $c^{\lfloor\theta\rfloor}(\pi_{st}^{\lfloor\theta\rfloor}) = \zeta^{\lfloor\theta\rfloor} \leq \zeta = \lfloor C\cdot\theta\rfloor \leq C\cdot\theta$. From $c^{\lfloor\theta\rfloor}(a) = \lfloor c(a)\cdot\theta\rfloor > c(a)\cdot\theta - 1$, it follows that $c(a) < \frac{c^{\lfloor\theta\rfloor}(a)}{\theta} + \frac{1}{\theta}$. Since $\pi_{st}^{\lfloor\theta\rfloor}$ has at most $n-1$ arcs, we conclude that

$$c_s(t) = \sum_{a\in\pi_{st}^*} c(a) < \sum_{a\in\pi_{st}^*} (\frac{c^{\lfloor\theta\rfloor}(a)}{\theta} + \frac{1}{\theta}) \leq \frac{c^{\lfloor\theta\rfloor}(\pi_{st}^*)}{\theta} + \frac{n-1}{\theta}. \tag{8}$$

From $c^{\lfloor\theta\rfloor}(a) \leq c^{\theta}(a) < c^{\lfloor\theta\rfloor}(a) + 1$, we conclude that $c^{\lfloor\theta\rfloor}(\pi_{st}^*) \leq c^{\theta}(\pi_{st}^*)$ and $c^{\theta}(\pi_{st}^{\lfloor\theta\rfloor}) < c^{\lfloor\theta\rfloor}(\pi_{st}^{\lfloor\theta\rfloor}) + (n-1)$. By the fact that $\pi_{st}^{\theta}$ is a $v_s$-to-$v_t$ CDP in $D^{\theta}$, we obtain $c^{\theta}(\pi_{st}^{\theta}) \leq c^{\theta}(\pi_{st}^{\lfloor\theta\rfloor})$. By the fact that a $v_s$-to-$v_t$ CDP in $D^{\theta}$ is equal to a $v_s$-to-$v_t$ CDP in $D$, we obtain $c^{\theta}(\pi_{st}^*) = c^{\theta}(\pi_{st}^{\theta})$. So,

$$c^{\lfloor\theta\rfloor}(\pi_{st}^*) \leq c^{\theta}(\pi_{st}^*) = c^{\theta}(\pi_{st}^{\theta}) \leq c^{\theta}(\pi_{st}^{\lfloor\theta\rfloor}) < c^{\lfloor\theta\rfloor}(\pi_{st}^{\lfloor\theta\rfloor}) + (n-1). \tag{9}$$

The combination of (8) and (9) yields that

$$c_s(t) < \frac{c^{\lfloor\theta\rfloor}(\pi_{st}^{\lfloor\theta\rfloor}) + (n-1)}{\theta} + \frac{n-1}{\theta} \leq \frac{C\cdot\theta}{\theta} + \frac{2(n-1)}{\frac{2(n-1)}{C\cdot\epsilon}} = (1+\epsilon)\cdot C.$$

Next, if the output is NO, then $c^{\lfloor\theta\rfloor}(\pi_{st}^{\lfloor\theta\rfloor}) = \zeta^{\lfloor\theta\rfloor} \geq \zeta + 1 = \lfloor C\cdot\theta\rfloor + 1 > C\cdot\theta$. From $c^{\lfloor\theta\rfloor}(a) = \lfloor c(a)\cdot\theta\rfloor \leq c(a)\cdot\theta$, it follows that $c(a) \geq \frac{c^{\lfloor\theta\rfloor}(a)}{\theta}$. Since $\pi_{st}^{\lfloor\theta\rfloor}$ is a $v_s$-to-$v_t$ CDP in $D^{\lfloor\theta\rfloor}$, we have $c^{\lfloor\theta\rfloor}(\pi_{st}^*) \geq c^{\lfloor\theta\rfloor}(\pi_{st}^{\lfloor\theta\rfloor})$. So,

$$c_s(t) = \sum_{a\in\pi_{st}^*} c(a) \geq \sum_{a\in\pi_{st}^*} \frac{c^{\lfloor\theta\rfloor}(a)}{\theta} = \frac{c^{\lfloor\theta\rfloor}(\pi_{st}^*)}{\theta} \geq \frac{c^{\lfloor\theta\rfloor}(\pi_{st}^{\lfloor\theta\rfloor})}{\theta} > \frac{C\cdot\theta}{\theta} = C.$$

□

### 3.3 FPTAS

First, we design $\mathsf{FPTAS}_1$ for CDPP in $D = (V, A, c, d, s, t)$ using $\mathsf{ALG}$ as a sub-procedure. The output solution of $\mathsf{FPTAS}_1$ is denoted as $\pi_{st}^{F_1}$, and its time cost is shown in Theorem 3. Let LB and UB be known lower and upper bounds on the optimal value, $c(\pi_{st}^*)$, of CDPP.

| $\mathsf{FPTAS}_1(D, \mathcal{B}, \epsilon, \text{LB}, \text{UB})$: FPTAS for CDPP. |
|---|
| **Input:** $D = (V, A, c, d, s, t)$, $\mathcal{B} = \{\mathbb{B}_1, \mathbb{B}_2, \ldots, \mathbb{B}_n\}$, $\text{LB}, \text{UB} > 0$, and $\epsilon \in (0, \frac{\text{UB}\cdot(n-1)}{\text{LB}}]$; **Output:** $\pi_{st}^{F_1}$. |
| $\theta \leftarrow \frac{n-1}{\text{LB}\cdot\epsilon}$; $c^{\lfloor\theta\rfloor}(a) \leftarrow \lfloor c(a)\cdot\theta \rfloor, \forall a \in A$; $\zeta \leftarrow \lfloor \text{UB}\cdot\theta \rfloor$; Use $\mathsf{ALG}(D^{\lfloor\theta\rfloor}, \mathcal{B}, \zeta)$ to obtain $\zeta^{\lfloor\theta\rfloor}$ and $\pi_{st}^{\lfloor\theta\rfloor}$; $\pi_{st}^{F_1} \leftarrow \pi_{st}^{\lfloor\theta\rfloor}$; Return $\pi_{st}^{F_1}$; |

**Theorem 3.** *Given $D = (V, A, c, d, s, t)$ with $|V| = n$ and $|A| = m$, a deadline set, $\mathcal{B} = \{\mathbb{B}_1, \mathbb{B}_2, \ldots, \mathbb{B}_n\}$, and three constants,* $\text{LB}, \text{UB} > 0$ *and* $0 < \epsilon \leq \frac{\text{UB}\cdot(n-1)}{\text{LB}}$, $\mathsf{FPTAS}_1(D, \mathcal{B}, \epsilon, \text{LB}, \text{UB})$ *can find a $(1+\epsilon)$-approximate $v_s$-to-$v_t$ CDP in $O(\frac{mn\cdot\text{UB}}{\epsilon\cdot\text{LB}})$ time.*

Theorem 3 shows that the time cost of $\mathsf{FPTAS}_1$ is closely related with $\frac{\text{UB}}{\text{LB}}$. So, we can reduce its time cost by reducing $\frac{\text{UB}}{\text{LB}}$. The common way is to first initialize LB and UB as easily computable values and then use the *binary* method with $\mathsf{TEST}$ as a sub-procedure to reduce $\frac{\text{UB}}{\text{LB}}$, see [3,10,15,19,20,24,25]. Here, LB is initialized as the $v_s$-to-$v_t$ *cheapest path cost*, denoted by $c(s,t)$, in $\widetilde{D} = (V, A, c, s, t)$ obtained by ignoring the delays on edges of $D$, and UB is initialized as $(n-1)\cdot \max_{a\in A} c(a)$.

Next, we use binary method to drive $\frac{\text{UB}}{\text{LB}}$ down to some number below $2(1+\epsilon)$, for any given $\epsilon > 0$. When $\frac{\text{UB}}{\text{LB}} > 2(1+\epsilon)$, the values of LB and UB are updated in the following way. Let $C = \sqrt{\frac{\text{UB}\cdot\text{LB}}{1+\epsilon}}$. If $\mathsf{TEST}(D, \mathcal{B}, C, \epsilon) = \mathsf{YES}$, then $(1+\epsilon)\cdot C$ becomes a new upper bound and LB is still the lower bound. If $\mathsf{TEST}(D, \mathcal{B}, C, \epsilon) = \mathsf{NO}$, then $C$ becomes a new lower bound and UB is still the upper bound. Such a process is called an *iteration*, and each iteration can be completed in a fully polynomial time of $O(\frac{mn}{\epsilon})$, due to Theorem 2. Incorporating the above iterations into $\mathsf{FPTAS}_1$ results in a faster FPTAS, named $\mathsf{FPTAS}_2$. The output solution of $\mathsf{FPTAS}_2$ is denoted as $\pi_{st}^{F_2}$ and the time cost of it is given in Theorem 4.

| $\mathsf{FPTAS}_2(D, \mathcal{B}, \epsilon)$: Faster FPTAS for CDPP. |
|---|
| **Input:** $D = (V, A, c, d, s, t)$, $\mathcal{B} = \{\mathbb{B}_1, \mathbb{B}_2, \ldots, \mathbb{B}_n\}$, $\epsilon > 0$.<br>**Output:** $\pi_{st}^{F_2}$. |
| Use Dijkstra's algorithm [2] to compute $c(s,t)$;<br>LB $\leftarrow c(s,t)$, UB $\leftarrow (n-1) \cdot \max_{a \in A} c(a)$;<br>**if** $\frac{\text{UB}}{\text{LB}} \leq 2(1+\epsilon)$ **then**<br>goto **Step 3**;<br>**else**<br>$C \leftarrow \sqrt{\frac{\text{LB} \cdot \text{UB}}{1+\epsilon}}$;<br>**if** $\mathsf{TEST}(D, \mathcal{B}, C, \epsilon) = \mathsf{YES}$ **then** UB $\leftarrow (1+\epsilon) \cdot C$;<br>**if** $\mathsf{TEST}(D, \mathcal{B}, C, \epsilon) = \mathsf{NO}$ **then** LB $\leftarrow C$; **endif**<br>goto **Step 2**;<br>**endif**<br>$\theta \leftarrow \frac{n-1}{\text{LB} \cdot \epsilon}$; $c^{\lfloor \theta \rfloor}(a) \leftarrow \lfloor c(a) \cdot \theta \rfloor, \forall a \in A$; $\zeta \leftarrow \lfloor \text{UB} \cdot \theta \rfloor$;<br>Use $\mathsf{ALG}(D^{\lfloor \theta \rfloor}, \mathcal{B}, \zeta)$ to obtain $\zeta^{\lfloor \theta \rfloor}$ and $\pi_{st}^{\lfloor \theta \rfloor}$;<br>$\pi_{st}^{F_2} \leftarrow \pi_{st}^{\lfloor \theta \rfloor}$; Return $\pi_{st}^{F_2}$; |

**Theorem 4.** *Given $D = (V, A, c, d, s, t)$ with $|V| = n$ and $|A| = m$, a deadline set, $\mathcal{B} = \{\mathbb{B}_1, \mathbb{B}_2, \ldots, \mathbb{B}_n\}$, and a constant, $\epsilon > 0$, $\mathsf{FPTAS}_2(D, \mathcal{B}, \epsilon)$ can find a $(1+\epsilon)$-approximate $v_s$-to-$v_t$ CDP in $O(\frac{mn}{\epsilon} \cdot \log_2 \mathcal{S})$ time, where $\mathcal{S}$ is the input size of the given instance.*

## References


1. Bernstein, A.: Near linear time $(1+\epsilon)$-approximation for restricted shortest paths in undirected graphs. In: Proceedings of 23rd SODA, pp. 189–201 (2012)
2. Dijkstra, E.W.: A note on two problems in connection with graphs. Numer. Math. **1**, 269–271 (1959)
3. Ding, W., Xue, G.: Minimum diameter cost-constrained Steiner trees. J. Comb. Optim. **27**(1), 32–48 (2013). https://doi.org/10.1007/s10878-013-9611-2
4. Ergun, F., Sinha, R., Zhang, L.: An improved FPTAS for restricted shortest path. Inf. Process. Lett. **83**(5), 287–291 (2002)
5. Fang, X., Yang, D.J., Xue, G.: MAP: multi-constrained anypath routing in wireless mesh networks. IEEE Trans. Mobile Comput. **12**(10), 1893–1906 (2013)
6. Feng, G., Korkmaz, T.: Finding multi-constrained multiple shortest paths. IEEE Trans. Comput. **64**(9), 2559–2572 (2015)
7. Garey, M.R., Johnson, D.S.: Computers and Intractability: A Guide to the Theory of NP-Completeness. Freeman, San Francisco (1979)
8. Goel, A., Ramakrishnan, K.G., Kataria, D., Logothetis, D.: Efficient computation of delay-sensitive routes from one source to all destinations. In: Proceedings of 20th INFOCOM, pp. 854–858 (2001)
9. Handler, G., Zang, I.: A dual algorithm for the constrained shortest path problem. Networks **10**(4), 293–309 (1980)
10. Hassin, R.: Approximation schemes for the restricted shortest path problem. Math. Oper. Res. **17**(1), 36–42 (1992)
11. Jaffe, J.M.: Algorithms for finding paths with multiple constraints. Networks **14**(1), 95–116 (1984)

12. Korkmaz, T., Krunz, M.: A randomized algorithm for finding a path subject to multiple QoS requirements. Comput. Netw. **36**(2/3), 251–268 (2001)
13. Liu, G., Ramakrishnan, K.G.: A*prune: an algorithm for finding K shortest paths subject to multiple constraints. In: Proceedings of 20th INFOCOM, pp. 743–749 (2001)
14. Lorenz, D., Raz, D.: A simple efficient approximation scheme for the restricted shortest path problem. Oper. Res. Lett. **28**(5), 213–219 (2001)
15. Misra, S., Xue, G., Yang, D.J.: Polynomial time approximations for multi-path routing with bandwidth and delay constraints. In: Proceedings of 28th INFOCOM, pp. 558–566 (2009)
16. Orda, A., Sprintson, A.: Precomputation schemes for QoS routing. IEEE/ACM Trans. Netw. **11**(4), 578–591 (2003)
17. Siachalou, S., Georgiadis, L.: Efficient QoS routing. Comput. Netw. **43**(3), 351–367 (2003)
18. Van Mieghem, P., Kuipers, F.A.: Concepts of exact QoS routing algorithms. IEEE/ACM Trans. Netw. **12**(5), 851–864 (2004)
19. Warburton, A.: Approximation of pareto optima in multiple-objective, shortest-path problems. Oper. Res. **35**(1), 70–79 (1987)
20. Xiao, Y., Thulasiraman, K., Fang, X., Yang, D.J., Xue, G.: Computing a most probable delay constrained path: NP-hardness and approximation schemes. IEEE Trans. Comput. **61**(5), 738–744 (2012)
21. Xiao, Y., Thulasiraman, K., Xue, G.: QoS routing in communication networks: approximation algorithms based on the primal simplex method of linear programming. IEEE Trans. Comput. **55**(7), 815–829 (2006)
22. Xue, G.: Minimum cost QoS multicast and unicast routing in communication networks. IEEE Trans. Commun. **51**(5), 817–824 (2003)
23. Xue, G., Makki, S.K.: Multi-constrained QoS routing: a norm approach. IEEE Trans. Comput. **56**(6), 859–863 (2007)
24. Xue, G., Sen, A., Zhang, W., Tang, J., Thulasiraman, K.: Finding a path subject to many additive QoS constraints. IEEE/ACM Trans. Netw. **15**(1), 201–211 (2007)
25. Xue, G., Xiao, W.: A polynomial time approximation scheme for minimum cost delay-constrained multicast tree under a Steiner topology. Algorithmica **41**(1), 53–72 (2005). https://doi.org/10.1007/s00453-004-1119-9
26. Xue, G., Zhang, W.: Multiconstrained QoS routing: greedy is good. In: Proceedings of GLOBECOM 2007, pp. 1866–1871 (2007)
27. Xue, G., Zhang, W., Tang, J., Thulasiraman, K.: Polynomial time approximation algorithms for multi-constrained QoS routing. IEEE/ACM Trans. Netw. **16**(3), 656–669 (2008)
28. Yuan, X.: Heuristic algorithms for multiconstrained quality-of-service routing. IEEE/ACM Trans. Netw. **10**(2), 244–256 (2002)

# Approximate the Lower-Bounded Connected Facility Location Problem

Lu Han[1], Chenchen Wu[2], and Yicheng Xu[3,4](✉)

[1] School of Science, Beijing University of Posts and Telecommunications, Beijing 100876, People's Republic of China
hl@bupt.edu.cn

[2] College of Science, Tianjin University of Technology, Tianjin 300384, People's Republic of China

[3] Shenzhen Institute of Advanced Technology, Chinese Academy of Sciences, Shenzhen 518055, People's Republic of China
yc.xu@siat.ac.cn

[4] Guangxi Key Laboratory of Cryptography and Information Security, Guilin 541004, People's Republic of China

**Abstract.** This paper studies the lower-bounded connected facility location (LB ConFL) problem, which extends the well-known connected facility location (ConFL) and lower-bounded facility location (LBFL) problems. In the LB ConFL, we are given a graph $G = (V, E)$, where $V$ and $E$ are all the vertices and edges, respectively. A facility set $\mathcal{F} \subseteq V$, a client set $\mathcal{D} \subseteq V$, a parameter $M \geq 1$, and an integer lower bound $L$ are also given. Each facility has an opening cost $f_i$, and each edge $e \in E$ has a connection cost $c_e$. Denote by $c_{uv}$ the shortest path with respect to the connection costs from vertex $u$ to $v$. Opening a facility $i$ incurs its opening cost. Assigning a client $j$ to some facility $i$ incurs a connection cost $c_{ij}$. Connecting a facility subset $S \subseteq \mathcal{F}$ by a Steiner tree $T$ incurs a cost of $M \sum_{e \in T} c_e$ called Steiner cost. The goal is to open some facilities $S \subseteq \mathcal{F}$, assign each client $j$ to some opened facility in $S$ and connect all the opened facilities $S$ by a Steiner tree, such that the number of clients connected to any opened facility is at least $L$, and the total incurred cost (i.e., the total opening, connection, and Steiner cost) is minimized.

As our main contribution, we propose two approximation algorithms for the LB ConFL with ratios of 696 and 169. The first algorithm is based on an intuitive idea that finding a suitable Steiner tree before considering the lower bound constraints may give a good solution. The second algorithm effectively avoids the shortcoming of the first one and successfully improves the approximation ratio. Moreover, we consider the general LB ConFL (GLB ConFL) problem, in which each facility $i$ has a non-uniform lower bound $L_i$. We give an approximation algorithm with a ratio related to the parameter $M$ for the GLB ConFL.

**Keywords:** Connected facility location · Lower bounds · Approximation algorithm


C.-Y. Chen et al. (Eds.): COCOON 2021, LNCS 13025, pp. 487–498, 2021.
https://doi.org/10.1007/978-3-030-89543-3_41

## 1 Introduction

The uncapacitated facility location (UFL) problem is famous in operations research for its wide range of applications. In a UFL instance, a set of facilities and a set of clients are given. Opening each facility incurs its own opening cost, and assigning a client to some facility incurs a connection cost. The aim is to open some facilities and assign each client to some opened facility, such that the total opening and connection cost is minimized. Usually assume that the connection costs in the UFL problem and its generalizations are non-negative, symmetric, and satisfy the triangle inequality. The UFL problem is NP-hard, and many approximation algorithms had been designed for it and its meaningful generalizations [2,7,9,10,14–16,19,20]. The currently best approximation ratio of 1.488 for the UFL problem is given by Li [14].

In reality, the opened facilities may want to have communications among them. This requirement in real-world scenarios can be captured as the generalization of the UFL called the connected facility location (ConFL) problem, which requires all the opened facilities must be connected by a Steiner tree. Karger and Minkoff [12] provide the first constant-factor approximation algorithm for the ConFL problem. Later, Gupta et al. [6] give an LP-rounding algorithm with a ratio of 10.66. Based on the technique of primal-dual, Swamy and Kumar [18] propose an 8.55-approximation algorithm. Hasan et al. [8] further improve the ratio to 8.29 via an LP-rounding algorithm. June et al. [11] offer a primal-dual 6.55-approximation algorithm. Eisenbrand et al. [3] give a 4-approximation algorithm by using the approximation algorithm for the UFL as a subroutine and randomly sampling the client to open the suitable facilities. Based on another random sampling, Grandoni and Rothvoß [4] present the currently best 3.19-approximation algorithm.

In a real situation, each facility wants to be connected by a certain number of clients for the sake of profit. This scenario can be captured as the lower-bounded facility location (LBFL) problem, in which an additional input of lower bound $L$ is given, and each opened facility is required to satisfy connecting by at least $L$ clients (i.e., the lower bound constraints). Guha et al. [5] and Karger and Minkoff [12] simultaneously give constant-factor bi-criteria approximation algorithms for the LBFL problem. These two algorithms approximately respect the lower bound constraints. Svitkina [17] reduces the LBFL problem to the capacitated facility location problem (CFL) and gives an approximation algorithm with a ratio of 488. Based on a more careful reduction, Ahmadian and Swamy [1] propose the currently best 82.6-approximation algorithm. For the general LBFL (GLBFL) problem, in which each facility $i$ has a non-uniform lower bound $L_i$, Li [13] also constructs a reduction to give its first and currently best 3926-approximation algorithm.

In this paper, we propose and study the lower-bounded connected facility location (LB ConFL) problem, generalizing both the ConFL and LBFL. In the LB ConFL problem, we require that all the opened facilities be connected by a Steiner tree, and each opened one should connect a certain uniform number of clients. When each facility is required to connect a non-uniform number of clients, the problem is called the general LB ConFL (GLB ConFL) problem. The following are our main contributions.

- Contribution 1: We propose the first constant-factor 696-approximation algorithm for the LB ConFL. The algorithm is based on an intuition that is concentrating on the requirement of connecting all the opened facility by a Steiner tree before thinking about the lower bound constraints may adequately work. To be specific, our first algorithm constructs and solves a ConFL instance then a LBFL instance.
- Contribution 2: Unfortunately, the ratio of the first algorithm is inevitably large. We then propose the currently best 169-approximation algorithm for the LB ConFL, which significantly improves the previous ratio. This algorithm avoids the weakness of the first algorithm by solving a constructed ConFL instance and a constructed LBFL instance simultaneously.
- Contribution 3: We study the GLB ConFL and given an $M$-related approximation algorithm, where $M$ is a given parameter. The idea of this algorithm is similar to the second algorithm for the LB ConFL. It uses the solutions of simultaneously constructed ConFL and GLBFL instances to construct a solution for the GLB ConFL instance. However, with non-uniform lower bounds, the construction process is a bit more complicated.

The remainder of the paper is structured as follows. Section 2 presents the algorithms for the LB ConFL, and Sect. 3 provides an algorithm for the GLB ConFL. Discussions are given in Sect. 4. Due to space constraints, all proofs are removed but will appear in a full version of this paper.

## 2 Dealing with Uniform Lower Bounds

In this section, we first give formal descriptions of the LB ConFL, ConFL, and LBFL. Then, we show how to use the approximation algorithms for the ConFL and LBFL as subroutines to solve the LB ConFL.

### 2.1 Problem Descriptions Related to the LB ConFL

In a LB ConFL instance $\mathcal{I}_{\text{LBCo}}$, a graph $G = (V, E)$ is given, where $V$ and $E$ are all the vertices and edges, respectively. A facility set $\mathcal{F} \subseteq V$, a client set $\mathcal{D} \subseteq V$, a parameter $M \geq 1$, and an integer lower bound $L$ are also given. Each facility $i \in \mathcal{F}$ has a non-negative opening cost $f_i$. Each edge $e \in E$ has a non-negative connection cost $c_e$. Denote by $c_{uv}$ the shortest path with respect to the connection costs from vertex $u$ to $v$. Opening a facility $i$ incurs its opening cost. Assigning a client $j$ to some facility $i$ incurs a connection cost $c_{ij}$. Connecting a facility subset $S \subseteq \mathcal{F}$ by a Steiner tree $T$ incurs a cost of $M \sum_{e \in T} c_e$ called Steiner cost. The goal is to open some facilities $S \subseteq \mathcal{F}$, assign each client $j$ to some opened facility $\sigma(j) \in S$ and connect all the opened facilities $S$ by a Steiner tree $T$, such that the number of clients connected to any opened facility is at least $L$, and the total incurred cost of $\sum_{i \in S} f_i + \sum_{j \in \mathcal{D}} c_{\sigma(j)j} + M \sum_{e \in T} c_e$ (i.e., the total opening, connection, and Steiner cost) is minimized.

The LB ConFL instance becomes a ConFL instance if we get rid of the lower bound constraints (i.e., set $L = 0$ in a LB ConFL instance). Specifically, in a ConFL instance $\mathcal{I}_{\text{Co}}$, a graph $G = (V, E)$, a facility set $\mathcal{F} \subseteq V$, a client set $\mathcal{D} \subseteq V$, and a parameter $M \geq 1$ are given. Each facility $i$ has a non-negative opening cost $f_i$. Each edge $e \in E$ has a non-negative connection cost $c_e$, and the connection cost between a facility $i$ and a client $j$ is $c_{ij}$, which is the shortest path with respect to the connection costs from $i$ to $j$. The aim is to open some facilities $S \subseteq \mathcal{F}$, assign each client $j$ to some opened facility $\sigma(j) \in S$ and connect all the opened facilities $S$ by a Steiner tree $T$, such that the total incurred cost of $\sum_{i \in S} f_i + \sum_{j \in \mathcal{D}} c_{\sigma(j)j} + M \sum_{e \in T} c_e$ (i.e., total opening, connection and Steiner cost) is minimized.

We use $(S, \sigma, T)$ to present a solution of a LB ConFL or ConFL instance. Here $S \subseteq \mathcal{F}$ is the set of open facilities, $\sigma : \mathcal{D} \rightarrow S$ is an assignment that maps each client $j$ to some opened facility in $S$, and $T$ is a tree. A solution $(S, \sigma, T)$ is feasible for the LB ConFL instance if $|\{j \in \mathcal{D} : \sigma(j) = i\}| \geq L$ for each $i \in S$, and the tree $T$ is a Steiner tree that spans all the vertices in $S$. A solution $(S, \sigma, T)$ is feasible for the ConFL instance if the tree is a Steiner tree that spans all the vertices in $S$. For the solution $(S, \sigma, T)$, denote by $cost_O(S, \sigma, T)$ its total opening cost, i.e.,

$$cost_O(S, \sigma, T) = \sum_{i \in S} f_i;$$

denote by $cost_C(S, \sigma, T)$ its total connection cost, i.e.,

$$cost_C(S, \sigma, T) = \sum_{j \in \mathcal{D}} c_{\sigma(j)j};$$

and denote by $cost_S(S, \sigma, T)$ its Steiner cost, i.e.,

$$cost_{St}(S, \sigma, T) = M \sum_{e \in T} c_e.$$

Let $cost(S, \sigma, T)$ be the total incurred cost of the solution $(S, \sigma, T)$. Therefore,

$$\begin{aligned} cost(S, \sigma, T) &= cost_O(S, \sigma, T) + cost_C(S, \sigma, T) + cost_{St}(S, \sigma, T) \\ &= \sum_{i \in S} f_i + \sum_{j \in \mathcal{D}} c_{\sigma(j)j} + M \sum_{e \in T} c_e. \end{aligned}$$

The LB ConFL instance becomes a LBFL instance if we remove the requirement that all the opened facilities be connected by a Steiner tree. Specifically, in a LBFL instance $\mathcal{I}_{\text{LB}}$, a facility set $\mathcal{F}$, a client set $\mathcal{D}$, and an integer lower bound $L$ are given. Each facility $i$ has a non-negative opening cost $f_i$. Each facility-client pair has a non-negative connection cost $c_{ij}$. The objective is to open some facilities $S \subseteq \mathcal{F}$, assign each client $j$ to some opened facility $\sigma(j) \in S$, such that the number of clients connected to any opened facility is at least $L$ (i.e., $|\{j \in \mathcal{D} : \sigma(j) = i\}| \geq L$ for each $i \in S$), and the total incurred cost of $\sum_{i \in S} f_i + \sum_{j \in \mathcal{D}} c_{\sigma(j)j}$ (i.e., the total opening and connection cost) is minimized.

We use $(S, \sigma)$ to denote a solution of a LBFL instance. Here $S \subseteq \mathcal{F}$ is the set of open facilities and $\sigma : \mathcal{D} \to S$ is an assignment which maps each client $j$ to some opened facility in $S$. A solution $(S, \sigma)$ is feasible for the LBFL instance if $|\{j \in \mathcal{D} : \sigma(j) = i\}| \geq L$ for each $i \in S$. For the solution $(S, \sigma)$, denote by $cost_O(S, \sigma)$ its total opening cost, i.e.,

$$cost_O(S, \sigma) = \sum_{i \in S} f_i;$$

denote by $cost_C(S, \sigma)$ its total connection cost, i.e.,

$$cost_C(S, \sigma) = \sum_{j \in \mathcal{D}} c_{\sigma(j)j}.$$

Let $cost(S, \sigma)$ be the total incurred cost of the solution $(S, \sigma)$. Therefore,

$$\begin{aligned} cost(S, \sigma) &= cost_O(S, \sigma) + cost_C(S, \sigma) \\ &= \sum_{i \in S} f_i + \sum_{j \in \mathcal{D}} c_{\sigma(j)j}. \end{aligned}$$

### 2.2 A Simple Algorithm for the LB ConFL

Our first approximation algorithm for the LB ConFL is based on a simple idea that finding a Steiner tree connecting all the possibly opened facilities before considering the lower bound constraints may give a good solution.

Here is a description of our simple algorithm. At the beginning of the algorithm, our primary concern is to find a suitable tree, and we pay no attention to the lower bound constraints. So for any LB ConFL instance $\mathcal{I}_{\text{LBCo}}$, we first get rid of the lower bound $L$ to yield a ConFL instance $\mathcal{I}_{\text{Co}}$. Use the currently best $\alpha$-approximation algorithm for the ConFL to solve $\mathcal{I}_{\text{Co}}$ and obtain a solution $(S_{\text{Co}}, \sigma_{\text{Co}}, T_{\text{Co}})$, where $\alpha = 3.19$. We view the tree $T_{\text{Co}}$ as our suitable tree. Then, we put our focus on satisfying the lower bound constraints. Since it is known that the facilities in $S_{\text{Co}}$ are already connected by the tree $T_{\text{Co}}$, it is advisable to open some facilities in $S_{\text{Co}}$ while considering the lower bound constraints. So we construct a LBFL instance as $\mathcal{I}_{\text{LB}} = (S_{\text{Co}}, \mathcal{D}, L, \{f_i\}_{i \in S_{\text{Co}}}, \{c_{ij}\}_{i \in S_{\text{Co}}, j \in \mathcal{D}})$ and use the currently best $\beta$-approximation algorithm for the LBFL problem to solve it in order to obtain a solution $(S_{\text{LB}}, \sigma_{\text{LB}})$, where $\beta = 82.6$. Last, we output $(S_{\text{LB}}, \sigma_{\text{LB}}, T_{\text{Co}})$ as our final solution. The simple algorithm is formally presented as Algorithm 1.

**Algorithm 1 :** A Simple Approximation Algorithm for the LB ConFL.

**Input:** A LB ConFL instance $\mathcal{I}_{\mathrm{LBCo}} = (V, E, \mathcal{F}, \mathcal{D}, M, L, \{f_i\}_{i\in\mathcal{F}}, \{c_e\}_{e\in E}, \{c_{ij}\}_{i\in\mathcal{F},j\in\mathcal{D}})$.
**Output:** A feasible solution $(S, \sigma, T)$ for instance $\mathcal{I}_{\mathrm{LBCo}}$.

**Step 1 Construct and solve a ConFL instance $\mathcal{I}_{\mathrm{Co}}$.**
For the LB ConFL instance $\mathcal{I}_{\mathrm{LBCo}}$, get rid of the lower bound $L$ to yield a ConFL instance $\mathcal{I}_{\mathrm{Co}} = (V, E, \mathcal{F}, \mathcal{D}, M, \{f_i\}_{i\in\mathcal{F}}, \{c_e\}_{e\in E}, \{c_{ij}\}_{i\in\mathcal{F},j\in\mathcal{D}})$. Use the currently best $\alpha$-approximation algorithm for the ConFL to solve $\mathcal{I}_{\mathrm{Co}}$ and obtain a solution $(S_{\mathrm{Co}}, \sigma_{\mathrm{Co}}, T_{\mathrm{Co}})$, where $\alpha = 3.19$.

**Step 2 Construct and solve a LBFL instance $\mathcal{I}_{\mathrm{LB}}$.**
Based on the opened facilities $S_{\mathrm{Co}}$ in the solution $(S_{\mathrm{Co}}, \sigma_{\mathrm{Co}}, T_{\mathrm{Co}})$, construct a LBFL instance as $\mathcal{I}_{\mathrm{LB}} = (S_{\mathrm{Co}}, \mathcal{D}, L, \{f_i\}_{i\in S_{\mathrm{Co}}}, \{c_{ij}\}_{i\in S_{\mathrm{Co}},j\in\mathcal{D}})$. Use the currently best $\beta$-approximation algorithm for the LBFL to solve $\mathcal{I}_{\mathrm{LB}}$ and obtain a solution $(S_{\mathrm{LB}}, \sigma_{\mathrm{LB}})$, where $\beta = 82.6$.

**Step 3 Construct a solution for the LB ConFL instance.**
Set $S := S_{\mathrm{LB}}$, $\sigma := \sigma_{\mathrm{LB}}$, and $T := T_{\mathrm{Co}}$. Output $(S, \sigma, T)$ as the solution for the LB ConFL instance $\mathcal{I}_{\mathrm{LBCo}}$.

Step 2 of Algorithm 1 ensures that the solution $(S, \sigma, T)$, where $S = S_{\mathrm{LB}}$, $\sigma = \sigma_{\mathrm{LB}}$, satisfies the lower bound constraints, and implies that $S_{\mathrm{LB}} \subseteq S_{\mathrm{Co}}$. Note that all the vertices in $S_{\mathrm{LB}}$ are connected by the tree $T_{\mathrm{Co}}$, since tree $T_{\mathrm{Co}}$ spans all the vertices in $S_{\mathrm{Co}}$. Therefore, the solution $(S, \sigma, T)$ is a feasible solution for the LB ConFL instance $\mathcal{I}_{\mathrm{LBCo}}$.

For any LB ConFL instance $\mathcal{I}_{\mathrm{LBCo}}$ and the corresponding ConFL instance $\mathcal{I}_{\mathrm{Co}}$ constructed from Step 1 in Algorithm 1, let $(S^*_{\mathrm{LBCo}}, \sigma^*_{\mathrm{LBCo}}, T^*_{\mathrm{LBCo}})$ and $(S^*_{\mathrm{Co}}, \sigma^*_{\mathrm{Co}}, T^*_{\mathrm{Co}})$ be the optimal solutions of them, and $OPT_{\mathrm{LBCo}}$ and $OPT_{\mathrm{Co}}$ be the total incurred cost of their optimal solutions.

**Lemma 1.** $OPT_{\mathrm{Co}} \leq OPT_{\mathrm{LBCo}}$.

**Lemma 2.** $cost_O(S, \sigma, T) \leq cost_O(S_{\mathrm{Co}}, \sigma_{\mathrm{Co}}, T_{\mathrm{Co}})$.

**Lemma 3.** $cost_{St}(S, \sigma, T) = cost_{St}(S_{\mathrm{Co}}, \sigma_{\mathrm{Co}}, T_{\mathrm{Co}})$.

**Lemma 4.** $cost_C(S, \sigma, T) \leq \beta \cdot cost_O(S_{\mathrm{Co}}, \sigma_{\mathrm{Co}}, T_{\mathrm{Co}}) + \beta \cdot (2 + \alpha) \cdot OPT_{\mathrm{LBCo}}$.

Combining Lemmas 1–4, we obtain the main result of Algorithm 1.

**Theorem 1.** *Algorithm 1 is a* 696*-approximation algorithm for the LB ConFL problem.*

## 2.3 An Improved Algorithm for the LB ConFL

The improved algorithm proposed in this subsection comes from a careful introspection of Algorithm 1 for the LB ConFL problem. Recall that Algorithm 1 first solves a ConFL instance and then deals with a LBFL instance. One can imagine that this sequential and rough idea will cause the approximation ratio at least at a scale of $O(\alpha)$ times $O(\beta)$, where $\alpha$ and $\beta$ are the currently best approximation ratios for the ConFL and LBFL. Now, we consider an improved idea which is construct and solve a ConFL instance and a LBFL instance simultaneously, and then use their solutions to construct a solution for the LB ConFL. It seems likely that this idea may lead to an approximation algorithm with a ratio at a scale of $O(\alpha)$ plus $O(\beta)$ and could significantly improve the previous ratio.

Here is a description of our improved algorithm. At the beginning of the algorithm, we focus on both finding a suitable tree and satisfying the lower bound constraints. For any LB ConFL instance $\mathcal{I}_{\text{LBCo}}$, get rid of the lower bound $L$ to yield a ConFL instance $\mathcal{I}_{\text{Co}}$; and get rid of the requirement that all the opened facilities must be connected by a Steiner tree to yield a LBFL instance $\mathcal{I}_{\text{LB}}$. Use the currently best $\alpha$-approximation algorithm for the ConFL problem to solve $\mathcal{I}_{\text{Co}}$ and obtain a solution $(S_{\text{Co}}, \sigma_{\text{Co}}, T_{\text{Co}})$. We look upon the tree $T_{\text{Co}}$ as a suitable tree. Use the currently best $\beta$-approximation algorithm for the LBFL problem to solve $\mathcal{I}_{\text{LB}}$ and obtain a solution $(S_{\text{LB}}, \sigma_{\text{LB}})$. Although the solution $(S_{\text{LB}}, \sigma_{\text{LB}})$ satisfies the lower bound constraints, the facilities in $S_{\text{LB}}$ may not be connected by the tree $T_{\text{Co}}$. Thus, we find the closest facility $i_c \in S_{\text{Co}}$ for each facility $i \in S_{\text{LB}}$, and then reconnect all the clients assigned to $i$ under the assignment of $\sigma_{\text{LB}}$ to the facility $i_c$. Denote by all the closest facilities in $S_{\text{Co}}$ as $S$. Denote by all the new assignments as $\sigma$. Therefore, it is not hard to see that $(S, \sigma, T_{\text{Co}})$ is a feasible solution for the LB ConFL instance $\mathcal{I}_{\text{LBCo}}$. The improved algorithm is formally shown as Algorithm 2.

---

**Algorithm 2** : An Improved Approximation Algorithm for the LB ConFL.

---

**Input:** A LB ConFL instance $\mathcal{I}_{\text{LBCo}} = (V, E, \mathcal{F}, \mathcal{D}, M, L, \{f_i\}_{i\in\mathcal{F}}, \{c_e\}_{e\in E}, \{c_{ij}\}_{i\in\mathcal{F}, j\in\mathcal{D}})$.
**Output:** A feasible solution $(S, \sigma, T)$ for instance $\mathcal{I}_{\text{LBCo}}$.

**Step 1 Construct and solve a ConFL instance $\mathcal{I}_{\text{Co}}$.**
It same as Step 1 in Algorithm. At the end of this step, we obtain a solution $(S_{\text{Co}}, \sigma_{\text{Co}}, T_{\text{Co}})$.

**Step 2 Construct and solve a LBFL instance $\mathcal{I}_{\text{LB}}$.**
For the LB ConFL instance $\mathcal{I}_{\text{LBCo}}$, get rid of the requirement, that all the opened facilities must be connected by a Steiner tree, to yield a LBFL instance $\mathcal{I}_{\text{LB}} = (\mathcal{F}, \mathcal{D}, L, \{f_i\}_{i\in\mathcal{F}}, \{c_{ij}\}_{i\in\mathcal{F}, j\in\mathcal{D}})$. Use the currently best $\beta$-approximation algorithm for the LBFL to solve $\mathcal{I}_{\text{LB}}$ and obtain a solution $(S_{\text{LB}}, \sigma_{\text{LB}})$, where $\beta = 82.6$.

**Step 3 Construct a solution for the LB ConFL instance.**
Initially, set $S := \emptyset$, $\sigma(j) := \sigma_{\text{LB}}(j)$ for any $j \in \mathcal{D}$.
**While** $S_{\text{LB}} \neq \emptyset$ **do**
Arbitrarily choose a facility $i \in S_{\text{LB}}$ and find the facility

$$i_c := \arg\min_{i' \in S_{\text{Co}}} c_{ii'}.$$

Update $S_{\text{LB}} = S_{\text{LB}} \setminus \{i\}$, $S := S \cup \{i_c\}$ and $\sigma(j) := i_c$ for any client $j$ with $\sigma(j) = i$.
Set $T := T_{\text{Co}}$. Output $(S, \sigma, T)$ as the solution for the LB ConFL instance $\mathcal{I}_{\text{LBCo}}$.

---

Recall that for any LB ConFL instance $\mathcal{I}_{\text{LBCo}}$, denote by $(S^*_{\text{LBCo}}, \sigma^*_{\text{LBCo}}, T^*_{\text{LBCo}})$ its optimal solutions and $OPT_{\text{LBCo}}$ the total incurred cost of the solution. For any corresponding LBFL instance $\mathcal{I}_{\text{LB}}$ constructed from Step 2 in Algorithm

2, let $(S^*_{\mathrm{LB}}, \sigma^*_{\mathrm{LB}})$ be the optimal solution, and $OPT_{\mathrm{LB}}$ be the total incurred cost of its optimal solution.

**Lemma 5.** $OPT_{\mathrm{LB}} \leq OPT_{\mathrm{LBCo}}$.

**Lemma 6.** $cost_O(S, \sigma, T) \leq cost_O(S_{\mathrm{Co}}, \sigma_{\mathrm{Co}}, T_{\mathrm{Co}})$.

**Lemma 7.** $cost_{St}(S, \sigma, T) = cost_{St}(S_{\mathrm{Co}}, \sigma_{\mathrm{Co}}, T_{\mathrm{Co}})$.

**Lemma 8.** $cost_C(S, \sigma, T) \leq cost_C(S_{\mathrm{Co}}, \sigma_{\mathrm{Co}}, T_{\mathrm{Co}}) + 2 \cdot cost_C(S_{\mathrm{LB}}, \sigma_{\mathrm{LB}})$.

Combining Lemmas 5–8, we obtain the main result of Algorithm 2.

**Theorem 2.** *Algorithm 2 is a* 169*-approximation algorithm for the LB ConFL problem.*

## 3 Dealing with Non-uniform Lower Bounds

In this section, we first describe the GLB ConFL and GLBFL. Then, we show that the approximation algorithms for the ConFL and GLBFL enable us to propose an approximation algorithm for the GLB ConFL.

### 3.1 Problem Descriptions Related to the GLB ConFL

In a GLB ConFL instance, each facility $i$ has its own non-uniform lower bound $L_i$. The goal is to open some facilities, assign each client to some opened facility and connect all the opened facilities by a Steiner tree, such that the number of clients connected to any opened facility $i$ is at least $L_i$, and the total incurred cost (i.e., the total opening, connection, and Steiner cost) is minimized.

Same as the LB ConFL, the GLB ConFL instance becomes a ConFL instance if we get rid of the lower bound constraints. We still use $(S, \sigma, T)$ to present a solution of a GLB ConFL instance. A solution $(S, \sigma, T)$ is feasible for the GLB ConFL instance if $|\{j \in \mathcal{D} : \sigma(j) = i\}| \geq L_i$ for each $i \in S$, and the tree $T$ is a Steiner tree that spans all the vertices in $S$. For the solution $(S, \sigma, T)$, still use the symbols of $cost_O(S, \sigma, T)$, $cost_C(S, \sigma, T)$, $cost_{St}(S, \sigma, T)$ and $cost(S, \sigma, T)$ to denote the total opening, total connection, Steiner, and total incurred cost, respectively.

The GLB ConFL instance becomes a GLBFL instance if we remove the requirement that all the opened facilities must be connected by a Steiner tree. We still use $(S, \sigma)$ to denote a solution of a GLBFL instance. A solution $(S, \sigma)$ is feasible for the GLBFL instance if $|\{j \in \mathcal{D} : \sigma(j) = i\}| \geq L_i$ for each $i \in S$. For the solution $(S, \sigma)$, still use the symbols of $cost_O(S, \sigma)$, $cost_C(S, \sigma)$ and $cost(S, \sigma)$ to denote the total opening, connection and incurred cost, respectively.

### 3.2 An Algorithm for the GLB ConFL

The idea of the algorithm proposed for the GLB ConFL problem is similar to Algorithm 2. We consider constructing and solving a ConFl instance and a GLBFL instance at the same time, and then use their solutions to construct a solution for the GLB ConFL. Since the lower bounds for the facilities could be different in the GLB ConFL problem, the step of constructing its solution is a bit more complicated than the one (i.e., Step 3) in Algorithm 2.

Here is a description of the algorithm for the GLB ConFL problem. The first two steps are similar to the ones in Algorithm 2. For any GLB ConFL instance $\mathcal{I}_{\mathrm{GLBCo}}$, get rid of the lower bounds $\{L_i\}_{i \in \mathcal{F}}$ to yield a ConFL instance $\mathcal{I}_{\mathrm{Co}}$, and get rid of the requirement that all the opened facilities must be connected by a Steiner tree to yield a GLBFL instance $\mathcal{I}_{\mathrm{GLB}}$. Use the currently best $\alpha$-approximation algorithm for the ConFL problem to solve $\mathcal{I}_{\mathrm{Co}}$ and obtain a solution $(S_{\mathrm{Co}}, \sigma_{\mathrm{Co}}, T_{\mathrm{Co}})$. We look upon the tree $T_{\mathrm{Co}}$ as a suitable tree for augmenting. Use the currently best $\eta$-approximation algorithm for the GLBFL problem to solve $\mathcal{I}_{\mathrm{GLB}}$ and obtain a solution $(S_{\mathrm{GLB}}, \sigma_{\mathrm{GLB}})$. Even though the solution $(S_{\mathrm{GLB}}, \sigma_{\mathrm{GLB}})$ satisfies the lower bound constraints, the facilities in $S_{\mathrm{GLB}}$ may not be connected by the tree $T_{\mathrm{Co}}$. Thus, we find the closest facility $i_c \in S_{\mathrm{Co}}$ for each facility $i \in S_{\mathrm{GLB}}$, and then reconnect all the clients assigned to $i$ under the assignment of $\sigma_{\mathrm{GLB}}$ to the facility $i_c$. We call the closest facility $i_c \in S_{\mathrm{Co}}$ to a facility $i \in S_{\mathrm{GLB}}$ its head, and call $i$ the tail of $i_c$. Denote by all the head facilities as $S_c$. Since the lower bounds are non-uniform, the new assignments cannot make sure that all the lower bound constraints of the heads are satisfied. Therefore, for each facility $i_c \in S_c$ find its closest tail facility $i_{cc} \in S_{\mathrm{GLB}}$ and reconnect all the clients assigned to $i_c$ to the facility $i_{cc}$. Denote by all the closest tail facilities as $S$. Denote by all the latest assignments as $\sigma$. Let $T$ be the tree obtained from augmenting the tree $T_{\mathrm{Co}}$ by adding the shortest connection path from each head facility to its closest tail facility. Therefore, it is not hard to see that $(S, \sigma, T)$ is a feasible solution for the GLB ConFL instance $\mathcal{I}_{\mathrm{GLBCo}}$, since the lower bound constraint for each tail facility must be satisfied. The algorithm is formally proposed as Algorithm 3.

**Algorithm 3** : An Approximation Algorithm for the GLB ConFL.

**Input:** An instance $\mathcal{I}_{\text{GLBCo}} = (V, E, \mathcal{F}, \mathcal{D}, M, \{L_i\}_{i\in\mathcal{F}}, \{f_i\}_{i\in\mathcal{F}}, \{c_e\}_{e\in E}, \{c_{ij}\}_{i\in\mathcal{F}, j\in\mathcal{D}})$.
**Output:** A feasible solution $(S, \sigma, T)$ for instance $\mathcal{I}_{\text{GLBCo}}$.

**Step 1 Construct and solve a ConFL instance $\mathcal{I}_{\text{Co}}$.**
For the GLB ConFL instance $\mathcal{I}_{\text{GLBCo}}$, get rid of the lower bounds $\{L_i\}_{i\in\mathcal{F}}$ to yield a ConFL instance $\mathcal{I}_{\text{Co}} = (V, E, \mathcal{F}, \mathcal{D}, M, \{f_i\}_{i\in\mathcal{F}}, \{c_e\}_{e\in E}, \{c_{ij}\}_{i\in\mathcal{F}, j\in\mathcal{D}})$. Use the currently best $\alpha$-approximation algorithm for the ConFL problem to solve $\mathcal{I}_{\text{Co}}$ and obtain a solution $(S_{\text{Co}}, \sigma_{\text{Co}}, T_{\text{Co}})$, where $\alpha = 3.19$.

**Step 2 Construct and solve a GLBFL instance $\mathcal{I}_{\text{GLB}}$.**
For the GLB ConFL instance $\mathcal{I}_{\text{GLBCo}}$, get rid of the requirement, that all the opened facilities must be connected by a Steiner tree, to yield a GLBFL instance $\mathcal{I}_{\text{GLB}} = (\mathcal{F}, \mathcal{D}, \{L_i\}_{i\in\mathcal{F}}, \{f_i\}_{i\in\mathcal{F}}, \{c_{ij}\}_{i\in\mathcal{F}, j\in\mathcal{D}})$. Use the currently best $\eta$-approximation algorithm for the GLBFL problem to solve $\mathcal{I}_{\text{GLB}}$ and obtain a solution $(S_{\text{GLB}}, \sigma_{\text{GLB}})$, where $\eta = 3926$.

**Step 3 Construct a solution for the GLB ConFL instance.**

**Step 3.1** Initially, set $S := \emptyset$, $\sigma(j) := \sigma_{\text{GLB}}(j)$ for any $j \in \mathcal{D}$. Set $S_1 := S_{\text{GLB}}$, $S_c := \emptyset$.

**Step 3.2 While $S_1 \neq \emptyset$ do**
Arbitrarily choose a facility $i \in S_1$ and find the facility

$$i_c := \arg \min_{i' \in S_{\text{Co}}} c_{ii'}.$$

Define $\sigma_c(i) := i_c$ for the facility $i$. Update $S_1 = S_1 \setminus \{i\}$, $S_c := S_c \cup \{i_c\}$ and $\sigma(j) := i_c$ for any client $j$ with $\sigma(j) = i$.

**Step 3.3 While $S_c \neq \emptyset$ do**
Arbitrarily choose a facility $i_c \in S_c$ and find the facility

$$i_{cc} := \arg \min_{i \in S_{\text{GLB}} : \sigma_c(i) = i_c} c_{ii_c}.$$

Update $S_c = S_c \setminus \{i_c\}$, $S := S \cup \{i_{cc}\}$ and $\sigma(j) := i_{cc}$ for any client $j$ with $\sigma(j) = i_c$.

**Step 3.4** Let $T$ be the tree obtained from augmenting the tree $T_{\text{Co}}$ by adding the shortest connection path from each facility $i_c \in S_c$ to its $i_{cc} \in S_{\text{GLB}}$. Output $(S, \sigma, T)$ as the solution for the GLB ConFL instance $\mathcal{I}_{\text{GLBCo}}$.

---

For any GLB ConFL instance $\mathcal{I}_{\text{GLBCo}}$, denote by $(S^*_{\text{GLBCo}}, \sigma^*_{\text{GLBCo}}, T^*_{\text{GLBCo}})$ its optimal solutions and $OPT_{\text{GLBCo}}$ the total incurred cost of the solution. For any corresponding ConFL instance $\mathcal{I}_{\text{Co}}$ constructed from Step 1 in Algorithm 3, still let $(S^*_{\text{Co}}, \sigma^*_{\text{Co}}, T^*_{\text{Co}})$ be the optimal solution of it, and $OPT_{\text{Co}}$ be the total incurred cost of its optimal solution. For any corresponding GLBFL instance $\mathcal{I}_{\text{GLB}}$ constructed from Step 2 in Algorithm 3, let $(S^*_{\text{GLB}}, \sigma^*_{\text{GLB}})$ be the optimal solutions of it, and $OPT_{\text{GLB}}$ be the total incurred cost of its optimal solution.

**Lemma 9.** $OPT_{\text{Co}} \leq OPT_{\text{GLBCo}}$.

**Lemma 10.** $OPT_{\text{GLB}} \leq OPT_{\text{GLBCo}}$.

**Lemma 11.** $cost_O(S, \sigma, T) \leq cost_O(S_{\text{GLB}}, \sigma_{\text{GLB}})$.

Let $cost_C^{\text{GLB}}$ and $cost_c^{\text{Co}}$ be $cost_C(S_{\text{GLB}}, \sigma_{\text{GLB}})$ and $cost_C(S_{\text{Co}}, \sigma_{\text{Co}}, T_{\text{Co}})$ for short.

**Lemma 12.** $cost_{St}(S, \sigma, T) \leq M \cdot cost_C^{\text{GLB}} + M \cdot cost_C^{\text{Co}} + cost_{St}(S_{\text{Co}}, \sigma_{\text{Co}}, T_{\text{Co}})$.

**Lemma 13.** $cost_C(S, \sigma, T) \leq 2cost_C(S_{\text{Co}}, \sigma_{\text{Co}}, T_{\text{Co}}) + 3 \cdot cost_C(S_{\text{GLB}}, \sigma_{\text{GLB}})$.

Combining Lemmas 9–13, we obtain the main result of Algorithm 3.

**Theorem 3.** *Algorithm 3 is an* $(\max\{(M+2) \cdot \alpha, (M+3) \cdot \eta\})$*-approximation algorithm for the LB ConFL problem, where* $\alpha = 3.19$ *and* $\eta = 3926$.

## 4 Discussions

We propose the first and currently best approximation algorithms for the LB ConFL in this paper. For the GLB ConFL, we give an $M$-related approximation algorithm. The reason for obtaining an unsatisfactory approximation ratio for the GLB ConFL is that the finally found tree in our algorithm is not good enough, since the augment process makes too much loss of the ratio. We do believe that by finding a better Steiner tree, the approximation ratio could be remarkably improved.

**Acknowledgments.** The first author is supported by National Natural Science Foundation of China (No. 12001523). The second author is supported by National Natural Science Foundation of China (No. 11971349). The third author is supported by Guangxi Key Laboratory of Cryptography and Information Security (No. GCIS202116) and National Natural Science Foundation of China (No. 11901558).

## References

1. Ahmadian, S., Swamy, C.: Improved approximation guarantees for lower-bounded facility location. In: Erlebach, T., Persiano, G. (eds.) WAOA 2012. LNCS, vol. 7846, pp. 257–271. Springer, Heidelberg (2013). https://doi.org/10.1007/978-3-642-38016-7_21
2. Arya, V., Garg, N., Khandekar, R., Meyerson, A., Munagala, K., Pandit, V.: Local search heuristics for $k$-median and facility location problems. SIAM J. Comput. **33**, 544–562 (2004)
3. Eisenbrand, F., Grandoni, F., Rothvoß, T., Schäfer, G.: Approximating connected facility location problems via random facility sampling and core detouring. In: Proceeding of 19th Annual ACM-SIAM Symposium, pp. 1174–1183 (2008)
4. Grandoni, F., Rothvoß, T.: Approximation algorithms for single and multi-commodity connected facility location. In: Günlük, O., Woeginger, G.J. (eds.) IPCO 2011. LNCS, vol. 6655, pp. 248–260. Springer, Heidelberg (2011). https://doi.org/10.1007/978-3-642-20807-2_20
5. Guha, S., Meyerson, A., Munagala, K.: Hierarchical placement and network design problems. In: Proceedings of the 41st Annual Symposium on Foundations of Computer Science, pp. 603–612 (2000)
6. Gupta, A., Kleinberg, J., Kumar, A., Rastogi, R. Yener, B.: Provisioning a virtual private network: a network design problem for multicommodity flow. In: Proceedings of the 33rd Annual ACM Symposium on Theory of Computing, pp. 389–398 (2001)

7. Han, L., Xu, D., Xu, Y., Zhang, D.: Approximating the $\tau$-relaxed soft capacitated facility location problem. J. Comb. Optim. **40**(3), 848–860 (2020). https://doi.org/10.1007/s10878-020-00631-y
8. Hasan, M.K., Jung, H., Chwa, K.Y.: Approximation algorithms for connected facility location problems. J. Comb. Optim. **16**, 155–172 (2008)
9. Jain, K., Mahdian, M., Markakis, E., Saberi, E., Vazirani, V.V.: Greedy facility location algorithms analyzed using dual fitting with factor-revealing LP. J. ACM **50**, 795–824 (2003)
10. Jain, K., Vazirani, V.V.: Approximation algorithms for metric facility location and $k$-median problems using the primal-dual schema and Lagrangian relaxation. J. ACM **48**, 274–296 (2001)
11. Jung, H., Hasan, M.K., Chwa, K.Y.: A 6.55 factor primal-dual approximation algorithm for the connected facility location problem. J. Combin. Optim. **18**, 258–271 (2009)
12. Karget, D.R., Minkoff, M.: Building Steiner trees with incomplete global knowledge. In: Proceedings of the 41st Annual Symposium on Foundations of Computer Science, pp. 613–623 (2000)
13. Li, S.: On facility location with general lower bounds. In: Proceedings of the 30th Annual ACM-SIAM Symposium on Discrete Algorithms, pp. 2279–2290 (2019)
14. Li, S.: A 1.488 approximation algorithm for the uncapacitated facility location problem. Inf. Comput. **222**, 45–58 (2013)
15. Shmoys, D.B., Tardos, É., Aardal, K.I.: Approximation algorithms for facility location problems. In: Proceedings of the 29th Annual ACM symposium on Theory of Computing, pp. 265–274 (1997)
16. Sviridenko, M.: An improved approximation algorithm for the metric uncapacitated facility location problem. In: Cook, W.J., Schulz, A.S. (eds.) IPCO 2002. LNCS, vol. 2337, pp. 240–257. Springer, Heidelberg (2002). https://doi.org/10.1007/3-540-47867-1_18
17. Svitkina, Z.: Lower-bounded facility location. ACM Trans. Algorithms **6**, 1–16 (2010)
18. Swamy, C., Kumar, A.: Primal-dual algorithms for connected facility location problems. Algorithmica **40**, 245–269 (2004)
19. Xu, Y., Xu, D., Du, D., Wu, C.: Improved approximation algorithm for universal facility location problem with linear penalties. Theoret. Comput. Sci. **774**, 143–151 (2019)
20. Xu, Y., Xu, D., Zhang, Y., Zou, J.: MpUFLP: universal facility location problem in the $p$-th power of metric space. Theoret. Comput. Sci. **838**, 58–67 (2020)

# Mechanism Design for Facility Location with Fractional Preferences and Minimum Distance

Longteng Duan[1], Zifan Gong[1], Minming Li[1], Chenhao Wang[2(✉)], and Xiaoying Wu[3]

[1] City University of Hong Kong, Kowloon Tong, Hong Kong
{longtduan2-c,zifangong2-c}@my.cityu.edu.hk, minming.li@cityu.edu.hk

[2] Kyushu University, Fukuoka, Japan
wangch@inf.kyushu-u.ac.jp

[3] AMSS, Chinese Academy of Sciences, Beijing, China
xywu@amss.ac.cn

**Abstract.** In this paper, we study the mechanism design for two-facility-location games with the fractional preferences of agents, in which each agent has private information including her location in an interval $[0, 1]$ and her fractional preference to indicate how much she prefers the two facilities. The decision maker needs to locate the two facilities to serve the agents, who has a utility equal to the interval length 1 minus the sum of weighted distances to both facilities. The facility locations are required to satisfy a minimum distance constraint, i.e., the distance of the two facilities must exceed a given number $d \in [0, 1]$. The goal is to design strategy-proof mechanisms to maximize the social/minimum utility among the agents. We propose a randomized strategy-proof mechanism, which is 2-approximation for both objectives of maximizing the social utility and minimum utility. We also propose a deterministic strategy-proof mechanism which has an approximation ratio of $\frac{4}{2-d}$ and 4 for the two objectives, respectively. Furthermore, we derive corresponding lower bounds on the approximation ratios of strategy-proof mechanisms.

**Keywords:** Facility location · Mechanism design · Approximation

## 1 Introduction

The facility location problem is a classic combinatorial optimization problem extensively studied in the communities of computer science, operations research, and economics, which aims at computing the optimal locations of facilities to minimize transportation costs for servicing customers.

More than one decade ago, Procaccia and Tennenholtz [14] propose a new perspective of *approximate mechanism design* for facility location problems,


C.-Y. Chen et al. (Eds.): COCOON 2021, LNCS 13025, pp. 499–511, 2021.
https://doi.org/10.1007/978-3-030-89543-3_42

using the most basic setting of locating facilities in a real line, which advocates the study of strategy-proof mechanisms through the lens of the *approximation ratio*, a celebrated notion in the field of theoretical computer science and approximation algorithms. In the setting studied by Procaccia and Tennenholtz [14], the agents/customers have private locations in a real line and need to report their locations to a mechanism; then the mechanism decides the locations of facilities to be built, under the objective of minimizing the total/maximum distance from the agents to the facilities. Each agent has a cost equal to her distance from the closest facility, and a mechanism is called *strategy-proof* if no agent can decrease her cost by misreporting. Later, lots of works [2,4,13,15] on mechanism design for facility location extend the classic setting in [14].

In this paper, we study the mechanism design for the two-facility-location model with *fractional preferences*, where each agent has a preference (weight) in the range $[0, 1]$ for each facility, to indicate how well she prefer this facility. The sum of an agent's preference for two facilities is equal to 1. The preferences are private information of agents, who can strategically report their preferences as well as locations to a mechanism. This fractional-preference model is proposed by Fong *et al.* [8], which models the scenario where the facilities are different but serve a similar purpose, such as the supermarket along with the convenience store, the hospital along with the clinic.

In addition, we study the *minimum distance constraint* between the facilities, which requires that the distance of the two facility locations should be at least a given value $d \in [0, 1]$, that is, the two facilities cannot be too close. The minimum distance constraint is first proposed by Duan *et al.* [5]. For the motivation, consider the first scenario that the social planner plans to deploy an Internet café and a primary school in a street, where all agents prefer living close for easy access to both internet surfing and education resource. In the worry that some pupils in the primary school may develop addiction to computer games after class, the two facilities should keep some distance away from each other.

### 1.1 Our Results

In this paper, we study the two-facility-location model with the fractional preferences of $n$ agents and the minimum distance constraint $d$. Suppose the underlying line segment is an interval $[0, 1]$. We derive upper and lower bounds on the approximation ratios of strategy-proof mechanisms under different objectives.

First, we note that there is no good approximation for the objective of minimizing the total cost, where the cost of each agent is the weighted total distance from the two facilities. Formally, every strategy-proof mechanism has an approximation ratio $\Omega(n^{\frac{1}{3}})$. Therefore, we turn to consider the utilities, where the utility of each agent is the interval length 1 minus her cost.

In Sect. 3, we study the objective of maximizing the social utility (total utility of all agents). After showing how to compute an optimal solution regardless of the strategy-proofness, we propose a simple randomized strategy-proof mechanism (Mechanism 1) with a proven approximation ratio 2, and a deterministic strategy-proof mechanism (Mechanism 2) with an approximation ratio $\frac{4}{2-d}$. Further, we derive a lower bound 2 for deterministic mechanisms, and a lower bound 1.06 for randomized mechanisms follows from the model with fractional preferences in [8].

In Sect. 4, we study the objective of maximizing the minimum utility. We prove that Mechanism 1 is 2-approximation, and Mechanism 2 is 4-approximation. Furthermore, we derive a lower bound 2 for all strategy-proof mechanisms.

Our results are summarized in the following Table 1, where UB and LB indicate upper bound and lower bound, respectively.

**Table 1.** A summary of our results.

| Objective | Deterministic | Randomized |
|---|---|---|
| Social utility | UB: $\frac{4}{2-d}$ (Theorem 2) | UB: 2 (Theorem 1) |
| | LB: 2 (Theorem 3) | LB: 1.06 [8] |
| Minimum utility | UB: 4 (Theorem 5) | UB: 2 (Proposition 3) |
| | LB: 2 (Theorem 6) | LB: 2 (Theorem 6) |

We compare our results with those for the problem with fractional preferences but without any distance constraint (i.e. $d = 0$). Regarding the social utility, for deterministic mechanisms, there is a 4-approximation for the case $d = 0$ [8], while we derive an improved upper bound $\frac{4}{2-d}$ even for our general problem. Regarding the minimum utility, for deterministic mechanisms, there is 2-approximation for the case $d = 0$ [8], while we can only have a 4-approximation for our problem.

### 1.2 Related Work

Procaccia and Tennenholtz [14] initiate the study of approximate mechanism design for facility location problems. For the setting of locating a single facility on a real line, they show that locating the facility at the median location of agents is strategy-proof and optimal for the objective of minimizing the social cost, and is 2-approximation for minimizing the maximum cost. For the two-facility setting, they show that the mechanism that places two facilities at the two extreme locations of agents is strategy-proof and $(n - 2)$-approximation for the social cost. Later, Lu *et al.* [12,13] derive lower bounds for strategy-proof mechanisms, and Fotakis and Tzamos [9] prove a tight lower bound $n-2$. Besides the classic setting, there are lots of variants, e.g., obnoxious facilities [4,19], multiple locations of agents [11], false-name manipulations [18], and different cost functions and spaces [7,10,16]. See an overview in [1].

*Preference Models.* A follow-up line of work considers the preference models, where the facilities are heterogeneous and agents may have different preferences or feelings to these facilities. Fong *et al.* [8] propose the *fractional preference* model, where each agent has a number between 0 and 1 for each facility to indicate how she likes this facility. This model can be regarded as a special case of our model by setting $d = 0$ (i.e., without the minimum distance constraint). The *dual preference* model is proposed independently by Zou and Li [21] and Feigenbaum and Sethuraman [6], in which each agent finds some facility either desirable or undesirable. A somewhat different model is the *optional preferences* [15,20], in which each agent is "interested" in either facility or both.

*Distance Constraints.* Zou and Li [21] initiate the study of distance constraint with two facilities. They consider the *maximum distance requirement*, namely that the distance between the two facilities cannot exceed a certain threshold. Later, Chen *et al.* [3,17] relaxed the maximum distance constraint by instead imposing a *penalty.* Analogously, Duan *et al.* [5] study the *minimum* distance requirement for heterogeneous two-facility locations, where the cost of an agent is the sum of her distances to the two facilities. The model in [5] can be regarded as a special case of our model by setting agents' preferences equally for all facilities.

## 2 Model

Let $N = \{1, 2, \cdots, n\}$ be the set of agents located on a line interval $[0, 1]$. We want to locate two facilities $F_1, F_2$ on this interval. Each agent $i$ has a profile $c_i = (x_i, p_i)$, where $x_i \in [0, 1]$ is her location, and $p_i = (p_{i,1}, p_{i,2})$ with $0 \leq p_{i,1}, p_{i,2} \leq 1$ and $p_{i,1} + p_{i,2} = 1$ is her preference on facilities $F_1, F_2$. Let $\mathbf{x} = (x_1, \ldots, x_n)$ be the location profile of agents, and $\mathbf{p} = (p_1, \ldots, p_n)$ be the preference profile. We denote an instance by $\mathbf{c} = (\mathbf{x}, \mathbf{p}) = (c_1, \ldots, c_n)$.

A *deterministic mechanism* $f$ is a function which maps an instance $\mathbf{c}$ to two facility locations $f(\mathbf{c}) = (y_1, y_2) \in [0, 1]^2$, where $y_j$ $(j = 1, 2)$ is the location of facility $F_j$. A *randomized mechanism* $f$ is a function which maps an instance $\mathbf{c}$ to a probability distribution over $[0, 1]^2$. We consider the *minimum distance constraint* with respect to the two facilities. That is, given a constant $d \in [0, 1]$, the facility locations $(y_1, y_2)$ must satisfy $|y_2 - y_1| \geq d$.

Given a deterministic outcome $f(\mathbf{c}) = (y_1, y_2)$, the *cost* of agent $i$ is

$$cost(f(\mathbf{c}), c_i) = |x_i - y_1| \cdot p_{i,1} + |x_i - y_2| \cdot p_{i,2},$$

that is, the weighted total distance from the two facilities. If $f(\mathbf{c})$ is a distribution returned by a randomized mechanism $f$, then the cost is defined as the weighted total distance in expectation:

$$cost(f(\mathbf{c}), c_i) = \mathbb{E}_{(y_1, y_2) \sim f(\mathbf{c})} \left[ |x_i - y_1| \cdot p_{i,1} + |x_i - y_2| \cdot p_{i,2} \right].$$

The *social cost* of a mechanism $f$ on an instance $\mathbf{c}$ is defined as the total cost of all $n$ agents:

$$SC(f(\mathbf{c}), \mathbf{c}) = \sum_{i \in N} cost(f(\mathbf{c}), c_i).$$

Now, we give the formal definition of strategy-proofness.

**Definition 1.** *A mechanism $f$ is* strategy-proof *if no agent can benefit from misreporting her profile $c_i$. Formally, for any agent $i \in N$, profile $\mathbf{c} = (c_i, \mathbf{c}_{-i})$ where $\mathbf{c}_{-i}$ is a collection of profile of the $n$ agents except agent $i$, and any misreported profile $c_i'$, it holds that*

$$cost(f(\mathbf{c}), c_i) \leq cost(f(c_i', \mathbf{c}_{-i}), c_i).$$

For the objective of minimizing the social cost, we say a mechanism $f$ has an approximation ratio $\alpha$ ($\alpha \geq 1$) or is $\alpha$-approximation, if for all instance $\mathbf{c}$, we have $\frac{SC(f(\mathbf{c}),\mathbf{c})}{OPT_{SC}(\mathbf{c})} \leq \alpha$, where $OPT_{SC}(\mathbf{c}) = \min_{(y_1,y_2)\in[0,1]^2} SC((y_1, y_2), \mathbf{c})$ is the optimal social cost. However, a special instance given in [8] shows that no mechanism can have a good approximation ratio.

**Proposition 1 (Theorem 1 of [8]).** *For the objective of minimizing social cost, if the agents can misreport preferences, the approximation ratio for any strategy-proof mechanism is $\Omega(n^{\frac{1}{3}})$.*

Therefore, we turn to consider the utility of agents instead of the cost. Given a deterministic outcome $f(\mathbf{c}) = (y_1, y_2)$, the *utility* of agent $i$ is defined as

$$u(f(\mathbf{c}), c_i) = 1 - |x_i - y_1| \cdot p_{i,1} - |x_i - y_2| \cdot p_{i,2}.$$

If $f$ is randomized then the utility is defined as the expectation. We consider two objectives: maximizing the social utility, and maximizing the minimum utility, where the *social utility* is $SU(f(\mathbf{c}), \mathbf{c}) = \sum_{i\in N} u(f(\mathbf{c}), c_i)$, and the *minimum utility* is $MU(f(\mathbf{c}), \mathbf{c}) = \min_{i\in N} u(f(\mathbf{c}), c_i)$.

For the $SU$ objective, we say a mechanism $f$ has an approximation ratio $\alpha$, if for any instance $\mathbf{c}$, we have $\frac{OPT_{SU}(\mathbf{c})}{SU(f(\mathbf{c}),\mathbf{c})} \leq \alpha$, where $OPT_{SU}(\mathbf{c})$ is the optimal social utility. For the $MU$ objective, the approximation ratio of a mechanism is defined similarly.

## 3 Maximizing the Social Utility

In this section, we study the objective of maximizing the social utility. We first show how to compute the optimal solution in the following, and then provide upper bounds and lower bounds on the approximation ratio of strategy-proof mechanisms.

### 3.1 Optimal Solution

We first study the optimization problem of maximizing the social utility, regardless of the strategy-proofness. Given instance $\mathbf{c}$, note that an optimal solution maximizes the social utility and minimizes the social cost simultaneously. So the optimal solution can be obtained by solving the following linear program:

$$\begin{aligned} &\min_{y_1,y_2} \; SC((y_1, y_2), \mathbf{c}) \\ &s.t. \; (y_1, y_2) \in D := \{(y_1, y_2) \in [0,1]^2 \mid d \leq |y_2 - y_1|\}. \end{aligned} \tag{1}$$

For $j = 1, 2$, we define $w_j = \sum_{i=1}^{n} p_{i,j}$, and $\tilde{y}_j = \arg\min_{x_k: k \in N}\{\sum_{i=1}^{k} p_{i,j} \geq \frac{w_j}{2}\}$. When the minimum distance constraint $|\tilde{y}_1 - \tilde{y}_2| \geq d$ is satisfied, $(\tilde{y}_1, \tilde{y}_2)$ is optimal for LP (1), because the point $\tilde{y}_j$, as a median, minimizes the total weighted distance from agents to $F_j$.

If $|\tilde{y}_1 - \tilde{y}_2| < d$, it is easy to see that the optimal social cost could be obtained within the boundary $\partial D = \{(y_1, y_2) \in [0,1]^2 \mid |y_2 - y_1| = d\}$. Therefore, it is equivalent to the following linear optimization problem:

$$\begin{aligned} \min_{y_1, y_2} \ & SC((y_1, y_2), \mathbf{c}) \\ s.t. \ & (y_1, y_2) \in \partial D. \end{aligned} \tag{2}$$

For any optimal solution $(y_1, y_2) \in \partial D$, there are two cases $y_2 = y_1 + d$ and $y_1 = y_2 + d$. We discuss them respectively.

**Case 1.** $y_2 = y_1 + d$. It is equivalent to solve

$$\min_{y_1 \in [0, 1-d]} \sum_{i=1}^{n} (p_{i,1}|y_1 - x_i| + p_{i,2}|y_1 + d - x_i|). \tag{3}$$

Consider the profile

$$L = ((p_{1,2}, x_1 - d), (p_{2,2}, x_2 - d), \ldots, (p_{n,2}, x_n - d), (p_{1,1}, x_1), (p_{2,1}, x_2), \ldots, (p_{n,1}, x_n)).$$

We rearrange the $2n$ elements in profile $L$ as $\tilde{L} = ((\tilde{p}_1, \tilde{x}_1), (\tilde{p}_2, \tilde{x}_2), \ldots, (\tilde{p}_{2n}, \tilde{x}_{2n}))$, such that $\tilde{x}_1 \leq \tilde{x}_2 \leq \cdots \leq \tilde{x}_{2n}$. Note that $\sum_{i=1}^{2n} \tilde{p}_i = \sum_{i \in N}(p_{i,1} + p_{i_2}) = n$. Let $k_1 = \arg\min_{k=1,\ldots,2n} \sum_{i=1}^{k} \tilde{p}_i \geq \frac{n}{2}$. We define $y_1^*$ as follows: $y_1^* = k_1$ if $k_1 \in [0, 1-d]$, $y_1^* = 0$ if $k_1 < 0$, and $y_1^* = 1 - d$ if $k_1 > 1 - d$. Then it is not hard to see that $y_1^*$ solves (3). Therefore, in this case, $(y_1^*, y_1^* + d)$ is optimal.

**Case 2.** $y_1 = y_2 + d$. It is equivalent to solve

$$\min_{y_2 \in [d, 1]} \sum_{i=1}^{n} (p_{i,1}|y_2 + d - x_i| + p_{i,2}|y_2 - x_i|). \tag{4}$$

Using a symmetric analysis, we can define $y_2^*$ that solves (4), and thus $(y_2^* + d, y_2^*)$ is an optimal solution in this case.

Now we can have the following proposition.

**Proposition 2.** *For maximizing the social utility, if $|\tilde{y}_1 - \tilde{y}_2| \geq d$, then $(\tilde{y}_1, \tilde{y}_2)$ is an optimal solution. Otherwise, the better one between $(y_1^*, y_1^* + d)$ and $(y_2^* + d, y_2^*)$ is optimal.*

### 3.2 Strategy-Proof Mechanisms

We first propose a simple randomized strategy-proof mechanism, and then derandomize it.

**Mechanism 1.** *Return* $(\frac{1-d}{2}, \frac{1+d}{2})$ *and* $(\frac{1+d}{2}, \frac{1-d}{2})$ *with half probability.*

**Theorem 1.** *Mechanism 1 is randomized and strategy-proof, and has an approximation ratio 2 for maximizing the social utility.*

*Proof.* The mechanism is trivially strategy-proof, because it does not take the agents' reports into consideration. The utility for each agent $i \in N$ is

$$\begin{aligned}
u(f(\mathbf{c}), c_i) &= 1 - \frac{1}{2}(|x_i - \frac{1-d}{2}| \cdot p_{i,1} + |x_i - \frac{1+d}{2}| \cdot p_{i,2}) \\
&\quad - \frac{1}{2}(|x_i - \frac{1+d}{2}| \cdot p_{i,1} + |x_i - \frac{1-d}{2}| \cdot p_{i,2}) \\
&= 1 - \frac{1}{2} \cdot |x_i - \frac{1-d}{2}| - \frac{1}{2} \cdot |x_i - \frac{1+d}{2}| \\
&\geq \frac{1}{2}
\end{aligned}$$

Note that the utility of each agent is at most 1. We have

$$\frac{OPT_{SU}(\mathbf{c})}{SU(f(\mathbf{c}), \mathbf{c})} \leq \frac{n}{n \cdot \frac{1}{2}} = 2,$$

indicating an approximation ratio of 2. □

Since Mechanism 1 outputs $(\frac{1-d}{2}, \frac{1+d}{2})$ and $(\frac{1+d}{2}, \frac{1-d}{2})$ randomly, an immediate idea is to output these two solutions based on the information of agents' locations and preferences. In the following we present such a deterministic mechanism. First we partition the agents into three types: we call agent $i \in N$ is *type-1* if $p_{i,1} > 0.5$, is *type-2* if $p_{i,2} > 0.5$, and is *type-3* if $p_{i,1} = p_{i,2} = 0.5$.

Define $L \subseteq N$ to be the set of those type-1 agents located in $[0, \frac{1}{2}]$, and those type-2 agents located in $(\frac{1}{2}, 1]$. Define $R \subseteq N$ to be the set of those type-2 agents located in $[0, \frac{1}{2}]$, and those type-1 agents located in $(\frac{1}{2}, 1]$.

**Mechanism 2.** *If* $|L| \geq |R|$, *output* $(\frac{1-d}{2}, \frac{1+d}{2})$, *otherwise, output* $(\frac{1+d}{2}, \frac{1-d}{2})$.

**Lemma 1.** *Mechanism 2 is strategy-proof.*

*Proof.* Given instance $\mathbf{c}$, assume w.l.o.g. that $|L| \geq |R|$, and the output is $(\frac{1-d}{2}, \frac{1+d}{2})$. First, all type-3 agents have no incentive to lie, because their utilities are invariant under these two outcomes. Second, for any agent in $L$, clearly solution $(\frac{1-d}{2}, \frac{1+d}{2})$ gives a larger utility than $(\frac{1+d}{2}, \frac{1-d}{2})$, implying that $i$ has no incentive to misreport. Finally, any agent in $R$ has no way to change the solution. Therefore, Mechanism 2 is strategy-proof. □

**Theorem 2.** *Mechanism 2 is a deterministic strategy-proof mechanism with an approximation ratio of* $\frac{4}{2-d}$ *for maximizing the social utility.*

*Proof.* The strategy-proofness is given by Lemma 1. For the approximation ratio, we first note that the optimal social utility is at most $n$, and it suffices to prove that the social utility of the mechanism for any instance $\mathbf{c}$ is at least $\frac{n}{2} - \frac{nd}{4}$. Assume w.l.o.g. that $|L| \geq |R|$, and the output is $s = (\frac{1-d}{2}, \frac{1+d}{2})$. For any type-1 agent $i \in L$ (i.e., $p_{i,1} > p_{i,2}$ and $x_i \in [0, \frac{1}{2}]$), the utility is

$$\begin{aligned}
u(s, c_i) &= 1 - p_{i,1}|x_i - \frac{1-d}{2}| - p_{i,2}|x_i - \frac{1+d}{2}| \\
&\geq 1 - \max\{p_{i,1}|0 - \frac{1-d}{2}| + p_{i,2}|0 - \frac{1+d}{2}|, p_{i,1}|\frac{1}{2} - \frac{1-d}{2}| + p_{i,2}|\frac{1}{2} - \frac{1+d}{2}|\} \\
&= 1 - \max\{\frac{1}{2} - \frac{d}{2}(p_{i,1} - p_{i,2}), \frac{d}{2}\} \\
&\geq \frac{1}{2}.
\end{aligned}$$

Similarly, any type-2 agent $i \in L$ (i.e., $p_{i,1} < p_{i,2}$ and $x_i \in (\frac{1}{2}, 1]$) has a utility

$$\begin{aligned}
u(s, c_i) &= 1 - p_{i,1}|x_i - \frac{1-d}{2}| - p_{i,2}|x_i - \frac{1+d}{2}| \\
&\geq 1 - \max\{p_{i,1}|1 - \frac{1-d}{2}| + p_{i,2}|1 - \frac{1+d}{2}|, p_{i,1}|\frac{1}{2} - \frac{1-d}{2}| + p_{i,2}|\frac{1}{2} - \frac{1+d}{2}|\} \\
&= 1 - \max\{\frac{1}{2} + \frac{d}{2}(p_{i,1} - p_{i,2}), \frac{d}{2}\} \\
&\geq \frac{1}{2}.
\end{aligned}$$

Note that the utility of any type-3 agent is at least $\frac{1}{2}$, and the number of agents in $N \backslash R$ is at least $\frac{n}{2}$. So the total utility of agents in $N \backslash R$ is at least $\frac{1}{2} \cdot \frac{n}{2} = \frac{n}{4}$. It remains to consider the agents in $R$. For any type-1 agent $i \in R$ (i.e., $p_{i,1} > p_{i,2}$ and $x_i \in (\frac{1}{2}, 1]$), the utility is

$$\begin{aligned}
u(s, c_i) &= 1 - p_{i,1}|x_i - \frac{1-d}{2}| - p_{i,2}|x_i - \frac{1+d}{2}| \\
&\geq 1 - \max\{\frac{1}{2} + \frac{d}{2}(p_{i,1} - p_{i,2}), \frac{d}{2}\} \\
&\geq \frac{1-d}{2}.
\end{aligned}$$

Similarly, for any type-2 agent $i \in R$ (i.e., $p_{i,1} < p_{i,2}$, $x_i \in [0, \frac{1}{2}]$), the utility is

$$\begin{aligned}
u(s, c_i) &= 1 - p_{i,1}|x_i - \frac{1-d}{2}| - p_{i,2}|x_i - \frac{1+d}{2}| \\
&\geq 1 - \max\{\frac{1}{2} - \frac{d}{2}(p_{i,1} - p_{i,2}), \frac{d}{2}\} \\
&\geq \frac{1-d}{2}.
\end{aligned}$$

Therefore, the social utility of the mechanism is at least

$$|n - R| \cdot \frac{1}{2} + |R| \cdot \frac{1-d}{2} \geq \frac{n}{2} \cdot \frac{1}{2} + \frac{n}{2} \cdot \frac{1-d}{2}$$
$$= \frac{n}{2} - \frac{nd}{4},$$

which completes the proof. □

We remark that the analysis in the proof of Theorem 2 is not necessarily tight, and Mechanism 2 may have a better approximation ratio than $\frac{4}{2-d}$.

Next, we provide a lower bound on the approximation ratio, and show the tightness of our mechanisms. Recall that in Definition 1 we say a mechanism is strategy-proof if no agent can gain by misreporting her location or preference. We can consider weaker definitions in which the agents are only allowed to misreport their preferences (i.e., locations are publicly known), or only allowed to misreport their locations (i.e., preferences are publicly known).

**Theorem 3.** *No deterministic strategy-proof mechanism has an approximation ratio less than* 2 *for maximizing the social utility, even if the agents can only misreport preferences.*

*Proof.* Suppose that $f$ is a deterministic strategy-proof mechanism with approximation ratio $2 - \delta$ for some $\delta > 0$. Consider an instance where $\frac{n}{3}$ agents are located at 0, and $\frac{2n}{3}$ agents are located at 1. The preference of each agent located at 0 is $p_{i1} = 1, p_{i2} = 0$, and the preference of each agent located at 1 is $p_{i1} = \frac{1}{2} + \epsilon, p_{i2} = \frac{1}{2} - \epsilon$, where $\epsilon > 0$ is a sufficiently small number. The minimum distance constraint is $d = 1$, and thus the only two possible solutions are $(0, 1)$ and $(1, 0)$. The optimal solution is $(0, 1)$, and the optimal social utility is $OPT = \frac{n}{3} + \frac{2n}{3} \cdot (\frac{1}{2} - \epsilon)$. On the other hand, the social utility of solution $(1, 0)$ is $\frac{2n}{3} \cdot (\frac{1}{2} + \epsilon) < \frac{OPT}{2-\delta}$, since $\epsilon$ is sufficiently small. By the approximation ratio, $f$ must output $(0, 1)$, and the utility of those agents located at 1 is $\frac{1}{2} - \epsilon$.

By [12], a strategy-proof mechanism must be *partial group strategy-proof*, which means that a group of agents with the same type (i.e., location and preference) cannot benefit even if they misreport type simultaneously. Next we consider the instance where the $\frac{n}{3}$ agents located as 0 still have a preference $p_{i1} = 1, p_{i2} = 0$, while the $\frac{2n}{3}$ agents located at 1 misreporting their preference to $p_{i1} = 1, p_{i2} = 0$. For this new instance, the optimal solution is $(1, 0)$, and the optimal social utility is $OPT' = \frac{2n}{3}$. On the other hand, the social utility of solution $(0, 1)$ is $\frac{n}{3} < \frac{OPT'}{2-\delta}$. By the approximation ratio, $f$ must output $(1, 0)$. Therefore, by misreporting their preferences, the agents located at 1 can increase the utility from $\frac{1}{2} - \epsilon$ to $\frac{1}{2} + \epsilon$, which contradicts the partial group strategy-proofness, and thus contradicts the strategy-proofness, □

**Theorem 4.** *No deterministic strategy-proof mechanism has an approximation ratio less than* 2 *for maximizing the social utility, even if the agents can only misreport locations.*

*Proof.* Suppose that $f$ is a deterministic strategy-proof mechanism with approximation ratio $2-\delta$ for some $\delta > 0$ Consider an instance where $\frac{n}{3}$ agents are located at 0, and $\frac{2n}{3}$ agents are located at $\frac{1}{2} - \epsilon$ for some sufficiently small $\epsilon < \frac{\delta}{n^2}$. The preference of each agent located at 0 is $p_{i1} = 1, p_{i2} = 0$, and the preference of each agent located at $\frac{1}{2} - \epsilon$ is $p_{i1} = 0, p_{i2} = 1$. The minimum distance constraint is $d = 1$, and thus the only two possible solutions are $(0, 1)$ and $(1, 0)$. The optimal solution is $(0, 1)$, and the optimal social utility is $OPT = \frac{n}{3} + \frac{2n}{3} \cdot (\frac{1}{2} - \epsilon)$. On the other hand, the social utility of solution $(1, 0)$ is $\frac{2n}{3} \cdot (\frac{1}{2} + \epsilon) < \frac{OPT}{2-\delta}$. By the approximation ratio, mechanism $f$ must output $(0, 1)$, and the utility of those agents located at $\frac{1}{2} - \epsilon$ is $\frac{1}{2} - \epsilon$.

Recall that a strategy-proof mechanism must be partial group strategy-proof. Next we consider the instance where the $\frac{2n}{3}$ agents located at $\frac{1}{2} - \epsilon$ misreport their locations as 0. For this new instance, the optimal solution is $(1, 0)$, and the optimal social utility is $OPT' = \frac{2n}{3}$. On the other hand, the social utility of solution $(0, 1)$ is $\frac{n}{3} < \frac{OPT'}{2-\delta}$. By the approximation ratio, mechanism $f$ must output $(1, 0)$. Therefore, by misreporting their locations as 0, the agents located at $\frac{1}{2} - \epsilon$ can increase the utility from $\frac{1}{2} - \epsilon$ to $\frac{1}{2} + \epsilon$, which contradicts the partial group strategy-proofness, and thus contradicts the strategy-proofness. □

For the lower bound of randomized strategy-proof mechanisms, we note that Theorem 9 of [8] provides a lower bound 1.06, in the case when all agents have the same preference $(0.5, 0.5)$ and the agents are only allowed to misreport locations. Because it is a special case of our model, this lower bound still holds for our model. Formally, no randomized strategy-proof mechanism has an approximation ratio less than 1.06, even if the agents can only misreport locations.

# 4 Maximizing the Minimum Utility

In this section, we study the objective of maximizing the minimum utility. We first consider Mechanism 1, which is randomized and strategy-proof. Recall from the proof of Theorem 1 that, for any instance $\mathbf{c}$, the utility of each agent $i \in N$ is at least $\frac{1}{2}$. Hence we have the following.

**Proposition 3.** *Mechanism 1 is a randomized strategy-proof mechanism with an approximation ratio of* 2 *for the minimum utility objective.*

Next, we consider Mechanism 2, which deterministically returns $(\frac{1-d}{2}, \frac{1+d}{2})$ if $|L| \geq |R|$, and $(\frac{1+d}{2}, \frac{1-d}{2})$ otherwise.

**Theorem 5.** *Mechanism 2 is a deterministic strategy-proof mechanism with an approximation ratio of* 4 *for the minimum utility objective.*

*Proof.* We prove the approximation ratio by constructing the worst-case instance for which the mechanism achieves the worst performance guarantee. Without loss of generality, assume that $|L| \geq |R|$ in the worst-case instance and the mechanism returns $(\frac{1-d}{2}, \frac{1+d}{2})$. Let $\epsilon > 0$ be a sufficiently small number. To construct the worst-case instance, noting that the solution is unfriendly for the agents in $R$, we want to ensure that an agent in $R$ achieves the minimum utility in the constructed instance. If $d < \frac{1}{2}$, the worst case is when $n-1$ agents located at $\frac{1}{2}+\epsilon$ have preference $(0,1)$, and one agent (say $j$) located at 1 has preference $(1,0)$. Clearly $j \in R$ and $|L| \geq |R|$. The utility of agent $j$ is $\frac{1-d}{2}$. The optimal solution is $(1, \frac{1}{2}+\epsilon)$, and the optimal minimum utility is 1. Thus the ratio is $\frac{1}{(1-d)/2} = \frac{2}{1-d} < 4$.

If $d \geq \frac{1}{2}$, the above instance is still the worst case. The utility of agent $j$ is still $\frac{1-d}{2}$. However, the optimal solution is $(1, 1-d)$, and the optimal minimum utility is $1 - [\frac{1}{2} + \epsilon - (1-d)] = 2 - d - \frac{1}{2} - \epsilon$. Thus the ratio is $\frac{2-d-1/2-\epsilon}{(1-d)/2} \leq \frac{1}{(1-d)/2} = \frac{2}{1-d} < 4$. □

We end this section by providing a lower bound 2 for randomized mechanisms, which implies the upper bound in Proposition 3 is tight, and thus Mechanism 1 is the best possible for the minimum utility objective. We also remark that this bound improves the lower bound 1.5 in [8] for the facility location problem with fractional preferences but without the minimum distance constraint.

**Theorem 6.** *No randomized strategy-proof mechanism has an approximation ratio less than 2 for maximizing the minimum utility, even if the agents can only misreport locations.*

*Proof.* Let $f$ be a randomized strategy-proof mechanism, and $d = 0$. We first consider an instance $\mathbf{c} = (\mathbf{x}, \mathbf{p})$, where the location profile of agents is $\mathbf{x} = (0, \frac{1}{2}, 1)$, and the preference profile is $\mathbf{p} = ((1,0),(1,0),(0,1))$. That is, two agents located at 0 and $\frac{1}{2}$ completely prefer $F_1$, and one agent located at 1 completely prefers $F_2$. Let $f(\mathbf{c}) = (y_1, y_2)$ and $y_1$ follows a probability distribution $P$ over $[0,1]$. It is easy to see that there exists $x_i$ for $i = 1, 2$ (without loss of generality suppose it is $x_2$), such that $\mathbb{E}_{y_1 \sim P}[|x_i - y_1|] \geq \frac{1}{4}$.

Now, consider the instance $\mathbf{c}' = (\mathbf{x}', \mathbf{p})$, where the location profile is $\mathbf{x}' = (0,1,1)$. Let $f(\mathbf{c}') = (y_1', y_2')$ be the output. By the strategy-proofness, the expected distance from $\frac{1}{2}$ to $y_1'$ must be at least $\frac{1}{4}$, otherwise in instance $\mathbf{c}$ agent 2 can gain by deviating from $x_2 = \frac{1}{2}$ to $x_2' = 1$. Therefore, the expected minimum utility between agents 1 and 2 is at most $1 - \frac{3}{4} = \frac{1}{4}$, while the optimal minimum utility is $\frac{1}{2}$, obtained by solution $(\frac{1}{2}, 1)$. It follows that the approximation ratio of $f$ is at least 2. □

## 5 Conclusion

In this paper, we studied the mechanism design for two-facility-location problem with fractional preferences and minimum distance constraint. For two objectives of maximizing the social utility and maximizing the minimum utility, we designed deterministic and randomized strategy-proof mechanisms with proven approximation ratios, and derived lower bounds on the approximation ratio of all strategy-proof mechanisms.

As extensions, we can study the problem of locating two *obnoxious* facilities, where each agent wants to stay as far away as possible, and has a utility equal to her total weighted distance from the two facilities. We note that the mechanism that returns $(0, 1)$ and $(1, 0)$ with equal probability is 2-approximation for the social utility, and there exists an 8-approximation in a deterministic way.

There are many other interesting future directions. For example, while we consider the minimum distance constraint, one can study the maximum distance constraint, where the two facilities cannot be located too far away. While this work is devoted to the fractional preferences, other types of preference model are worth studying, e.g., optional preferences and dual preferences.

**Acknowledgement.** Minming Li was partially supported by NSFC under Grant No. 11771365, and by Project No. CityU 11200518 from Research Grants Council of HKSAR.

## References

1. Chan, H., Filos-Ratsikas, A., Li, B., Li, M., Wang, C.: Mechanism design for facility location problems: a survey. In: IJCAI (2021)
2. Chen, X., Hu, X., Jia, X., Li, M., Tang, Z., Wang, C.: Mechanism design for two-opposite-facility location games with penalties on distance. In: Deng, X. (ed.) SAGT 2018. LNCS, vol. 11059, pp. 256–260. Springer, Cham (2018). https://doi.org/10.1007/978-3-319-99660-8_24
3. Chen, X., Hu, X., Tang, Z., Wang, C.: Tight efficiency lower bounds for strategy-proof mechanisms in two-opposite-facility location game. Inf. Process. Lett. **168**, 106098 (2021)
4. Cheng, Y., Yu, W., Zhang, G.: Strategy-proof approximation mechanisms for an obnoxious facility game on networks. Theoret. Comput. Sci. **497**, 154–163 (2013)
5. Duan, L., Li, B., Li, M., Xu, X.: Heterogeneous two-facility location games with minimum distance requirement. In: AAMAS, pp. 1461–1469 (2019)
6. Feigenbaum, I., Sethuraman, J.: Strategyproof mechanisms for one-dimensional hybrid and obnoxious facility location models. In: Workshops at the Twenty-Ninth AAAI Conference on Artificial Intelligence (2015)
7. Feigenbaum, I., Sethuraman, J., Ye, C.: Approximately optimal mechanisms for strategyproof facility location: minimizing $l_p$ norm of costs. Math. Oper. Res. **42**(2), 434–447 (2017)
8. Fong, K.K.C., Li, M., Lu, P., Todo, T., Yokoo, M.: Facility location games with fractional preferences. In: AAAI (2018)
9. Fotakis, D., Tzamos, C.: On the power of deterministic mechanisms for facility location games. ACM Trans. Econ. Comput. (TEAC) **2**(4), 1–37 (2014)
10. Fotakis, D., Tzamos, C.: Strategyproof facility location for concave cost functions. Algorithmica **76**(1), 143–167 (2016)
11. Hossain, S., Micha, E., Shah, N.: The surprising power of hiding information in facility location. In: AAAI, vol. 34, pp. 2168–2175 (2020)
12. Lu, P., Sun, X., Wang, Y., Zhu, Z.A.: Asymptotically optimal strategy-proof mechanisms for two-facility games. In: EC, pp. 315–324 (2010)
13. Lu, P., Wang, Y., Zhou, Y.: Tighter bounds for facility games. In: Leonardi, S. (ed.) WINE 2009. LNCS, vol. 5929, pp. 137–148. Springer, Heidelberg (2009). https://doi.org/10.1007/978-3-642-10841-9_14
14. Procaccia, A.D., Tennenholtz, M.: Approximate mechanism design without money. In: EC, pp. 177–186 (2009)

15. Serafino, P., Ventre, C.: Truthful mechanisms without money for non-utilitarian heterogeneous facility location. In: AAAI, pp. 1029–1035 (2015)
16. Tang, P., Yu, D., Zhao, S.: Characterization of group-strategyproof mechanisms for facility location in strictly convex space. In: EC, pp. 133–157 (2020)
17. Tang, Z., Wang, C., Zhang, M., Zhao, Y.: Mechanism design for facility location games with candidate locations. In: Wu, W., Zhang, Z. (eds.) COCOA 2020. LNCS, vol. 12577, pp. 440–452. Springer, Cham (2020). https://doi.org/10.1007/978-3-030-64843-5_30
18. Wada, Y., Ono, T., Todo, T., Yokoo, M.: Facility location with variable and dynamic populations. In: AAMAS, pp. 336–344 (2018)
19. Ye, D., Mei, L., Zhang, Y.: Strategy-proof mechanism for obnoxious facility location on a line. In: Xu, D., Du, D., Du, D. (eds.) COCOON 2015. LNCS, vol. 9198, pp. 45–56. Springer, Cham (2015). https://doi.org/10.1007/978-3-319-21398-9_4
20. Yuan, H., Wang, K., Fong, C.K.K., Zhang, Y., Li, M.: Facility location games with optional preference. In: ECAI, pp. 1520–1527 (2016)
21. Zou, S., Li, M.: Facility location games with dual preference. In: AAMAS, pp. 615–623 (2015)

15. [illegible] mechanisms [illegible] for [illegible] detection [illegible] (2015)
16. [illegible] mechanism [illegible] (2020)
17. [illegible] design [illegible] In: [illegible] ECCV 2020 [illegible] Springer, Cham (2020) [illegible]
18. [illegible] a camera [illegible] (2016)
19. [illegible] for the formation [illegible] ECCV [illegible] (2018) [illegible]
20. [illegible]
21. [illegible]

# Online Algorithm and Streaming Algorithms

# On the Hardness of Opinion Dynamics Optimization with $L_1$-Budget on Varying Susceptibility to Persuasion

T.-H. Hubert Chan(✉) and Chui Shan Lee

Department of Computer Science, The University of Hong Kong, Pok Fu Lam, Hong Kong
hubert@cs.hku.hk, leechuishan@connect.hku.hk

**Abstract.** Recently, Abebe et al. (KDD 2018) and Chan et al. (WWW 2019) have considered an opinion dynamics optimization problem that is based on a popular model for social opinion dynamics, in which each agent has some fixed innate opinion, and a resistance that measures the importance it places on its innate opinion; moreover, the agents influence one another's opinions through an iterative process. Under certain conditions, this iterative process converges to some equilibrium opinion vector. Previous works gave an efficient local search algorithm to solve the unbudgeted variant of the problem, for which the goal is to modify the resistance of any number of agents (within some given range) such that the sum of the equilibrium opinions is minimized. On the other hand, it was proved that the $L_0$-budgeted variant is NP-hard, where the $L_0$-budget is a restriction given upfront on the number of agents whose resistance may be modified.

Inspired by practical situations in which the effort to modify an agent's resistance increases with the magnitude of the change, we propose the $L_1$-budgeted variant, in which the $L_1$-budget is a restriction on the sum of the magnitudes of the changes over all agents' resistance parameters. In this work, we show that the $L_1$-budgeted variant is NP-hard via a reduction from the vertex cover problem. However, contrary to the $L_0$-budgeted variant, a very technical argument is needed to show that the optimal solution can be achieved by focusing the given $L_1$-budget on as small a number of agents as possible, as opposed to spreading the budget over a large number of agents.

## 1 Introduction

The process of social influence is a significant basis for opinion formation, decision making and the shaping of an individual's identity. It drives many social phenomenon ranging from the emergence of trends, diffusion of rumor, and the shaping of public views about social issues. An opinion formation model was

This work is partially supported by the Hong Kong RGC grant 17201220. The full version of this paper is available at [3].


C.-Y. Chen et al. (Eds.): COCOON 2021, LNCS 13025, pp. 515–527, 2021.
https://doi.org/10.1007/978-3-030-89543-3_43

introduced by the works of DeGroot [5] and Friedkin and Johnsen [7], which considered how agents' influence one another's opinions in discrete time steps. In this model, each agent $i$ has some *innate opinion* $s_i$ in $[0, 1]$, which reflects the intrinsic position of the agent on a certain topic. The expressed opinion of an agent is updated in each iteration according to the weighted average of other agents' opinions (according to the *interaction matrix*) and its innate opinion. The weight that an agent assigns to its own innate opinion is captured by a *resistance parameter* $\alpha_i \in [0, 1]$, where a higher value for the resistance parameter means that the agent is less susceptible to persuasion by the opinions of other agents. Under very mild conditions, the expressed opinions of the agents converge to an equilibrium, which is a vector-valued function of the innate opinions, the interaction matrix between the agents and the agents' resistance parameters.

Recent works by Abebe et al. [1] and Chan et al. [4] have considered the opinion dynamics optimization problem, in which the innate opinions and the interaction matrix between agents are given as the input, and the goal is to minimize the average equilibrium opinion by varying the agents' resistance parameters. As mentioned in their works, the motivation of the problem has been inspired by empirical works in social psychology that studied people's susceptibility to persuasion, and more references to related work and applications are given in [1,4].

Restrictions on how the agents' resistance parameters may be modified lead to different variants of the problem with different hardness. At the trivial end of the spectrum, if one can choose any $\alpha_i \in [0, 1]$ for every agent $i$, then the trivial solution to minimize the average equilibrium is to set $\alpha_i = 1$ for the agent with the minimum innate opinion and set the resistance of all other agents to 0, provided that the interaction matrix among the agents is *irreducible*, in the sense that every agent has some direct or indirect influence over every other agent. If the resistance of each agent $i$ must be chosen from some restricted interval $[l_i, u_i] \subseteq [0, 1]$, then an efficient local search method is given in [4] such that the minimum average equilibrium can be achieved by setting each agent's resistance parameter to either its lower $l_i$ or upper $u_i$ bound. In addition to the restriction intervals, the problem gets harder if one places further restrictions on the number of agents whose resistance parameters may be modified. The $L_0$-budgeted variant has some initial resistance vector $\widehat{\alpha}$ for all agents and some budget $k$, and the algorithm is allowed to change the resistance parameters of at most $k$ agents. Indeed, it is shown in [1] that the $L_0$-budgeted variant is NP-hard via a reduction from the vertex cover problem.

Intuitively, in the reduction construction for proving the NP-hardness of the $L_0$-budgeted variant, the set of agents whose resistance parameters are modified corresponds to a set of vertices to be considered as a candidate as a vertex cover in some graph. Hence, it seems that the binary nature of the choice for each agent contributes to the hardness of the problem. A natural question is whether the problem becomes easier if one is allowed to make a "fractional" decision for each agent. From a practical point of view, the $L_0$-budget uses the implicit assumption that modifying the resistance parameter of an agent a little takes the same effort as modifying it a lot. However, it is reasonable to assume that the effort it takes to modify an agent's resistance should be proportional to the magnitude of the change.

With such motivations in mind, we propose the $L_1$-budgeted variant in this work. Similar to the $L_0$-budgeted variant in which an initial resistance vector $\widehat{\alpha}$ and a budget $k$ is given, the goal is to minimize the average equilibrium opinion by choosing a vector $\alpha$ (satisfying any restriction interval placed on each agent) such that $\|\alpha - \widehat{\alpha}\|_1 \leq k$.

### 1.1 Our Contributions

At first sight, the efficient local search techniques in [4] and the fractional nature of the $L_1$-budget suggest that the problem might be solved optimally by some gradient method or mathematical program. Indeed, there are examples in which the optimal solution is achieved by assigning the budget to modify the agents' resistance partially, i.e., the resistance parameter of an agent does not reach its specified lower or upper bound. However, it turns out that this variant is also NP-hard.

**Theorem 1.** *The $L_1$-budgeted variant of the opinion dynamics optimization problem is NP-hard.*

*Hardness Intuition.* Given an $L_1$-budget, whether one should spread the budget among many agents or focus it on a small number of agents depends on the interaction matrix among the agents. When we tried to understand the structure of the problem by studying various examples, we discovered that if two agents have little direct or indirect influence on each other, then the $L_1$-budget should be shared among them. However, if all agents are well-connected such as in the case of a clique, then the budget should be as focused as possible on a small number of agents. Hence, intuitively, the problem should be hard if the underlying interaction matrix among the agents resembles a well-connected graph.

The difficulty here is that we do not yet know how to quantify the well-connectedness of the interaction matrix in relation to this *budget-focus* effect. Furthermore, the hardness still relies on the reduction from the vertex cover problem, whose hardness has not been extensively studied for graphs with different connectivity. Our solution to the reduction construction is to consider an interaction matrix that is a convex combination of a clique and some given graph $G$ that is supposed to be an instance of the vertex cover problem.

The high-level argument is that as long as the weight of the clique is large enough, the aforementioned budget-focus effect should be in place. However, as long as there is a non-zero weight of the graph $G$ on the interaction matrix, the existence of a vertex cover for $G$ of a certain size will have a quantifiable effect on the optimal average equilibrium opinion given a certain $L_1$-budget. Even though the general approach is not too complicated, combining these ideas requires quite technical calculations.

*Outline.* In Sect. 2, we will introduce the notation and formally recall various variants of the problem. In Sect. 3, we will give our reduction construction and explain the intuition behind the proofs. In Sect. 4, we give an outline of our technical proofs, while the most technical details are deferred to the full version [3].

## 2 Preliminaries

We recall the problem setting as described in [1,4]. Consider a set $N$ of *agents*, where each agent $i \in N$ is associated with an *innate opinion* $s_i \in [0, 1]$ (where higher values correspond to more favorable opinions towards a given topic) and a parameter measuring an agent's susceptibility to persuasion $\alpha_i \in [0, 1]$ (where higher values signify agents who are less susceptible to changing their opinions). We call $\alpha_i$ the *resistance parameter.*

The agents interact with one another in discrete time steps. The interaction matrix captures the relationship between agents and is simply a row stochastic matrix[1] $P \in [0, 1]^{N \times N}$ (i.e., each entry of $P$ is non-negative and every row sums to 1, but $P$ needs not be symmetric). We denote $A = \mathrm{Diag}(\alpha)$ as the diagonal matrix with $A_{ii} = \alpha_i$, and $I$ as the identity matrix. Starting from some arbitrary initial expressed opinion vector $z^{(0)} \in [0, 1]^N$, the expressed opinion vector is updated in each time step according to the following equation:

$$z^{(t+1)} := As + (I - A)Pz^{(t)}. \tag{1}$$

Equating $z^{(t)}$ with $z^{(t+1)}$, one can see that the equilibrium opinion vector is given by $z = [I - (I - A)P]^{-1}As$, which exists under very mild conditions such as the following.

**Fact 1 (Convergence Assumption).** *Suppose $P$ is irreducible and at least one $i \in N$ has $\alpha_i > 0$. Then, Eq. (1) converges to a unique equilibrium $\lim_{t \to \infty} z^{(t)}$.*

The opinion susceptibility problem is defined below. Intuitively, the objective is to choose a resistance vector $\alpha$ to minimize the sum of equilibrium opinions $\langle \mathbf{1}, z \rangle = \mathbf{1}^\top z$, i.e., the goal is to drive the average opinion towards 0. Observe that one can also consider maximizing the sum of equilibrium opinions. To see that the minimization and maximization problems are equivalent, consider the transformation $x \mapsto 1 - x$ on the opinion space $[0, 1]$ that is applied to the innate opinions and expressed opinions in every time step. In this paper, we will focus on the minimization problem as follows.

---

[1] Given sets $U$ and $W$, we use the notation $U^W$ to denote the collection of all functions from $W$ to $U$. Each such function can also be interpreted as a vector (or a matrix if $W$ itself is a Cartesian product), where each coordinate is labeled by an element in $W$ and takes a value in $U$. As an example, a member of $[0, 1]^{N \times N}$ is a matrix whose rows and columns are labeled by elements of $N$. The alternative notation $[0, 1]^{n \times n}$ implicitly assumes a linear ordering on $N$, which does not have any importance in our case and would simply be an artefact of the notation.

**Definition 1 (Opinion Susceptibility Problem (Unbudgeted Variant)).** *Given a set $N$ of agents with innate opinions $s \in [0,1]^N$ and interaction matrix $P \in [0,1]^{N \times N}$, suppose for each $i \in N$, its resistance is restricted to some interval $\mathcal{I}_i := [l_i, u_i] \subseteq [0,1]$ where we assume*[2] *that $0 \leq l_i \leq u_i \leq 1$.*

*The objective is to choose $\alpha \in \mathcal{I}_N := \times_{i \in N} \mathcal{I}_i \subseteq [0,1]^N$ such that the objective function, $f(\alpha) := \mathbf{1}^\top [I - (I - A)P]^{-1} A s$, where $A = \mathrm{Diag}(\alpha)$ is the diagonal matrix with $A_{ii} = \alpha_i$, is minimized. Observe that the assumption in Fact 1 ensures that the above inverse exists.*

*Budgeted Variants.* To describe different types of budgets, we use the following norms. Given $x \in \mathbb{R}^N$, we denote its $L_0$-norm $\|x\|_0 := |\{i \in N : x_i \neq 0\}|$ and its $L_1$-norm $\|x\|_1 := \sum_{i \in N} |x_i|$. For $b \in \{0,1\}$, the $L_b$-budgeted variant of the problem also has some initial resistance vector $\widehat{\alpha} \in \mathcal{I}_N$ and a given budget $k > 0$. The goal is to find $\alpha \in \mathcal{I}_N$ to minimize $f(\alpha)$ subject to $\|\alpha - \widehat{\alpha}\|_b \leq k$.

*Hardness of the Various Variants.* A polynomial-time algorithm is given for the unbudgeted variant in [4], while the $L_0$-budgeted variant is shown to be NP-hard in [1] via a reduction from the vertex cover problem. The main result of this work is to show that the $L_1$-budgeted variant is also NP-hard.

# 3 Hardness of $L_1$-Budgeted Variant

As in [1], we shall prove the NP-hardness of the $L_1$-budgeted variant via reduction from the vertex cover problem on regular graphs [6], where a graph is $d$-regular if every vertex has degree $d$.

**Fact 2 (Vertex Cover on Regular Graphs).** *Given a $d$-regular undirected graph $G = (V, E)$ and some $k > 0$, it is NP-hard (even for $d = 3$) to decide if $G$ has a vertex cover $T$ of size $k$, where $T \subseteq V$ is a vertex cover for $G$ if every edge in $E$ has at least one end-point in $T$.*

## 3.1 Warmup: Reduction for $L_0$-Budget

Before we give our final reduction construction for $L_1$-budget, we give a simplified reduction for $L_0$-budget, which will offer some intuition on why the $L_1$-budget reduction is more complicated. The reduction here for $L_0$-budget is similar to the one given in [1], but is even simpler because we allow different agents $i$ to have different ranges $[l_i, u_i]$ for their resistance parameters.

Recall that an instance of the vertex cover problem consists of a $d$-regular graph $G = (V, E)$ with $n = |V|$ and some target vertex cover size $k$.

*Reduction Construction.* In addition to the original vertices in $V$, we create one extra agent 0 to form $N := V \cup \{0\}$. For the innate opinions, $s_0 = 1$ and $s_i = 0$ for $i \neq 0$; for the initial resistances, $\widehat{\alpha}_0 = 1$ and $\widehat{\alpha}_i = 0$ for $i \neq 0$. For the

[2] In view of Fact 1, we assume that for at least one $i$, $l_i > 0$.

range of resistance parameter, we restrict[3] $\mathcal{I}_0 = \{1\}$ and $\mathcal{I}_i = [0, 1]$ for $i \neq 0$; in other words, agent 0 will always remain the most stubborn, while agents in $V$ have 0 resistance initially, but their resistance parameters could be increased to 1 subject to the budget constraint. The $L_0$-budget is $k$, which is the same as the target cover size. Our final construction for the $L_1$-budget will also share the above parameter settings.

*Interaction Matrix for $L_0$-Budget Reduction.* We next describe the interaction matrix $P$. For agent 0, we will always have resistance $\alpha_0 = 1$, and so the corresponding row in $P$ is irrelevant, but we could set $P_{0i} = \frac{1}{n}$ for $i \in V$ to be concrete. For $i \neq 0$, let $P_{ij} = \frac{1}{d+1}$ if $j = 0$ or $\{i, j\} \in E$ is an edge in $G$, recalling that each node $i \in V$ has degree $d$ in $G$. As we shall see, the reduction for $L_1$-budget will have a different interaction matrix.

*Intuition.* For the $L_0$-budgeted variant, if we wish to change the resistance parameter of an agent $i \in V$ (who has innate opinion $s_i = 0$), we might as well set it to $\alpha_i = 1$, because the goal is to minimize the expressed opinion. To complete the reduction proof, it suffices to give a threshold $\vartheta$ for the objective function $f$ that can distinguish between the YES and NO instances of the vertex cover problem. The following two lemmas complete the reduction argument.

**Lemma 1 (YES Instance).** *Let $\vartheta := 1 + \frac{n-k}{d+1}$. Suppose $G = (V, E)$ has a vertex cover $T$ of size $k$. Then, by changing the resistance parameters to $\alpha_i = 1$ for all $i \in T$ (while those for other agents are not changed), we can achieve $f(\alpha) = \vartheta$.*

*Proof.* We compute the equilibrium expressed opinion of each agent. For agent 0, we have $z_0 = 1$; for $i \in T$ in the vertex cover, we have $z_i = 0$.

For $i \in V \backslash T$, we still have $\alpha_i = 0$. Since all its neighbors in $V$ are in $T$ and $i$ is influenced by agent 0, we have $z_i = \frac{1}{d+1}$.

Therefore, $f(\alpha) = \sum_{i \in N} z_i = 1 + |T| \cdot 0 + |V \backslash T| \cdot \frac{1}{d+1} = \vartheta$, as required.

**Lemma 2 (NO Instance).** *Suppose $G = (V, E)$ has no vertex cover of size $k$. Then, for any $\alpha \in \mathcal{I}_N$ such that $\|\alpha - \widehat{\alpha}\|_0 \leq k$, $f(\alpha) \geq \vartheta + \frac{2}{d(d+1)}$.*

*Proof.* Since the goal is to minimize $f$, the minimum can be achieved by using all of the $L_0$-budget $k$. Moreover, as each $i \in V$ has innate opinion $s_i = 0$, if we change its resistance parameter, we should set it to $\alpha_i = 1$. Hence, we can assume that there is some $T \subseteq V$ of size $|T| = k$ such that $\alpha_i = 1$ for $i \in T$ and $\alpha_i = 0$ for $i \in V \backslash T$. We remark that to reach the same conclusion for the $L_1$-budget reduction later will require a lot more technical details.

[3] Observe that we could make every agent have the same range of resistance parameter. For instance, we can replace agent 0 with a large enough clique of initially stubborn agents with innate opinions 1. However, this will make the presentation and calculation more cumbersome, and it is quite obvious that the hardness of the problem is not due to difference in the range of resistance parameters among different agents.

As in Lemma 1, we can conclude $z_0 = 1$ and $z_i = 0$ for $i \in T$.

For $i \in V \backslash T$, we now only have the inequality $z_i \geq \frac{1}{d+1}$. However, we can achieve a stronger lower bound because $T$ is not a vertex cover for $G$.

Let $\gamma$ be the minimum over all $z_i$ such that $i$ is incident on an edge in $E$ that is not covered by $T$. Suppose $i \in V \backslash T$ is a vertex that attains $\gamma$ and the edge $\{i, j\}$ is not covered by $T$. Then, we have $z_j \geq \gamma$. Hence, we have $\gamma = z_i \geq \frac{z_0 + z_j}{d+1} \geq \frac{1+\gamma}{d+1}$, which implies that $\gamma \geq \frac{1}{d} = \frac{1}{d+1} + \frac{1}{d(d+1)}$.

Since there is at least one edge in $E$ that is not covered by $T$, we have $f(\alpha) \geq \vartheta + \frac{2}{d(d+1)}$, as required.

### 3.2 Reduction for $L_1$-Budget

We first describe the main challenge for adapting the reduction proof for $L_0$-budget to $L_1$-budget. The issue is that to adapt Lemma 2 for the NO instances of the vertex cover problem, we need a desirable structural property on an optimal solution for the $L_1$-budgeted variant of the opinion optimization problem. Specifically, we would like to argue that to minimize the objective function $f$ with an integral budget $k$, one should pick a subset $T \subseteq V$ of exactly $k$ agents on whom to use the budget, as opposed to spreading the budget fractionally over more than $k$ agents.

Unfortunately, this is not true for the reduction construction given in Sect. 3.1. Indeed, we have discovered that for two agents $i$ and $j$ that are somehow not "well-connected" in $G$, if some fixed $L_1$-budget of less than 2 is assigned to them, spreading the budget fractionally among the two agents would yield a lower objective value than biasing the budget towards one agent. On the other hand, we discovered that the desirable structural property holds in some cases where all the vertices in $G$ are "well-connected", for instance, if $G$ is a clique on the $n$ vertices. However, since there is no connectivity assumption on the given instance $G$ of vertex cover, we consider the following interaction matrix.

*Interaction Matrix for $L_1$-Budget Reduction.* Recall that we are given a $d$-regular graph $G = (V, E)$ with $n = |V|$, and $N = V \cup \{0\}$. Let $C \in [0, 1]^{N \times N}$ be a row-stochastic matrix such that for $i \neq j$, $C_{ij} = \frac{1}{n}$, recalling that $|N| = n + 1$; in other words, $C$ behaves like a clique on $N$. Let $R \in [0, 1]^{N \times N}$ be a row-stochastic matrix such that $R_{00} = 1$, and $R_{ij} = \frac{1}{d}$ *iff* $\{i, j\} \in E$, where all other entries of $R$ are 0; in other words, $R$ is the normalized adjacency matrix of $G$ with an additional isolated vertex 0. For some appropriate $\delta \in (0, 1)$ (that depends only on $d$ and $n$), we consider the following interaction matrix $P^{(\delta)} := (1 - \delta)C + \delta R$.

Recall that the innate opinions and initial resistance parameters are the same as in Sect. 3.1, i.e., $s_0 = \widehat{\alpha}_0 = 1$ and $s_i = \widehat{\alpha}_i = 0$ for $i \in V$. Moreover, the $L_1$-budget can be used to change the resistance parameters of only the agents in $V$, i.e., $\alpha_0 = 1$ must remain.

The lemma for YES instance is similar to that in Sect. 3.1. We define the threshold $\vartheta := 1 + \frac{(1-\delta)(n-k)}{n-(1-\delta)(n-k-1)}$.

**Lemma 3 (YES Instance).** *Suppose $G = (V, E)$ has a vertex cover $T$ of size $k$. Then, by changing the resistance parameters to $\alpha_i = 1$ for all $i \in T$ (while those for other agents are not changed), we can achieve $f(\alpha) = \vartheta$.*

*Proof.* As in Lemma 1, the equilibrium expressed opinions are $z_0 = 1$ and $z_i = 0$ for $i \in T$.

For $j \in V \backslash T$, recall that $\alpha_j = 0$ and we exploit the symmetry in $P^{(\delta)} = (1-\delta)C + \delta R$ to analyze the value of $z_j$. Note that with respect to $C$, agent $j$ observes that agent 0 has $z_0 = 1$ and the $k$ agents $i \in T$ have $z_i = 0$, and there are $n-k-1$ other agents like itself; with respect to $R$, agent $j$ observes that all its $d$ neighbors in $G$ are in $T$, and so has $z_i = 0$.

Therefore, every agent $j \in V \backslash T$ has this same observation, and we can conclude that $z_j$'s have some common value $\gamma$ for all $j \in V \backslash T$ satisfying:

$$\gamma = (1-\delta) \cdot \frac{1}{n} \cdot (1 + k \cdot 0 + (n-k-1) \cdot \gamma).$$

This gives $\gamma = \frac{1-\delta}{n-(1-\delta)(n-k-1)}$, and so $f(\alpha) = 1 + k \cdot 0 + (n-k) \cdot \gamma = \vartheta$.

The following structural property is needed for the analysis of NO instances. Its proof is technical and is deferred to full version [3].

**Lemma 4 (Structural Property of Optimal Solution).** *Suppose we set $\delta := \frac{d^3(2d-1)^{3n-3}}{(n+1)^6(2d+1)^{3n}}$ to define the interaction matrix $P^{(\delta)}$ above, and an $L_1$-budget of $k$ is given to change the resistance parameters of agents in $V$. Then, the objective function $f$ can be minimized by picking some $T \subseteq V$ of size $k$, and setting $\alpha_i = 1$ for $i \in T$.*

**Lemma 5 (NO Instance).** *Suppose $G = (V, E)$ has no vertex cover of size $k$, and $\delta \in (0, 1)$ is chosen to satisfy Lemma 4. Then, for any $\alpha \in \mathcal{I}_N$ such that $\|\alpha - \widehat{\alpha}\|_1 \leq k$, $f(\alpha) \geq \vartheta + \frac{\delta}{dn}$.*

*Proof.* Because of Lemma 4, we can assume that the minimum $f(\alpha)$ is achieved by picking some $T \subseteq V$ of size $k$ and set $\alpha_i = 1$ for $i \in T$ and $\alpha_j = 0$ remains for $j \in V \backslash T$.

Again, the equilibrium expressed opinions satisfy $z_0 = 1$ and $z_i = 0$ for $i \in T$.

Let $\gamma := \min_{j \in V \backslash T} z_j$. Then, a similar argument as in Lemma 3 gives the inequality $\gamma \geq \gamma_0 := \frac{1-\delta}{n-(1-\delta)(n-k-1)}$.

We can get a stronger lower bound because $T$ is not a vertex cover for $G = (V, E)$. Let $\widehat{\gamma}$ be the minimum $z_j$ among $j \in V \backslash T$ such that $j$ is an end-point of an edge not covered by $T$.

Then, we have the inequality $\widehat{\gamma} \geq (1-\delta) \cdot \frac{1}{n} \cdot (1 + k \cdot 0 + (n-k-1) \cdot \gamma_0) + \delta \cdot \frac{\widehat{\gamma}}{d} = \gamma_0 + \widehat{\gamma} \cdot \frac{\delta}{d}$.

Hence, we have $\widehat{\gamma} \geq (1 - \frac{\delta}{d})^{-1} \cdot \gamma_0 \geq \gamma_0 + \frac{\delta \gamma_0}{d}$.

Since there is at least one edge in $E$ that is not covered by $T$ (and such an edge has two end-points), we have $f(\alpha) \geq \vartheta + \frac{2\delta\gamma_0}{d} \geq \vartheta + \frac{\delta}{dn}$, as required.

**Corollary 1.** *It is NP-hard to solve the $L_1$-budgeted variant with additive error at most $\frac{\delta}{dn}$ on the objective function $f$.*

*Proof.* Observe that from Lemmas 3 and 5, it suffices to use a precision of $\Theta(\frac{\delta}{dn})$ on the objective function, which can be achieved using $O(\log \frac{dn}{\delta}) = \texttt{poly}(n)$ number of bits.

# 4 Technical Proofs for Well-Connected Interaction Matrices

To complete our hardness proof in Sect. 3.2, it suffices to prove Lemma 4. Recall that the goal is to pick some $\delta \in (0,1)$ to define the interaction matrix $P^{(\delta)} = (1-\delta)C + \delta R$ such that given an $L_1$-budget $k \in \mathbb{Z}$, the objective function $f(\alpha)$ can be minimized by allocating the budget to exactly $k$ agents in $V$. One way to achieve this is to pick two arbitrary agents $i, j \in V$ and show that if some fixed $L_1$-budget $b < 2$ is assigned for $i$ and $j$ (while the other agents are not modified), then the objective function $f(\alpha)$ can be minimized by prioritizing the budget to either $i$ or $j$ as much as possible. Denoting $e_i \in [0,1]^N$ as the unit vector with $i \in N$ as the only non-zero coordinate, we shall prove the following.

**Formal Goal.** By setting $\delta = \frac{d^3(2d-1)^{3n-3}}{(n+1)^6(2d+1)^{3n}}$, we will show that if $i, j \in V$ such that $0 \leq \alpha_i, \alpha_j < 1$, then the second derivative of $f$ in the direction $e_i - e_j$ satisfies:

$$(e_i - e_j)^\top \nabla f(\alpha) = 0 \implies (e_i - e_j)^\top \nabla^2 f(\alpha)(e_i - e_j) < 0. \qquad (2)$$

Statement (2) implies that if two agents in $V$ both receive non-zero fractional budget to change their resistance parameters, then it will not increase the objective function $f$ by biasing the budget towards one of them.

We will outline the proof ideas to achieve the above formal goal. The complete technical proofs are given in the full version [3].

**Notation Recap.** Recall that given $\alpha \in [0,1]^N$, we write $A := \text{Diag}(\alpha)$. Moreover, we denote $X = X(\alpha) := I - (I-A)P^{(\delta)}$. Under conditions such as Fact 1, we write $M = M(\alpha) := X^{-1}$ and the equilibrium vector $z = z(\alpha) := MAs$, where $s \in [0,1]^N$ is the innate opinion vector from Sect. 3.2 such that $s_0 = 1$ and $s_i = 0$ for $i \in V$. Finally, the objective function is $f(\alpha) := \mathbf{1}^\top z(\alpha)$. Note that the quantities have a dependence on $\delta$, and we will use a superscript such as $f^{(\delta)}$ when we wish to emphasize this dependence.

**Fact 3 (Technical Calculations [4]).** *Whenever the above quantities are well-defined, we have*

- $M \geq 0$ *and* $M_{ii} \geq 1$ *for all* $i \in N$.
- *If* $\alpha_i \neq 1$, $(PM)_{ii} = \frac{M_{ii}-1}{1-\alpha_i}$ *and* $(PM)_{ij} = \frac{M_{ij}}{1-\alpha_i}$ *for* $j \neq i$.
- *For* $i, j \in N$, $\frac{\partial z_i(\alpha)}{\partial \alpha_j} = \frac{s_j - z_j(\alpha)}{1-\alpha_j} \cdot M_{ij}$ *and* $e_i^\top \nabla f(\alpha) = \frac{\partial f}{\partial \alpha_i} = \frac{s_i - z_i(\alpha)}{1-\alpha_i} \cdot \mathbf{1}^\top M e_i$.

Statement (2) inspires us to analyze the following quantities.

**Lemma 6.** *Suppose $i, j \in V$ such that $0 \leq \alpha_i, \alpha_j < 1$ and $(e_i - e_j)^\top \nabla f(\alpha) = 0$. Then, we have:*

$$(e_i - e_j)^\top \nabla^2 f(\alpha)(e_i - e_j) = \frac{2}{(1-\alpha_i)(1-\alpha_j)} \cdot \{z_i(\alpha) y_{ij}(\alpha) + z_j(\alpha) y_{ji}(\alpha)\},$$

*where $y_{ij}(\alpha) := \mathbf{1}^\top M e_i \cdot (M_{jj} - 1) - \mathbf{1}^\top M e_j \cdot M_{ji}$.*

In view of Lemma 6, it suffices to analyze the quantity $y_{ij}$. Since every $i \in V$ with $\alpha_i < 1$ is influenced by agent 0 (with $z_0 = 1$) in $P^{(\delta)}$, we have $z_i(\alpha) > 0$ for all $i \in V$. Hence, to show statement (2), it remains to show that $y_{ij}(\alpha) < 0$. We next analyze $y_{ij}$ as functions of $\alpha_j$ and $P$ respectively in the next two lemmas.

**Lemma 7 (Monotonicity).** *Given $\alpha \in [0,1]^n$, let $A := \mathrm{Diag}(\alpha)$ and $P$ be a row-stochastic matrix such that $M := [I - (I - A)P]^{-1}$ exists. For $i \neq j \in V$, we fix $\alpha_k$ for every $k \neq j$ and all entries of $P$ and consider the following quantity as a function of $\alpha_j$:*

$$y_{ij}(\alpha_j) := \mathbf{1}^\top M e_i \cdot (M_{jj} - 1) - \mathbf{1}^\top M e_j \cdot M_{ji}.$$

*Then, $y_{ij}(\alpha_j)$ is a strictly monotone or constant function of $\alpha_j$ on $[0,1]$, i.e., it is either strictly increasing, strictly decreasing, or constant. In addition, $y_{ij}(1) = 0$.*

*Proof.* When $\alpha_j = 1$, note that the $j$-th row of the matrix $I - (I - A)P$ is equal to $e_j^\top$. By considering the $j$-th row of the equation $[I - (I - A)P]M = I$, we have $M_{jj} = 1$ and $M_{jk} = 0$ for any $k \neq j$. Hence, $M_{jj} - 1 = M_{ji} = 0$ and so $y_{ij}(1) = 0$.

We will now show that $y_{ij}(\alpha_j)$ is a strictly monotone function of $\alpha_j$. Notice that $y_{ij}$ is a continuous function of $\alpha_j$ since $y_{ij}$ is a continuous function of $M$, $M$ is a continuous function of $\alpha_j$ (because of the continuity of matrix inversion), and a composition of continuous functions is continuous. In addition, we know that $\frac{\partial B^{-1}}{\partial t} = -B^{-1} \frac{\partial B}{\partial t} B^{-1}$ for any invertible matrix $B$. Applying the above result with $B = I - (I - A)P$ and $t = \alpha_j$, we get $\frac{\partial M}{\partial \alpha_j} = -M e_j e_j^\top P M$.

Hence, when $\alpha_j \neq 1$, the partial derivative of $y_{ij}$ with respect to $\alpha_j$ is

$$\begin{aligned}
\frac{\partial y_{ij}}{\partial \alpha_j} &= \frac{\partial}{\partial \alpha_j}\left[1^\top M e_i \cdot (e_j^\top M e_j - 1) - 1^\top M e_j \cdot e_j^\top M e_i\right] \\
&= 1^\top \frac{\partial M}{\partial \alpha_j} e_i \cdot (e_j^\top M e_j - 1) + 1^\top M e_i \cdot e_j^\top \frac{\partial M}{\partial \alpha_j} e_j - \\
&\quad 1^\top \frac{\partial M}{\partial \alpha_j} e_j \cdot e_j^\top M e_i - 1^\top M e_j \cdot e_j^\top \frac{\partial M}{\partial \alpha_j} e_i \\
&= -1^\top M e_j e_j^\top P M e_i \cdot (e_j^\top M e_j - 1) - 1^\top M e_i \cdot e_j^\top M e_j e_j^\top P M e_j + \\
&\quad 1^\top M e_j e_j^\top P M e_j \cdot e_j^\top M e_i + 1^\top M e_j \cdot e_j^\top M e_j e_j^\top P M e_i \\
&= -1^\top M e_j \cdot \frac{M_{ji}}{1-\alpha_j} \cdot (M_{jj} - 1) - 1^\top M e_i \cdot M_{jj} \cdot \frac{M_{jj}-1}{1-\alpha_j} + \\
&\quad 1^\top M e_j \cdot \frac{M_{jj}-1}{1-\alpha_j} \cdot M_{ji} + 1^\top M e_j \cdot M_{jj} \cdot \frac{M_{ji}}{1-\alpha_j} \qquad \text{(Fact 3)} \\
&= 1^\top M e_j \cdot M_{jj} \cdot \frac{M_{ji}}{1-\alpha_j} - 1^\top M e_i \cdot M_{jj} \cdot \frac{M_{jj}-1}{1-\alpha_j} \\
&= -\frac{M_{jj}}{1-\alpha_j}\left[1^\top M e_i \cdot (M_{jj} - 1) - 1^\top M e_j \cdot M_{ji}\right] \\
&= -\frac{M_{jj}}{1-\alpha_j} y_{ij},
\end{aligned}$$

where $M_{jj} \geq 1 > 0$ by Fact 3. Observe that $M_j j$ is a function of $\alpha_j$, and we denote this by $M(\cdot)$.

Rewriting $g(t) = -y_{ij}(1-t)$, we have an alternative form $\frac{dg}{dt} = \frac{M(1-t)\cdot g}{t}$, where $M(\cdot) \geq 1$ and $g(0) = -y_{ij}(1) = 0$. It follows that if $g$ is a continuous function, then either $g$ stays 0 in [0,1] or $g$ is strictly monotone.

Recall that our goal is to choose some $\delta > 0$ to define the interaction matrix $P^{(\delta)} := (1-\delta)C + \delta R$ such that we can prove that the quantity $y_{ij}^{(\delta)} < 0$. The next lemma shows that for the special case $\delta = 0$, we can argue that $y_{ij}^{(0)} < 0$ for $\alpha_j < 1$.

**Lemma 8.** *For $\delta = 0$, consider $P = P^{(0)} = C$, whose diagonal entries are 0 and every other entry is $\frac{1}{n}$. For $\alpha \in [0,1]^N$ (with $\alpha_0 = 1$), we have $A := \mathrm{Diag}(\alpha)$ and $M = [I - (I-A)C]^{-1}$. Fix some $i \neq j \in V$, and consider*

$$y_{ij}(\alpha_j) := \mathbf{1}^\top M e_i \cdot (M_{jj} - 1) - \mathbf{1}^\top M e_j \cdot M_{ji}.$$

*as a function of $\alpha_j$; when $\alpha_j = 0$, $y_{ij}(0) \leq -\frac{1}{n+1}$. Moreover, $\mathbf{1}^\top M \mathbf{1} \leq n$.*

Lemma 8 says that $y_{ij}^{(0)}|_{\alpha_j=0} \leq -\frac{1}{n+1}$. The hope is that if $\delta > 0$ is small enough, then $P^{(\delta)}$ would still be close to $P^{(0)} = C$, and so $y_{ij}^{(\delta)}|_{\alpha_j=0}$ will stay negative. Hence, we next analyze the quantity $y_{ij}$ as a function of the interaction matrix $P$.

**Lemma 9.** *Fixing some* $\alpha \in [0,1]^N$*, let* $A := \mathrm{Diag}(\alpha)$ *and let* $P$ *be a row-stochastic matrix such that* $M = M(P) := [I - (I - A)P]^{-1}$ *exists. For any distinct* $i, j \in V$*, denote*

$$y_{ij}(P) := \mathbf{1}^\top M e_i \cdot (M_{jj} - 1) - \mathbf{1}^\top M e_j \cdot M_{ji}$$

*as a function of* $P$*. Then,* $\sum_{k,l \in N} \left| \frac{\partial y_{ij}}{\partial P_{kl}} \right| \leq 4(\mathbf{1}^\top M \mathbf{1})^3$.

In view of Lemma 9, we wish to bound the entries of $M^{(\delta)}$ for small $\delta > 0$. The next fact was given in Alfa et al. [2] that gives two-sided bounds to the inverse of a perturbed nonsingular diagonally dominant matrix. We use the operator $|\cdot|$ on a matrix to denote the matrix with the same dimension by taking absolute values entrywise.

**Fact 4 (Entrywise Bounds for Diagonally Dominant Matrix Inverse [2]).** *Suppose* $X$ *and* $\tilde{X}$ *are nonsingular matrices of the form* $I - B$*, where* $B \geq 0$ *and has spectral norm strictly less than 1, and each row of* $B$ *sums to at most 1.*

*Let* $0 \leq \varepsilon < 1$ *such that* $|X_{ij} - \tilde{X}_{ij}| \leq \varepsilon |X_{ij}|$ *for* $i \neq j$ *and* $|X\mathbf{1} - \tilde{X}\mathbf{1}| \leq \varepsilon |X\mathbf{1}|$*. Then,*

$$\frac{(1-\varepsilon)^n}{(1+\varepsilon)^{n-1}} X^{-1} \leq \tilde{X}^{-1} \leq \frac{(1+\varepsilon)^n}{(1-\varepsilon)^{n-1}} X^{-1}.$$

Recall that $R$ is a row-stochastic matrix that represents a normalized adjacency matrix of a $d$-regular graph $G = (V, E)$ with the insertion of an isolated vertex 0; also, recall that $C = \frac{1}{n}(J - I)$ represents a clique on $N$.

**Lemma 10.** *Let* $\alpha \in [0,1]^N$ *such that* $\alpha_0 = 1$*, and* $A := \mathrm{Diag}(\alpha)$*. For* $0 \leq \delta < \frac{d}{n}$*, define* $P^{(\delta)} := (1-\delta)C + \delta R$ *and* $M^{(\delta)} := [I - (I - A)P^{(\delta)}]^{-1}$*. Then,*

$$\mathbf{1}^\top M^{(\delta)} \mathbf{1} \leq \frac{n(1+\varepsilon)^n}{(1-\varepsilon)^{n-1}} \quad \textit{where } \varepsilon = \frac{\delta n}{d}.$$

With the previous preparation, the next lemma proposes an exact universal perturbation parameter $\delta$ that guarantees the negativity of $y_{ij}$ when $\alpha_j \in [0, 1)$, for any distinct $i, j \in V$.

**Lemma 11.** *Let* $\alpha \in [0,1]^N$ *and* $M^{(t)} := [I - (I - A)P^{(t)}]^{-1}$ *for* $0 \leq t < \frac{d}{n}$ *be as defined in Lemma 10. For any distinct* $i, j \in V$*, define:*

$$y_{ij}^{(t)} := \mathbf{1}^\top M^{(t)} e_i \cdot (M_{jj}^{(t)} - 1) - \mathbf{1}^\top M^{(t)} e_j \cdot M_{ji}^{(t)}.$$

*Let* $\delta = \frac{d^3(2d-1)^{3n-3}}{(n+1)^6(2d+1)^{3n}}$*. If* $\alpha_j \in [0, 1)$*, then* $y_{ij}^{(\delta)} < 0$.

Finally, we are ready to prove the main result of this section.

**Proof of Lemma 4**

As mentioned before, the formal goal is to show statement (2). Lemma 6 says that it suffices to show that $y_{ij}(\alpha) < 0$ when $\alpha_j < 1$. Finally, Lemma 11 says that by choosing $\delta = \frac{d^3(2d-1)^{3n-3}}{(n+1)^6(2d+1)^{3n}}$ to define $P^{(\delta)} = (1-\delta)C + \delta R$, we have $y_{ij}^{(\delta)}(\alpha) < 0$ for $\alpha_j < 1$, as required. □

## References

1. Abebe, R., Kleinberg, J.M., Parkes, D.C., Tsourakakis, C.E.: Opinion dynamics with varying susceptibility to persuasion. In: KDD, pp. 1089–1098. ACM (2018). DOI: https://doi.org/10.1145/3219819.3219983
2. Alfa, A., Xue, J., Ye, Q.: Entrywise perturbation theory for diagonally dominant M-matrices with applications. Numer. Math. **90**, 401–414 (2002). https://doi.org/10.1007/s002110100289
3. Chan, T.H., Lee, C.S.: On the hardness of opinion dynamics optimization with $l_1$-budget on varying susceptibility to persuasion. CoRR abs/2105.04105 (2021)
4. Chan, T.H., Liang, Z., Sozio, M.: Revisiting opinion dynamics with varying susceptibility to persuasion via non-convex local search. In: WWW, pp. 173–183. ACM (2019). https://doi.org/10.1145/3308558.3313509
5. DeGroot, M.H.: Reaching a consensus. J. Am. Stat. Assoc. **69**(345), 118–121 (1974). http://www.jstor.org/stable/2285509
6. Feige, U.: Vertex cover is hardest to approximate on regular graphs. Manuscript (2003)
7. Friedkin, N.E., Johnsen, E.C.: Social influence networks and opinion change. Adv. Group Process. **16**, 1–19 (1999)

# Symmetric Norm Estimation and Regression on Sliding Windows

Vladimir Braverman[1], Viska Wei[1], and Samson Zhou[2](✉)

[1] Johns Hopkins University, Baltimore, USA
vova@cs.jhu.edu, swei20@jhu.edu

[2] Carnegie Mellon University, Pittsburgh, USA

**Abstract.** The sliding window model generalizes the standard streaming model and often performs better in applications where recent data is more important or more accurate than data that arrived prior to a certain time. We study the problem of approximating symmetric norms (a norm on $\mathbb{R}^n$ that is invariant under sign-flips and coordinate-wise permutations) in the sliding window model, where only the $W$ most recent updates define the underlying frequency vector. Whereas standard norm estimation algorithms for sliding windows rely on the smooth histogram framework of Braverman and Ostrovsky (FOCS 2007), analyzing the *smoothness* of general symmetric norms seems to be a challenging obstacle. Instead, we observe that the symmetric norm streaming algorithm of Braverman *et al.* (STOC 2017) can be reduced to identifying and approximating the frequency of heavy-hitters in a number of substreams. We introduce a heavy-hitter algorithm that gives a $(1+\epsilon)$-approximation to each of the reported frequencies in the sliding window model, thus obtaining the first algorithm for general symmetric norm estimation in the sliding window model. Our algorithm is a universal sketch that simultaneously approximates all symmetric norms in a parametrizable class and also improves upon the smooth histogram framework for estimating $L_p$ norms, for a range of large $p$. Finally, we consider the problem of overconstrained linear regression problem in the case that loss function that is an Orlicz norm, a symmetric norm that can be interpreted as a scale-invariant version of $M$-estimators. We give the first sublinear space algorithms that produce $(1+\epsilon)$-approximate solutions to the linear regression problem for loss functions that are Orlicz norms in both the streaming and sliding window models.

**Keywords:** Streaming algorithms · Symmetric norms · Sliding window model · Linear regression

## 1 Introduction

The efficient estimation of norms is a fundamental problem in the *streaming model*, which implicitly defines an underlying frequency vector through a series of sequential updates to coordinates of the vector, but each update may only be


C.-Y. Chen et al. (Eds.): COCOON 2021, LNCS 13025, pp. 528–539, 2021.
https://doi.org/10.1007/978-3-030-89543-3_44

observed once. For example, the $L_2$ and entropy norms are frequently used to detect network anomalies [16,28,34], while the $L_1$ norm is used to monitor network traffic [23] and perform low-rank approximation and linear regression [24], and the top-$k$ and Ky Fan norms are commonly used in matrix optimization problems [39]. These norms all have the property that they are invariant to permutations and sign flips of the coordinates of the underlying vectors:

**Definition 1 (Symmetric norm).** *A norm $\ell : \mathbb{R}^n \to \mathbb{R}$ is a* symmetric norm *if for all $x \in \mathbb{R}^n$ and any $n \times n$ permutation matrix $P$, we have $\ell(x) = \ell(Px)$ and $\ell(x) = \ell(|x|)$, where $|x|$ is the coordinate-wise absolute value of $x$.*

Symmetric norms include the $L_p$, entropy, top-$k$, $k$-support, and box norms, and many other examples that we detail in Sect. 3.1. Braverman *et al.* [5] show that a symmetric norm $\ell$ can be approximated using space roughly $\text{mmc}(\ell)^2$, where mmc is the *maximum modulus of concentration* of the norm $\ell$, whose formal definition we will defer to Sect. 3.1. Informally, $\text{mmc}(\ell)$ is roughly the ratio of the maximum value $\ell$ achieves on a unit ball compared to the meidan value of $\ell$ on the unit ball.

*Sliding Window Model.* Unfortunately, the streaming model does not prioritize recent data that is considered more accurate and important than data that arrived prior to a certain time. Thus for a number of time-sensitive applications [4,29,32,35], the streaming model has inferior performance compared to the *sliding window model*, in which the underlying dataset consists of only the $W$ most recent updates in the stream. The fixed parameter $W > 0$ represents the window size for the active data and the goal is to process information about the dataset using space sublinear in $W$. Note that the sliding window model is a generalization of the streaming model, e.g., when the stream length $m$ is at most $W$. The sliding window model is especially relevant in time-dependent settings such as network monitoring [18–20], event detection in social media [31], data summarization [17,22], and has been also studied in a number of additional settings [6–8,10–13,15,21,26,37].

*Problem Statement.* Formally, the model is as follows. Given a symmetric norm $\ell : \mathbb{R}^n \to \mathbb{R}$, we receive updates $u_1, \ldots, u_m$ to the coordinates of an underlying frequency vector $f$. Each update with $i \in [m]$ satisfies $u_i \in [n]$ so that the $i$-th update effectively increments the $u_i$-th coordinate of $f$. However, in the sliding window model, only the last $W$ updates define $f$ so that for each $j \in [n]$, we have $f_j = |\{i : u_i = j, i \geq m - W + 1\}|$. The goal is to approximate $\ell(f)$ at the end of the stream, but $m$ is not given in advance so we cannot simply maintain a sketch of the last $W$ elements because we do not know the value of $m - W + 1$ a priori.

The main challenge of the sliding window model is that updates to $f$ expire implicitly. Thus we cannot apply linear sketching techniques, which forms the backbone of many streaming algorithms. For example, we do not know that the update $u_{m-W}$ does not affect the value of $f$ until the very last update. Thus if we maintain a sketch of the updates that includes $u_{m-W}$, we must "undo" the inclusion of $u_{m-W}$ at time $m$; however at that time, it may be too late to remember the value of $u_{m-W}$.

### 1.1 Our Results

In this paper, we give the first generic framework that can approximate any symmetric norm of an underlying frequency vector in the sliding window model.

**Theorem 1.** *Given an accuracy parameter $\epsilon > 0$ and a symmetric norm $\ell$, there exists a sliding window algorithm that outputs a $(1+\epsilon)$-approximation to the $\ell$-norm of the underlying frequency vector with probability $\frac{2}{3}$ and uses space $mmc(\ell)^2 \cdot poly\left(\frac{1}{\epsilon}, \log n\right)$.*

Our framework has specific implications to the well-studied $L_p$ norms and the top-$k$ norm that is used in matrix optimization, as well as the $k$-support, box, and more generally, $Q'$-norms that are frequently used to regularize sparse recovery problems in machine learning. We summarize these applications in Fig. 1 and provide additional detail on these norms in Sect. 3.1. In particular for sufficiently large $p > 2$, our $L_p$ norm sliding window algorithm improves upon the $\tilde{\mathcal{O}}\left(\frac{1}{\epsilon^{p+2}} n^{1-2/p}\right)$ space algorithm by [13]. Our framework not only uses near-optimal space complexity for these applications, but is also a *universal sketch* that suffices to simultaneously approximate all symmetric norms in a wide parametrizable class.

**Theorem 2.** *Given an accuracy parameter $\epsilon > 0$ and a space parameter $S$, there exists a sliding window algorithm that uses space $S \cdot poly\left(\frac{1}{\epsilon}, \log n\right)$ and outputs a $(1+\epsilon)$-approximation to* any *symmetric norm $\ell$ with $mmc(\ell) \leq \sqrt{S}$, with probability $\frac{2}{3}$.*

| Problem | Space Complexity | Reference |
|---|---|---|
| Symmetric norm $\ell$ | $\text{mmc}(\ell)^2 \cdot \text{poly}\left(\frac{1}{\epsilon}, \log n\right)$ | Theorem 1 |
| $L_p$ norm, $p \in [1,2]$ | $\text{poly}\left(\frac{1}{\epsilon}, \log n\right)$ | Corollary 1 |
| $L_p$ norm, $p > 2$ | $\text{poly}\left(\frac{1}{\epsilon}, \log n\right) \cdot n^{1-2/p}$ | Corollary 2 |
| $k$-support norm | $\text{poly}\left(\frac{1}{\epsilon}, \log n\right)$ | Corollary 1 |
| $Q'$ norm | $\text{poly}\left(\frac{1}{\epsilon}, \log n\right)$ | Corollary 1 |
| Box norm | $\text{poly}\left(\frac{1}{\epsilon}, \log n\right)$ | Corollary 1 |
| Top-$k$ norm | $\frac{n}{k} \cdot \text{poly}\left(\frac{1}{\epsilon}, \log n\right)$ | Corollary 3 |

**Fig. 1.** Summary of our sliding window algorithms

The general approach to sliding window algorithms is to use the smooth histogram framework by Braverman and Ostrovsky [13]. The smooth histogram framework requires the desired objective to be smooth, where given adjacent substreams $A$, $B$, and $C$, a smooth function states that $(1-\eta)f(A \cup B) \leq f(B)$ implies $(1-\epsilon)f(A \cup B \cup C) \leq f(B \cup C)$ for some constants $0 < \eta \leq \epsilon < 1$. Intuitively, once a suffix of a data stream becomes a $(1 \pm \eta)$-approximation for a smooth function, then it is *always* a $(1 \pm \epsilon)$-approximation, regardless of the subsequent updates that arrive in the stream. Since the resulting space complexity depends on $\eta$, this approach requires analyzing the smoothness of each symmetric norm and it is not clear how these parameters relate to $\text{mmc}(\ell)$ or whether there is a general parametrization for each norm.

Instead, we observe that [5] effectively reduces the problem to computing a $(1+\nu)$-approximation to the frequency of all $\eta$-heavy hitters for a number of various substreams.

**Definition 2 ($\nu$-approximate $\eta$-heavy hitters).** *Given any accuracy parameter $\nu$, a threshold parameter $\eta$, and a frequency vector $f$, an algorithm $\mathcal{A}$ is said to solve the $\nu$-approximate $\eta$-heavy hitters problem if it outputs a set $H$ and a set of approximations $\widehat{f_i}$ for all $i \in H$ such that:*

*(1) If $f_i \geq \eta \|f\|_2$ for any $i \in [n]$, then $i \in H$. That is, $H$ contains all $\eta$-heavy hitters of $f$.*
*(2) There exists an absolute constant $C > 0$ so that if $f_i \leq \frac{C\eta}{2} \|f\|_2$ for any $i \in [n]$, then $i \notin H$. That is, $H$ does not contain any item that is not an $\frac{C\eta}{2}$-heavy hitter of $f$.*
*(3) If $i \in H$, then $\mathcal{A}$ reports a value $\widehat{f_i}$ such that $(1-\nu)f_i \leq \widehat{f_i} \leq (1+\nu)f_i$. That is, $\mathcal{A}$ outputs a $(1 \pm \nu)$-approximation to the frequency $f_i$, for all $i \in H$.*

Thus to approximate a symmetric norm on the active elements, it suffices to find $\nu$-approximate $\eta$-heavy hitters for a number of substreams. Whereas the sliding window heavy-hitter algorithms [9,10] optimize for space complexity and only output constant factor approximations to the frequencies of the reported elements, we give a simple modification to their ideas to output $\nu$-approximate $\eta$-heavy hitters.

**Theorem 3.** *Let $f$ be a frequency vector on $[n]$ induced by the active window of an insertion-only data stream. For any accuracy parameter $\nu \in \left(0, \frac{1}{4}\right)$ and threshold $\eta \in (0,1)$, there exists a one-pass streaming algorithm that outputs a list that includes all $\eta$-heavy hitters and no element that is not a $\frac{\eta}{8}$-heavy hitter. Moreover, the algorithm reports a $(1+\nu)$-approximation to the frequency $f_i$ of all reported items $i$. The algorithm uses $\mathcal{O}\left(\frac{1}{\nu^3\eta^2}\log^3 n\right)$ bits of space and succeeds with high probability.*

In summary, our main conceptual contribution is the existence of a $(1+\epsilon)$-approximation algorithm for general symmetric norms in the sliding window model. Our technical contributions include an overall framework that incorporates *any* symmetric norm in a plug-and-play manner as well as a heavy-hitter subroutine that may be of independent interest. Finally, we perform a number of empirical evaluations comparing our algorithms to uniform sampling on large-scale real-world datasets.

*Subsequent Related Work.* Subsequent to our work, [27] has given a framework for subadditive functions that extends beyond the smooth histogram approach of [13]. In particular, their framework gives a $(2+\epsilon)$-approximation for symmetric norms in the sliding window model. By comparison, our algorithm achieves a $(1+\epsilon)$-approximation for symmetric norms on sliding windows. Their techniques are based on black-boxing the streaming algorithm of [5] that approximates the symmetric norm and initializing various instances of the algorithm as the stream progresses. We open up the black box by instead introducing a new heavy-hitter algorithm in the sliding window model and using properties of heavy-hitters and level sets to enable a finer approximation to the symmetric norm, e.g., [14,25,36,38].

*Symmetric Norm Regression.* As a further application of our work, we consider the fundamental overconstrained linear regression problem in the case that loss function that is a symmetric norm, which includes many standard loss functions such as $L_p$ norms, top-$k$ norms, and $Q'$-norms. Specifically, given a data matrix $\mathbf{A} \in \mathbb{R}^{n \times d}$ and a response vector $\mathfrak{b} \in \mathbb{R}^n$ with $n \gg d$, we aim to minimize the optimization problem $\min_{\mathbf{x} \in \mathbb{R}^d} \mathcal{L}(\mathbf{Ax} - \mathfrak{b})$, where $\mathcal{L} : \mathbb{R}^n \to \mathbb{R}$ is a loss function. When $\mathcal{L}$ is a symmetric norm, then the loss function places emphasis on the magnitude of the incorrect coordinates rather than their specific indices. In particular, we consider the general case where $\mathcal{L}$ is an Orlicz norm, which can be interpreted as a scale-invariant version of $M$-estimators. Embeddings for $(1+\epsilon)$-approximate solutions to the linear regression problem for loss functions that are Orlicz norms in the central model, where complete access to $\mathbf{A}$ is given, was recently studied by [2,33]. We give the first algorithms that produce $(1+\epsilon)$-approximate solutions to the linear regression problem for loss functions that are Orlicz norms in both the streaming and sliding window models. Our algorithms are parametrized by a constant $\Delta$, which represents the aspect ratio of the dataset under the norm. We defer the following result to the full version of the paper.

**Theorem 4.** *Given an accuracy $\epsilon > 0$ and a matrix $\mathbf{A} \in \mathbb{R}^{W \times d}$ whose rows $\mathbf{a}_1, \ldots, \mathbf{a}_W$ arrive sequentially in a stream $\mathbf{r}_1, \ldots, \mathbf{r}_n$ with condition number at most $\kappa$, there exists both a streaming algorithm and a sliding window algorithm that outputs a $(1+\epsilon)$ embedding for an Orlicz norm with high probability. The algorithms sample $\frac{d^2 \Delta}{\epsilon^2} \log \kappa \operatorname{polylog} n$ rows, with high probability.*

## 2 Approximate Heavy-Hitters in the Sliding Window Model

In this section, we describe our $\nu$-approximate $\eta$-heavy hitters algorithm that appears in Algorithm 1, slightly perturbing constants for the ease of discussion. Our starting point is the $L_2$ norm estimation algorithm FreqEst in [13].

FREQEST maintains a number of timestamps $\{t_i\}$ throughout the data stream, along with a separate streaming algorithm for each $t_i$ that stores a sketch of the $L_2$ norm of the elements in the stream after $t_i$. [13] observes that it suffices for $\{t_i\}$ to maintain the invariant that the sketches of at most two timestamps produce values that are within 2 of each other, since it can be shown that they would always output values that are within 2 afterwards. Hence, if the length of the stream $m$ is polynomially bounded in $n$, then the number of total timestamps is $\mathcal{O}(\log n)$. Moreover, two of these timestamps will sandwich the starting point of the sliding window and provide a 2-approximation to the $L_2$ norm of the active elements and more generally, there exists an algorithm FREQEST that outputs a 2-approximation to *any* suffix of the stream.

To transition from $L_2$ norm estimation to $\eta$-heavy hitters, [9,10] simultaneously run instances of the COUNTSKETCH heavy-hitter algorithm starting at each of the timestamps $t_i$. Any $\eta$-heavy hitter of the active elements must be a $\frac{\eta}{2}$-heavy hitter of the stream starting at some timestamp, since one of these timestamps $t_i$ contains the active elements but has $L_2$ norm at most 2 times the $L_2$ norm of the active elements. Hence, all $\eta$-heavy hitters will be reported by the corresponding COUNTSKETCH starting at $t_i$. However, it can also report elements that do not appear in the window at all, e.g., the elements after $t_i$ but before $m-W+1$. Thus, [9,10] also maintains a constant factor approximation to the frequency of each item reported by COUNTSKETCH as a final check, through comparison with the estimated $L_2$ norm from FREQEST. These parameters are insufficient to obtain $\nu$-approximate $\eta$-heavy hitters, since 1) a constant factor approximation to each frequency cannot give a $(1+\nu)$-approximation and 2) if COUNTSKETCH only reports elements once they are $\eta$-heavy, then it is possible that a constant fraction of the frequency is missed, e.g., if the frequency is $2\eta \cdot \|f\|_2$. To address these issues, we apply two simple fixes in Fig. 2.

(1) Find a superset of the possible heavy-hitters of the active window by taking heavy-hitters of a superset of the active window, but with a lower threshold, i.e. $\mathcal{O}(\nu\eta)$ rather than $\eta$.
(2) For each possible heavy-hitter, maintain a $(1+\mathcal{O}(\nu))$-approximation to its frequency.
(3) Report the items with sufficiently high estimated frequency.

**Fig. 2.** Crude outline of $\nu$-approximate $\eta$-heavy hitter sliding window algorithm.

First, we maintain a $(1+\mathcal{O}(\nu))$-approximation to the frequency of each item reported by COUNTSKETCH. However, we note that we only track the frequency of an item once it is reported by COUNTSKETCH and thus the second issue still prevents our algorithm from reporting a $(1+\nu)$-approximation for sufficiently small $\nu$ because a constant fraction of the frequency can still be missed before being reported by COUNTSKETCH. Thus the second idea is to report items once they are $\frac{\nu\eta}{32}$-heavy hitters, so that only a $\mathcal{O}(\nu)$ fraction of the frequency can be missed before each heavy-hitter is tracked. We give a crude outline of our approach in Fig. 2 and the algorithm in full in Algorithm 1.

**Algorithm 1.** Algorithm for $\eta$-heavy hitters in sliding window model, with $(1+\nu)$-approximation to frequency of reported items.

**Input:** A stream of elements $u_1, \ldots, u_m \in [n]$, a window parameter $W > 0$, threshold $\eta \in (0,1)$, accuracy parameter $\nu \in \left(0, \frac{1}{4}\right)$

**Output:** A list that contains all $\eta$-heavy hitters and no element that is not a $\frac{\eta}{2}$-heavy hitters, along with a $(1+\nu)$ to the frequency of all items.

1: Run an instance of FreqEst on the stream.
2: $\mathcal{T} \leftarrow \emptyset$
3: **for** each update $u_t \in [n]$ with $t \in [m]$ **do**
4: $\quad \mathcal{T} \leftarrow \mathcal{T} \cup \{t\}$
5: $\quad$ Initialize CountSketch$_t$ with threshold $\frac{\nu\eta}{32}$. ▷Identify a superset of the heavy-hitters
6: $\quad X_a \leftarrow$ estimated $L_2$ norm of the frequency vector from time $a \in \mathcal{T}$ to $t$ by FreqEst.
7: $\quad$ **while** exist $b < c \in \mathcal{T}$ with $c < t - W + 1$ or $a < b < c \in \mathcal{T}$ with $X_a \leq \frac{17}{16} X_c$ **do**
8: $\quad\quad$ Delete $b$ from $\mathcal{T}$ and CountSketch$_b$.
9: $\quad H_a \leftarrow$ heavy-hitters reported by CountSketch$_a$ from time $a \in \mathcal{T}$ to $t$.
10: $\quad F \leftarrow$ estimated $L_2$ norm of the frequency vector from time $\min(1, t - W + 1)$ to $t$ by FreqEst.
11: $\quad$ **for** all $a \in \mathcal{T}$ and $i \in H_a$ **do**
12: $\quad\quad$ Use Counter for $i$, starting at time $a$. ▷$(1+\frac{\nu}{4})$-accuracy
13: $\quad\quad \widehat{f_i} \leftarrow$ any underestimate to the frequency of $i$ in the last $W$ updates by Counter.
14: $\quad\quad$ **if** $\widehat{f_i} \geq \frac{\eta}{2} \cdot F$ **then**
15: $\quad\quad\quad$ Report $i$, with estimated frequency $\widehat{f_i}$

We first show that Algorithm 1 does not output any items with sufficiently low frequency.

**Lemma 1 (Low frequency items are not reported).** *Let $f$ be the frequency vector induced by the active window. For each $i \in [n]$, if $f_i \leq \frac{\eta}{8} \|f\|_2$, then Algorithm 1 does not report $i$.*

Next we show that not only are the heavy-hitters reported, but the estimated frequency for each reported item is also a $(1+\nu)$ approximation to the true frequency.

**Lemma 2 (Heavy-hitters are reported accurately).** *Let $f$ be the frequency vector induced by the active window. For each $i \in [n]$, if $f_i \geq \eta \cdot \|f\|_2$, then Algorithm 1 reports $i$. Moreover, $\widehat{f_i} \leq f_i \leq (1+\nu)\widehat{f_i}$ for any item $i$ reported by Algorithm 1.*

Theorem 3 then follows from Lemma 1 and Lemma 2 and an analysis of the space complexity.

## 3 Symmetric Norms

In this section, we formalize our symmetric norm sliding window algorithm and give a number of applications. We first require the following preliminary definitions that quantify specific properties of symmetric norms.

**Definition 3 (Modulus of concentration).** *Let $X \in \mathbb{R}^n$ be a random variable uniformly distributed on the $L_2$-unit sphere $S^{n-1}$. The median of a symmetric norm $\ell$ is the unique value $M_\ell$ such that $\mathbf{Pr}\left[\ell(X) \geq M_\ell\right] \geq \frac{1}{2}$ and $\mathbf{Pr}\left[\ell(X) \leq M_\ell\right] \geq \frac{1}{2}$. Then if $\mathfrak{b}_\ell$ denotes the maximum value of $\ell(x)$ over $x \in S^{n-1}$, then the ratio $mc(\ell) := \frac{\mathfrak{b}_\ell}{M_\ell}$ is called the* modulus of concentration *of the norm $\ell$.*

The modulus of concentration characterizes the average behavior of the norm $\ell$ on $\mathbb{R}^n$. However, even if $\ell$ is well-behaved on average, more difficult norms can be embedded and hidden in a lower-dimensional subspace. For example, [5] observes that $\text{mc}(\ell) = \mathcal{O}(1)$ for the $L_1$ norm $\ell$, but when $x$ has fewer than $\sqrt{n}$ nonzero coordinates, the norm $\ell(x) = \max(L_\infty(x), L_1(x)/\sqrt{n})$ on the unit ball becomes identically $L_\infty(x)$, which requires $\Omega(\sqrt{n})$ space [1]. Thus, we instead consider the modulus of concentration over all lower dimensions.

**Definition 4 (Maximum modulus of concentration).** *For every $k \leq n$, the norm $\ell : \mathbb{R}^n \to \mathbb{R}$ induces a norm on $\mathbb{R}^k$ by setting $\ell^{(k)}((x_1, \ldots, x_k)) = \ell((x_1, \ldots, x_k, 0, \ldots, 0))$. The* maximum modulus of concentration *of the norm $\ell$ is defined as $mmc(\ell) := \max_{k \leq n} mc(\ell^{(k)}) = \max_{k \leq n} \frac{\mathfrak{b}_{\ell^{(k)}}}{M_{\ell^{(k)}}}$.*

We now reduce the problem of approximating a symmetric norm $\ell$ to the $\nu$-approximate $\eta$-heavy hitters problem.

**Lemma 3 (Symmetric norm approximation through heavy-hitters).** [5] *Let $\ell$ be any symmetric norm, $\epsilon > 0$ and $\nu := \mathcal{O}\left(\frac{\epsilon^2}{\log n}\right)$ be fixed accuracy parameters, and $\eta := \mathcal{O}\left(\frac{\epsilon^{5/2}}{mmc(\ell) \log^{5/2} n}\right)$ be a fixed threshold. Let $R = \Theta\left(\frac{\log^{10} n}{\epsilon^5}\right)$ and for each $i \in [\log n]$ and $r \in [R]$, let $j \in [n]$ be sampled into $S_{i,r}$ with probability $\frac{1}{2^i}$. Let $f$ be a frequency vector (possibly implicitly) defined on $[n]$ and for each $i \in [\log n]$, let $g_{i,r}$ be the frequency vector induced by setting all coordinates $j \in [n]$ of $f$ with $j \notin S_i$.*

*Suppose there exists an algorithm that outputs $\nu$-approximate $\eta$-heavy hitters $H_{i,r}$ for each $g_{i,r}$. There exists a recovery function* Estimate *that recovers a $(1+\epsilon)$-approximation to $\ell(f)$ using $\{H_{i,r}\}$. The running time of* Estimate *is polynomial in $\frac{1}{\epsilon}$ and $n$ and the working space of* Estimate *is the space used to store $\{H_i\}$.*

Informally, Lemma 3 states that to obtain a $(1+\epsilon)$-approximation to any symmetric norm $\ell$ of an underlying frequency vector, it suffices to use a $\nu$-approximate $\eta$ heavy-hitter algorithm. Here, $\eta$ and $\nu$ are parameters dependent on the norm $\ell$. We give additional intuition into Lemma 3 and its proof by [5] in the full version of the paper.

### 3.1 Applications

In this section, we demonstrate the application of Theorem 1 and Theorem 2 to a number of symmetric norms. We summarize our results in Fig. 1.

*Q'-Norms.* We first that a $(1+\epsilon)$-approximation of any $Q'$-norm, i.e., quadratic norm, in the sliding window model only requires polylogarithmic space, using the maximum modulus of concentration characterization of $Q$-norms by [5].

**Definition 5 ($Q$-norm and $Q'$-norm).** *A norm $\ell : \mathbb{R}^n \to \mathbb{R}$ is a $Q$-norm if there exists a symmetric norm $L : \mathbb{R}^n \to \mathbb{R}$ such that for all $x \in \mathbb{R}^n$, we have $\ell(x) = L(x^2)^{1/2}$, where $x^2$ denotes the coordinate-wise square power of $x$. Then a norm $\ell' : \mathbb{R}^n \to \mathbb{R}$ is a $Q'$-norm if its dual norm is a $Q$-norm.*

$Q'$-norms includes the $L_p$ norms for $1 \le p \le 2$. [5] also notes that multiple $Q'$-norms have been proposed to regularize sparse recovery problems in machine learning. For example, [3] shows that the $k$-support norm, whose unit ball is the convex hull of the set $\{x \in \mathbb{R}^n : \|x\|_0 \le k \text{ and } \ell_2(x) \le 1\}$, is a $Q'$-norm that has a tighter relaxation than elastic nets and can thus be more effective for sparse prediction. The box norm [30], defined for $\Theta = \{\theta \in [a,b]^n : \ell_1(x) \le c\}$, given parameters $0 < a < b \le c$, as $\ell_\Theta(x) = \min_{\theta\in\Theta} \left(\sum_{i=1}^n x_i^2/\theta_i\right)^{1/2}$, is a $Q'$-norm that is also a generalization of the $k$-support norm. The box norm has been used to further optimize algorithms for the sparse prediction problem specifically in the context of multitask clustering [30].

**Corollary 1.** *Given $\epsilon > 0$, there exists a sliding window algorithm that uses poly $\left(\frac{1}{\epsilon}, \log n\right)$ bits of space and outputs a $(1+\epsilon)$-approximation to the $Q'$-norm.*

*$L_p$ norms.* Since $Q'$-norms include $L_p$ norms for $p \in [1,2]$, we now consider the approximation of $L_p$ norms for $p > 2$.

**Corollary 2.** *Given $\epsilon > 0$ and $p > 2$, there exists a sliding window algorithm that uses poly $\left(\frac{1}{\epsilon}, \log n\right) \cdot n^{1-2/p}$ bits of space and outputs a $(1+\epsilon)$-approximation to the $L_p$-norm.*

In particular, since the exponents of $\epsilon$ and $\log n$ are fixed, then for sufficiently large $p$, Corollary 2 improves on the results of [13], who give an algorithm using space $\frac{1}{\epsilon^{p+2}}$ polylog$n \cdot n^{1-2/p}$.

*Top-k Norms.* We now show that a $(1+\epsilon)$-approximation of any top-$k$ norm in the sliding window model only requires sublinear space, for sufficiently large $k$.

**Definition 6 (Top-$k$ norm).** *The top-$k$ norm for a vector $x \in \mathbb{R}^n$ is the sum of the largest $k$ coordinates of $|x|$.*

The top-$k$ norm is a special case of the Ky Fan $k$-norm [39] when the vector $x$ represents the entries in a diagonal matrix. Thus the top-$k$ norm is often used to understand the Ky Fan $k$-norm, which is used to regularize optimization problems in numerical linear algebra.

**Corollary 3.** *Given $\epsilon > 0$, there exists a sliding window algorithm that uses $\frac{n}{k} \cdot \text{poly}\left(\frac{1}{\epsilon}, \log n\right)$ bits of space and outputs a $(1+\epsilon)$-approximation to the top-$k$ norm.*

## References


1. Alon, N., Matias, Y., Szegedy, M.: The space complexity of approximating the frequency moments. J. Comput. Syst. Sci. **58**(1), 137–147 (1999)
2. Andoni, A., Lin, C., Sheng, Y., Zhong, P., Zhong, R.: Subspace embedding and linear regression with Orlicz norm. In: Proceedings of the 35th International Conference on Machine Learning, ICML, pp. 224–233 (2018)
3. Argyriou, A., Foygel, R., Srebro, N.: Sparse prediction with the $k$-support norm. In: Advances in Neural Information Processing Systems 25: Annual Conference on Neural Information Processing Systems, pp. 1466–1474 (2012)
4. Babcock, B., Babu, S., Datar, M., Motwani, R., Widom, J.: Models and issues in data stream systems. In: Proceedings of the Twenty-first ACM SIGACT-SIGMOD-SIGART Symposium on Principles of Database Systems, pp. 1–16 (2002)
5. Blasiok, J., Braverman, V., Chestnut, S.R., Krauthgamer, R., Yang, L.F.: Streaming symmetric norms via measure concentration. In: Proceedings of the 49th Annual ACM SIGACT Symposium on Theory of Computing, STOC, pp. 716–729 (2017)
6. Borassi, M., Epasto, A., Lattanzi, S., Vassilvitskii, S., Zadimoghaddam, M.: Better sliding window algorithms to maximize subadditive and diversity objectives. In: Proceedings of the 38th ACM SIGMOD-SIGACT-SIGAI Symposium on Principles of Database Systems, PODS, pp. 254–268 (2019)
7. Borassi, M., Epasto, A., Lattanzi, S., Vassilvitskii, S., Zadimoghaddam, M.: Sliding window algorithms for k-clustering problems. In: Advances in Neural Information Processing Systems 33: Annual Conference on Neural Information Processing Systems, NeurIPS (2020)
8. Braverman, V., et al.: Near optimal linear algebra in the online and sliding window models. In: 61st IEEE Annual Symposium on Foundations of Computer Science, FOCS, pp. 517–528 (2020)
9. Braverman, V., Gelles, R., Ostrovsky, R.: How to catch $L_2$-heavy-hitters on sliding windows. Theor. Comput. Sci. **554**, 82–94 (2014)
10. Braverman, V., Grigorescu, E., Lang, H., Woodruff, D.P., Zhou, S.: Nearly optimal distinct elements and heavy hitters on sliding windows. In: Approximation, Randomization, and Combinatorial Optimization. Algorithms and Techniques, APPROX/RANDOM, pp. 7:1–7:22 (2018)

11. Braverman, V., Lang, H., Levin, K., Monemizadeh, M.: Clustering on sliding windows in polylogarithmic space. In: 35th IARCS Annual Conference on Foundation of Software Technology and Theoretical Computer Science, FSTTCS, pp. 350–364 (2015)
12. Braverman, V., Lang, H., Levin, K., Monemizadeh, M.: Clustering problems on sliding windows. In: Proceedings of the Twenty-Seventh Annual ACM-SIAM Symposium on Discrete Algorithms, SODA, pp. 1374–1390 (2016)
13. Braverman, V., Ostrovsky, R.: Smooth histograms for sliding windows. In: Proceedings of 48th Annual IEEE Symposium on Foundations of Computer Science (FOCS), pp. 283–293 (2007)
14. Braverman, V., Ostrovsky, R., Roytman, A.: Zero-one laws for sliding windows and universal sketches. In: Approximation, Randomization, and Combinatorial Optimization. Algorithms and Techniques, APPROX/RANDOM, pp. 573–590 (2015)
15. Braverman, V., Ostrovsky, R., Zaniolo, C.: Optimal sampling from sliding windows. J. Comput. Syst. Sci. **78**(1), 260–272 (2012)
16. Chakrabarti, A., Ba, K.D., Muthukrishnan, S.: Estimating entropy and entropy norm on data streams. Internet Math. **3**(1), 63–78 (2006)
17. Chen, J., Nguyen, H.L., Zhang, Q.: Submodular maximization over sliding windows. CoRR, abs/1611.00129 (2016)
18. Cormode, G.: The continuous distributed monitoring model. SIGMOD Rec. **42**(1), 5–14 (2013)
19. Cormode, G., Garofalakis, M.N.: Streaming in a connected world: querying and tracking distributed data streams. In: EDBT 2008, Proceedings of 11th International Conference on Extending Database Technology, p. 745 (2008)
20. Cormode, G., Muthukrishnan, S.: What's new: finding significant differences in network data streams. IEEE/ACM Trans. Netw. **13**(6), 1219–1232 (2005)
21. Datar, M., Motwani, R.: The sliding-window computation model and results. In: Garofalakis, M., Gehrke, J., Rastogi, R. (eds.) Data Stream Management. DSA, pp. 149–165. Springer, Heidelberg (2016). https://doi.org/10.1007/978-3-540-28608-0_7
22. Epasto, A., Lattanzi, S., Vassilvitskii, S., Zadimoghaddam, M.: Submodular optimization over sliding windows. In: Proceedings of the 26th International Conference on World Wide Web, WWW, pp. 421–430 (2017)
23. Feigenbaum, J., Kannan, S., Strauss, M., Viswanathan, M.: An approximate l1-difference algorithm for massive data streams. SIAM J. Comput. **32**(1), 131–151 (2002)
24. Feldman, D., Monemizadeh, M., Sohler, C., Woodruff, D.P.: Coresets and sketches for high dimensional subspace approximation problems. In: Proceedings of the Twenty-First Annual ACM-SIAM Symposium on Discrete Algorithms, SODA, pp. 630–649 (2010)
25. Indyk, P., Woodruff, D.P.: Optimal approximations of the frequency moments of data streams. In: Proceedings of the 37th Annual ACM Symposium on Theory of Computing, pp. 202–208 (2005)
26. Jayaram, R., Woodruff, D.P., Zhou, S.: Truly perfect samplers for data streams and sliding windows. CoRR, abs/2108.12017 (2021)
27. Krauthgamer, R., Reitblat, D.: Almost-smooth histograms and sliding-window graph algorithms. CoRR, abs/1904.07957 (2019)
28. Krishnamurthy, B., Sen, S., Zhang, Y., Chen, Y.: Sketch-based change detection: methods, evaluation, and applications. In: Proceedings of the 3rd ACM SIGCOMM Internet Measurement Conference, IMC, pp. 234–247 (2003)

29. Manku, G.S., Motwani, R.: Approximate frequency counts over data streams. PVLDB **5**(12), 1699 (2012)
30. McDonald, A.M., Pontil, M., Stamos, D.: Spectral k-support norm regularization. In: Advances in Neural Information Processing Systems 27: Annual Conference on Neural Information Processing Systems, pp. 3644–3652 (2014)
31. Osborne, M., et al.: Real-time detection, tracking and monitoring of automatically discovered events in social media. In: Proceedings of the 52nd Annual Meeting of the Association for Computational Linguistics (2014)
32. Papapetrou, O., Garofalakis, M., Deligiannakis, A.: Sketching distributed sliding-window data streams. VLDB J. **24**(3), 345–368 (2015). https://doi.org/10.1007/s00778-015-0380-7
33. Song, Z., Wang, R., Yang, L.F., Zhang, H., Zhong, P.: Efficient symmetric norm regression via linear sketching. In: Advances in Neural Information Processing Systems 32: Annual Conference on Neural Information Processing Systems, pp. 828–838 (2019)
34. Thorup, M., Zhang, Y.: Tabulation based 4-universal hashing with applications to second moment estimation. In Proceedings of the Fifteenth Annual ACM-SIAM Symposium on Discrete Algorithms, SODA, pp. 615–624 (2004)
35. Wei, Z., Liu, X., Li, F., Shang, S., Du, X., Wen, J.-R.: Matrix sketching over sliding windows. In: Proceedings of the 2016 International Conference on Management of Data, SIGMOD Conference, pp. 1465–1480 (2016)
36. Woodruff, D.P., Zhang, Q.: Distributed statistical estimation of matrix products with applications. In: Proceedings of the 37th ACM SIGMOD-SIGACT-SIGAI Symposium on Principles of Database Systems, pp. 383–394 (2018)
37. Woodruff, D.P., Zhou, S.: Tight bounds for adversarially robust streams and sliding windows via difference estimators. CoRR, abs/2011.07471 (2020)
38. Woodruff, D.P., Zhou, S.: Separations for estimating large frequency moments on data streams. In: 48th International Colloquium on Automata, Languages, and Programming, ICALP, pp. 112:1–112:21 (2021)
39. Bin, W., Ding, C., Sun, D., Toh, K.-C.: On the Moreau-Yosida regularization of the vector k-norm related functions. SIAM J. Optim. **24**(2), 766–794 (2014)

# Single-Pass Streaming Algorithms to Partition Graphs into Few Forests

Cheng-Hung Chiang and Meng-Tsung Tsai(✉)

Institute of Information Science, Academia Sinica, Taipei, Taiwan
{timmychiang,mttsai}@iis.sinica.edu.tw

**Abstract.** We devise a single-pass $O(n)$-space deterministic streaming algorithm to partition any $n$-node undirected simple graph $G$ into $O(\alpha \log n)$ forests where $\alpha$ is the minimum number of forests which $G$ can be partitioned into. We then apply this result to obtain single-pass streaming algorithms for other graph problems, including low outdegree orientation, partitioning graphs into few planar subgraphs, and finding small dominating sets.

**Keywords:** Greedy · Arboricity · Orientation · Thickness · Dominating set

## 1 Introduction

The ***arboricity*** $a(G)$ of an undirected simple graph $G = (V, E)$ is defined to be the minimum number of forests into which the edge set $E$ can be partitioned. On a RAM, the arboricity can be computed exactly in $O(n^3 \log n)$ time [12], and $(1 + \varepsilon)$-approximated in $O(\varepsilon^{-1} m \log n)$ time for any $\varepsilon > 0$ [4], where $n = |V|$ denotes the number of nodes and $m = |E|$ denotes the number of edges. In the model of streaming, property testing, and distributed computation, the exact value of $a(G)$ may be impossible to obtain, and only algorithms that approximate $a(G)$ are known [7,9,14]. Instead of computing the exact or an approximate value of $a(G)$, we consider the problem of partitioning $E$ into $t$ forests so that the approximation ratio $t/a(G)$ is small.

Our model of computation is the ***semi-streaming model*** [19,21], which is a variant of the streaming model frequently used for the computation of graph problems [2,3,6,13,16,20]. It allows the edges of an $n$-node input graph to be read sequentially in $p$ passes using $O(n \operatorname{polylog} n)$ space. Some multi-pass streaming algorithms are known to partition graphs into few forests [9,10], but it is still unknown whether this problem can be approximated well by single-pass algorithms. We show that there exists a single-pass $O(n)$-space deterministic streaming algorithm that can partition a given $n$-node undirected simple graph $G$ into $O(a(G) \cdot \min\{\log n, a(G) + 2\})$ forests, formally stated in Theorem 1. We

This research was supported in part by the Ministry of Science and Technology of Taiwan under contract MOST grant 109-2221-E-001-025-MY3.


C.-Y. Chen et al. (Eds.): COCOON 2021, LNCS 13025, pp. 540–552, 2021.
https://doi.org/10.1007/978-3-030-89543-3_45

show also that the space usage can be further reduced to $s$ for any $s = o(n)$, at the cost of increasing the approximation factor to $O(n/\sqrt{s})$.

**Theorem 1.** *Let $e_1, e_2, \dots, e_m$ be the edges of a given $n$-node undirected simple graph $G$ in an arbitrary order. There exists a single-pass deterministic streaming algorithm that runs in $O(m \log n)$ time and uses $O(n)$ space to output a sequence of tuples $(e'_1, c_1), (e'_2, c_2), \dots, (e'_m, c_m)$ so that the following three conditions simultaneously hold.*

*(a)* $e'_1, e'_2, \dots, e'_m$ *is a permutation of* $e_1, e_2, \dots, e_m$.
*(b) There exists an integer* $t = O(a(G) \cdot \min\{\log n, a(G)+2\})$ *so that* $1 \le c_i \le t$ *for every* $i \in \{1, 2, \dots, m\}$.
*(c) For every* $i \in \{1, 2, \dots, t\}$, $\{e_j : 1 \le j \le m, c_j = i\}$ *is a forest of* $G$.

In addition to the algorithmic results, we prove some lower bounds on the space usage for any streaming algorithm that solves this problem. We get:

**Theorem 2.** *For any $p$-pass randomized streaming algorithm $\mathcal{A}$ that partitions a given $n$-node undirected simple graph $G$ into at most $t$ forests with success probability at least 2/3,*

- *if* $p = 1$, $t = a(G)$, *the space usage of* $\mathcal{A}$ *is at least* $\Omega(n^2)$ *bits;*
- *if* $p = O(1)$, $t < 2a(G)$, *the space usage of* $\mathcal{A}$ *is at least* $\Omega(n)$ *bits.*

***Applications.*** Partitioning the edges of a given graph into few forests has several known applications to other graph problems [11,17,18]. In Sect. 4, we will show how to apply Theorem 1 to devise a single-pass streaming algorithm to obtain $O(\log^2 n)$-approximation for the lowest outdegree orientation, $O(\log n)$-approximation for partitioning graphs into the minimum number of planar subgraphs, and $O(a^2(G) \log^2 n)$-approximation for the minimum dominating set. For the minimum dominating set problem, we need to assume that the input stream are organized in the vertex-arrival order; that is, the input stream is a concatenation of adjacency lists of the input graph.

***Notation.*** The input graph $G = (V, E)$ is an $n$-node $m$-edge undirected simple graph whose arboricity is $a(G)$. We use $\alpha$ as an abbreviation of $a(G)$ when the context is clear. By $\{u, v\}$ we denote an undirected edge incident to nodes $u$ and $v$, and by $(u, v)$ we denote a directed edge from node $u$ to node $v$. We define $[n] := \{1, 2, \dots, n\}$ and $[a, b] := \{a, a+1, \dots, b\}$. For any two graphs $G = (V, E)$ and $H = (V, F)$ that share the same node set, $G \cup H = (V, E \cup F)$ and $G \backslash H = (V, E \backslash F)$. A maximal forest $F$ of $G$ is a forest that has the same number of connected components as $G$. When we say $O(s)$ space, we mean the space that can store $O(s)$ edges, i.e. $O(s \log n)$ bits.

***Organization.*** In Sect. 2, we devise single-pass streaming algorithms that partition a given graph into few forests. Then, we prove lower bounds on the space usage of any streaming algorithm that solves this problem in Sect. 3. Finally, in Sect. 4, we apply our algorithmic results to devise single-pass streaming algorithms for some applications.

## 2 Streaming Algorithms

In this section, we devise two single-pass streaming algorithms to partition a given $n$-node undirected simple graph into few forests. One algorithm uses space linear in $n$, and the other uses space sublinear in $n$.

### 2.1 A Linear-Space Algorithm

A key observation for our linear-space algorithm is that, for any maximal forest $F$ of an undirected graph $G$, $F$ contains a large fraction of edges in $G$. The formal statement is given in Lemma 3. Hence, if one iteratively removes a maximal forest from $G$, then the number of iterations can not be very large, compared to the arboricity $\alpha$ of $G$.

**Lemma 3.** *For any undirected graph $G$ whose arboricity is $\alpha$, if $F$ is a maximal forest of an $m'$-edge subgraph $H$ of $G$, i.e. adding any edge in $H \backslash F$ to $F$ makes the resulting $F$ cyclic, then $F$ has at least $\lceil m'/\alpha \rceil$ edges.*

*Proof.* Since $H$ is a subgraph of $G$, the $m'$ edges in $H$ can be partitioned into $\alpha$ forests. By an average argument, we get that some forest $F_{avg}$ of $H$ contains at least $\lceil m'/\alpha \rceil$ edges. Observe that $F_{avg}$ cannot have more edges than $F$. Here is why. Suppose for the contradiction that $F_{avg}$ has more edges than $F$, then it contains fewer connected components than $F$. By the pigeonhole principle, there exist two nodes $x, y \in H$ so that $x, y$ are contained in different connected components in $F$ but in the same connected component in $F_{avg}$. Thus, $F_{avg}$ contains a path $P$ from $x$ and $y$. Not every edge on $P$ connects two nodes in the same connected component in $F$; otherwise, $x$ and $y$ are connected in $F$. Hence, some edge on $P$ can be added to $F$ while retaining $F$ acyclic, contradicting with the maximality of $F$. □

Our algorithm works as follows. Let $G_0 = G$. For every $i \geq 1$, let $F_i$ be some maximal forest of $G_{i-1}$ to be determined later and $G_i$ be the subgraph of $G_{i-1}$ obtained from the removal of edges in $F_i$. Let $G$ be an $n$-node $m$-edge undirected simple graph with arboricity $\alpha$. By Lemma 3, the number of edges in $F_1$ is at least $m/\alpha$, so the number of edges in $G_1$ is at most $m(1 - 1/\alpha)$. Extend this argument for every $i > 1$, we have that the number of edges in $G_i$ is at most

$$m\left(1 - \frac{1}{\alpha}\right)^i \leq me^{-i/\alpha}.$$

Hence, let $t = \alpha\lceil \log m \rceil + 1$, $G_t$ has no edge. In other words, this greedy approach yields a partition of $G$ into $F_1, F_2, \ldots, F_t$ so that $F_i$ for every $i \in [t]$ is a forest. Since $G$ has arboricity $\alpha$, we have:

> The greedy procedure is an $O(\log n)$-approximation algorithm for partitioning an $n$-node undirected simple graph into forests.

**Algorithm 1:** A streaming implementation of the greedy procedure.

```
1 A_i ← ∅ for every i ≥ 1;
2 foreach incoming edge e do
3 |   k ← 1;
4 |   while A_k ∪ {e} contains a cycle do
5 |   |   k ← k + 1;
6 |   end
7 |   A_k ← A_k ∪ {e};
8 |   output (e, k);
9 end
```

The above greedy procedure can be implemented in the semi-streaming model, as shown in Algorithm 1, if the space usage of $A_1, A_2, \ldots, A_t$ can be bounded in $O(n \operatorname{polylog} n)$. For each incoming edge $e \in G_0$, Algorithm 1 attempts to add $e$ to the ***latest*** $A_1$ (i.e. the $A_1$ at the moment when Algorithm 1 processes all the edges preceding $e$ in the arrival order but not yet processes $e$). If the union of the latest $A_1$ and $\{e\}$ does not contain a cycle, then $e$ is added to the latest $A_1$. Note that if $e$ cannot be added to the latest $A_1$, then it cannot be added to any super set of the latest $A_1$, in particular the ***final*** $A_1$, i.e. the $A_1$ at the moment when Algorithm 1 is completely executed. Hence, the final $A_1$ is a maximal forest of $G_0$. So we can define $F_1$ to be the final $A_1$ without violating the requirements. The edges that Algorithm 1 attempts to add to $A_2$ are those in $G_0$ but not in the final $A_1$, or equivalently $G_1$. By a similar argument, the final $A_2$ is a maximal forest of $G_1$, so we can define $F_2$ to be the final $A_2$. Repeating this argument for every $A_i$ $(i > 2)$, we get:

After Algorithm 1 is completely executed, $A_i$ is a maximal forest of $G_{i-1}$ for every $i \in [t]$. Hence, we can define $F_i$ to be the final $A_i$ for every $i \in [t]$ without violating the requirements.

It is impossible to store $A_1, A_2, \ldots, A_t$ entirely using $O(n \operatorname{polylog} n)$ space. Hence, we output the edges in $A_i$ for $i \in [t]$ on the fly without keeping them in memory (Line 8 in Algorithm 1). To execute Algorithm 1 without accessing the entire $A_1, A_2, \ldots, A_t$, we maintain an $O(n \log n)$-bit data structure that supports the following two operations. Unlike the data structure that organizes edges for low-arboricity graphs due to Brodal and Fagerberg [5], our data structure cannot support membership queries, so its space usage can be sublinear in $m$.

1. ACYCLIC$(i, e)$: a query operation that returns TRUE if the union of the latest $A_i$ and edge $e$ does not contain a cycle, or FALSE otherwise. This operation implements Line 4 in Algorithm 1.
2. INSERT$(i, e)$: an update operation that inserts edge $e$ to the latest $A_i$. This operation implements Line 7 in Algorithm 1.

Let $e_1, e_2, \ldots, e_m$ be the arrival order of edges in $G$. For every $s \geq 1$, define $A_{s,i}$ to be the $A_i$ at the moment when Algorithm 1 processes all the edges preceding $e_s$ but not yet processes $e_s$. Let $M_s$ be a maximum-weight spanning forest of the union $U_s$ of $A_{s,1}, A_{s,2}, \ldots, A_{s,t}$ whose edge weights $\omega(e)$ for every edge $e$ in $U_s$ is set as $i$ if $e \in A_{s,i}$. Hence, $M_s$ contains at most $n-1$ edges. Every edge uses $O(\log n)$ bits to store the indices of its end-nodes and $O(\log m)$ bits to store its weight. In total, the space usage of $M_s$ is $O(n \log n)$ bits for every $s \geq 1$. By Lemma 4, one can utilize $M_s$ to implement ACYCLIC$(i, e)$. By Lemma 5, for every $i, s \geq 1$, INSERT$(i, e_s)$ can be performed given access to $M_s$. Thus, we get:

The space usage of Algorithm 1 can be bounded by $O(n \log n)$ bits.

**Lemma 4.** *For every* $s \geq 1, i \in [t]$, *let* $M_{s,i} = \{e' \in M_s : \omega(e') \geq i\}$,

$$A_{s,i} \cup \{e_s\} \text{ contains a cycle iff } M_{s,i} \cup \{e_s\} \text{ contains a cycle.}$$

*Proof.* ($\Rightarrow$) Let the end-nodes of $e_s$ be $x$ and $y$. Since $A_{s,i} \cup \{e_s\}$ contains a cycle, there is a path $P$ in $A_{s,i}$ from $x$ to $y$. We claim that, for every edge $\{u, v\}$ in $P$, nodes $u, v$ are connected in $M_{s,i}$. Suppose that $u, v$ are not connected in $M_{s,i}$, since $M_{s,i}$ is a maximal forest of $\bigcup_{j \geq i} A_{s,j}$, $M_{s,i}$ contains $\{u, v\}$, a contradiction. Thus, there is a path $Q$ from $x$ to $y$ in $M_{s,i}$. $Q \cup \{e_s\}$ gives a cycle in $M_{s,i}$.

($\Leftarrow$) Let the end-nodes of $e_s$ be $x$ and $y$. Since $M_{s,i} \cup \{e_s\}$ contains a cycle, there is a path $P$ in $M_{s,i}$ from $x$ to $y$. We claim that, for every edge $\{u, v\}$ in $P$, nodes $u, v$ are connected in $A_{s,i}$. Suppose that $u, v$ are not connected in $A_{s,i}$, then $\{u, v\}$ is added to $A_{s,j}$ for some $j > i$. This contradicts with that $A_{s,i}$ is a maximal forest of $\bigcup_{j \geq i} A_{s,j}$. Thus, there is a path $Q$ in $A_{s,i}$ from $x$ to $y$. $Q \cup \{e_s\}$ forms a cycle in $A_{s,i}$, as desired. □

**Lemma 5.** *For every* $s \geq 1, i \in [t]$, *a maximum-weight spanning forest of* $M_s \cup \{e_s\}$ *is also a maximum-weight spanning forest of* $\{e_s\} \cup \bigcup_{j \in [t]} A_{s,j}$ *where* $\omega(e_s)$ *is set as* $i$.

*Proof.* This lemma is a special case of Lemma 4.1 in [8]. □

Simple implementations of ACYCLIC$(i, e)$ and INSERT$(i, e)$ take $O(n)$ time for each invocation. By Lemma 4, for every $s \geq 1$ finding the least index $k_s$ so that $A_{s,k_s} \cup \{e_s\}$ is acyclic can be reduced to finding the edge $\ell_s$ with the least weight on the cycle in $M_s \cup \{e_s\}$ and letting $k_s = \omega(\ell_s) + 1$. If such a cycle does not exist, we define $k_s$ as 1. By Lemma 5, the operation INSERT$(k_s, e_s)$ is equivalent to replacing $\ell_s$ in $M_s$ with $e_s$. These operations can be realized in $O(\log n)$ time per edge by the dynamic tree due to Sleator and Tarjan [23]. We remark that:

In total, the running time of Algorithm 1 can be bounded by $O(m \log n)$.

***Our Analysis of the Approximation Factor is Tight.*** We prove this statement by constructing an $n$-node graph $\Gamma_n$ for every $n \geq 1$ whose arboricity $\leq 2$ but the greedy procedure partitions it into $\Omega(\log n)$ forests. A small example $\Gamma_9$ is depicted in Fig. 1.

For general $n$, construct $\Gamma_n$ as follows. Let the nodes be $v_1, v_2, \ldots, v_n$. Connect nodes $v_x$ and $v_y$ with an edge if $x - y = 1$. Let $B_0$ (resp. $R_0$) be the set of the edges added by this rule whose $x$ is even (resp. odd). For each integer $k \geq 1$, connect nodes $v_x$ and $v_y$ with an edge if $x - y = 2^k$ and $x \equiv y \equiv 1 \pmod{2^k}$ and let $B_k$ ($R_k$) be the set of the edges added by this rule whose $y \equiv 1 \pmod{2^{k+1}}$ (resp. $x \equiv 1 \pmod{2^{k+1}}$).

The greedy procedure may output $B_1 \cup R_1$, $B_2 \cup R_2$, $\ldots$, $B_{\log n} \cup R_{\log n}$ as the forests because, for every $i \geq 1$, $B_i \cup R_i$ is a maximal forest in $\bigcup_{j \geq i} B_j \cup R_j$. The remaining to show is that $\bigcup_{i \geq 0} B_i$ is a forest and $\bigcup_{i \geq 0} R_i$ is another, so $\Gamma_n$ has arboricity $\leq 2$. Suppose for contradiction that $\bigcup_{i \geq 0} B_i$ contains a cycle $C$. $C \cap B_0 = \emptyset$ because all edges in $B_0$ have an end-node whose degree in $\bigcup_{i \geq 0} B_i$ is 1. By the same argument, we get $C \cap B_i = \emptyset$ for every $i \geq 1$, a contradiction. Similarly, $\bigcup_{i \geq 0} R_i$ does not contain a cycle.

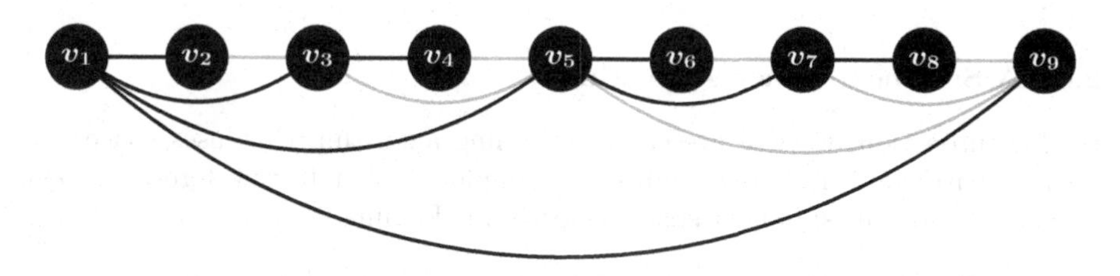


**Fig. 1.** A graph with arboricity 2, in which the set of edges with any one color forms a forest. The greedy procedure may partition this graph into 4 forests because the edges in every row is a maximal forest when the edges in the previous rows are removed.

***Bounding the Approximation Factor by $\alpha + 1$.*** We show that the approximation factor can be bounded by $\alpha + 1$ if the greedy procedure is slightly modified, as below. For $O(1)$-arboricity graphs, the modified algorithm yields an $O(1)$-approximation using $O(n)$ space. This complements the hardness result, Lemma 9, stating that any single-pass $r$-approximation for $r < 2$ streaming algorithm requires $\Omega(n)$-bit space.

We claim that there exists an index $\Delta$ that simultaneously satisfies the following two conditions.

1. For every $i \leq \Delta$, the number of edges in the forest $F_i$ is at least $n/\alpha$.
2. The total number of edges in $F_i$ for all $i > \Delta$ is at most $n$.

Here is why. Since $F_i$ is a maximal forest of $\bigcup_{j\geq i} F_j$, the number of edges in $F_i$ is no less than that in $F_j$ for any $j > i$. The first condition holds. By Lemma 3, if the total number of edges in $F_i$ for $i \in [r,t]$ exceeds $n$, then $F_r$ contains at least $\lceil n/\alpha \rceil$ edges. The second condition holds.

If we output $F_1, F_2, \ldots, F_\Delta$ followed by an optimal partition of $\bigcup_{i>\Delta} F_i$ into forests, then this yields an $(\alpha+1)$-approximation. To see why, observe that $\Delta \leq \alpha^2$ because $G$ contains at most $\alpha(n-1)$ edges, and that any subgraph of $G$ can be partitioned into $\alpha$ forests. Consequently, the approximation factor is $(\alpha^2+\alpha)/\alpha = \alpha+1$. This modification can be realized in the streaming model by buffering the output of the edges in $\bigcup_{i>\Delta} F_i$ (Line 8 in Algorithm 1) and computing an optimal partition for them in memory. It worth noting that this modification can be made without knowing $\Delta$ in advance. One can underestimate $\Delta$ at the beginning of the execution of Algorithm 1 and increase $\Delta$ by 1 every time the $O(n)$-space buffer gets full. To bound the running time in $O(m \log n)$, we use the $(1+\varepsilon)$-approximation algorithm [4] to partition $\bigcup_{j>\Delta} F_j$ into $(1+\varepsilon)\alpha$ forests. Thus, we get:

A modification of Algorithm 1 partitions $\alpha$-arboricity graphs into $(\alpha+1+\varepsilon)\alpha$ forests for any constant $\varepsilon > 0$ in $O(m \log n)$ time using $O(n)$ space.

The highlighted statements in this section together give a proof of Theorem 1.

### 2.2 A Sublinear-Space Algorithm

In this subsection, we will present a streaming algorithm that uses space $s = o(n)$ to partition an $n$-node undirected simple graph into few forests. A key observation for our sublinear-space algorithm is Lemma 6.

**Lemma 6.** *Every $n$-node $m$-edge undirected simple graph $G$ contains a forest of at least $\lceil \sqrt{m} \rceil$ edges.*

*Proof.* Let $C_1, C_2, \ldots, C_t$ be the connected components in $G$. Let $n_i \geq 1$ denote the number of nodes in $C_i$ for every $i \in [t]$. Because $C_i$ consists of $n_i$ nodes and no more than $\binom{n_i}{2}$ edges for every $i \in [t]$, we have:

$$\sum_{i\in[t]} n_i = n \text{ and } \sum_{i\in[t]} \frac{n_i^2 - n_i}{2} \geq m. \tag{1}$$

Our goal is to prove that $G$ has a forest $F$ of many edges. Since $C_i$ for each $i \in [t]$ can contribute $n_i - 1$ edges to $F$, it is equivalent to ***minimize*** $\mu := \sum_{i\in[t]} n_i - 1$ subject to (1).

$$\mu^2 \geq \sum_{i\in[t]} (n_i - 1)^2 = \left(\sum_{i\in[t]} n_i^2 - n_i\right) - \left(\sum_{i\in[t]} n_i - 1\right) \geq 2m - \mu$$

Since $\mu$ is a non-negative integer, $2\mu^2 \geq \mu^2 + \mu \geq 2m$. As a result $\mu \geq \sqrt{m}$. In every case, the number of edges in $F$ is no less than the minimum possible $\sqrt{m}$. $\square$

For every integer $s \in [n]$, we present a streaming algorithm to partition a given graph into forests using $O(s)$ space as follows. Initialize a counter $c$ as 1. The counter $c$ will not exceed $m$ and $m \leq n(n-1)$ for any $n$-node simple graph, so $c$ occupies $O(1)$ space. Fill the memory space with the first $s$ edges in the input stream. By Lemma 6, some of the $s$ edges form a forest $F$ of at least $\lceil \sqrt{s} \rceil$ edges. For each edge $e$ in $F$, output $(e, c)$. Free the space occupied by the edges in $F$. Increase the counter $c$ by 1. Refill the memory space with the subsequent edges in the input stream. This algorithm partitions the input graph into at most $\lceil m/\sqrt{s} \rceil + \alpha$ forests. Since any partition of $G$ consists of at least $\lceil m/n \rceil$ forests, we get an $O(n/\sqrt{s})$-approximation.

## 3 Space Lower Bounds

In this section, we prove two space lower bounds. One is for any single-pass streaming algorithm that partitions $n$-node undirected simple graphs into the minimum number of forests, and the other is for any $p$-pass streaming algorithm that yields a 2-approximation. Both lower bounds are proven by a reduction from some 2-player communication game to the targeted problem. They together gives a proof of Theorem 2.

**Lemma 7.** *Any single-pass streaming algorithm that partitions an $n$-node graph $G$ into $k$ forests so that $k$ equals the minimum possible (i.e. the arboricity of $G$) requires $\Omega(n^2)$-bit space.*

*Proof.* Our proof is a reduction from the INDEX problem, defined below, to partitioning a graph into the minimum number of forests in the streaming model. We assume w.l.o.g. that $n$ is a multiple of 8.

- Input: Alice has a private subset $A$ of $[U]$, and Bob has a private integer $k$ in $[U]$ where $U$ is set as $\binom{n/4}{2}$ in this proof.
- Goal: Determine whether $k \in A$.

Ablayev shows in [1] that:

**Theorem 8. (Ablayev [1]).** *For any 1-way, randomized protocol from Alice to Bob that solves INDEX with success probability at least 2/3, Alice has to send $\Omega(U)$ bits to Bob.*

Here is our reduction. Given $A$, construct an $n/4$-node graph $G_A$ so that $G_A$ contains edge $e$ if and only if $f(e) \in A$ where $f$ is any one-to-one mapping from the $\binom{n/4}{2}$ distinct edges of $G_A$ to $[\binom{n/4}{2}]$. Given $k$, construct $G_B$ as a $3n/4$-node graph whose arboricity is $n/4 + 1$, and that adding any other edge to $G_B$ increases the arboricity by 1. Such a $G_B$ exists because an $3n/4$-node empty graph has arboricity 0, a $3n/4$-node complete graph has arboricity $3n/8$ (due to the Nash-Williams Theorem [22]), and adding an edge to a graph increases its arboricity by at most 1. Let $x, y$ be the end-nodes of $f^{-1}(k)$, and $G_A, G_B$ have only these two nodes in common. We claim that $G_A$ contains the edge $f^{-1}(k)$ if and only if $G_A \cup G_B$ has arboricity $\geq n/4 + 2$. By the definition of $G_B$, one direction ($\Rightarrow$) clearly holds. In what follows, we prove the other direction ($\Leftarrow$). We partition $G_A$ into $A_1, A_2, \ldots, A_{n/4+1}$ so that:

- $A_1$ contains all the edges in $G_A$ incident to node $y$.
- $A_2, A_3, \ldots, A_{n/4+1}$ is any partition of the remaining edges into forests. Such a partition exists because $G_A$ has arboricity at most $n/8$ by the Nash-Williams Theorem.

Then, we partition all edges in $G_B$ into forests $B_1, B_2, \ldots, B_{n/4+1}$. For every $i \in [n/4+1]$, $A_i, B_i$ are acyclic. If $A_i \cup B_i$ contains a cycle, the cycle contains both $x$ and $y$. If the cycle contains $y$, then it has a node of degree $\leq 1$ because edge $\{x, y\}$ does not exist. Hence, $G_A \cup G_B$ has arboricity at most $n/4 + 1$.

To complete the reduction, we place the edges of $G_A$ in the first half of the input stream and the edges of $G_B$ in the second half. Let $\mathcal{A}$ be any single-pass streaming algorithm that uses $s$-bit space. When $\mathcal{A}$ transits between the two halves while scanning over the input stream, at most $s$-bit information is communicated. If $\mathcal{A}$ can partition $G_A \cup G_B$ into the minimum number of forests, then it determines whether $k \in A$, so by Theorem 8 $s = \Omega(n^2)$. □

**Lemma 9.** *Any $p$-pass streaming algorithm that partitions an $n$-node graph $G$ into $k$ forests so that $k$ approximates the minimum possible to within a factor $< 2$ (i.e. $k$ is less than twice the arboricity of $G$) requires $\Omega(n/p)$ bits.*

*Proof.* Our proof is a reduction from the SetDisjontness problem to partitioning a graph into few forests in the streaming model. The SetDisjontness problem is a two-player communication game, defined as follows. We assume w.l.o.g. that $n$ is a multiple of 3.

- Input: Alice has a private $(\alpha U)$-size subset $A$ of $[U]$, and Bob has another private $(\alpha U)$-size subset of $[U]$ where $\alpha$ is some positive constant $< 1/2$ and $U = n/3$ in this proof.
- Goal: Determine whether the intersection $A \cap B$ is an empty set.

Kalyanasundaram and Schnitger in [15] show that:

**Theorem 10.** *No matter which 2-way, multi-round protocol Alice and Bob use to solve the disjointness problem, they have to communicate at least $\Omega(U)$ bits to succeed with probability greater than $2/3$.*

Let $H = (X \cup Y \cup Z, E_A \cup E_B)$ be a tripartite graph so that $X, Y, Z$ are disjoint node subsets and each contains $n/3$ nodes. Let $X = \{x_1, x_2, \ldots, x_{n/3}\}$, $Y = \{y_1, y_2, \ldots, y_{n/3}\}$, and $Z = \{z_1, z_2, \ldots, z_{n/3}\}$. Initially, $E_A = E_B = \emptyset$. For every $i \in A$, add two edges $\{x_i, y_i\}, \{y_i, z_i\}$ to $E_A$. For every $i \in B$, add an edge $\{x_i, z_i\}$ to $E_B$. If $A \cap B = \emptyset$, then $H$ is a union of disjoint paths, so $H$ has arboricity 1. Otherwise $A \cap B \neq \emptyset$, then $H$ contains some triangle, so $H$ has arboricity 2. Hence, any protocol that approximates the arboricity to within a factor of smaller than 2 answers the SETDISJONTESS problem. By Theorem 10, we are done. □

# 4 Applications

In this section, we show how to apply Theorem 1 to devise single-pass streaming algorithms for some applications of partitioning graphs into few forests.

## 4.1 Low Outdegree Orientation

In [17], Kowalik devises $(1+\varepsilon)$-approximation algorithms for any $\varepsilon > 0$ for the lowest outdegree orientation problem for any given $n$-node undirected graph $G$; that is, assign a direction to each edge in $G$ so as to minimize the maximum outdegree in the resulting directed graph. By using Algorithm 1 as a building block, we show that the lowest outdegree orientation problem can be approximated to within a factor of $O(\log^2 n)$. Formally, our result is:

**Corollary 11.** *Let $e_1, e_2, \ldots, e_m$ be the edges of a given $n$-node $m$-edge undirected graph $G$ in an arbitrary order. There exists a single-pass deterministic algorithm that runs in $O(m \log n)$ time and uses $O(n)$ space to output the assigned directions to the $m$ edges so that the maximum outdegree of the directed graph comprised of these directed edges is at most $O(\log^2 n)$ times the minimum possible.*

*Proof.* Algorithm 1 is a single-pass $O(n)$-space deterministic streaming algorithm to partition the edges of $G$ into $t$ forests $F_1, F_2, \ldots, F_t$ so that $t$ is at most $O(\log n)$ times the minimum possible. For every $i \in [t]$, if we root $F_i$ and orient the edges from descendants to ancestors, then for each node $x$ at most one edge in $F_i$ leaves $x$ in the orientation. By [17], this approach gives an $O(\log n)$-approximation for the lowest outdegree orientation problem. However, it cannot be directly implemented in the streaming model. The reason is that edges of $F_i$ for each $i \in [t]$ are outputted on the fly, so Algorithm 1 does not have access to all edges of $F_i$ for each $i \in [t]$. Instead, Algorithm 1 has access to $M_s$. To remedy, we assign the directions to the edges as follows. For every $s \geq 1$, for some $i \in [t]$, let $e_s = \{u, v\}$ and $C_{u,i}$ (resp. $C_{v,i}$) be the connected components in $M_{s,i}$ that contains node $u$ (resp. $v$). When $e$ is added to $M_s$ to join the connected components $C_{u,i}, C_{v,i}$, we assign $e_s$ a direction from $u$ to $v$ iff $C_{u,i}$ has fewer nodes than $C_{v,i}$. In this way, every time an edge in $F_i$ is assigned to a direction that leaves node $x$, the connected component in $M_{s,i}$ that contains $x$ doubles its size. Thus, for any node $x$, for any $i \in [t]$, the orientation in $F_i$ contributes at most $O(\log n)$ out-going edges to $x$. This gives an $O(\log^2 n)$-approximation as desired. □

### 4.2 Partitioning Graphs into Few Planar Subgraphs

The key observation for partitioning graphs into few forests (i.e. Lemma 3) can be extended to other kinds of subgraphs, such as planar subgraphs, bipartite subgraphs, etc. For the subgraphs other than forests, the $O(n)$-space data structure $M_s$, which we use to decide the maximal forest that edge $e_s$ belongs to, may not have a replacement. However, by Algorithm 1 we can still get an analogous result for planar subgraphs:

**Corollary 12.** *Let $e_1, e_2, \ldots, e_m$ be the edges of a given $n$-node undirected simple graph $G$ in an arbitrary order. There exists a single-pass deterministic streaming algorithm that runs in $O(m \log n)$ time and uses $O(n)$-space to output a sequence of tuples $(e'_1, c_1), (e'_2, c_2), \ldots, (e'_m, c_m)$ so that the following three conditions simultaneously hold.*

*(a) $e'_1, e'_2, \ldots, e'_m$ is a permutation of $e_1, e_2, \ldots, e_m$.*
*(b) There exists an integer $t = O(\theta \log n)$, for every $i \in [m]$, $c_i$ is an integer in $[t]$, where $\theta$ denotes the thickness of $G$.*
*(c) For every $i \in [t]$, $\{e_j : j \in [m], c_j = i\}$ is a planar subgraph of $G$.*

*Proof.* Every forest is a planar graph, so the output of Algorithm 1 is a feasible solution for partitioning graphs into planar subgraphs. Because every planar graph has arboricity at most 3 [11], we know $\alpha \leq 3\theta$. Hence, a partition of a graph into $O(\alpha \log n)$ forests is a partition of the graph into $O(\theta \log n)$ planar subgraphs. □

### 4.3 Finding Small Dominating Sets

In [18], Lenzen and Wattenhofer devise an $O(a^2(G))$-approximation algorithm for the minimum dominating set problem. By using Algorithm 1 as a building block, we show that the minimum dominating set can be approximated to within a factor of $O(a^2(G) \log^2 n)$. We have:

**Corollary 13.** *Let $e_1, e_2, \ldots, e_m$ be the edges of a given $n$-node undirected simple graph $G$ in the* **vertex-arrival order**. *There exists a single-pass deterministic streaming algorithm that runs in $O(m \log n)$ time and uses $O(n)$-space to output a dominating set of size $O(a(G)^2 \log^2 n)$ times the minimum possible.*

*Proof.* As the algorithm in the proof of Corollary 11, one can assign a direction to each edge so that the maximum outdegree in the orientation is at most $O(a(G) \log n)$. Since the edges arrive in the vertex-arrival order, for each incoming adjacency list, say of node $x$, if $x$ is not yet dominated, we are able to add every node $y$ that has a directed edge from $x$ to $y$ in the orientation to the dominating set. If no such a node exists, add $x$ to the dominating set. By [18], this yields an $O(a^2(G) \log^2 n)$-approximation. □

**Acknowledgements.** We thank the anonymous reviewers for their helpful comments.

## References


1. Ablayev, F.M.: Lower bounds for one-way probabilistic communication complexity and their application to space complexity. Theor. Comput. Sci. **157**(2), 139–159 (1996)
2. Ahn, K.J., Guha, S., McGregor, A.: Analyzing graph structure via linear measurements. In: SODA, pp. 459–467 (2012)
3. Assadi, S., Chen, Y., Khanna, S.: Sublinear algorithms for $(\Delta + 1)$ vertex coloring. In: SODA, pp. 767–786 (2019)
4. Blumenstock, M., Fischer, F.: A constructive arboricity approximation scheme. In: Chatzigeorgiou, A., et al. (eds.) SOFSEM 2020. LNCS, vol. 12011, pp. 51–63. Springer, Cham (2020). https://doi.org/10.1007/978-3-030-38919-2_5
5. Brodal, G.S., Fagerberg, R.: Dynamic representations of sparse graphs. In: Dehne, F., Sack, J.-R., Gupta, A., Tamassia, R. (eds.) WADS 1999. LNCS, vol. 1663, pp. 342–351. Springer, Heidelberg (1999). https://doi.org/10.1007/3-540-48447-7_34
6. Chang, Y., Farach-Colton, M., Hsu, T., Tsai, M.: Streaming complexity of spanning tree computation. In: STACS. LIPIcs, vol. 154, pp. 34:1–34:19. Schloss Dagstuhl - Leibniz-Zentrum für Informatik (2020)
7. Eden, T., Levi, R., Ron, D.: Testing bounded arboricity. ACM Trans. Algorithms **16**(2), 18:1–18:22 (2020)
8. Eppstein, D., Galil, Z., Italiano, G.F., Nissenzweig, A.: Sparsification - a technique for speeding up dynamic graph algorithms. J. ACM **44**(5), 669–696 (1997)
9. Farach-Colton, M., Tsai, M.-T.: Computing the degeneracy of large graphs. In: Pardo, A., Viola, A. (eds.) LATIN 2014. LNCS, vol. 8392, pp. 250–260. Springer, Heidelberg (2014). https://doi.org/10.1007/978-3-642-54423-1_22
10. Farach-Colton, M., Tsai, M.-T.: Tight approximations of degeneracy in large graphs. In: Kranakis, E., Navarro, G., Chávez, E. (eds.) LATIN 2016. LNCS, vol. 9644, pp. 429–440. Springer, Heidelberg (2016). https://doi.org/10.1007/978-3-662-49529-2_32
11. Chartrand, G., Kronk, H.V., Wall, C.E.: The point-arboricity of a graph. Israel J. Math. **6**, 169–175 (1968)
12. Gabow, H.N., Westermann, H.H.: Forests, frames, and games: algorithms for matroid sums and applications. Algorithmica **7**(5 & 6), 465–497 (1992)
13. Guha, S., McGregor, A., Tench, D.: Vertex and hyperedge connectivity in dynamic graph streams. In: PODS, pp. 241–247 (2015)
14. Harris, D.G., Su, H., Vu, H.T.: On the locality of Nash-Williams forest decomposition and star-forest decomposition. CoRR abs/2009.10761 (2020)
15. Kalyanasundaram, B., Schnitger, G.: The probabilistic communication complexity of set intersection. SIAM J. Discret. Math. **5**(4), 545–557 (1992)
16. Kapralov, M., Lee, Y.T., Musco, C., Musco, C., Sidford, A.: Single pass spectral sparsification in dynamic streams. SIAM J. Comput. **46**(1), 456–477 (2017)
17. Kowalik, Ł: Approximation scheme for lowest outdegree orientation and graph density measures. In: Asano, T. (ed.) ISAAC 2006. LNCS, vol. 4288, pp. 557–566. Springer, Heidelberg (2006). https://doi.org/10.1007/11940128_56
18. Lenzen, C., Wattenhofer, R.: Minimum dominating set approximation in graphs of bounded arboricity. In: Lynch, N.A., Shvartsman, A.A. (eds.) DISC 2010. LNCS, vol. 6343, pp. 510–524. Springer, Heidelberg (2010). https://doi.org/10.1007/978-3-642-15763-9_48
19. McGregor, A.: Graph stream algorithms: a survey. SIGMOD Rec. **43**(1), 9–20 (2014)

20. McGregor, A., Tench, D., Vorotnikova, S., Vu, H.T.: Densest subgraph in dynamic graph streams. In: Italiano, G.F., Pighizzini, G., Sannella, D.T. (eds.) MFCS 2015. LNCS, vol. 9235, pp. 472–482. Springer, Heidelberg (2015). https://doi.org/10.1007/978-3-662-48054-0_39
21. Muthukrishnan, S.: Data streams: algorithms and applications. Found. Trends Theor. Comput. Sci. **1**(2), 117–236 (2005)
22. Nash-Williams, C.S.J.A.: Decomposition of finite graphs into forests. J. Lond. Math. Soc. **39**(1), 12 (1964)
23. Sleator, D.D., Tarjan, R.E.: A data structure for dynamic trees. J. Comput. Syst. Sci. **26**(3), 362–391 (1983)

# The Secretary Problem with Reservation Costs

Elisabet Burjons, Matthias Gehnen(✉), Henri Lotze, Daniel Mock, and Peter Rossmanith

RWTH Aachen University, Aachen, Germany
{burjons,gehnen,lotze,mock,rossmani}@cs.rwth-aachen.de

**Abstract.** We introduce two variants of the secretary problem, where a reservation fee can be paid to keep candidates on a short-list instead of rejecting them on the spot. In the first model, the fee has to be paid only once and keeps the reservation forever. In the second model, the fee has to be paid in every round as long as the reservation is kept. We analyze the competitive ratio for both variants and present optimal, relatively simple strategies.

## 1 Introduction

The secretary problem has been extensively researched in the areas of online algorithms and stopping theory. One way to state the problem is the following: You are presented with $n$ numbers in a random order and have to choose one. What is the best strategy to choose the highest number among them? It is important to know that you have to decide whether to choose a number as soon as it arrives without seeing the following numbers and that you cannot take back your decision. This classic problem was first solved by Lindley [23] and Dynkin [20]. The solution basically is as follows: Look at the first $n/e$ numbers and after that choose the first number that is the biggest one of all numbers seen so far. Then the probability of getting the highest number is asymptotically $1/e$, which is the best possible. Online algorithms are usually analyzed in terms of the competitive ratio, which is the worst case factor between the gain of the optimal solution and the gain of the algorithm over all instances [24], see also [15]. In the secretary problem the gain is either 1 or 0, depending on whether the algorithm has chosen the highest number or not.

### 1.1 Related Work

Today the secretary problem is very well-known and many variants have been considered. For example, instead of hiring just one person we might be looking for $k$ persons. Ideally, the highest numbers will be chosen, but the gain is defined as the number of persons chosen that are among the best $k$ ones. Here—not surprisingly—the competitive ratio will get better and better as $k$ grows. This


C.-Y. Chen et al. (Eds.): COCOON 2021, LNCS 13025, pp. 553–564, 2021.
https://doi.org/10.1007/978-3-030-89543-3_46

so-called $k$-secretary problem was introduced by Kleinberg [21] and has seen a lot of attention, too [1–3,16,18].

While the competitive ratio is the predominant way to analyze online computations, there are also alternative ways to judge such an algorithm. Regret minimization [17,19] was arguably motivated by the fact that the competitive ratio had been criticized for being too pessimistic: Here we minimize the difference between the optimal reward and the actual reward. Another alternative is advice complexity of online algorithms, where we analyze how much information of the future is needed to compensate for the online setting or to achieve a given competitive ratio [4,5,7–14,22]. Another idea to address the problem that an online setting is usually too unrealistic is to change the model. For example, in real life we can often guard ourselves against the unknowns of the future by, e.g., hedging, insurances, or reservations. If we reserve an item for a fee we might not need it after all and forfeit the reservation fee, but it cannot happen anymore that the item is not available because some competitor took it before us. This model has been used, for example, for the online knapsack problem, where items of different sizes have to packed into an knapsack of limited capacity in an online fashion. While the classical knapsack problem is not competitive as an online problem, introducing the possibility of reserving items for a fee makes a big difference even if the reservation fee is relatively high [6].

### 1.2 Our Results

In this paper we are applying the reservation model to the secretary problem. Instead of being forced to hire a person on the spot we have the alternative to ask for a call-back, but we have to pay a fee for this privilege that will be deducted from the gain.

We distinguish two natural variants: In the *reservation per item*-model we assume that a fee reserves an item forever. In the *reservation per step*-model a fee has to be paid to reserve an item for each step in which we like to keep the reservation alive. We analyze both models precisely and prove matching upper and lower bounds for the best possible competitive ratio. Let us say that the reservation cost is $c = \alpha/n$ per item. It turns out that in the *reservation per item*-model, the competitive ratio is basically the same as without reservation if $c$ is asymptotically bigger than $1/e$. Otherwise the competitive ratio beats $1/e$ and gets smaller and smaller with diminishing reservation costs. Theorems 1 and 2 contain the detailed results and Fig. 1 shows a plot. An optimal algorithm is relatively simple and works in three phases: In the first phase it just watches the first $\lfloor \alpha \rfloor$ items.

In the second phase, which takes a certain number of steps, it reserves every item that is the biggest seen so far. Finally, in the third phase, the algorithm looks at the remaining items. As soon as one arrives that is at least as big as the biggest seen so far it chooses that item immediately and stops. Otherwise, at the end of the sequence, the largest reserved item is chosen.

The *reservation per step*-model is different. Here a reservation fee has to be paid for every step in which an item is reserved. A reservation can be dropped

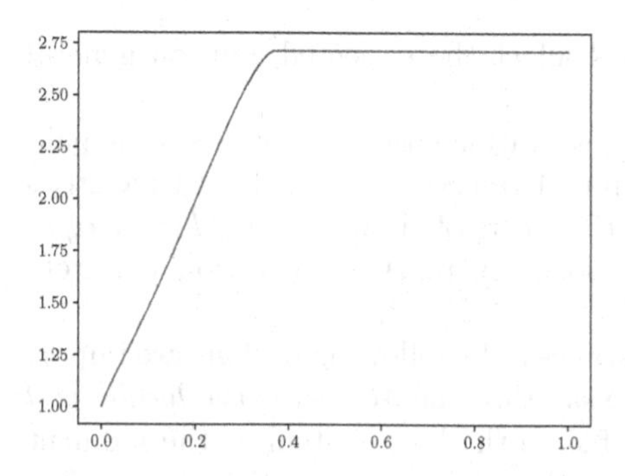


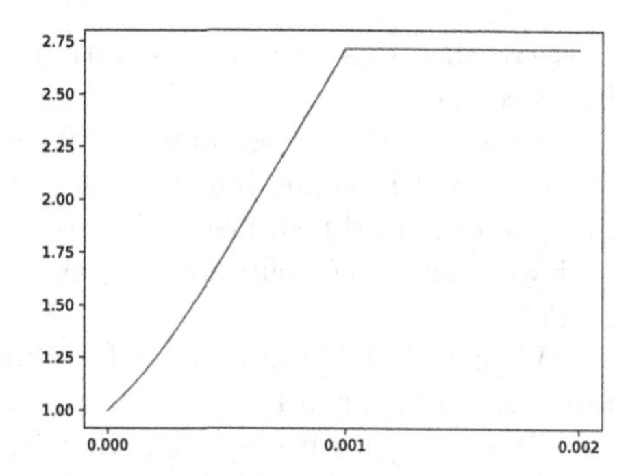


**Fig. 1.** Left: competitive ratio for $n = 1000$ in the reservation per item model. The $x$-axis contains the cost per reservation and the $y$-axis the resulting best possible competitive ratio. Right: the same for the reservation per step model.

anytime and no costs occur from this point onward. Let us call the reservation cost per step $c$. It turns out that for $c > 1/n$ again the classical algorithm without reservations is optimal. Otherwise, an optimal strategy is to ignore a certain number of items (depending on $c$) and then reserving all items that are the best up to this point in time until the last item arrives. Of course, only one reservation is kept at any time. The last item is chosen, if it is the best, and otherwise the reserved item, if it exists. Theorems 4 and 5 contain the details and Fig. 1 an example.

## 1.3 Preliminaries

An instance $I$ of size $n$ for the secretary problem consists of $n$ items $(x_1, \dots, x_n)$ defined by real numbers, such that item $x_i$ is larger than item $x_j$ if $x_i > x_j$. An online algorithm solving the secretary problem wants to choose the maximal item of an instance. The gain of an algorithm $A$ on instance $I$ is $\text{gain}_A(I) = 1$ if the item selected by $A$ is maximal and 0 otherwise.

A general instance $\mathcal{I}$ for the secretary problem consists of $n$ items $\mathcal{I} = \{x_1, \dots, x_n\}$, which are then presented to the algorithm in a random order. We can think of $\mathcal{I}$ as the set of all instances containing the same $n$ items. The gain of an algorithm $A$ on instance $\mathcal{I}$ is the expected gain over all instances in $I$ that are one possible ordering of the items in $\mathcal{I}$. Formally,

$$\text{gain}_A(\mathcal{I}) = \frac{1}{n!} \sum_{\pi \in S_n} \text{gain}_A(x_{\pi(1)}, \dots, x_{\pi(n)}),$$

where $S_n$ is the set of all permutations of the set $\{1, \dots, n\}$.

In the reservation per item model, we consider the reservation cost $c = \alpha/n$ to be paid per reserved item, if an algorithm $A$ reserves $r$ items on an instance $I$, the reservation costs of $A$ on $I$ are $r_A(I) = r \cdot \alpha/n$ and the gain is defined as $\text{gain}_A(I) = 1 - r_A(I)$ if the item selected by $A$ is maximal and $-r_A(I)$ otherwise.

Observe that this modified definition will affect the expected gain on general instances as well.

In the reservation per step model the reservation cost $c = \alpha/n$ has to be paid per step, which means that if an algorithm $A$ reserves an item for $t$ time steps in total for a fixed instance $I$, the reservation costs of $A$ on $I$ are $r_A(I) = t \cdot \alpha/n$, and we adapt the definition of gain analogously to the reservation per item model.

When we talk about items in fixed instances, the following notions are important. On an instance $I = (x_1, \ldots, x_n)$, we say that the item $x_i$ is the *locally best item* if $x_i \geq x_j$ for every $j < i$, that is, if $x_i$ is the largest item at the moment when it is presented. Analogously, we say that an item $x_i$ is the *globally best item* (gbi) if $x_i \geq x_j$ for every $1 \leq j \leq n$. Finally, we say that an item $x_i$ is the *locally best item after* $x_k$ if $x_i \geq x_j$ for every $j \leq i$ and $i > k$. If an item is the *first locally best item after* $x_k$, it additionally means that there is an item $x_l$ with $l \leq k$, such that $x_l \geq x_j$ for every $k < j < i$.

## 2 Reservation per Item

First, we make two key observations that are well known. They can be found for example in the first introduction of the marriage problem [23].

After that we will define an algorithm $A$ and then prove that $A$ is optimal for the reservation per item model.

**Lemma 1.** *Let $I = \{x_1, \ldots, x_n\}$ be an instance for the online secretary problem with $n$ items ordered at random. An algorithm that accepts the first occurring locally best item after the $k$-th request will accept the globally best item with a probability of*

$$F(k) := \frac{k}{n} \sum_{i=k}^{n-1} \frac{1}{i} = \frac{k}{n} \ln\left(\frac{n}{k}\right) + O(1/n).$$

**Lemma 2.** *The probability $F(k) = \frac{k}{n} \sum_{i=k}^{n-1} \frac{1}{i}$ is maximal if $k$ is the biggest integer for which $\sum_{i=k}^{n-1} \frac{1}{i} > 1$. Let $m$ be the integer for which $F(m)$ is maximized, then $F(m) \geq \frac{m}{n}$, $F(m) = \frac{m}{n} + O(1/n)$, and $m = n/e + O(1)$.*

Now we are able to present four necessary conditions that an optimal algorithm must fulfill. Note that every item must be accepted or reserved or neither of them. In the latter case we will also write that an item is ignored. In any case, the algorithm remembers the value of the item.

**Lemma 3 (Condition 1).** *An optimal algorithm must ignore an item if it is neither a locally best item nor the last item.*

*Proof.* Assume that an algorithm $A$ reserves items that are not locally best items on some instance $I$. We can define an algorithm $A'$ with a better expected gain as follows. $A'$ simulates $A$ but does not reserve or accept any item if a better item has been presented, unless the item presented is the last one. Let us look at the gain function. Because $A'$ follows the same strategy as $A$, it accepts exactly the same item, or the last item if the item accepted by $A$ was not optimal, thus if we ignore the reservation costs $r(A)$ and $r(A')$,

$$\mathrm{gain}_A(I) + r(A) = \mathrm{gain}_{A'}(I) + r(A') ,$$

but by definition of $A'$, $r(A') < r(A)$, as it reserves fewer items. Thus, $\mathrm{gain}_A(I) < \mathrm{gain}_{A'}(I)$ on any instance $I$, and $A$ is not optimal. □

**Lemma 4 (Condition 2).** *An optimal algorithm must ignore a locally best item $x_k$, if $k < \alpha$ and $k < m$, where $m$ is the integer for which $F(m)$ is maximized.*

*Proof.* Assume that an algorithm $A$ reserves an item $x_k$ with $k < \alpha$ and $k < m$. The probability that $k$ is the globally best one is $k/n$, which is less than the reservation cost $\alpha/n$. Therefore, the expected gain of reserving this particular item is $k/n - \alpha/n < 0$.

We define algorithm $A'$ that just differs from $A$ by ignoring any reservations for element $x_k$, when we look at the overall gain, if we take into account the linearity of expectations,

$$\mathrm{gain}_{A'}(I) = \mathrm{gain}_A(I) - \frac{k}{n} + \frac{\alpha}{n} > \mathrm{gain}_A(I) .$$

Thus, algorithm $A$ is not optimal.

If $A$ takes item $x_k$ instead, the expected gain is still $k/n$ which is lower than an algorithm that took the locally best item after $m$ with expected gain $F(m) > m/n$ by Lemma 2. Thus the optimal online algorithm for the secretary problem without reservation has a better gain, which means that $A$ is not optimal. □

Let $r$ denote the biggest integer such that $F(r) > c$. Note that $r$ is only well defined if there is such an $r$, which is not always the case. Asymptotically, if $c$ is smaller than $n/e$ then $r$ exists, and otherwise it does not.

**Lemma 5 (Condition 3).** *An optimal algorithm must accept a locally best item $x_k$, if $k > r$ and $k > m$, or, if $r$ is not well defined, if $k > m$.*

*Proof.* This can be proven by induction on the placement of item $x_k$ with the second to last item being the base case. Assume that $x_{n-1}$ is a locally best item, and $r < n-1$, which means that $c < F(n-1) = 1/n$. The probability that $x_{n-1}$ is also the gbi is $n - 1/n$.

An algorithm $A$ accepting $x_{n-1}$ given that it is the locally best item on instance $I$, would have an expected gain

$$\mathrm{gain}_A(I) = \frac{n-1}{n} - r_A(I).$$

An algorithm $A'$ behaving identically as $A$ but reserving $x_{n-1}$ would have an expected gain

$$\mathrm{gain}_{A'}(I) = 1 - r_A(I) - c.$$

Observe that $\frac{n-1}{n} > 1 - c$ when $c > \frac{1}{n}$, thus $A$ is a better algorithm.

An algorithm $A''$ behaving identically as $A$ but rejecting $x_{n-1}$ would have an expected gain

$$\text{gain}_{A''}(I) = \frac{1}{n} - r_A(I).$$

Here it is clear that $\frac{n-1}{n} > \frac{1}{n}$, thus $A$ is a better algorithm.

Assume that an item $x_k$ with $k < n-1$ is a locally best item and either $k > r$ or, $r$ is not defined but $k > m$. As an induction hypothesis assume that an optimal algorithm must accept any locally best item $x_{k'}$ with $k' > k$.

The probability that item $x_k$ is the gbi given that it is a locally best item is $k/n$. If an algorithm $A$ accepts $x_k$, its gain is

$$\text{gain}_A(I) = \frac{k}{n} - r_A(I).$$

An algorithm $A'$ behaving identically as $A$ but reserving $x_k$ would need to take the locally best item after $x_k$ to be optimal, but the probability that the best item is the locally best item arriving after $k$ is $F(k)$. Hence, $A'$ would have an expected gain

$$\text{gain}_{A'}(I) = \frac{k}{n} + F(k) - r_A(I) - c.$$

Observe that $F(k) < c$ because $k < r$ and $k < m$ and thus $A$ is a better algorithm.

An algorithm $A''$ behaving identically as $A$ but rejecting $x_k$ would still have to accept the next locally best item by induction hypothesis, thus it would have an expected gain

$$\text{gain}_{A''}(I) = F(k) - r_A(I).$$

Here it is clear that $F(k) < \frac{k}{n}$, as $k > m$, thus $A$ is, yet again, a better algorithm.

So the best an algorithm can do is to accept the current item. □

**Lemma 6 (Condition 4).** *If $r$ exists, an optimal algorithm must reserve an item $x_k$ if it is a locally best item and $k > \alpha$ or if $k > m$ and $k \leq r$.*

*Proof.* First, we take a look on the case that $\alpha < k < m$.

Let $x_k$ be a locally best item with $\alpha < k < m$. Let us consider an algorithm $A$ that ignores $x_k$ and has a gain of $\text{gain}_A(I)$. An algorithm $A'$ that proceeds identically but reserves $x_k$ will have a gain of $\text{gain}_{A'}(I) = \text{gain}_A(I) + \frac{k}{n} - c$, as the probability of the globally best item being $x_k$—given that it is the locally best item—is $k/n$. Thus $A$ is not optimal

Consider now an algorithm $A$ that accepts $x_k$. Its gain is $\text{gain}_A(I) = k/n - r_A(I)$. We know that the optimal online algorithm for the secretary problem achieves an expected gain of $F(m) > m/n$ by Lemma 2. Thus, $A$ is not optimal.

Now consider the case $m \leq k \leq r$ and that $x_k$ is a locally best item.

An algorithm $A$ that accepts $x_k$ has a gain of $\text{gain}_A(I) = k/n - r_A(I)$. An algorithm $A'$ reserving the item and accepting the next locally best item will have a gain of $\text{gain}_{A'}(I) = k/n + F(k) - c - r_A(I)$. But by construction $F(k) \geq c$, thus $A'$ performs better than $A$. Lastly, an algorithm $A''$ that proceeds exactly as $A'$ but ignores $x_k$ has a gain of $\text{gain}_{A''}(I) = F(k) - c - r_A(I)$. Trivially $A'$ is a better algorithm. □

**Theorem 1 (Optimal strategy).** *If $c > F(m)$, then the classic strategy without reservation is optimal.*

*If $c \leq F(m)$, then the following strategy is optimal:*

- *If an item $x_k$ is either at a position $k < \alpha$ or not a locally best item, it will be ignored.*
- *If an item $x_k$ is a locally best item and $\alpha \leq k \leq r$, it will be reserved.*
- *If an item is at a position $k > r$ and a locally best item, it will be accepted.*

*If no item was accepted, the strategy either accepts the best reserved item (if there is one) or otherwise the last item.*

*Proof.* Observe that if $c > F(m)$, $r$ is not defined. Thus Conditions 1 to 3 describe exactly the classic online strategy. If $c \leq F(m)$ then $r$ is defined, and the second algorithm follows Conditions 1 to 4. Moreover, the conditions describe strict boundaries for a strategy. If an algorithm proceeds differently it will break one of the conditions. □

**Theorem 2 (Expected gain).** *Given a size $n$ and reservation costs $c = \alpha/n$, an optimal algorithm has the following properties:*

- *If $c > F(m)$, then the expected gain is $F(m)$.*
- *If $c \leq F(m)$, then the expected gain is $\frac{r - \lceil\alpha\rceil + 1}{n} - \sum_{i=\lceil\alpha\rceil}^{r} \frac{c}{i} + F(r)$.*

*Proof.* The expected gain for the optimal strategy is the following:

If $c > F(m)$, we are in the classical online case, the algorithm will pick the locally best item after item $m$, thus the expected gain is $F(m)$.

If $c \leq F(m)$, then the strategy will reserve items. The expected gain can be divided in three parts: If the globally best item is $x_k$ then, if $k < \alpha$ the gain is 0, if $\alpha \leq k \leq r$ the gain is 1, and this happens with probability $(r - \lceil\alpha\rceil + 1)/n$ and with an expected reservation cost of $c \cdot \sum_{i=\lceil\alpha\rceil}^{r-1} \frac{1}{i}$. Finally, if $k > r$ the expected gain is $F(r)$. Thus, the total expected gain is

$$\text{gain}_A(\mathcal{I}) = 0 + \frac{r - \lceil\alpha\rceil + 1}{n} - c \cdot \sum_{i=\lceil\alpha\rceil}^{r-1} \frac{1}{i} + F(r). \tag{1}$$

□

Theorem 2 expresses the expected gain as an exact, but complicated and not closed formula. If the reservation costs are small enough, the following theorem provides a much nicer estimate.

**Theorem 3.** *The expected gain for a given* $c \leq F(m)$ *is*

$$e^{-c} + c\ln(c) + O(1/n) + O(c^2).$$

*In particular, for* $c = 1/n$ *the gain is* $1 - \ln(n)/n + O(1/n)$.

*Proof.* The only complicated term in the sum presented in Theorem 2 is $r$. It is defined via

$$c = \frac{r}{n}(H_n - H_r) = \frac{r}{n}(\ln(n) - \ln(r) + O(1/r)) = \frac{r}{n}\ln\left(\frac{n}{r}\right) + O(1/n).$$

With the Lambert $W$ function we can solve this equation for $r/n$:

$$\frac{r}{n} = e^{W(-c+O(1/n))} = e^{-c+O(c^2)+O(1/n)} = e^{-c}(1 + O(c^2) + O(1/n))$$

We get rid of the $W$-function by a Taylor approximation with $W'(0) = 1$. Our goal is to approximate the gain found in (1) with an additive error term of $O(1/n)$. In addition to that we have $\lceil \alpha \rceil / n = c + O(1/n)$, the sum is $\ln(r) - \ln(a)$ and $\ln(r) = -c + \ln(n) + O(1/n)$, and finally $F(r) = c + O(1/n)$. Note that $\ln(n) - \ln(\alpha) = \ln(c)$. Altogether this yields the bound stated in the theorem. □

## 3 Reserving Cost for Every Step

Here we have the possibility to reserve an item for one step from the position $k$ to $k+1$. In every step, we can decide if we want to reserve the item for one more step. The reservation costs are $c$ for every reservation.

Again, the optimal strategy can be given by showing that every deviation cannot be optimal:

**Lemma 7 (Condition 1).** *An optimal algorithm must ignore an item, if it is neither a locally best item nor the last item.*

*Proof.* Analogous to Lemma 3. □

**Lemma 8 (Condition 2).** *If* $c > 1/n$, *then no optimal algorithm reserves any item. Therefore, the optimal strategy is the same as in the basic secretary problem without reservation costs.*

*Proof.* We prove this by induction over the items in descending order as in Lemma 5.

If $x_n$ is the globally best item, it is clear that only accepting the current or reserved item is optimal.

For the induction step, consider a locally best item $x_k$. An algorithm $A$ accepting $x_k$ gets the globally best item with probability $\frac{k}{n}$, thus it has a gain of $\text{gain}_A(I) = \frac{k}{n}$. An alternative algorithm $A'$ reserving it would cause additional costs of $c$. From the induction hypothesis we know that in the next step the item will be accepted. Hence, $\text{gain}_{A'}(I) = \frac{k+1}{n} - c$ and $A'$ is not optimal. □

**Lemma 9 (Condition 3).** *If $c < 1/n$, an optimal algorithm does not accept an item before the last one.*

*Proof.* An algorithm $A$ accepting a locally best item at position $k$ has $\text{gain}_A(I) = \frac{k}{n} - r_A(I)$. An algorithm $A'$ reserving this item and accepting in the next step has $\text{gain}_{A'}(I) = \frac{k+1}{n} - c - r_A(I) > \frac{k}{n} - r_A(I)$. □

**Lemma 10 (Condition 4).** *An optimal algorithm with $c \leq 1/n$ reserves the current locally best item from a point $\hat{k}(c, n)$ until the last step.*

*Proof.* There are only three strategies that optimal algorithms can follow at this point. An algorithm $A$ can reserve the current locally best item from a specific point $k$ until the end. Alternatively, an algorithm $A'$ will reserve from point $k_1$ until point $k_2$ and then reject the reserved item and then perhaps at a future point $k_3$ reserve the next locally best item until the end. Finally, an algorithm $A''$ may refuse to reserve items.

Algorithm $A''$ cannot be optimal because, by Lemma 9, $A''$ cannot accept any item before the last step. Thus, $\text{gain}_{A''}(I) = 1/n$, which is not optimal.

Algorithm $A'$ on the other hand, cannot be optimal because an equivalent algorithm that only reserves items from $k_3$ on will have the exact same success chance and lower reservation costs.

Thus, we are left to analyze the gain of algorithm $A$. The probability that the globally best item occurs after step $k$ is $(n-k)/n$ and the expected cost of reserving the next locally best item until the end is

$$c \cdot \sum_{i=k+1}^{n} \Pr(i \text{ is first lbi after } k) \cdot (n-i) = c \cdot \sum_{i=k+1}^{n} \frac{n-i}{i} \frac{k}{i-1}$$
$$= c\left(n - k - k \ln\left(\frac{n}{k}\right) + O(1/k)\right). \quad (2)$$

This means that $A$ would have an expected gain of

$$\text{gain}_A(\mathcal{I}) = \frac{n-k}{n} - c \cdot \sum_{i=k+1}^{n} \frac{n-i}{i} \frac{k}{i-1}. \quad (3)$$

This expression is maximized for one value $k$ (with a given $n$ and $c$), since $\frac{n-k}{n}$ decreases linearly with growing $k$, while the sum increases with growing $k$.

We can define $\hat{k}(n, c)$ to be the integer $k$ that maximizes $\text{gain}_A(\mathcal{I})$. Because all other strategies are worse, this algorithm must be the optimal one. □

From these four conditions, we get an optimal online algorithm for the secretary problem in the reservation per step model.

**Theorem 4 (Optimal strategy for reservation per step).** *If $c > 1/n$ then the strategy for the basic secretary problem without reservation is optimal. If $c \leq 1/n$ there exists a $\hat{k}(c, n)$ such that the optimal strategy is to keep the locally best item after $x_{\hat{k}(c,n)}$ reserved and accept only after the last step.*

*Proof.* In case of $c > 1/n$, an optimal strategy cannot reserve items (Condition 2). Therefore the best strategy without a reservation is optimal here.

If $c \leq 1/n$, an optimal strategy will start to reserve items at some point and will not accept an item until the end (Condition 3 and 4). The optimal point for reserving items is $\hat{k}$ (Condition 4). Every item that is not a locally optimal item will be ignored (Condition 1). □

**Theorem 5 (Expected gain).** *If $c > 1/n$, the expected gain is $F(m) = 1/e + O(1/n)$. If $c \leq 1/n$, the expected gain is*

$$\frac{n-\hat{k}}{n} - c \cdot \sum_{i=\hat{k}+1}^{n} \frac{n-i}{i} \frac{\hat{k}}{i-1}$$

*where $\hat{k}$ is chosen such that the expression gets maximal.*

*Proof.* If $c > 1/n$, the optimal algorithm does not reserve any item. So the strategy and the expected gain is the same as in the case in Sect. 2 where reserving items was not optimal.

If $c \leq 1/n$, then the expected gain is calculated as in Condition 4. □

**Theorem 6.** *The expected gain for a given $c \leq \frac{1}{n}$ is*

$$1 - cn(1 - e^{-1/cn}) + O(1/n).$$

*In particular, for $c = 1/n$ the gain becomes $1/e + O(1/n)$.*

*Proof.* Using methods from calculus we find that the optimal $\hat{k}$ is $\hat{k} = ne^{-1/cn}(1 + O(1/\hat{k}))$. As $e^{-1/cn}$ is between $1/e$ and 1, we know that $\hat{k} = \Theta(n)$ and therefore $\hat{k} = ne^{-1/cn}(1+O(1/n))$. Inserting this $\hat{k}$ into the estimate for the gain using a combination of (2) and (3) yields the result.

# 4 Conclusion

We found two optimal online algorithms for the basic secretary model with reservations. It seems that the reservation model might also make sense for other variations of the secretary problem. You might, for example, want to look at the variant where you also get more than zero gain if you pick the second or third best candidate.

**Acknowledgement.** We like to thank our anonymous referees for their useful comments that helped to improve the exposition of the paper.

# References

1. Albers, S., Ladewig, L.: New results for the $k$-secretary problem. In: 30th International Symposium on Algorithms and Computation, ISAAC. LIPIcs, vol. 149, pp. 18:1–18:19 (2019). https://doi.org/10.4230/LIPIcs.ISAAC.2019.18

2. Babaioff, M., Immorlica, N., Kempe, D., Kleinberg, R.: A knapsack secretary problem with applications. In: Charikar, M., Jansen, K., Reingold, O., Rolim, J.D.P. (eds.) APPROX/RANDOM -2007. LNCS, vol. 4627, pp. 16–28. Springer, Heidelberg (2007). https://doi.org/10.1007/978-3-540-74208-1_2
3. Babaioff, M., Immorlica, N., Kempe, D., Kleinberg, R.: Online auctions and generalized secretary problems. SIGecom Exch. **7**(2), (2008). https://doi.org/10.1145/1399589.1399596
4. Bianchi, M.P., Böckenhauer, H.-J., Brülisauer, T., Komm, D., Palano, B.: Online minimum spanning tree with advice. Int. J. Found. Comput. Sci. **29**(4), 505–527 (2018). https://doi.org/10.1142/S0129054118410034
5. Bianchi, M.P., Böckenhauer, H.-J., Hromkovič, J., Krug, S., Steffen, B.: On the advice complexity of the online $L(2,1)$-coloring problem on paths and cycles. Theor. Comput. Sci. **554**, 22–39 (2014). https://doi.org/10.1016/j.tcs.2014.06.027
6. Böckenhauer, H.-J., Burjons, E., Hromkovič, J., Lotze, H., Rossmanith, P.: Online simple knapsack with reservation costs. In: 38th International Symposium on Theoretical Aspects of Computer Science, STACS 2021. LIPIcs, vol. 187, pp. 16:1–16:18 (2021). https://doi.org/10.4230/LIPIcs.STACS.2021.16
7. Böckenhauer, H.-J., Dobson, R., Krug, S., Steinhöfel, K.: On energy-efficient computations with advice. In: Xu, D., Du, D., Du, D. (eds.) COCOON 2015. LNCS, vol. 9198, pp. 747–758. Springer, Cham (2015). https://doi.org/10.1007/978-3-319-21398-9_58
8. Böckenhauer, H.-J., Fuchs, J., Unger, W.: Exploring sparse graphs with advice (extended abstract). In: Epstein, L., Erlebach, T. (eds.) WAOA 2018. LNCS, vol. 11312, pp. 102–117. Springer, Cham (2018). https://doi.org/10.1007/978-3-030-04693-4_7
9. Böckenhauer, H.-J., Hromkovič, J., Komm, D., Královič, R., Rossmanith, P.: On the power of randomness versus advice in online computation. In: Bordihn, H., Kutrib, M., Truthe, B. (eds.) Languages Alive. LNCS, vol. 7300, pp. 30–43. Springer, Heidelberg (2012). https://doi.org/10.1007/978-3-642-31644-9_2
10. Böckenhauer, H.-J., Hromkovič, J., Komm, D., Krug, S., Smula, J., Sprock, A.: The string guessing problem as a method to prove lower bounds on the advice complexity. Theor. Comput. Sci. **554**, 95–108 (2014). https://doi.org/10.1016/j.tcs.2014.06.006
11. Böckenhauer, H.-J., Hromkovič, J., Krug, S., Unger, W.: On the advice complexity of the online dominating set problem. Theor. Comput. Sci. **862**, 81–96 (2021). https://doi.org/10.1016/j.tcs.2021.01.022
12. Böckenhauer, H.-J., Komm, D., Královič, R., Královič, R.: On the advice complexity of the k-server problem. In: Aceto, L., Henzinger, M., Sgall, J. (eds.) ICALP 2011. LNCS, vol. 6755, pp. 207–218. Springer, Heidelberg (2011). https://doi.org/10.1007/978-3-642-22006-7_18
13. Böckenhauer, H.-J., Komm, D., Královič, R., Královič, R., Mömke, T.: Online algorithms with advice: the tape model. Inf. Comput. **254**, 59–83 (2017). https://doi.org/10.1016/j.ic.2017.03.001
14. Böckenhauer, H.-J., Komm, D., Královič, R., Rossmanith, P.: The online knapsack problem: advice and randomization. Theor. Comput. Sci. **527**, 61–72 (2014). https://doi.org/10.1016/j.tcs.2014.01.027
15. Borodin, A., El-Yaniv, R.: Online Computation and Competitive Analysis. Cambridge University Press, Cambridge (1998)
16. Buchbinder, N., Jain, K., Singh, M.: Secretary problems via linear programming. Math. Oper. Res. **39**(1), 190–206 (2014). https://doi.org/10.1287/moor.2013.0604

17. Cesa-Bianchi, N., Lugosi, G.: Prediction, Learning, and Games. Cambridge University Press, Cambridge (2006). https://doi.org/10.1017/CBO9780511546921
18. Hubert Chan, T.-H., Chen, F., Jiang, S.H.-C.: Revealing optimal thresholds for generalized secretary problem via continuous LP: impacts on online $k$-item auction and bipartite $k$-matching with random arrival order. In: Proceedings of the Twenty-Sixth Annual ACM-SIAM Symposium on Discrete Algorithms, SODA 2015, San Diego, CA, USA, 4–6 January 2015, pp. 1169–1188. SIAM (2015). https://doi.org/10.1137/1.9781611973730.78
19. Daniely, A., Mansour, Y.: Competitive ratio vs regret minimization: Achieving the best of both worlds. In: Algorithmic Learning Theory, ALT 2019, 22–24 March 2019, Chicago, Illinois, USA. Proceedings of Machine Learning Research, vol. 98, pp. 333–368. PMLR (2019)
20. Dynkin, E.B.: The optimum choice of the instant for stopping a Markov process. Sov. Math. **4**, 627–629 (1963)
21. Kleinberg, R.D.: A multiple-choice secretary algorithm with applications to online auctions. In: Proceedings of the Sixteenth Annual ACM-SIAM Symposium on Discrete Algorithms, SODA 2005, Vancouver, British Columbia, Canada, 23–25 January 2005, pp. 630–631. SIAM (2005)
22. Komm, D.: An Introduction to Online Computation - Determinism, Randomization, Advice. Texts in Theoretical Computer Science. An EATCS Series, Springer, Cham (2016). https://doi.org/10.1007/978-3-319-42749-2
23. Lindley, D.V.: Dynamic programming and decision theory. J. R. Stat. Soc. Ser. C (Appl. Stat.) **10**(1), 39–51 (1961). http://www.jstor.org/stable/2985407
24. Sleator, D.D., Tarjan, R.E.: Amortized efficiency of list update and paging rules. Commun. ACM **28**(2), 202–208 (1985). https://doi.org/10.1145/2786.2793

# Online Ride-Hitching in UAV Travelling

Songhua Li[1(✉)], Minming Li[1], Lingjie Duan[2], and Victor C. S. Lee[3]

[1] City University of Hong Kong, Kowloon, Hong Kong SAR, China
songhuali3-c@my.cityu.edu.hk, minming.li@cityu.edu.hk

[2] Singapore University of Technology and Design, Singapore, Singapore
lingjie_duan@sutd.edu.sg

[3] The University of Hong Kong, Pok Fu Lam, Hong Kong SAR, China
csvlee@eee.hku.hk

**Abstract.** The unmanned aerial vehicle (UAV) has emerged as a promising solution to provide delivery and other mobile services to customers rapidly, yet it drains its stored energy quickly when travelling on the way and (even if solar-powered) it takes time for charging power on the way before reaching the destination. To address this issue, existing works focus more on UAV's path planning with designated system vehicles providing charging service. However, in some emergency cases and rural areas where system vehicles are not available, public trucks can provide more feasible and cost-saving services and hence a silver lining. In this paper, we explore how a single UAV can save flying distance by exploiting public trucks, to minimize the travel time of the UAV. We give the first theoretical work studying online algorithms for the problem, which guarantees a worst-case performance. We first consider the offline problem knowing future truck trip information far ahead of time. By delicately transforming the problem into a graph satisfying both time and power constraints, we present a shortest-path algorithm that outputs the optimal solution of the problem. Then, we proceed to the online setting where trucks appear in real-time and only inform the UAV of their trip information some certain time $\Delta t$ beforehand. As a benchmark, we propose a well-constructed lower bound that an online algorithm could achieve. We propose an online algorithm MyopicHitching that greedily takes truck trips and an improved algorithm $\Delta t$-Adaptive that further tolerates a waiting time in taking a ride. Our theoretical analysis shows that $\Delta t$-Adaptive is asymptotically optimal in the sense that its ratio approaches the proposed lower bounds as $\Delta t$ increases.

**Keywords:** Ride-hitching · Energy efficiency · Online algorithm

## 1 Introduction

As technologies in navigation and control progress, the application of unmanned aerial vehicle (UAV) in package delivery is proved to be a promising approach. For example, the e-commerce giant Amazon has been pushing to deliver packages to its millions of customers by drones. DHL applies drones to provide fully


C.-Y. Chen et al. (Eds.): COCOON 2021, LNCS 13025, pp. 565–576, 2021.
https://doi.org/10.1007/978-3-030-89543-3_47

autonomous loading and offloading in the last-mile delivery [8], which provides a silver lining in special situations (e.g., the COVID-19 pandemic) with social-distancing. However, due to the nature of the UAV/drone in both the low energy storage and high-rate flying consumption, it quickly drains its stored energy, which limits the delivery range and affects the service effectiveness remarkably. Although the UAV can be solar-powered by utilizing solar radiation as energy [1], this is not sufficient since the charging-rate is not high enough. Fortunately, the UAV is able to dock with road vehicles automatically [16], which makes it possible for UAV to team up with trucks spontaneously and instantaneously for reducing the transportation cost. This is inspired by ride-sharing platforms, for example, GrabHitch allows passengers to hitch rides on the way. However, it is still not clear what ramifications of online ride-hitching are on UAV's energy saving. This paper aims to provide theoretical foundations for online ride-hitching in UAV travelling. We note that the UAV may not catch a truck which is far away from its current location (*spatial issue*), nor wait for a truck for too long (*temporal issue*) as time efficiency is critical in UAV's delivery service.

**Related Works.** We survey relevant researches along two threads. *The first thread* studies UAV's energy-efficiency problem with either routing or speed scheduling optimization. For example, [4] proposed the *looking before crossing* algorithm, which is proved to be optimal for the offline speed scheduling problem under a practical flight energy model. [5] considers an energy-aware path planning algorithm that minimizes energy consumption while satisfying coverage and resolution constraints. Please refer to [2] for a survey work. In contrast, we focus on theoretical issues of the problem especially when truck trips are released in an online fashion. We aim to unveil the adaptability of the "ride-sharing" in UAV travelling in the worst-case scenario, which is usually measured by online algorithms and competitive ratio [7]. *The second thread* focuses on classical combinatorial optimization problems. The $k$-server problem aims to efficiently move $k$ servers to serve a batch of online requests of the metric space [10] such that the total moving distance of all servers is minimized. The famous work function algorithm achieves a competitive ratio of $2k-1$ on general metrics and hence is optimal for 1-server problem on the line [11]. When the server moves in constant velocity, the work function algorithm is optimal in achieving minimum completion time of serving all the online requests. A variant of the 1-server problem is the online repairman problem [9] which asks for a tour that visits a set of online cities in the metric space such that the weighted sum of completion times of the cities is minimized. [9] proposes a $(1+\sqrt{2})^2$-competitive deterministic online algorithm for the general metric spaces and [12] gave an improved 5.429-competitive algorithm for line metrics. A generalized version of the $k$-server problem is the $k$-taxi problem in which each request is represented by a pair $(s,t)$ of two points (including the start point $s$ and end point $t$) instead [13]. In the gas station problem [3], a vehicle with a given tank capacity $U$ and an initial amount $\mu_s$ of gas, can purchase gas at each vertex of a complete graph at a certain price. And the objective of [3] is to find the cheapest way from a given start node $s$ to a given target node $t$ of the graph. Note that the gas station problem involves neither the time constraint on vertices nor restrictions in the set of visited vertices. Another related problem is the online maximum $k$-interval coverage problem which aims to select $k$ online sub-intervals (i.e., truck trips) such that the total covered length of a target interval (i.e., the UAV's path) is maximized [14]. In contrast, the problem studied in this paper is more complicated since one has to face both the *spatial* and the *temporal issues* simultaneously and the power constraint of UAV travelling is further involved.

Main contributions of this paper are summarized below.

- We are the first to study UAV's traveling problem by hitching truck rides in an online setting, for the purpose of minimizing the UAV's travel time. Comprehensively, we investigate different cases according to how early (i.e., $\Delta t$) a truck should inform the UAV of its trip before the departure.
- For the offline version of the problem, we give a graph-based optimal solution. Since it is intricate to capture both power and time constraints in mapping truck trips to nodes of a graph, we delicately construct the graph by screening unnecessary trucks iteratively, which is on top of some characteristics of the problem. Based on this, we find an optimal solution in $O(n^2)$ time.
- As a benchmark, we construct lower bounds on the competitive ratio for any online algorithms, by considering different time gap $\Delta t$ between the start time and release time.
- We show that a simple myopic algorithm (where the UAV flies forward constantly by default until using up its stored energy, and myopically accepts as many rides as possible halfway) has a defect, which can be easily exploited by the adversary, leading to negligible energy saved from taking rides. To fix this defect, we propose a $\Delta t$-ADAPTIVE algorithm by tolerating a waiting time at most $\frac{\Delta t}{2}$ in taking each ride, which achieves a provable competitive ratio very close to the lower bound.

Due to space constraints, some results and proofs are deferred to the full version [15].

## 2 Problem Formulation

We consider the following problem: the UAV, which is at its origin $O$, is supposed to move to its destination $A$ as early as possible, in which the path length $|OA| = a$. The UAV has a low charging rate $\alpha$ per unit time and a high power-consuming rate $\beta$ ($> \alpha$) per unit time. Initially at time 0, the UAV stores an amount $P_0$ of energy (which is small and could be zero) and it flies at its maximum velocity[1] $v_0$. To avoid the trivial case that the UAV directly flies to the destination by charging on the way, we assume the UAV has insufficient energy $P_0$ ($< \frac{(\beta-\alpha)a}{v_0}$). That is, the UAV needs to hitch truck rides to save flying distance or charge for sufficiently large amount of time to fly to the destination.

[1] Considering the dominant travel time, we assume the flying velocity as a constant as in [5,6].

Along the path $OA$ of the UAV, a sequence $\mathbb{V} = \{V_1, V_2, ..., V_n\}$ of $n \in \mathbb{N}^+$ trucks will be released one by one to potentially offer rides to the UAV. Each truck $V_i = (r_i, t_i, o_i, d_i, v_i) \in \mathbb{V}$ releases its trip information to the UAV at time $r_i$ and departs from its origin $o_i$ at time $t_i$ ($\geq r_i$) with a constant velocity $v_i$ to its destination $d_i$. Further, we denote $\Delta t_i = t_i - r_i$ as the time gap between the start time $t_i$ and the release time $r_i$ of a truck ride $V_i$. The UAV is not informed of each ride $V_i \in \mathbb{V}$ until its release time $r_i$ (when the truck's schedule is determined) and needs to determine whether to accept/catch or reject $V_i$ irrevocably at $r_i$. The *objective* is to minimize the UAV's travel time to the destination $A$ (or equivalent, the arrival time at $A$) by using online truck rides to save energy halfway. Key notations of this paper are given below in Table 1.

**Table 1.** Notations in this paper.

| Notations | Physical meanings |
|---|---|
| $[0, a]$ | The line segment representing the UAV's route |
| $\mathbb{V}_i = \{V_1, V_2, ..., V_i\}$ | The sequence of the first $i$ rides released, particularly, $\mathbb{V}_n = \mathbb{V}$ |
| $V_i = (r_i, t_i, o_i, d_i, v_i)$ | The $i$th released ride, with its release time $r_i$, departure time $t_i$, start location $o_i$, end location $d_i$ and flying velocity $v_i$ |
| $\Delta t_i = t_i - r_i$ | The time gap between the start time $t_i$ and the release time $r_i$ of a ride $V_i$ |
| $v = \min_{V_i \in \mathbb{V}} v_i$ | The smallest possible velocity of a truck/ride |
| $P_0$ | The initial power that the UAV contains at time $t_0 = 0$ |
| $P_i$ | The power that the UAV contains at the start time of the $i$th ride taken by the UAV |
| $\beta$ | The UAV's power-consumption-rate for flying |
| $\alpha$ | The UAV's recharging-rate, which satisfies $\alpha < \beta$; |
| $\xi(U)$ and $\xi(V_{\lvert U\rvert})$ | The arrival time (or the overall travel time) of the UAV to the target $A$, given the set $U$ of rides taken by the UAV and the last ride $V_{\lvert U\rvert}$ taken by the UAV, respectively, as formally defined later in (9) and (10) |

Suppose that $U = (V_1, \cdots, V_{|U|})$ is the sequence of rides to be taken by the UAV. Denote $P_i$ as the power that the UAV contains at the start time $t_i$ of the $i$th taken ride $V_i$. Now, we formally formulate *our model* as the following mathematical problem (2)–(8). where the objective (2) is to minimize the UAV's travel time from $O$ to $A$, which is according to the following Proposition 1 where we discuss the physical meaning of the two terms of (2) and why we take the maximizing operation between the two. Constraints (3)–(4) indicate the power $P_{i+1}$ that the UAV contains at the start time $t_{i+1}$ of $V_{i+1}$ by transferring from $V_i$, which can be calculated by the following *power transfer function* in (1).

$$\mathrm{PTF}(P_i, V_i, V_{i+1}) = P_i + (t_{i+1} - t_i) \cdot \alpha - \frac{|o_{i+1} - d_i|\beta}{v_0} \tag{1}$$

After leaving each truck $V_i$, for $i \in \{1, \cdots, |U|-1\}$, note that the UAV needs to have enough energy to fly to the following truck $V_{i+1}$'s start location, which leads to the *power compatibility* constraint (5) as $\text{PTF}(P_i, V_i, V_{i+1}) \geq 0$; and it also needs to catch $V_{i+1}$'s start time, which is reflected by the *time compatibility* constraints (6)–(7). When the inequality in constraint (7) holds strictly, the UAV needs to stop at the roadside to wait for $V_{i+1}$'s departure[2] after leaving $V_i$. Constraint (8) indicates the total flying distance of the UAV.

$$\min_{U \subseteq \mathbb{V}} \max\{\lambda \frac{\beta}{\alpha v_0} - \frac{P_0}{\alpha}, t_{|U|} + \frac{d_{|U|} - o_{|U|}}{v_{|U|}} + \frac{a - d_{|U|}}{v_0}\} \tag{2}$$

$$\text{s.t.} \quad P_1 = P_0 + t_1 \cdot \alpha - \frac{o_1}{v_0}\beta \tag{3}$$

$$P_{i+1} = \text{PTF}(P_i, V_i, V_{i+1}) \tag{4}$$

$$\text{PTF}(P_i, V_i, V_{i+1}) \geq 0 \tag{5}$$

$$t_1 \geq \frac{o_1}{v_0} \tag{6}$$

$$t_{i+1} \geq t_i + \frac{d_i - o_i}{v_i} + \frac{|o_{i+1} - d_i|}{v_0} \tag{7}$$

$$\lambda = (o_1 + \sum_{i=1}^{|U|-1} |o_{i+1} - d_i| + a - d_{|U|}) \tag{8}$$

For analytical tractability, the overall travel time of the UAV in objective (2) is converted to (9) in the following Proposition 1. Intuitively, when the UAV contains enough power to fly constantly to $A$ after completing the last-taken ride $V_{|U|}$, UAV's arrival time only corresponds to $V_{|U|}$; otherwise, UAV's arrival time to $A$ only corresponds to the UAV's overall flying distance. Thus, we have

**Proposition 1.** *Given the sequence $U = (V_1, \cdots, V_{|U|})$ of rides taken by the UAV, the UAV's arrival time $\xi(U)$ at the target $A$ is given by*

$$\xi(U) = \max\{\underbrace{(o_1 + \sum_{i=1}^{|U|-1} |o_{i+1} - d_i| + a - d_{|U|})}_{\text{i.e., UAV's overall flying distance}} \frac{\beta}{\alpha v_0} - \frac{P_0}{\alpha}, t_{|U|} + \frac{d_{|U|} - o_{|U|}}{v_{|U|}} + \frac{a - d_{|U|}}{v_0}\} \tag{9}$$

**Competitive Ratio.** Online algorithms are typically measured by the competitive ratio [7]. Given a sequence $\mathbb{V}$ of online trucks that can offer rides to the UAV, denote by $\xi_{\text{ALG}}(\mathbb{V})$ and $\xi_{\text{OPT}}(\mathbb{V})$ the UAV's arrival time to the destination by an online algorithm (ALG) and the optimal offline solution (OPT) where complete information of $\mathbb{V}$ is given beforehand, respectively. Then, the competitive ratio $\rho$ of the problem is defined as $\rho = \max_{\mathbb{V}} \frac{\xi_{\text{ALG}}(\mathbb{V})}{\xi_{\text{OPT}}(\mathbb{V})}$. When a number $\theta \geq 1$ satisfies $\theta \leq \rho$ for all deterministic online algorithms, we say $\theta$ is a lower bound on the competitive ratio of the problem.

[2] We use stop-and-recharge time and waiting time interchangeably in this paper, to refer to the time that the UAV stops at roadside.

## 3 Offline Problem and Algorithm Design

We present an optimal solution, named OPTIMALHITCHING, for the offline problem in this section, in which the idea behind is to map the offline rides to nodes in a graph, and further the taking sequence in achieving the earliest arrival time is converted to a minimum-weight path in the constructed graph. In a graph-based solution, we find it is difficult to map the UAV's arrival time to edge weight in the graph directly by using $\xi(U)$. This is because the arrival time $\xi(U)$ corresponds to multi nodes/rides in the set $U$. Hence, we transform the objective (9) to the following (10), which only corresponds to the last ride $V_{|U|}$ taken by the UAV and the power $P_{|U|}$ that the UAV contains at the start time $t_{|U|}$ of $V_{|U|}$.

$$\xi(V_{|U|}) = t_{|U|} + \frac{d_{|U|} - o_{|U|}}{v_{|U|}} + \frac{a - d_{|U|}}{v_0} + \underbrace{\frac{\max\{\frac{a-d_{|U|}}{v_0}(\beta - \alpha) - (P_{|U|} + \frac{d_{|U|}-o_{|U|}}{v_{|U|}}), 0\}}{\alpha}}_{\text{the possible stop-and-recharge time of the UAV}} \tag{10}$$

Before going into the details of our offline algorithm, we first give the following definitions together with some preliminary results.

**Definition 1 (Adjacent).** *Given the set $U \subset \mathbb{V}$ of rides accepted by the UAV, two trips $V_i$ and $V_j$ in $U$ are called adjacent rides if and only if $V_j \in \{\arg\min_{V_x \in U, o_x > o_i}\{o_x - o_i\}, \arg\max_{V_x \in U, o_x < o_i}\{o_x - o_i\}\}$. Specifically, $V_j$ is regarded as the prior one (resp. the following one) of the two if $o_j < o_i$ (resp. $o_j > o_i$).*

**Definition 2 (Sequentially-taken).** *Given the set $U \subset \mathbb{V}$ of rides taken by the UAV, we say $V_i$ and $V_j$ in $U$ are sequentially-taken if they are neighbors in the taking sequence $U$ of rides. For example, when the UAV transfers from $V_i$ to $V_j$ in the sequence, we call $V_i$ and $V_j$ the prior and the following ride of the two sequentially-taken rides respectively.*

To verify whether two rides can be taken together by the UAV or not, we have Proposition 2, which is summarized from (2)–(8) and helps to determine whether two nodes/rides should be connected/compatible in the constructed graph.

**Proposition 2 (Compatible condition).** *Given two rides $V_j$ and $V_i$ with $o_j < o_i$ and the power $P_j$ of the UAV at time $t_j$, we say they are compatible only when the UAV is able to take both rides by transferring from $V_j$ to $V_i$. Specifically, they satisfy the following compatible condition (11).*

$$\begin{aligned} &a) \text{ [time compatibility] } t_i \geq t_j + \frac{d_j - o_j}{v_j} + \frac{|o_i - d_j|}{v_0} \\ &b) \text{ [power compatibility] } \mathrm{PTF}(P_j, V_j, V_i) \geq 0 \end{aligned} \tag{11}$$

The moment while the UAV is either landing-on or flying-off a truck, note that the time and space dimensions keep in consistency between the UAV and the truck. Given two sequentially-taken rides $V_i$ (the prior one) and $V_j$, the UAV is supposed to land on $V_j$ right at time $t_j$ to catch the start of $V_j$ without reducing its remaining power at time $t_j$. This is because the *power transfer function* in (1) is independent from when the UAV departs in transferring between $V_i$ and $V_j$. Or, the UAV can stop-and-recharge at the end location $d_i$ of $V_i$ until it can fly constantly to the start $o_j$ of $V_j$ (right at $t_j$). This helps us to better understand the location of the UAV while transferring between two rides. Theorem 1 shows the taking sequence of the UAV in a given set of accepted rides.

**Theorem 1.** *Given the set $U = \{V_1, \cdots, V_{|U|}\}$ of rides that are accepted by the optimal solution (OPT), OPT takes all rides in $U$ following the increasing order of the rides' start locations, i.e. the smaller $o_i$ is, the earlier $V_i \in U$ is taken.*

**Offline Algorithm.** At the high level, OPTIMALHITCHING first constructs a graph by screening unnecessary truck rides iteratively, which is on top of some characteristics of the problem. Afterwards, the optimal solution of the offline problem in this paper is converted to a minimum-weight shortest path of the graph. In the constructed graph, two virtual nodes $V_0$ and $V_{n+1}$ are introduced to represent the origin $O$ and the destination $A$ respectively, while the other nodes in $\{V_1, \cdots, V_n\}$ are constructed to represent the taking sequence of rides in $\mathbb{V}$ respectively due to Theorem 1. Each node of $\{V_1, \cdots, V_n\}$ maintains a weight of the maximum power that the UAV could remain at the moment transferring to this node/ride, and connects to the previously constructed node from which the UAV transfers to the new node and remains that maximum power. In other words, node weights in the graph are only used for checking the power compatibility of rides taken by the UAV. Due to (10), all edges of the graph are set as zero weight except for those connecting to $V_{n+1}$. In this way, the weight of a path connecting $V_0/O$ and $V_{n+1}/A$ indicates the arrival time of the UAV at $A$ taking those rides on the path. Below gives details of OPTIMALHITCHING.

1. *First*, sort offline rides in $\mathbb{V}_n$, by increasing order of their start locations as $(V_1, V_2, \cdots, V_n)$; create virtual nodes $V_0 = (0, 0, 0, 0, 0)$ and $V_{n+1}$ representing the origin $O$ and the destination $A$ respectively.
2. *Then*, construct a graph $G = (N, E, w)$ with the weight function $w$ applying to both nodes in $N$ and edges in $E$, in an iterative way:
   (a) Include $V_0$ in $N$. Check in sequence $(V_1, V_2, \cdots, V_n)$ the rides one by one. When $V_i \in (V_1, V_2, \cdots, V_n)$ is compatible with at least one ride in $N$ by (11), denote $\Phi(V_i)$ as the set of rides in $N$ that are compatible with $V_i$:
      i. find in $\Phi(V_i)$ the node/ride, denoted by $V_*(i)$, to which the most power remains to the UAV on arrival at the start of $V_i$ by transferring from a ride in $\Phi(V_i)$;
      ii. include $V_i$ in $N$ and set weight $w(V_i) = \mathrm{PTF}(w(V_*(i)), V_*(i), V_i)$; include $(V_*(i), V_i)$ in $E$ and set edge weight $w(V_*(i), V_i) = 0$;
   (b) Include $V_{n+1}$ in $N$. For each $V_j \in N$, add $(V_j, V_{n+1})$ to $E$ with weight $w(V_j, V_{n+1}) = \xi(V_j)$ that representing the UAV's arrival time to $A$ with taking $V_j$ as the last ride.
3. *Finally*, find in $N$ the node $\overline{V} = \arg\min_{V \in N - \{V_{n+1}\}} w(V, V_{n+1})$ that has the minimum-weight edge connecting with $V_{n+1}$, and further find backwards

(from those nodes joining in $N$ earlier than $\overline{V}$) the node that connects with $\overline{V}$.[3] Repeat this step backwards until node $V_0$ is reached. Output those nodes found by OPTIMALHITCHING in the sequence of their joining time in $N$.

Note that the running time of OPTIMALHITCHING is dominated by graph construction steps, which is in $O(n^2)$. We have the following Theorem 2.

**Theorem 2.** OPTIMALHITCHING *runs in* $O(n^2)$*-time and outputs the sequence of rides that are taken by an optimal offline solution.*

# 4 Lower Bounds

**Table 2.** Derived notations for bound analysis

| Derived notations | Physical meanings |
|---|---|
| $T_{ru} := \frac{\beta-\alpha}{\alpha \cdot v_0}$ | The time duration for recharging the amount of power to fly a unit distance constantly |
| $T_{mu} := \frac{\beta}{\alpha \cdot v_0}$ | The time duration for moving a unit distance in the case that the UAV contains no power at the beginning |
| $T_{f0} := \frac{P_0}{\beta-\alpha}$ | The time duration that the UAV flies constantly from time $t_0$ on |
| $l_f := \frac{P_0 v_0}{\beta-\alpha}$ | The furthest location to which the UAV can reach by flying constantly from time $t_0$ on, note that $l_f = T_{f0} \cdot v_0$ |
| $\xi(\varnothing) := \frac{\beta}{\alpha}\frac{a}{v_0} - \frac{P_0}{\alpha}$ | The earliest arrival time of the UAV to $A$ when no ride is taken, by (10) with $V_{\|U\|} = (0, 0, 0, 0, v)$ and $P_{\|U\|} = P_0$ |
| $T_{ra} := \frac{\beta-\alpha}{\alpha}\frac{a}{v_0} - \frac{P_0}{\alpha}$ | The least time duration of the UAV to stop-and-recharge, to reach the target $A$ without taking rides, $T_{ra} = \xi(\varnothing) - \frac{a}{v_0}$ |
| $L_{\min} := \frac{(a\beta - a\alpha - P_0 v_0)v}{v_0\alpha - v\alpha + v\beta}$ | The minimum length to be saved by rides (with velocity $v$) to reach the target $A$ without stop-and-recharge, by Lemma 1 |
| $Len(T) := \max\{\frac{(a\beta - a\alpha - P_0 v_0 - v_0 T\alpha)v}{v_0\alpha - v\alpha + v\beta}, 0\}$ | The least amount of length that the UAV needs to save for avoiding more time in stop-and-recharge afterwards, given that the UAV already stop-and-recharges for a total amount $T$ of time |

We present the lower bounds of the UAV's travel time by first releasing a hook ride, and then releasing rides that are not compatible to ALG by either power or

[3] Step 2.b.i ensures only one node joining in $N$ earlier than $\overline{V}$ that connects with $\overline{V}$.

time constraint if ALG rejects/accepts the hook ride, since taking a ride helps to save more energy in an early stage but moves more slowly. For bound analysis, we further derive some notations as summarized in Table 2.

Notice that the UAV does not need a ride $V_i$ with $\Delta t_i \geq T_{ra} + T_{f0}$. This is because the moment when $V_i$ starts, the UAV already contains at least $(\Delta t_i - T_{f0})\alpha$ of power by stop-and-recharging, which enables the UAV to fly constantly to $A$. Thus, we have the following Proposition 3. Then, Lemma 1 is given to better figure out the minimum travel time that an OPT could achieve, based on which a lower bound is presented in Theorem 3 as a benchmark for further online algorithm design.

**Proposition 3.** *$\Delta t_i = t_i - r_i$ of each ride $V_i \in \mathbb{V}$ ranges in $[0, T_{ra} + T_{f0}]$.*

**Lemma 1.** *The UAV needs to save an overall distance of at least $L_{\min} = \frac{(a\beta - a\alpha - P_0 v_0)v}{v_0\alpha - v\alpha + v\beta}$ by taking rides, in order to avoid stop-and-recharge halfway.*

**Theorem 3.** *For the problem with flexible $\Delta t_i \in [0, T_{ra} + T_{f0}]$, no online deterministic algorithm can achieve a competitive ratio better than*

$$\frac{\xi(\varnothing) - T_{mu}}{\frac{\lceil L_{\min} \rceil}{v} + \frac{a+1-\lceil L_{\min} \rceil}{v_0}} \tag{12}$$

## 5 Online Algorithms with Competitive Analysis

We propose a myopic algorithm MyopicHitching and a near-optimal algorithm $\Delta t$-Adaptive respectively, both of which inherit notations from Table 2.

### 5.1 MyopicHitching Algorithm

We first present the MyopicHitching algorithm under fixed $\Delta t$. By some small changes in the following accepting conditions (i-ii), one can easily extend it to the flexible $\Delta t$. MyopicHitching follows the route by default to fly forward constantly until the first time it runs out of power. Afterwards, the UAV stops-and-recharge until containing enough power to fly constantly to the target $A$. The by-default action possibly changes only when a ride is accepted. Denote $P(V)$ as the power that the UAV remains at the start time of an accepted ride $V$ in the current solution, and $U$ as the set of rides accepted by MyopicHitching. A new ride $V_i \in \mathbb{V}$ is accepted only when $V_i$ meets the following two conditions together, i.e., $l_{\text{rc}} \cdot l_{\text{aa}} = 1$. Accordingly, update both $P(V_i)$ and power attributes of those rides in $U$ that depart after $V_i$ by the power-transfer-function in (1).

**i)** *ride-compatible*[4]: the new ride should meet (11) with $V_{\text{left}} = \underset{\{V_j \in U | o_j < o_i\}}{\arg\min} \{o_i - o_j\}$ which is the only ride in $U$ that is sequentially-taken with $V_i$, i.e., the following indicator should be equal to 1.

[4] Under fixed $\Delta t$, online algorithm must accept rides departing after previously accepted rides due to Observation 1. But this is not the case under flexible $\Delta t$, a newly accepted ride can depart between two rides in $U$, and the potential $V_{\text{right}} = \underset{\{V_j \in U | o_j > o_i\}}{\arg\min}\, o_j - o_i$ should also be compatible with $V_i$.

$$1_{\rm rc} = \begin{cases} 1, & t_i \geq t_{\rm left} + \frac{d_{\rm left} - o_{\rm left}}{v_{\rm left}} + \frac{|o_i - d_{\rm left}|}{v_0} \\ & \textbf{and } \mathrm{PTF}(P_{\rm left}, V_{\rm left}, V_i) \geq 0 \\ 0, & \text{otherwise} \end{cases}$$

**Observation 1.** *For the problem with fixed $\Delta t$, the earlier a ride is released, the earlier the ride departs.*

ii) *arrival-ahead*: the UAV will reduce its overall travel time when taking the new ride, i.e., the following indicator $1_{aa}$ (9) should be equal to 1.

$$1_{\rm aa} = \begin{cases} 1, & \xi(U \cup \{V_i\}) \leq \xi(U) \\ 0, & \text{otherwise} \end{cases}$$

To take each accepted ride in $U$, MyopicHitching guides the UAV to reach the origin of the ride right at its start time. Whenever the UAV contains enough power to fly constantly towards the target $A$, i.e., (13) is satisfied, it stops accepting new ride and flies directly to $A$ after taking the last-accepted ride.

$$\begin{cases} u_{\rm time} \geq \max\{(\frac{\beta}{\alpha} - 1)\frac{a - d_{|U|}}{v_0} - \frac{\widetilde{P}(V_{|U|})}{\alpha}, t_{|U|} + \frac{d_{|U|} - o_{|U|}}{v_{|U|}}\} \text{ or} \\ \widetilde{P}(V_{|U|}) \geq \frac{a - d_{|U|}}{v_0}(\beta - \alpha) \ \& \ u_{\rm time} < t_{|U|} + \frac{d_{|U|} - o_{|U|}}{v_{|U|}} \text{ or} \\ U = \varnothing \ \& \ u_{\rm time} = T_{ra} \end{cases} \tag{13}$$

in which $\widetilde{P}(V_{|U|}) = P(V_{|U|}) + \frac{\alpha(d_{|U|} - o_{|U|})}{v_{|U|}}$ indicates the power remaining to the UAV on completing $V_{|U|}$, while $u_{\rm time}$ indicates the real time in the execution.

**Lemma 2.** *For the problem with fixed $\Delta t$,* MyopicHitching *always takes rides in increasing order of their start locations.*

Notice in the following example that MyopicHitching has a *defect* which can be exploited by an adversary leading to a very bad competitive ratio: suppose $V_1 = (\frac{1}{2v_0}, \frac{1}{2v_0} + \Delta t, \varepsilon, 1 + \varepsilon, v)$ is the first released ride with a small $\Delta t$. At the release time $\frac{1}{2v_0}$ of $V_1$, the UAV is at location $\frac{1}{2}$ and contains power $\overline{P} = P_0 + (\alpha - \beta)\frac{1}{2v_0}$. We note that the *arrival-ahead* condition (ii) implies a ride released at an early stage could be accepted when the ride could help the UAV to save more energy. Since the UAV can save a small amount $\frac{2\varepsilon}{v_0}\beta$ of power by taking $V_1$, $V_1$ is accepted by accepting conditions (i)–(ii). Notice that MyopicHitching actually costs $t_{\rm wait} = \Delta t + \frac{1}{v_1} + \frac{\frac{1}{v_0} - 1 - 2\varepsilon}{v_0}$ of waiting time. The *defect* appears when $\Delta t < t_{\rm wait}$ since the adversary could further releases rides making MyopicHitching violate constraints in (11). When $\varepsilon \to 0$, we get the following Theorem 4.

**Theorem 4.** *For the problem with fixed $\Delta t$,* MyopicHitching *achieves a competitive ratio no worse than*

$$\frac{\xi(\varnothing)}{(a - \lceil \mathrm{L}_{\min} \rceil)T_{mu} - \frac{P_0}{\alpha}} \tag{14}$$

Particularly, in the scenario where the UAV already exhausts its power (i.e., $P_0 = 0$) at the very beginning, MyopicHitching achieves a competitive ratio no worse than $\frac{a}{a - \lceil \mathrm{L}_{\min} \rceil}$.

### 5.2 $\Delta t$-Adaptive Algorithm

We fix the *defect* of MyopicHitching by leveraging adaptability of $\Delta t$ in MyopicHitching and present the $\Delta t$-Adaptive algorithm by including the following *conditional-start* condition (i.e., $1_{\text{cs}} = 1$) in the accepting conditions. Specifically, $\Delta t$-Adaptive only replaces the accepting conditions $1_{\text{rc}} \cdot 1_{\text{aa}} == 1$ in MyopicHitching by $1_{\text{rc}} \cdot 1_{\text{aa}} \cdot 1_{\text{cs}} == 1$, but keeps the remaining the same.

**iii)** *conditional-start* condition: the following indicator $1_{cs}$ should be one.

$$1_{\text{cs}} = \begin{cases} 1, & \begin{array}{l} o_i \geq \frac{\Delta t v_0}{2} + l(u_{\text{time}}) \text{ for } u_{\text{time}} \in [0, \frac{l_f}{v_0}], \\ \text{or, } o_i \geq l_f + \frac{\Delta t v_0}{2} \text{ for } u_{\text{time}} \in (\frac{l_f}{v_0}, \xi(\varnothing)] \end{array} \\ 0, & \quad\quad\quad \text{otherwise} \end{cases}$$

**Theorem 5.** *For the problem with fixed $\Delta t$, $\Delta t$-Adaptive algorithm achieves a competitive ratio no worse than*

$$\frac{\xi(\varnothing)}{\frac{\Delta t}{2} + \frac{a - Len(\frac{\Delta t}{2})}{v_0} + \frac{Len(\frac{\Delta t}{2})}{v}} \tag{15}$$

Recall that rides with $\Delta t \geq T_{ra} + T_{f0}$ are not worth taking (see Lemma 3). By Theorem 5, we know $\Delta t$-Adaptive prefers larger $\Delta t$ since its competitive ratio decreases as $\Delta_t$ increases. Comparing (15) and (14), we find that $\Delta t$-Adaptive algorithm outperforms *MyopicHitching* especially when $\Delta t$ is large, this is because $\Delta t$-Adaptive guarantees that OPT has to cost some waiting time of at least $\frac{\Delta t}{2}$ in the case when all released truck rides are not compatible to $\Delta t$-Adaptive. What's more, $\Delta t$-Adaptive algorithm actually achieves near-optimal performance compared to the best possible online algorithm since the latter can save at most one ride while the OPT has to pay a waiting time of $\Delta t$.

## 6 Concluding Remarks

In this paper, we give the first theoretical work on the problem of online ride-hitching in UAV travelling. By mapping truck trips to nodes in a graph in an iterative way, we give a shortest-path-like solution for the offline version of this problem where truck trips are all known in advance. As a benchmark, we present lower bounds on the competitive ratio of the problem, respectively, for different settings. Then, we show that a greedy algorithm which accepts as many rides as possible has a defect. To fix the defect, we propose the $\Delta t$-Adaptive algorithm, achieving near-optimal performance in terms of the competitive ratio.

**Acknowledgement.** We thank the anonymous referees for their helpful feedback. Part of this work was done while Songhua LI was visiting the Singapore University of Technology and Design. Minming LI is also from City University of Hong Kong Shenzhen Research Institute, Shenzhen, P.R. China. The work described in this paper was sponsored by Project 11771365 supported by NSFC.

## References


1. Wu, J., Wang, H., Li, N., Yao, P., Huang, Y., Yang, H.: Path planning for solar-powered UAV in urban environment. Neurocomputing **275**, 2055–2065 (2018)
2. Chung, S.H., Sah, B., Lee, J.: Optimization for drone and drone-truck combined operations: a review of the state of the art and future directions. Comput. Oper. Res. **123**, 105004 (2020)
3. Khuller, S., Malekian, A., Mestre, J.: To fill or not to fill: the gas station problem. ACM Trans. Algorithms **7**(3), 1–16 (2011)
4. Shan, F., Luo, J., Xiong, R., Wu, W., Li, J.: Looking before crossing: an optimal algorithm to minimize UAV energy by speed scheduling with a practical flight energy model. In: IEEE INFOCOM 2020-IEEE Conference on Computer Communications, pp. 1758–1767. IEEE (2020)
5. Di Franco, C., Buttazzo, G.: Energy-aware coverage path planning of UAVs. In: 2015 IEEE International Conference on Autonomous Robot Systems and Competitions, pp. 111–117 (2015)
6. Henchey, M.J., Batta, R., Karwan, M., Crassidis, A.: A flight time approximation model for unmanned aerial vehicles: estimating the effects of path variations and wind. In: Operations Research for Unmanned Systems, pp. 95–117 (2016)
7. Borodin, A., El-Yaniv, R.: Online Computation and Competitive Analysis. Cambridge University Press, Cambridge (2005)
8. DHL Express. https://www.dhl.com/tw-en/home/press/press-archive/2019/dhl-express-launches-its-first-regular-fully-automated-and-intelligent-urban-drone-delivery-service. Accessed August 2021
9. Krumke, S.O., De Paepe, W.E., Poensgen, D., Stougie, L.: News from the online traveling repairman. Theoret. Comput. Sci. **295**(1–3), 279–294 (2003)
10. Koutsoupias, E.: The k-server problem. Comput. Sci. Rev. **3**(2), 105–118 (2009)
11. Koutsoupias, E., Papadimitriou, C.H.: On the k-server conjecture. J. ACM (JACM) **42**(5), 971–983 (1995)
12. Bienkowski, M., Liu, H.H.: An improved online algorithm for the traveling repairperson problem on a line. In: 44th International Symposium on Mathematical Foundations of Computer Science. Schloss Dagstuhl-Leibniz-Zentrum fuer Informatik (2019)
13. Coester, C., Koutsoupias, E.: The online $k$-taxi problem. In: Proceedings of the 51st Annual ACM SIGACT Symposium on Theory of Computing, pp. 1136–1147 (2019)
14. Li, S., Li, M., Duan, L., Lee, V.C.S.: Online maximum $k$-interval coverage problem. In: Wu, W., Zhang, Z. (eds.) COCOA 2020. LNCS, vol. 12577, pp. 455–470. Springer, Cham (2020). https://doi.org/10.1007/978-3-030-64843-5_31
15. Li, S., Li, M., Duan, L., Lee, V.C.S.: Online ride-hitching in UAV travelling. arXiv preprint. http://arxiv.org/abs/2108.09606 (2021)
16. Bokeno, E.T., Bort, T.M., Burns, S.S., Rucidlo, M., Wei, W., Wires, D.L.: U.S. Patent No. 9,915,956. U.S. Patent and Trademark Office, Washington, DC (2018)

# Parameterized Complexity and Algorithms

# Disconnected Matchings

Guilherme C. M. Gomes[1], Bruno P. Masquio[2](✉), Paulo E. D. Pinto[2], Vinicius F. dos Santos[1], and Jayme L. Szwarcfiter[2,3]

[1] Departamento de Ciência da Computação, Universidade Federal de Minas Gerais (UFMG), Belo Horizonte, MG, Brazil
{gcm.gomes,viniciussantos}@dcc.ufmg.br

[2] Instituto de Matemática e Estatística, Universidade do Estado do Rio de Janeiro (UERJ), Rio de Janeiro, RJ, Brazil
{brunomasquio,pauloedp}@ime.uerj.br

[3] Instituto de Matemática e PESC/COPPE, Universidade Federal do Rio de Janeiro (UFRJ), Rio de Janeiro, RJ, Brazil
jayme@nce.ufrj.br

**Abstract.** In 2005, Goddard, Hedetniemi, Hedetniemi and Laskar [Generalized subgraph-restricted matchings in graphs, Discrete Mathematics, 293 (2005) 129 – 138] asked the computational complexity of determining the maximum cardinality of a matching whose vertex set induces a disconnected graph. In this paper we answer this question. In fact, we consider the generalized problem of finding *c-disconnected matchings*; such matchings are ones whose vertex sets induce subgraphs with at least $c$ connected components. We show that, for every fixed $c \geq 2$, this problem is NP-complete even if we restrict the input to bounded diameter bipartite graphs. For the case when $c$ is part of the input, we show that the problem is NP-complete for chordal graphs while being solvable in polynomial time for interval graphs, FPT when parameterized by treewidth, and XP for graphs with a polynomial number of minimal separators, when parameterized by $c$.

**Keywords:** Algorithms · Complexity · Induced subgraphs · Matchings

## 1 Introduction

Matchings are a widely studied subject both in structural and algorithmic graph theory [10,15–20]. A matching is a subset $M \subseteq E$ of edges of a graph $G = (V, E)$ that do not share any endpoint. A $\mathscr{P}$-matching is a matching such that $G[M]$, the subgraph of $G$ induced by the endpoints of edges of $M$, satisfies property $\mathscr{P}$. The complexity of deciding whether or not a graph admits a $\mathscr{P}$-matching has been investigated for many different properties $\mathscr{P}$ over the years. One of the most well

G. C. M. Gomes—Partially supported by CAPES
V. F. dos Santos—Partially supported by FAPEMIG and CNPq
J. L. Szwarcfiter—Partially supported by CNPq.


C.-Y. Chen et al. (Eds.): COCOON 2021, LNCS 13025, pp. 579–590, 2021.
https://doi.org/10.1007/978-3-030-89543-3_48

known examples is the NP-completeness of the INDUCED MATCHING problem [5], where $\mathscr{P}$ is the property of being a 1-regular graph. Other NP-hard problems include ACYCLIC MATCHING [13], $k$-DEGENERATE MATCHING [2], deciding if the subgraph induced by a matching contains a unique maximum matching [14], and LINE-COMPLETE MATCHING[1] [6]; the latter was originally named CONNECTED MATCHING, but we adopt the more recent meaning of CONNECTED MATCHING given by Goddard et al. [13], where we want the subgraph induced by the matching to be connected. We summarize the above results in Table 1.

**Table 1.** $\mathscr{P}$-matchings and some of its complexity results. Entries marked with a † are presented in this paper.

| $\mathscr{P}$-matching | Property $\mathscr{P}$ | Complexity |
|---|---|---|
| INDUCED MATCHING | 1-regular | NP-complete [5] |
| ACYCLIC MATCHING | acyclic | NP-complete [13] |
| $k$-DEGENERATE MATCHING | $k$-degenerate | NP-complete [2] |
| UNIQUELY RESTRICTED MATCHING | has a unique maximum matching | NP-complete [14] |
| CONNECTED MATCHING | connected | Polynomial [13] Same as MAXIMUM MATCHING† |
| $c$-DISCONNECTED MATCHING, for each $c \geq 2$ | has $c$ connected components | NP-complete for bipartite graphs† |
| DISCONNECTED MATCHING, with $c$ as part of the input | has $c$ connected components | NP-complete for chordal graphs† |

Motivated by a question posed by Goddard et al. [13] about the complexity of finding a matching that induces a disconnected graph, in this paper we study the DISCONNECTED MATCHING problem, which we define as follows:

> DISCONNECTED MATCHING
> *Instance*: A graph $G$ and two integers $k$ and $c$.
> *Question*: Is there a matching $M$ with at least $k$ edges such that $G[M]$ has at least $c$ connected components?

Our first result is an alternative proof for the polynomial time solvability of CONNECTED MATCHING. We then answer Goddard et al.'s question by showing that DISCONNECTED MATCHING is NP-complete for $c = 2$. Indeed, we show that the problem remains NP-complete even on bipartite graphs of diameter three,

[1] A line-complete matching $M$ is a matching such that every pair of edges of $M$ has a common adjacent edge.

for every fixed $c \geq 2$; we denote this version of the problem by $c$-DISCONNECTED MATCHING. Note that, while INDUCED MATCHING is the particular case of DISCONNECTED MATCHING when $c = k$, our result is much more general since we decouple these two parameters. Then, we turn our attention to the complexity of this problem on graph classes and parameterized complexity. We begin by showing that, unlike INDUCED MATCHING, DISCONNECTED MATCHING remains NP-complete even when restricted to chordal graphs of diameter 3; in this case, however, $c$ is part of the input, and we also prove that, for every fixed $c$, we can solve the problem in XP time. Afterwards, we present a polynomial time dynamic programming algorithm for interval graphs and an FPT algorithm parameterized by treewidth. We summarize our results in Table 2.

**Table 2.** Complexity results for DISCONNECTED MATCHING restricted to some input scopes. We denote by $\eta_i$ the $i$-th Bell number, which corresponds to the number of distinct partitions of a set of $i$ elements.

| Graph class | $c$ | Complexity | Proof |
|---|---|---|---|
| General | $c = 1$ | Same as MAXIMUM MATCHING | Theorem 2 |
| Bipartite | Fixed $c \geq 2$ | NP-complete | Theorem 4 |
| Chordal | Input | XP and NP-complete | Theorems 6 and 7 |
| Interval | Input | $\mathcal{O}\left(\lvert V\rvert^2 c \max\{\lvert V\rvert c, \lvert E\rvert\sqrt{\lvert V\rvert}\}\right)$ | Theorem 8 |
| Treewidth $t$ | Input | $\mathcal{O}(8^t \eta_{t+1}^3 \lvert V\rvert^2)$ | Theorem 9 |

**Preliminaries.** For an integer $k$, we define $[k] = \{1, \dots, k\}$. For a set $S$, we say that $A, B \subseteq S$ partition $S$ if $A \cap B = \emptyset$ and $A \cup B = S$; we denote a partition of $S$ in $A$ and $B$ by $A \dot{\cup} B = S$. A parameterized problem $(\Pi, k)$ is said to be XP when parameterized by $k$ if it admits an algorithm running in $f(k)n^{g(k)}$ time for computable functions $f, g$; it is said to be FPT when parameterized by $k$ if $g \in \mathcal{O}(1)$. For more on parameterized complexity, we refer to [8]. We use standard graph theory notation and nomenclature as in [3,4]. Let $G = (V, E)$ be a graph, $W \subseteq V(G)$, $M \subseteq E(G)$, and $V(M)$ to be the set of endpoints of edges of $M$, which are also called $M$-saturated vertices. We denote by $G[W]$ the subgraph of $G$ induced by $W$; in an abuse of notation, we define $G[M] = G[V(M)]$. A matching is said to be maximum if no other matching of $G$ has more edges than $M$, and perfect if $V(M) = V(G)$. Also, $M$ is said to be connected if $G[M]$ is connected and $c$-disconnected if $G[M]$ has at least $c$ connected components. A graph $G$ is $H$-free if $G$ has no copy of $H$ as an induced subgraph; $G$ is chordal if it has no induced cycle with more than three edges. A graph is an interval graph if it is the intersection graph of intervals on a line. In $G$, we denote by $\beta(G)$ the number of edges in a maximum matching, by $\beta_c(G)$ the cardinality of a maximum connected matching, by $\beta_*(G)$ the size of a maximum induced matching, and by $\beta_{d,i}(G)$ the size of a maximum $i$-disconnected matching. If $G$ is connected, note that:

1. Every maximum induced matching $M^*$ is a $\beta_*(G)$-disconnected matching, since each connected component of $G[M^*]$ is an edge.
2. Since $\beta_*(G)$ is the maximum number of components that $G[M]$ can have with any matching $M$, there exists no $c$-disconnected matching for $c > \beta_*(G)$.
3. Every matching is a 1-disconnected matching.
4. As shown in [13], $\beta(G) = \beta_c(G)$.

Consequently, we have that both Theorem 1 and the following bounds hold:

$$\beta = \beta_c = \beta_{d,1} \geq \beta_{d,2} \geq \beta_{d,3} \geq \dots \geq \beta_{d,\beta_*} \geq \beta_*$$

**Theorem 1.** Disconnected Matching *is* NP-complete *for every graph class for which the* Induced Matching *is* NP-complete.

This paper is organized as follows. In Sect. 2, we give an alternative proof to the fact that Connected Matching is in P and present an algorithm for Maximum Connected Matching, then present a construction used to show that $c$-Disconnected Matching is NP-complete for every fixed $c \geq 2$ on bipartite graphs of diameter three. In Sect. 3, we prove our final negative result, that Disconnected Matching is NP-complete on chordal graphs. We show, in Sect. 4.1, that the previous proof cannot be strengthened to fixed $c$ by giving an XP algorithm for Disconnected Matching parameterized by $c$ on graphs with a polynomial number of minimal separators. Finally, in Sects. 4.2 and 4.3, we present polynomial time algorithms for Disconnected Matching in interval and bounded treewidth graphs. We present our concluding remarks and directions for future work in Sect. 5.

## 2 Complexity of $c$-Disconnected Matching

### 2.1 1-disconnected and Connected Matchings

We consider that the input graph has at least one edge and is connected. Otherwise, the solution is trivial or we can solve the problem independently for each connected component. Recall that 1-Disconnected Matching allows its solution to have any number of connected components. Consequently, any matching with at least $k$ edges is a valid solution to an instance $(G, k, 1)$, which leads to Theorem 2.

**Theorem 2.** 1-Disconnected Matching *is in* P.

Note that if $M$ is a solution to an instance $(G, k)$ of Connected Matching, then it is also a solution to the instance $(G, k, 1)$ of 1-Disconnected Matching. Our next Theorem shows that the converse is also true and, using the former theorem, that Maximum Matching and Connected Matching are also related. The algorithm is based on the proof of Goddard et al. [13] that $\beta(G) = \beta_c(G)$, and can be implemented in linear time.

**Theorem 3.** *Given a maximum matching, a maximum connected matching can be found in linear time.*

**Corollary 1.** Maximum Connected Matching *has the same time complexity of* Maximum matching.

### 2.2 2-disconnected Matchings

Next, we show that 2-Disconnected Matching is NP-complete for bipartite graphs with bounded diameter. Our reduction is from the NP-hard problem 1-in-3 3SAT [11]; in this problem, we are given a set of $m$ clauses $I$ with exactly three literals in each clause, and asked if there is a truth assignment of the variables such that only one literal of each clause resolves to true. We consider that each variable must be present in at least one clause and that a variable is not repeated in the same clause. This follows from the original NP-completeness reduction [22].

**Input transformation in** 1-in-3 3SAT**.** We use $k = 12m$ and build a bipartite graph $G = (V_1 \dot{\cup} V_2, E)$ from a set of clauses $I$ as follows.

(I) For each clause $c_i$, generate a subgraph $B_i$ as described below.
- $V(B_i) = \{l_{ij}, r_{ij} \mid j \in \{1, \dots, 9\}\}$
- $E(B_i)$ is as shown in Fig. 1.

(II) For each variable $x$ present in two clauses $c_i$ and $c_j$, being the $q$-th literal of $c_i$ and the $t$-th literal of $c_j$, add two edges. If $x$ is negated in exactly one of the clauses, add the set of edges $\{r_{iq}l_{jt}, l_{i(q+3)}r_{j(t+3)}\}$. Otherwise, add $\{l_{i(q+3)}r_{jt}, r_{iq}l_{j(t+3)}\}$.

(III) Generate two complete bipartite subgraphs $H_1$ and $H_2$, both isomorphic to $K_{3m,3m}$, $V(H_1) = V(U_1) \dot{\cup} V(U_2)$ and $V(H_2) = V(U_3) \dot{\cup} V(U_4)$.

(IV) For each $u_2 \in V(U_2)$ and clause $c_i$, add the edge $\{u_2 l_{ij} \mid j \in \{1, \dots, 6\}\}$.

(V) For each $u_3 \in V(U_3)$ and clause $c_i$, add the edge $\{u_3 r_{ij} \mid j \in \{1, \dots, 6\}\}$.

Besides $G$ being bipartite, it is possible to observe that its diameter is 5, regardless of the set of clauses and its cardinality. This holds due to the distance between, for example, $u_1$ and $u_4$, $u_1 \in V(U_1)$, $u_4 \in V(U_4)$, as well as $l_{i7}$ and $r_{j7}$, $i, j \in [m]$, distinct, such that the clauses $i$ and $j$ do not have literals related to the same variable. Also, consider $G_i^+ = G[V(B_i) \cup V(H_1) \cup V(H_2)]$. Note that $|V(G)| = \mathcal{O}(m)$.

**Properties of Disconnected Matchings in the Generated Graphs.** We now prove some properties of the disconnected matching with cardinality at least $k$ in a graph $G$ generated by the transformation described.

Initially, we show, from Lemmas 1 and 2, that a subgraph induced by the saturated vertices of such matching has exactly two connected components, one containing vertices of $H_1$ and the other, vertices of $H_2$. Afterwards, Lemma 3 shows the possible sets of edges contained in the matching.

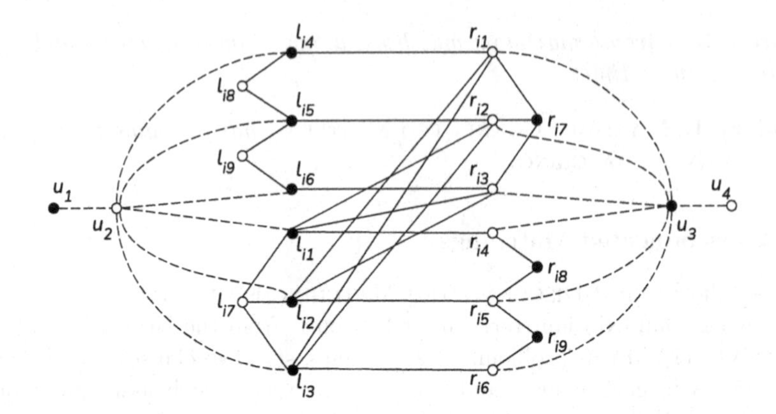


**Fig. 1.** The subgraph $G[V(B_i) \cup \{u_1, u_2, u_3, u_4\}]$, $u_1 \in V(U_1)$, $u_2 \in V(U_2)$, $u_3 \in V(U_3)$ and $u_4 \in V(U_4)$. The solid edges are the edges of $E(B_i)$ and the bold vertices represent a bipartition of $G$.

**Lemma 1.** *If $M$ is a disconnected matching with cardinality $k \geq 12m$, then there exists two saturated vertices $h_1 \in V(H_1)$ and $h_2 \in V(H_2)$.*

**Lemma 2.** *If $M$ is a disconnected matching with cardinality $k \geq 12m$, then $G[M]$ has exactly two connected components.*

**Lemma 3.** *Let $M$ be a disconnected matching with cardinality $k \geq 12m$ and $B_i$ be a clause subgraph. There are exactly 6 edges saturated by $M$ in $G[V(B_i)]$ and, moreover, there are exactly 3 sets of edges that satisfy this constraint.*

**Transforming a Disconnected Matching into a Variable Assignment.** First, we define, starting from a 2-disconnected matching $M$, $|M| = 12m$, a variable assignment $R$ and, in sequence, we present Lemma 4, proving that $R$ is a 1-IN-3 3SAT solution.

(I) For each clause $c_i$, where $x_{ij}$ corresponds to the $j$-th literal of $c_i$, generate the following assignments.
  - If $l_{ij}$ is $M$-saturated, then assign $x_{ij} = T$.
  - Else, assign $x_{ij} = F$.

Note that, analyzing the generated graph, the pair of saturated vertices $l_{ij}$ and $r_{ij}$, $j \in \{1, 2, 3\}$ define that the $j$-th literal is the true of the clause $c_i$. Similarly, each pair of saturated vertices $l_{iq}$ and $r_{iq}$, $q \in \{4, 5, 6\}$, $q \neq j + 3$, defines that the $(q - 3)$-th literal is false.

**Lemma 4.** *Let $M$ be a 2-disconnected matching with cardinality $k = 12m$ in a graph generated from a input $I$ of* 1-IN-3 3SAT. *It is possible to generate in polynomial time an assignment to variables in $I$ that solves* 1-IN-3 3SAT.

**Transforming a Variable Assignment into a Disconnected Matching.** Finally, we define a 2-disconnected matching $M$, obtained from a solution of 1-IN-3 3SAT. Then, Lemma 5 proves that $M$ is a 2-disconnected matching with the desired cardinality $12m$.

(I) For each clause $c_i$, whose true literal is the $j$-th, add to $M$ the edge set defined as $\{l_{ij}l_{i7}, r_{ij}r_{i7}, l_{iq}l_{i8}, r_{iq}r_{i8}, r_{it}l_{i9}, r_{it}r_{i9} \mid q \in \{4,5\}, t \in \{5,6\}, q \neq j+3 \neq t \neq q\}$.

(II) For $H_1$, add to the matching $M$ any $3m$ disjoint edges. Repeat the process for $H_2$.

**Lemma 5.** *Let $R$ be a variable assignment of an input $I$ from* 1-IN-3 3SAT*. It is possible, in polynomial time, to generate a disconnected matching with cardinality $k = 12m$ from $I$ in a graph generated by the transformation described below.*

Note that for any graph with diameter $d \leq 1$ the answer to DISCONNECTED MATCHING is always NO. We may assume that there are no isolated vertices as their removal does not change the solution. If the graph is disconnected, the problem can be solved in polynomial time by finding a maximum matching $M$ and checking if $|M| \geq k$. These statements are used in the proof of Theorem 4, which has a slight modification of the above construction, but allows us to reduce the diameter of the graph to 3.

**Lemma 6.** *Let $G = (V_1 \dot\cup V_2, E)$ be the bipartite graph from the transformation mentioned and $G' = (V_1' \dot\cup V_2', E')$ so that $V(G') = V(G) \cup \{w_1, w_2\}$ and $E(G') = E(G) \cup \{w_1w_2\} \cup \{vw_1 \mid v \in V(V_1)\} \cup \{vw_2 \mid v \in V(V_2)\}$. If $M$ is a 2-disconnected matching in $G'$, $|M| \geq k$, so $M$ is also a 2-disconnected matching in $G$.*

Combining the previous results, we obtain Theorem 4.

**Theorem 4.** 2-DISCONNECTED MATCHING *is* NP-complete *even if the input is restricted to bipartite graphs with diameter* 3.

These results imply the following dichotomies, in terms of diameter.

**Corollary 2.** *For bipartite graphs with diameter $\leq d$,* DISCONNECTED MATCHING *is* NP-complete *if $d$ is at least* 3 *and belongs to P otherwise.*

**Corollary 3.** *For graphs with diameter $\leq d$,* DISCONNECTED MATCHING *is* NP-complete *if $d$ is at least* 2 *and belongs to P otherwise.*

## 2.3 NP-completeness for Any Fixed $c$

We now generalize our hardness proof to $c$-DISCONNECTED MATCHING for every fixed $c > 2$. We begin by setting the number of edges in the matching $k = 12m + c - 2$, defining $G'$ to be the graph obtained in our hardness proof for 2-DISCONNECTED MATCHING, and $H$ to be the graph with $c - 2$ isolated edges $\{v_{i1}v_{i2} \mid i \in [c-2]\}$. To obtain our input graph to $c$-DISCONNECTED MATCHING, we make $w_1 \in V(G')$ adjacent to $v_{i1}$ and $w_2 \in V(G')$ adjacent $v_{i2}$, for every $i \in [c-2]$, where $w_1$ and $w_2$ are as defined in Lemma 6. This proves that the problem is NP-complete on bipartite graphs of diameter three. Note that, if we identify $w_1$ and $w_2$, we may reason as before, but now conclude that $c$-DISCONNECTED MATCHING is NP-complete on general graphs of diameter 2. We summarize this discussion as Theorem 5.

**Theorem 5.** *Let $c \geq 1$. The $c$-DISCONNECTED MATCHING problem belongs to P if $c = 1$. Otherwise, it is NP-complete even for bipartite graphs of diameter 3 or for general graphs of diameter 2.*

## 3 NP-completeness for Chordal Graphs

In this section, we prove that DISCONNECTED MATCHING is NP-complete even for chordal graphs with diameter 3. In order to prove it, we describe a reduction from the NP-complete problem EXACT COVER BY 3-SETS [11]. This problem consists in, given two sets $X$, $|X| = 3q$, and $C$, $|C| = m$ of 3-element subsets of $X$, decide if there exists a subset $C' \subseteq C$ such that every element of $X$ occurs in exactly one member of $C'$.

For the reduction, we define $c = m - q + 1$, $k = m + 3q$ and build the chordal graph $G = (V, E)$ from the sets $C$ and $X$ as follows.

(I) For each 3-element set $c_i = (x, y, z)$, $c_i \in C$, generate a complete subgraph $H_i$ isomorphic to $K_5$ and label its vertices as $W_i = \{w_{ix}, w_{iy}, w_{iz}, w_i^+, w_i^-\}$.
(II) For each pair of 3-element sets $c_i = (x, y, z)$ and $c_j = (a, b, c)$ such that $c_i, c_j \in C$, add all edges between vertices of $\{w_{ix}, w_{iy}, w_{iz}\}$ and $\{w_{ja}, w_{jb}, w_{jc}\}$.
(III) For each element $x \in X$, generate a vertex $v_x$ and the edges $v_x w_{ix}$ for every $i$ such that $c_i$ contains the element $x$.

An example of the reduction and its corresponding $c$-disconnected matching is presented in Fig. 2. For easier visualization, the edges from rule (II) are omitted.

In Lemmas 7 and 8, we define the polynomial transformation between a $(m - q + 1)$-disconnected matching $M$, $|M| \geq m + 3q$, and a subset $C'$ that solves the EXACT COVER BY 3-SETS. Then, Theorem 6 concludes the NP-completeness for chordal graphs.

**Lemma 7.** *Let $(C, X)$ be an input of* EXACT COVER BY 3-SETS *with $|C| = m$, $|X| = 3q$ and a solution $C'$. A $(m - q + 1)$-disconnected matching $M$, $|M| = m + 3q$, can be built in the transformation graph $G$ in polynomial time.*

**Lemma 8.** *Let $(C, X)$ be an input of* EXACT COVER BY 3-SETS *with $|C| = m$, $|X| = 3q$. Given a $(m - q + 1)$-disconnected matching $M$, $|M| = m + 3q$, in the transformation graph $G$ described, a solution $C'$ to* EXACT COVER BY 3-SETS *can be built in polynomial time.*

**Theorem 6.** DISCONNECTED MATCHING *is NP-complete even for chordal graphs with diameter 3.*

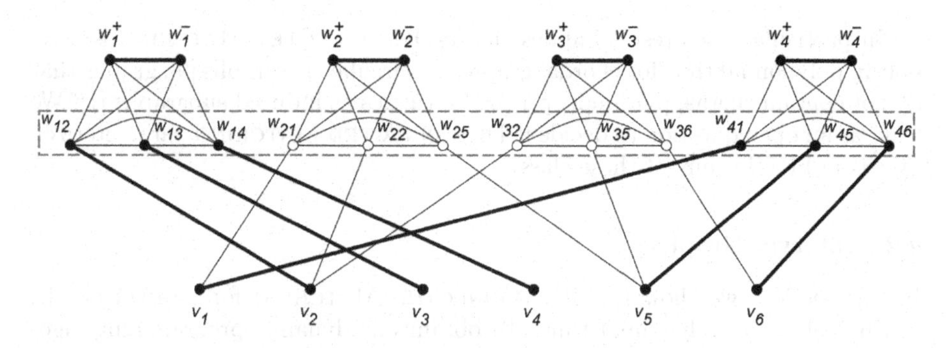


**Fig. 2.** An example of reduction for the input $X = \{1, 2, 3, 4, 5, 6\}$ and $C = \{\{2, 3, 4\}, \{1, 2, 5\}, \{2, 5, 6\}, \{1, 5, 6\}\}$. The subgraph induced by the vertices inside the dotted rectangle is complete and the matching in bold corresponds the solution $C' = \{\{2, 3, 4\}, \{1, 5, 6\}\}$.

# 4 Polynomial Time Algorithms

For our final contributions, we turn our attention to positive results, showing that the problem is efficiently solvable in some graph classes.

## 4.1 Minimal Separators and Disconnected Matchings

It is not surprising that minimal separators play a role when looking for $c$-disconnected matchings. In fact, for $c = 2$, Goddard et al. [13] showed how to find 2-disconnected matchings in graphs with a polynomial number of minimal separators. We generalize their result by showing that Disconnected Matching parameterized by the number $c$ of connected components is in XP; note that we do not need to assume that the family of minimal separators is part of the input, as it was shown in [23] it can be constructed in polynomial time.

**Theorem 7.** Disconnected Matching *parameterized by the number of connected components is in* XP *for graphs with a polynomial number of minimal separators.*

*Proof.* Note that if a matching $M$ is a maximum $c$-disconnected matching of $G = (V, E)$, then there is a family $\mathcal{S}$ of at most $c - 1$ minimal separators such that $V(G) - V(M)$ contains $\bigcup_{S \in \mathcal{S}} S$. Therefore, if we find such $\mathcal{S}$ that maximizes a maximum matching $M$ in $G[V - (\bigcup_{S \in \mathcal{S}} S)]$ and $M$ is $c$-disconnected, then $M$ is a maximum $c$-disconnected matching. Considering that $G$ has $|V|^{\mathcal{O}(1)}$ many minimal separators, the number of possible candidates for $\mathcal{S}$ is bounded by $|V|^{\mathcal{O}(c)}$. Computing a maximum matching can be done in polynomial time and checking whether $G[M]$ has $c$ components can be done in linear time. Therefore, the whole procedure takes $|V|^{\mathcal{O}(c)}$ time and finds a maximum $c$-disconnected matching.

In particular, this result implies that $c$-DISCONNECTED MATCHING is solvable in polynomial time for chordal graphs [7], circular-arc graphs [9], graphs that do not contain thetas, pyramids, prisms, or turtles as induced subgraphs [1]. We leave as an open question to decide if DISCONNECTED MATCHING parameterized by $c$ is in FPT for any of these classes.

### 4.2 Interval Graphs

In this section, we show that DISCONNECTED MATCHING for interval graphs can be solved in polynomial time. To obtain our dynamic programming algorithm, we rely on the ordering property of interval graphs [12]; that is, there is an ordering $\mathcal{Q} = \langle Q_1, \dots Q_p \rangle$ of the $p$ maximal cliques of $G$ such that each vertex of $G$ occurs in consecutive elements of $\mathcal{Q}$ and, moreover the intersection $S_i = Q_i \cap Q_{i-1}$ between two consecutive cliques is a minimal separator of $G$. Our algorithm builds a table $f(i, j, c')$, where $i, j \in \{1, \dots, p\}$ and $c' \in \{1, \dots, c\}$, and is equal to $q$ if and only if the largest $c'$-disconnected matching of $G\left[\bigcup_{i \leq \ell \leq j} Q_i \setminus (S_i \cup S_{j+1})\right]$ has $q$ edges; that is, $(G, k, c)$ is a positive instance if and only if $f(1, p, c) \geq k$.

**Theorem 8.** DISCONNECTED MATCHING *can be solved in polynomial time on interval graphs.*

### 4.3 Treewidth

A *tree decomposition* of a graph $G$ is a pair $\mathbb{T} = (T, \mathcal{B} = \{B_j \mid j \in V(T)\})$, where $T$ is a tree and $\mathcal{B} \subseteq 2^{V(G)}$ is a family where: $\bigcup_{B_j \in \mathcal{B}} B_j = V(G)$; for every edge $uv \in E(G)$ there is some $B_j$ such that $\{u, v\} \subseteq B_j$; for every $i, j, q \in V(T)$, if $q$ is in the path between $i$ and $j$ in $T$, then $B_i \cap B_j \subseteq B_q$. Each $B_j \in \mathcal{B}$ is called a *bag* of the tree decomposition. $G$ has treewidth has most $t$ if it admits a tree decomposition such that no bag has more than $t$ vertices. For further properties of treewidth, we refer to [21]. After rooting $T$, $G_x$ denotes the subgraph of $G$ induced by the vertices contained in any bag that belongs to the subtree of $T$ rooted at node $x$. Our final result is a standard dynamic programming algorithm on tree decompositions; we omit the proof and further discussions on how to construct the dynamic programming table for brevity.

**Theorem 9.** DISCONNECTED MATCHING *can be solved in* FPT *time when parameterized by treewidth.*

## 5 Conclusions and Future Works

We have presented $c$-disconnected matchings and the corresponding decision problem, which we named DISCONNECTED MATCHING. They generalize the well studied induced matchings and the problem of recognizing graphs that admit a sufficiently large induced matching. Our results show that, when the number of connected components $c$ is fixed, $c$-Disconnected Matching is solvable in polynomial time if $c = 1$ but NP-complete even on bipartite graphs if $c \geq 2$. We also proved that, unlike Induced Matching, Disconnected Matching remains NP-complete on chordal graphs. On the positive side, we show that the problem can be solved in polynomial time for interval graphs, in XP time for graphs with a polynomial number of minimal separators when parameterized by the number of connected components $c$, and in FPT time when parameterized by treewidth.

Possible directions for future work include determining the complexity of the problem on different graph classes. In particular, we would like to know the complexity of Disconnected Matching for strongly chordal graphs; we note that the reduction presented in Sect. 3 has many induced subgraphs isomorphic to a sun graph. We are also interested in the parameterized complexity of the problem. Our results show that, when parameterized by $c$, the problem is paraNP-hard; on the other hand, it is W[1]-hard parameterized by the number of edges in the matching since Induced Matching is W[1]-hard under this parameterization [19]. A first question of interest is whether chordal graphs admit an FPT algorithm when parameterized by $c$; while the algorithm presented in Sect. 4.1 works for all classes with a polynomial number of minimal separators, chordal graphs offer additional properties that may aid in the proof of an FPT algorithm. Another research direction would be the investigation of other structural parameterizations, such as vertex cover and cliquewidth; while the former yields a fixed-parameter tractable algorithm due to Theorem 9, we would like to know if we can find a single exponential time algorithm under this weaker parameterization. On the other hand, cliquewidth is a natural next step, as graphs of bounded treewidth have bounded cliquewidth, but the converse does not hold. Finally, we would like to study Disconnected Matching from the kernelization point of view and settle for which parameterizations we can obtain polynomial kernels and for which no such kernels exist, unless $\mathsf{NP} \subseteq \mathsf{coNP/poly}$.

## References

1. Abrishami, T., Chudnovsky, M., Dibek, C., Thomassé, S., Trotignon, N., Vušković, K.: Graphs with polynomially many minimal separators. arXiv preprint arXiv:2005.05042 (2020)
2. Baste, J., Rautenbach, D.: Degenerate matchings and edge colorings. Discret. Appl. Math. **239**, 38–44 (2018). https://doi.org/10.1016/j.dam.2018.01.002
3. Bondy, J.A., Murty, U.S.R.: Graph Theory. Springer, Heidelberg (2008). https://www.springer.com/gp/book/9781846289699
4. Brandstädt, A., Le, V.B., Spinrad, J.P.: Graph Classes: A Survey. Society for Industrial and Applied Mathematics, Philadelphia, PA, USA (1999)
5. Cameron, K.: Induced matchings. Discret. Appl. Math. **24**(1), 97–102 (1989). https://doi.org/10.1016/0166-218X(92)90275-F
6. Cameron, K.: Connected matchings. In: Jünger, M., Reinelt, G., Rinaldi, G. (eds.) Combinatorial Optimization — Eureka, You Shrink! LNCS, vol. 2570, pp. 34–38. Springer, Heidelberg (2003). https://doi.org/10.1007/3-540-36478-1_5

7. Chandran, L.S.: A linear time algorithm for enumerating all the minimum and minimal separators of a chordal graph. In: Wang, J. (ed.) COCOON 2001. LNCS, vol. 2108, pp. 308–317. Springer, Heidelberg (2001). https://doi.org/10.1007/3-540-44679-6_34
8. Cygan, M., et al.: Parameterized Algorithms, vol. 3. Springer, Heidelberg (2015)
9. Deogun, J.S., Kloks, T., Kratsch, D., Müller, H.: On the vertex ranking problem for trapezoid, circular-arc and other graphs. Discret. Appl. Math. **98**(1), 39–63 (1999). https://doi.org/10.1016/S0166-218X(99)00179-1
10. Edmonds, J.: Paths, trees, and flowers. Can. J. Math. **17**, 449–467 (1965). https://doi.org/10.4153/CJM-1965-045-4
11. Garey, M.R., Johnson, D.S.: Computers and Intractability: A Guide to the Theory of NP-Completeness. W. H. Freeman & Co., New York (1979)
12. Gilmore, P.C., Hoffman, A.J.: A characterization of comparability graphs and of interval graphs. Can. J. Math. **16**, 539–548 (1964). https://doi.org/10.4153/CJM-1964-055-5
13. Goddard, W., Hedetniemi, S.M., Hedetniemi, S.T., Laskar, R.: Generalized subgraph-restricted matchings in graphs. Discret. Math. **293**(1), 129–138 (2005). https://doi.org/10.1016/j.disc.2004.08.027
14. Golumbic, M.C., Hirst, T., Lewenstein, M.: Uniquely restricted matchings. Algorithmica **31**(2), 139–154 (2001). https://doi.org/10.1007/s00453-001-0004-z
15. Kobler, D., Rotics, U.: Finding maximum induced matchings in subclasses of claw-free and p5-free graphs, and in graphs with matching and induced matching of equal maximum size. Algorithmica **37**(4), 327–346 (2003). https://doi.org/10.1007/s00453-003-1035-4
16. Lozin, V.V.: On maximum induced matchings in bipartite graphs. Inf. Process. Lett. **81**(1), 7–11 (2002). https://doi.org/10.1016/S0020-0190(01)00185-5
17. Masquio, B.P.: Emparelhamentos desconexos. Master's thesis, Universidade do Estado do Rio de Janeiro (2019). http://www.bdtd.uerj.br/handle/1/7663
18. Micali, S., Vazirani, V.V.: An $O(\sqrt{|V|}|E|)$ algorithm for finding maximum matching in general graphs. In: 21st Annual Symposium on Foundations of Computer Science, pp. 17–27, October 1980. https://doi.org/10.1109/SFCS.1980.12
19. Moser, H., Sikdar, S.: The parameterized complexity of the induced matching problem. Discret. Appl. Math. **157**(4), 715–727 (2009). https://doi.org/10.1016/j.dam.2008.07.011
20. Panda, B.S., Chaudhary, J.: Acyclic matching in some subclasses of graphs. In: Gąsieniec, L., Klasing, R., Radzik, T. (eds.) IWOCA 2020. LNCS, vol. 12126, pp. 409–421. Springer, Cham (2020). https://doi.org/10.1007/978-3-030-48966-3_31
21. Robertson, N., Seymour, P.D.: Graph minors. II. Algorithmic aspects of tree-width. J. Algorithms **7**(3), 309–322 (1986). https://doi.org/10.1016/0196-6774(86)90023-4
22. Schaefer, T.J.: The complexity of satisfiability problems. In: Proceedings of the Tenth Annual ACM Symposium on Theory of Computing. STOC 1978, pp. 216–226. Association for Computing Machinery, New York (1978). https://doi.org/10.1145/800133.804350
23. Shen, H., Liang, W.: Efficient enumeration of all minimal separators in a graph. Theoret. Comput. Sci. **180**(1–2), 169–180 (1997)

# On the $d$-Claw Vertex Deletion Problem

Sun-Yuan Hsieh[1], Van Bang Le[2], and Sheng-Lung Peng[3](✉)

[1] Department of Computer Science and Information Engineering, National Cheng Kung University, Tainan City, Taiwan
hsiehsy@mail.ncku.edu.tw

[2] Institut für Informatik, Universität Rostock, Rostock, Germany
van-bang.le@uni-rostock.de

[3] Department of Creative Technologies and Product Design, National Taipei University of Business, Taipei, Taiwan
slpeng@ntub.edu.tw

**Abstract.** For an integer $d \geq 1$, let *d-claw* stand for $K_{1,d}$, the complete bipartite graph with 1 and $d$ vertices on each part. The $d$-claw vertex deletion problem, $d$-CLAW-VD, asks for a given graph $G$ and an integer $k$ whether one can delete at most $k$ vertices from $G$ such that the resulting graph has no $d$-claw as an induced subgraph. Thus, 1-CLAW-VD and 2-CLAW-VD are just the famous VERTEX COVER problem and the CLUSTER-VD problem, respectively.

In this paper, we show that CLUSTER-VD remains NP-complete when restricted to bipartite graphs of maximum degree 3. Moreover, for every $d \geq 3$, we show that $d$-CLAW-VD is NP-complete even when restricted to bipartite graphs of maximum degree $d$. These hardness results are optimal with respect to degree constraint. We also show that, for every $d \geq 3$, $d$-CLAW-VD is NP-complete when restricted to split graphs without $d+1$-claws. On the positive side, we prove that $d$-CLAW-VD is polynomially solvable on what we call $d$-block graphs, a class that properly contains all block graphs.

## 1 Introduction

Graph modification problems are a very extensively studied topic in graph algorithm. One important class of graph modification problems is the following. Let $H$ be a fixed graph. The $H$ VERTEX DELETION ($H$-VD) problem takes as input a graph $G$ and an integer $k$. The question is whether it is possible to delete a vertex set $S$ of most $k$ vertices from $G$ such that the resulting graph is $H$-*free*, i.e., $G - S$ contains no induced subgraphs isomorphic to $H$. The optimization version asks for such a vertex set $S$ of minimum size, and is denoted by MIN $H$ VERTEX DELETION (MIN $H$-VD).

The case $H$ is the 2-vertex path is the famous VERTEX COVER problem, one of the basic NP-complete problems. The case $H$ is 3-vertex path is well known under the name CLUSTER VERTEX DELETION (CLUSTER-VD for short). Very recently, the COCOON 2020 paper [1] addresses the case $H$ is the *claw*


C.-Y. Chen et al. (Eds.): COCOON 2021, LNCS 13025, pp. 591–603, 2021.
https://doi.org/10.1007/978-3-030-89543-3_49

$K_{1,3}$, the complete bipartite graph with 1 and 3 vertices in each part, thus the CLAW-VD problem.

For any integer $d > 0$, let *d-claw* stand for $K_{1,d}$, the complete bipartite graph with 1 and $d$ vertices on each part. In this paper, we go on with the CLAW-VD problem by considering the $d$-CLAW-VD problem for any given integer $d > 0$:

> $d$-CLAW-VD
> *Instance:* A graph $G = (V, E)$ and an integer $k < |V|$.
> *Question:* Is there a subset $S \subset V$ of size at most $k$ such that $G - S$ is $d$-claw free?

Thus, 1-CLAW-VD and 2-CLAW-VD are just the well known NP-complete problems VERTEX COVER and CLUSTER-VD, respectively, and 3-CLAW-VD is the CLAW-VD problem addressed in the recent paper [1] mentioned above.

While 1-CLAW-VD is polynomially solvable when restricted to perfect graphs (including chordal and bipartite graphs) [6], $d$-CLAW-VD is NP-complete for any $d \geq 2$ even when restricted to bipartite graphs [11,17]. When restricted to chordal graphs, it is shown in [1] that 3-CLAW-VD remains NP-complete even on split graphs. The computational complexity of 2-CLAW-VD on chordal graphs is still unknown [2,3]. Both 2-CLAW-VD and 3-CLAW-VD can be solved in polynomial time on block graphs [1,3], a proper subclass of chordal graphs containing all trees.

It is well known that the classical NP-complete problem VERTEX COVER remains hard when restricted to planar graphs of maximum degree 3 and arbitrary large girth. It is also known that, assuming ETH, VERTEX COVER admits no subexponential-time algorithm in the vertex number [12] and, assuming UGC, MIN VERTEX COVER cannot be approximated within to a factor better then 2 [9].

By the standard bounded search tree technique, $d$-CLAW-VD admits a parameterized algorithm running in $O^*((d+1)^k)$ time.[1] The current fastest parameterized algorithm for VERTEX COVER and CLUSTER-VD has runtime $O^*(1.2738^k)$ [5] and $O^*(1.811^k)$ [16], respectively. By the greedy algorithm, $d$-CLAW-VD can be approximated within a factor $d + 1$ but there is no polynomial-time approximation scheme [13]. From the results in [10] it is known that, for any $d \geq 2$, MIN$d$-CLAW-VD admits a $d$-approximation algorithm on bipartite graphs. This result was improved later by a result in [7], where the related problem $d$-CLAW-TRANSVERSAL was considered. Given a graph $G$, this problem asks to find a smallest vertex set $S \subseteq V(G)$ such that $G - S$ does not contain a $d$-claw as a (not necessarily induced) subgraph. In [7], it was shown that, in contrast to our MIN$d$-CLAW-VD problem, $d$-CLAW-TRANSVERSAL can be approximated within a factor of $O(\log(d + 1))$. Since $d$-CLAW-VD and $d$-CLAW-TRANSVERSAL coincide when restricted to bipartite graphs, $d$-CLAW-VD admits an $O(\log(d + 1))$-approximation on bipartite graphs. The case 2-CLAW-TRANSVERSAL is known under the name $P_3$ VERTEX COVER; see, e.g., [4,15].

In this paper, we first derive some hardness results by a simple reduction from VERTEX COVER to $d$-CLAW-VD, stating that $d$-CLAW-VD does not admit

[1] The $O^*$ notation hides polynomial factors.

a subexponential-time algorithm in the vertex number unless the Exponential-Time Hypothesis (ETH) fails, and that $d$-CLAW-VD remains NP-complete when restricted to planar graphs of maximum degree $d+1$ and arbitrary large girth.

We then revisit the case of bipartite input graphs by showing that CLUSTER-VD remains NP-complete on bipartite graphs of maximum degree 3, and for $d \geq 3$, $d$-CLAW-VD remains NP-complete on bipartite graphs of maximum degree $d$ and on bipartite graphs of diameter 3. These hardness results for $d$-CLAW-VD are optimal with respect to degree and diameter constraints, and improve the corresponding hardness results for $d$-CLAW-VD, $d \geq 2$, on bipartite graphs in [17].

Further, we extend the hardness results in [1] for CLAW-VD to $d$-CLAW-VD for every $d \geq 3$. We show that $d$-CLAW-VD is NP-complete even when restricted to split graphs without $d+1$-claws and, assuming the Unique Game Conjecture (UGC), it is hard to approximate MIN$d$-CLAW-VD to within a factor better than $d-1$.

We complement the negative results by showing that $d$-CLAW-VD is polynomial-time solvable on what we call $d$-block graphs, a class that contains all block graphs. As block graphs are 2-block graphs, and $d$-block graphs are $d+1$-block graphs but not the converse, our positive result extends the polynomial-time algorithm for 2-CLAW-VD on block graphs in [3] to $d$-CLAW-VD for all $d \geq 2$, and for 3-CLAW-VD on block graphs in [1] to 3-block graphs.

## 2 Preliminaries

Let $G = (V, E)$ be a graph with vertex set $V(G) := V$ and edge set $E(G) := E$. The neighborhood of a vertex $v$ in $G$, denoted by $N_G(v)$, is the set of all vertices in $G$ adjacent to $v$; if the context is clear, we simply write $N(v)$. Set $\deg(v) := |N(v)|$, the degree of the vertex $v$. We call a vertex *universal* if it is adjacent to all other vertices. Vertices of degree 1 are called *leaves*. The *distance* between two vertices in $G$ is the length of a shortest path connecting the two vertices, the *diameter* is the maximal distance between any two vertices, the *girth* is the length of a shortest cycle in $G$ (if any).

A *center vertex* of a $d$-claw $H$ is a universal vertex of $H$; if $d \geq 2$, the center of $d$-claws are unique. We say that a $d$-claw is *centered at vertex* $v$ if $v$ is a center vertex of that $d$-claw.

An *independent set* (a *clique*) in a graph $G = (V, E)$ is a set of pairwise non-adjacent (adjacent) vertices. $G$ is a *split graph* if its vertex set $V$ can be partitioned into an independent set and a clique.

For a subset $S \subseteq V$, $G[S]$ is the subgraph of $G$ induced by $S$, and $G - S$ stands for $G[V \setminus S]$. Let $H$ be a fixed graph. An *$H$-deletion set* is a vertex set $S \subseteq V(G)$ such that $G - S$ is $H$-free. A $K_{1,1}$-deletion set and a $K_{1,2}$-deletion set are known as *vertex cover* and *cluster deletion set*, respectively.

A *hypergraph* $G = (V, E)$ consists of a vertex set $V$ and an edge set $E$ where each edge $e \in E$ is a subset of $V$. Let $r \geq 2$ be an integer. A hypergraph is *$r$-uniform* if each of its edges is an $r$-element set. (Thus, a 2-uniform hypergraph is a graph in usual sense.) A *vertex cover* in a hypergraph $G = (V, E)$ is a vertex set $S \subseteq V$ such that $S \cap e \neq \emptyset$ for any edge $e \in E$. The $r$-HYPERGRAPH VERTEX COVER ($r$-HVC for short) problem asks, for a given $r$-uniform hypergraph $G = (V, E)$ and an integer $k < |V|$, whether $G$ has a vertex cover $S$ of size at most $k$. The optimization version asks for such a vertex set $S$ of minimum size and is denoted by MIN$r$-HVC. Note that 2-HVC and MIN2-HVC are the famous VERTEX COVER problem and MIN VERTEX COVER problem, respectively. It is known that $r$-HVC is NP-complete and MIN$r$-HVC is UGC-hard to approximate within a factor better than $r$ [9].

## 3 Hardness Results

We begin with two simple observations.

**Observation 1.** *$d$-CLAW-VD remains NP-complete on graphs of diameter 2.*

*Proof.* Given an instance $(G, k)$ for $d$-CLAW-VD, let $G'$ be obtained from $G$ by adding a $d$-claw with center vertex $v$ and joining $v$ to all vertices in $G$. Then $v$ is a universal vertex in $G'$ and hence $G'$ has diameter 2. Moreover, $(G, k) \in$ $d$-CLAW-VD if and only if $(G', k + 1) \in d$-CLAW-VD. □

We remark that the graph $G'$ in the proof above is a split graph whenever $G$ is a split graph, and $G'$ has only one vertex of unbounded degree whenever $G$ has bounded maximum degree. The bipartite version of Observation 1 is:

**Observation 2.** *For any $d \geq 2$, $d$-CLAW-VD remains NP-complete on bipartite graphs of diameter 3.*

*Proof.* Let $(G, k)$ be an instance for $d$-CLAW-VD, where $G = (X, Y, E)$ is a bipartite graph. Let $G'$ be the bipartite graph obtained from $G$ by adding two $d$-claws with center vertices $x$ and $y$, respectively, and joining $x$ to all vertices in $Y \cup \{y\}$ and $y$ to all vertices in $X \cup \{x\}$. Then $G'$ has diameter 3. Moreover, $(G, k) \in d$-CLAW-VD if and only if $(G', k + 2) \in d$-CLAW-VD. □

We remark that in the bipartite graph $G'$ in the proof above has only two vertices of unbounded degree whenever $G$ has bounded maximum degree.

We now describe a simple reduction from VERTEX COVER to $d$-CLAW-VD and some implications for the hardness of $d$-CLAW-VD. Let $d \geq 2$. Given a graph $G = (V, E)$, construct a graph $G' = (V', E')$ as follows.

- for each $v \in V$ let $I(v)$ be an independent set of $d - 1$ new vertices;
- $V' = V \cup \bigcup_{v \in V} I(v)$;
- $E' = E \cup \bigcup_{v \in V} \{vx \mid x \in I(v)\}$.

Thus, $G'$ is obtained from $G$ by attaching to each vertex $v$ a set $I(v)$ of $d - 1$ leaves.

**Fact 1.** *If $S$ is a vertex cover in $G$, then $S$ is a $d$-claw-deletion set in $G'$.*

*Proof.* This follows immediately from the construction of $G'$. Indeed, since $G-S$ is edgeless, every connected component of $G'-S$ is a single vertex (from $I(v)$ for some $v \in S$) or a $(d-1)$-claw (induced by $v$ and $I(v)$ for some $v \notin S$). Thus, $S$ is a $d$-claw-deletion set in $G'$. □

**Fact 2.** *If $S'$ is a d-claw-deletion set in $G'$, then $G$ has a vertex cover $S$ with $|S| \leq |S'|$.*

*Proof.* Let $S'$ be a $d$-claw-deletion set in $G'$. We may assume that, for every $v \in V(G)$, $S'$ contains no vertex in $I(v)$. Otherwise $(S' \setminus I(v)) \cup \{v\}$ is also a $d$-claw-deletion set in $G'$ of size at most $|S'|$. Thus $S' \subseteq V(G)$ and $S = S'$ is a vertex cover in $G$. For, if $uv$ is an edge in $G-S$, then $v$ and $\{u\} \cup I(v)$ induced a $d$-claw in $G'-S = G'-S'$, a contradiction. □

We now derive hardness results for $d$-CLAW-VD from the above reduction.

**Theorem 1.** *Assuming ETH, there is no $O^*(2^{o(n)})$ time algorithm for $d$-CLAW-VD on $n$-vertex graphs, even on graphs of diameter 2.*

*Proof.* By Facts 1 and 2, and the known fact that, assuming ETH, there is no $O^*(2^{o(n)})$ time algorithm for VERTEX COVER on $n$-vertex graphs [12]. Since the graph $G'$ in the construction has $|V'| = |V| + (d-1)|V| = O(|V|)$ vertices, we obtain that there is no $O^*(2^{o(n)})$ time algorithm for $d$-CLAW-VD, too. By (the proof of) Observation 1, the statement also holds for graphs of diameter 2. □

**Theorem 2.** *Let $d \geq 2$ be a fixed integer.*

*(i) $d$-CLAW-VD is NP-complete, even when restricted to planar graphs of maximum degree $d+1$ and arbitrary large girth.*
*(ii) $d$-CLAW-VD is NP-complete, even when restricted to diameter-2 graphs with only one vertex of unbounded degree.*

*Proof.* It is known (and it can be derived, e.g., from [8,14]) that VERTEX COVER remains NP-complete on planar graphs $G$ of maximum degree 3 and arbitrary large girth, and in which the neighbors of any vertex of degree 3 in $G$ have degree 2.

Given such a graph $G$, let $G'$ be obtained from $G$ by attaching, for every vertex $v$ of degree 2, $d-1$ leaves to $v$. Then $G'$ is planar, has maximum degree $d+1$ and arbitrary large girth. Moreover, similarly to Facts 1 and 2, it can be seen that $G$ has a vertex cover of size at most $k$ if and only if $G'$ has a $d$-claw deletion set of size at most $k$. This proves (i). Part (ii) follows from (i) and the reduction in the proof of Observation 1. □

We remark that the hardness result stated in Theorem 2 (ii) is optimal in the sense that graphs of bounded diameter and bounded vertex degree have bounded size, hence $d$-CLAW-VD is trivial when restricted to such graphs.

Note that $d$-CLAW-VD is trivial on graph of maximum degree less than $d$ (because such graphs contain no $d$-claws). Moreover, CLUSTER-VD is easily solvable on graphs of maximum degree 2. Thus, with Theorem 2 (i), the computational complexity of $d$-CLAW-VD, $d \geq 3$, on graphs of maximum degree $d$ remains to discuss. We will show in the next subsection that the problem is still hard even on bipartite graphs of maximum degree $d$.

### 3.1 Bipartite Graphs of Bounded Degree

Recall that VERTEX COVER is polynomially solvable on bipartite graphs, hence previous results reported above cannot be stated for bipartite graphs. In this subsection, we first give a polynomial reduction from POSITIVE 1-IN-3 3-SAT to CLUSTER-VD showing that CLUSTER-VD is NP-complete even when restricted to bipartite graphs of degree 3.

Then, we will modify this reduction to obtain a reduction from POSITIVE 1-IN-3 3-SAT to $d$-CLAW-VD showing that, for any $d \geq 3$, $d$-CLAW-VD is NP-complete even when restricted to bipartite graphs of maximum degree $d$. Thus, we obtain an interesting dichotomy for all $d \geq 3$: $d$-CLAW-VD is polynomial-time solvable on graphs of maximum degree less than $d$ and NP-complete otherwise.

Recall that an instance for POSITIVE 1-IN-3 3-SAT is a 3-CNF formula $F = C_1 \wedge C_2 \wedge \cdots \wedge C_m$ over $n$ variables $x_1, x_2, \ldots, x_n$, in which each clause $C_j$ consists of three distinct variables. The problem asks whether there is a truth assignment of the variables such that every clause in $F$ has exactly one true variable. Such an assignment is called *1-in-3 assignment.* It is well known that POSITIVE 1-IN-3 3-SAT is NP-complete.

Our reduction is inspired by a reduction from NAE 3-SAT to CLUSTER-VD on bipartite graphs in [17].

Let $F = C_1 \wedge C_2 \wedge \cdots \wedge C_m$ over $n$ variables $x_1, x_2, \ldots, x_n$, in which each clause $C_j$ consists of three distinct variables. We may assume that $m$ is even. We construct an instance $(G, k)$ for CLUSTER-VD as follows.

**Variable Gadget.** For each variable $x_i$ we introduce $m$ *variable vertices* $x_{ij}$ one for each clause $C_j$, $1 \leq j \leq m$, as follows. First, take a cycle with $m$ vertices $x_{i1}$, $x_{i2}$, ..., $x_{im}$ and edges $x_{i1}x_{i2}$, $x_{i2}x_{i3}$, ..., $x_{i(m-1)}x_{im}$ and $x_{i1}x_{im}$. Then subdivide every edge with 2 new vertices to obtain a cycle on $3m$ vertices. Finally, attach to every vertex $d-2$ leaves, and, in case $d > 2$, label an arbitrary leaf of each variable vertex $x_i^j$ by $x'_{ij}$. The obtained graph is denoted by $G(x_i)$.

The following property of $G(x_i)$ can be verified immediately:

**Fact 3.** *Any d-claw deletion set for $G(x_i)$ contains at least $m$ vertices, and every independent set of $m$ non-leaf vertices, two of which have no common neighbors, is a d-claw deletion set. Moreover, the set of all variable vertices $\{x_{i1}, x_{j2}, \ldots, x_{jm}\}$ is a d-claw deletion set, and any d-claw deletion set of size $m$ contains all or none variable vertices.*

**Clause Gadget.** First, take a triangle with *clause vertices* $c_{j1}$, $c_{j2}$ and $c_{j3}$. Then subdivide every edge with one new vertex to obtain a 6-cycle. Finally, attach to every vertex $d-2$ leaves, and, in case $d > 2$, label an arbitrary leaf of each clause vertex $c_{jk}$ by $c'_{jk}$. The obtained graph is denoted by $G(C_j)$.

The following property of $G(C_j)$ can be verified immediately:

**Fact 4.** *Any d-claw deletion set for $G(C_j)$ contains at least 2 vertices, and two non-leaf vertices at distance three form a d-claw deletion set. In particular, every 2-element d-claw deletion set for $G(C_j)$ contains exactly one of the clause vertices $c_{j1}, c_{j2}$ and $c_{j3}$.*

Finally, the graph $G$ is obtained by connecting the variable and clause gadgets as follows.

**Case** $d = 2$**:** connect the variable vertex $x_{ij}$ in $G(x_i)$ to a clause vertex in $G(C_j)$ by an edge whenever $x_i$ appears in clause $C_j$, i.e., $x_i = c_{jk}$ for some $k \in \{1, 2, 3\}$.

**Case** $d \geq 3$**:** connect the leaf $x'_{ij}$ of the variable vertex $x_{ij}$ in $G(x_i)$ to the labeled leaf of a clause vertex in $G(C_j)$ by an edge whenever $x_i$ appears in clause $C_j$, i.e., if $x_i = c_{jk}$ for some $k \in \{1, 2, 3\}$, then the there is an edge between $x'_{ij}$ and $c'_{jk}$.

It follows from construction, that $G$ has maximum degree 3 and is bipartite.

Set $k = nm + 2m$. We now show that $F \in positive1 - in - 33 - sat$ if and only if $(G, k) \in d$-CLAW-VD. First, assume that there is a 1-in-3 assignment for $F$. Then a $d$-claw deletion set $S$ of size $k = nm + 2m$ for $G$ is as follows, according to Facts 3 and 4.

**Case** $d = 2$**:**
- in each $G(C_j)$, put the true clause vertex $c_{jk}$ and the vertex at distance three to $c_{jk}$ into $S$,
- in each $G(x_i)$, if $x_i$ is *false* then put all $m$ variable vertices $x_{ij}$, $1 \leq j \leq m$, into $S$. Otherwise, if $x_i$ is true then put other $m$ vertices, which form a cluster deletion set in $G(x_i)$ and none of them is a variable vertex, into $S$.

**Case** $d \geq 3$**:**
- in each $G(C_j)$, put the true clause vertex $c_{jk}$ and the *non-leaf* vertex at distance three to $c_{jk}$ into $S$,
- in each $G(x_i)$, if $x_i$ is *true* then put all $m$ variable vertices $x_{ij}$, $1 \leq j \leq m$, into $S$. Otherwise, if $x_i$ is true then put other $m$ non-leaf vertices, which form a $d$-claw deletion set in $G(x_i)$ and none of them is a variable vertex, into $S$.

Second, if $S$ is a $d$-claw deletion set of $G$ with $|S| \leq mn + 2m$ then $S$ contains exactly 2 vertices from each $G(C_j)$. Hence, by Fact 4, $|S \cap \{c_{j1}, c_{j2}, c_{j3}\}| = 1$ for each $1 \leq j \leq m$. Thus, by Fact 3, defining $x_i = c_{jk}$ be `true` if $c_{jk} \in S$ and `false` if $c_{jk} \notin S$ we obtain a 1-in-3 assignment for $F$.

Since $G$ is bipartite and has maximum degree $d$, we obtain:

**Theorem 3.** CLUSTER-VD *is* NP*-complete even when restricted to bipartite graphs of maximum degree* 3*. For any* $d \geq 3$*,* $d$-CLAW-VD *is* NP*-complete even when restricted to bipartite graphs of maximum degree* $d$*.*

From Theorems 3 and (the proof of) Observation 2 we conclude:

**Theorem 4.** *For any* $d \geq 2$*,* $d$-CLAW-VD *is* NP*-complete even when restricted to bipartite graphs of diameter* 3 *with only two unbounded vertices.*

We remark that MIN$d$-CLAW-VD is polynomially solvable on bipartite graphs of diameter at most two. This can be seen as follows. Let $G = (X, Y, E)$ be a bipartite of diameter $\leq 2$; such a bipartite graph is complete bipartite. Note first that $X$ and $Y$ are $d$-claw deletion sets for $G$. We will see that any optimal $d$-claw deletion set is $X$ or $Y$ or is of the form $(X \setminus X') \cup (Y \setminus Y')$ for some $d-1$-element sets $X' \subseteq X$ and $Y' \subseteq Y$. (In particular, all optimal $d$-claw deletion sets can be found in $O(n^{d-1})$ time.) Indeed, let $S$ be an optimal $d$-claw deletion set. If $X \subseteq S$, then by the optimality of $S$, $S = X$. Similarly, if $Y \subseteq S$, then $S = Y$. So, let $X' = X \setminus S \neq \emptyset$ and $Y' = Y \setminus S \neq \emptyset$. Then $|X'| \leq d-1$ and $|Y'| \leq d-1$: if $|X'| \geq d$, say, then any vertex in $Y'$ and $d$ vertices in $X'$ together would induce a $d$-claw in $G - S$. Thus, by the optimality of $S$, $|X'| = |Y'| = d - 1$.

Unfortunately, we have to leave open the complexity of $d$-CLAW-VD on bipartite graphs of diameter 3 with only one vertex of unbounded degree.

### 3.2 Split Graphs

In this subsection, we show that, for any $d \geq 3$, $d$-CLAW-VD is NP-hard even when restricted to split graphs. Note that split graphs have diameter 3. By Observation 1, however, $d$-CLAW-VD is hard even on split graphs of diameter 2. Recall that 1-CLAW-VD and 2-CLAW-VD are solvable in polynomial time on split graphs.

Let $d \geq 3$ be a fixed integer. We reduce $(d-1)$-HVC to $d$-CLAW-VD. Our reduction is inspired by the reduction from VERTEX COVER to 3-CLAW-VD in [1]. Let $G = (V, E)$ be a $(d-1)$-uniform hypergraph with $n = |V|$ vertices and $m = |E|$ edges. We may assume that for any hyperedge $e \in E$ there is another hyperedge $f \in E$ such that $e \cap f = \emptyset$. For otherwise, $G$ has a vertex cover of size $\leq |e| = d-1$ and therefore $d-1$-HVC is polynomially solvable on such inputs $G$. We construct a split graph $G' = (V', E')$ with $V' = Q \cup I$, where $Q$ is a clique and $I$ is an independent set, as follows.

- $I = \{v' \mid v \in V\}$;
- for each edge $e \in E$, let $Q(e)$ be a clique of size $n$;
- all sets $I$ and $Q(e)$, $e \in E$, are pairwise disjoint;
- make $\bigcup_{e \in E} Q(e)$ to clique $Q$;
- for each $v' \in I$ and $e \in E$, connect $v'$ to all vertices in $Q(e)$ if and only if $v \in e$.

The description of the split graph $G'$ is complete. Note that $G'$ has $nm + n$ vertices and $O(n^2m^2)$ edges and can be constructed in $O(n^2m^2)$ time.

For each $e \in E$, write $e' = \{v' \in I \mid v \in e\}$. By construction, every vertex in $Q(e)$ has exactly $d-1$ neighbors in $I$, namely the vertices in $e'$. Hence, every induced $d$-claw in $G'$ is formed by a center vertex $x \in Q(e)$ for some $e \in E$ and $e' \cup \{y\}$, where $y$ is any vertex in $Q(f)$, $f \in E$, such that $f \cap e = \emptyset$. It follows that $G'$ contains no induced $(d+1)$-claws.

**Fact 5.** *If $S$ is a vertex cover in the hypergraph $G$, then $S' = \{v' \mid v \in S\}$ is a $d$-claw-deletion set in the split graph $G'$.*

*Proof.* If $C$ is a $d$-claw in $G'$ with center vertex $x \in Q(e)$ for some $e \in E$ such that $C \cap S' = \emptyset$, then $e' \cap S' = \emptyset$. This means $e \cap S = \emptyset$, contradicting the fact that $S$ is a vertex cover of the hypergraph $G$. □

**Fact 6.** *If $S'$ is a $d$-claw-deletion set in the split graph $G'$ of size $< n$, then $S = \{v \mid v' \in S'\}$ is a vertex cover in the hypergraph $G$.*

*Proof.* First, for each $e \in E$, $S' \cap e' \neq \emptyset$. For otherwise let $S' \cap e' = \emptyset$ for some $e \in E$. Since $|S'| < n$, there is a vertex $x \in Q(e) \setminus S'$ and a vertex $y \in Q(f) \setminus S'$ with $f \cap e = \emptyset$. Then $x, y$ and $e'$ induce a $d$-claw in $G' - S'$, a contradiction. We have seen that, for each $e \in E$, $S' \cap e' \neq \emptyset$. Then, with $S = \{v \mid v' \in S'\}$, we have $S \cap e \neq \emptyset$ for all $e \in E$. That is, $S$ is a vertex cover of the hypergraph $G$. □

**Fact 7.** *The size of a smallest vertex cover of $G$, $\text{OPT}_{\text{VC}}(G)$, and the size of a smallest $d$-claw-deletion set in $G'$, $\text{OPT}_{d\text{-CLAW-VD}}(G')$, are equal.*

*Proof.* By Fact 5, $\text{OPT}_{d\text{-CLAW-VD}}(G) \leq \text{OPT}_{\text{VC}}(G')$. Let $S'$ be a smallest $d$-claw-deletion set in $G'$. Then $|S'| < n$ because $I$ minus an arbitrary vertex is a $d$-claw-deletion set in $G'$ with $n - 1$ vertices. Hence, by Fact 6, $S = \{v \mid v' \in S'\}$ is a vertex cover in the hypergraph $G$ with $|S| \leq |S'|$. Thus, $\text{OPT}_{\text{VC}}(G') \leq \text{OPT}_{d\text{-CLAW-VD}}(G)$. □

We now derive hardness results for $d$-CLAW-VD and MIN$d$-CLAW-VD from the above reduction.

**Theorem 5.** *For any fixed $d \geq 3$, $d$-CLAW-VD is NP-complete, even when restricted to*

*(i) split graphs without induced $(d+1)$-claws, and*
*(ii) split graphs of diameter 2.*

*Proof.* Part (i) follows from Facts 5 and 6, and the fact that the split graph $G'$ contains no induced $(d+1)$-claws. Part (ii) follows from (i) and Observation 1. □

We remark that both hardness results in Theorem 5 are optimal in the sense that $d$-CLAW-VD is trivial for graphs without induced $d$-claws, in particular for graphs of diameter 1, i.e., complete graphs. We also remark that Theorem 5 implies, in particular, that $d$-CLAW-VD is NP-complete on chordal graphs for any $d \geq 3$, while the complexity of 2-CLAW-VD on chordal graphs is still open (cf. [2,3]).

Since it is UGC-hard to approximate MIN$(d-1)$-HVC to within a factor $(d-1) - \epsilon$ for any $\epsilon > 0$ [9], Fact 7 implies:

**Theorem 6.** *Let $d \geq 3$ be a fixed integer. Assuming the UGC, there is no approximation algorithm for* MIN$d$-CLAW-VD *within a factor better than $d - 1$, even when restricted to split graphs without induced $(d+1)$-claws.*

We remark that for triangle-free graphs, in particular bipartite graphs, MIN$d$-CLAW-VD and $d$-CLAW-TRANSVERSAL coincide, hence a result in [7] implies that MIN$d$-CLAW-VD admit an $O(\log(d+1))$-approximation when restricted to bipartite graphs.

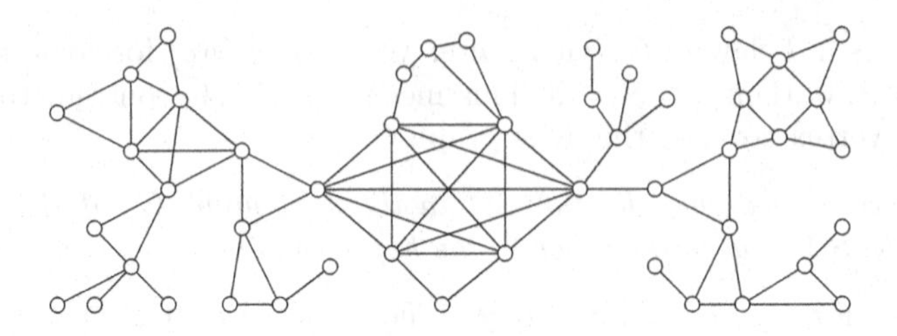

**Fig. 1.** A 3-block graph.

# 4 A Polynomially Solvabe Case

In this section, we will show a polynomial-time algorithm solving MIN$d$-CLAW-VD for what we call $d$-block graphs. As $d$-block graphs generalize block graphs, this result extends the polynomial-time algorithm for 2-CLAW-VD on block graphs given in [3] to $d$-CLAW-VD for all $d \geq 2$ on block graphs, and improves the polynomial-time algorithm for MIN3-CLAW-VD given in [1] on block graphs to 3-block graphs.

Recall that a *block* in a graph is a maximal biconnected subgraph. *Block graphs* are those in which every block is a clique. For each integer $d \geq 2$, the $d$-block graphs defined below generalize block graphs.

**Definition 1.** *Let $d \geq 2$ be an integer. A graph $G$ is $d$-*block graph *if, for every block $B$ of $G$,*

- *$B$ is $d$-claw free,*
- *for every cut vertex $v$ of $G$, $N(v) \cap B$ is a clique, and*
- *the cut vertices of $G$ in $B$ induce a clique.*

Note that block graphs are exactly the 2-block graphs and $d$-block graphs are $(d+1)$-block graphs, but not the converse. Note also that, for $d \geq 3$, $d$-block graphs need not be chordal; they may contain arbitrary long induced cycles. An example of a 3-block graph is shown in Fig. 1.

Let $d \geq 2$ and let $G$ be a $d$-block graph. Recall that a block in $G$ is an *endblock* if it contains at most one cut vertex. Vertices that are not cut vertices are called *endvertices*. Thus, if $u$ is an endvertex then the block containing $u$ (which may or may not be an endblock) is unique. We call a vertex $u$ a *pseudo-endvertex* if $u$ is an endvertex or $u$ belongs to at most $d-2$ endblocks and exactly one non-endblock. Thus, for a pseudo-endvertex $u$, we say that $B$ is the unique block containing $u$, meaning that in case $u$ is a cut vertex, $B$ is the unique non-endblock that contains $u$.

In computing an optimal $d$-claw deletion set for $G$, we will use the following facts.

**Fact 8.** *Let $u$ be a pseudo-endvertex. Then any $d$-claw $C$ containing $u$, if any, is centered at a cut vertex $v \neq u$. Moreover,*

– *if $B$ is the unique block containing $u$, then $C \cap B = \{u, v\}$;*
– *if $u$ is a cut vertex and $B'$ is an endblock containing $u$, then $C \cap B' = \{u\}$.*

*Proof.* This is because every block is $d$-claw-free and the neighbors of any cut vertex in any block induce a clique. □

An optimal $d$-claw deletion set for $G$ is also called *solution.*

**Fact 9.** *There is a solution that contains no pseudo-endvertices.*

*Proof.* Let $S$ be a solution for $G$ and assume that $S$ contains a pseudo-endvertex $u$. Let $B$ be the unique block of $G$ containing $u$. Since $S - u$ is not a $d$-claw deletion set, there is some $d$-claw $C$ of $G$ outside $S \setminus \{u\}$. Then, of course,

$$C \cap S = \{u\}. \tag{1}$$

By Fact 8, the center $v$ of $C$ is a cut vertex of $G$ in $B$, and $C \cap B = \{u, v\}$. Thus, for every $w \in N(v) \cap B$, $C - u + w$ is a $d$-claw, and by (1), $w \in S$. Hence

$$N(v) \cap B \subseteq S. \tag{2}$$

We now claim that $S' = S - u + v$ is a $d$-claw deletion set (and thus $S'$ is a solution). Indeed, let $C'$ be an arbitrary $d$-claw. If $u \notin C'$ or $v \in C'$ then $C' \cap S' \neq \emptyset$. So let us consider the case in which $u \in C'$ and $v \notin C'$. Then, by Fact 8, the center $v'$ of $C'$ is a cut vertex of $G$ in $B$. Hence $v'$ and $v$ are adjacent, and by (2), $v' \in S'$. □

We remark that Fact 9 is best possible in the sense that a cut vertex $u$ belonging to exactly two non-endblocks may be contained in any solution; take the $d$-block graph that consists of two $d$-claws with exactly one common leaf $u$.

**Fact 10.** *Let $v$ be a cut vertex and let $B$ be a block containing $v$. If every vertex in $N(v) \cap B$ is a cut vertex, then $B = N[v] \cap B$. In particular, $B$ is a clique.*

*Proof.* Suppose the contrary that $B \setminus N[v] \neq \emptyset$. Then, as $B - v$ is connected, there is an edge connecting a vertex $w \in N(v) \cap B$ and a vertex $w' \in B \setminus N[v]$. Now, as $w$ is a cut vertex, $N(w) \cap B$ is a clique, implying $vw'$ is an edge. This is a contradiction, hence $B = N[v] \cap B$. □

**Lemma 1.** *If $G$ has at most one block that is not an endblock, then a solution for $G$ can be computed in polynomial time.*

**Theorem 7.** MIN$d$-CLAW-VD *is polynomially solvable on $d$-block graphs.*

*Proof.* Let $T$ be the block-cut vertex tree of $G$. Nodes in $T$ corresponding to blocks in $G$ are *block nodes*; for a block node $u$ we use $B(u)$ to denote the corresponding block in $G$. Nodes in $T$ corresponding to cut vertices in $G$ are *cut nodes* and are denoted by the same labels.

Choose a node $r$ of $T$ and root $T$ at $r$. For a node $x \neq r$ of $T$, let $p(x)$ denote the parent of $x$ in $T$. Note that all leaves of $T$ are block nodes, the parent of a block node is a cut node and the parent of a cut node is a block node.

Let $u$ be a leaf of $T$ on the lowest level and let $v = p(u)$ be the parent of $u$. Note that all children of $v$ correspond to endblocks in $G$, and if $r = p(v)$, then Lemma 1 is applicable. So, assume $r \neq p(u)$ and write $u' = p(v)$, $v' = p(u')$. Note that by the choice of $u$, $B' = B(u')$ is the unqiue non-endblock containing vertices in $B' - v'$.

If $v$ has at most $d-2$ children, then $v$ is a pseudo-endvertex in $G$. By Fact 9, we remove $v$ and all children of $v$ from $T$.

If $v$ has at least $d$ children, or $v$ has exactly $d-1$ children and some vertex in $N_G(v) \cap B'$ is a pseudo-endvertex, then put $v$ into the solution $S$ and remove $v$ and all children of $v$ from $T$. Correctness follows again from Fact 9.

It remains the case that $v$ have exactly $d-1$ children and no vertex in $N_G(v) \cap B'$ is a pseudo-endvertex. In particular, every vertex in $N_G(v) \cap B'$ is a cut vertex, hence $B'$ is a clique by Fact 10. Moreover, every vertex in $N_G(v) \cap (B' - v')$ belongs to at least $d-1$ endblocks. Now, that all $d$-claws in $G$ containing $v$ contain a vertex in $B' - v$, and every solution for $G$ not containing pseudo-endvertices must contain at least $|B'| - 1$ vertices in $B'$. Thus, we put $B' - v$ into solution $S$ and remove the subtree rooted at $v'$ from $T$, and for each other child $u_i \neq u$ of $v'$, we solve the problem on the subgraph induced by $B(u_i)$ and its children. Note that, by the choice of $u$, all these subgraphs satisfy the condition of Lemma 1. □

## 5 Conclusion

This paper considers the $d$-claw vertex deletion problem, $d$-CLAW-VD, which generalizes the famous VERTEX COVER (that is 1-CLAW-VD) and the CLUSTER-VD (that is 2-CLAW-VD) problems and goes on with the recent study [1] on claw vertex deletion problem, 3-CLAW-VD. It is shown that CLUSTER-VD remains NP-complete on bipartite graphs of maximum degree 3 and, for each $d \geq 3$, $d$-CLAW-VD remains NP-complete on bipartite graphs of degree $d$, and thus a complexity dichotomy with respect to degree constraint. It is also shown that $d$-CLAW-VD remains NP-complete when restricted to split graphs of diameter 2 and to bipartite graphs of diameter 3 (with only two vertices of unbounded degree) and polynomially solvable on bipartite graphs of diameter 2, and thus another dichotomy with respect to diameter. We show that $d$-CLAW-VD is solvable in polynomial time on $d$-block graphs, a class that contains all block graphs, extending the algorithm for CLUSTER-VD on block graphs in [3] to $d$-CLAW-VD, and improving the algorithm for (unweighted) 3-CLAW-VD on block graphs in [1] to 3-block graphs.

## References

1. Bonomo-Braberman, F., Nascimento, J.R., Oliveira, F.S., Souza, U.S., Szwarcfiter, J.L.: Linear-time algorithms for eliminating claws in graphs. In: Kim, D., Uma, R.N., Cai, Z., Lee, D.H. (eds.) COCOON 2020. LNCS, vol. 12273, pp. 14–26. Springer, Cham (2020). https://doi.org/10.1007/978-3-030-58150-3_2
2. Cao, Y., Ke, Y., Otachi, Y., You, J.: Vertex deletion problems on chordal graphs. In: 37th IARCS Annual Conference on Foundations of Software Technology and Theoretical Computer Science, FSTTCS 2017, volume 93 of LIPIcs, pp. 22:1–22:14 (2017). https://doi.org/10.4230/LIPIcs.FSTTCS.2017.22
3. Cao, Y., Ke, Y., Otachi, Y., You, J.: Vertex deletion problems on chordal graphs. Theoret. Comput. Sci. **745**, 75–86 (2018). https://doi.org/10.1016/j.tcs.2018.05.039
4. Chang, M.-S., Chen, L.-H., Hung, L.-J., Rossmanith, P., Ping-Chen, S.: Fixed-parameter algorithms for vertex cover $p_3$. Discrete Optim. **19**, 12–22 (2016). https://doi.org/10.1016/j.disopt.2015.11.003
5. Chen, J., Kanj, I.A., Xia, G.: Improved upper bounds for vertex cover. Theoret. Comput. Sci. **411**(40–42), 3736–3756 (2010). https://doi.org/10.1016/j.tcs.2010.06.026
6. Grötschel, M., Lovász, L., Schrijver, A.: Geometric Algorithms and Combinatorial Optimization. Springer, Heidelberg (1988). https://doi.org/10.1007/978-3-642-97881-4
7. Guruswami, V., Lee, E.: Inapproximability of h-transversal/packing. SIAM J. Discrete Math. **31**(3), 1552–1571 (2017). https://doi.org/10.1137/16M1070670
8. Joseph Douglas Horton and Kyriakos Kilakos: Minimum edge dominating sets. SIAM J. Discrete Math. **6**(3), 375–387 (1993). https://doi.org/10.1137/0406030
9. Khot, S., Regev, O.: Vertex cover might be hard to approximate to within 2-epsilon. J. Comput. Syst. Sci. **74**(3), 335–349 (2008). https://doi.org/10.1016/j.jcss.2007.06.019
10. Kumar, M., Mishra, S., Safina Devi, N., Saurabh, S.: Approximation algorithms for node deletion problems on bipartite graphs with finite forbidden subgraph characterization. Theoret. Comput. Sci. **526**, 90–96 (2014)
11. Lewis, J.M., Yannakakis, M.: The node-deletion problem for hereditary properties is NP-complete. J. Comput. Syst. Sci. **20**(2), 219–230 (1980). https://doi.org/10.1016/0022-0000(80)90060-4
12. Lokshtanov, D., Marx, D., Saurabh, S.: Lower bounds based on the exponential time hypothesis. Bull. EATCS **105**, 41–72 (2011). http://eatcs.org/beatcs/index.php/beatcs/article/view/92
13. Lund, C., Yannakakis, M.: The approximation of maximum subgraph problems. In: Lingas, A., Karlsson, R., Carlsson, S. (eds.) ICALP 1993. LNCS, vol. 700, pp. 40–51. Springer, Heidelberg (1993). https://doi.org/10.1007/3-540-56939-1_60
14. Murphy, O.J.: Computing independent sets in graphs with large girth. Discrete Appl. Math. **35**(2), 167–170 (1992). https://doi.org/10.1016/0166-218X(92)90041-8
15. Tsur, D.: Parameterized algorithm for 3-path vertex cover. Theoret. Comput. Sci. **783**, 1–8 (2019). https://doi.org/10.1016/j.tcs.2019.03.013
16. Tsur, D.: Faster parameterized algorithm for cluster vertex deletion. Theory Comput. Syst. **65**(2), 323–343 (2020). https://doi.org/10.1007/s00224-020-10005-w
17. Yannakakis, M.: Node-deletion problems on bipartite graphs. SIAM J. Comput. **10**(2), 310–327 (1981). https://doi.org/10.1137/0210022

# Constrained Hitting Set Problem with Intervals

Ankush Acharyya[1], Vahideh Keikha[1], Diptapriyo Majumdar[2], and Supantha Pandit[3](✉)

[1] Institute of Computer Science, The Czech Academy of Sciences, Prague, Czech Republic
{acharyya,keikha}@cs.cas.cz
[2] Royal Holloway, University of London, Egham, UK
diptapriyo.majumdar@rhul.ac.uk
[3] Dhirubhai Ambani Institute of Information and Communication Technology, Gandhinagar, India

**Abstract.** We study a constrained version of the *Geometric Hitting Set* problem where we are given a set of points, partitioned into disjoint subsets, and a set of intervals. The objective is to hit all the intervals with a minimum number of points such that if we select a point from a subset then we must select all the points from that subset. In general, when the intervals are disjoint, we prove that the problem is in FPT, when parameterized by the size of the solution. We also complement this result by giving a lower bound in the size of the kernel for disjoint intervals, and we also provide a polynomial kernel when the size of all subsets is bounded by a constant.

Next, we consider two special cases of the problem where each subset can have at most 2 and 3 points. If each subset contains at most 2 points and the intervals are disjoint, we show that the problem admits a polynomial-time algorithm. However, when each subset contains at most 3 points and intervals are disjoint, we prove that the problem is NP-Hard and we provide two constant factor approximations for the problem.

## 1 Introduction

The *Hitting Set* problem is a well-studied problem in theoretical computer science, especially in combinatorics, computational geometry, operation research, complexity theory, etc. In the classical setup of the *Hitting Set* problem, a universe of elements $U$ and a collection $\mathcal{F} \subseteq 2^U$ are given. The goal is to find the smallest subset $S \subseteq U$ that intersects every set in $\mathcal{F}$. The decision version of the *Hitting Set* problem is known to be NP-Complete, whereas the optimization version of the problem is NP-Hard [16]. Significant attention is also given to the geometric version of the *Hitting Set* problem due to its practical importance. In this version, $U$ is considered to be a set of points and $\mathcal{F}$ is a set of geometric

A. Acharyya and V. Keikha—The author is supported by the Czech Science Foundation, grant number GJ19-06792Y, and by institutional support RVO: 67985807.


C.-Y. Chen et al. (Eds.): COCOON 2021, LNCS 13025, pp. 604–616, 2021.
https://doi.org/10.1007/978-3-030-89543-3_50

objects (such as intervals, disks, boxes, etc.). Due to the underlying geometric structure of these objects, different *Geometric Hitting Set* problems are shown to be polynomial-time solvable, however many problems remain NP-Hard [15,16].

We study a constrained variation of the *Geometric Hitting Set* problem, the *Constrained Hitting Set* problem with intervals, defined as follows:

**Constrained Hitting Set with Intervals (*CHSI*):** We are given a set of closed intervals, $\mathcal{I}$ and a set $P$ of $n$ points in $\mathbb{R}$ partitioned into $d$ nonempty subsets $P_1, P_2, \ldots, P_d$, such that $\bigcup_{i=1}^{d} P_i = P$ and $P_i \cap P_j = \emptyset$ for all $i \neq j$, $i, j \in \{1, 2, \ldots, d\}$. The objective is to find a subset $P' \subseteq P$ of minimum number of points such that each interval in $\mathcal{I}$ is hit[a] and for each $p \in P'$, if $p \in P_i$ for some $i \in \{1, \ldots, d\}$ then $P_i \subseteq P'$.

[a] An interval $I$ is said to be *hit* by a point $p$ if and only if $p \in I$.

To be precise, we are interested in the following variations of the *CHSI* problem based on the size (number of points) of the subsets and the underlying structure of the intervals. We define the *CHSI-tD* (resp. *CHSI-tO*) problem as the *CHSI* problem with intervals where each subset $P_i$ is of size at most $t$ and the given intervals are disjoint (resp. overlapping). Note that for $t = 1$, the *CHSI-tO* problem is the standard *Hitting Set* problem with intervals on the real line and can be solved in $O(n \log n)$ time [20]. When the size of the subsets is not bounded by any fixed number, then we call this variant as *CHSI-D* problem. We also consider a variation where we minimize the number of subsets of points, instead of the total number of points. We call such a variation as *Weak Constrained Hitting Set* with intervals (*WCHSI-D* problem).

We denote the decision version of the *CHSI-D* problem as the *DCHSI-D* problem where one additional parameter $k$ is given as part of the input with usual input of the *CHSI-D* problem and the objective is to decide whether there is a set of at most $k$ points that satisfy the constraints and hit all the intervals. The total number of points in the solution is at most $k$. Similarly, we denote the decision versions of the variations *CHSI-tD*, *CHSI-tO* as *DCHSI-tD*, *DCHSI-tO* problems. Further, we denote the decision version of the *WCHSI-D* problem as the *DWCHSI-D* problem.

A possible application of the *CHSI* problem is to provide efficient project management system. To satisfy the requirement of a project with a set of skills like developing, programming, visualizing, etc., the workload needs to be divided among the employees with proficiency in programming, data analysis, design, etc. The requirements of the project can be modeled as intervals and the expertise of employees as the set of points. To manage all the requirements of the project, we need all the employees to have the required expertise. Then the objective is to assign each of the projects to a set of employees satisfying the project requirements, and identify a number of smallest possible resources to complete it. Another possible application of a special case of the *CHSI* problem where the intervals are disjoint is as follows. Suppose that there is a number of working sites where a number of workers work. These sites need to be supervised by a collection of supervisors during the working hours of a day. The total working hour is divided into small chunks of time windows. A supervisor needs to visit many sites as assigned to him/her during the start of a day. Now for a particular site a number of supervisors visit in different time windows. During each time window a supervisor needs to be present. This problem can be modelled as the *CHSI-D* problem, where time windows are represented as intervals, visiting a particular site in a time window by a supervisor represents a point in that time window, and visiting the site by a supervisor in different time windows represents a subset of points hitting a collection of intervals (corresponds to the time windows). Now minimizing the number of supervisors visiting a particular site is same as minimizing the number of subsets that hit all the disjoint intervals.

### 1.1 Related Work

A rich body of work has been done for the classical version of the *Hitting Set* problem that is equivalent to the classical *Set Cover* problem [5]. There is a well-known greedy algorithm for the *Hitting Set* problem that gives an $O(\log n)$-factor approximation [16,17] and we can not get an $o(\log n)$-factor approximation unless P=NP [13]. However, exploiting the underlying geometry, the*Hitting Set* problem on some geometric objects can be solved in polynomial-time or some NP-Hard problems have better approximation factors [4,6,7,22]. More specifically, both *Set Cover* and *Hitting Set* problems with intervals on the real line can be solved in polynomial time using greedy algorithms [20]. In one dimension, the *Geometric Hitting Set* (also *Set Cover*) problem on different objects remains NP-Complete [3], however, for a restricted class of objects they can be solved efficiently [16]. The *Constrained Hitting Set* problem was introduced by Cornet and Lafornet [9] on general graphs. They provided various computational hardness status and approximation algorithms for different problems, such as *Vertex Cover*, *Connected Vertex Cover*, *Dominating Set*, *Total Dominating Set*, *Independent Dominating Set*, *Spanning Tree*, *Connected Minimum Weighted Spanning Graph*, *Matching*, and *Hamiltonian Path* problems. These vertex deletion problems on graphs with obligation can be interpreted as variants of the *Constrained Implicit Hitting Set* problems on graphs. On the contrary, the *conflict-free* versions of *Implicit Hitting Set* problems on graphs have also been studied [2,8,19,28]. In the conflict-free version, a different conflict graph with the same input vertex set is provided as part of the input. The goal is to find a set of size at most $k$ that forms a corresponding implicit hitting set in the original input graph, but an independent set in the conflict graph.

From the perspective of parameterized complexity, *Hitting Set* and the *Set Cover* problems are W[2]-hard [11] parameterized by solution size. However, when all sets in the input have at most $d$ (for some constant $d$), elements, the *d-Hitting Set* problem admits a polynomial kernel [1] parameterized by solution size. Computing a kernel of smaller size is also studied in [24]. Also, polynomial-sized kernels for hitting set for a fixed $d$ have already been presented in [14]. On the other hand, if the *Set Cover* problem is parameterized by the number of elements in the universe, then it is FPT and does not admit a polynomial kernel unless $\mathsf{NP} \subseteq \mathsf{coNP/poly}$ [10]. Jacob et al. [18] studied the conflict-free version of the Set Cover problem with parameterized complexity and kernelization perspective. Related problems of [18] are also studied in [26]. See also [25] and the references therein.

### 1.2 Our Contribution

➢ We show that the *DCHSI-D* problem admits an algorithm taking $O^*(2^k)$-time, where $k$ is the total number of points in the solution. We also prove that the *DCHSI-tD* problem admits a kernel with $k$ intervals and $O(t^2k^t)$ points.
➢ We prove that the *DWCHSI-D* problem parameterized by the number of intervals does not admit a polynomial kernel unless $\mathsf{NP} \subseteq \mathsf{coNP/poly}$. We also give an algorithmic lower bound of the *DWCHSI-D* problem based on the Set Cover Conjecture.
➢ The *CHSI-2D* problem admits a polynomial-time algorithm.
➢ The *CHSI-3D* problem is NP-Hard. We present two constant factor approximations for this problem.

Due to lack of space, some proofs are omitted; they will be provided in the full version of the paper.

## 2 Parameterized Complexity for Disjoint Intervals

### 2.1 Preliminaries

A parameterized problem is $\Pi \subseteq \Sigma^* \times \mathbb{N}$ for some finite alphabet $\Sigma$. An instance of a parameterized problem is $(x, k) \in \Sigma^* \times \mathbb{N}$ where $k$ is called the parameter and $x$ is the input. We assume that $k$ is given in unary and without loss of generality $k \leq |x|$, and $|x|$ is the input length. A parameterized problem $\Pi \subseteq \Sigma \times \mathbb{N}$ is said to be *fixed-parameter tractable* (or *FPT*) if there exists an algorithm that runs in $f(k)|x|^c$ time where $f : \mathbb{N} \to \mathbb{N}$ is a computable function and $c$ is a constant.

Kernelization in parameterized complexity is a polynomial-time preprocessing algorithm. Formally, given an instance $(x, k)$ of a parameterized problem $\Pi \subseteq \Sigma^* \times \mathbb{N}$, *kernelization* is a polynomial-time algorithm that transforms the input instance $(x, k)$ to $(x', k')$ such that **(i)** $(x, k) \in \Pi$ if and only if $(x', k') \in \Pi$, and **(ii)** $|x'| + k' \leq f(k)$ for some function $f : \mathbb{N} \to \mathbb{N}$ depending only on $k$. If $f(k)$ is $k^{O(1)}$, then we say that $\Pi$ has a *polynomial kernel.* Informally speaking, kernelization is a collection of *reduction rules* that have to be applied in sequence to reduce the original instance into an equivalent instance. A reduction rule that replaces input instance $(x, k)$ by $(x', k')$ is *safe* if $(x, k)$ is a yes-instance if and only if $(x', k')$ is a yes-instance. It is well-known that, a parameterized problem is FPT if and only if it admits a kernelization [11]. Another type of polynomial-time preprocessing in parameterized complexity is called a “compression”. Formally, given an instance $(x, k)$ of parameterized problem $\Pi \subseteq \Sigma^* \times \mathbb{N}$, *compression* transforms $(x, k)$ to an equivalent instance $x' \in \Sigma^*$ of an unparameterized problem $L \subseteq \Sigma^*$ in polynomial-time such that $x'$ can be represented by $f(k)$ bits. If $f(k)$ is in $k^{O(1)}$, then we say that $\Pi$ admits a *polynomial compression.* Informally speaking, polynomial compression is a polynomial-time preprocessing algorithm that transforms the input instance of a parameterized problem to an input instance of a possibly different unparameterized problem with a polynomial number of bits.

Let $\Pi_1$ and $\Pi_2$ be two parameterized problems. If there exists a polynomial-time reduction that given an instance $(x, k)$ of $\Pi_1$, constructs an instance $(x', k')$ such that $k' = O(k^{O(1)})$, then we say that there exists a *polynomial parameter transformation (PPT)* from $\Pi_1$ to $\Pi_2$.

### 2.2 Fixed-Parameter Tractability for Disjoint Intervals

We show that the *DCHSI-D* problem with disjoint intervals is fixed parameter tractable parameterized by the size of the solution. *DCHSI-D* is NP-Hard when there are subsets of points that are of size at least 3 (see Sect. 4). Our kernel lower bound results also prove that the *DWCHSI-D* problem is NP-Hard.

We apply the following reduction rules in sequence.

**Reduction Rule 1.** If the number of intervals is more than $k$, then the given instance is a “NO” instance.

**Reduction Rule 2.** If there are two subsets $P_i, P_j$ in $P$ such that $|P_i| = |P_j|$ and both of them hit the same set of intervals, then we remove $P_i$ from the input.

**Reduction Rule 3.** If there exists a subset $P_i$ that does not hit any interval, i.e. $P_i \cap \mathcal{I} = \emptyset$, then we simply remove that subset $P_i$ from the input.

**Reduction Rule 4.** If any subset $P_i$ contains more than $k$ points, we can remove $P_i$ from the input. Such a subset only makes the size of the solution more than $k$. Thus we have the following lemma.

**Lemma 1.** *Reduction Rules 1, 2, 3, and 4 are safe, and can be implemented in polynomial-time. Thus the DCHSI-D problem admits a kernel of size $O(2^k k)$ and an FPT algorithm with $O^*(2^{k^2})$ running time.*

**Dynamic Programming:** Now, we describe an improved $O^*(2^k)$ time algorithm by using *dynamic programming* over subsets of intervals $(\mathcal{I})$ where the set of points are $P = P_1 \cup P_2 \cup \ldots \cup P_d$, and $P_i \cap P_j = \emptyset$ for all $i \neq j$. For every $P_i$, we denote $w(P_i) = |P_i|$ (number of points in $P_i$). Since Reduction Rule 1 is not applicable, $|\mathcal{I}| \leq k$. We fix an arbitrary ordering $P_1, P_2, \ldots, P_d$. For every $i \in [d]$, we use $c(P_i)$ to denote the set of intervals hit by the points in $P_i$.

For every subsets of intervals $\mathcal{X} \subseteq \mathcal{I}$, for every $i \in \{1, \ldots, d\}$, we define $\mathsf{B}[\mathcal{X}, i]$, the weight of a smallest subset $\mathcal{P} \subseteq \{P_1, \ldots, P_i\}$ such that $\mathcal{X} \subseteq$

$\bigcup_{P \in \mathcal{P}} c(P)$. Informally speaking, in the table entry $\mathsf{B}[\mathcal{X}, i]$, we store the weight of a smallest subset $\mathcal{P} \subseteq \{P_1, \ldots, P_i\}$ that hits all intervals in $\mathcal{X}$. Since no point is required to hit $\emptyset \subseteq \mathcal{I}$, for all $i \in \{1, 2, \ldots, d\}$, we initialize $\mathsf{B}[\emptyset, i] = 0$. For $\mathcal{X} \neq \emptyset$, if $\mathcal{X} \subseteq c(P_1)$, then $\mathsf{B}[\mathcal{X}, 1] = |P_1| = w(P_1)$. Otherwise, when $\mathcal{X} \not\subset c(P_1)$, we denote $\mathsf{B}[\mathcal{X}, 1] = \infty$. For all other $\mathcal{X} \subseteq \mathcal{I}$, and for all $i \in [d]$, we initialize $\mathsf{B}[\mathcal{X}, i] = \infty$. We use the following recurrence relation. For every $\mathcal{X} \subseteq \mathcal{I}$ such that $\mathcal{X} \neq \emptyset$, and for every $i \geq 2$, we denote $\mathsf{B}[\mathcal{X}, i] = \min\{|P_i| + \mathsf{B}[\mathcal{X} \setminus c(P_i), i-1], \mathsf{B}[\mathcal{X}, i-1]]\}$. Thus we have the following.

**Theorem 1.** *The DCHSI-D problem can be solved in $O^*(2^k)$ time.*

### 2.3 Kernelization and FPT Lower Bound for Disjoint Intervals:

We prove that the *DWCHSI-D* problem admits no polynomial compression unless $\mathsf{NP} \subseteq \mathsf{coNP/poly}$. In this variant, we aim to minimize the number of distinct subsets of points rather than the total number of points in the solution. We also prove a lower bound based on *Set Cover Conjecture* for the same problem.

We give a reduction from the *Set Cover* as follows. The input to a *Set Cover* is a universe $U = \{1, 2, \ldots, n\} = [n]$, and a family $\mathcal{F} \subseteq 2^U$, and an integer $k$. The objective is to find a subfamily $\mathcal{F}' \subseteq \mathcal{F}$ such that $|\mathcal{F}'| \leq k$ and $U = \bigcup_{A \in \mathcal{F}'} A$.

**Lemma 2** ([12])**.** *The Set Cover problem parameterized by $|U|$ admits no polynomial compression unless $\mathsf{NP} \subseteq \mathsf{coNP/poly}$.*

**Conjecture 1 (Set Cover Conjecture** [10]**).** *The Set Cover problem cannot be solved in $O^*((2-\epsilon)^{|U|})$ time.*

**Lemma 3** ([11])**.** *Let $\Pi_1, \Pi_2$ be two parameterized problems and suppose that there exists a polynomial parameter transformation from $\Pi_1$ to $\Pi_2$. Then, if $\Pi_1$ does not admits a polynomial compression, neither does $\Pi_2$.*

**Reduction:** Let $(U, \mathcal{F}, k)$ be an instance of the *Set Cover* problem such that $U = \{x_1, x_2, \ldots, x_n\}$ and $\mathcal{F} = \{S_1, \ldots, S_m\}$. For every $i \in [n]$, let us denote $\mathsf{oc}(x_i) = \{j \in [m] | x_i \in S_j\}$ and let $\delta = \max\{|\mathsf{oc}(x_i)| : i \in [n]\}$. Informally speaking, $\delta$ is the maximum number of sets at which an element of the universe can occur. For every $x_i \in U$, we arrange the indices of $\mathsf{oc}(x_i)$ in increasing order. **(i)** We construct the set of intervals $\mathcal{I} = \{[(i-1)\delta+1, i\delta] : i \in \{1, 2, \ldots, n\}\}$, **(ii)** we construct the point set $P$ and its partition into $m$ nonempty sets $P_1, \ldots, P_m$ as follows, Observe that for every $x_i \in U$, the set $\mathsf{oc}(x_i)$ denotes the increasing order at which $x_i$ occurs across several sets in $\mathcal{F}$. For every $j \in [m]$, we create $P_j$ as follows. Consider an arbitrary $x_i \in S_j$. If $j$ is the $r$'th occurrence of $x_i$ in $\mathsf{oc}(x_i)$, then we add the point $(i-1)\delta + r$ into $P_j$. In other words, every occurrence of an element is represented by a unique point in a specific subset of points, **(iii)** finally, we denote $P = P_1 \cup \ldots \cup P_m$.

Observe that by construction, every element in $U$ has its corresponding interval in $\mathcal{I}$. Also observe that the point $(i-1)\delta + r$ (corresponding to $x_i \in U$) hits the interval $[(i-1)\delta, i\delta]$ since $r \leq \delta$. Hence, the sets $P_j$'s are pairwise disjoint.

Because the occurrence of an element $x_i$ across distinct sets in the family is represented by various points in the same interval $[(i-1)\delta, i\delta]$. Also observe that for all $x_i \in U$, $\mathsf{oc}(x_i) \neq \emptyset$.

Next consider $\mathcal{F}' = \{S_{j_1}, \ldots, S_{j_k}\}$ be a subfamily of size at most $k$ that covers $U$. Then, $P' = P_{j_1} \cup \ldots \cup P_{j_k}$ is the solution to *DWCHSI-D* instance. The idea is that the corresponding interval $[(i-1)\delta+1, i\delta]$ will be hit by $r$'th ($r \leq \delta$) occurrence of $x_i$ in $P_{j_r}$. Therefore, there are $k$ subsets points $P_{j_1} \cup \ldots \cup P_{j_k}$ that hit all intervals and satisfy the constraints. On the other hand, let $P' = P_{j_1} \cup \ldots \cup P_{j_k}$ be $k$ subsets of points that hit all the intervals. If interval $[(i-1)\delta+1, i\delta]$ is hit by a point in $P_{j_r}$, then, the element $x_i$ has $t$'th occurrence ($t \leq \delta$) in $S_{j_r}$. Hence, $\mathcal{F}' = \{S_{j_1}, \ldots, S_{j_k}\}$ covers $U$. This leads to the following lemma.

**Lemma 4.** *For a Set Cover instance $(U, \mathcal{F}, k)$, the instance $(U, \mathcal{F}, k)$ has a feasible solution of size at most $k$ if and only if there are $k$ subsets of points that hit all intervals in DWCHSI-D instance we created by the reduction.*

Lemmas 2, 4, and Conjecture 1 lead the following theorem.

**Theorem 2.** *The DWCHSI-D problem parameterized by $|\mathcal{I}|$ admits no polynomial compression unless* NP $\subseteq$ coNP/poly. *Moreover, unless Conjecture 1 fails, the DWCHSI-D problem cannot be solved in $O^*((2-\epsilon)^{|\mathcal{I}|})$ time.*

### 2.4 Polynomial Kernel for Subsets of Size at Most *t* for Fixed *t*

We provide a polynomial kernel for the *DCHSI-tD* problem parameterized by solution size ($k$). Recall that all subsets of points in the input instance has size at most $t$. Thus we have the following:

**Theorem 3.** *When Reduction Rules 1, 2, 3, and 4 are not applicable, then an instance of the DCHSI-tD problem has $k$ intervals and at most $O(t^2k^t)$ points. Hence, the DCHSI-tD problem parameterized by solution size ($k$) admits a kernel with $k$ intervals and $O(t^2k^t)$ points.*

## 3 Subset Size at Most 2, Disjoint Intervals

In this section, we show that the *CHSI-2D* problem can be solved in polynomial-time. We first convert the *CHSI-2D* problem to an equivalent problem, where the size of each subset is exactly 2. We call it as *CHSI-2D-exact* problem. Next, we reduce the *CHSI-2D-exact* problem to the edge cover problem[1] in a graph.
***CHSI-2D-exact* problem instance construction:**

Let $\mathcal{I} = \{i_1, i_2, \ldots, i_\gamma\}$ be a set of pairwise disjoint intervals and $P$ be a set of points partitioned into subsets $\{P_1, P_2, \ldots, P_d\}$, where $2d \geq \gamma$. Note that, in the given instance each $P_i$, $1 \leq i \leq d$, contains at most 2 points. Without loss of generality, we assume that each point hits at least one interval in the set $\mathcal{I}$,

[1] The edge cover problem defined on a graph finds the set of edges of a minimum size such that every vertex of the graph is incident to at least one edge of the set.

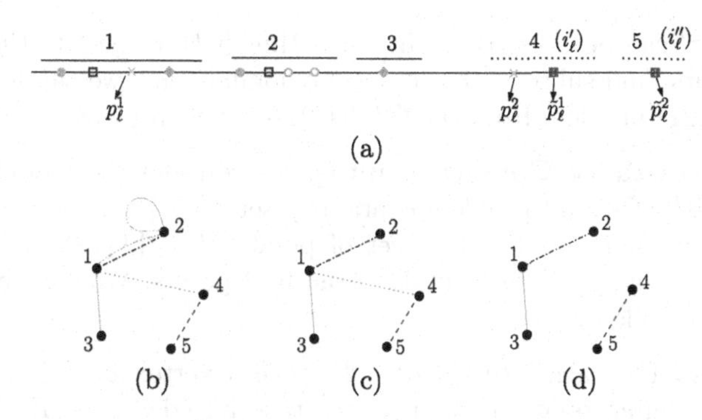


**Fig. 1.** (a) From the *CHSI-2D* problem to the *CHSI-2D-exact* problem: $p^2_\ell$ is a dummy point for the single point $p^1_\ell$, $\tilde{p}_\ell{}^1$ and $\tilde{p}_\ell{}^2$ are the dummy points for the dummy intervals $i'_\ell$ and $i''_\ell$. (b) converting into an equivalent graph $G_\tau$ (c) removing self-loops and parallel edges (d) corresponding edge-cover of $G_\tau$ (for interpretation of the references to color in this figure legend, the reader is referred to the web version).

otherwise, we can do the following. If any set (having one or two points) does not hit any interval then we delete that set from $P$. If only point $\alpha$ of a set hits an interval $\tilde{i}$ (the other point does not) then the following cases can happen:

(i) $\tilde{i}$ is only hit by $\alpha$, then we include the set containing $\alpha$ into our solution and delete the interval from set $\mathcal{I}$,
(ii) $\tilde{i}$ is hit by other points also apart from $\alpha$, then remove the set containing $\alpha$ from our consideration. Next, for each subset of $P_\ell \in P$ that contains exactly one point, say $p^1_\ell$, we do the following, as illustrated in Fig. 1(a) :

Take one extra (dummy) point $p^2_\ell \in P_\ell$, take two extra (dummy) points $\tilde{p}^1_\ell$ and $\tilde{p}^2_\ell$ that belongs to a single new set, say $\tilde{P}_\ell$, and take two additional (dummy) disjoint intervals $i'_\ell$ and $i''_\ell$. We place the intervals $i'_\ell$ and $i''_\ell$ to the extreme right of the current configuration such that these two intervals do not overlap with the existing configuration. The points $p^2_\ell$ and $\tilde{p}^1_\ell$ hit the interval $i'_\ell$ and the point $\tilde{p}^2_\ell$ hit the interval $i''_\ell$.

Let $\tau$ be an original instance of the *CHSI-2D* problem with exactly one set $P_\ell$ that contains exactly one point and $\tau^*$ be the instance constructed above. We have the following lemma.

**Lemma 5.** *The instance $\tau$ has a solution of size $s$ if and only if either (i) $\tau^*$ has a solution of size $s+3$ if $P_\ell$ is in the solution of $\tau$ or (ii) $\tau^*$ has a solution of size $s+2$ if $P_\ell$ is not in the solution of $\tau$.*

We repeat the above procedure for each subset of $P$ one by one that contains exactly 1 point. Therefore, in the final instance, say $\tau'$ all the subsets contain exactly 2 points. By repeated application of Lemma 5 we ensure that finding a solution of the *CHSI-2D* problem is equivalent to finding a solution to the *CHSI-2D-exact* problem. Observe that $\tau'$ can contain at most $3\gamma$ number of intervals and at most $4d$ number of points partitioned into at most $2d$ subsets.

The instance $\tau'$ can be constructed in linear time with respect to the number of intervals, points, and subsets. Hence, in polynomial-time, we can get a solution of the *CHSI-2D* problem from the *CHSI-2D-exact* problem.

**Edge Cover Instance Construction:** Let us consider the modified instance $\tau'$ of the *CHSI-2D-exact* problem contains a set $\mathcal{I}' = \{i_1, i_2, \ldots, i_{\gamma'}\}$, $\gamma' \leq 3\gamma$, of pairwise disjoint intervals and a set of points $P = \{P_1, P_2, \ldots, P_{d'}\}$ where each $P_i$, $1 \leq i \leq d'$, $d' \leq 2d$, contains exactly 2 points. We construct a graph $G_{\tau'} = (V, E)$ as follows:

**Construction:** For each interval $i_l \in \mathcal{I}'$, take a vertex $v_l \in V$ and for each subset $P_j$ containing points $p_j^1$ and $p_j^2$, we take an edge $e_j \in E$. The edge $e_j$ connects the vertices $v_{l'}$ and $v_{l''}$ if and only if the interval $i_{l'}$ corresponding to $v_{l'}$ contains the point $p_j^1$ and the interval $i_{l''}$ corresponding to $v_{l''}$ contains the point $p_j^2$. Note that, if a single interval $i_l$ contains both the points $p_j^1$ and $p_j^2$ then the edge $e_j$ is a self loop on the vertex $v_l$. If both intervals $i_{l'}$ and $i_{l''}$ are hit by the two points of the subset $P_{j'}$ as well as by the two points of the subset $P_{j''}$, then there are parallel edges between $v_{l'}$ and $v_{l''}$.

We now process (removing redundant and trivial edges) the graph $G_{\tau'}$ without affecting the size of the solution. If there are parallel edges between two vertices of $G_{\tau'}$, then we keep exactly one edge between them and remove the remaining edges. Note that this modification does not affect the size of the optimal solution, since the subsets corresponding to the parallel edges hit the same set of intervals, and hence only one of them can be selected in the optimal solution. Let the resultant graph be $\tilde{G_\tau}$. Next, we remove all the self-loops from $\tilde{G_\tau}$. Let the resultant graph be $\tilde{G}'$ (Fig. 1(c)). Let $v$ be a vertex in $\tilde{G}'$. Here two cases may arise based on the number of the loops and edges incident on $v$; Case (i) only loops ($\geq 1$) are incident on $v$ and Case (ii) loops as well as some other edges are incident on $v$. In Case (i), the interval corresponding to $v$ covers the subsets (both points) corresponding to the loops. We arbitrarily choose one loop and insert the corresponding subset into our solution $P'$ and remove $v$ from the graph $\tilde{G_\tau}$. However, in Case (ii), the interval corresponding to $v$ covers the subsets (both points) corresponding to the loops and at least one subset that has exactly one point hit the interval. In this case, we delete the self-loops incident on $v$, because choosing an edge as opposed to choosing a self-loop incident on $v$ does not worsen the size of the solution. After processing the parallel edges and self-loops let the resultant graph is $G' = (V', E')$.

Next, we find an edge cover (see Fig. 1(d)) in the graph $G'$ using the maximum matching algorithm and then greedily add a minimum number of edges such that all the vertices are covered. We add all the points corresponding to those edges to our solution $P'$. Given a graph $G'$ of $n$ vertices and $m$ edges then we can find its edge cover in $O(mn)$-time [16,21]. Thus we have the following:

**Theorem 4.** *The CHSI-2D-exact problem (hence the CHSI-2D problem) can be solved in $O(n \log n + \gamma n)$ time, where $\gamma$ denotes the number of intervals in $\mathcal{I}$.*

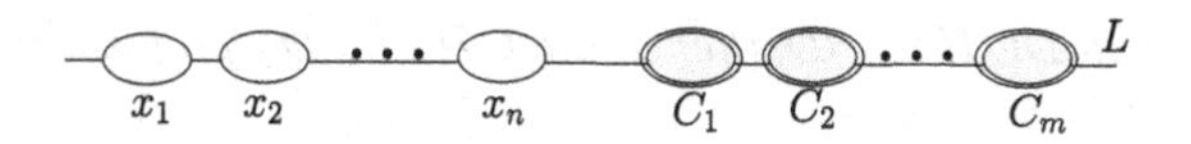


**Fig. 2.** Overall structure

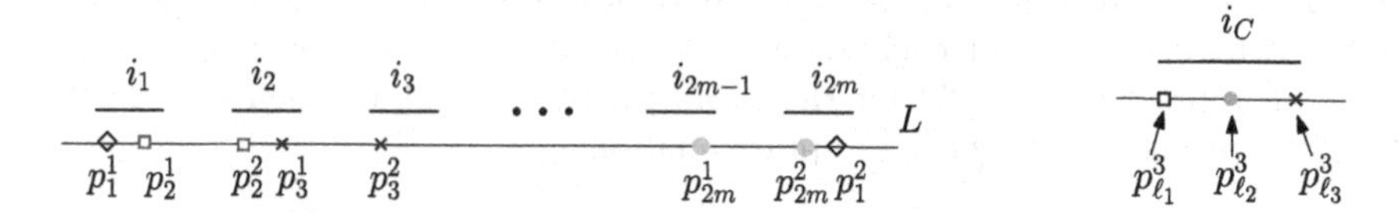


**Fig. 3.** A variable gadget.

**Fig. 4.** A clause gadget.

## 4 Subset Size at Most 3, Disjoint Intervals

We now prove that the *CHSI*-3*D* problem is NP-hard. We give a reduction from the *Positive-1-in-3-SAT* problem that is known to be NP-complete [16,23].

***Positive-1-in-3-SAT*** [16,23]**:** We are given a 3-SAT formula $\phi$ with $m$ clauses and $n$ variables such that each clause contains exactly three positive literals, the objective is to decide whether there exists an assignment of truth values to the variables of $\phi$ such that exactly one literal is true in each clause of $\phi$.

**Reduction:** We create an instance $I_\phi$ of the *DCHSI*-3*D* problem from an instance $\phi$ of the *Positive-1-in-3-SAT* problem as follows.

**Overall Structure:** We place the variable and clause gadgets one by one from left to right on a line $L$ (see Fig. 2 for a schematic diagram). To the left, place the variable gadgets, and after that place the clause gadgets one after another.

**Variable Gadget:** For each variable, we take $2m$ subsets of points and each subset $P_i$ contains two points $p_i^1$ and $p_i^2$, for $1 \leq i \leq 2m$. The points are ordered left-to-right on a real line $L$ as $p_1^1, p_2^1, p_2^2, \dots, p_{2m}^1, p_{2m}^2, p_1^2$. We also take $2m$ unit intervals $\{i_1, i_2, \dots, i_{2m}\}$ such that interval $i_1$ is hit by points $p_1^1$ and $p_2^1$, $i_{2m}$ is hit by points $p_{2m}^2$ and $p_1^2$, and for $2 \leq j \leq 2m-1$, interval $i_j$ is hit by points $p_j^2$ and $p_{j+1}^1$. See the construction in Fig. 3. Observe that there are exactly two optimal solutions that hit the intervals: either $G_1 = \{P_1, P_3, \dots, P_{2m-1}\}$ or $G_2 = \{P_2, P_4, \dots, P_{2m}\}$, each solution contains $2m$ points (Fig. 4).

**Clause Gadget:** Let $C$ be a clause that contains the three positive literals $x_i$, $x_j$, and $x_k$. Also let $x_i$ is the $l_1$-th, $x_j$ is the $l_2$-th, and $x_k$ is the $l_3$-th occurrences in the formula $\phi$. For $C$, the gadget consists of three points $p_{l_1}^3$, $p_{l_2}^3$, and $p_{l_3}^3$ and one interval $i_C$ that is hit by these three points. The point $p_{l_1}^3$ is in the subset $P_{l_1}$ of the gadget of $x_i$. Similarly, the points $p_{l_2}^3$ and $p_{l_3}^3$ is in the subsets $P_{l_2}$ and $P_{l_3}$ of the gadget of $x_j$ and $x_k$ respectively.

This completes the construction that can be done in polynomial-time with respect to the number of variables and clauses in $\phi$. We have the following lemma.

**Lemma 6.** *Exactly one literal is true in every clause of $\phi$ if and only if the intervals in $I_\phi$ are hit by $2mn + m$ points.*

**Theorem 5.** *The DCHSI-3D (hence the CHSI-3D) problem is NP-hard.*

### 4.1 Approximation Algorithms

The *CHSI-kD* problem can be reduced to the standard weighted set cover problem where the size of each set is bounded by $k$. Thus, we can obtain a $H_k$ factor approximation [27] for the problem. In particular, when $k = 3$ (the *CHSI-3D* problem), we get a $\frac{11}{6}$ approximation.

**Lemma 7.** *The CHSI-3D problem can be approximated by a* $H_3 = \frac{11}{6}$*-factor.*

**A $\frac{5}{3}$-factor approximation algorithm:** We now propose an improved $\frac{5}{3}$ factor approximation algorithm for the *CHSI-3D* problem in Algorithm 1.

For each subset $P_i$ we define its $\rho$ value as $\left\lceil \frac{\text{number of intervals hit by the points in } P_i}{\text{number of points in } P_i} \right\rceil$.

**Algorithm 1.** $\frac{5}{3}$-factor approximation algorithm

---

**Input:** $\mathcal{I}$: set of intervals, $P = \{P_1 \cup \ldots \cup P_d\}$: set of points where $|P_i| \leq 3$, $P_i \cap P_j = \emptyset$.
**Output:** *A subset $P'$ of $P$ that hits all the intervals of $\mathcal{I}$.*
1: $P' \leftarrow \emptyset$;
2: **while** all the intervals are not "hit" **do**
3: Sort the $\rho$ values of the subsets;
4: $P' \leftarrow P' \cup$ set having largest $\rho$;
5: Remove that subset and the intervals those are "hit";
6: update the corresponding $\rho$ values of remaining subsets;
7: **return** $P'$;

Algorithm 1 picks a subset with the highest $\rho$ value in each iteration to $P'$ and also updates the $\rho$ values of the subsets in $P \setminus P'$ after removing the intervals those are hit by $P'$. As we have disjoint intervals and each subset contains at most 3 points, the possible $\rho$ values are $1, \frac{2}{3}, \frac{1}{2}, \frac{1}{3}$. We select the subsets with respect to the non-decreasing order of their $\rho$ values. It can be justified that Algorithm 1 returns a $\frac{5}{3}$-approximate solution, by ensuring that at each iteration, our algorithm chooses at most $\frac{5}{3}$ points compared to the optimum solution for those set of intervals hit till that step. Thus, we conclude the following theorem.

**Theorem 6.** *The CHSI-3D problem can be approximated within a factor of* $\frac{5}{3}$ *in* $O(n \log n)$ *time.*

## 5 Conclusion

We study a constrained version of the *Geometric Hitting Set* problem where the intervals are either disjoint (*CHSI-tD* problem) or overlapping (*CHSI-tO* problem). We show that the *DCHSI-D* problem is in FPT. We also prove that the *CHSI-tD* problem is NP-Hard for $t = 3$ while the *CHSI-tD* problem is polynomial-time solvable for $t = 2$ and gave a $\frac{5}{3}$-factor approximation algorithm for *CHSI-3D*. It would be interesting to investigate whether the approximation can be generalized for any $t$. The computational complexity of the *CHSI-tO* problem for $t = 2$, the parameterized complexity and approximation algorithm for the *CHSI-tO* problem, for any $t \geq 2$ also remains interesting open questions.

## References


1. Abu-Khzam, F.N.: A kernelization algorithm for $d$-hitting set. J. Comput. Syst. Sci. **76**(7), 524–531 (2010)
2. Agrawal, A., Jain, P., Kanesh, L., Saurabh, S.: Parameterized complexity of conflict-free matchings and paths. Algorithmica **82**(7), 1939–1965 (2020). https://doi.org/10.1007/s00453-020-00681-y
3. Aho, A.V., Hopcroft, J.E., Ullman, J.D.: The Design and Analysis of Computer Algorithms. Addison-Wesley, Boston (1974)
4. Aronov, B., Ezra, E., Sharir, M.: Small-size $\epsilon$-nets for axis-parallel rectangles and boxes. SIAM J. Comput. **39**(7), 3248–3282 (2010)
5. Ausiello, G., D'Atri, A., Protasi, M.: Structure preserving reductions among convex optimization problems. J. Comput. Syst. Sci. **21**(1), 136–153 (1980)
6. Brönnimann, H., Goodrich, M.T.: Almost optimal set covers in finite VC-dimension. Discrete Comput. Geom. **14**(4), 463–479 (1995). https://doi.org/10.1007/BF02570718
7. Clarkson, K.L., Varadarajan, K.R.: Improved approximation algorithms for geometric set cover. Discrete Comput. Geom. **37**(1), 43–58 (2007). https://doi.org/10.1007/s00454-006-1273-8
8. Cornet, A., Laforest, C.: Total domination, connected vertex cover and Steiner tree with conflicts. Discrete Math. Theoret. Comput. Sci. **19**(3) (2017)
9. Cornet, A., Laforest, C.: Graph problems with obligations. In: Kim, D., Uma, R.N., Zelikovsky, A. (eds.) COCOA 2018. LNCS, vol. 11346, pp. 183–197. Springer, Cham (2018). https://doi.org/10.1007/978-3-030-04651-4_13
10. Cygan, M., et al.: On problems as hard as CNF-SAT. ACM Trans. Algorithms **12**(3), 41:1-41:24 (2016)
11. Cygan, M., et al.: Parameterized Algorithms. Springer, Cham (2015). https://doi.org/10.1007/978-3-319-21275-3
12. Dom, M., Lokshtanov, D., Saurabh, S.: Kernelization lower bounds through colors and ids. ACM Trans. Algorithms **11**(2), 13:1-13:20 (2014)
13. Feige, U.: A threshold of $\ln n$ for approximating set cover. J. ACM **45**(4), 634–652 (1998)
14. Flum, J., Grohe, M.: Parameterized Complexity Theory. TTCSAES, Springer, Heidelberg (2006). https://doi.org/10.1007/3-540-29953-X
15. Fowler, R.J., Paterson, M., Tanimoto, S.L.: Optimal packing and covering in the plane are NP-Complete. Inf. Process. Lett. **12**(3), 133–137 (1981)
16. Garey, M.R., Johnson, D.S.: Computers and Intractability: A Guide to the Theory of NP-Completeness. W. H. Freeman, New York (1979)
17. Goldschmidt, O., Hochbaum, D.S., Yu, G.: A modified greedy heuristic for the set covering problem with improved worst case bound. Inf. Process. Lett. **48**(6), 305–310 (1993)

18. Jacob, A., Majumdar, D., Raman, V.: Parameterized complexity of conflict free set cover. Theory Comput. Syst. **65**(3), 515–540 (2021). https://doi.org/10.1007/s00224-020-10022-9
19. Jain, P., Kanesh, L., Misra, P.: Conflict free version of covering problems on graphs: classical and parameterized. Theory Comput. Syst. **64**(6), 1067–1093 (2020). https://doi.org/10.1007/s00224-019-09964-6
20. Kleinberg, J.M., Tardos, É.: Algorithm Design. Addison-Wesley, Boston (2006)
21. Lawler, E.L.: Combinatorial Optimization: Networks and Matroids. Courier Corporation, Chelmsford (2001)
22. Mustafa, N.H., Raman, R., Ray, S.: Quasi-polynomial time approximation scheme for weighted geometric set cover on pseudodisks and halfspaces. SIAM J. Comput. **44**(6), 1650–1669 (2015)
23. Schaefer, T.J.: The complexity of satisfiability problems. In: Proceedings of the 10th Annual ACM Symposium on Theory of Computing, pp. 216–226. ACM (1978)
24. Van Bevern, R.: Towards optimal and expressive kernelization for $d$-hitting set. Algorithmica **70**(1), 129–147 (2014). https://doi.org/10.1007/s00453-013-9774-3
25. van Bevern, R., Smirnov, P.V.: Optimal-size problem kernels for $d$-hitting set in linear time and space. Inf. Process. Lett. **163**, 105998 (2020)
26. van Bevern, R., Tsidulko, O.Y., Zschoche, P.: Representative families for matroid intersections, with applications to location, packing, and covering problems. Discrete Appl. Math. **298**, 110–128 (2021)
27. Vazirani, V.V.: Approximation Algorithms. Springer, Heidelberg (2001)
28. Yinnone, H.: On paths avoiding forbidden pairs of vertices in a graph. Discrete Appl. Math. **74**(1), 85–92 (1997)

# Exact Algorithms for Maximum Weighted Independent Set on Sparse Graphs (Extended Abstract)

Sen Huang[1], Mingyu Xiao[1](✉), and Xiaoyu Chen[2]

[1] University of Electronic Science and Technology of China, Chengdu, China
[2] Nanjing University, Nanjing, China

**Abstract.** The maximum independent set problem is one of the most important problems in graph algorithms and has been extensively studied in the line of research on the worst-case analysis of exact algorithms for NP-hard problems. In the weighted version, each vertex in the graph is associated with a weight and we are going to find an independent set of maximum total vertex weight. In this paper, we design several reduction rules and a fast exact algorithm for the maximum weighted independent set problem, and use the measure-and-conquer technique to analyze the running time bound of the algorithm. Our algorithm works on general weighted graphs and it has a good running time bound on sparse graphs. If the graph has an average degree at most 3, our algorithm runs in $O^*(1.1443^n)$ time and polynomial space, improving previous running time bounds for the problem in cubic graphs using polynomial space.

**Keywords:** Maximum weighted independent set · Exact algorithms · Measure-and-Conquer · Graph algorithms · Reduction rules

## 1 Introduction

The Maximum Independent Set problem on unweighted graphs belongs to the first batch of 21 NP-hard problems proved by Karp [12]. In the line of research on the worst-case analysis of exact algorithms for NP-hard problems, Maximum Independent Set, as one of the most fundamental problems, is used to test the efficiency of new techniques of exact algorithms. There is a long list of contributions to exact algorithms for Maximum Independent Set in unweighted graphs [2,8,11,13,16,17]. Now it can be solved in $O^*(1.1996^n)$ time and polynomial space [21]. If the maximum degree of the graph is 3, the running time bound can be improved to $O^*(1.0836^n)$ [20].

In this paper, we will consider the weighted version of Maximum Independent Set, called Maximum Weighted Independent Set, where each vertex in the graph has a nonnegative weight and we are asked to find an independent

The work is supported by the National Natural Science Foundation of China, under grant 61972070.


C.-Y. Chen et al. (Eds.): COCOON 2021, LNCS 13025, pp. 617–628, 2021.
https://doi.org/10.1007/978-3-030-89543-3_51

set with the maximum total vertex weight. It has many applications in various real-world problems. For example, the dynamic map labeling problem [1,15] can be naturally encoded as MAXIMUM WEIGHTED INDEPENDENT SET. Some experimental algorithms, such as the algorithms in [14,19] have been developed to solve instances from real world and known benchmarks. These algorithms run fast even on large scale sparse instances but lack running time analysis. For running time bounds, most known results were obtained via two counting problems: COUNTING MAXIMUM WEIGHTED INDEPENDENT SET and COUNTING WEIGHTED 2-SAT. Most of these counting algorithms can also list out all independent sets and then we can find a maximum one by increasing only a polynomial factor. Dahllöf et al. [4] presented an $O^*(1.3247^n)$-time algorithm for COUNTING MAXIMUM WEIGHTED INDEPENDENT SET. Later, the running time bound was improved to $O^*(1.2431^n)$ by Fomin et al. [6]. COUNTING MAXIMUM WEIGHTED INDEPENDENT SET can also be reduced to COUNTING WEIGHTED 2-SAT, preserving the exponential part of the running time. For COUNTING WEIGHTED 2-SAT, the running time bound was improved from $O^*(1.2561^n)$ [5] to $O^*(1.2461^n)$ [9] and then to $O^*(1.2377^n)$ [18]. Wahlström [18] also showed that the running time bound could be further improved to $O^*(1.1499^n)$ and $O^*(1.2117^n)$ if the maximum degree of the variables or the vertices in the graph is bounded by 3 and 4, respectively. Most of the above algorithms use only polynomial space. If exponential space is allowed, dynamic programming algorithms based on tree decompositions, by using the treewidth bound on degree-3 graphs in [7], may achieve a better running time bound $O^*(1.1225^n)$.

In this paper, we will focus on exact algorithms specifying for MAXIMUM WEIGHTED INDEPENDENT SET. We develop structural properties and design reduction rules for the problem, and then design a fast exact algorithm based on them. By using the measure-and-conquer technique, we can prove that the algorithm runs in $O^*(1.1443^{(0.624x-0.872)n})$ time and polynomial space, where $x$ is the average degree of the graph. For some sparse graphs, our result beats the known bounds. For example, the running time bound of our algorithm in graphs with the average degree at most three is $O^*(1.1443^n)$, which improves the previously known bound of $O^*(1.1499^n)$ using polynomial space [18]. For graphs with the average degree at most 3.68, the running time of our algorithm is strictly better than the running time bound $O^*(1.2117^n)$ for MAXIMUM WEIGHTED INDEPENDENT SET in degree-4 graphs [18].

Due to the limited space, the proofs of lemmas marked with (*) were omitted, which can be found in the full version of this paper [10].

## 2 Preliminaries

Let $G = (V, E, w)$ denote an undirected vertex-weighted graph with $|V| = n$ vertices and $|E| = m$ edges, where each vertex $v \in V$ is associated with a positive weight $w(v)$. Although our graphs are undirected, we may use an arc to denote the relation of the weights of the two endpoints of an edge. An *arc* $\overrightarrow{uv}$ from vertex $u$ to vertex $v$ means that there is an edge between $u$ and $v$ and it holds that $w(u) \geq w(v)$.

Let $V' \subseteq V$ be a vertex subset. We let $w(V') = \sum_{v \in V'} w(v)$, and $N(V')$ denote the set of vertices not in $V'$ but adjacent to at least one vertex in $V'$. We also denote $d(V') = |N(V')|$ and $N[V'] = N(V') \cup V'$. We use $G[V']$ to denote the subgraph of $G$ induced by $V'$ and use $G - V'$ to denote $G[V \setminus V']$. For a graph $G'$, we use $\mathcal{C}(G')$ to denote the set of connected components of $G'$. A *chain* is an induced path such that the degree of each vertex except the two endpoints of the path is exactly 2. One vertex is a *chain-neighbor* of another vertex if they are connected by a chain. For a vertex-weighted graph, a *maximum weighted independent set* is an independent set $S$ such that $w(S)$ is maximized among all independent sets in the graph. We use $S(G)$ to denote a maximum weighted independent set in graph $G$ and $\alpha(G)$ to denote the total vertex weight of $S(G)$. Our problem is defined below.

---

MAXIMUM WEIGHTED INDEPENDENT SET (MWIS)
**Input**: An undirected vertex-weighted graph $G = (V, E, w)$.
**Output**: the weight of a maximum weighted independent set in $G$., i.e., $\alpha(G)$.

---

### 2.1 Measure-and-Conquer

Our algorithm is a branch-and-search algorithm. We will use a measure to evaluate the time complexity. For a branching operation, if the measure decreases by at least $a_i$ in the $i$-th substance, then we say the *branching vector* of the operation is $[a_1, a_2, \ldots, a_l]$. The largest root of the function $f(x) = 1 - \sum_{i=1}^{l} x^{-a_i}$ is called the *branching factor* of the recurrence.

The measure-and-conquer technique, introduced in [8], is a powerful tool to analyze branch-and-search algorithms. The main idea is to use a non-traditional measure to evaluate the running time. Let $n_i$ denote the number of vertices of degree $i$ in the graph. We associate a cost $\delta_i \geq 0$ for each degree-$i$ vertex in the graph. Our measure $p$ is defined as follows:

$$p := \sum_{i=0}^{n} n_i \delta_i. \tag{1}$$

The cost $\delta_i$ in this paper is given by

$$\delta_i = \begin{cases} 0 & \text{if } i \leq 1 \\ 0.376 & \text{if } i = 2 \\ 1 & \text{if } i = 3 \\ 1 + 0.624(i-3) & \text{if } i \geq 4. \end{cases} \tag{2}$$

We also define $\delta_i^{<-k>} := \delta_i - \delta_{i-k}$ for each integer $k \geq 0$. In our analysis, we may use the following inequalities and equalities to simplify some arguments: $\delta_i^{<-1>} = \delta_3^{<-1>}$ for $i \geq 4$; $\delta_3 \geq 2.5\delta_2$; $3\delta_2 \geq \delta_3$.

With the above setting, we know that when $p \leq 0$, the instance contains only degree-0 and degree-1 vertices and can be solved directly. We will design an algorithm with running time bound $O^*(c^p)$ for some constant $c$. If the initial graph has degree at most 3, then we have that $p \leq n$ and then the running time bound of the algorithm is $O^*(c^n)$. In general, if we have $p \leq f(n)$ for some function $f$ on $n$, then we can get a running time bound of $O^*(c^{f(n)})$. We have the following lemma for the relation between $p$ and $n$.

**Lemma 1.** *(*) For a graph of $n$ vertices, if the average degree of the graph is at most $x$, then the measure $p$ of the graph is at most* $(0.624x - 0.872)n$.

## 3 Reduction Rules

We first introduce reduction rules that will be applied to reduce the instance directly by eliminating some local structures of the graph. Some reduction rules may include a set $S$ of vertices in the solution set directly. We use $M_c$ to store the weight of the vertices that have been included in the solution set. When a set $S$ of vertices is included in the solution set, we will remove $N[S]$ from the graph and update $M_c$ by adding $w(S)$.

### 3.1 General Reductions for Some Special Structures

We use several reduction rules based on unconfined vertices, twins, vertices with a clique neighborhood, and heavy vertices. Some of these reduction rules were introduced in [14] and [19].

**Unconfined Vertices.** A vertex $v$ in $G$ is called *removable* if $\alpha(G) = \alpha(G - v)$, i.e., there is a maximum weighted independent set in $G$ that does not contain $v$. We can say that a vertex $v$ is removable if a contradiction is obtained from the assumption that every maximum weighted independent set in $G$ contains $v$. A sufficient condition for a vertex to be removable in unweighted graphs has been studied in [20]. We extend this concept to weighted graphs.

For an independent set $S$ of $G$, a vertex $u \in N(S)$ is called a *child* of $S$ if $w(u) \geq w(S \cap N(u))$. A child $u$ is called an *extending child* if it holds that $|N(u) \setminus N[S]| = 1$, and the only vertex $v \in N(u) \setminus N[S]$ is called a *satellite* of $S$.

**Lemma 2.** *(*) Let $S$ be an independent set that is contained in any maximum weighted independent set in $G$. Then every maximum weighted independent set contains at least one vertex in $N(u) \setminus N[S]$ for each child $u$ of $S$.*

We introduce a method based on Lemma 2 to find possible removable vertices. Let $v$ be an arbitrary vertex in the graph. After starting with $S := \{v\}$, we repeat (1) until (2) or (3) holds:

(1) If $S$ has some extending child in $N(S)$, then let $S'$ be the set of satellites. Update $S$ by letting $S := S \cup S'$.
(2) If $S$ is not an independent set or there is a child $u$ such that $N(u) \setminus N[S] = \emptyset$, then halt and conclude that $v$ is *unconfined.*
(3) If $|N(u) \setminus N[S]| \geq 2$ for all children $u \in N(S)$, then halt and return $S_v = S$.

Obviously, the procedure can be executed in polynomial time for any starting set $S$ of a vertex. If the procedure halts in (2), we say vertex $v$ *unconfined.* If the procedure halts in (3), then we say that the set $S_v$ returned in (3) *confines* vertex $v$ and vertex $v$ is also called *confined.* Note that the set $S_v$ confining $v$ is uniquely determined by the procedure with starting set $S := \{v\}$. It is easy to observe the following lemma.

**Lemma 3.** *(*) Any unconfined vertex is removable.*

**Reduction Rule 1 (R1).** *If a vertex $v$ is unconfined, remove $v$ from $G$.*

**Twins.** A set $A = \{u, v\}$ of two non-adjacent vertices is called a *twin* if they have the same neighbor set, i.e., $N(u) = N(v)$.

**Reduction Rule 2 (R2)** [14]**.** *If there is a twin $A = \{u, v\}$, delete $v$ and update the weight of $u$ by letting $w(u) := w(u) + w(v)$.*

**Clique Neighborhood.** A vertex $v$ has a *clique neighborhood* if the graph $G[N(v)]$ induced by the open neighbor set of $v$ is a clique, which was introduced as isolated vertices in [14].

**Reduction Rule 3 (R3)** [14]**.** *If there is a vertex $v$ having a clique neighborhood and $w(v) < w(u)$ holds for all $u \in N(v)$, then remove $v$ from the graph, update the weight $w(u) := w(u) - w(v)$ for all $u \in N_G(v)$, and add $w(v)$ to $M_c$.*

**Heavy Vertices.** A vertex $v$ is called a *heavy vertex* if its weight is not less the weight of the maximum weighted independent set in subgraph induced by the open neighborhood of it, i.e., $w(v) \geq \alpha(G[N(v)])$.

**Reduction Rule 4 (R4).** *If there is a heavy vertex $v$ of degree at most 5, then delete $N[v]$ from the graph and add $w(v)$ to $M_c$.*

It is an effective rule that has been used in some experimental algorithms [14, 19]. In this paper, we will only check heavy vertices of degree bounded by 5 and then it can be done in polynomial time. Note that degree-0 vertices will be reduced as heavy vertices in this step.

### 3.2 Reductions Based on Degree-2 Vertices

For unweighted graphs, we have good reduction rules to deal with all degree-2 vertices (see the reduction rule in [3]). However, for weighted graphs, it becomes much more complicated. The following R5 is generalized from the concept of folding degree-2 vertices in unweighted graphs in [3], which has been also used in some experimental algorithms [14,19]. We also consider more reduction rules for degree-2 vertices in some complicated structures.

**Reduction Rule 5 (R5).** *If there is a degree-2 vertex $v$ with two neighbors $\{u_1, u_2\}$ such that $w(u_1) + w(u_2) > w(v) \geq max\{w(u_1), w(u_2)\}$, then delete $\{v, u_1, u_2\}$ from the graph $G$, introduce a new vertex $v'$ adjacent to $N_G(\{v, u_1, u_2\})$ with weight $w(v') := w(u_1) + w(u_2) - w(v)$, and add $w(v)$ to $M_c$.*

**Reduction Rule 6 (R6)** ([19]). *If there is a path $v_1v_2v_3v_4$ such that $d_G(v_2) = d_G(v_3) = 2$ and $w(v_1) \geq w(v_2) \geq w(v_3) \geq w(v_4)$, then remove $v_2$ and $v_3$ from the graph, add an edge $v_1v_4$ if it does not exist, update the weight of $v_1$ by letting $w(v_1) := w(v_1) + w(v_3) - w(v_2)$, and add $w(v_2)$ to $M_c$.*

**Reduction Rule 7 (R7)** ([19]). *If there is a 4-cycle $v_1v_2v_3v_4$ such that $d_G(v_2) = d_G(v_3) = 2$ and $w(v_1) \geq w(v_2) \geq w(v_3)$, then remove $v_2$ and $v_3$, update the weight of $v_1$ by letting $w(v_1) := w(v_1) + w(v_3) - w(v_2)$, and add $w(v_2)$ to $M_c$.*

**Reduction Rule 8 (R8).** *If there is a 4-path $v_1v_2v_3v_4v_5$ such that $d_G(v_2) = d_G(v_3) = d_G(v_4) = 2$ and $w(v_1) \geq w(v_2) \geq w(v_3) \leq w(v_4) \leq w(v_5)$, then remove $v_2$ and $v_4$, add edges $v_1v_3$ and $v_3v_5$, update the weight of $v_1$ by letting $w(v_1) := w(v_1) + w(v_3) - w(v_2)$ and the weight of $v_5$ by letting $w(v_5) := w(v_5) + w(v_3) - w(v_4)$, and add $w(v_2) + w(v_4) - w(v_3)$ to $M_c$.*

**Reduction Rule 9 (R9).** *For a 5-cycle $v_1v_2v_3v_4v_5$ such that $d_G(v_2) = d_G(v_3) = d_G(v_5) = 2$, $\min\{d(v_1), d(v_4)\} \geq 3$, and $w(v_1) \geq w(v_2) \geq w(v_3) \leq w(v_4)$,*

*(1) if $w(v_3) > w(v_5)$, then remove $v_5$, update the weight of $v_i$ by letting $w(v_i) := w(v_i) - w(v_5)$ for $i = 1, 2, 3, 4$, and add $2w(v_5)$ to $M_c$.*
*(2) if $w(v_3) \leq w(v_5)$, then remove $v_2$ and $v_3$, update the weight of $v_1$ by letting $w(v_1) := w(v_1) - w(v_2)$, the weight of $v_4$ by letting $w(v_4) := w(v_4) - w(v_3)$ and the weight of $v_5$ by letting $w(v_5) := w(v_5) - w(v_3)$, and add $w(v_2) + w(v_3)$ to $M_c$.*

**Reduction Rule 10 (R10).** *For a 6-cycle $v_1v_2v_3v_4v_5v_6$ such that $d_G(v_2) = d_G(v_3) = d_G(v_5) = d_G(v_6) = 2$, $w(v_1) \geq \max\{w(v_2), w(v_6)\}$, $w(v_4) \geq \max\{w(v_3), w(v_5)\}$, and $w(v_6) \geq w(v_5)$,*

*(1) if $w(v_2) \geq w(v_3)$, then remove $v_5$ and $v_6$, and update the weight of $v_2$ by letting $w(v_2) := w(v_2) + w(v_6)$ and the weight of $v_3$ by letting $w(v_3) := w(v_3) + w(v_5)$;*
*(2) if $w(v_2) < w(v_3)$, then remove $v_6$, add edge $v_1v_5$, and update the weight of $v_2$ by letting $w(v_2) := w(v_2) + w(v_6)$, the weight of $v_3$ by letting $w(v_3) := w(v_3) + w(v_5)$, and the weight of $v_5$ by letting $w(v_5) := w(v_6) + w(v_3) - \max\{w(v_2) + w(v_6), w(v_3) + w(v_5)\}$.*

### 3.3 Reductions Based on Small Cuts

We also have some reduction rules to deal with vertex-cuts of size one or two, which can even be used to design a polynomial-time divide-and-conquer algorithm. However, a graph may not always have vertex-cuts of small size.

**Reduction Rule 11 (R11).** *For a vertex-cut* $\{u\}$ *with a connected component* $G^*$ *in* $G - u$ *such that* $2\delta_3 - \delta_2 \leq \sum_{v \in G^*} \delta_{d_G(v)} \leq 10$,

*(1) if* $w(u) + \alpha(G^* - N[u]) \leq \alpha(G^*)$, *then remove* $G^*$ *and* $\{u\}$ *from* $G$ *and add* $\alpha(G^*)$ *to* $M_c$;
*(2) if* $w(u) + \alpha(G^* - N[u]) > \alpha(G^*)$, *then remove* $G^*$ *from* $G$, *update the weight of* $u$ *by letting* $w(u) := w(u) + \alpha(G^* - N[u]) - \alpha(G^*)$, *and add* $\alpha(G^*)$ *to* $M_c$.

**Lemma 4.** *(*) Let* $\{u, u'\}$ *be a vertex-cut of size two in* $G$ *and* $G^*$ *be a connected component in* $G - \{u, u'\}$, *where we assume w.l.o.g. that* $\alpha(G^* - N[u]) \geq \alpha(G^* - N[u'])$. *We construct a new graph* $G'$ *from* $G$ *as follows: remove* $G^*$; *add three new vertices* $\{v_1, v_2, v_3\}$ *with weight* $w(v_1) = \alpha(G^* - N[u']) - \alpha(G^* - N[\{u, u'\}])$, $w(v_2) = \alpha(G^* - N[u]) - \alpha(G^* - N[\{u, u'\}])$ *and* $w(v_3) = \alpha(G^*) - \alpha(G^* - N[u])$, *and add five new edges* $uv_1$, $v_1v_2$, $v_2u'$, $uv_3$ *and* $u'v_3$. *It holds that*

$$\alpha(G) = \alpha(G') + \alpha(G^* - N[\{u, u'\}]).$$

**Reduction Rule 12 (R12).** *For a vertex-cut* $\{u, u'\}$ *of size two with a connected component* $G^*$ *in* $G - \{u, u'\}$ *such that* $2\delta_3 + \delta_2 \leq \sum_{v \in G^*} \delta_{d_G(v)} \leq 10$, *we construct the graph* $G'$ *in Lemma 4, replace* $G$ *with* $G'$, *and add* $\alpha(G^* - N[\{u, u'\}])$ *to* $M_c$.

### 3.4 Analyzing Reduction Rules

It is easy to see that each application of our reduction rules can be executed in polynomial time. We also show that

**Lemma 5.** *The measure* $p$ *will not increase after applying any reduction rule.*

**Definition 1.** *An instance is* reduced, *if no reduction rule can be applied.*

**Lemma 6.** *(*) In a reduced instance, any two degree-2 vertices in different chains have at most one common chain-neighbor of degree at least 3, and each cycle contains at least three vertices of degree* $\geq 3$.

**Lemma 7.** *(*) For a triangle* $C$ *in a reduced instance, each vertex in* $C$ *is a vertex of degree* $\geq 3$ *and it has a chain-neighbor of degree at least* 3 *not in* $C$.

## 4 Branching Rules

### 4.1 Two Branching Rules

We have two branching rules. The first branching rule is to branch on a vertex $v$ by considering two cases: (i) there is a maximum weighted independent set in $G$ which does not contain $v$; (ii) every maximum weighted independent set in $G$ contains $v$. For the former case, we simply delete $v$ from the graph. For the latter case, by Lemma 2 we know that we can include the set $S_v$ confining $v$ in the independent set. So we delete $N[S_v]$ from the graph.

**Branching Rule 1 (Branching on a vertex).** *Branch on a vertex $v$ to generate two sub instances by either deleting $v$ from the graph or deleting $N[S_v]$ from the graph and adding $w(S_v)$ to $M_c$.*

Since each independent set contains at most two vertices in each 4-cycle, we have the second rule.

**Branching Rule 2 (Branching on a 4-cycle).** *Branch on a 4-cycle $v_1v_2v_3v_4$ to generate two sub instances by deleting either $\{v_1, v_3\}$ or $\{v_2, v_4\}$ from $G$.*

### 4.2 The Analysis and Some Properties

The hardest part is to analyze how much we can decrease the measure in each sub-branch of a branching operation. Usually, we need to deeply analyze the local graph structure and use case-analysis. Here we try to summarize some common properties. The following notations will be frequently used in the whole paper.

Let $S$ be a vertex subset in a reduced graph $G$. We use $G_{-S}$ to denote the graph after deleting $S$ from $G$ and iteratively applying R1 to R4 until none of them can be applied. We use $R_S$ to denote the set of deleted vertices during applying R1 to R4 on $G - S$. Then $G_{-S} = G - (S \cup R_S)$. We also use $e_S$ to denote the number of edges between $S \cup R_S$ and $V \setminus (S \cup R_S)$ in $G$. We have the following lemmas for some bounds on $p(G) - p(G_{-S})$. Note that $G_{-S}$ may not be a reduced graph because of reduction rules from R5 to R12 and we may further apply reduction rules to further decrease the measure $p$.

**Lemma 8.** *(*) It holds that*

$$p(G) - p(G_{-S}) \geq \sum_{u \in S \cup R_S} \delta_{d_G(u)} + e_S \delta_3^{<-1>}. \tag{3}$$

In some cases, we can not use the bound in (3) directly, since we may not know the vertex set $R_S$. So we also consider some special cases and relaxed bounds.

**Lemma 9.** *(*) Let $S = \{v\}$ be a set of a vertex of degree $\geq 3$. We have that*

$$p(G) - p(G_{-S}) \geq \delta_{d(v)} + \sum_{u \in N(v)} \delta_{d(u)}^{<-1>} + q_2 \delta_3^{<-1>},$$

*where $q_2$ is the number of degree-2 vertices in $N(v)$.*

**Lemma 10.** *(*) If $S \cup R_S$ contains $N[v]$ for some vertex $v$ of degree $\geq 3$, then we have that*

$$p(G) - p(G_{-S}) \geq \sum_{u \in N[v]} \delta_{d(u)} + q_2 \delta_3^{<-1>},$$

*where $q_2$ is the number of degree-2 vertices in $N(v)$.*

Recall that we use $\mathcal{C}(G')$ to denote the set of connected components of the graph $G'$. We can easily observe the following lemma, which will be used to prove several bounds on $p(G) - p(G_{-S})$.

**Lemma 11.** *Let $S$ be a vertex subset. Let $S'$ be a subset of $S \cup R_S$ and $R' = S \cup R_S \setminus S'$. The number of edges between $S \cup R_S$ and $V \setminus (S \cup R_S)$ is $e_S$, and the number of edges between $S'$ and $V \setminus S'$ is $k$. For any component $H \in \mathcal{C}(G[R'])$, the number of edges between $S'$ and $H$ is $l_H$ and the number of edges between $H$ and $N(S \cup R_S)$ is $r_H$. We have that*

$$k - e_S = \sum_{H \in \mathcal{C}(G[R'])} (l_H - r_H).$$

*Furthermore, for any component $H \in \mathcal{C}(G[R'])$ containing only degree-2 vertices, it holds that $l_H - r_H = 0$ or 2.*

**Lemma 12.** *(*) For any subset $S' \subseteq S \cup R_S$ with $k$ edges between $S'$ and $V \setminus S'$, it holds that*

$$p(G) - p(G_{-S}) \geq \sum_{u \in S'} \delta_{d_G(u)} + e_S \delta_3^{<-1>} + \begin{cases} 0, & k - e_S \leq 0 \\ \delta_3, & k - e_S = 1 \\ \delta_2, & k - e_S = 2 \\ \delta_3, & k - e_S = 3 \\ 2\delta_2, & k - e_S > 3. \end{cases}$$

**Lemma 13.** *(*) Assume that a reduced graph $G$ has a maximum degree 3 and has no 3 or 4-cycles. For any subset $S' \subseteq S \cup R_S$ with $k$ edges between $S'$ and $V \setminus S'$, if the diameter of the induced graph $G[S']$ is 2, then it holds that either $p(G) - p(G_{-S}) > 10$ or*

$$p(G) - p(G_{-S}) \geq \sum_{u \in S'} \delta_{d_G(u)} + 3\delta_3^{<-1>} + \begin{cases} 0, & k \leq 3 \\ \delta_3^{<-1>}, & k = 4 \\ 2\delta_2, & k = 5 \\ \delta_2 + \delta_3, & k = 6. \end{cases}$$

**Lemma 14.** *(*) Assume that a reduced graph $G$ has a maximum degree 3, and each cycle $C$ in it contains at least five vertices, where at least four vertices are degree-3 vertices. For any subset $S' \subseteq S \cup R_S$ with $k$ edges between $S'$ and $V \setminus S'$, if each path $P$ in the induced graph $G[S']$ contains either at most three vertices or at most two degree-3 vertices, then it holds either $p(G) - p(G_{-S}) > 10$ or*

$$p(G) - p(G_{-S}) \geq \sum_{u \in S'} \delta_{d_G(u)} + \begin{cases} k\delta_3^{<-1>}, & k \leq 5 \\ \delta_3 + 2\delta_2 + 3\delta_3^{<-1>}, & k = 6. \end{cases}$$

## 5 The Algorithm

Now we describe the main steps of the algorithm. When the algorithm executes one step, we assume that all previous steps can not be applied.

**Step 1 (Applying Reductions).** *If the instance is not reduced, iteratively apply reduction rules in order, i.e., when one reduction rule is applied, no reduction rule with a smaller index can be applied on the graph.*

**Step 2 (Solving Small Components).** *If there is a connected component $G^*$ of $G$ such that $p(G^*) \leq 10$, solve the component $G^*$ directly and return $\alpha(G - G^*) + \alpha(G^*)$.*

**Step 3 (Branching on Vertices of Degree $\geq 5$).** *If there is a vertex $v$ with degree $d(v) \geq 5$, then branch on $v$ with Branching Rule 1 by either excluding $v$ from the independent set or including $S_v$ in the independent set.*

**Lemma 15.** *(*) Step 3 followed by applications of reduction rules creates a branching vector covered by* $[5.368, 7.248]$.

**Step 4 (Branching on 4-Cycles with Chords).** *If there is a 4-cycle $C = v_1v_2v_3v_4$ with a chord $v_1v_3 \in E$, then branch on the 4-cycle with Branching Rule 2 by excluding either $\{v_1, v_3\}$ or $\{v_2, v_4\}$ from the independent set.*

**Lemma 16.** *(*) Step 4 followed by applications of reduction rules creates a branching vector covered by one of* $[3\delta_4 + \delta_3^{<-1>}, 4\delta_4 + 2\delta_3^{<-1>}] = [5.496, 7.744]$ *and* $[4\delta_4, 2\delta_4 + 2\delta_3 + 2\delta_2] = [6.496, 6]$.

**Step 5 (Branching on Degree-4 Vertices).** *If there is a degree-4 vertex $v$, then branch on it with Branching Rule 1 by either excluding $v$ from the independent set or including $S_v$ in the independent set.*

**Lemma 17.** *(*) Step 5 followed by applications of reduction rules creates a branching vector covered by one of* $[5.624, 5.624]$, $[5.248, 6]$, $[4.872, 6.624]$, $[4.496, 7.248]$, *and* $[4.12, 7.872]$.

**Step 6 (Branching on Other 4-Cycles).** *If there is a 4-cycle $C = v_1v_2v_3v_4$, then branch on the 4-cycle with Branching Rule 2 by excluding either $\{v_1, v_3\}$ or $\{v_2, v_4\}$ from the independent set.*

**Lemma 18.** *(*) Step 6 followed by applications of reduction rules creates a branching vector covered by* $[6\delta_3 - 2\delta_2, 6\delta_3 - 2\delta_2] = [5.248, 5.248]$.

**Step 7 (Branching on Triangles).** *If there is a triangle $C = v_1v_2v_3$, where we assume without loss of generality that $w(v_1) \geq \max\{w(v_2), w(v_3)\}$ and $v_1$ is chain-adjacent to a degree-3 vertex $u \neq v_2, v_3$, then branch on $u$ with Branching Rule 1.*

**Lemma 19.** *(*) Step 7 followed by applications of reduction rules creates a branching vector covered by one of* $[6\delta_3 - 3\delta_2, 7\delta_3 + \delta_2] = [4.872, 7.376]$ *and* $[6\delta_3 - 2\delta_2, 5\delta_3 + 2\delta_2] = [5.248, 5.752]$.

**Step 8 (Branching on Cycles Containing Three Degree-3 Vertices).** *If there is a cycle $C$ containing exactly three degree-3 vertices $\{v_1, v_2, v_3\}$, where we assume without loss of generality that $v_1$ is chain-adjacent to a degree-3 vertex $u \neq v_2, v_3$, then branch on $u$ with Branching Rule 1.*

**Lemma 20.** *(*) Step 8 followed by applications of reduction rules can create a branching vector covered by one of* $[6\delta_3 - 4\delta_2, 8\delta_3 - 2\delta_2] = [4.496, 7.248]$, $[6\delta_3 - 3\delta_2, 6\delta_3 - \delta_2] = [4.872, 5.624]$, *and* $[6\delta_3 - 2\delta_2, 6\delta_3 - 2\delta_2] = [5.248, 5.248]$.

**Step 9 (Branching on Degree-3 Vertices with Two Degree-2 Neighbors).** *If there is degree-3 vertex $u$ having two degree-2 neighbors and one degree-3 neighbor $v$, then branch on $v$ with Branching Rule 1.*

**Lemma 21.** *(*) Step 9 followed by applications of reduction rules creates a branching vector covered by one of* $[4\delta_3 - \delta_2, 8\delta_3 - \delta_2] = [3.624, 7.624]$, $[4\delta_3, 8\delta_3 - 4\delta_2] = [4, 6.496]$, *and* $[4\delta_3 + \delta_2, 6\delta_3] = [4.376, 6]$.

**Step 10 (Branching on Degree-3 Vertices of a Mixed Case).** *If a degree-3 vertex $u$ without degree-3 neighbors is chain-adjacent to a degree-3 vertex $v$ with exactly two degree-3 neighbors, then branch on $v$ with Branching Rule 1.*

**Lemma 22.** *(*) Step 10 followed by applications of reduction rules creates a branching vector covered by* $[4\delta_3, 8\delta_3 - 2\delta_2] = [4, 7.248]$.

**Step 11 (Branching on Degree-3 Vertices With At Least Two Degree-3 Neighbors).** *If there is a connected component $H$ containing a degree-3 vertex with at least two degree-3 neighbors, we let $u$ be the vertex of the maximum weight in $H$ and let $v$ be a degree-3 neighbor of $u$, and branch on $v$ with Branching Rule 1.*

**Lemma 23.** *(*) Step 11 followed by applications of reduction rules creates a branching vector covered by one of* $[4\delta_3 - \delta_2, 8\delta_3 - \delta_2] = [3.624, 7.624]$, *and* $[4\delta_3, 8\delta_3 - 4\delta_2] = [4, 6.496]$.

**Step 12 (Branching on Other Degree-3 Vertices).** *Pick up an arbitrary degree-3 vertex $v$ and branch on it with Branching Rule 1.*

**Lemma 24.** *(*) Step 12 followed by applications of reduction rules creates a branching vector covered by* $[4\delta_3 + 6\delta_2, 4\delta_3 + 6\delta_2] = [6.256, 6.256]$.

It is easy to see that above steps cover all the cases. Among all the branching vectors, the bottleneck ones are $[4\delta_3, 8\delta_3 - 4\delta_2] = [4, 6.496]$ in Lemma 21, $[4\delta_3 + \delta_2, 6\delta_3] = [4.376, 6]$ in Lemma 21, and $[4\delta_3, 8\delta_3 - 4\delta_2] = [4, 6.496]$ in Lemma 23. All of them have a branching factor of 1.14427. So we get that

**Theorem 1.** Maximum Weighted Independent Set *can be solved in* $O^*(1.1443^p)$ *time and polynomial space.*

By Lemma 1 and Theorem 1, we get that

**Corollary 1.** Maximum Weighted Independent Set *in graphs with average degree $x$ can be solved in* $O^*(1.1443^{(0.624x-0.872)n})$ *time and polynomial space.*

Let $x = 3$ in Lemma 1, we get that $p \leq n$ and the following result.

**Theorem 2.** Maximum Weighted Independent Set *in graphs with the average degree at most 3 can be solved in* $O^*(1.1443^n)$ *time and polynomial space.*

## References


1. Been, K., Daiches, E., Yap, C.K.: Dynamic map labeling. IEEE Trans. Visual Comput. Graph. **12**(5), 773–780 (2006)
2. Bourgeois, N., Escoffier, B., Paschos, V.T., van Rooij, J.M.M.: Fast algorithms for max independent set. Algorithmica **62**(1), 382–415 (2012)
3. Chen, J., Kanj, I.A., Jia, W.: Vertex cover: further observations and further improvements. J. Algorithms **41**(2), 280–301 (2001)
4. Dahllöf, V., Jonsson, P.: An algorithm for counting maximum weighted independent sets and its applications. In: SODA 2002, pp. 292–298 (2002)
5. Dahllöf, V., Jonsson, P., Wahlström, M.: Counting models for 2SAT and 3SAT formulae. Theor. Comput. Sci. **332**(1–3), 265–291 (2005)
6. Fomin, F.V., Gaspers, S., Saurabh, S.: Branching and treewidth based exact algorithms. In: ISAAC 2006. LNCS, vol. 4288, pp. 16–25 (2006)
7. Fomin, F.V., Gaspers, S., Saurabh, S., Stepanov, A.A.: On two techniques of combining branching and treewidth. Algorithmica **54**(2), 181–207 (2009)
8. Fomin, F.V., Grandoni, F., Kratsch, D.: A measure & conquer approach for the analysis of exact algorithms. J. ACM **56**(5), 25:1–25:32 (2009)
9. Fürer, M., Kasiviswanathan, S.P.: Algorithms for counting 2-SAT solutions and colorings with applications. In: AAIM 2007, pp. 47–57 (2007)
10. Huang, S., Xiao, M., Chen, X.: Exact algorithms for maximum weighted independent set on sparse graphs. CoRR abs/2108.12840 (2021)
11. Jian, T.: An $O(2^{0.304n})$ algorithm for solving maximum independent set problem. IEEE Trans. Comput. **35**(9), 847–851 (1986)
12. Karp, R.M.: Reducibility among combinatorial problems. In: Proceedings of a Symposium on the Complexity of Computer Computations, pp. 85–103. The IBM Research Symposia Series (1972)
13. Kneis, J., Langer, A., Rossmanith, P.: A fine-grained analysis of a simple independent set algorithm. In: Proceedings of IARCS Annual Conference on Foundations of Software Technology and Theoretical Computer Science. LIPIcs, vol. 4, pp. 287–298 (2009)
14. Lamm, S., Schulz, C., Strash, D., Williger, R., Zhang, H.: Exactly solving the maximum weight independent set problem on large real-world graphs. In: Proceedings of Algorithm Engineering and Experiments, pp. 144–158 (2019)
15. Liao, C.S., Liang, C.W., Poon, S.H.: Approximation algorithms on consistent dynamic map labeling. Theor. Comput. Sci. **640**, 84–93 (2016)
16. Robson, J.M.: Algorithms for maximum independent sets. J. Algorithms **7**(3), 425–440 (1986)
17. Tarjan, R.E., Trojanowski, A.E.: Finding a maximum independent set. SIAM J. Comput. **6**(3), 537–546 (1977)
18. Wahlström, M.: A tighter bound for counting max-weight solutions to 2SAT instances. In: Proceedings of Third International Workshop on Parameterized and Exact Computation. LNCS, vol. 5018, pp. 202–213 (2008)
19. Xiao, M., Huang, S., Zhou, Y., Ding, B.: Efficient reductions and a fast algorithm of maximum weighted independent set. In: WWW 2021: The Web Conference 2021, pp. 3930–3940 (2021)
20. Xiao, M., Nagamochi, H.: Confining sets and avoiding bottleneck cases: a simple maximum independent set algorithm in degree-3 graphs. Theor. Comput. Sci. **469**, 92–104 (2013)
21. Xiao, M., Nagamochi, H.: Exact algorithms for maximum independent set. Inf. Comput. **255**, 126–146 (2017)

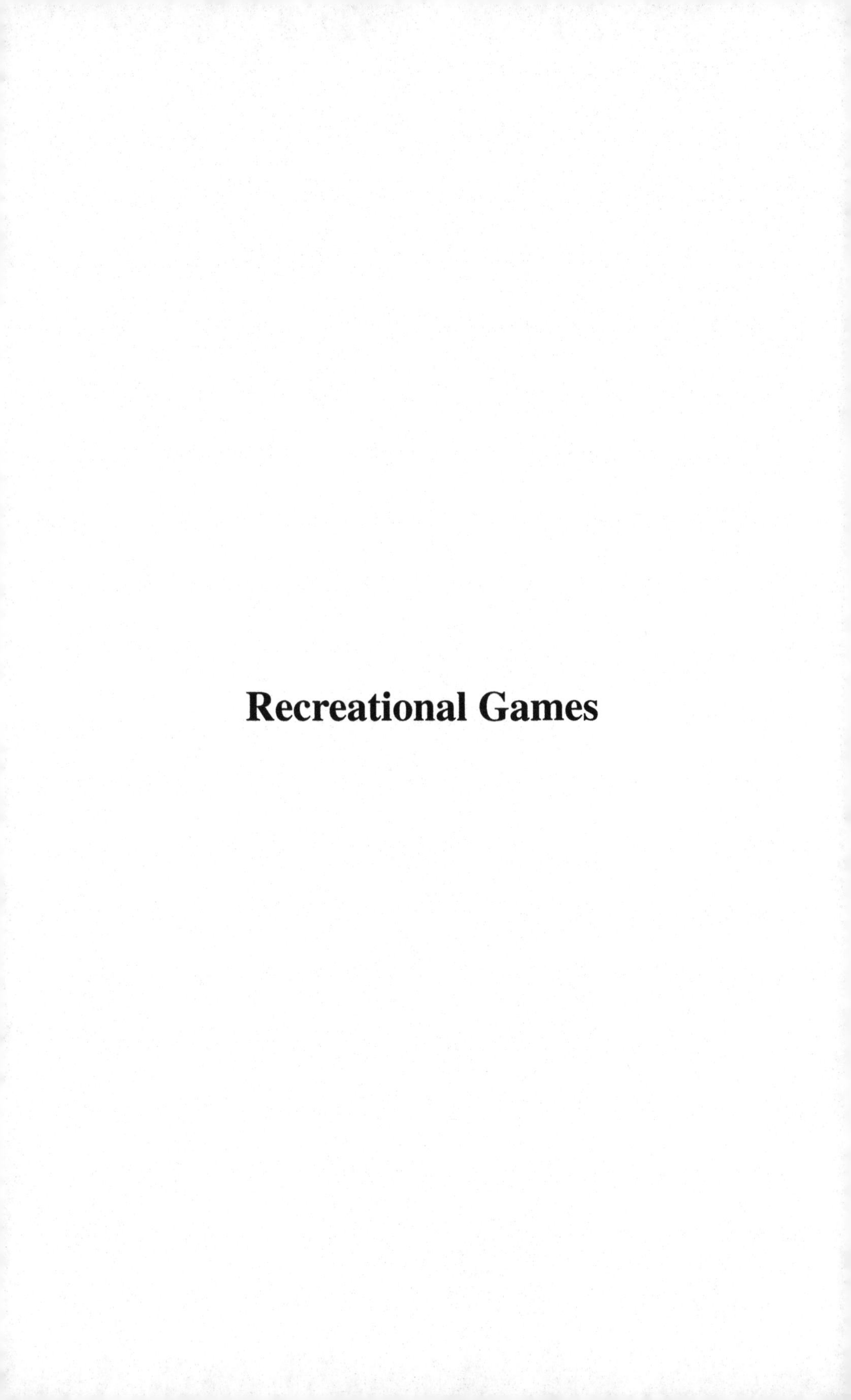

# Recreational Games

# Two Standard Decks of Playing Cards Are Sufficient for a ZKP for Sudoku

Suthee Ruangwises(✉)

Department of Mathematical and Computing Science,
Tokyo Institute of Technology, Tokyo, Japan

**Abstract.** Sudoku is a logic puzzle with an objective to fill a number between 1 and 9 into each empty cell of a $9 \times 9$ grid such that every number appears exactly once in each row, each column, and each $3 \times 3$ block. In 2020, Sasaki et al. proposed a physical zero-knowledge proof (ZKP) protocol for Sudoku using 90 cards, which allows a prover to physically show that he/she knows a solution without revealing it. However, their protocol requires nine identical copies of some cards, which cannot be found in a standard deck of playing cards (with 52 different cards and two jokers). Therefore, nine identical decks are actually required in order to perform that protocol. In this paper, we propose a new ZKP protocol for Sudoku that can be performed using only two standard decks of playing cards. In general, we develop the first ZKP protocol for an $n \times n$ Sudoku that can be performed using a deck of all different cards.

**Keywords:** Zero-knowledge proof · Card-based cryptography · Sudoku · Puzzle

## 1 Introduction

*Sudoku* is one of the world's most popular logic puzzles. A Sudoku puzzle consists of a $9 \times 9$ grid divided into nine blocks of size $3 \times 3$. Some of the cells in the grid are already filled with numbers between 1 and 9. The player has to fill a number into each empty cell such that every number from 1 to 9 appears exactly once in each row, each column, and each $3 \times 3$ block [18] (see Fig. 1). There is also a generalized version of Sudoku where the grid has size $n \times n$ and is divided into $n$ blocks of size $\sqrt{n} \times \sqrt{n}$, where $n$ is a perfect square. The generalized Sudoku is known to be NP-complete [25].

### 1.1 Zero-Knowledge Proof

We want to construct a *zero-knowledge proof (ZKP)* for Sudoku, which allows a prover $P$ to convince a verifier $V$ that he/she knows a solution of the puzzle without revealing any information about it. Formally, a ZKP is an interactive

A full version of this paper is available at https://arxiv.org/abs/2106.13646.


C.-Y. Chen et al. (Eds.): COCOON 2021, LNCS 13025, pp. 631–642, 2021.
https://doi.org/10.1007/978-3-030-89543-3_52

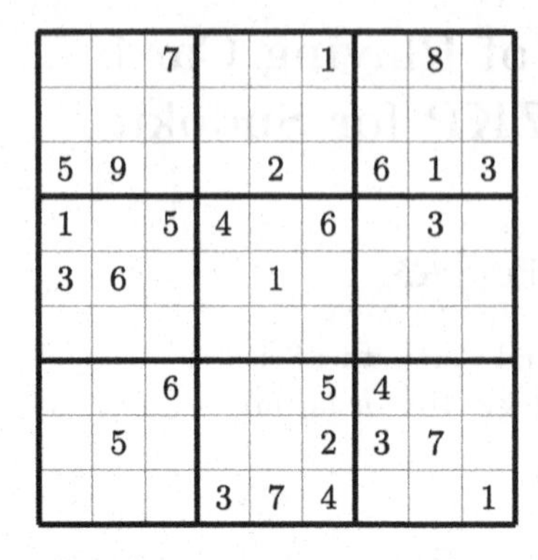

| 6 | 2 | 7 | 5 | 3 | 1 | 9 | 8 | 4 |
|---|---|---|---|---|---|---|---|---|
| 8 | 1 | 3 | 6 | 4 | 9 | 2 | 5 | 7 |
| 5 | 9 | 4 | 8 | 2 | 7 | 6 | 1 | 3 |
| 1 | 7 | 5 | 4 | 9 | 6 | 8 | 3 | 2 |
| 3 | 6 | 9 | 2 | 1 | 8 | 7 | 4 | 5 |
| 2 | 4 | 8 | 7 | 5 | 3 | 1 | 9 | 6 |
| 7 | 3 | 6 | 1 | 8 | 5 | 4 | 2 | 9 |
| 4 | 5 | 1 | 9 | 6 | 2 | 3 | 7 | 8 |
| 9 | 8 | 2 | 3 | 7 | 4 | 5 | 6 | 1 |

**Fig. 1.** An example of a $9 \times 9$ Sudoku puzzle (left) and its solution (right)

proof between $P$ and $V$ where both of them are given a computational problem $x$, but only $P$ knows a solution $w$. A ZKP with perfect completeness and perfect soundness must satisfy the following properties.

1. **Perfect Completeness:** If $P$ knows $w$, then $V$ always accepts.
2. **Perfect Soundness:** If $P$ does not know $w$, then $V$ always rejects.
3. **Zero-knowledge:** $V$ does not obtain any information about $w$. Formally, there exists a probabilistic polynomial time algorithm $S$ (called a *simulator*) that does not know $w$, and the outputs of $S$ follow the same probability distribution as the outputs of the actual protocol.

The concept of a ZKP was first introduced by Goldwasser et al. [5]. Recently, many results have been focusing on constructing physical ZKPs using objects found in everyday life such as a deck of cards. These protocols have a benefit that they do not require electronic devices, and also have didactic values since they are easy to understand and verify the correctness, even for non-experts in cryptography.

## 2 Previous Protocols

The first ZKP protocols for Sudoku were developed by Gradwohl et al. [6] in 2009. However, each of their six proposed protocols either has a nonzero soundness error or requires special tools such as scratch-off cards. In 2020, Sasaki et al. [24] proposed the improved ZKP protocols for Sudoku that have perfect soundness without using special tools.

### 2.1 Uniqueness Verification Protocol

Before showing the protocol of Sasaki et al., we first explain the following subprotocol, which was also developed by the same authors [24]. This protocol allows the prover $P$ to convince the verifier $V$ that a sequence $\sigma$ of $n$ face-down cards is a permutation of different cards $a_1, a_2, ..., a_n$ in some order, without revealing their orders. It also preserves the orders of the cards in $\sigma$ (so that the sequence can be later used in other protocols).

Let $x_1, x_2, ..., x_n$ be another set of $n$ different cards. $P$ performs the following steps.

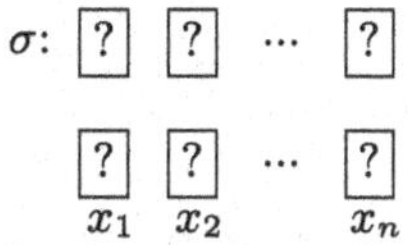

$$\sigma: \boxed{?}\ \boxed{?}\ \cdots\ \boxed{?}$$

$$\underset{x_1}{\boxed{?}}\ \underset{x_2}{\boxed{?}}\ \cdots\ \underset{x_n}{\boxed{?}}$$

**Fig. 2.** A $2 \times n$ matrix constructed in Step 1

1. Publicly place face-down cards $x_1, x_2, ..., x_n$ below the face-down sequence $\sigma$ in this order from left to right to form a $2 \times n$ matrix of cards (see Fig. 2).
2. Rearrange all columns of the matrix by a uniformly random permutation. This can be performed in real world by putting both cards in each column into an envelope and scrambling all envelopes together.
3. Turn over all cards in the top row. $V$ verifies that the sequence is a permutation of $a_1, a_2, ..., a_n$. Otherwise, $V$ rejects.
4. Turn over all face-up cards. Rearrange all columns of the matrix by a uniformly random permutation.
5. Turn over all cards in the bottom row. Rearrange the columns such that the cards in the bottom rows are $x_1, x_2, ..., x_n$ in this order from left to right. The sequence in the top row now returns to its original state.

### 2.2 Protocol of Sasaki et al.

Sasaki et al. [24] proposed three protocols to verify a solution of an $n \times n$ Sudoku puzzle. Here we will show only the first protocol, which is the one using the least number of cards.

Each card used in this protocol has a positive number on the front side $\boxed{1}$, $\boxed{2}$, ...; all cards have identical back sides $\boxed{?}$. On each cell already having a number $j$, $P$ publicly places a face-down $\boxed{j}$. On each empty cell having a number $j$ is $P$'s solution, $P$ secretly places a face-down $\boxed{j}$.

$P$ then applies the uniqueness verification protocol to verify that every row, column, and block contains a permutation of $\boxed{1}$, $\boxed{2}$, ..., $\boxed{n}$.

In total, this protocol uses $n^2+n$ cards: $n$ identical copies of $\boxed{1}$, $\boxed{2}$, ..., $\boxed{n}$ (to encode the numbers in the grid), and another set of $n$ different cards (to use in the uniqueness verification protocol). For a standard $9 \times 9$ puzzle, the protocol uses 90 cards, which is less than the number of cards in two standard decks; however, it requires nine identical copies of $\boxed{1}$, $\boxed{2}$, ..., $\boxed{9}$. As a standard deck consists of 54 different cards (including two different jokers), nine identical decks are actually required in order to perform this protocol, which are too many to be practical. Another choice is to use a different kind of deck (e.g. cards from board games) that includes several identical copies of some cards, but these decks are more difficult to find in everyday life.

Considering a drawback of this protocol, we aim to develop a more practical ZKP protocol for a $9 \times 9$ Sudoku that can be performed using only two standard decks of playing cards.

### 2.3 Related Work

After the discovery of the physical ZKP protocols for Sudoku, physical ZKP protocols for other popular logic puzzles have been proposed as well, including Nonogram [3], Akari [1], Takuzu [1,14], Kakuro [1,15], KenKen [1], Makaro [2], Norinori [4], Slitherlink [12], Juosan [14], Numberlink [21], Suguru [20], Ripple Effect [22], Nurikabe [19], Hitori [19], and Bridges [23].

Besides verifying solutions of logic puzzles, card-based protocols have also been extensively studied in secure multi-party computation, a setting where multiple parties want to jointly compute a function of their secret inputs without revealing the input of any party. The vast majority of work in this area, however, also uses identical copies of $\boxed{\clubsuit}$ and $\boxed{\heartsuit}$ in the protocols. The only exceptions are [9,11,16,17] which introduced AND, XOR, and copy protocols using a standard deck, and [13] which introduced a Yao's millionaire protocol using a standard deck. In [11], the authors also posed an open problem to develop ZKP protocols for logic puzzles using a standard deck.

Pratically, a standard deck of playing cards consists of 54 different cards (including two different jokers). Theoretically, it is also a challenging problem to develop a protocol that uses a deck of all different cards, so we also study the setting where the deck consists of $\boxed{1}$, $\boxed{2}$, ... where each card can have an arbitrarily large number on it.

## 3 Our Contribution

In this paper, we propose a new ZKP protocol for a generalized $n \times n$ Sudoku puzzle with perfect completeness and soundness using a set of all different cards.

There are two slightly different methods to implement our protocol. The first one uses $n^2 + n\sqrt{n} + n + \sqrt{n}$ cards and $4n\sqrt{n}$ shuffles. The second one uses $n^2 + 2n + 3\sqrt{n}$ cards and at most $2n^2(\sqrt{n} - 1) + 2$ shuffles (see Table 1).

In particular, for a standard $9 \times 9$ Sudoku puzzle, our protocol (with the second method of implementation) uses 108 cards and can be performed using two standard decks of playing cards, regardless of whether the two decks are the same or different types (see Table 2).

Theoretically, this work is an important step in card-based cryptography as it is the first ZKP protocol for any logic puzzle that can be performed using a deck of all different cards, an open problem posed in [11].

**Table 1.** The number of required cards and shuffles for an $n \times n$ Sudoku

| Protocol | Standard Deck? | #Cards | #Shuffles |
|---|---|---|---|
| Sasaki et al. [24] | No | $n^2 + n$ | $5n$ |
| Ours (Sect. 5.1) | Yes | $n^2 + n\sqrt{n} + n + \sqrt{n}$ | $4n\sqrt{n}$ |
| Ours (Sect. 5.2) | Yes | $n^2 + 2n + 3\sqrt{n}$ | $2n^2(\sqrt{n} - 1)$ for even $n$<br>$2n^2(\sqrt{n} - 1) + 2$ for odd $n > 9$ |

**Table 2.** The number of required cards and shuffles for a $9 \times 9$ Sudoku

| Protocol | Standard Deck? | #Cards | #Shuffles |
|---|---|---|---|
| Sasaki et al. [24] | No | 90 | 45 |
| Ours (Sect. 5.1) | Yes | 120 | 108 |
| Ours (Sect. 5.2) | Yes | 108 | 322 |

# 4 Preliminaries

At first, we assume that all cards used in our protocols have different front sides and identical back sides (although we will later show that some pairs of cards can have identical front sides or different back sides, and our protocol still works correctly).

## 4.1 Marked Matrix

Suppose we have a $k \times \ell$ matrix of face-down cards (we call these cards *encoding cards*). Let Row $i$ denote an $i$-th topmost row and let Column $j$ denote a $j$-th leftmost column. To the left of Column 1, publicly place face-down cards $p_1, p_2, ..., p_k$ in this order from top to bottom; this new column is called Column 0. Analogously, above Row 1, publicly place face-down cards $q_1, q_2, ..., q_\ell$ in this order from left to right; this new row is called Row 0.

We call this structure a $k \times \ell$ *marked matrix* (see Fig. 3), and we call the cards in Row 0 and Column 0 *marking cards*.

## 4.2 Shuffle Operations

Suppose we have a $k \times \ell$ marked matrix. For a set $S \subseteq \{1, 2, ..., k\}$, an operation $\texttt{row\_shuf}(S)$ rearranges the rows in $S$ (including marking cards in Column 0) by a uniformly random permutation. For example, $\texttt{row\_shuf}(\{1, 3, 4\})$ rearranges Row 1, Row 3, and Row 4 of the matrix by a uniformly random permutation. This can be performed in real world by putting all cards in each row in $S$ into an envelope and scrambling all envelopes together.

Analogously, for a set $S \subseteq \{1, 2, ..., \ell\}$, an operation $\texttt{col\_shuf}(S)$ rearranges the columns in $S$ (including marking cards in Row 0) by a uniformly random permutation.

| Row \ Column | 0 | 1 | 2 | 3 | 4 | 5 |
|---|---|---|---|---|---|---|
| 0 | | $q_1$ ? | $q_2$ ? | $q_3$ ? | $q_4$ ? | $q_5$ ? |
| 1 | $p_1$ ? | ? | ? | ? | ? | ? |
| 2 | $p_2$ ? | ? | ? | ? | ? | ? |
| 3 | $p_3$ ? | ? | ? | ? | ? | ? |
| 4 | $p_4$ ? | ? | ? | ? | ? | ? |

**Fig. 3.** An example of a $4 \times 5$ marked matrix

### 4.3 Rearrangement Protocol

After applying some shuffle operations to a marked matrix, a *rearrangement protocol* reverts the matrix back to its original state. Slightly different variants of this protocol with the same idea has been used in previous work [2,7,8,21,22,24].

Suppose we have a $k \times \ell$ marked matrix $M$ with marking cards $p_1, p_2, ..., p_k$ in Column 0 and $q_1, q_2, ..., q_\ell$ in Row 0. We perform the following steps.

1. Apply $\texttt{row\_shuffle}(\{1, 2, ..., k\})$ and $\texttt{col\_shuffle}(\{1, 2, ..., \ell\})$ to $M$.
2. Turn over all marking cards in Column 0 and Row 0. Rearrange the rows of $M$ such that the marking cards in Column 0 are $p_1, p_2, ..., p_k$ in this order from top to bottom. Rearrange the columns of $M$ such that the marking cards in Row 0 are $q_1, q_2, ..., q_\ell$ in this order from left to right.

### 4.4 Standard Deck Chosen Cut Protocol

Given a $k \times \ell$ marked matrix $M$, this protocol allows the prover $P$ to choose a card located at Row $i$ and Column $j$ of $M$ he/she wants without revealing $i$ or $j$. It was modified from an original chosen cut protocol of Koch and Walzer [10] (which uses identical copies of ♣ and ♡) so that it can be performed using a standard deck. $P$ performs the following steps.

1. Secretly stack a face-down card $x_1$ on a card located at Row $i$ and Column $j$.
2. On each of the remaining $k\ell - 1$ cards in the matrix, secretly stack each of face-down cards $x_2, x_3, ..., x_{k\ell}$ in a uniformly random order. The cards $x_1, x_2, ..., x_{k\ell}$ are called *helper cards.*
3. Apply $\texttt{row\_shuffle}(\{1, 2, ..., k\})$ and $\texttt{col\_shuffle}(\{1, 2, ..., \ell\})$ to $M$.
4. Turn over all helper cards. Locate the position of $x_1$. The encoding card in that stack is the one originally located at Row $i$ and Column $j$ as desired.
5. Remove all helper cards. Apply the rearrangement protocol to revert $M$ to its original state.

This protocol will be implicitly used in our main protocol, with Step 3 being replaced by equivalent operations.

## 5 Main Protocol

For simplicity, we will show a protocol for a standard $9 \times 9$ Sudoku puzzle. Our protocol can be straightforwardly generalized to an $n \times n$ puzzle.

We use the following cards in our protocol.

- encoding cards $a_j, b_j, c_j, d_j, e_j, f_j, g_j, h_j, i_j$ ($j = 1, 2, ..., 9$)
- marking cards $p_j$ ($j = 1, 2, 3$) and $q_j$ ($j = 1, 2, ..., 9$)
- helper cards $x_j, y_j, z_j$ ($j = 1, 2, ..., 9$)

Suppose the grid is divided into blocks $A, B, ..., I$ (see Fig. 4). We use a card $a_j$ ($j = 1, 2, ..., 9$) to encode a number $j$ in Block $A$. Analogously, we use cards $b_j, c_j, ..., i_j$ ($j = 1, 2, ..., 9$) to encode numbers $j$ in blocks $B, C, ..., I$, respectively.

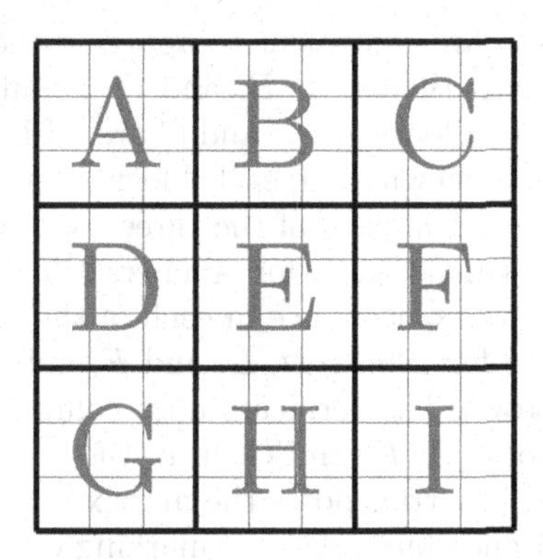


**Fig. 4.** Blocks $A, B, C, D, E, F, G, H$, and $I$ in the grid

On each cell already having a number, $P$ publicly places a face-down corresponding card (e.g. places a card $b_3$ on a cell with a number 3 in Block $B$). On each empty cell, $P$ secretly places a face-down corresponding card according to his/her solution.

Apply the uniqueness verification protocol in Sect. 2.1 to verify that Block $A$ consists of cards $a_1, a_2, ..., a_9$ in some order. Do the same for Blocks $B, C, ..., I$. Now $V$ is convinced that every number from 1 to 9 appears exactly once in each block.

Next, we will show two methods to verify that every number from 1 to 9 appears exactly once in each row and column.

### 5.1 Method A

First, $P$ performs the following steps to verify that a number 1 appears exactly once in each of the three topmost rows.

1. Take the cards from the three topmost rows to form a $3 \times 9$ matrix and publicly place marking cards $p_1, p_2, p_3$ in Column 0 and $q_1, q_2, ..., q_9$ in Row 0 to create a marked matrix $M$.

2. Secretly stack face-down cards $x_1$, $y_1$, and $z_1$ on $a_1$, $b_1$, and $c_1$, respectively.
3. On each of the remaining 8 cards in Block $A$, secretly stack each of face-down cards $x_2, x_3, ..., x_9$ in a uniformly random order. Do the same for cards $y_2, y_3, ..., y_9$ in Block $B$ and $z_2, z_3, ..., z_9$ in Block $C$.
4. Apply `row_shuffle`($\{1,2,3\}$), `col_shuffle`($\{1,2,3\}$), `col_shuffle`($\{4,5,6\}$), and `col_shuffle`($\{7,8,9\}$) to $M$.
5. Turn over all helper cards. Locate the positions of $x_1$, $y_1$, and $z_1$. Turn over the encoding cards in these three stacks to show that they are $a_1$, $b_1$, and $c_1$, respectively, and that they are all located at different rows. Otherwise, $V$ rejects.
6. Remove all helper cards and turn all encoding cards face-down. Apply the rearrangement protocol in Sect. 4.3 to revert $M$ to its original state.

Note that Steps 2 to 6 are equivalent to applying the standard deck chosen cut protocol in Sect. 4.4 to Blocks $A$, $B$, and $C$, simultaneously. These steps ensure that the three 1s in Blocks $A$, $B$, and $C$ are all located at different rows. Since it has already been shown that each block contains exactly one 1, this implies there is exactly one 1 in each of the three topmost rows.

$P$ performs these steps analogously for numbers 2, 3, ..., 9. Now $V$ is convinced that every number appears exactly once in each of the three topmost rows.

$P$ then does the same for Blocks $D$, $E$, and $F$ and for Blocks $G$, $H$, and $I$ to verify the rest of the rows. The verification for columns works analogously ($P$ takes the cards from Blocks $A$, $D$, and $G$, from Blocks $B$, $E$, and $H$, and from Blocks $C$, $F$, and $I$, and just transposes the matrix).

This method uses 81 encoding cards, 12 marking cards, and 27 helper cards, resulting in the total of 120 cards, slightly more than the number of cards in two standard decks, and uses 342 shuffles.[1] We aim to further reduce the number of required cards as a trade-off between the numbers of cards and shuffles.

### 5.2 Method B

In Method A, we verify that the three 1s in Blocks $A$, $B$, and $C$ are all located at different rows by verifying these three blocks at the same time, which requires a lot of marking and helper cards. Instead, we can first verify that the two 1s in Blocks $A$ and $B$ are located at different rows, then do the same for Blocks $A$ and $C$, and for Blocks $B$ and $C$. This leads to the same conclusion that the three 1s in Blocks $A$, $B$, and $C$ are all located at different rows. The formal steps for verifying Blocks $A$ and $B$ are shown below.

1. Take the cards from blocks $A$ and $B$ to form a $3 \times 6$ matrix and publicly place marking cards $p_1, p_2, p_3$ in Column 0 and $q_1, q_2, ..., q_6$ in Row 0 to create a marked matrix $M$.
2. Secretly stack face-down cards $x_1$ and $y_1$ on $a_1$ and $b_1$, respectively.

[1] The number of shuffles can be reduced to 108 after optimization. See the full version.

3. On each of the remaining 8 cards in Block $A$, secretly stack each of face-down cards $x_2, x_3, ..., x_9$ in a uniformly random order. Do the same for cards $y_2, y_3, ..., y_9$ in Block $B$.
4. Apply `row_shuffle`($\{1,2,3\}$), `col_shuffle`($\{1,2,3\}$), and `col_shuffle`($\{4,5,6\}$) to $M$.
5. Turn over all helper cards. Locate the positions of $x_1$ and $y_1$. Turn over the encoding cards in both stacks to show that they are $a_1$ and $b_1$, respectively, and that they are located at different rows. Otherwise, $V$ rejects.
6. Remove all helper cards and turn all encoding cards face-down. Apply the rearrangement protocol in Sect. 4.3 to revert $M$ to its original state.

We say that two cards are from the same set if they are denoted by the same letter with different indices (e.g. $d_2$ and $d_5$ are from the same set). Notice that in both methods, cards from different sets never get mixed together. Therefore, cards from different sets can have identical front sides or different back sides (or even different sizes) and our protocol still works correctly. The only requirement is that all cards from the same set must have different front sides and identical back sides.

This method uses 81 encoding cards, nine marking cards, and 18 helper cards, resulting in the total of 108 cards, which is exactly the number of cards from two standard decks (including jokers), and uses 828 shuffles.[2] We can, for example, use 54 cards from the first deck in the sets $a_j, b_j, ..., f_j$ and 54 cards from the second deck in the remaining sets. The protocol works correctly regardless of whether the two decks are identical or different, since it allows cards from different sets to have identical front sides (in case of identical decks) or different back sides or sizes (in case of different decks). Note that in some decks, the two jokers are identical; in that case, we just need to make sure that the two jokers are in different sets.

### 5.3 Generalization

Our protocol can be straightforwardly generalized to an $n \times n$ puzzle.

Method A uses $n^2$ encoding cards, $n + \sqrt{n}$ marking cards, and $n\sqrt{n}$ helper cards, resulting in the total of $n^2 + n\sqrt{n} + n + \sqrt{n}$ cards. It uses $4n\sqrt{n}$ shuffles (after the optimization).

Method B uses $n^2$ encoding cards, $3\sqrt{n}$ marking cards, and $2n$ helper cards, resulting in the total of $n^2 + 2n + 3\sqrt{n}$ cards. It uses at most $2n^2(\sqrt{n} - 1) + 2$ shuffles (after the optimization).

## 6 Proof of Correctness and Security

We will prove the perfect completeness, perfect soundness, and zero-knowledge properties of our protocol.

[2] The number of shuffles can be reduced to 322 after optimization. See the full version.

**Lemma 1 (Perfect Completeness).** *If $P$ knows a solution of the Sudoku puzzle, then $V$ always accepts.*

*Proof.* Suppose $P$ knows a solution and places cards on the grid accordingly. Every number from 1 to 9 will appear exactly once in each row, each column, and each block. Hence, the uniqueness verification protocol will pass for every block. Also, the same numbers from different blocks are always located at different rows and columns, so both Methods A and B will pass. Therefore, $V$ always accepts. □

**Lemma 2 (Perfect Soundness).** *If $P$ does not know a solution of the Sudoku puzzle, then $V$ always rejects.*

*Proof.* Suppose $P$ does not know a solution. There will be a number that appears at least twice in the same row, column, or block. If it appears twice in a block, the uniqueness verification protocol for that block will fail. If it appears twice in different blocks in the same row (resp. column), Method A will fail when verifying the three blocks containing that row (resp. column); also, method B will fail when verifying the two blocks where these two numbers appear. Therefore, $V$ always rejects. □

**Lemma 3 (Zero-Knowledge).** *During the verification, $V$ learns nothing about $P$'s solution.*

*Proof.* It is sufficient to show that all distributions of cards that are turned face-up can be simulated by a simulator $S$ that does not know $P$'s solution.

- In Steps 3 and 5 of the uniqueness verification protocol in Sect. 2.1, the orders of the $n$ cards are uniformly distributed among all $n!$ permutations. Hence, it can be simulated by $S$.
- In Step 2 of the rearrangement protocol in Sect. 4.3, the orders of $p_1, p_2, ..., p_k$ and $q_1, q_2, ..., q_\ell$ are uniformly distributed among all $k!$ permutations and $\ell!$ permutations, respectively. Hence, it can be simulated by $S$.
- In Step 5 of Method A in Sect. 5.1, the rows where $x_1$, $y_1$, and $z_1$ are located are uniformly distributed among all $3! = 6$ permutations of the first three rows; the columns where they are located are uniformly distributed among all $3^3 = 27$ combinations of three columns from Blocks $A$, $B$, and $C$. Also, the orders of $x_2, x_3, ..., x_9$ are uniformly distributed among all $8!$ permutations of the remaining cards in Block $A$; the same goes for $y_2, y_3, ..., y_9$ in Block $B$ and $z_2, z_3, ..., z_9$ in Block $C$. Hence, it can be simulated by $S$.
- In Step 5 of Method B in Sect. 5.2, the rows where $x_1$ and $y_1$ are located are uniformly distributed among all $\frac{3!}{1!} = 6$ 2-permutations of the first three rows; the columns where they are located are uniformly distributed among all $3^2 = 9$ combinations of two columns from Blocks $A$ and $B$. Also, the orders of $x_2, x_3, ..., x_9$ are uniformly distributed among all $8!$ permutations of the remaining cards in Block $A$; the same goes for $y_2, y_3, ..., y_9$ in Block $B$. Hence, it can be simulated by $S$. □

## 7 Future Work

We developed the first ZKP protocol for Sudoku, and also the first one for any logic puzzle, that uses a deck of all different cards. Our protocol for a standard $9 \times 9$ Sudoku can be performed using two standard decks of playing cards. However, the drawback of our protocol is that it uses a large number of shuffles, which makes it impractical. A possible future work is to develop an equivalent protocol for Sudoku that uses asymptotically less number of shuffles. Other challenging future work includes developing ZKP protocols for other logic puzzles (e.g. Kakuro, Numberlink) that uses a deck of all different cards.

## References

1. Bultel, X., Dreier, J., Dumas, J.G., Lafourcade, P.: Physical zero-knowledge proofs for akari, takuzu, kakuro and kenken. In: Proceedings of the 8th International Conference on Fun with Algorithms (FUN), pp. 8:1–8:20 (2016)
2. Bultel, X., et al.: Physical zero-knowledge proof for makaro. In: Izumi, T., Kuznetsov, P. (eds.) SSS 2018. LNCS, vol. 11201, pp. 111–125. Springer, Cham (2018). https://doi.org/10.1007/978-3-030-03232-6_8
3. Chien, Y.-F., Hon, W.-K.: Cryptographic and physical zero-knowledge proof: from sudoku to nonogram. In: Boldi, P., Gargano, L. (eds.) FUN 2010. LNCS, vol. 6099, pp. 102–112. Springer, Heidelberg (2010). https://doi.org/10.1007/978-3-642-13122-6_12
4. Dumas, J.-G., Lafourcade, P., Miyahara, D., Mizuki, T., Sasaki, T., Sone, H.: Interactive physical zero-knowledge proof for norinori. In: Du, D.-Z., Duan, Z., Tian, C. (eds.) COCOON 2019. LNCS, vol. 11653, pp. 166–177. Springer, Cham (2019). https://doi.org/10.1007/978-3-030-26176-4_14
5. Goldwasser, S., Micali, S., Rackoff, C.: The knowledge complexity of interactive proof systems. SIAM J. Comput. **18**(1), 186–208 (1989)
6. Gradwohl, R., Naor, M., Pinkas, B., Rothblum, G.N.: Cryptographic and physical zero-knowledge proof systems for solutions of sudoku puzzles. Theory Comput. Syst. **44**(2), 245–268 (2009)
7. Hashimoto, Y., Shinagawa, K., Nuida, K., Inamura, M., Hanaoka, G.: Secure grouping protocol using a deck of cards. IEICE Trans. Fundam. Electron. Commun. Comput. Sci. **101.A**(9), 1512–1524 (2018)
8. Ibaraki, T., Manabe, Y.: A more efficient card-based protocol for generating a random permutation without fixed points. In: Proceedings of the 3rd International Conference on Mathematics and Computers in Sciences and Industry (MCSI), pp. 252–257 (2016)
9. Koch, A., Schrempp, M., Kirsten, M.: Card-based cryptography meets formal verification. New Gener. Comput. **39**(1), 115–158 (2021)
10. Koch, A., Walzer, S.: Foundations for actively secure card-based cryptography. In: Proceedings of the 10th International Conference on Fun with Algorithms (FUN), pp. 17:1–17:23 (2020)
11. Koyama, H., Miyahara, D., Mizuki, T., Sone, H.: A secure three-input and protocol with a standard deck of minimal cards. In: Proceedings of the 16th International Computer Science Symposium in Russia (CSR), pp. 242–256 (2021)

12. Lafourcade, P., Miyahara, D., Mizuki, T., Sasaki, T., Sone, H.: A physical ZKP for slitherlink: how to perform physical topology-preserving computation. In: Heng, S.-H., Lopez, J. (eds.) ISPEC 2019. LNCS, vol. 11879, pp. 135–151. Springer, Cham (2019). https://doi.org/10.1007/978-3-030-34339-2_8
13. Miyahara, D., Hayashi, Y., Mizuki, T., Sone, H.: Practical card-based implementations of Yao's millionaire protocol. Theor. Comput. Sci. **803**, 207–221 (2020)
14. Miyahara, D., et al.: Card-based ZKP protocols for takuzu and juosan. In: Proceedings of the 10th International Conference on Fun with Algorithms (FUN), pp. 20:1–20:21 (2020)
15. Miyahara, D., Sasaki, T., Mizuki, T., Sone, H.: Card-based physical zero-knowledge proof for kakuro. IEICE Trans. Fundam. Electron. Commun. Comput. Sci. **E102.A**(9): 1072–1078 (2019)
16. Mizuki, T.: Efficient and secure multiparty computations using a standard deck of playing cards. In: Proceedings of the 15th International Conference on Cryptology and Network Security (CANS), pp. 484–499 (2016)
17. Niemi, V., Renvall, A.: Solitaire zero-knowledge. Fundam. Inform. **38**(1,2), 181–188 (1999)
18. Nikoli: Sudoku. https://www.nikoli.co.jp/en/puzzles/sudoku.html
19. Robert, L., Miyahara, D., Lafourcade, P., Mizuki, T.: Interactive physical ZKP for connectivity: applications to nurikabe and hitori. In: Proceedings of the 17th Conference on Computability in Europe (CiE), pp. 373–384 (2021)
20. Robert, L., Miyahara, D., Lafourcade, P., Mizuki, T.: Physical zero-knowledge proof for suguru puzzle. In: Devismes, S., Mittal, N. (eds.) SSS 2020. LNCS, vol. 12514, pp. 235–247. Springer, Cham (2020). https://doi.org/10.1007/978-3-030-64348-5_19
21. Ruangwises, S., Itoh, T.: Physical zero-knowledge proof for numberlink puzzle and $k$ vertex-disjoint paths problem. New Gener. Comput. **39**(1), 3–17 (2021)
22. Ruangwises, S., Itoh, T.: Physical zero-knowledge proof for ripple effect. In: Proceedings of the 15th International Conference and Workshops on Algorithms and Computation (WALCOM), pp. 296–307 (2021)
23. Ruangwises, S., Itoh, T.: Physical ZKP for connected spanning subgraph: applications to bridges puzzle and other problems. In: Proceedings of the 19th International Conference on Unconventional Computation and Natural Computation (UCNC) (2021, in press)
24. Sasaki, T., Miyahara, D., Mizuki, T., Sone, H.: Efficient card-based zero-knowledge proof for Sudoku. Theor. Comput. Sci. **839**, 135–142 (2020)
25. Yato, T., Seta, T.: Complexity and completeness of finding another solution and its application to puzzles. IEICE Trans. Fundam. Electron. Commun. Comput. Sci. **86.A**(5), 1052–1060 (2003)

# Token Shifting on Graphs

Win Hlaing Hlaing Myint(✉), Ryuhei Uehara, and Giovanni Viglietta

School of Information Science, Japan Advanced Institute of Science and Technology (JAIST), Nomi, Japan
{winhlainghlaingmyint,uehara,johnny}@jaist.ac.jp

**Abstract.** We investigate a new variation of a token reconfiguration problem on graphs using the cyclic shift operation. A colored or labeled token is placed on each vertex of a given graph, and a "move" consists in choosing a cycle in the graph and shifting tokens by one position along its edges. Given a target arrangement of tokens on the graph, our goal is to find a shortest sequence of moves that will re-arrange the tokens as in the target arrangement. The novelty of our model is that tokens are allowed to shift along any cycle in the graph, as opposed to a given subset of its cycles. We first discuss the problem on special graph classes: we give efficient algorithms for optimally solving the 2-Colored Token Shifting Problem on complete graphs and block graphs, as well as the Labeled Token Shifting Problem on complete graphs and variants of barbell graphs. We then show that, in the 2-Colored Token Shifting Problem, the shortest sequence of moves is NP-hard to approximate within a factor of $2 - \varepsilon$, even for grid graphs. The latter result settles an open problem posed by Sai et al.

**Keywords:** Reconfiguration problem · Cyclic shift · Barbell graph · Block graph · NP-hard

## 1 Introduction

Reconfiguration arises in countless problems that involve movement and change, including problems in computational geometry such as morphing graph drawings and polygons, and problems relating to games and puzzles, such as the 15-puzzle, a topic of research since 1879 [5]. The general questions that are considered in reconfiguration problems are: can any arrangement be reconfigured to any other; what is the worst-case number of steps required; and what is the complexity of computing the minimum number of steps required to get from one given configuration to another given configuration [5]. These questions can be rephrased in terms of the *configuration graph*, which is the graph whose vertices are all possible configurations, and whose edges represent feasible moves: is the configuration graph connected; what is its diameter; how efficiently can one compute distances between vertices in this graph? Previously studied token reconfiguration problems include the Token Swapping Problem, where pairs of tokens can be swapped along the edges of a graph. The Token Swapping Problem


C.-Y. Chen et al. (Eds.): COCOON 2021, LNCS 13025, pp. 643–654, 2021.
https://doi.org/10.1007/978-3-030-89543-3_53

is proved to be NP-complete, and there are many special classes of graphs on which the Token Swapping Problem can be solved exactly by polynomial-time algorithms, including complete graphs, paths, cycles, stars, brooms, complete bipartite graphs, and complete split graphs (see, e.g., [2] for comprehensive surveys).

Recently, the Token Shifting Problem was introduced by Sai et al. in [6], inspired by puzzles based on cyclic shift operations. The input of the problem is a graph with a distinguished set of cycles $\mathcal{C}$, and an initial and a final arrangement of colored tokens on the vertices of the graph. The basic operation is called "shift" along a cycle $C \in \mathcal{C}$, and it moves each token located on a vertex of $C$ into the next vertex along $C$. The problem asks for a sequence of shift operations that transforms the initial configuration into the final configuration. We can further distinguish between the Labeled Token Shifting Problem, where all tokens are distinct, and the $k$-Colored Token Shifting Problem, where tokens come in $k$ different colors, and same-colored tokens are indistinguishable.

It was shown in [6] that the Labeled Token Shifting Problem is solvable in polynomial time on a large class of graphs, while solving the $k$-Colored Token Shifting Problem in the minimum number of moves is NP-hard, even for $k = 2$.

In this paper, we study a variation of the Token Shifting Problem where the set of cycles $\mathcal{C}$ consists of *all* cycles in the graph (as opposed to a subset of them). On one hand, our choice makes the problem's description more natural and compact; on the other hand, proving hardness results is now more challenging. Indeed, previous NP-hardess proofs for variations of the Token Shifting Problem crucially relied on the fact that only shifts along certain cycles were allowed.

In Sect. 3, we give linear-time algorithms for the shortest shift sequence for both the 2-Colored and the Labeled Token Shifting Problem for complete graphs. In Sect. 4, we discuss the shortest shift sequence for the Labeled Token Shifting Problem on standard barbell graphs, and then on generalized barbell graphs with more than one connecting edge. In Sect. 5, we study the 2-Colored Token Shifting Problem for block graphs. Finally, in Sect. 6 we prove that, in the 2-Colored Token Shifting Problem, the shortest sequence of moves is NP-hard to approximate within a factor of $2 - \varepsilon$, even for planar graphs with a maximum degree of 4.

Notably, our NP-hardness result settles a problem left open in [6], which asked whether the Token Shifting Problem remains NP-hard when restricted to planar graphs or graphs of constant maximum degree. We remark that in [1], Amano et al. proved that a 2-Colored Token Shifting Problem called *Torus Puzzle* is NP-hard to solve in the minimum number of shifts. This puzzle consists of two arrays of horizontal and vertical cycles arranged in a grid, which yields a planar graph of maximum degree 4. However, in this puzzle the number of moves is measured in a different way: any number $k > 0$ of consecutive shifts along the same cycle is counted as only one move, while in our model (as well as in [6]) we count them as $k$ moves. Because of this, the NP-hardness reduction in [1] does not work in our model. In addition, the majority of cycles in the graph of the Torus Puzzle are forbidden from shifting (such is, for example, the 4-cycle determined by any cell in the grid). However, as already remarked, in our model we insist on allowing shifts along any cycle.

## 2 Preliminaries

Let $G = (V, E)$ be an undirected connected graph, where $V$ is the vertex set and $E$ is the edge set, and let $\mathrm{Col} = \{1, 2, \ldots, c\}$ be the color set for tokens, where $c$ is constant. A *token arrangement* (or *configuration*) is a function $f \colon V \to \mathrm{Col}$, where $f(v)$ represents the color of the token located on the vertex $v \in V$.

The *token shift operation* can be defined as follows. Let $C = (v_1, v_2, \ldots, v_k)$ be a cycle of $k > 1$ distinct vertices of $G = (V, E)$, where $\{v_i, v_{i+1}\} \in E$ for all $1 \leq i < k$ and $\{v_k, v_1\} \in E$. Then, a token shift along $C$ will transform any arrangement $f$ into the arrangement $f'$, which coincides with $f$ on all vertices except the ones in $C$. Specifically, for $v_i \in \{v_1, v_2, \ldots, v_{k-1}\}$, we have $f'(v_{i+1}) = f(v_i)$, and $f'(v_1) = f(v_k)$. All cycles in $G$ are eligible for token shift, and the length of the cycle can range from 2 to $|V|$. Note that we consider each edge of $G$ as a cycle of length 2; in this case, the result of the shift operation will be equivalent to a token swap along that edge.

The *Token Shifting Problem* takes as input a connected graph $G = (V, E)$, a color set Col, an initial arrangement $f_0$, and a final arrangement $f_t$. The problem asks to determine a shortest sequence of shift operations OPT that transforms $f_0$ into $f_t$, assuming that such a sequence exists.

Note that, since swaps along edges are allowed, it is possible to transform $f_0$ into $f_t$ if and only if they have the same number of tokens of each color, which is checkable in linear time given $f_0$ and $f_t$. Thus, without loss of generality, we may assume that there is always a sequence of shift operations that transforms $f_0$ into $f_t$, and our goal is to find the shortest one. Furthermore, it is easy to prove that $|\mathrm{OPT}| \leq |V|(|V| - 1)/2$ (this bound is obtained by using swap operations only; cf. [7, Theorem 1]). Since we have a polynomial upper bound of the number of shift operations, the Token Shifting Problem is in NP.

We distinguish between the *$k$-Colored Token Shifting Problem*, where the size of Col is a fixed constant $k$, and the *Labeled Token Shifting Problem*, where $\mathrm{Col} = V$, and $f_0$ and $f_t$ are permutations of $V$ (that is, all tokens have distinct labels). In this paper, we will mostly focus on the 2-Colored Token Shifting Problem (i.e., where $\mathrm{Col} = \{c_1, c_2\}$) and the Labeled Token Shifting Problem.

## 3 Token Shifting on Complete Graphs

### 3.1 2-Colored Token Shifting on Complete Graphs

In this section, we show that for the 2-Colored Token Shifting Problem on complete graphs, an optimal shift sequence can be constructed in linear time.

**Theorem 1.** *The 2-Colored Token Shifting Problem on a complete graph $G = (V, E)$ can be solved in linear time by a single shift operation.*

*Proof.* Let $\mathrm{Col} = \{c_1, c_2\}$ be the color set and let $f_0$ and $f_t$ be the initial and target token arrangements, respectively. We can construct two sets $V_1$ and $V_2$ of vertices as follows:

$$V_1 = \{v \in V \mid f_0(v) = c_1 \text{ and } f_t(v) = c_2\} \text{ and}$$

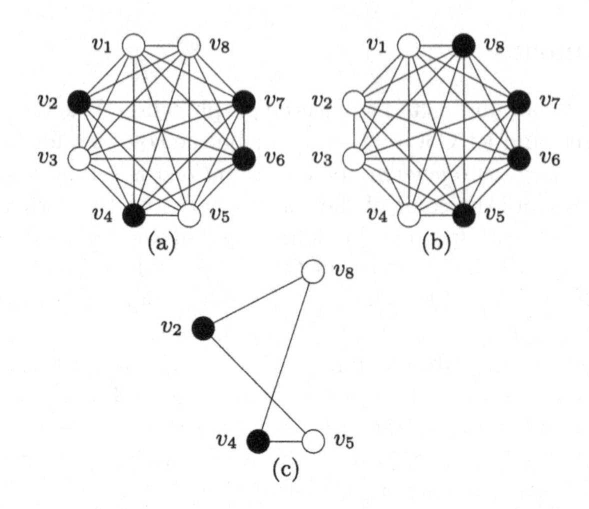


**Fig. 1.** 2-colored token shifting on a complete graph: (a) an initial token arrangement $f_0$, (b) a target token arrangement $f_t$, and (c) an optimal shift cycle

$$V_2 = \{v \in V |\, f_0(v) = c_2 \text{ and } f_t(v) = c_1\}.$$

Given that $f_0$ is re-configurable to $f_t$, $|V_1| = |V_2| = m$ for a complete graph with $2m$ misplaced tokens. Thus, we can construct a cycle of length $2m$ that visits each vertex in $V_1$ and $V_2$ alternately. For $V_1 = \{x_1, x_2, \ldots, x_m\}$ and $V_2 = \{y_1, y_2, \ldots, y_m\}$, the shift $(x_1, y_1, x_2, y_2, \ldots, x_m, y_m)$ transforms $f_0$ into $f_t$. □

For example, in Fig. 1, $V_1 = \{v_5, v_8\}$ and $V_2 = \{v_2, v_4\}$. From $V_1$ and $V_2$ the shift cycle $(v_2, v_5, v_4, v_8)$ can be constructed, which transforms $f_0$ into $f_t$.

### 3.2 Labeled Token Shifting on Complete Graphs

In this section, we show that the Labeled Token Shifting Problem on a complete graph can be solved by at most two shift operations.

**Theorem 2.** *The Labeled Token Shifting Problem on a complete graph $G = (V, E)$ can be solved with a minimum shift sequence $|\mathrm{OPT}| \leq 2$ in linear time.*

*Proof.* Let $f_0$ and $f_t$ be the initial and target token arrangements, respectively. We define the *conflict graph* $D(f_a, f_b) = (V', E')$ for two arrangements $f_a$ and $f_b$ as follows [7]:

$$V' = \{v \in V |\, f_a(v) \neq f_b(v)\} \text{ and}$$

$$E' = \{e = (v_i, v_j) |\, f_a(v_i) = f_b(v_j) \text{ and } v_i, v_j \in V'\}.$$

$D(f_0, f_t)$ is a digraph that includes vertices that hold different tokens in the initial and target token arrangements and there is an arc from $v_i$ to $v_j$ if the token on $v_i$ needs to be moved to $v_j$. A simple example is given in Fig. 2. One way to transform $f_0$ to $f_t$ would be to perform a token shift along each directed cycle

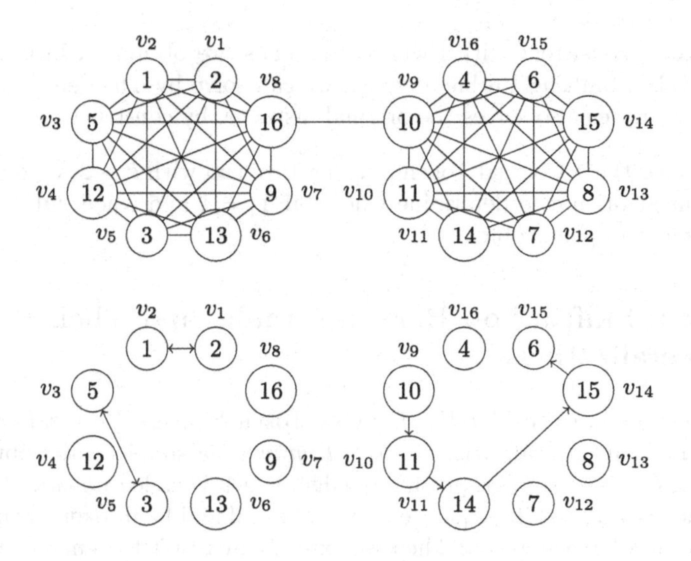


**Fig. 2.** (a) An initial token arrangement $f_0$, (b) the conflict graphs $D_A(f_0, f_t)$ and $D_B(f_0, f_t)$

in $D(f_0, f_t)$; if there are only 1 or 2 cycles, this strategy is optimal. However, it is not optimal when the number of cycles is greater than 2.

We consider the disjoint cycles in $D(f_0, f_t)$ as permutation cycles. For example, in Fig. 2(c) we have the three disjoint cycles $(v_1, v_4)$, $(v_2, v_6, v_3, v_7)$, and $(v_5, v_8)$, which collectively correspond to the permutation (14)(2637)(58).

We will use the following general fact: let us be given $m$ disjoint cyclic permutations involving $n$ elements in total; the product of these $m$ disjoint cycles and a length-$m$ cycle consisting of one element from each disjoint cycle is a single length-$n$ cycle that includes all $n$ elements. For example, $(1\mathbf{4})(\mathbf{2}637)(58)(\mathbf{521}) = (18563724)$. Equivalently, $(\mathbf{14})(\mathbf{2}637)(\mathbf{5}8) = (18563724)(\mathbf{125})$. In other words, we can express the product of any set of $m > 2$ disjoint cyclic permutations as the product of only two cycles.

Therefore, we construct a first cycle including one vertex from each cycle in $D(f_0, f_t)$, and we shift along this cycle once. This will result in an arrangement $f_1$ whose conflict graph $D(f_1, f_t)$ consists of a single directed cycle (see Fig. 2(d)). We can then perform a single shift along this cycle to obtain the target token arrangement $f_t$. □

**Corollary 1.** *For the k-Colored Token Shifting Problem on a complete graph $G = (V, E)$, we have $|\mathrm{OPT}| \leq 2$.*

*Proof.* Let $f_0$ and $f_t$ be the initial and final arrangements, respectively. Let $\mathrm{Col}' = V$, and let us define $f'_0$ as an arbitrary bijection $f'_0 \colon V \to \mathrm{Col}'$. We then define $f'_t \colon V \to \mathrm{Col}'$ as a bijection that, for all $v_i, v_j \in V$, satisfies $f'_0(v_i) = f'_t(v_j) \implies f_0(v_i) = f_t(v_j)$. Essentially, we assign unique labels to tokens in a way that is consistent with their colors. Thus, we obtain an instance of the Labeled Token Shifting Problem, which we can solve by Theorem 2. The same sequence of moves also solves the original instance, by construction. □

Note that, for the $k$-Colored Token Shifting Problem with $k > 2$, we do not have an efficient algorithm to determine when $|\mathrm{OPT}| = 1$ and when $|\mathrm{OPT}| = 2$. We leave this as an open problem.

# 4 Token Shifting on Barbell Graphs and Their Generalizations

In this section, we consider the Labeled Token Shifting Problem on barbell graphs and their generalization. A *barbell graph* is a simple graph obtained by connecting two complete graphs by an edge, which is called its *bar*. Our goal is to find the minimum shift sequence between initial and final token arrangements $f_0$ and $f_t$ on a barbell graph. Then we extend our result to generalized barbell graphs that have two or more bars.

## 4.1 Token Shifting on Barbell Graphs

We first show that we can find the minimum shift sequence on a barbell graph in linear time. Let $G$ be a barbell graph composed of two cliques $A$ and $B$, each of size $n$, connected by a single edge: the bar.

The two cliques $A$ and $B$ contain $n$ vertices each, from $v_1$ to $v_n$ and from $v_{n+1}$ to $v_{2n}$, respectively. The two vertices joined by the bar will be referred as *gate* vertices. Furthermore, we subdivide the tokens into two types, based on their matching vertices in the target arrangement: *local* tokens and *foreign* tokens, as follows. Tokens on vertices in a clique whose target vertices are in the other clique are referred to as *foreign* tokens. Let foreign$(A)$ be the set of foreign tokens in $A$ in $f_0$ and foreign$(B)$ be the set of foreign tokens in $B$ in $f_0$, as follows:

$$\mathrm{foreign}(A) = \{v_i \in V \,|\, f_0(v_i) = f_t(v_j) \text{ where } v_i \in A \text{ and } v_j \in B\},$$

$$\mathrm{foreign}(B) = \{v_i \in V \,|\, f_0(v_i) = f_t(v_j) \text{ where } v_i \in B \text{ and } v_j \in A\}.$$

Let $F = |\mathrm{foreign}(A)| = |\mathrm{foreign}(B)|$. In the following, we will prove that $3F - 2 \leq |\mathrm{OPT}| \leq 3F + 4$. Note that $|\mathrm{foreign}(A)| = |\mathrm{foreign}(B)| = F$ must hold in order for $f_0$ to be re-configurable to $f_t$. Let $S_F$ be a shortest sequence of shifts that moves all $2F$ foreign tokens to their matching vertices. Note that this may still leave some non-foreign tokens on incorrect vertices; we will deal with re-configuring these tokens later.

**Lemma 1.** *In the Labeled Token Shifting Problem on a barbell graph, we have* $3F - 2 \leq |S_F| \leq 3F + 2$.

*Proof.* To transform $f_0$ to $f_t$, it is required for every foreign token on $A$ and $B$ to cross the bar at least once. Note that we can move two foreign tokens by performing a token exchange across the bar. In the worst case, a foreign token needs to be moved three times: from the current vertex to the nearest gate vertex, then across the bar to the gate vertex of the target clique, and then to the target vertex. Firstly, a foreign token on each clique must be moved to the gate vertex of that clique, which takes 2 shifts in total. Then, the actual exchange of tokens on gate vertices in a shift cycle $(v_n, v_{n+1})$ of length 2 occurs. Next, in each clique, the token on the gate vertex, say $v_n$, is moved to its target vertex $v_i$, while a new foreign token is moved from $v_j$ to the gate. This is done with the single cycle $(v_n, v_i, v_j)$. After the $F$th exchange, we need one more shift in each clique to move the token from the gate vertex to its target vertex. Therefore, in the worst case we do $F$ exchanging shifts and $2F + 2$ local shifts, which is $3F + 2$ shifts in total. However, we also need to consider the following special cases.

**Condition 1.** *A gate vertex already holds a foreign token in the initial arrangement $f_0$.*

If a gate vertex already holds a foreign token in the initial arrangement, then the initial shift for moving a foreign token to that gate vertex is not necessary. Hence, in the cases where $A$ or $B$ (or both) satisfy Condition 1, we need one (or two) fewer shift than $2F + 2$.

**Condition 2.** *The target token of a gate vertex (i.e., the token that is on a gate vertex in $f_t$) is in the opposite clique in $f_0$.*

If this condition is satisfied, we can move that gate's final token across the bar in the $F$th exchange. This way it is already in place when it enters the clique, and we can spare the final shift in that clique. Thus, in the extreme case where both gate vertices satisfy Conditions 1 and 2, and only $3F - 2$ shifts are necessary. □

As for the local tokens, their target vertices are within the same clique. Hence, by Theorem 2, at most 2 shifts are necessary to solve the problem in each clique. We can now present this section's main result

**Theorem 3.** *The Labeled Token Shifting Problem on a barbell graph $G = (V, E)$ can be solved with an optimal shift sequence in linear time, satisfying $3F - 2 \leq |\mathrm{OPT}| \leq 3F + 4$.* □

### 4.2 Token Shifting on Generalized Barbell Graphs with Two Bars

In this section, we extend our previous result to generalized barbell graphs. That is, we join two cliques by two bars instead of one, and this allows us to more effectively exploit the cyclic shift operation.

Let $G$ be a generalized barbell graph with $2n$ vertices, with cliques $A$ and $B$ consisting of vertices from $v_1$ to $v_n$ and $v_{n+1}$ to $v_{2n}$, respectively. Two *bars* $e_1$ and $e_2$ connect $A$ and $B$ such that $e_1$ is incident to $v_n$ and $v_{n+1}$ and $e_2$ is

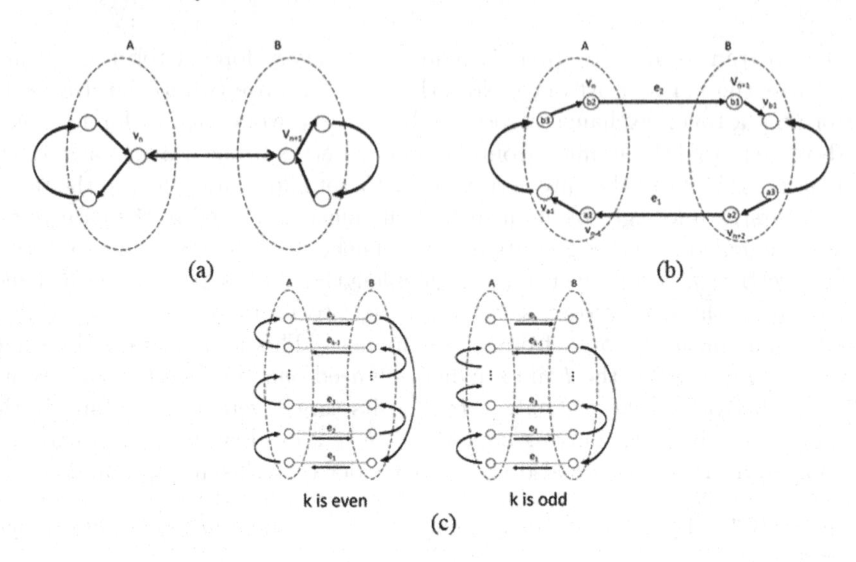


**Fig. 3.** Representation of token shifting on (a) a barbell graph, (b) a generalized barbell graph with 2 bars, and (c) a generalized barbell graph with $k > 2$ bars

incident to $v_{n-1}$ and $v_{n+2}$. Let $F = |\text{foreign}(A)| = |\text{foreign}(B)|$, defined as in the previous section. We can combine the two steps into one by exchanging foreign tokens and bringing the foreign tokens to the gate vertices for the next exchange in a single shift.

**Theorem 4.** *The Labeled Token Shifting Problem on a generalized barbell graph $G = (V, E)$ with 2 bars can be solved with an optimal shift sequence in linear time, satisfying $F \leq |\text{OPT}| \leq F + 4$.* □

### 4.3 Token Shifting on Generalized Barbell Graphs with $k \geq 2$ Bars

For the next step, we discuss the Labeled Token Shifting Problem on generalized barbell graphs with $k > 2$ bars. Here, $G$ is a graph consisting of two equal cliques $A$ and $B$ connected by $k$ edges, called *bars*, such that no two bars are incident to the same vertex. Let $F = \text{foreign}(A) = \text{foreign}(B)$, defined as usual.

**Theorem 5.** *The Labeled Token Shifting Problem on a generalized barbell graph $G = (V, E)$ with $k \geq 2$ bars can be solved with an optimal shift sequence that satisfies $F/\lfloor k/2 \rfloor \leq |\text{OPT}| \leq F/\lfloor k/2 \rfloor + 4$.*

*Proof.* In the previous section, we proved that token shifting on a barbell graph with 2 connecting edges for $2F$ foreign tokens uses $F + 4$ shifts: 2 local shifts for moving foreign tokens on gate vertices at the start, $F$ shifts for exchanging foreign tokens between cliques, and 2 local shifts to rearrange tokens within cliques. Now, while the number of local shifts remains the same, the number of exchanging shifts decreases as $k$ increases.

Half of the $k$ edges can be used to move the foreign tokens from $A$ to $B$ and another half of the $k$ edges can be used to move foreign tokens from $B$ to $A$. In one shift, we can exchange $k$ tokens for even $k$ and $k-1$ tokens for odd $k$ (see Fig. 3(c)). Thus, for $F$ tokens, we only need $F/\lfloor k/2 \rfloor$ shifts. □

## 5 2-Colored Token Shifting on Block Graphs

In this section, we discuss the 2-Colored Token Shifting Problem on block graphs. A block graph (or a clique tree) is a graph in which every bi-connected component (block) is a clique (see Fig. 4).

**Definitions.** In order to state this section's result, we need some definitions. Given a block graph $G = (V, E)$, where a block is a maximal clique, an *articulation point* is a vertex that belongs to more than one block. Let $P \subseteq V$ be the set of articulation points of $G$, and let $K$ be the set of blocks of $G$. We define the *tree representation* of $G$ (see [3]) as the undirected graph $T(G) = (V', E')$, where $V' = P \cup K$ and

$$E' = \{\{k, p\} | \text{ the articulation point } p \in P \text{ lies in the block } k \in K\}.$$

When referring to $T(G)$, the nodes in $P$ are called *articulation nodes*, and the nodes in $K$ are called *clique nodes*. Figure 4(c) shows an example of a tree representation. For a clique node $k \in K$, we write $I(k)$ to indicate the vertices of $G$ that are in the block $k$ but are not articulation points, i.e., $I(k) = k \setminus P$. Note that $I(k)$ induces a (possibly empty) clique in $G$.

Now, let $G = (V, E)$ be a block graph with $n$ vertices, let $\mathrm{Col} = \{c_1, c_2\}$ be the color set, and let $f_0$ and $f_t$ be the initial and target token arrangements on $G$. We say that an articulation node $p \in P$ *holds* color $c \in \mathrm{Col}$ if $f_0(p) = c$. Also, if $f$ is an arbitrary arrangement, we write $n_c(f(p)) = 1$ if $f(p) = c$, and $n_c(f(p)) = 0$ otherwise. Similarly, for a clique node $k \in K$, let $n_c(f(k))$ be the number of $c$-colored tokens in $I(k) \subseteq V$ in the arrangement $f$. Then, we say that a clique node $k$ of $T(G)$ *holds* color $c$ if $n_c(f_0(k)) > n_c(f_t(k))$.

For each node $x$ in $T(G)$, $x$ has a *value* of $n_{c_1}(f_0(x)) - n_{c_1}(f_t(x))$. For each edge $e$ in $E'$ connecting two nodes $k \in K$ and $p \in P$, we define the number $\mathrm{diff}(e)$ as follows (cf. [8]). Let $T_k$ be the subtree including node $k$ resulted by the removal of $e$ from $T(G)$. $n_{c_1}(f(T'))$ is the number of $c_1$ tokens on the set of vertices of $G$ represented by $T'$ in arrangement $f$. Then, $\mathrm{diff}(e) = n_{c_1}(f_t(T_k)) - n_{c_1}(f_0(T_k))$, i.e., the difference in number of $c_1$ tokens on $T'$ between $f_0$ and $f_t$. For simplicity, $\mathrm{diff}(e)$ can be defined as the number of $c_1$ tokens (and, symmetrically, also $c_2$ tokens) that we must move along $e$ to transform $f_0$ into $f_t$. If $\mathrm{diff}(e) = d > 0$, it means we need to move $d$ tokens of color $c_1$ to $k$. If $\mathrm{diff}(e) = -d < 0$, it means we need to move $d$ tokens of color $c_2$ to $k$.

Finally, we define $E'_k \subseteq E'$ to be the set of edge of $T(G)$ that are incident to the clique node $k$.

**Theorem 6.** *For the 2-Colored Token Shifting Problem on a block graph $G = (V, E)$, we have*

$$\sum_{k \in K} \max_{e \in E'_k} \{|\mathrm{diff}(e)|\} \le |\mathrm{OPT}| \le \sum_{k \in K} \max \left\{ \sum_{\substack{e \in E'_k \\ \mathrm{diff}(e)>0}} \mathrm{diff}(e), \sum_{\substack{e \in E'_k \\ \mathrm{diff}(e)<0}} |\mathrm{diff}(e)|, 1 \right\},$$

*and a shift sequence within these bounds can be computed in $O(n^2)$ time.*

*Proof.* For the upper bound, we will give a procedure for finding a shift sequence. We first construct the tree representation $T(G)$ in $O(n^2)$ time. From $T(G)$, we determine the sequence of shifts by deciding on which clique the shift must be performed in each step (note that, in a block graph, every cycle is included in a single clique).

For a clique $k$ with an excess of $c_1$ tokens connected to an articulation vertex $p$, some $c_1$ tokens in $k$ must be moved out and some $c_2$ tokens must be moved in through $p$. We need to perform a shift that moves the extra $c_1$ token in $k$ to the articulation vertex $p$ and the $c_2$ tokens on $p$ to the target vertex in $k$. On $T(G)$, it will be a token exchange between a clique node $k$ that holds color $c_1$ and the articulation node $p$ that holds color $c_2$ along the edge $e = \{k, p\} \in E'$. This exchange will decrease $|\mathrm{diff}(e)|$ and change the color of $p$ to $c_1$. However, in the case where the $p$ holds the same color $c_1$ as $k$, it is pointless to perform a shift between them. The same goes for a clique with $n_{c_2}(f_0(k)) > n_{c_2}(f_t(k))$. If $\mathrm{diff}(e) = 0$, no token needs to be moved across $e$, and $e$ can be removed from $T(G)$. For $G$ to achieve the target arrangement $f_t$, all the edges in $T(G)$ must be removed. Thus, we can construct the shift sequence for $G$ from $T(G)$ by determining the clique nodes for an exchange in each step.

We now discuss how to choose a feasible clique node for token exchange. There are three types of clique nodes in $T(G)$.

A leaf node is a clique node with an articulation node, the removal of which will disconnect the clique node from the other clique nodes in $T(G)$. When we look for a clique for token exchange, we start with the leaf nodes and go up the tree $T(G)$. A leaf node $k$ connected to node $p$ by edge $e$ is feasible for an exchange if $k$ and $p$ hold different colors and $|\mathrm{diff}(e)| > 0$.

Non-leaf nodes are those with multiple articulation nodes connecting them to other clique nodes in $T(G)$. In non-leaf nodes, we can exchange one or more pairs of different color tokens in one shift. For a non-leaf node $k$ with $m$ articulation nodes $p_1, p_2, \ldots, p_m$, $k$ is feasible for an exchange (1) if there are one or more edges $e = (k, p)$ with $|\mathrm{diff}(e)| > 0$, and $k$ and $p$ hold different colors, where $p \in \{p_1, p_2, \ldots, p_m\}$ and $k$ has non-zero value or (2) if $k$ is connected to one or more pairs of articulation nodes $p_i$ and $p_j \in \{p_1, p_2, \ldots, p_m\}$ where $p_i$ and $p_j$ hold different colors, and $\mathrm{diff}(e_i = \{k, p_i\})$ and $\mathrm{diff}(e_j = \{k, p_j\})$ have opposite sign (one positive, one negative).

An isolated node is already disconnected from other clique nodes in $T(G)$ and the amount of both $c_1$ and $c_2$ tokens in it is the same for $f_0$ and $f_t$. For each

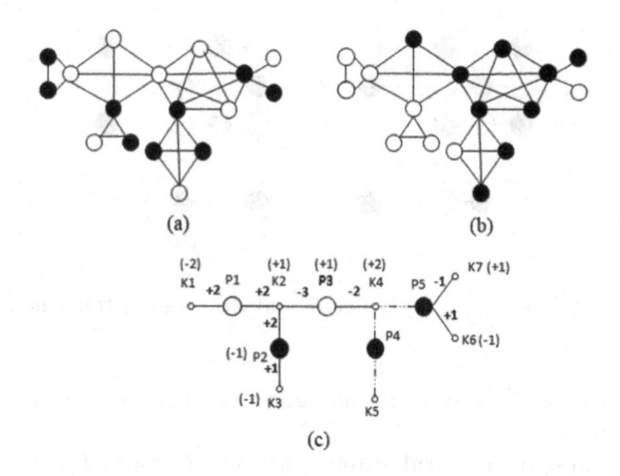


**Fig. 4.** (a) Initial arrangement $f_0$, (b) target arrangement $f_t$, and (c) tree representation $T(G)$ of block graph $G$ with positive values over nodes that need black tokens, negative values over nodes that need white tokens, diff$(e)$ values over each edge $e$, and dotted lines for removed edges

isolated node $k$ with no edge in $T(G)$, if $f_0(k) \neq f_t(k)$, then one shift suffices to reach the target arrangement as $n_c(f_0(k)) = n_c(f_t(k))$, $c \in \{c_1, c_2\}$.

As for the lower bound, we observe that, for each clique node $k$, we can only move one token to or from each articulation point in a shift and decrease the $|\text{diff}(e)|$ of each edge by one. Therefore, if $k$ is incident to an edge $e$ with $|\text{diff}(e)| = d$, then at least $d$ shifts must be performed in the clique corresponding to $k$. Thus, to remove all the edges incident to a clique node $k$ in $T(G)$, at least $\max_{e \in E'_k} \{|\text{diff}(e)|\}$ shifts are necessary. □

## 6 Hardness of 2-Colored Token Shifting

In this section, we show that a shortest shift sequence for the 2-Colored Token Shifting Problem is not only NP-hard to compute, but also NP-hard to approximate within a factor of $2 - \varepsilon$, for any $\varepsilon > 0$. This is true even if the graph $G$ is a grid graph, hence planar and with maximum degree 4. We will prove it by a reduction from the NP-complete problem of deciding if a grid graph has a Hamiltonian cycle, i.e., a cycle involving all vertices [4].

**Theorem 7.** *The optimal shifting sequence for the 2-Colored Token Shifting Problem is NP-hard to approximate within a factor of* $2 - \varepsilon$*, for any* $\varepsilon > 0$*, even for grid graphs.*

*Proof.* Let $G = (V, E)$ be a connected grid graph (i.e., a vertex-induced finite subgraph of the infinite grid), and let a *checkered arrangement* be an arrangement of two-colored tokens on $G$ such that tokens on any two adjacent vertices

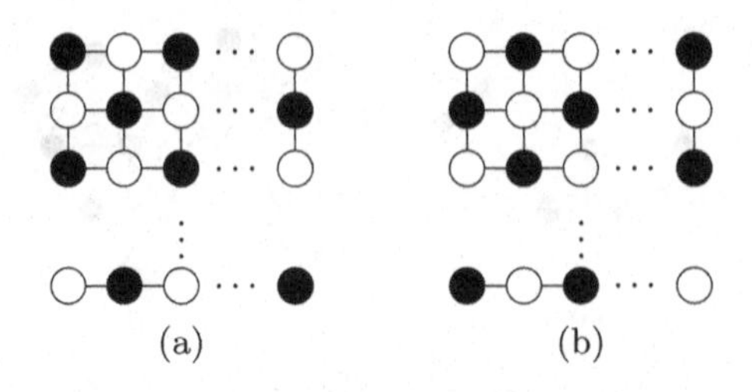


**Fig. 5.** (a) Initial arrangement $f_0$ and (b) target arrangement $f_t$

have different colors. Note that, for any given $G$, there are exactly two different checkerboard arrangements.

Our reduction maps the grid graph $G$ to the 2-Colored Token Shifting Problem on the same graph $G$, where the initial arrangement $f_0$ and the target arrangement $f_t$ are the two distinct checkerboard arrangements (see Fig. 5).

Observe that $f_0(v) \neq f_t(v)$ for all $v \in V$, and thus a sequence of shift operations that transforms $f_0$ into $f_t$ must move every token at least once. More precisely, $f_t$ is reached if and only if every token takes part in an odd number of shift operations. If $G$ has a Hamiltonian cycle $C$, then the shift operation along $C$ immediately transforms $f_0$ into $f_t$, and hence $|\text{OPT}| = 1$. Conversely, if $|\text{OPT}| = 1$, the single shift operation that transforms $f_0$ into $f_t$ must involve every vertex, and thus it must be a Hamiltonian cycle.

We have proved that, if $G$ has a Hamiltonian cycle, then $|\text{OPT}| = 1$, and that if $G$ does not have a Hamiltonian cycle, then $|\text{OPT}| \geq 2$. Thus, if we could compute an approximation of $|\text{OPT}|$ within a factor of $2 - \varepsilon$ in polynomial time, we would also be able to decide if $G$ has a Hamiltonian cycle. Since the latter problem is NP-hard [4], then so is the former problem. □

## References


1. Amano, K., et al.: How to solve the torus puzzle. Algorithms **5**(1), 18–29 (2012)
2. Biniaz, A., et al.: Token swapping on trees. arXiv preprint arXiv:1903.06981 (2019)
3. Brimkov, B., Hicks, I.V.: Memory efficient algorithms for cactus graphs and block graphs. Discret. Appl. Math. **216**, 393–407 (2017)
4. Itai, A., Papadimitriou, C.H., Szwarcfiter, J.L.: Hamilton paths in grid graphs. SIAM J. Comput. **11**(4), 676–686 (1982)
5. Nishimura, N.: Introduction to reconfiguration. Algorithms **11**(4), 52 (2018)
6. Sai, K.K., Uehara, R., Viglietta, G.: Cyclic shift problems on graphs. In: Uehara, R., Hong, S.-H., Nandy, S.C. (eds.) WALCOM 2021. LNCS, vol. 12635, pp. 308–320. Springer, Cham (2021). https://doi.org/10.1007/978-3-030-68211-8_25
7. Yamanaka, K., et al.: Swapping labeled tokens on graphs. Theor. Comput. Sci. **27**, 81–94 (2015)
8. Yamanaka, K., et al.: Swapping colored tokens on graphs. In: Dehne, F., Sack, J.-R., Stege, U. (eds.) WADS 2015. LNCS, vol. 9214, pp. 619–628. Springer, Cham (2015). https://doi.org/10.1007/978-3-319-21840-3_51

# Computational Complexity of Jumping Block Puzzles

Masaaki Kanzaki[1], Yota Otachi[2], and Ryuhei Uehara[1(✉)]

[1] School of Information Science, Japan Advanced Institute of Science and Technology, Asahidai, Nomi, Ishikawa 923-1292, Japan
{kanzaki,uehara}@jaist.ac.jp

[2] Department of Mathematical Informatics, Graduate School of Informatics, Nagoya University, Furocho, Chikusa-ku, Nagoya, Aichi 464-8601, Japan
otachi@nagoya-u.jp

**Abstract.** In combinatorial reconfiguration, the reconfiguration problems on a vertex subset (e.g., an independent set) are well investigated. In these problems, some tokens are placed on a subset of vertices of the graph, and there are three natural reconfiguration rules called "token sliding," "token jumping," and "token addition and removal". In the context of computational complexity of puzzles, the sliding block puzzles play an important role. Depending on the rules and set of pieces, the sliding block puzzles characterize the computational complexity classes including P, NP, and PSPACE. The sliding block puzzles correspond to the token sliding model in the context of combinatorial reconfiguration. On the other hand, a relatively new notion of jumping block puzzles is proposed in puzzle society. This is the counterpart to the token jumping model of the combinatorial reconfiguration problems in the context of block puzzles. We investigate several variants of jumping block puzzles and determine their computational complexities.

**Keywords:** Combinatorial reconfiguration · Computational complexity · Jumping block puzzle · Sliding block puzzle · Token jumping

## 1 Introduction

Recently, the *reconfiguration problems* attracted the attention in theoretical computer science. These problems arise when we need to find a step-by-step transformation between two feasible solutions of a problem such that all intermediate results are also feasible and each step abides by a fixed reconfiguration rule, that is, an adjacency relation defined on feasible solutions of the original problem. The reconfiguration problems have been studied extensively for several well-known problems, including INDEPENDENT SET [12,14], SATISFIABILITY [8,16], SET COVER, CLIQUE, MATCHING [12], VERTEX COLORING [2–4], and SHORTEST PATH [13].


C.-Y. Chen et al. (Eds.): COCOON 2021, LNCS 13025, pp. 655–667, 2021.
https://doi.org/10.1007/978-3-030-89543-3_54

In the reconfiguration problems that consist in transforming a vertex subset (e.g., an independent set), there are three natural reconfiguration rules called "token sliding," "token jumping," and "token addition and removal" [14]. In these rules, a vertex subset is represented by the set of *tokens* placed on each of the vertices in the set. In the token sliding model, we can slide a token on a vertex to another along an edge joining these vertices. On the other hand, a token can jump to any other vertex in the token jumping model.

In the puzzle society, tons of puzzles have been invented which can be seen as realizations of some reconfiguration problems. Among them, the family of *sliding block puzzles* has been playing an important role bridging recreational mathematics and theoretical computer science. A classic puzzle is called the Dad puzzle; it consists of rectangle pieces in a rectangle frame, and the goal is to slide a specific piece (e.g., the largest one) to the goal position. The Dad puzzle was invented in 1909 and the computational complexity of this puzzle was open since Martin Gardner mentioned it. After almost four decades, Hearn and Demaine proved that the puzzle is PSPACE-complete in general (a comprehensive survey can be found in [9]). When all pieces are unit squares, we obtain another famous puzzle called the *15-puzzle*; in this puzzle, we slide each unit square using an empty area and arrange the pieces in order. From the viewpoint of combinatorial reconfiguration, this puzzle has remarkable properties in the general form of the $(n^2 - 1)$-puzzle. For given initial and final arrangements of pieces, the decision problem asks if we can transform from the initial arrangement to the final arrangement. Then the decision problem can be solved in linear time, and for a yes-instance, while a feasible solution can be found in $O(n^3)$ time, finding a shortest solution is NP-complete (see [5,18]).

To see how the computational complexity of a reconfiguration problem depends on its reconfiguration rule, it is natural to consider the "jumping block" variant as a counterpart of sliding block puzzles. In fact, in the puzzle society, some realizations of *jumping block puzzles* have been invented. As far as the authors know, "Flying Block" was the first jumping block puzzle, which was designed by Dries de Clercq and popularized at International Puzzle Party by Dirk Weber in 2008.[1] The Flying Block consists of four polyominoes (see [7] for the notion of polyominoes) within a rectangle frame. The goal is similar to the Dad puzzle; moving a specific piece to the goal position. In one move, we first pick up a piece, rotate it if necessary, and then put it back into the frame. A key feature is that each piece has a small tab for picking up. Because of this feature, flipping a piece is inhibited when we put it into the frame. In this framework, several puzzles were designed by Hideyuki Ueyama, and some of them can be found in [1] (see Concluding Remarks for further details).

Later, Fujio Adachi invented another jumping block puzzle which is called *Flip Over* (or Turn Over). It was invented in 2016 and popularized at Inter-

[1] You can find "Flying Block", "Flying Block II", and "Flying Block III" at http://www.robspuzzlepage.com/sliding.htm (accessed in June 2021).

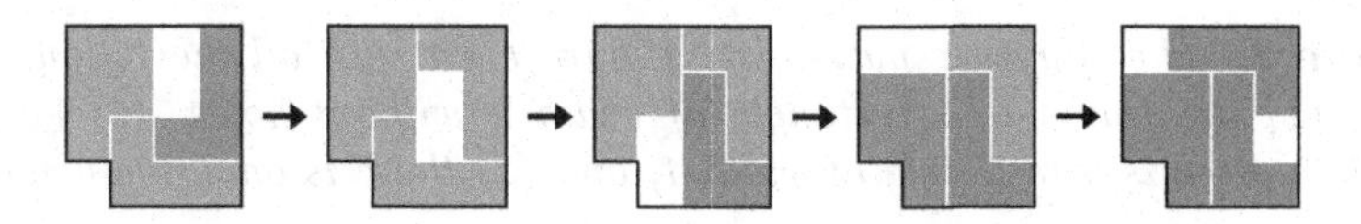

**Fig. 1.** An example of the flip over puzzle. Each piece has its front side and back side, and the goal is to make them back-side up.

national Puzzle Party by Naoyuki Iwase in 2019.[2] This puzzle consists of four polyominoes in an orthogonal frame, which is not a rectangle. The operation is a bit different from Flying Block; you can *flip* the piece in addition to the rotation if you like. Each piece has its front side and back side (distinguished by their colors). The goal of this puzzle is to make all pieces back-side up. A simple example is given in Fig. 1.

As far as the authors know, the jumping block puzzles have never been investigated in the context of the reconfiguration problems. In this paper, we first investigate the jumping block puzzle under the model of Flip Over. That is, the *flip over puzzle* problem is formalized as follows:

FLIP OVER PUZZLE
**Input:** A set of polyominoes in an orthogonal frame.
**Operation:** Pick up a piece, rotate and flip it if desired, and put it back into the frame.
**Goal:** To make every piece back-side up in the frame.

We first observe that each piece can be flipped in-place if it is line symmetric. That is, when all polyominoes are line symmetric, it is a yes-instance and all pieces can be flipped in a trivial (shortest) way. In contrast with that, the existence of one asymmetric piece changes the computational complexity of this puzzle. The first result in this paper is the following theorem:

**Theorem 1.** *The flip over puzzle is PSPACE-complete even with all the following conditions: (1) the frame is a rectangle without holes, (2) every line symmetric piece is a rectangle of size* $1 \times k$ *with* $1 \leq k \leq 3$*, (3) there is only one asymmetric piece of size 4, and (4) there is only one vacant unit square.*

We note that all polyominoes of size at most 3 are line symmetric (i.e., monomino, domino, I-tromino, and L-tromino in terms of polyomino). Therefore, Theorem 1 is tight since it contains only one minimum asymmetric piece.

We show Theorem 1 by a reduction from the Nondeterministic Constraint Logic (NCL). It is known that the NCL is PSPACE-complete even if the given input NCL graph has constant bandwidth [20]. Using the result in [20], we will have the claim in Theorem 1 even if the frame is a rectangle of constant width. On the other hand, if the frame is a rectangle of size $1 \times m$, the problem is trivial: each piece should be a rectangle of width 1 and hence it can be flipped to back side in-place in one step. Interestingly, the problem is intractable when the frame has width 2.

[2] Some commercial products can be found at http://www.puzzlein.com/ (in Japanese; accessed in June 2021).

**Theorem 2.** *The flip over puzzle is NP-hard even with all the following conditions: (1) the frame is a rectangle of width 2 without holes, (2) every line symmetric piece is a rectangle of width 1, and (3) there is only one asymmetric piece.*

In our reduction for proving Theorem 2, the constructed instances admit a sequence of flips, if any exists, of length polynomial in $n$. However, we do not know whether the flip over puzzle on a frame of width $w$ is in NP or not for some small constant $w \geq 2$. On the other hand, when the number of flips of each piece is bounded above by a constant, we can observe that this problem is in NP. Under this assumption, we show that the flip over puzzle is NP-complete even if we have some combination of natural conditions.

Next we turn to the *flying block puzzle* problem, which is formalized as follows:

FLYING BLOCK PUZZLE

**Input:** A set of polyominoes in an orthogonal frame, a specific piece $P$ in the set, and a goal position of $P$ in the frame.

**Operation:** Pick up a piece, rotate it if desired, and put it back into the frame.

**Goal:** To move $P$ to the goal position.

In our results of the flip over puzzle, the unique asymmetric piece plays an important role and flips of rectangles are not essential. In some cases, we can show corresponding results for the flying block puzzle by modifying the unique asymmetric piece in the flip over puzzle. Using the idea, we first show natural counterparts of Theorems 1 and 2. However, while the goal of the flip over puzzle is to flip *all* pieces, the goal of the flying block puzzle is to arrange the *specific* piece to the goal position. This difference requires different techniques, and some counterparts of the NP-completeness results of the flip over puzzle remain open. Intuitively, throughout these counterparts, the flying block puzzle seems to be more difficult than the flip over puzzle since the hardness results hold under stronger restrictions. In fact, the computational complexity of the flying block puzzle is not trivial even if the frame is a rectangle of width 1, while it is trivial in the flip over puzzle since all pieces are rectangles of width 1. We show weakly NP-completeness of the flying block puzzle even if the frame is of width 1 and each piece can be moved at most once. On the other hand, we show a nontrivial polynomial time algorithm when we can move each piece any number of times in the frame of width 1.

## 2 Preliminaries

A *polyomino* is a polygon formed by joining one or more unit squares edge to edge. A polyomino is also called a *k-omino* if it consists of $k$ unit squares. When $k = 1, 2, 3, 4$, we sometimes call it *monomino*, *domino*, *tromino*, and *tetromino*, respectively. An instance of the *jumping block puzzle* consists of a set $\mathcal{P}$ of polyominoes $P_1, P_2, \ldots, P_n$ and a polyomino $F$. Each polyomino $P_i \in \mathcal{P}$ is called a *piece*, and $F$ is called a *frame*. Each piece $P_i$ has its *front side* and *back side* (in the figures in this paper, a bright color and a dark color indicate front side and back side, respectively).

A *feasible packing* of $\mathcal{P}$ to $F$ is an arrangement of all pieces of $\mathcal{P}$ into $F$ such that each piece is placed in $F$, no pair of pieces overlaps (except at edges and vertices), every vertex of pieces of $\mathcal{P}$ is placed on a grid point in $F$, and each edge of pieces of $\mathcal{P}$ is parallel or perpendicular to the edges of $F$.

For a feasible packing of $\mathcal{P}$ to $F$, a *flip* of $P_i$ is an operation that consists of the following steps: (1) pick up $P_i$ from $F$, (2) translate, rotate, and flip $P_i$ if necessary, and (3) put $P_i$ back into $F$ so that the resulting arrangement is a feasible packing. On the other hand, a *fly* of $P_i$ is an operation that consists of the following steps: (1) pick up $P_i$ from $F$, (2) translate and rotate $P_i$ if necessary, and (3) put $P_i$ back into $F$ so that the resulting arrangement is a feasible packing.

For a given feasible packing $X$ of $\mathcal{P}$ to $F$, the *flip over puzzle* asks whether $X$ can be reconfigured by a sequence of flips to a feasible packing in which all pieces are back-side up in $F$. In contrast, in the flying block puzzle, the input consists of three tuples; a feasible packing of $\mathcal{P}$ to $F$, a specific piece, say $P_n$, and the goal position of $P_n$ in $F$. That is, the *flying block puzzle* asks if we can move the specific piece $P_n$ to the goal position starting from the feasible packing by a sequence of flies.

In order to determine the computational complexities of the jumping block puzzles, we use the notion of the *constraint logic* which was introduced by Hearn and Demaine [10] (see also [9]). A *constraint graph* $G = (V, E, w)$ is an edge-weighted undirected 3-regular graph such that (1) each edge $e \in E$ is weighted 1 or 2, (we sometimes describe the values 1 by *red* and 2 by *blue*, respectively) and (2) each vertex is either an *AND vertex* or an *OR vertex* such that (2a) an AND vertex is incident to two red edges and one blue edge, and (2b) an OR vertex is incident to three blue edges. A *configuration* of a constraint graph is an orientation of the edges in the graph. A configuration is *legal* if the total weight of the edges pointing to each vertex is at least 2. The problem *NCL* on the constraint logic is defined as follows.

**Input:** A constraint graph $G = (V, E, w)$, a legal configuration $C_0$ for $G$, and an edge $e_t \in E$.

**Question:** Is there a sequence of legal configurations $(C_0, C_1, \ldots)$ such that (1) $C_i$ is obtained by reversing the direction of a single edge in $C_{i-1}$ with $i \geq 1$, and (2) the last configuration is obtained by reversing the direction of $e_t$?

The *Bounded NCL* is a variant of the NCL that requires one additional restriction that every edge can be reversed at most once. For these two problems, the following theorem is known:

**Theorem 3** ([9]). *(1) NCL is PSPACE-complete even on planar graphs and (2) Bounded NCL is NP-complete even on planar graphs.*

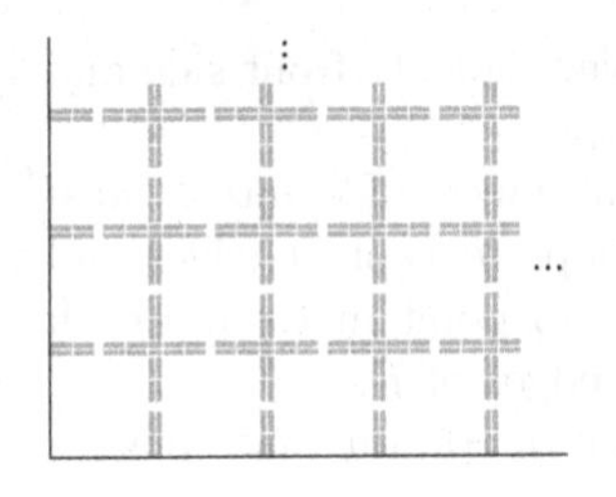

**Fig. 2.** The cell border, which is a framework of embedding gadgets.

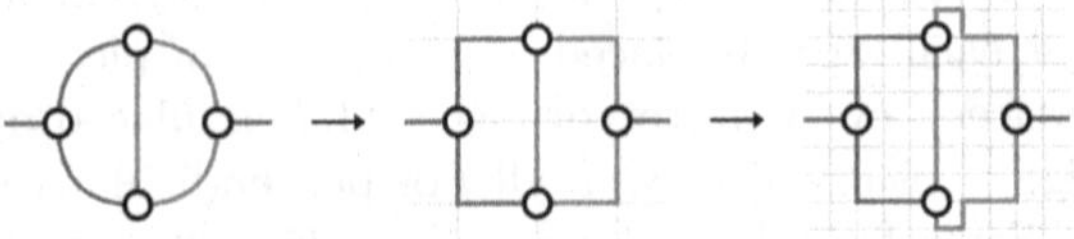

**Fig. 3.** A grid drawing of the graph in Step (1). Two red edges incident to an AND vertex are collinear.

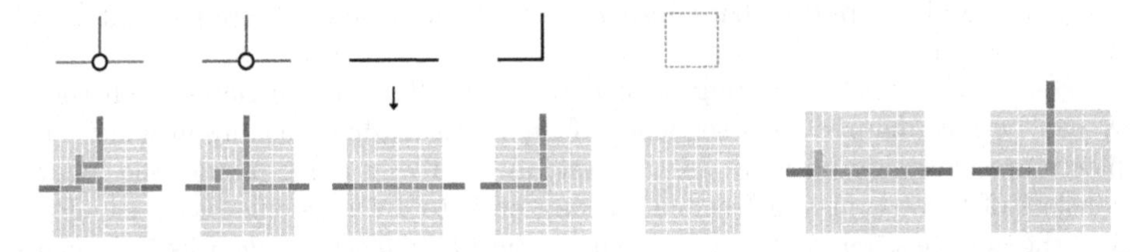

**Fig. 4.** Gadgets of size $13 \times 13$ for an AND vertex, an OR vertex, a wire, a corner, and an empty cell.

**Fig. 5.** Two edge gadgets for $e_t$ for embedding the L-tetromino.

## 3 Flip over Puzzles

In this section, we focus on the flip over puzzles. We first show that this puzzle is PSPACE-complete even on quite restricted input. Next, we show that it is NP-hard even if the frame $F$ is a rectangle of width 2. Lastly, we show that it is NP-complete if the number of flips of each piece is $O(1)$.

### 3.1 PSPACE-completeness in General

In this section, we give the outline of the proof of Theorem 1. We give a polynomial-time reduction from the NCL problem to the flip over puzzle. Let $G = (V, E, w)$, $e_t$, and $C_0$ be the input of the NCL problem. That is, $G$ is an edge-weighted planar graph and $C_0$ is a legal configuration of $G$ and $e_t$ is an edge in $E$. The framework of the reduction is similar to the reduction of the NCL problem to the sliding block puzzle shown in [9, Sec. 9.3]: The frame $F$ is a big rectangle, and it is filled with a regular grid of gate gadgets, within a "cell border" construction. The internal construction of the gates is such that none of the cell-border blocks may move, thus providing overall integrity to the configuration. In our reduction, the cell border has width 2 (Fig. 2), and each gadget in a cell will be designed of size $13 \times 13$. We construct an instance of the flip over puzzle as follows.

We first compute a rectilinear embedding of $G$ (see [15] for the definition). To simplify the construction, we bend some edges (by refining the grid) to make two red edges incident to an AND vertex collinear (Fig. 3). We then define the frame $F$ so that each *cell* of size $13 \times 13$ contains one of an AND vertex, an OR vertex, a unit straight segment of an edge (*wire*), and a unit turning segment of an edge (*corner*). The vertices, unit segments of edges, and empty cells are replaced by the gadgets shown in Fig. 4.

Pick up any unit segment of the edge $e_t$, and embed the left gadget in Fig. 5 if it is a straight, and the right gadget in Fig. 5 if it is a turn. We note that each of them contains the unique asymmetric polyomino of size 4 in L-shape (which is called the *L-tetromino*), and this L-tetromino can be flipped when the leftmost I-tromino moves to left. Then we put polyominoes of unit size (which are called *monominoes*) to fill all vacant unit squares except one. (This one vacant unit square can be arbitrary.) This is the end of our reduction.

It is clear that this reduction can be done in polynomial time, the flip over puzzle constructed from $G$ satisfies the conditions (1) to (4) in Theorem 1, and the flip over puzzle is in PSPACE. We show that $G$ is a yes instance of NCL if and only if the instance of the flip over puzzle constructed from $G$ is a yes instance. The basic idea is the same as one used in [9, Sec. 9.3]; when an edge $e$ changes the direction and points to a vertex $v$, the corresponding *I-tromino* (of size $1 \times 3$ and in green in the figure) moves out one unit. Using the same arguments in [9, Sec. 9.3], we can show that the flip over puzzle simulates the movements of the NCL and vice versa.

We note that the instance of the flip over puzzle constructed here has only one vacant unit square. However, by jumping monominoes, we can move the vacant unit square to anywhere we need as long as it is occupied by a monomino. Thus, if we can eventually reverse $e_t$, we can make a vacant unit square next to the L-tetromino, and then we can make it back-side up. This is the outline of the proof of Theorem 1.

It is known that NCL is still PSPACE-complete even if the input NCL graph is planar and have constant bandwidth and it is given with a rectilinear embedding of constant height (see [20, Thm. 11] for the details). Thus we have the following corollary.

**Corollary 1.** *The flip over puzzle is PSPACE-complete even with the conditions in Theorem 1, and the frame is of constant width.*

## 3.2 NP-hardness on a Frame of Width 2

In this section, we give a proof of Theorem 2. We reduce the following 3-PARTITION problem to our problem.

**Input:** Positive integers $a_1, a_2, a_3, \ldots, a_{3m}$ such that $\sum_{i=1}^{3m} a_i = mB$ for some positive integer $B$ and $B/4 < a_i < B/2$ for $1 \leq i \leq 3m$.

**Output:** Determine whether we can partition $\{1, 2, \ldots, 3m\}$ into $m$ subsets $A_1, A_2, \ldots, A_m$ so that $\sum_{i \in A_j} a_i = B$ for $1 \leq j \leq m$.

It is well known that the 3-PARTITION problem is strongly NP-complete [6]. Without loss of generality, we assume that $m < B/2$ (otherwise, we multiply

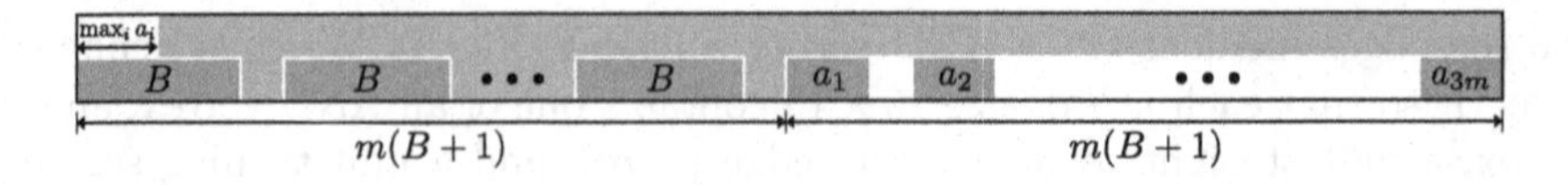


**Fig. 6.** Reduction for NP-hardness.

each $a_i$ by $m'$ for some $m' > m$). For a given instance $\langle a_1, a_2, \ldots, a_{3m} \rangle$ of 3-PARTITION, we construct an instance of the jumping block puzzle as follows. It consists of a frame $F$ and a set of pieces $\mathcal{P} = \{P_1, P_2, \ldots, P_{4m+1}\}$. The frame $F$ is a $2 \times (2m(B+1))$ rectangle. For $1 \le i \le 3m$, the piece $P_i$ is a $1 \times a_i$ rectangle. The pieces $P_{3m+1}, P_{3m+2}, \ldots P_{4m}$ are $1 \times B$ rectangles. The unique asymmetric piece $P_{4m+1}$ is drawn in Fig. 6. It almost covers the whole frame $F$ except (1) one rectangle of size $1 \times \max_i a_i$, (2) $m$ rectangles of size $1 \times B$ in the left side, and (3) one rectangle of size $1 \times m(B+1)$ in the right side. The initial feasible packing of $\mathcal{P}$ to $F$ is also described in Fig. 6. All pieces are placed front-side up. The asymmetric piece $P_{4m+1}$ is placed in $F$ as shown in Fig. 6, and $m$ rectangles of size $1 \times B$ are occupied by $m$ pieces $P_{3m+1}, P_{3m+2}, \ldots, P_{4m}$ of size $1 \times B$. The other $3m$ polyominoes $P_1, P_2, \ldots, P_{3m}$ are put in the rectangle of size $1 \times m(B+1)$ in the right side of $P_{4m+1}$ in arbitrary order. We assume that the piece $P_{3m+i}$ occupies the $i$-th vacant space of size $1 \times B$ from the center of $P_{4m+1}$.

Then we can show that all pieces in $\mathcal{P}$ can be flipped back-side up if and only if the original instance of 3-PARTITION is a yes instance.

### 3.3 NP-completeness with Constant Flips

In the proof of Theorem 2, we reduced the 3-PARTITION problem to the flip over puzzle. We saw that the instances constructed there had polynomial-length yes-witnesses, however, we do not know whether the flip over puzzle on a frame of width 2 is in NP or not. In this section, we focus on the flip over puzzle with constant flips. In this model, we restrict ourselves that the number of flips for each piece is bounded above by a constant. We assume that each piece can be moved at most $c$ times for some constant $c$. Then, if an instance of the puzzle has a solution, the solution can be represented by a sequence of moves, and the number of moves is bounded above by $cn$, where $n$ is the number of pieces. Therefore, each yes-instance has a witness of polynomial length, which implies that the puzzle with the restriction is in NP. In this section, we show that the flip over puzzle is still NP-complete even if each piece can be flipped at most once with some additional restrictions.

**Theorem 4.** *The flip over puzzle with constant flips is NP-complete even if it satisfies all the following conditions (0), (1), (2), and (3). (0) Each piece can be moved at most once, (1) the frame is a rectangle without holes, (2) every line symmetric piece is a rectangle of size* $1 \times k$ *with* $1 \le k \le 3$*, and (3) there is only one asymmetric piece of size 4.*

*Proof.* (Outline) The claim can be proved by following the same strategy of the proof of Theorem 1 in Sect. 3.1. □

**Theorem 5.** *The flip over puzzle with constant flips is NP-complete even if it satisfies all the following conditions (0), (1), (2), (3), and (4). (0) Each piece can be moved at most once, (1) the frame is a rectangle without holes, (2) every line symmetric piece is a rectangle of size* $1 \times k$ *with* $1 \leq k \leq 3$*, (3) all asymmetric pieces are of size 4, and (4) there is only one vacant unit square.*

*Proof.* (Outline) In order to prove NP-hardness, we give a polynomial-time reduction from the *Hamiltonian cycle problem* on 3-regular planar digraphs where each vertex has indegree 1 or 2 which is a classic well-known NP-complete problem [17]. We use an idea similar to the one in [19]: For a given 3-regular planar digraph $G$, we replace the two types of vertices by the two types of gadgets, respectively, and join them by the same wire gadget in Sect. 3.1. □

**Theorem 6.** *The flip over puzzle with constant flips is NP-complete even with all the following conditions: (0) Each piece can be moved at most once, (1) the frame is a rectangle of width* 2 *without holes, (2) all pieces are rectangles of width 1 except 2, and (3) there is only one asymmetric piece.*

*Proof.* (Outline) The basic idea is similar to the proof of Theorem 2. We reduce the 3-PARTITION problem to our problem. □

## 4 Flying Block Puzzles

In this section, we turn to the flying block puzzles. In flying block puzzles, we can translate and rotate pieces but cannot flip them. In the flip over puzzle in Sect. 3, almost all pieces are rectangles except a few asymmetric pieces. Since a rectangle does not change by a flip, we can inherit most of the results for flip over puzzles in Sect. 3 to ones for flying block puzzles by changing some special pieces. However, while the goal of the flip over puzzle is to flip *all* pieces, the goal of the flying block puzzle is to arrange a *specific* piece to the goal position. This difference requires different techniques.

The counterpart of Theorem 1 is as follows:

**Theorem 7.** *The flying block puzzle is PSPACE-complete even with all the following conditions: (1) the frame is a rectangle without holes, (2) every piece is a rectangle of size* $1 \times k$ *with* $1 \leq k \leq 3$*, and (3) there is only one vacant unit square.*

*Proof.* (Outline) In the proof of Theorem 1, we use an L-tetromino as the unique asymmetric piece. In the flying block puzzle, we do not need to use this trick. Instead of that, the goal of the resulting flying block puzzle is just to move any specific I-tromino in the wire or corner gadget in Fig. 4 corresponding to $e_t$, which plays the same role of the flip of the L-tetromino in the proof of Theorem 1. □

The counterpart of Theorems 2 and 6 is as follows:

**Theorem 8.** *The flying block puzzle is NP-complete even with all the following conditions: (1) the frame is a rectangle of width 2 without holes, (2) every line symmetric piece is a rectangle of width 1, and (3) there is only one non-rectangle piece, which is line symmetric.*

*Proof.* (Outline) The basic idea is the same as the proof of Theorem 2: We reduce from the 3-PARTITION problem. □

We note that the claim in the proof of Theorem 8 still holds even if each piece can be moved at most once. In this sense, Theorem 8 is the counterpart of both Theorems 2 and 6.

The counterpart of Theorem 4 is as follows:

**Theorem 9.** *The flying block puzzle with constant flies is NP-complete even if it satisfies the following conditions (0), (1), (2), and (3). (0) Each piece can be moved at most once, (1) the frame is a rectangle without holes, and (2) every piece is a rectangle of size* $1 \times k$ *with* $1 \leq k \leq 3$.

*Proof.* (Outline) We apply the same technique of the proof of Theorem 7 to Theorem 4. □

Interestingly, the reduction of the proof of Theorem 5 from the Hamiltonian cycle cannot be extended to the flying block puzzle since we do not need to "visit" every piece to flip. Therefore, the computational complexity of the flying block puzzle with the conditions in Theorem 5 is open. We now turn to the flying block puzzles with the frame of width 1. While this case is trivially "yes" in any instance of the flip over puzzle since every piece is a rectangle of width 1, we show weak NP-hardness of the flying block puzzle if the frame is of width 1 and each piece can be moved at most once. On the other hand, when we can move the pieces any number of times, we have nontrivial polynomial time algorithm to solve it.

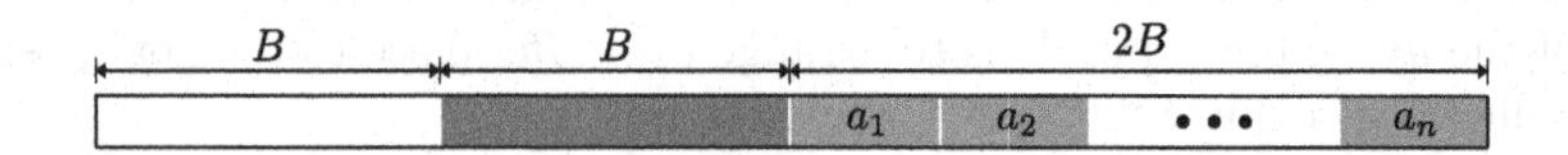


**Fig. 7.** The reduction from PARTITION.

**Theorem 10.** *The flying block puzzle with constant flies is weakly NP-complete even if each piece can be moved at most once and the frame is a rectangle of width* 1.

*Proof.* (Outline) We present a polynomial-time reduction from PARTITION [6], which asks, given $n$ positive integers $a_1, \ldots, a_n$ with $\sum_{i=1}^{n} a_i = 2B$, whether there exists a partition of $\{1, \ldots, n\}$ into two sets $A$ and $A'$ such that $\sum_{i \in A} a_i = \sum_{i \in A'} a_i = B$. This problem is weakly NP-complete. For a given instance

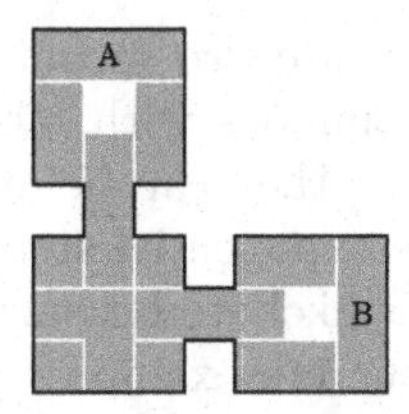


**Fig. 8.** A "simple" example of the flying block puzzle designed by Hideyuki Ueyama in [1] (with permission): It requires 256 steps to exchange A and B.

$\langle a_1, a_2, \ldots, a_n \rangle$ of PARTITION, we construct an instance of the flying block puzzle by a set of pieces $\mathcal{P} = \{P_1, P_2, \ldots, P_{n+1}\}$, where the piece $P_i$ is a $1 \times a_i$ rectangle for $1 \leq i \leq n$, and the frame $F$ which is a $1 \times (4B)$ rectangle. The special piece $P_{n+1}$ is a $1 \times B$ rectangle. The initial feasible packing of $\mathcal{P}$ to $F$ is given in Fig. 7. The goal is to move $P_{n+1}$ to the right end of $F$. The further details are omitted. □

**Theorem 11.** *The flying block puzzle can be solved in polynomial time when the frame is a rectangle of width* 1.

*Proof.* Omitted. □

## 5 Concluding Remarks

In this paper, we investigate the computational complexities of the jumping block puzzles which form the token-jumping counterpart of the sliding block puzzles in the context of reconfiguration problems. The other well-studied model in the field of combinatorial reconfiguration would allow "removals and additions" of blocks, which would be future work. Even in the jumping block puzzles, we still have many variants.

One natural variant in the context of the combinatorial reconfiguration is that the input consists of the initial feasible packing and the target feasible packing. In this reconfiguration problem, we have two observations (proofs are omitted):

**Observation 12.** *The reconfiguration problem of the jumping block puzzle is tractable if the frame $F$ is of width* 1.

**Observation 13.** *The reconfiguration problem of the jumping block puzzle is tractable if all pieces are rectangles of size* $1 \times 2$.

Extension of them to general cases, e.g., blocks of size $1 \times 3$, seems to be interesting (cf. [11]).

We may allow the frame to have holes (or fixed obstacles). Moreover, we may distinguish congruent pieces. One interesting example can be found in [1]: the puzzle in Fig. 8 was designed by Hideyuki Ueyama. It consists of 12 rectangle pieces and one L-tromino. There are two vacant unit squares. The goal of this puzzle is to exchange two I-trominoes with labels A and B. At a glance, it seems to be impossible. However, they can be exchanged in 256 steps. In such a case, it seems that it requires exponential steps if we have a few vacant unit squares. Intuitively, the puzzle is likely to become easier if we have many vacant unit squares. (In fact, some concrete examples can be found in [1].) It would be interesting to ask whether there is an algorithm that runs faster when $k$ is larger, where $k$ is the number of vacant unit squares.

## References

1. Akiyama, H.: Board Puzzle Reader (2009). (in Japanese). Shin Kigen Sha
2. Bonamy, M., Johnson, M., Lignos, I., Patel, V., Paulusma, D.: On the diameter of reconfiguration graphs for vertex colourings. Electron. Notes Discret. Math. **38**, 161–166 (2011)
3. Bonsma, P., Cereceda, L.: Finding Paths between graph colourings: PSPACE-completeness and superpolynomial distances. Theor. Comput. Sci. **410**(50), 5215–5226 (2009). http://dx.doi.org/10.1016/j.tcs.2009.08.023
4. Cereceda, L., van den Heuvel, J., Johnson, M.: Finding paths between 3-colourings. J. Graph Theory **67**, 69–82 (2011)
5. Demaine, E.D., Rudoy, M.: A simple proof that the $(n^2 - 1)$-puzzle is hard. Theor. Comput. Sci. **732**, 80–84 (2018)
6. Garey, M.R., Johnson, D.S.: Computers and Intractability – A Guide to the Theory of NP-Completeness. Freeman (1979)
7. Golomb, S.W.: Polyominoes. Princeton University Press, Princeton (1994)
8. Gopalan, P., Kolaitis, P.G., Maneva, E.N., Papadimitriou, C.H.: The connectivity of Boolean satisfiability: computational and structural dichotomies. SIAM J. Comput. **38**, 2330–2355 (2009)
9. Hearn, R.A., Demaine, E.D.: Games, Puzzles, and Computation. A K Peters Ltd. (2009)
10. Hearn, R.A., Demaine, E.D.: PSPACE-completeness of sliding-block puzzles and other problems through the nondeterministic constant logic model of computation. Theor. Comput. Sci. **343**(1–2), 72–96 (2005)
11. Horiyama, T., Ito, T., Nakatsuka, K., Suzuki, A., Uehara, R.: Packing trominoes is np-complete, #P-hard and ASP-complete. In: The 24th Canadian Conference on Computational Geometry (CCCG 2012), pp. 219–224 (2012)
12. Ito, T., et al.: On the complexity of reconfiguration problems. Theor. Comput. Sci. **412**, 1054–1065 (2011)
13. Kamiński, M., Medvedev, P., Milanič, M.: Shortest paths between shortest paths. Theor. Comput. Sci. **412**, 5205–5210 (2011)
14. Kamiński, M., Medvedev, P., Milanič, M.: Complexity of independent set reconfigurability problems. Theor. Comput. Sci. **439**, 9–15 (2012)
15. Liu, Y., Morgana, A., Simeone, B.: A linear algorithm for 2-bend embeddings of planar graphs in the two-dimensional grid. Discret. Appl. Math. **81**(1), 69–91 (1998)
16. Makino, K., Tamaki, S., Yamamoto, M.: An exact algorithm for the Boolean connectivity problem for $k$-CNF. Theor. Comput. Sci. **412**, 4613–4618 (2011)
17. Plesník, J.: The NP-completeness of the Hamiltonian cycle problem in planar digraphs with degree bound two. Inf. Process. Lett. **8**(4), 199–201 (1979)

18. Ratner, D., Warmuth, M.: The $(n^2 - 1)$-puzzle and related relocation problems. J. Symb. Comput. **10**, 111–137 (1990)
19. Uehara, R., Iwata, S.: Generalized Hi-Q is NP-Complete. Trans. IEICE **E73**(2), 270–273 (1990). http://www.jaist.ac.jp/~uehara/pdf/phd7.ps.gz
20. van der Zanden, T.C.: Parameterized complexity of graph constraint logic. In: IPEC 2015, pp. 282–293. LIPIcs, Vol. 43, Dagsthul (2015). https://doi.org/10.4230/LIPIcs.IPEC.2015.282

# A Card-Minimal Three-Input AND Protocol Using Two Shuffles

Raimu Isuzugawa[1(✉)], Kodai Toyoda[1], Yu Sasaki[1], Daiki Miyahara[2,3], and Takaaki Mizuki[1,2]

[1] Tohoku University, Sendai, Japan
raimu.isuzugawa.q6@dc.tohoku.ac.jp, mizuki+lncs@tohoku.ac.jp
[2] National Institute of Advanced Industrial Science and Technology, Tokyo, Japan
[3] The University of Electro-Communications, Tokyo, Japan

**Abstract.** Card-based cryptography typically uses a physical deck comprising black and red cards to perform secure computations, where a one-bit value is encoded using a pair of cards with different colors such that the order of black to red represents 0 and red to black represents 1. One of the most fundamental classes of card-based protocols is the class of "card-minimal" $n$-input AND protocols, which require $2n$ face-down cards as input to securely evaluate the AND value after applying a number of shuffles; here, the $2n$ cards are minimally required to describe an $n$-bit input. The best $n$-input AND protocols currently known use two shuffles for $n = 2$, five shuffles for $n = 3$, and $n + 1$ shuffles for $n > 3$. These upper bounds on the numbers of shuffles have not been improved for several years. In this work, we present a better upper bound for the $n = 3$ case by designing a new card-minimal three-input AND protocol using only two shuffles. Therefore, our proposed protocol reduces the number of required shuffles from five to two; we believe that this is a significant improvement.

## 1 Introduction

Many card-based protocols have been designed in the history of *card-based cryptography* to perform secure computations using a deck of physical cards. Typically, card-based protocols work on two-colored decks comprising black ♣ and red ♡ cards whose backs are denoted by ? and indistinguishable. These cards are used to represent Boolean values as follows:

$$\clubsuit\,\heartsuit = 0, \quad \heartsuit\,\clubsuit = 1. \tag{1}$$

When two face-down cards represent a bit $x \in \{0, 1\}$ according to the above encoding rule (1), we call them a *commitment* to $x$, denoted as

C.-Y. Chen et al. (Eds.): COCOON 2021, LNCS 13025, pp. 668–679, 2021.
https://doi.org/10.1007/978-3-030-89543-3_55

### 1.1 Card-Minimal AND Protocols

In 1989, Den Boer [2] designed the first card-based protocol, called the "five-card trick," which takes two commitments to $x_1, x_2 \in \{0,1\}$ and a helping card $\boxed{\heartsuit}$ as input to securely evaluate the AND value $x_1 \wedge x_2$ (without revealing any information about $x_1, x_2$ more than necessary) through a series of actions, such as shuffling and turning over cards:

$$\underbrace{\boxed{?}\boxed{?}}_{x_1}\ \underbrace{\boxed{?}\boxed{?}}_{x_2}\ \boxed{\heartsuit} \rightarrow \cdots \rightarrow x_1 \wedge x_2.$$

Thus, this is a two-input AND protocol using one helping card.

Since then, a challenging open problem had been to construct a two-input AND protocol without any helping card. More generally, could we construct an $n$-input AND protocol using only $2n$ cards?

$$\underbrace{\boxed{?}\boxed{?}}_{x_1}\ \underbrace{\boxed{?}\boxed{?}}_{x_2} \cdots \underbrace{\boxed{?}\boxed{?}}_{x_n} \rightarrow \cdots \rightarrow x_1 \wedge x_2 \wedge \cdots \wedge x_n.$$

Because $2n$ cards are necessary for arranging the $n$ input commitments to the values $x_1, x_2, \ldots, x_n \in \{0,1\}$ (as long as we obey the encoding rule (1)), this type of AND protocol (using exactly $2n$ cards) is said to be *card-minimal.* This study addresses the class of card-minimal AND protocols.

### 1.2 Known Results

The first card-minimal AND protocol was proposed by Kumamoto et al. in 2012 [12]:

$$\underbrace{\boxed{?}\boxed{?}}_{x_1}\ \underbrace{\boxed{?}\boxed{?}}_{x_2} \xrightarrow{\text{2 RBCs}} \cdots \rightarrow \begin{cases} x_1 \wedge x_2 = 0 & \text{if } \boxed{?}\boxed{\clubsuit}\boxed{?}\boxed{\heartsuit} \\ x_1 \wedge x_2 = 0 & \text{if } \boxed{\clubsuit}\boxed{\heartsuit}\boxed{?}\boxed{?} \\ x_1 \wedge x_2 = 1 & \text{if } \boxed{?}\boxed{\clubsuit}\boxed{?}\boxed{\clubsuit} \\ x_1 \wedge x_2 = 1 & \text{if } \boxed{\heartsuit}\boxed{\heartsuit}\boxed{?}\boxed{?}. \end{cases}$$

That is, using only four cards, a two-input AND protocol was constructed. This protocol uses two shuffles; more precisely, it applies a "random bisection cut (RBC)" twice, which is a kind of shuffling operation (explained later in Sect. 2.3). See the first protocol listed in Table 1.

How about $n$-input AND protocols for $n \geq 3$? This open problem was solved in 2016 [11]. Specifically, for the case of $n = 3$, a card-minimal three-input AND protocol was proposed by Mizuki [11]:

$$\underbrace{\boxed{?}\boxed{?}}_{x_1}\ \underbrace{\boxed{?}\boxed{?}}_{x_2}\ \underbrace{\boxed{?}\boxed{?}}_{x_3} \xrightarrow{\text{3 RBCs \& 2 RCs}} \cdots \rightarrow \begin{cases} x_1 \wedge x_2 \wedge x_3 = 1 & \text{if } \boxed{\clubsuit}\boxed{\clubsuit}\boxed{\clubsuit}\cdots \\ x_1 \wedge x_2 \wedge x_3 = 1 & \text{if } \boxed{\heartsuit}\boxed{\heartsuit}\boxed{\heartsuit}\cdots \\ x_1 \wedge x_2 \wedge x_3 = 0 & \text{otherwise.} \end{cases}$$

**Table 1.** Existing card-minimal AND protocols and our proposed protocol (using random cuts and/or random bisection cuts)

| | #Inputs | #Cards | #Shuffles |
|---|---|---|---|
| Kumamoto et al. [12] | 2 | 4 | 2 |
| Mizuki [11] | 3 | 6 | 5 |
| Mizuki [11] | $n\,(\geq 4)$ | $2n$ | $n{+}1$ |
| **This paper** | **3** | **6** | **2** |

Thus, this six-card protocol uses three random bisection cuts along with two "random cuts (RCs)," for a total of five shuffles; the random cut is another common type of shuffling operation, which will be explained in Sect. 2.2. See the second protocol listed in Table 1.

For the case of $n \geq 4$, Mizuki [11] also proposed a card-minimal $n$-input AND protocol:

$$\underbrace{\boxed{?}\boxed{?}}_{x_1}\ \underbrace{\boxed{?}\boxed{?}}_{x_2}\ \underbrace{\boxed{?}\boxed{?}}_{x_3}\ \underbrace{\boxed{?}\boxed{?}}_{x_4}\ \cdots\ \underbrace{\boxed{?}\boxed{?}}_{x_n} \xrightarrow{n+1 \text{ RBCs}} \cdots \rightarrow \begin{cases} x_1 \wedge x_2 \wedge \cdots \wedge x_n = 0 & \text{if } \boxed{\clubsuit}\cdots \\ x_1 \wedge x_2 \wedge \cdots \wedge x_n = 1 & \text{if } \boxed{\heartsuit}\cdots. \end{cases}$$

This protocol takes $n$ input commitments (such that $n \geq 4$) and uses $n+1$ random bisection cuts to securely evaluate their AND value. See the third protocol shown in Table 1.

### 1.3 Contribution

In this work, we focus on the number of required shuffles: From Table 1, it is observed that the number of shuffles used in the second protocol, i.e., 5 in the three-input AND protocol [11], is somewhat large. Actually, the three-input protocol [11] is elaborate but rather complicated, and it seems difficult for lay-people to execute practically. Therefore, our goal is to improve this existing three-input AND protocol [11].

Specifically, we will construct a new card-minimal three-input AND protocol using only two shuffles, namely one random bisection cut and one random cut:

$$\underbrace{\boxed{?}\boxed{?}}_{x_1}\ \underbrace{\boxed{?}\boxed{?}}_{x_2}\ \underbrace{\boxed{?}\boxed{?}}_{x_3} \xrightarrow{1 \text{ RBC } \& \text{ 1 RC}} \cdots \rightarrow \begin{cases} x_1 \wedge x_2 \wedge x_3 = 1 & \text{if } \boxed{\clubsuit}\boxed{\clubsuit}\boxed{\clubsuit}\cdots \\ x_1 \wedge x_2 \wedge x_3 = 1 & \text{if } \boxed{\heartsuit}\boxed{\heartsuit}\boxed{\heartsuit}\cdots \\ x_1 \wedge x_2 \wedge x_3 = 0 & \text{otherwise.} \end{cases}$$

The performance of this approach is shown in Table 1 in the last row.

Figure 1 shows the numbers of required shuffles for all the protocols listed in Table 1. As seen from this figure, the previous three-input protocol [11] requires five shuffles while our proposed three-input protocol uses only two shuffles, thereby successfully reducing the number of required shuffles significantly from

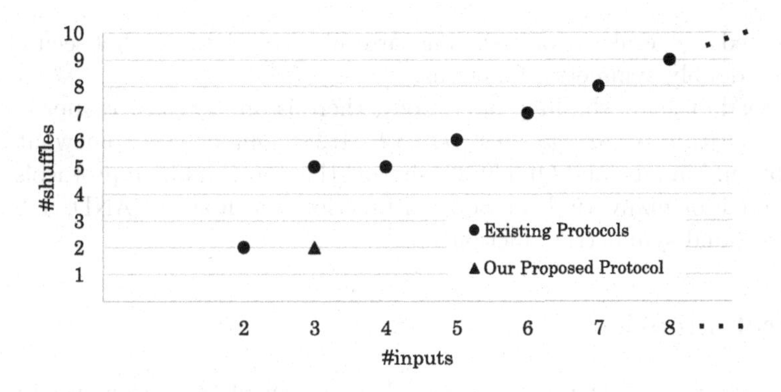


**Fig. 1.** Numbers of shuffles used in the existing card-minimal $n$-input AND protocols (for all $n \geq 2$) and our proposed protocol

five to two. As shown later in Sect. 3, our designed protocol is simple enough for lay-people to execute practically. Therefore, we believe that our new protocol is important from both theoretical and practical points of view.

### 1.4 Related Work

All the protocols mentioned thus far are not *committed-format* AND protocols because their AND values are not obtained as commitments. Contrarily, there are committed-format protocols, such as committed-format two-input AND protocols, which produce commitments to the AND values:

$$\underbrace{\boxed{?}\boxed{?}}_{x_1}\ \underbrace{\boxed{?}\boxed{?}}_{x_2} \rightarrow \cdots \rightarrow \underbrace{\boxed{?}\boxed{?}}_{x_1 \wedge x_2}.$$

Since this output is a (hidden) commitment, it can be used as input to another computation; thus, by repeatedly executing a committed-format AND protocol $n-1$ times, we can perform an $n$-input AND computation. Therefore, card-minimal committed-format two-input AND protocols are considered useful. Unfortunately however, such known AND protocols [4,8,23] require nonuniform or nonclosed shuffles, which are difficult to implement manually (cf. [17,18,25]); furthermore, Kastner et al. [3] proved that there exist no card-minimal two-input AND protocols that use only uniform closed shuffles. It should be noted that both random cuts and random bisection cuts (which all the protocols listed in Table 1 rely on) are uniform closed shuffles, which are easy to implement (as shown in Sects. 2.2 and 2.3).

If we allow the use of helping cards, we have a six-card committed-format AND protocol [15] and a five-card committed-format AND protocol [1] that rely only on random cuts and/or random bisection cuts; however, of course, they are not card-minimal.

Apart from the AND computation, because there is a card-minimal committed-format two-input XOR protocol [15], we can construct a card-minimal $n$-input XOR protocol for any $n \geq 2$. Recently, Ruangwises and Itoh [24] constructed a general way of designing card-minimal protocols that securely compute any doubly symmetric functions.

In stead of using shuffling operations, there is an alternative approach that relies on *private operations* (e.g., [16,19–21,28]); under this somewhat strong assumption, Manabe and Ono [9,22] showed that card-minimal protocols can be constructed for many kinds of Boolean functions, such as the AND, half-adder, full-adder, and symmetric functions.

## 2 Preliminaries

In this section, we first present a formal treatment of the actions used in card-based protocols (which has been developed in [3,5,13,14]). Then, we formally introduce two shuffling operations, namely a random cut and a random bisection cut.

### 2.1 Actions in Card-Based Protocols

In card-based protocols, the following three main actions are applied to a sequence of cards; below, we assume a sequence of $m$ cards.

**Permute.** This is denoted by $(\mathsf{perm}, \pi)$, where $\pi$ is a permutation applied to the sequence of cards as follows:

$$\overset{1}{\boxed{?}}\overset{2}{\boxed{?}}\cdots\overset{m}{\boxed{?}} \xrightarrow{(\mathsf{perm},\pi)} \overset{\pi^{-1}(1)}{\boxed{?}}\;\overset{\pi^{-1}(2)}{\boxed{?}}\;\cdots\;\overset{\pi^{-1}(m)}{\boxed{?}}\;.$$

**Turn.** This is denoted by $(\mathsf{turn}, T)$, where $T$ is a set of indexes, indicating that for every $t \in T$, the $t$-th card is turned over as follows:

$$\overset{1}{\boxed{?}}\overset{2}{\boxed{?}}\cdots\overset{t\in T}{\boxed{?}}\cdots\overset{m}{\boxed{?}} \xrightarrow{(\mathsf{turn},T)} \overset{1}{\boxed{?}}\overset{2}{\boxed{?}}\cdots\overset{t\in T}{\boxed{\clubsuit}}\cdots\overset{m}{\boxed{?}}.$$

**Shuffle.** This is denoted by $(\mathsf{shuf}, \Pi, \mathcal{F})$, where $\Pi$ is a permutation set and $\mathcal{F}$ is a probability distribution on $\Pi$, indicating that $\pi \in \Pi$ is drawn according to $\mathcal{F}$ and applied to the sequence of cards as follows:

$$\overset{1}{\boxed{?}}\overset{2}{\boxed{?}}\cdots\overset{m}{\boxed{?}} \xrightarrow{(\mathsf{shuf},\Pi,\mathcal{F})} \overset{\pi^{-1}(1)}{\boxed{?}}\;\overset{\pi^{-1}(2)}{\boxed{?}}\;\cdots\;\overset{\pi^{-1}(m)}{\boxed{?}}\;.$$

Here, the permutation in $\Pi$ that is applied remains unknown. If the distribution $\mathcal{F}$ is uniform, then its description can be omitted.

### 2.2 Random Cut

A *random cut* (RC) is the simplest and most easy-to-implement shuffle in card-based cryptography, denoted by $\langle\cdot\rangle$, which shifts a sequence of cards cyclically and randomly. If a random cut is applied to a sequence of $m$ cards, then the resulting sequence becomes one of the following $n$ sequences, each of which occurs with a probability of $1/m$:

$$\left\langle \overset{1}{\boxed{?}}\overset{2}{\boxed{?}}\overset{3}{\boxed{?}} \cdots \overset{m-1}{\boxed{?}}\overset{m}{\boxed{?}} \right\rangle \rightarrow \begin{cases} \overset{1}{\boxed{?}}\overset{2}{\boxed{?}}\overset{3}{\boxed{?}} \cdots \overset{m-1}{\boxed{?}}\overset{m}{\boxed{?}}, \\ \overset{2}{\boxed{?}}\overset{3}{\boxed{?}}\overset{4}{\boxed{?}} \cdots \overset{m}{\boxed{?}}\overset{1}{\boxed{?}}, \\ \qquad\vdots \\ \overset{m-1}{\boxed{?}}\overset{m}{\boxed{?}}\overset{1}{\boxed{?}} \cdots \overset{m-3}{\boxed{?}}\overset{m-2}{\boxed{?}}, \\ \overset{m}{\boxed{?}}\overset{1}{\boxed{?}}\overset{2}{\boxed{?}} \cdots \overset{m-2}{\boxed{?}}\overset{m-1}{\boxed{?}}. \end{cases}$$

This random cut is formally described as

$$(\mathsf{shuf}, \{\mathsf{id}, \pi, \pi^2, \ldots, \pi^{m-1}\})$$

for a cyclic permutation $\pi = (1\,2\,3\,\cdots\,m)$, where $\mathsf{id}$ denotes the identity permutation.

Hereinafter, we use $\mathsf{RC}_{1,2,\ldots,m}$ to represent $\{\mathsf{id}, \pi, \pi^2, \ldots, \pi^{m-1}\}$. For example, $(\mathsf{shuf}, \mathsf{RC}_{1,2,3,4,5,6})$ is a random cut to a sequence of six face-down cards:

$$\left\langle \boxed{?}\boxed{?}\boxed{?}\boxed{?}\boxed{?}\boxed{?} \right\rangle.$$

A random cut can be easily performed manually; a secure implementation called the Hindu cut is a well-known instance [27].

## 2.3 Random Bisection Cut

A *random bisection cut* (RBC) is another major shuffle action, which was invented in 2009 [15]. This shuffle, denoted by $[\cdot\,|\,\cdot]$, bisects a sequence of $2m$ cards and randomly swaps the two halves; the resulting sequence becomes one of the following two sequences, with a probability of $1/2$:

$$\left[ \overset{1}{\boxed{?}} \cdots \overset{m}{\boxed{?}} \,\middle|\, \overset{m+1}{\boxed{?}} \cdots \overset{2m}{\boxed{?}} \right] \rightarrow \begin{cases} \overset{1}{\boxed{?}} \cdots \overset{m}{\boxed{?}}\overset{m+1}{\boxed{?}} \cdots \overset{2m}{\boxed{?}}, \\ \overset{m+1}{\boxed{?}} \cdots \overset{2m}{\boxed{?}}\overset{1}{\boxed{?}} \cdots \overset{m}{\boxed{?}}. \end{cases}$$

That is, the resulting sequence either remains unchanged compared with the original or is obtained such that the two halves are swapped, with a probability of $1/2$. The random bisection cut can be expressed as follows:

$$(\mathsf{shuf}, \{\mathsf{id}, (1\ m{+}1)(2\ m{+}2)\cdots(m\ 2m)\}).$$

Secure implementations of a random bisection cut were shown in [26].

# 3 Our Proposed Protocol

In this section, we present the new card-minimal three-input AND protocol:

$$\underbrace{\boxed{?}\boxed{?}}_{x_1}\ \underbrace{\boxed{?}\boxed{?}}_{x_2}\ \underbrace{\boxed{?}\boxed{?}}_{x_3} \xrightarrow{\text{1 RBC \& 1 RC}} \cdots \rightarrow \begin{cases} x_1 \wedge x_2 \wedge x_3 = 1 & \text{if } \boxed{\clubsuit}\boxed{\clubsuit}\boxed{\clubsuit}\cdots \\ x_1 \wedge x_2 \wedge x_3 = 1 & \text{if } \boxed{\heartsuit}\boxed{\heartsuit}\boxed{\heartsuit}\cdots \\ x_1 \wedge x_2 \wedge x_3 = 0 & \text{otherwise.} \end{cases}$$

This protocol uses one random bisection cut and one random cut.

We present the description of our protocol in Sect. 3.1 as well as its pseudocode in Sect. 3.2. We also present an intuitive explanation in Sect. 3.3 as to why the proposed protocol works correctly. Formal proofs of the correctness and security of our proposed protocol (based on the so-called KWH-tree [8]) are omitted owing to length limitations.

## 3.1 Description

Our card-minimal three-input AND protocol proceeds, as follows.

1. Apply $(\mathsf{shuf}, \{\mathsf{id}, (1\,2)(3\,6)\})$ by performing operations (a)–(c) noted below.
   (a) Swap the second and third cards as well as the fourth and sixth cards:
   $$\overset{1}{\boxed{?}}\overset{2}{\boxed{?}}\overset{3}{\boxed{?}}\overset{4}{\boxed{?}}\overset{5}{\boxed{?}}\overset{6}{\boxed{?}} \rightarrow \overset{1}{\boxed{?}}\overset{3}{\boxed{?}}\overset{2}{\boxed{?}}\overset{6}{\boxed{?}}\overset{5}{\boxed{?}}\overset{4}{\boxed{?}}.$$
   (b) Apply a random bisection cut to the four cards on the extreme left:
   $$\left[\boxed{?}\boxed{?}\middle|\boxed{?}\boxed{?}\right]\ \boxed{?}\boxed{?} \rightarrow \boxed{?}\boxed{?}\boxed{?}\boxed{?}\boxed{?}\boxed{?}.$$
   (c) Swap the second and third cards as well as the fourth and sixth cards again:
   $$\overset{1}{\boxed{?}}\overset{2}{\boxed{?}}\overset{3}{\boxed{?}}\overset{4}{\boxed{?}}\overset{5}{\boxed{?}}\overset{6}{\boxed{?}} \rightarrow \overset{1}{\boxed{?}}\overset{3}{\boxed{?}}\overset{2}{\boxed{?}}\overset{6}{\boxed{?}}\overset{5}{\boxed{?}}\overset{4}{\boxed{?}}.$$
2. Turn over the first card to check its color. If the card is $\boxed{\clubsuit}$, swap the first and second cards as well as the third and sixth cards:
   $$\overset{1}{\boxed{\clubsuit}}\overset{2}{\boxed{?}}\overset{3}{\boxed{?}}\overset{4}{\boxed{?}}\overset{5}{\boxed{?}}\overset{6}{\boxed{?}} \rightarrow \overset{2}{\boxed{?}}\overset{1}{\boxed{\clubsuit}}\overset{6}{\boxed{?}}\overset{4}{\boxed{?}}\overset{5}{\boxed{?}}\overset{3}{\boxed{?}}.$$
   If the card color is $\boxed{\heartsuit}$, proceed to Step 3 directly.
3. After turning over the revealed card in a face-down manner, apply a random cut to the entire sequence:
   $$\left\langle\boxed{?}\boxed{?}\boxed{?}\boxed{?}\boxed{?}\boxed{?}\right\rangle \rightarrow \boxed{?}\boxed{?}\boxed{?}\boxed{?}\boxed{?}\boxed{?}.$$

4. Turn over the first, third, and fifth cards. If these are $\boxed{\heartsuit}\boxed{\clubsuit}\boxed{\heartsuit}$ or $\boxed{\clubsuit}\boxed{\heartsuit}\boxed{\clubsuit}$ (apart from cyclic rotation), then $x_1 \wedge x_2 \wedge x_3 = 0$; if the cards are $\boxed{\heartsuit}\boxed{\heartsuit}\boxed{\heartsuit}$ or $\boxed{\clubsuit}\boxed{\clubsuit}\boxed{\clubsuit}$, then $x_1 \wedge x_2 \wedge x_3 = 1$:

$$\boxed{\heartsuit}\boxed{?}\boxed{\clubsuit}\boxed{?}\boxed{\heartsuit}\boxed{?} \quad \boxed{\clubsuit}\boxed{?}\boxed{\heartsuit}\boxed{?}\boxed{\clubsuit}\boxed{?}$$
$$\boxed{\heartsuit}\boxed{?}\boxed{\heartsuit}\boxed{?}\boxed{\clubsuit}\boxed{?} \quad \boxed{\clubsuit}\boxed{?}\boxed{\clubsuit}\boxed{?}\boxed{\heartsuit}\boxed{?}$$
$$\boxed{\clubsuit}\boxed{?}\boxed{\heartsuit}\boxed{?}\boxed{\heartsuit}\boxed{?} \quad \boxed{\heartsuit}\boxed{?}\boxed{\clubsuit}\boxed{?}\boxed{\clubsuit}\boxed{?}$$
$$x_1 \wedge x_2 \wedge x_3 = 0,$$

$$\boxed{\heartsuit}\boxed{?}\boxed{\heartsuit}\boxed{?}\boxed{\heartsuit}\boxed{?} \quad \boxed{\clubsuit}\boxed{?}\boxed{\clubsuit}\boxed{?}\boxed{\clubsuit}\boxed{?}$$
$$x_1 \wedge x_2 \wedge x_3 = 1.$$

## 3.2 Pseudocode

The pseudocode for our protocol is depicted in Algorithm 1, where $(\mathsf{result}, i, j, k)$ specifies the output positions.

**Algorithm 1.** Our proposed protocol

input set:

$$\Big\{\Big(\tfrac{?}{\clubsuit},\tfrac{?}{\heartsuit},\tfrac{?}{\clubsuit},\tfrac{?}{\heartsuit},\tfrac{?}{\clubsuit},\tfrac{?}{\heartsuit}\Big),\Big(\tfrac{?}{\clubsuit},\tfrac{?}{\heartsuit},\tfrac{?}{\clubsuit},\tfrac{?}{\heartsuit},\tfrac{?}{\heartsuit},\tfrac{?}{\clubsuit}\Big),\Big(\tfrac{?}{\clubsuit},\tfrac{?}{\heartsuit},\tfrac{?}{\heartsuit},\tfrac{?}{\clubsuit},\tfrac{?}{\clubsuit},\tfrac{?}{\heartsuit}\Big),$$
$$\Big(\tfrac{?}{\clubsuit},\tfrac{?}{\heartsuit},\tfrac{?}{\heartsuit},\tfrac{?}{\clubsuit},\tfrac{?}{\heartsuit},\tfrac{?}{\clubsuit}\Big),\Big(\tfrac{?}{\heartsuit},\tfrac{?}{\clubsuit},\tfrac{?}{\clubsuit},\tfrac{?}{\heartsuit},\tfrac{?}{\clubsuit},\tfrac{?}{\heartsuit}\Big),\Big(\tfrac{?}{\heartsuit},\tfrac{?}{\clubsuit},\tfrac{?}{\clubsuit},\tfrac{?}{\heartsuit},\tfrac{?}{\heartsuit},\tfrac{?}{\clubsuit}\Big),$$
$$\Big(\tfrac{?}{\heartsuit},\tfrac{?}{\clubsuit},\tfrac{?}{\heartsuit},\tfrac{?}{\clubsuit},\tfrac{?}{\clubsuit},\tfrac{?}{\heartsuit}\Big),\Big(\tfrac{?}{\heartsuit},\tfrac{?}{\clubsuit},\tfrac{?}{\heartsuit},\tfrac{?}{\clubsuit},\tfrac{?}{\heartsuit},\tfrac{?}{\clubsuit}\Big)\Big\}$$

1. $(\mathsf{shuf}, \{\mathsf{id}, (1\ 2)(3\ 6)\})$
2. $(\mathsf{turn}, \{1\})$
3. **if** visible sequence $= (\clubsuit, ?, ?, ?, ?, ?)$ **then**
4. $\quad(\mathsf{perm}, (1\ 2)(3\ 6))$
5. $\quad(\mathsf{turn}, \{2\})$
6. **else if** visible sequence $= (\heartsuit, ?, ?, ?, ?, ?)$ **then**
7. $\quad(\mathsf{turn}, \{1\})$
8. $(\mathsf{shuf}, \mathsf{RC}_{1,2,3,4,5,6})$
9. $(\mathsf{result}, 1, 3, 5)$

## 3.3 Why Our Protocol Works Correctly

Herein, we intuitively explain why our proposed protocol works correctly. Note that the input sequence has eight possibilities depending on the input values $(x_1, x_2, x_3) \in \{0,1\}^3$, as shown in the second column of Table 2. The idea behind our protocol is to assign each possible input to one of two sequence patterns without leaking the input value. One of the two patterns is an alternating pattern

**Table 2.** Resulting sequences after applying (perm,(1 2)(3 6)) and Step 2

| Input | Input Sequence | Apply (perm,(1 2)(3 6)) | After Step 2 |
|---|---|---|---|
| (0,0,0) | ♣♡♣♡♣♡ | ♡♣♡♡♣♣ | ♡♣♡♡♣♣ |
| (0,0,1) | ♣♡♣♡♡♣ | ♡♣♣♡♡♣ | ♡♣♣♡♡♣ |
| (0,1,0) | ♣♡♡♣♣♡ | ♡♣♡♣♣♡ | ♡♣♡♣♣♡ |
| (0,1,1) | ♣♡♡♣♡♣ | ♡♣♣♣♡♡ | ♡♣♣♣♡♡ |
| (1,0,0) | ♡♣♣♡♣♡ | ♣♡♡♡♣♣ | ♡♣♣♡♣♡ |
| (1,0,1) | ♡♣♣♡♡♣ | ♣♡♣♡♡♣ | ♡♣♣♡♡♣ |
| (1,1,0) | ♡♣♡♣♣♡ | ♣♡♡♣♣♡ | ♡♣♡♣♣♡ |
| (1,1,1) | ♡♣♡♣♡♣ | ♣♡♣♣♡♡ | ♡♣♡♣♡♣ |

of ♣ and ♡, i.e., either ♣♡♣♡♣♡ or ♡♣♡♣♡♣, which is a possible input sequence when $(x_1, x_2, x_3) = (0, 0, 0), (1, 1, 1)$; we call this an *alternating* sequence. The other pattern corresponds to the remaining sequences, which we call *non-alternating* sequences.

Suppose that we apply the permutation $(1\,2)(3\,6)$ to the input sequence; this appears in the shuffle applied in Step 1 and permutation in Step 2. The resulting sequence is the one shown in the third column of Table 2. Note that the second column (i.e., input sequence) and third column of Table 2 are equivalent to the transition possibilities after applying $(\mathsf{shuf}, \{\mathsf{id}, (1\,2)(3\,6)\})$ to the input sequence in Step 1. In the third column of Table 2, among the four sequences from the top (which are obtained when $x_1 = 0$), the sequences corresponding to $(0, 0, 1)$, $(0, 1, 0)$, and $(0, 1, 1)$ are still non-alternating, and the sequence corresponding to $(0, 0, 0)$ is converted to a non-alternating sequence. If we combine the four sequences from the top in the third column with the four sequences from the bottom in the second column, we obtain the eight sequences shown in the fourth column of Table 2, where only the sequence corresponding to $(1, 1, 1)$ is alternating. To achieve this, we apply the permutation $(1\,2)(3\,6)$ when the first cards in the sequences in the second and third columns of Table 2 are ♣. This is the reason behind performing $(\mathsf{shuf}, \{\mathsf{id}, (1\,2)(3\,6)\})$ in Step 1 and $(\mathsf{perm}, (1\,2)(3\,6))$ in Step 2 if the first card revealed is ♣.

By applying a random cut to the sequence of cards in Step 3, the resulting sequence is one among the following four sequences (up to cyclic rotation):

(a) ♡♡♣♣♡♣, if $(0, 0, 0)$ or $(1, 0, 0)$;
(b) ♣♣♡♡♣♡, if $(0, 0, 1)$, $(0, 1, 0)$, $(1, 0, 1)$, or $(1, 1, 0)$;
(c) ♡♡♡♣♣♣, if $(0, 1, 1)$;
(d) ♡♣♡♣♡♣, if $(1, 1, 1)$.

Here, when the input is either $(0, 0, 0)$ or $(1, 0, 0)$, the resulting sequence is equal to that in (a); when the input is $(0, 0, 1)$, $(0, 1, 0)$, $(1, 0, 1)$, or $(1, 1, 0)$, the sequence is equal to that in (b); when the input is $(0, 1, 1)$, the sequence is equal to that in (c); when the input is $(1, 1, 1)$, the sequence is equal to that in (d). Now, if we revealed all the cards in the sequence, then we could obtain the value of $x_1 \wedge x_2 \wedge x_3$; however, information about the input would be leaked because the revealed sequence depends on the input values. Therefore, we need an alternative approach to obtain the output value.

Let us focus on the cards at the odd-numbered positions. If we reveal only these cards, then the revealed cards in every sequence of (a), (b), and (c) will have the same pattern, i.e., either ♡ ♣ ♡ or ♣ ♡ ♣ up to cyclic rotation. By contrast, if (d) occurs, then the revealed cards will be either ♡ ♡ ♡ or ♣ ♣ ♣. Hence, the sequences of (a), (b), and (c) become indistinguishable, and we can obtain only the value of $x_1 \wedge x_2 \wedge x_3$.

## 4 Conclusion

In this work, we proposed a card-minimal three-input AND protocol using only two shuffles. The minimality means that the protocol uses exactly six cards. The shuffles used in our protocol are one random cut and one random bisection cut. Since the existing three-input protocol [11] requires five shuffles, our proposed protocol successfully reduces the number of required shuffles from five to two. We believe that this is a significant improvement and that our protocol is simple enough for easy execution by lay-people.

An interesting open problem is improving the numbers of shuffles for card-minimal $n$-input AND computations for $n \geq 4$. It is also an intriguing problem to seek lower bounds on the numbers of shuffles using the "formal method approach" recently developed by Koch, Schrempp, and Kirsten [6,7].

This work considers the number of shuffles as the quality metric for evaluating a protocol because the shuffle action is the most time-consuming step (cf. [10]). However, considering the other actions for a more fine-grained analysis would be an interesting line of investigation in the future.

**Acknowledgements.** We thank the anonymous referees, whose comments have helped us improve the presentation of the paper. We would like to thank Hideaki Sone for his cooperation in preparing a Japanese draft version at an earlier stage of this work. This work was supported in part by JSPS KAKENHI Grant Numbers JP19J21153 and JP21K11881.

## References

1. Abe, Y., Hayashi, Y., Mizuki, T., Sone, H.: Five-card AND computations in committed format using only uniform cyclic shuffles. New Gener. Comput. **39**(1), 97–114 (2021). https://doi.org/10.1007/s00354-020-00110-2

2. Boer, B.: More efficient match-making and satisfiability *the five card trick*. In: Quisquater, J.-J., Vandewalle, J. (eds.) EUROCRYPT 1989. LNCS, vol. 434, pp. 208–217. Springer, Heidelberg (1990). https://doi.org/10.1007/3-540-46885-4_23
3. Kastner, J., et al.: The minimum number of cards in practical card-based protocols. In: Takagi, T., Peyrin, T. (eds.) ASIACRYPT 2017. LNCS, vol. 10626, pp. 126–155. Springer, Cham (2017). https://doi.org/10.1007/978-3-319-70700-6_5
4. Koch, A.: The landscape of optimal card-based protocols. Cryptology ePrint Archive, Report 2018/951 (2018). https://eprint.iacr.org/2018/951
5. Koch, A.: Cryptographic protocols from physical assumptions. Ph.D. thesis, Karlsruhe Institute of Technology (2019). https://doi.org/10.5445/IR/1000097756
6. Koch, A., Schrempp, M., Kirsten, M.: Card-based cryptography meets formal verification. In: Galbraith, S.D., Moriai, S. (eds.) ASIACRYPT 2019. LNCS, vol. 11921, pp. 488–517. Springer, Cham (2019). https://doi.org/10.1007/978-3-030-34578-5_18
7. Koch, A., Schrempp, M., Kirsten, M.: Card-based cryptography meets formal verification. New Gener. Comput. **39**(1), 115–158 (2021). https://doi.org/10.1007/s00354-020-00120-0
8. Koch, A., Walzer, S., Härtel, K.: Card-based cryptographic protocols using a minimal number of cards. In: Iwata, T., Cheon, J.H. (eds.) ASIACRYPT 2015. LNCS, vol. 9452, pp. 783–807. Springer, Heidelberg (2015). https://doi.org/10.1007/978-3-662-48797-6_32
9. Manabe, Y., Ono, H.: Card-based cryptographic protocols for three-input functions using private operations. In: Flocchini, P., Moura, L. (eds.) IWOCA 2021. LNCS, vol. 12757, pp. 469–484. Springer, Cham (2021). https://doi.org/10.1007/978-3-030-79987-8_33
10. Miyahara, D., Ueda, I., Hayashi, Y., Mizuki, T., Sone, H.: Evaluating card-based protocols in terms of execution time. Int. J. Inf. Secur. **20**(5), 729–740 (2021). https://doi.org/10.1007/s10207-020-00525-4
11. Mizuki, T.: Card-based protocols for securely computing the conjunction of multiple variables. Theor. Comput. Sci. **622**(C), 34–44 (2016). https://doi.org/10.1016/j.tcs.2016.01.039
12. Mizuki, T., Kumamoto, M., Sone, H.: The five-card trick can be done with four cards. In: Wang, X., Sako, K. (eds.) ASIACRYPT 2012. LNCS, vol. 7658, pp. 598–606. Springer, Heidelberg (2012). https://doi.org/10.1007/978-3-642-34961-4_36
13. Mizuki, T., Shizuya, H.: A formalization of card-based cryptographic protocols via abstract machine. Int. J. Inf. Secur. **13**(1), 15–23 (2014). https://doi.org/10.1007/s10207-013-0219-4
14. Mizuki, T., Shizuya, H.: Computational model of card-based cryptographic protocols and its applications. IEICE Trans. Fundam. **E100.A**(1), 3–11 (2017). https://doi.org/10.1587/transfun.E100.A.3
15. Mizuki, T., Sone, H.: Six-card secure AND and four-card secure XOR. In: Deng, X., Hopcroft, J.E., Xue, J. (eds.) FAW 2009. LNCS, vol. 5598, pp. 358–369. Springer, Heidelberg (2009). https://doi.org/10.1007/978-3-642-02270-8_36
16. Nakai, T., Misawa, Y., Tokushige, Y., Iwamoto, M., Ohta, K.: How to solve millionaires' problem with two kinds of cards. New Gener. Comput. **39**(1), 73–96 (2021). https://doi.org/10.1007/s00354-020-00118-8
17. Nishimura, A., Hayashi, Y., Mizuki, T., Sone, H.: An implementation of non-uniform shuffle for secure multi-party computation. In: Proceedings of the 3rd ACM International Workshop on ASIA Public-Key Cryptography, AsiaPKC 2016, pp. 49–55. ACM, New York (2016). https://doi.org/10.1145/2898420.2898425,https://doi.acm.org/10.1145/2898420.2898425

18. Nishimura, A., Nishida, T., Hayashi, Y., Mizuki, T., Sone, H.: Card-based protocols using unequal division shuffles. Soft Comput. **22**(2), 361–371 (2018). https://doi.org/10.1007/s00500-017-2858-2
19. Ono, H., Manabe, Y.: Efficient card-based cryptographic protocols for the millionaires' problem using private input operations. In: Asia Joint Conference on Information Security (AsiaJCIS), pp. 23–28 (2018). https://doi.org/10.1109/AsiaJCIS.2018.00013
20. Ono, H., Manabe, Y.: Card-based cryptographic protocols with the minimum number of cards using private operations. In: Zincir-Heywood, N., Bonfante, G., Debbabi, M., Garcia-Alfaro, J. (eds.) FPS 2018. LNCS, vol. 11358, pp. 193–207. Springer, Cham (2019). https://doi.org/10.1007/978-3-030-18419-3_13
21. Ono, H., Manabe, Y.: Card-based cryptographic protocols with the minimum number of rounds using private operations. In: Pérez-Solà, C., Navarro-Arribas, G., Biryukov, A., Garcia-Alfaro, J. (eds.) DPM/CBT -2019. LNCS, vol. 11737, pp. 156–173. Springer, Cham (2019). https://doi.org/10.1007/978-3-030-31500-9_10
22. Ono, H., Manabe, Y.: Card-based cryptographic logical computations using private operations. New Gener. Comput. **39**(1), 19–40 (2021). https://doi.org/10.1007/s00354-020-00113-z
23. Ruangwises, S., Itoh, T.: AND protocols using only uniform shuffles. In: van Bevern, R., Kucherov, G. (eds.) CSR 2019. LNCS, vol. 11532, pp. 349–358. Springer, Cham (2019). https://doi.org/10.1007/978-3-030-19955-5_30
24. Ruangwises, S., Itoh, T.: Securely computing the n-variable equality function with 2n cards. Theor. Comput. Sci. (2021, in press). https://doi.org/10.1016/j.tcs.2021.07.007
25. Saito, T., Miyahara, D., Abe, Y., Mizuki, T., Shizuya, H.: How to implement a non-uniform or non-closed shuffle. In: Martín-Vide, C., Vega-Rodríguez, M.A., Yang, M.-S. (eds.) TPNC 2020. LNCS, vol. 12494, pp. 107–118. Springer, Cham (2020). https://doi.org/10.1007/978-3-030-63000-3_9
26. Ueda, I., Miyahara, D., Nishimura, A., Hayashi, Y., Mizuki, T., Sone, H.: Secure implementations of a random bisection cut. Int. J. Inf. Secur. **19**(4), 445–452 (2020). https://doi.org/10.1007/s10207-019-00463-w
27. Ueda, I., Nishimura, A., Hayashi, Y., Mizuki, T., Sone, H.: How to implement a random bisection cut. In: Martín-Vide, C., Mizuki, T., Vega-Rodríguez, M.A. (eds.) TPNC 2016. LNCS, vol. 10071, pp. 58–69. Springer, Cham (2016). https://doi.org/10.1007/978-3-319-49001-4_5
28. Yasunaga, K.: Practical card-based protocol for three-input majority. IEICE Trans. Fundam. **E103.A**(11), 1296–1298 (2020). https://doi.org/10.1587/transfun.2020EAL2025

# Spy Game: FPT-Algorithm and Results on Graph Products

Eurinardo Rodrigues Costa[1], Nicolas Almeida Martins[2], and Rudini Sampaio[3](✉)

[1] Universidade Federal do Ceará, Campus Russas, Russas, Brazil
eurinardo@ufc.br

[2] Univ. Integração Internacional Lusofonia Afrobrasileira Unilab, Redenção, Brazil
nicolasam@unilab.edu.br

[3] Departamento de Computação, Universidade Federal do Ceará, Fortaleza, Brazil
rudini@dc.ufc.br

**Abstract.** In the $(s, d)$-spy game over a graph $G$, $k$ guards and one spy occupy some vertices of $G$ and, at each turn, the spy may move with speed $s$ (along at most $s$ edges) and each guard may move along one edge. The spy and the guards may occupy the same vertices. The spy wins if she reaches a vertex at distance more than the surveilling distance $d$ from every guard. This game was introduced by Cohen et al. in 2016 and is related to two well-studied games: Cops and robber game and Eternal Dominating game. The guard number $gn_{s,d}(G)$ is the minimum number of guards such that the guards have a winning strategy (of controlling the spy) in the graph $G$. In 2018, it was proved that deciding if the spy has a winning strategy is NP-hard for every speed $s \geq 2$ and distance $d \geq 0$. In this paper, we initiate the investigation of the guard number in grids and in graph products. We obtain a strict upper bound on the strong product of two general graphs and obtain examples with King grids that match this bound and other examples for which the guard number is smaller. We also obtain the exact value of the guard number in the lexicographical product of two general graphs for any distance $d \geq 2$. From the algorithmic point of view, we prove a positive result: if the number $k$ of guards is fixed, the spy game is solvable in polynomial XP time $O(n^{3k+2})$ for every speed $s \geq 2$ and distance $d \geq 0$. This XP algorithm is used to obtain an FPT algorithm on the $P_4$-fewness of the graph. As a negative result, we prove that the spy game is W[2]-hard even in bipartite graphs when parameterized by the number of guards, for every speed $s \geq 2$ and distance $d \geq 0$, extending the hardness result of Cohen et al. in 2016.

**Keywords:** Games on graphs · Spy game · Graph products · XP algorithm · Parameterized complexity

## 1 Introduction

Given a graph $G = (V, E)$ and $v \in V$, let $N(v) = \{w \mid vw \in E\}$ denote the set of neighbors of $v$ and let $N[v] = N(v) \cup \{v\}$.


C.-Y. Chen et al. (Eds.): COCOON 2021, LNCS 13025, pp. 680–691, 2021.
https://doi.org/10.1007/978-3-030-89543-3_56

Given integers $s \geq 2$, $d \geq 0$ and $k \geq 1$ (respectively the spy speed, the surveillance distance and the number of guards), the $(s,d)$-spy game is a two-player game on a finite graph $G$ with $k$ guards and one spy occupying vertices of $G$. The guards and even the spy may occupy the same vertex. One player controls the guards and the other player controls the spy. It is a full information game: any player has full knowledge of the positions of the other player. Initially, the spy is placed at some vertex of $G$ and then the $k$ guards are placed at some vertices of $G$. The game proceeds turn-by-turn: first the spy may move along at most $s$ edges and then each guard may move along one edge. The spy wins if, after a finite number of turns (after the guards' move), she reaches a vertex at distance greater than $d$ from every guard. Otherwise, the guards win the game: there is always at least one guard at distance at most $d$ from the spy. From the classical Zermelo-von Neumann theorem [28], one of the two players has a winning strategy, since it is a perfect information finite game without draw. Here we can consider the spy game as a finite game since the number of possible configurations of the spy and the guards is finite in a finite graph $G$. For example, we may consider that the guards win the game if the spy repeats a game configuration (after her move).

The *guard-number* $gn_{s,d}(G)$ is the minimum number of guards such that the guards have a winning strategy in the $(s,d)$-spy game. From the definition, only one guard is always sufficient if the spy speed $s = 1$. For this reason, we consider $s \geq 2$.

The spy game was introduced by Cohen et al. in 2016 [13] and is closely related to the well known *Cops and robber* game [7,26]. In this game, first $k$ cops occupy some vertices of the graph and then one robber occupies a vertex. Turn-by-turn, each player may move (the cops first and then the robber) along one edge. The cops win if one cop occupies the same vertex of the robber after a finite number of turns. The *cop-number* $cn(G)$ of a graph $G$ is the minimum number of cops required to win in $G$ [1].

There are many generalizations of the Cops and robber game [2,6,12,18,19]. For example, allowing a faster robber with speed $s \geq 2$. In this variant, the exact number of cops with speed one required to capture a robber with speed two is unknown even in 2-dimensional grids [5,17]. In 2010, Bonato et al. [6] introduced other variant of the cops and robbers in which the game is over if a cop occupies a vertex at distance at most a given integer $d$ from the robber. This is equivalent to the spy game with speed $s = 1$ when the spy is placed after the guards. For speed $s > 1$, the equivalence is not true and the games are significantly different. Therefore, we only consider the speed $s \geq 2$ for the spy in the spy game.

Other related well know game is the *eternal domination* game [20,21,23,24]. A set of $k$ *defenders* occupy some vertices of a graph $G$. At each turn, an *attacker* chooses a vertex $v \in V$ and the defenders may move along an edge in such a way that at least one defender is at distance at most a given integer $d$ from $v$. There are some variants of the eternal domination game allowing more defenders to move at each turn and to occupy the same vertex [21,23,24], which is equivalent to the spy game when the spy speed is at least the diameter of the graph.

Regarding results in the spy game, Cohen et al. [14] proved in 2018 that the guard number is NP-hard in general graphs and a directed version is PSPACE-hard even in DAGs. The authors left open the question of the PSPACE-hardness on the undirected case and the question of the spy game on grids: "*Many open questions remain such as the characterization of the guard-number in other graph classes, e.g., in grids*". Moreover, in 2020, it was proved [15] that the guard number is computable in polynomial time for trees by using Linear Programming and a fractional relaxation of the game. The authors also obtained an upper bound for the fractional guard number on the square grid $G_{n\times n}$, a parameter different from the guard number, which was proved to be equal in trees.

### 1.1 Our Contribution

In this paper, we initiate the investigation of the spy game guard number in grids. Specifically, we study the King grid (the strong product of two path graphs) and, more generally, we obtain a strict upper bound on the strong product $G_1 \boxtimes G_2$ of general graphs $G_1$ and $G_2$. We prove that $gn_{s,d}(G_1 \boxtimes G_2) \leq gn_{s_d}(G_1) \times gn_{s,d}(G_2)$ and obtain examples that match this upper bound and other examples for which the guard number is smaller. We also show that this upper bound of the strong product does not work in the cartesian product nor the lexicographical product of two graphs. Regarding the lexicographical product, we obtain the exact value of the guard number for any distance $d \geq 2$.

Regarding algorithmic results, we obtain positive and negative results in the spy game decision problem (the problem of deciding if the spy has a winning strategy in the $(s, d)$-spy game with $k$ guards in a finite graph $G$). We prove that, if the number $k$ of guards is fixed, the spy game decision problem is $O(n^{3k+2})$-time solvable for every speed $s \geq 2$ and distance $d \geq 0$. In other words, the spy game decision problem is XP when parameterized by the number $k$ of guards. This XP algorithm is very important in order to obtain an FPT algorithm for the spy game on the $P_4$-fewness of the graph, which, in our opinion, is one of our most important results, since it solves a game on graphs for many graph classes.

As a negative result, we prove that the spy game decision problem is W[2]-hard even in bipartite graphs when parameterized by the number $k$ of guards, for every speed $s \geq 2$ and distance $d \geq 0$. This hardness result is a generalization of the W[2]-hardness result of the spy game in general graphs [14] and follows a similar (but significantly different) structure of the reduction from Set Cover in [14]. However, the extension to bipartite graphs brings much more technical difficulties to the reduction, making this extension a relevant and non-trivial result.

## 2 Spy Game on King Grids and Graph Products

Given graphs $G_1$ and $G_2$, the strong product $G_1 \boxtimes G_2$ is the graph with vertex set $V(G_1 \boxtimes G_2) = V(G_1) \times V(G_2)$ (cartesian product) such that distinct vertices $(u_1, u_2)$ and $(v_1, v_2)$ are adjacent in $G_1 \boxtimes G_2$ if and only if (a) $u_1 = v_1$ and

$u_2v_2 \in E(G_2)$, or (b) $u_2 = v_2$ and $u_1v_1 \in E(G_1)$, or (c) $u_1v_1 \in E(G_1)$ and $u_2v_2 \in E(G_2)$. We say that a graph $G$ is a King grid if it is the strong product of two path graphs $G = P_n \boxtimes P_m$. Figure 1 shows the King grids $P_3 \boxtimes P_3$ and $P_5 \boxtimes P_5$. The name of the King grid is due to the king's moves in the chess game (for example, in the chess board $P_8 \boxtimes P_8$). There are many papers investigating King grids in several different problems (for example, [8,16,27]).

Our first result is a general upper bound on the guard number $gn_{s,d}(G_1 \boxtimes G_2)$ for any two graphs $G_1$ and $G_2$ and any speed $s \geq 2$ and surveillance distance $d \geq 0$.

**Theorem 1.** *Let $s \geq 2$ and $d \geq 0$. Given two graphs $G_1$ and $G_2$, the guard number of the strong product of $G_1$ and $G_2$ satisfies the following inequality. Moreover, the equality holds if $gn_{s,d}(G_1) = 1$ or $gn_{s,d}(G_2) = 1$.*

$$gn_{s,d}(G_1 \boxtimes G_2) \leq gn_{s,d}(G_1) \times gn_{s,d}(G_2).$$

From this upper bound on the guard number on the strong product of two general graphs, the natural question about the equality or the looseness of the bound arises when the guard numbers of both graphs are at least 2. In the following, we present two lemmas with general examples in King grids when the guard numbers of both graphs are exactly 2: the first with the strict inequality $gn_{s,d}(P_{2d+3} \boxtimes P_{2d+3}) \leq 2 < 4 = gn_{s,d}(P_{2d+3})^2$ and the second with the equality $gn_{s,d}(P_{2d+4} \boxtimes P_{2d+4}) = 4 = gn_{s,d}(P_{2d+4})^2$.

First we show that $gn_{s,d}(P_{2d+3} \boxtimes P_{2d+3}) \leq 2$ for any $d \geq 0$ and $s \geq 2$. Notice that $gn_{s,d}(P_{2d+3}) = 2$ for any $d \geq 0$ and $s \geq 2d + 2$, since a spy in the first vertex $v_0$ of the path $P_{2d+3}$ can go to the last vertex $v_{2d+2}$ of the path, but a guard surveilling the spy in the vertex $v_d$ of the path $P_{2d+3}$ cannot jump to the vertex $v_{d+2}$ to keep surveilling the spy. Moreover, two guards in the vertices $v_d$ and $v_{d+2}$ are sufficient to surveil the spy. That is, $gn_{s,d}(P_{2d+3})^2 = 4$.

**Lemma 1.** *Let $d \geq 0$ and $s \geq 2$ be fixed. Then $gn_{s,d}(P_{2d+3} \boxtimes P_{2d+3}) \leq 2$.*

*Proof (Sketch of the proof).* See Fig. 1. □

Next we show that $gn_{s,d}(P_{2d+4} \boxtimes P_{2d+4}) = 4$ for any $d \geq 0$ and $s \geq d + 2$. Notice that $gn_{s,d}(P_{2d+4}) = 2$ for any $d \geq 0$ and $s \geq d + 2$, since a spy in the first vertex $v_0$ of the path $P_{2d+4}$ can go in two time steps to the last vertex $v_{2d+3}$ of the path, but a guard surveilling the spy in the vertex $v_d$ of the path $P_{2d+3}$ cannot go to the vertex $v_{d+3}$ in two time steps to keep surveilling the spy. Moreover, two guards in the vertices $v_d$ and $v_{d+3}$ are sufficient to surveil the spy. That is, $gn_{s,d}(P_{2d+4})^2 = 4$.

**Lemma 2.** *Let $d \geq 0$ be a fixed integer and let $s \geq d + 2$. Then $gn_{s,d}(P_{2d+4} \boxtimes P_{2d+4}) = 4$.*

*Proof (Sketch of the proof).* See Fig. 2. □

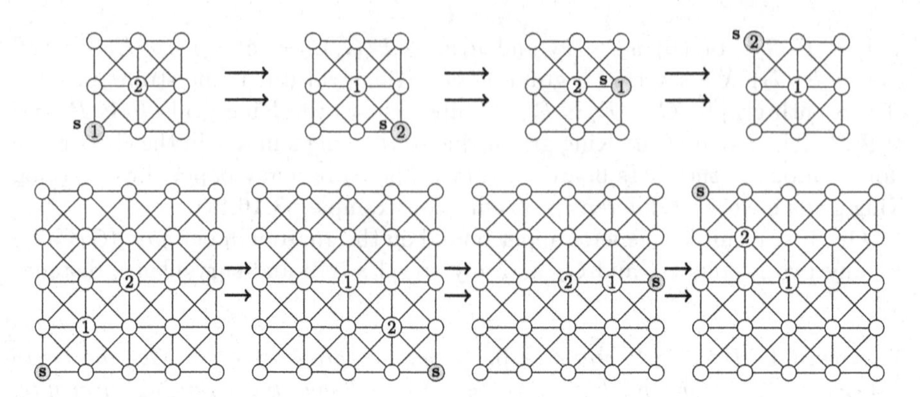

**Fig. 1.** Guards winning strategies with 2 guards in the King grids $P_3 \boxtimes P_3$ and $P_5 \boxtimes P_5$ with $(s, d) = (2, 0)$ and $(s, d) = (4, 1)$, respectively. The spy is represented by $s$ and the guards by 1 and 2.

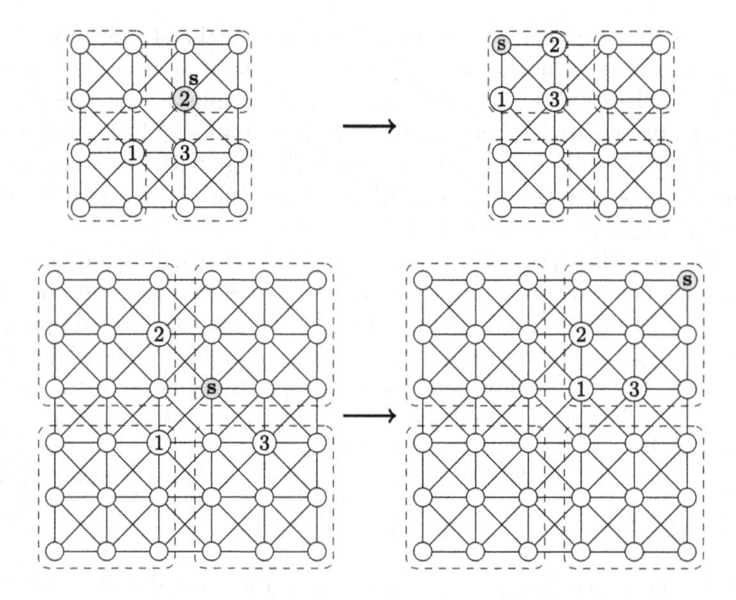

**Fig. 2.** Spy winning strategies with 3 guards in the King grids $P_4 \boxtimes P_4$ and $P_6 \boxtimes P_6$ with $(s, d) = (2, 0)$ and $(s, d) = (3, 1)$, respectively. The spy is represented by $s$ and the guards by 1, 2 and 3.

The previous lemmas showed examples with few guards. We show that $gn_{s,d}(P_n \boxtimes P_n)$ can be very close to $gn_{s,d}(P_n)^2$, when many guards are necessary. More specifically, we show that $gn_{s,d}(P_n \boxtimes P_n) \geq (gn_{s,d}(P_n) - 1)^2$ in many cases where $gn_{s,d}(P_n)$ can be any positive integer.

**Lemma 3.** *Let $d \geq 0$ and $2 \leq k \leq 2d + 2$ be fixed integers and let $s \geq (k-1)(2d+3)$. Then $gn_{s,d}(P_{k(2d+3)}) = k + 1$ and $k^2 \leq gn_{s,d}(P_{k(2d+3)} \boxtimes P_{k(2d+3)}) \leq (k+1)^2$.*

There are other well studied graph products on the vertex set $V(G_1) \times V(G_2)$, such as the cartesian product $G_1 \square G_2$ and the lexicographical products $G_1 \cdot G_2$ and $G_2 \cdot G_1$ (see [22] for a reference). In the cartesian product $G_1 \square G_2$, $(u_1, u_2)$ and $(v_1, v_2)$ are adjacent if and only if (a) $u_1 = v_1$ and $u_2 v_2 \in E(G_2)$ or (b) $u_2 = v_2$ and $u_1 v_1 \in E(G_1)$. In the lexicographical product $G_1 \cdot G_2$, $(u_1, u_2)$ and $(v_1, v_2)$ are adjacent if and only if (a) $u_1 = v_1$ and $u_2 v_2 \in E(G_2)$ or (b) $u_1 v_1 \in E(G_1)$.

It is easy to see, from the definitions, that the cartesian product $G_1 \square G_2$ is a subgraph of the strong product $G_1 \boxtimes G_2$, which is a subgraph of the lexicographical products $G_1 \cdot G_2$ and $G_2 \cdot G_1$. It is also easy to find examples in which the upper bound of Theorem 1 fails in the cartesian product. For example, $gn_{2,0}(P_2 \square P_2) = gn_{2,0}(C_4) = 2 > 1 = gn_{2,0}(P_2)^2$. Regarding the lexicographical product, failing examples are not so easy to find. The next lemma presents an example in which the upper bound of Theorem 1 fails in both cartesian product and lexicographical product.

**Lemma 4.** *$gn_{2,1}(P_5 \square P_5) > gn_{2,1}(P_5)^2$ and $gn_{2,1}(P_5 \cdot P_5) > gn_{2,1}(P_5)^2$.*

*Proof (Sketch of the proof).* See Fig. 3. □

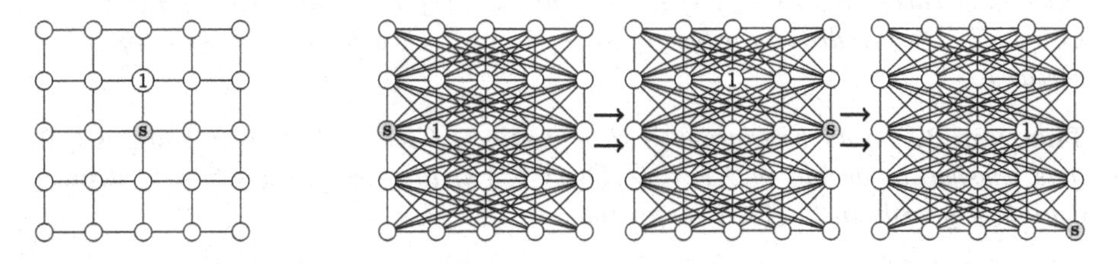


**Fig. 3.** The cartesian product $P_5 \square P_5$ and the lexicographical product $P_5 \cdot P_5$ with a spy winning strategy with 1 guard and $(s, d) = (2, 1)$. The spy is represented by $s$ and the guard by 1.

Nevertheless, regarding the lexicographical product, it is possible to prove a better general upper bound if the surveilling distance $d \geq 2$. In fact, we obtain the following stronger result.

**Theorem 2.** *Let $s \geq 2$, $d \geq 2$ and let $G_1$ and $G_2$ be two graphs. If $G_1$ has no isolated vertex, then the guard number of the lexicographical product of $G_1$ and $G_2$ with surveilling distance $d \geq 2$ is the guard number of $G_1$:*

$$gn_{s,d}(G_1 \cdot G_2) \quad = \quad gn_{s,d}(G_1).$$

*Otherwise, it is the maximum among the guard numbers of $G_1$ and $G_2$:*

$$gn_{s,d}(G_1 \cdot G_2) \quad = \quad \max \{ \ gn_{s,d}(G_1), \ gn_{s,d}(G_2) \ \}.$$

There are still some open questions related to graph products. For example, determine the exact value of the guard number in the lexicographical product for $d \in \{0, 1\}$ and in the strong product and the cartesian product of two general graphs.

## 3 Spy Game Is XP Parameterized by the Number of Guards

In this section, we prove that, if the number $k$ of guards is fixed, the spy game is solvable in polynomial time $O(n^{3k+2})$ for any speed $s \geq 2$ and distance $d \geq 0$. For this, we first define spy and guard configurations.

**Definition 1.** *Given $k \geq 1$, $s \geq 2$, $d \geq 0$ and a graph $G$, let a* configuration *in $G$ be a possible scenario of the spy game, with all $k$ guards and the spy occupying vertices. We define two types of configurations: a* spy configuration *(before the spy's move) and a* guard configuration *(before the guards' move). A spy configuration may be identical to a guard configuration (the only difference is that the spy is the next to move). It is easy to see that there are exactly $2n^{k+1}$ configurations (2 possibilities for the next to move, $n$ possible vertices for the spy and all $k$ guards). We say that a spy configuration $C_1$* leads to *a guard configuration $C_2$ if $C_2$ can be obtained from $C_1$ by moving the spy along at most $s$ edges. We say that a guard configuration $C_1$* leads to *a spy configuration $C_2$ if $C_2$ can be obtained from $C_1$ by moving the guards along at most one edge each. Let the* digraph *$D^*$ defined as follows: for every spy or guard configuration $C$, create an associated vertex $v_C$ in $D^*$. If a configuration $C_1$ leads to a configuration $C_2$, we add in $D^*$ the directed edge from $v_{C_1}$ to $v_{C_2}$. Clearly, $D^*$ has exactly $2n^{k+1}$ vertices.*

**Theorem 3.** *Let $G$ be a graph with $n$ vertices. Given $k \geq 1$, $s \geq 2$ and $d \geq 0$, it is possible to decide in XP time $O(n^{3k+2})$ if the spy has a winning strategy against $k$ guards in the $(s,d)$-spy game.*

*Proof (Sketch of the Proof).* First consider the original $(s,d)$-spy game. The algorithm is as follows. First, we define a *spy winning configuration* as a spy configuration such that the spy is at distance more than $d$ from every guard. Mark all vertices of $D^*$ associated to spy winning configurations. Next, repeat the following until no more vertices of $D^*$ are marked. For every guard configuration $C$, mark the vertex $v_C$ in $D^*$ if all out-neighbors of $v_C$ are marked (in words, any guards' move will lead to a spy winning configuration). Moreover, for every spy configuration $C$, mark the vertex $v_C$ in $D^*$ if there exists at least one marked out-neighbor of $v_C$ (in words, there is a spy's move which leads to a spy winning configuration). Finally, at the end, if there exists a vertex $u$ of $G$ such that, for every spy configuration $C$ with the spy occupying $u$, $v_C$ is marked in $D^*$, then the spy has a winning strategy (by occupying vertex $u$ first). Otherwise, the guards have a winning strategy. By applying breadth-first searchs in each vertex of $G$ as a preprocessing to check distances, we can obtain all out-neighbors of a vertex $v_C$ in $D^*$ in time $O(n^3)$ and consequently $D^*$ can be constructed in time $O(n^{k+4})$. The algorithm will make at most $2n^{k+1}$ iterations, since at least one vertex will be marked in each iteration. Moreover, since each spy configuration leads to at most $n$ guard configurations and each guard configuration leads to at most $n^k$ spy configurations, every iteration takes time $O(n^{k+1}n^k)$. Thus the whole algorithm takes time $O(n^{3k+2})$. □

A graph $G$ is a $(q, q-4)$-graph for some integer $q \geq 4$ if every subset of at most $q$ vertices induces at most $q-4$ distinct $P_4$'s. For instance, cographs and $P_4$-sparse graphs are exactly the $(q, q-4)$-graphs for $q = 4$ and $q = 5$, respectively. The $P_4$-fewness $q(G)$ of a graph $G$ is the minimum $q \geq 4$ such that $G$ is a $(q, q-4)$-graph [9–11,25]. These graphs have received a lot of attention in the literature (under the expression "graphs with few $P_4$'s") and have a nice recursive decomposition based on unions, joins, spiders and small separable p-components [4].

**Theorem 4.** *The spy game decision problem is FPT on the $P_4$-fewness $q(G)$ of the graph $G$, with time $O(m + q^{3q+3} \cdot n)$, where $q = q(G)$ and $m$ and $n$ are the number of edges and vertices of $G$, respectively.*

*Proof (Sketch of the Proof).* From the primeval decomposition of $(q, q-4)$-graphs and the XP algorithm of Theorem 3. □

However, despite these positive results, it was proved in [14] that the spy game decision problem is W[2]-hard in general graphs when parameterized by the number $k$ of guards. In the next section, we extend this hardness result to bipartite graphs for any speed $s \geq 2$ and surveilling distance $d \geq 0$.

## 4 Spy Game Is W[2]-Hard in Bipartite Graphs

In this section, we prove that the spy game decision problem with speed $s$ and distance $d$ is W[2]-hard for any $s \geq 2$ and $d \geq 0$ even in bipartite graphs.

As mentioned in the introduction, we obtain an FPT-reduction from the Set Cover problem (parameterized by the size of the solution) structurally similar to the reduction of [14] for general graphs. However, it turns out that the extension to bipartite graphs brings much more technical difficulties to the reduction and the constructed bipartite graph is significantly different from the one of [14]. In the reduction of [14], the constructed graph is highly non-bipartite: there is a very large clique (representing the sets in the Set Cover problem), which is the main place for the guards. With this, in one time step, any guard can replace the main position of the guard who is surveilling the spy. This is an important recurrent argument in the proof of [14], since a fast spy could force some guard to abandon his main position and later run to a vertex that should be surveilled by this guard. By removing the edges of this main clique in the reduction of [14], we obtain a bipartite graph. However, with this, a guard may help other guard in two time steps only, favoring the spy. Also increasing the number of guards favors the guards in some cases and the balance is not always easy to maintain. This brings some tricky timing problems, which are solved in the following by changing significant parts of the reduction of [14].

**Theorem 5.** *Let $s \geq 2$, $d \geq 0$ and $k \geq 1$ be fixed. The $(s, d)$-spy game decision problem with $k$ guards is NP-hard and W[2]-hard (parameterized by the number of guards). Morever, the $(s, d)$-spy game problem of minimizing the number of guards is Log-APX-hard and $(1 - \varepsilon) \ln n$-inapproximable in polynomial time for any constant $0 < \varepsilon < 1$, unless $P = NP$.*

The reduction is from the SET COVER Problem and is divided in three cases: (i) $s \geq 2d+2$, (ii) $d+1 < s < 2d+2$ and (iii) $s \leq d+1$. Proofs are omitted due to space restrictions.

An instance of the SET COVER Problem is a family $\mathcal{S} = \{S_1, \ldots, S_m\}$ of sets and an integer $c$, and the objective is to decide if there exists a subfamily $\mathcal{C} = \{S_{i_1}, \ldots, S_{i_c}\} \subseteq \mathcal{S}$ such that $|\mathcal{C}| \leq c$ and $S_{i_1} \cup \ldots \cup S_{i_c} = U$, where $U = S_1 \cup \ldots \cup S_m$ (we say that $\mathcal{C}$ is a set cover of $U$). Given an instance $(\mathcal{S}, c)$ of Set Cover, we construct a graph $G = G_{s,d}(\mathcal{S}, c)$ and an integer $K = K_{s,d}(\mathcal{S}, c)$ such that there exists a cover $\mathcal{C} \subseteq \mathcal{S}$ of $U$ with size at most $c$ if and only if $g_{s,d}(G) \leq K$. Note that the reductions presented below are actually FPT-reductions and preserve approximation ratio. Therefore, since the SET COVER Problem is W[2]-hard (when parameterized by the size $c$ of the set cover) and has no $\alpha \ln(n)$ approximation algorithm for some constant $0 < \alpha < 1$ (unless P = NP) [3], we not only prove the NP-hardness but also the fact that the problem is W[2]-hard (when parameterized by the number of guards) and cannot be approximated in polynomial time up to some logarithmic ratio (unless $P = NP$).

**Definition 2.** *Given $s \geq 2$ and $d \geq 0$, let $p = p(s,d) = d + \left\lceil \frac{d+1}{s-1} \right\rceil$ and $q = q(s,d)$ be 0 (if $d+1 < s < 2d+2$), or $d+1+\left\lceil \frac{d}{s-1} \right\rceil$ (if $s \leq d+1$), or d+1, otherwise. Let $(\mathcal{S}, c)$ an instance of Set Cover, where $\mathcal{S} = \{S_1, \ldots, S_m\}$, and let $U = S_1 \cup \ldots \cup S_m = \{u_1, \ldots, u_n\}$. Let the number $K = K_{s,d}(\mathcal{S}, c)$ of guards be $c$ (if $d+1 < s < 2d+2$), or $c+2$, otherwise. Let $G = G_{s,d}(\mathcal{S}, c)$ be defined as follows: for every set $S_j \in \mathcal{S}$, create a new vertex $S_j$ in $G$ and, for every element $u_i \in U$, create a path $U_i$ with $p$ vertices $u_{i,1}, \ldots, u_{i,p}$. If $u_i \in S_j$, add the edge $u_{i,1}S_j$ in $G$. Create a new vertex $z_0$ and add all possible edges between $z_0$ and $\{S_1, \ldots, S_m\}$ in $G$. Finally, if $q > 0$, create a path $Z$ with $q$ vertices $z_1, \ldots, z_q$, and add the edge $z_0z_1$. Moreover, if $s \leq d+1$ or $s \geq 2d+2$, then create a path $Z'$ with $q$ vertices $z'_1, \ldots, z'_q$ and add the edge $z_0z'_1$.*

See Fig. 4, 5 and 6 for examples.

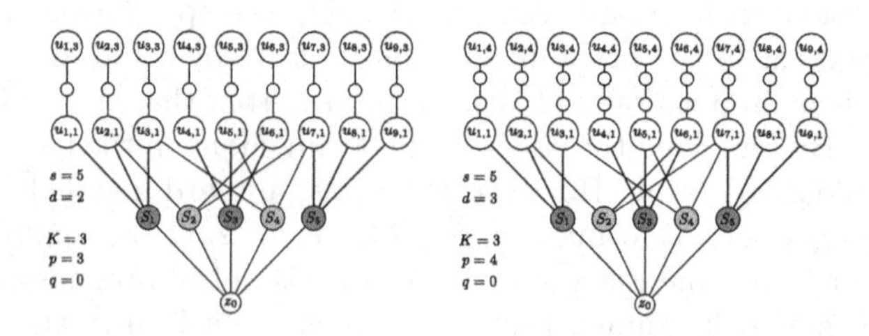


**Fig. 4.** Reduction from Set Cover instance $(\mathcal{S}, c)$, where $c = 3$, $\mathcal{S} = \{S_1, S_2, S_3, S_4, S_5\}$, $S_1 = \{1,2,3\}$, $S_2 = \{2,6,7\}$, $S_3 = \{4,5,6\}$, $S_4 = \{3,5,7\}$, $S_5 = \{7,8,9\}$ and $U = \{1,2,3,4,5,6,7,8,9\}$. Cases for speed $s = 5$ and distance $d = 2, 3$. Illustration of the proof of Lemma 5.

**Lemma 5.** *Given a graph $G$ and an integer $K > 0$, deciding if $g_{s,d}(G) \leq K$ is NP-hard for every $s, d \geq 0$ such that $d+1 < s < 2d+2$.*

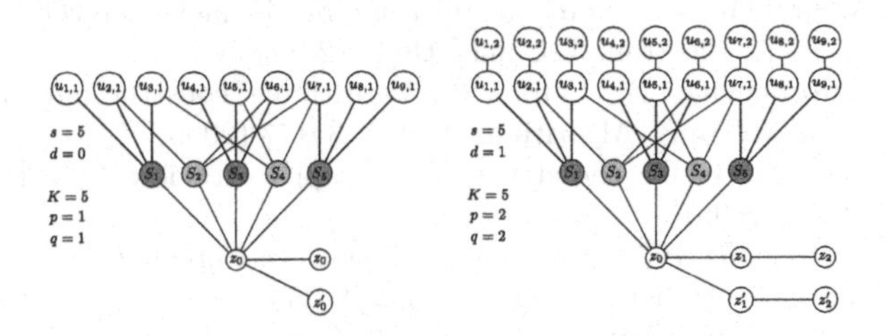


**Fig. 5.** Reduction from Set Cover instance $(\mathcal{S}, c)$, where $c = 3$, $\mathcal{S} = \{S_1, S_2, S_3, S_4, S_5\}$, $S_1 = \{1, 2, 3\}$, $S_2 = \{2, 6, 7\}$, $S_3 = \{4, 5, 6\}$, $S_4 = \{3, 5, 7\}$, $S_5 = \{7, 8, 9\}$ and $U = \{1, 2, 3, 4, 5, 6, 7, 8, 9\}$. Cases for speed $s = 5$ and distance $d \in \{0, 1\}$. Illustration of the proof of Lemma 6.

**Lemma 6.** *Given a graph $G$ and an integer $K$, deciding if $g_{s,d}(G) \leq K$ is NP-hard for every $s, d \geq 0$ such that $s \geq 2d + 2$.*

**Lemma 7.** *Given a graph $G$ and an integer $K$, deciding if $g_{s,d}(G) \leq K$ is NP-hard for every $d > 0$ and $2 \leq s \leq d + 1$.*

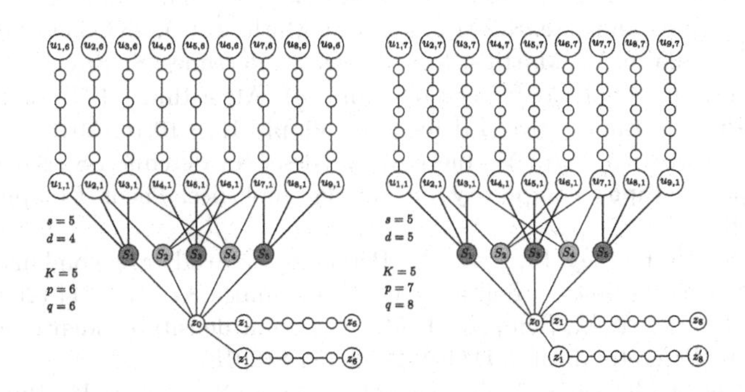


**Fig. 6.** Reduction from Set Cover instance $(\mathcal{S}, c)$, where $c = 3$, $\mathcal{S} = \{S_1, S_2, S_3, S_4, S_5\}$, $S_1 = \{1, 2, 3\}$, $S_2 = \{2, 6, 7\}$, $S_3 = \{4, 5, 6\}$, $S_4 = \{3, 5, 7\}$, $S_5 = \{7, 8, 9\}$ and $U = \{1, 2, 3, 4, 5, 6, 7, 8, 9\}$. Cases for speed $s = 5$ and distance $d \in \{4, 5\}$. Illustration of the proof of Lemma 7.

**Acknowledgments.** This research was supported by Funcap [4543945/2016] Pronem, CNPq [314031/2018-9] Produtividade, CNPq Universal [425297/2016-0] and [437841/2018-9], CAPES [88887.143992/2017-00] DAAD Probral and CAPES [88881.197438/2018-01] STIC AmSud.

## References


1. Aigner, M., Fromme, M.: A game of cops and robbers. Discrete Appl. Math. **8**, 1–12 (1984)
2. Alon, N., Mehrabian, A.: On a generalization of Meyniel's conjecture on the cops and robbers game. Electron. J. Comb. **18**(1) (2011)
3. Alon, N., Moshkovitz, D., Safra, S.: Algorithmic construction of sets for $k$-restrictions. ACM Trans. Algorithms **2**(2), 153–177 (2006)
4. Babel, L., Olariu, S.: On the structure of graphs with few $P_4$'s. Discrete Appl. Math. **84**, 1–13 (1998)
5. Balister, P., Bollobás, B., Narayanan, B., Shaw, A.: Catching a fast robber on the grid. J. Comb. Theory Ser. A **152**, 341–352 (2017)
6. Bonato, A., Chiniforooshan, E., Pralat, P.: Cops and robbers from a distance. Theor. Comput. Sci. **411**(43), 3834–3844 (2010)
7. Bonato, A., Nowakovski, R.: The Game of Cops and Robber on Graphs. American Mathematical Society (2011)
8. Bouznif, M., Darlay, J., Moncel, J., Preissmann, M.: Exact values for three domination-like problems in circular and infinite grid graphs of small height. Discrete Math. Theor. Comput. Sci. **21**(3) (2019)
9. Campos, V., Klein, S., Sampaio, R., Silva, A.: Fixed-parameter algorithms for the cocoloring problem. Discrete Appl. Math. **167**, 52–60 (2014)
10. Campos, V., Linhares-Sales, C., Sampaio, R., Maia, A.K.: Maximization coloring problems on graphs with few P4's. Discrete Appl. Math. **164**, 539–546 (2014)
11. Campêlo, M., Huiban, C., Sampaio, R., Wakabayashi, Y.: Hardness and inapproximability of convex recoloring problems. Theor. Comput. Sci. **533**, 15–25 (2014)
12. Chalopin, J., Chepoi, V., Nisse, N., Vaxès, Y.: Cop and robber games when the robber can hide and ride. SIAM J. Discrete Math. **25**(1), 333–359 (2011)
13. Cohen, N., Hilaire, M., Martins, N.A., Nisse, N., Pérennes, S.: Spy-game on graphs. In: 8th International Conference on Fun with Algorithms (FUN 2016). Leibniz Intern. Proc. in Informatics (LIPIcs), vol. 49, pp. 10:1–10:16 (2016)
14. Cohen, N., Martins, N.A., Mc Inerney, F., Nisse, N., Pérennes, S., Sampaio, R.M.: Spy-game on graphs: complexity and simple topologies. Theor. Comput. Sci. **725**, 1–15 (2018)
15. Cohen, N., Mc Inerney, F., Nisse, N., Pérennes, S.: Study of a combinatorial game in graphs through linear programming. Algorithmica **82**, 212–244 (2020)
16. Dantas, R., Havet, F., Sampaio, R.M.: Minimum density of identifying codes of king grids. Discrete Math. **341**(10), 2708–2719 (2018)
17. Fomin, F.V., Golovach, P.A., Kratochvíl, J., Nisse, N., Suchan, K.: Pursuing a fast robber on a graph. Theor. Comput. Sci. **411**(7–9), 1167–1181 (2010)
18. Fomin, F.V., Golovach, P.A., Lokshtanov, D.: Cops and robber game without recharging. In: Kaplan, H. (ed.) SWAT 2010. LNCS, vol. 6139, pp. 273–284. Springer, Heidelberg (2010). https://doi.org/10.1007/978-3-642-13731-0_26
19. Fomin, F., Golovach, P.A., Pralat, P.: Cops and robber with constraints. SIAM J. Discrete Math. **26**(2), 571–590 (2012)
20. Goddard, W., Hedetniemi, S., Hedetniemi, S.: Eternal security in graphs. J. Comb. Math. Comb. Comput. **52**, 169–180 (2005)
21. Goldwasser, J.L., Klostermeyer, W.: Tight bounds for eternal dominating sets in graphs. Discrete Math. **308**(12), 2589–2593 (2008)
22. Hammack, R., Imrich, W., Klavžar, S.: Handbook of Product Graphs. CRC Press, Boca Raton (2011)

23. Klostermeyer, W., MacGillivray, G.: Eternal dominating sets in graphs. J. Comb. Math. Comb. Comput. **68** (2009)
24. Klostermeyer, W., Mynhardt, C.: Graphs with equal eternal vertex cover and eternal domination numbers. Discrete Math. **311**(14), 1371–1379 (2011)
25. Linhares-Sales, C., Maia, A.K., Martins, N., Sampaio, R.: Restricted coloring problems on graphs with few P4's. Anna. Oper. Res. **217**, 385–397 (2014)
26. Nowakowski, R.J., Winkler, P.: Vertex-to-vertex pursuit in a graph. Discrete Math. **43**, 235–239 (1983)
27. Pelto, M.: Optimal $(r, \leq 3)$-locating–dominating codes in the infinite king grid. Discrete Appl. Math. **161**(16), 2597–2603 (2013)
28. Zermelo, E.: Über eine anwendung der mengenlehre auf die theorie des schachspiels. In: Proceedings of the Fifth International Congress of mathematicians, pp. 501–504 (1913)

# Author Index

Printed in the United States
by Baker & Taylor Publisher Services